COLLINS FAMILY ATLAS OF THE WORLD

Collins Family Atlas of the World

Collins
An Imprint of HarperCollins*Publishers*
77-85 Fulham Palace Road, Hammersmith, London W6 8JB

First published 1998

Maps © HarperCollins*Publishers*

All rights reserved. No part of this publication may be reproduced, stored in a retrieval system, or transmitted, in any form or by any means, electronic, mechanical, photocopying, recording, or otherwise, without the prior written permission of the publisher and copyright owners.

The contents of this edition of the Collins Family Atlas of the World are believed correct at the time of printing. Nevertheless the publisher can accept no responsibility for errors or omissions, changes in the detail given or for any expense or loss thereby caused.

Printed in Italy

ISBN 0 00 448610 2

Photo Credits
Jacket: Galen Rowell/Corbis
p.14: Pictor International - London
p.15: Pictor International - London
p.16: Pictor International - London
p.17: Pictor International - London
p.18: Pictor International - London (topx2)
The Stock Market Photo Agency (centre)
p.19: M. Ashworth
p.20: M. Ashworth (centre)
Pictor International - London (bottomx2)
p.21: Science Photo Library (topx2)
Pictor International - London (bottom)
p.22: Pictor International - London
p.23: Pictor International - London (centre)
Science Photo Library (bottomx2)
p.26: Pictor International - London
p.27: Jim Holmes, Axiom
p.36: Science Photo Library

Data Credits
pp.36-37: Telegeography Inc, Washington DC, www.telegeography.com
Petroleum Economist Ltd., London, www.petroleum-economist.com
Network Wizards, www.nw.com

KH9297 Imp 001

COLLINS FAMILY ATLAS OF THE WORLD

HarperCollins*Publishers*

CONTENTS

THE WORLD

EUROPE

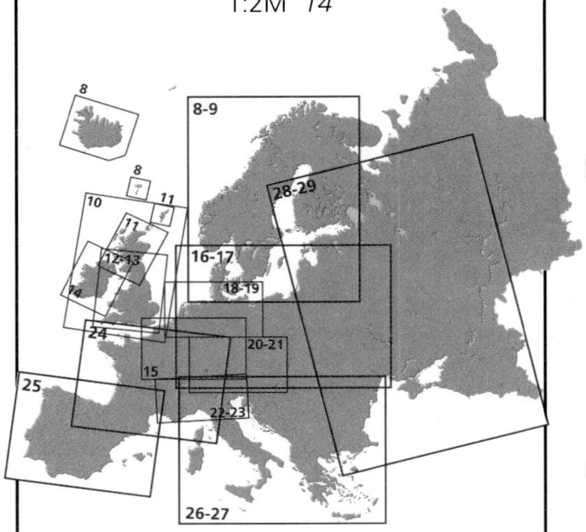

ASIA

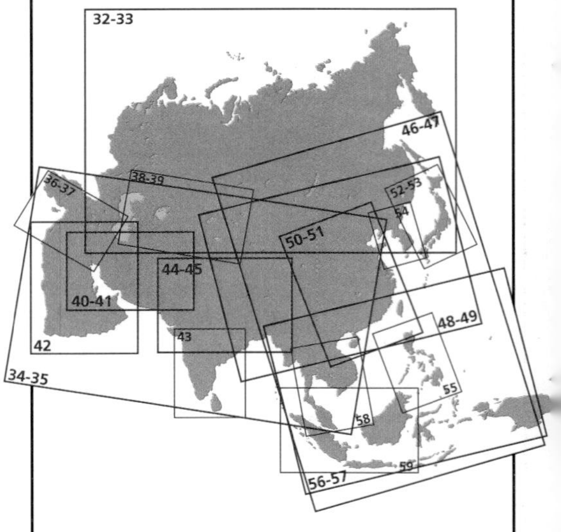

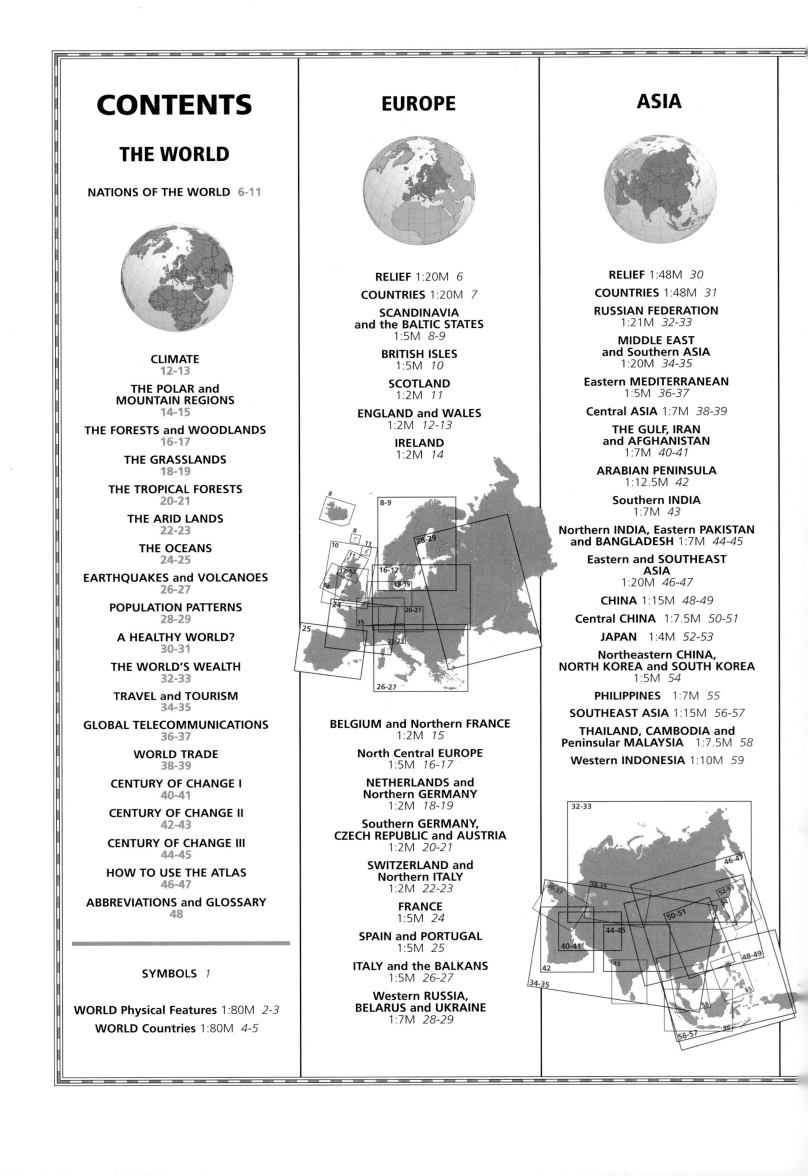

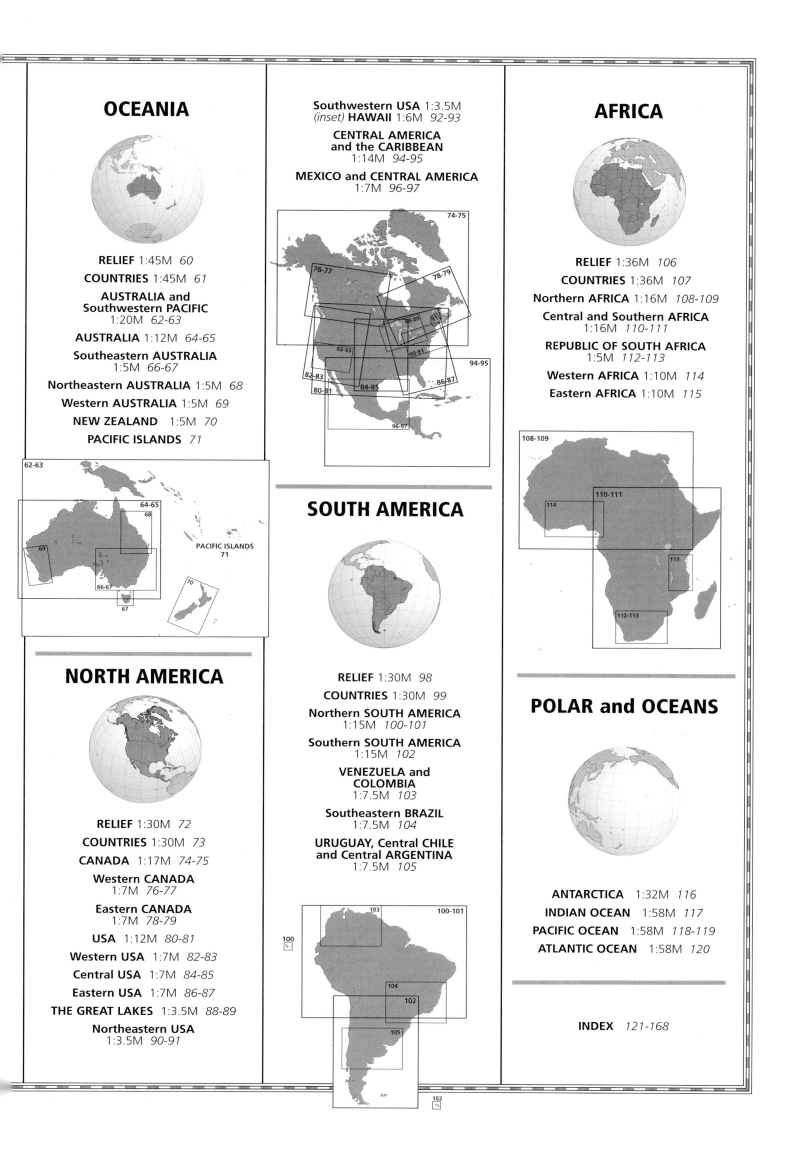

NATIONS OF THE WORLD

EUROPE

COUNTRY	AREA sq ml	AREA sq km	POPULATION TOTAL	DENSITY PER sq ml	DENSITY PER sq km	CITY CAPITAL	LANGUAGES	RELIGIONS	CURRENCY
ALBANIA	11 100	28 748	3 414 000	308	119	Tirana	Albanian	Muslim, Orthodox, Roman Catholic	Lek
ANDORRA	180	465	65 000	362	140	Andorra la Vella	Catalan, Spanish, French	Roman Catholic	Fr franc, Sp peseta
AUSTRIA	32 377	83 855	8 031 000	248	96	Vienna	German, Serbo-Croat	Roman Catholic, Protestant	Schilling
AZORES	868	2 247	237 800	274	106	Ponta Delgada	Turkish Portuguese	Roman Catholic, Protestant	Port. escudo
BELARUS	80 155	207 600	10 355 000	129	50	Minsk	Belorussian, Russian, Ukrainian	Orthodox, Roman Catholic	Rouble
BELGIUM	11 784	30 520	10 080 000	855	330	Brussels	Dutch (Flemish), French, German	Roman Catholic, Protestant	Franc
BOSNIA-HERZEGOVINA	19 741	51 130	4 459 000	226	87	Sarajevo	Serbo-Croat	Muslim, Orthodox, Roman Catholic, Protestant	Dinar
BULGARIA	42 855	110 994	8 443 000	197	76	Sofia	Bulgarian, Turkish	Orthodox, Muslim	Lev
CHANNEL ISLANDS	75	195	149 000	1987	764	St Helier, St Peter Port	English, French	Protestant, Roman Catholic	Pound
CROATIA	21 829	56 538	4 777 000	219	84	Zagreb	Serbo-Croat	Roman Catholic, Orthodox, Muslim	Kuna
CZECH REPUBLIC	30 450	78 864	10 336 000	339	131	Prague	Czech, Moravian, Slovak	Roman Catholic, Protestant	Koruna
DENMARK	16 631	43 075	5 205 000	313	121	Copenhagen	Danish	Protestant, Roman Catholic	Krone
ESTONIA	17 452	45 200	1 499 000	86	33	Tallinn	Estonian, Russian	Protestant, Orthodox	Kroon
FAROE ISLANDS	540	1 399	47 000	87	34	Tórshavn	Danish, Faeroese	Protestant	Danish krone
FINLAND	130 559	338 145	5 088 000	39	15	Helsinki	Finnish, Swedish	Protestant, Orthodox	Markka
FRANCE	210 026	543 965	58 375 000	278	107	Paris	French, French dialects, Arabic,German (Alsatian)	Roman Catholic, Protestant, Muslim	Franc
GERMANY	138 174	357 868	81 912 000	593	229	Berlin	German	Protestant, Roman Catholic,	Mark
GIBRALTAR	3	7	28 000	11157	4308	Gibraltar	English, Spanish	Roman Catholic, Protestant,	Pound
GREECE	50 949	131 957	10 426 000	205	79	Athens	Greek	Greek Orthodox	Drachma
HUNGARY	35 919	93 030	10 261 000	286	110	Budapest	Hungarian	Roman Catholic, Protestant	Forint
ICELAND	39 699	102 820	266 000	7	3	Reykjavik	Icelandic	Protestant, Roman Catholic	Króna
ISLE OF MAN	221	572	73 000	331	128	Douglas	English	Protestant, Roman Catholic	Pound
ITALY	116 311	301 245	57 193 000	492	190	Rome	Italian, Italian dialects	Roman Catholic	Lira
LATVIA	24 595	63 700	2 548 000	104	40	Riga	Latvian, Russian	Protestant, Roman Catholic, Orthodox	Lat
LIECHTENSTEIN	62	160	31 000	502	194	Vaduz	German	Roman Catholic, Protestant	Swiss franc
LITHUANIA	25 174	65 200	3 721 000	148	57	Vilnius	Lithuanian, Russian, Polish	Roman Catholic, Protestant, Orthodox	Litas
LUXEMBOURG	998	2 586	404 000	405	156	Luxembourg	Letzeburgish, Portuguese	Roman Catholic, Protestant	Franc
MACEDONIA, Former Yugoslavian Republic of	9 928	25 713	2 142 000	216	83	Skopje	Macedonian	Orthodox, Muslim, Roman Catholic	Denar
MADEIRA	307	794	253 000	825	319	Funchal	Portuguese	Roman Catholic, Protestant	Port. escudo
MALTA	122	316	373 000	3057	1180	Valletta	Maltese, English	Roman Catholic	Lira
MOLDOVA	13 012	33 700	4 350 000	334	129	Chişinău	Romanian, Russian, Ukrainian	Moldovan, Orthodox,	Leu
MONACO	1	2	32 821	32821	16410	Monaco	French, Monegasque, Italian	Roman Catholic	French franc
NETHERLANDS	16 033	41 526	15 517 000	968	374	Amsterdam	Dutch, Frisian	Roman Catholic, Protestant	Guilder
NORWAY	125 050	323 878	4 325 000	35	13	Oslo	Norwegian	Protestant, Roman Catholic	Krone
POLAND	120 728	312 683	38 544 000	319	123	Warsaw	Polish, German	Roman Catholic, Orthodox	Zæoty
PORTUGAL	34 340	88 940	9 902 000	288	111	Lisbon	Portuguese	Roman Catholic, Protestant	Escudo
REPUBLIC OF IRELAND	27 136	70 282	3 571 000	132	51	Dublin	English, Irish	Roman Catholic, Protestant	Punt
ROMANIA	91 699	237 500	22 731 000	248	96	Bucharest	Romanian	Orthodox	Leu
RUSSIAN FEDERATION	6 592 849	17 075 400	147 739 000	22	9	Moscow	Russian, Tatar, Ukrainian, local languages	Orthodox, Muslim, other Christian, Jewish	Rouble
RUSSIAN FEDERATION (In Europe)	1 527 343	3 955 800	105 984 000	69	27				
SAN MARINO	24	61	25 000	1061	410	San Marino	Italian	Roman Catholic	Ital. lira
SLOVAKIA	18 933	49 035	5 347 000	282	109	Bratislava	Slovak, Hungarian, Czech	Roman Catholic, Protestant	Koruna
SLOVENIA	7 819	20 251	1 989 000	254	98	Ljubljana	Slovene, Serbo-Croat	Roman Catholic, Protestant	Tólar
SPAIN	194 897	504 782	39 143 000	201	78	Madrid	Spanish, Catalan, Galician, Basque	Roman Catholic	Peseta
SWEDEN	173 732	449 964	8 781 000	51	20	Stockholm	Swedish	Protestant, Roman Catholic	Krona
SWITZERLAND	15 943	41 293	6 995 000	439	169	Bern	German, French, Italian, Romansch	Roman Catholic, Protestant	Franc
UNITED KINGDOM	94 241	244 082	58 144 000	617	238	London	English, South Indian languages, Welsh, Gaelic	Protestant, Roman Catholic, Muslim	Pound
UKRAINE	233 090	603 700	51 910 000	223	86	Kiev	Ukrainian, Russian, regional languages	Orthodox, Roman Catholic	Karbovanets
VATICAN CITY		0.44	1000		2273		Italian	Roman Catholic	Ital. lira
YUGOSLAVIA	39 449	102 173	10 516 000	267	103	Belgrade	Serbo-Croat, Albanian, Hungarian	Serbian and Montenegrin Orthodox, Muslim	Dinar

ASIA

COUNTRY	AREA sq ml	AREA sq km	POPULATION TOTAL	DENSITY PER sq ml	DENSITY PER sq km	CITY CAPITAL	LANGUAGES	RELIGIONS	CURRENCY
AFGHANISTAN	251 825	652 225	18 879 000	75	29	Kābul	Dari, Pushtu, Uzbek	Sunni & Shi'a Muslim	Afghani
ARMENIA	11 506	29 800	3 548 000	308	119	Yerevan	Armenian, Azeri, Russian	Orthodox, Roman Catholic, Muslim	Dram
AZERBAIJAN	33 436	86 600	7 472 000	223	86	Baku	Azeri, Armenian, Russian,	Shi'a & Sunni Muslim,	Manat
BAHRAIN	267	691	599 000	2244	867	Al Manāmah	Arabic, English	Shi'a & Sunni Muslim, Christian	Dinar
BANGLADESH	55 598	143 998	120 073 000	2160	834	Dhaka	Bengali	Muslim, Hindu	Taka
BHUTAN	18 000	46 620	1 614 000	90	35	Thimphu	Dzongkha, Nepali, Assamese	Buddhist, Hindu, Muslim	Ngultrum
BRUNEI	2 226	5 765	280 000	126	49	Bandar Seri Begawan	Malay, English, Chinese	Muslim, Buddhist, Christian	Dollar (ringgit)
CAMBODIA	69 884	181 000	9 568 000	137	53	Phnum Penh	Khmer	Buddhist, Muslim	Riel
CHINA	3 691 899	9 560 900	1 232 083 000	334	129	Beijing	Chinese, regional languages	Confucian, Taoist, Buddhist	Yuan
CYPRUS	3 572	9 251	726 000	203	78	Nicosia	Greek, Turkish, English	Greek Orthodox, Muslim	Pound
GEORGIA	26 911	69 700	5 450 000	203	78	T'bilisi	Georgian, Russian, Armenian	Orthodox, Muslim	Lari
INDIA	1 269 219	3 287 263	944 580 000	744	287	New Delhi	Hindi, English, regional languages	Hindu, Muslim, Sikh, Christian	Rupee
INDONESIA	741 102	1 919 445	196 813 000	266	103	Jakarta	Indonesian, local languages	Muslim, Protestant, Roman Catholic	Rupiah
IRAN	636 296	1 648 000	61 128 000	96	37	Tehrān	Farsi, Azeri, Kurdish	Shi'a & Sunni Muslim	Rial
IRAQ	169 235	438 317	19 925 000	118	45	Baghdād	Arabic, Kurdish	Shi'a & Sunni Muslim	Dinar
ISRAEL	8 019	20 770	5 399 000	673	260	Jerusalem	Hebrew, Arabic	Jewish, Muslim, Christian	Shekel
JAPAN	145 841	377 727	125 761 000	862	333	Tōkyō	Japanese	Shintoist, Buddhist	Yen
JORDAN	34 443	89 206	5 198 000	151	58	'Ammān	Arabic	Sunni & Shi'a Muslim	Dinar
KAZAKSTAN	1 049 155	2 717 300	17 027 000	16	6	Almaty	Kazakh, Russian	Muslim, Orthodox, Protestant	Tanga
KUWAIT	6 880	17 818	1 620 000	235	91	Kuwait	Arabic	Sunni & Shi'a Muslim, Christian	Dinar
KYRGYZSTAN	76 641	198 500	4 473 000	58	23	Bishkek	Kirghiz, Russian, Uzbek	Muslim, Orthodox	Som
LAOS	91 429	236 800	4 742 000	52	20	Vientiane	Lao, local languages	Buddhist, traditional beliefs	Kip
LEBANON	4 036	10 452	2 915 000	722	279	Beirut	Arabic, French, Armenian	Shi'a & Sunni Muslim, Protestant, Roman Catholic	Pound
MACAU	7	17	440 000	62857	25882	Macau	Chinese, Portuguese	Buddhist	Pataca
MALAYSIA	128 559	332 965	20 097 000	156	60	Kuala Lumpur	Malay, English, Chinese,	Muslim, Buddhist, Roman Catholic, Christian, trad. beliefs	Ringgit
MALDIVES	115	298	263 015	2287	883	Male	Divehi (Maldivian)	Muslim	Rufiyaa
MONGOLIA	604 250	1 565 000	2 363 000	4	2	Ulaanbaatar	Khalka (Mongolian), Kazakh	Buddhist, Muslim, traditional beliefs	Tugrik
MYANMAR	261 228	676 577	43 922 000	168	65	Yangon	Burmese, Shan, Karen	Buddhist, Muslim, Protestant,	Kyat
NEPAL	56 827	147 181	21 360 000	376	145	Kathmandu	Nepali, Maithili, Bhojpuri	Hindu, Buddhist	Rupee
NORTH KOREA	46 540	120 538	23 483 000	505	195	P'yŏngyang	Korean	Trad. beliefs, Chondoist	Won
OMAN	105 000	271 950	2 096 000	20	8	Muscat	Arabic, Baluchi	Muslim	Rial
PAKISTAN	310 403	803 940	134 146 000	432	167	Islamabad	Urdu, Punjabi, Sindhi, Pushtu	Muslim, Christian, Hindu	Rupee
PALAU	192	497	17 000	89	34	Koror	Palauan, English	Roman Catholic, Protestant	US dollar
PHILIPPINES	115 831	300 000	71 899 000	621	240	Manila	Filipino, Cebuano, local languages	Roman Catholic, Aglipayan, Muslim	Peso
QATAR	4 416	11 437	593 000	134	52	Doha	Arabic, Indian lang.	Muslim, Christian	Riyal
RUSSIAN FEDERATION	6 592 849	17 075 400	147 739 000	22	9	Moscow	Russian, Tatar, Ukrainian, local languages	Orthodox, other Christian, Muslim, Jewish	Rouble
RUSSIAN FEDERATION (In Asia)	5 065 506	13 119 600	41 755 000	8	3				
SAUDI ARABIA	849 425	2 200 000	17 451 000	21	8	Riyadh	Arabic	Sunni & Shi'a Muslim	Riyal
SINGAPORE	247	639	3 044 000	12324	4764	Singapore	Chinese, English, Malay, Tamil	Buddhist, Taoist, Muslim, Christian	Dollar
SOUTH KOREA	38 330	99 274	45 547 000	1188	450	Seoul	Korean	Buddhist, Protestant, Roman Catholic	Won
SRI LANKA	25 332	65 610	17 865 000	705	272	Colombo	Sinhalese, Tamil, English	Buddhist, Hindu, Muslim	Rupee
SYRIA	71 498	185 180	13 844 000	194	75	Damascus	Arabic, Kurdish	Muslim, Christian	Pound
TAIWAN	13 969	36 179	21 212 000	1518	586	T'ai-pei	Chinese, local languages	Buddhist, Taoist, Confucian,	Dollar
TAJIKISTAN	55 251	143 100	5 933 000	107	41	Dushanbe	Tajik, Uzbek, Russian	Muslim	Rouble
THAILAND	198 115	513 115	60 003 000	303	117	Bangkok	Thai, Lao, Chinese	Buddhist, Muslim	Baht
TURKEY	300 948	779 452	62 697 000	208	80	Ankara	Turkish, Kurdish	Sunni & Shi'a Muslim	Lira
TURKMENISTAN	188 456	488 100	4 010 000	21	8	Ashgabat	Turkmen, Russian	Muslim	Manat
UNITED ARAB EMIRATES	30 000	77 700	1 861 000	62	24	Abu Dhabi	Arabic	Sunni & Shi'a Muslim, Christian	Dirham
UZBEKISTAN	172 742	447 400	22 633 000	131	51	Tashkent	Uzbek, Russian, Tajik, Kazakh	Muslim, Orthodox	Som
VIETNAM	127 246	329 565	75 181 000	591	228	Ha Nôi	Vietnamese	Buddhist, Roman Catholic	Dong
YEMEN	203 850	527 968	12 672 000	62	24	Şan'ā	Arabic	Sunni & Shi'a Muslim	Dinar, rial

NATIONS OF THE WORLD

OCEANIA

COUNTRY	AREA sq ml	AREA sq km	POPULATION TOTAL	DENSITY PER sq ml	DENSITY PER sq km	CITY CAPITAL	LANGUAGES	RELIGIONS	CURRENCY
AMERICAN SAMOA	76	197	55 000	723	279	Pago Pago	Samoan, English	Protestant, Roman Catholic	US dollar
AUSTRALIA	2 966 153	7 682 300	17 838 000	6	2	Canberra	English, Aboriginal languages	Protestant, Roman Catholic, Aboriginal beliefs	Dollar
FIJI	7 077	18 330	784 000	111	43	Suva	English, Fijian, Hindi	Christian, Hindu, Muslim	Dollar
FRENCH POLYNESIA	1 261	3 265	215 000	171	66	Papeete	French, Polynesian languages	Protestant, Roman Catholic,	Pacific franc
GUAM	209	541	146 000	699	270	Agana	Chamorro, English	Roman Catholic	US dollar
KIRIBATI	277	717	77 000	278	107	Bairiki	I-Kiribati (Gilbertese), English	Roman Catholic, Protestant	Austr. dollar
MARSHALL ISLANDS	70	181	54 000	773	298	Dalap-Uliga-Darrit	Marshallese, English	Protestant, Roman Catholic	US dollar
FED. STATES OF MICRONESIA	271	701	104 000	384	148	Palikir	English, Trukese, Pohnpeian, local languages	Protestant, Roman Catholic	US dollar
NAURU	8	21	11 000	1357	524	Yaren	Nauruan, Gilbertese, English	Protestant, Roman Catholic	Austr. dollar
NEW CALEDONIA	7 358	19 058	184 000	25	10	Nouméa	French, local languages	Roman Catholic, Protestant	Pacific franc
NEW ZEALAND	104 454	270 534	3 493 000	33	13	Wellington	English, Maori	Protestant, Roman Catholic	Dollar
NORTH. MARIANA IS.	184	477	47 000	255	99	Saipan	English, Chamorro, Tagalog	Roman Catholic, Protestant	US dollar
PAPUA NEW GUINEA	178 704	462 840	3 997 000	22	9	Port Moresby	English, Tok Pisin	Protestant, Roman Catholic	Kina
SOLOMON ISLANDS	10 954	28 370	366 000	33	13	Honiara	English, Pidgin	Protestant, Roman Catholic	Dollar
TONGA	289	748	98 000	339	131	Nuku'alofa	Tongan, English	Protestant, Roman Catholic, Mormon	Pa'anga
TUVALU	10	25	10 000	1000	400	Fongafale	Tuvaluan, English	Protestant	Dollar
VANUATU	4 707	12 190	165 000	35	14	Port Vila	English, Bislama, French	Protestant, Roman Catholic	Vatu
WALLIS AND FUTUNA	106	274	14 000	132	51	Mata-Utu	French, Polynesian	Roman Catholic	Pacific franc
WESTERN SAMOA	1 093	2 831	164 000	150	58	Apia	Samoan, English	Protestant, Roman Catholic	Tala

ECONOMIC GROUPS

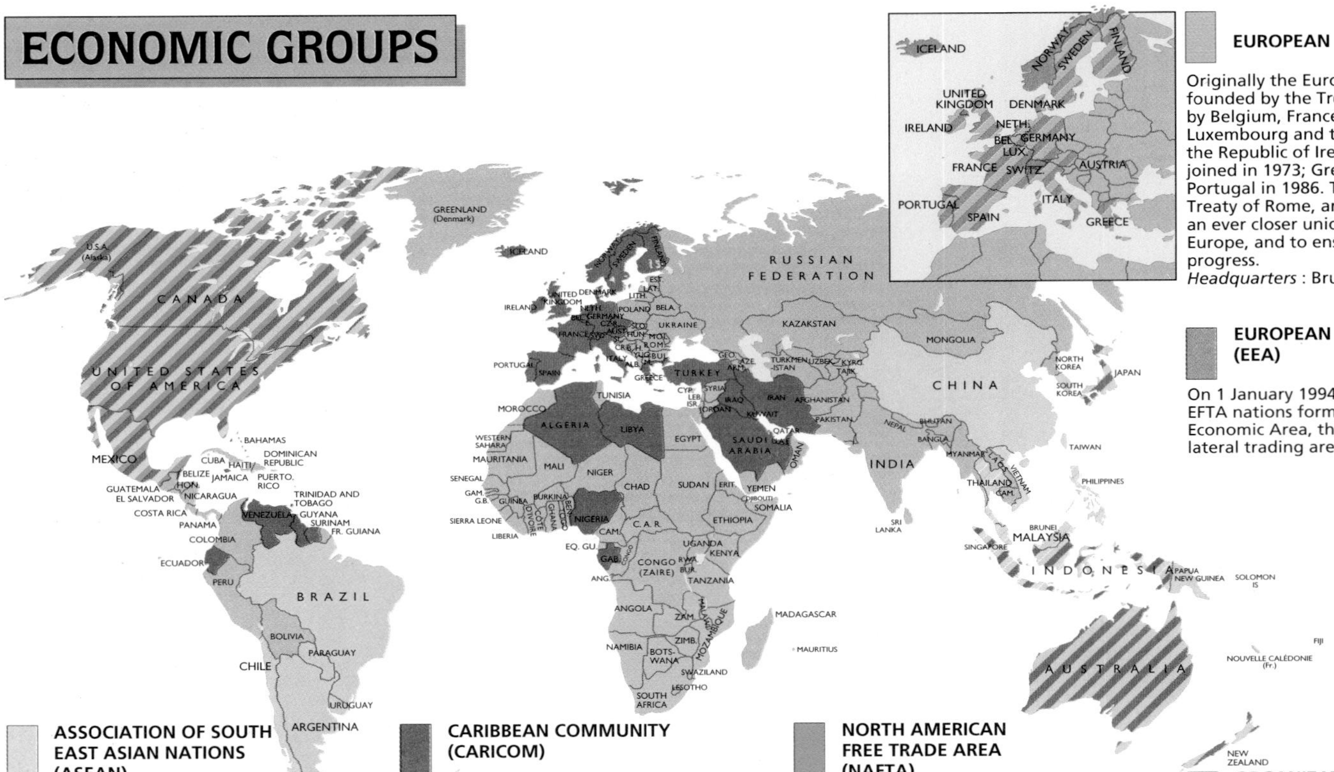

EUROPEAN UNION (EU)

Originally the European Economic Community, founded by the Treaty of Rome in 1957, signed by Belgium, France, West Germany, Italy, Luxembourg and the Netherlands. Denmark, the Republic of Ireland and the United Kingdom joined in 1973; Greece in 1981 and Spain and Portugal in 1986. The objectives, under the Treaty of Rome, are to lay the foundations of an ever closer union among the peoples of Europe, and to ensure economic and social progress.
Headquarters : Brussels, Belgium

EUROPEAN ECONOMIC AREA (EEA)

On 1 January 1994 the EU nations and the EFTA nations formed the European Economic Area, the World's largest multi-lateral trading area.

ASSOCIATION OF SOUTH EAST ASIAN NATIONS (ASEAN)

Established in 1967, the objectives of ASEAN are to promote economic, political and social co-operation. The founder members were Indonesia, Malaysia, the Philippines, Singapore and Thailand; Brunei joined in 1984 and Vietnam in 1995. Cambodia, Laos and Myanmar have applied for membership.
Headquarters : Jakarta, Indonesia

ASIA PACIFIC ECONOMIC CO-OPERATION FORUM (APEC)

Formed in 1989 to promote trade and economic co-operation, with the long term aim of the creation of a Pacific free trade area. The original members were Australia, Brunei, Canada, Indonesia, Japan, Malaysia, New Zealand, the Philippines, Singapore, South Korea, Thailand and U.S.A.. China, Hong Kong and Taiwan joined in 1991, Mexico and Papua New Guinea in 1993, and Chile in 1994.
Headquarters : Singapore

CARIBBEAN COMMUNITY (CARICOM)

CARICOM was established in 1973. The original members were Barbados, Guyana, Jamaica and Trinidad and Tobago; in May 1974, Belize, Dominica, Grenada, Montserrat, St Lucia, and St Vincent joined followed by Antigua and St Kitts-Nevis in August 1974, the Bahamas in 1984 and Surinam in 1995. The objectives of CARICOM are to foster co-operation, co-ordinate foreign policy, and to formulate and carry out common policies on health, education and culture, communications and industrial relations.
Headquarters : Georgetown, Guyana

MERCADO COMMUN DEL SUR (Southern Common Market MERCOSUR)

Established by a treaty signed in Paraguay in 1991 by Argentina, Brazil, Paraguay and Uruguay, Mercosur's objective is to establish a regional common market.
Headquarters : Mercosur's headquarters rotate between member states' capitals.

NORTH AMERICAN FREE TRADE AREA (NAFTA)

NAFTA grew out of a 1988 free trade agreement between U.S.A. and Canada, which was extended to include Mexico in 1992. The accord came into force in January 1994.

ORGANISATION FOR ECONOMIC CO-OPERATION AND DEVELOPMENT (OECD)

Established in 1961, the OECD's objective is to promote economic and social welfare throughout the OECD area. It does this by assisting member governments in the formulation and co-ordination of policies to meet this objective; it also aims to stimulate and harmonise members' efforts in favour of developing countries.
Headquarters : Paris, France

ORGANIZATION OF PETROLEUM EXPORTING COUNTRIES (OPEC)

Established in 1960 to co-ordinate the price and supply policies of oil-producing states, and to provide member countries with economic and technical aid. Member countries are Algeria, Ecuador, Gabon, Indonesia, Iran, Iraq, Kuwait, Libya, Nigeria, Qatar, Saudi Arabia, U.A.E., and Venezuela.
Headquarters : Vienna, Austria

SOUTHERN AFRICAN DEVELOPMENT COMMUNITY (SADC)

The founder members of SADC were Angola, Botswana, Lesotho, Malawi, Mozambique, Swaziland, Tanzania, Zambia and Zimbabwe. Namibia joined in 1990, South Africa in 1994 and Mauritius in 1995. The objectives are deeper economic co-operation and integration and the promotion of political and social values, human rights and the alleviation of poverty.
Headquarters : Gaborone, Botswana

NORTH AMERICA

COUNTRY	AREA sq ml	AREA sq km	POPULATION TOTAL	DENSITY PER sq ml	DENSITY PER sq km	CITY CAPITAL	LANGUAGES	RELIGIONS	CURRENCY
ANGUILLA	60	155	8 000	134	52	The Valley	English	Protestant, Roman Catholic	E. Carib. dollar
ANTIGUA & BARBUDA	171	442	65 000	381	147	St John's	English, Creole	Protestant, Roman Catholic	E. Carib. dollar
BAHAMAS	5 382	13 939	272 000	51	20	Nassau	English, Creole, French Creole	Protestant, Roman Catholic	Dollar
BARBADOS	166	430	264 000	1590	614	Bridgetown	English, Creole (Bajan)	Protestant, Roman Catholic	Dollar
BELIZE	8 867	22 965	211 000	24	9	Belmopan	English, Creole, Spanish, Mayan	Roman Catholic, Protestant	Dollar
BERMUDA	21	54	64 000	3048	1185	Hamilton	English	Protestant, Roman Catholic	Dollar
CANADA	3 849 674	9 970 610	29 251 000	8	3	Ottawa	English, French, Amerindian languages, Inuktitut (Eskimo)	Roman Catholic, Protestant	Dollar
CAYMAN ISLANDS	100	259	31 000	310	120	George Town	English	Protestant, Roman Catholic	Dollar
COSTA RICA	19 730	51 100	3 071 000	156	60	San José	Spanish	Roman Catholic, Protestant	Colón
CUBA	42 803	110 860	10 960 000	256	99	Havana	Spanish	Roman Catholic, Protestant	Peso
DOMINICA	290	750	71 000	245	95	Roseau	English, French Creole	Roman Catholic, Protestant	E. Carib. dollar,
DOMINICAN REPUBLIC	18 704	48 442	7 769 000	415	160	Santo Domingo	Spanish, French Creole	Roman Catholic, Protestant	Peso
EL SALVADOR	8 124	21 041	5 641 000	694	268	San Salvador	Spanish	Roman Catholic, Protestant	Colón
GREENLAND	840 004	2 175 600	55 000			Nuuk	Greenlandic, Danish	Protestant	Danish krone
GRENADA	146	378	92 000	630	243	St George's	English, Creole	Roman Catholic, Protestant	E. Carib. dollar
GUADELOUPE	687	1 780	421 000	613	237	Basse Terre	French, French Creole	Roman Catholic	French franc
GUATEMALA	42 043	108 890	10 322 000	246	95	Guatemala	Spanish, Mayan languages	Roman Catholic, Protestant	Quetzal
HAITI	10 714	27 750	7 041 000	657	254	Port-au-Prince	French, French Creole	Roman Catholic, Protestant	Gourde
HONDURAS	43 277	112 088	5 770 000	133	51	Tegucigalpa	Spanish, Amerindian languages	Roman Catholic, Protestant	Lempira
JAMAICA	4 244	10 991	2 429 000	572	221	Kingston	English, Creole	Protestant, Roman Catholic	Dollar
MARTINIQUE	417	1 079	375 000	900	348	Fort-de-France	French, French Creole	Roman Catholic	French franc
MEXICO	761 604	1 972 545	96 578 000	127	49	México	Spanish, Amerindian languages	Roman Catholic	Peso
MONTSERRAT	39	100	11 000	285	110	Plymouth	English	Protestant, Roman Catholic	E. Carib. dollar
NICARAGUA	50 193	130 000	4 401 000	88	34	Managua	Spanish, Amerindian languages	Roman Catholic, Protestant	Córdoba
PANAMA	29 762	77 082	2 583 000	87	34	Panamá	Spanish, English Creole, Amerindian languages	Roman Catholic	Balboa
PUERTO RICO	3 515	9 104	3 736 000	1063	410	San Juan	Spanish, English	Roman Catholic, Protestant	US dollar
ST KITTS & NEVIS	101	261	41 000	407	157	Basseterre	English, Creole	Protestant, Roman Catholic	E. Carib. dollar
ST LUCIA	238	616	141 000	593	229	Castries	English, French Creole	Roman Catholic, Protestant	E. Carib. dollar
ST VINCENT & THE GRENADINES	150	389	111 000	739	285	Kingstown	English, Creole	Protestant, Roman Catholic	E. Carib. dollar
TURKS & CAICOS ISLANDS	166	430	14 000	84	33	Cockburn Town	English	Protestant	US dollar
USA	3 787 425	9 809 386	266 557 000	70	27	Washington	English, Spanish, Amerindian languages	Protestant, Roman Catholic	Dollar
VIRGIN ISLANDS (UK)	59	153	18 000	305	118	Road Town	English	Protestant, Roman Catholic	US dollar
VIRGIN ISLANDS (USA)	136	352	104 000	765	295	Charlotte Amalie	English, Spanish	Protestant, Roman Catholic	US dollar

INTERNATIONAL ORGANIZATIONS

ARAB LEAGUE

Founded in 1945 in Cairo, by Egypt, Syria, Iraq, Lebanon, Jordan, Saudi Arabia and Yemen. The membership has been extended to include 14 other countries in the region.
Headquarters : Cairo, Egypt

THE COMMONWEALTH

The status and relationship of members of the Commonwealth, which grew out of the British Empire was defined in 1926 and enshrined in the 1931 Statute of Westminster. There are 53 members of the Commonwealth.

COMMONWEALTH OF INDEPENDENT STATES (CIS)

Established by the Minsk agreement and the Alma-Ata Declaration in 1991 following the collapse of the U.S.S.R..
Headquarters : Minsk, Belarus

ORGANIZATION OF AMERICAN STATES (OAS)

The OAS claims to be the oldest regional organization in the world, tracing its origins back to 1826. The charter of the present OAS came into force in 1951. There are 34 member states spread throughout North and South America.
Headquarters : Washington, U.S.A.

THE UNITED NATIONS

The United Nations is the largest international group of countries. Formed in 1945 to promote world peace and co-operation between nations. The 185 members regularly meet in a General Assembly to settle disputes and agree on common policies to world problems. The work of the United Nations is carried out through its various agencies which include:

Headquarters : New York, U.S.A.

Agency:	Responsibility:
UNESCO	Science, education and culture.
UNICEF	Children's welfare.
UNDRO	Disaster relief.
UNHCR	Aid to refugees.
WHO	Health.
FAO	Food & agriculture.
UNEP	Environment.
UNDP	Development programme.

ORGANIZATION OF AFRICAN UNITY (OAU)

The OAU grew out of the Union of Africa states which was founded in 1961. All continental African countries are members together with Cape Verde, the Comoros, São Tomé and Príncipe, and Seychelles.
Headquarters : Addis Ababa, Ethiopia

NATIONS OF THE WORLD

SOUTH AMERICA

COUNTRY	AREA sq ml	AREA sq km	POPULATION TOTAL	DENSITY PER sq ml	DENSITY PER sq km	CITY CAPITAL	LANGUAGES	RELIGIONS	CURRENCY
ARGENTINA	1 068 302	2 766 889	34 180 000	32	12	Buenos Aires	Spanish, Amerindian languages	Roman Catholic	Peso
ARUBA	75	193	69 000	926	358	Oranjestad	Dutch, Papiamento,	Roman Catholic, Protestant	Florin
BOLIVIA	424 164	1 098 581	7 237 000	17	7	La Paz	Spanish, Quechua, Aymara	Roman Catholic	Boliviano
BRAZIL	3 286 488	8 511 965	157 872 000	48	19	Brasília	Portuguese, Italian, Amerindian languages	Roman Catholic	Real
CHILE	292 258	756 945	13 994 000	48	18	Santiago	Spanish, Amerindian languages	Roman Catholic	Peso
COLOMBIA	440 831	1 141 748	34 520 000	78	30	Bogotá	Spanish, Amerindian languages	Roman Catholic	Peso
ECUADOR	105 037	272 045	11 221 000	107	41	Quito	Spanish, Amerindian languages	Roman Catholic	Sucre
FALKLAND ISLANDS	4 699	12 170	2 000			Stanley	English	Protestant, Roman Catholic	Pound
FRENCH GUIANA	34 749	90 000	141 000	4	2	Cayenne	French, French Creole	Roman Catholic, Protestant	French franc
GUYANA	83 000	214 969	825 000	10	4	Georgetown	English, Creole, Hindi, Amerindian languages	Protestant, Hindu, Roman Catholic, Sunni Muslim	Dollar
NETH. ANTILLES	283	732	158 206	560	216	Willemstad	Dutch, Papiamento	Roman Catholic, Protestant	Guilder
PARAGUAY	157 048	406 752	4 700 000	30	12	Asunción	Spanish, Guaraní	Roman Catholic	Guaraní
PERU	496 225	1 285 216	23 088 000	47	18	Lima	Spanish, Quechua	Roman Catholic	Sol
SURINAME	63 251	163 820	418 000	7	3	Paramaribo	Dutch, Surinamese	Hindu, Roman Catholic, Protestant, Muslim	Guilder
TRINIDAD AND TOBAGO	1 981	5 130	1 250 000	631	244	Port Of Spain	English, Creole, Hindi	Roman Catholic, Hindu, Protestant	Dollar
URUGUAY	68 037	176 215	3 167 000	47	18	Montevideo	Spanish	Roman Catholic, Protestant	Peso
VENEZUELA	352 144	912 050	21 177 000	60	23	Caracas	Spanish, Amerindian languages	Roman Catholic	Bolívar

AFRICA

COUNTRY	AREA sq ml	AREA sq km	POPULATION TOTAL	DENSITY PER sq ml	DENSITY PER sq km	CITY CAPITAL	LANGUAGES	RELIGIONS	CURRENCY
ALGERIA	919 595	2 381 741	27 561 000	30	12	Algiers	Arabic, French, Berber	Muslim	Dinar
ANGOLA	481 354	1 246 700	10 674 000	22	9	Luanda	Portuguese	Roman Catholic, Protestant, traditional beliefs	Kwanza
BENIN	43 483	112 620	5 387 000	124	48	Porto Novo	French, local languages	Trad. beliefs, Roman Catholic	CFA franc
BOTSWANA	224 468	581 370	1 443 000	6	2	Gaborone	English, Setswana	Traditional beliefs, Protestant, Roman Catholic	Pula
BURKINA	105 869	274 200	9 889 000	93	36	Ouagadougou	French, local languages	Traditional beliefs, Muslim, Roman Catholic	CFA franc
BURUNDI	10 747	27 835	6 134 000	571	220	Bujumbura	Kirundi, French	Roman Catholic, Protestant	Franc
CAMEROON	183 569	475 442	12 871 000	70	27	Yaoundé	French, English	Trad. beliefs, Roman Catholic, Muslim, Protestant	CFA franc
CAPE VERDE	1 557	4 033	381 000	245	94	Praia	Portuguese	Roman Catholic	Escudo
CENTRAL AFRICAN REPUBLIC	240 324	622 436	3 235 000	13	5	Bangui	French, Sango	Protestant, Roman Catholic, traditional beliefs	CFA franc
CHAD	495 755	1 284 000	6 214 000	13	5	Ndjamena	Arabic, French	Muslim, traditional beliefs, Roman Catholic	CFA franc
COMOROS	719	1 862	630 000	876	338	Moroni	Comorian, French, Arabic	Muslim	Franc
CONGO	132 047	342 000	2 516 000	19	7	Brazzaville	French, local languages	Roman Catholic, Protestant	CFA franc
CONGO (ZAIRE)	905 568	2 345 410	42 552 000	47	18	Kinshasa	French, local languages	Roman Catholic, Protestant	Zaïre
CÔTE D'IVOIRE	124 504	322 463	13 695 000	110	42	Yamoussoukro	French, local languages	Traditional beliefs, Muslim, Roman Catholic	CFA franc
DJIBOUTI	8 958	23 200	566 000	63	24	Djibouti	French, Arabic	Muslim	Franc
EGYPT	386 199	1 000 250	60 603 000	157	61	Cairo	Arabic, French	Muslim, Coptic Christian	Pound
EQUATORIAL GUINEA	10 831	28 051	389 000	36	14	Malabo	Spanish	Roman Catholic	CFA franc
ERITREA	45 328	117 400	3 437 000	76	29	Asmara	Tigrinya, Arabic, Tigre, English	Muslim, Coptic Christian	Ethiopian birr
ETHIOPIA	437 794	1 133 880	58 506 000	134	52	Addis Ababa	Amharic, local languages	Ethiopian Orthodox, Muslim, traditional beliefs	Birr
GABON	103 347	267 667	1 283 000	12	5	Libreville	French, local languages	Roman Catholic, Protestant	CFA franc
GAMBIA	4 361	11 295	1 081 000	248	96	Banjul	English	Muslim	Dalasi
GHANA	92 100	238 537	16 944 000	184	71	Accra	English, local languages	Protestant, Roman Catholic, Muslim, traditional beliefs	Cedi
GUINEA	94 926	245 857	6 501 000	68	26	Conakry	French, local languages	Muslim	Franc
GUINEA-BISSAU	13 948	36 125	1 050 000	75	29	Bissau	Portuguese, local languages	Traditional beliefs, Muslim	Peso
KENYA	224 961	582 646	29 292 000	130	50	Nairobi	Swahili, English	Roman Catholic, Protestant, traditional beliefs	Shilling
LESOTHO	11 720	30 355	1 996 000	170	66	Maseru	Sesotho, English	Roman Catholic, Protestant	Loti
LIBERIA	43 000	111 369	2 700 000	63	24	Monrovia	English, local languages	Muslim, Christian	Dollar
LIBYA	679 362	1 759 540	4 899 000	7	3	Tripoli	Arabic, Berber	Muslim	Dinar
MADAGASCAR	226 658	587 041	14 303 000	63	24	Antananarivo	Malagasy, French	Traditional beliefs, Roman Catholic, Protestant	Franc

COUNTRY	AREA sq ml	AREA sq km	POPULATION TOTAL	DENSITY PER sq ml	DENSITY PER sq km	CITY CAPITAL	LANGUAGES	RELIGIONS	CURRENCY
MALAWI	45 747	118 484	9 461 000	207	80	Lilongwe	English, Chichewa	Protestant, Roman Catholic, traditional beliefs, Muslim	Kwacha
MALI	478 821	1 240 140	10 462 000	22	8	Bamako	French, local languages	Muslim, traditional beliefs	CFA franc
MAURITANIA	397 955	1 030 700	2 211 000	6	2	Nouakchott	Arabic, local languages	Muslim	Ouguiya
MAURITIUS	788	2 040	1 134 000	1439	556	Port Louis	English	Hindu, Roman Catholic, Muslim	Rupee
MOROCCO	172 414	446 550	26 590 000	154	60	Rabat	Arabic	Muslim	Dirham
MOZAMBIQUE	308 642	799 380	16 614 000	54	21	Maputo	Portuguese, local languages	Traditional beliefs,Roman Catholic, Muslim	Metical
NAMIBIA	318 261	824 292	1 500 000	5	2	Windhoek	English, Afrikaans, German, Ovambo	Protestant, Roman Catholic	Dollar
NIGER	489 191	1 267 000	8 846 000	18	7	Niamey	French, local languages	Muslim, traditional beliefs	CFA franc
NIGERIA	356 669	923 768	115 020 000	322	126	Abuja	English, Hausa, Yoruba, Ibo, Fulani	Muslim, Protestant, Roman Catholic, traditional beliefs	Naira
RÉUNION	985	2 551	644 000	654	252	St-Denis	French	Roman Catholic	French franc
RWANDA	10 169	26 338	7 750 000	762	294	Kigali	Kinyarwanda, French	Roman Catholic, traditional beliefs	Franc
SÃO TOMÉ AND PRÍNCIPE	372	964	125 000	336	130	São Tomé	Portuguese	Roman Catholic, Protestant	Dobra
SENEGAL	75 954	196 720	8 102 000	107	41	Dakar	French, local languages	Muslim	CFA franc
SEYCHELLES	176	455	74 000	421	163	Victoria	Seychellois, English	Roman Catholic, Protestant	Rupee
SIERRA LEONE	27 699	71 740	4 402 000	159	61	Freetown	English, local languages	Traditional beliefs, Muslim	Leone
SOMALIA	246 201	637 657	9 077 000	37	14	Mogadishu	Somali, Arabic	Muslim	Shilling
SOUTH AFRICA	470 689	1 219 080	40 436 000	86	33	Pretoria/Cape Town	Afrikaans, English, local languages	Protestant, Roman Catholic	Rand
SUDAN	967 500	2 505 813	28 947 000	30	12	Khartoum	Arabic, local languages	Muslim, traditional beliefs	Dinar
SWAZILAND	6 704	17 364	879 000	131	51	Mbabane	Swazi, English	Protestant, Roman Catholic, traditional beliefs	Emalangeni
TANZANIA	364 900	945 087	28 846 000	79	31	Dodoma	Swahili, English, local languages	Christian, Muslim, traditional beliefs	Shilling
TOGO	21 925	56 785	3 928 000	179	69	Lomé	French, local languages	Traditional beliefs, Roman Catholic, Muslim	CFA franc
TUNISIA	63 379	164 150	8 814 000	139	54	Tunis	Arabic	Muslim	Dinar
UGANDA	93 065	241 038	20 621 000	222	86	Kampala	English, Swahili	Roman Catholic, Protestant	Shilling
ZAMBIA	290 586	752 614	9 196 000	32	12	Lusaka	English, local languages	Christian, traditional beliefs	Kwacha
ZIMBABWE	150 873	390 759	11 150 000	74	29	Harare	English, Shona, Ndebele	Protestant, Roman Catholic, traditional beliefs	Dollar

TIME ZONES

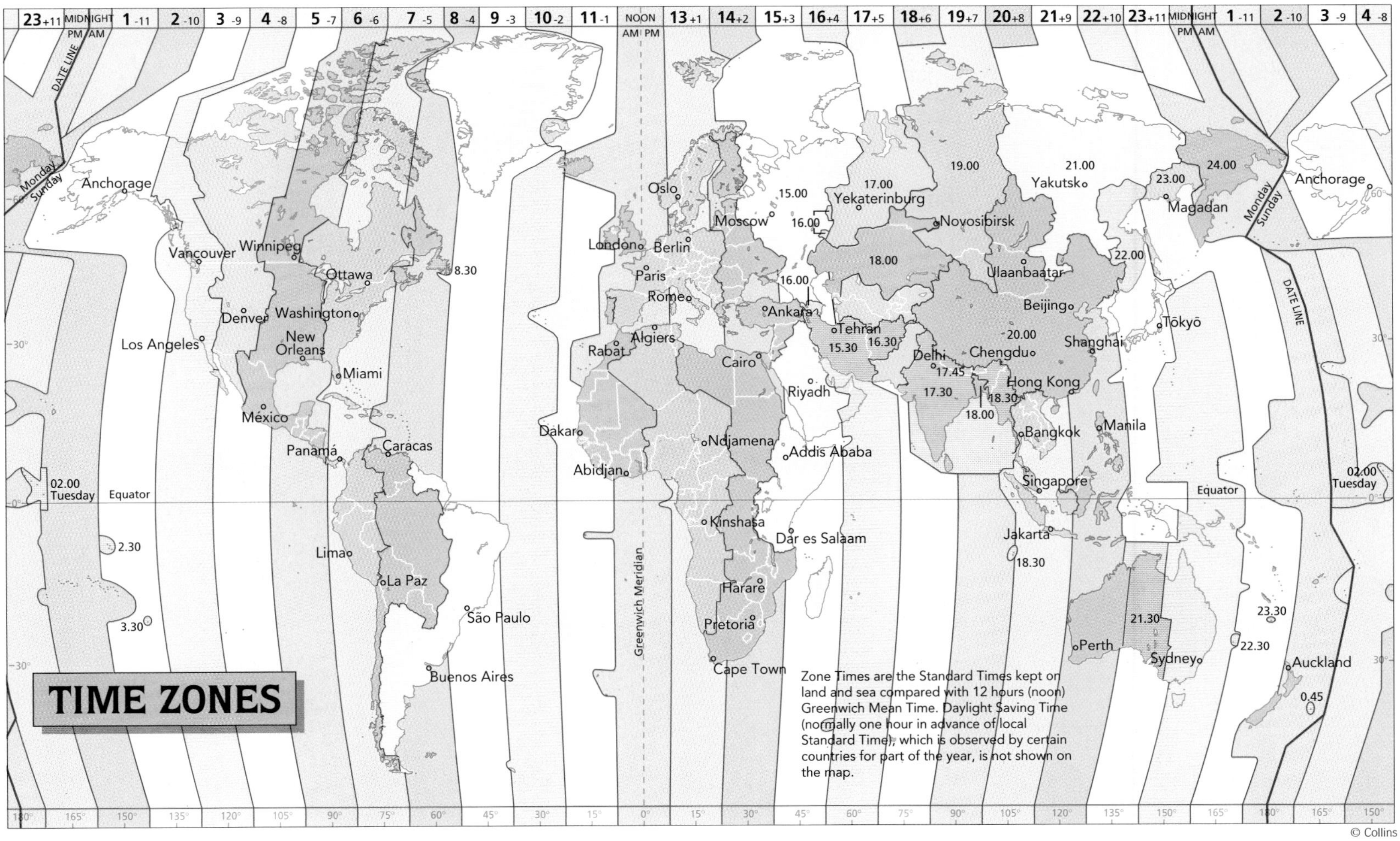

Zone Times are the Standard Times kept on land and sea compared with 12 hours (noon) Greenwich Mean Time. Daylight Saving Time (normally one hour in advance of local Standard Time), which is observed by certain countries for part of the year, is not shown on the map.

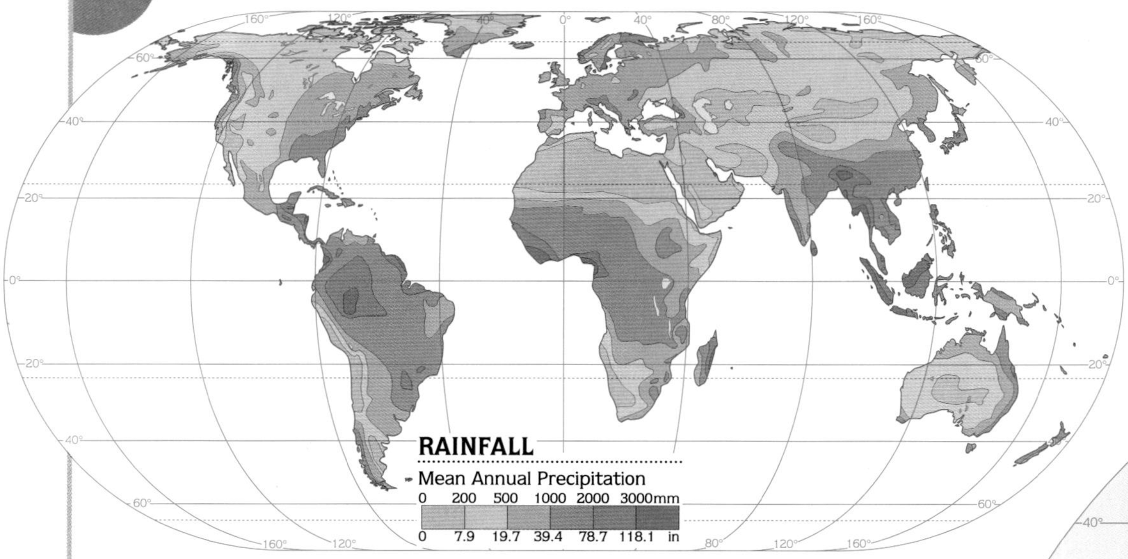

CLIMATE TYPES

- Ice cap
- **Tundra**: warmest month below 10°C
- **Subarctic**: rainy with severe cold winters and less than 4 months over 10°C

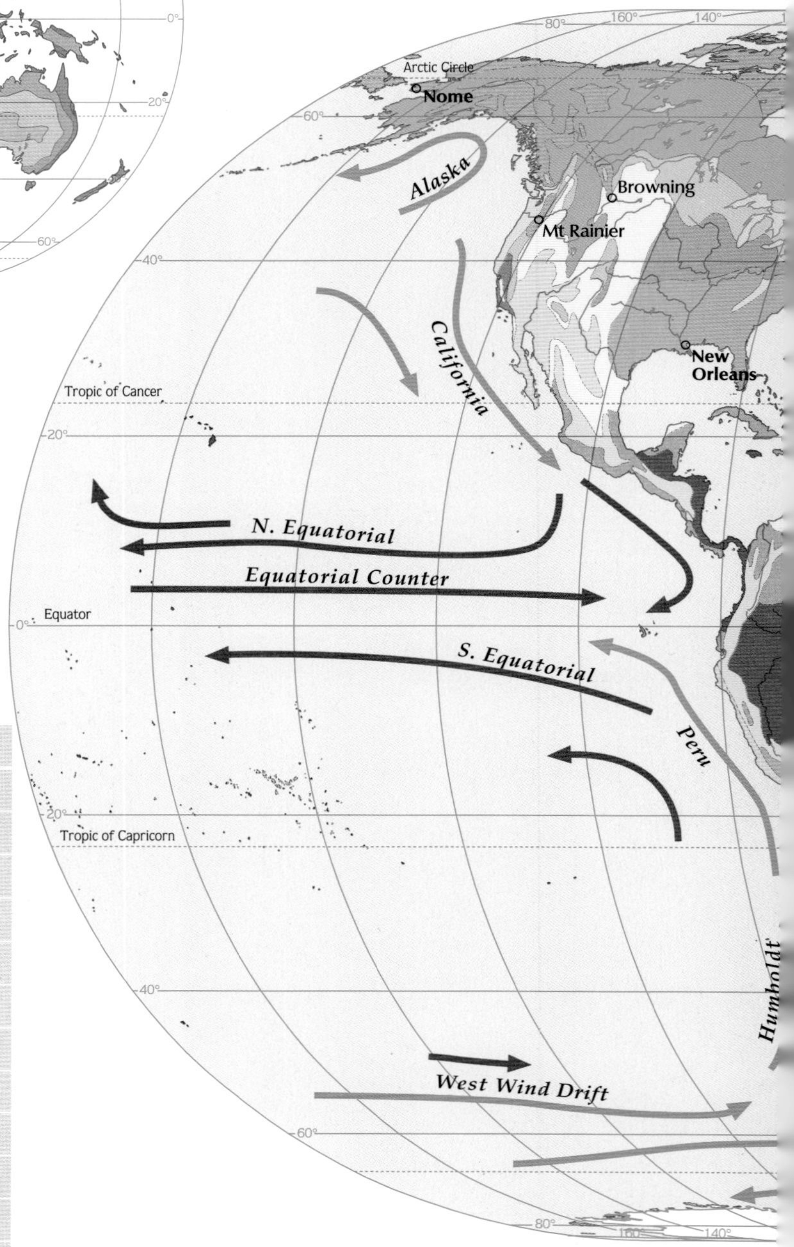

CLIMATE

WEATHER EXTREMES

Hottest annual average	**Dalol,** Ethiopia	**94°F/34.4°C** 1960-1966
Coldest annual average	**Pole of Inaccessibility,** Antarctica	**-72°F/-57.8°C**
Hottest location	**Al 'Azīzīyah,** Libya	**136°F/57.8°C** 13 September 1992
Coldest location	**Vostok Station,** Antarctica	**-128.6°F/-89.2°C** 21 July 1983
Highest annual average rainfall	**Meghalaya,** India	**467.5in/11 874.5mm**
Greatest measured annual rainfall	**Cherrapunji,** India	**1 041.75in/26 461.7mm** 1 August 1860 - 31 July 1861
Greatest 24hr rainfall	**Chilaos, Réunion,** Indian Ocean	**73.5in/1 869.9mm** 15 March 1952
Driest location	**Atacama Desert,** Chile	**0.003in/0.08mm** (annual rainfall)
Greatest 24hr temperature change	**Browning,** USA	**100°F/55.6°C** (From 44°F/6.7°C to -56°F/-49°C) 23-24 January 1916
Greatest annual snowfall	**Mt Rainier,** USA	**1 224.5in/31 102mm** 19 February 1971 - 8 February 1972
Largest hailstone	**Gopalganj,** Bangladesh	**2.25lbs/1.02kg** 14 April 1986
Highest measured wind gust	**Mt Washington,** USA	**231mph/372kph** 12 April 1934

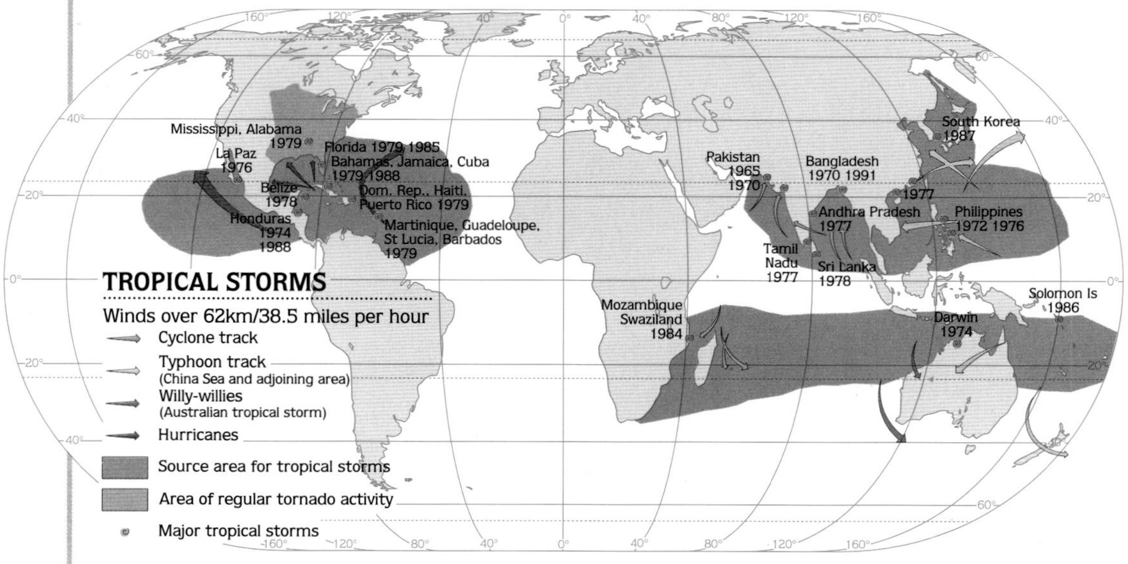

CLIMATE GRAPHS

The graphs show the average monthly temperature and the average monthly rainfall; the colour relates to the Climate Type shown on the map and key above.

The climate stations are shown on the map in bold type; the names in light type on the map show the locations of the Weather Extremes on the chart on the left.

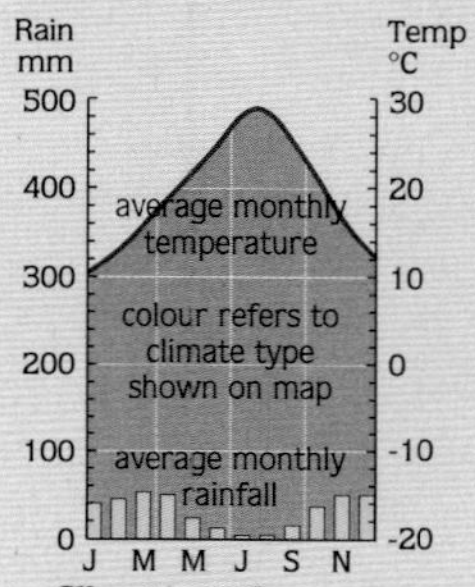

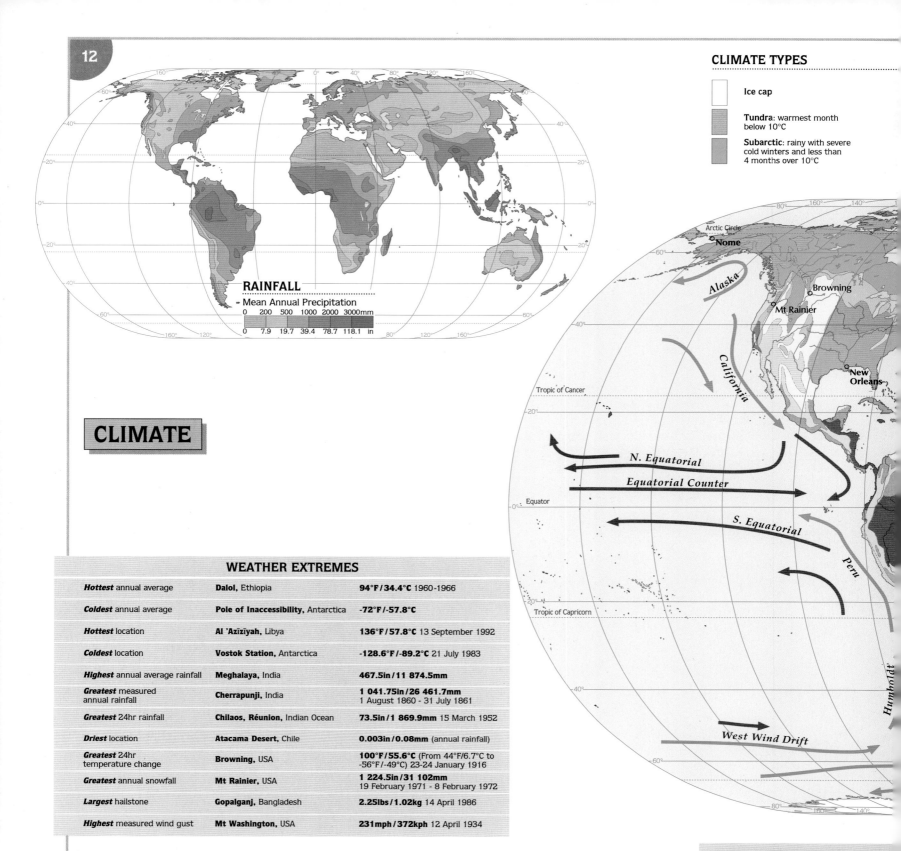

OCEAN CURRENTS

Warm Currents

Cold Currents

Continental cool summer: rainy with warmest month below 22°C

Continental warm summer: with warmest month above 20°C

Temperate: rainy with mild winters and coolest month above 0°C

Humid subtropical: coolest month above 0°C and warmest month above 22°C

Mediterranean: rainy with mild wet winters and dry summers

Steppe: semi-arid, dry

Desert

Savanna: rainy tropical climate

Rain forest: rainy tropical climate, constantly wet throughout the year

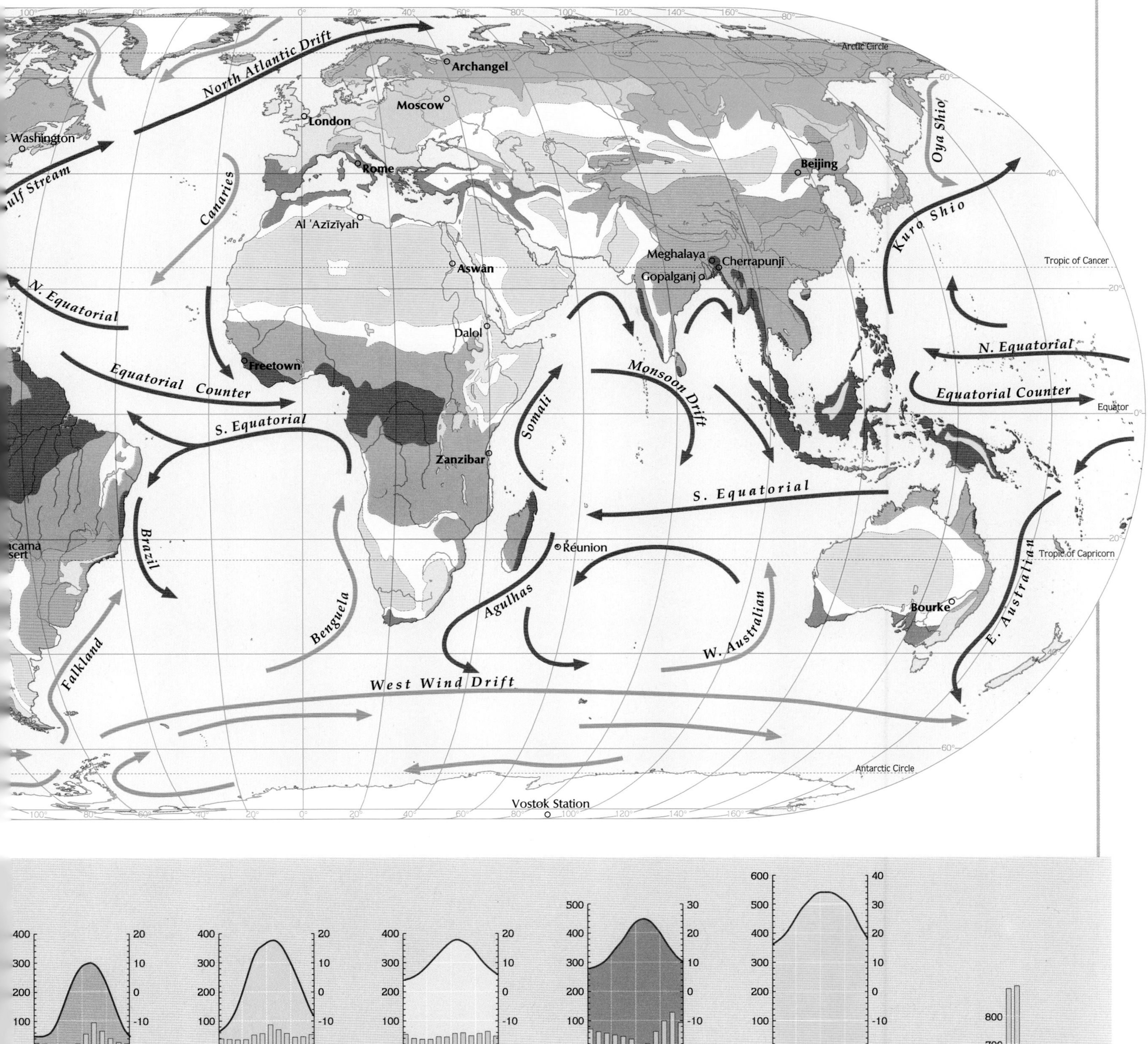

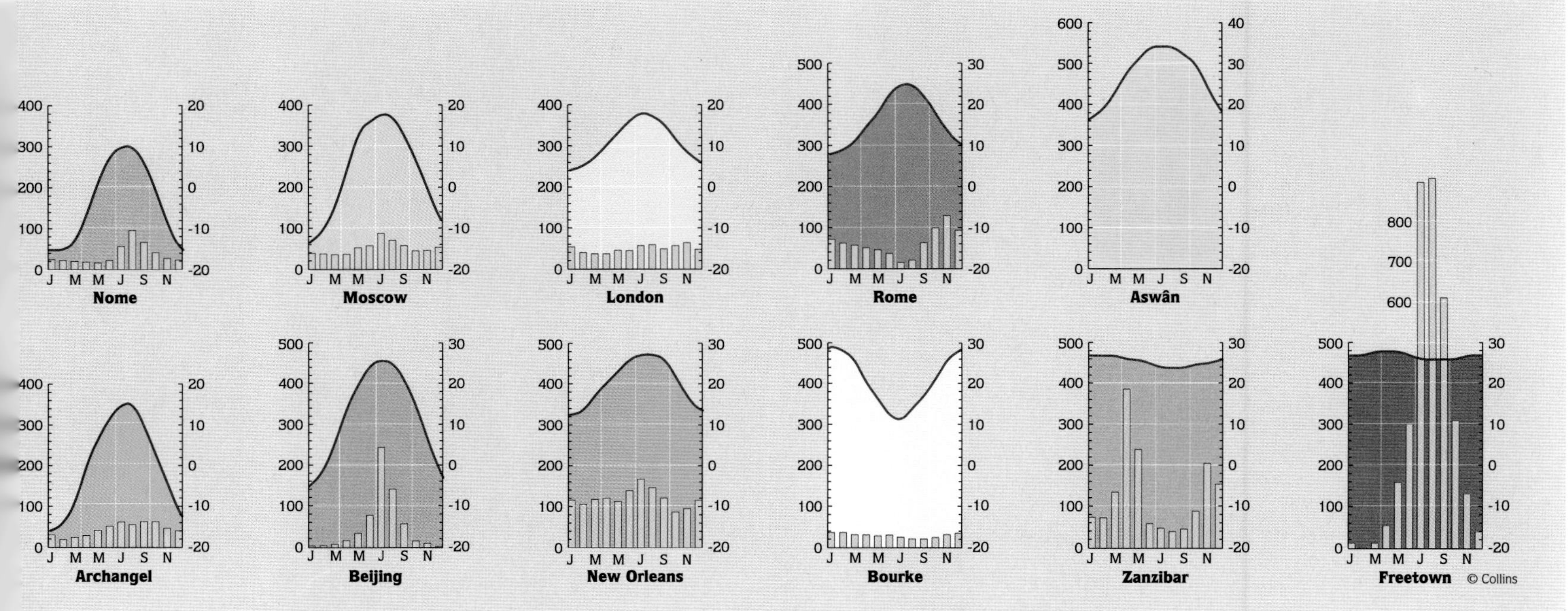

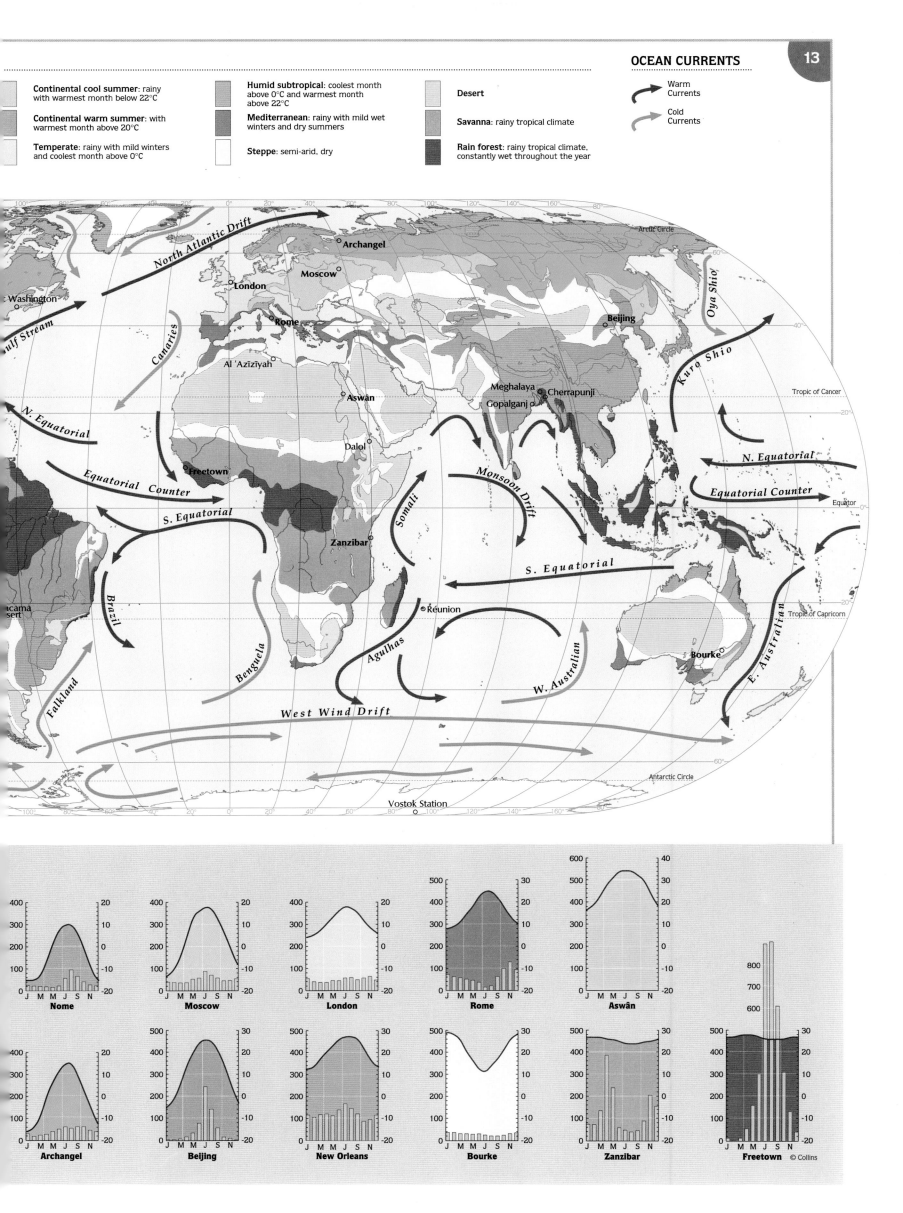

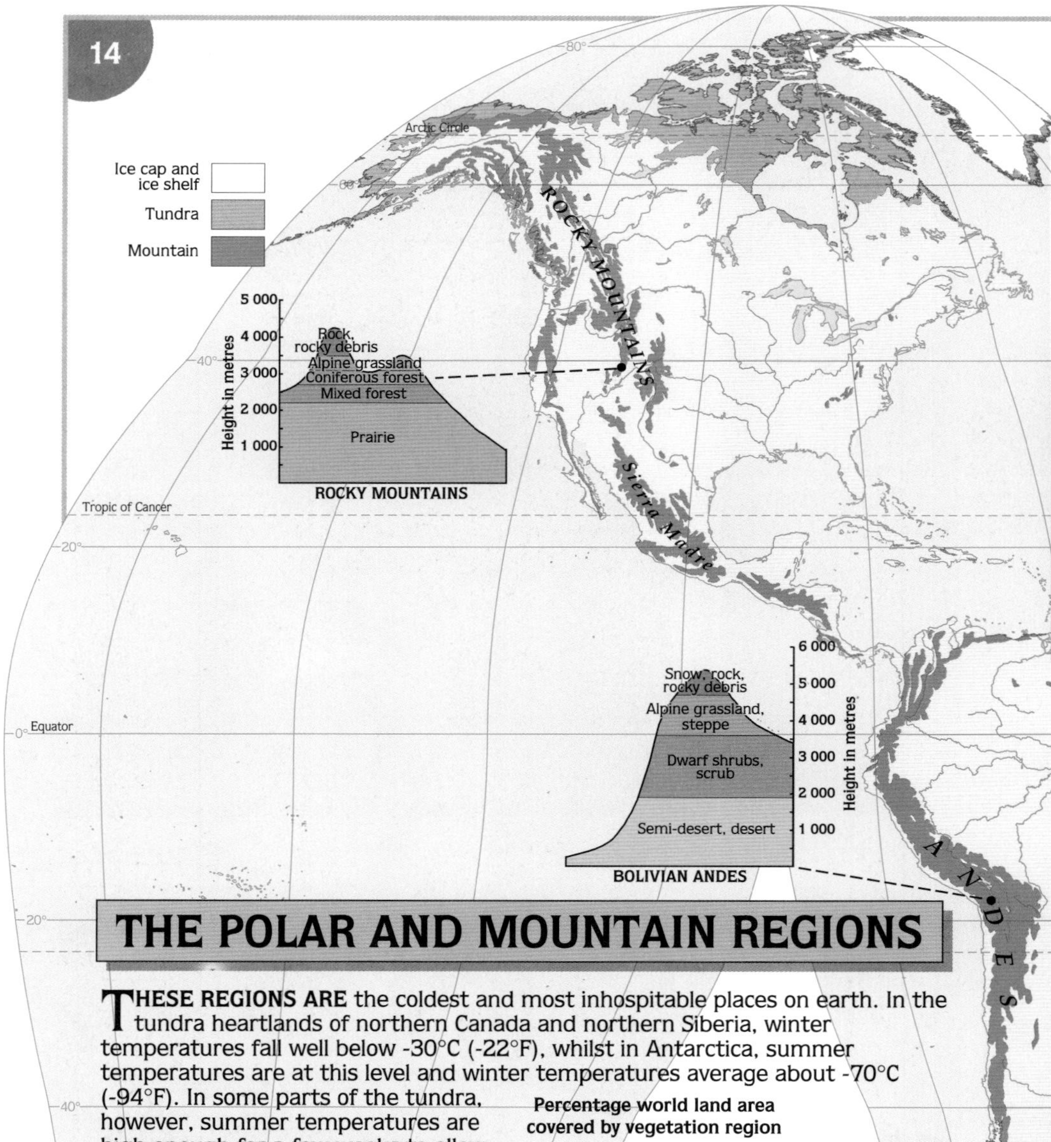

Ice cap and ice shelf
Tundra
Mountain
Arctic Circle
Tropic of Cancer
Equator
Antarctic Circle
ROCKY MOUNTAINS
Sierra Madre
ANDES
Atlas Mounta
5 000
4 000
3 000
2 000
1 000
Height in metres
Rock, rocky debris
Alpine grassland
Coniferous forest
Mixed forest
Prairie
ROCKY MOUNTAINS
6 000
Snow, rock, rocky debris
Alpine grassland, steppe
Dwarf shrubs, scrub
Semi-desert, desert
BOLIVIAN ANDES
Snow, ice, rock, rocky debris
Alpine grassland
Dwarf shrubs, tree heaths
Tropical montane forest, bamboo
Savanna
KILIMANJARO

TUNDRA

Sub-arctic areas which are usually frozen, but during the short summers the top layer of soil thaws, creating vast marshes. Vegetation is low growing to allow it to be insulated by snow in the winter. Plant species include mosses, lichens, rushes, grasses and flowering herbs; animals include the reindeer and arctic fox (see photograph).

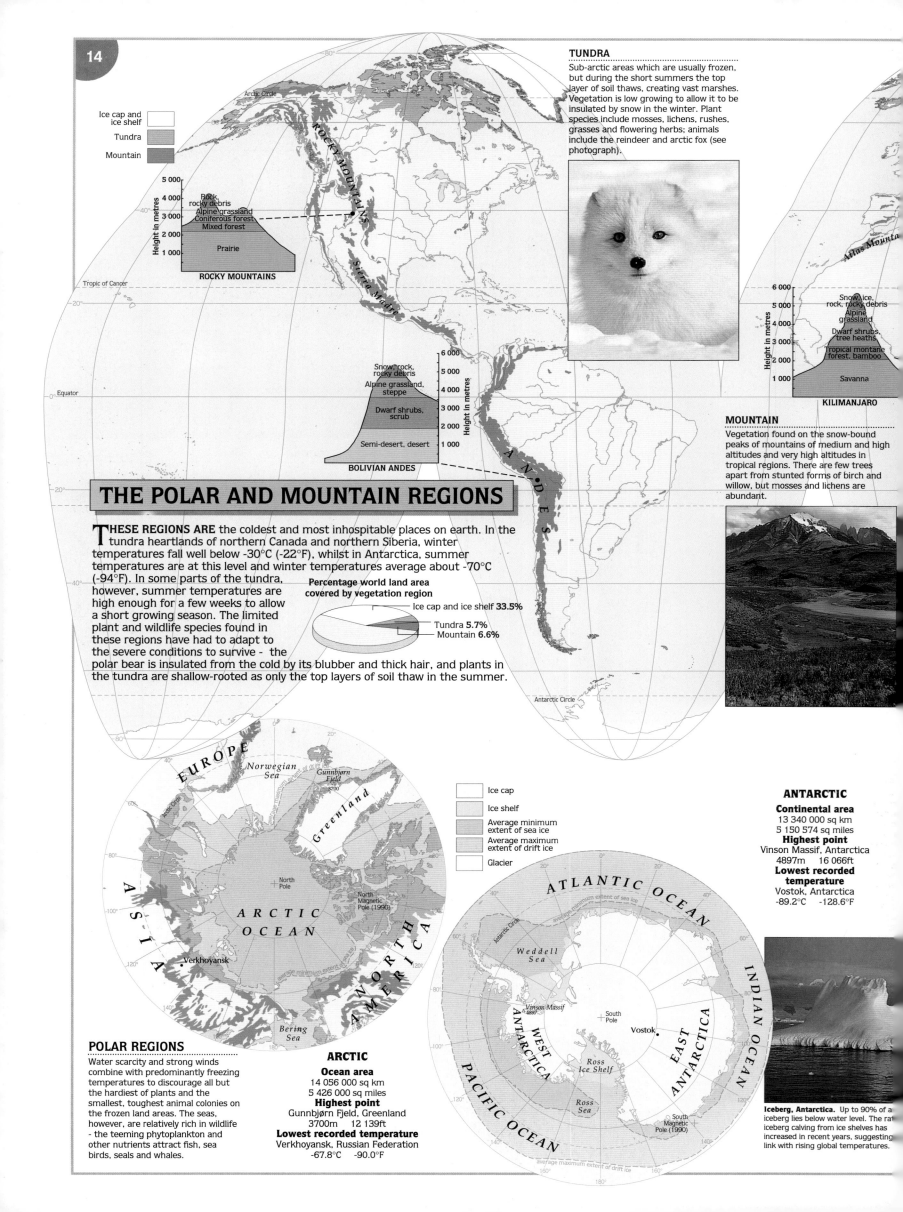

MOUNTAIN

Vegetation found on the snow-bound peaks of mountains of medium and high altitudes and very high altitudes in tropical regions. There are few trees apart from stunted forms of birch and willow, but mosses and lichens are abundant.

THE POLAR AND MOUNTAIN REGIONS

THESE REGIONS ARE the coldest and most inhospitable places on earth. In the tundra heartlands of northern Canada and northern Siberia, winter temperatures fall well below -30°C (-22°F), whilst in Antarctica, summer temperatures are at this level and winter temperatures average about -70°C (-94°F). In some parts of the tundra, however, summer temperatures are high enough for a few weeks to allow a short growing season. The limited plant and wildlife species found in these regions have had to adapt to the severe conditions to survive - the polar bear is insulated from the cold by its blubber and thick hair, and plants in the tundra are shallow-rooted as only the top layers of soil thaw in the summer.

Percentage world land area covered by vegetation region

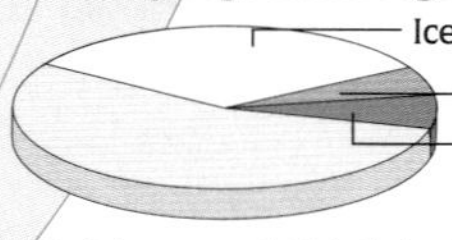

Ice cap and ice shelf 33.5%
Tundra 5.7%
Mountain 6.6%

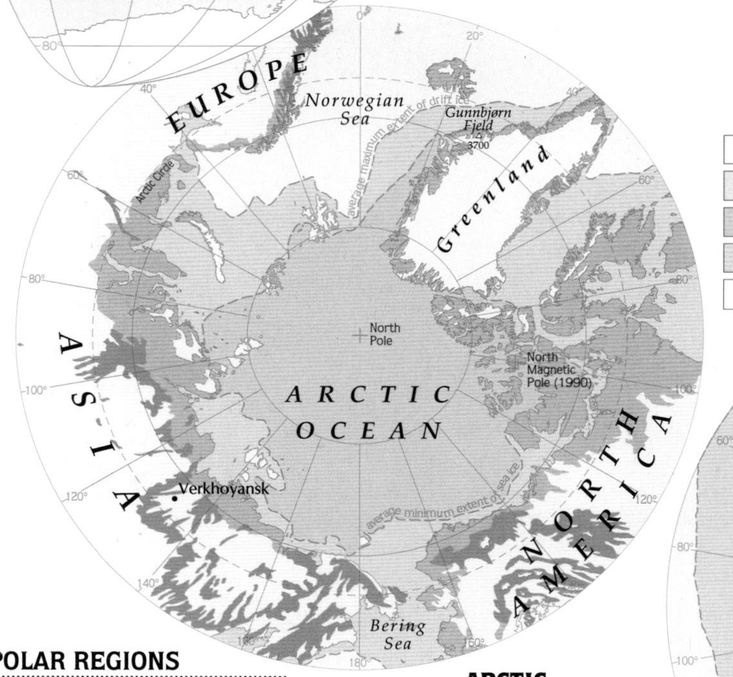

EUROPE
Norwegian Sea
Gunnbjørn Fjeld 3700
Greenland
Arctic Circle
average maximum extent of drift ice
North Pole
North Magnetic Pole (1990)
ASIA
ARCTIC OCEAN
NORTH AMERICA
Verkhoyansk
average minimum extent of sea ice
Bering Sea

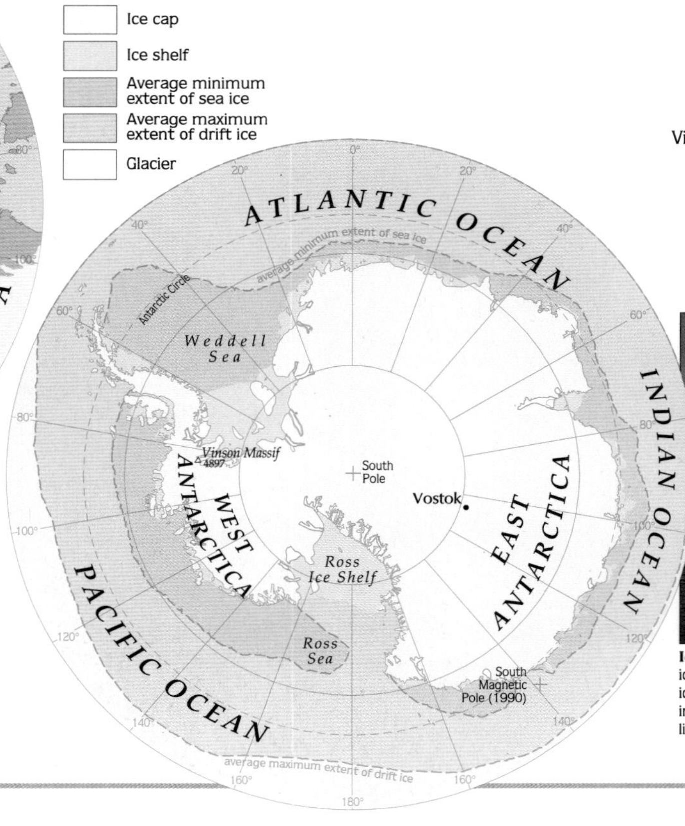

Ice cap
Ice shelf
Average minimum extent of sea ice
Average maximum extent of drift ice
Glacier
ATLANTIC OCEAN
average minimum extent of sea ice
Antarctic Circle
Weddell Sea
Vinson Massif 4897
South Pole
Vostok
WEST ANTARCTICA
EAST ANTARCTICA
INDIAN OCEAN
Ross Ice Shelf
Ross Sea
South Magnetic Pole (1990)
PACIFIC OCEAN
average maximum extent of drift ice

POLAR REGIONS

Water scarcity and strong winds combine with predominantly freezing temperatures to discourage all but the hardiest of plants and the smallest, toughest animal colonies on the frozen land areas. The seas, however, are relatively rich in wildlife - the teeming phytoplankton and other nutrients attract fish, sea birds, seals and whales.

ARCTIC

Ocean area
14 056 000 sq km
5 426 000 sq miles
Highest point
Gunnbjørn Fjeld, Greenland
3700m 12 139ft
Lowest recorded temperature
Verkhoyansk, Russian Federation
-67.8°C -90.0°F

ANTARCTIC

Continental area
13 340 000 sq km
5 150 574 sq miles
Highest point
Vinson Massif, Antarctica
4897m 16 066ft
Lowest recorded temperature
Vostok, Antarctica
-89.2°C -128.6°F

Iceberg, Antarctica. Up to 90% of a[n] iceberg lies below water level. The ra[te of] iceberg calving from ice shelves has increased in recent years, suggesting [a] link with rising global temperatures.

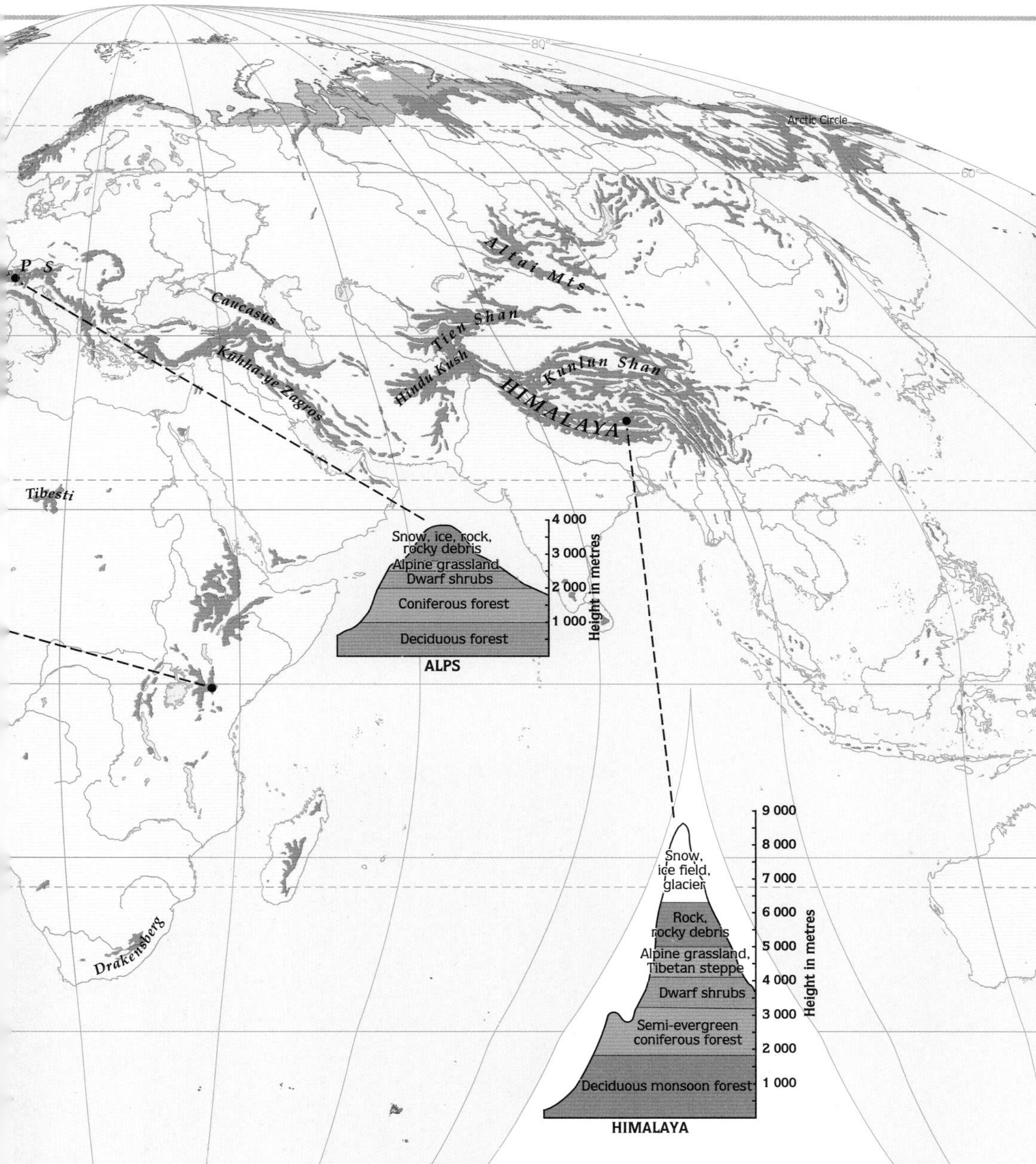
80°
Arctic Circle
60°
P S
Altai Mts
Caucasus
Tien Shan
Kuhha-ye Zagros
Hindu Kush
Kunlun Shan
HIMALAYA
Tibesti
Tropic of Cancer
20°
Equator
0°
20°
Tropic of Capricorn
Drakensberg
40°
Snow, ice, rock, rocky debris
Alpine grassland
Dwarf shrubs
Coniferous forest
Deciduous forest
4 000
3 000
2 000
1 000
Height in metres
ALPS
Snow, ice field, glacier
Rock, rocky debris
Alpine grassland, Tibetan steppe
Dwarf shrubs
Semi-evergreen coniferous forest
Deciduous monsoon forest
9 000
8 000
7 000
6 000
5 000
4 000
3 000
2 000
1 000
Height in metres
HIMALAYA

ICE CAP and ICE SHELF

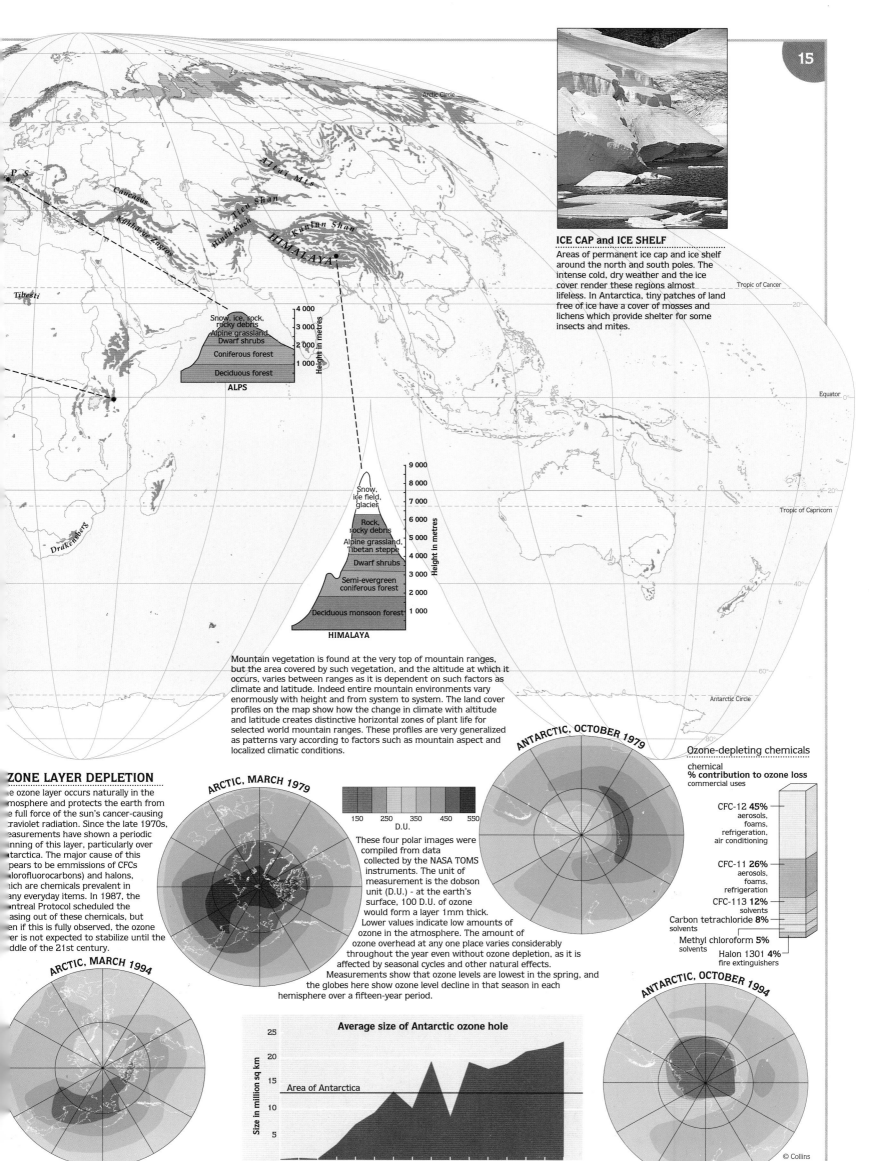

Areas of permanent ice cap and ice shelf around the north and south poles. The intense cold, dry weather and the ice cover render these regions almost lifeless. In Antarctica, tiny patches of land free of ice have a cover of mosses and lichens which provide shelter for some insects and mites.

Mountain vegetation is found at the very top of mountain ranges, but the area covered by such vegetation, and the altitude at which it occurs, varies between ranges as it is dependent on such factors as climate and latitude. Indeed entire mountain environments vary enormously with height and from system to system. The land cover profiles on the map show how the change in climate with altitude and latitude creates distinctive horizontal zones of plant life for selected world mountain ranges. These profiles are very generalized as patterns vary according to factors such as mountain aspect and localized climatic conditions.

60°
Antarctic Circle
80°

ZONE LAYER DEPLETION

e ozone layer occurs naturally in the mosphere and protects the earth from e full force of the sun's cancer-causing traviolet radiation. Since the late 1970s, easurements have shown a periodic inning of this layer, particularly over ntarctica. The major cause of this pears to be emmissions of CFCs lorofluorocarbons) and halons, ich are chemicals prevalent in any everyday items. In 1987, the ontreal Protocol scheduled the asing out of these chemicals, but en if this is fully observed, the ozone er is not expected to stabilize until the ddle of the 21st century.

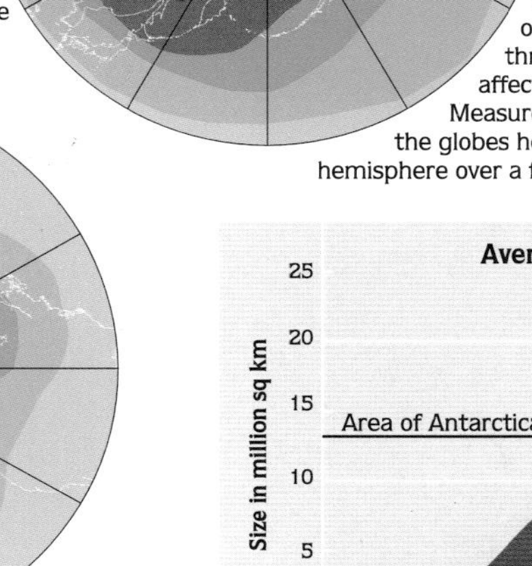
ARCTIC, MARCH 1979

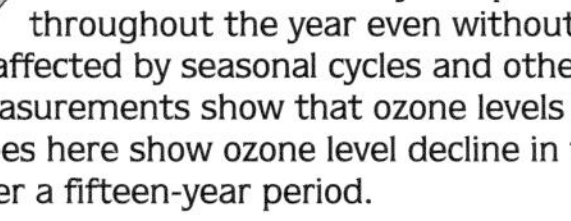
150
250
350
450
550
D.U.

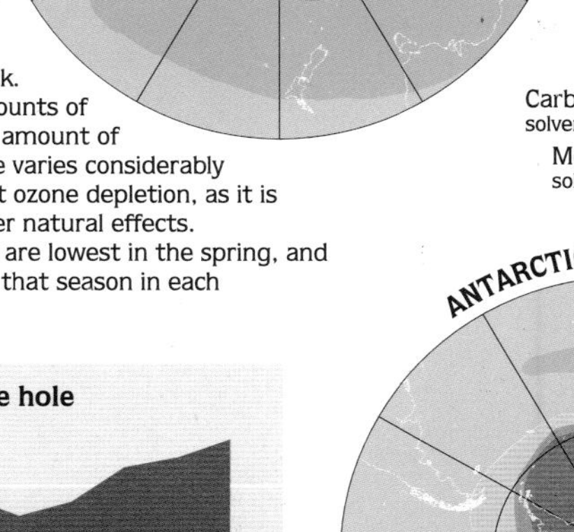
ANTARCTIC, OCTOBER 1979

These four polar images were compiled from data collected by the NASA TOMS instruments. The unit of measurement is the dobson unit (D.U.) - at the earth's surface, 100 D.U. of ozone would form a layer 1mm thick. Lower values indicate low amounts of ozone in the atmosphere. The amount of ozone overhead at any one place varies considerably throughout the year even without ozone depletion, as it is affected by seasonal cycles and other natural effects. Measurements show that ozone levels are lowest in the spring, and the globes here show ozone level decline in that season in each hemisphere over a fifteen-year period.

Ozone-depleting chemicals

chemical
% contribution to ozone loss
commercial uses
CFC-12 45%
aerosols, foams, refrigeration, air conditioning
CFC-11 26%
aerosols, foams, refrigeration
CFC-113 12%
solvents
Carbon tetrachloride 8%
solvents
Methyl chloroform 5%
solvents
Halon 1301 4%
fire extinguishers

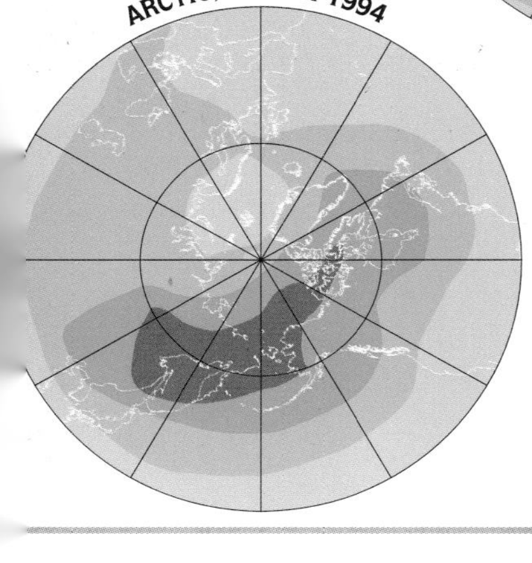
ARCTIC, MARCH 1994

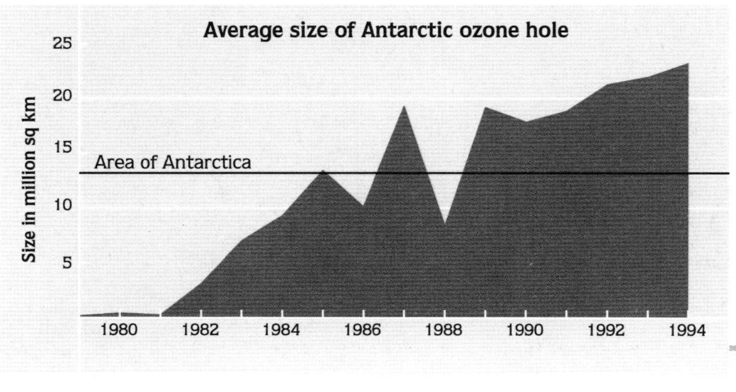
Average size of Antarctic ozone hole
Size in million sq km
25
20
15
10
5
Area of Antarctica
1980
1982
1984
1986
1988
1990
1992
1994

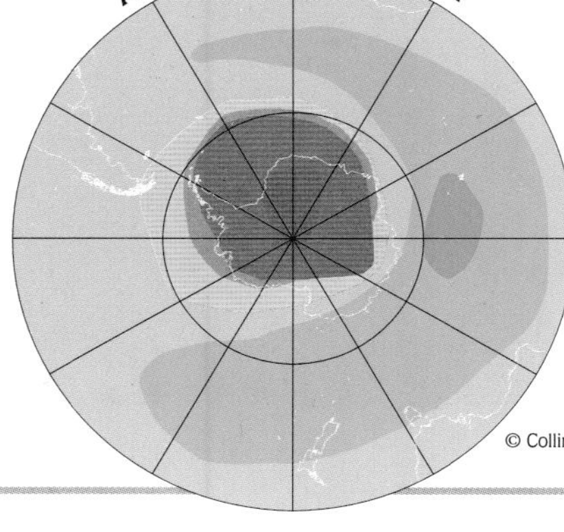
ANTARCTIC, OCTOBER 1994

IMPACT OF POPULATION ON THE NATURAL LANDSCAPE

Over the centuries much of the natural forest and woodland area has disappeared, exploited by humankind for fuel and materials, and cleared for cultivation, industry and settlement. The map shows how much of these naturally wooded areas now support the highest population densities and largest cities in North America, Europe and eastern Asia.

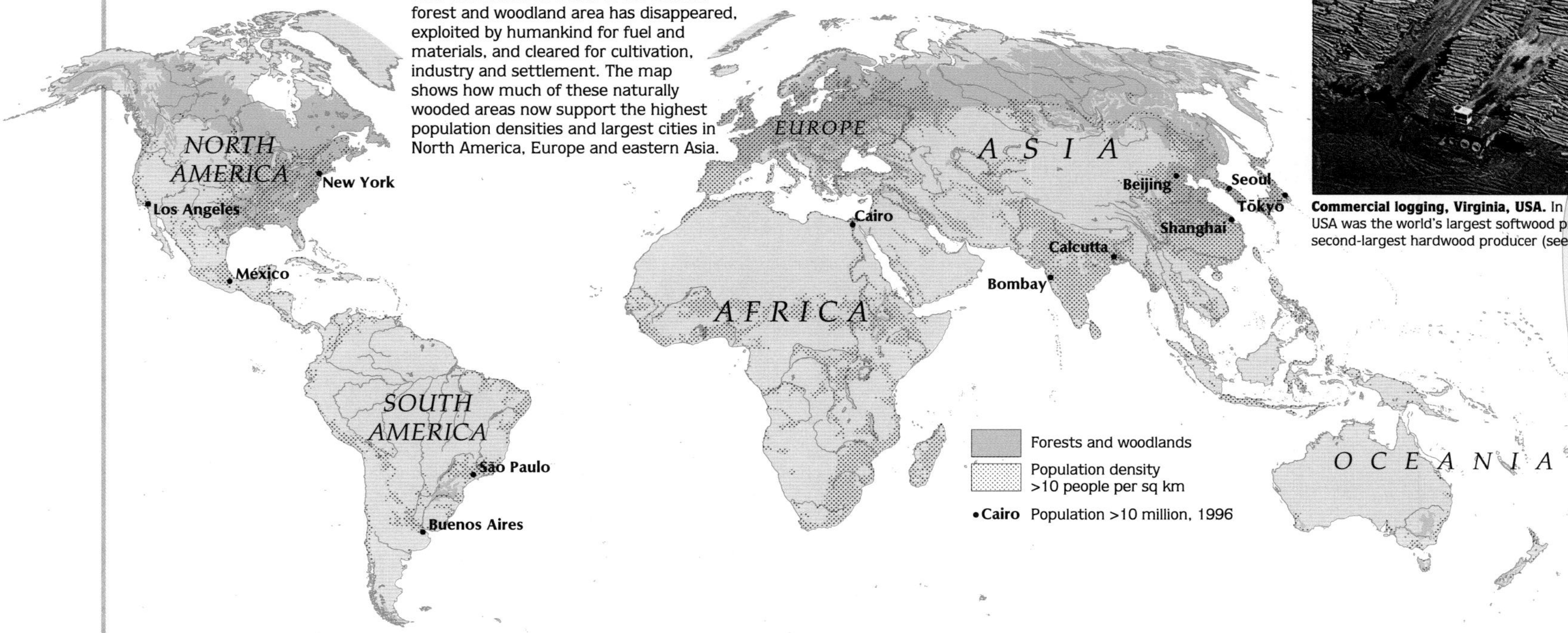

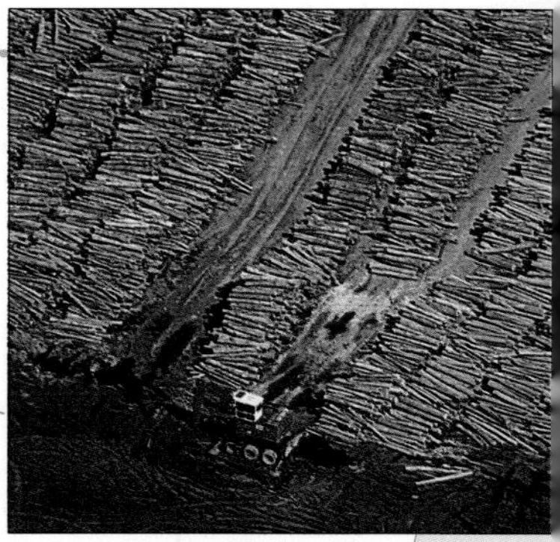

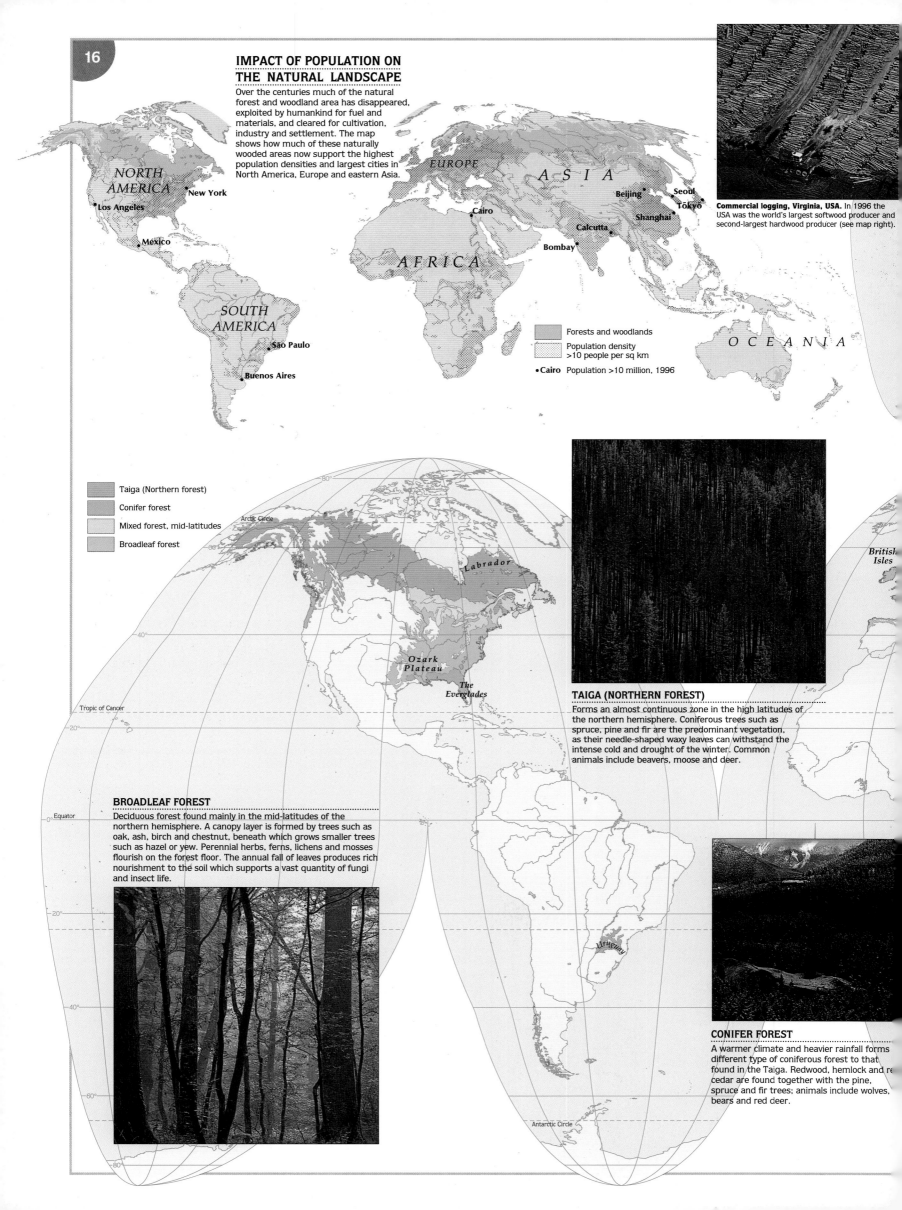

Commercial logging, Virginia, USA. In 1996 the USA was the world's largest softwood producer and second-largest hardwood producer (see map right).

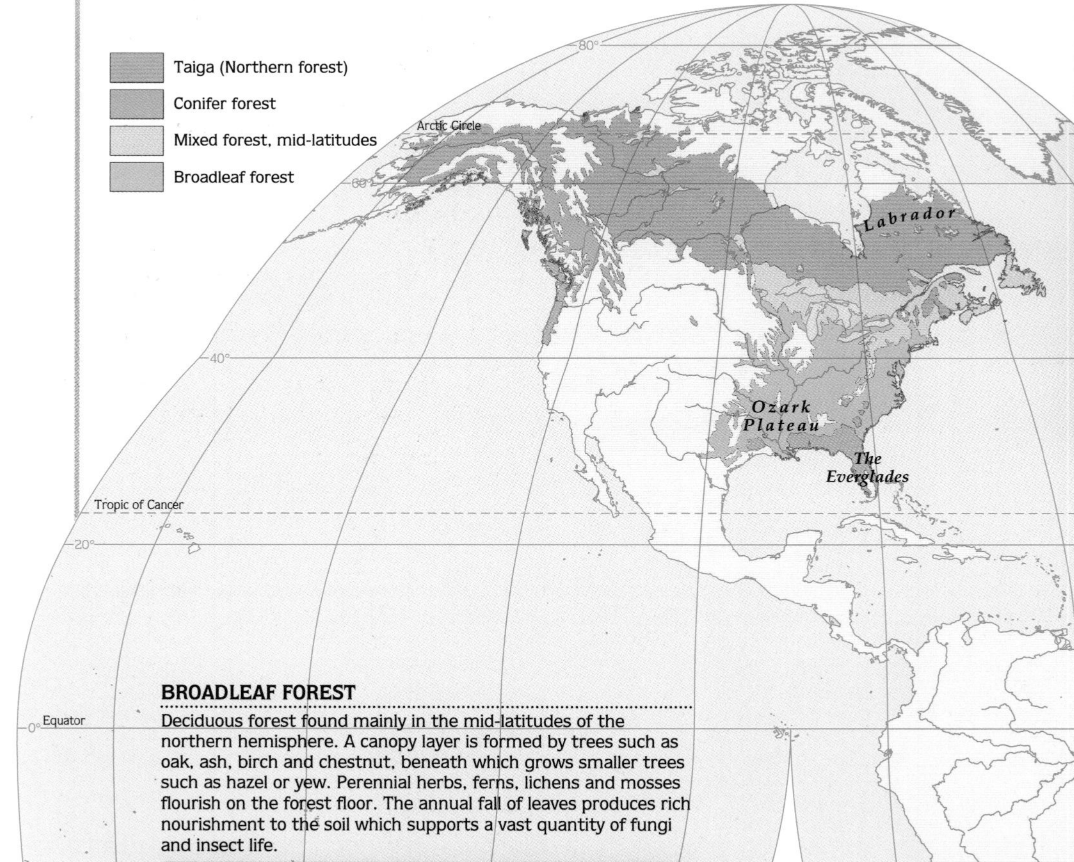

TAIGA (NORTHERN FOREST)

Forms an almost continuous zone in the high latitudes of the northern hemisphere. Coniferous trees such as spruce, pine and fir are the predominant vegetation, as their needle-shaped waxy leaves can withstand the intense cold and drought of the winter. Common animals include beavers, moose and deer.

BROADLEAF FOREST

Deciduous forest found mainly in the mid-latitudes of the northern hemisphere. A canopy layer is formed by trees such as oak, ash, birch and chestnut, beneath which grows smaller trees such as hazel or yew. Perennial herbs, ferns, lichens and mosses flourish on the forest floor. The annual fall of leaves produces rich nourishment to the soil which supports a vast quantity of fungi and insect life.

CONIFER FOREST

A warmer climate and heavier rainfall forms different type of coniferous forest to that found in the Taiga. Redwood, hemlock and re cedar are found together with the pine, spruce and fir trees; animals include wolves, bears and red deer.

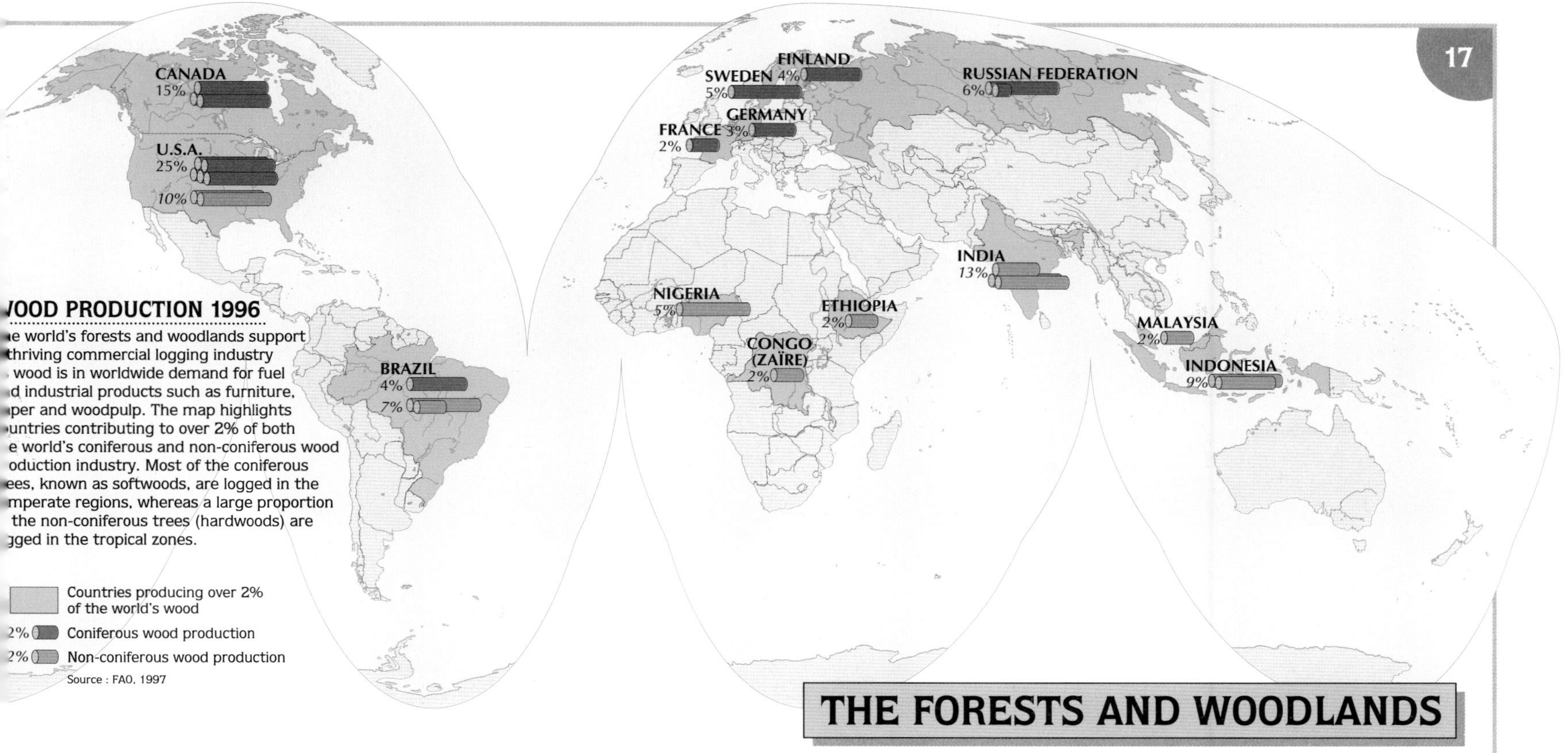

/OOD PRODUCTION 1996

ɪe world's forests and woodlands support thriving commercial logging industry wood is in worldwide demand for fuel d industrial products such as furniture, per and woodpulp. The map highlights untries contributing to over 2% of both e world's coniferous and non-coniferous wood oduction industry. Most of the coniferous ees, known as softwoods, are logged in the mperate regions, whereas a large proportion the non-coniferous trees (hardwoods) are gged in the tropical zones.

Countries producing over 2% of the world's wood

2% Coniferous wood production

2% Non-coniferous wood production

Source : FAO, 1997

THE FORESTS AND WOODLANDS

FORESTS IN THE mid-latitudes are described as temperate because there are no excesses in climate - rainfall is always adequate for tree growth (averaging 800mm/32 inches per annum), and is usually evenly spread throughout the year. Summers are warm (averaging 20°C/68°F) but winters can be cold, with temperatures dropping below freezing in some areas. The Taiga (Northern forest) is outside this zone as its northern location means that it has great seasonal variation in climate, with warm summers and extremely cold winters. Precipitation patterns also differ from those in temperate regions, being low, falling to less than 150mm (6 inches) per annum in the far north.

Human exploitation and clearance of the temperate forest zone has had such an impact on the landscape that today little of the natural cover remains.

Percentage world land area covered by vegetation region

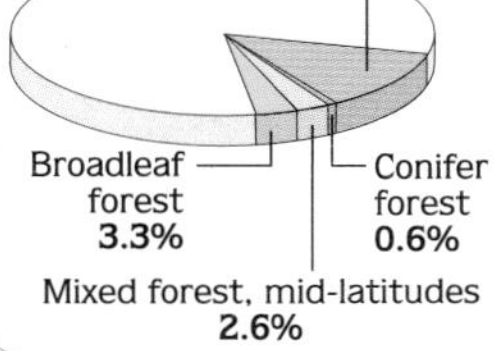

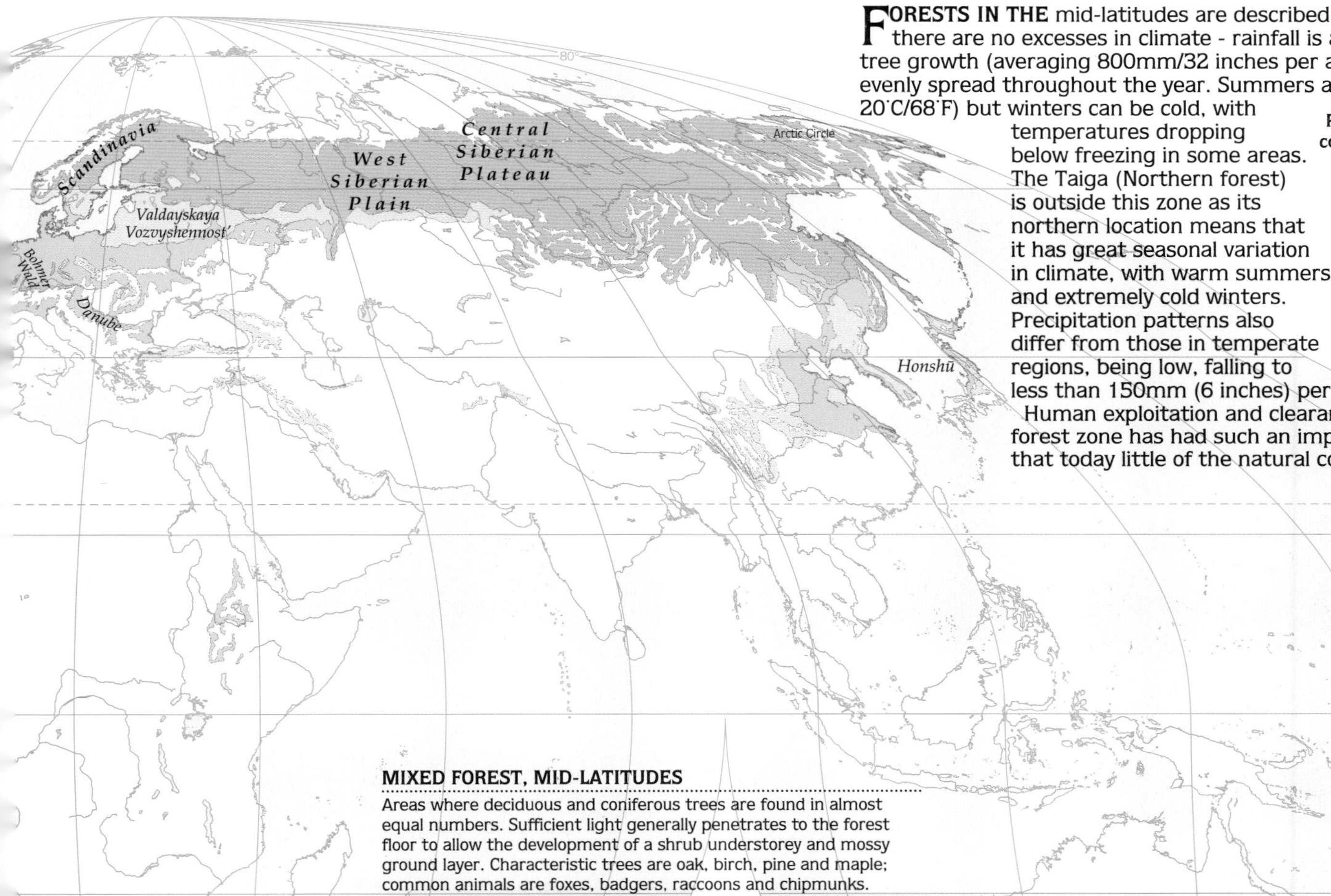

MIXED FOREST, MID-LATITUDES

Areas where deciduous and coniferous trees are found in almost equal numbers. Sufficient light generally penetrates to the forest floor to allow the development of a shrub understorey and mossy ground layer. Characteristic trees are oak, birch, pine and maple; common animals are foxes, badgers, raccoons and chipmunks.

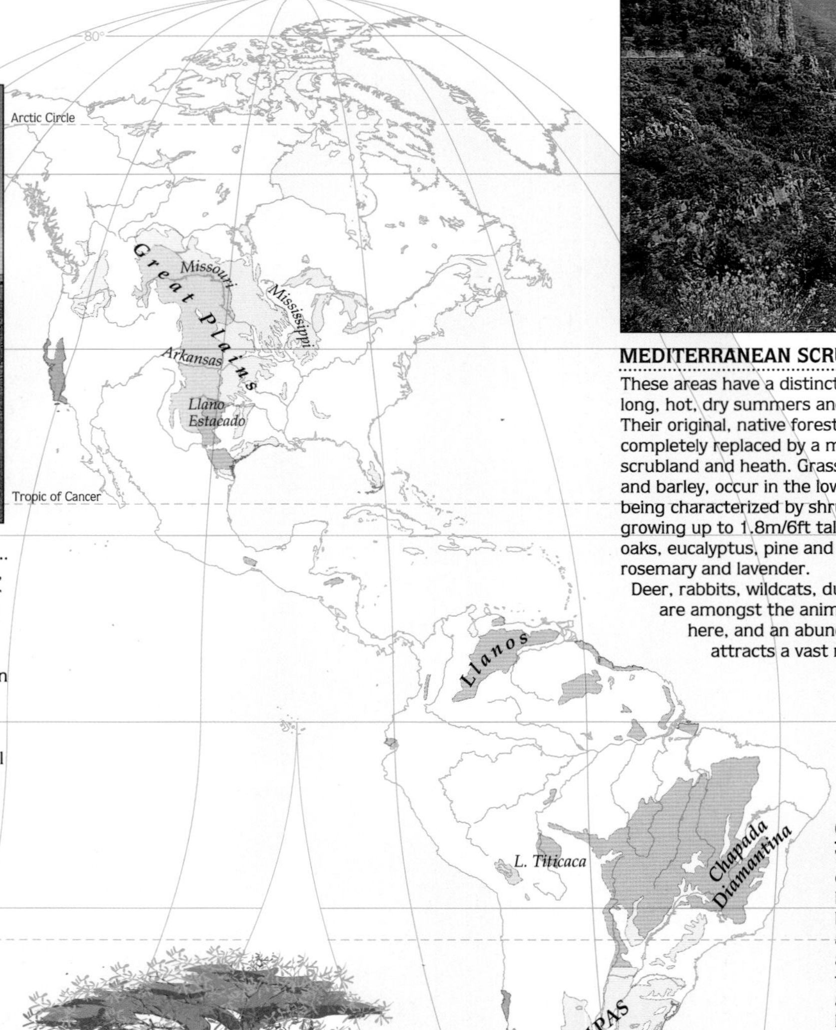

PRAIRIE

Similar in many ways to steppe, this temperate, mid-latitude grassland has a semi-moist, rather than semi-arid, climate with hot summers, cold winters and moderate rainfall. Prairie has continuous grass cover, mainly blue grama and buffalo grasses, up to 2.1m/7ft high - taller than the steppe.

Large areas are now cultivated for grain production with little of the natural vegetation remaining. Cultivation has put the native animal species, including bison, antelopes, prairie dogs and wolves, under pressure.

MEDITERRANEAN SCRUB

These areas have a distinct semi-arid climate of long, hot, dry summers and short, mild winters. Their original, native forest has been almost completely replaced by a mixture of open grassland, scrubland and heath. Grasses, mainly wild wheat and barley, occur in the lowlands, the hilly areas being characterized by shrubs and stunted trees growing up to 1.8m/6ft tall, including evergreen oaks, eucalyptus, pine and herbs such as rosemary and lavender.

Deer, rabbits, wildcats, duiker and kangaroos are amongst the animal species found here, and an abundance of insects attracts a vast range of birdlife.

Game parks

The savanna and steppe of Africa support an enormous range of wildlife which has always been in demand - both by game hunters and by tourists on safari. The variety of grasses and scattered trees support herds of grazing animals such as zebra, antelopes and buffalo; the larger herbivores giraffe and elephants; and their predators - lion, leopard, cheetah, hyena, jackals and hunting dogs. To protect and preserve the wildlife and the landscapes, reserves and game parks have been established across the continent.

Height and structure of grasslands

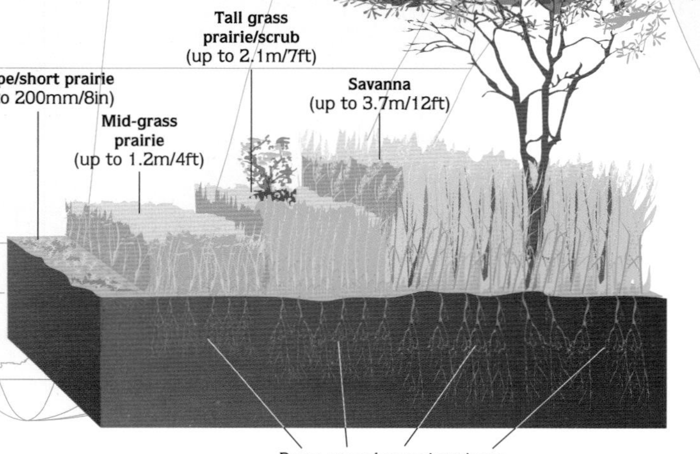

Deep, complex root systems assist quick recovery from grazing and fire.

THE GRASSLANDS

THE TWO MAIN sub-divisions within this group are determined largely by latitude: tropical grasslands - savanna and temperate grasslands - prairie and steppe. Mediterranean scrub is a distinct vegetation type with some elements in common with savanna.

The grasslands lie largely between forest belts and arid regions. Rainfall is too light for substantial tree growth, and yet is more regular, and therefore more productive, than in the deserts. Local patterns of rainfall determine the variations between the different types - from the treeless plains of the prairie and steppe, to the sparsely wooded savanna and stunted trees and shrubs of the mediterranean regions.

All the areas are widely grazed and are often affected by fire - both natural and man-made. Grasses, with their low growth points and deep roots, are more able to recover from these effects than other vegetation types.

Percentage world land area covered by vegetation region

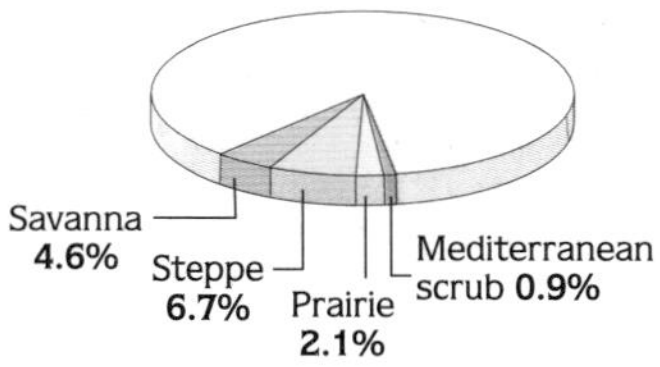

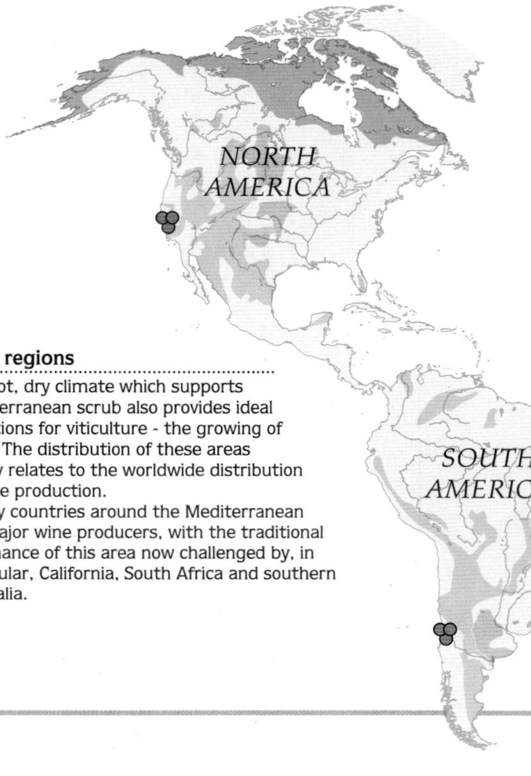

Wine regions

The hot, dry climate which supports mediterranean scrub also provides ideal conditions for viticulture - the growing of vines. The distribution of these areas closely relates to the worldwide distribution of wine production.

Many countries around the Mediterranean are major wine producers, with the traditional dominance of this area now challenged by, in particular, California, South Africa and southern Australia.

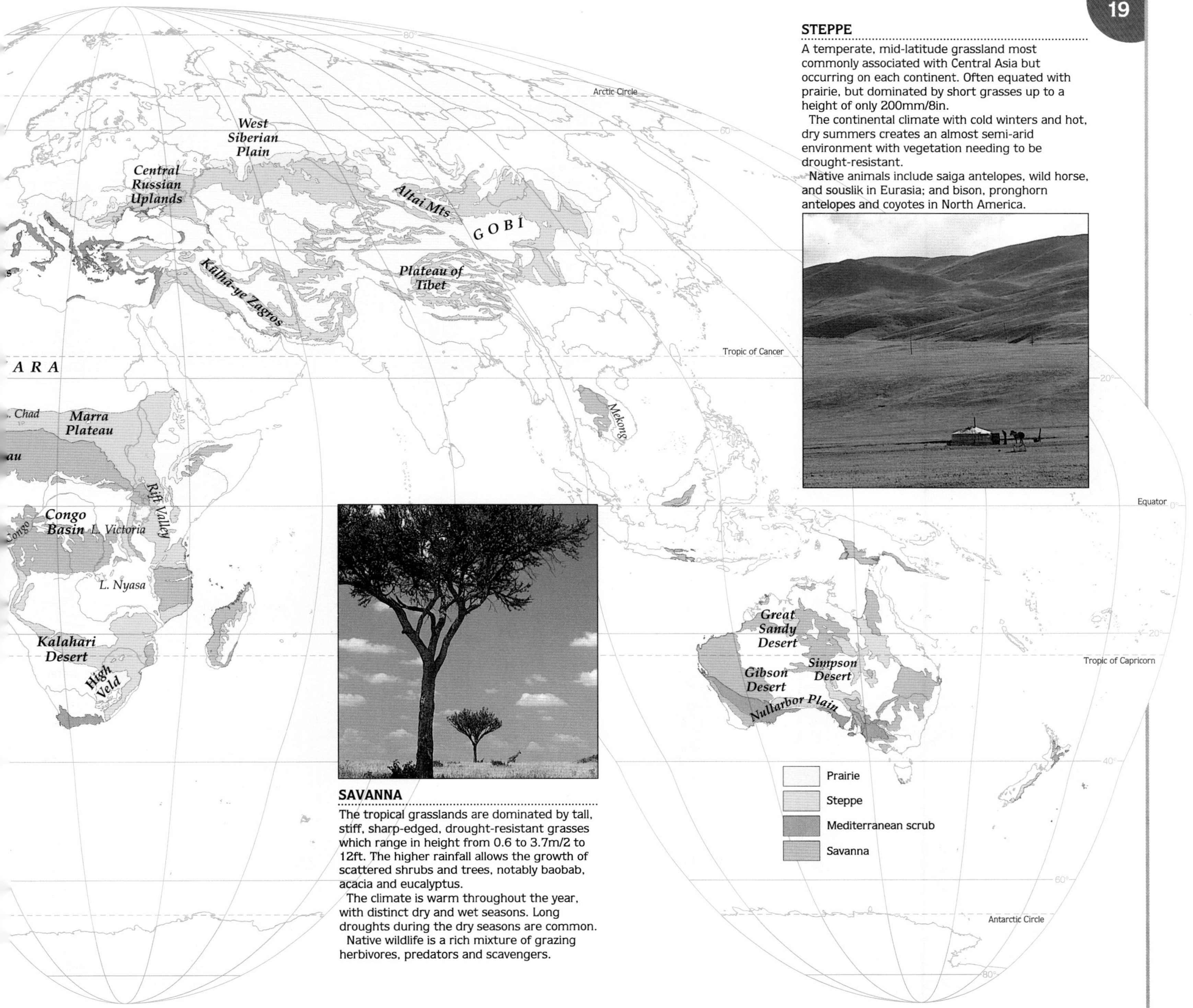

STEPPE

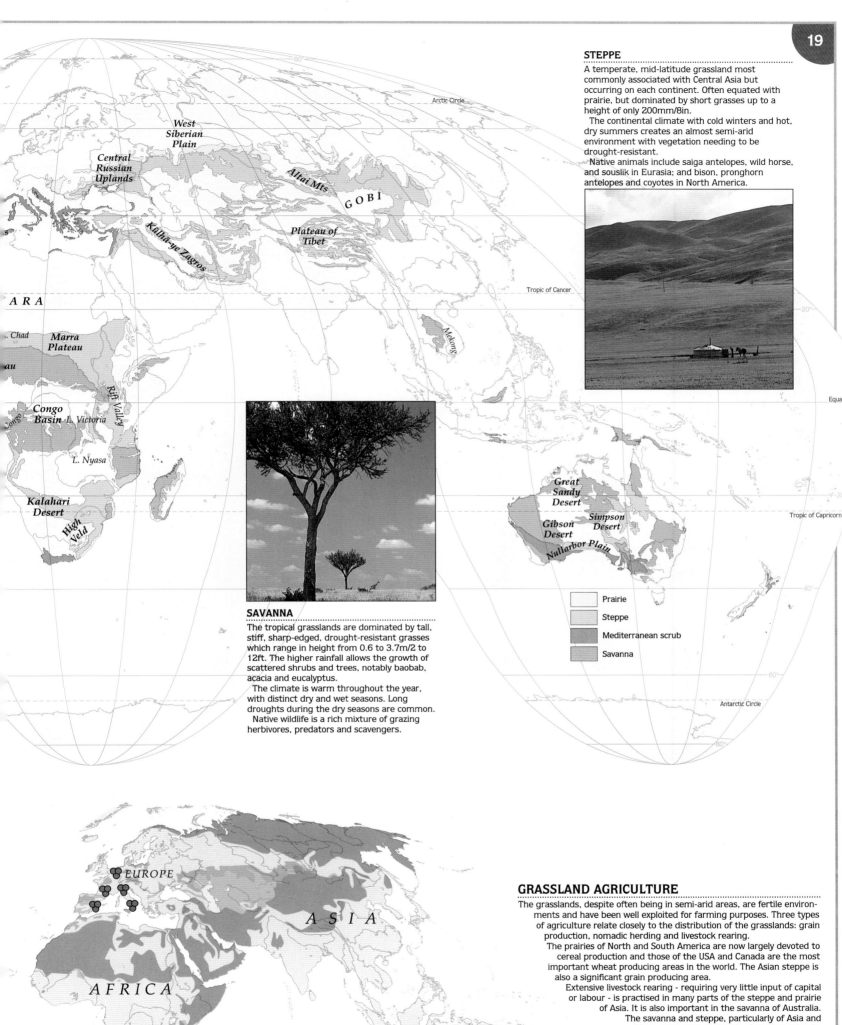

A temperate, mid-latitude grassland most commonly associated with Central Asia but occurring on each continent. Often equated with prairie, but dominated by short grasses up to a height of only 200mm/8in.

The continental climate with cold winters and hot, dry summers creates an almost semi-arid environment with vegetation needing to be drought-resistant.

Native animals include saiga antelopes, wild horse, and souslik in Eurasia; and bison, pronghorn antelopes and coyotes in North America.

SAVANNA

The tropical grasslands are dominated by tall, stiff, sharp-edged, drought-resistant grasses which range in height from 0.6 to 3.7m/2 to 12ft. The higher rainfall allows the growth of scattered shrubs and trees, notably baobab, acacia and eucalyptus.

The climate is warm throughout the year, with distinct dry and wet seasons. Long droughts during the dry seasons are common.

Native wildlife is a rich mixture of grazing herbivores, predators and scavengers.

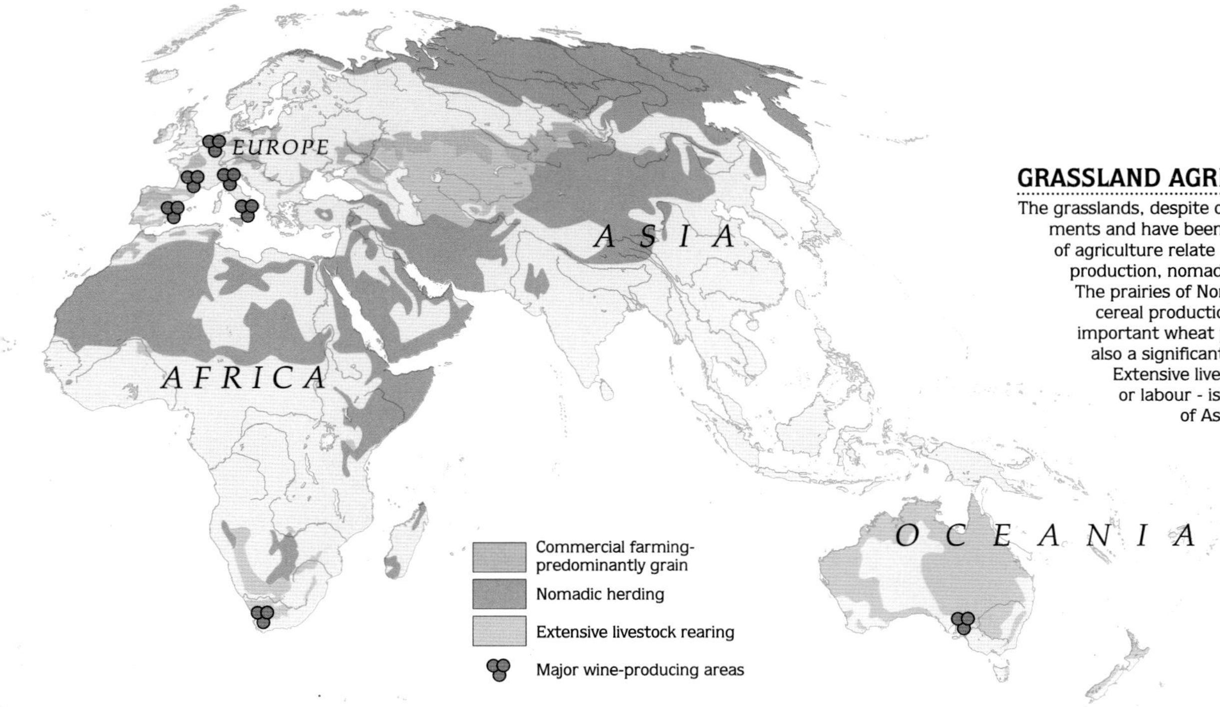

GRASSLAND AGRICULTURE

The grasslands, despite often being in semi-arid areas, are fertile environments and have been well exploited for farming purposes. Three types of agriculture relate closely to the distribution of the grasslands: grain production, nomadic herding and livestock rearing.

The prairies of North and South America are now largely devoted to cereal production and those of the USA and Canada are the most important wheat producing areas in the world. The Asian steppe is also a significant grain producing area.

Extensive livestock rearing - requiring very little input of capital or labour - is practised in many parts of the steppe and prairie of Asia. It is also important in the savanna of Australia.

The savanna and steppe, particularly of Asia and Africa, support large herds of grazing animals. Outside the deserts and polar regions, these are the most important areas of nomadic herding.

A notable exception to these patterns is that of the southern savanna belt of Africa. This area is infested by the tsetse fly - transmitter of sleeping sickness in humans and nagana in animals - and cannot support any livestock production.

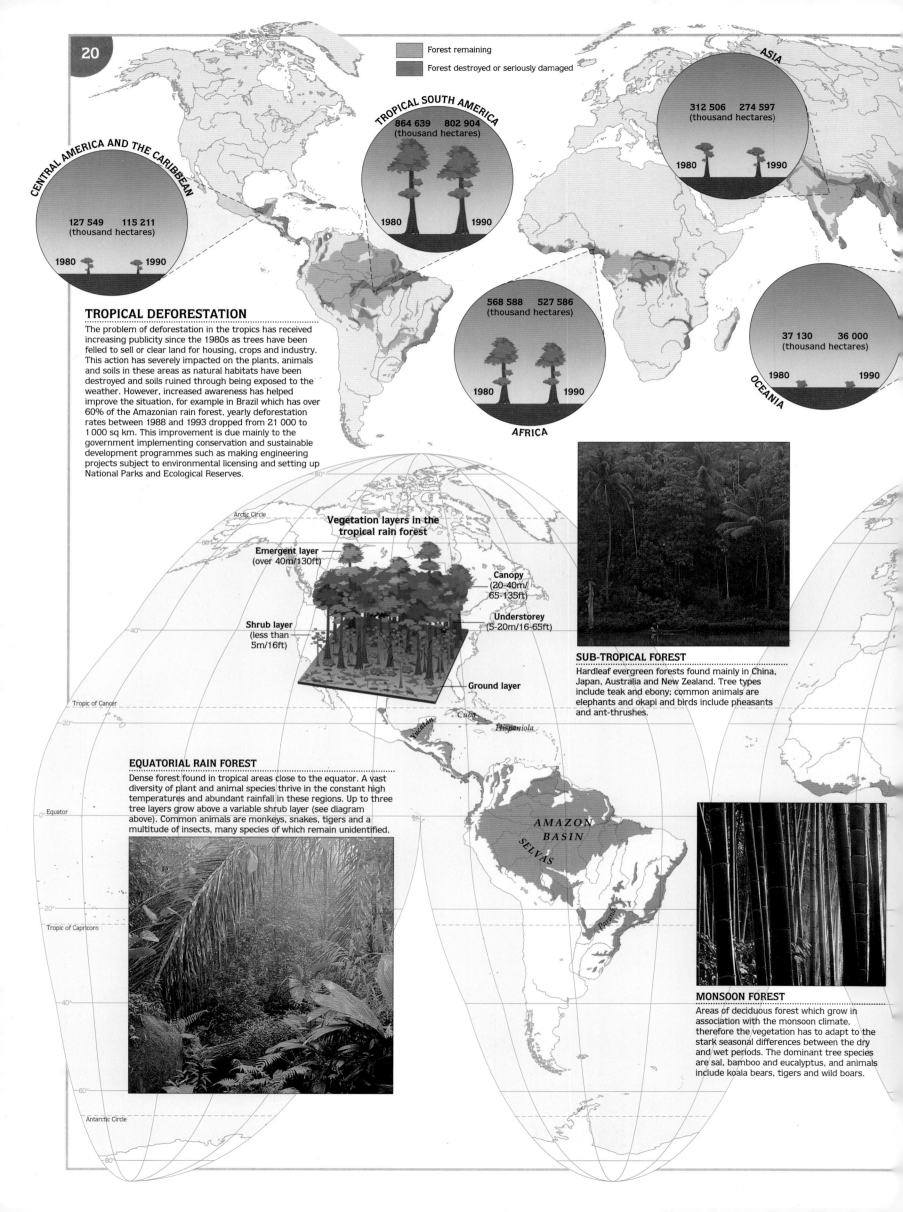

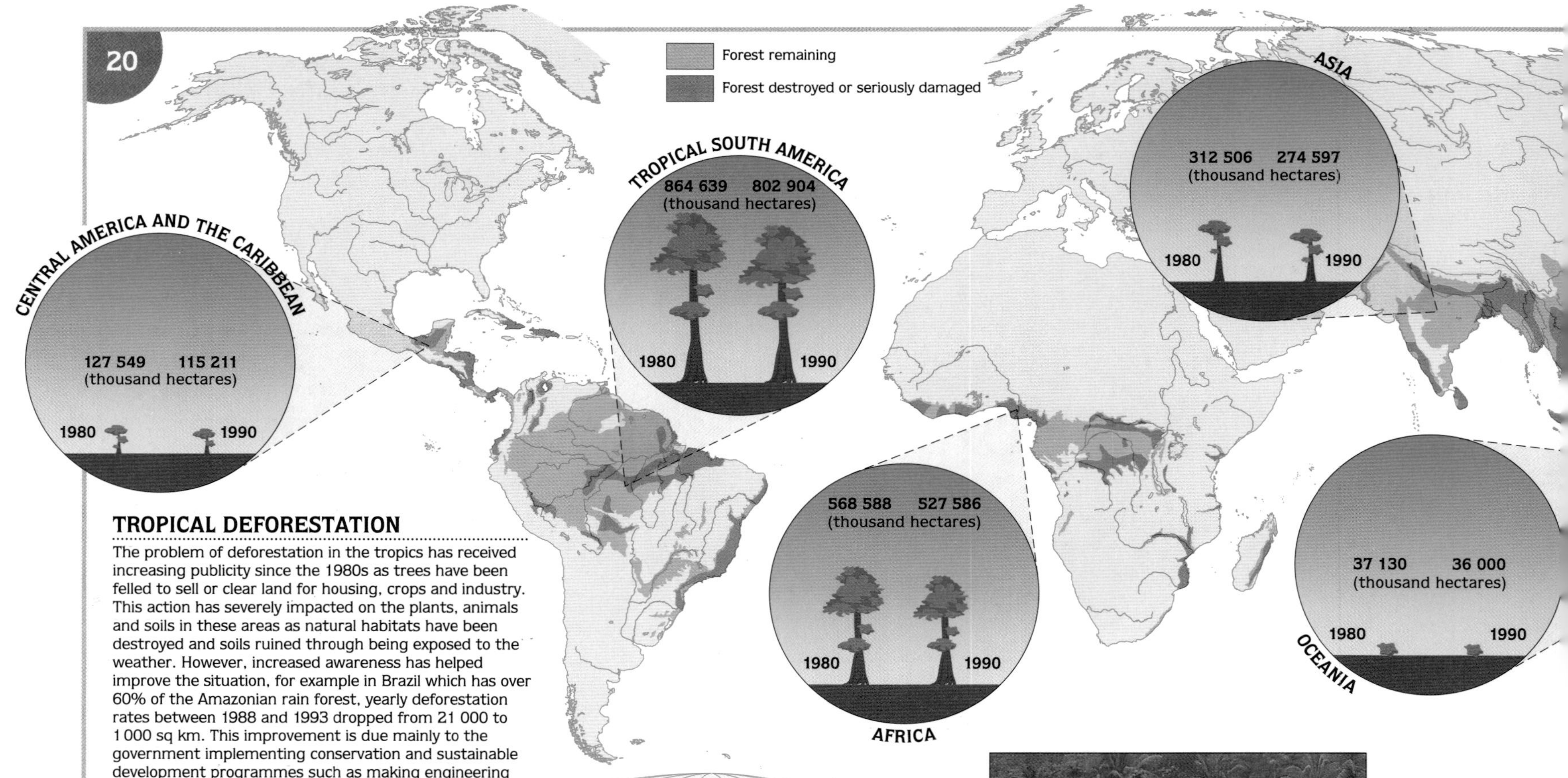

Forest remaining
Forest destroyed or seriously damaged
CENTRAL AMERICA AND THE CARIBBEAN
127 549 115 211
(thousand hectares)
1980 1990
TROPICAL SOUTH AMERICA
864 639 802 904
(thousand hectares)
1980 1990
ASIA
312 506 274 597
(thousand hectares)
1980 1990
AFRICA
568 588 527 586
(thousand hectares)
1980 1990
OCEANIA
37 130 36 000
(thousand hectares)
1980 1990

TROPICAL DEFORESTATION

The problem of deforestation in the tropics has received increasing publicity since the 1980s as trees have been felled to sell or clear land for housing, crops and industry. This action has severely impacted on the plants, animals and soils in these areas as natural habitats have been destroyed and soils ruined through being exposed to the weather. However, increased awareness has helped improve the situation, for example in Brazil which has over 60% of the Amazonian rain forest, yearly deforestation rates between 1988 and 1993 dropped from 21 000 to 1 000 sq km. This improvement is due mainly to the government implementing conservation and sustainable development programmes such as making engineering projects subject to environmental licensing and setting up National Parks and Ecological Reserves.

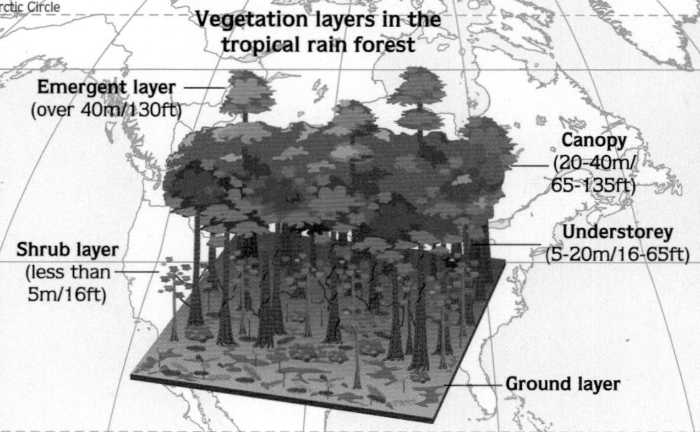

Vegetation layers in the tropical rain forest
Emergent layer (over 40m/130ft)
Canopy (20-40m/ 65-135ft)
Understorey (5-20m/16-65ft)
Shrub layer (less than 5m/16ft)
Ground layer

SUB-TROPICAL FOREST

Hardleaf evergreen forests found mainly in China, Japan, Australia and New Zealand. Tree types include teak and ebony; common animals are elephants and okapi and birds include pheasants and ant-thrushes.

EQUATORIAL RAIN FOREST

Dense forest found in tropical areas close to the equator. A vast diversity of plant and animal species thrive in the constant high temperatures and abundant rainfall in these regions. Up to three tree layers grow above a variable shrub layer (see diagram above). Common animals are monkeys, snakes, tigers and a multitude of insects, many species of which remain unidentified.

MONSOON FOREST

Areas of deciduous forest which grow in association with the monsoon climate, therefore the vegetation has to adapt to the stark seasonal differences between the dry and wet periods. The dominant tree species are sal, bamboo and eucalyptus, and animals include koala bears, tigers and wild boars.

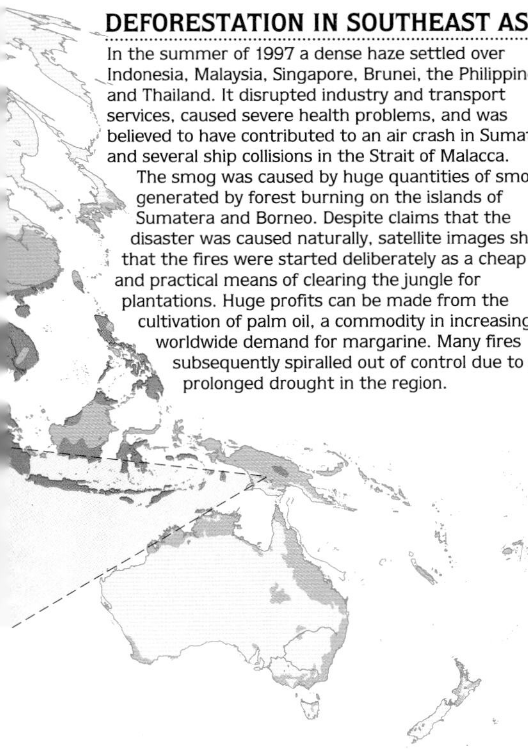

DEFORESTATION IN SOUTHEAST ASIA

In the summer of 1997 a dense haze settled over Indonesia, Malaysia, Singapore, Brunei, the Philippines and Thailand. It disrupted industry and transport services, caused severe health problems, and was believed to have contributed to an air crash in Sumatera and several ship collisions in the Strait of Malacca.

The smog was caused by huge quantities of smoke generated by forest burning on the islands of Sumatera and Borneo. Despite claims that the disaster was caused naturally, satellite images show that the fires were started deliberately as a cheap and practical means of clearing the jungle for plantations. Huge profits can be made from the cultivation of palm oil, a commodity in increasing worldwide demand for margarine. Many fires subsequently spiralled out of control due to a prolonged drought in the region.

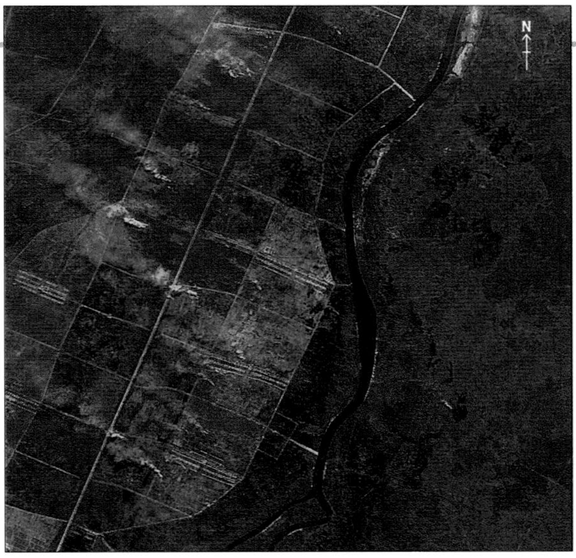
N

Kalimantan, Indonesia, 29th July 1997. Red areas show lush vegetation, green/brown is bare ground, black is water and white is smoke on these infrared SPOT satellite images. To the west of the main river, irrigation canals mark out the fields of a plantation. The many smoke plumes show that further areas of jungle are being burned to clear land for the expanding plantation, and these contributed to the formation of a vast smog cloud over Southeast Asia.

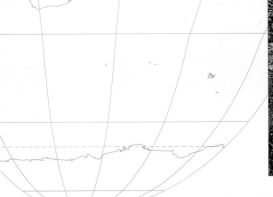

Kalimantan, Indonesia, 8th September 1997. After less than six weeks this SPOT image shows that further forest has been destroyed for cultivation in the region. The field areas are now almost totally cleared, and are expanding to the west and east. Large areas of jungle to the east of the main river have also been removed and the smoke plumes indicate that this action is continuing.

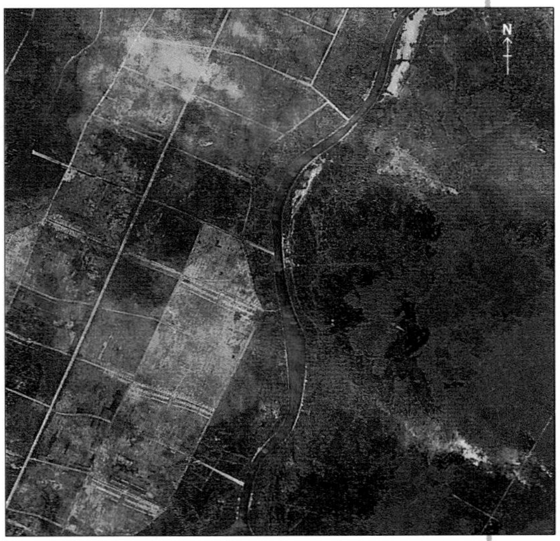
N

THE TROPICAL FORESTS

IN THE TROPICAL forest regions temperatures are consistently high (above 20°C/68°F) but precipitation patterns vary - in the equatorial forests rainfall is abundant (over 1500mm/59 inches) throughout the year, but it becomes more seasonal away from the equator and decreases to periods of drought (about 255mm/10 inches) in the dry tropical forests.

The tropical forests are the most diverse of all vegetation groups - up to two hundred tree and plant species per hectare have been recorded in some areas of rain forest, compared with around fifteen per hectare in forests at higher latitudes. This variety provides a consistent abundant food supply which supports a diversity of animals, birds and insects.

Percentage world land area covered by vegetation region

Sub tropical forest 0.8%
Dry tropical forest 1.3%
Monsoon forest 1.1%
Equatorial rain forest 5.0%

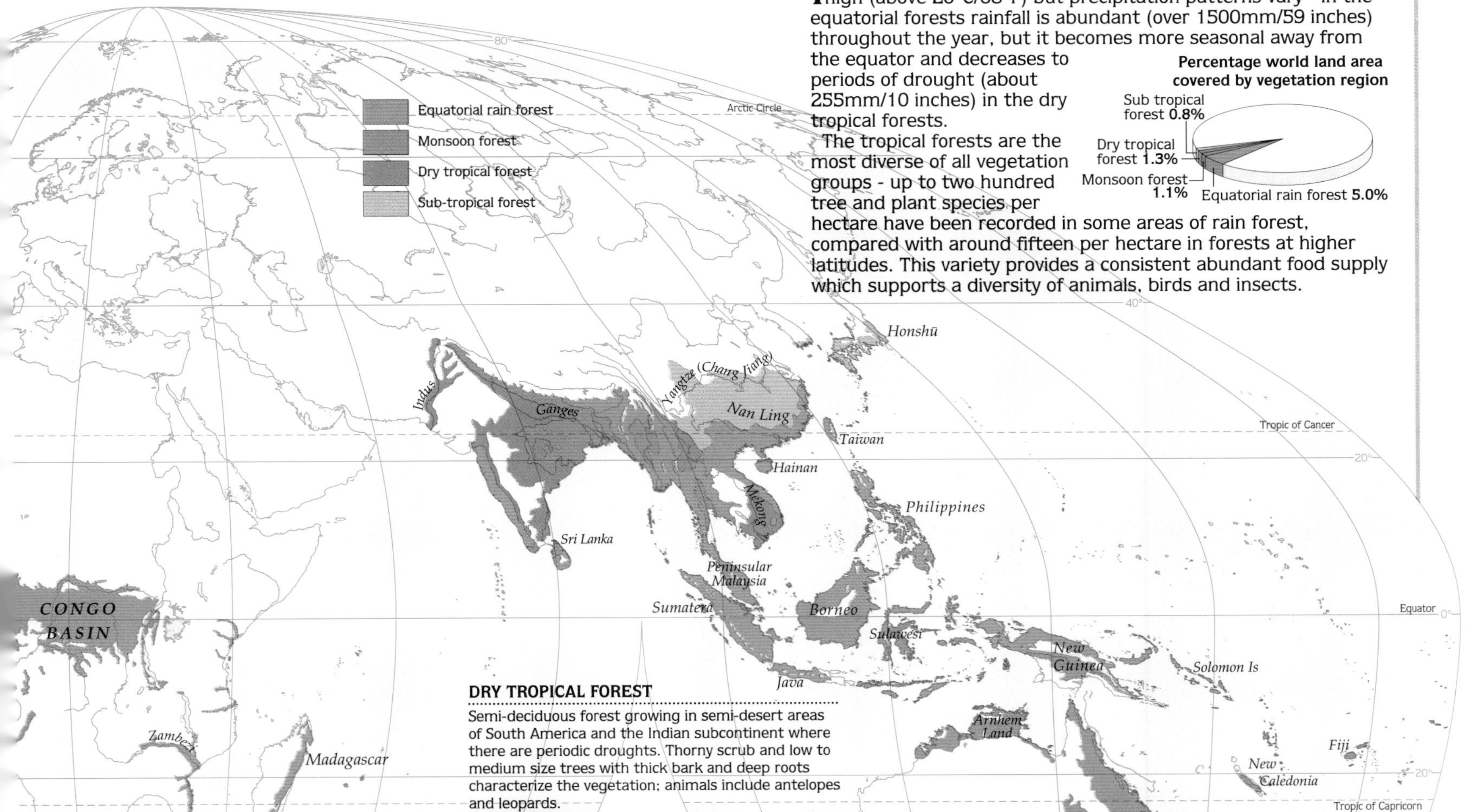
Equatorial rain forest
Monsoon forest
Dry tropical forest
Sub-tropical forest
Arctic Circle
80°
40°
20°
0°
60°
Tropic of Cancer
Equator
Tropic of Capricorn
Antarctic Circle
Honshū
Indus
Ganges
Yangtze (Chang Jiang)
Nan Ling
Taiwan
Hainan
Mekong
Philippines
Sri Lanka
Peninsular Malaysia
CONGO BASIN
Sumatera
Borneo
Sulawesi
New Guinea
Solomon Is
Java
Zambezi
Madagascar
Arnhem Land
Fiji
New Caledonia
New Zealand
Tasmania

DRY TROPICAL FOREST

Semi-deciduous forest growing in semi-desert areas of South America and the Indian subcontinent where there are periodic droughts. Thorny scrub and low to medium size trees with thick bark and deep roots characterize the vegetation; animals include antelopes and leopards.

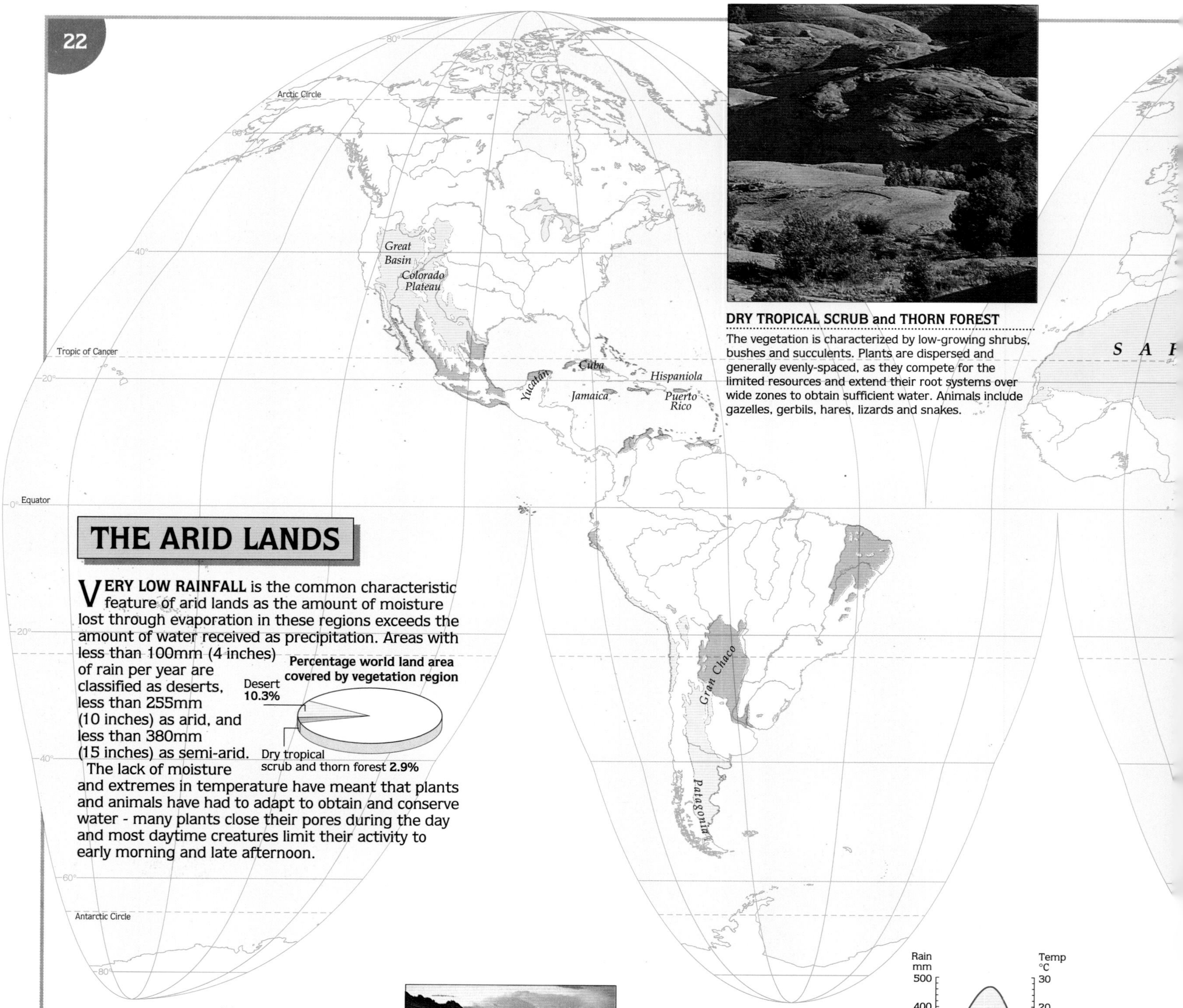

DRY TROPICAL SCRUB and THORN FOREST

The vegetation is characterized by low-growing shrubs, bushes and succulents. Plants are dispersed and generally evenly-spaced, as they compete for the limited resources and extend their root systems over wide zones to obtain sufficient water. Animals include gazelles, gerbils, hares, lizards and snakes.

THE ARID LANDS

VERY LOW RAINFALL is the common characteristic feature of arid lands as the amount of moisture lost through evaporation in these regions exceeds the amount of water received as precipitation. Areas with less than 100mm (4 inches) of rain per year are classified as deserts, less than 255mm (10 inches) as arid, and less than 380mm (15 inches) as semi-arid.

The lack of moisture and extremes in temperature have meant that plants and animals have had to adapt to obtain and conserve water - many plants close their pores during the day and most daytime creatures limit their activity to early morning and late afternoon.

The photograph above illustrates the sculpted canyon surface of the Colorado Plateau, USA, whilst that below left shows the gravel surface in the northern Namib Desert, Namibia.

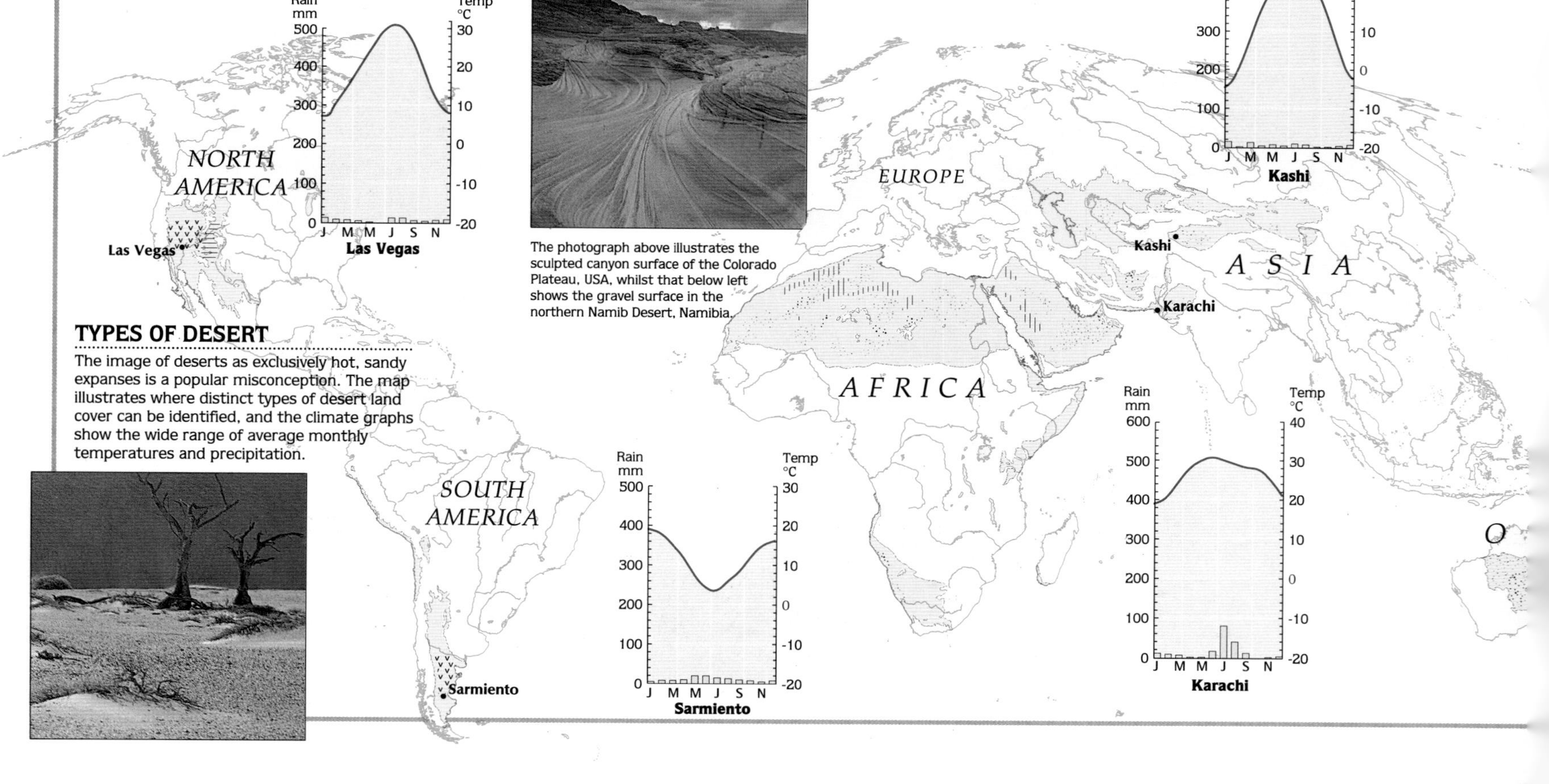

TYPES OF DESERT

The image of deserts as exclusively hot, sandy expanses is a popular misconception. The map illustrates where distinct types of desert land cover can be identified, and the climate graphs show the wide range of average monthly temperatures and precipitation.

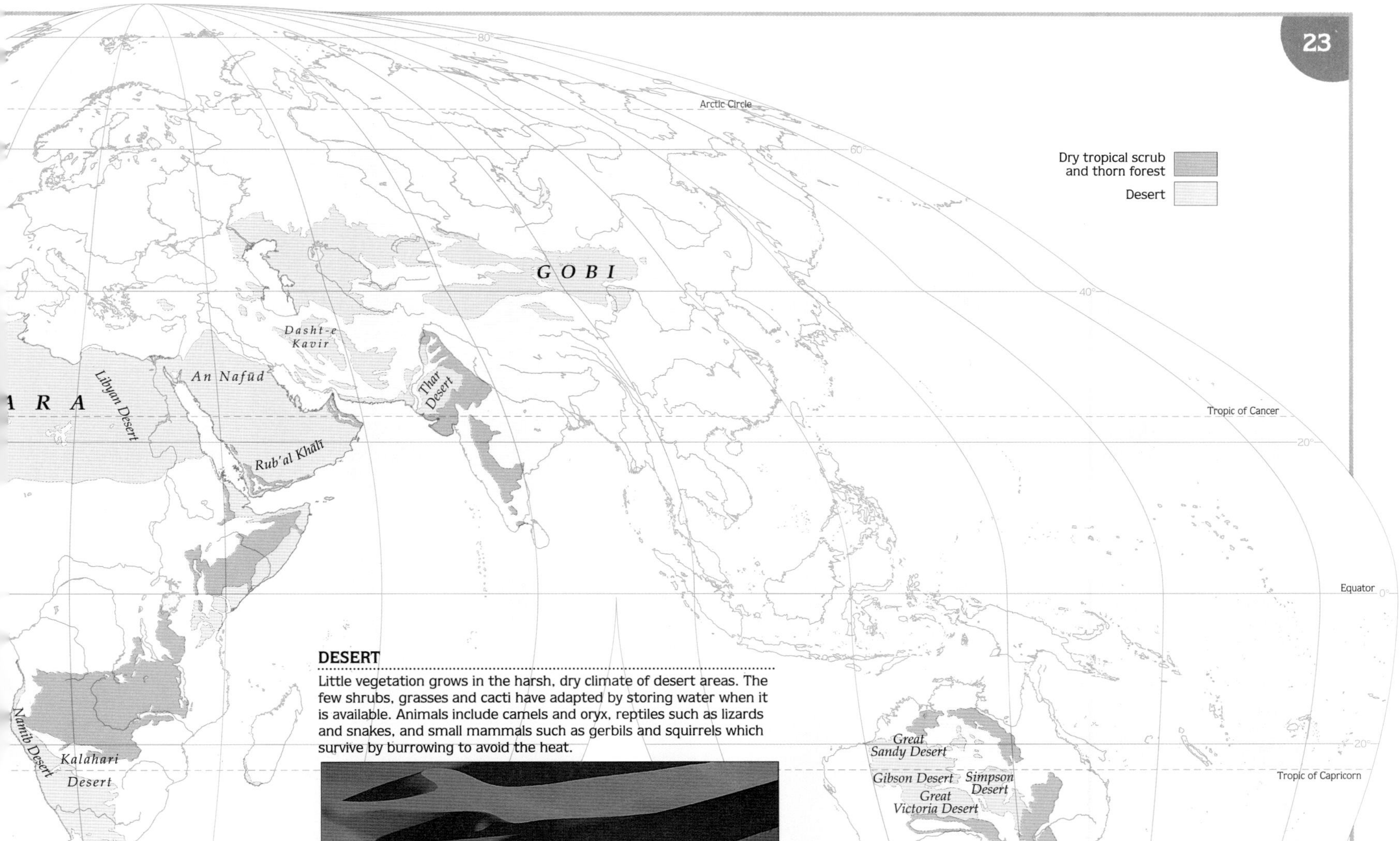
Arctic Circle
80°
60°
40°
Dry tropical scrub and thorn forest
Desert
GOBI
Dasht-e Kavir
An Nafūd
ARA
Libyan Desert
Thar Desert
Rub'al Khali
Tropic of Cancer
20°
Equator
0°
Namib Desert
Kalahari Desert
Great Sandy Desert
Gibson Desert
Simpson Desert
Great Victoria Desert
Tropic of Capricorn
20°
40°
60°
Antarctic Circle
80°

DESERT

Little vegetation grows in the harsh, dry climate of desert areas. The few shrubs, grasses and cacti have adapted by storing water when it is available. Animals include camels and oryx, reptiles such as lizards and snakes, and small mammals such as gerbils and squirrels which survive by burrowing to avoid the heat.

- Sandy (Erg)
- Gravel (Reg)
- Rocky (Hamada)
- Canyons
- Sparse Vegetation

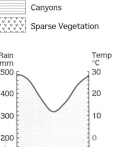

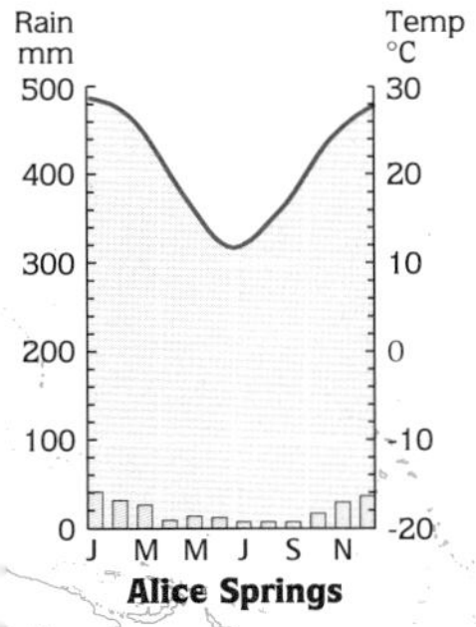
Rain mm
Temp °C
500
400
300
200
100
0
30
20
10
0
-10
-20
J M M J S N
Alice Springs

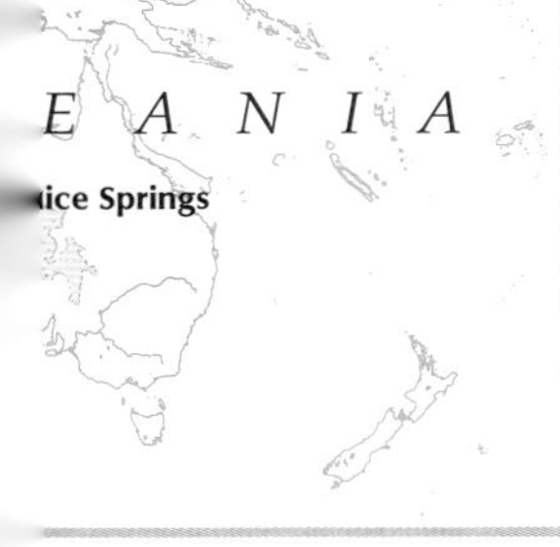
EANIA
ice Springs

Centre-pivot irrigation, Colorado, USA. Each circular field is irrigated by water sprinklers set upon a long metal frame which revolves around the central point of the field.

IRRIGATION

Irrigation is the artificial application of water to land to promote crop growth. Its development has greatly increased world food production. Today approximately 15% of all world land under cultivation is irrigated and, because the volume and timing of water supply can be firmly controlled, yields are often over twice that from non-irrigated fields.

There are four main types of irrigation currently in use worldwide: the first, flood irrigation is used for crops such as rice where fields are level and there is plenty of water. Second, furrow irrigation, utilised on row crops such as cotton, uses parallel furrows to water fields which are too irregular to flood. Third, drip irrigation uses narrow plastic tubes to provide small amounts of moisture to the roots of each plant when water is needed. Finally, sprinkler irrigation uses sprinklers to spray water in a continuous circle. Centre-pivot irrigation uses long lines of sprinklers which move around circular fields - the satellite image and photograph illustrate the impact of this on the landscape.

The success of irrigation techniques depends on a stable water source and the ability to store and distribute water efficiently. The minimal rainfall in deserts means that water has to be obtained from other sources, such as from below ground using wells, or from rivers which originate outside the arid zone (e.g. Nile, Indus). Water from rivers outside the arid zone can be transferred to drier areas by a system of dams, aqueducts and canals.

The main problem with irrigation is the salt content of the water, which collects in the upper layers of soil in poorly drained areas and affects plant growth. This means that more advanced techniques will have to be employed to maintain the efficiency of currently irrigated land, and deal with the challenge of extending irrigated areas.

Desert irrigation, Saudi Arabia. This SPOT-1 satellite image shows hundreds of irrigated circular fields, each watered by a structure of radius 300 m (984 ft). Vegetation appears brown and red - in the northeast a town can be identified, and in the southeast there is a barren, rocky ridge.

N

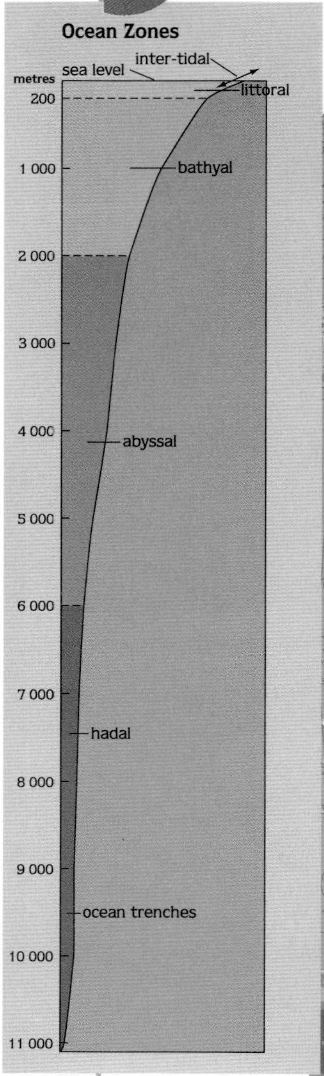
Ocean Zones
metres
sea level
inter-tidal
littoral
200
1 000
bathyal
2 000
3 000
4 000
abyssal
5 000
6 000
7 000
hadal
8 000
9 000
ocean trenches
10 000
11 000

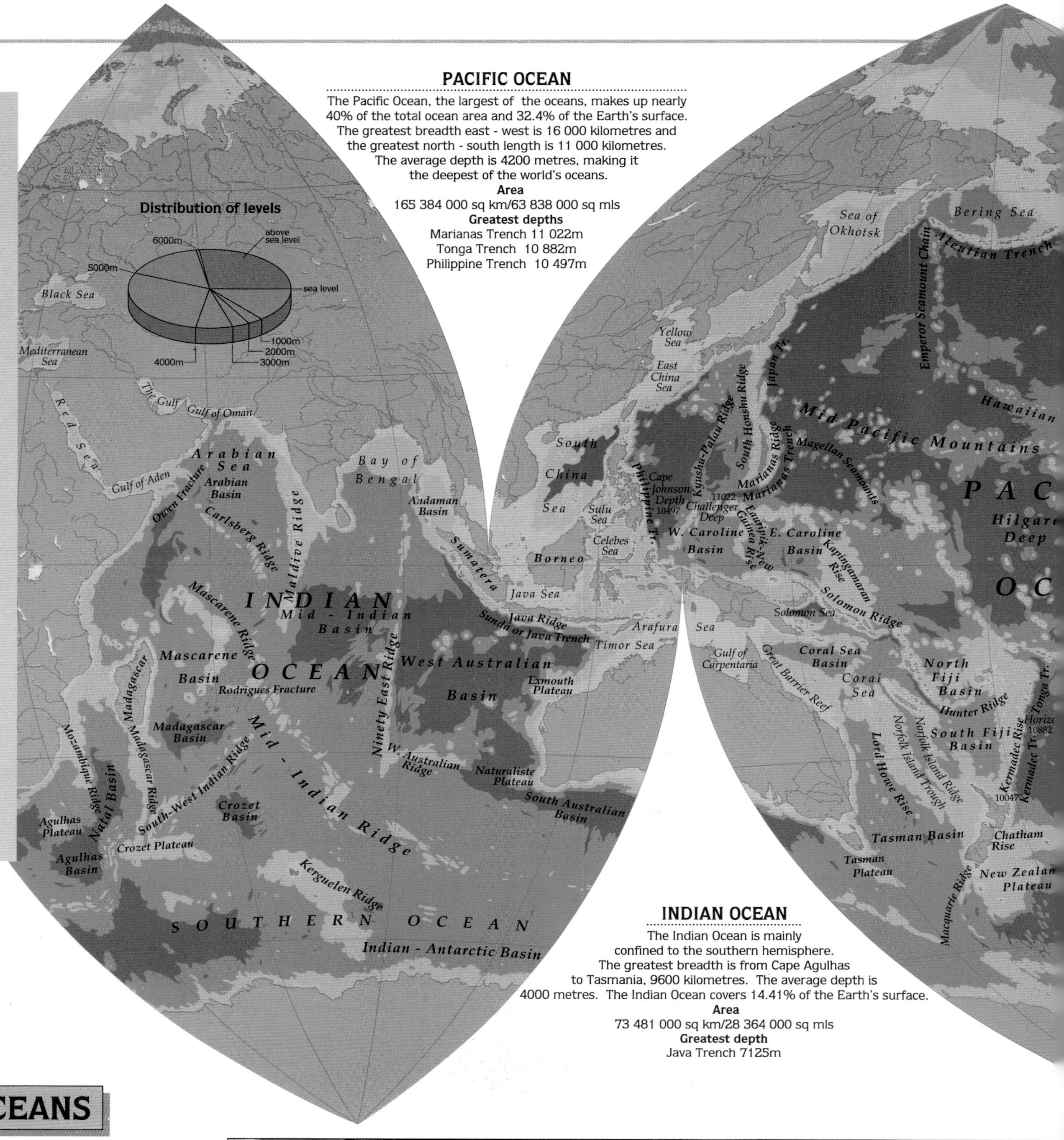
Distribution of levels
6000m
above sea level
5000m
sea level
1000m
2000m
3000m
4000m
Black Sea
Mediterranean Sea
Red Sea
The Gulf
Gulf of Oman
Gulf of Aden
Arabian Sea
Arabian Basin
Owen Fracture
Carlsberg Ridge
Maldive Ridge
Bay of Bengal
Andaman Basin
Sumatera
INDIAN
Mid - Indian Basin
OCEAN
Mascarene Ridge
Mascarene Basin
Rodrigues Fracture
Madagascar
Madagascar Basin
Madagascar Ridge
Mozambique Ridge
Natal Basin
Agulhas Plateau
Agulhas Basin
Crozet Plateau
Crozet Basin
South-West Indian Ridge
Mid - Indian Ridge
Ninety East Ridge
West Australian Basin
W. Australian Ridge
Exmouth Plateau
Naturaliste Plateau
South Australian Basin
Kerguelen Ridge
SOUTHERN OCEAN
Indian - Antarctic Basin
Java Sea
Java Ridge
Sunda or Java Trench
Arafura Sea
Timor Sea
Gulf of Carpentaria
Great Barrier Reef
Coral Sea Basin
Coral Sea
Sea of Okhotsk
Bering Sea
Aleutian Trench
Emperor Seamount Chain
Yellow Sea
East China Sea
Japan Tr.
South Honshu Ridge
Kyushu-Palau Ridge
Mid Pacific Mountains
Magellan Seamounts
Hawaiian
South China Sea
Philippine Tr.
Cape Johnson Depth 10497
11022 Challenger Deep
Marianas Ridge
Marianas Trench
Sulu Sea
Celebes Sea
Borneo
W. Caroline Basin
E. Caroline Basin
Eauripik-New Guinea Rise
Kapingamarangi Rise
PAC
Hilgard Deep
OC
Solomon Sea
Solomon Ridge
North Fiji Basin
Hunter Ridge
South Fiji Basin
Tonga Tr.
Horizon 10882
Kermadec Rise
Kermadec Tr.
10047
Norfolk Island Trough
Norfolk Island Ridge
Lord Howe Rise
Tasman Basin
Tasman Plateau
Chatham Rise
New Zealand Plateau
Macquarie Ridge

PACIFIC OCEAN

The Pacific Ocean, the largest of the oceans, makes up nearly 40% of the total ocean area and 32.4% of the Earth's surface. The greatest breadth east - west is 16 000 kilometres and the greatest north - south length is 11 000 kilometres. The average depth is 4200 metres, making it the deepest of the world's oceans.

Area
165 384 000 sq km/63 838 000 sq mls

Greatest depths
Marianas Trench 11 022m
Tonga Trench 10 882m
Philippine Trench 10 497m

INDIAN OCEAN

The Indian Ocean is mainly confined to the southern hemisphere. The greatest breadth is from Cape Agulhas to Tasmania, 9600 kilometres. The average depth is 4000 metres. The Indian Ocean covers 14.41% of the Earth's surface.

Area
73 481 000 sq km/28 364 000 sq mls

Greatest depth
Java Trench 7125m

THE OCEANS

THE OCEANS FORM part of the Earth's surface known as the hydro-sphere (the surface waters) and cover 361 million square kilometres, over 70 per cent of the total surface area of the earth.

The largest ocean is the Pacific Ocean and its associated seas, followed by the Atlantic which includes the Arctic basin, and finally the Indian Ocean. The name Southern Ocean is often given to the vast ocean area surrounding Antarctica. As the northern edge of the ocean is not well defined by land masses its area is combined into the areas of the other oceans. The ocean floor can be defined by zones based on water depth and the natural fauna. Between the high and low water marks, is the littoral or inter-tidal zone. The sublittoral zone, or continental platform is the relatively narrow zone that slopes gently from the continental coasts to the shelf break at around 200 metres, where it merges into the bathyal zone. The boundary between the bathyal zone and the abyssal zone varies between depths of 1000 and 3000 metres; the hadal zone is the ocean trenches below 6000 metres.

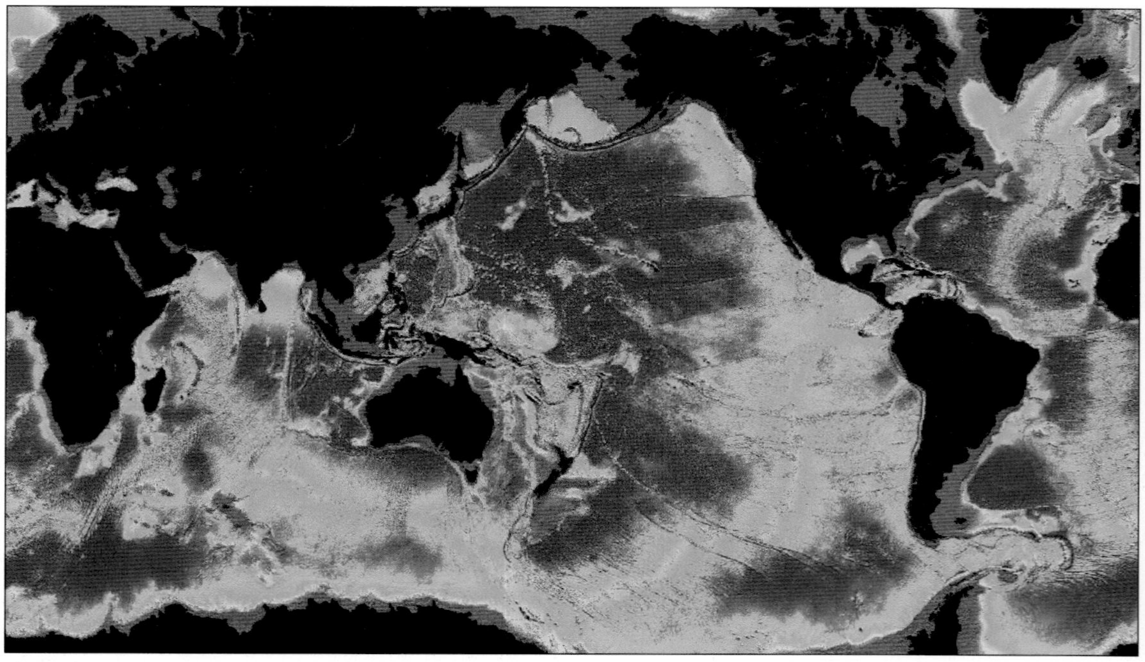

Global seafloor topography. Modern exploration of the oceans is mainly by 'remote' means. This image shows the shape of the seafloor as measured and estimated by two methods - depth soundings and gravity data obtained from satellite altimetry.
(Image courtesy of WHF Smith, U.S. National Oceanic and Atmospheric Administration.)

ATLANTIC OCEAN

Sometimes divided into the North Atlantic and the South Atlantic. The greatest breadth in the North is 7200 kilometres (Florida to Morocco) and in the South, 9600 kilometres (Brazil to Guinea). The average depth is 3600 metres, and it makes up 16.12 % of the Earth's surface area.

Area

82 217 000 sq km/31 737 000 sq mls

Greatest depths

Puerto Rico Trench 9220m
South Sandwich Trench 8325m
Romanche Gap 7728m

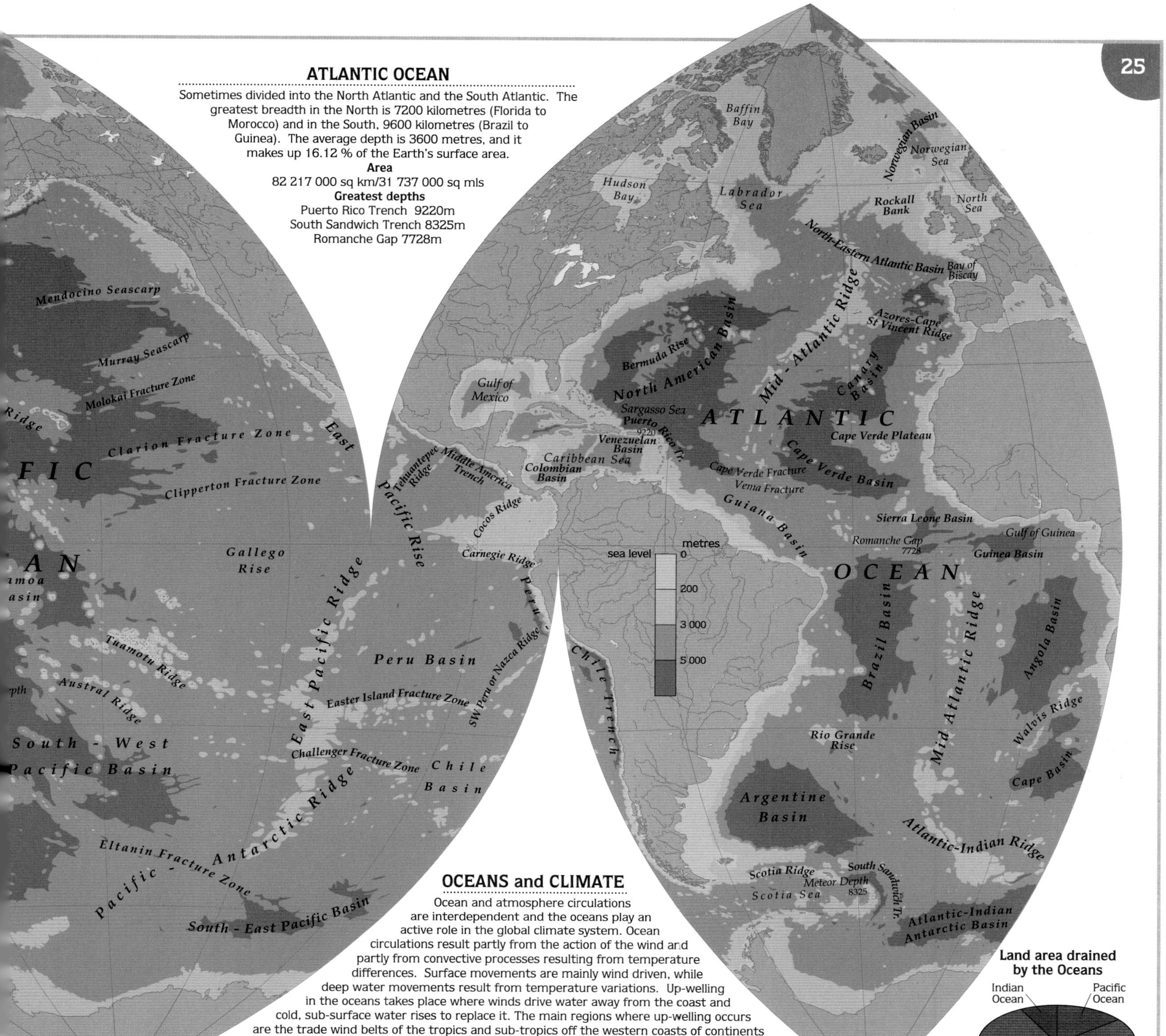

Baffin Bay
Hudson Bay
Labrador Sea
Norwegian Basin
Norwegian Sea
Rockall Bank
North Sea
North-Eastern Atlantic Basin
Bay of Biscay
Mendocino Seascarp
Murray Seascarp
Molokai Fracture Zone
Ridge
Clarion Fracture Zone
East
FIC
Clipperton Fracture Zone
Gulf of Mexico
Bermuda Rise
North American Basin
Mid - Atlantic Ridge
Azores-Cape St Vincent Ridge
Canary Basin
Sargasso Sea
Puerto Rico Tr.
9220
ATLANTIC
Cape Verde Plateau
Venezuelan Basin
Caribbean Sea
Colombian Basin
Tehuantepec Ridge
Middle America Trench
Pacific Rise
Cocos Ridge
Cape Verde Fracture
Vema Fracture
Cape Verde Basin
Guiana Basin
Sierra Leone Basin
Romanche Gap
7728
Gulf of Guinea
Guinea Basin
Gallego Rise
AN
Carnegie Ridge
metres
sea level
0
200
3 000
5 000
OCEAN
Peru
Tuamotu Ridge
Austral Ridge
Peru Basin
SW Peru or Nazca Ridge
Easter Island Fracture Zone
East Pacific Ridge
Chile Trench
Brazil Basin
Mid Atlantic Ridge
Angola Basin
Walvis Ridge
South - West Pacific Basin
Challenger Fracture Zone
Chile Basin
Rio Grande Rise
Cape Basin
Argentine Basin
Pacific - Antarctic Ridge
Eltanin Fracture Zone
Atlantic-Indian Ridge
Scotia Ridge
South Sandwich Tr.
Meteor Depth
8325
Scotia Sea
South - East Pacific Basin
Atlantic-Indian Antarctic Basin

OCEANS and CLIMATE

Ocean and atmosphere circulations are interdependent and the oceans play an active role in the global climate system. Ocean circulations result partly from the action of the wind and partly from convective processes resulting from temperature differences. Surface movements are mainly wind driven, while deep water movements result from temperature variations. Up-welling in the oceans takes place where winds drive water away from the coast and cold, sub-surface water rises to replace it. The main regions where up-welling occurs are the trade wind belts of the tropics and sub-tropics off the western coasts of continents and around Antarctica. The surface waters of the southern Pacific Ocean off Peru and Chile are particularly cold for their latitude, and this is due to up-welling.

Land area drained by the Oceans

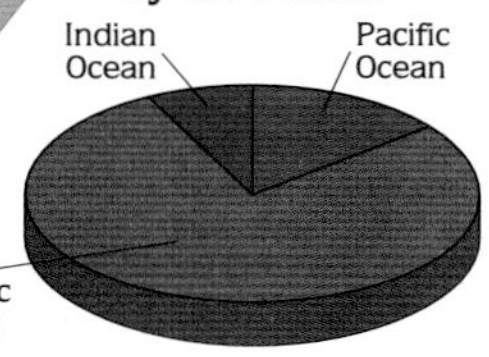

Indian Ocean
Pacific Ocean
Atlantic Ocean

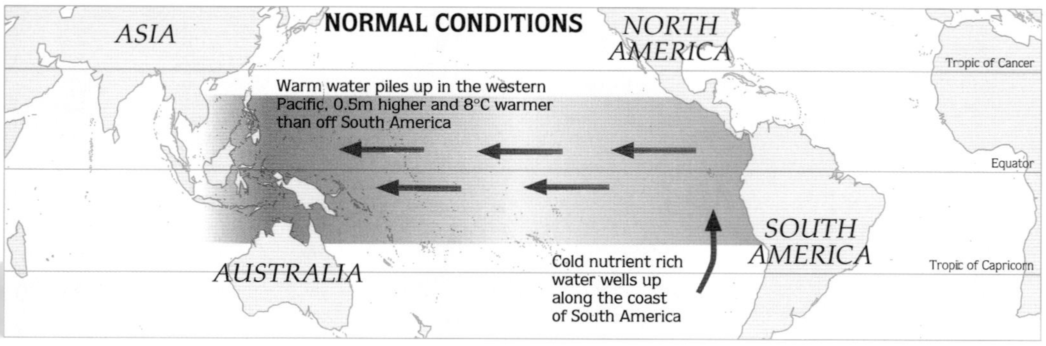

NORMAL CONDITIONS
ASIA
NORTH AMERICA
Tropic of Cancer
Warm water piles up in the western Pacific, 0.5m higher and 8°C warmer than off South America
Equator
SOUTH AMERICA
AUSTRALIA
Cold nutrient rich water wells up along the coast of South America
Tropic of Capricorn

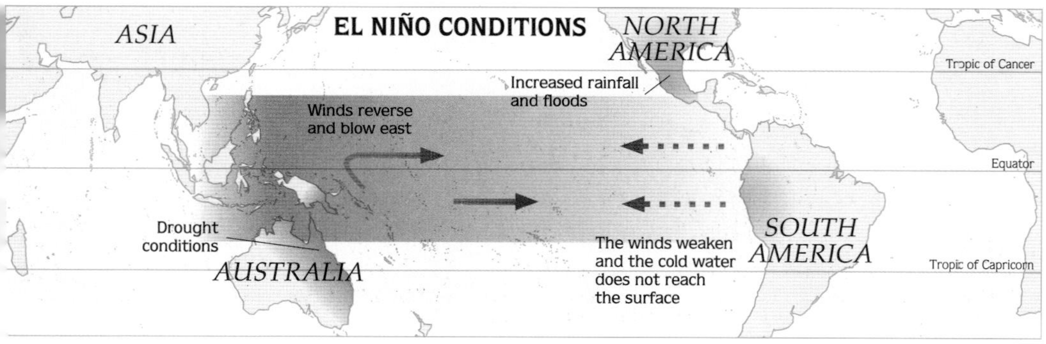

EL NIÑO CONDITIONS
ASIA
NORTH AMERICA
Tropic of Cancer
Increased rainfall and floods
Winds reverse and blow east
Equator
Drought conditions
AUSTRALIA
The winds weaken and the cold water does not reach the surface
SOUTH AMERICA
Tropic of Capricorn

EL NIÑO

The natural oceanic phenomenon, El Niño, and the associated atmospheric changes known as the Southern Oscillation appear to have been happening for millennia. There have been about twenty five this century and each cycle lasts for three to six years with the most intense El Niño effect lasting around a year. Though no two El Niños are alike, recent cycles have been more frequent, the temperature changes greater, or they have lasted longer than previously.

In normal conditions, with no El Niño effect, the trade winds blow west across the tropical Pacific toward southeast Asia and northern Australia. These winds blow warm surface water west, which causes nutrient-rich colder water to well up in the eastern Pacific along the tropical coast of South America. As the cycle progresses and the El Niño conditions develop, the increased winds and rainstorms over the warmer waters in the west affect atmospheric heating which in turn decreases the trade winds in the western and central Pacific. This reduces the up-welling of the cold waters in the eastern Pacific, the cold water no longer reaches the surface, resulting in increased sea-surface temperatures. The reduction in up-welling also means a reduction of the nutrient level of the surface waters which, in turn, has a devastating effect on the fishing on the west coast of South America.

Climatic effects include flooding in Peru and drought in Indonesia and Australia, but, although the areas most affected by El Niño conditions are the countries around the tropical Pacific, El Niño affects the weather around the globe.

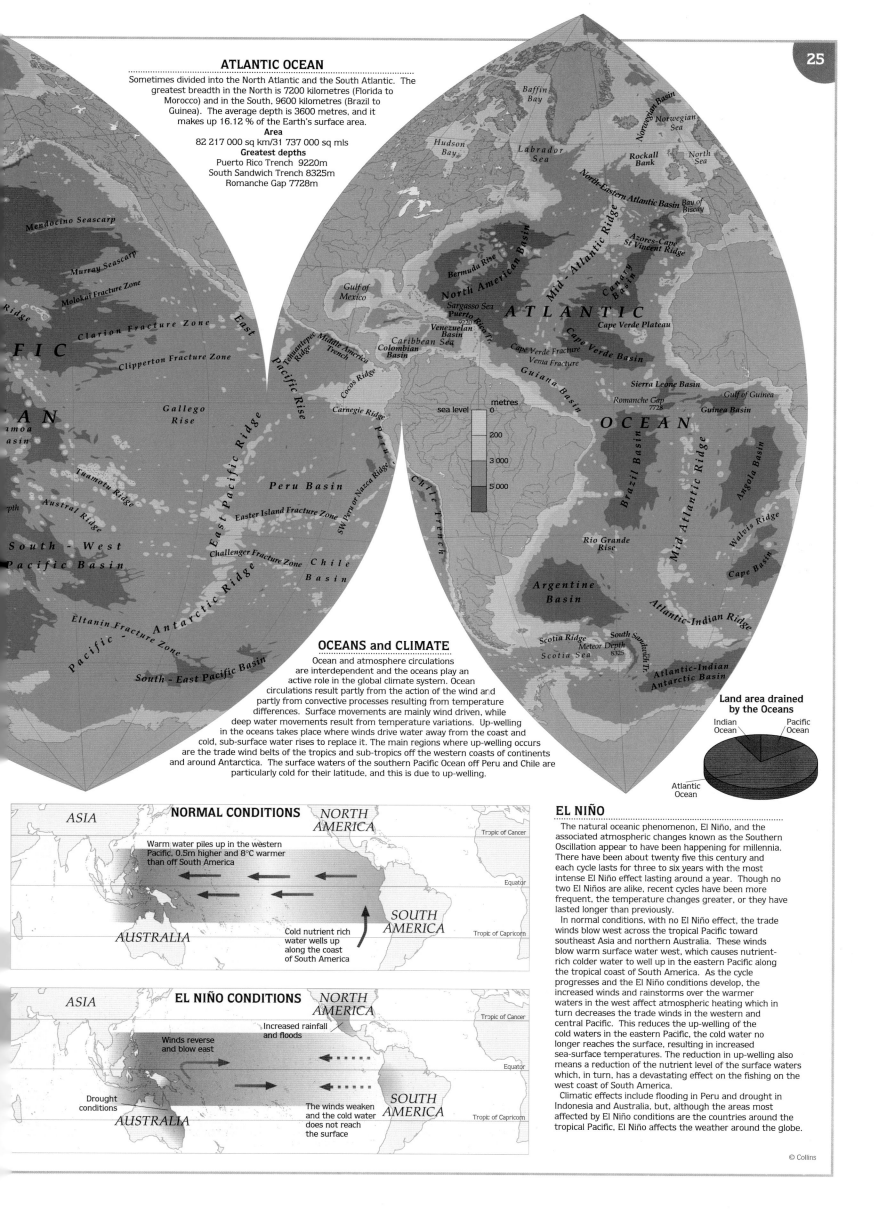

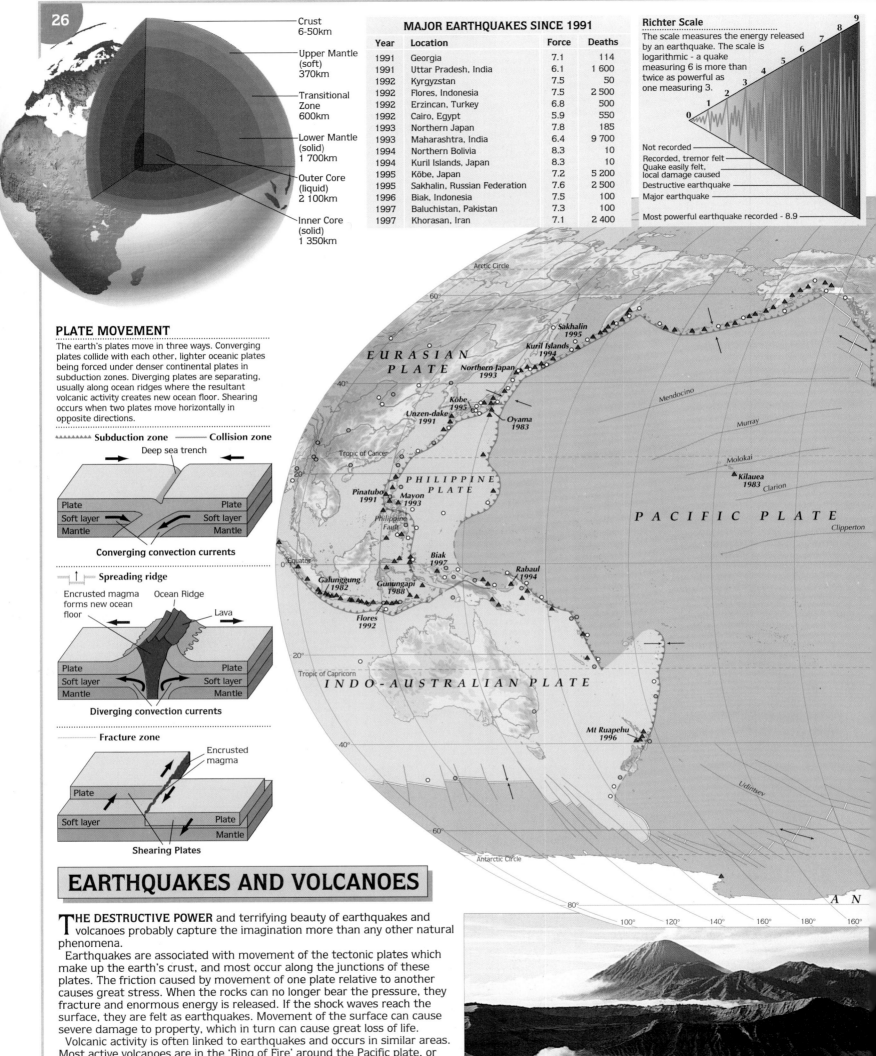

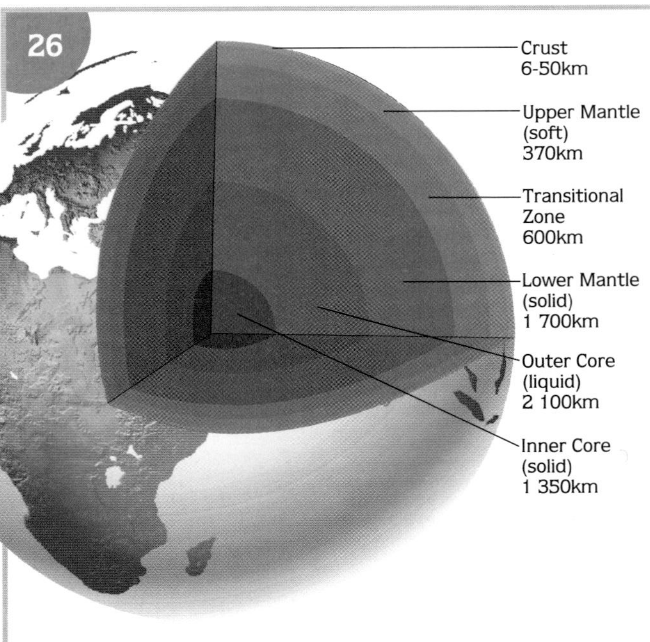
Crust
6-50km
Upper Mantle
(soft)
370km
Transitional
Zone
600km
Lower Mantle
(solid)
1 700km
Outer Core
(liquid)
2 100km
Inner Core
(solid)
1 350km

MAJOR EARTHQUAKES SINCE 1991

Year	Location	Force	Deaths
1991	Georgia	7.1	114
1991	Uttar Pradesh, India	6.1	1 600
1992	Kyrgyzstan	7.5	50
1992	Flores, Indonesia	7.5	2 500
1992	Erzincan, Turkey	6.8	500
1992	Cairo, Egypt	5.9	550
1993	Northern Japan	7.8	185
1993	Maharashtra, India	6.4	9 700
1994	Northern Bolivia	8.3	10
1994	Kuril Islands, Japan	8.3	10
1995	Kōbe, Japan	7.2	5 200
1995	Sakhalin, Russian Federation	7.6	2 500
1996	Biak, Indonesia	7.5	100
1997	Baluchistan, Pakistan	7.3	100
1997	Khorasan, Iran	7.1	2 400

Richter Scale

The scale measures the energy released by an earthquake. The scale is logarithmic - a quake measuring 6 is more than twice as powerful as one measuring 3.

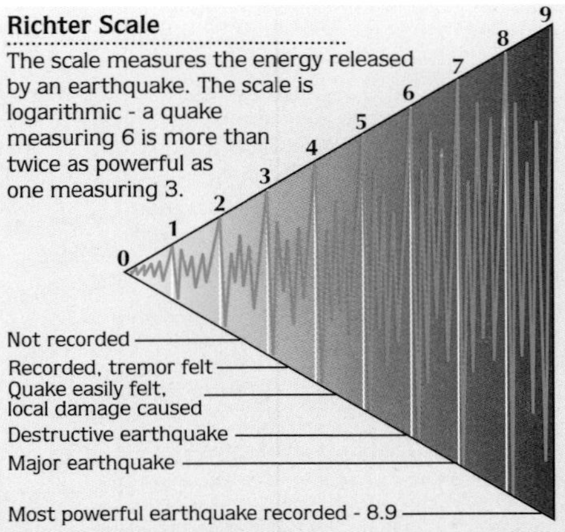
0
1
2
3
4
5
6
7
8
9
Not recorded
Recorded, tremor felt
Quake easily felt, local damage caused
Destructive earthquake
Major earthquake
Most powerful earthquake recorded - 8.9

PLATE MOVEMENT

The earth's plates move in three ways. Converging plates collide with each other, lighter oceanic plates being forced under denser continental plates in subduction zones. Diverging plates are separating, usually along ocean ridges where the resultant volcanic activity creates new ocean floor. Shearing occurs when two plates move horizontally in opposite directions.

Subduction zone **Collision zone**

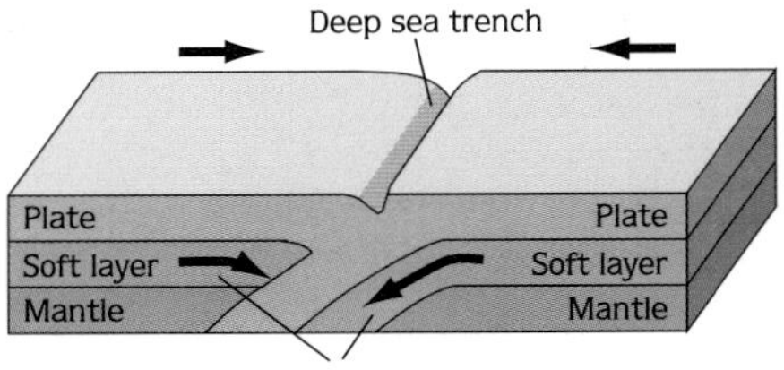
Deep sea trench
Plate
Plate
Soft layer
Soft layer
Mantle
Mantle

Converging convection currents

Spreading ridge

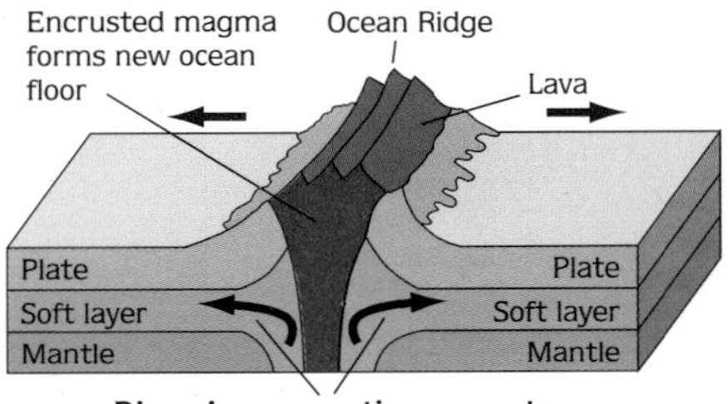
Encrusted magma forms new ocean floor
Ocean Ridge
Lava
Plate
Plate
Soft layer
Soft layer
Mantle
Mantle

Diverging convection currents

Fracture zone

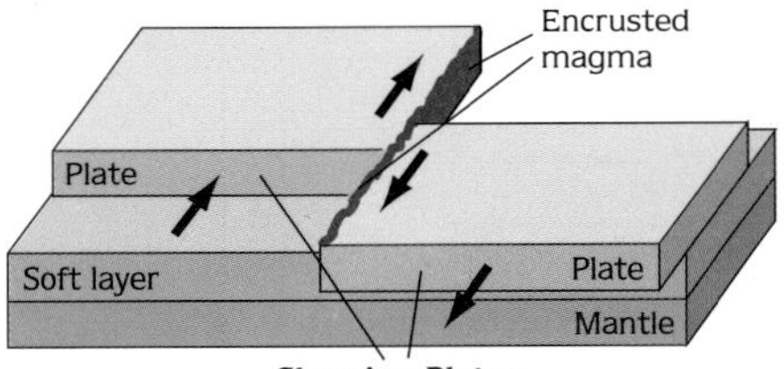
Encrusted magma
Plate
Soft layer
Plate
Mantle

Shearing Plates

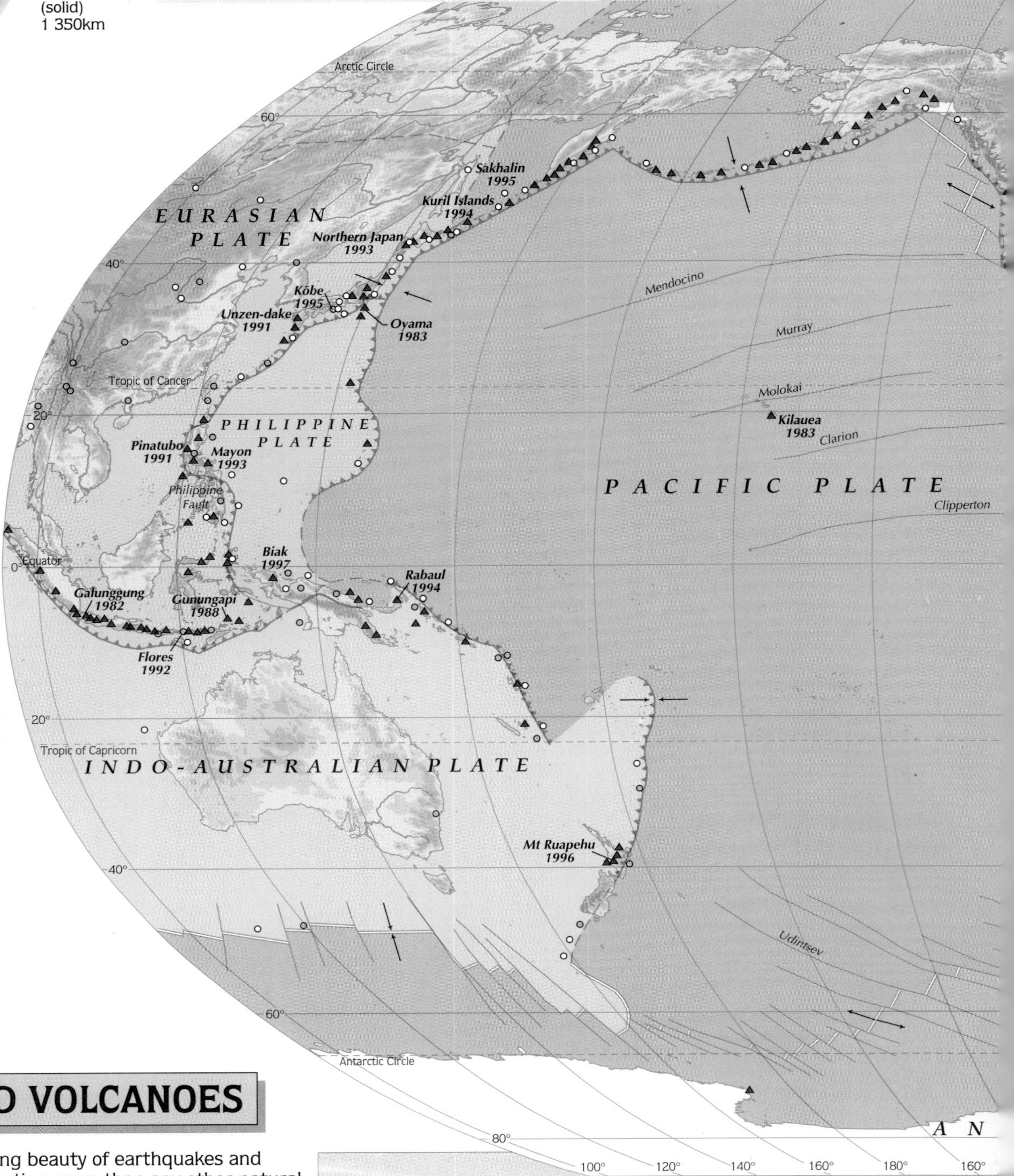
Arctic Circle
60°
40°
20°
0°
20°
40°
60°
80°
100°
120°
140°
160°
180°
160°
EURASIAN PLATE
Sakhalin 1995
Kuril Islands 1994
Northern Japan 1993
Kōbe 1995
Unzen-dake 1991
Oyama 1983
Mendocino
Murray
Tropic of Cancer
Molokai
Kilauea 1983
Clarion
PHILIPPINE PLATE
Pinatubo 1991
Mayon 1993
Philippine Fault
PACIFIC PLATE
Clipperton
Equator
Biak 1997
Rabaul 1994
Galunggung 1982
Gunungapi 1988
Flores 1992
Tropic of Capricorn
INDO-AUSTRALIAN PLATE
Mt Ruapehu 1996
Udintsev
Antarctic Circle
A N

EARTHQUAKES AND VOLCANOES

THE DESTRUCTIVE POWER and terrifying beauty of earthquakes and volcanoes probably capture the imagination more than any other natural phenomena.

Earthquakes are associated with movement of the tectonic plates which make up the earth's crust, and most occur along the junctions of these plates. The friction caused by movement of one plate relative to another causes great stress. When the rocks can no longer bear the pressure, they fracture and enormous energy is released. If the shock waves reach the surface, they are felt as earthquakes. Movement of the surface can cause severe damage to property, which in turn can cause great loss of life.

Volcanic activity is often linked to earthquakes and occurs in similar areas. Most active volcanoes are in the 'Ring of Fire' around the Pacific plate, or along the southern edge of the Eurasian plate. The most violent eruptions occur when two plates collide. The force involved can cause the surrounding rock to melt to form magma. The heat generated creates upward pressure and the magma is forced through weaknesses in the rock. If the pressure is great enough, material breaks through the surface as a volcano.

Semeru Peak, Java, Indonesia. Java has over one hundred and twenty volcanoes, thirty of which are active. Indonesia as a whole has 34% of the world's active volcanoes.

EARTHQUAKES

- ○ **High magnitude (over 7.8 Richter scale)**
- ◉ **Lesser magnitude**

Approximately 5000 earthquakes, of varying intensity, are recorded each year in Japan - a country which lies at the very active junction of the Eurasian, North American, Philippine and Pacific plates.

In January 1995 the city of Kōbe, a major port on the southern coast of Honshū (left), was struck by an earthquake measuring 7.2 on the Richter scale. The result was extensive damage and over 5000 deaths. The cause was lateral movement along a relatively minor fault line in an area of very complex geology. The Philippine and Eurasian plates are colliding at an oblique angle and this creates stress in areas away from the main plate junction. This can cause movement along secondary faults such as that causing the Kōbe disaster.

Earthquake mechanisms

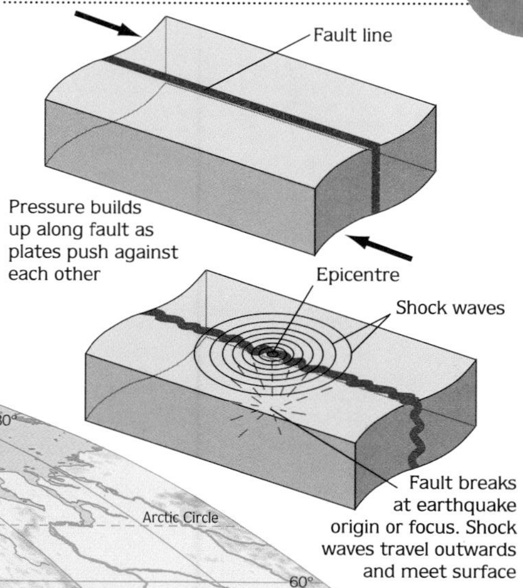

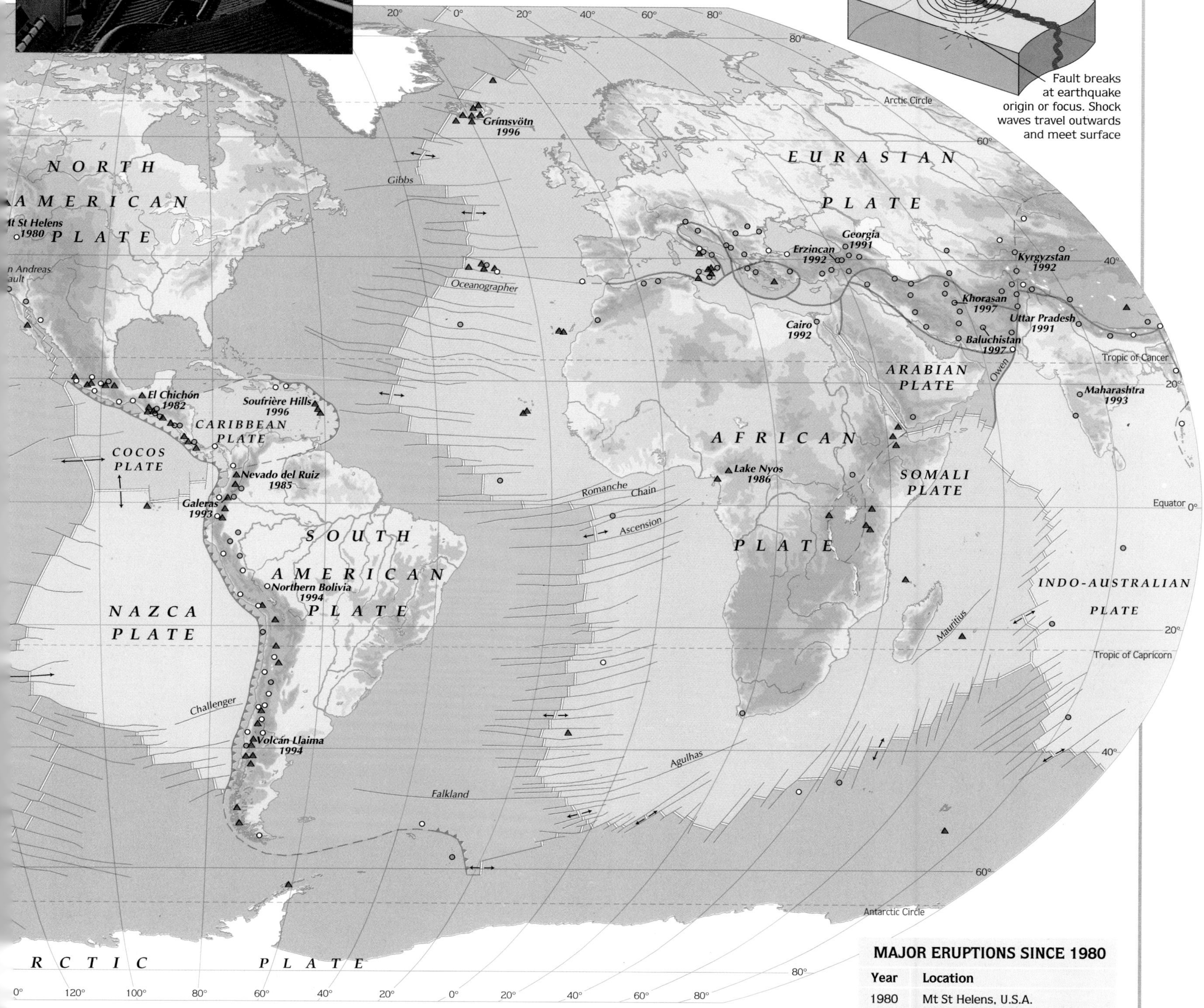

Soufrière Hills, Montserrat

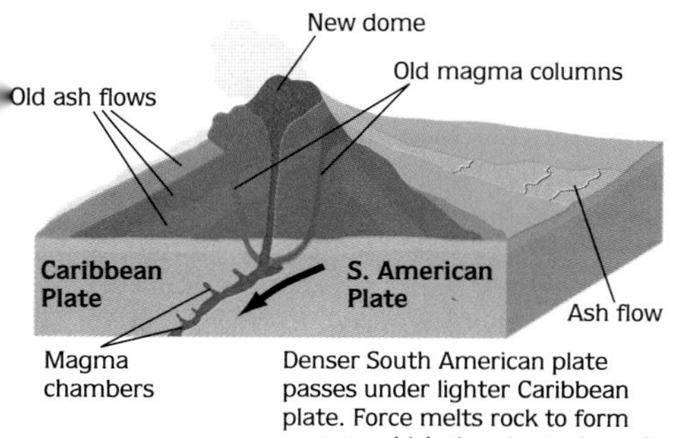

Denser South American plate passes under lighter Caribbean plate. Force melts rock to form magma which rises towards surface.

VOLCANOES

- ▲ **Active Volcano**

Most volcanoes occur along subduction zones where two plates collide with one passing underneath the other. One area prone to such activity is the Caribbean, at the junction of the North American, South American and Caribbean plates. The American plates are moving towards and beneath the Caribbean plate at a rate of approximately 2cm per year. It was this movement which created the Lesser Antilles and the same forces are still at work today.

Major eruptions of the Soufrière Hills volcano (left), on the island of Montserrat, occurred in 1996 and 1997. The whole range of volcanic material was emitted - rocks, lava, ash, gases - and the danger was so great that almost the entire population of the island was evacuated.

MAJOR ERUPTIONS SINCE 1980

Year	Location
1980	Mt St Helens, U.S.A.
1982	El Chichón, Mexico
1982	Galunggung, Indonesia
1983	Kilauea, Hawaii
1983	Oyama, Japan
1985	Nevado del Ruiz, Colombia
1986	Lake Nyos, Cameroon
1988	Gunungapi, Indonesia
1991	Pinatubo, Philippines
1991	Unzen-dake, Japan
1993	Mayon, Philippines
1993	Galeras, Colombia
1994	Volcán Llaima, Chile
1994	Rabaul, PNG
1996	Soufrière Hills, Montserrat
1996	Mt Ruapehu, New Zealand
1996	Grímsvötn, Iceland

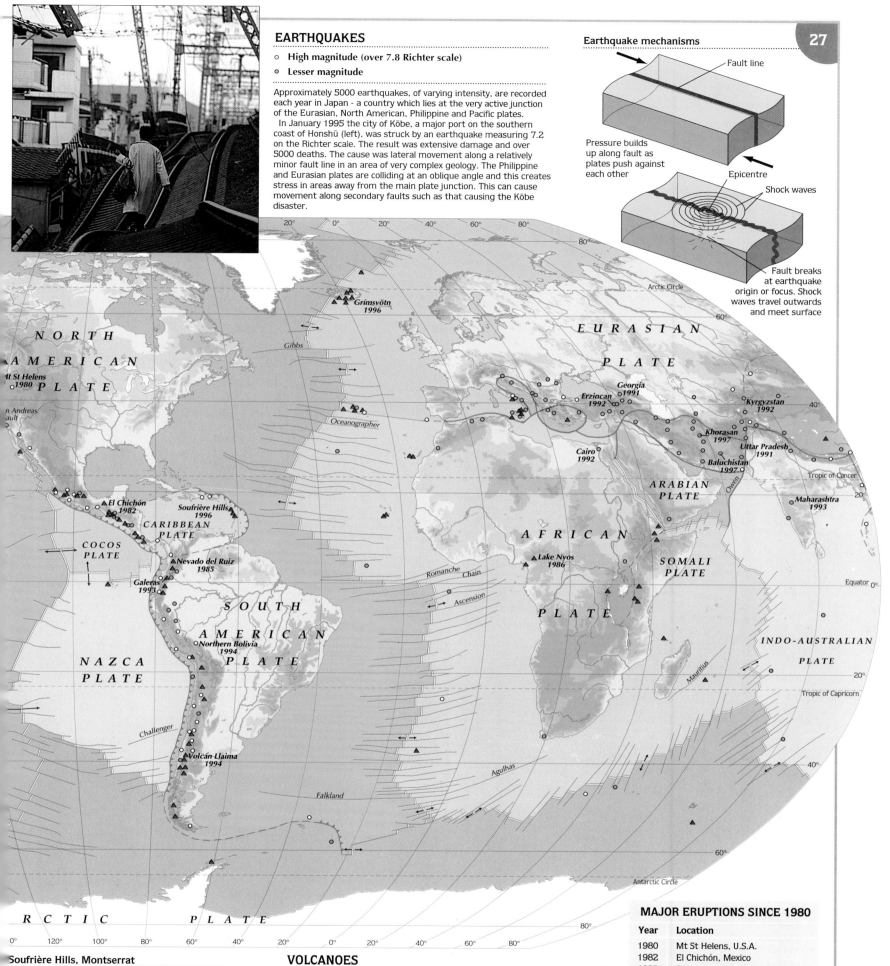

WORLD POPULATION RANKINGS

Rank	Mid 1996 population		Population density 1996 (persons per square kilometre)	
1	CHINA	1 232 083 000	MACAU	25882.4
2	INDIA	944 580 000	MONACO	16410.3
3	U.S.A.	266 557 000	SINGAPORE	4763.7
4	INDONESIA	196 813 000	GIBRALTAR	4307.7
5	BRAZIL	157 872 000	VATICAN CITY	2272.7
6	RUSSIAN FEDERATION	147 739 000	BERMUDA	1185.2
7	PAKISTAN	134 146 000	MALTA	1180.4
8	JAPAN	125 761 000	MALDIVES	882.6
9	BANGLADESH	120 073 000	BAHRAIN	866.9
10	NIGERIA	115 020 000	BANGLADESH	833.9
11	MEXICO	96 578 000	CHANNEL ISLANDS	764.1
12	GERMANY	81 912 000	BARBADOS	607.0
13	VIETNAM	75 181 000	TAIWAN	586.3
14	PHILIPPINES	71 899 000	MAURITIUS	555.9
15	TURKEY	62 697 000	NAURU	523.8
16	IRAN	61 128 000	SOUTH KOREA	458.8
17	EGYPT	60 603 000	PUERTO RICO	410.4
18	THAILAND	60 003 000	SAN MARINO	409.8
19	ETHIOPIA	58 506 000	TUVALU	400.0
20	FRANCE	58 375 000	NETHERLANDS	373.7

Urban population as a percentage of the total population

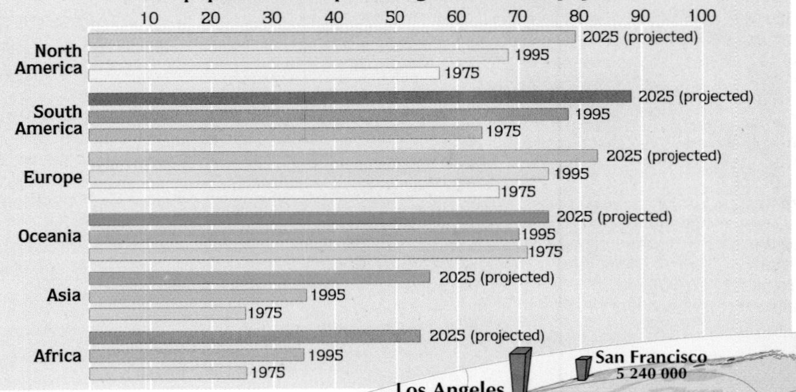
10
20
30
40
50
60
70
80
90
100
North America
2025 (projected)
1995
1975
South America
2025 (projected)
1995
1975
Europe
2025 (projected)
1995
1975
Oceania
2025 (projected)
1995
1975
Asia
2025 (projected)
1995
1975
Africa
2025 (projected)
1995
1975

POPULATION PATTERNS

THE WORLD'S POPULATION in mid-1996 totalled 5.8 billion, over half of which live in six countries: China, India, USA, Indonesia, Brazil and the Russian Federation. 80% of the world's population live in developing countries - 95% of people added to the world total are born in the developing world.

The total is still rising, but there are signs that worldwide growth is slowly coming under control. Growth rates and fertility rates are declining, although there are great regional variations which still cause concern. The average annual growth rate in the developed world is 0.4% per annum, whilst in the less developed world it is 1.8%, reaching as high as 2.8% in Africa. Developed regions also have lower fertility rates - an average of 1.7 children per woman, below the 'replacement level' target of 2. In the developing world the rate is 3.4 and can reach 5.6 in the poorest countries.

Until growth is brought under tighter control, the developing world in particular will continue to face enormous problems of supporting a rising population.

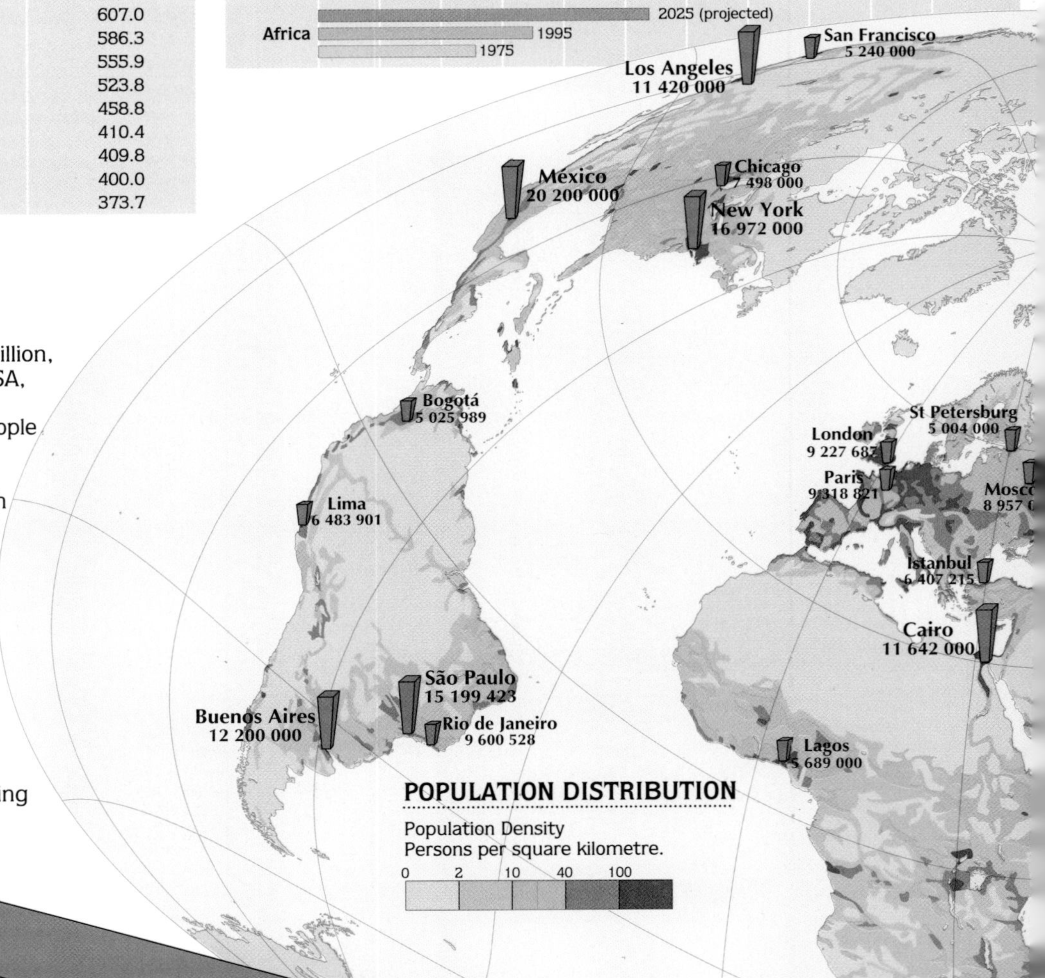
Los Angeles
11 420 000
San Francisco
5 240 000
México
20 200 000
Chicago
7 498 000
New York
16 972 000
Bogotá
5 025 989
Lima
6 483 901
London
9 227 687
St Petersburg
5 004 000
Paris
9 318 821
Istanbul
6 407 215
Cairo
11 642 000
São Paulo
15 199 423
Buenos Aires
12 200 000
Rio de Janeiro
9 600 528
Lagos
5 689 000

POPULATION DISTRIBUTION

Population Density
Persons per square kilometre.
0 2 10 40 100

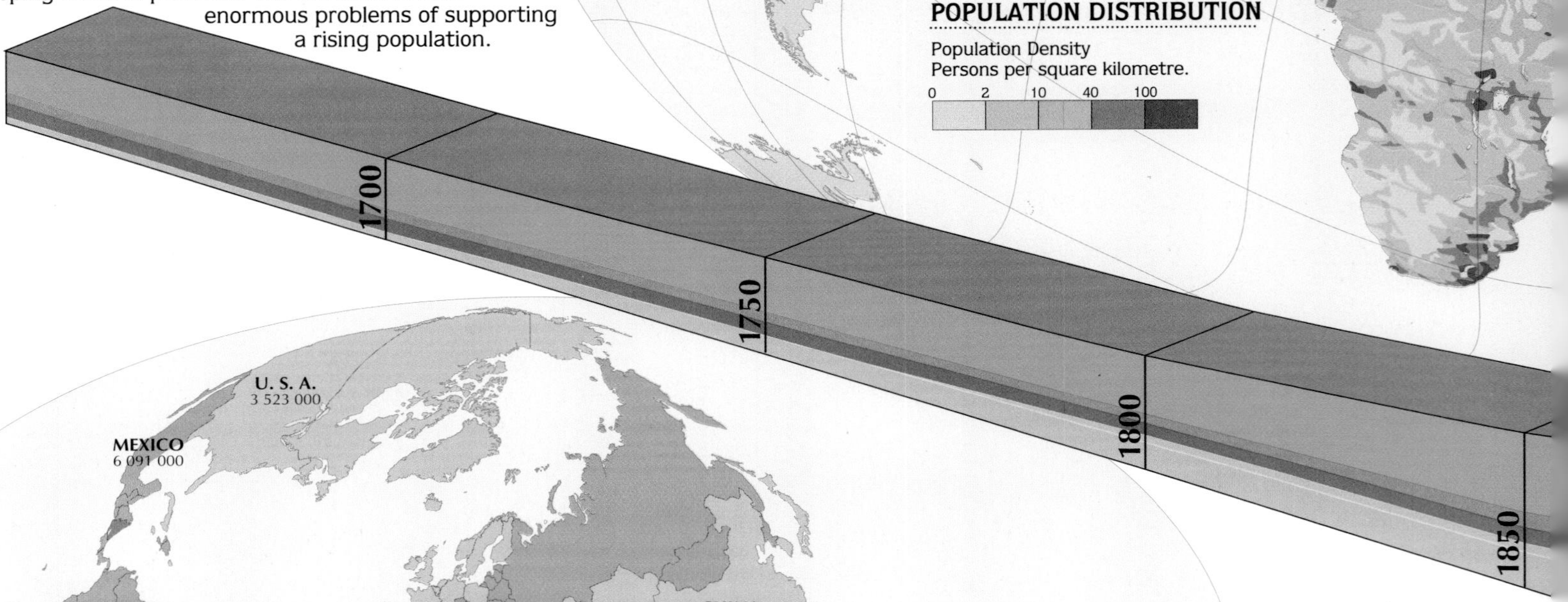
1700
1750
1800
1850

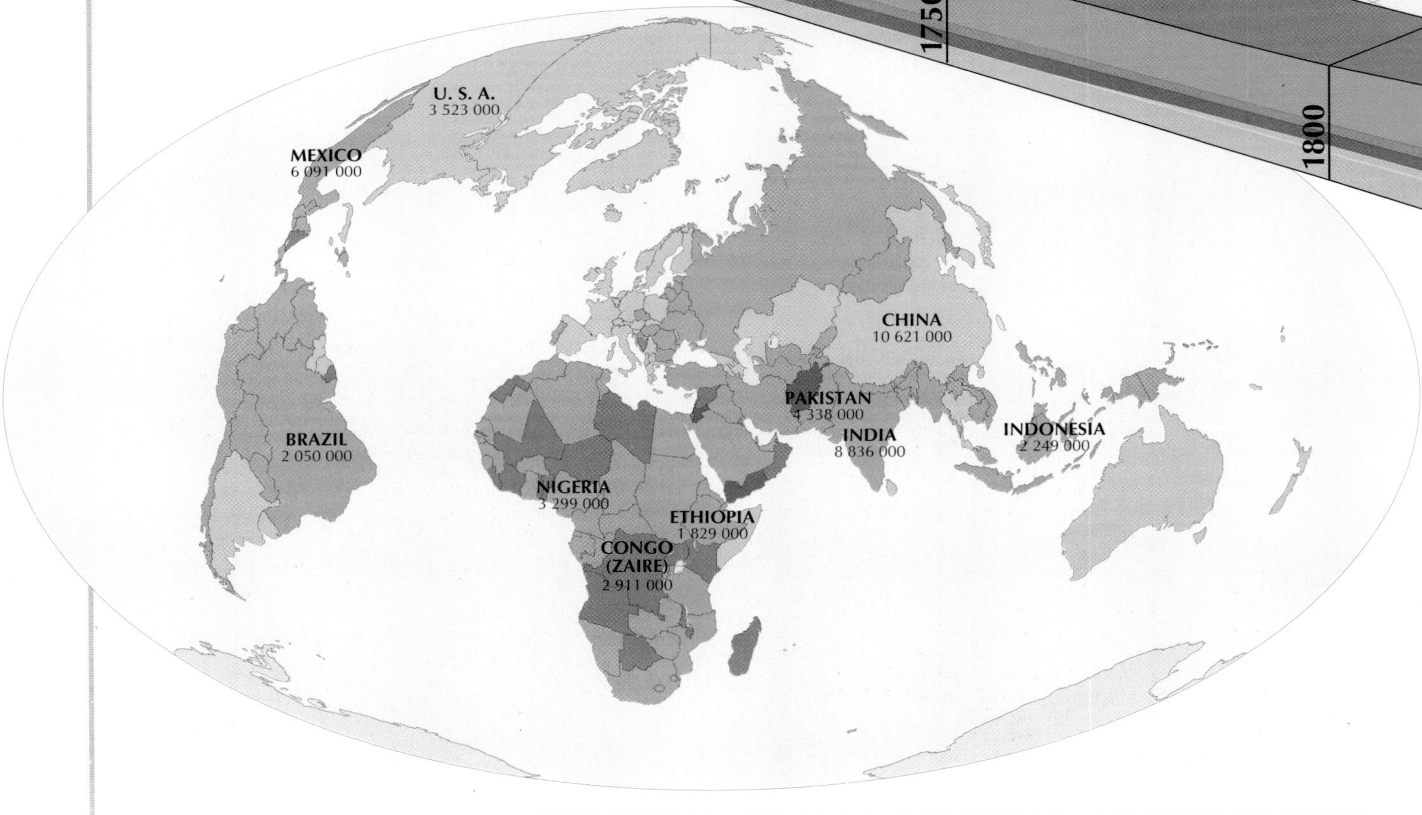
U. S. A.
3 523 000
MEXICO
6 091 000
CHINA
10 621 000
BRAZIL
2 050 000
PAKISTAN
4 338 000
INDIA
8 836 000
INDONESIA
2 249 000
NIGERIA
3 299 000
ETHIOPIA
1 829 000
CONGO
(ZAIRE)
2 911 000

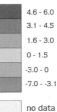

Population change

Average annual population change 1990-1995 (%)

- 4.6 - 6.0
- 3.1 - 4.5
- 1.6 - 3.0
- 0 - 1.5
- -3.0 - 0
- -7.0 - -3.1
- no data

Countries named are those with the 10 greatest population increases between 1995 and 1996

URBANIZATION

One of the dominant themes of world population is that of urbanization - the movement of people from the countryside to towns and cities. In 1995 43% of the world's population lived in urban areas and the number of urban dwellers is expected to double between 1990 and 2025 to over 5 billion.

The degree of urbanization varies between regions. The populations of Europe, North America and South America are over 70% urban compared to only 30-35% in Africa and Asia. It is the developing regions, however, that are experiencing the fastest growth of urban populations. Urban growth rates reach over 7% per year in some of the poorest countries of the world, including Burkina and Mozambique.

One side effect of urbanization is that fertility rates tend to decrease as a country's population becomes more urbanized. Decreasing fertility rates are a crucial factor in controlling population growth. It seems ironic that increased urbanization, with all its problems, may ultimately help the overall situation

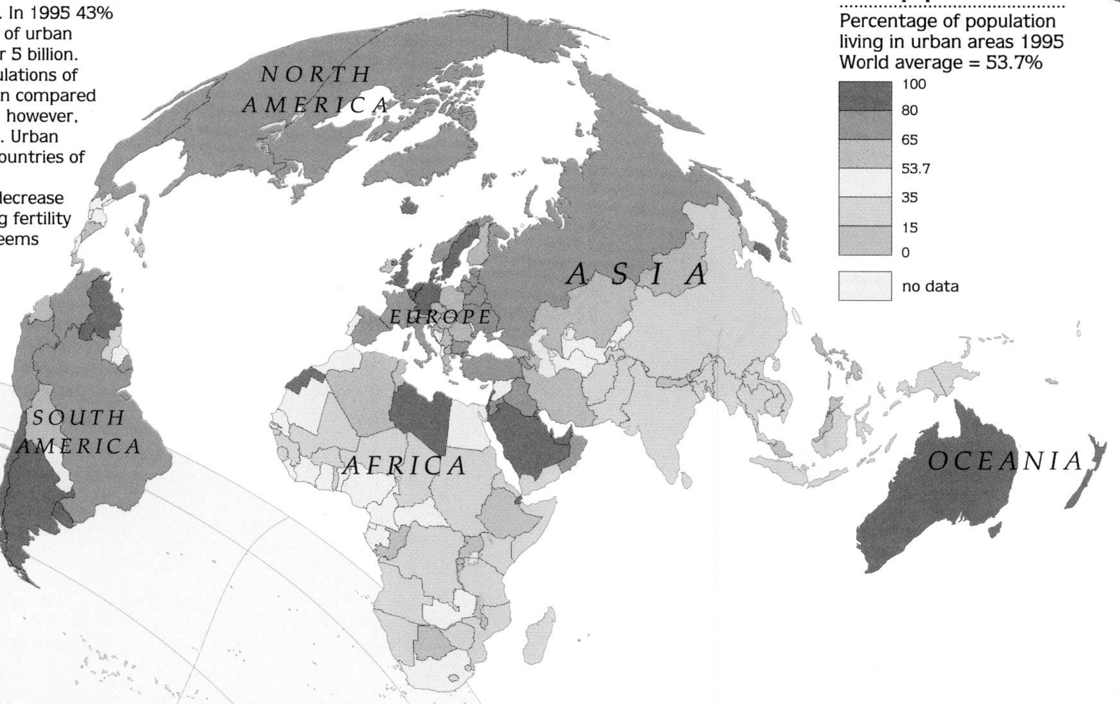

Urban population
Percentage of population living in urban areas 1995
World average = 53.7%
100
80
65
53.7
35
15
0
no data
NORTH AMERICA
SOUTH AMERICA
EUROPE
ASIA
AFRICA
OCEANIA

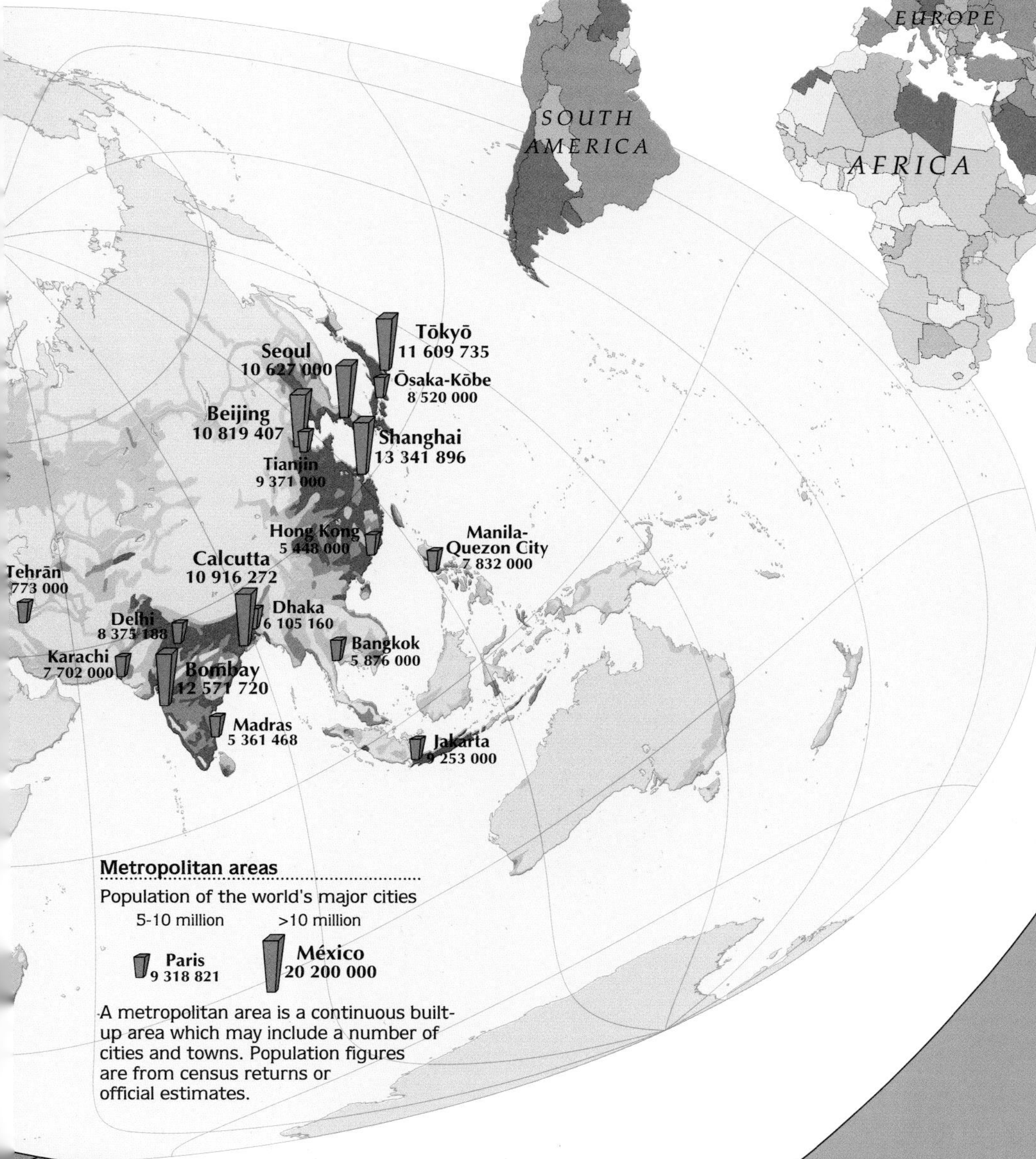

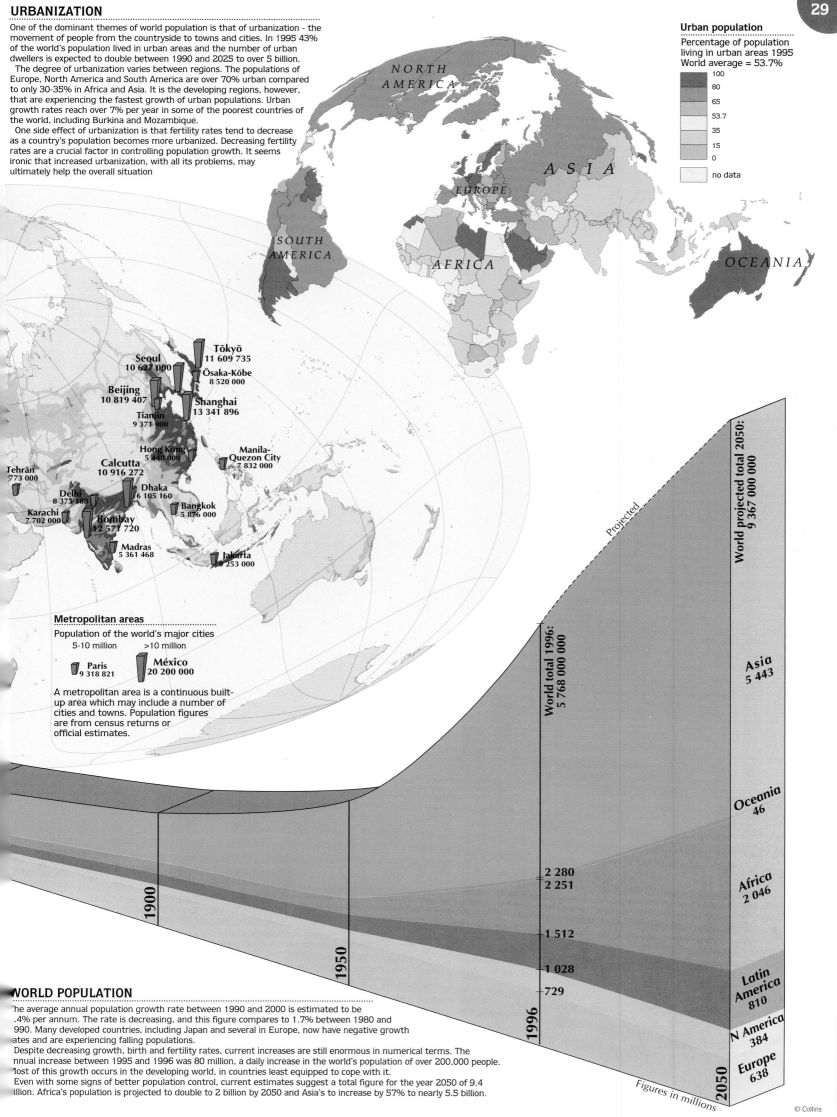

Tōkyō
11 609 735
Seoul
10 627 000
Ōsaka-Kōbe
8 520 000
Beijing
10 819 407
Shanghai
13 341 896
Tianjin
9 371 000
Hong Kong
5 448 000
Manila-Quezon City
7 832 000
Tehrān
773 000
Calcutta
10 916 272
Dhaka
6 105 160
Delhi
8 375 188
Karachi
7 702 000
Bangkok
5 876 000
Bombay
12 571 720
Madras
5 361 468
Jakarta
9 253 000
Metropolitan areas
Population of the world's major cities
5-10 million
>10 million
Paris
9 318 821
México
20 200 000
A metropolitan area is a continuous built-up area which may include a number of cities and towns. Population figures are from census returns or official estimates.

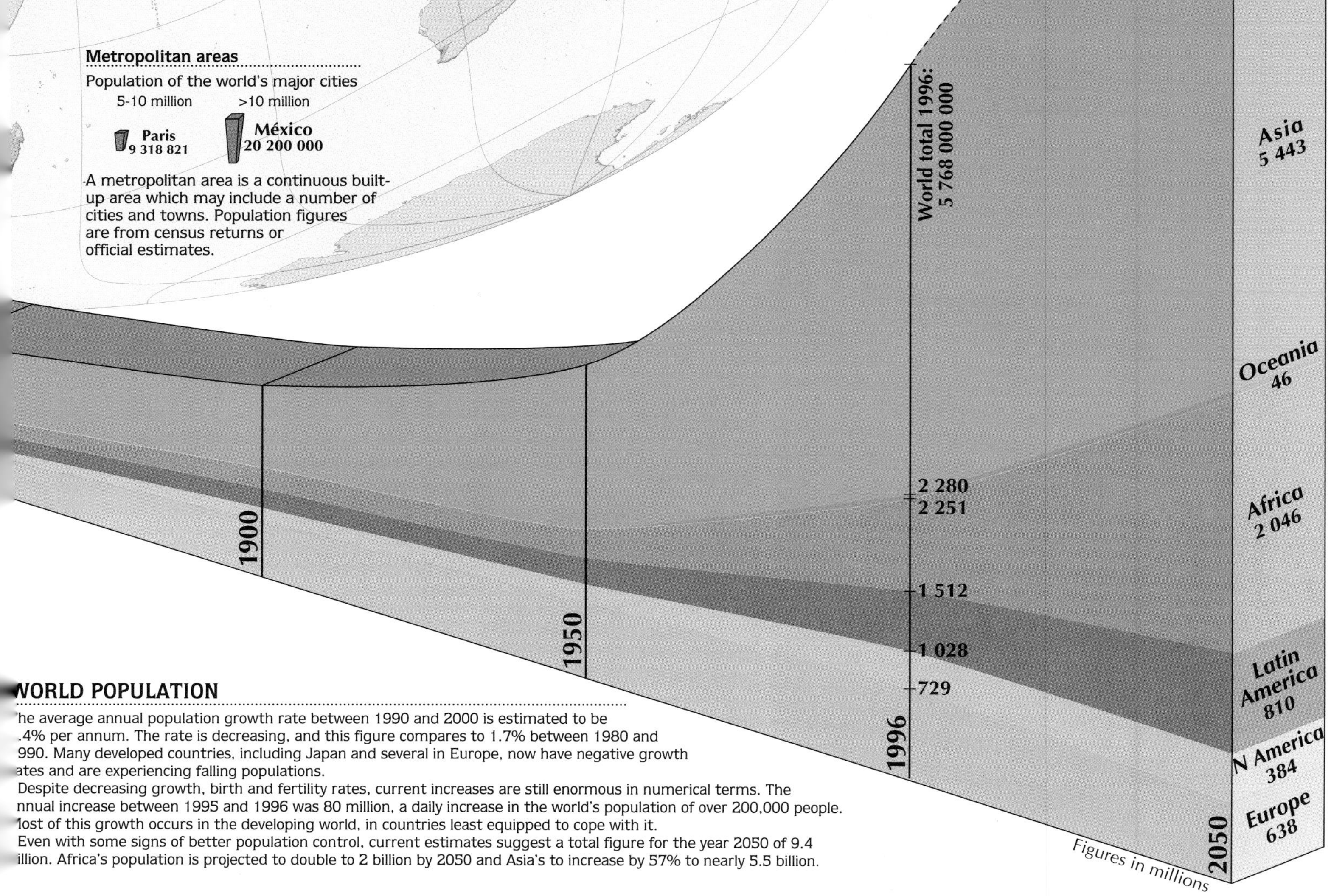

Projected
World projected total 2050: 9 367 000 000
World total 1996: 5 768 000 000
Asia
5 443
Oceania
46
2 280
2 251
Africa
2 046
1 512
1 028
729
Latin America
810
N America
384
Europe
638
1900
1950
1996
2050
Figures in millions

WORLD POPULATION

The average annual population growth rate between 1990 and 2000 is estimated to be 1.4% per annum. The rate is decreasing, and this figure compares to 1.7% between 1980 and 1990. Many developed countries, including Japan and several in Europe, now have negative growth rates and are experiencing falling populations.

Despite decreasing growth, birth and fertility rates, current increases are still enormous in numerical terms. The annual increase between 1995 and 1996 was 80 million, a daily increase in the world's population of over 200,000 people. Most of this growth occurs in the developing world, in countries least equipped to cope with it.

Even with some signs of better population control, current estimates suggest a total figure for the year 2050 of 9.4 billion. Africa's population is projected to double to 2 billion by 2050 and Asia's to increase by 57% to nearly 5.5 billion.

Health Indicators
(Selected Countries)

	Life Expectancy 1995	Country	Infant Mortality Rate 1995
BEST	80	Japan	4
	79	Sweden	4
	79	Iceland	5
	78	Switzerland	6
	78	Canada	6
	78	Netherlands	6
	78	Italy	7
	78	Australia	7
	78	Spain	8
	77	U.K.	6
WORST	48	Somalia	125
	47	Mali	117
	47	Mozambique	158
	46	Gambia	80
	46	Guinea	128
	45	Guinea-Bissau	134
	45	Malawi	138
	45	Afghanistan	165
	44	Uganda	111
	40	Sierra Leone	164

Infant mortality

Deaths before first birthday per 1000 live births 1995

- 0-9
- 10-49
- 50-99
- 100-149
- >149
- no data

A HEALTHY WORLD?

STANDARDS OF HEALTH vary widely throughout the world. Richer countries enjoy higher standards than poorer countries and there are similar inequalities within countries.

Great progress is being made in all aspects of health and there have been significant improvements in the two main indicators of health levels over the last few decades. The world average life expectancy increased from 48 to 65 years between 1955 and 1995; and infant mortality rates have fallen - deaths among children under five declined from 19 million in 1960 to 11 million in 1996.

Longer life and a greater chance of survival do not necessarily mean a healthy life. There is still a need to ensure a freedom from additional years of ill-health and poverty - quality of life is as important as quantity.

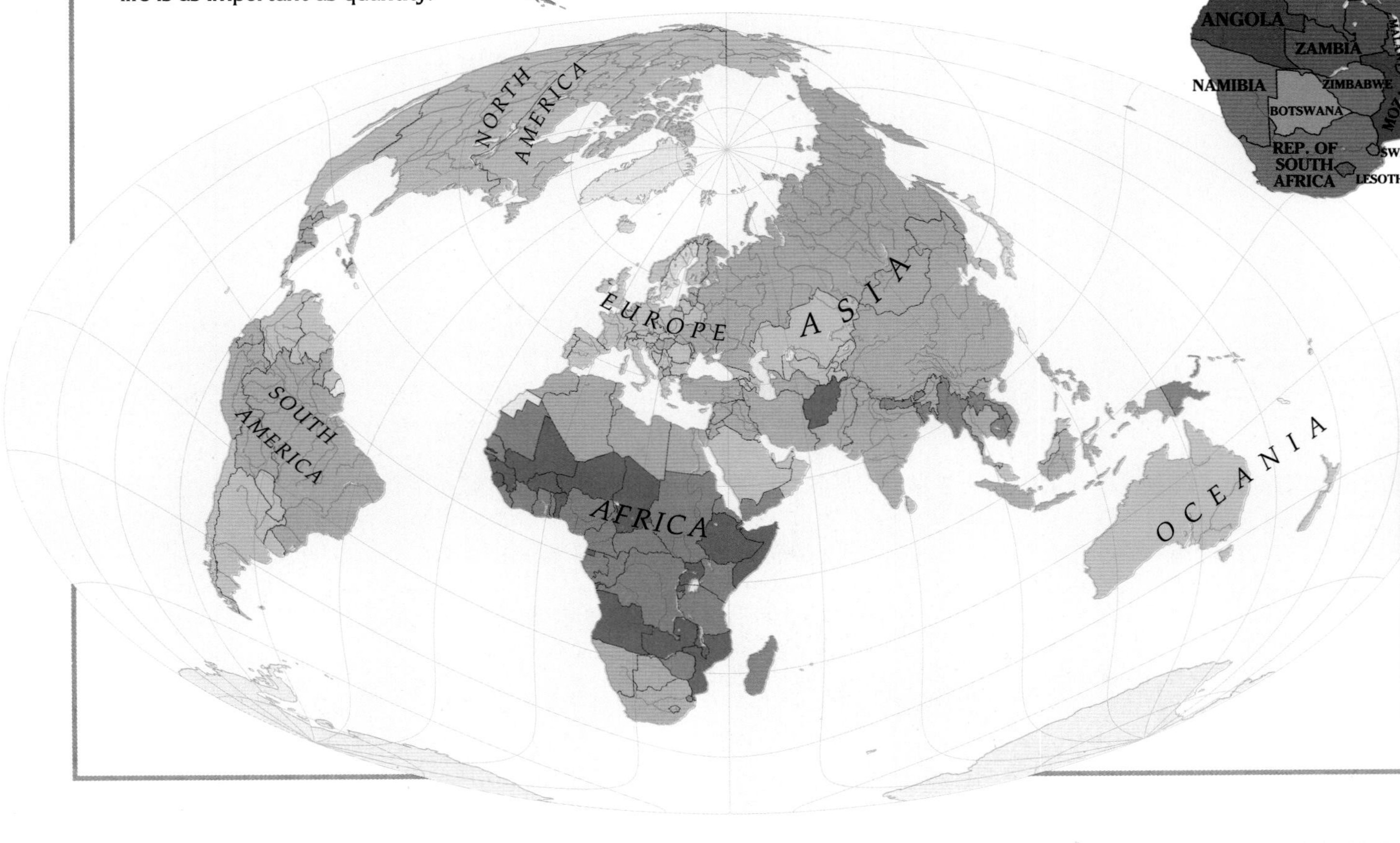

Life expectancy

Number of years a new born ch[ild] can expect to survive

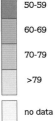

- <50
- 50-59
- 60-69
- 70-79
- >79
- no data

CAUSES OF DEATH

The dominant causes of death in the developing world are infectious and parasitic diseases, in particular lower respiratory infection, tuberculosis, diarrhoeal diseases and malaria. By contrast, people in the richer, developed countries suffer more from circulatory diseases and cancers - illnesses generally occurring later in life and often associated with life-style.

This pattern is gradually changing. As living standards and life expectancy increase throughout the world there is a corresponding increase in the risks from diseases prevalent in the developed world. This provides the developing world with the 'double burden' of coping with existing high rates of infectious diseases and increasing rates of chronic illness.

Patterns in causes of death again reflect relative wealth. Infectious and parasitic diseases are the easiest to prevent and eradicate but, despite great successes in the eradication of smallpox and the imminent demise of polio, the resources are so often lacking in the areas suffering most.

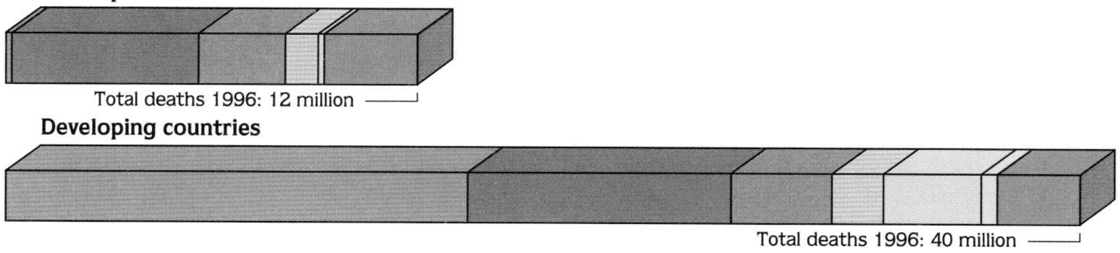

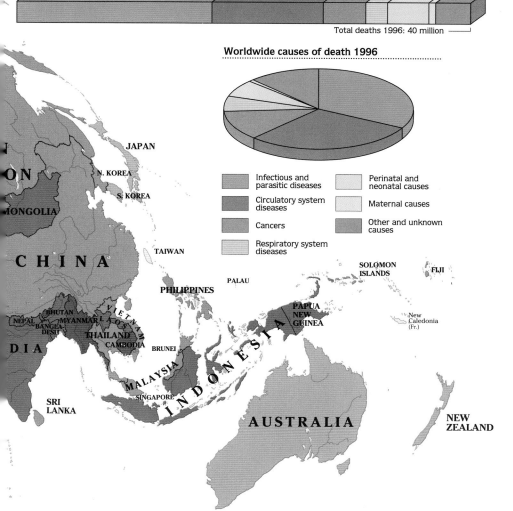

Worldwide causes of death 1996

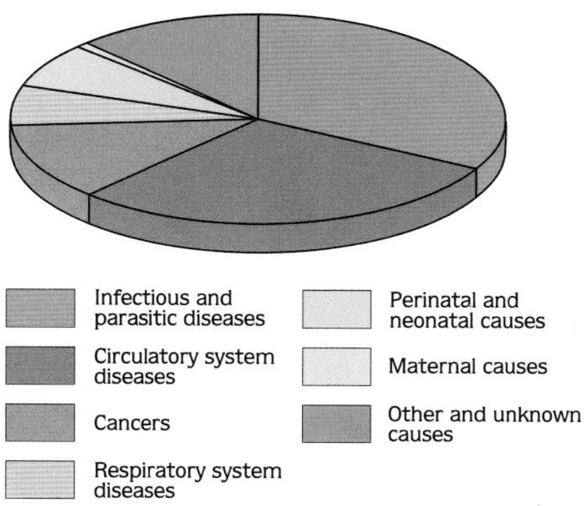

HEALTH PROVISION

Easy access to appropriate health services are taken for granted in the developed world where 100% access for the population can often be assumed. In developing countries, however, such access is often far more limited, as the graph below shows. Social and economic conditions in a country can greatly influence the chances of a healthy life.

The provision of conditions for good health, including trained personnel, medical facilities and equipment as well as those of safe water, food and sanitation, obviously costs money. Some countries are in a much better position to meet these costs than others. This is reflected in the overall differences in standards of health between the developed and developing world. The richer countries not only enjoy better health, but also spend proportionately more of their Gross National Product on health provision. The poorer countries of the world are spending a smaller proportion of much less money on vital facilities.

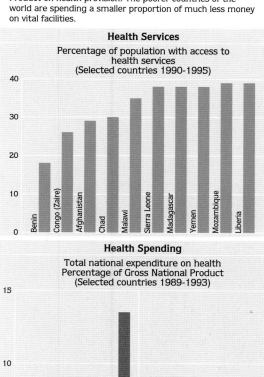

Health Services

Percentage of population with access to health services (Selected countries 1990-1995)

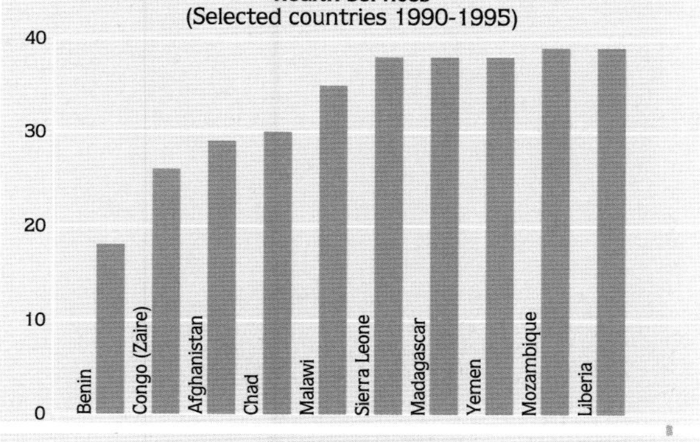

Health Spending

Total national expenditure on health Percentage of Gross National Product (Selected countries 1989-1993)

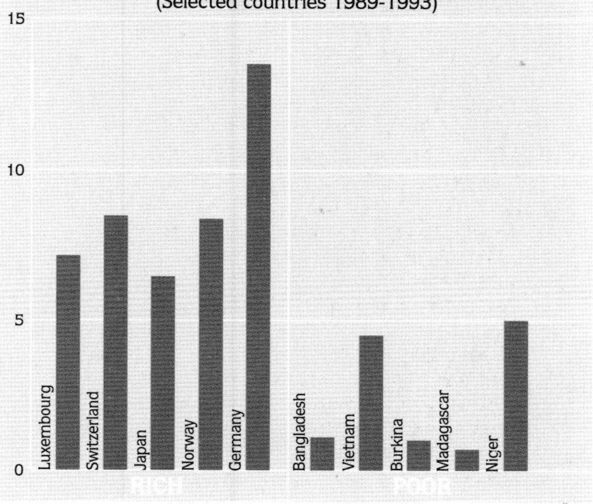

Access to safe water

Percentage of population with reasonable access to sufficient safe water 1995

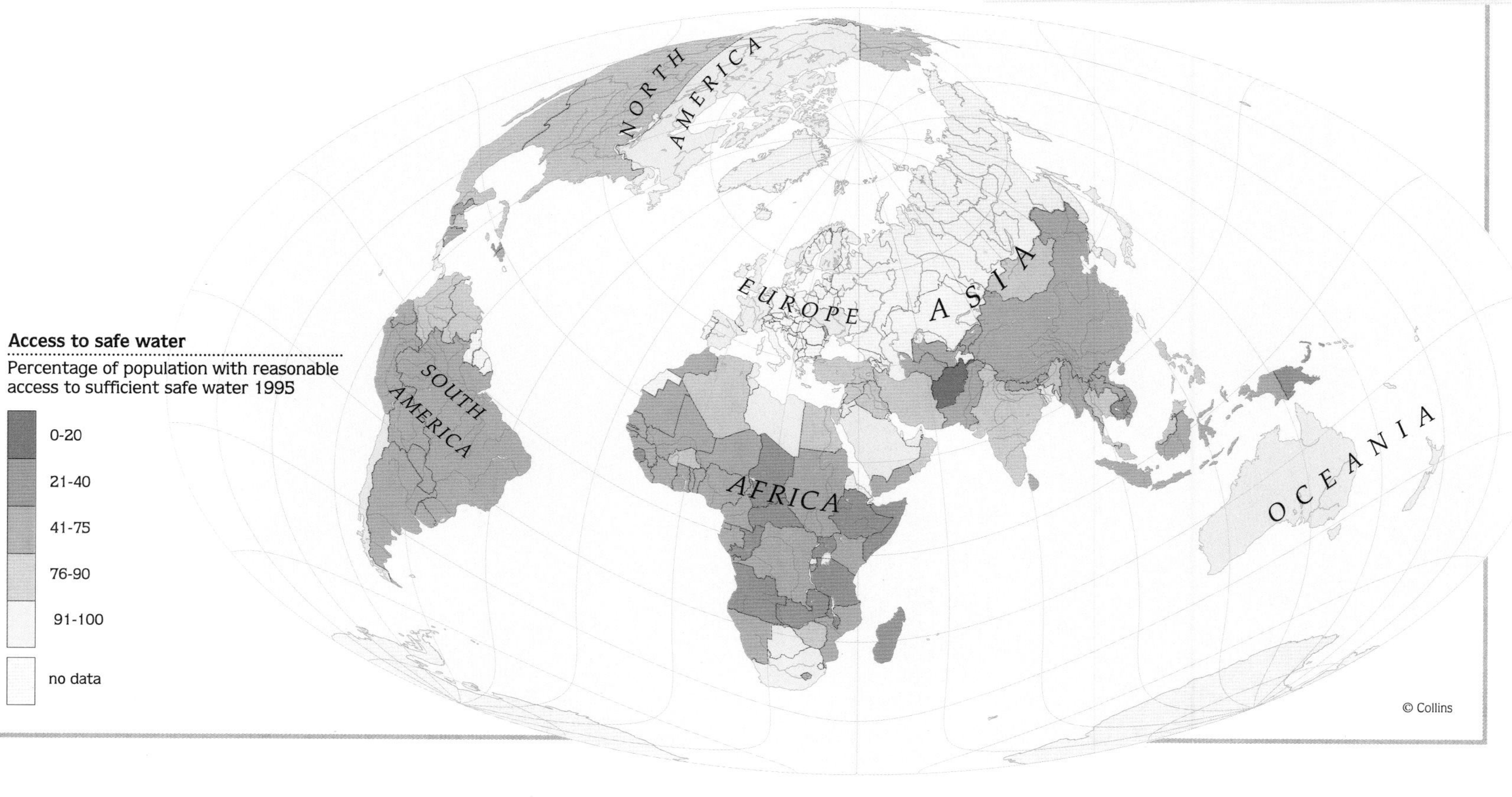

MEASURING THE WORLD'S WEALTH

A commonly used measure of wealth is Gross National Product (GNP). This is the total value of goods and services produced by a country in any one year, including income from investments abroad. If the total GNP figure is divided by the country's population, the average wealth per person is provided as GNP per capita.

GNP per capita statistics provide only an average figure and give no indication of the relative distribution of wealth within the country. They show neither the great inequalities which can exist, nor the relative numbers of people living in poverty. Also, GNP is usually based on valuations in US$ and not in local currencies. Purchasing Power Parity (PPP), measured in International Dollars, is another means of measuring individual wealth which uses local exchange rates and takes account of cost of living differences between countries (see table below).

No method provides a perfect picture of the world's complex ecomomy, but whichever method is used clear patterns of wealth and poverty - the rich and poor of the world - emerge.

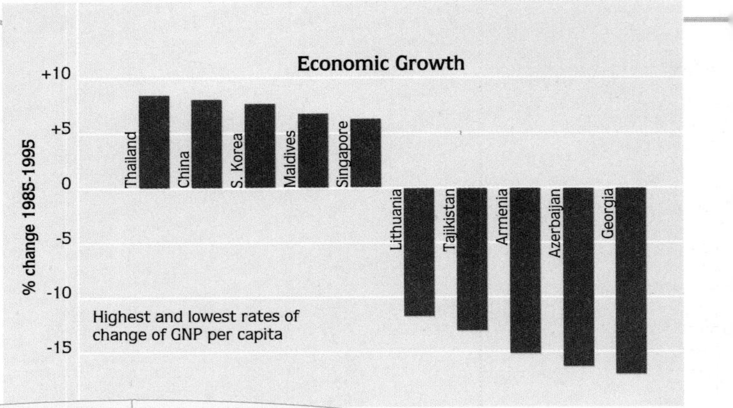

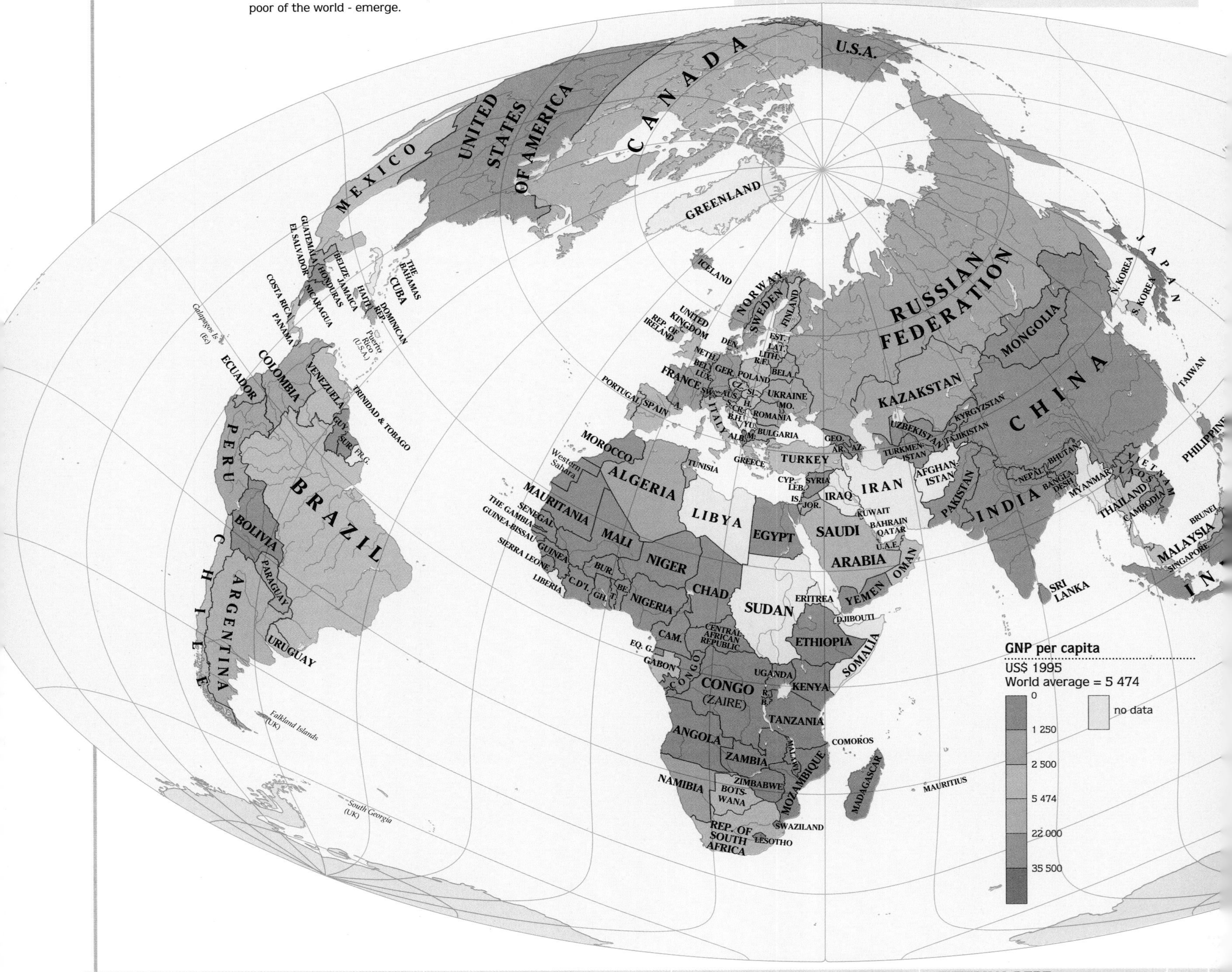

	Rank	GROSS NATIONAL PRODUCT (GNP) (US$ Millions 1995)	GNP PER CAPITA (US$ 1995)	PURCHASING POWER PARITY(PPP) ($ International 1995)	EXTERNAL DEBT (% of GNP 1995) Low- and middle-income economies	Rank
HIGHEST	1	U.S.A. (7 100 007)	LUXEMBOURG (41 210)	LUXEMBOURG (37 930)	NICARAGUA (520)	1
	2	JAPAN (4 963 587)	SWITZERLAND (40 630)	U.S.A. (26 980)	MOZAMBIQUE (333)	2
	3	GERMANY (2 252 343)	JAPAN (39 640)	SWITZERLAND (25 860)	CONGO (325)	3
	4	FRANCE (1 451 051)	NORWAY (31 250)	KUWAIT (23 790)	ANGOLA (260)	4
	5	UNITED KINGDOM (1 094 734)	DENMARK (29 890)	SINGAPORE (22 770)	GUINEA-BISSAU (235)	5
	6	ITALY (1 088 085)	GERMANY (27 510)	JAPAN (22 110)	CÔTE D'IVOIRE (185)	6
	7	CHINA (744 890)	U.S.A. (26 980)	NORWAY (21 940)	MAURITANIA (166)	7
	8	BRAZIL (579 787)	AUSTRIA (26 890)	BELGIUM (21 660)	TANZANIA (148)	8
	9	CANADA (573 695)	SINGAPORE (26 730)	AUSTRIA (21 250)	ZAMBIA (139)	9
	10	SPAIN (532 347)	FRANCE (24 990)	DENMARK (21 230)	VIETNAM (138)	10
LOWEST	10	COMOROS (237)	YEMEN (260)	NIGER (750)	KYRGYZSTAN (15)	10
	9	DOMINICA (218)	GUINEA-BISSAU (250)	CHAD (700)	ARMENIA (14)	9
	8	MICRONESIA (215)	HAITI (250)	MADAGASCAR (640)	BOTSWANA (13)	8
	7	ST KITTS & NEVIS (212)	MALI (250)	TANZANIA (640)	UKRAINE (10)	7
	6	VANUATU (202)	BANGLADESH (240)	BURUNDI (630)	LITHUANIA (9)	6
	5	WESTERN SAMOA (184)	UGANDA (240)	SIERRA LEONE (580)	AZERBAIJAN (8)	5
	4	TONGA (170)	VIETNAM (240)	MALI (550)	LATVIA (7)	4
	3	EQUATORIAL GUINEA (152)	BURKINA (230)	RWANDA (540)	UZBEKISTAN (7)	3
	2	KIRIBATI (73)	MADAGASCAR (230)	CONGO (ZAIRE) (490)	BELARUS (6)	2
	1	SÃO TOMÉ & PRÍNCIPE (45)	NIGER (220)	ETHIOPIA (450)	ESTONIA (6)	1

FOREIGN DEBT

A critical problem facing many of the world's poorest countries is that of foreign debt. To assist them in development programmes in the past, many countries borrowed huge amounts from such agencies as the World Bank. Changes in the world's economy have created conditions in which it is virtually impossible for these countries to repay their loans.

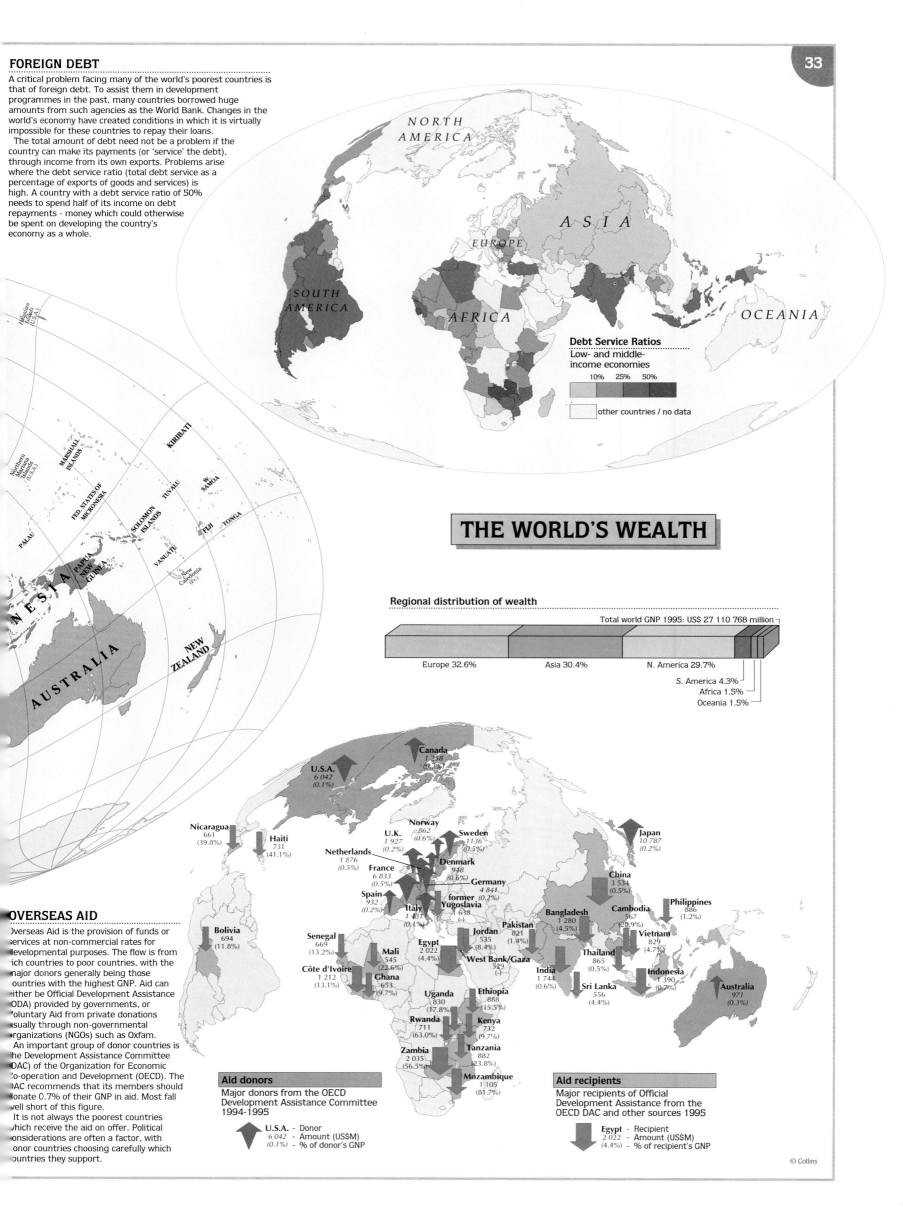

The total amount of debt need not be a problem if the country can make its payments (or 'service' the debt), through income from its own exports. Problems arise where the debt service ratio (total debt service as a percentage of exports of goods and services) is high. A country with a debt service ratio of 50% needs to spend half of its income on debt repayments - money which could otherwise be spent on developing the country's economy as a whole.

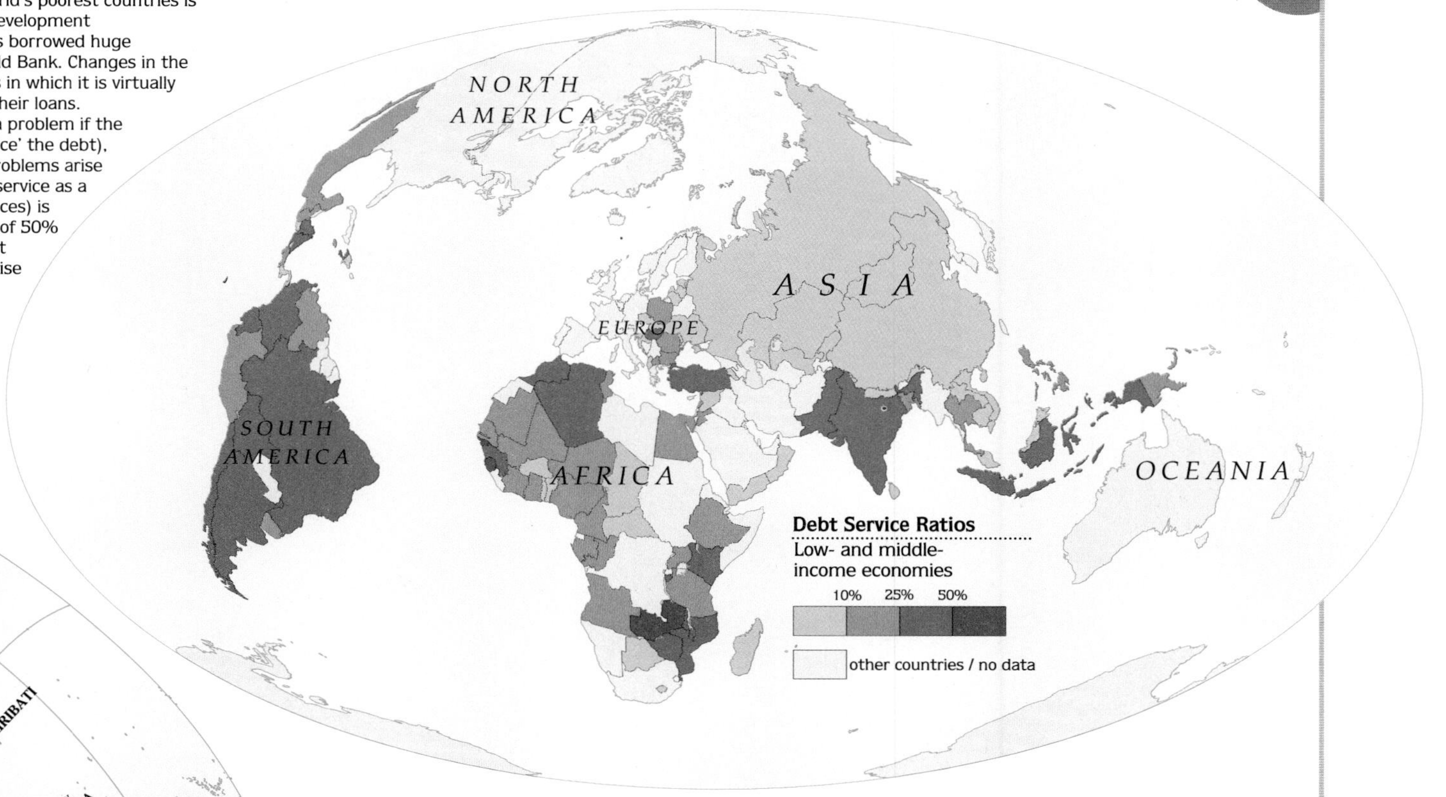

THE WORLD'S WEALTH

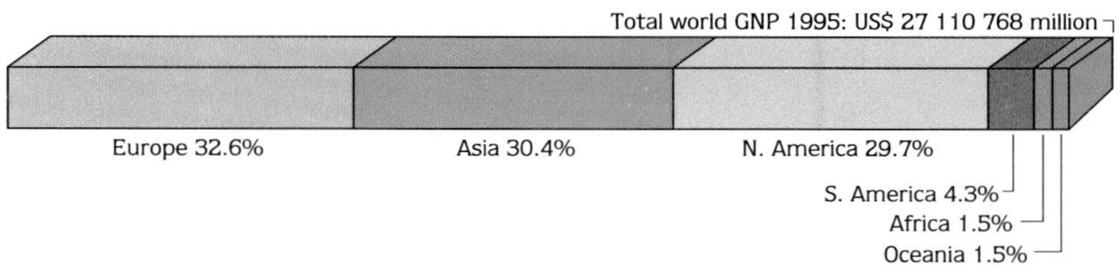

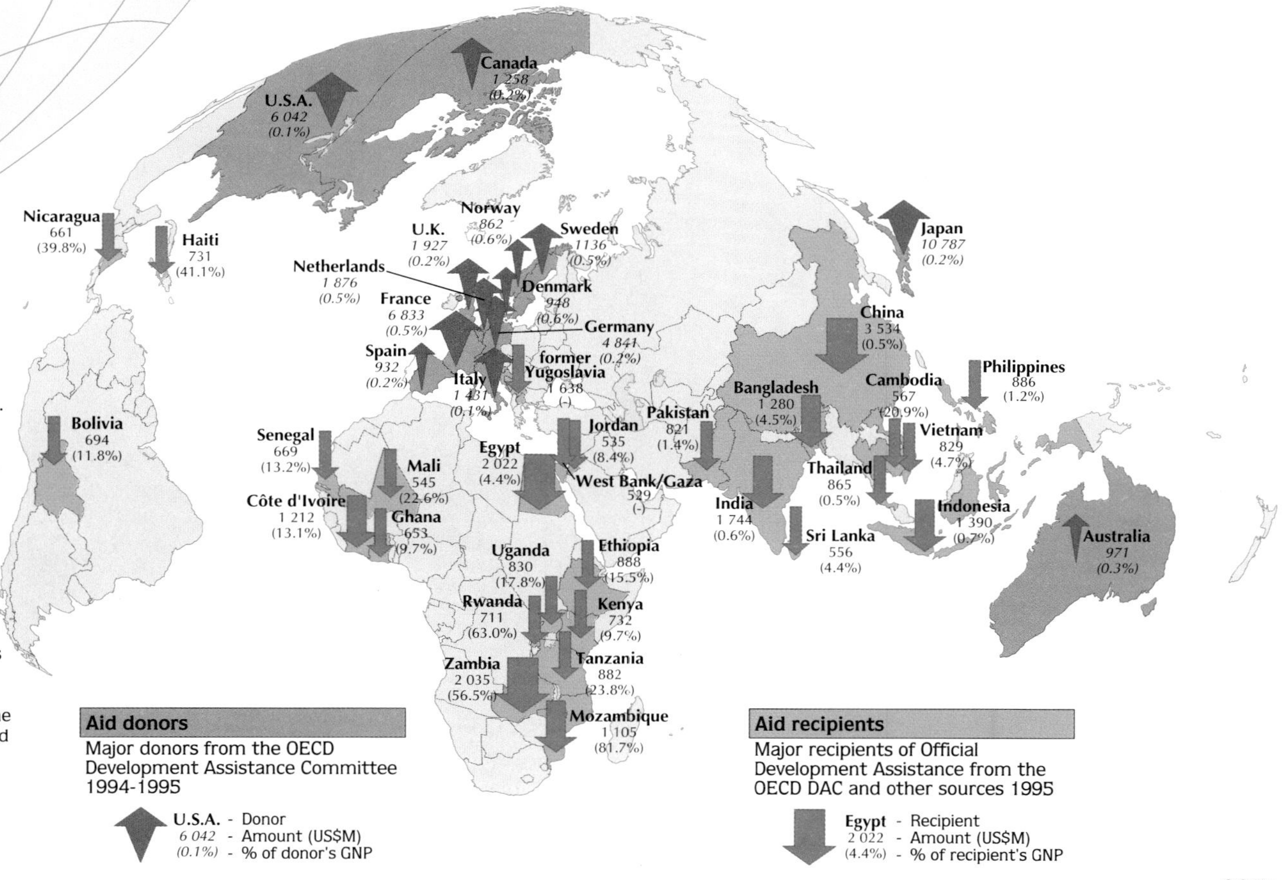

OVERSEAS AID

)verseas Aid is the provision of funds or services at non-commercial rates for developmental purposes. The flow is from rich countries to poor countries, with the major donors generally being those countries with the highest GNP. Aid can either be Official Development Assistance (ODA) provided by governments, or Voluntary Aid from private donations usually through non-governmental organizations (NGOs) such as Oxfam.

An important group of donor countries is the Development Assistance Committee (DAC) of the Organization for Economic Co-operation and Development (OECD). The DAC recommends that its members should donate 0.7% of their GNP in aid. Most fall well short of this figure.

It is not always the poorest countries which receive the aid on offer. Political considerations are often a factor, with donor countries choosing carefully which countries they support.

INTERNATIONAL TRAVEL

All parts of the world are experiencing steady growth in air travel and tourism. Worldwide, airline traffic for passengers and freight grew by 7% in 1995 and the rise is expected to continue at a similar rate into the next century. Rates of growth vary between regions, with the dominant region being East Asia and the Pacific which is expected to sustain a growth rate in passenger traffic of approximately 12%.

New international airports have recently been completed in Denver, USA and Macau; and Hong Kong's new Chek Lap Kok airport is due to open in 1998 - further indicators of the strength of the aviation business.

Healthy economic conditions, particularly in the developed world, have encouraged recent increases in international tourism. Both tourist arrivals and receipts increased significantly in 1995 with the fastest-growing region again being that of East Asia and the Pacific.

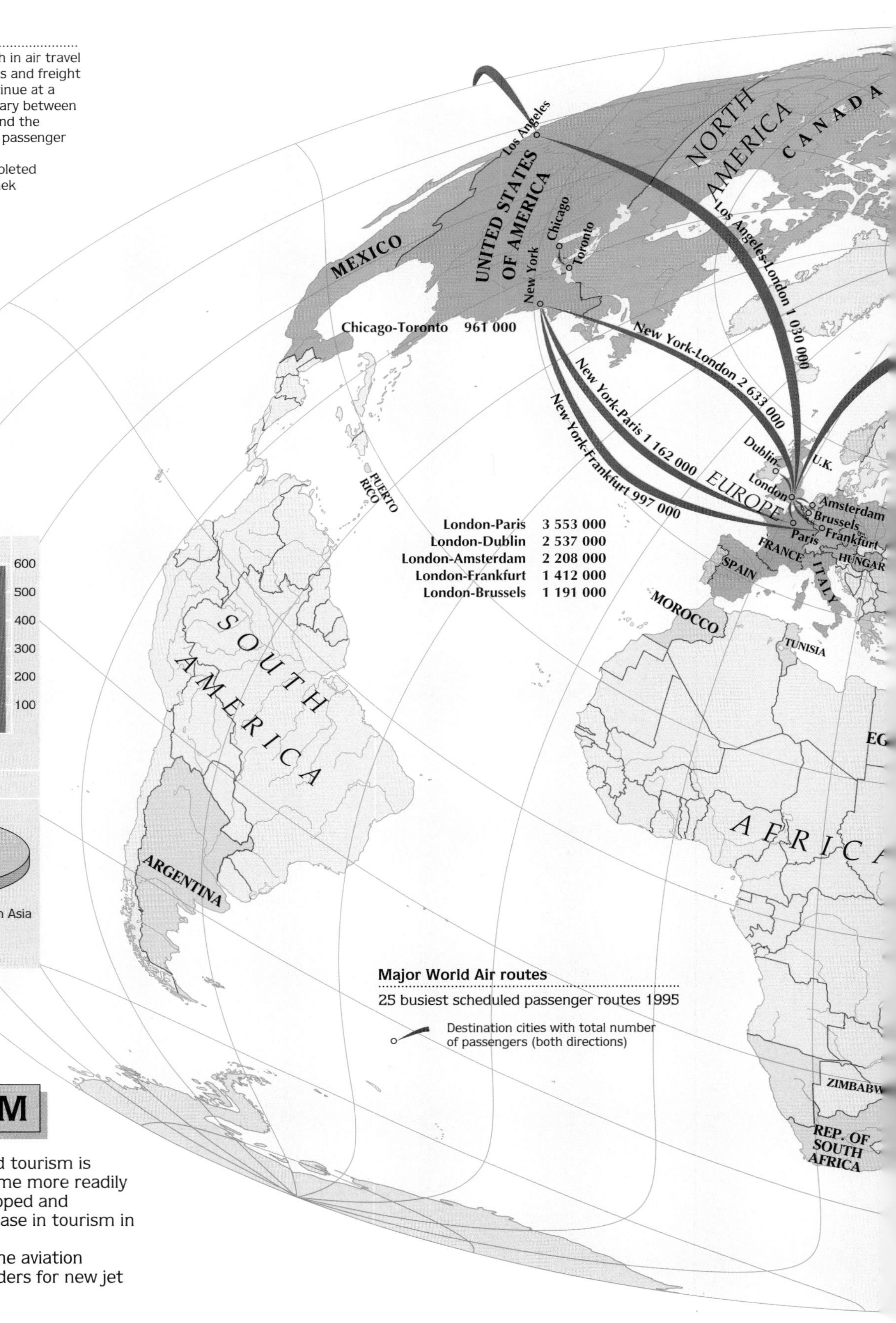

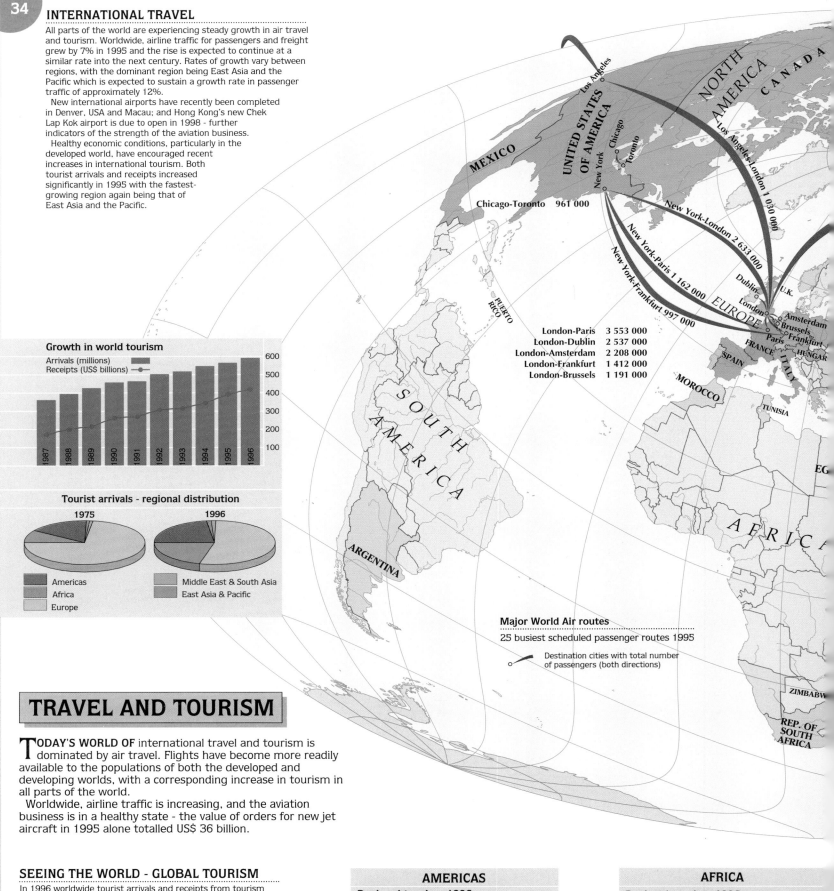

MEXICO
UNITED STATES OF AMERICA
NORTH AMERICA
CANADA
Los Angeles
Chicago
Toronto
New York
Chicago-Toronto 961 000
Los Angeles-London 1 030 000
New York-London 2 633 000
New York-Paris 1 162 000
New York-Frankfurt 997 000
Dublin
U.K.
London
EUROPE
Amsterdam
Brussels
Paris
Frankfurt
FRANCE
SPAIN
ITALY
HUNGARY
MOROCCO
TUNISIA
PUERTO RICO
London-Paris 3 553 000
London-Dublin 2 537 000
London-Amsterdam 2 208 000
London-Frankfurt 1 412 000
London-Brussels 1 191 000
SOUTH AMERICA
ARGENTINA
AFRICA
ZIMBABWE
REP. OF SOUTH AFRICA

Major World Air routes

25 busiest scheduled passenger routes 1995

Destination cities with total number of passengers (both directions)

Growth in world tourism

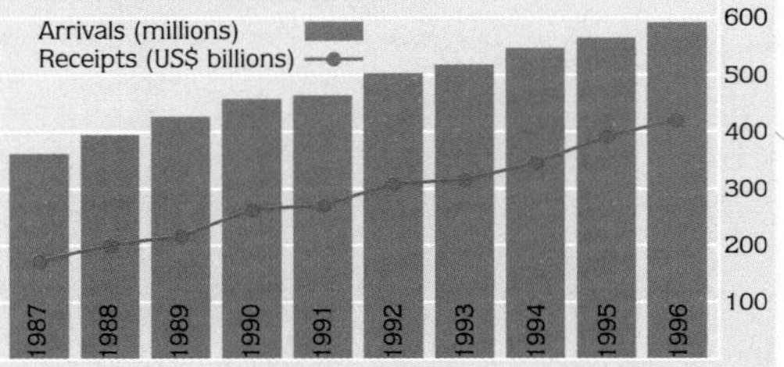

Arrivals (millions)
Receipts (US$ billions)
600
500
400
300
200
100
1987
1988
1989
1990
1991
1992
1993
1994
1995
1996

Tourist arrivals - regional distribution

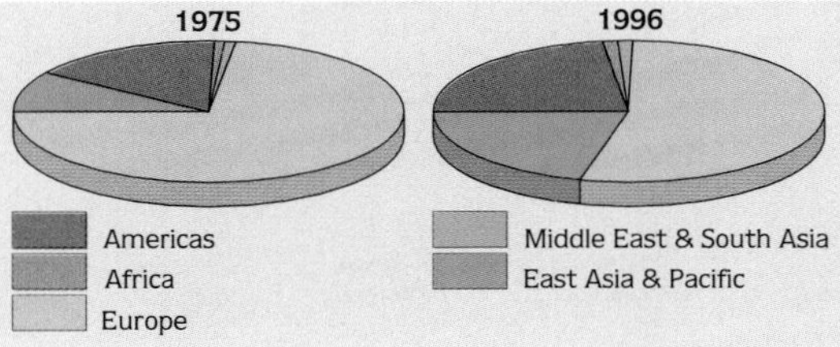

1975
1996
Americas
Africa
Europe
Middle East & South Asia
East Asia & Pacific

TRAVEL AND TOURISM

TODAY'S WORLD OF international travel and tourism is dominated by air travel. Flights have become more readily available to the populations of both the developed and developing worlds, with a corresponding increase in tourism in all parts of the world.

Worldwide, airline traffic is increasing, and the aviation business is in a healthy state - the value of orders for new jet aircraft in 1995 alone totalled US$ 36 billion.

SEEING THE WORLD - GLOBAL TOURISM

In 1996 worldwide tourist arrivals and receipts from tourism increased by 4.5% and 7.6% respectively. Growth is expected to continue, largely because of increasing numbers of short-duration overseas visits by travellers from the developed world. Foreign travel from within the developing regions is also increasing steadily.

Europe is the dominant region in terms of arrivals and receipts, but its relative share is decreasing. Its market share has fallen by over 10% since 1975, compared with a corresponding rise of 11.3% in the East Asia and Pacific region. This shift reflects an overall increase in long-haul flights, from Europe in particular, to tourist destinations such as China (including Hong Kong), Malaysia and Thailand.

AMERICAS

Regional tourism 1996

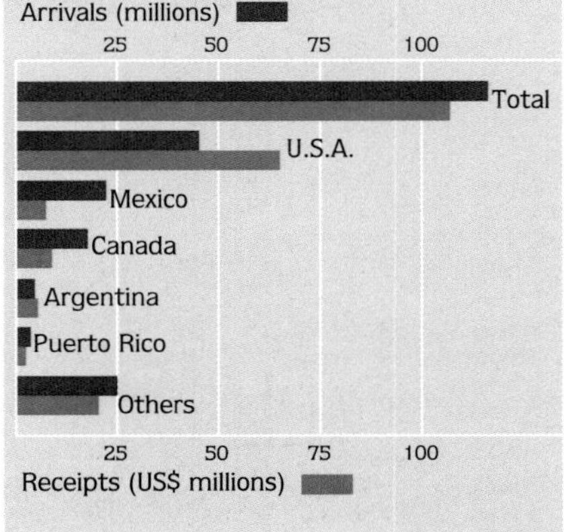

Arrivals (millions)
25
50
75
100
Total
U.S.A.
Mexico
Canada
Argentina
Puerto Rico
Others
Receipts (US$ millions)

AFRICA

Regional tourism 1996

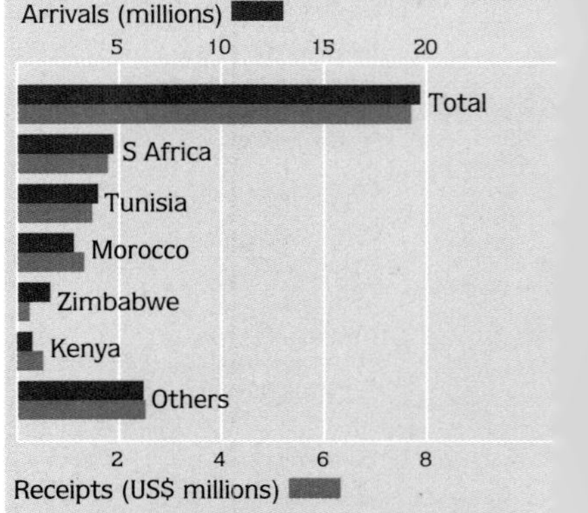

Arrivals (millions)
5
10
15
20
Total
S Africa
Tunisia
Morocco
Zimbabwe
Kenya
Others
2
4
6
8
Receipts (US$ millions)

THE WORLD'S BUSIEST AIRPORTS

City/Airport	Code	Passengers (1996) Total	International
Chicago O'Hare	ORD	69 153 528	7 218 461
Atlanta Hartsfield	ATL	63 303 171	3 060 173
Dallas /Fort Worth	DFW	58 034 503	3 341 007
Los Angeles	LAX	57 974 559	14 032 531
London Heathrow	LHR	56 037 798	48 257 080
Tokyo Haneda	HND	46 631 475	843 644
San Francisco	SFO	39 251 942	6 644 354
Frankfurt/Main	FRA	38 761 174	31 016 550
Seoul Kimpo	SEL	34 706 158	14 705 015
Miami	MIA	33 504 579	14 913 477

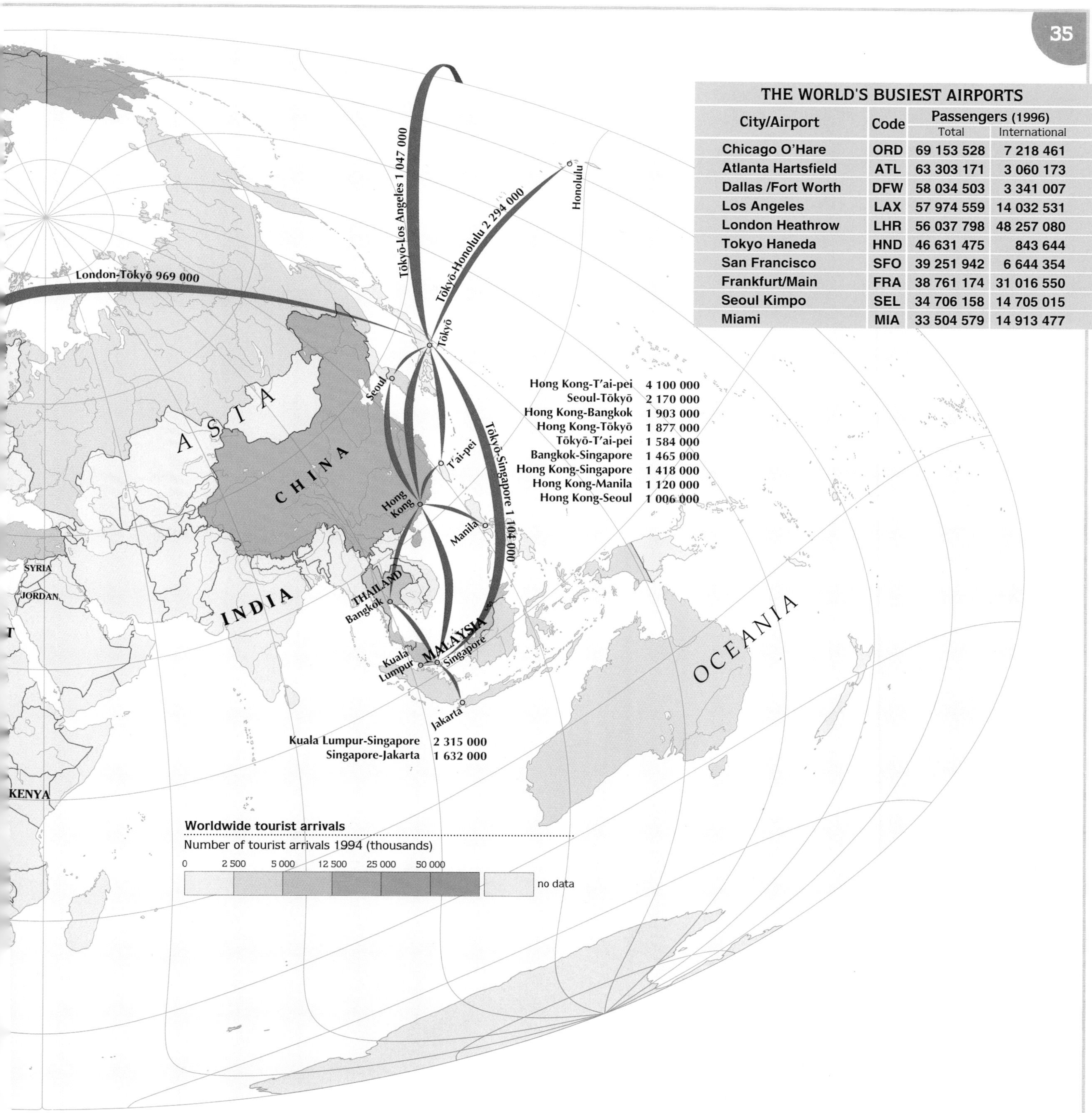
London-Tōkyō 969 000
Tōkyō-Los Angeles 1 047 000
Tōkyō-Honolulu 2 294 000
Honolulu
Tōkyō
Seoul
T'ai-pei
Hong Kong
Manila
Tōkyō-Singapore 1 104 000
Bangkok
THAILAND
MALAYSIA
Kuala Lumpur
Singapore
Jakarta
ASIA
CHINA
INDIA
OCEANIA
SYRIA
JORDAN
KENYA
Hong Kong-T'ai-pei 4 100 000
Seoul-Tōkyō 2 170 000
Hong Kong-Bangkok 1 903 000
Hong Kong-Tōkyō 1 877 000
Tōkyō-T'ai-pei 1 584 000
Bangkok-Singapore 1 465 000
Hong Kong-Singapore 1 418 000
Hong Kong-Manila 1 120 000
Hong Kong-Seoul 1 006 000
Kuala Lumpur-Singapore 2 315 000
Singapore-Jakarta 1 632 000
Worldwide tourist arrivals
Number of tourist arrivals 1994 (thousands)
0 2 500 5 000 12 500 25 000 50 000
no data

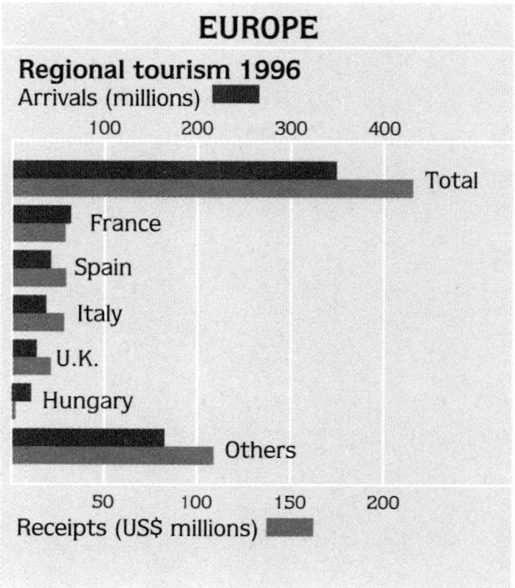
EUROPE
Regional tourism 1996
Arrivals (millions)
100 200 300 400
Total
France
Spain
Italy
U.K.
Hungary
Others
50 100 150 200
Receipts (US$ millions)

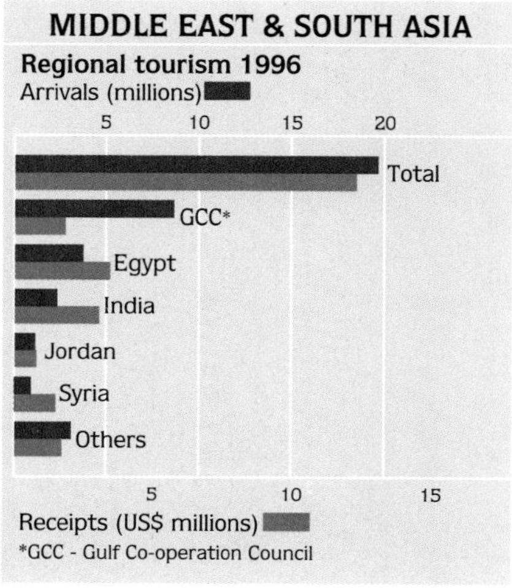
MIDDLE EAST & SOUTH ASIA
Regional tourism 1996
Arrivals (millions)
5 10 15 20
Total
GCC*
Egypt
India
Jordan
Syria
Others
5 10 15
Receipts (US$ millions)
*GCC - Gulf Co-operation Council

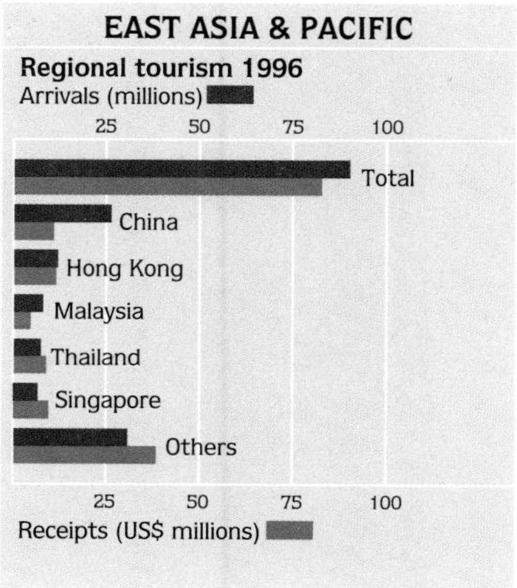
EAST ASIA & PACIFIC
Regional tourism 1996
Arrivals (millions)
25 50 75 100
Total
China
Hong Kong
Malaysia
Thailand
Singapore
Others
25 50 75 100
Receipts (US$ millions)

Fibre-optic cable

COMMUNICATIONS TECHNOLOGY

Satellites are used by earth stations to receive and amplify information which is in the form of high-powered, high-frequency signals, and retransmit it back to stations in another part of the world.

Fibre-optic cables have been developed to overcome the limited capacity of the copper wires traditionally used for communication. Information is encoded into beams of laser light and sent down fine fibres of coated glass which can carry signals over large distances with little loss of quality.

Improving cost:capacity ratios have meant that fibre-optic cables are overtaking satellites in the transmission of point-to-point communications (e.g. telephone calls) but satellites remain the prime carrier of point-to-multi point signals (e.g. television broadcasts).

Satellite transmission basics

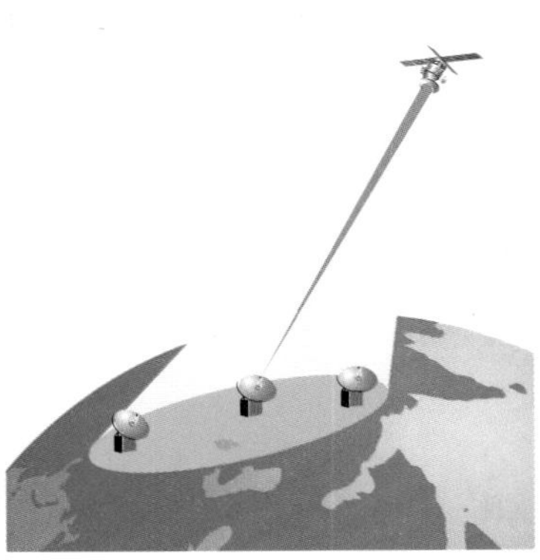

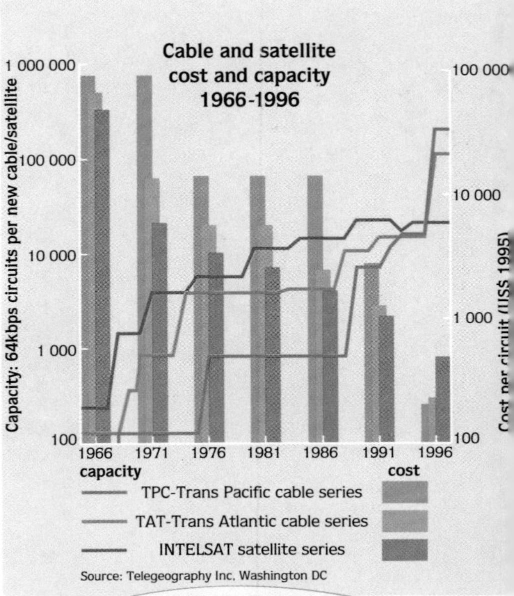

GLOBAL TELECOMMUNICATIONS

DEVELOPMENTS IN TECHNOLOGY have improved the speed and extent of voice, image and text communication to a level previously unimaginable, now making possible almost instant connection between people throughout the world by means of telephone, television, facsimile, and personal computer.

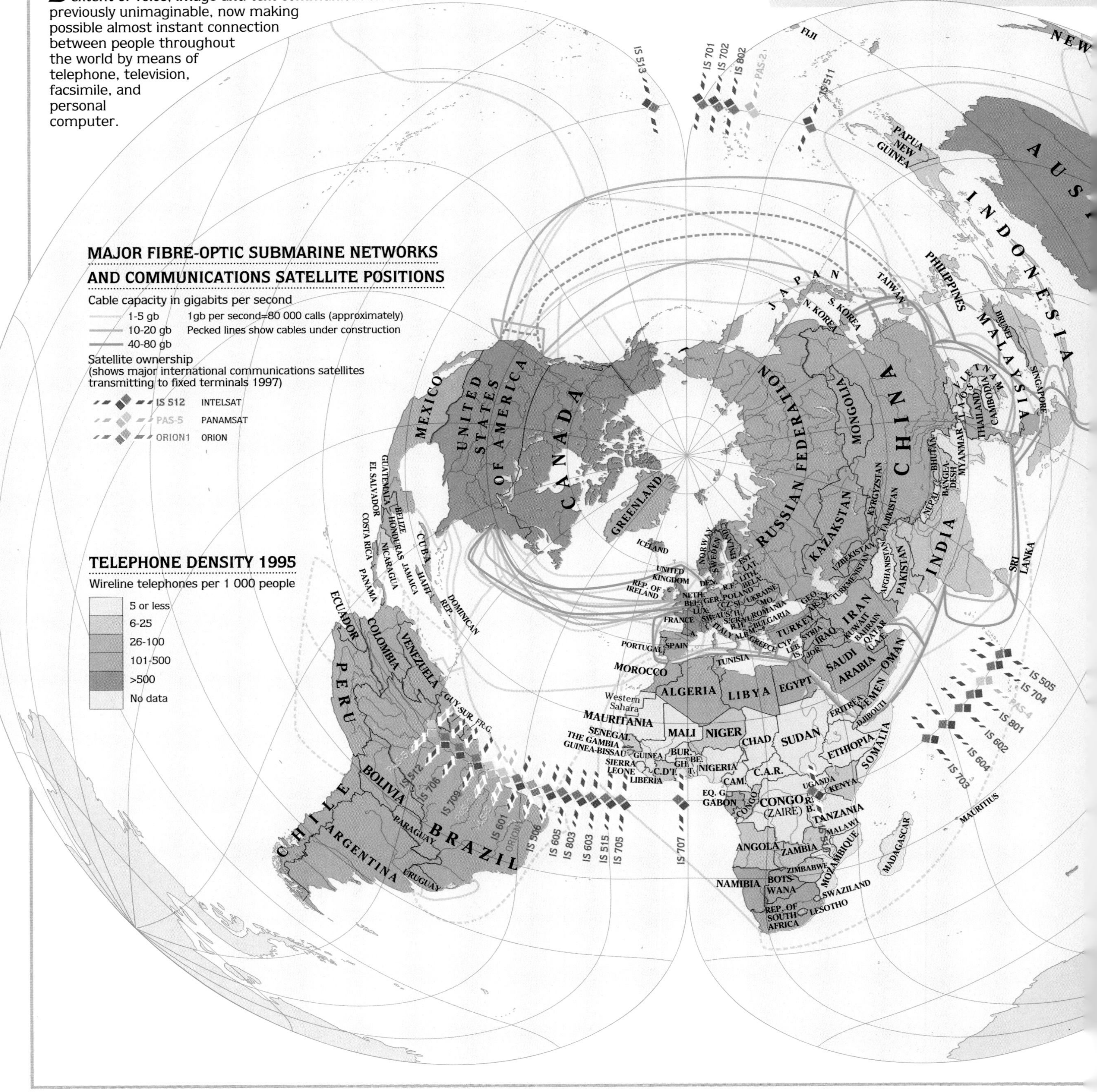

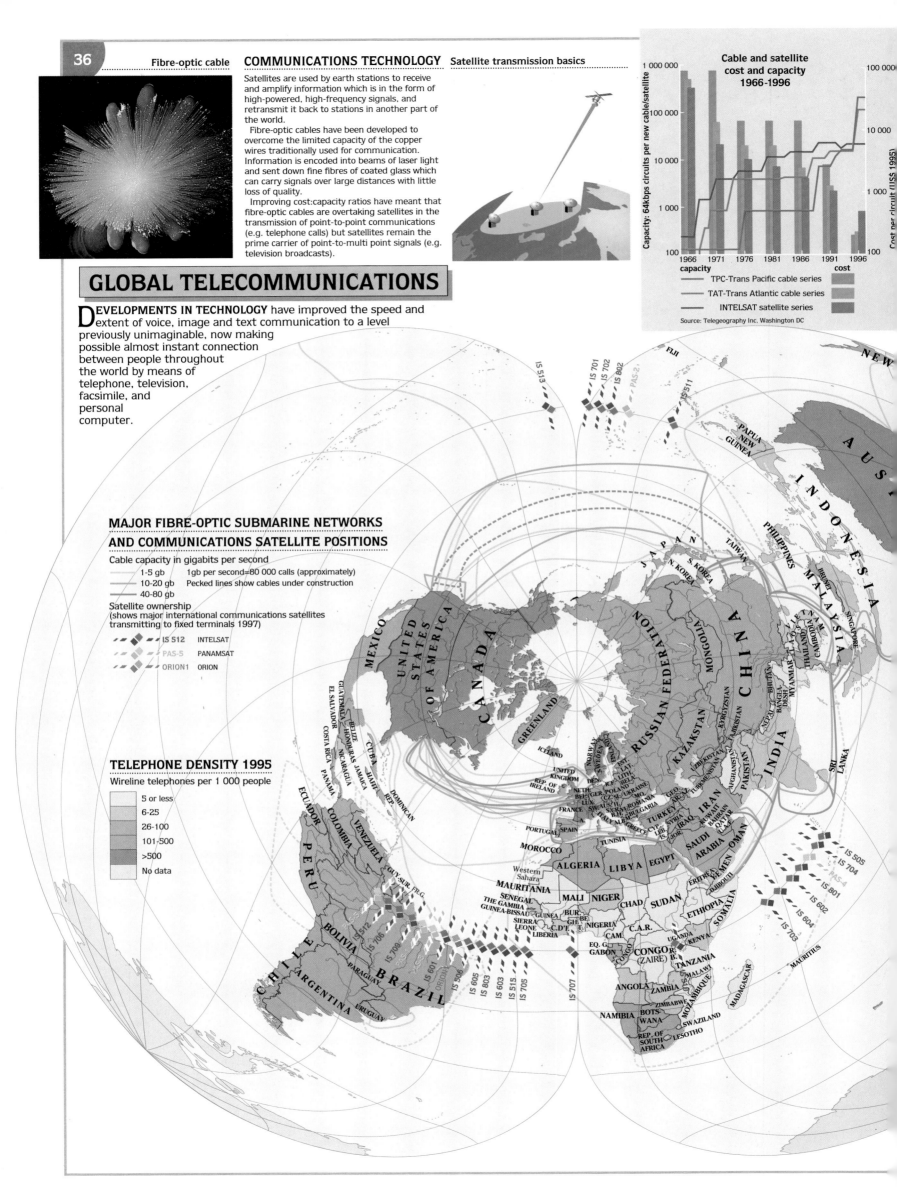

WORLD COMMUNICATIONS EQUIPMENT

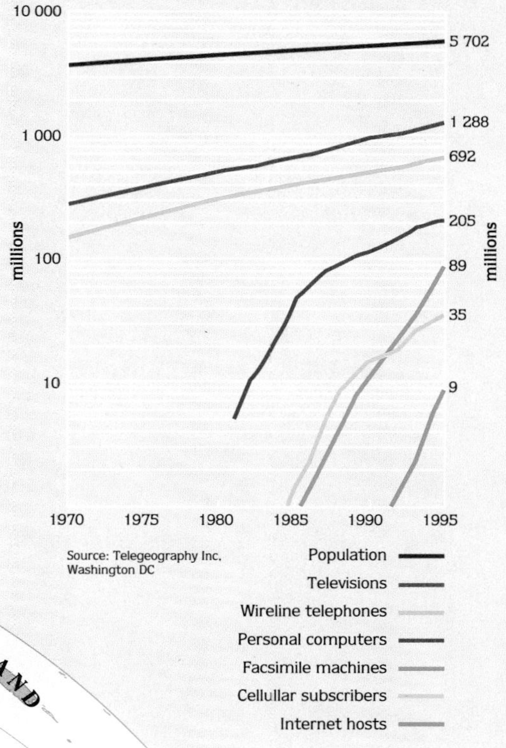

10 000
1 000
100
10
millions
5 702
1 288
692
205
89
35
9
millions
1970
1975
1980
1985
1990
1995
Source: Telegeography Inc, Washington DC
Population
Televisions
Wireline telephones
Personal computers
Facsimile machines
Cellullar subscribers
Internet hosts

INTERCONTINENTAL TELECOMMUNICATIONS TRAFFIC 1995

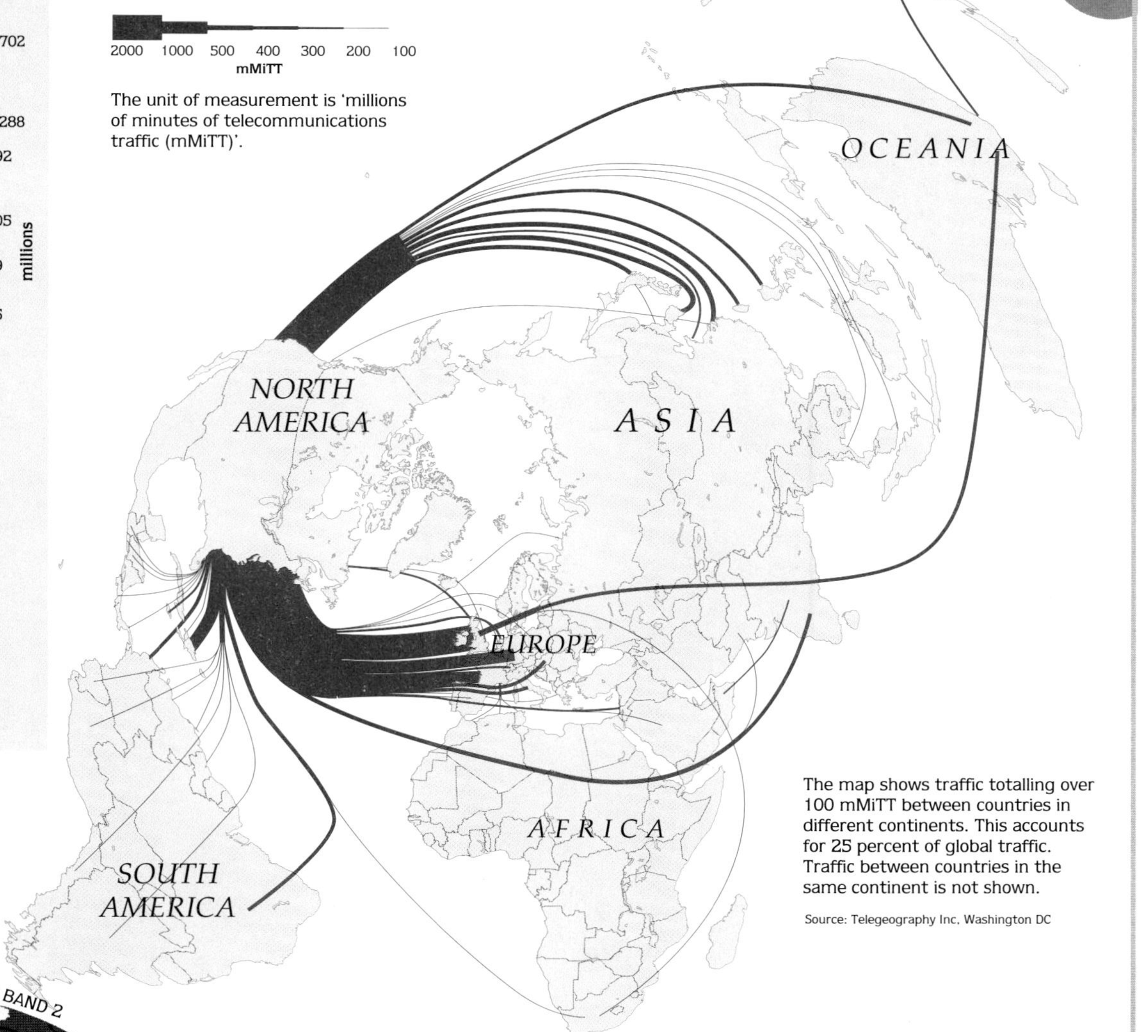

2000 1000 500 400 300 200 100
mMiTT
The unit of measurement is 'millions of minutes of telecommunications traffic (mMiTT)'.
OCEANIA
NORTH AMERICA
ASIA
EUROPE
AFRICA
SOUTH AMERICA

The map shows traffic totalling over 100 mMiTT between countries in different continents. This accounts for 25 percent of global traffic. Traffic between countries in the same continent is not shown.

Source: Telegeography Inc, Washington DC

ZEALAND
ALIA

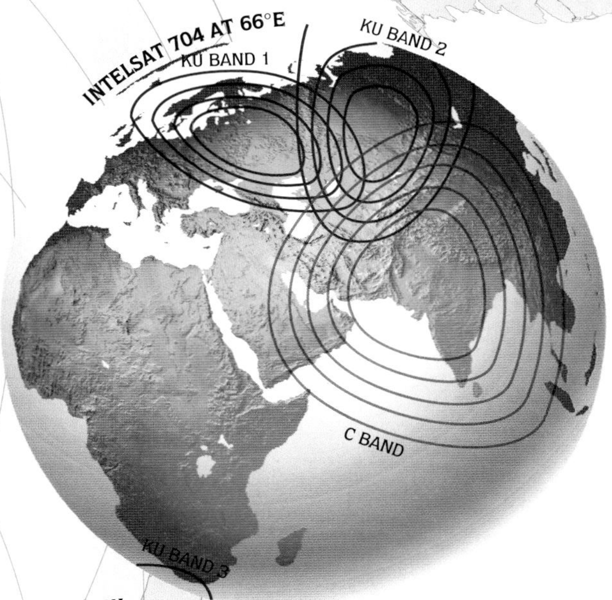

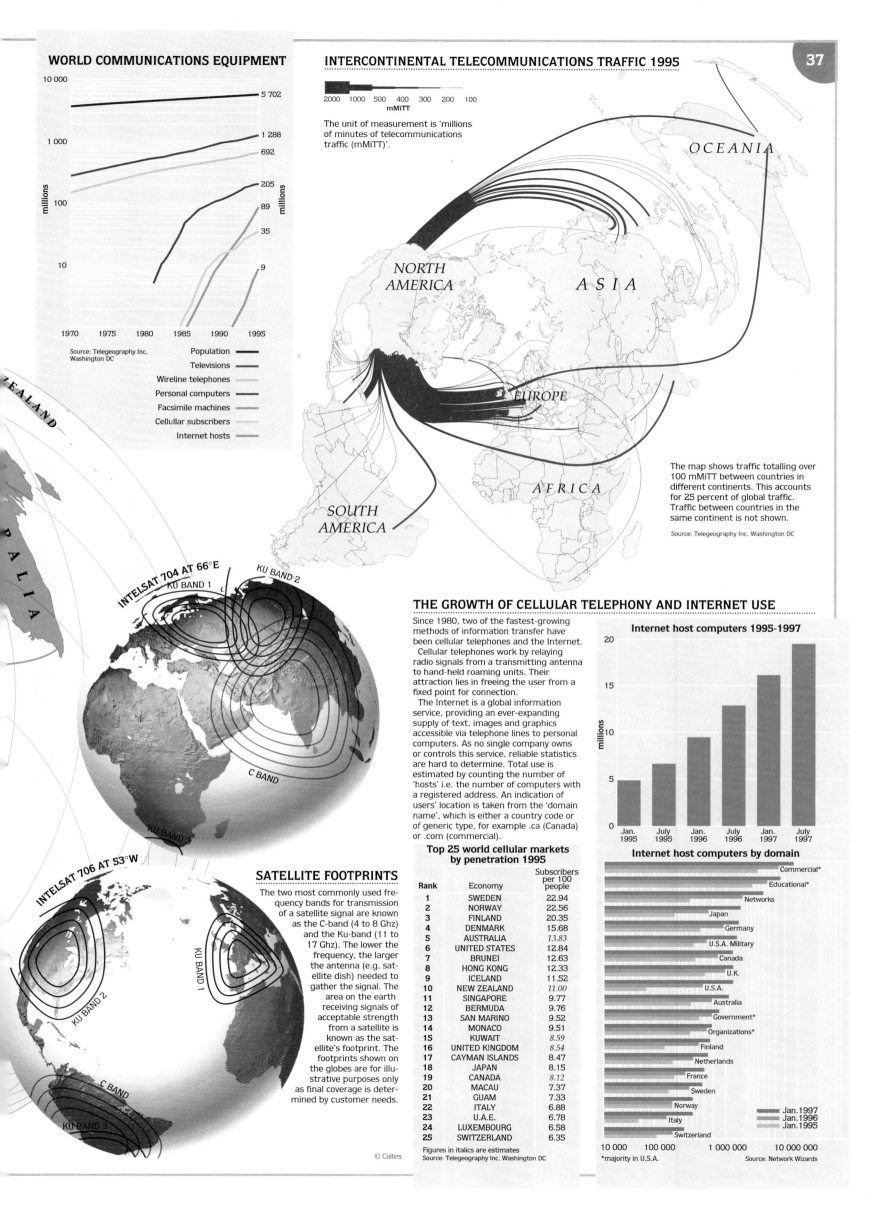

INTELSAT 704 AT 66°E
KU BAND 1
KU BAND 2
C BAND
KU BAND 3

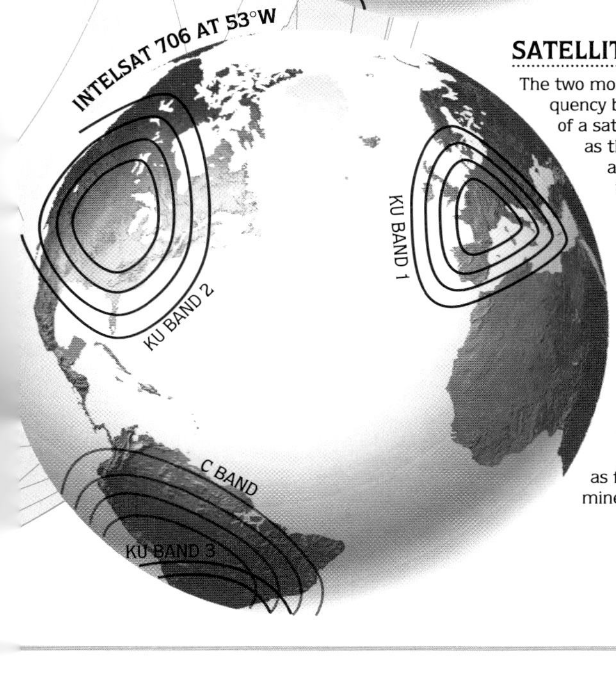

INTELSAT 706 AT 53°W
KU BAND 1
KU BAND 2
C BAND
KU BAND 3

SATELLITE FOOTPRINTS

The two most commonly used frequency bands for transmission of a satellite signal are known as the C-band (4 to 8 Ghz) and the Ku-band (11 to 17 Ghz). The lower the frequency, the larger the antenna (e.g. satellite dish) needed to gather the signal. The area on the earth receiving signals of acceptable strength from a satellite is known as the satellite's footprint. The footprints shown on the globes are for illustrative purposes only as final coverage is determined by customer needs.

THE GROWTH OF CELLULAR TELEPHONY AND INTERNET USE

Since 1980, two of the fastest-growing methods of information transfer have been cellular telephones and the Internet.

Cellular telephones work by relaying radio signals from a transmitting antenna to hand-held roaming units. Their attraction lies in freeing the user from a fixed point for connection.

The Internet is a global information service, providing an ever-expanding supply of text, images and graphics accessible via telephone lines to personal computers. As no single company owns or controls this service, reliable statistics are hard to determine. Total use is estimated by counting the number of 'hosts' i.e. the number of computers with a registered address. An indication of users' location is taken from the 'domain name', which is either a country code or of generic type, for example .ca (Canada) or .com (commercial).

Top 25 world cellular markets by penetration 1995

Rank	Economy	Subscribers per 100 people
1	SWEDEN	22.94
2	NORWAY	22.56
3	FINLAND	20.35
4	DENMARK	15.68
5	AUSTRALIA	*13.83*
6	UNITED STATES	12.84
7	BRUNEI	12.63
8	HONG KONG	12.33
9	ICELAND	11.52
10	NEW ZEALAND	*11.00*
11	SINGAPORE	9.77
12	BERMUDA	9.76
13	SAN MARINO	9.52
14	MONACO	9.51
15	KUWAIT	*8.59*
16	UNITED KINGDOM	*8.54*
17	CAYMAN ISLANDS	8.47
18	JAPAN	8.15
19	CANADA	*8.12*
20	MACAU	7.37
21	GUAM	7.33
22	ITALY	6.88
23	U.A.E.	6.78
24	LUXEMBOURG	6.58
25	SWITZERLAND	6.35

Figures in italics are estimates
Source: Telegeography Inc, Washington DC

Internet host computers 1995-1997

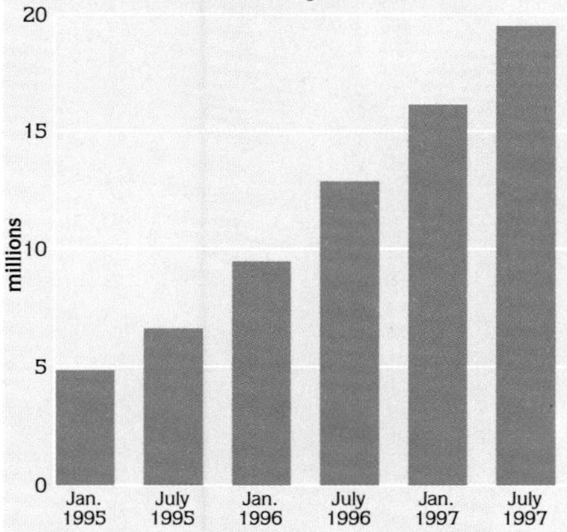

20
15
10
5
0
millions
Jan. 1995
July 1995
Jan. 1996
July 1996
Jan. 1997
July 1997

Internet host computers by domain

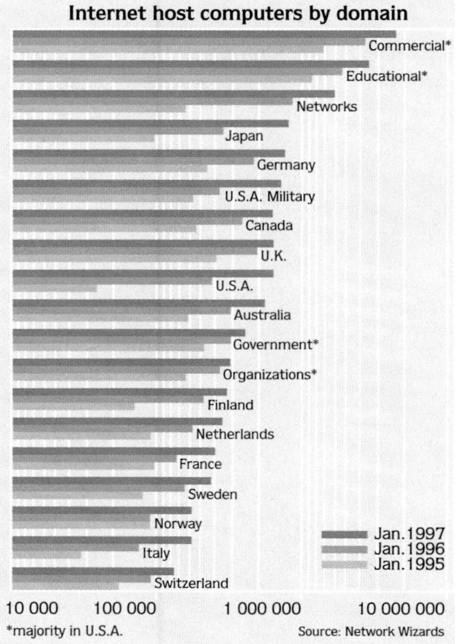

Commercial*
Educational*
Networks
Japan
Germany
U.S.A. Military
Canada
U.K.
U.S.A.
Australia
Government*
Organizations*
Finland
Netherlands
France
Sweden
Norway
Italy
Switzerland
Jan.1997
Jan.1996
Jan.1995
10 000
100 000
1 000 000
10 000 000

*majority in U.S.A.

Source: Network Wizards

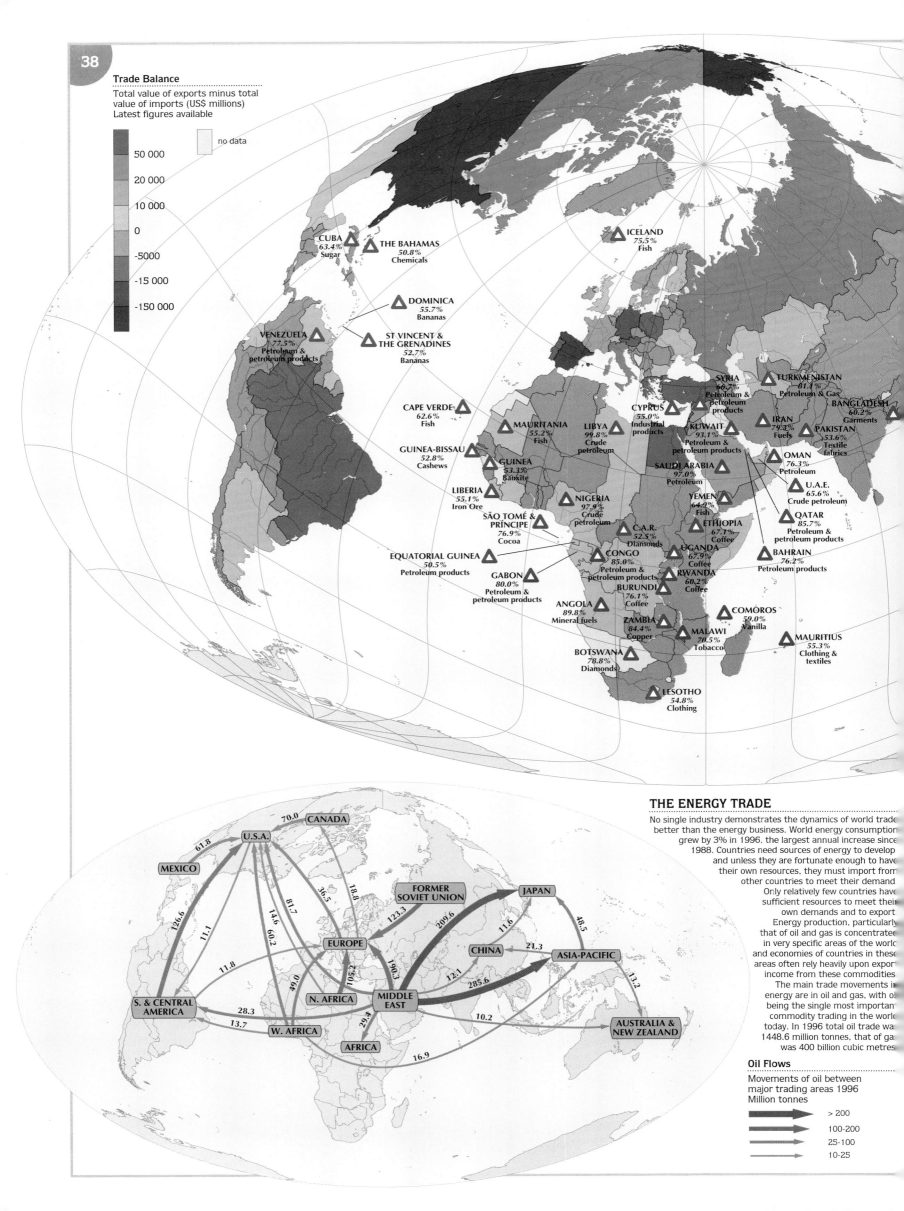

Trade Balance

Total value of exports minus total value of imports (US$ millions)
Latest figures available

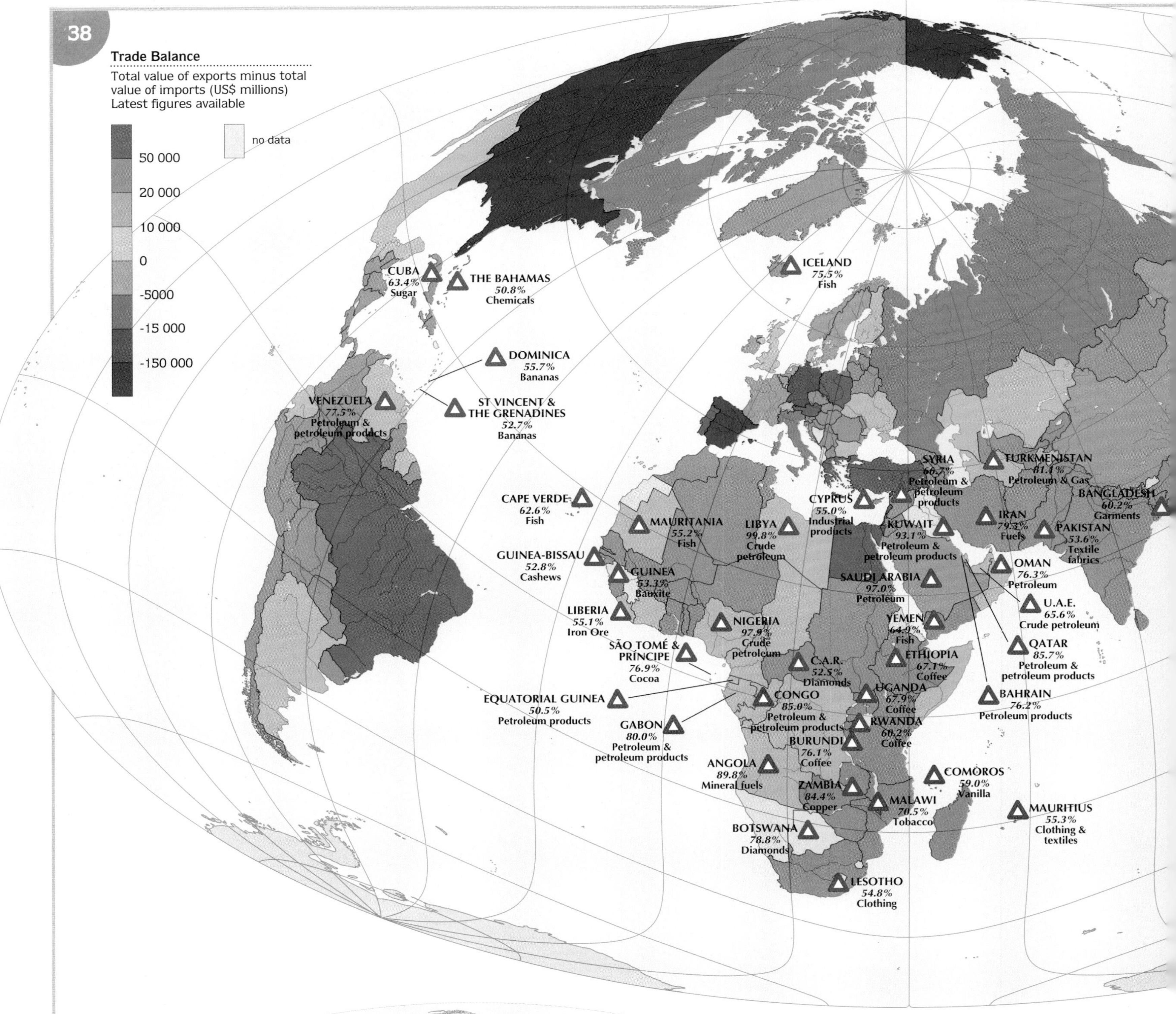
no data
50 000
20 000
10 000
0
-5000
-15 000
-150 000
CUBA
63.4%
Sugar
THE BAHAMAS
50.8%
Chemicals
ICELAND
75.5%
Fish
DOMINICA
55.7%
Bananas
ST VINCENT &
THE GRENADINES
52.7%
Bananas
VENEZUELA
77.5%
Petroleum &
petroleum products
CAPE VERDE
62.6%
Fish
MAURITANIA
55.2%
Fish
LIBYA
99.8%
Crude
petroleum
CYPRUS
55.0%
Industrial
products
SYRIA
66.7%
Petroleum &
petroleum
products
TURKMENISTAN
81.1%
Petroleum & Gas
BANGLADESH
60.2%
Garments
KUWAIT
93.1%
Petroleum &
petroleum products
IRAN
79.3%
Fuels
PAKISTAN
53.6%
Textile
fabrics
GUINEA-BISSAU
52.8%
Cashews
GUINEA
53.3%
Bauxite
SAUDI ARABIA
97.0%
Petroleum
OMAN
76.3%
Petroleum
U.A.E.
65.6%
Crude petroleum
LIBERIA
55.1%
Iron Ore
NIGERIA
97.9%
Crude
petroleum
YEMEN
64.9%
Fish
SÃO TOMÉ &
PRÍNCIPE
76.9%
Cocoa
C.A.R.
52.5%
Diamonds
ETHIOPIA
67.1%
Coffee
QATAR
85.7%
Petroleum &
petroleum products
EQUATORIAL GUINEA
50.5%
Petroleum products
CONGO
85.0%
Petroleum &
petroleum products
UGANDA
67.9%
Coffee
BAHRAIN
76.2%
Petroleum products
GABON
80.0%
Petroleum &
petroleum products
RWANDA
60.2%
Coffee
BURUNDI
76.1%
Coffee
ANGOLA
89.8%
Mineral fuels
COMOROS
59.0%
Vanilla
ZAMBIA
84.4%
Copper
MALAWI
70.5%
Tobacco
MAURITIUS
55.3%
Clothing &
textiles
BOTSWANA
78.8%
Diamonds
LESOTHO
54.8%
Clothing

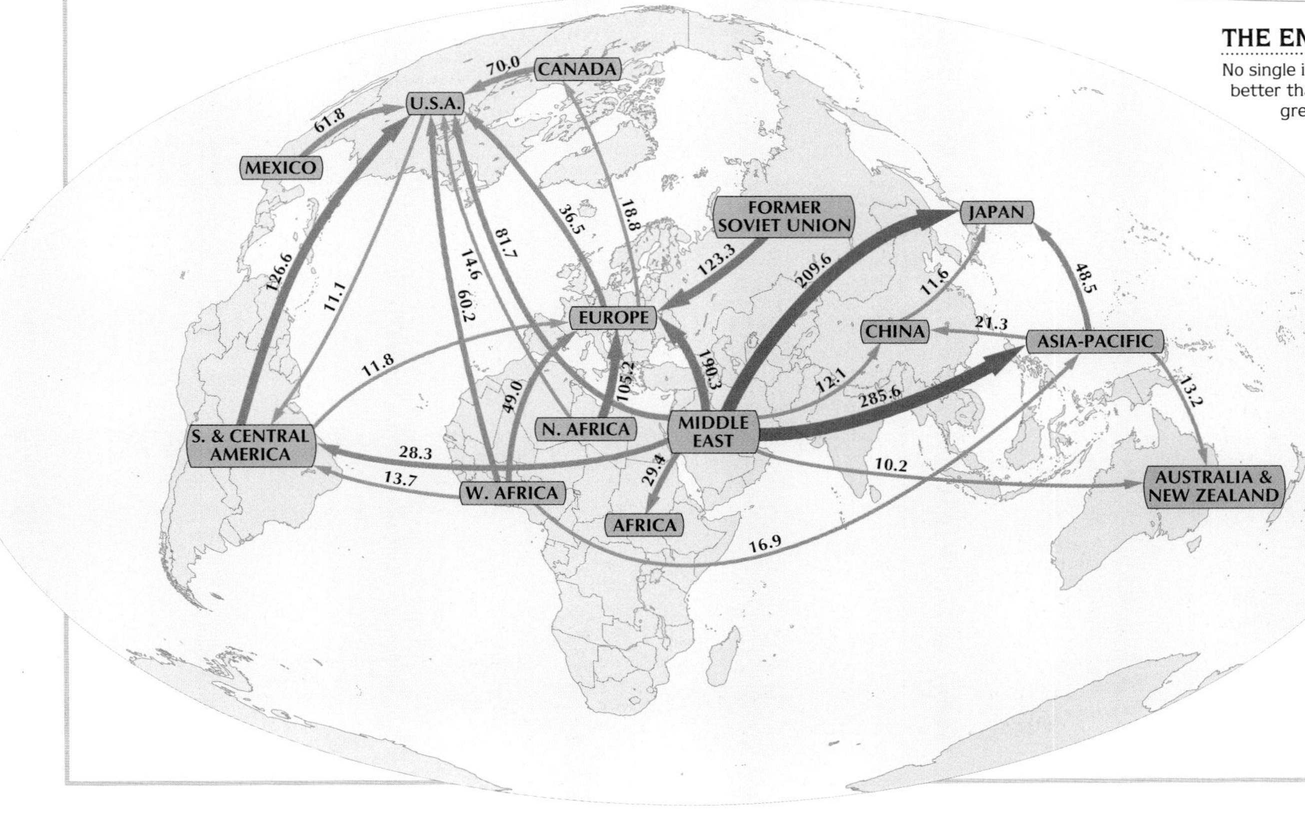
CANADA
U.S.A.
MEXICO
FORMER
SOVIET UNION
JAPAN
EUROPE
CHINA
ASIA-PACIFIC
S. & CENTRAL
AMERICA
N. AFRICA
MIDDLE
EAST
W. AFRICA
AFRICA
AUSTRALIA &
NEW ZEALAND
70.0
61.8
126.6
11.1
36.5
18.8
14.6
81.7
60.2
123.3
209.6
11.6
48.5
21.3
11.8
49.0
105.2
190.3
12.1
285.6
13.2
28.3
13.7
29.4
10.2
16.9

THE ENERGY TRADE

No single industry demonstrates the dynamics of world trade better than the energy business. World energy consumption grew by 3% in 1996, the largest annual increase since 1988. Countries need sources of energy to develop and unless they are fortunate enough to have their own resources, they must import from other countries to meet their demand. Only relatively few countries have sufficient resources to meet their own demands and to export.

Energy production, particularly that of oil and gas is concentrated in very specific areas of the world and economies of countries in these areas often rely heavily upon export income from these commodities.

The main trade movements in energy are in oil and gas, with oil being the single most important commodity trading in the world today. In 1996 total oil trade was 1448.6 million tonnes, that of gas was 400 billion cubic metres.

Oil Flows

Movements of oil between major trading areas 1996
Million tonnes

- > 200
- 100-200
- 25-100
- 10-25

WORLD TRADE

THE WORLD ECONOMY is enormously dependent upon international trade - the movement of goods and services between countries. The total annual value of such movements is over five million million US dollars.

World trade is dominated by the richer countries. As countries develop they are able to become more efficient in producing goods which are in demand elsewhere. Their own demand for commodities from other countries and therefore their imports, also increase. The richer they become, the more they trade; the more they trade the richer they become.

Commodity dependence

△ Countries with over 50% of exports in a single commodity (latest figures)

OMAN - Exporting country
76.3% - % of exports
Petroleum - Main export commodity

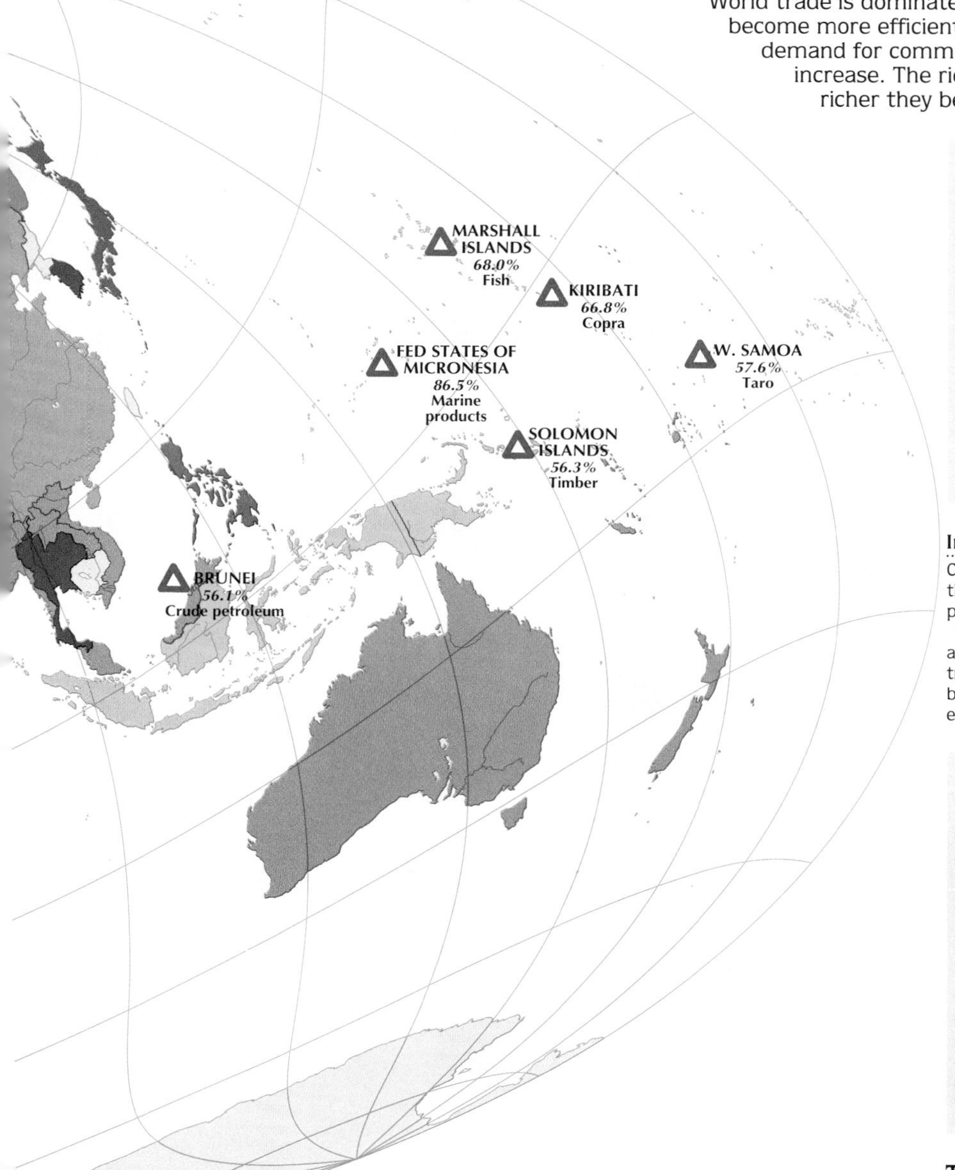

MARSHALL ISLANDS
68.0%
Fish
KIRIBATI
66.8%
Copra
FED STATES OF MICRONESIA
86.5%
Marine products
W. SAMOA
57.6%
Taro
SOLOMON ISLANDS
56.3%
Timber
BRUNEI
56.1%
Crude petroleum

TRADE COMPOSITION

Type of exports of richest and poorest countries

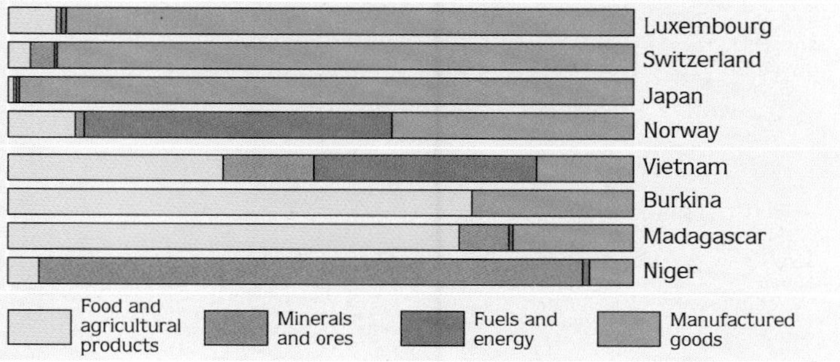

RICH
POOR
Luxembourg
Switzerland
Japan
Norway
Vietnam
Burkina
Madagascar
Niger
Food and agricultural products
Minerals and ores
Fuels and energy
Manufactured goods

Imports and Exports

Countries cannot produce for themselves all the goods they need. They must import the required goods from countries willing to sell, and must seek to generate income by producing and exporting goods in demand elsewhere.

The state of a country's economy is reflected by its balance of payments. This accounts for both 'visible' trade - goods passing through customs points - and 'invisible' trade - financial transactions and services. The most important single element of the balance of payments is the trade balance - the difference between earnings from exports and spending on imports.

TOP TRADERS

Rank	Imports			Exports		
	Country	Total (US$M)	% of world total	Country	Total (US$M)	% of world total
1	U.S.A.	817795	15.6	U.S.A.	624528	12.1
2	GERMANY	455925	8.7	GERMANY	521008	10.1
3	JAPAN	349173	6.6	JAPAN	410924	8.0
4	CHINA	337504	6.4	CHINA	331942	6.4
5	FRANCE	274073	5.2	FRANCE	289351	5.6
6	UNITED KINGDOM	261713	5.0	UNITED KINGDOM	270826	5.2
7	ITALY	204047	3.9	ITALY	231331	4.5
8	NETHERLANDS	176426	3.4	CANADA	201636	3.9
9	CANADA	174962	3.3	NETHERLANDS	195912	3.8
10	BELGIUM	152791	2.9	BELGIUM	165807	3.2

TRADE DEPENDENCY

Because demand for goods, and relationships between countries are not equal, complex patterns of trade have developed throughout the world. Trading patterns are often highly regional, with countries interacting more with their immediate neighbours than with other parts of the world, but there are many less straightforward relationships. Individual countries may concentrate on developing trading links with specific partners for political as well as economic reasons.

There are dangers in a country becoming too dependent upon either a single importing trading partner or a single export commodity. Economic and political conditions in a partner country may change rapidly, putting the trade relationship at risk; and natural disasters or economic events (in particular falling world prices) can have a devastating effect on an economy over-dependent on a single product.

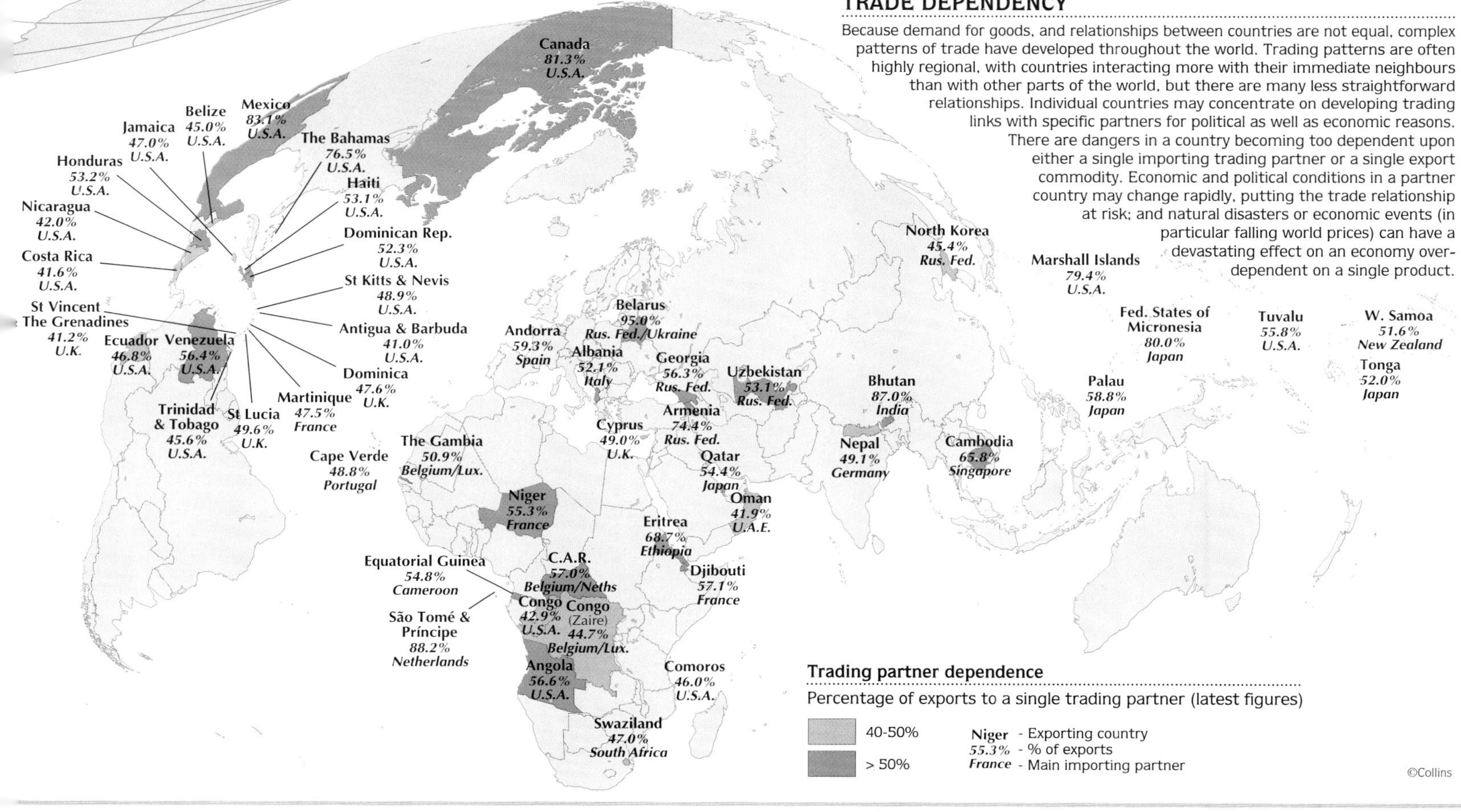

Canada
81.3%
U.S.A.
Mexico
83.1%
U.S.A.
Belize
45.0%
U.S.A.
Jamaica
47.0%
U.S.A.
Honduras
53.2%
U.S.A.
Nicaragua
42.0%
U.S.A.
Costa Rica
41.6%
U.S.A.
St Vincent
The Grenadines
41.2%
U.K.
Ecuador
46.8%
U.S.A.
Venezuela
56.4%
U.S.A.
Trinidad & Tobago
45.6%
U.S.A.
St Lucia
49.6%
U.K.
Martinique
47.5%
France
The Bahamas
76.5%
U.S.A.
Haiti
53.1%
U.S.A.
Dominican Rep.
52.3%
U.S.A.
St Kitts & Nevis
48.9%
U.S.A.
Antigua & Barbuda
41.0%
U.S.A.
Dominica
47.6%
U.K.
Cape Verde
48.8%
Portugal
The Gambia
50.9%
Belgium/Lux.
Andorra
59.3%
Spain
Belarus
95.0%
Rus. Fed./Ukraine
Albania
52.1%
Italy
Georgia
56.3%
Rus. Fed.
Armenia
74.4%
Rus. Fed.
Uzbekistan
53.1%
Rus. Fed.
Cyprus
49.0%
U.K.
Qatar
54.4%
Japan
Oman
41.9%
U.A.E.
Niger
55.3%
France
Eritrea
68.7%
Ethiopia
Djibouti
57.1%
France
Equatorial Guinea
54.8%
Cameroon
C.A.R.
57.0%
Belgium/Neths
Congo
42.9%
U.S.A.
Congo
(Zaire)
44.7%
Belgium/Lux.
São Tomé & Príncipe
88.2%
Netherlands
Angola
56.6%
U.S.A.
Comoros
46.0%
U.S.A.
Swaziland
47.0%
South Africa
Bhutan
87.0%
India
Nepal
49.1%
Germany
Cambodia
65.8%
Singapore
North Korea
45.4%
Rus. Fed.
Marshall Islands
79.4%
U.S.A.
Fed. States of Micronesia
80.0%
Japan
Palau
58.8%
Japan
Tuvalu
55.8%
U.S.A.
W. Samoa
51.6%
New Zealand
Tonga
52.0%
Japan

Trading partner dependence

Percentage of exports to a single trading partner (latest figures)

40-50%
> 50%

Niger - Exporting country
55.3% - % of exports
France - Main importing partner

OUTBREAK OF THE FIRST WORLD WAR 1914

In 1912 the desire for independence and territory motivated the people in the Balkan peninsula to go to war, first with their Ottoman rulers, and then amongst themselves. These actions had repercussions for the other European powers who saw their security threatened by the instability of the frontier area between three great empires. By this time, Europe had divided itself into two alliance blocs, the Central Powers (Germany, Austria-Hungary, Italy) and the Entente Powers (France, Russia, Britain) whose opposing interests prevented them resolving the Balkan dispute.

The crisis in the Balkans intensified the growing rivalry between the old dynastic empires of Russia and Austria-Hungary, into which their respective allies were inextricably drawn. The turning point came in June 1914 when Austria blamed Serbia for the assassination of the heir to their throne, Franz Ferdinand, and Russia pledged to defend their Balkan ally in the event of war. These actions triggered the wider system of alliances, whose members had considerably reinforced their armies, and so by August the major states of Europe found themselves at war.

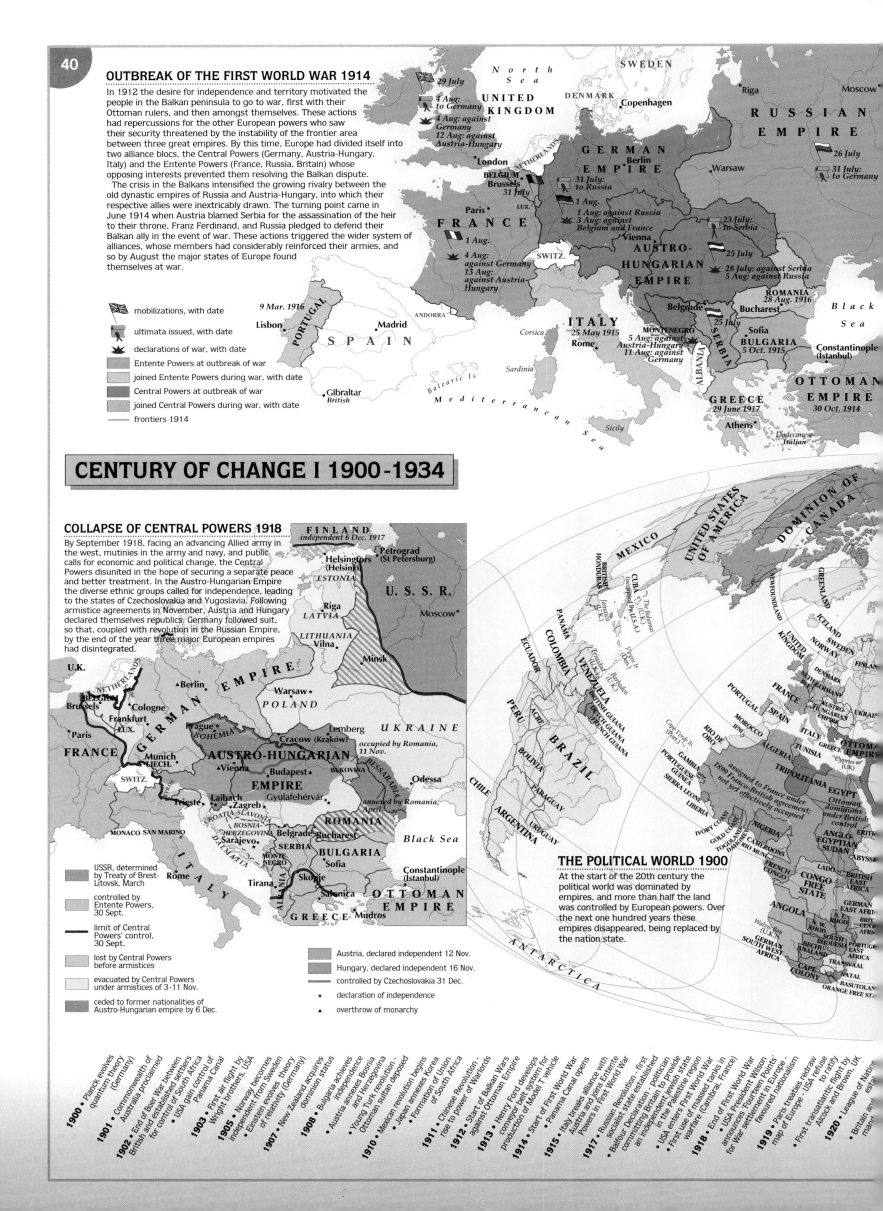

- mobilizations, with date
- ultimata issued, with date
- declarations of war, with date
- Entente Powers at outbreak of war
- joined Entente Powers during war, with date
- Central Powers at outbreak of war
- joined Central Powers during war, with date
- frontiers 1914

CENTURY OF CHANGE I 1900-1934

COLLAPSE OF CENTRAL POWERS 1918

By September 1918, facing an advancing Allied army in the west, mutinies in the army and navy, and public calls for economic and political change, the Central Powers disunited in the hope of securing a separate peace and better treatment. In the Austro-Hungarian Empire the diverse ethnic groups called for independence, leading to the states of Czechoslovakia and Yugoslavia. Following armistice agreements in November, Austria and Hungary declared themselves republics. Germany followed suit, so that, coupled with revolution in the Russian Empire, by the end of the year three major European empires had disintegrated.

- USSR, determined by Treaty of Brest-Litovsk, March
- controlled by Entente Powers, 30 Sept.
- limit of Central Powers' control, 30 Sept.
- lost by Central Powers before armistices
- evacuated by Central Powers under armistices of 3-11 Nov.
- ceded to former nationalities of Austro-Hungarian empire by 6 Dec.
- Austria, declared independent 12 Nov.
- Hungary, declared independent 16 Nov.
- controlled by Czechoslovakia 31 Dec.
- ★ declaration of independence
- ▲ overthrow of monarchy

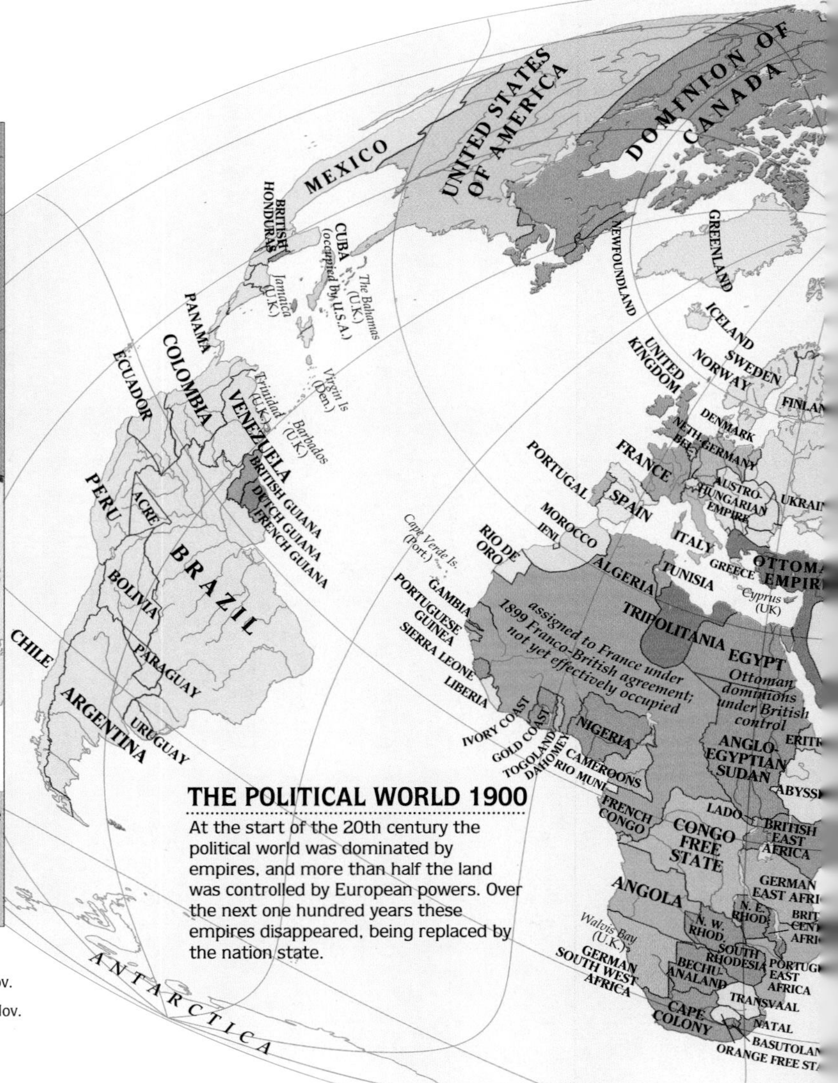

THE POLITICAL WORLD 1900

At the start of the 20th century the political world was dominated by empires, and more than half the land was controlled by European powers. Over the next one hundred years these empires disappeared, being replaced by the nation state.

1900 • Planck evolves quantum theory (Germany)
1901 • Commonwealth of Australia proclaimed
1902 • End of Boer War between British and established settlers for control of South Africa
• USA gain control of Panama Canal
1903 • First air flight by Wright brothers, USA
1905 • Norway becomes independent from Sweden
• Einstein evolves theory of relativity (Germany)
1907 • New Zealand acquires dominion status
1908 • Bulgaria achieves independence
• Austria annexes Bosnia and Herzegovina
• Young Turk revolution - Ottoman sultan deposed
1910 • Mexican revolution begins
• Japan annexes Korea
• Formation of Union of South Africa
1911 • Chinese Revolution - rise to power of Warlords
1912 • Start of Balkan Wars against Ottoman Empire
1913 • Henry Ford develops conveyor belt system for production of Model T vehicle
1914 • Start of First World War
• Panama Canal opens
1915 • Italy breaks alliance with Austria and joins Entente Powers in First World War
1917 • Russian Revolution - first socialist state established
• Balfour Declaration - politician committing Britain to provide an independent jewish state in the Palestine region
• USA enters First World War
• First use of massed tanks in warfare (Cambrai, France)
1918 • End of First World War
• USA President Wilson announces 'Fourteen Points' for War settlement in Europe - favoured nationalism
1919 • Paris treaties redraw map of Europe - USA refuse to ratify
• First transatlantic flight by Alcock and Brown, UK
1920 • League of Natio… esta…
• Britain and … mand…

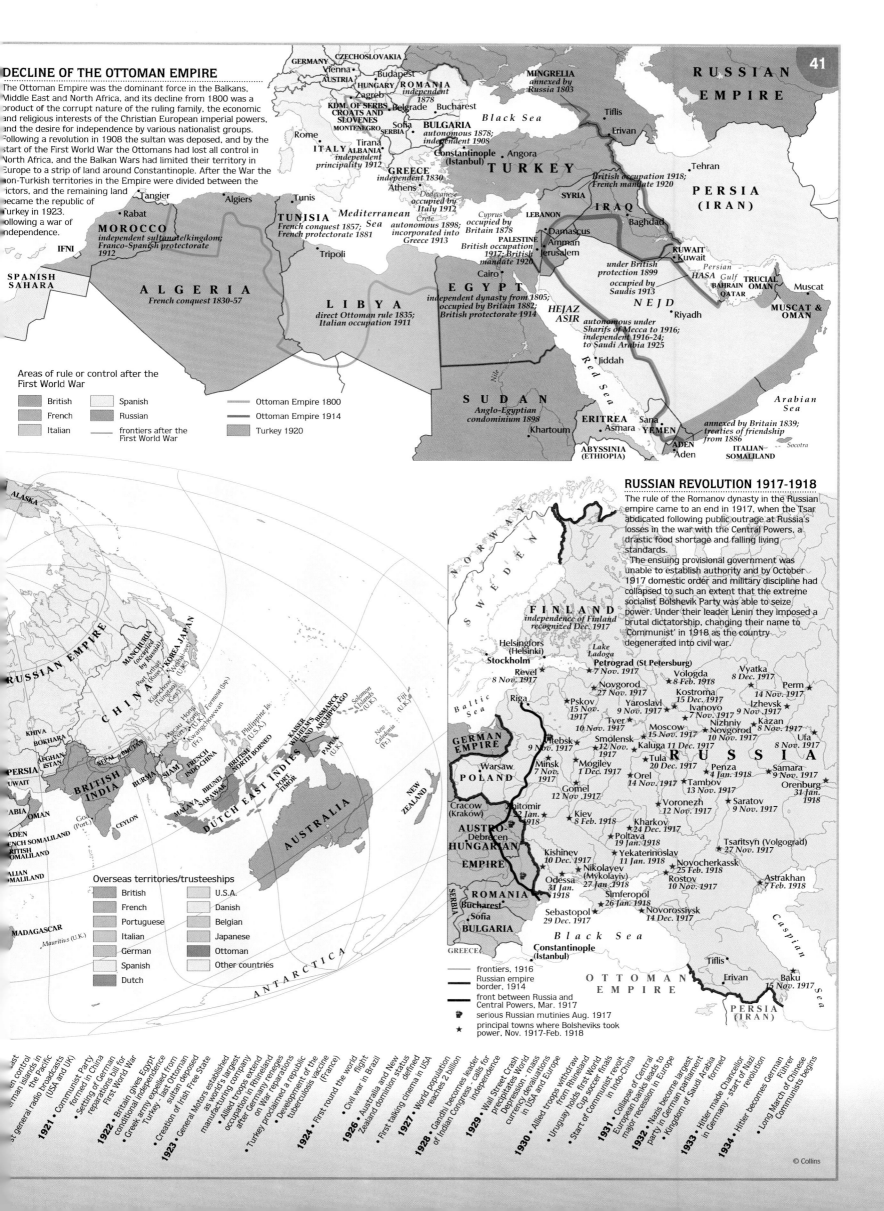

DECLINE OF THE OTTOMAN EMPIRE

The Ottoman Empire was the dominant force in the Balkans, Middle East and North Africa, and its decline from 1800 was a product of the corrupt nature of the ruling family, the economic and religious interests of the Christian European imperial powers, and the desire for independence by various nationalist groups. Following a revolution in 1908 the sultan was deposed, and by the start of the First World War the Ottomans had lost all control in North Africa, and the Balkan Wars had limited their territory in Europe to a strip of land around Constantinople. After the War the non-Turkish territories in the Empire were divided between the victors, and the remaining land became the republic of Turkey in 1923, following a war of independence.

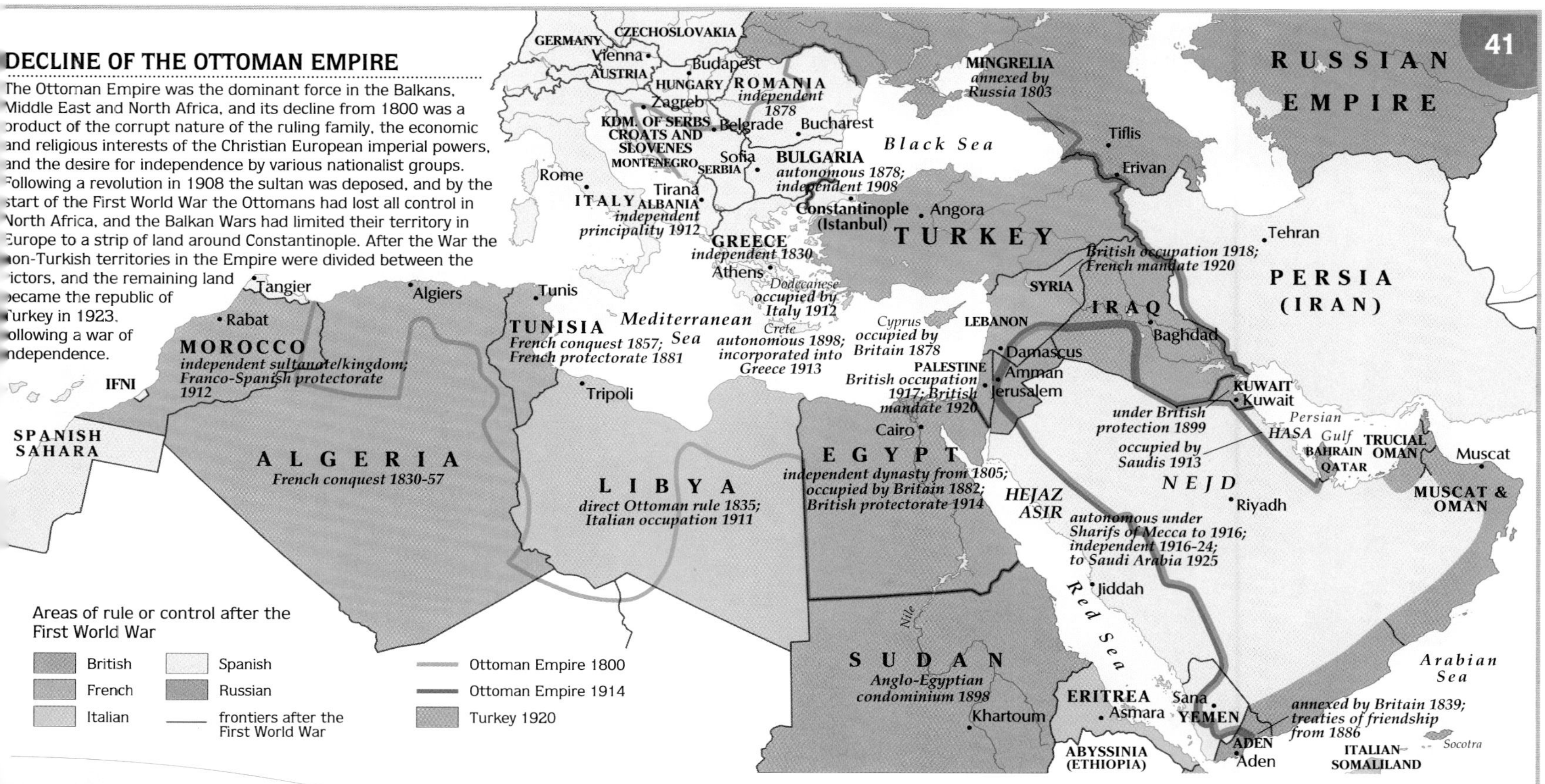

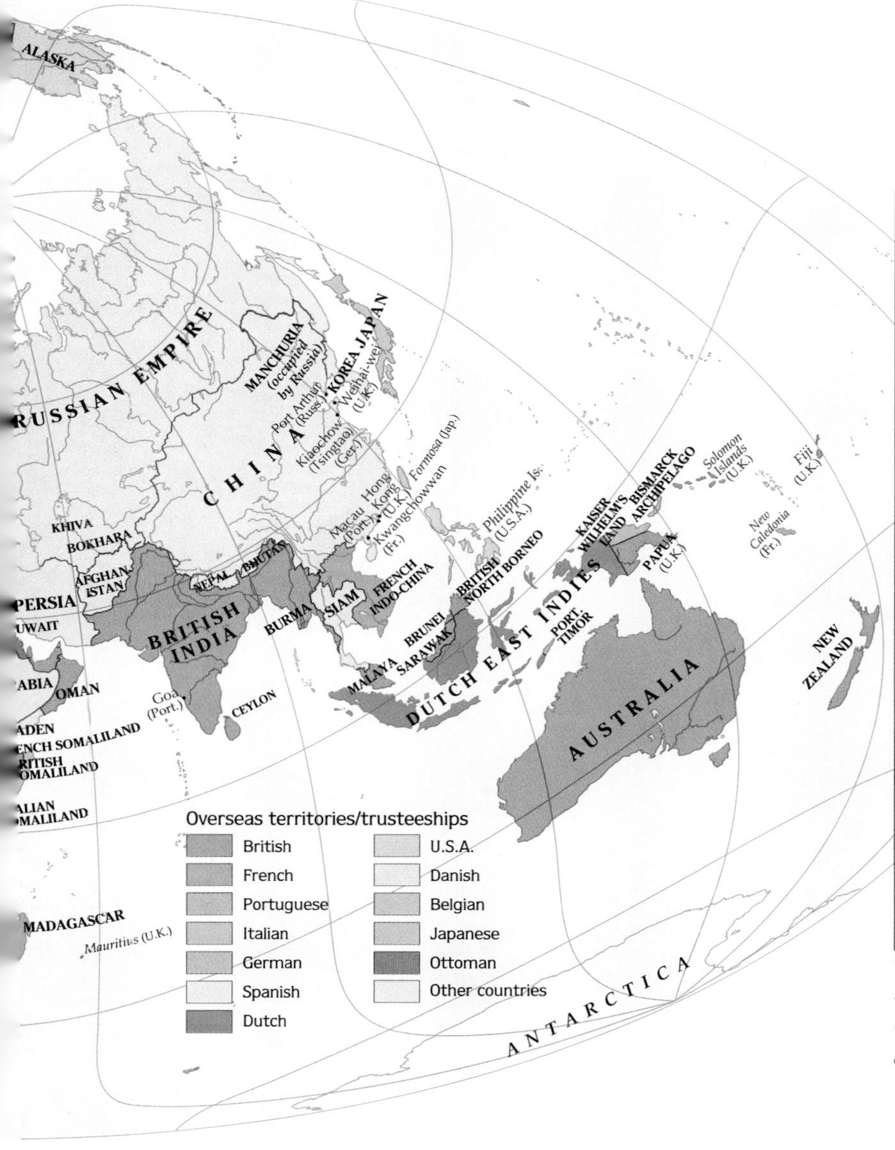

RUSSIAN REVOLUTION 1917-1918

The rule of the Romanov dynasty in the Russian empire came to an end in 1917, when the Tsar abdicated following public outrage at Russia's losses in the war with the Central Powers, a drastic food shortage and falling living standards.

The ensuing provisional government was unable to establish authority and by October 1917 domestic order and military discipline had collapsed to such an extent that the extreme socialist Bolshevik Party was able to seize power. Under their leader Lenin they imposed a brutal dictatorship, changing their name to 'Communist' in 1918 as the country degenerated into civil war.

...ast
...en control
...rman islands in the Pacific
...st general radio broadcasts (USA and UK)

1921 • Communist Party formed in China
• Setting of German reparations bill for First World War

1922 • Britain gives Egypt conditional independence
• Greek army expelled from Turkey - last Ottoman sultan deposed
• Creation of Irish Free State

1923 • General Motors established as world's largest manufacturing company
• Allied troops extend occupation in Rhineland after Germany reneges on War reparations
• Turkey proclaimed a republic
• Development of the tuberculosis vaccine (France)

1924 • First round the world flight
• Civil war in Brazil

1926 • Australia and New Zealand dominion status defined
• First talking cinema in USA

1927 • World population reaches 2 billion

1928 • Gandhi becomes leader of Indian Congress - calls for independence

1929 • Wall Street Crash precipitates World Depression - mass currency devaluations in USA and Europe

1930 • Allied troops withdraw from Rhineland
• Uruguay holds first World Cup soccer finals
• Start of Communist revolt in Indo-China

1931 • Collapse of Central European banks leads to major recession in Europe

1932 • Nazis become largest party in German parliament
• Kingdom of Saudi Arabia formed

1933 • Hitler made Chancellor in Germany - start of Nazi revolution

1934 • Hitler becomes German Führer
• Long March of Chinese Communists begins

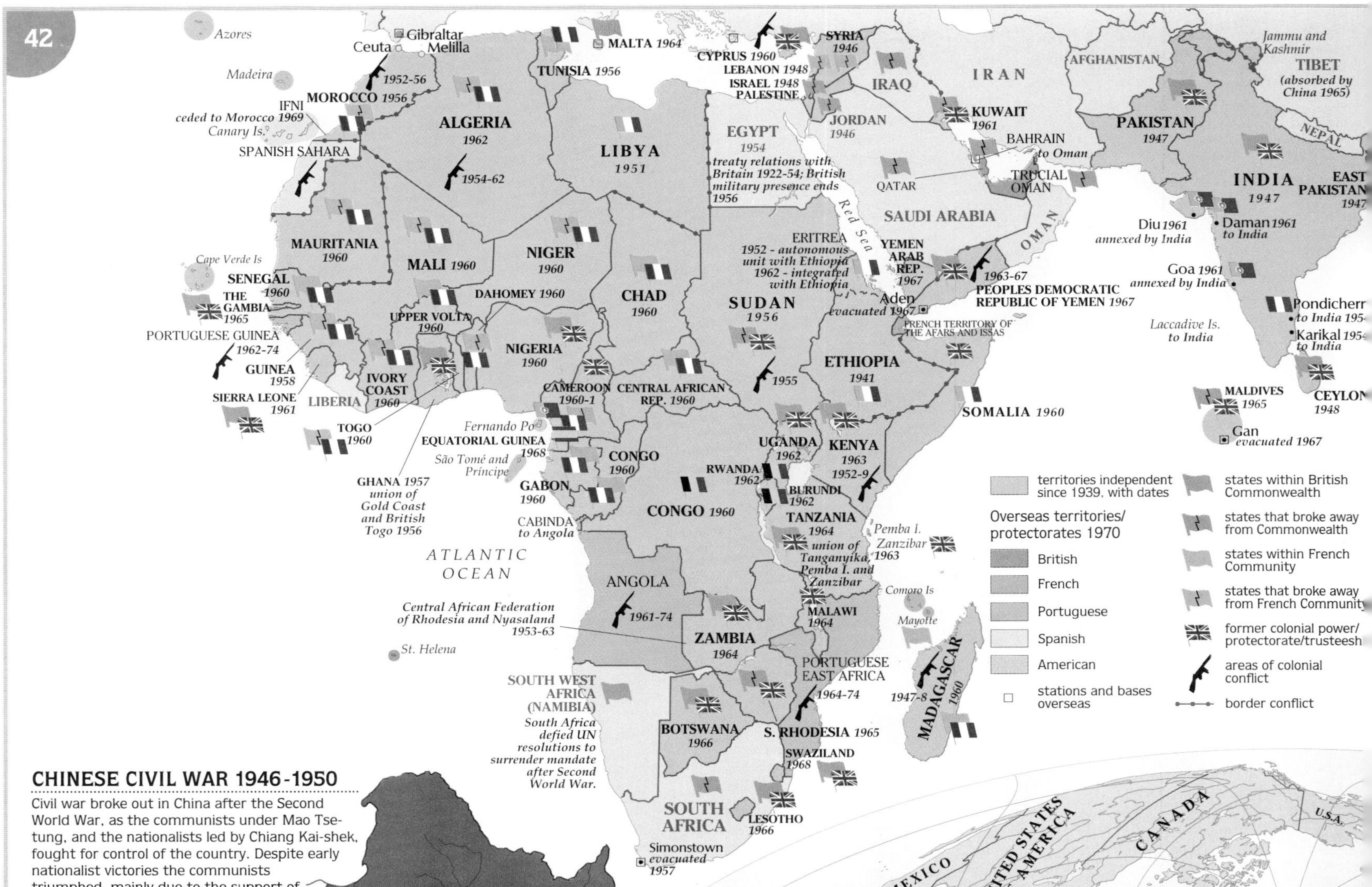

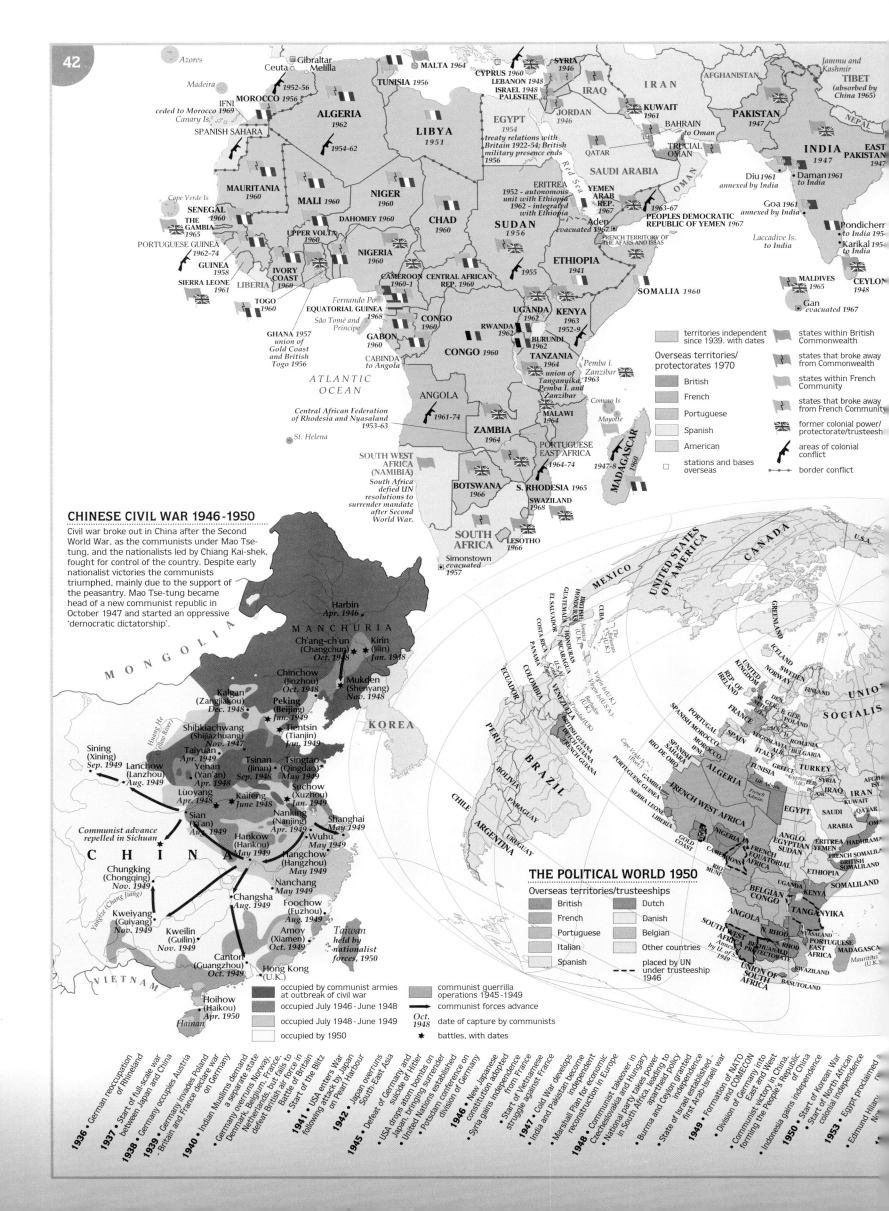

CHINESE CIVIL WAR 1946-1950

Civil war broke out in China after the Second World War, as the communists under Mao Tse-tung, and the nationalists led by Chiang Kai-shek, fought for control of the country. Despite early nationalist victories the communists triumphed, mainly due to the support of the peasantry. Mao Tse-tung became head of a new communist republic in October 1947 and started an oppressive 'democratic dictatorship'.

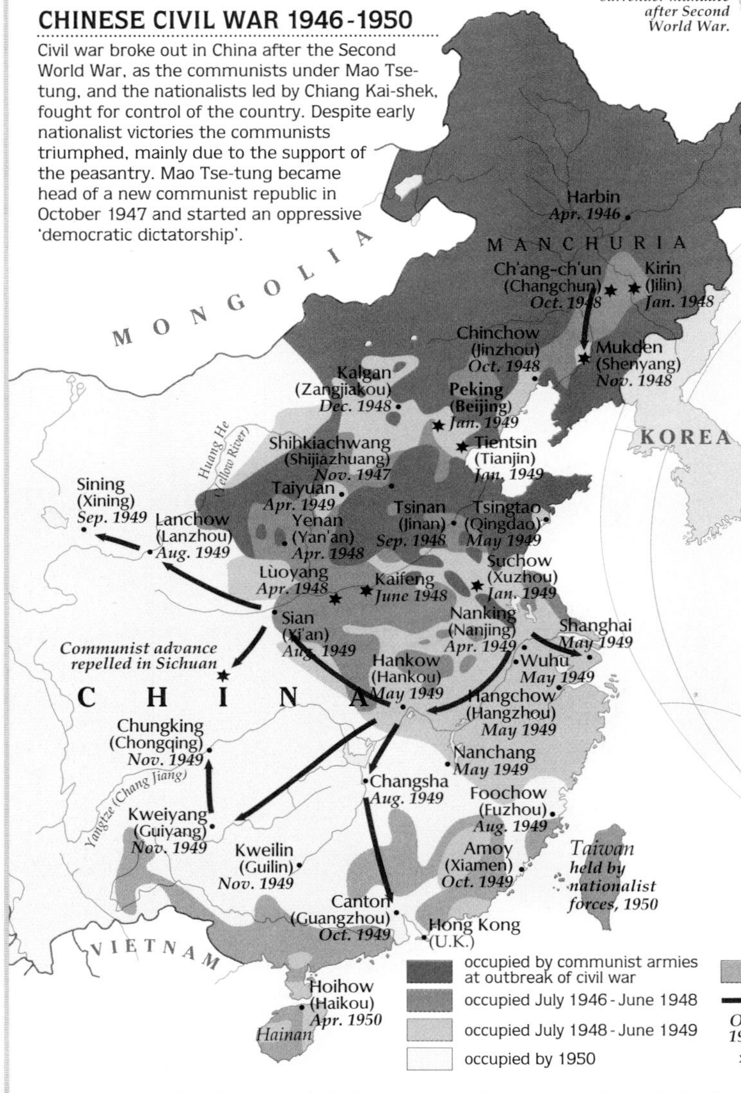

1936 • German reoccupation of Rhineland
1937 • Start of full-scale war between Japan and China
1938 • Germany occupies Austria
1939 • Germany invades Poland - Britain and France declare war on Germany
1940 • Indian Muslims demand a separate state
• Germany overruns Norway, Denmark, Belgium, France, Netherlands, but fails to defeat British air force in Battle of Britain
• Start of the Blitz
1941 • USA enters War following attack by Japan on Pearl Harbour
1942 • Japan overruns South-East Asia
1945 • Defeat of Germany and suicide of Hitler
• USA drops atom bombs on Japan, bringing surrender
• United Nations established
• Potsdam conference on division of Germany
1946 • New Japanese constitution adopted
• Syria gains independence from France
• Start of Vietnamese struggle against France
1947 • Cold War develops
• India and Pakistan become independent
• Marshall Plan for economic reconstruction in Europe
1948 • Communist takeover in Czechoslovakia and Hungary
• National party takes power in South Africa, leading to apartheid policy
• Burma and Ceylon granted independence
• State of Israel established - first Arab-Israeli war
1949 • Formation of NATO and COMECON
• Division of Germany into East and West
• Communist victory in China, forming the People's Republic of China
• Indonesia gains independence
1950 • Start of Korean War
• Start of North African colonial independence
1953 • Egypt proclaimed a re[...]
• Edmund Hilary [...] No[...]

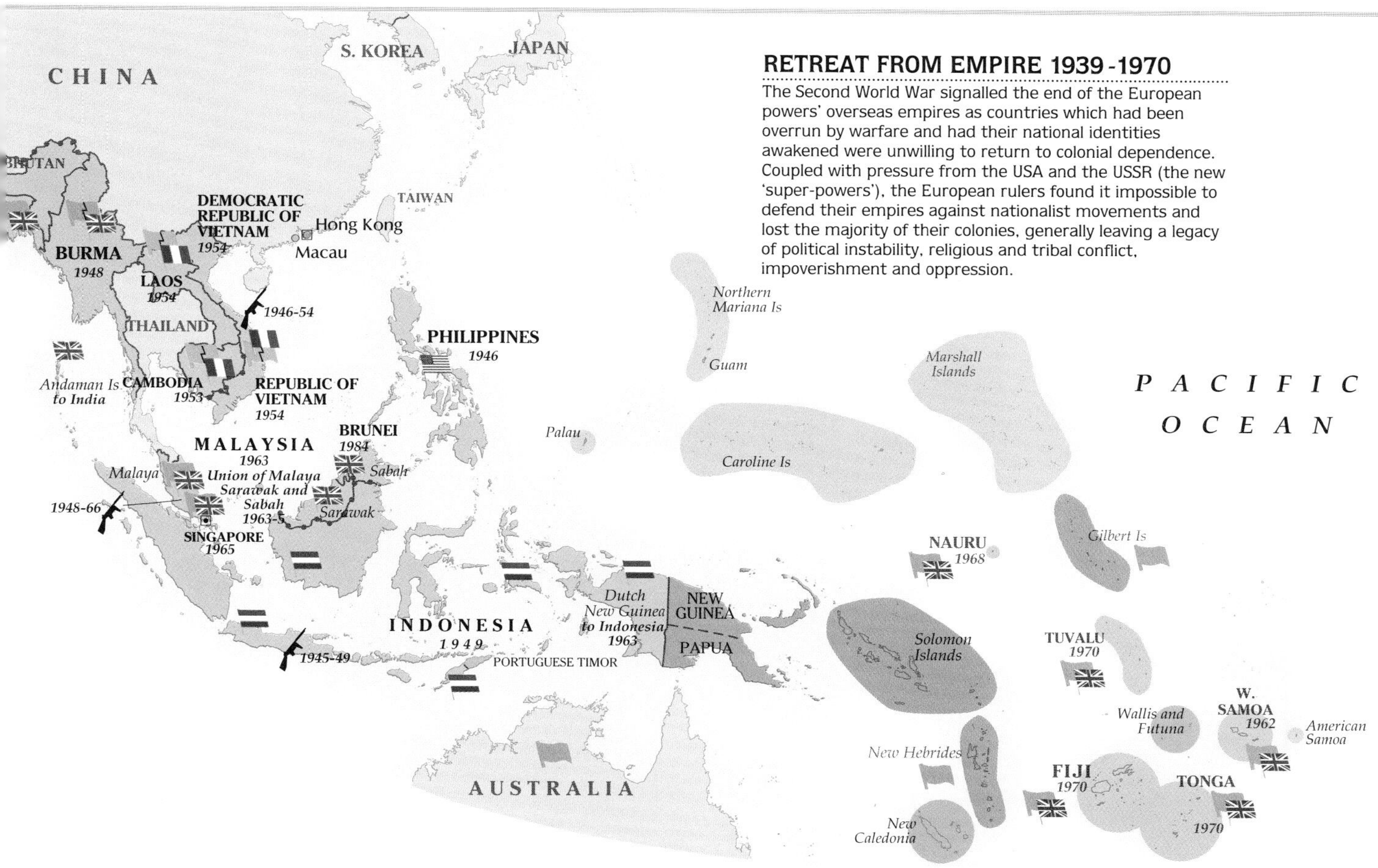
CHINA
S. KOREA
JAPAN
BHUTAN
DEMOCRATIC REPUBLIC OF VIETNAM 1954
TAIWAN
Hong Kong
Macau
BURMA 1948
LAOS 1954
1946-54
THAILAND
PHILIPPINES 1946
Andaman Is to India
CAMBODIA 1953
REPUBLIC OF VIETNAM 1954
MALAYSIA 1963
BRUNEI 1984
Malaya
Union of Malaya Sarawak and Sabah 1963-5
Sabah
Sarawak
1948-66
SINGAPORE 1965
INDONESIA 1949
1945-49
PORTUGUESE TIMOR
Dutch New Guinea to Indonesia 1963
NEW GUINEA
PAPUA
AUSTRALIA
Northern Mariana Is
Guam
Palau
Marshall Islands
Caroline Is
PACIFIC OCEAN
NAURU 1968
Gilbert Is
Solomon Islands
TUVALU 1970
Wallis and Futuna
W. SAMOA 1962
American Samoa
New Hebrides
FIJI 1970
TONGA 1970
New Caledonia

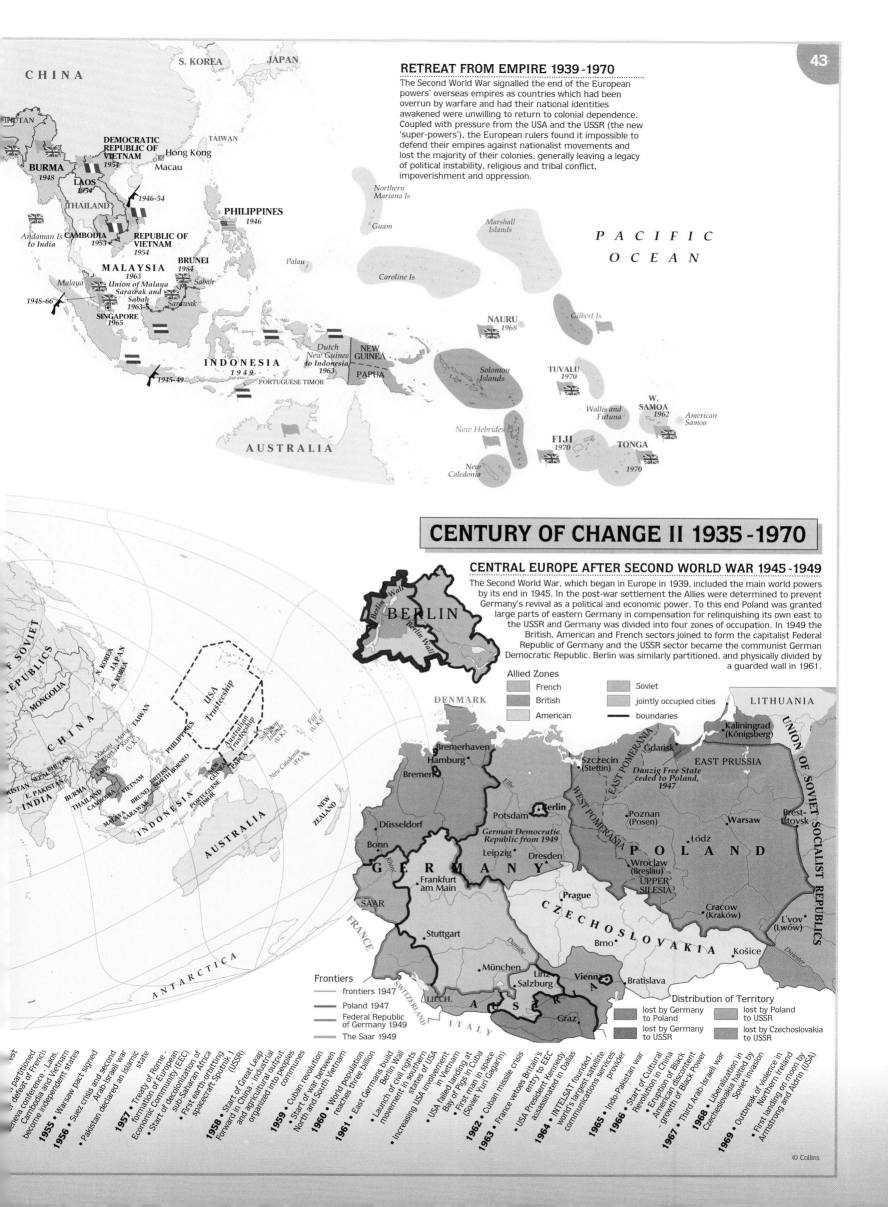

RETREAT FROM EMPIRE 1939-1970

The Second World War signalled the end of the European powers' overseas empires as countries which had been overrun by warfare and had their national identities awakened were unwilling to return to colonial dependence. Coupled with pressure from the USA and the USSR (the new 'super-powers'), the European rulers found it impossible to defend their empires against nationalist movements and lost the majority of their colonies, generally leaving a legacy of political instability, religious and tribal conflict, impoverishment and oppression.

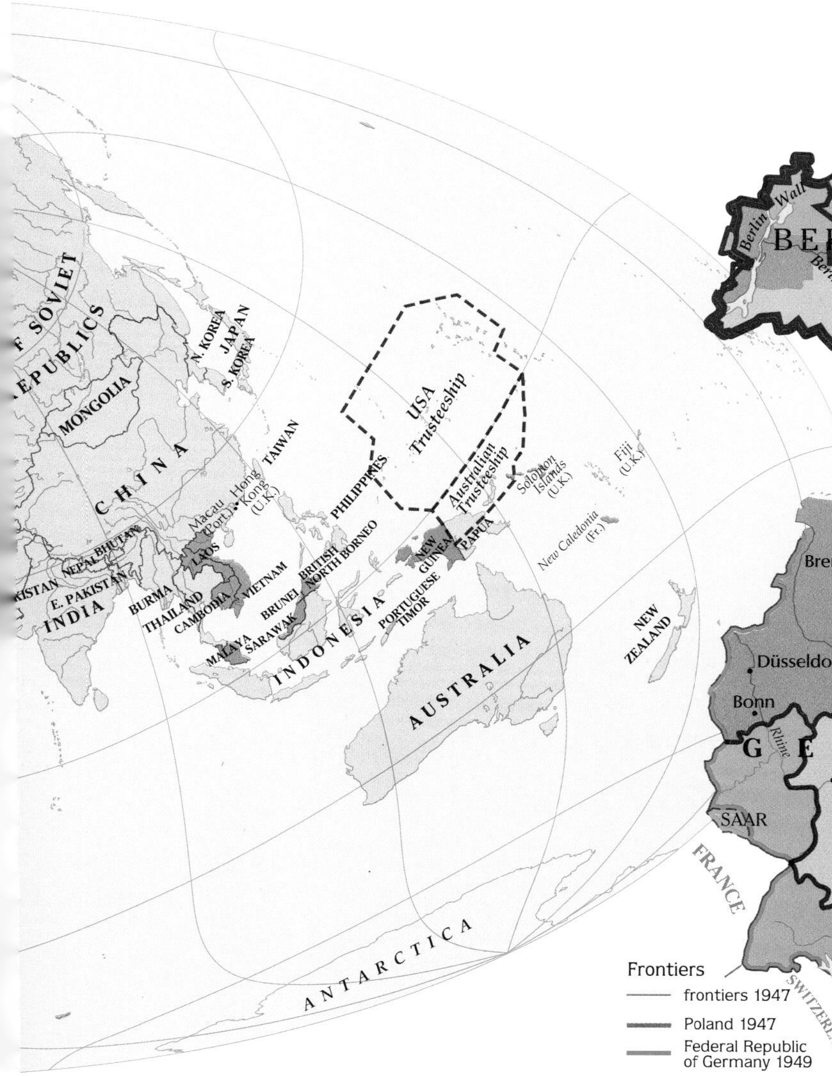
UNION OF SOVIET SOCIALIST REPUBLICS
MONGOLIA
CHINA
N. KOREA
JAPAN
S. KOREA
TAIWAN
Hong Kong (U.K.)
Macau (Port.)
PHILIPPINES
USA Trusteeship
Australian Trusteeship
Solomon Islands (U.K.)
Fiji (U.K.)
New Caledonia (Fr.)
NEPAL
BHUTAN
E. PAKISTAN
INDIA
BURMA
THAILAND
LAOS
CAMBODIA
VIETNAM
MALAYA
BRUNEI
SARAWAK
BRITISH NORTH BORNEO
INDONESIA
PORTUGUESE TIMOR
NEW GUINEA
PAPUA
AUSTRALIA
NEW ZEALAND
ANTARCTICA

CENTURY OF CHANGE II 1935-1970

CENTRAL EUROPE AFTER SECOND WORLD WAR 1945-1949

BERLIN
Berlin Wall
Berlin Wall

The Second World War, which began in Europe in 1939, included the main world powers by its end in 1945. In the post-war settlement the Allies were determined to prevent Germany's revival as a political and economic power. To this end Poland was granted large parts of eastern Germany in compensation for relinquishing its own east to the USSR and Germany was divided into four zones of occupation. In 1949 the British, American and French sectors joined to form the capitalist Federal Republic of Germany and the USSR sector became the communist German Democratic Republic. Berlin was similarly partitioned, and physically divided by a guarded wall in 1961.

Allied Zones
French
British
American
Soviet
jointly occupied cities
boundaries
DENMARK
LITHUANIA
UNION OF SOVIET SOCIALIST REPUBLICS
Kaliningrad (Königsberg)
Gdańsk
EAST PRUSSIA
Danzig Free State ceded to Poland, 1947
Bremerhaven
Hamburg
Bremen
Szczecin (Stettin)
EAST POMERANIA
WEST POMERANIA
Elbe
Berlin
Potsdam
Poznan (Posen)
Warsaw
Brest-Litovsk
Düsseldorf
German Democratic Republic from 1949
Łódź
Bonn
Leipzig
Dresden
POLAND
GERMANY
Rhine
Wroclaw (Breslau)
UPPER SILESIA
Frankfurt am Main
Prague
SAAR
CZECHOSLOVAKIA
Cracow (Kraków)
L'vov (Lwów)
FRANCE
Stuttgart
Brno
Danube
Košice
Dniester
München
Linz
Salzburg
Vienna
Bratislava
AUSTRIA
SWITZERLAND
LIECH.
Graz
ITALY
Frontiers
frontiers 1947
Poland 1947
Federal Republic of Germany 1949
The Saar 1949
Distribution of Territory
lost by Germany to Poland
lost by Poland to USSR
lost by Germany to USSR
lost by Czechoslovakia to USSR

...est ... partitioned ... defeat of French ... Geneva conference - Laos, Cambodia and Vietnam become independent states

1955 • Warsaw pact signed

1956 • Suez crisis and second Arab-Israeli war • Pakistan declared an Islamic state

1957 • Treaty of Rome - formation of European Economic Community (EEC) • Start of decolonization of sub-Saharan Africa • First earth-orbitting spacecraft Sputnik I (USSR)

1958 • Start of Great Leap Forward in China - industrial and agricultural output organized into peoples communes

1959 • Cuban revolution • Start of war between North and South Vietnam

1960 • World population reaches three billion

1961 • East Germans build Berlin Wall • Launch of civil rights movement in southern states of USA • Increasing USA involvement in Vietnam • USA failed landing at Bay of Pigs in Cuba • First man in space (Soviet Yuri Gagarin)

1962 • Cuban missile crisis

1963 • France vetoes Britain's entry to EEC • USA President Kennedy assassinated in Dallas

1964 • INTELSAT founded - world's largest satellite communications services provider

1965 • Indo-Pakistan war

1966 • Start of Cultural Revolution in China • Eruption of Black American discontent - growth of Black Power

1967 • Third Arab-Israeli war

1968 • Liberalization in Czechoslovakia halted by Soviet invasion

1969 • Outbreak of violence in Northern Ireland • First landing on moon by Armstrong and Aldrin (USA)

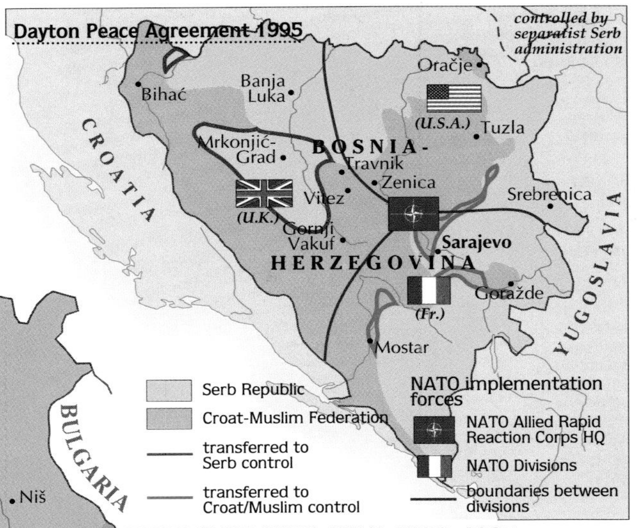

YUGOSLAV CIVIL WAR 1990-1995

In 1990 multi-party elections in Yugoslavia uncovered the deep ethnic divisions throughout the country. Over the next three years, withstanding initial opposition from federal troops and some fierce inter-ethnic warfare, the six Yugoslav republics declared their independence as separate states. There was bitter conflict in multi-ethnic Bosnia-Herzegovina where civil war broke out following independence. Muslims were caught between Croats and Serbs in the fight over territory. After four years of combat which devastated the area, a ceasefire was negotiated in October 1995. A month later in Dayton, Ohio, a settlement was agreed in which Bosnia-Herzegovina was divided into two separately-administered areas, a Serb Republic and a Muslim-Croat Federation, with NATO troops maintaining order.

CENTURY OF CHANGE III 1971-1997

ISRAELI-PALESTINIAN AGREEMENT 1993-1996

Claims on territory between Palestinian Arabs and Israelis in the Middle East have manifested themselves in numerous terrorist and fundamentalist attacks, violent rebellions and four wars since 1948. In 1993 and 1995 the Israeli prime minister, Yitzhak Rabin, negotiated settlements with Yasser Arafat (leader of the Palestinian Liberation Organization) involving a phased withdrawal of Israeli troops and limited Palestinian self-rule until 1999 in Gaza and parts of the West Bank. The status of other West Bank areas, particularly those with Jewish settlers, and the future of Jerusalem remained unresolved. The course to peace became increasingly uncertain following the election of right-wing Binyamin Netanyahu as prime minister after Rabin's assassination in November 1995, and in the face of continued terrorism by extremist groups.

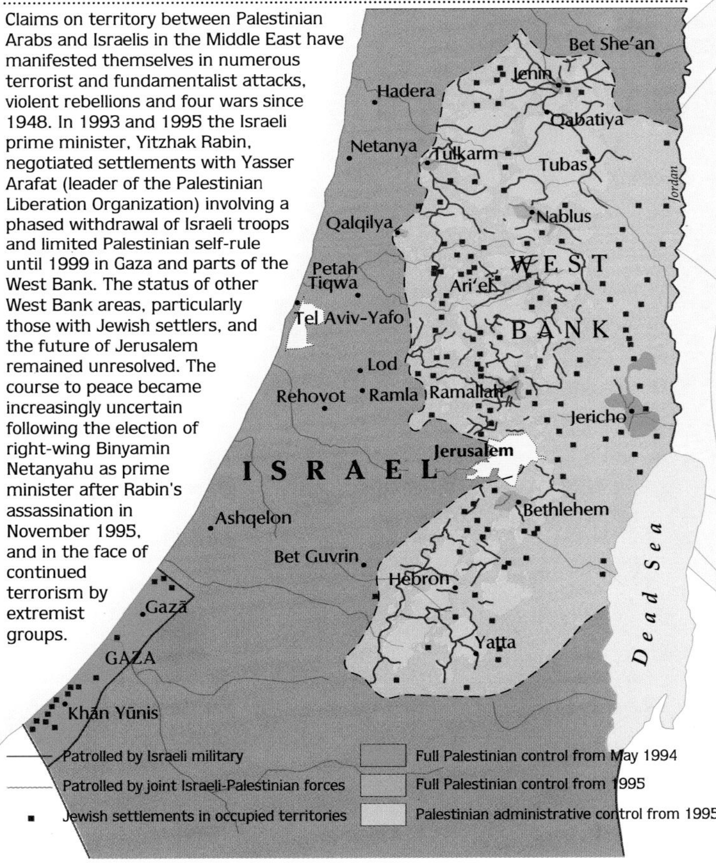

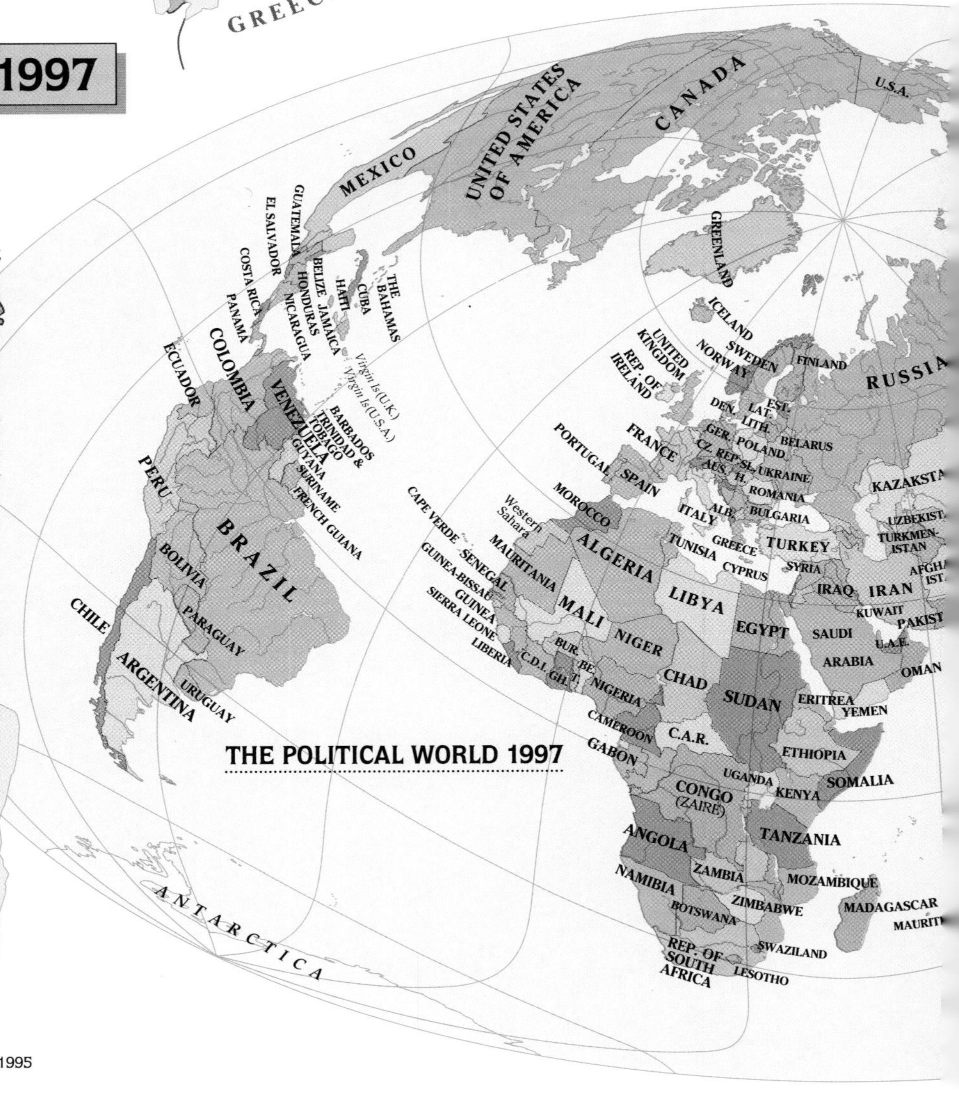

1971 • Indo-Pakistan war leads to breakaway of East Pakistan (Bangladesh)

1973 • Oil crisis caused by OPEC countries trebling price of oil
• Fourth Arab-Israeli war
• USA withdraws from Vietnam

1974 • USA Watergate scandal - President Nixon resigns
• Turkey invades Cyprus

1975 • Communists take over in Vietnam, Laos and Cambodia

1976 • Morocco and Mauritania partition Spanish Sahara
• First supersonic transatlantic passenger service begins with Concorde

1977 • Egypt-Israel peace talks
• Last recorded case of smallpox (Somalia) - eradicated worldwide following vaccination programme by World Health Organization

1979 • Civil war in Nicaragua
• Civil war in El Salvador
• Fall of Shah in Iran - creation of Islamic republic under Ayatolla Khomeini
• Vietnam invades Cambodia, expelling Khmer Rouge Government under Pol Pot
• USSR invades Afghanistan

1980 • Black majority rule established in Zimbabwe
• Outbreak of Iran-Iraq war

1981 • USA launch Space Shuttle - first re-usable spacecraft

1982 • Argentina invades South Georgia and Falkland Is - surrenders to UK forces after two month war

1985 • Civil unrest in South Africa - state of emergency declared
• Gorbachev comes to power in USSR - initiates glasnost (openness) and perestroika (restructuring)

1986 • Explosion at Chernobyl nuclear plant (USSR)
• Development of fibre-optic cables, capable of transmitting vast amounts of digital information over long distances
• USSR launch first manned earth-orbitting space station (Mir)
• Halley's Comet enters inner solar system

1987 • 'Intifada' revolt begins in Gaza and West Bank
• World population reaches five billion

1988 • PLO recogni

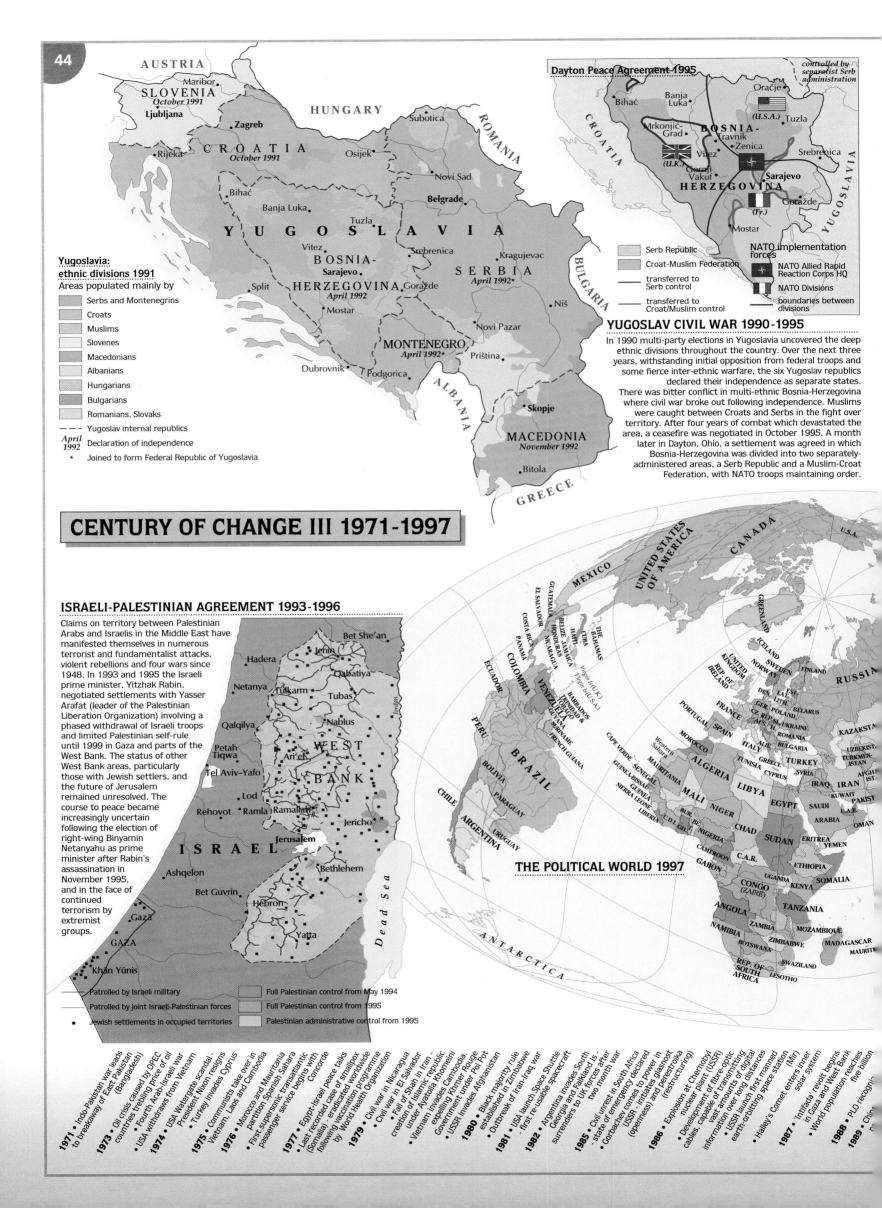

1989 • Chi

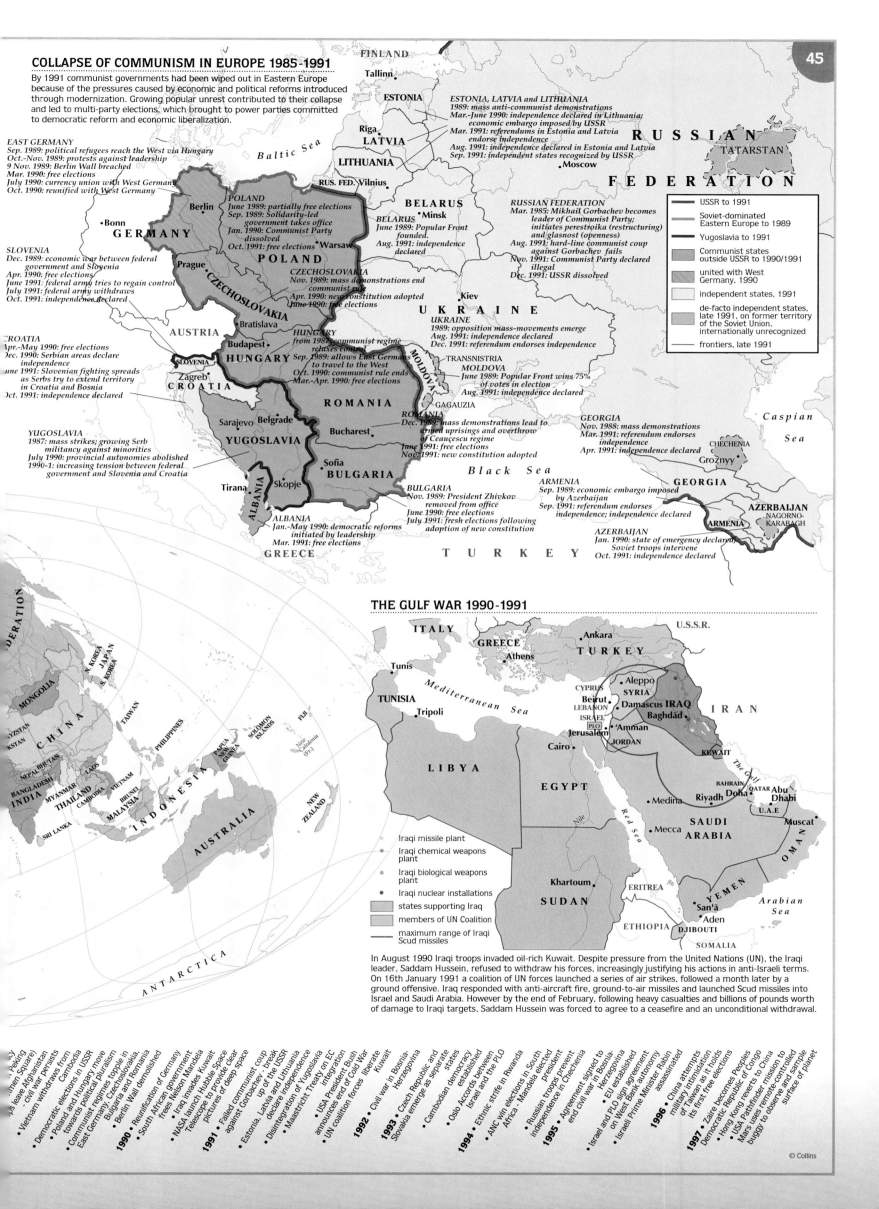

COLLAPSE OF COMMUNISM IN EUROPE 1985-1991

By 1991 communist governments had been wiped out in Eastern Europe because of the pressures caused by economic and political reforms introduced through modernization. Growing popular unrest contributed to their collapse and led to multi-party elections, which brought to power parties committed to democratic reform and economic liberalization.

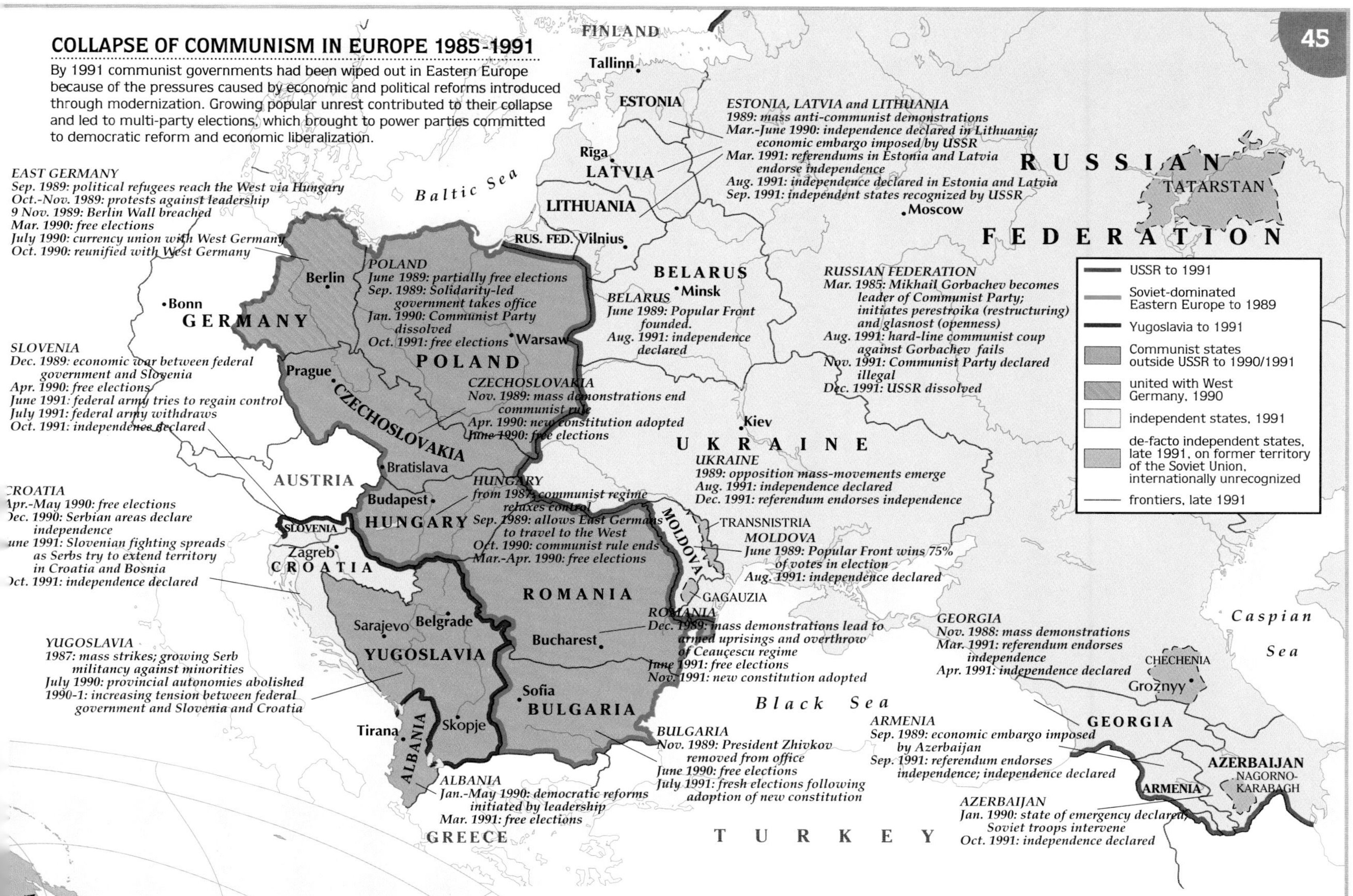
EAST GERMANY
Sep. 1989: political refugees reach the West via Hungary
Oct.-Nov. 1989: protests against leadership
9 Nov. 1989: Berlin Wall breached
Mar. 1990: free elections
July 1990: currency union with West Germany
Oct. 1990: reunified with West Germany
SLOVENIA
Dec. 1989: economic war between federal government and Slovenia
Apr. 1990: free elections
June 1991: federal army tries to regain control
July 1991: federal army withdraws
Oct. 1991: independence declared
CROATIA
Apr.-May 1990: free elections
Dec. 1990: Serbian areas declare independence
June 1991: Slovenian fighting spreads as Serbs try to extend territory in Croatia and Bosnia
Oct. 1991: independence declared
YUGOSLAVIA
1987: mass strikes; growing Serb militancy against minorities
July 1990: provincial autonomies abolished
1990-1: increasing tension between federal government and Slovenia and Croatia
POLAND
June 1989: partially free elections
Sep. 1989: Solidarity-led government takes office
Jan. 1990: Communist Party dissolved
Oct. 1991: free elections
CZECHOSLOVAKIA
Nov. 1989: mass demonstrations end communist rule
Apr. 1990: new constitution adopted
June 1990: free elections
HUNGARY
from 1987: communist regime relaxes control
Sep. 1989: allows East Germans to travel to the West
Oct. 1990: communist rule ends
Mar.-Apr. 1990: free elections
ROMANIA
Dec. 1989: mass demonstrations lead to armed uprisings and overthrow of Ceauçescu regime
June 1991: free elections
Nov. 1991: new constitution adopted
BULGARIA
Nov. 1989: President Zhivkov removed from office
June 1990: free elections
July 1991: fresh elections following adoption of new constitution
ALBANIA
Jan.-May 1990: democratic reforms initiated by leadership
Mar. 1991: free elections
ESTONIA, LATVIA and LITHUANIA
1989: mass anti-communist demonstrations
Mar.-June 1990: independence declared in Lithuania; economic embargo imposed by USSR
Mar. 1991: referendums in Estonia and Latvia endorse independence
Aug. 1991: independence declared in Estonia and Latvia
Sep. 1991: independent states recognized by USSR
BELARUS
June 1989: Popular Front founded.
Aug. 1991: independence declared
RUSSIAN FEDERATION
Mar. 1985: Mikhail Gorbachev becomes leader of Communist Party; initiates perestroika (restructuring) and glasnost (openness)
Aug. 1991: hard-line communist coup against Gorbachev fails
Nov. 1991: Communist Party declared illegal
Dec. 1991: USSR dissolved
UKRAINE
1989: opposition mass-movements emerge
Aug. 1991: independence declared
Dec. 1991: referendum endorses independence
MOLDOVA
June 1989: Popular Front wins 75% of votes in election
Aug. 1991: independence declared
GEORGIA
Nov. 1988: mass demonstrations
Mar. 1991: referendum endorses independence
Apr. 1991: independence declared
ARMENIA
Sep. 1989: economic embargo imposed by Azerbaijan
Sep. 1991: referendum endorses independence; independence declared
AZERBAIJAN
Jan. 1990: state of emergency declared, Soviet troops intervene
Oct. 1991: independence declared
USSR to 1991
Soviet-dominated Eastern Europe to 1989
Yugoslavia to 1991
Communist states outside USSR to 1990/1991
united with West Germany, 1990
independent states, 1991
de-facto independent states, late 1991, on former territory of the Soviet Union, internationally unrecognized
frontiers, late 1991
FINLAND
Tallinn
ESTONIA
Riga
LATVIA
LITHUANIA
RUS. FED.
Vilnius
Baltic Sea
RUSSIAN FEDERATION
TATARSTAN
Moscow
BELARUS
Minsk
Berlin
Bonn
GERMANY
POLAND
Warsaw
Prague
CZECHOSLOVAKIA
Bratislava
Kiev
UKRAINE
AUSTRIA
Budapest
HUNGARY
SLOVENIA
Zagreb
CROATIA
MOLDOVA
TRANSNISTRIA
GAGAUZIA
ROMANIA
Bucharest
Sarajevo
Belgrade
YUGOSLAVIA
Sofia
BULGARIA
Skopje
Tirana
ALBANIA
GREECE
Black Sea
TURKEY
Caspian Sea
CHECHENIA
Groznyy
GEORGIA
ARMENIA
AZERBAIJAN
NAGORNO-KARABAGH

MONGOLIA
CHINA
N. KOREA
JAPAN
S. KOREA
TAIWAN
PHILIPPINES
NEPAL
BHUTAN
BANGLADESH
INDIA
SRI LANKA
MYANMAR
THAILAND
LAOS
CAMBODIA
VIETNAM
BRUNEI
MALAYSIA
INDONESIA
PAPUA NEW GUINEA
SOLOMON ISLANDS
FIJI
AUSTRALIA
NEW ZEALAND
ANTARCTICA

THE GULF WAR 1990-1991

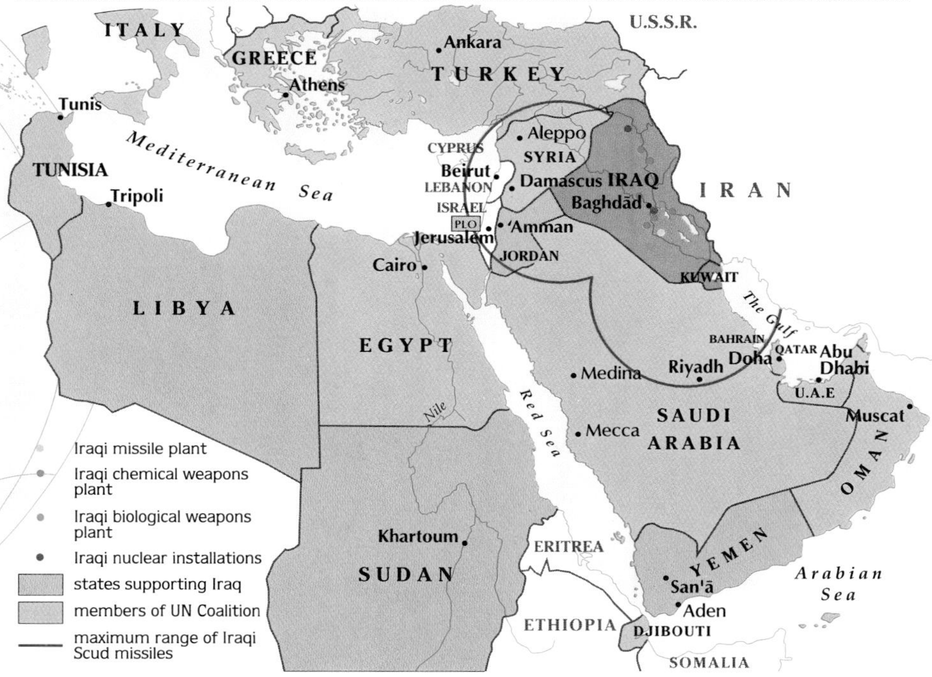
ITALY
GREECE
Athens
Ankara
TURKEY
U.S.S.R.
Tunis
TUNISIA
Tripoli
Mediterranean Sea
CYPRUS
Aleppo
SYRIA
Beirut
LEBANON
Damascus
IRAQ
ISRAEL
PLO
Jerusalem
'Amman
JORDAN
Baghdād
IRAN
Cairo
LIBYA
EGYPT
KUWAIT
The Gulf
BAHRAIN
QATAR
Doha
Abu Dhabi
U.A.E.
Riyadh
Medina
Mecca
SAUDI ARABIA
Muscat
OMAN
Nile
Red Sea
Khartoum
SUDAN
ERITREA
YEMEN
San'ā
Aden
Arabian Sea
ETHIOPIA
DJIBOUTI
SOMALIA
Iraqi missile plant
Iraqi chemical weapons plant
Iraqi biological weapons plant
Iraqi nuclear installations
states supporting Iraq
members of UN Coalition
maximum range of Iraqi Scud missiles

In August 1990 Iraqi troops invaded oil-rich Kuwait. Despite pressure from the United Nations (UN), the Iraqi leader, Saddam Hussein, refused to withdraw his forces, increasingly justifying his actions in anti-Israeli terms. On 16th January 1991 a coalition of UN forces launched a series of air strikes, followed a month later by a ground offensive. Iraq responded with anti-aircraft fire, ground-to-air missiles and launched Scud missiles into Israel and Saudi Arabia. However by the end of February, following heavy casualties and billions of pounds worth of damage to Iraqi targets, Saddam Hussein was forced to agree to a ceasefire and an unconditional withdrawal.

…cy
…Peking
…men Square)
…ps leave Afghanistan - civil war persists
• Vietnam withdraws from Cambodia
• Democratic elections in USSR
• Poland and Hungary move towards political pluralism
• Communist regimes topple in East Germany, Czechoslovakia, Bulgaria and Romania
• Berlin Wall demolished

1990 • Reunification of Germany
• South African government frees Nelson Mandela
• Iraq invades Kuwait
• NASA launch Hubble Space Telescope to provide clear pictures of deep space

1991 • Failed communist coup against Gorbachev - break up of the USSR
• Estonia, Latvia and Lithuania declare independence
• Disintegration of Yugoslavia
• Maastricht Treaty on EC integration
• USA President Bush announces end of Cold War
• UN coalition forces liberate Kuwait

1992 • Civil war in Bosnia-Herzegovina

1993 • Czech Republic and Slovakia emerge as separate states
• Cambodian democracy established
• Oslo Accords between Israel and the PLO

1994 • Ethnic strife in Rwanda
• ANC win elections in South Africa - Mandela elected president
• Russian troops prevent independence in Chechenia

1995 • Agreement signed to end civil war in Bosnia-Herzegovina
• EU established
• Israel and PLO sign agreement on West Bank autonomy
• Israeli Prime Minister Rabin assassinated

1996 • China attempts military intimidation of Taiwan as it holds its first free elections

1997 • Zaire becomes Democratic Republic of Congo
• Hong Kong reverts to China
• USA Pathfinder mission to Mars uses remote-controlled buggy to observe and sample surface of planet

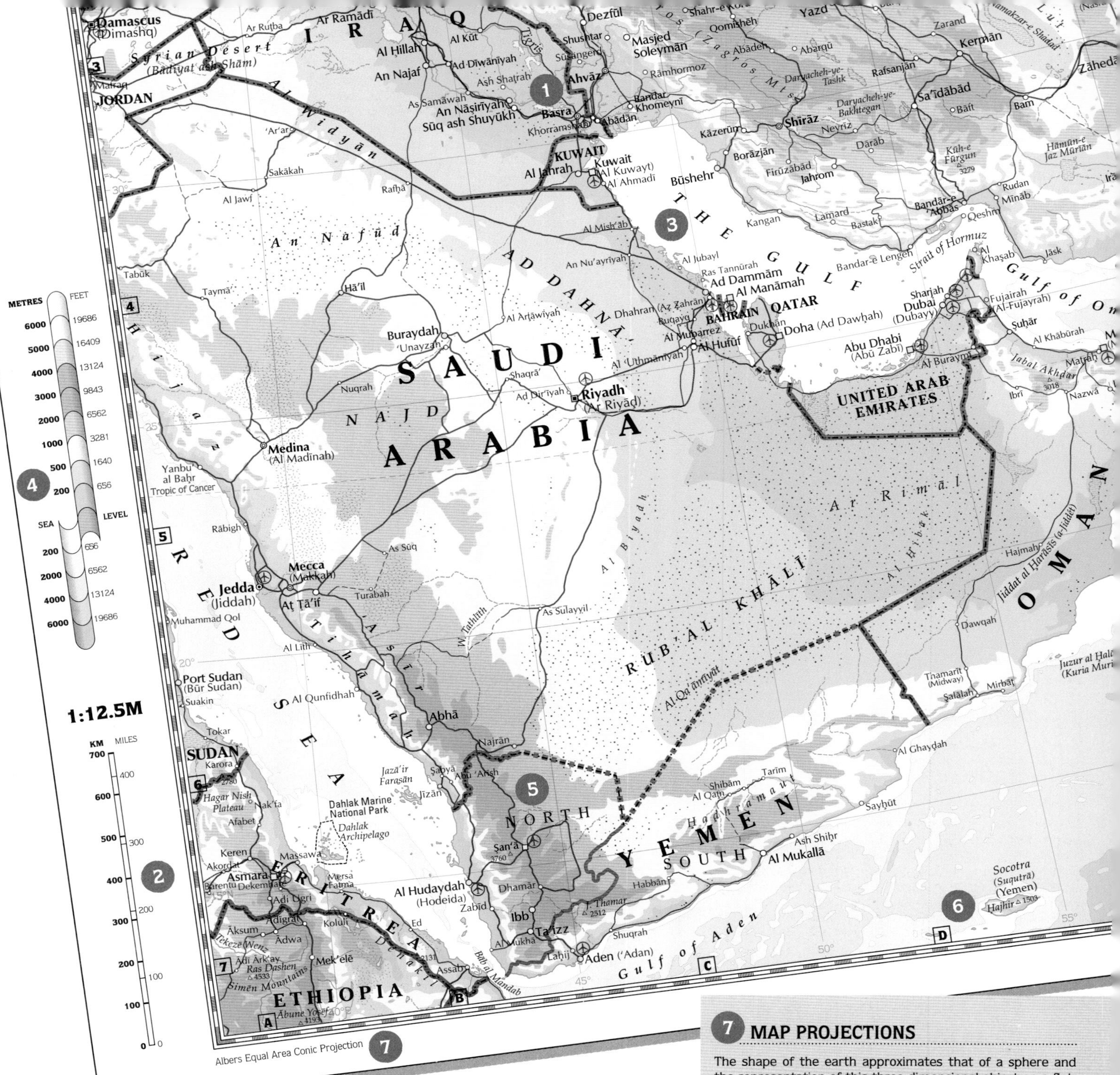

HOW TO USE THE ATLAS

THE ATLAS CONTAINS several types of map. The preliminary section includes a varied selection of thematic, statistical, and historical mapping. These vary greatly in their content and carry individual keys and supplementary information to help in their interpretation. The main Atlas of the World section contains topographic reference mapping covering all parts of the world, designed to a standard specification. The 'HOW TO USE THE ATLAS' section helps the reader to interpret these reference maps. It explains the main features shown on the mapping and the policies adopted in deciding what to show and how to show it.

The databases used to create the maps provide the freedom to select the best map coverage for each part of the world. Maps are arranged on a continental basis, with each continent being introduced by maps of the political situation and the main physical features. Maps of Antarctica and the world's oceans complete the extensive worldwide coverage.

1 SYMBOLS & GENERALIZATION

Maps show information by using signs, or symbols, which are designed to reflect the features on the earth that they represent. Symbols can be in the form of points - such as those used to show towns, mountain summits and airports; lines - used to represent roads, rivers and boundaries; or areas - such as lakes, marsh and urban areas. Variations in the size, shape and colour of the symbols used allow a great range of information to be shown.

The symbols used in this atlas are explained on the Symbols page or 'key' on the first page of the Atlas of the World section. A full range of symbols described above are used to depict communications, physical features, relief (see 4 below), political entities (see 5 below), settlements and miscellaneous features.

Not all features on the ground can be shown, nor can all characteristics of a feature be depicted. Much has to be generalized to be clearly shown on the maps. The degree of generalization is determined largely by the scale of the map (see 2 below). As map scale decreases, fewer features can be shown, and their depiction becomes less detailed.

The most common generalization techniques are selection and simplification. Selection is the inclusion of some features and the omission of others of less importance. Smaller scale maps can show fewer features than larger scales, and therefore only the more important features will be selected. Simplification is the process of smoothing lines, combining areas, or slightly displacing symbols to add clarity. Smaller scale maps require more simplification. All the techniques are carried out in such a way that the overall character of the area mapped is retained.

2 SCALE

The amount of detail shown on a map is determined by its scale - the relationship between the size of an area shown on the map and the actual size of the area on the ground. Larger scales show more detail, smaller scales require more generalization (see 1 above) and show less. The scale can be used to measure the distance between two points and to calculate comparative areas, though the projection of the

7 MAP PROJECTIONS

The shape of the earth approximates that of a sphere and the representation of this three-dimensional object on a flat, two-dimensional map presents a problem for the cartographer. For maps to be constructed, the shape of the earth needs to be 'projected' onto the flat sheet. There are several ways of doing this which have given rise to a wide variety of 'map projections'. Whichever method is chosen, transferring information from a sphere to a flat surface is impossible without distortion. It is the pattern of this distortion, and the question of how best to control and limit it, that must be considered when choosing a map projection. The area covered by the map and its overall purpose will also influence the decision.

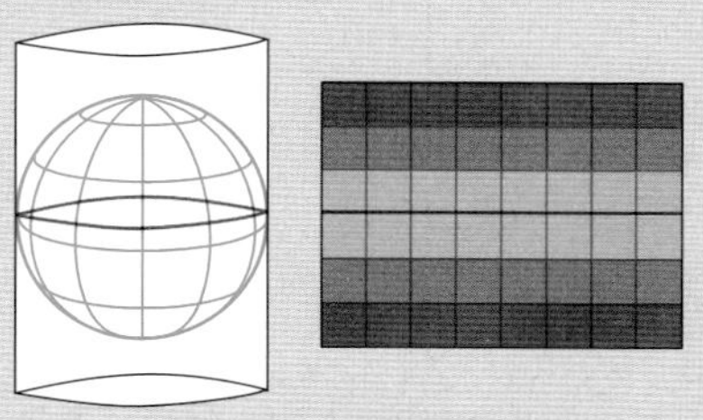

CYLINDRICAL

The four basic properties of any projection are area, shape, distance (scale) and direction. Any one of these can be controlled but only at the expense of the others. For example, a projection can retain true distances, but if it does, the shape of the areas shown will not be accurate. Projections are commonly described in terms of these three properties. Those with accurate representation of area are known as 'equal area' or 'equivalent projections'; those with true shape are known as 'conformal' and those with correct distances from a specific point are 'equidistant'. True directions (bearings) are only possible on an 'azimuthal' projection.

The geometric characteristics or methods of construction of projections are also used to describe them. The three

map must also be taken into account when taking measurements (see panel below).

The scale is shown in the margin of each map in two ways. The representative fraction (1:12.5M on the extract opposite) tells us that a distance of 1 mm on the map actually measures 12,500,000 mm (or 12.5 kilometres) on the ground. Any unit of measure can be used to apply this principle. The larger the denominator of the representative fraction (12.5M in the example above) the smaller the scale of the map. The second method used to depict scale - the linear scale, or scale bar - converts the representative fraction into easily measurable units.

Scales used for the main maps in this atlas range from 1:2M (large scale) to 1:80M (small scale world maps). Insets are used to show areas of the world of particular interest or which cannot be included in the page layouts in their true position. They are at larger scales to allow the inclusion of more detail.

3 GEOGRAPHICAL NAMES

Place names are, to an extent, a mirror for the changes that continue to transform the political globe, and their spelling on maps is a complex problem for the cartographer. There is no single standard way of spelling names or of converting them from one alphabet, or symbol set, to another.

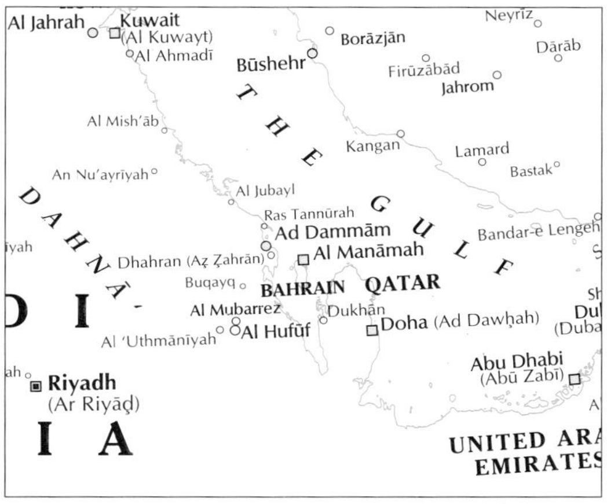

Instead, conventional ways of spelling have evolved, and the results often differ significantly from the original name in the local language. Familiar examples in English include Munich (München in German), Florence (Firenze in Italian) and Moscow (Moskva from Russian). Other factors also stand in the way of achieving a single standard. In many countries different languages are in use in different regions, or side-by-side in the same region. A worldwide trend towards national, regional and ethnic self-determination is operating at the same time as an inevitable pressure towards more international standardization.

All these factors, and any changes in official languages have to be taken into account when creating maps and databases, and policies need to be established for the spelling of names on individual atlases and maps. Such policies must take account of the local official position, international conventions or traditions, and the purpose of the atlas or map. The policy in this atlas is to use local name forms which are officially recognized by the governments of the countries concerned. This is a basic principle laid down by the Permanent Committee on Geographical Names (PCGN) - the body responsible for determining official UK government policy on place names around the world. PCGN rules, and those of the equivalent US agency, the US Board on Geographical Names (BGN), are also applied to the conversion of non-roman alphabet names, for example in the Russian Federation, into the roman alphabet used in English.

However, the overall policy of using local names is varied slightly, with English conventional name forms being used for the most well-known places. In these cases, the local form is included in brackets on the map and is referred to as an 'alternative' name. It also appears as a cross-reference in the index. Examples of this policy on the above names extract include:

ENGLISH FORM	LOCAL FORM
Kuwait	**Al Kuwayt**
Riyadh	**Ar Riyāḍ**
Doha	**Ad Dawḥah**

Other examples in the atlas include:

Moscow	**Moskva**
Vienna	**Wien**
Crete	**Kriti**
Serbia	**Srbija**

Other alternative names, such as well-known historical names or those in other languages, may also be included in brackets and as index cross-references. Another variation from the above policy is that all country names and those for international physical features (stretching over more than one country) appear in their English forms.

Names relating to different types of feature are distinguished by the use of different type styles. This helps to quickly identify the feature on a map to which a name relates. The full range of type styles used is shown on the Symbols page.

Abbreviations used in place names on the maps and in the index are explained in the Abbreviations and Glossary.

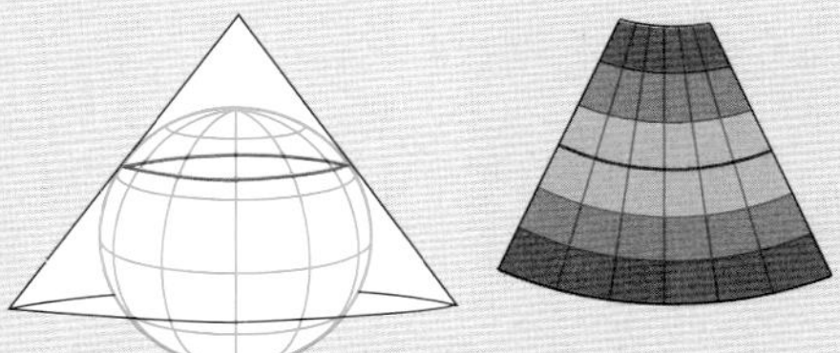

CONIC

main methods of construction are illustrated here. In each one a globe representing the earth is brought into contact with a flat surface (a map). The flat surface can be made into a cylinder or cone, or it can remain flat. The shape of the earth is 'projected' from the globe onto the map to create cylindrical, conical and azimuthal projections respectively. The different results are illustrated by the pattern of the graticule - the lines of latitude and longitude used to define position on the earth. Distortions increase away from the point of contact between the globe and the map surface. To minimize distortion, the point of contact is chosen in accordance with the shape of the area being mapped, and can be further reduced if the map surface is made to cut through the globe rather than just touch it.

A variety of projections are used for the reference mapping, each chosen to retain the most important characteristics for the scale and area involved. The projection used is indicated in the margin of each map, using terms such as those explained above.

AZIMUTHAL

increasing distortion

4 REPRESENTATION OF RELIEF

One important element of mapping the earth is the depiction of relief - the 'shape' of the land. The reader needs to be made immediately aware of which areas are high, which are low, of where the main mountain ranges and plains are. This presents a problem to the cartographer who has to show the earth's three-dimensional surface on the two-dimensional page. The maps in this atlas use three methods of relief representation.

Hypsometric layers, or layer tints, use variations in colour to distinguish areas of land which lie in specific altitude or relief bands. The colours give an immediate impression of the height and shape of the land and are indicated on the altitude bar in the margin of each map. This method applies also to sea and ocean areas. The bathymetry, or depth of the

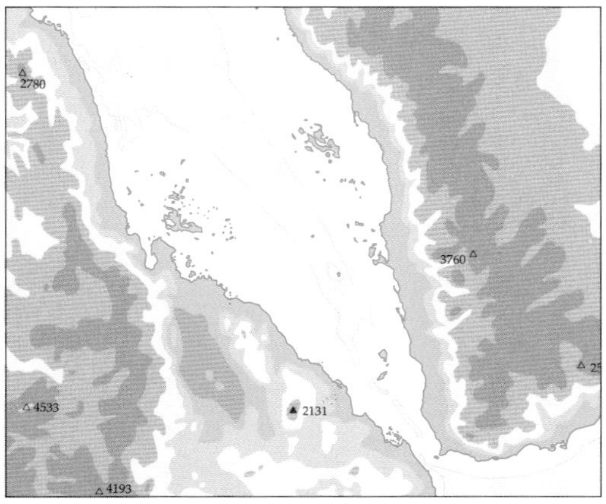

sea-floor is indicated by layers representing different depth bands. These are shown on all the reference maps, including those of the oceans. This method of relief representation was pioneered by our predecessor John Bartholomew of Edinburgh, and is now widely established as the most commonly used in atlas mapping.

The height of the land at selected points is shown by summit symbols or spot heights which indicate the height at that point in metres above sea level. On the maps of the oceans, small dots show the depth of the ocean below sea level in metres.

A third method of relief depiction is used on the maps of the oceans and Antarctica - hill or relief shading simulates the effect of a light shining across the landscape, providing an impression of the shape of the land.

5 BOUNDARIES

The status of nations and their boundaries, and the names associated with them, are shown in this atlas as they are in reality at the time of going to press, as far as can be ascertained. All recent changes of the status of nations and their boundaries have been taken into account.

Where international boundaries are the subject of dispute it may be that no portrayal of them will meet with the approval of the countries involved, but it is not seen as the function of this atlas to try to adjudicate between the rights and wrongs of political issues. The aim in these cases is to take a strictly neutral viewpoint based on advice from expert consultants. Although atlas reference mapping is not a suitable medium for indicating the claims of the many separatist movements of the world, or of one country's claims against another, every reasonable attempt is made to show where an active territorial dispute exists. It is also important to show where there is a

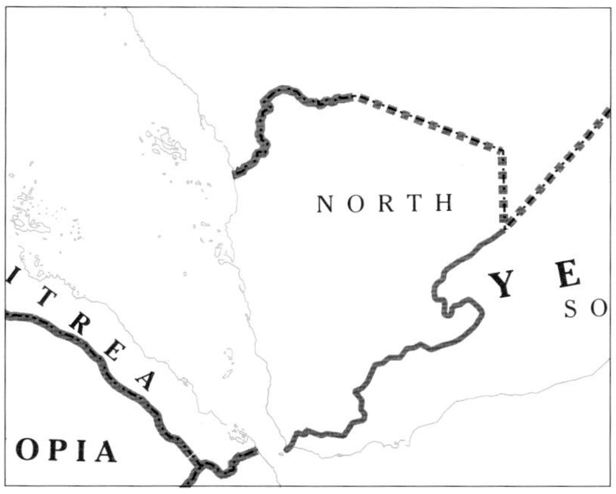

difference between 'de facto' (existing on the ground) and 'de jure' (according to law) boundaries. Generally, prominence is given to the former situation. The depiction of boundaries and their current status varies accordingly, as illustrated on the above extract. The full range of methods of boundary depiction is shown on the Symbols page.

International boundaries are shown on all the reference maps, and those of a large enough scale also include internal administrative boundaries of selected countries. The delineation of international boundaries in the sea is often a very contentious issue, and in many cases an official alignment is not defined. Boundaries in the sea are generally only shown where they are required to clarify the ownership of specific islands or island groups.

6 INDEXING

All names appearing on the reference maps are included in the index and can be easily found from the information included in the index entry. In addition to the page number, each entry includes a grid reference which relates to the alpha-numeric boxes on the frame of each map. Entries for names on inset maps do not include grid references, but are marked with a symbol (□) and the number of the inset on the page (where applicable).

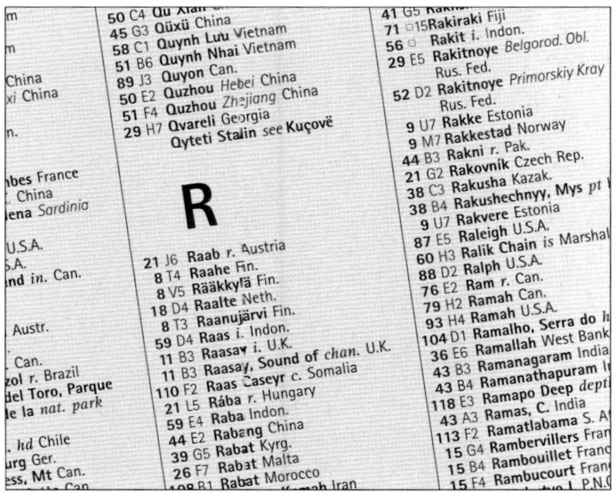

45 G3 **Qüxü** China
58 C1 **Quynh Luu** Vietnam
51 B6 **Quynh Nhai** Vietnam
89 J3 **Quyon** Can.
50 E2 **Quzhou** *Hebei* China
51 F4 **Quzhou** *Zhejiang* China
29 H7 **Qvareli** Georgia
Qyteti Stalin *see* **Kuçovë**

R

21 J6 **Raab** *r.* Austria
8 T4 **Raahe** Fin.
8 V5 **Rääkkylä** Fin.
18 D4 **Raalte** Neth.
8 T3 **Raanujärvi** Fin.
59 D4 **Raas** *i.* Indon.
11 B3 **Raasay** *i.* U.K.
11 B3 **Raasay, Sound of** *chan.* U.K.
110 F2 **Raas Caseyr** *c.* Somalia
21 L5 **Rába** *r.* Hungary
59 E4 **Raba** Indon.
44 E2 **Rabang** China
39 G5 **Rabat** Kyrg.
26 F7 **Rabat** Malta

71 □15 **Rakiraki** Fiji
56 □ **Rakit** *i.* Indon.
29 E5 **Rakitnoye** *Belgorod. Obl.* Rus. Fed.
52 D2 **Rakitnoye** *Primorskiy Kray* Rus. Fed.
9 U7 **Rakke** Estonia
9 M7 **Rakkestad** Norway
44 B3 **Rakni** *r.* Pak.
21 G2 **Rakovník** Czech Rep.
38 C3 **Rakusha** Kazak.
9 U7 **Rakvere** Estonia
87 E5 **Raleigh** U.S.A.
88 D2 **Ralph** U.S.A.
76 E2 **Ram** *r.* Can.
79 H2 **Ramah** Can.
93 H4 **Ramah** U.S.A.
36 E6 **Ramallah** West Bank
43 B3 **Ramanagaram** India
43 A3 **Ramas, C.** India
113 F2 **Ramatlabama** S. A
15 G4 **Rambervillers** Fran
15 B4 **Rambouillet** Franc
15 F4 **Rambucourt** Fran

The country in which the feature falls is included and if two similar features of the same name fall in one country, the administrative division is also shown. For physical features, descriptors are used to clarify the type of feature to which a name refers, where this is not obvious from the name itself; for example, r. for river, l. for lake, mt for mountain. Abbreviations used in the index are explained in the Abbreviations and Glossary overleaf. Alternative names (see 3 above) appear in the index as cross-references and direct the reader to the main entry to which the name relates.

To allow easier and quicker searching of the index, the order of names of physical features, in particular lakes and mountains, are often adjusted to ensure that the main part of the name appears first. This 'permuting' saves the reader from searching through many hundreds of names beginning with the same term. Examples include: Lake Superior being indexed as 'Superior, Lake' and Mt Everest as 'Everest, Mt'.

ABBREVIATIONS AND GLOSSARY

A.	Alp Alpen Alpi *alp*
	Alt *upper*
A.C.T.	Australian Capital Territory
Afgh.	Afghanistan
Afr.	Africa African
Aig.	Aiguille *peak*
AK	Alaska
AL	Alabama
Alg.	Algeria
Alta	Alberta
Ant.	Antarctica
AR	Arkansas
Arch.	Archipelago
Arg.	Argentina
Arr.	Arrecife *reef*
Atl.	Atlantic
Austr.	Australia
AZ	Arizona
Azer.	Azerbaijan
B.	Bad *spa*
	Ban *village*
	Bay
Bangla.	Bangladesh
B.C.	British Columbia
Bg	Berg *mountain*
Bge.	Barragem *reservoir*
Bgt	Bight Bugt *bay*
Bj	Burj *hills*
Bol.	Bolivia
Bos.-Herz.	Bosnia Herzegovina
Br.	Burun Burnu *point, cape*
Bt	Bukit *bay*
Bü.	Büyük *big*
Bulg.	Bulgaria
C.	Cape
	Col *high pass*
Ç.	Çay *river*
CA	California
Cabo	Cabeço *summit*
Can.	Canada
	Canal Canale *canal*
	Cañon Canyon *canyon*
C.A.R.	Central African Republic
Cat.	Cataract
	Catena *mountains*
Cd	Ciudad *town city*
Ch.	Chaung *stream*
	Chott *salt lake, marsh*
Chan.	Channel
Che	Chaîne *mountain chain*
Cma	Cima *summit*
Cno	Corno *peak*
Co	Cerro *hill, peak*
CO	Colorado
Co.	County
Col.	Colombia
Cord.	Cordillera *mountain chain*
Cr.	Creek
CT	Connecticut
Cuch.	Cuchilla *mountain chain*
Czo	Cozzo *mountain*
D.	Da *big, river*
	Dag Dagh Dağı Dağlari *mountain, mountains*
-d.	-dake *peak*
DC	District of Columbia
DE	Delaware
Des.	Desert
Div.	Division
Dj.	Djebel *mountain*
Dom. Rep.	Dominican Republic
Eil.	Eiland *island*
	Eilanden *islands*
Emb.	Embalse *reservoir*
Eng.	England
Equat.	Equatorial
Esc.	Escarpment
Est.	Estuary
Eth.	Ethiopia
Etg	Etang *lake, lagoon*
F.	Firth
Fin.	Finland
Fj.	Fjell *mountain*
	Fjord Fjördur *fjord*
Fl.	Fleuve *river*
FL	Florida
Fr. Guiana	French Guiana
G.	Gebel *mountain*
	Göl Gölö Göl *lake*
	Golfe Golfo Gulf *gulf, bay*
	Góra *mountain*
	Gunung *mountain*
-g.	-gawa *river*
GA	Georgia
Gd	Grand *big*
Gde	Grande *big*
Geb.	Gebergte *mountain range*
	Gebirge *mountains*
Gl.	Glacier
Ger.	Germany
Gr.	Graben *trench, ditch*
	Gross Grosse Grande *big*
Grp	Group
Gt	Great Groot Groote *big*
Gy	Góry Gory *mountains*
H.	Hawr *lake*
	Hill Hills
	Hoch *high*
	Hora *mountain*
	Hory *mountains*
Halv.	Halvøy *peninsula*
Harb.	Harbour
Hd	Head Headland
Hg.	Hegység *mountains*
Hgts	Heights
HI	Hawaii
Ht	Haut *high*
Hte	Haute *high*
I.	Île Ilha Insel Isla
	Island Isle *island, isle*
	Isola Isole *island*
IA	Iowa
ID	Idaho
IL	Illinois
IN	Indiana
In.	Inlet
Indon.	Indonesia
Is	Islas Îles Ilhas
	Islands Isles *islands, isles*
Isr.	Israel
Isth.	Isthmus
J.	Jabal Jebel *mountain*
	Jibāl *mountains*
	Jrvi Jaure Jezero
	Jezioro *lake*
	Jökull *glacier*
K.	Kaap Kap Kapp *cape*
	Kaikyō *strait*
	Kato Káto *lower*
	Kiang *river or stream*
	Ko *island, lake, inlet*
	Koh Kūh Kūhha *island*
	Kolpos *gulf*
	Kopf *hill*
	Kuala *estuary*
	Kyst *coast*
	Küçük *small*
Kan.	Kanal Kanaal *canal*
Kazak.	Kazakstan
Kep.	Kepulauan *archipelago, islands*
Kg	Kampong *village*
Khr	Khrebet *mountain range*
Kl.	Klein Kleine *small*
Kör.	Körfez Körfezi *bay, gulf*
KS	Kansas
KY	Kentucky
Kyrg.	Kyrgyzstan
L.	Lac Lago Lake
	Liqen Loch Lough *lake, loch*
	Lam *stream*
LA	Louisiana
Lag.	Lagoon Laguna
	Lagôa *lagoon*
Lith.	Lithuania
Lux.	Luxembourg
M.	Mae *river*
	Me *great, chief, mother*
	Meer *lake, sea*
	Muang *kingdom, province, town*
	Muong *town*
	Mys *cape*
	Maloye *small*
MA	Massachusetts
Madag.	Madagascar
Man.	Manitoba
Maur.	Mauritania
MD	Maryland
ME	Maine
Mex.	Mexico
Mf	Massif *mountains, upland*
Mgna	Montagna *mountain*
Mgne	Montagne *mountain*
Mgnes	Montagnes *mountains*
MI	Michigan
MN	Minnesota
MO	Missouri
Mon.	Monasterio Monastery *monastery*
	Monument *monument*
Moz.	Mozambique
MS	Mississippi
Mt	Mont Mount Mountain *mountain*
MT	Montana
Mte	Monte *mountain*
Mtes	Montes *mountains*
Mti	Monti Munţi *mountains*
Mtii	Munţii *mountains*
Mth	Mouth
Mths	Mouths
Mtn	Mountain
Mts	Monts Mountains
N.	Nam *south(ern), river*
	Neu Ny *new*
	Nevado *peak*
	Nudo *mountain*
	Noord Nord Nörre
	Nørre North *north(ern)*
	Nos *spit, point*
Nac.	Nacional *national*
Nat.	National
N.B.	New Brunswick
NC	North Carolina
ND	North Dakota
NE	Nebraska
Neth.	Netherlands
Neth. Ant.	Netherlands Antilles
Nfld	Newfoundland
NH	New Hampshire
Nic.	Nicaragua
Nizh.	Nizhneye Nizhniy
	Nizhnyaya *lower*
Nizm.	Nizmennost' *lowland*
NJ	New Jersey
NM	New Mexico
N.O.	Noord Oost Nord Ost *northeast*
Nov.	Novyy Novaya
	Noviye Novoye *new*
N.S.	Nova Scotia
N.S.W.	New South Wales
N.T.	Northern Territory
NV	Nevada
Nva	Nueva *new*
N.W.T.	Northwest Territories
NY	New York
N.Z.	New Zealand
O.	Oost Ost *east*
	Ostrov *island*
Ø	Østre *east*
Ob.	Ober *upper, higher*
Oc.	Ocean
Ode	Oude *old*
Ogl.	Oglat *well*
OH	Ohio
OK	Oklahoma
Ont.	Ontario
Or.	Óri Óros Ori *mountains*
	Oros *mountain*
OR	Oregon
Orm.	Ormos *bay*
O-va	Ostrova *islands*
Ot	Olet *mountain*
Öv.	Över Övre *upper*
Oz.	Ozero *lake*
	Ozera *lakes*
P.	Pass
	Pic Pico Piz *peak*
	Pou *mountain*
	Pulau *island*
PA	Pennsylvania
Pac.	Pacific
Pak.	Pakistan
Para.	Paraguay
Pass.	Passage
Peg.	Pegunungan *mountain range*
P.E.I.	Prince Edward Island
Pen.	Peninsula Penisola *peninsula*
Per.	Pereval *pass*
Phil.	Philippines
Phn.	Phnom *hill, mountain*
Pgio	Poggio *hill*
Pk	Peak Park
Pl.	Planina Planinski *mountain(s)*
Pla	Playa *beach*
Plat.	Plateau
Plosk.	Ploskogor'ye *plateau*
P.N.G.	Papua New Guinea
Pno	Pantano *reservoir, swamp*
Pol.	Poland
Por.	Porog *rapids*
Port.	Portugal
P-ov	Poluostrov *peninsula*
P.P.	Pulau-pulau *islands*
Pr.	Proliv *strait*
	Przylądek *cape*
Presq.	Presqu'île *peninsula*
Prom.	Promontory
Prov.	Province Provincial
Psa	Presa *dam*
Pso	Passo *dam*
Pt	Point
	Pont *bridge*
	Petit *small*
Pta	Ponta Punta *cape, point*
	Puerta *narrow pass*
Pte	Pointe *cape, point*
	Ponte Puente *bridge*
Pto	Porto Puerto *harbour, port*
Pzo	Pizzo *mountain peak, mountain*
Qld.	Queensland
Que.	Quebec
R.	Reshteh *mountain range*
	Rūd River *river*
Ra.	Range
Rca	Rocca *rock, fortress*
Reg.	Region
Rep.	Republic
Res.	Reserve
Resr	Reservoir
Resp.	Respublika *republic*
Rf	Reef
Rge	Ridge
RI	Rhode Island
Riba	Ribeira *coast, bottom of the river valley*
Rte	Route
Rus. Fed.	Russian Federation
S.	Salar Salina *salt pan*
	San São *saint*
	See *lake*
	Seto *strait, channel*
	Sjö *lake*
	Sör Süd Sud Syd South *south(ern)*
Sa	Serra Sierra *mountain range*
S.A.	South Australia
Sab.	Sabkhat *salt flat*
Sask.	Saskatchewan
S. Arabia	Saudi Arabia
SC	South Carolina
Sc.	Scoglio *rock, reef*
Sd	Sound Sund *sound*
SD	South Dakota
Seb.	Sebjet Sebkhat Sebkra *salt flat*
Serr.	Serrania *mountain range*
Sev.	Severnaya Severnyy *north(ern)*
Sh.	Shā'ib *watercourse*
	Shaţţ *river (-mouth)*
	Shima *island*
	Shankou *pass*
Si	Sidi *lord, master*
Sing.	Singapore
Sk.	Shuiku *reservoir*
Skt	Sankt *saint*
Smt	Seamount
Snra	Senhora *Mrs, lady*
Snro	Senhoro *Mr, gentleman*
Sp.	Spain Spanish
	Spitze *peak*
Sr	Sönder Sønder *southern*
Sr.	Sredniy Srednyaya *middle*
St	Saint Sint
	Staryy *old*
St.	Stor Store *big*
	Stung *river*
Sta	Santa *saint*
Ste	Sainte *saint*
	Store *big*
Sto	Santo *saint*
Str.	Strait Stretta *strait*
Sv.	Sväty Sveti *holy, saint*
Switz.	Switzerland
T.	Tal *valley*
	Tall Tell *hill*
	Tepe Tepesi *hill, peak*
Tajik.	Tajikistan
Tanz.	Tanzania
Tas.	Tasmania
Terr.	Territory
Tg	Tanjung Tanjong *cape, point*
Thai.	Thailand
Tk	Teluk *bay*
Tmt	Tablemount
TN	Tennessee
Tr.	Trench Trough
Tre	Torre *tower, fortress*
Tte	Teniente *lieutenant*
Turk.	Turkmenistan
TX	Texas
U.A.E.	United Arab Emirates
Ug	Ujung *point, cape*
U.K.	United Kingdom
Ukr.	Ukraine
Unt.	Unter *lower*
Upr	Upper
Uru.	Uruguay
U.S.A.	United States of America
UT	Utah
Uzbek.	Uzbekistan
V.	Val Valle Valley *valley*
	Väster Vest Vester *west(ern)*
	Vatn *lake*
	Ville *town*
Va	Vila *small town*
VA	Virginia
Venez.	Venezuela
Vic.	Victoria
Volc.	Volcán Volcan
	Volcano *volcano*
Vdkhr.	Vodokhranilishche *reservoir*
Vdskh.	Vodoskhovshche
	Vodaskhovishcha *reservoir*
Vel.	Velikiy Velikaya
	Velikiye *big*
Verkh.	Verkhniy Verkhneye
	Verkhne *upper*
	Verkhnyaya *upper*
Vost.	Vostochnyy *eastern*
Vozv.	Vozvyshennost' *hills, upland*
VT	Vermont
W.	Wadi *watercourse*
	Wald *forest*
	Wan *bay*
	Water *water*
WA	Washington
W.A.	Western Australia
WI	Wisconsin
Wr	Wester
WV	West Virginia
WY	Wyoming
-y	-yama *mountain*
Y.T.	Yukon Territory
Yt.	Ytre Ytter Ytri *outer*
Yugo.	Yugoslavia
Yuzh.	Yuzhnaya Yuzhno
	Yuzhnyy *southern*
Zal.	Zaliv *bay*
Zap.	Zapadnyy Zapadnaya
	Zapadno Zapadnoye *western*
Zem.	Zemlya *land*

RELIEF

Contour intervals used in layer colouring

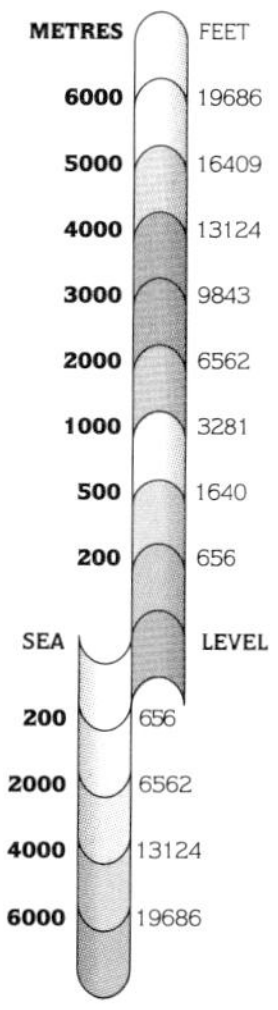

METRES		FEET
6000		19686
5000		16409
4000		13124
3000		9843
2000		6562
1000		3281
500		1640
200		656
SEA		LEVEL
200		656
2000		6562
4000		13124
6000		19686

Additional bathymetric contour layers are shown at scales greater than 1:2 million. These are labelled on an individual basis.

213 Summit
height in metres

PHYSICAL FEATURES

Freshwater lake

Seasonal freshwater lake

Saltwater lake *or* Lagoon

Seasonal saltwater lake

Dry salt lake *or* Saltpan

Marsh

River

Waterfall

Dam *or* Barrage

Seasonal river *or* Wadi

Canal

Flood dyke

Reef

Volcano

Lava field

Sandy desert

Rocky desert

Oasis

Escarpment

Mountain pass
height in metres
923

Ice cap *or* Glacier

COMMUNICATIONS

Motorway

Motorway
under construction

Motorway tunnel

Motorways are classified separately at scales greater than 1:5 million. At smaller scales motorways are classified with main roads.

Main road

Main road
under construction

Main road tunnel

Other road

Other road
under construction

Other road tunnel

Track

Main railway

Main railway
under construction

Main railway tunnel

Other railway

Other railway
under construction

Other railway tunnel

Main airport

Other airport

BOUNDARIES

International

International
disputed

Ceasefire line

Main administrative (U.K.)

Main administrative

Main administrative
through water

OTHER FEATURES

National park

Reserve

Ancient wall

Historic *or* Tourist site

Urban area

SETTLEMENTS

POPULATION	NATIONAL CAPITAL	ADMINISTRATIVE CAPITAL	CITY OR TOWN
Over 5 million	▣ **Beijing**	◉ **Tianjin**	◉ **New York**
1 to 5 million	▣ **Seoul**	◉ **Lagos**	◉ **Barranquilla**
500000 to 1 million	▣ **Bangui**	◎ **Douala**	◎ **Memphis**
100000 to 500000	▢ Wellington	○ Mansa	○ Mara
50000 to 100000	▢ Port of Spain	○ Lubango	○ Arecibo
10000 to 50000	▫ Malabo	○ Chinhoyi	○ El Tigre
Less than 10000	▫ Roseau	○ Áti	○ Soledad

STYLES OF LETTERING

COUNTRY NAME	MAIN ADMINISTRATIVE NAME	AREA NAME	MISCELLANEOUS NAME	PHYSICAL NAME
CANADA	XINJIANG UYGUR ZIZHIQU	*PATAGONIA*	Charles de Gaulle Airport	*Long Island*
SUDAN	MAHARASHTRA	*KALIMANTAN*	Rocky Mountains Forest Reserve	*LAKE ERIE*
TURKEY	KENTUCKY	*ARTOIS*	Disneyland Paris	*ANDES*
LIECHTENSTEIN	BRANDENBURG	*PENINSULAR MALAYSIA*	Great Wall	*Rio Grande*

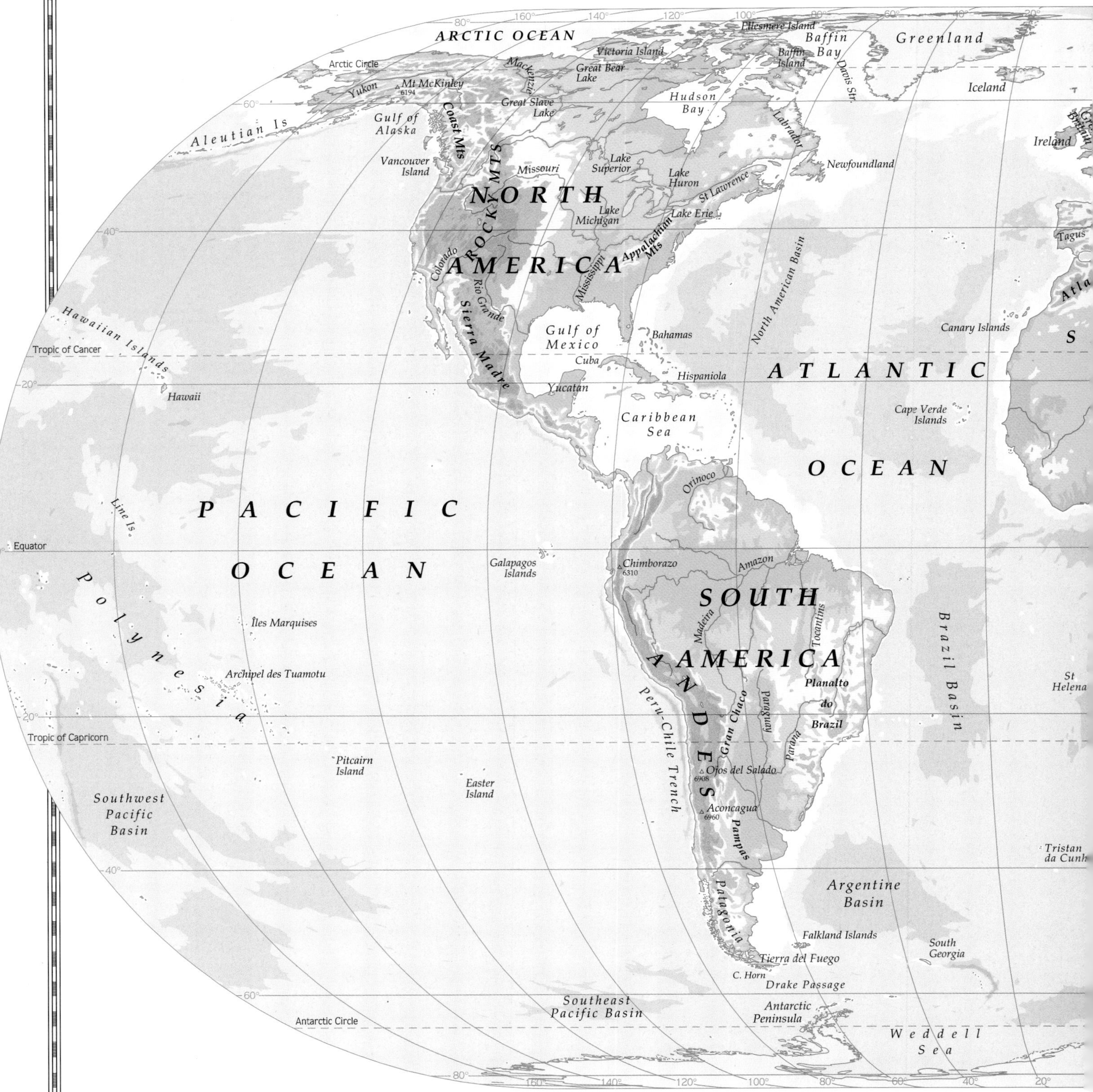
ARCTIC OCEAN
Ellesmere Island
Victoria Island
Baffin Bay
Greenland
Arctic Circle
Baffin Island
Davis Str.
Mackenzie
Great Bear Lake
Yukon
Mt McKinley 6194
Great Slave Lake
Hudson Bay
Iceland
Aleutian Is
Gulf of Alaska
Coast Mts
Labrador
Ireland
Vancouver Island
Missouri
Lake Superior
Lake Huron
Newfoundland
NORTH AMERICA
ROCKY MTS
Lake Michigan
St Lawrence
Lake Erie
Appalachian Mts
Tagus
Colorado
Rio Grande
Mississippi
North American Basin
Hawaiian Islands
Sierra Madre
Gulf of Mexico
Bahamas
Canary Islands
Tropic of Cancer
Cuba
Hispaniola
ATLANTIC
Hawaii
Yucatan
Cape Verde Islands
Caribbean Sea
OCEAN
Orinoco
PACIFIC
Line Is.
Equator
Galapagos Islands
Chimborazo 6310
Amazon
OCEAN
SOUTH AMERICA
Polynesia
Îles Marquises
Madeira
Tocantins
Brazil Basin
Archipel des Tuamotu
ANDES
Planalto do Brazil
St Helena
Peru-Chile Trench
Gran Chaco
Paraguay
Paraná
Tropic of Capricorn
Pitcairn Island
Ojos del Salado 6908
Easter Island
Aconcagua 6960
Southwest Pacific Basin
Pampas
Tristan da Cunha
Argentine Basin
Patagonia
Falkland Islands
South Georgia
Tierra del Fuego
C. Horn
Drake Passage
Southeast Pacific Basin
Antarctic Peninsula
Antarctic Circle
Weddell Sea

sq km	CONTINENTS and OCEANS	sq miles
45 036 492	Asia	17 388 590
30 343 578	Africa	11 715 655
25 680 331	North America	9 529 076
17 815 420	South America	6 878 534
13 340 000	Antarctica	5 150 574
9 908 599	Europe	3 825 710
8 504 241	Oceania	3 283 487
165 384 000	Pacific Ocean	63 838 000
82 217 000	Atlantic Ocean	31 736 000
73 481 000	Indian Ocean	28 364 000
14 056 000	Arctic Ocean	5 426 000

metres	MOUNTAINS	feet
8 848	Mt Everest (Nepal/China)	29 028
8 611	K2 (India/China)	28 251
8 598	Kangchenjunga (Nepal/India)	28 210
6 960	Aconcagua (Argentina)	22 834
6 908	Ojos del Salado (Arg./Chile)	22 664
6 310	Chimborazo (Ecuador)	20 703
6 194	Mt McKinley (USA)	20 321
5 895	Kilimanjaro (Tanzania)	19 340
5 642	Elbrus (Russian Federation)	18 510
5 199	Kirinyaga (Kenya)	17 057
5 030	Puncak Jaya (Indonesia)	16 503
4 808	Mt Blanc (France/Italy)	15 774

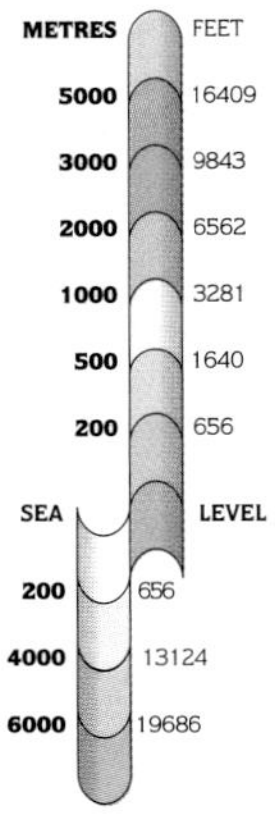
METRES
FEET
5000
16409
3000
9843
2000
6562
1000
3281
500
1640
200
656
SEA
LEVEL
200
656
4000
13124
6000
19686

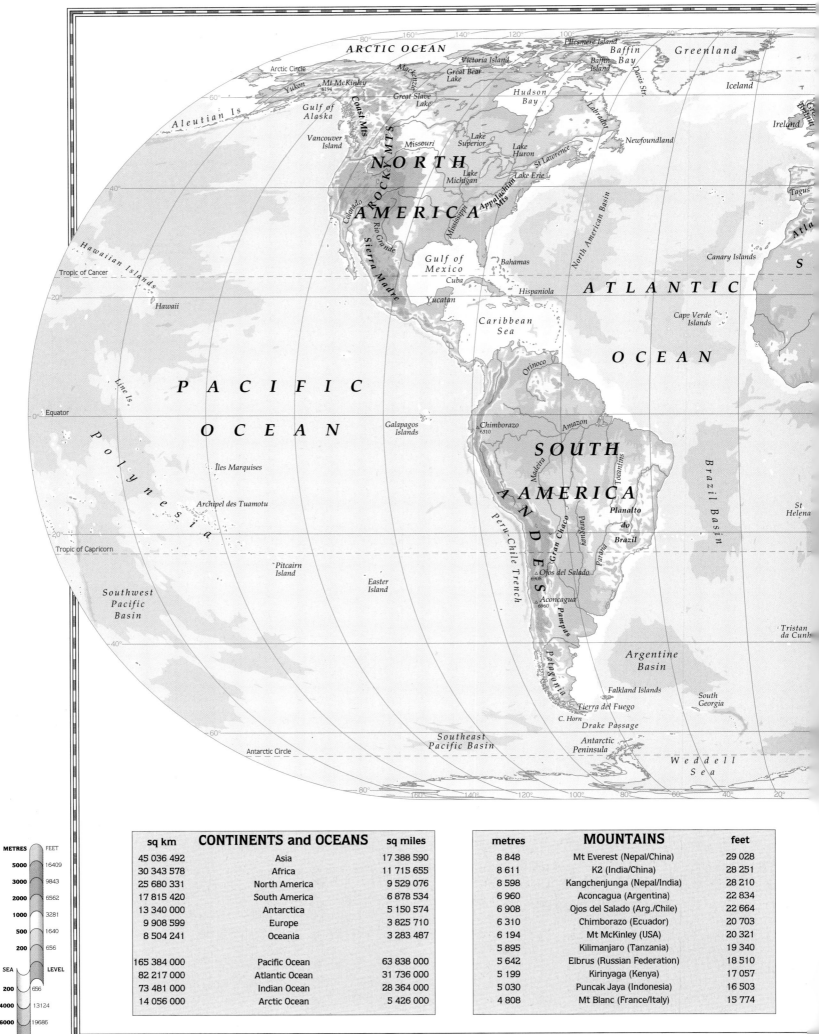

Eckert IV Projection

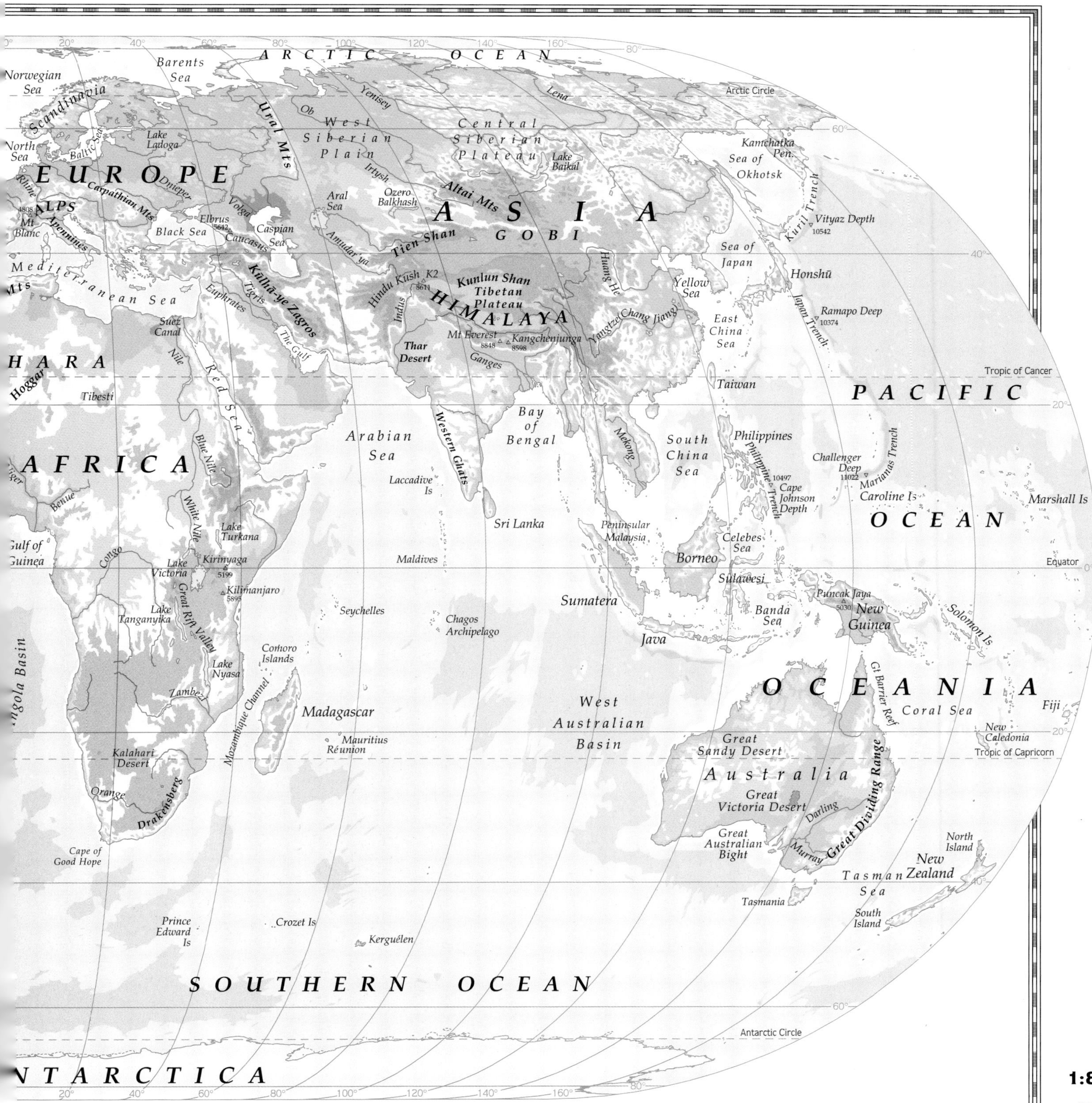

1:80M

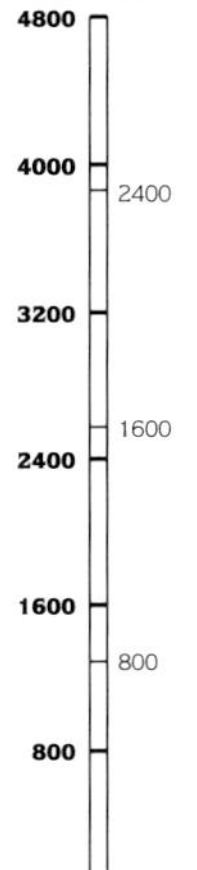

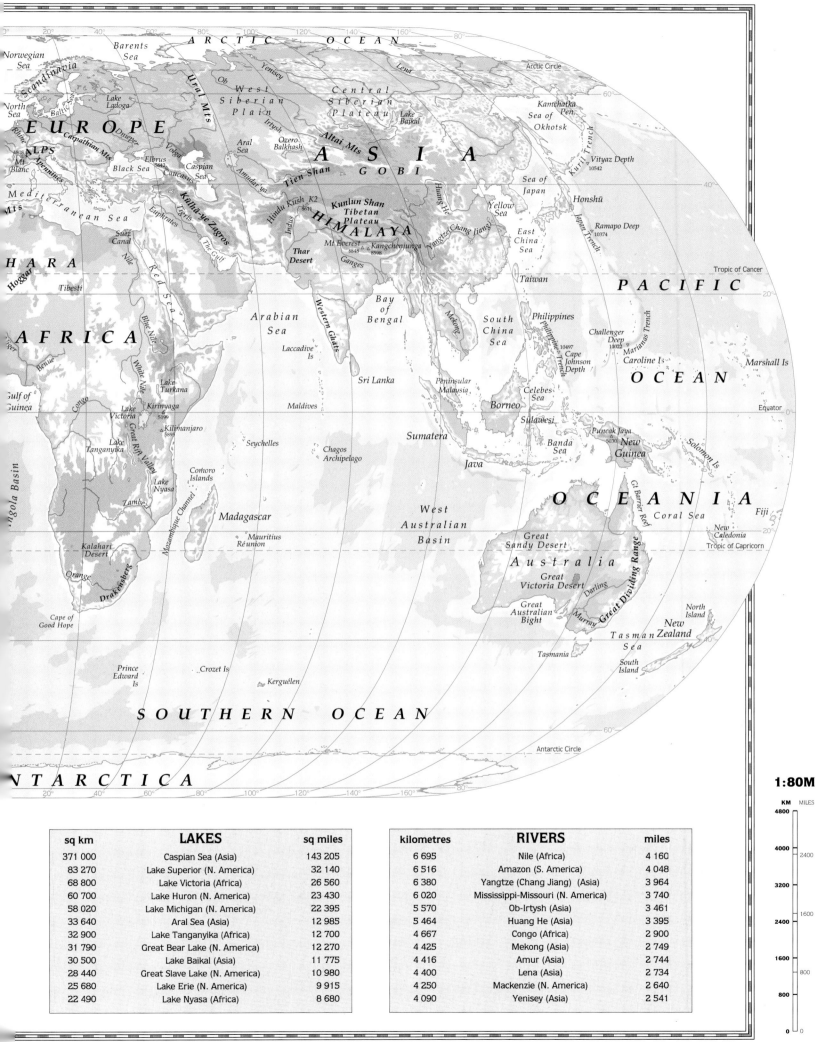

sq km	LAKES	sq miles
371 000	Caspian Sea (Asia)	143 205
83 270	Lake Superior (N. America)	32 140
68 800	Lake Victoria (Africa)	26 560
60 700	Lake Huron (N. America)	23 430
58 020	Lake Michigan (N. America)	22 395
33 640	Aral Sea (Asia)	12 985
32 900	Lake Tanganyika (Africa)	12 700
31 790	Great Bear Lake (N. America)	12 270
30 500	Lake Baikal (Asia)	11 775
28 440	Great Slave Lake (N. America)	10 980
25 680	Lake Erie (N. America)	9 915
22 490	Lake Nyasa (Africa)	8 680

kilometres	RIVERS	miles
6 695	Nile (Africa)	4 160
6 516	Amazon (S. America)	4 048
6 380	Yangtze (Chang Jiang) (Asia)	3 964
6 020	Mississippi-Missouri (N. America)	3 740
5 570	Ob-Irtysh (Asia)	3 461
5 464	Huang He (Asia)	3 395
4 667	Congo (Africa)	2 900
4 425	Mekong (Asia)	2 749
4 416	Amur (Asia)	2 744
4 400	Lena (Asia)	2 734
4 250	Mackenzie (N. America)	2 640
4 090	Yenisey (Asia)	2 541

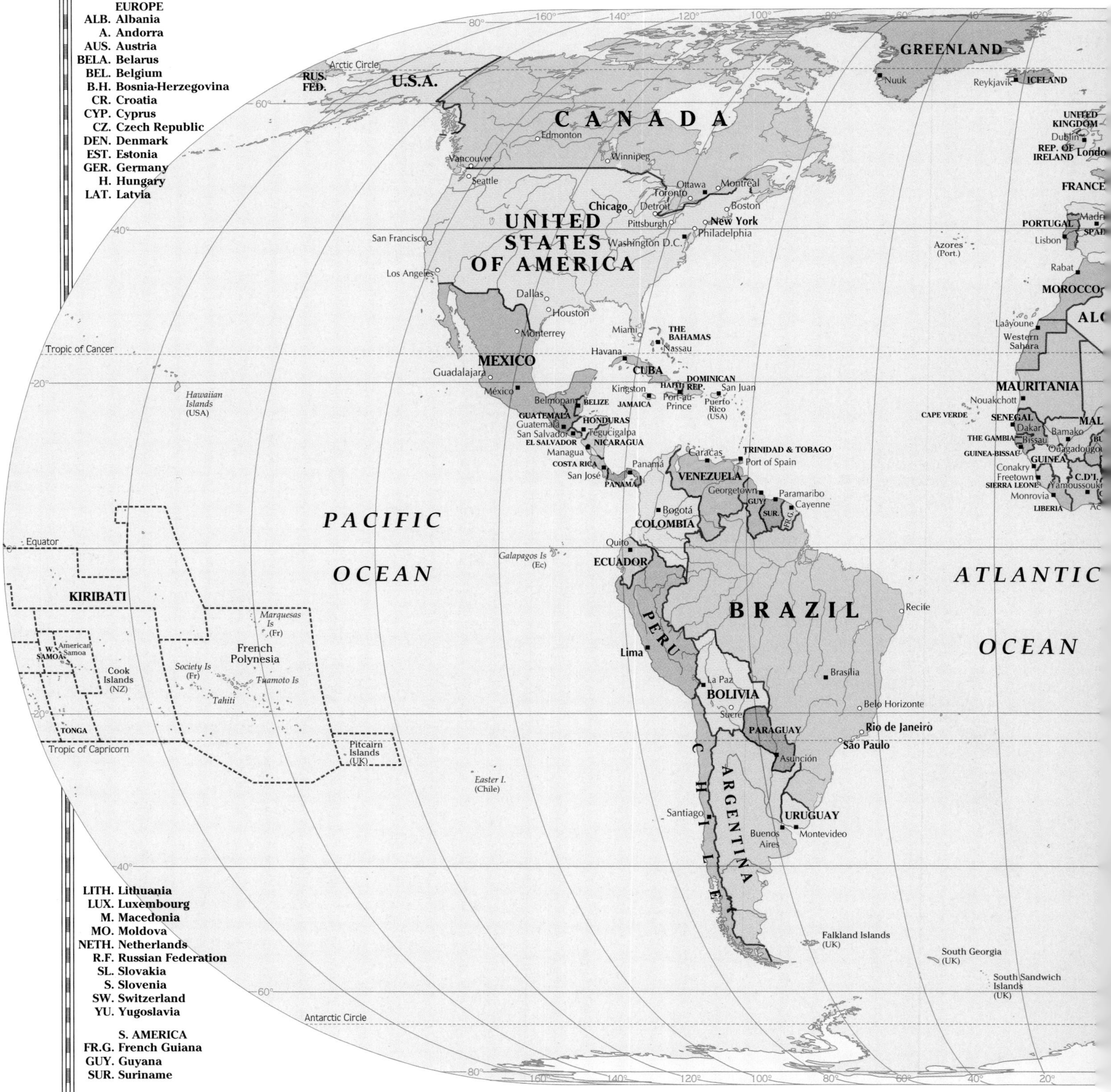

METROPOLITAN AREA POPULATIONS

A metropolitan area is a built-up zone containing a number of cities and towns. The total combined population for a selection of these is given below, either as an estimate or from census returns.

NORTH and CENTRAL AMERICA

20,200,000	México *Mexico*
16,972,000	New York *U.S.A.*
11,420,000	Los Angeles *U.S.A.*
7,498,000	Chicago *U.S.A.*
5,240,000	San Francisco *U.S.A.*
4,941,000	Philadelphia *U.S.A.*
4,497,000	Boston *U.S.A.*
4,293,000	Washington D.C. *U.S.A.*
4,285,000	Detroit *U.S.A.*
4,135,000	Dallas-Fort Worth *U.S.A.*
3,893,000	Toronto *Canada*
3,437,000	Houston *U.S.A.*
3,264,000	Miami *U.S.A.*
3,127,000	Montréal *Canada*
3,051,000	Atlanta *U.S.A.*
2,846,720	Guadalajara *Mexico*
2,583,000	Minneapolis-St Paul *U.S.A.*
2,549,000	San Diego *U.S.A.*
2,521,697	Monterrey *Mexico*
2,507,000	St Louis *U.S.A.*
2,414,000	Baltimore *U.S.A.*
2,404,000	Pittsburg *U.S.A.*
2,099,000	Havana *Cuba*
2,055,000	Santo Domingo *Dominican Republic*
1,603,000	Vancouver *Canada*
1,522,126	San Salvador *El Salvador*
1,402,000	Port-au-Prince *Haiti*
1,390,000	San Juan *Puerto Rico*
1,132,730	Guatemala *Guatemala*
1,012,000	Managua *Nicaragua*
920,857	Ottawa *Canada*

SOUTH AMERICA

15,199,423	São Paulo *Brazil*
12,200,000	Buenos Aires *Argentina*
9,600,528	Rio de Janeiro *Brazil*
6,483,901	Lima *Peru*
5,025,989	Bogotá *Colombia*
4,628,320	Santiago *Chile*
4,092,000	Caracas *Venezuela*
3,461,905	Belo Horizonte *Brazil*
2,859,469	Recife *Brazil*
1,596,274	Brasília *Brazil*
1,383,660	Montevideo *Uruguay*
1,234,000	La Paz *Bolivia*
1,100,847	Quito *Ecuador*

EUROPE

9,318,821	Paris *France*
9,227,687	London *U.K.*
8,957,000	Moscow *Rus. Fed.*
6,407,215	Istanbul *Turkey*
5,004,000	St Petersburg *Rus. Fed.*
3,447,916	Berlin *Germany*
3,097,000	Athens *Greece*
2,909,792	Madrid *Spain*
2,723,327	Rome *Italy*
2,720,400	Essen-Dortmund *Germany*
2,616,000	Kiev *Ukraine*
2,578,900	Manchester *U.K.*
2,350,984	Bucharest *Romania*
2,329,600	Birmingham *U.K.*
1,992,343	Budapest *Hungary*
1,742,000	Lisbon *Portugal*
1,669,840	Stockholm *Sweden*
1,669,000	Hamburg *Germany*
1,655,700	Warsaw *Poland*
1,633,600	Minsk *Belarus*
1,565,800	Vienna *Austria*
1,342,679	Copenhagen *Denmark*
1,221,000	Sofia *Bulgaria*
1,214,174	Prague *Czech Rep.*
1,174,512	Zagreb *Croatia*
1,168,454	Belgrade *Yugoslavia*
1,091,338	Amsterdam *Neth.*
976,883	Helsinki *Finland*
954,045	Brussels *Belgium*
915,516	Dublin *R. of Ireland*
915,000	Rīga *Latvia*

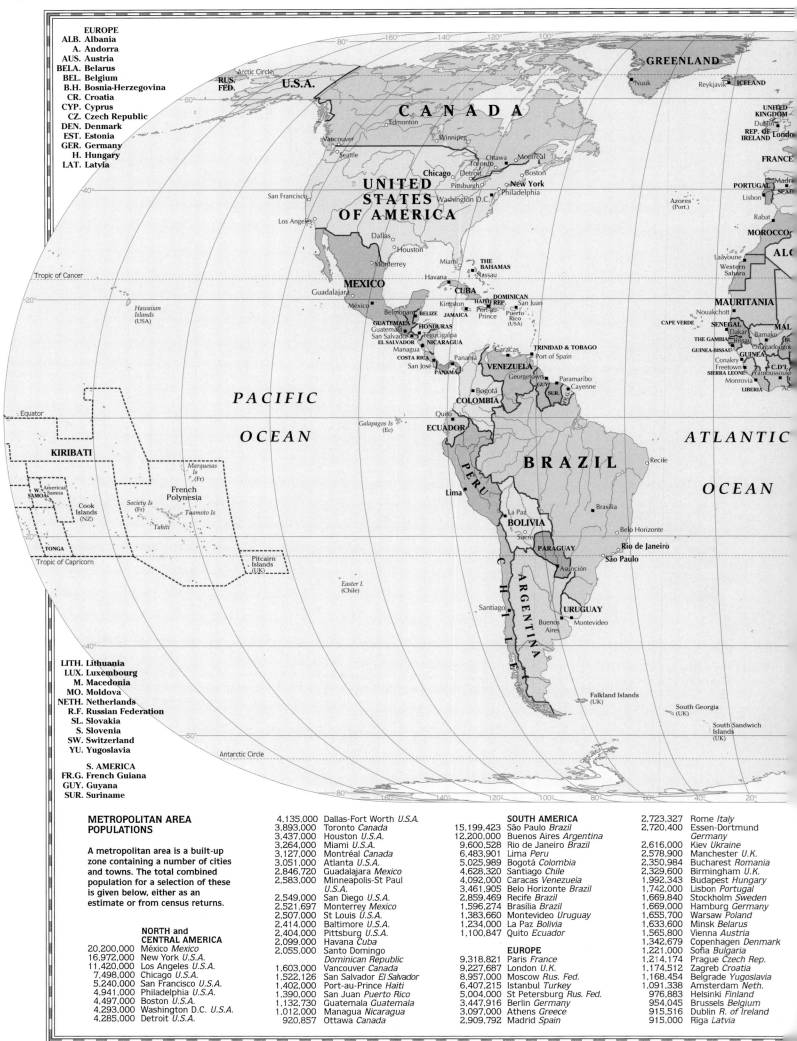

Eckert IV Projection

ARCTIC OCEAN
RUSSIAN FEDERATION
KAZAKSTAN
MONGOLIA
CHINA
INDIA
IRAN
SAUDI ARABIA
TURKEY
LIBYA
EGYPT
NIGER
CHAD
SUDAN
NIGERIA
ETHIOPIA
SOMALIA
KENYA
CONGO (ZAIRE)
TANZANIA
ANGOLA
ZAMBIA
MOZAMBIQUE
ZIMBABWE
NAMIBIA
BOTSWANA
REP. OF SOUTH AFRICA
MADAGASCAR
YEMEN
OMAN
AFGHANISTAN
PAKISTAN
MYANMAR
THAILAND
MALAYSIA
INDONESIA
PHILIPPINES
JAPAN
N. KOREA
S. KOREA
TAIWAN
PAPUA NEW GUINEA
AUSTRALIA
NEW ZEALAND
PACIFIC OCEAN
INDIAN OCEAN
SOUTHERN OCEAN
ANTARCTICA
Arctic Circle
Tropic of Cancer
Equator
Tropic of Capricorn
Antarctic Circle
Northern Mariana Islands (USA)
MARSHALL ISLANDS
FEDERATED STATES OF MICRONESIA
PALAU
NAURU
KIRIBATI
SOLOMON ISLANDS
TUVALU
VANUATU
FIJI
New Caledonia (Fr.)
MALDIVES
SEYCHELLES
COMOROS
MAURITIUS
Réunion (Fr.)
British Indian Ocean Territory
French Southern and Antarctic Lands
Kerguelen (Fr)
ASIA
AR. Armenia
AZ. Azerbaijan
GEO. Georgia
IS. Israel
JOR. Jordan
LEB. Lebanon
U.A.E. United Arab Emirates
AFRICA
BE. Benin
BUR. Burkina
B. Burundi
CAM. Cameroon
C.D'I. Côte d'Ivoire
EQ. G. Equatorial Guinea
GH. Ghana
R. Rwanda
T. Togo

758,949 Oslo *Norway*
582,000 Vilnius *Lithuania*
499,183 Tallinn *Estonia*

ASIA

13,341,896 Shanghai *China*
12,571,720 Bombay *India*
11,609,735 Tōkyō *Japan*
10,916,272 Calcutta *India*
10,819,407 Beijing *China*
10,627,000 Seoul *S. Korea*
9,371,000 Tianjin *China*
9,253,000 Jakarta *Indonesia*
8,520,000 Ōsaka-Kōbe *Japan*
8,375,188 Delhi *India*
7,832,000 Manila-Quezon City *Philippines*
7,702,000 Karachi *Pakistan*
6,773,000 Tehrān *Iran*
6,105,160 Dhaka *Bangladesh*
5,876,000 Bangkok *Thailand*
5,448,000 Hong Kong *China*
5,361,468 Madras *India*
4,763,000 Shenyang *China*
4,280,261 Hyderabad *India*
4,092,000 Lahore *Pakistan*
4,086,548 Bangalore *India*
4,044,000 Baghdād *Iraq*
3,924,435 Hô Chi Minh *Vietnam*
3,921,000 Wuhan *China*
3,797,566 Pusan *S. Korea*
3,671,000 Guangzhou *China*
3,297,655 Ahmadabad *India*
3,295,000 Yangon *Myanmar*
3,250,548 Yokohama *Japan*
3,151,000 Chongqing *China*
3,022,236 Ankara *Turkey*
3,004,000 Chengdu *China*
2,966,000 Harbin *China*
2,913,000 Damascus *Syria*
2,874,000 Singapore *Singapore*
2,859,000 Xi'an *China*
2,768,000 Aleppo *Syria*
2,720,000 T'ai-pei *Taiwan*
2,665,105 Izmir *Turkey*
2,543,000 Dalian *China*
2,485,014 Pune *India*
2,473,272 Surabaya *Indonesia*
2,265,000 Nanjing *China*
2,230,000 P'yŏngyang *N. Korea*
2,214,000 Changchun *China*
2,094,000 Tashkent *Uzbekistan*
2,000,000 Kābul *Afghanistan*
1,711,000 Kuala Lumpur *Malaysia*
1,500,000 Beirut *Lebanon*
1,500,000 Riyadh *Saudi Arabia*
1,442,000 Novosibirsk *Rus. Fed.*
1,400,000 Tbilisi *Georgia*
1,272,000 'Ammān *Jordan*
1,200,000 Yerevan *Armenia*
1,151,300 Almaty *Kazakstan*
1,056,146 Ha Nôi *Vietnam*
616,000 Colombo *Sri Lanka*
549,900 Jerusalem *Israel*
537,000 Islamabad *Pakistan*
200,000 Kuwait *Kuwait*

AFRICA

11,642,000 Cairo *Egypt*
5,689,000 Lagos *Nigeria*
3,505,000 Kinshasa *Congo (Zaire)*
3,380,000 Alexandria *Egypt*
3,210,000 Casablanca *Morocco*
3,033,000 Algiers *Algeria*
2,350,157 Cape Town *S. Africa*
1,947,000 Khartoum *Sudan*
1,891,000 Addis Ababa *Ethiopia*
1,717,000 Luanda *Angola*
1,636,000 Tunis *Tunisia*
1,503,000 Nairobi *Kenya*
1,500,000 Tripoli *Libya*
1,492,000 Dakar *Senegal*
1,472,000 Rabat *Morocco*
1,098,000 Maputo *Mozambique*
1,000,000 Harare *Zimbabwe*
523,900 Abuja *Nigeria*

OCEANIA

3,700,000 Sydney *Australia*
3,178,000 Melbourne *Australia*
1,386,000 Brisbane *Australia*
1,215,000 Perth *Australia*
1,065,000 Adelaide *Australia*
896,200 Auckland *New Zealand*
325,700 Wellington *New Zealand*
310,000 Canberra *Australia*

1:80M

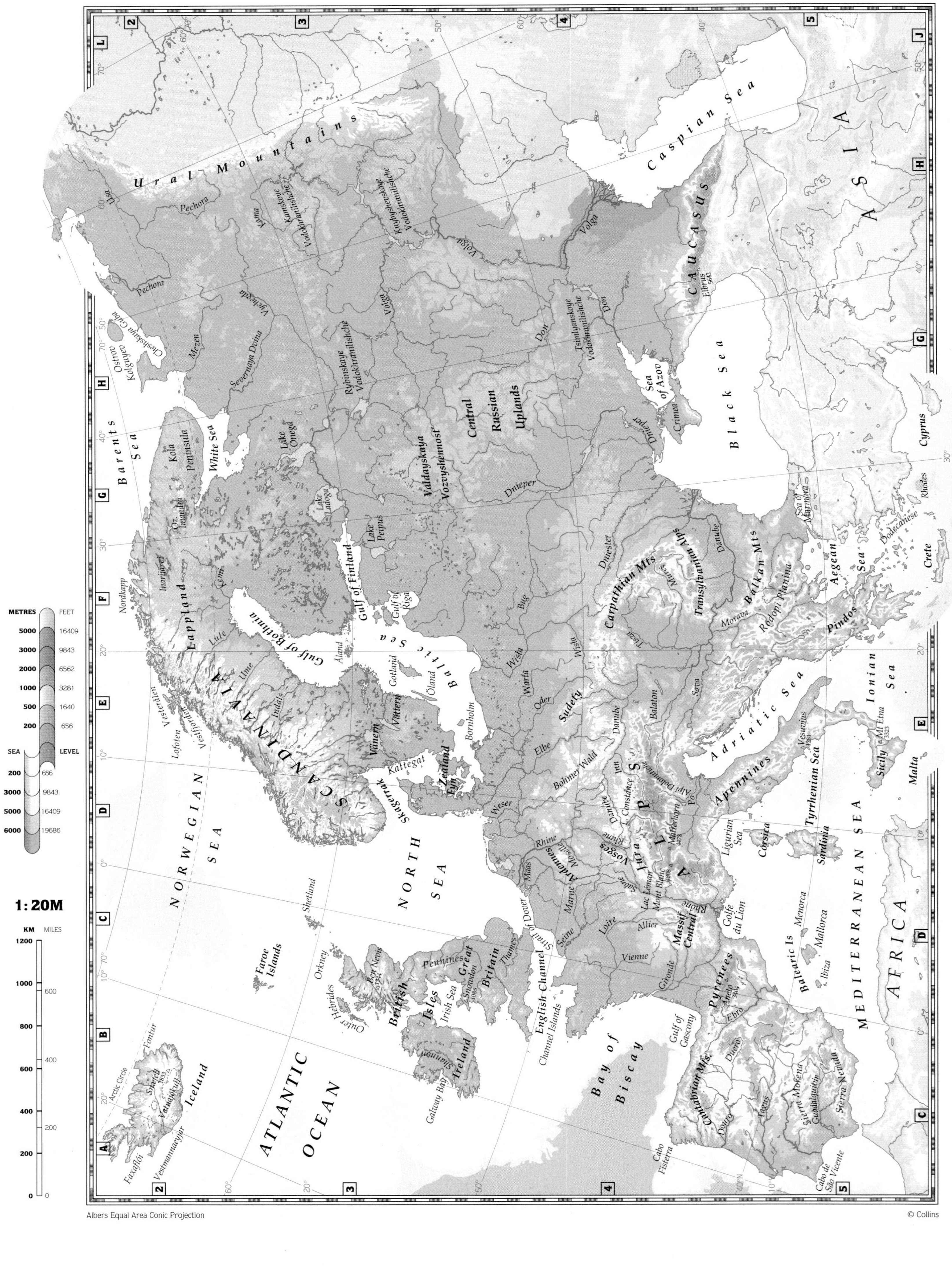
METRES
FEET
5000
16409
3000
9843
2000
6562
1000
3281
500
1640
200
656
SEA
LEVEL
200
656
3000
9843
5000
16409
6000
19686
1:20M
KM
MILES
1200
1000
800
600
400
200
0
600
400
200
0
Ural Mountains
Caspian Sea
ASIA
CAUCASUS
Elbrus
Black Sea
Sea of Azov
Crimea
Central Russian Uplands
Valdayskaya Vozvyshennost
Barents Sea
White Sea
Kola Peninsula
Lake Onega
Lake Ladoga
Lake Peipus
Gulf of Finland
Gulf of Riga
Gulf of Bothnia
Baltic Sea
Lappland
SCANDINAVIA
Pechora
Mezen
Severnaya Dvina
Vychegda
Kama
Volga
Don
Dnieper
Dniester
Bug
Wisła
Warta
Oder
Elbe
Weser
Rhine
Danube
Carpathian Mts
Transylvanian Alps
Balkan Mts
Rodopi Planina
Pindos
Aegean Sea
Sea of Marmara
Dodecanese
Rhodes
Crete
Cyprus
Ionian Sea
Adriatic Sea
Apennines
Tyrrhenian Sea
Sicily
Malta
Sardinia
Corsica
Ligurian Sea
Golfe du Lion
Balearic Is
Mallorca
Menorca
Ibiza
MEDITERRANEAN SEA
AFRICA
ALPS
Jura
Vosges
Ardennes
Massif Central
Pyrenees
Cantabrian Mts.
Sierra Morena
Sierra Nevada
Bay of Biscay
Gulf of Gascony
English Channel
Channel Islands
Strait of Dover
NORTH SEA
Skagerrak
Kattegat
Zealand
Fyn
Bornholm
Gotland
Öland
Vänern
Vättern
NORWEGIAN SEA
Lofoten
Vestfjorden
Nordkapp
Shetland
Orkney
Faroe Islands
Outer Hebrides
British Isles
Great Britain
Pennines
Ben Nevis
Snowdon
Irish Sea
Ireland
Shannon
Galway Bay
Thames
Seine
Loire
Marne
Allier
Vienne
Garonne
Rhône
Ebro
Douro
Tagus
Guadalquivir
Cabo Fisterra
Cabo de São Vicente
ATLANTIC OCEAN
Iceland
Vatnajökull
Faxaflói
Arctic Circle

Albers Equal Area Conic Projection

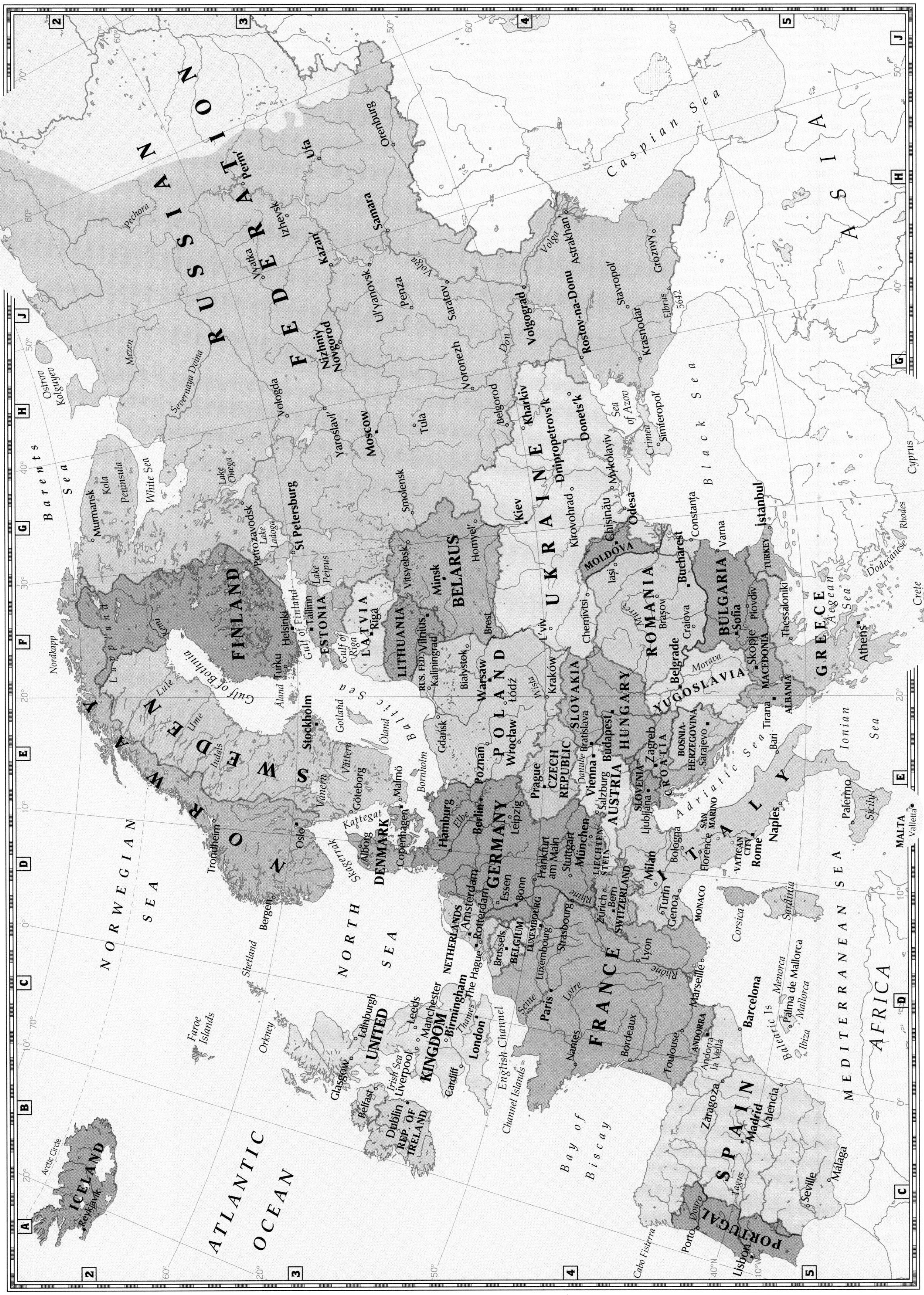

1:20M

KM 0 200 400 600 800 1000 1200

MILES 0 200 400 600

Albers Equal Area Conic Projection

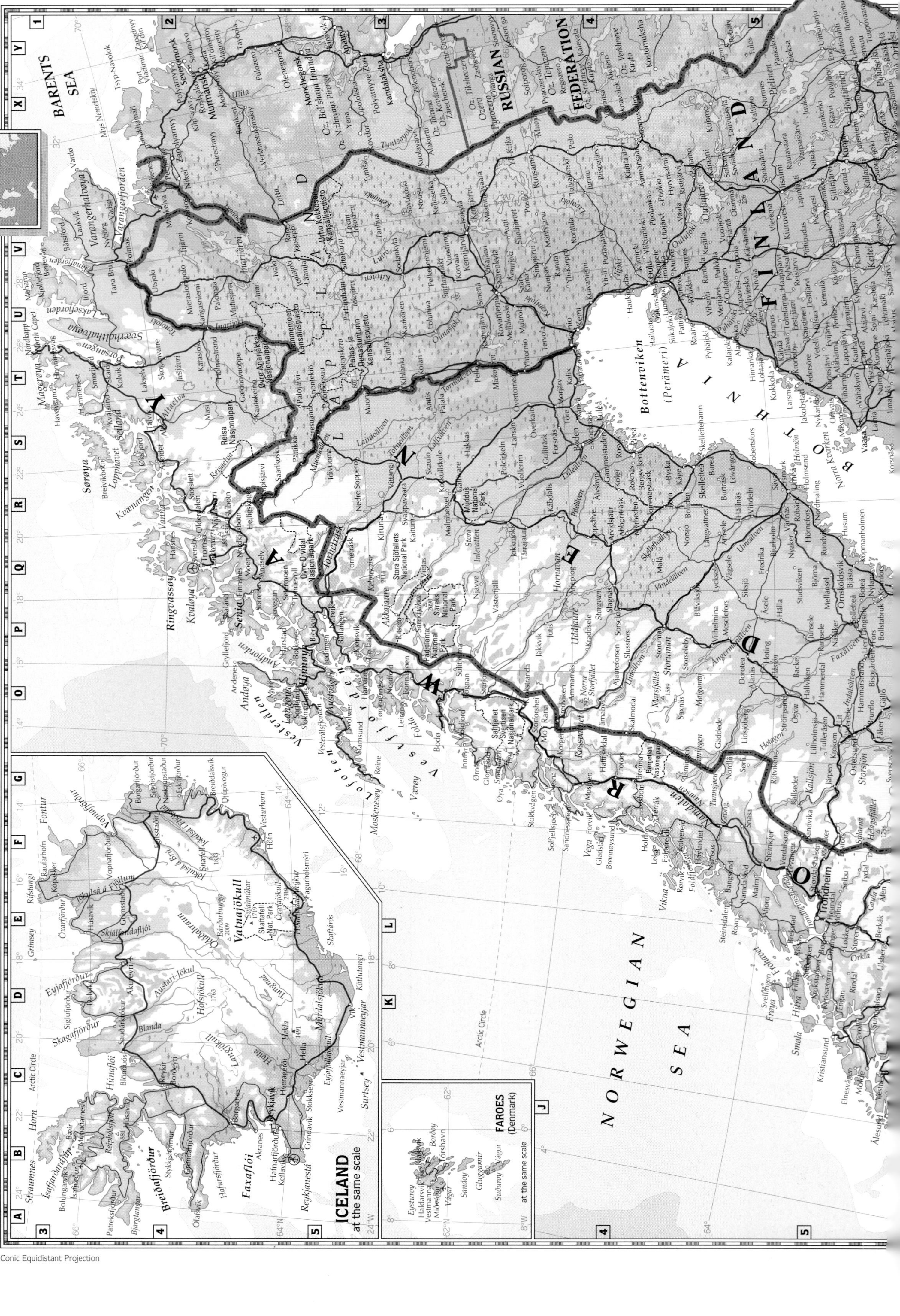
BARENTS SEA
RUSSIAN FEDERATION
FINLAND
SWEDEN
NORWAY
NORWEGIAN SEA
LAPLAND
Bottenviken
(Perämeri)
BOTHNIA
Lofoten
Vestfjorden
Vesterålen
Arctic Circle
ICELAND
at the same scale
FAROES
(Denmark)
at the same scale
Vatnajökull
Reykjavík

METRES	FEET
6000	19686
5000	16409
4000	13124
3000	9843
2000	6562
1000	3281
500	1640
200	656
SEA	LEVEL
200	656
2000	6562
4000	13124
6000	19686

Conic Equidistant Projection

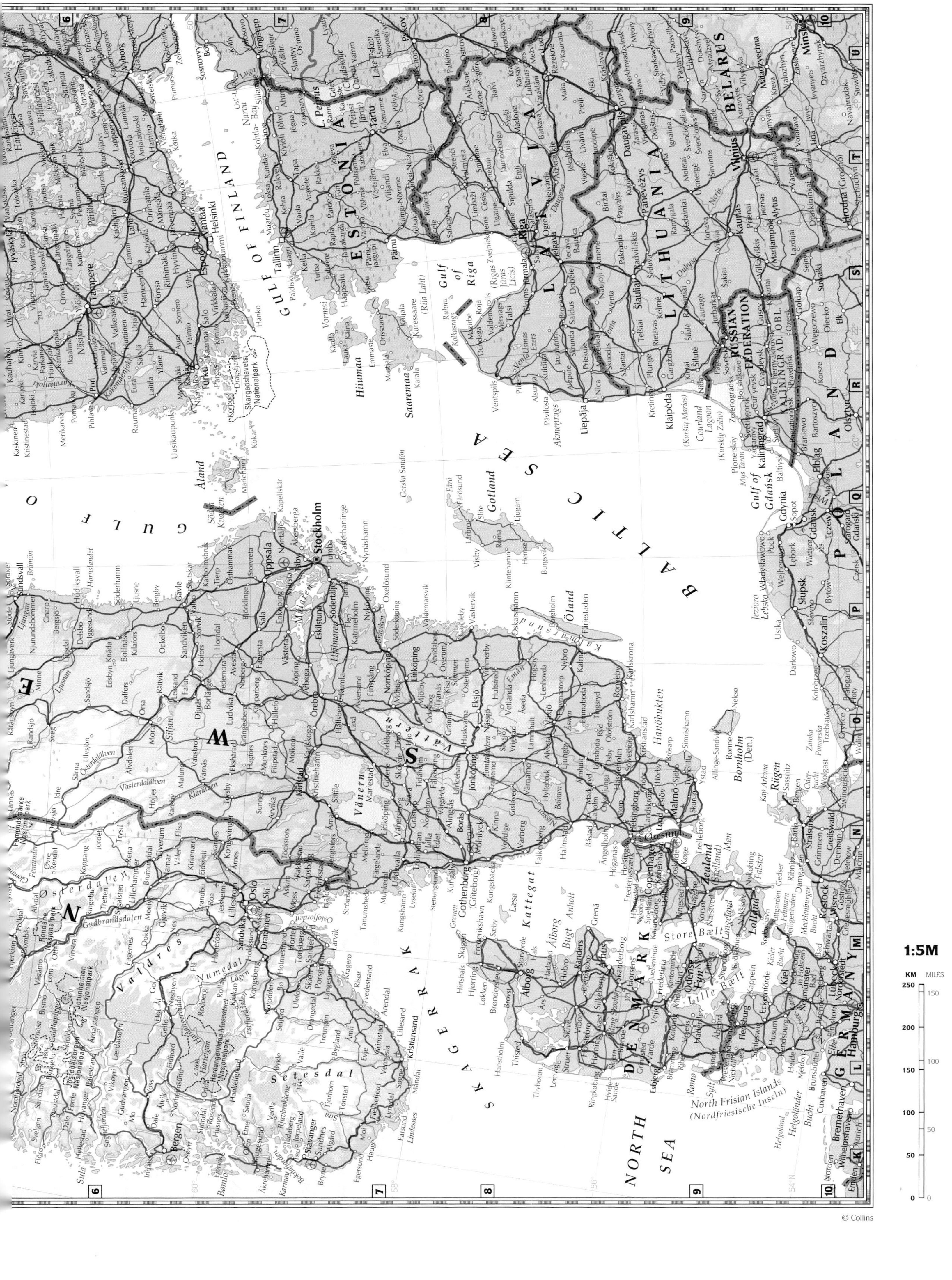
GULF OF FINLAND
ESTONIA
LATVIA
LITHUANIA
BELARUS
RUSSIAN FEDERATION
KALININGRAD OBL.
POLAND
GERMANY
DENMARK
BALTIC SEA
SKAGERRAK
Kattegat
NORTH SEA
Gulf of Riga
Gulf of Gdańsk
Gotland
Öland
Åland
Hiiumaa
Saaremaa
Bornholm (Den.)
Rügen
Sjælland (Zealand)
Fyn
North Frisian Islands (Nordfriesische Inseln)
Stockholm
Helsinki
Tallinn
Rīga
Vilnius
Minsk
Oslo
Copenhagen (København)
Gothenburg (Göteborg)
Malmö
Hamburg
Gdańsk
Kaliningrad
Bergen
Stavanger
1:5M
KM
MILES

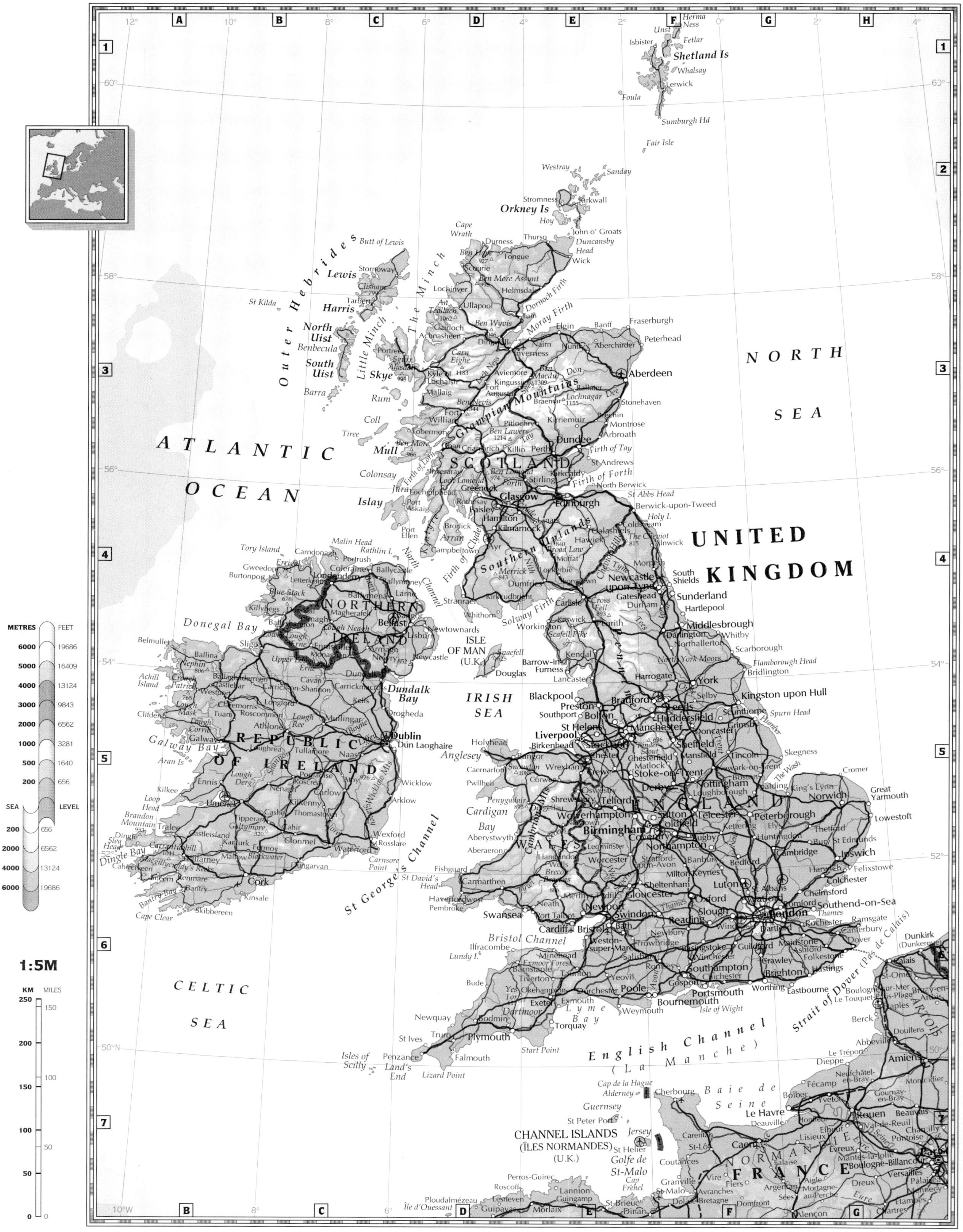
NORTH SEA
ATLANTIC OCEAN
CELTIC SEA
IRISH SEA
UNITED KINGDOM
SCOTLAND
NORTHERN IRELAND
REPUBLIC OF IRELAND
ENGLAND
WALES
FRANCE
ISLE OF MAN (U.K.)
CHANNEL ISLANDS (ÎLES NORMANDES) (U.K.)
English Channel (La Manche)
Strait of Dover (Pas de Calais)
St George's Channel
Bristol Channel
North Channel
The Minch
Little Minch
Outer Hebrides
Shetland Is
Orkney Is
Grampian Mountains
Southern Uplands
Cambrian Mts
Pennines
Moray Firth
Firth of Forth
Firth of Clyde
Solway Firth
Galway Bay
Donegal Bay
Dingle Bay
Cardigan Bay
Lyme Bay
Baie de Seine
Golfe de St-Malo
Lewis
Harris
North Uist
South Uist
Skye
Mull
Islay
Arran
Anglesey
Isles of Scilly
Jersey
Guernsey
Alderney
Aberdeen
Inverness
Dundee
Edinburgh
Glasgow
Belfast
Londonderry
Dublin
Cork
Limerick
Galway
Newcastle upon Tyne
Sunderland
Middlesbrough
Leeds
Bradford
York
Kingston upon Hull
Manchester
Liverpool
Sheffield
Nottingham
Leicester
Birmingham
Coventry
Stoke-on-Trent
Norwich
Cambridge
Oxford
Reading
London
Southend-on-Sea
Brighton
Southampton
Portsmouth
Bournemouth
Bristol
Cardiff
Swansea
Plymouth
Dover
Calais
Dunkirk
Le Havre
Rouen
Caen
Cherbourg
Amiens
Paris
METRES
FEET
6000
5000
4000
3000
2000
1000
500
200
SEA
LEVEL
19686
16409
13124
9843
6562
3281
1640
656
1:5M
KM
MILES
250
200
150
100
50
0

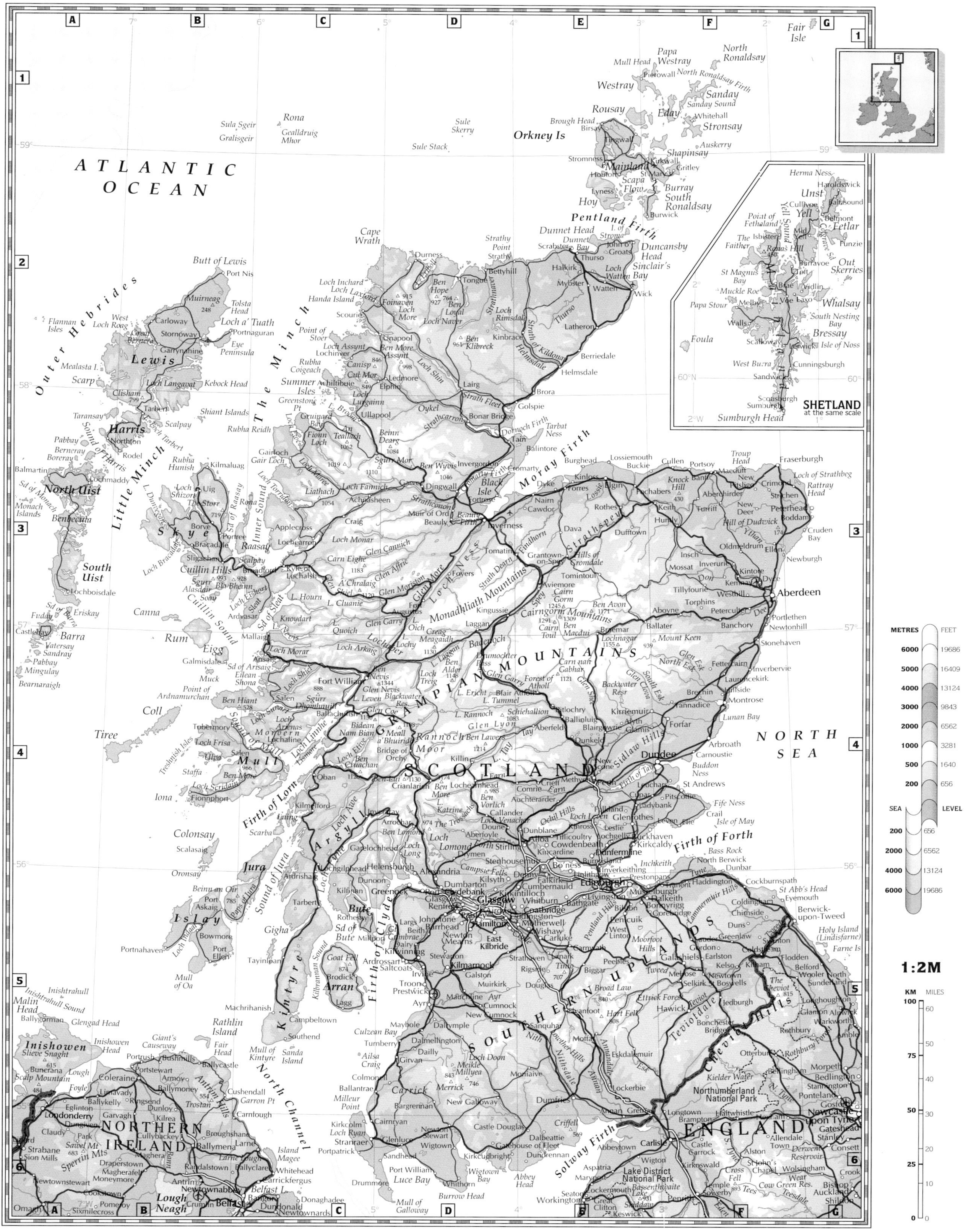
ATLANTIC OCEAN
NORTH SEA
SCOTLAND
ENGLAND
NORTHERN IRELAND
Orkney Is
SHETLAND
at the same scale
Outer Hebrides
The Minch
Little Minch
Lewis
Harris
North Uist
South Uist
Barra
Skye
Mull
Jura
Islay
Arran
Kintyre
Argyll
GRAMPIAN MOUNTAINS
SOUTHERN UPLANDS
Cairngorm Mountains
Monadhliath Mountains
Moray Firth
Firth of Forth
Firth of Tay
Firth of Clyde
Firth of Lorn
Solway Firth
Pentland Firth
North Channel
Cheviot Hills
Sidlaw Hills
Ochil Hills
Pentland Hills
Lammermuir Hills
Inner Sound
Sound of Jura
Loch Ness
Loch Lomond
Loch Linnhe
Glen More
Strathspey
Inverness
Aberdeen
Dundee
Perth
Stirling
Edinburgh
Glasgow
Paisley
Kilmarnock
Ayr
Dumfries
Stranraer
Fort William
Oban
Ullapool
Thurso
Wick
Kirkwall
Stornoway
Lerwick
Carlisle
Newcastle upon Tyne
Belfast
Londonderry
Cape Wrath
Butt of Lewis
Ben Nevis
1344
Mull of Kintyre
Mull of Galloway
Fair Isle
Rona
Sula Sgeir
Sule Skerry
Sule Stack
St Kilda
Tiree
Coll
Colonsay
Rum
Eigg
Canna
Iona
Staffa
Bute
Lake District National Park
Northumberland National Park
Lough Neagh
Inishowen
A
B
C
D
E
F
G
1
2
3
4
5
6
7°
6°
5°
4°
3°
2°
59°
58°
57°
56°
METRES
FEET
6000
19686
5000
16409
4000
13124
3000
9843
2000
6562
1000
3281
500
1640
200
656
SEA
LEVEL
1:2M
KM
MILES
100
75
50
25
0

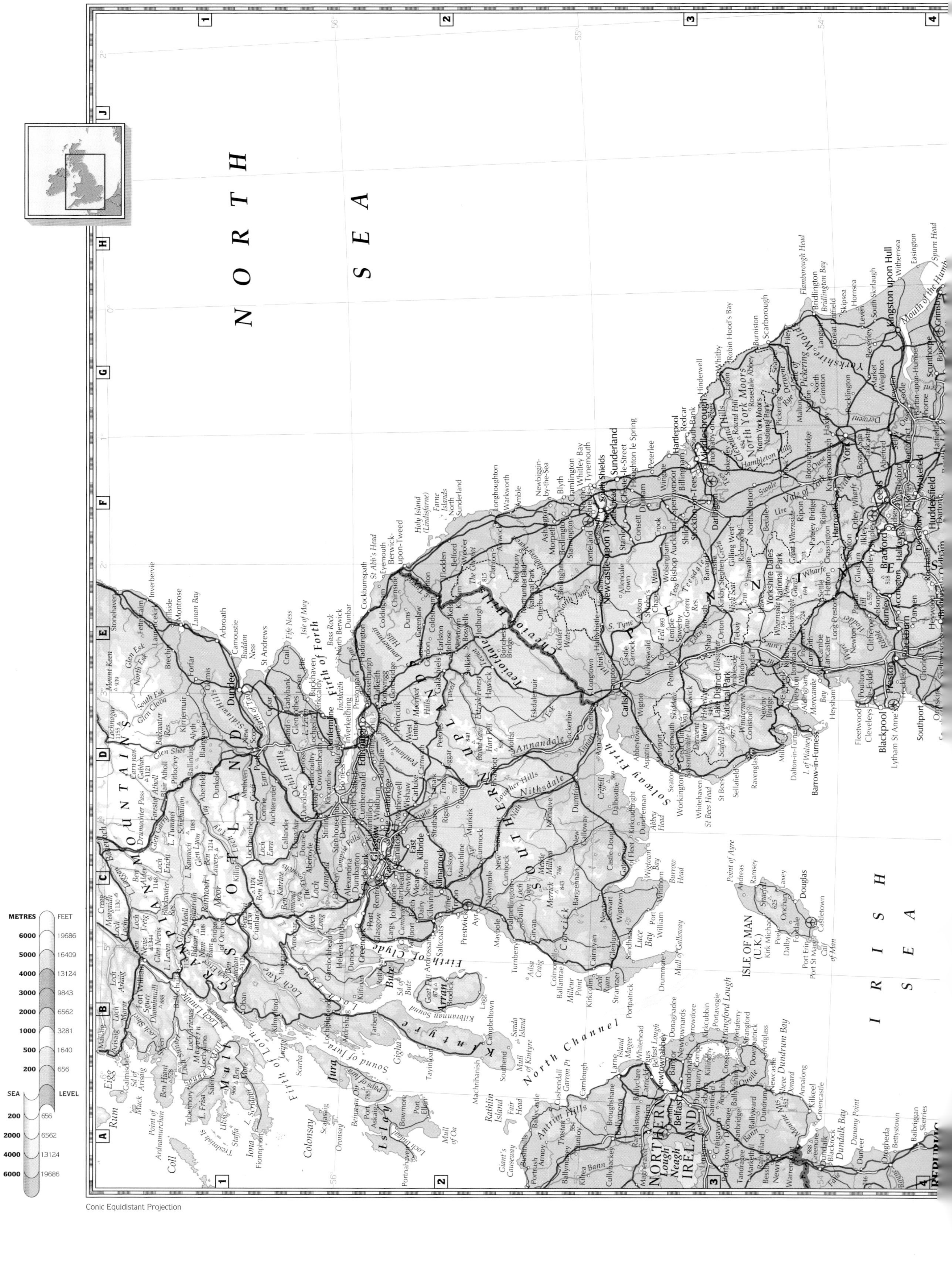

Conic Equidistant Projection

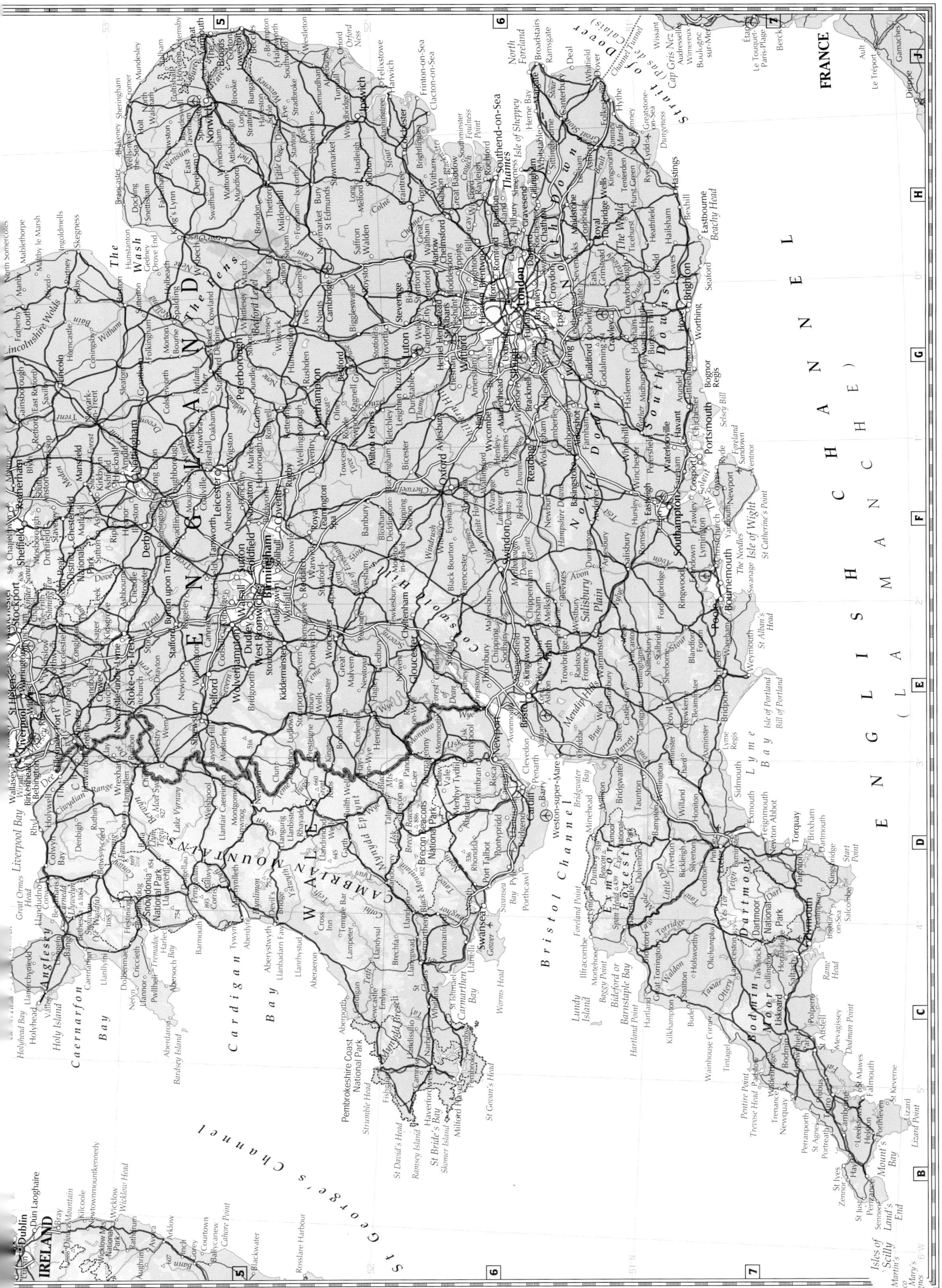

1:2M

KM MILES

100 60

75 50

40

50 30

20

25

10

0 0

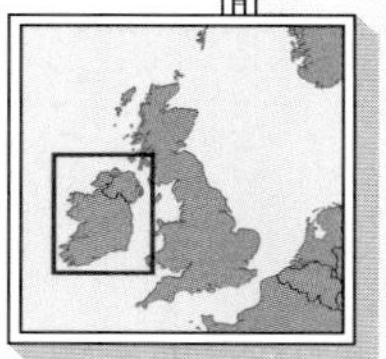

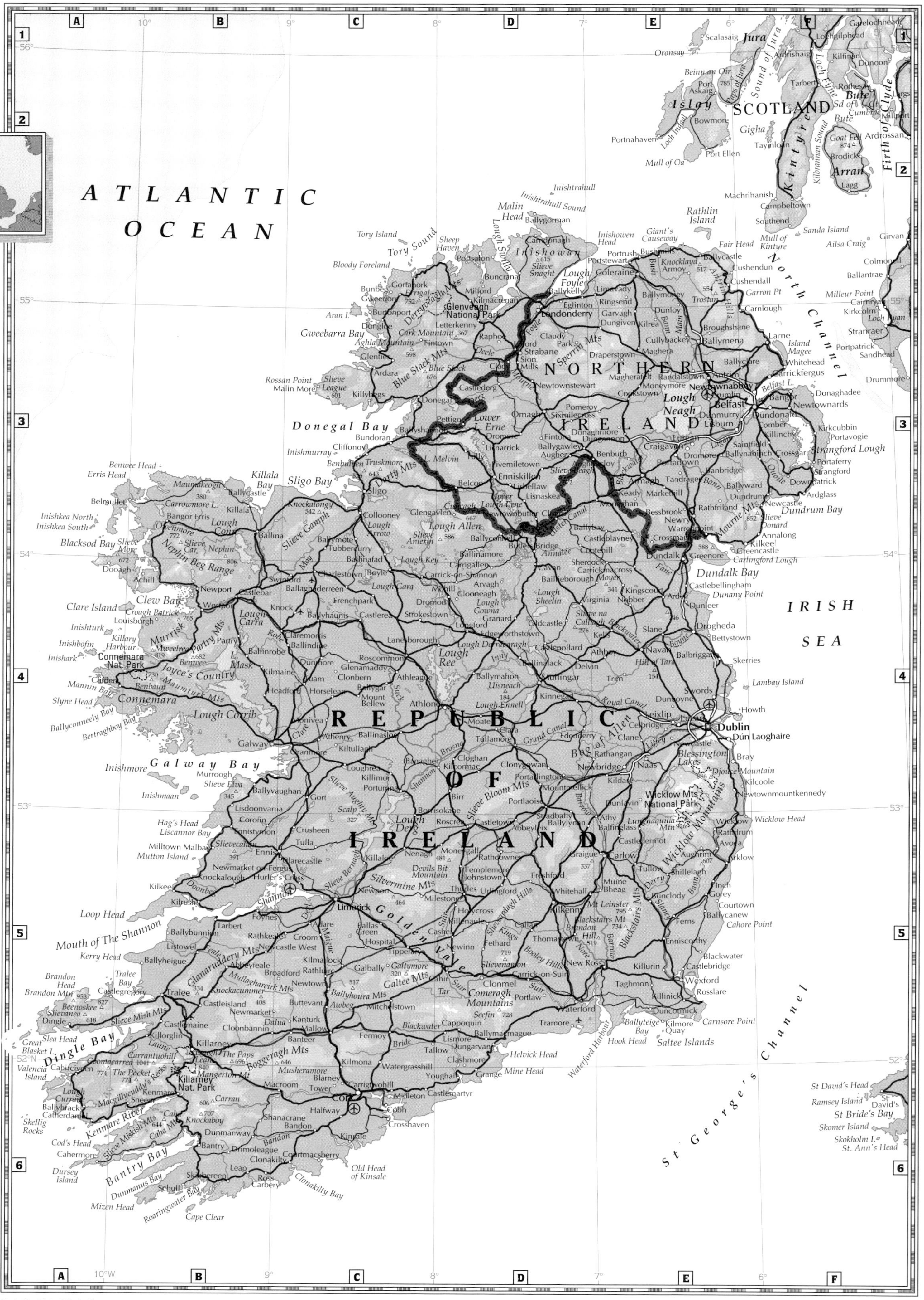

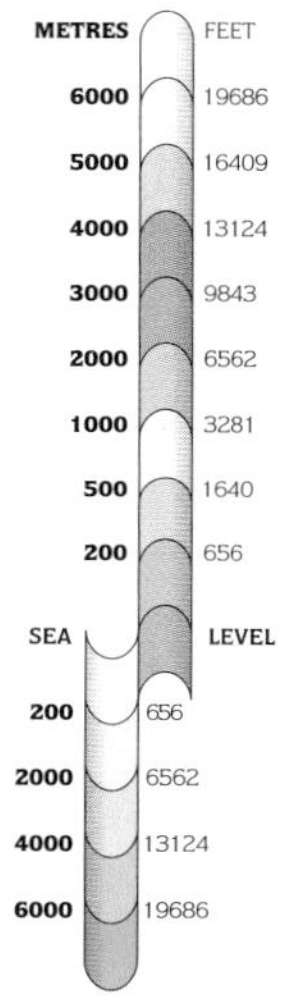

1:2M

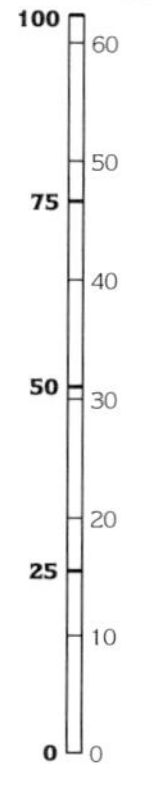

Conic Equidistant Projection

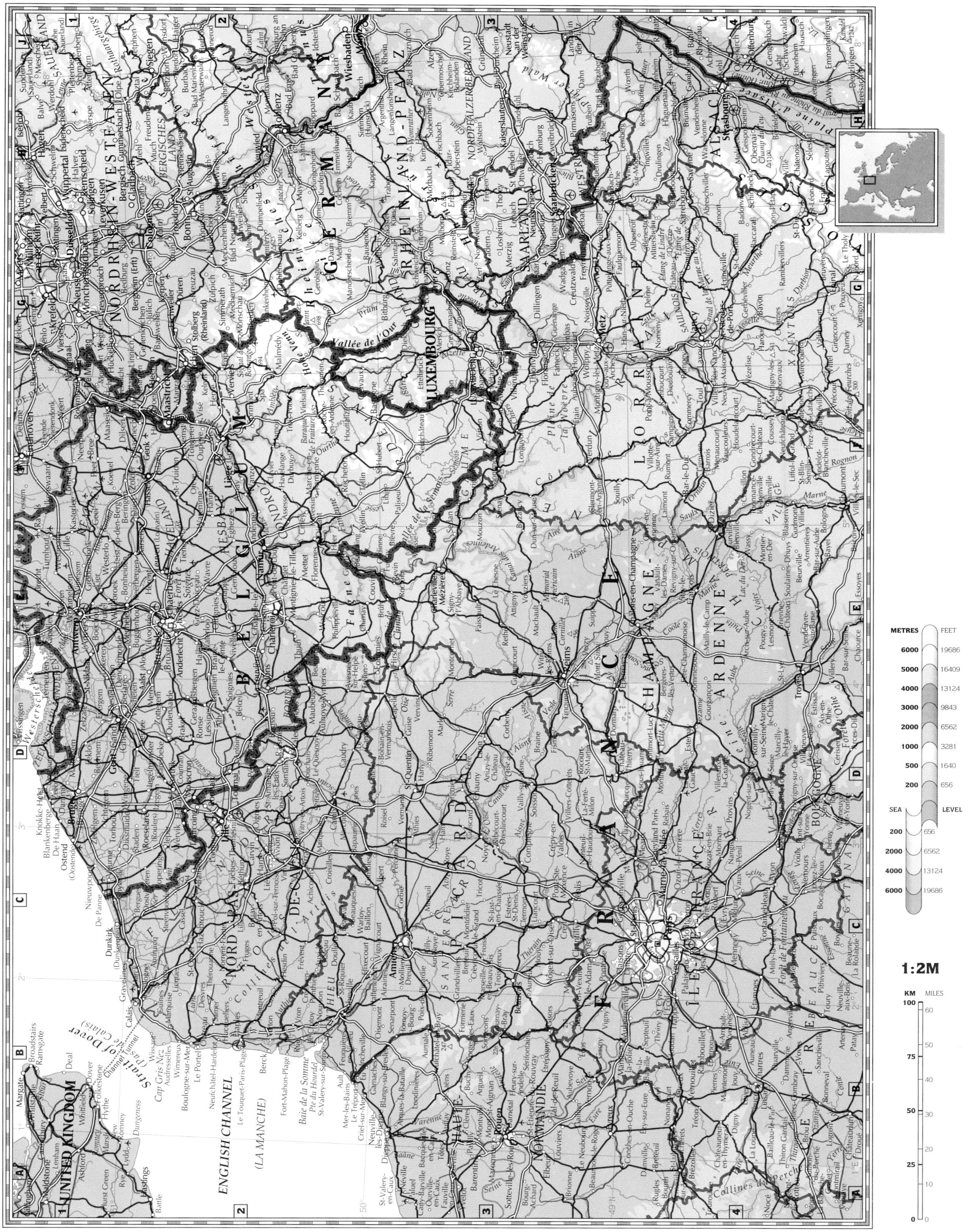

Conic Equidistant Projection

NORWAY
SWEDEN
DENMARK
NETHERLANDS
BELGIUM
LUXEMBOURG
GERMANY
FRANCE
SWITZERLAND
LIECHTENSTEIN
AUSTRIA
CZECH REPUBLIC
POL
NORTH SEA
SKAGERRAK
Kattegat
BALTIC
North Frisian Islands (Nordfriesische Inseln)
East Frisian Islands
Waddeneilanden
Waddenzee
Helgoländer Bucht
Kieler Bucht
Mecklenburger Bucht
Ålborg Bugt
Hanöbukten
Øresund
Store Bælt
Lille Bælt
Bornholm (Den.)
Öland
Rügen
Fehmarn
Zealand (Sjælland)
Fyn
Lolland
Falster
Møn
Langeland
Sylt
Læsø
Vättern
Kalmarsund
Kristiansand
Arendal
Grimstad
Lillesand
Mandal
Lindesnes
Farsund
Gothenburg (Göteborg)
Halmstad
Helsingborg
Malmö
Lund
Trelleborg
Ystad
Karlskrona
Kalmar
Växjö
Jönköping
Linköping
Norrköping
Nyköping
Borås
Varberg
Falkenberg
Kristianstad
Ålborg
Århus
Randers
Esbjerg
Odense
Copenhagen (København)
Roskilde
Helsingør
Fredericia
Vejle
Kolding
Frederikshavn
Hjørring
Skagen
Flensburg
Kiel
Lübeck
Hamburg
Bremen
Bremerhaven
Cuxhaven
Rostock
Stralsund
Greifswald
Schwerin
Wismar
Hannover
Braunschweig
Wolfsburg
Magdeburg
Berlin
Potsdam
Frankfurt an der Oder
Cottbus
Dresden
Leipzig
Halle
Chemnitz (Karl-Marx-Stadt)
Zwickau
Jena
Weimar
Erfurt
Gotha
Eisenach
Kassel
Göttingen
Bielefeld
Münster
Osnabrück
Dortmund
Essen
Duisburg
Düsseldorf
Wuppertal
Cologne (Köln)
Bonn
Aachen
Koblenz
Wiesbaden
Mainz
Frankfurt am Main
Offenbach am Main
Darmstadt
Mannheim
Ludwigshafen am Rhein
Heidelberg
Saarbrücken
Karlsruhe
Stuttgart
Pforzheim
Heilbronn
Würzburg
Nürnberg
Fürth
Erlangen
Bamberg
Bayreuth
Regensburg
Ingolstadt
Augsburg
Ulm
München (Munich)
Passau
Freiburg im Breisgau
Konstanz
Amsterdam
Rotterdam
The Hague
Utrecht
Haarlem
Leiden
Groningen
Eindhoven
Arnhem
Nijmegen
Antwerp
Brussels
Liège
Namur
Charleroi
Luxembourg
Metz
Nancy
Strasbourg
Colmar
Mulhouse
Belfort
Besançon
Basel
Zürich
Bern
Luzern
Lausanne
Genève (Geneva)
Lac Léman
Innsbruck
Salzburg
Linz
Vienna (Wien)
Graz
Klagenfurt
Villach
Prague (Praha)
Plzeň
Brno
Liberec
Ústí nad Labem
Karlovy Vary
Szczecin
Koszalin
Słupsk
Poznań
Wrocław
Legnica
Zielona Góra
Gorzów Wielkopolski
Bydgoszcz
Opole
Bratislava
Győr
Sopron
Szombathely
Maribor
Bolzano
Merano
Trento
ALPS
Niedere Tauern
Hohe Tauern
Bayerischer Wald
Böhmer Wald
Šumava
Odenwald
Schwäbische Alb
Elbe
Oder
Rhein
Weser
Danube (Donau)
Conic Equidistant Projection

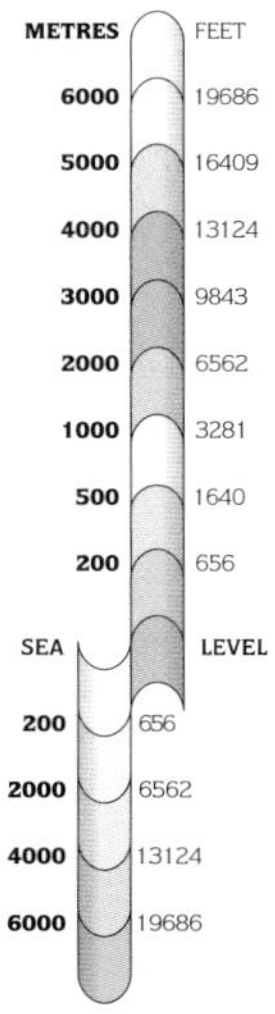
METRES
FEET
6000
19686
5000
16409
4000
13124
3000
9843
2000
6562
1000
3281
500
1640
200
656
SEA
LEVEL
200
656
2000
6562
4000
13124
6000
19686

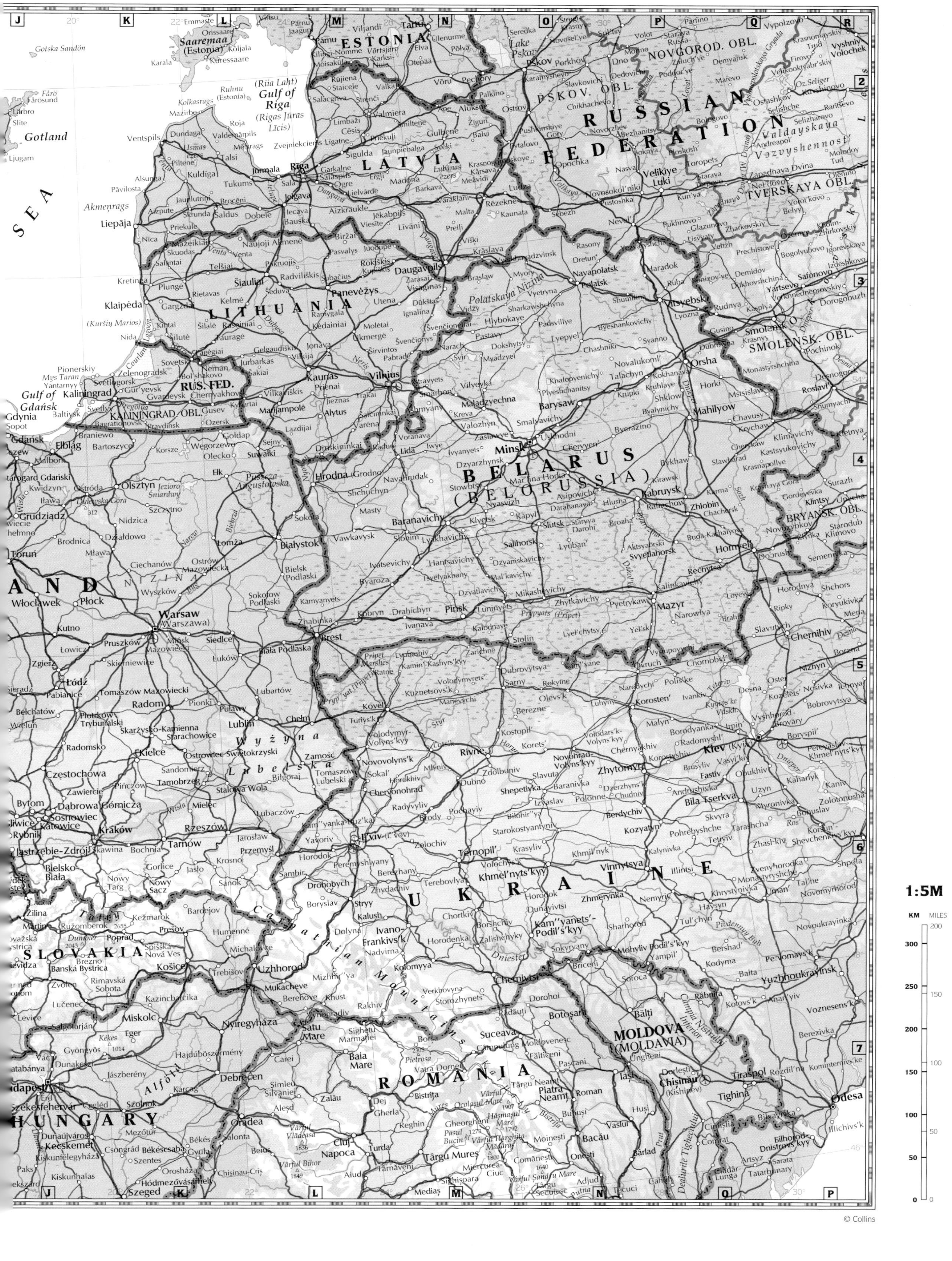
ESTONIA
LATVIA
LITHUANIA
RUS. FED.
KALININGRAD OBL.
BELARUS
(BELORUSSIA)
RUSSIAN FEDERATION
PSKOV. OBL.
NOVGOROD. OBL.
TVERSKAYA OBL.
SMOLENSK. OBL.
BRYANSK. OBL.
UKRAINE
MOLDOVA
(MOLDAVIA)
ROMANIA
HUNGARY
SLOVAKIA
SEA
Gulf of Riga
(Riias Laht)
(Rigas Jūras Līcis)
Gulf of Gdańsk
Gotland
Saaremaa (Estonia)
Carpathian Mountains
Wyżyna Lubelska
Polatskaya Nizina
Riga
Tallinn
Tartu
Pskov
Vilnius
Kaunas
Klaipėda
Liepāja
Daugavpils
Kaliningrad
Minsk
Hrodna (Grodno)
Brest
Homyel'
Mahilyow
Vitsyebsk
Smolensk
Warsaw (Warszawa)
Łódź
Lublin
Kraków
Kiev (Kyiv)
L'viv (L'vov)
Chişinău (Kishinev)
Odesa
Debrecen
Budapest
Cluj-Napoca
Iaşi
1:5M
KM
MILES
300
250
200
150
100
50
0
200
150
100
50
0

Conic Equidistant Projection

1:2M

KM / MILES: 100, 75, 50, 25, 0 / 60, 50, 40, 30, 20, 10, 0

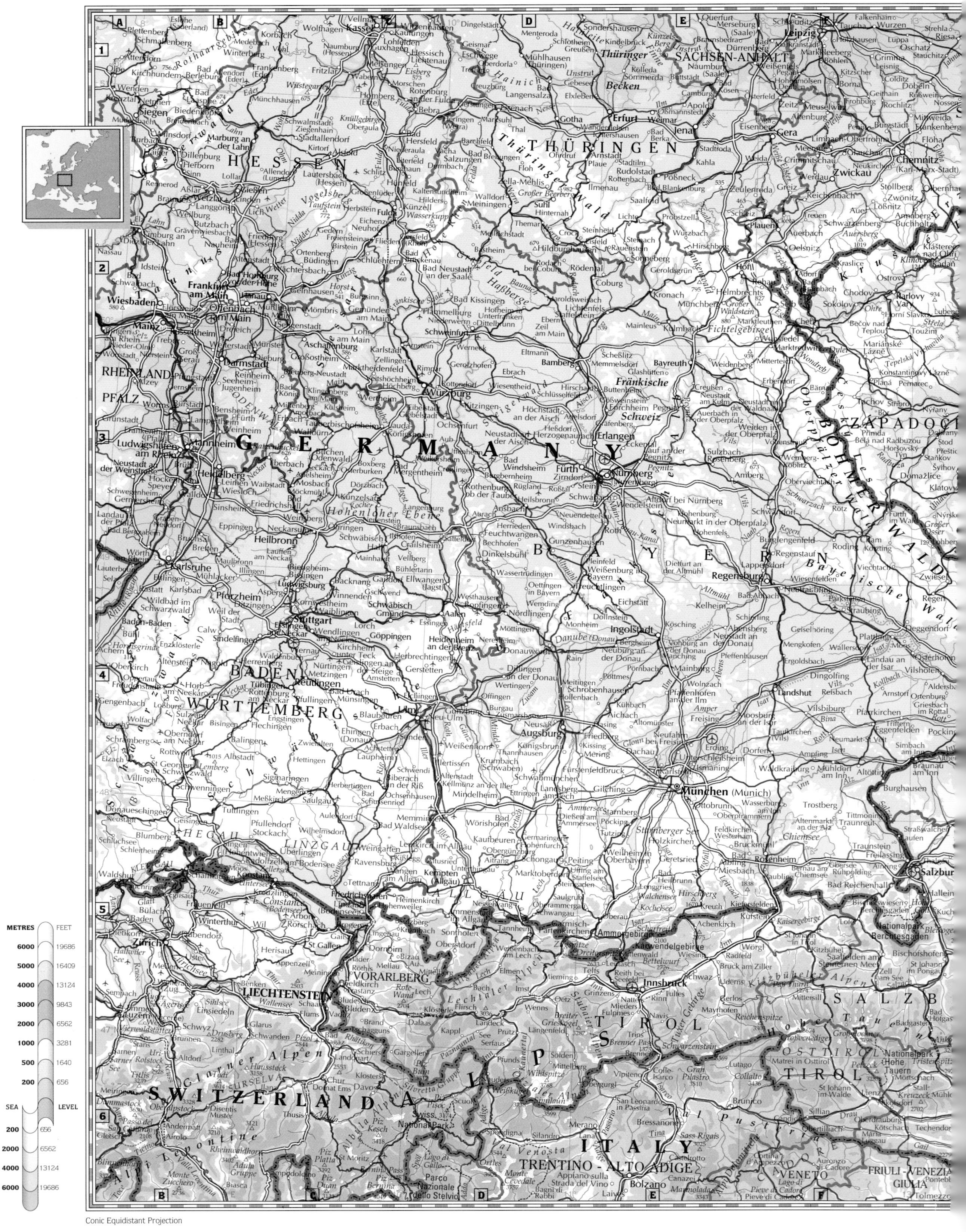

Conic Equidistant Projection

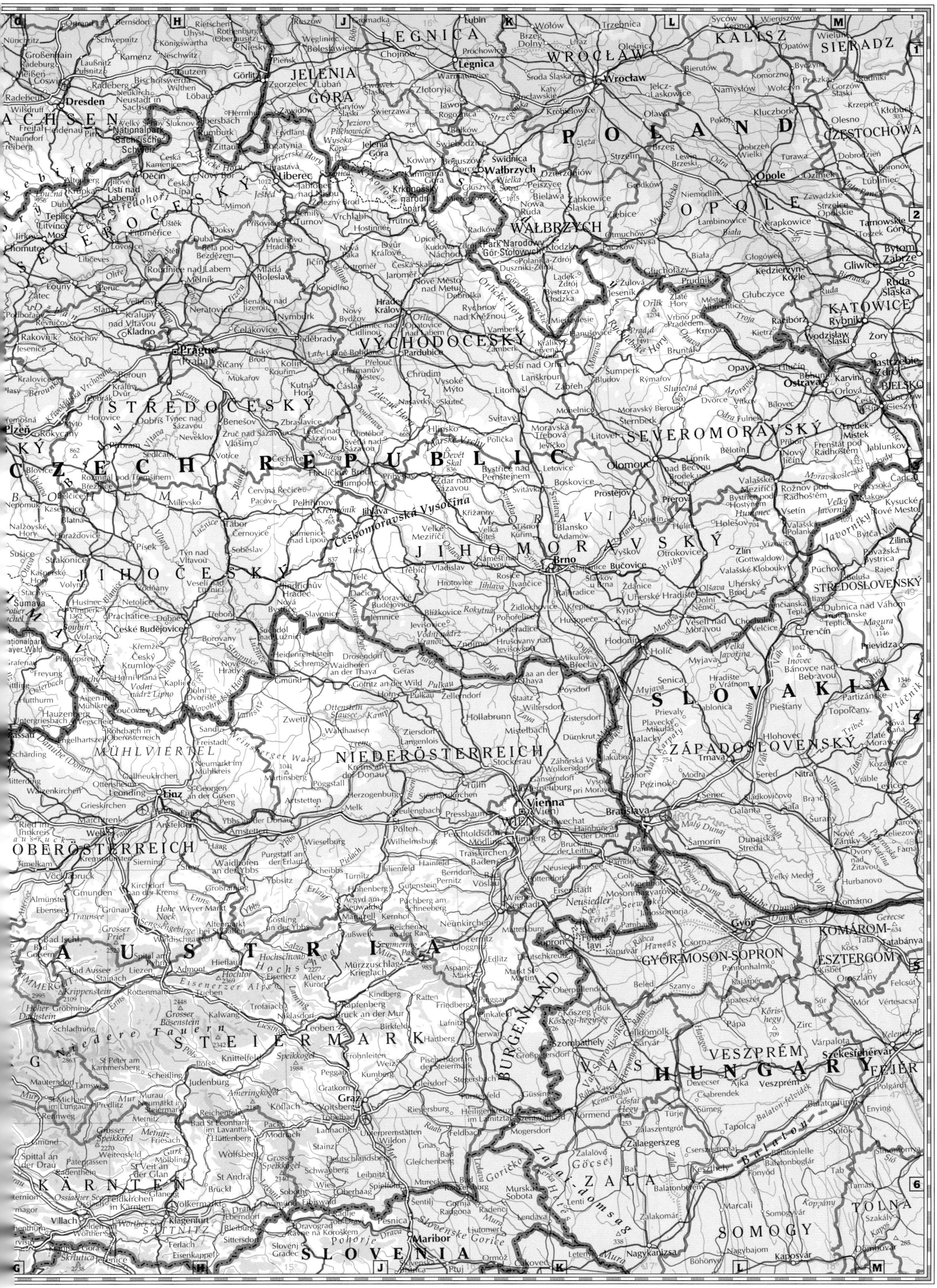

1:2M

KM MILES

100 60
50
75 40
30
50
20
25
10
0 0

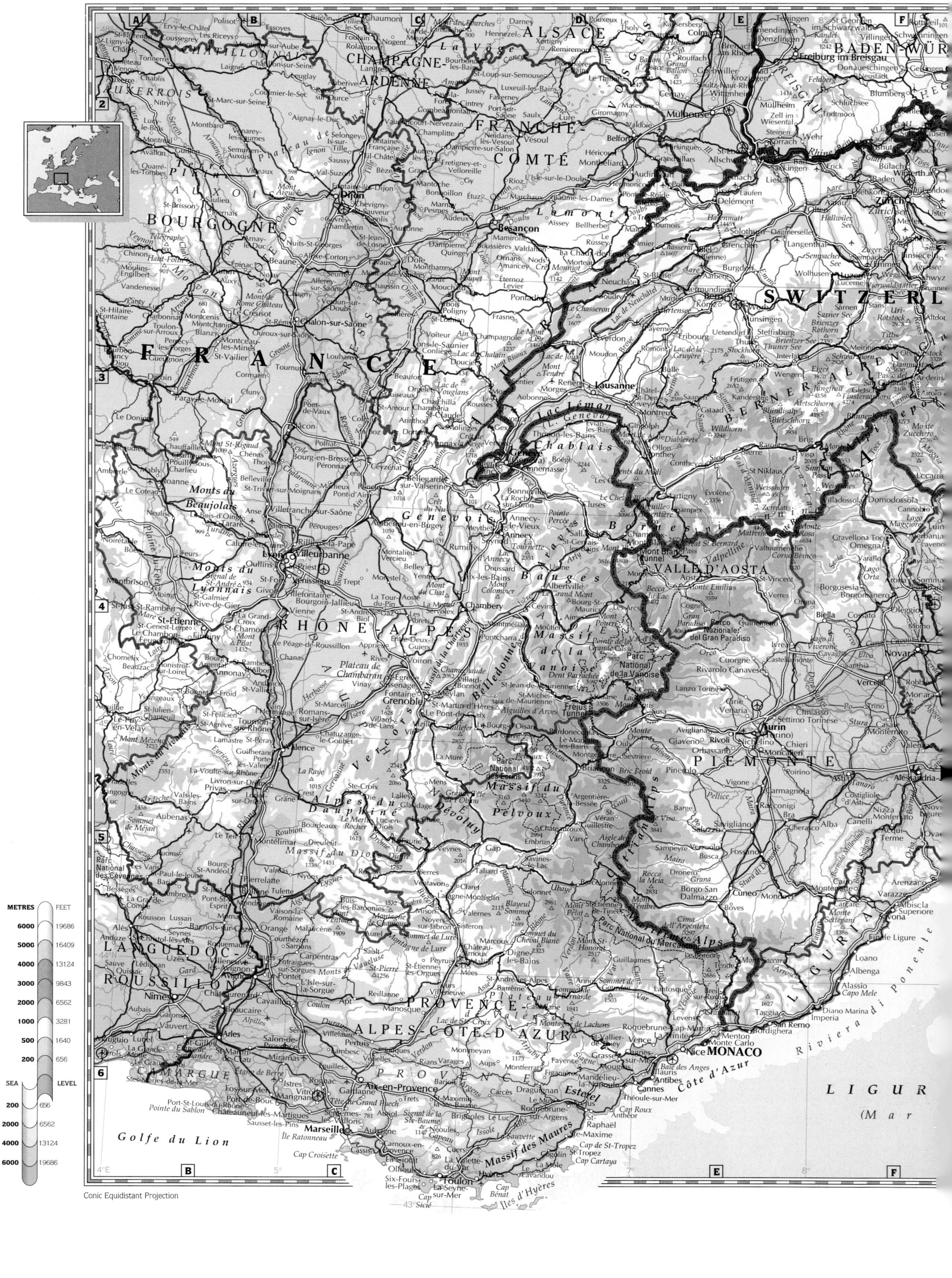

Conic Equidistant Projection

1:2M

KM 100, 75, 50, 25, 0 — MILES 60, 50, 40, 30, 20, 10, 0

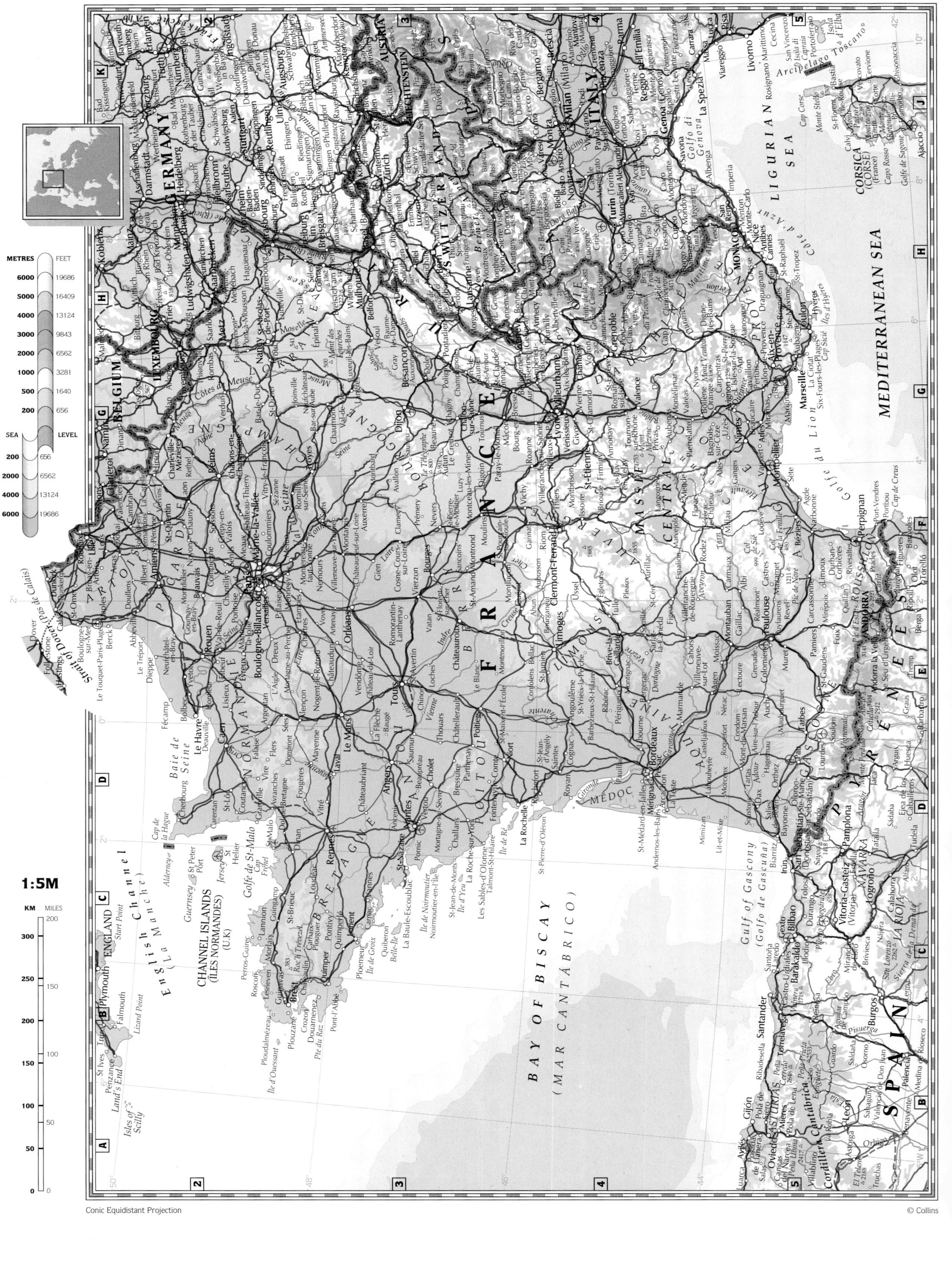

Conic Equidistant Projection

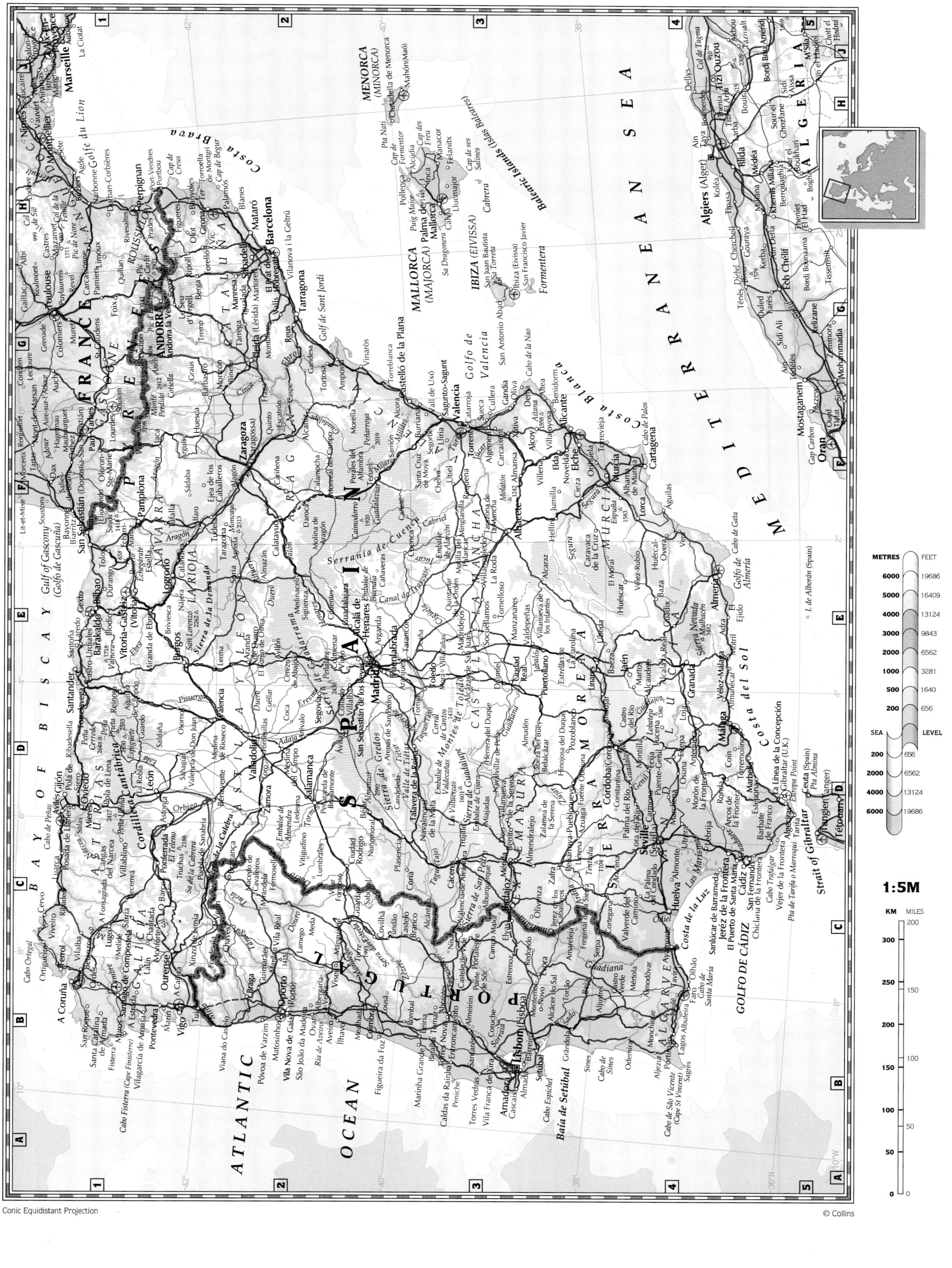

Conic Equidistant Projection

Conic Equidistant Projection

HUNGARY
ROMANIA
MOLDOVA
UKRAINE
YUGOSLAVIA
SERBIA
MONTENEGRO (CRNA GORA)
KOSOVO
VOJVODINA
BULGARIA
MACEDONIA (F.Y.R.O.M.)
ALBANIA
GREECE
TURKEY
BLACK SEA
AEGEAN SEA
Sea of Marmara (Marmara Denizi)
Ionian Islands (Ionioi Nisoi)
Strait of Otranto
CARPATHIAN MTS
Transylvanian Alps (Carpaţii Meridionali)
BALKAN MTS (STARA PLANINA)
Rhodope Mts
Pindus Mts
Cyclades (Kyklades)
Dodecanese (Dodekanisos)
Krytiko Pelagos
Mirtoö Pelagos
Thrakiko Pelagos
Thermaïkos Kolpos
CRETE (KRITI)
Bucharest (Bucureşti)
Belgrade (Beograd)
Sofia (Sofiya)
Athens (Athina)
Istanbul
Skopje
Tirana (Tiranë)
Sarajevo
Thessaloniki
Izmir
Constanţa
Varna
Burgas
Odesa
Corfu (Kerkyra)
Kefallonia (Cephalonia)
Lesbos (Lesvos)
Chios
Samos
Rhodes (Rodos)
Naxos
Limnos
EVVOIA (EUBOEA)
Kythira
Karpathos (Scarpanto)
1:5M
KM
MILES
250
200
150
100
50
0

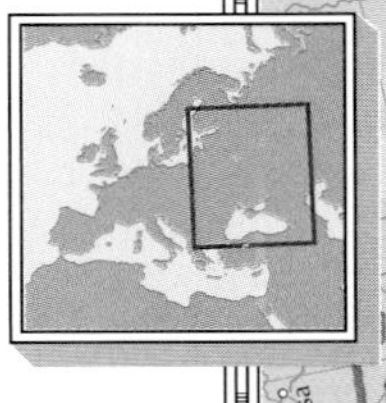

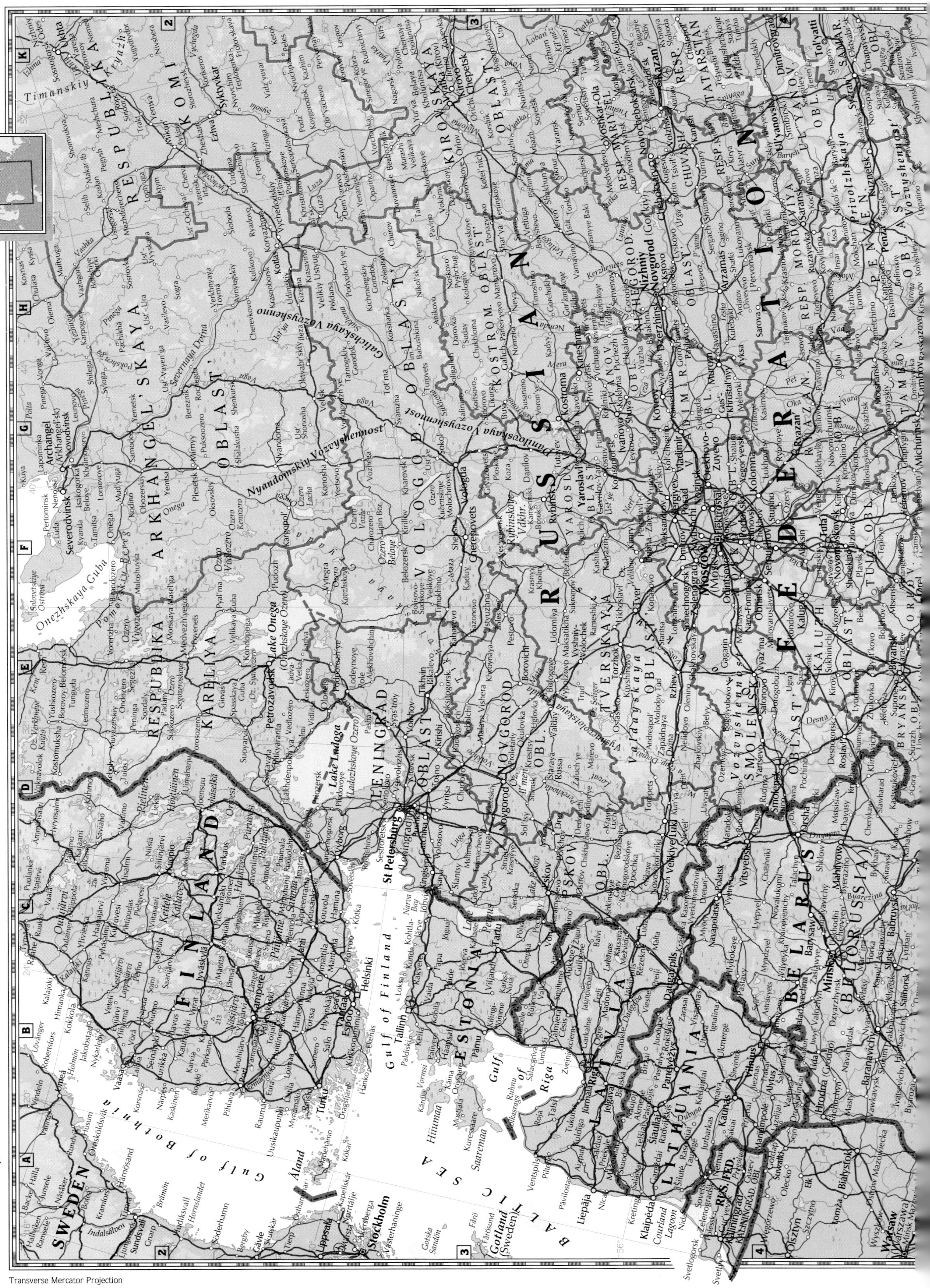

METRES	FEET
6000	19686
5000	16409
4000	13124
3000	9843
2000	6562
1000	3281
500	1640
200	656
SEA	LEVEL
200	656
2000	6562
4000	13124
6000	19686

Transverse Mercator Projection

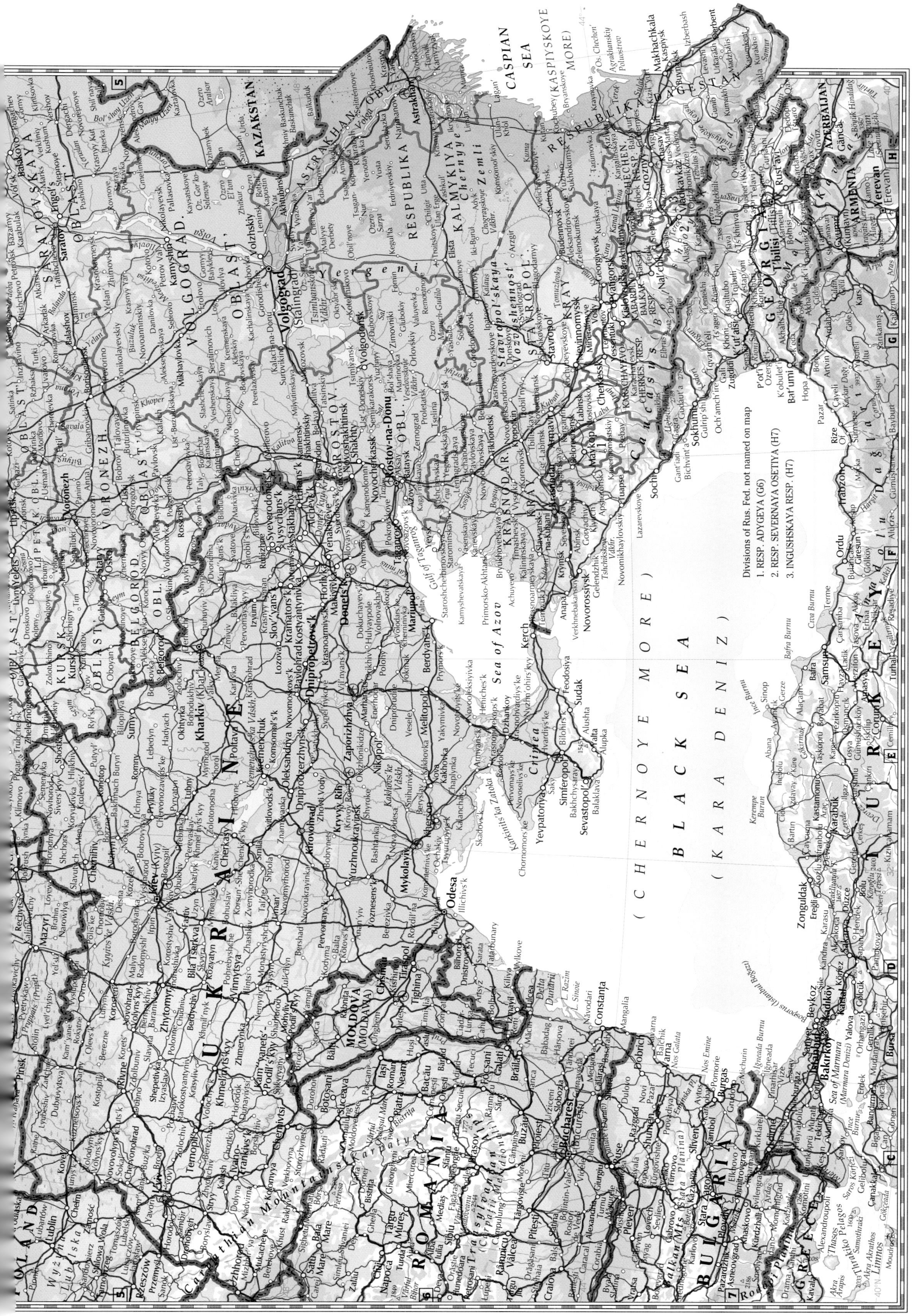
BLACK SEA
(CHERNOYE MORE)
(KARA DENIZ)
Sea of Azov
CASPIAN SEA
(KASPIYSKOYE MORE)
Divisions of Rus. Fed. not named on map
1. RESP. ADYGEYA (G6)
2. RESP. SEVERNAYA OSETIYA (H7)
3. INGUSHSKAYA RESP. (H7)
UKRAINE
ROMANIA
MOLDOVA
BULGARIA
GEORGIA
ARMENIA
AZERBAIJAN
KAZAKHSTAN
TURKEY
GREECE
POLAND
Crimea
Caucasus

1:7M

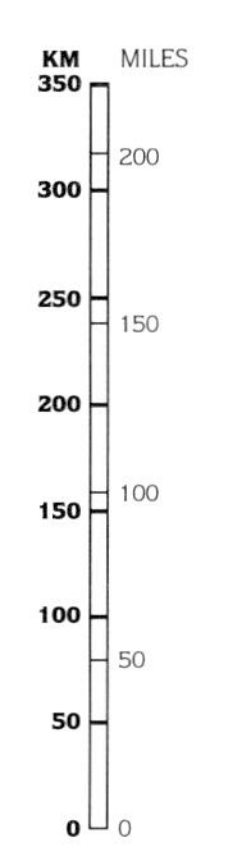
KM
MILES
350
300
250
200
150
100
50
0
200
150
100
50
0

METRES	FEET
5000	16409
3000	9843
2000	6562
1000	3281
500	1640
200	656
SEA	LEVEL
200	656
3000	9843
5000	16409
6000	19686

1:48M

KM	MILES
2500	1500
2000	
1500	1000
1000	
500	500
0	0

Lambert Azimuthal Equal Area Projection

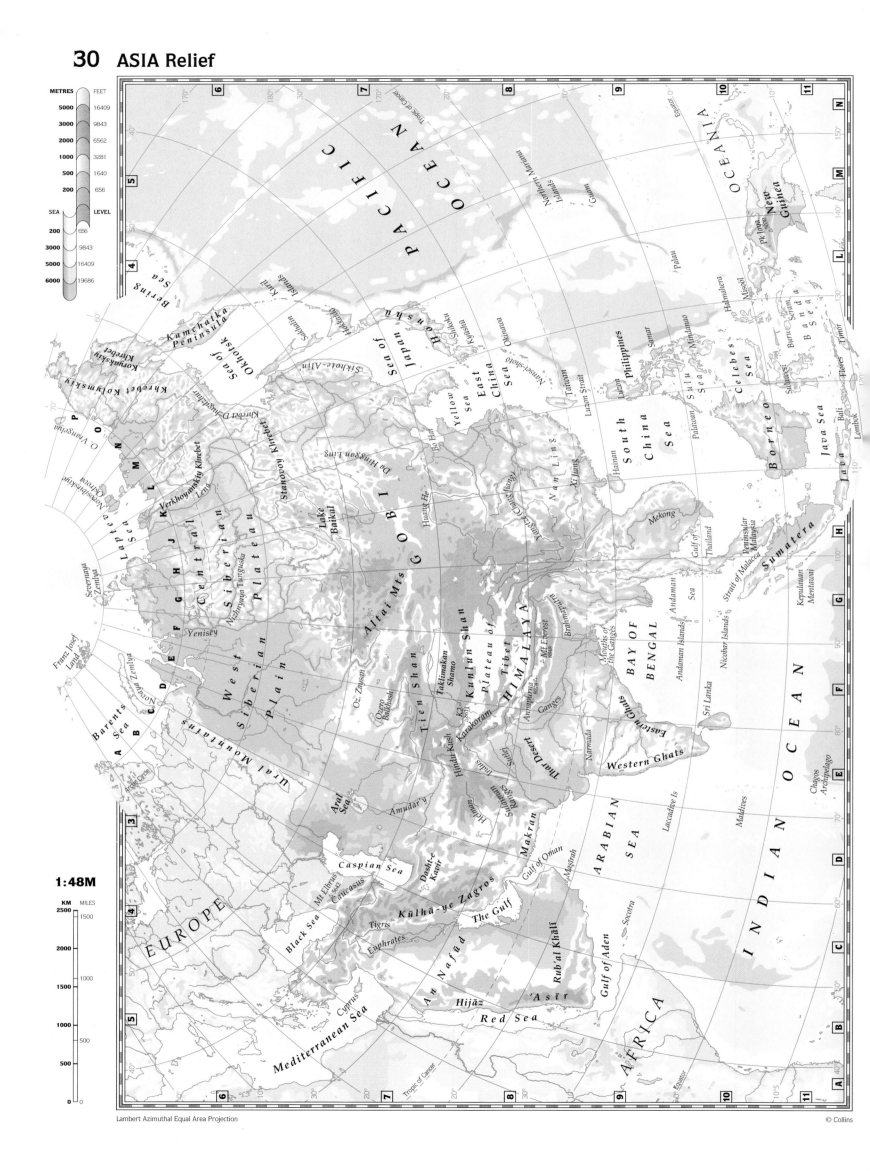

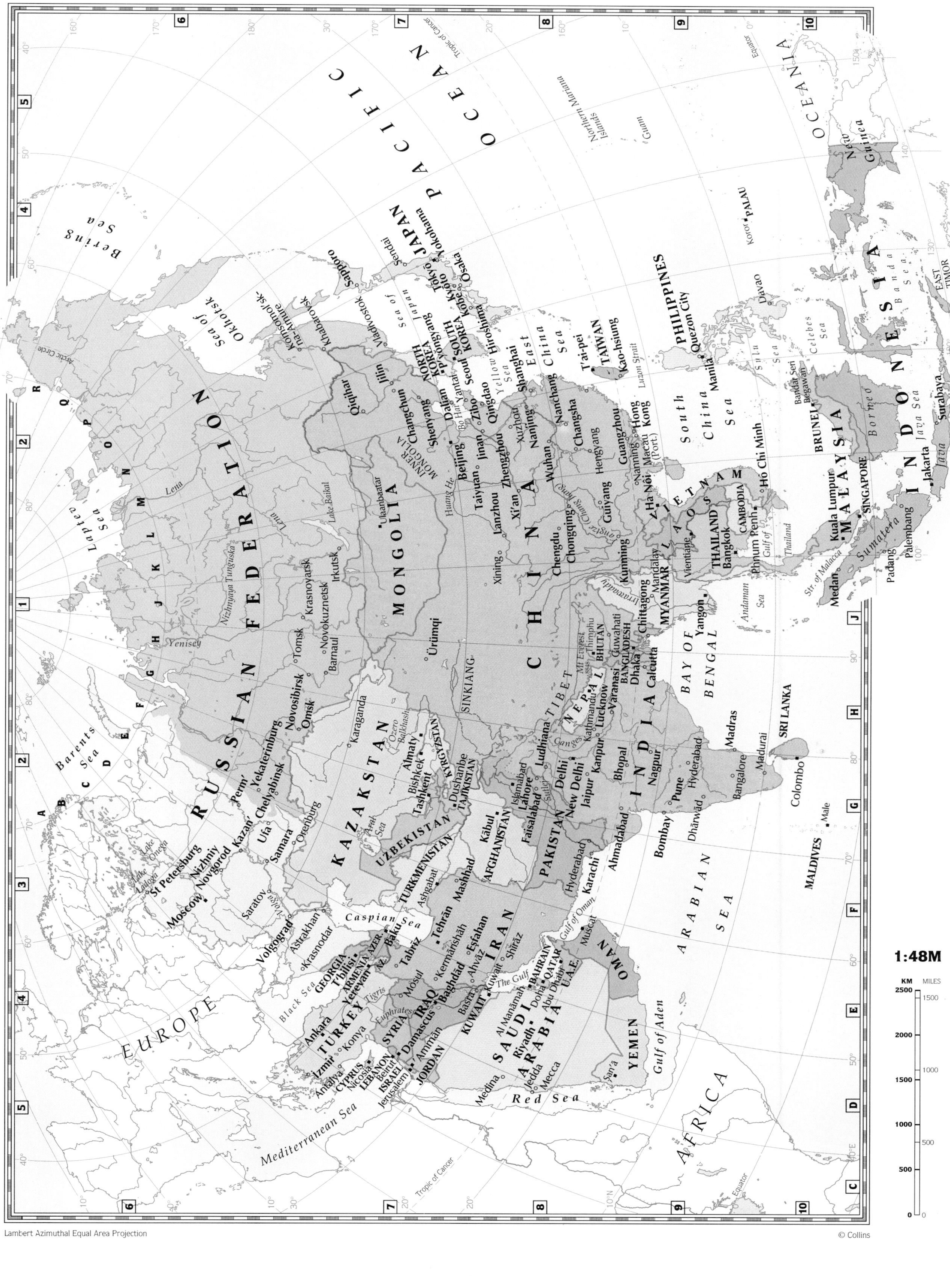
PACIFIC OCEAN
OCEANIA
EUROPE
AFRICA
RUSSIAN FEDERATION
KAZAKSTAN
MONGOLIA
CHINA
INDIA
JAPAN
NORTH KOREA
SOUTH KOREA
PHILIPPINES
TAIWAN
VIETNAM
LAOS
THAILAND
CAMBODIA
MYANMAR
MALAYSIA
SINGAPORE
BRUNEI
INDONESIA
EAST TIMOR
PALAU
SRI LANKA
MALDIVES
BANGLADESH
BHUTAN
NEPAL
PAKISTAN
AFGHANISTAN
TAJIKISTAN
KYRGYZSTAN
UZBEKISTAN
TURKMENISTAN
IRAN
IRAQ
SYRIA
TURKEY
GEORGIA
ARMENIA
AZER.
CYPRUS
LEBANON
ISRAEL
JORDAN
KUWAIT
BAHRAIN
QATAR
U.A.E.
OMAN
YEMEN
SAUDI ARABIA
TIBET
SINKIANG
INNER MONGOLIA
Bering Sea
Sea of Okhotsk
Sea of Japan
East China Sea
South China Sea
Yellow Sea
Laptev Sea
Barents Sea
Caspian Sea
Black Sea
Mediterranean Sea
Red Sea
Gulf of Aden
Gulf of Oman
The Gulf
Arabian Sea
Bay of Bengal
Andaman Sea
Gulf of Thailand
Sulu Sea
Celebes Sea
Banda Sea
Java Sea
Aral Sea
Lake Baikal
Arctic Circle
Tropic of Cancer
Equator
Lambert Azimuthal Equal Area Projection
1:48M
KM
MILES
2500
2000
1500
1000
500
0
1500
1000
500
0

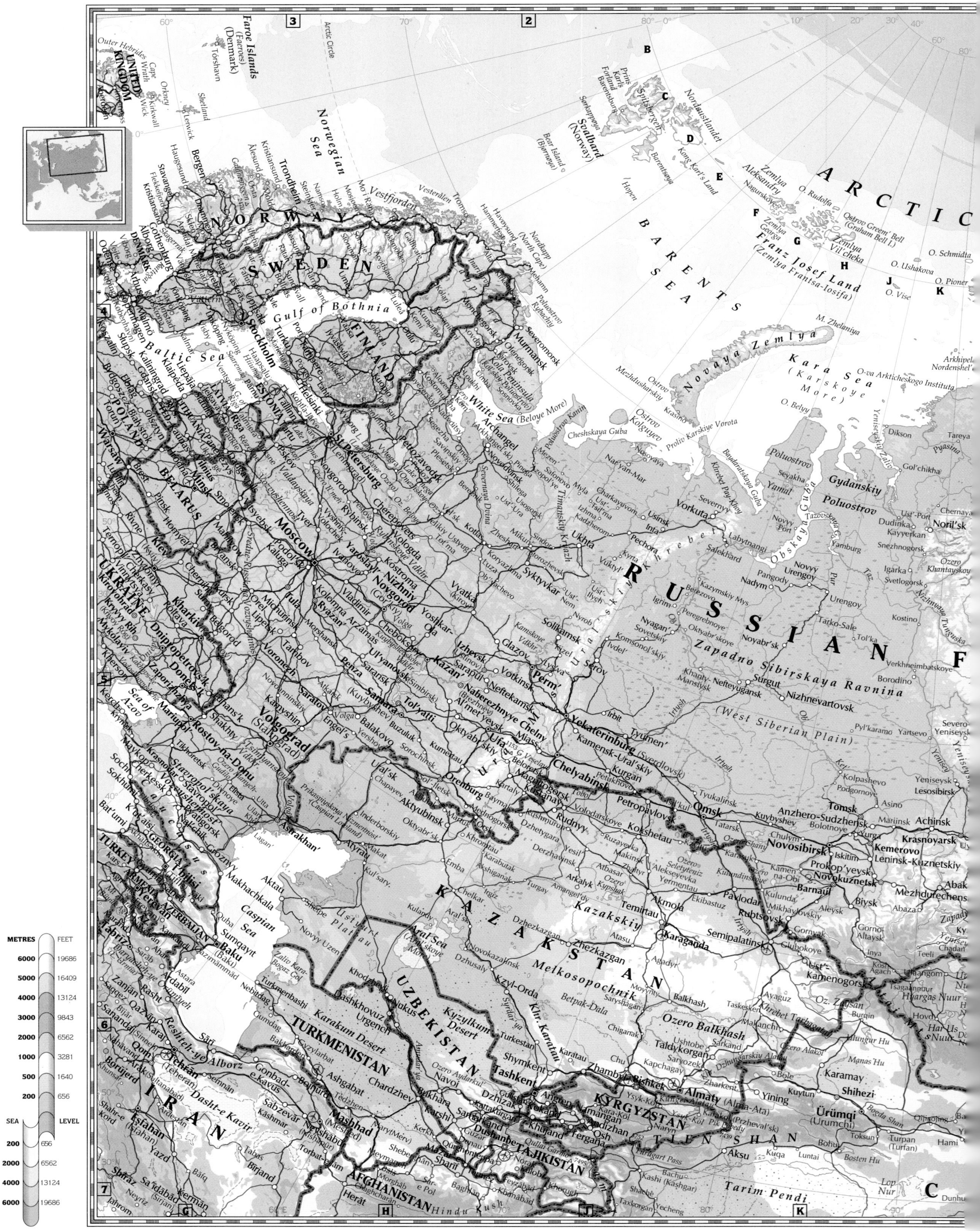

Conic Equidistant Projection

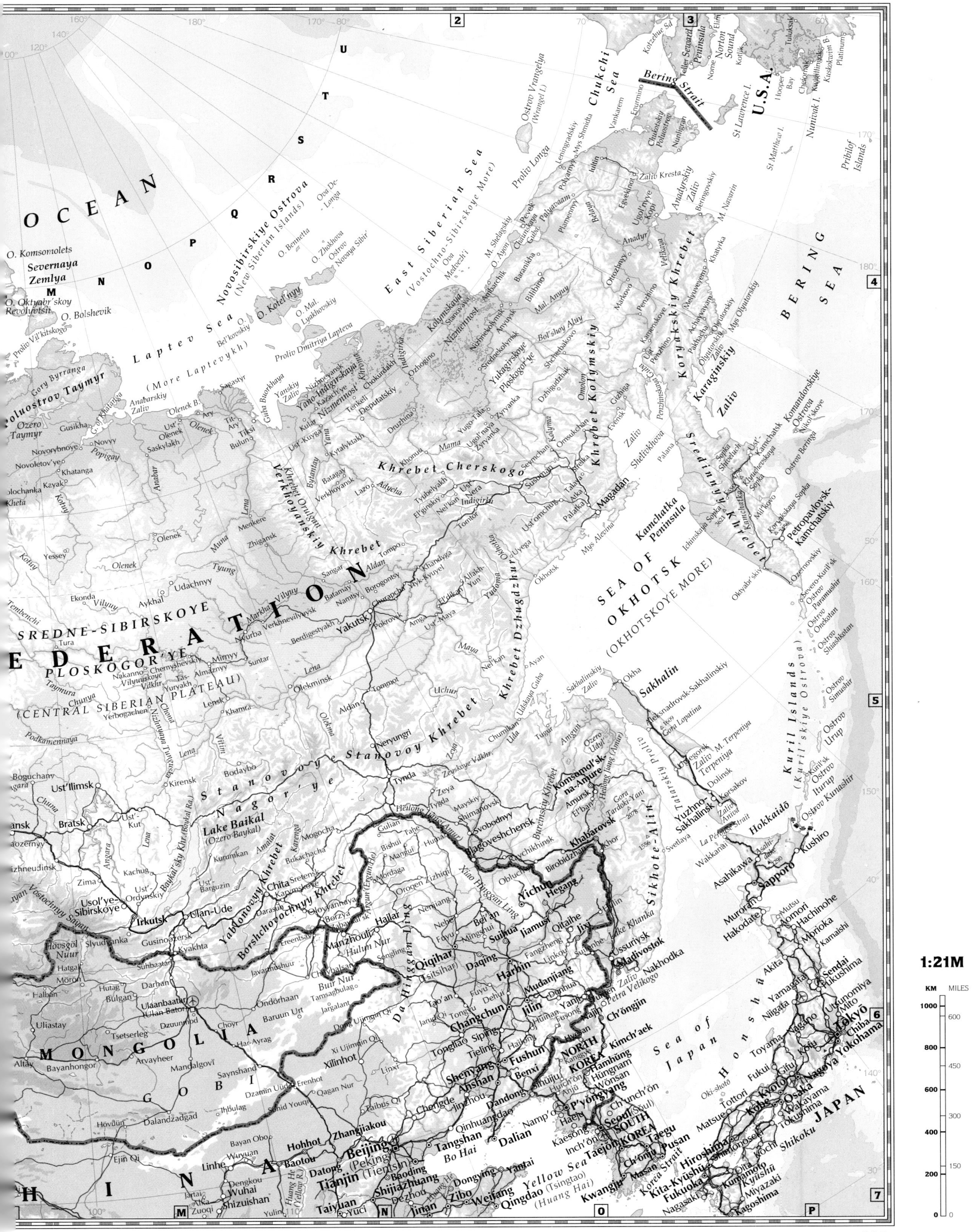
OCEAN
Chukchi Sea
Bering Strait
U.S.A.
St Lawrence I.
Norton Sound
Nome
Seward Peninsula
Pribilof Islands
Ostrov Vrangelya (Wrangel I.)
East Siberian Sea (Vostochno-Sibirskoye More)
Novosibirskiye Ostrova (New Siberian Islands)
O. Kotel'nyy
O. Komsomolets
Severnaya Zemlya
O. Oktyabr'skoy Revolyutsii
O. Bolshevik
Laptev Sea (More Laptevykh)
Poluostrov Taymyr
Ozero Taymyr
Zaliv Kresta
Anadyrskiy Zaliv
Anadyr'
BERING SEA
Koryakskiy Khrebet
Karaginskiy Zaliv
Komandorskiye Ostrova
Sredinnyy Khrebet
Kamchatka Peninsula
Petropavlovsk-Kamchatskiy
SEA OF OKHOTSK (OKHOTSKOYE MORE)
Magadan
Khrebet Kolymskiy
Khrebet Cherskogo
Verkhoyanskiy Khrebet
Tiksi
Yakutsk
Khrebet Dzhugdzhur
Okhotsk
Sakhalin
Kuril Islands (Kuril'skiye Ostrova)
SREDNE-SIBIRSKOYE PLOSKOGOR'YE (CENTRAL SIBERIAN PLATEAU)
FEDERATION
Stanovoy Khrebet
Stanovoye Nagor'ye
Lake Baikal (Ozero Baykal)
Bratsk
Ust'-Ilimsk
Irkutsk
Ulan-Ude
Chita
Yablonovyy Khrebet
Neryungri
Tynda
Blagoveshchensk
Komsomol'sk-na-Amure
Khabarovsk
Sikhote-Alin'
Tatarskiy Proliv
Yuzhno-Sakhalinsk
La Pérouse Strait
Hokkaidō
Sapporo
Asahikawa
Kushiro
Hakodate
Aomori
Sendai
Vladivostok
Nakhodka
Harbin
Qiqihar
Hailar
Manzhouli
Jilin
Changchun
Shenyang
Anshan
Fushun
Dalian
NORTH KOREA
P'yŏngyang
Seoul
SOUTH KOREA
Taejŏn
Taegu
Pusan
Kwangju
Sea of Japan
Honshū
JAPAN
Tōkyō
Yokohama
Nagoya
Ōsaka
Kyōto
Kōbe
Hiroshima
Shikoku
Kyūshū
Fukuoka
Kagoshima
Beijing (Peking)
Tianjin (Tientsin)
Qingdao (Tsingtao)
Yellow Sea (Huang Hai)
Bo Hai
Hohhot
Baotou
Zhangjiakou
Shijiazhuang
Taiyuan
Jinan
MONGOLIA
Ulaanbaatar (Ulan Bator)
GOBI
CHINA
1:21M
KM
MILES
1000
800
600
400
200
0
600
450
300
150
0

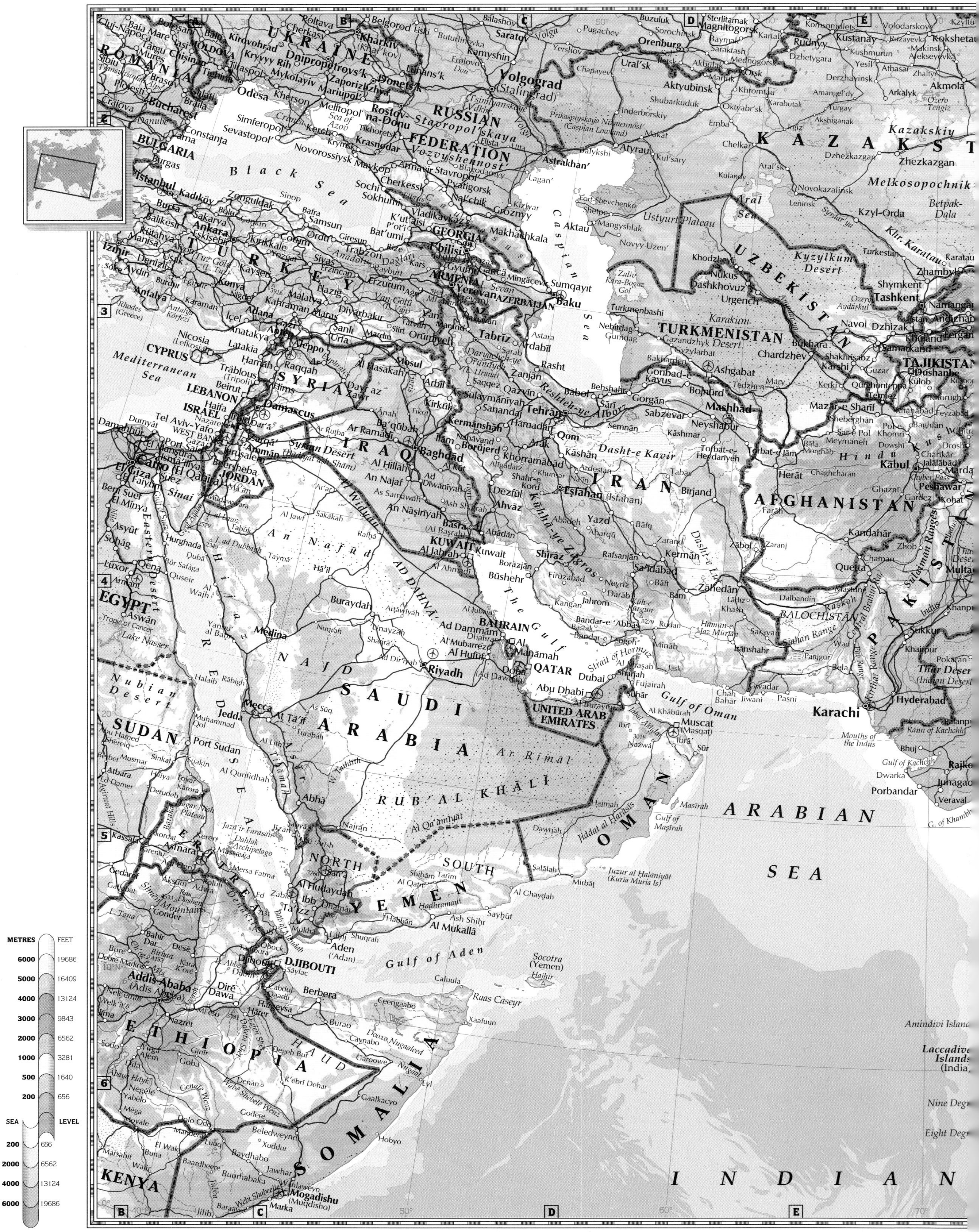

Albers Equal Area Conic Projection

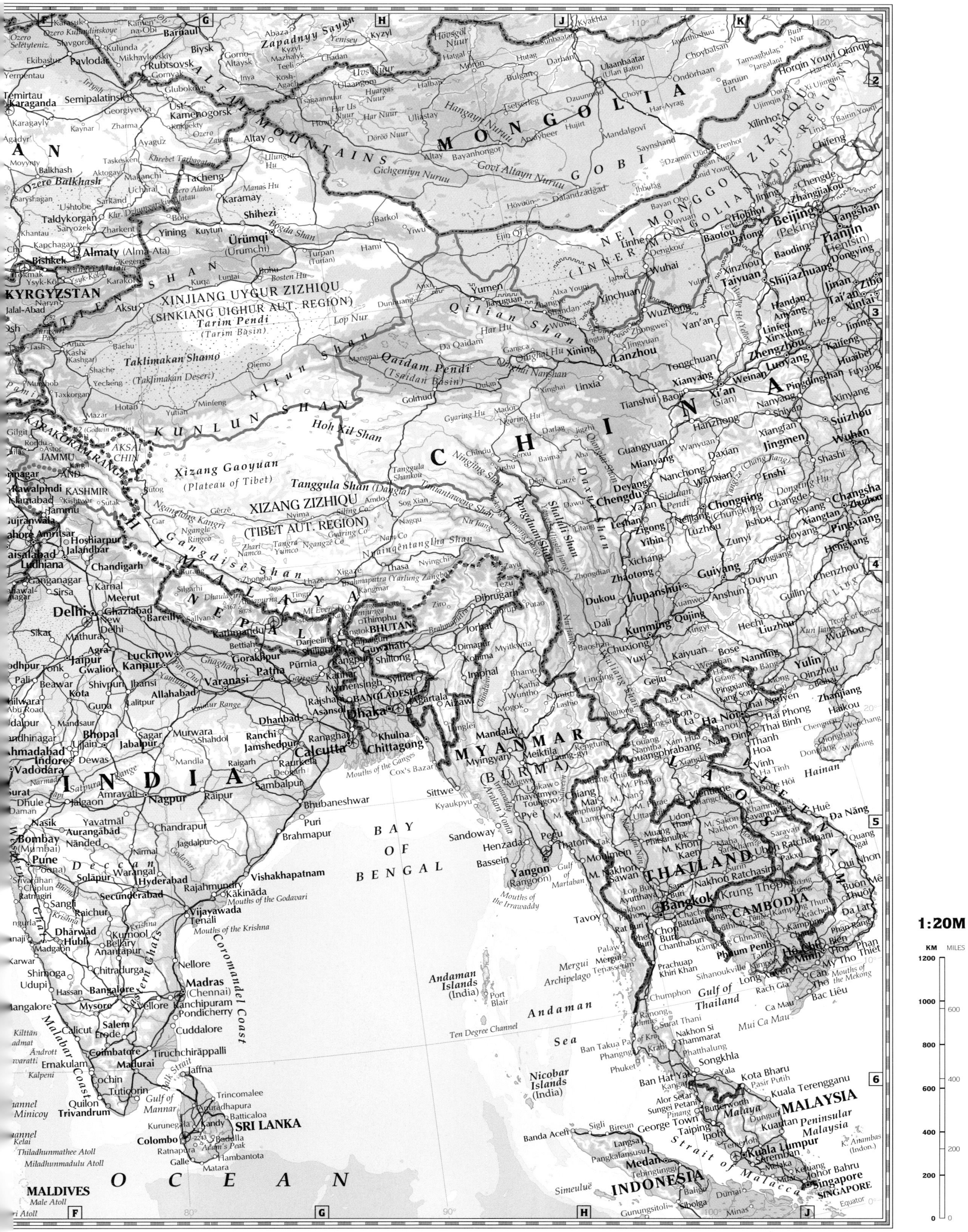

MONGOLIA
CHINA
INDIA
NEPAL
BHUTAN
BANGLADESH
MYANMAR
(BURMA)
THAILAND
LAOS
CAMBODIA
VIETNAM
MALAYSIA
SINGAPORE
INDONESIA
SRI LANKA
MALDIVES
KYRGYZSTAN
ALTAI MOUNTAINS
GOBI
HIMALAYA
KUNLUN SHAN
XINJIANG UYGUR ZIZHIQU
(SINKIANG UIGHUR AUT. REGION)
XIZANG ZIZHIQU
(TIBET AUT. REGION)
NEI MONGGOL ZIZHIQU
(INNER MONGOLIAN AUT. REGION)
Xizang Gaoyuan
(Plateau of Tibet)
Taklimakan Shamo
(Taklimakan Desert)
Tarim Pendi
(Tarim Basin)
Qaidam Pendi
(Tsaidam Basin)
Tanggula Shan
Qilian Shan
Altun Shan
Hoh Xil Shan
Gangdisê Shan
Nyainqêntanglha Shan
Hengduan Shan
Daxue Shan
Deccan
Eastern Ghats
Western Ghats
Coromandel Coast
Malabar Coast
BAY OF BENGAL
Andaman Sea
Andaman Islands (India)
Nicobar Islands (India)
Gulf of Thailand
Gulf of Martaban
Gulf of Mannar
Strait of Malacca
Peninsular Malaysia
Hainan
OCEAN
Ten Degree Channel
Mouths of the Ganges
Mouths of the Irrawaddy
Mouths of the Godavari
Mouths of the Krishna
Mouths of the Mekong
Beijing (Peking)
Tianjin (Tientsin)
Ulaanbaatar (Ulan Bator)
Ürümqi (Urumchi)
Almaty (Alma-Ata)
Bishkek
Lhasa
Delhi
New Delhi
Calcutta
Dhaka
Kathmandu
Thimphu
Bombay (Mumbai)
Madras (Chennai)
Hyderabad
Bangalore
Colombo
Yangon (Rangoon)
Bangkok (Krung Thep)
Vientiane
Ha Nôi (Hanoi)
Phnom Penh
Hô Chi Minh
Kuala Lumpur
Xi'an (Sian)
Chengdu
Chongqing (Chungking)
Wuhan
Kunming
Lanzhou
Xining
Changsha
Nagpur
Ahmadabad
Jaipur
Lucknow
Kanpur
Varanasi
Patna
Chittagong
Mandalay
Vishakhapatnam
Vijayawada
Pune (Poona)
Madurai
Jaffna
Kandy
Medan
George Town
1:20M
KM
MILES
1200
1000
800
600
400
200
0

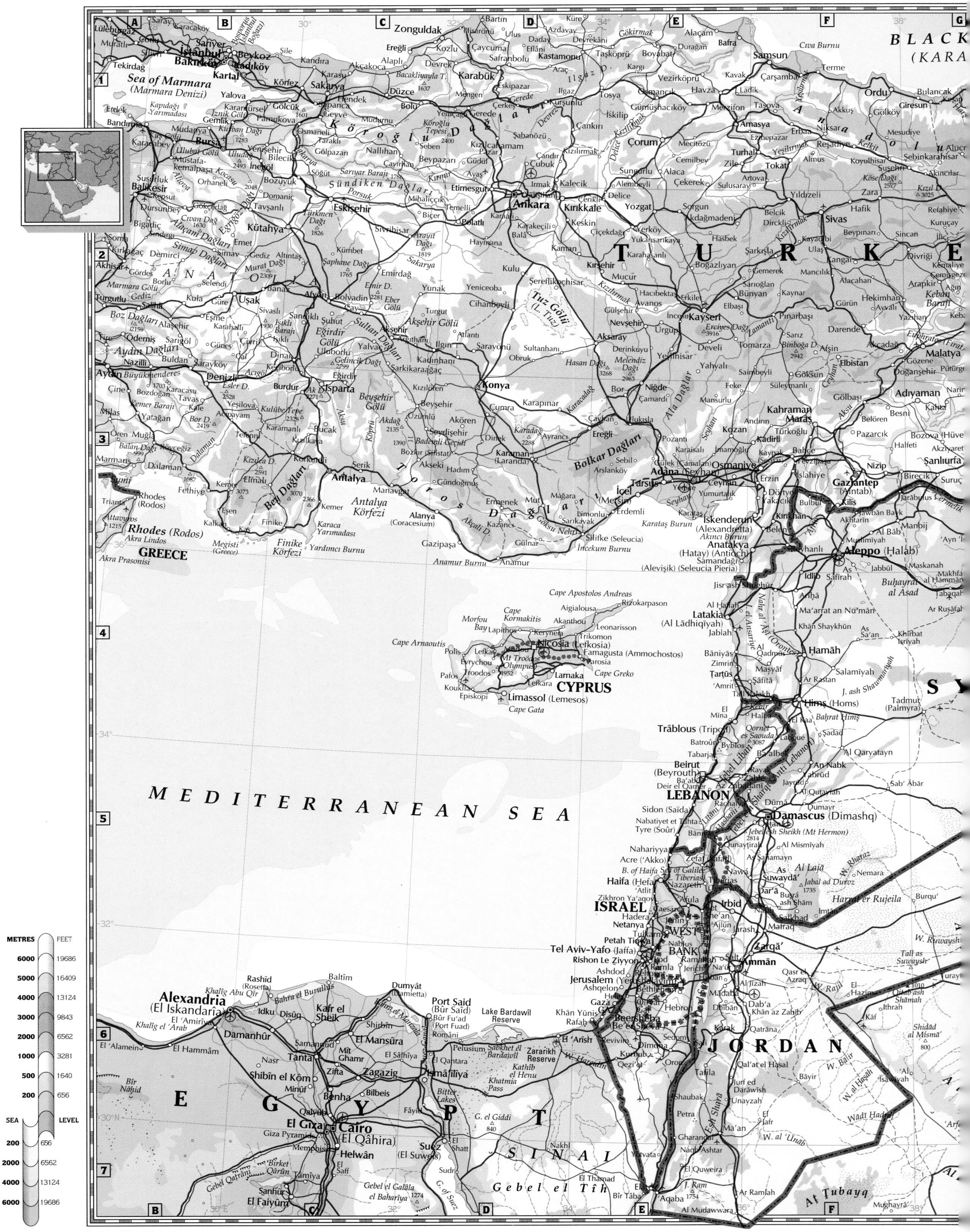

Conic Equidistant Projection

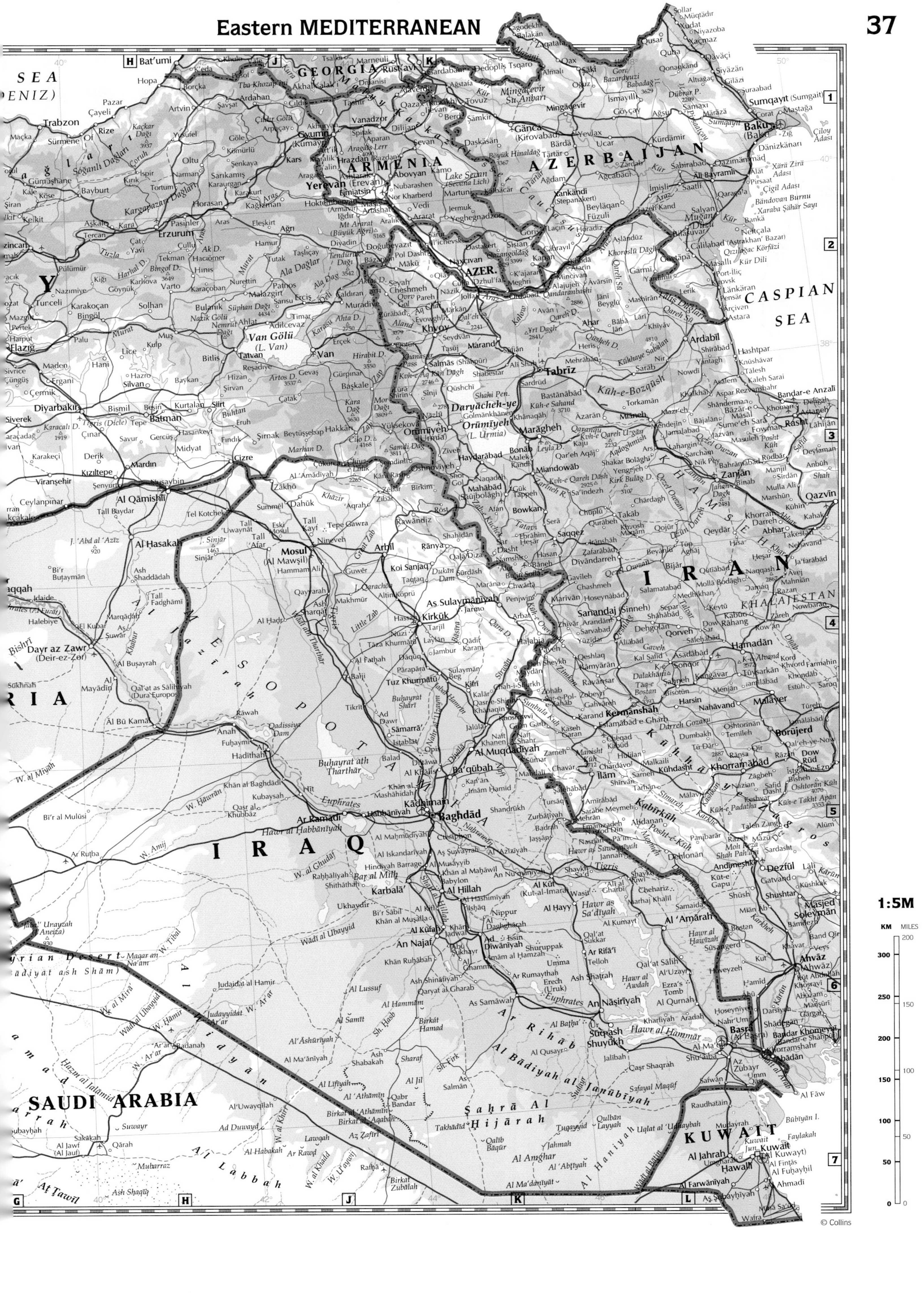
GEORGIA
ARMENIA
AZERBAIJAN
AZER.
IRAN
IRAQ
KUWAIT
SAUDI ARABIA
CASPIAN SEA
MESOPOTAMIA
KHAMSEH
KHALAJESTAN
Tbilisi
Yerevan
Baku (Bakı)
Tabrīz
Mosul (Al Mawşil)
Arbīl
Kirkūk
As Sulaymānīyah
Baghdād
Karbalā'
An Najaf
Al Kūt
Al Ḩillah
Al 'Amārah
An Nāşirīyah
Başrah (Al Başrah)
Ābādān
Kuwait
Al Jahrah
Hamadān
Kermānshāh
Sanandaj (Sinneh)
Orūmīyeh (Urmia)
Daryācheh-ye Orūmīyeh (L. Urmia)
Van Gölü (L. Van)
Van
Erzurum
Trabzon
Diyarbakır
Mardin
Batman
Al Qāmishlī
Al Ḩasakah
Dayr az Zawr (Deir-ez-Zor)
Ar Ramādī
Qazvīn
Rasht
Ardabīl
Zanjān
Ahvāz (Ahwāz)
Dezfūl
Khorramābād
Borūjerd
Gäncä (Kirovabad)
Sumqayıt (Sumgait)
Lake Sevan (Sevana Lich)
Mt Ararat (Büyük Ağrı)
Tigris
Euphrates
Ash Shām
Al Widyān
Al Bādiyah al Janūbīyah
1:5M
KM
MILES

Conic Equidistant Projection

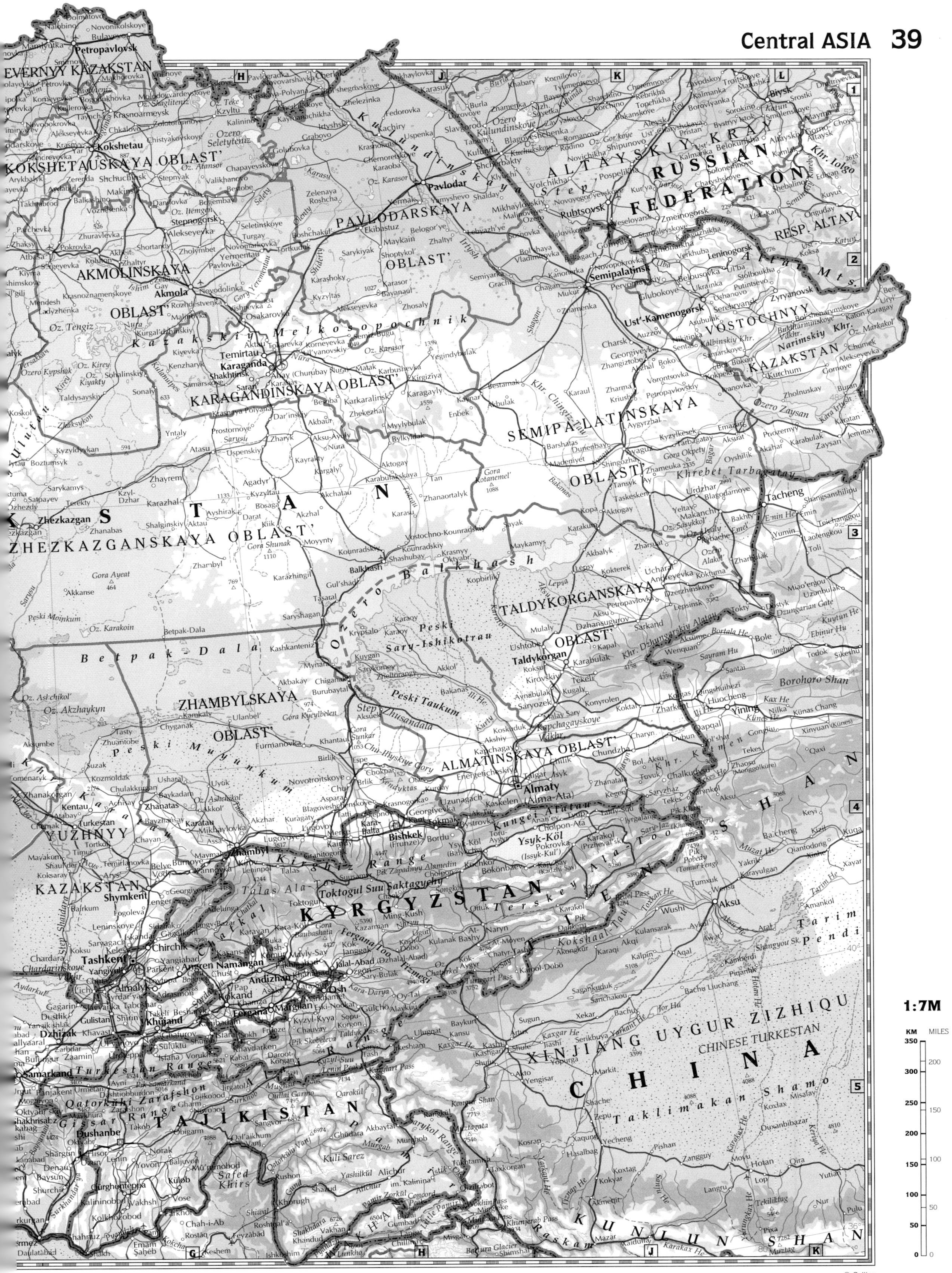
KAZAKSTAN
RUSSIAN FEDERATION
KYRGYZSTAN
TAJIKISTAN
CHINA
XINJIANG UYGUR ZIZHIQU
CHINESE TURKESTAN
KOKSHETAUSKAYA OBLAST'
PAVLODARSKAYA OBLAST'
ALTAYSKIY KRAY
RESP. ALTAY
AKMOLINSKAYA OBLAST'
KARAGANDINSKAYA OBLAST'
SEMIPALATINSKAYA OBLAST'
VOSTOCHNYY KAZAKSTAN
ZHEZKAZGANSKAYA OBLAST'
TALDYKORGANSKAYA OBLAST'
ZHAMBYLSKAYA OBLAST'
ALMATINSKAYA OBLAST'
YUZHNYY KAZAKSTAN
Kazakskiy Melkosopochnik
Kulundinskaya Step'
Betpak-Dala
Ozero Balkhash
Peski Taukum
Peski Muyunkum
Ozero Zaysan
Khrebet Tarbagatay
Altai Mts.
Tien Shan
Kunlun Shan
Taklimakan Shamo
Tarim Pendi
Kirghiz Range
Talas Ala-Too
Terskey Ala-Too
Kungei Ala-Too
Ysyk-Köl
Turkestan Range
Pamir
Petropavlovsk
Kokshetau
Akmola
Pavlodar
Ekibastuz
Semipalatinsk
Ust'-Kamenogorsk
Rubtsovsk
Biysk
Karaganda
Temirtau
Zhezkazgan
Balkhash
Taldykorgan
Almaty (Alma-Ata)
Bishkek (Frunze)
Zhambyl
Shymkent
Tashkent
Chirchik
Andizhan
Namangan
Osh
Fergana
Kokand
Khujand
Samarkand
Dushanbe
Kulob
Khorugh
Tacheng
Yining
Aksu
Kashi (Kashgar)
Shache
Yecheng
Hotan
1:7M
KM
MILES
350
300
250
200
150
100
50
0
200
150
100
50
0

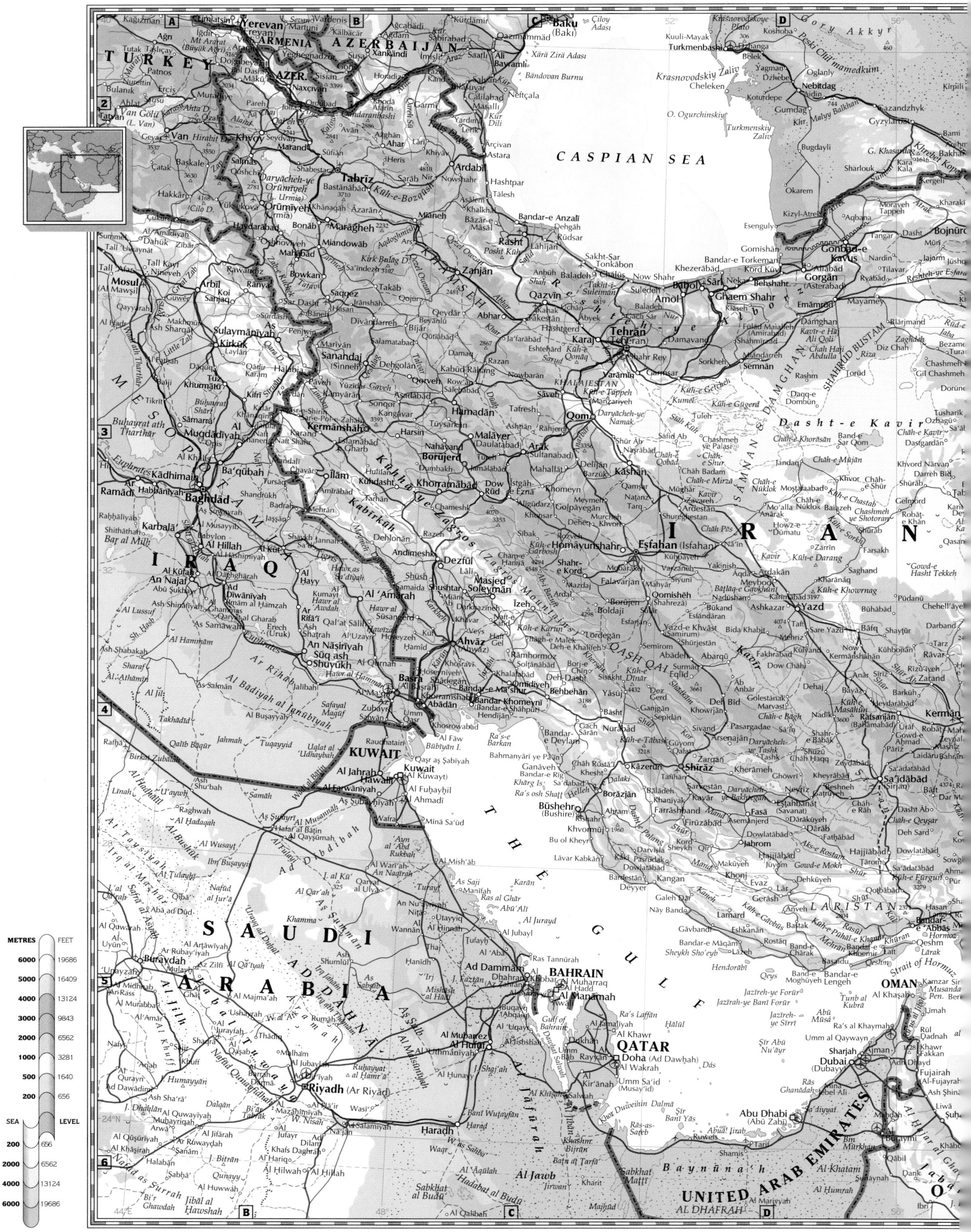

Conic Equidistant Projection

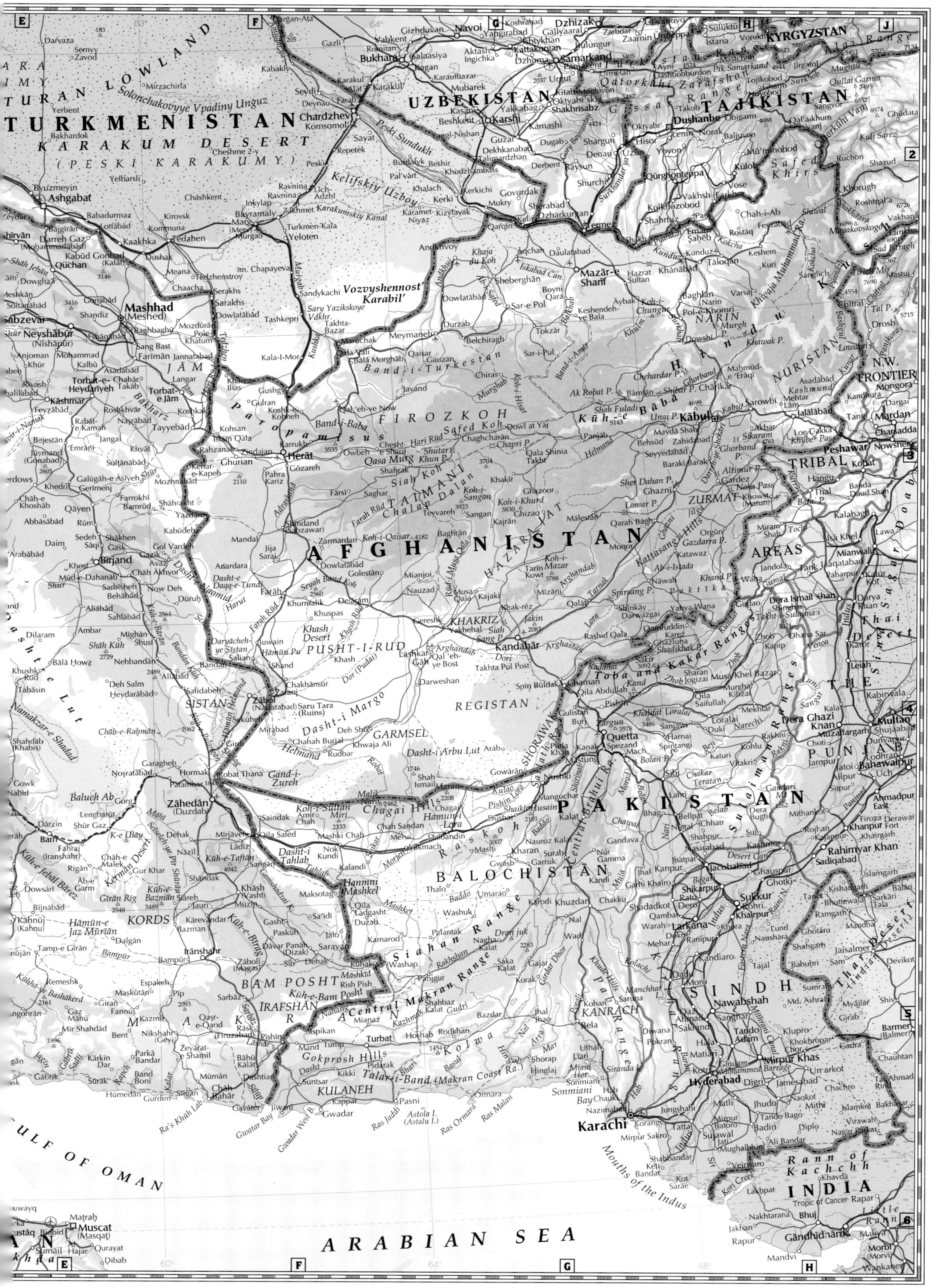

1:7M

KM 350 300 250 200 150 100 50 0

MILES 200 150 100 50 0

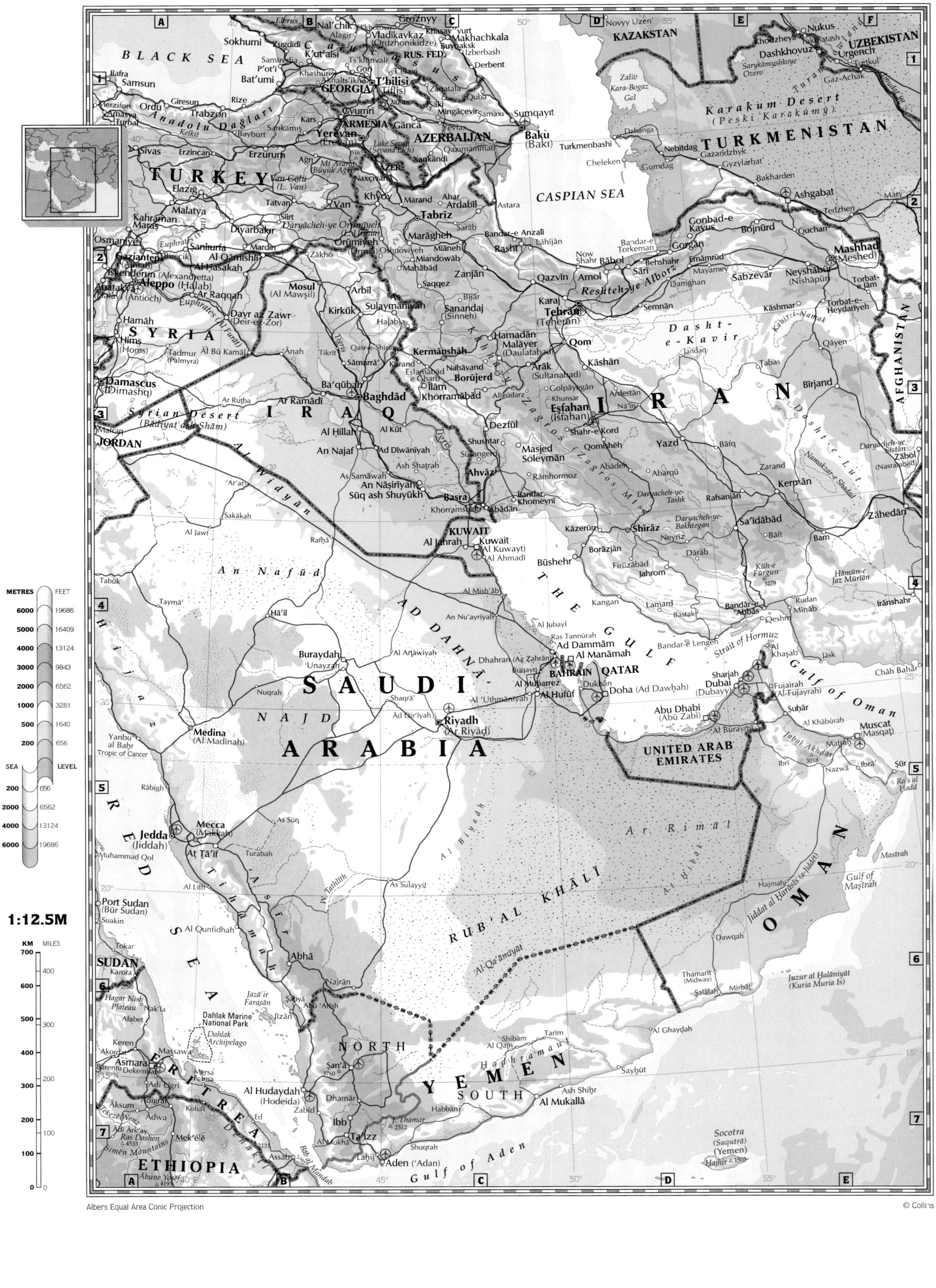

Albers Equal Area Conic Projection

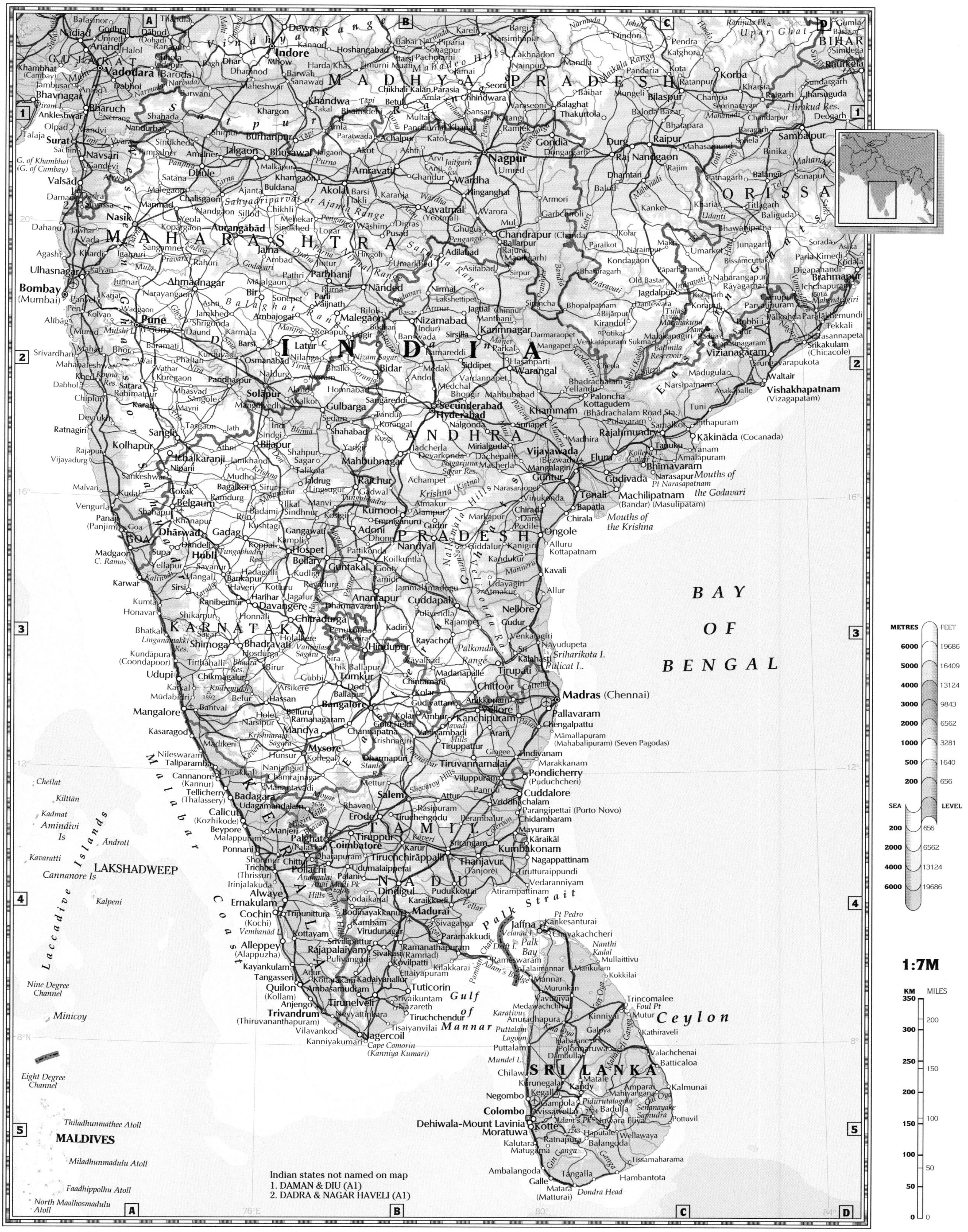
MADHYA PRADESH
MAHARASHTRA
GUJARAT
ORISSA
BIHAR
ANDHRA PRADESH
KARNATAKA
KERALA
TAMIL NADU
INDIA
GOA
LAKSHADWEEP
SRI LANKA
MALDIVES
BAY OF BENGAL
Ceylon
Gulf of Mannar
Palk Strait
Palk Bay
Laccadive Islands
Malabar Coast
Nine Degree Channel
Eight Degree Channel
Bombay (Mumbai)
Madras (Chennai)
Hyderabad
Secunderabad
Bangalore
Colombo
Nagpur
Indore
Pune
Surat
Vadodara
Ahmadnagar
Aurangābād
Solāpur
Hubli
Mysore
Coimbatore
Madurai
Trivandrum
(Thiruvananthapuram)
Cochin
Calicut
Mangalore
Vishakhapatnam
(Vizagapatam)
Vijayawada
Kākināda
Pondicherry
Jaffna
Kandy
Trincomalee
Tiruchchirāppalli
Salem
Raipur
Bilaspur
Korba
Sambalpur
Brahmapur
Gulbarga
Bijapur
Belgaum
Kolhapur
Nellore
Guntur
Warangal
Kurnool
Ongole
Tuticorin
Nagercoil
Indian states not named on map
1. DAMAN & DIU (A1)
2. DADRA & NAGAR HAVELI (A1)
METRES
FEET
6000
19686
5000
16409
4000
13124
3000
9843
2000
6562
1000
3281
500
1640
200
656
SEA
LEVEL
200
656
2000
6562
4000
13124
6000
19686
1:7M
KM
MILES
350
300
250
200
150
100
50
0
200
150
100
50
0

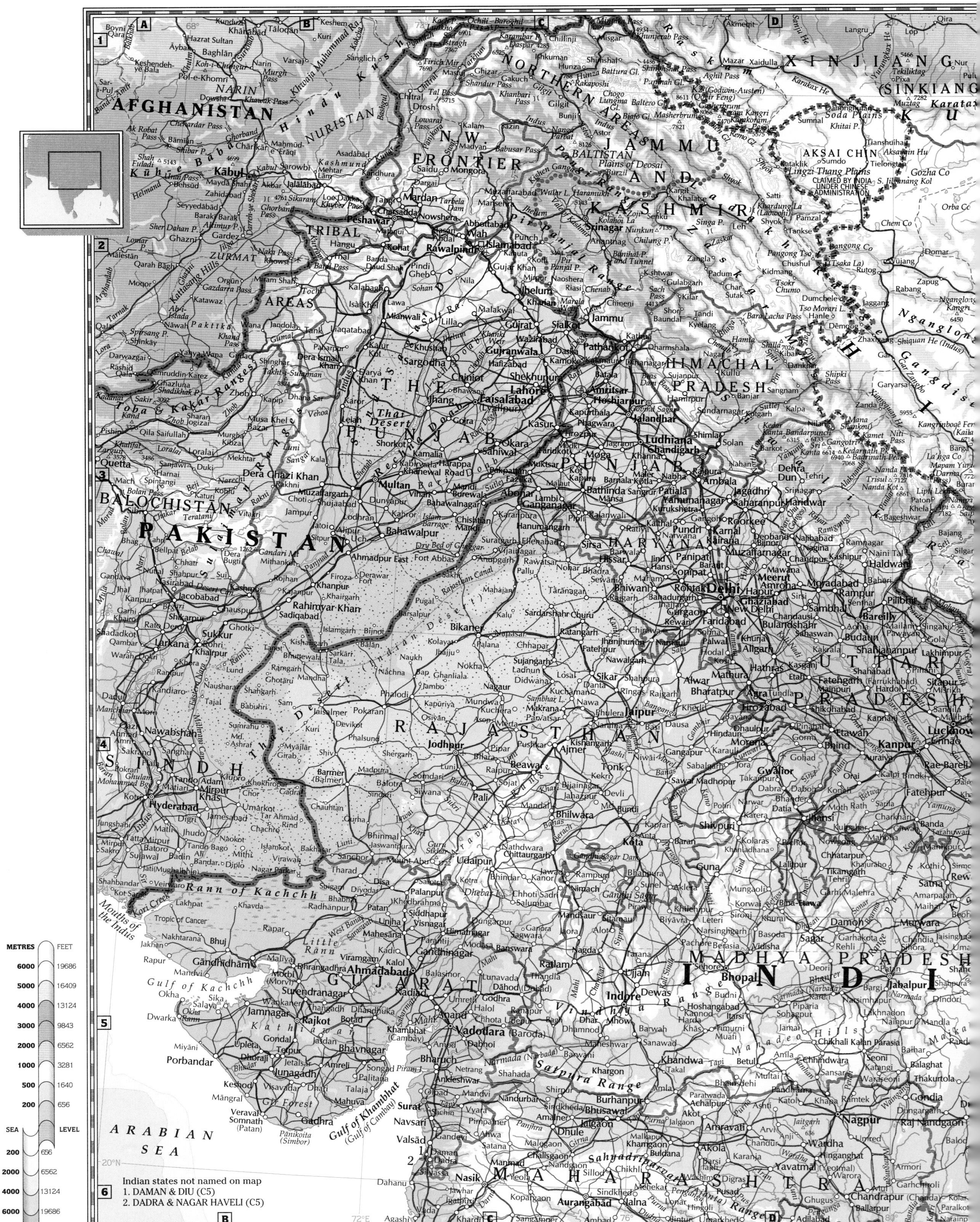

Conic Equidistant Projection

UYGUR ZIZHIQU
UIGHUR AUT. REGION
KUNLUN SHAN
Hoh Xil Shan
XIZANG GAOYUAN
(PLATEAU OF TIBET)
QINGHAI
Tanggula (Dangla) Shan
Tanintawweng Shan
CHINA
XIZANG ZIZHIQU
(TIBET AUT. REGION)
NYAINQÊNTANGLHA SHAN
Kailas Range
HIMALAYA
NEPAL
Kathmandu
Mt Everest
SIKKIM
BHUTAN
Thimphu
ARUNACHAL PRADESH
ASSAM
NAGALAND
MEGHALAYA
Shillong
MANIPUR
Imphal
MIZORAM
Aizawl
TRIPURA
Agartala
BANGLADESH
Dhaka (Dacca)
Chittagong
Khulna
Lhasa
Brahmaputra (Yarlung Zangbo)
BIHAR
Patna
Varanasi
Gorakhpur
Ranchi
Jamshedpur
WEST BENGAL
Calcutta
ORISSA
Cuttack
Bhubaneshwar
Puri
Mouths of the Ganges
BAY OF BENGAL
MYANMAR (BURMA)
Sittwe (Akyab)
1:7M
KM
MILES
© Collins

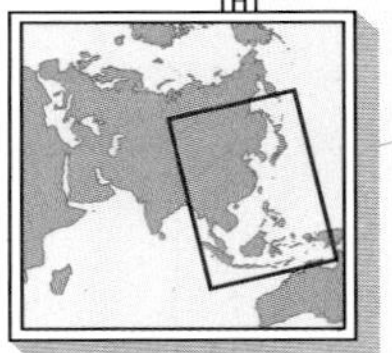

METRES	FEET
6000	19686
5000	16409
4000	13124
3000	9843
2000	6562
1000	3281
500	1640
200	656
SEA	LEVEL
200	656
2000	6562
4000	13124
6000	19686

Conic Equidistant Projection

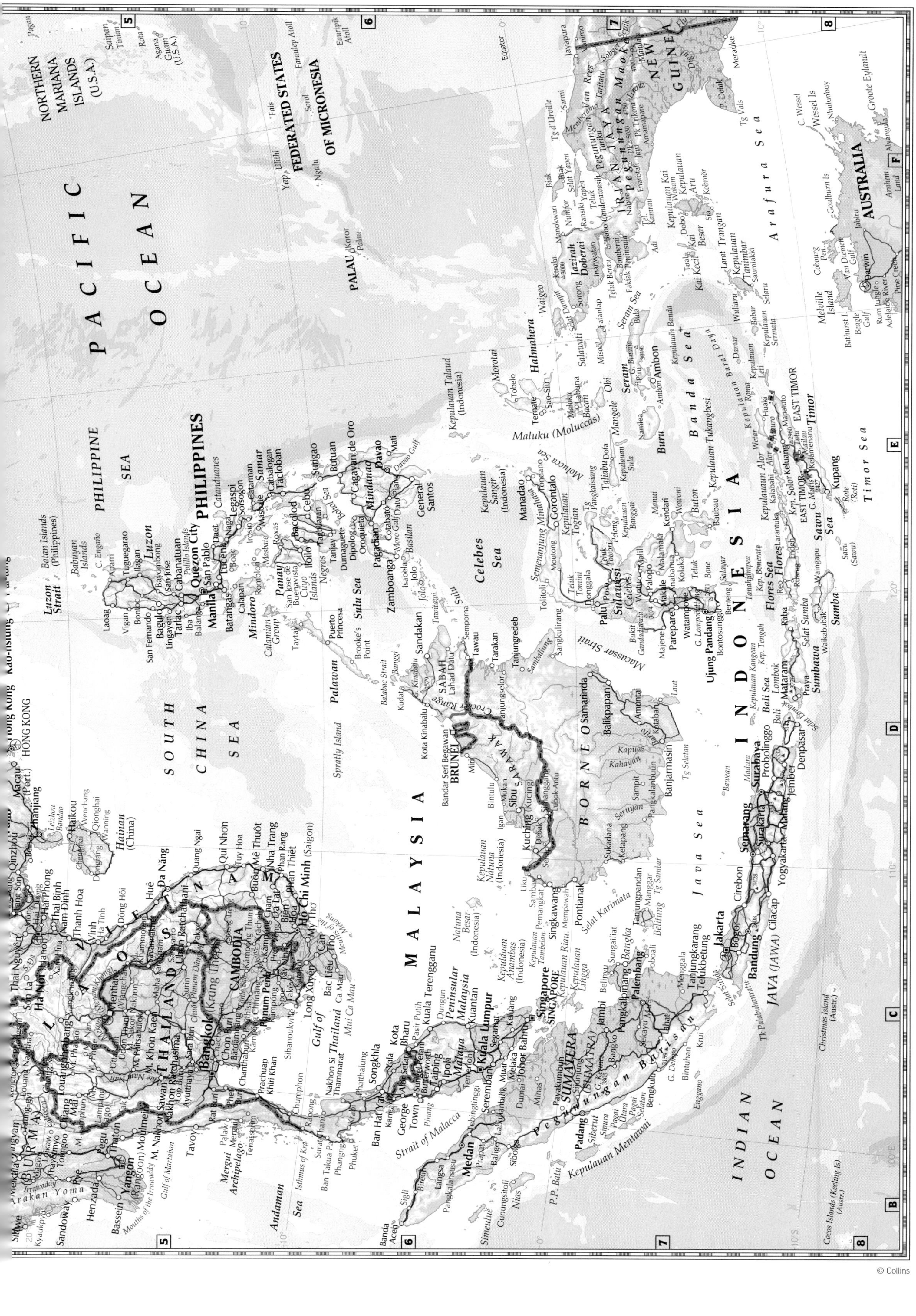

1:20M

KM 0 200 400 600 800 1000 1200

MILES 0 200 400 600

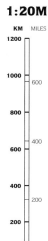

Albers Equal Area Conic Projection

Sea of Japan
Yellow Sea
(Huang Hai)
EAST CHINA SEA
SOUTH CHINA SEA
PACIFIC OCEAN
Bo Hai
Korea Bay
Korea Strait
Taiwan Strait
Luzon Strait
NORTH KOREA
SOUTH KOREA
JAPAN
PHILIPPINES
TAIWAN (FORMOSA)
MONGGOL ZIZHIQU (INNER MONGOLIAN AUT. REGION)
HEILONGJIANG
JILIN
LIAONING
HEBEI
SHANXI
SHANDONG
HENAN
JIANGSU
ANHUI
HUBEI
ZHEJIANG
JIANGXI
HUNAN
FUJIAN
GUANGDONG
HAINAN
Hainan (China)
Beijing (Peking)
Tianjin
Shanghai
Harbin
Changchun
Shenyang
Dalian
Qingdao (Tsingtao)
Nanjing (Nanking)
Wuhan
Hangzhou
Guangzhou (Canton)
Hong Kong
Macau (Port.)
Xiamen (Amoy)
Fuzhou
P'yŏngyang
Seoul (Sŏul)
Pusan
Taegu
T'ai-pei
Kao-hsiung
Tōkyō
Yokohama
Ōsaka
Nagoya
Kyōto
Kōbe
Hiroshima
Fukuoka
Sapporo
Sendai
Hokkaidō
Honshū
Shikoku
Kyūshū
Sakhalin
Vladivostok
Khabarovsk
Blagoveshchensk
Chita
Okinawa
Nansei-shotō (Japan)
Sakishima-Guntō (Japan)
Batan Islands (Philippines)
Babuyan Islands
Tropic of Cancer
K L M N O P Q R
2 3 4 5 6 7 8

1:15M

KM | MILES
900 | 600
750 | 450
600 | 300
450 | 150
300 |
150 |
0 | 0

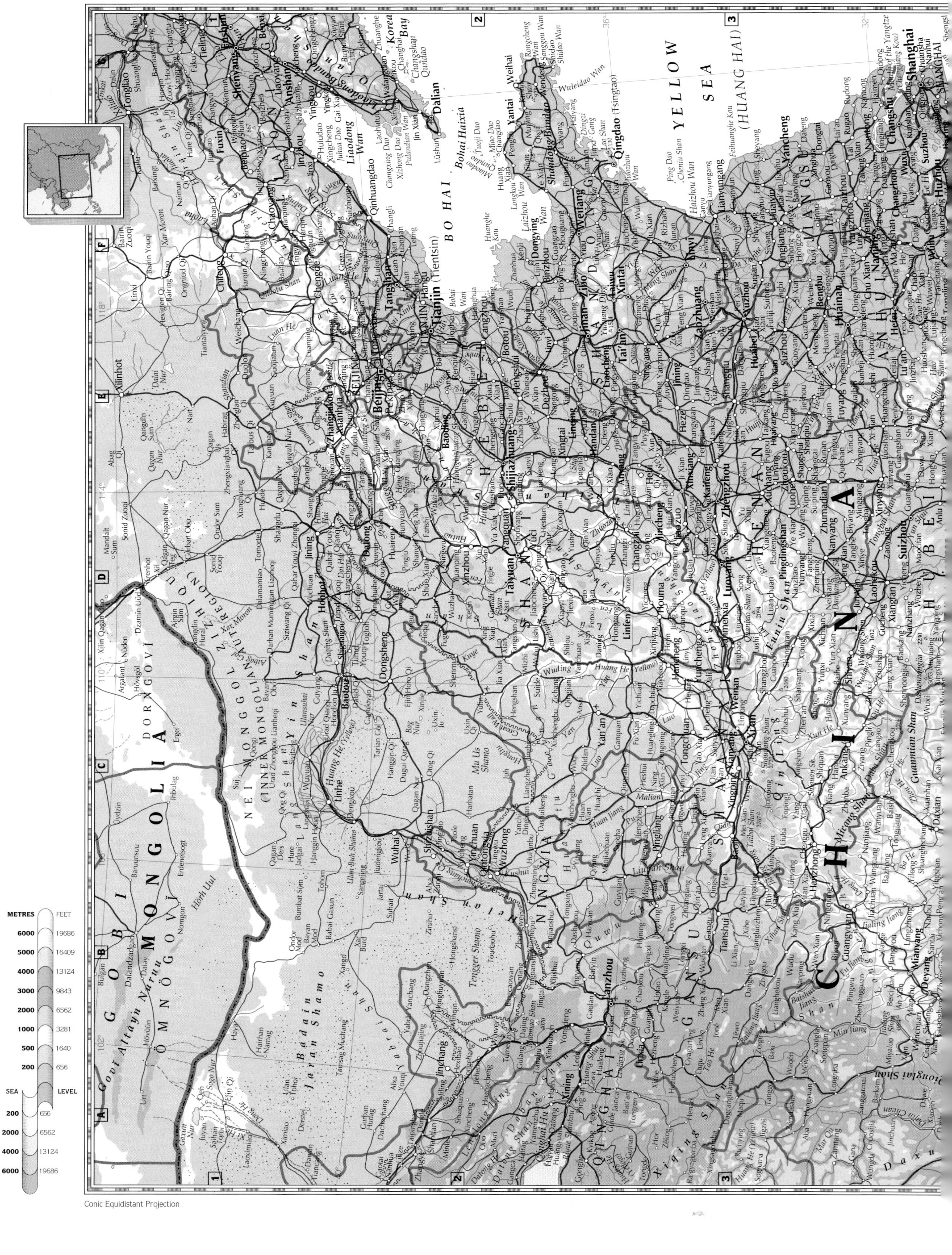

Conic Equidistant Projection

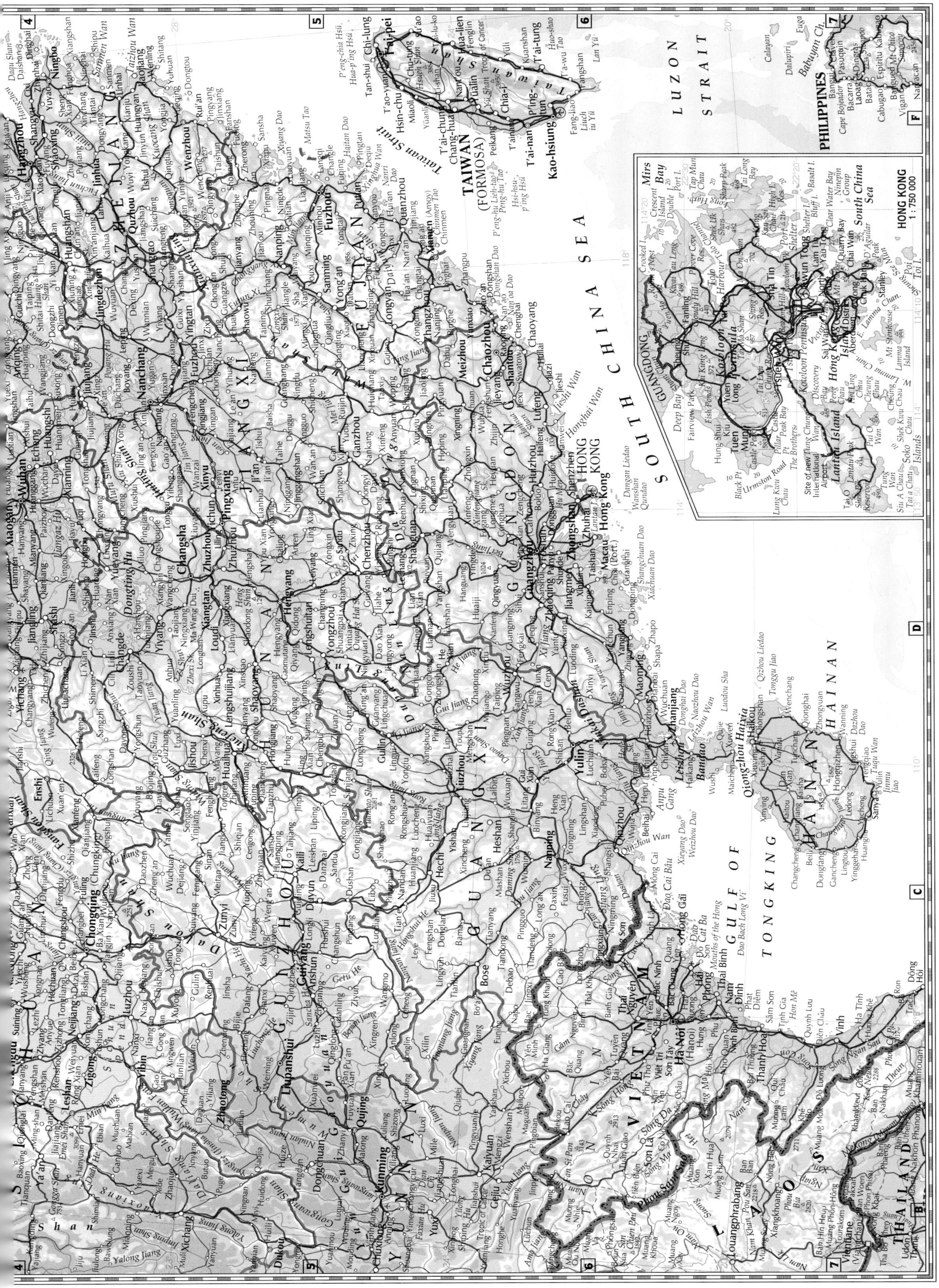
SOUTH CHINA SEA
LUZON STRAIT
GULF OF TONGKING
Taiwan Strait
TAIWAN (FORMOSA)
PHILIPPINES
HONG KONG
1 : 750 000
South China Sea
HAINAN
Qiongzhou Haixia
GUANGDONG
GUANGXI
FUJIAN
JIANGXI
HUNAN
GUIZHOU
YUNNAN
ZHEJIANG
VIETNAM
LAOS
THAILAND
Hong Kong
Macau
Guangzhou
Shenzhen
Xiamen
Fuzhou
Nanchang
Changsha
Wuhan
Chongqing (Chungking)
Guiyang
Kunming
Nanning
Hà Nôi
Haikou
Kao-hsiung
T'ai-pei

1:7.5M

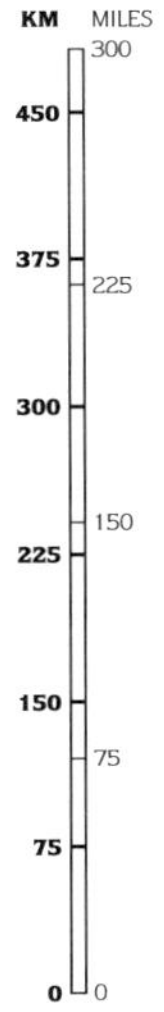
KM
MILES
450
375
300
225
150
75
0
300
225
150
75
0

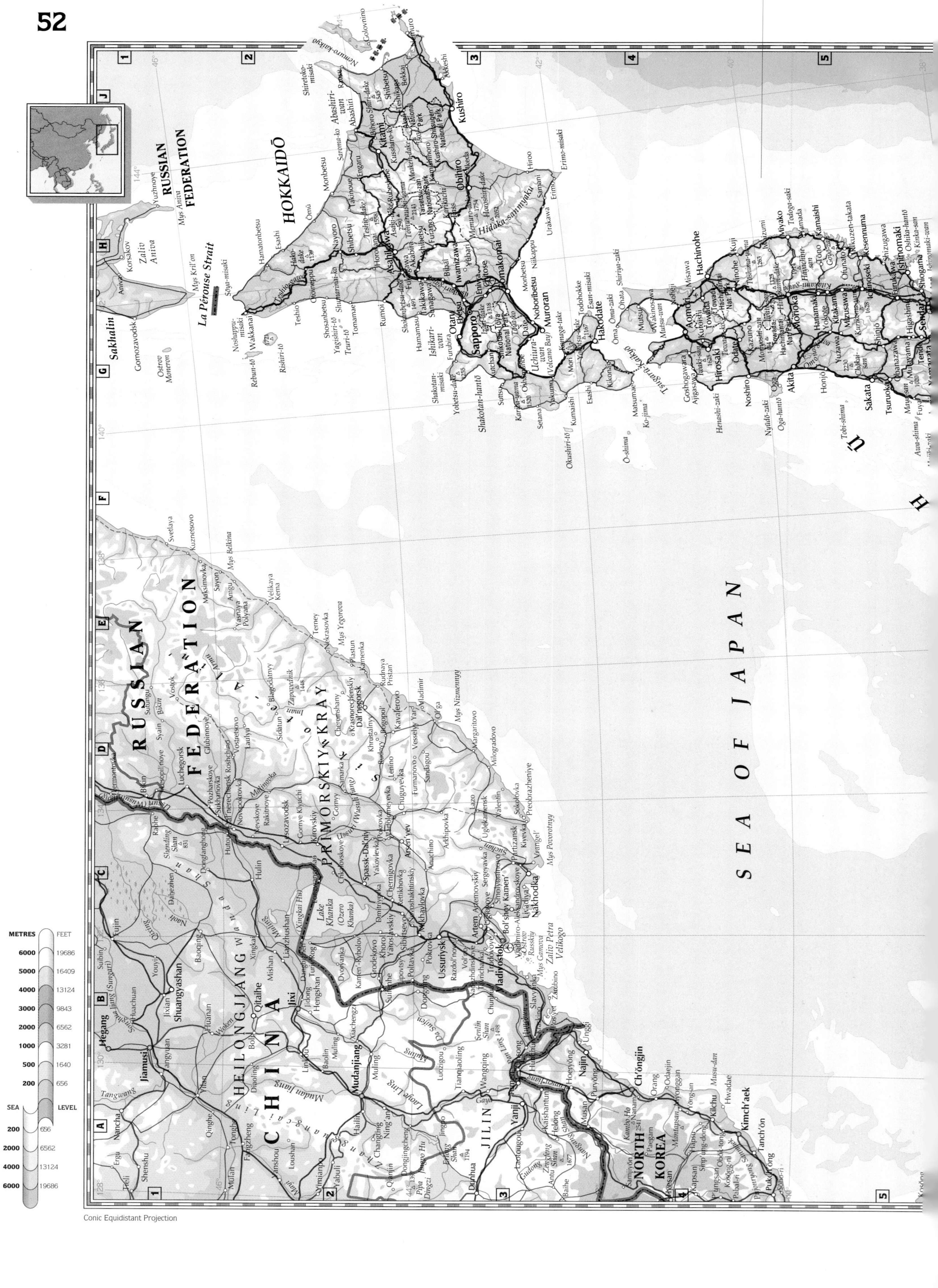

Conic Equidistant Projection

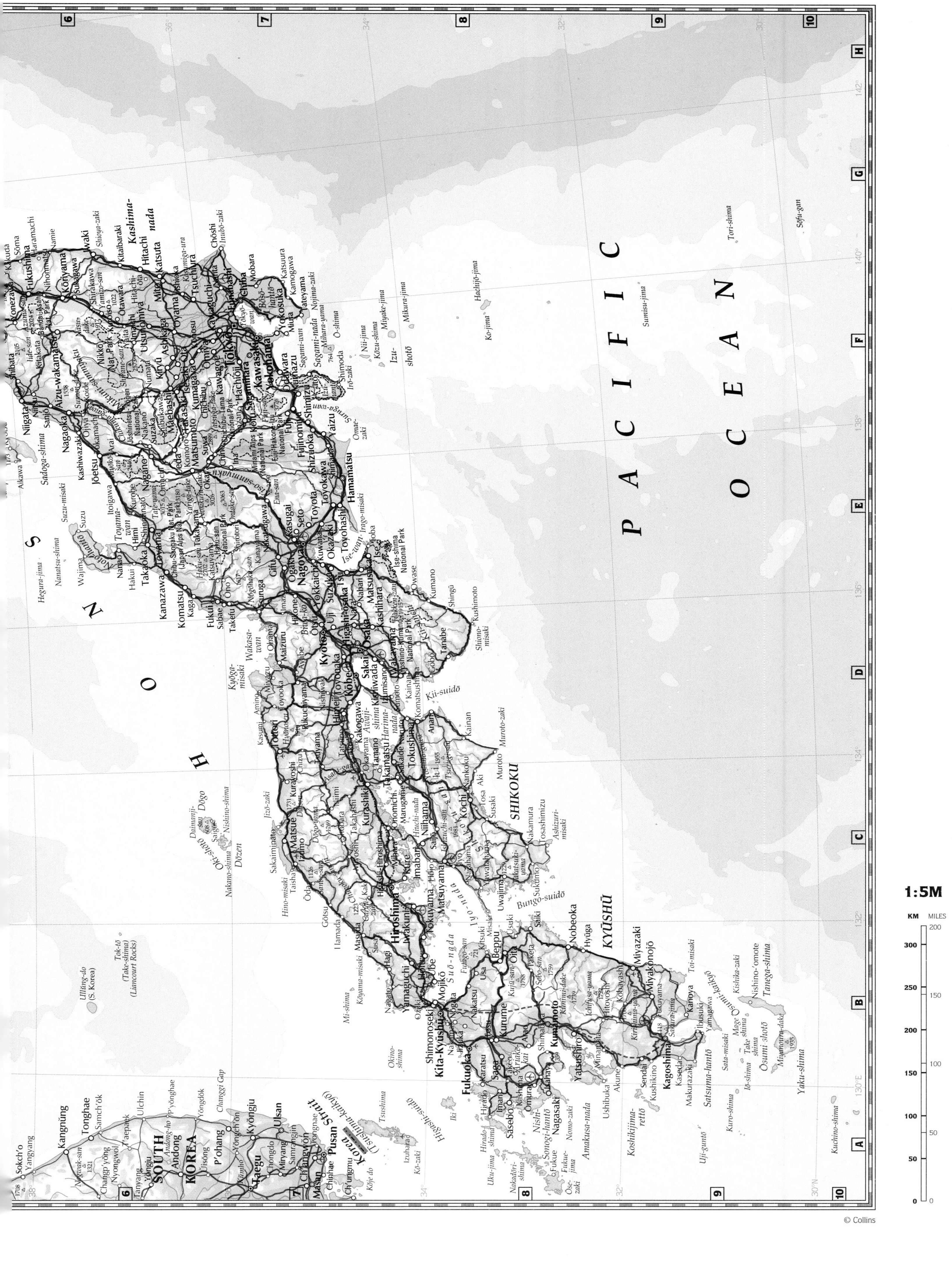
PACIFIC
OCEAN
SHIKOKU
KYŪSHŪ
SOUTH KOREA
Korea Strait
1:5M
KM
MILES
300
250
200
150
100
50
0
200
150
100
50
0

Northeastern CHINA, NORTH KOREA and SOUTH KOREA

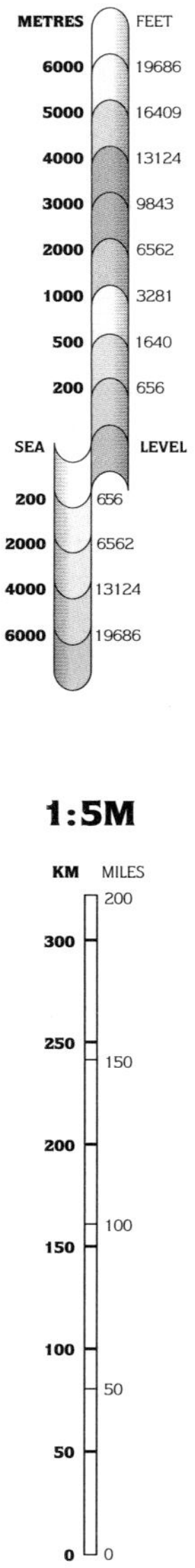

Conic Equidistant Projection

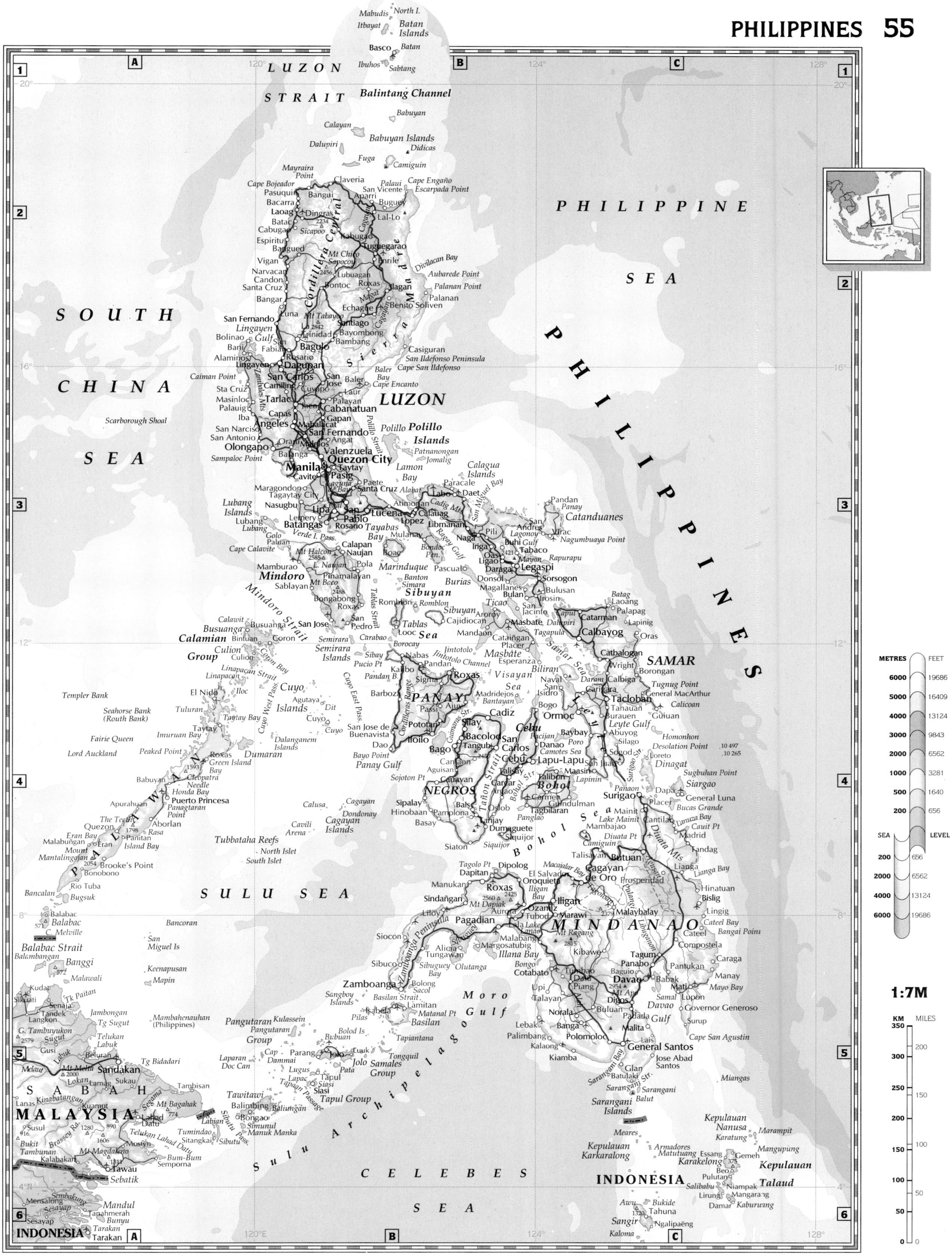
LUZON STRAIT
Balintang Channel
Batan Islands
Mabudis
North I.
Itbayat
Basco
Batan
Ibuhos
Sabtang
Babuyan
Calayan
Babuyan Islands
Dalupiri
Didicas
Fuga
Camiguin
Mayraira Point
Cape Bojeador
Claveria
Palaui
Cape Engaño
Escarpada Point
San Vicente
Aparri
Pasuquin
Bangui
Buguey
Bacarra
Laoag
Dingras
Lal-Lo
Batac
Cabugao
Sicapoo
Espiritu
Kabugao
Tuguegarao
Bangued
Mt Chico
Sapocoy
Enrile
Vigan
Narvacan
Candon
Santa Cruz
Cordillera Central
Bontoc
Lubuagan
Roxas
Ilagan
Sierra Madre
Divilacan Bay
Aubarede Point
Palanan Point
Palanan
Benito Soliven
Echague
Bangar
Luna
San Fernando
Mt Tabayoc
Santiago
Lingayen Gulf
Bolinao
Bani
Alaminos
Trinidad
Bayombong
Bambang
Baguio
San Fabian
Rosario
Casiguran
San Ildefonso Peninsula
Cape San Ildefonso
Lingayen
Dagupan
San Carlos
Camiling
San Jose
Baler
Baler Bay
Cape Encanto
Caiman Point
Sta Cruz
Masinloc
Palauig
Zambales Mts
Cuyapo
Laur
Tarlac
Capas
Jaen
Cabanatuan
LUZON
Palayan
Gapan
Angeles
Mabalacat
Iba
San Narciso
San Antonio
Olongapo
Orani
San Fernando
Angat
Polillo Strait
Polillo
Polillo Islands
Patnanongan
Jomalig
Malolos
Valenzuela
Quezon City
Balanga
Manila
Cavite
Taytay
Pasig
Laguna de Bay
Lamon Bay
Sampaloc Point
Scarborough Shoal
SOUTH CHINA SEA
PHILIPPINE SEA
PHILIPPINES
Maragondon
Tagaytay City
Paete
Santa Cruz
Alabat
Calagua Islands
Paracale
Labo
Daet
San Miguel Bay
Atimonan
Lubang Islands
Lubang
Nasugbu
Lemery
Lipa
San Pablo
Lucena
Lopez
Calauag
Cadig Mts
Pandan
Panay
Catanduanes
Batangas
Rosario
Tayabas Bay
Libmanan
Naga
Pili
San Andres
Lagonoy
Virac
Nagumbuaya Point
Golo
Verde I. Pass.
Mulanay
Ragay Gulf
Bondoc Pen.
Iriga
Buhi
Lagonoy Gulf
Tabaco
Cape Calavite
Paluan
Calapan
Naujan
Boac
Oas
Ligao
Mayon
Rapurapu
Mt Halcon
L. Naujan
Pola
Daraga
Legaspi
Mamburao
Mindoro
Sablayan
Mt Baco
Pinamalayan
Marinduque
Pascual
Burias
Donsol
Sorsogon
Banton
Simara
Magallanes
Bulan
Bulusan
Irosin
Bongabong
Roxas
Sibuyan
Romblon
Tablas Strait
Ticao
Batag
Laoang
Palapag
Mindoro Strait
Calawit
Busuanga
Calamian Group
Coron
Binluan
Culion
Linapacan Strait
Linapacan
San Jose
San Pedro
Semirara
Semirara Islands
Carabao
Tablas
Looc
Sibuyan Sea
Cajidiocan
Aroroy
Mandaon
Masbate
San Jacinto
Capul
Dalupiri
Catarman
Calbayog
Lapinig
Oras
Boracay
Jintotolo
Jintotolo Channel
Cataingan
Placer
Esperanza
Tagapula
Samar Sea
Catbalogan
Wright
Borongan
SAMAR
Masbate
Cuyo East Pass.
Sibay
Pucio Pt
Nabas
Pandan
Kalibo
Pandan B.
Sigma
Roxas
Visayan Sea
Biliran
Naval
Daram
Calbiga
Tugnug Point
El Nido
Iloc
Cuyo West Pass.
Cuyo
Cuyo Islands
Agutaya
Dit
Barboza
PANAY
Madridejos
Bantayan
San Isidro
Caricara
Tacloban
General MacArthur
Calicoan
Templer Bank
Seahorse Bank (Routh Bank)
Tuluran
Taytay Bay
Taytay
Cuyo
Passi
Ajuy
Cadiz
Bogo
Ormoc
Tanauan
Burauen
Guiuan
Leyte Gulf
Fairie Queen
Imuruan Bay
Dalanganem Islands
Cordillera Range
Guimaras Str.
Potolan
San Jose de Buenavista
Silay
Bacolod
Iloilo
Cebu
Pacijan
Baybay
Abuyog
Homonhon
Desolation Point
Lord Auckland
Peaked Point
Roxas
Green Island Bay
Dumaran
Dao
Bago
Tangub
San Carlos
Camotes Sea
Poro
Danao
Lapu-Lapu
Silago
Sogod
Loreto
Dinagat
Cleopatra Needle
Bayo Point
Panay Gulf
Cebu
Tañon Strait
Talisay
Maasin
San Juan
Surigao Str.
Sugbuhan Point
Siargao
Babuyan
Honda Bay
PALAWAN
Puerto Princesa
Panagtaran Point
Sojoton Pt
Candoni
Aguisan
Cauayan
NEGROS
Carcar
Argao
Bohol Str.
Talibon
Bohol
Lapinin
Panaon
Dapa
Surigao
Placer
General Luna
Bucas Grande
Apurahuan
Calusa
Cagayan
Dondonay
Cagayan Islands
Sipalay
Hinobaan
Bais
Pamplona
Carmen
Guindulman
Tagbilaran
Mainit
Lake Mainit
Cantilan
Lanuza Bay
Cauit Pt
The Teeth
Quezon
Aborlan
Rasa
Cavili
Arena
Basay
Dalob
Tanjay
Panglao
Dumaguete
Siquijor
Mambajao
Madrid
Eran Bay
Malabuñgan
Panitan
Eran
Island Bay
Tubbataha Reefs
North Islet
South Islet
Bohol Sea
Diuata Pt
Camiguin
Diuata Mts
Mount Mantalingajan
Brooke's Point
Bonobono
Siaton
Siquijor
Tandag
Talisayan
Butuan
Tagolo Pt
Dapitan
Dipolog
El Salvador
Macajalar Bay
Cagayan de Oro
Lianga
Lianga Bay
Bancalan
Rio Tuba
Bugsuk
SULU SEA
Manukan
Roxas
Oroquieta
Iligan Bay
Prosperidad
Hinatuan
Bislig
Sindañgan
Mt Dapiak
Aurora
Ozamiz
Iligan
Marawi
Malaybalay
Lingig
Balabac
Melville
Bancoran
Liloy
Pagadian
Lake Lanao
Tubod
MINDANAO
Mt Ragang
Cateel Bay
Bangai Point
Cateel
Balabac Strait
Balambangan
Banggi
San Miguel Is
Siocon
Zamboanga Peninsula
Sibuguey
Aliciag
Tungawan
Margosatubig
Malabang
Illana Bay
Kibawe
Compostela
Tagum
Panabo
Pantukan
Caraga
Malawali
Keenapusan
Mapin
Sibuco
Sibuguey Bay
Olutanga
Bongo
Cotabato
Baguio
Davao
Babak
Mati
Manay
Mayo Bay
Kudat
Sikuati
Tk Paitan
Tandek
Senaja
Langkon
Jambongan
Tg Sugut
Mambahenauhan (Philippines)
Zamboanga
Bolong
Sacol
Sangboy Islands
Basilan Strait
Isabela
Lamitan
Matanal Pt
Basilan
Moro Gulf
Upi
Talayan
Tumbao
Dalo
Piang
Digos
Mt Apo
Samal
Davao Gulf
Lupon
Governor Generoso
Surup
G. Tambuyukon
Sugut
Telukan Labuk
Gusi
Labuk
Beluran
Pangutaran Group
Pangutaran
Kulassein
Bolod Is
Bubuan
Tapiantana
Lebak
Palimbang
Kalaong
Norala
Banga
Polomolok
Buluan
Malita
Lais
Cape San Agustin
Melaut
Mt Meliau
Sandakan
Tg Bidadari
Laparan
Doc Can
Cap
Dammai
Parang
Jolo
Luuk
Tongquil
Samales Group
Kiamba
General Santos
Jose Abad Santos
Glan
Batulaki
Sarangani Str.
Sarangani
Balut
Sarangani Bay
Miangas
SABAH
Lokan
Lamag
Sukau
Tambisan
Lanas
Kinabatangan
Kuamut
Tawitawi
Lugus
Lapac
Siasi
Tapul
Tapaan Passage
Tapul Group
Sulu Archipelago
Sarangani Islands
Kepulauan Nanusa
Karatung
Marampit
MALAYSIA
Segama
Mt Bagahak
Lahad Datu
Sibutu Passage
Labian
Bongao
Balimbing
Bilatan
Simunul
Manuk Manka
Sibutu
Tumindao
Sitangkai
Susul
Bukit Tambunan
Brassey Ra.
Telukan Lahad Datu
Mostyn
Bum-Bum
Semporna
Kalabakan
Mt Magdalena
Tawau
Sebatik
CELEBES SEA
Meares
Kepulauan Karkaralong
Armadores
Matutuang
Essang
Karakelong
Gemeh
Mangupung
Beo
Kepulauan Talaud
Pulutan
Salibabu
Niampak
Lirung
Mangarang
Damar
Kaburuang
INDONESIA
Mensalong
Sembakung
Sesayap
Mandul
Tanahmerah
Bunyu
Tarakan
Awu
Bukide
Tahuna
Sangir
Ngalipaëng
Kaloma
INDONESIA
120°E
124°
128°
20°
16°
12°
8°
4°N
A
B
C
1
2
3
4
5
6
METRES
FEET
6000
19686
5000
16409
4000
13124
3000
9843
2000
6562
1000
3281
500
1640
200
656
SEA
LEVEL
200
656
2000
6562
4000
13124
6000
19686
1:7M
KM
MILES
350
300
250
200
150
100
50
0

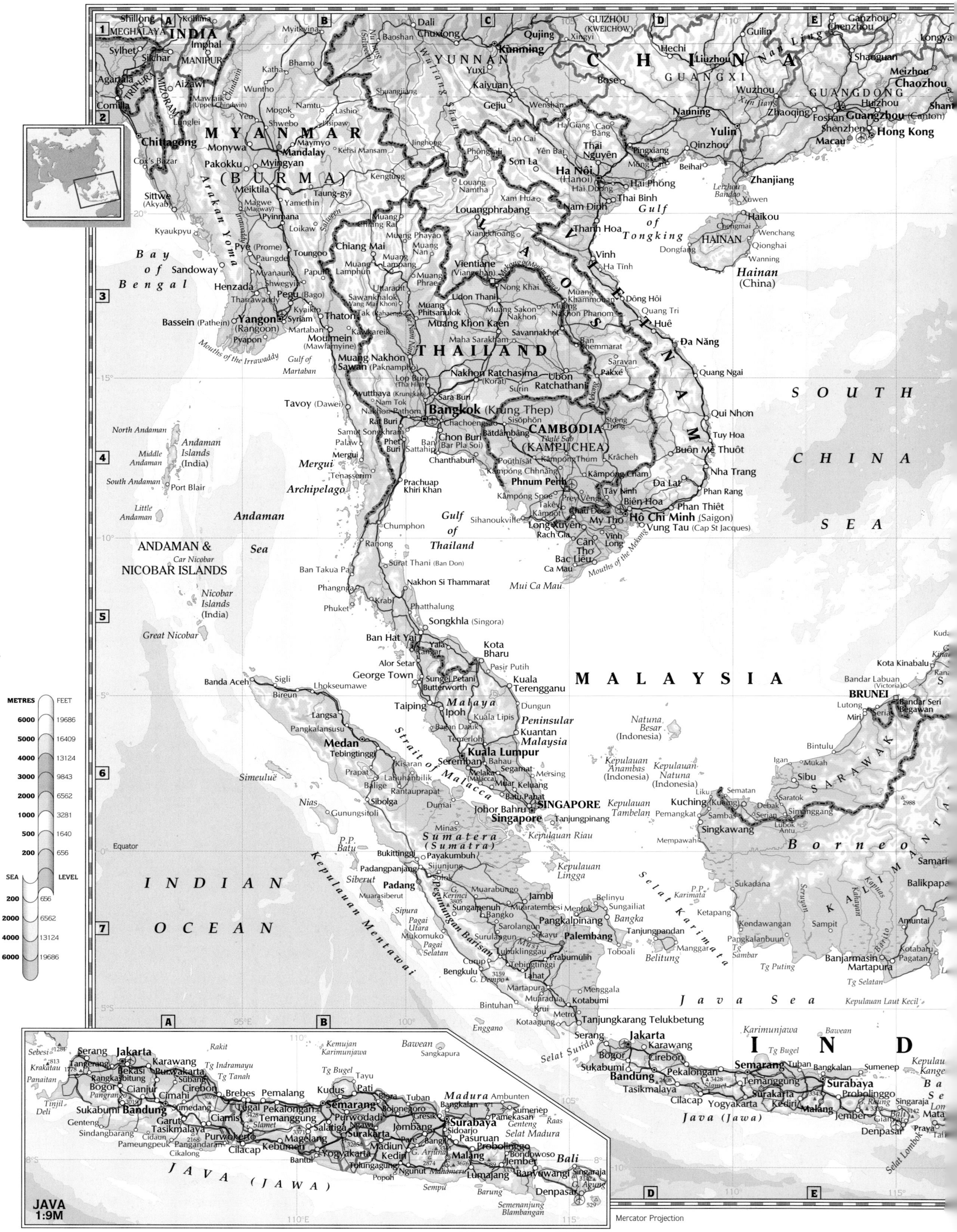
INDIA
MEGHALAYA
MANIPUR
TRIPURA
MIZORAM
Shillong
Kohima
Imphal
Sylhet
Silchar
Agartala
Aizawl
Comilla
Chittagong
Cox's Bazar
Sittwe (Akyab)
Kyaukpyu
Sandoway
Bay of Bengal
MYANMAR (BURMA)
Myitkyina
Bhamo
Katha
Wuntho
Mawlaik (Upper Chindwin)
Mogok
Namtu
Lashio
Hsipaw
Yeu
Shwebo
Monywa
Maymyo
Mandalay
Pakokku
Myingyan
Meiktila
Taung-gyi
Magwe (Majway)
Yamethin
Pyinmana
Loikaw
Arakan Yoma
Pye (Prome)
Paungde
Toungoo
Myanaung
Shwegyin
Henzada
Tharrawaddy
Pegu (Bago)
Kyaikto
Syriam
Bassein (Pathein)
Yangon (Rangoon)
Pyapon
Thaton
Martaban
Moulmein (Mawlamyine)
Kawkareik
Mouths of the Irrawaddy
Gulf of Martaban
Tavoy (Dawei)
Mergui
Tenasserim
Mergui Archipelago
North Andaman
Middle Andaman
South Andaman
Andaman Islands (India)
Port Blair
Little Andaman
Andaman Sea
ANDAMAN & NICOBAR ISLANDS
Car Nicobar
Nicobar Islands (India)
Great Nicobar
INDIAN OCEAN
Dali
Chuxiong
Baoshan
YUNNAN
Kunming
Qujing
Yuxi
Kaiyuan
Gejiu
Wenshan
Shuangjiang
Jinghong
Kengtung
GUIZHOU (KWEICHOW)
Xingyi
CHINA
GUANGXI
Hechi
Liuzhou
Bose
Nanning
Wuzhou
Guilin
Yulin
Qinzhou
Beihai
Zhanjiang
GUANGDONG
Ganzhou
Shaoguan
Meizhou
Chaozhou
Huizhou
Zhaoqing
Foshan
Guangzhou (Canton)
Shenzhen
Hong Kong
Macau
Haikou
HAINAN
Hainan (China)
Wenchang
Qionghai
Wanning
Dongfang
Gulf of Tongking
Lao Cai
Ha Giang
Cao Bang
Yen Bai
Thai Nguyen
Son La
Ha Noi (Hanoi)
Hai Duong
Hai Phong
Thai Binh
Nam Dinh
Thanh Hoa
Vinh
Ha Tinh
Dong Hoi
Quang Tri
Hue
Da Nang
Quang Ngai
Qui Nhon
Tuy Hoa
Buon Me Thuot
Nha Trang
Phan Rang
Da Lat
Phan Thiet
Bien Hoa
Ho Chi Minh (Saigon)
Vung Tau (Cap St Jacques)
My Tho
Vinh Long
Can Tho
Bac Lieu
Ca Mau
Mui Ca Mau
Mouths of the Mekong
VIETNAM
LAOS
Phongsali
Louang Namtha
Louangphrabang
Xam Hua
Xiangkhoang
Vientiane (Viangchan)
Nong Khai
Savannakhet
Saravan
Pakxe
THAILAND
Chiang Rai
Chiang Mai
Lamphun
Muang Nan
Muang Phrae
Uttaradit
Sawankhalok
Tak
Muang Phitsanulok
Muang Khon Kaen
Muang Sakon Nakhon
Nakhon Phanom
Maha Sarakham
Nakhon Sawan
Lop Buri
Nakhon Ratchasima (Korat)
Ubon Ratchathani
Surin
Sara Buri
Ayutthaya
Bangkok (Krung Thep)
Nakhon Pathom
Rat Buri
Chachoengsao
Samut Songkhram
Phet Buri
Chon Buri
Chanthaburi
Prachuap Khiri Khan
Chumphon
Ranong
Surat Thani (Ban Don)
Nakhon Si Thammarat
Phatthalung
Songkhla (Singora)
Ban Hat Yai
Phuket
Phangnga
Krabi
Ban Takua Pa
Gulf of Thailand
CAMBODIA (KAMPUCHEA)
Sisophon
Batdambang
Tonle Sap
Pouthisat
Kampong Thum
Kracheh
Kampong Chhnang
Kampong Cham
Phnum Penh
Kampong Spoe
Takev
Kampot
Sihanoukville
Chau Doc
Long Xuyen
Rach Gia
MALAYSIA
Kota Bharu
Pasir Puteh
Alor Setar
George Town
Sungei Petani
Butterworth
Kuala Terengganu
Dungun
Taiping
Ipoh
Kuala Lipis
Peninsular Malaysia
Kuantan
Bagan Datuk
Temerloh
Kuala Lumpur
Seremban
Bahau
Segamat
Mersing
Melaka (Malacca)
Muar
Keluang
Batu Pahat
Johor Bahru
SINGAPORE
Singapore
Strait of Malacca
Tanjungpinang
Kepulauan Riau
Natuna Besar (Indonesia)
Kepulauan Anambas (Indonesia)
Kepulauan Natuna (Indonesia)
Kepulauan Tambelan
SOUTH CHINA SEA
Kota Kinabalu
Bandar Labuan (Victoria)
BRUNEI
Bandar Seri Begawan
Lutong
Miri
Bintulu
Sibu
SARAWAK
Kuching
Sematan
Pemangkat
Sambas
Singkawang
Mempawah
Borneo
KALIMANTAN
Sukadana
Ketapang
Kendawangan
Sampit
Pangkalanbuun
Tg Puting
Banjarmasin
Martapura
Balikpapan
Samarinda
Amuntai
Kotabaru
Pagatan
Tg Selatan
Kepulauan Laut Kecil
Banda Aceh
Sigli
Bireun
Lhokseumawe
Langsa
Pangkalansusu
Medan
Tebingtinggi
Kisaran
Prapat
Labuhanbilik
Balige
Rantauprapat
Sibolga
Simeulue
Nias
Gunungsitoli
Dumai
P.P. Batu
Bukittinggi
Payakumbuh
Padangpanjang
Sijunjung
Siberut
Padang
Muarasiberut
Sumatera (Sumatra)
Kepulauan Mentawai
Sipura
Pagai Utara
Mukomuko
Pagai Selatan
Pegunungan Barisan
Muarabungo
Jambi
Sungaipenuh
Muaratembesi
Bangko
Sarolangun
Surulangun
Sekayu
Pangkalpinang
Bangka
Palembang
Lubuklinggau
Prabumulih
Tebingtinggi
Lahat
Bengkulu
Curup
G. Dempo
Martapura
Muaradua
Menggala
Kotabumi
Bintuhan
Krui
Metro
Kotaagung
Tanjungkarang Telukbetung
Enggano
Kepulauan Lingga
Belitung
Tanjungpandan
Manggar
Selat Karimata
Selat Sunda
Java Sea
Serang
Jakarta
Karawang
Bogor
Cirebon
Sukabumi
Bandung
Pekalongan
Tasikmalaya
Semarang
Tuban
Bangkalan
Sumenep
Surabaya
Probolinggo
Surakarta
Kediri
Malang
Jember
Yogyakarta
Cilacap
Java (Jawa)
Karimunjawa
Bawean
Tg Bugel
INDONESIA
Denpasar
Singaraja
Selat Lombok
Equator
METRES FEET
6000 19686
5000 16409
4000 13124
3000 9843
2000 6562
1000 3281
500 1640
200 656
SEA LEVEL
200 656
2000 6562
4000 13124
6000 19686
JAVA 1:9M
Sebesi
Krakatau
Panaitan
Tanjung Lesung
Rangkasbitung
Bekasi
Purwakarta
Subang
Tg Indramayu
Tg Tanah
Cianjur
Cimahi
Sumedang
Garut
Ciamis
Sindangbarang
Cidaun
Pameungpeuk
Pangandaran
Cikalong
Brebes
Tegal
Slamet
Pemalang
Purwokerto
Kebumen
Magelang
Salatiga
Kudus
Pati
Purwodadi
Blora
Bojonegoro
Ngawi
Madiun
Jombang
Gresik
Sidoarjo
Pasuruan
Madura
Ambunten
Pamekasan
Genteng
Raas
Selat Madura
Bondowoso
Lumajang
Tulungagung
Blitar
Ngunut
Popoh
Sempu
Banyuwangi
Bali
Barung
Semenanjung Blambangan
Mercator Projection

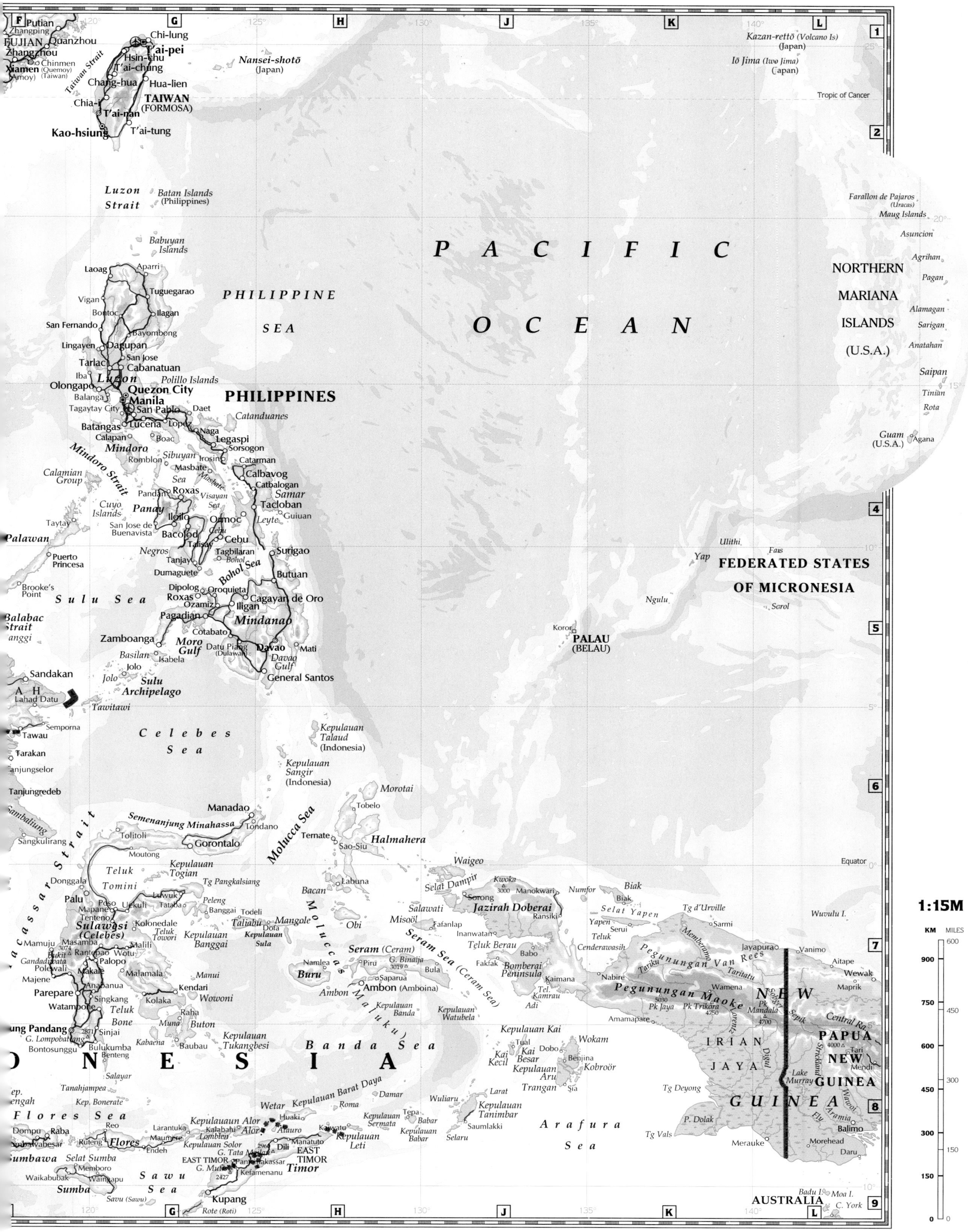

PACIFIC OCEAN
PHILIPPINE SEA
PHILIPPINES
NORTHERN MARIANA ISLANDS (U.S.A.)
FEDERATED STATES OF MICRONESIA
PALAU (BELAU)
TAIWAN (FORMOSA)
FUJIAN
Putian
Zhangping
Quanzhou
Zhangzhou
Chinmen (Quemoy) (Taiwan)
Xiamen (Amoy)
Taiwan Strait
Chi-lung
T'ai-pei
Hsin-chu
T'ai-chung
Chang-hua
Hua-lien
Chia-i
T'ai-nan
Kao-hsiung
T'ai-tung
Nansei-shotō (Japan)
Kazan-rettō (Volcano Is) (Japan)
Iō Jima (Iwo Jima) (Japan)
Tropic of Cancer
Luzon Strait
Batan Islands (Philippines)
Babuyan Islands
Farallon de Pajaros (Uracas)
Maug Islands
Asuncion
Agrihan
Pagan
Alamagan
Sarigan
Anatahan
Saipan
Tinian
Rota
Guam (U.S.A.)
Agana
Laoag
Aparri
Vigan
Tuguegarao
Bontoc
Ilagan
San Fernando
Bayombong
Lingayen
Dagupan
San Jose
Tarlac
Cabanatuan
Iba
Luzon
Polillo Islands
Olongapo
Quezon City
Balanga
Manila
Tagaytay City
San Pablo
Daet
Batangas
Lucena
Lopez
Catanduanes
Calapan
Naga
Legaspi
Boac
Sorsogon
Mindoro
Romblon
Sibuyan
Irosin
Catarman
Mindoro Strait
Masbate
Calamian Group
Sea
Calbayog
Catbalogan
Samar
Cuyo Islands
Pandan
Roxas
Visayan Sea
Panay
Tacloban
Iloilo
Ormoc
Leyte
Guiuan
Taytay
San Jose de Buenavista
Bacolod
Cebu
Palawan
Talisay
Negros
Tagbilaran
Surigao
Bohol
Puerto Princesa
Tanjay
Bohol Sea
Dumaguete
Butuan
Brooke's Point
Dipolog
Oroquieta
Sulu Sea
Roxas
Cagayan de Oro
Ozamiz
Iligan
Balabac Strait
Pagadian
Mindanao
Cotabato
Zamboanga
Moro Gulf
Mati
Datu Piang (Dulawan)
Davao
Basilan
Isabela
Davao Gulf
Jolo
Jolo
General Santos
Sulu Archipelago
Sandakan
Lahad Datu
Tawitawi
Semporna
Tawau
Celebes Sea
Tarakan
Tanjungselor
Tanjungredeb
Kepulauan Talaud (Indonesia)
Kepulauan Sangir (Indonesia)
Morotai
Tobelo
Manadao
Semenanjung Minahassa
Tondano
Molucca Sea
Tolitoli
Ternate
Halmahera
Sangkulirang
Gorontalo
Sao-Siu
Makassar Strait
Moutong
Kepulauan Togian
Teluk Tomini
Tg Pangkalsiang
Donggala
Labuna
Bacan
Waigeo
Palu
Luwuk
Selat Dampir
Kwoka 3000
Manokwari
Numfor
Biak
Poso
Uekuli
Tataba
Peleng
Sorong
Biak
Mapane
Banggai
Todeli
Salawati
Jazirah Doberai
Ransiki
Selat Yapen
Tg d'Urville
Tentena
Taliabu
Dofa
Mangole
Misoöl
Fafanlap
Yapen
Serui
Sarmi
Wuvulu I.
Sulawesi (Celebes)
Kolonedale
Teluk Towori
Kepulauan Banggai
Kepulauan Sula
Obi
Inanwatan
Teluk
Mamuju
Masamba
Malili
Seram (Ceram)
Seram Sea (Ceram Sea)
Teluk Berau
Babo
Cenderawasih
Jayapura
Vanimo
Pegunungan Van Rees
Rantepao
Wotu
Namlea
Piru
G. Binaija 3019
Fakfak
Bomberai Peninsula
Tariku
Taritatu
Aitape
Wewak
Gandadiwata
Palopo
Polewali
Bula
Kaimana
Nabire
Maprik
Majene
Makale
Malamala
Manui
Buru
Saparua
Ambon (Amboina)
Pegunungan Maoke
Wamena
Anabanua
Kendari
Tel. Kamrau
Parepare
Singkang
Kolaka
Wowoni
Ambon
Kepulauan Banda
Adi
Watampone
Teluk Bone
Raha
Kepulauan Watubela
Pk Jaya 5030
Pk Trikora 4750
Pk Mandala 4700
NEW
Amamapare
Central Ra.
Ujung Pandang
2871
Sinjai
Muna
Buton
Kepulauan Kai
Wokam
G. Lompobatang
Kabaena
Baubau
Kepulauan Tukangbesi
Banda Sea
Tual
Dobo
IRIAN JAYA
PAPUA NEW GUINEA
4000
Tari
Mendi
Bontosunggu
Bulukumba
Benteng
Kai Kecil
Kai Besar
Kepulauan Aru
Benjina
Kobroör
Lake Murray
Salayar
INDONESIA
Tanahjampea
Kep. Bonerate
Wetar
Kepulauan Barat Daya
Damar
Larat
Trangan
Sia
Tg Deyong
Digul
Fly
Strickland
Sepik
Sobger
Aramia
Wawoi
Flores Sea
Roma
Wuliaru
Kepulauan Tanimbar
GUINEA
Huaki
Tepa
Babar
Kepulauan Sermata
Saumlakki
Arafura Sea
P. Dolak
Balimo
Dompu
Raba
Reo
Larantuka
Kalabahi
Alor
Kepulauauan Alor
Lomblen
Atauro
Kisar
Kepulauan Babar
Selaru
Tg Vals
Merauke
Morehead
Daru
Flores
Maumere
Ende
Kepulauan Solor
G. Tata Mailau 2963
Dili
Manatuto
EAST TIMOR
Kepulauan Leti
Sumbawa
Selat Sumba
EAST TIMOR
Pantai Makassar
Timor
Memboro
G. Mutis 2427
Kefamenanu
Waikabubak
Waingapu
Sawu Sea
Sumba
Sawu (Sawu)
Kupang
Rote (Roti)
Badu I.
Moa I.
AUSTRALIA
C. York
Ulithi
Fais
Yap
Ngulu
Sorol
Koror
Equator
1:15M
KM
MILES
900
750
600
450
300
150
0
600
450
300
150
0

© Collins

THAILAND, CAMBODIA and Peninsular MALAYSIA

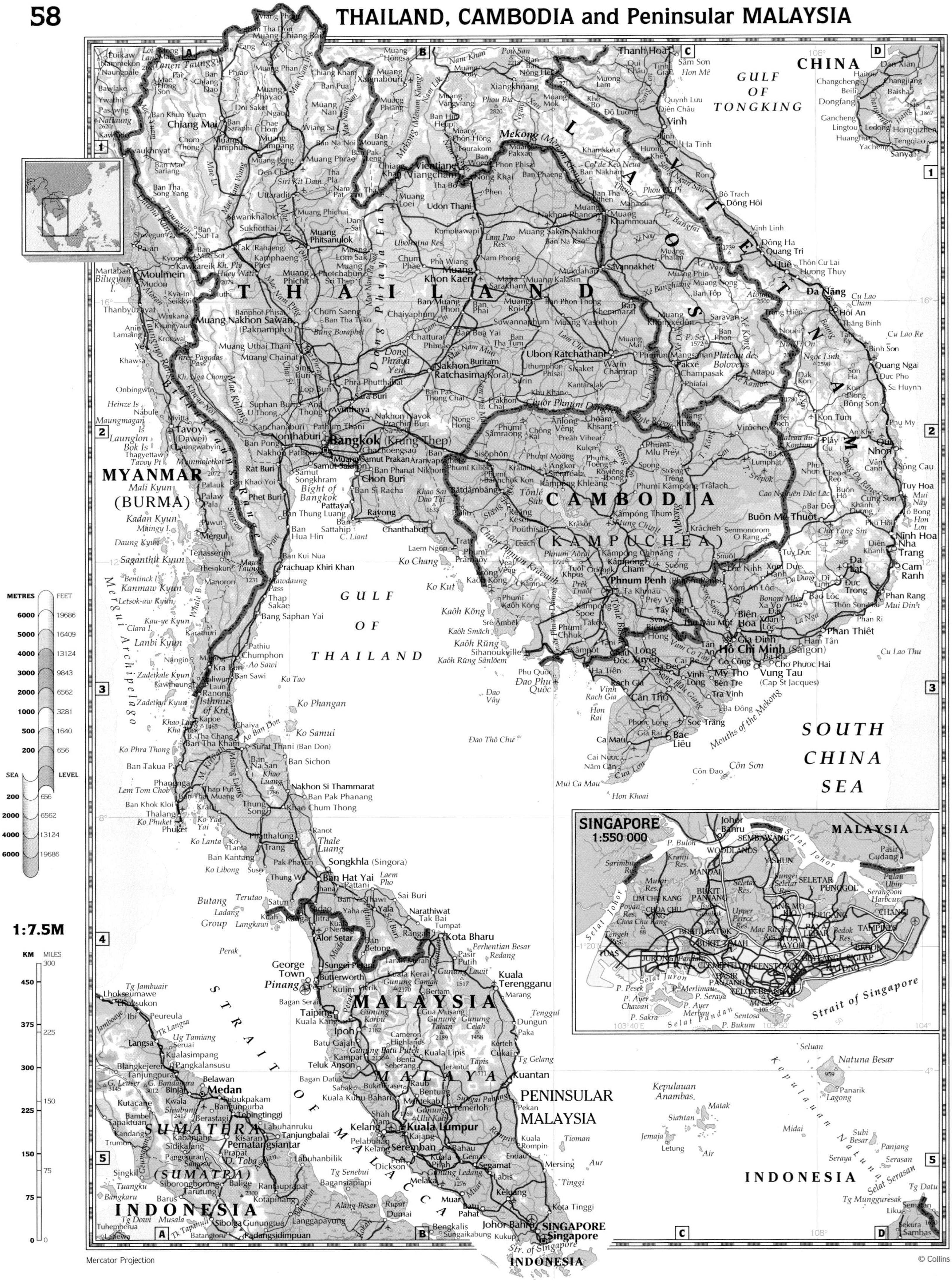

Mercator Projection

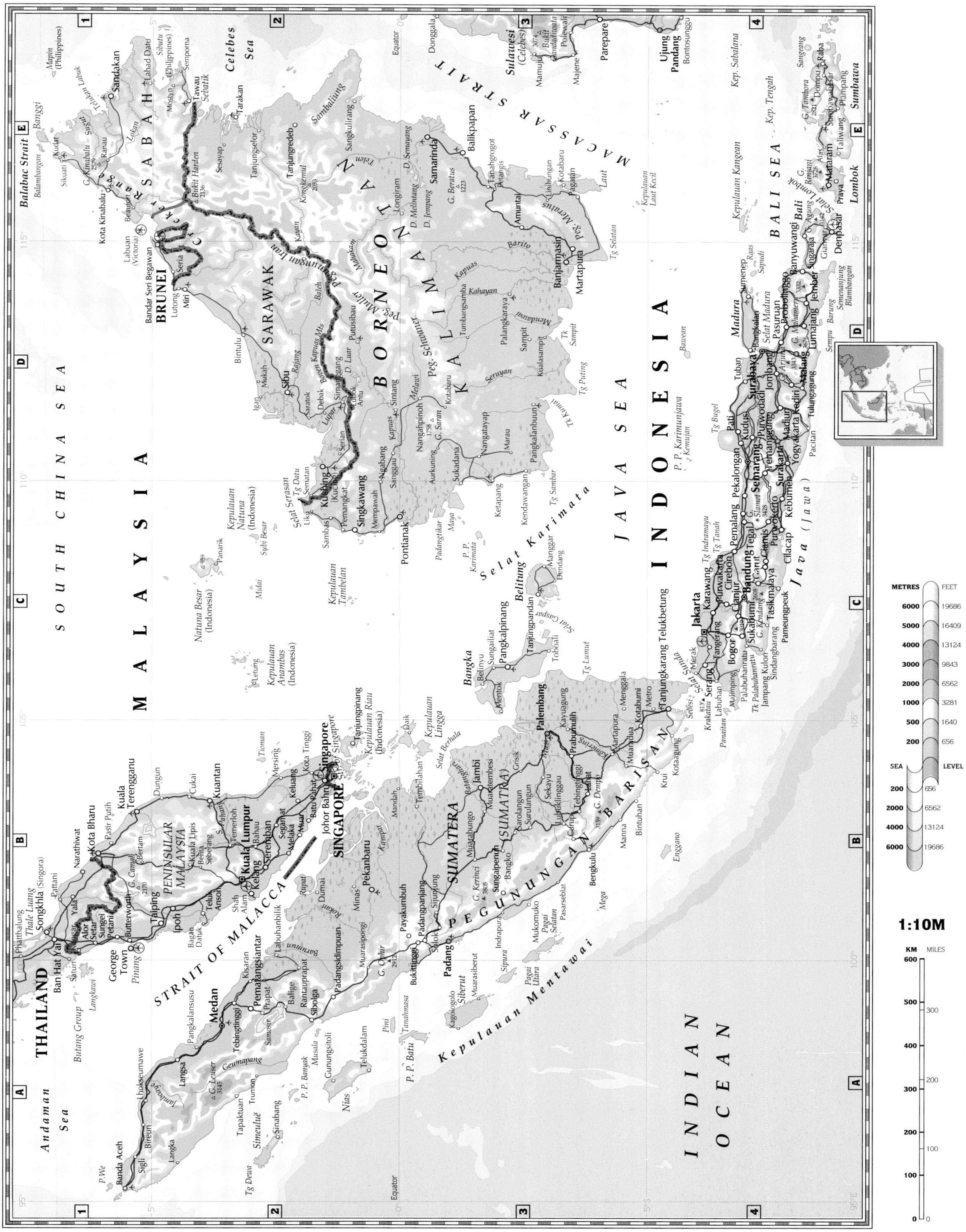

Mercator Projection

METRES	FEET
5000	16409
3000	9843
2000	6562
1000	3281
500	1640
200	656
SEA	LEVEL
200	656
3000	9843
5000	16409
6000	19686

1:45M

KM	MILES
2500	1500
2000	
1500	1000
1000	500
500	
0	0

Lambert Azimuthal Equal Area Projection

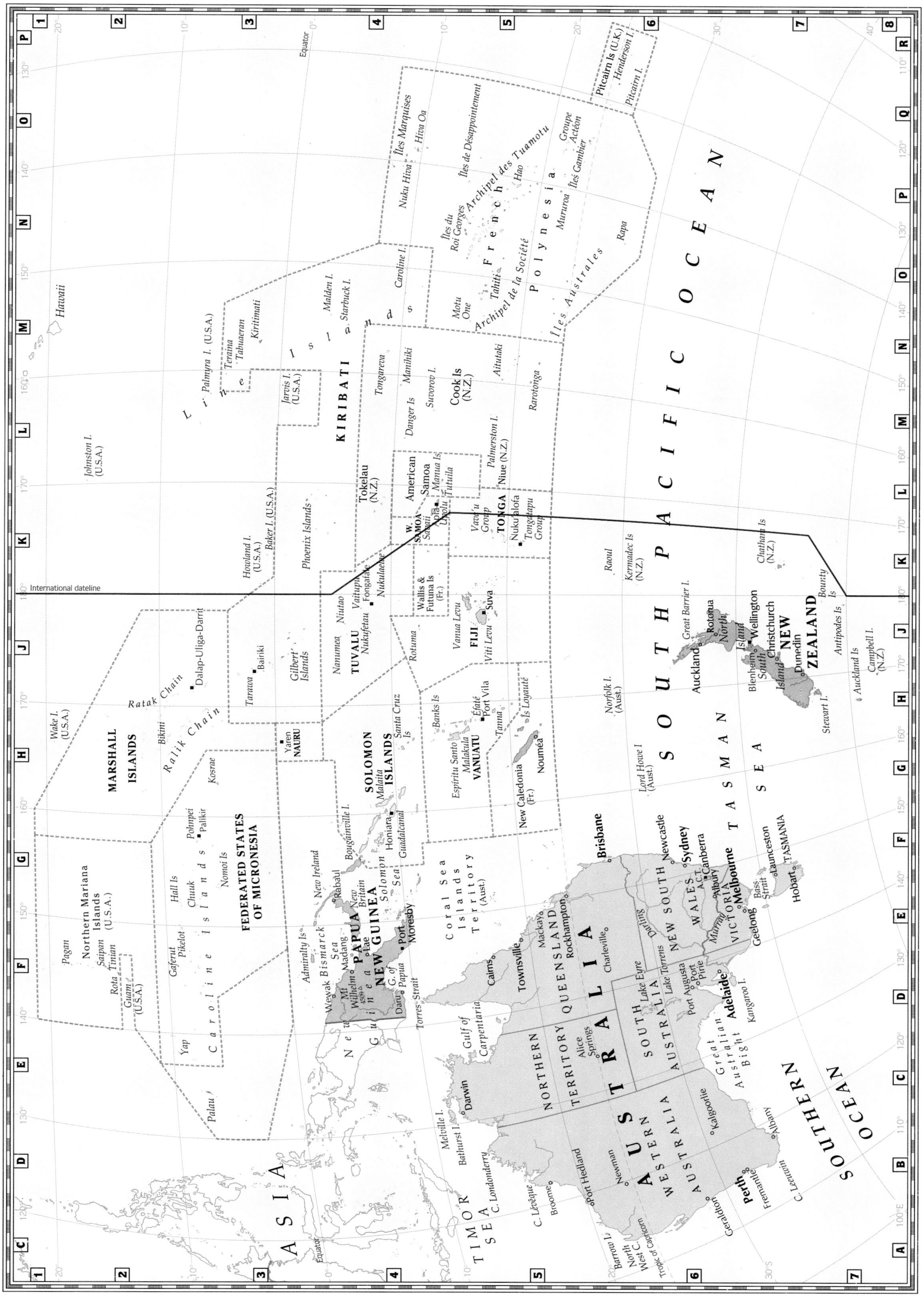

1:45M

KM	MILES
2500	1500
2000	1000
1500	500
1000	0
500	
0	

Lambert Azimuthal Equal Area Projection

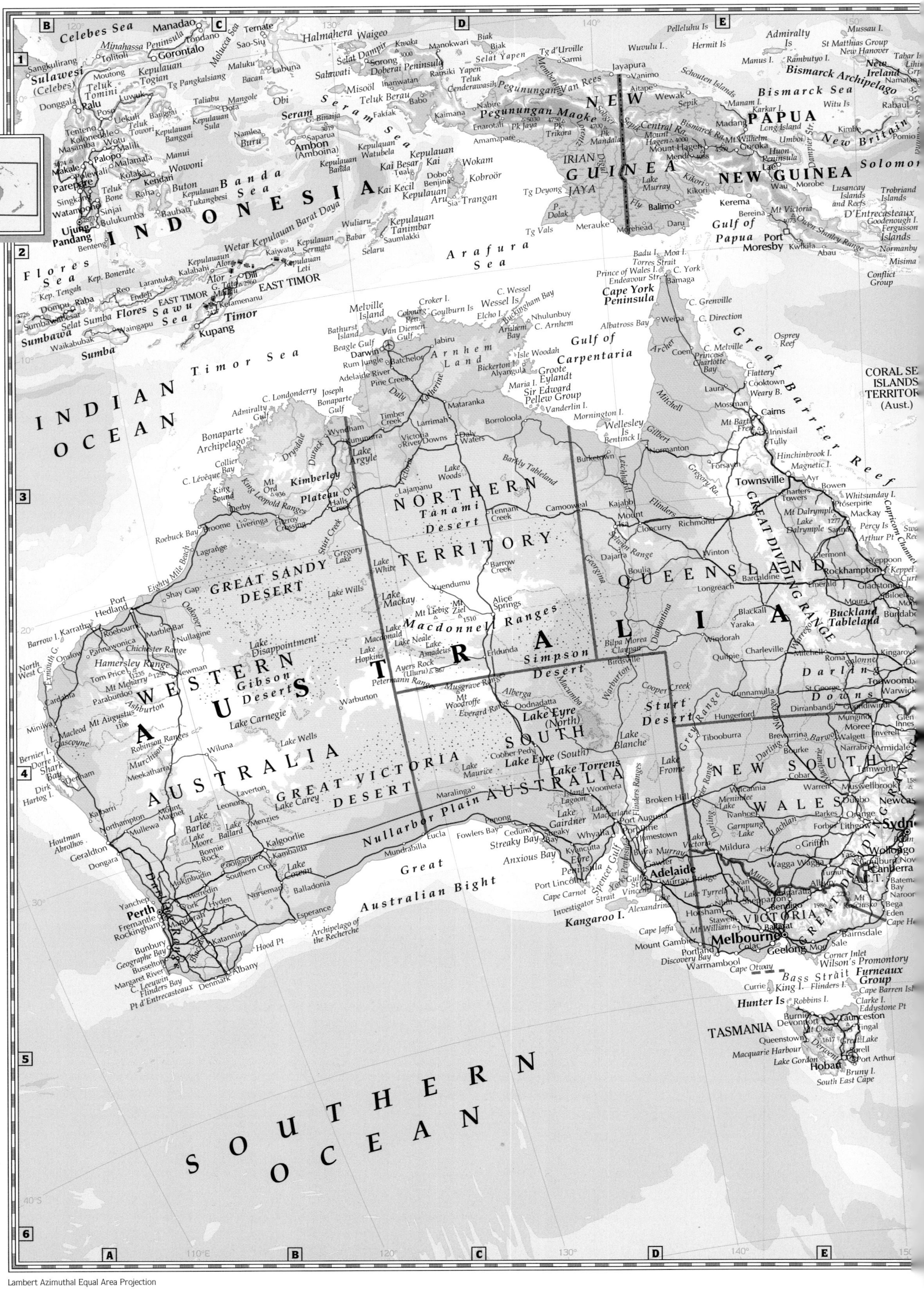

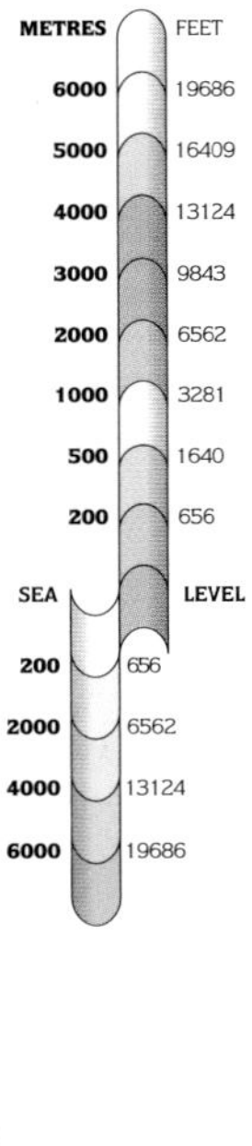

Lambert Azimuthal Equal Area Projection

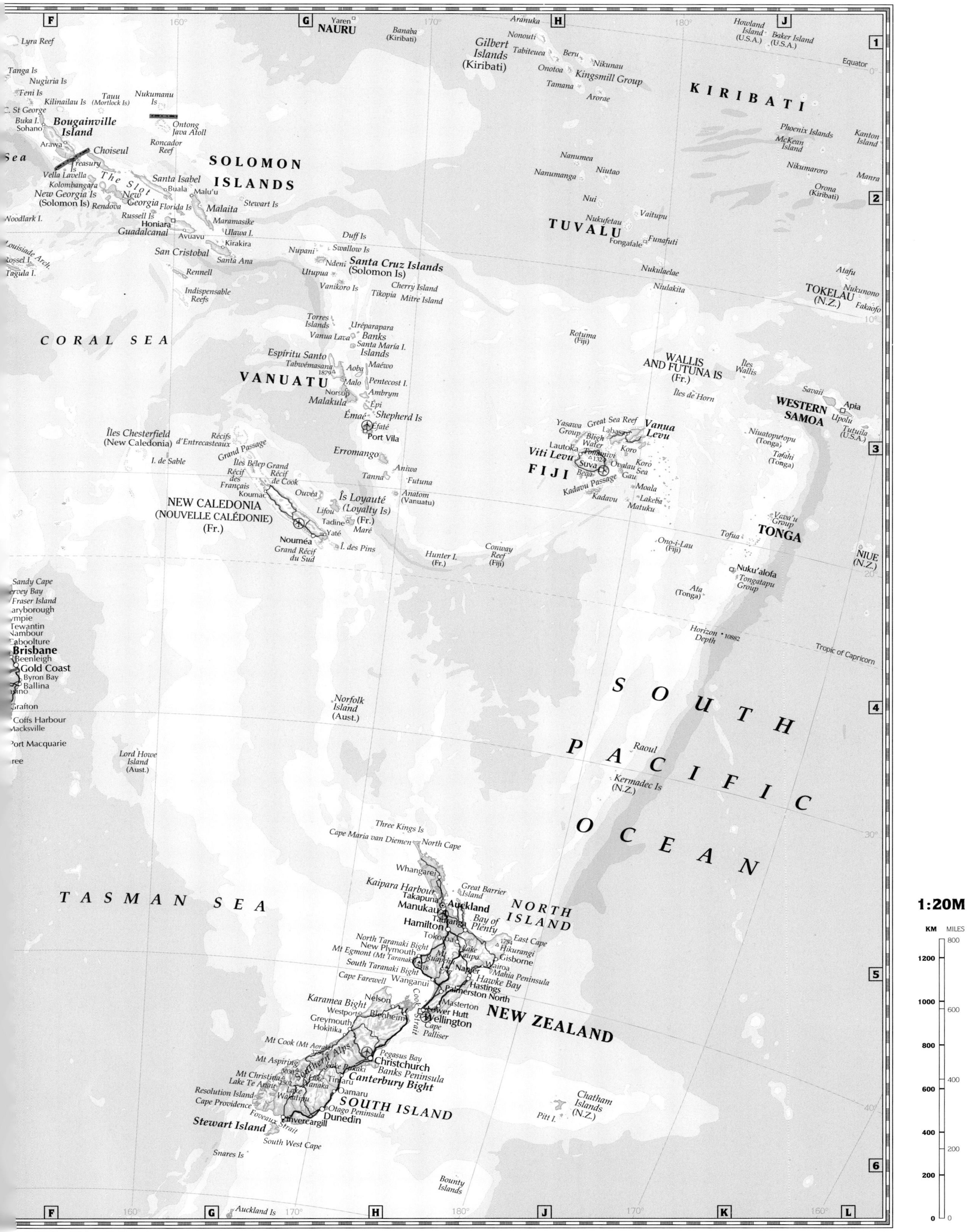
NAURU
Yaren
Banaba (Kiribati)
Gilbert Islands (Kiribati)
Aranuka
Nonouti
Tabiteuea
Beru
Nikunau
Onotoa
Kingsmill Group
Tamana
Arorae
Howland Island (U.S.A.)
Baker Island (U.S.A.)
Equator
KIRIBATI
Phoenix Islands
McKean Island
Kanton Island
Nikumaroro
Orona (Kiribati)
Manra
Lyra Reef
Tanga Is
Nuguria Is
Feni Is
Kilinailau Is
Tauu (Mortlock Is)
Nukumanu Is
St George
Buka I.
Sohano
Bougainville Island
Ontong Java Atoll
Arawa
Roncador Reef
Choiseul
Sea
SOLOMON ISLANDS
Treasury Is
Vella Lavella
Kolombangara
New Georgia Is (Solomon Is)
Rendova
The Slot
New Georgia
Santa Isabel
Buala
Malu'u
Florida Is
Malaita
Stewart Is
Russell Is
Honiara
Guadalcanal
Maramasike
Ulawa I.
Avuavu
Kirakira
San Cristobal
Santa Ana
Woodlark I.
Louisiade Arch.
Rossel I.
Tagula I.
Rennell
Indispensable Reefs
Duff Is
Swallow Is
Nupani
Ndeni
Santa Cruz Islands (Solomon Is)
Utupua
Vanikoro Is
Cherry Island
Tikopia
Mitre Island
Nanumea
Nanumanga
Niutao
Nui
Vaitupu
Nukufetau
TUVALU
Fongafale
Funafuti
Nukulaelae
Niulakita
Atafu
Nukunono
TOKELAU (N.Z.)
Fakaofo
CORAL SEA
Torres Islands
Uréparapara
Vanua Lava
Banks Islands
Santa Maria I.
Espiritu Santo
Tabwémasana 1879
Aoba
Maéwo
VANUATU
Malo
Pentecost I.
Norsup
Ambrym
Malakula
Épi
Émaé
Shepherd Is
Éfaté
Port Vila
Erromango
Rotuma (Fiji)
WALLIS AND FUTUNA IS (Fr.)
Îles Wallis
Îles de Horn
Savaii
Apia
WESTERN SAMOA
Upolu
Tutuila (U.S.A.)
Niuatoputapu (Tonga)
Tafahi (Tonga)
Yasawa Group
Great Sea Reef
Vanua Levu
Labasa
Bligh Water
Lautoka
Viti Levu
Koro
FIJI
Suva
Ovalau
Koro Sea
Beqa
Gau
Kadavu Passage
Moala
Kadavu
Lakeba
Matuku
Îles Chesterfield (New Caledonia)
Récifs d'Entrecasteaux
Grand Passage
Îles Belep
Grand Récif de Cook
Récif des Français
I. de Sable
Koumac
Ouvéa
NEW CALEDONIA (NOUVELLE CALÉDONIE) (Fr.)
Lifou
Îs Loyauté (Loyalty Is) (Fr.)
Tadine
Maré
Yaté
Nouméa
Grand Récif du Sud
I. des Pins
Tanna
Aniwa
Futuna
Anatom (Vanuatu)
Hunter I. (Fr.)
Conway Reef (Fiji)
Vava'u Group
Tofua
TONGA
Ono-i-Lau (Fiji)
NIUE (N.Z.)
Nuku'alofa
Tongatapu Group
Ata (Tonga)
Horizon Depth 10882
Tropic of Capricorn
Sandy Cape
Hervey Bay
Fraser Island
Maryborough
Gympie
Tewantin
Nambour
Caboolture
Brisbane
Beenleigh
Gold Coast
Byron Bay
Ballina
Lismore
Grafton
Coffs Harbour
Macksville
Port Macquarie
Taree
Norfolk Island (Aust.)
Lord Howe Island (Aust.)
SOUTH PACIFIC OCEAN
Raoul
Kermadec Is (N.Z.)
Three Kings Is
Cape Maria van Diemen
North Cape
Whangarei
Kaipara Harbour
Great Barrier Island
Takapuna
Auckland
Manukau
NORTH ISLAND
Hamilton
Tauranga
Bay of Plenty
Tokoroa
East Cape
TASMAN SEA
North Taranaki Bight
New Plymouth
Mt Egmont (Mt Taranaki) 2518
Lake Taupo
Mt Ruapehu 2797
Hikurangi 1754
Gisborne
South Taranaki Bight
Wairoa
Napier
Mahia Peninsula
Cape Farewell
Wanganui
Hawke Bay
Hastings
Palmerston North
Karamea Bight
Nelson
Cook Strait
Masterton
Westport
Blenheim
Lower Hutt
Wellington
NEW ZEALAND
Greymouth
Hokitika
Cape Palliser
Mt Cook (Mt Aoraki) 3754
Southern Alps
Pegasus Bay
Christchurch
Mt Aspiring 3030
Lake Pukaki
Banks Peninsula
Mt Christina 2502
Lake Tekapo
Lake Te Anau
Lake Wanaka
Canterbury Bight
Resolution Island
Lake Wakatipu
Oamaru
Cape Providence
SOUTH ISLAND
Otago Peninsula
Dunedin
Invercargill
Foveaux Strait
Stewart Island
South West Cape
Snares Is
Chatham Islands (N.Z.)
Pitt I.
Bounty Islands
Auckland Is
F G H J K L
1 2 3 4 5 6
160° 170° 180° 170° 160°
0° 10° 20° 30° 40°
1:20M
KM MILES
1200 1000 800 600 400 200 0
800 600 400 200 0

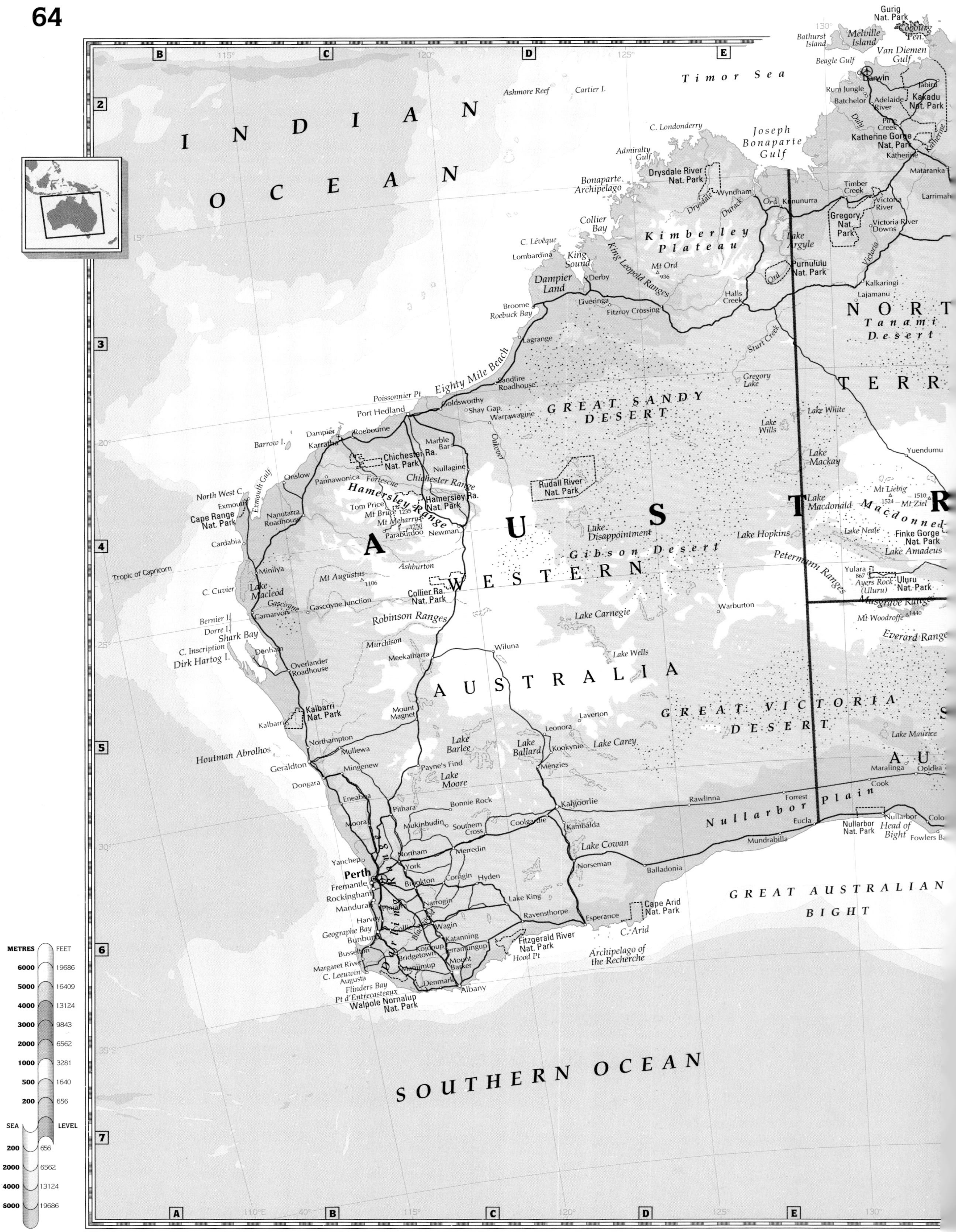
INDIAN OCEAN
Timor Sea
Ashmore Reef
Cartier I.
Gurig Nat. Park
Melville Island
Bathurst Island
Coburg Pen.
Van Diemen Gulf
Beagle Gulf
Darwin
Jabiru
Rum Jungle
Batchelor
Adelaide River
Kakadu Nat. Park
Pine Creek
Daly
Katherine Gorge Nat. Park
Katherine
Mataranka
Larrimah
Timber Creek
Victoria River
Gregory Nat. Park
Victoria River Downs
C. Londonderry
Joseph Bonaparte Gulf
Admiralty Gulf
Drysdale River Nat. Park
Bonaparte Archipelago
Wyndham
Kununurra
Ord
Collier Bay
Kimberley Plateau
Lake Argyle
Purnululu Nat. Park
C. Lévêque
Lombardina
King Sound
Mt Ord 936
Dampier Land
Derby
King Leopold Ranges
Halls Creek
Kalkaringi
Lajamanu
Broome
Roebuck Bay
Liveringa
Fitzroy Crossing
Tanami Desert
NORT
Lagrange
Sturt Creek
Eighty Mile Beach
Sandfire Roadhouse
Gregory Lake
TERR
Poissonnier Pt
Port Hedland
Goldsworthy
Shay Gap
Warrawagine
GREAT SANDY DESERT
Lake White
Lake Wills
Dampier
Roebourne
Karratha
Barrow I.
Marble Bar
Oakover
Yuendumu
Chichester Ra. Nat. Park
Nullagine
Lake Mackay
Onslow
Pannawonica
Fortescue
Chichester Range
Rudall River Nat. Park
North West C.
Exmouth
Exmouth Gulf
Hamersley Range
Hamersley Ra. Nat. Park
Mt Liebig
Mt Ziel 1524
1510
Cape Range Nat. Park
Tom Price
Mt Bruce 1235
Mt Meharry 1250
Lake Macdonald
Macdonnell
Nanutarra Roadhouse
Paraburdoo
Newman
Lake Disappointment
Lake Hopkins
Lake Neale
Finke Gorge Nat. Park
Cardabia
AUSTRALIA
WESTERN AUSTRALIA
Gibson Desert
Lake Amadeus
Minilya
Ashburton
Petermann Ranges
Tropic of Capricorn
Mt Augustus 1106
Yulara
Uluru
867
Ayers Rock (Uluru)
Uluru Nat. Park
C. Cuvier
Lake Macleod
Collier Ra. Nat. Park
Gascoyne
Gascoyne Junction
Musgrave Ranges
Carnarvon
Warburton
Mt Woodroffe 1440
Bernier I.
Robinson Ranges
Lake Carnegie
Dorre I.
Shark Bay
Everard Range
C. Inscription
Dirk Hartog I.
Denham
Murchison
Wiluna
Meekatharra
Lake Wells
Overlander Roadhouse
GREAT VICTORIA DESERT
Kalbarri Nat. Park
Kalbarri
Mount Magnet
Laverton
Leonora
Lake Maurice
Northampton
Lake Barlee
Lake Ballard
Kookynie
Lake Carey
Houtman Abrolhos
Mullewa
Geraldton
Mingenew
Payne's Find
Menzies
Maralinga
Ooldea
Dongara
Lake Moore
Cook
Eneabba
Bonnie Rock
Kalgoorlie
Rawlinna
Forrest
Nullarbor Plain
Pithara
Moora
Mukinbudin
Southern Cross
Coolgardie
Kambalda
Eucla
Mundrabilla
Nullarbor Nat. Park
Nullarbor
Head of Bight
Fowlers Ba
Colo
Yanchep
Northam
Merredin
Lake Cowan
Perth
York
Norseman
Balladonia
Fremantle
Rockingham
Brookton
Corrigin
Hyden
GREAT AUSTRALIAN BIGHT
Mandurah
Narrogin
Lake King
Cape Arid Nat. Park
Harvey
Geographe Bay
Collie
Wagin
Ravensthorpe
Esperance
Bunbury
Katanning
C. Arid
Busselton
Kojonup
Fitzgerald River Nat. Park
Bridgetown
Hood Pt
Archipelago of the Recherche
Margaret River
C. Leeuwin
Augusta
Manjimup
Mount Barker
Denmark
Albany
Flinders Bay
Pt d'Entrecasteaux
Walpole Nornalup Nat. Park
SOUTHERN OCEAN
B C D E
A B C D E
2 3 4 5 6 7
115° 120° 125° 130°
110°E 115° 120° 125° 130°
15° 20° 25° 30° 35°S
METRES FEET
6000 19686
5000 16409
4000 13124
3000 9843
2000 6562
1000 3281
500 1640
200 656
SEA LEVEL
200 656
2000 6562
4000 13124
6000 19686

Lambert Azimuthal Equal Area Projection

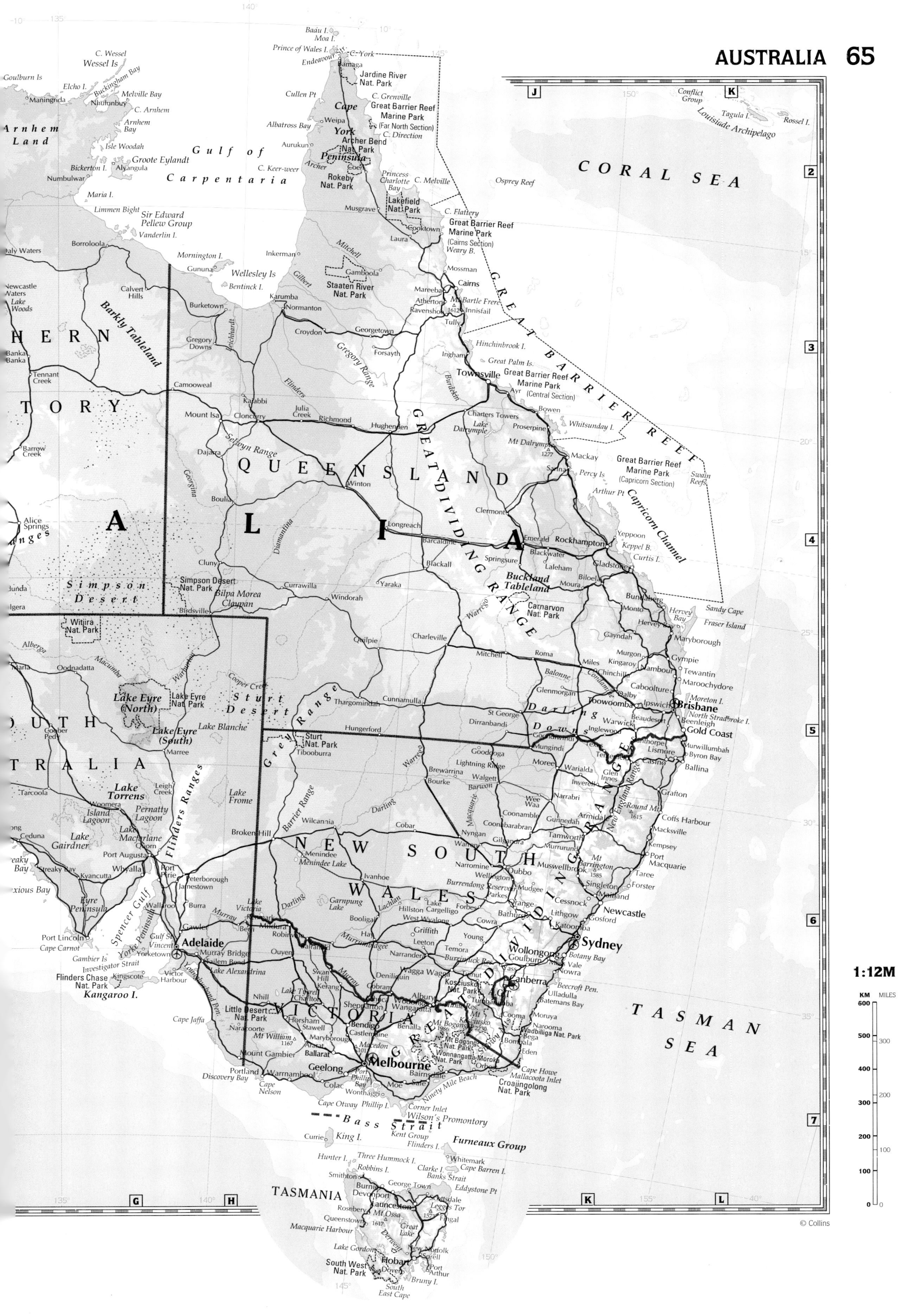

CORAL SEA
TASMAN SEA
GREAT BARRIER REEF
Gulf of Carpentaria
QUEENSLAND
NEW SOUTH WALES
VICTORIA
TASMANIA
GREAT DIVIDING RANGE
Arnhem Land
Cape York Peninsula
Simpson Desert
Sturt Desert
Barkly Tableland
Buckland Tableland
Darling Downs
Bass Strait
Furneaux Group
Great Barrier Reef Marine Park (Far North Section)
Great Barrier Reef Marine Park (Cairns Section)
Great Barrier Reef Marine Park (Central Section)
Great Barrier Reef Marine Park (Capricorn Section)
Capricorn Channel
Louisiade Archipelago
Osprey Reef
Whitsunday I.
Fraser Island
Moreton I.
North Stradbroke I.
King I.
Flinders I.
Kangaroo I.
Lake Eyre (North)
Lake Eyre (South)
Lake Torrens
Lake Frome
Lake Gairdner
Spencer Gulf
Gulf St Vincent
Eyre Peninsula
York Peninsula
Flinders Ranges
Grey Range
Barrier Range
Selwyn Range
Gregory Range
Darling
Murray
Murrumbidgee
Lachlan
Warrego
Cooper Creek
Diamantina
Georgina
Flinders
Mitchell
Gilbert
Burdekin
Mt Kosciuszko Nat. Park
Wilson's Promontory
Ninety Mile Beach
Croajingolong Nat. Park
South West Nat. Park
Carnarvon Nat. Park
Lakefield Nat. Park
Staaten River Nat. Park
Simpson Desert Nat. Park
Sturt Nat. Park
Witjira Nat. Park
Cairns
Townsville
Mackay
Rockhampton
Gladstone
Bundaberg
Maryborough
Gympie
Toowoomba
Brisbane
Gold Coast
Mount Isa
Cloncurry
Longreach
Charleville
Roma
Bourke
Broken Hill
Dubbo
Tamworth
Armidale
Coffs Harbour
Port Macquarie
Newcastle
Sydney
Wollongong
Canberra
Wagga Wagga
Albury
Mildura
Bendigo
Ballarat
Geelong
Melbourne
Adelaide
Port Augusta
Port Pirie
Port Lincoln
Mount Gambier
Warrnambool
Portland
Launceston
Devonport
Hobart
Queenstown
Alice Springs
1:12M
KM
MILES
© Collins

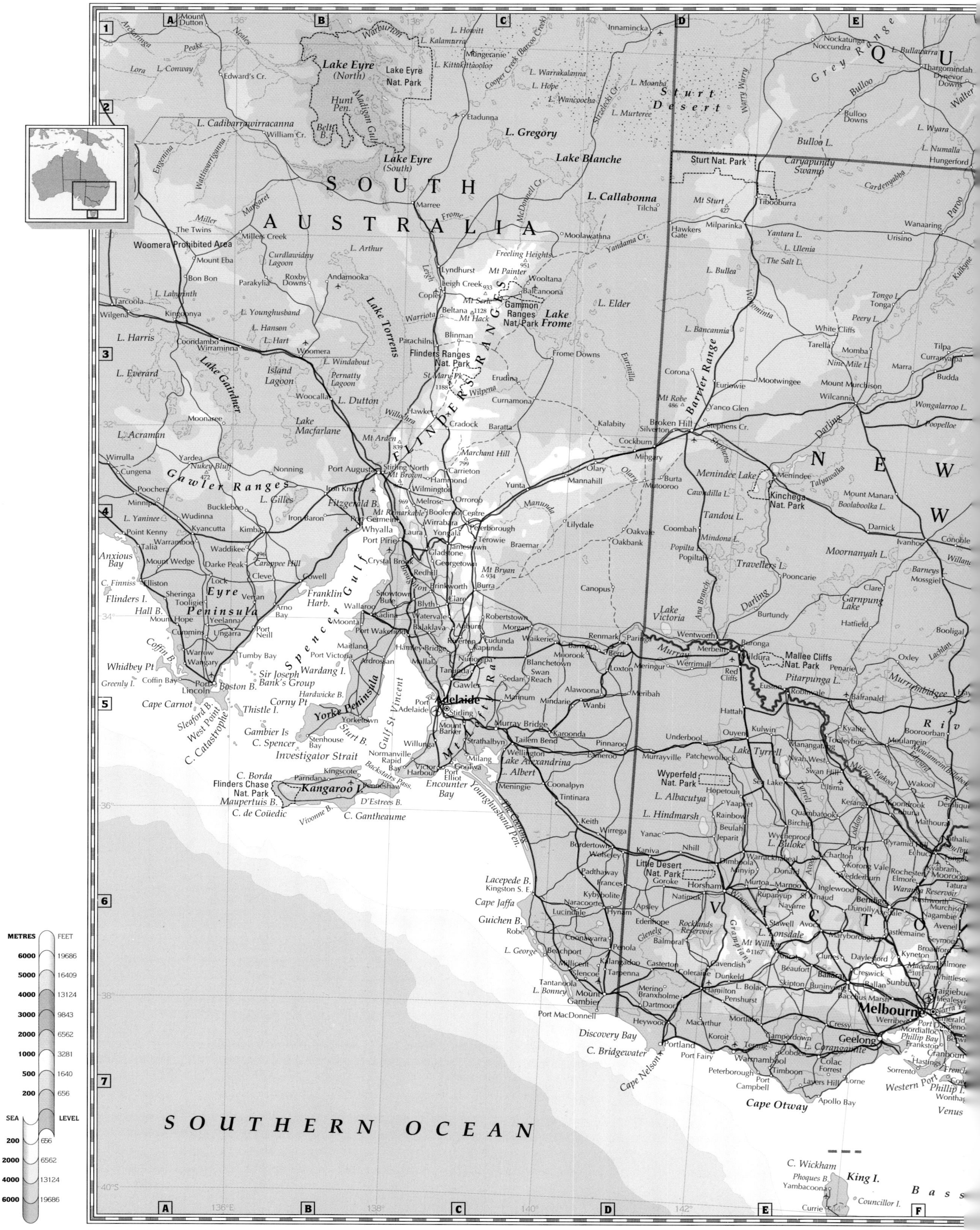

Lambert Azimuthal Equal Area Projection

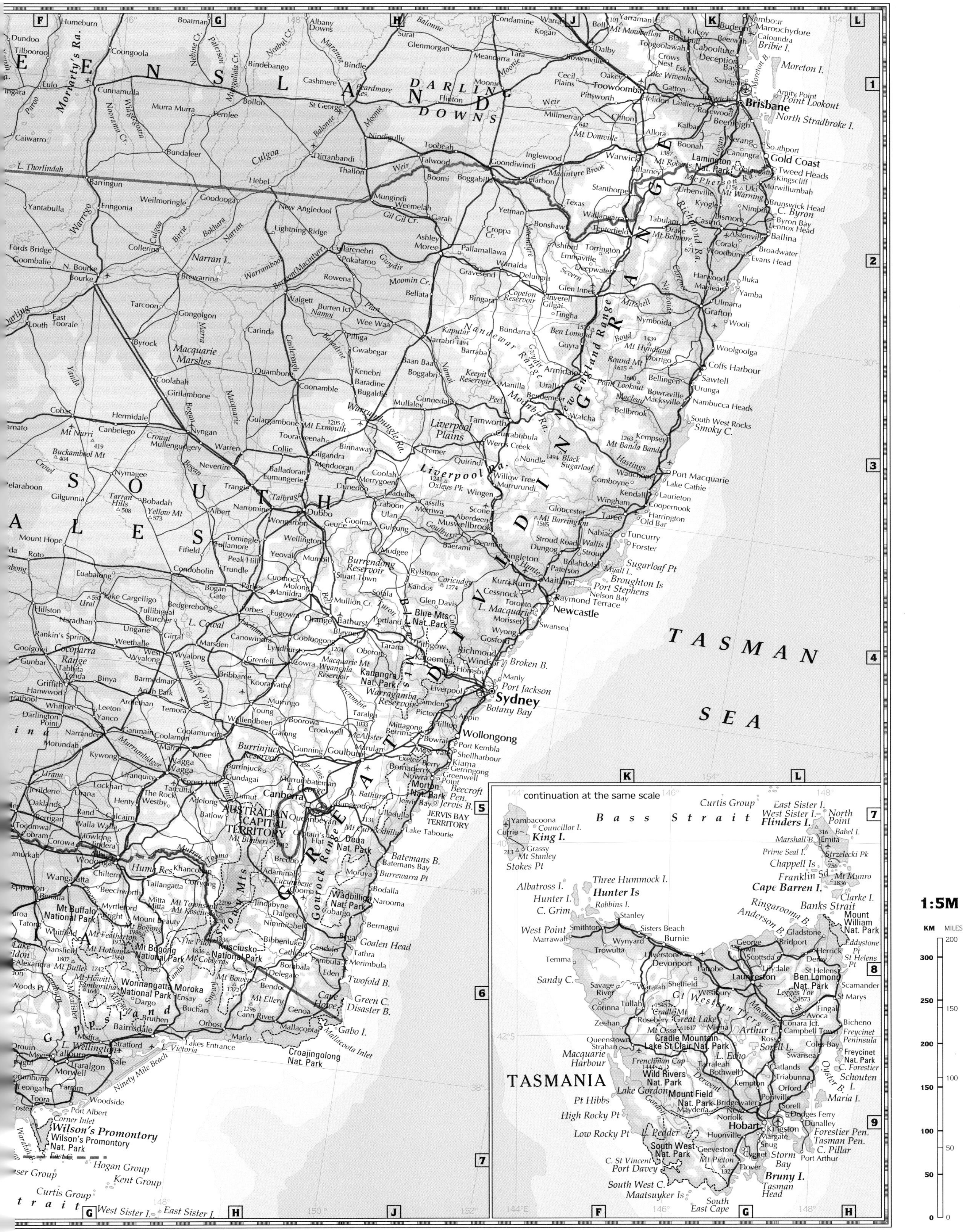
QUEENSLAND
DARLING DOWNS
NEW SOUTH WALES
GREAT DIVIDING RANGE
NEW ENGLAND RANGE
TASMAN SEA
Brisbane
Gold Coast
Toowoomba
Moreton I.
North Stradbroke I.
Point Lookout
Tweed Heads
C. Byron
Byron Bay
Lennox Head
Ballina
Lismore
Casino
Grafton
Coffs Harbour
Armidale
Tamworth
Port Macquarie
Kempsey
Newcastle
Maitland
Cessnock
Port Stephens
Sydney
Port Jackson
Botany Bay
Wollongong
Canberra
AUSTRALIAN CAPITAL TERRITORY
JERVIS BAY TERRITORY
Jervis B.
Batemans B.
Dubbo
Orange
Bathurst
Wagga Wagga
Albury
Goulburn
Bourke
Moree
Narrabri
Gunnedah
Mudgee
Lithgow
Katoomba
Blue Mts Nat. Park
Kosciusko National Park
Snowy Mts
Mt Kosciuszko
Cape Howe
Gabo I.
Croajingolong Nat. Park
Ninety Mile Beach
Lakes Entrance
Wilson's Promontory
Wilson's Promontory Nat. Park
Hogan Group
Kent Group
Curtis Group
West Sister I.
East Sister I.
Lamington Nat. Park
Liverpool Plains
Liverpool Ra.
Warrumbungle Ra.
Nandewar Range
Macquarie Marshes
Burrinjuck Reservoir
Burrendong Reservoir
L. Macquarie
L. Cowal
Narran L.
Sugarloaf Pt
Smoky C.
continuation at the same scale
TASMANIA
Bass Strait
King I.
Flinders I.
Cape Barren I.
Hunter Is
Three Hummock I.
Hobart
Launceston
Devonport
Burnie
Queenstown
Macquarie Harbour
Cradle Mountain Lake St Clair Nat. Park
Great Lake
Ben Lomond Nat. Park
Freycinet Nat. Park
Maria I.
Tasman Pen.
Bruny I.
South West Nat. Park
Mount Field Nat. Park
Wild Rivers Nat. Park
South West C.
South East Cape
1:5M
KM
MILES

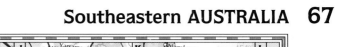

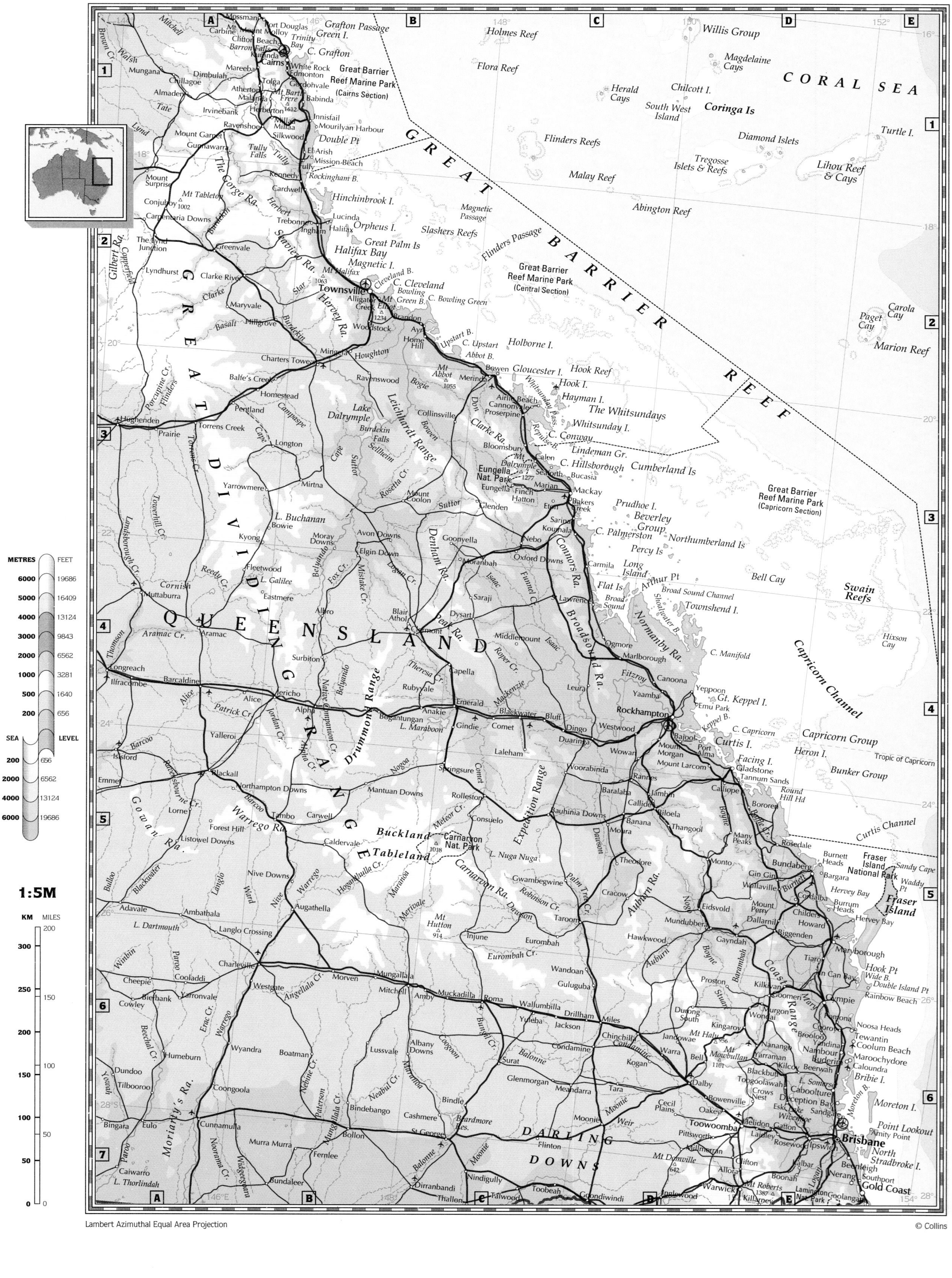
METRES FEET
6000 19686
5000 16409
4000 13124
3000 9843
2000 6562
1000 3281
500 1640
200 656
SEA LEVEL
200 656
2000 6562
4000 13124
6000 19686
1:5M
KM MILES
CORAL SEA
GREAT BARRIER REEF
Great Barrier Reef Marine Park (Cairns Section)
Great Barrier Reef Marine Park (Central Section)
Great Barrier Reef Marine Park (Capricorn Section)
QUEENSLAND
GREAT DIVIDING RANGE
DARLING DOWNS
Cairns
Townsville
Mackay
Rockhampton
Bundaberg
Toowoomba
Brisbane
Gold Coast
Holmes Reef
Flora Reef
Willis Group
Magdelaine Cays
Herald Cays
Chilcott I.
South West Island
Coringa Is
Flinders Reefs
Diamond Islets
Turtle I.
Tregosse Islets & Reefs
Lihou Reef & Cays
Malay Reef
Abington Reef
Magnetic Passage
Slashers Reefs
Flinders Passage
Carola Cay
Paget Cay
Marion Reef
Holborne I.
Hook Reef
Hook I.
Hayman I.
The Whitsundays
Whitsunday I.
Lindeman Gr.
Cumberland Is
Prudhoe I.
Beverley Group
Northumberland Is
Percy Is
Bell Cay
Swain Reefs
Broad Sound Channel
Townshend I.
Capricorn Channel
Hixson Cay
Capricorn Group
Heron I.
Bunker Group
Tropic of Capricorn
Curtis Channel
Fraser Island
Moreton I.
North Stradbroke I.
Hinchinbrook I.
Halifax Bay
Magnetic I.
Grafton Passage
Hughenden
Charters Towers
Longreach
Barcaldine
Emerald
Charleville
Roma
Gladstone
Maryborough
Gympie
Dalby
Warwick
Goondiwindi
St George
Cunnamulla
Lake Dalrymple
L. Buchanan
Buckland Tableland
Carnarvon Nat. Park
Eungella Nat. Park
Lambert Azimuthal Equal Area Projection
© Collins

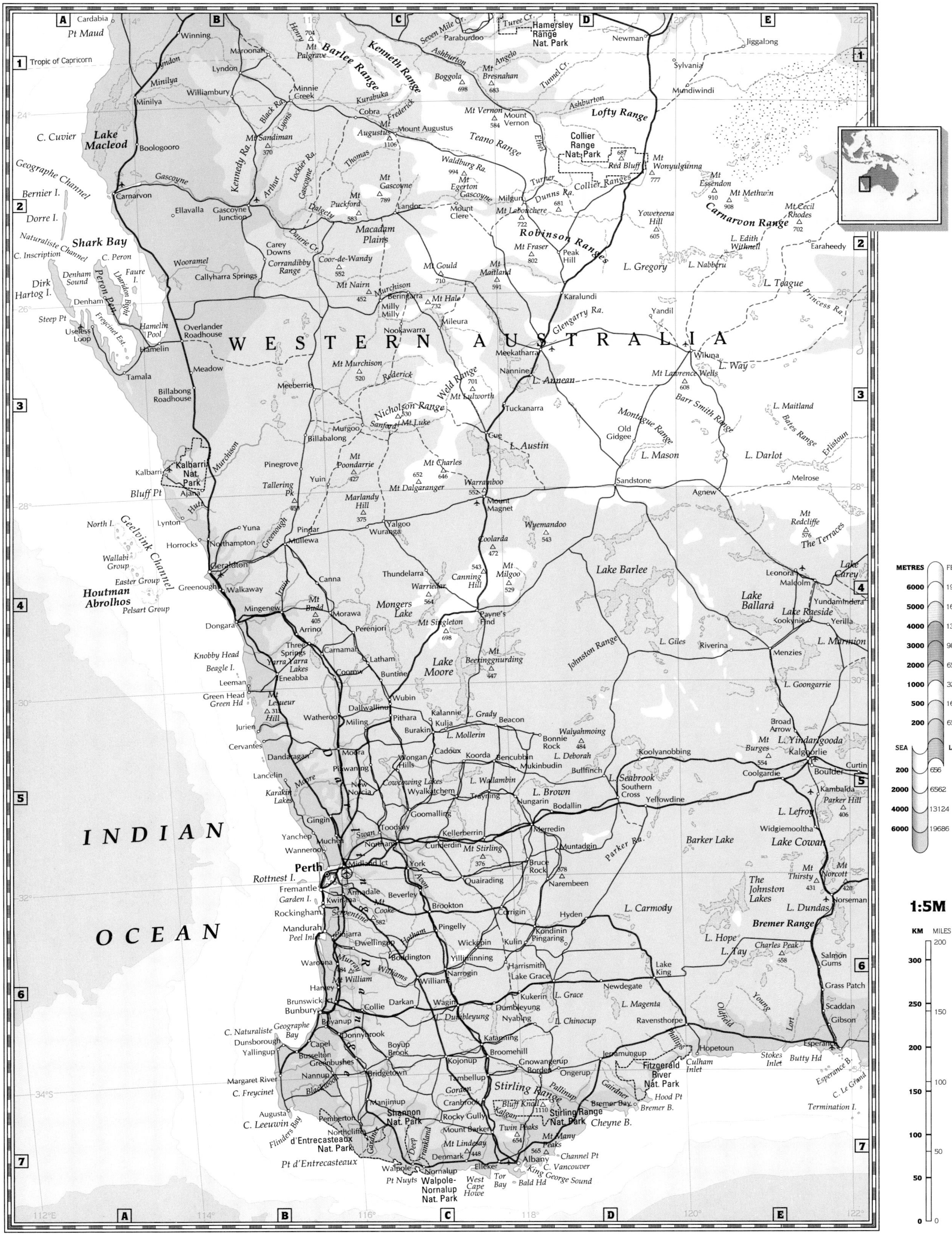
WESTERN AUSTRALIA
INDIAN OCEAN
Tropic of Capricorn
Cardabia
Pt Maud
Winning
Maroonah
Lyndon
Minilya
Williambury
Minnie Creek
Henry
Mt Palgrave 704
Barlee Range
Kenneth Range
Seven Mile Cr.
Paraburdoo
Turee Cr.
Hamersley Range Nat. Park
Newman
Jiggalong
Sylvania
Mundiwindi
Ashburton
Mt Angelo
Mt Bresnahan 683
Boggola 698
Tunnel Cr.
Kurabuka
Cobra
Mt Frederick
Mt Vernon 584
Mount Vernon
Lofty Range
Mt Augustus 1106
Mount Augustus
Teano Range
Collier Range Nat. Park
Red Bluff 687
Mt Wonyulgunna 777
Mt Essendon 910
Mt Methwin 908
Carnarvon Range
Mt Cecil Rhodes 702
C. Cuvier
Lake Macleod
Boolgooro
Kennedy Ra.
Mt Sandiman 370
Black Ra.
Lyons
Lockier Ra.
Thomas
Waldburg Ra. 994
Mt Egerton
Gascoyne
Milgun
Turner
Dunns Ra. 681
Collier Ranges
Ethel
Geographe Channel
Bernier I.
Dorre I.
Carnarvon
Ellavalla
Gascoyne Junction
Arthur
Mt Puckford 583
Mt Gascoyne 789
Landor
Mount Clere
Mt Labouchere 722
Yowereena Hill 605
Shark Bay
Naturaliste Channel
C. Inscription
C. Peron
Carey Downs
Dalgety
Daurie Cr.
Macadam Plains
Robinson Ranges
Mt Fraser 802
Peak Hill
L. Edith Withnell
Earaheedy
Dirk Hartog I.
Denham Sound
Faure I.
Peron Pen.
Uhridon Bight
Denham
Steep Pt
Useless Loop
Freycinet Est.
Wooramel
Callyharra Springs
Corrandibby Range
Coor-de-Wandy 552
Mt Nairn 452
Murchison
Mt Gould 710
Mt Maitland 591
L. Gregory
L. Nabberu
L. Teague
Princess Ra.
Beringarra
Mt Hale 732
Milly Milly
Mileura
Karalundi
Glengarry Ra.
Yandil
Hamelin Pool
Hamelin
Overlander Roadhouse
Nookawarra
Meekatharra
Wiluna
L. Way
Tamala
Meadow
Mt Murchison 520
Roderick
Weld Range 701
Nannine
L. Annean
Mt Lawrence Wells 608
Billabong Roadhouse
Meeberrie
Nicholson Range
Mt Lulworth
Tuckanarra
Barr Smith Range
L. Maitland
Bates Range
Montague Range
Murgoo
Sanford
Mt Luke 530
Billabalong
Cue
L. Austin
Old Gidgee
Eriistoun
Kalbarri
Kalbarri Nat. Park
Pinegrove
Mt Poondarrie 427
Mt Charles 646 652
L. Mason
L. Darlot
Tallering Pk 459
Yuin
Mt Dalgaranger
Warramboo 552
Sandstone
Melrose
Bluff Pt
Ajana
Hutt
Marlandy Hill 375
Mount Magnet
Agnew
North I.
Geelvink Channel
Lynton
Yuna
Greenough
Yalgoo
Wuranga
Pindar
Wyemandoo 543
Mt Redcliffe 576
The Terraces
Horrocks
Northampton
Mullewa
Coolarda 472
Wallabi Group
Easter Group
Geraldton
Canna
Thundelarra
Canning Hill 543
Mt Milgoo 529
Lake Barlee
Leonora
Lake Carey
Houtman Abrolhos
Greenough
Walkaway
Warriedar 564
Malcolm
Pelsart Group
Mingenew
Mt Budd 405
Morawa
Mongers Lake
Mt Singleton 698
Payne's Find
Lake Ballard
Yundamindera
Lake Raeside
Kookynie
Yerilla
Dongara
Arrino
Perenjori
Johnston Range
L. Giles
Riverina
L. Marmion
Knobby Head
Beagle I.
Three Springs
Yarra Yarra Lakes
Carnamah
Latham
Coorow
Mt Beringbnurding 447
Lake Moore
Menzies
Leeman
Eneabba
Buntine
L. Goongarrie
Green Head
Mt Lesueur 311
Hill
Wubin
Jurien
Watheroo
Dalwallinu
Miling
Pithara
Kalannie
Kulja
L. Grady
Beacon
Broad Arrow
Burakin
L. Mollerin
Walyahmoing 484
Bonnie Rock
L. Deborah
Mt Burges 554
L. Yindarlgooda
Kalgoorlie
Cervantes
Dandaragan
Moora
Cadoux
Koorda
Bencubbin
Mukinbudin
Koolyanobbing
Curtin
Boulder
Lancelin
Wongan Hills
Bullfinch
L. Seabrook
Coolgardie
Karakin Lakes
Moore
Dandaragan
Piawaning
Cowcowing Lakes
L. Wallambin
Southern Cross
New Norcia
Wyalkatchem
Trayning
L. Brown
Nungarin
Yellowdine
Kambalda
Parker Hill 406
Gingin
Goomalling
Bodallin
L. Lefroy
Widgiemooltha
Yanchep
Wanneroo
Muchea
Swan
Toodyay
Northam
Kellerberrin
Cunderdin
Merredin
Mt Stirling 376
Muntadgin
Parker Ra.
Barker Lake
Lake Cowan
Perth
Midland Jct
York
Bruce Rock 378
Mt Thirsty 431
Mt Norcott 420
Rottnest I.
Fremantle
Garden I.
Armadale
Kwinana
Avon
Beverley
Quairading
Narembeen
The Johnston Lakes
Norseman
Rockingham
Serpentine
Mt Cooke 582
Brookton
Corrigin
Hyden
L. Carmody
L. Dundas
Bremer Range
Mandurah
Peel Inlet
Pinjarra
Pingelly
Dwellingup
Hotham
Wickepin
Kulin
Kondinin
Pingaring
L. Hope
L. Tay
Charles Peak 558
Salmon Gums
Waroona
Boddington
Murray
Yilliminning
Harrismith
Lake Grace
Lake King
Narrogin
Williams
Harvey
Mt William
Darling Range
Brunswick
Bunbury
Collie
Darkan
Williams
Wagin
Kukerin
L. Grace
Newdegate
L. Magenta
Oldfield
Young
Lort
Grass Patch
Scaddan
Dumbleyung
L. Dumbleyung
Nyabing
L. Chinocup
Ravensthorpe
Gibson
C. Naturaliste
Geographe Bay
Dunsborough
Yallingup
Boyanup
Donnybrook
Capel
Busselton
Greenbushes
Boyup Brook
Katanning
Broomehill
Jerramungup
Phillips
Hopetoun
Culham Inlet
Esperance
Stokes Inlet
Butty Hd
Esperance B.
Margaret River
C. Freycinet
Nannup
Bridgetown
Blackwood
Kojonup
Tambellup
Gnowangerup
Borden
Ongerup
Fitzgerald River Nat. Park
Gairdner
Hood Pt
Bremer Bay
Bremer B.
C. Le Grand
Termination I.
Augusta
C. Leeuwin
Flinders Bay
Manjimup
Pemberton
Shannon Nat. Park
Gordon
Cranbrook
Stirling Range
Pallinup
Bluff Knoll 1110
Rocky Gully
Kalgan
Stirling Range Nat. Park
Cheyne B.
Northcliffe
d'Entrecasteaux Nat. Park
Gardner
Deep
Frankland
Mount Barker
Twin Peaks 654
Mt Lindesay 448
Mt Many Peaks 565
Channel Pt
Albany
C. Vancouver
King George Sound
Denmark
Elleker
Pt d'Entrecasteaux
Walpole
Nornalup
Pt Nuyts
Walpole-Nornalup Nat. Park
West Cape Howe
Tor Bay
Bald Hd
112°E
114°
116°
118°
120°
122°
24°S
26°
28°
30°
32°
34°S
A
B
C
D
E
1
2
3
4
5
6
7
METRES
FEET
6000
19686
5000
16409
4000
13124
3000
9843
2000
6562
1000
3281
500
1640
200
656
SEA
LEVEL
200
656
2000
6562
4000
13124
6000
19686
1:5M
KM
MILES
300
250
200
150
100
50
0
200
150
100
50
0

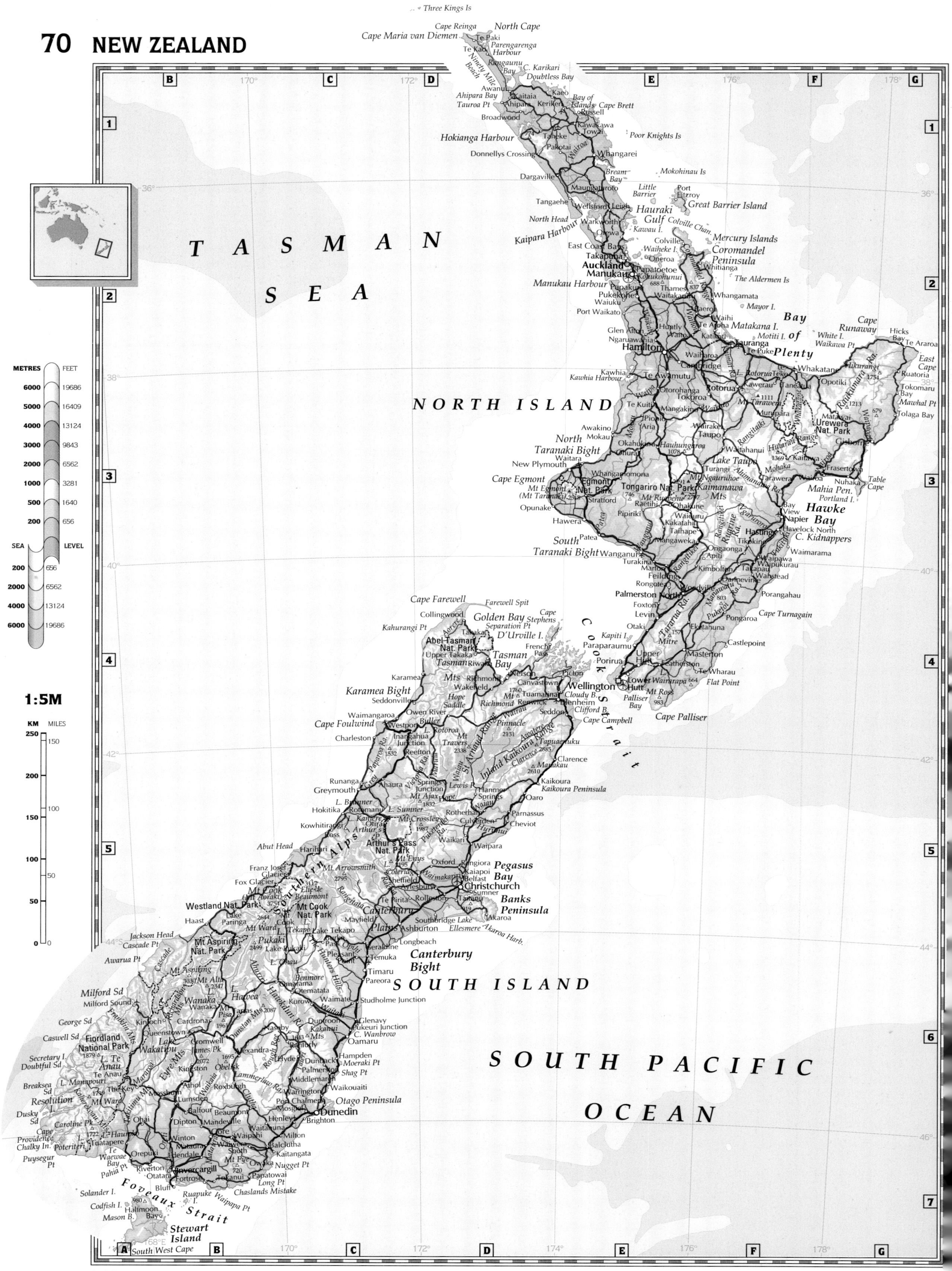

TASMAN SEA
SOUTH PACIFIC OCEAN
NORTH ISLAND
SOUTH ISLAND
Three Kings Is
Cape Reinga
North Cape
Cape Maria van Diemen
Parengarenga Harbour
Ninety Mile Beach
Rangaunu Bay
Doubtless Bay
Bay of Islands
Cape Brett
Kaitaia
Kerikeri
Russell
Kawakawa
Hokianga Harbour
Poor Knights Is
Whangarei
Donnellys Crossing
Dargaville
Bream Bay
Mokohinau Is
Little Barrier
Great Barrier Island
Hauraki Gulf
Kaipara Harbour
Mercury Islands
Coromandel Peninsula
Auckland
Manukau
Manukau Harbour
Thames
The Aldermen Is
Mayor I.
Bay of Plenty
Matakana I.
Motiti I.
White I.
Cape Runaway
Hicks Bay
East Cape
Tauranga
Te Puke
Hamilton
Whakatane
Opotiki
Rotorua
Kawhia Harbour
Te Awamutu
Te Kuiti
Taupo
Lake Taupo
Urewera Nat. Park
Gisborne
Tolaga Bay
North Taranaki Bight
New Plymouth
Cape Egmont
Mt Egmont (Mt Taranaki)
Egmont Nat. Park
Tongariro Nat. Park
Mt Ruapehu
Kaimanawa Mts
Mahia Pen.
Table Cape
Hawke Bay
Napier
Hastings
C. Kidnappers
South Taranaki Bight
Wanganui
Palmerston North
Cape Turnagain
Castlepoint
Masterton
Wellington
Lower Hutt
Upper Hutt
Cape Palliser
Palliser Bay
Cook Strait
Kapiti I.
Cape Farewell
Farewell Spit
Golden Bay
Cape Stephens
D'Urville I.
Abel Tasman Nat. Park
Tasman Bay
Kahurangi Pt
Nelson
Picton
Blenheim
Cloudy B.
Clifford B.
Cape Campbell
Karamea Bight
Cape Foulwind
Westport
Greymouth
Hokitika
Kaikoura
Kaikoura Peninsula
Arthur's Pass Nat. Park
Southern Alps
Pegasus Bay
Christchurch
Banks Peninsula
Lake Ellesmere
Westland Nat. Park
Mt Cook
Mt Cook Nat. Park
Mt Aspiring Nat. Park
Canterbury Plains
Ashburton
Canterbury Bight
Timaru
Oamaru
Jackson Head
Cascade Pt
Awarua Pt
Milford Sd
George Sd
Caswell Sd
Fiordland National Park
Queenstown
L. Wakatipu
L. Te Anau
L. Manapouri
Dunedin
Otago Peninsula
Nugget Pt
Invercargill
Bluff
Foveaux Strait
Stewart Island
South West Cape
Solander I.
Puysegur Pt
Chalky In.
Dusky Sd
Breaksea Sd
Doubtful Sd
METRES
FEET
6000 19686
5000 16409
4000 13124
3000 9843
2000 6562
1000 3281
500 1640
200 656
SEA LEVEL
200 656
2000 6562
4000 13124
6000 19686
1:5M
KM MILES

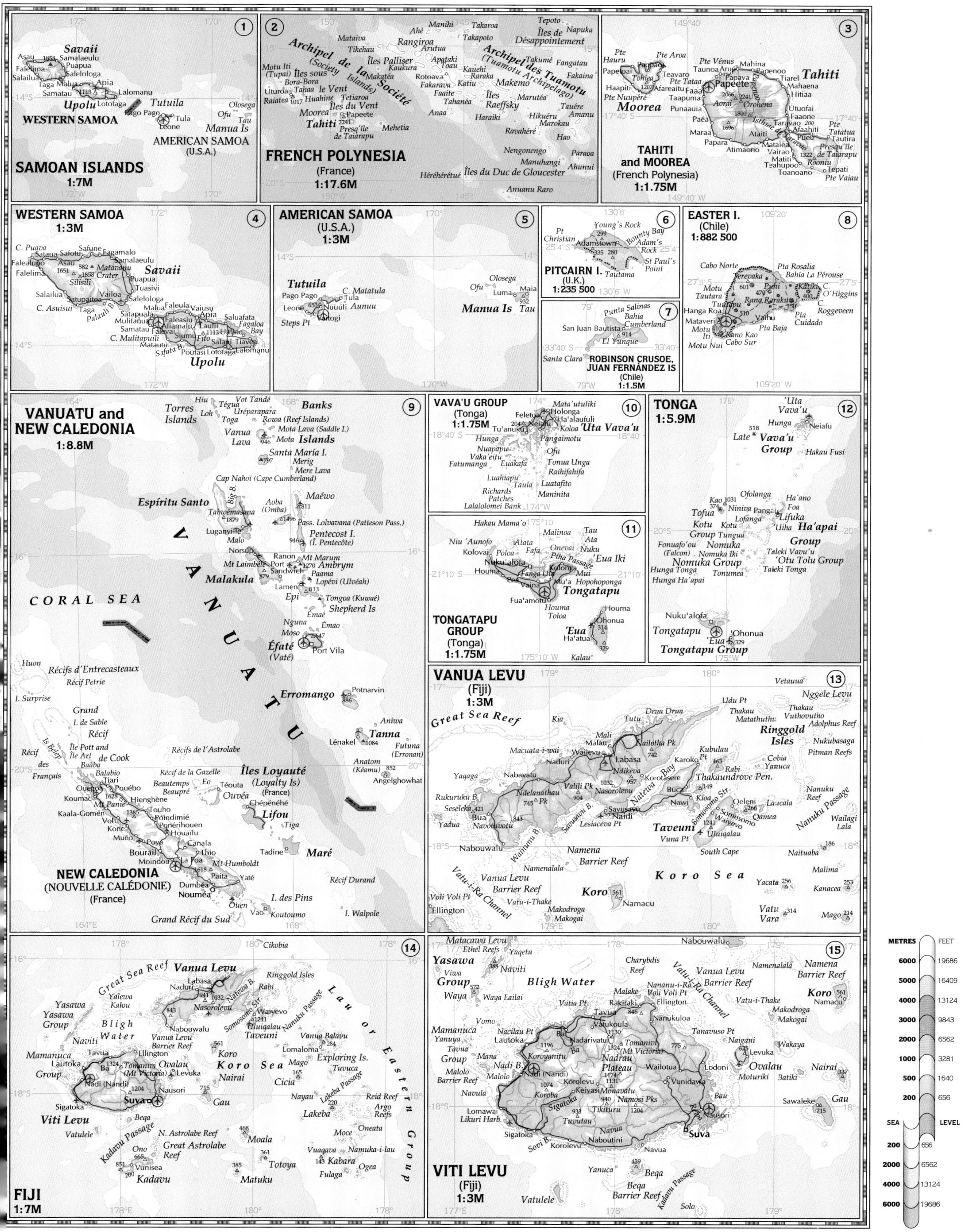
SAMOAN ISLANDS
1:7M
Savaii
Upolu
WESTERN SAMOA
Tutuila
Manua Is
AMERICAN SAMOA
(U.S.A.)
FRENCH POLYNESIA
(France)
1:17.6M
Archipel de la Société
(Society Islands)
Archipel des Tuamotu
(Tuamotu Archipelago)
Tahiti
TAHITI and MOOREA
(French Polynesia)
1:1.75M
Moorea
Papeete
WESTERN SAMOA
1:3M
AMERICAN SAMOA
(U.S.A.)
1:3M
PITCAIRN I.
(U.K.)
1:235 500
ROBINSON CRUSOE,
JUAN FERNÁNDEZ IS
(Chile)
1:1.5M
EASTER I.
(Chile)
1:882 500
VANUATU and
NEW CALEDONIA
1:8.8M
CORAL SEA
VANUATU
NEW CALEDONIA
(NOUVELLE CALÉDONIE)
(France)
Îles Loyauté
(Loyalty Is)
(France)
VAVA'U GROUP
(Tonga)
1:1.75M
TONGATAPU
GROUP
(Tonga)
1:1.75M
TONGA
1:5.9M
VANUA LEVU
(Fiji)
1:3M
Koro Sea
FIJI
1:7M
VITI LEVU
(Fiji)
1:3M
Bligh Water
Suva
METRES
FEET
6000
19686
5000
16409
4000
13124
3000
9843
2000
6562
1000
3281
500
1640
200
656
SEA
LEVEL
200
656
2000
6562
4000
13124
6000
19686

NORTH AMERICA Relief

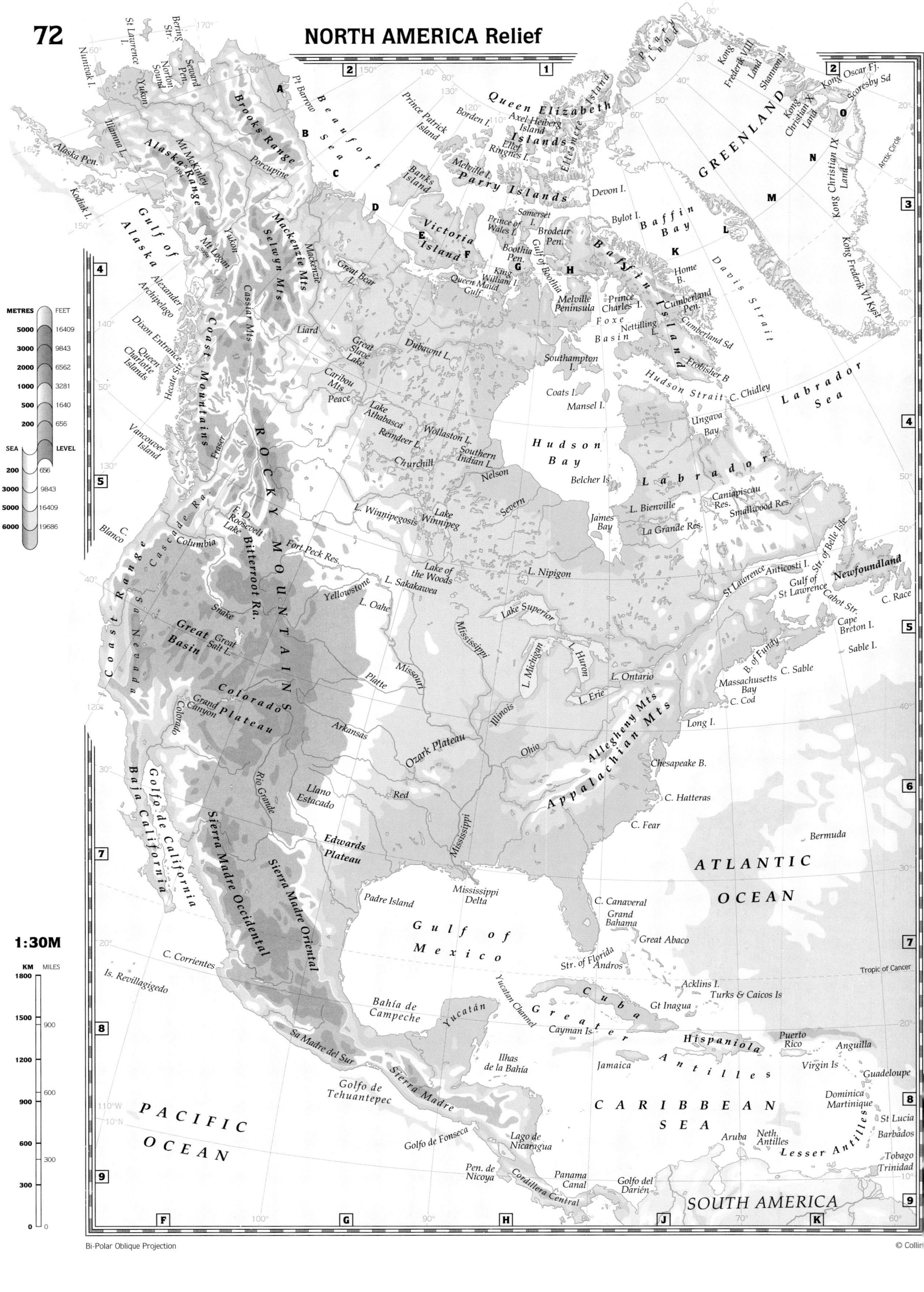
METRES
FEET
5000
16409
3000
9843
2000
6562
1000
3281
500
1640
200
656
SEA
LEVEL
200
656
3000
9843
5000
16409
6000
19686
1:30M
KM
MILES
1800
1500
1200
900
600
300
0
900
600
300
0
Brooks Range
Beaufort Sea
Pt Barrow
Alaska Range
Mt McKinley
Mt Logan
Alaska Pen.
Kodiak I.
Iliamna L.
Yukon
Norton Sound
Seward Pen.
Bering Str.
St Lawrence I.
Nunivak I.
Gulf of Alaska
Alexander Archipelago
Dixon Entrance
Queen Charlotte Islands
Hecate Str.
Vancouver Island
Coast Mountains
Mackenzie Mts
Selwyn Mts
Cassiar Mts
Mackenzie
Porcupine
Great Bear L.
Great Slave Lake
Liard
Peace
Caribou Mts
Lake Athabasca
Reindeer L.
Wollaston L.
Southern Indian L.
Churchill
Nelson
Dubawnt L.
Prince Patrick Island
Borden I.
Queen Elizabeth Islands
Axel Heiberg Island
Ellef Ringnes I.
Ellesmere Island
Melville I.
Parry Islands
Banks Island
Victoria Island
Devon I.
Somerset I.
Prince of Wales I.
Boothia Pen.
Brodeur Pen.
Gulf of Boothia
King William I.
Queen Maud Gulf
Melville Peninsula
Prince Charles I.
Bylot I.
Baffin Bay
Baffin Island
Home B.
Cumberland Pen.
Cumberland Sd
Nettilling L.
Foxe Basin
Frobisher B
Southampton I.
Coats I.
Mansel I.
Hudson Strait
C. Chidley
Labrador Sea
Davis Strait
Ungava Bay
Hudson Bay
Belcher Is
James Bay
Labrador
L. Bienville
Caniapiscau Res.
Smallwood Res.
La Grande Res.
Severn
Peary Land
GREENLAND
Kong Frederik VIII Land
Shannon I.
Kong Oscar Fj.
Scoresby Sd
Kong Christian X Land
Kong Christian IX Land
Kong Frederik VI Kyst
Arctic Circle
ROCKY MOUNTAINS
Fraser
Cascade Ra.
Coast Range
Columbia
F. D. Roosevelt Lake
Bitterroot Ra.
Fort Peck Res.
Yellowstone
L. Sakakawea
L. Oahe
Snake
Great Basin
Great Salt L.
Sierra Nevada
C. Blanco
Colorado Plateau
Grand Canyon
Colorado
Platte
Missouri
Arkansas
Lake Winnipeg
L. Winnipegosis
Lake of the Woods
L. Nipigon
Lake Superior
Mississippi
L. Michigan
L. Huron
L. Ontario
L. Erie
Illinois
Ohio
Ozark Plateau
Allegheny Mts
Appalachian Mts
St Lawrence
Anticosti I.
Gulf of St Lawrence
Str. of Belle Isle
Newfoundland
Cabot Str.
C. Race
Cape Breton I.
Sable I.
B. of Fundy
C. Sable
Massachusetts Bay
C. Cod
Long I.
Chesapeake B.
C. Hatteras
C. Fear
Bermuda
ATLANTIC OCEAN
Baja California
Golfo de California
Rio Grande
Llano Estacado
Red
Edwards Plateau
Sierra Madre Occidental
Sierra Madre Oriental
Mississippi Delta
Padre Island
Gulf of Mexico
C. Canaveral
Grand Bahama
Great Abaco
Str. of Florida
Andros
Acklins I.
Turks & Caicos Is
Gt Inagua
Cuba
Yucatan Channel
Cayman Is
Greater Antilles
Hispaniola
Puerto Rico
Anguilla
Virgin Is
Guadeloupe
Dominica
Martinique
St Lucia
Barbados
Tobago
Trinidad
Jamaica
Lesser Antilles
Neth. Antilles
Aruba
CARIBBEAN SEA
Tropic of Cancer
C. Corrientes
Is. Revillagigedo
Bahía de Campeche
Yucatán
Ilhas de la Bahía
Sa Madre del Sur
Sierra Madre
Golfo de Tehuantepec
Golfo de Fonseca
Lago de Nicaragua
Pen. de Nicoya
Cordillera Central
Panama Canal
Golfo del Darién
SOUTH AMERICA
PACIFIC OCEAN

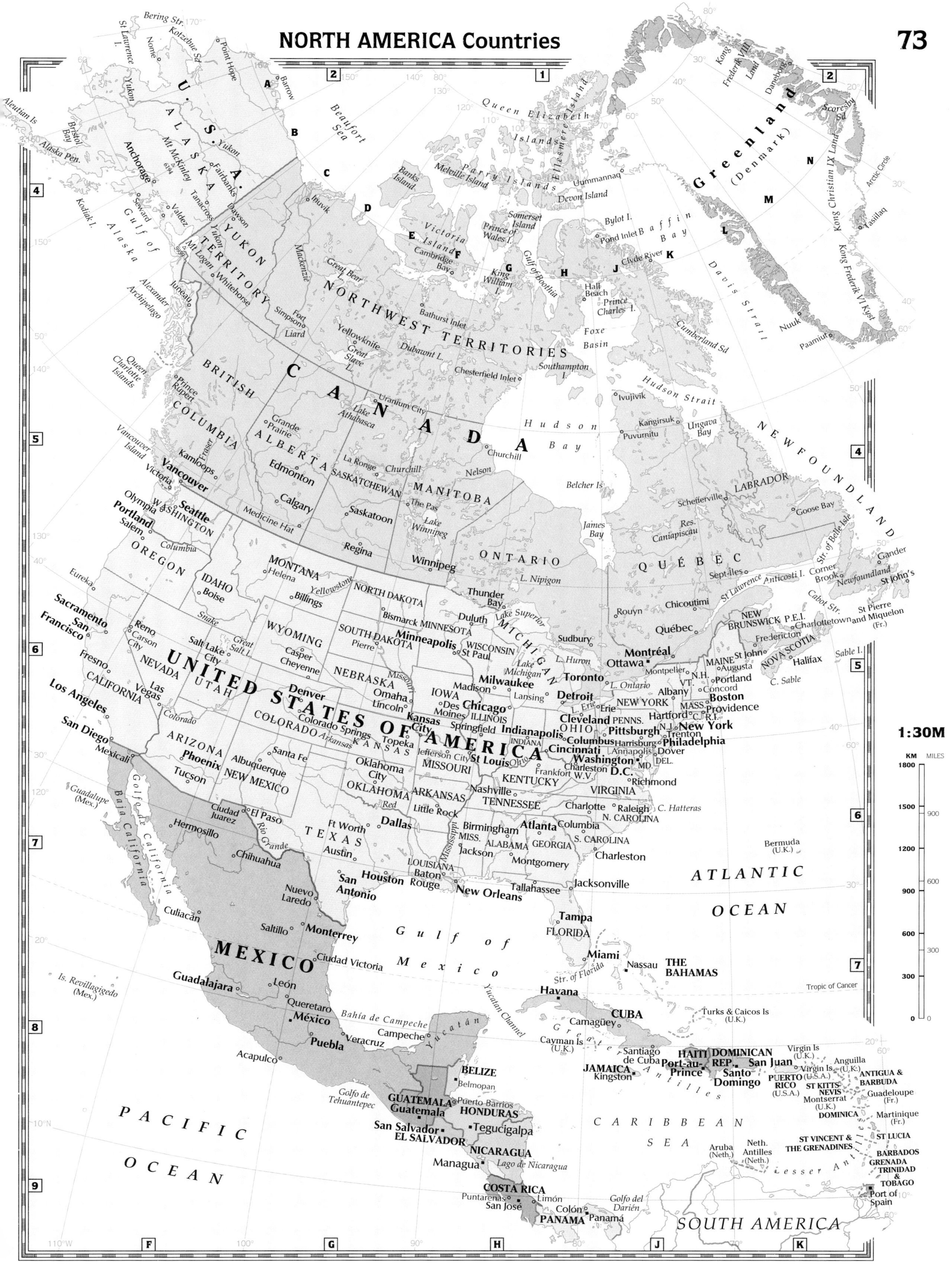
1:30M
KM MILES
ATLANTIC OCEAN
PACIFIC OCEAN
CARIBBEAN SEA
Gulf of Mexico
Hudson Bay
Baffin Bay
Davis Strait
Hudson Strait
Beaufort Sea
Gulf of Alaska
Foxe Basin
James Bay
Ungava Bay
Greenland (Denmark)
CANADA
UNITED STATES OF AMERICA
U.S.A.
ALASKA
MEXICO
THE BAHAMAS
CUBA
JAMAICA
HAITI
DOMINICAN REP.
PUERTO RICO (U.S.A.)
BELIZE
GUATEMALA
HONDURAS
EL SALVADOR
NICARAGUA
COSTA RICA
PANAMA
SOUTH AMERICA
YUKON TERRITORY
NORTHWEST TERRITORIES
BRITISH COLUMBIA
ALBERTA
SASKATCHEWAN
MANITOBA
ONTARIO
QUÉBEC
NEWFOUNDLAND
LABRADOR
NEW BRUNSWICK
NOVA SCOTIA
P.E.I.
Anchorage
Fairbanks
Juneau
Whitehorse
Inuvik
Yellowknife
Churchill
Vancouver
Victoria
Edmonton
Calgary
Saskatoon
Regina
Winnipeg
Thunder Bay
Toronto
Ottawa
Montréal
Québec
Halifax
St John's
Seattle
Portland
San Francisco
Sacramento
Los Angeles
San Diego
Las Vegas
Salt Lake City
Denver
Phoenix
Dallas
Houston
San Antonio
New Orleans
Chicago
Detroit
Minneapolis
St Louis
Kansas City
Atlanta
Miami
Washington D.C.
Philadelphia
New York
Boston
Havana
Nassau
Kingston
Port-au-Prince
Santo Domingo
San Juan
México
Guadalajara
Monterrey
Puebla
Guatemala
San Salvador
Tegucigalpa
Managua
San José
Panamá
Port of Spain
Tropic of Cancer
Arctic Circle

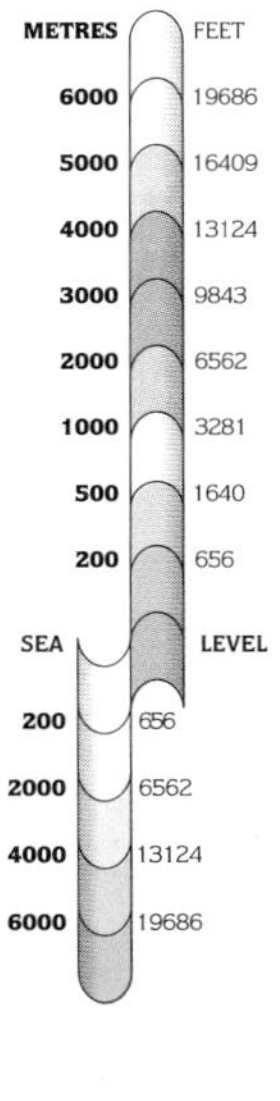

Chamberlin Trimetric Projection

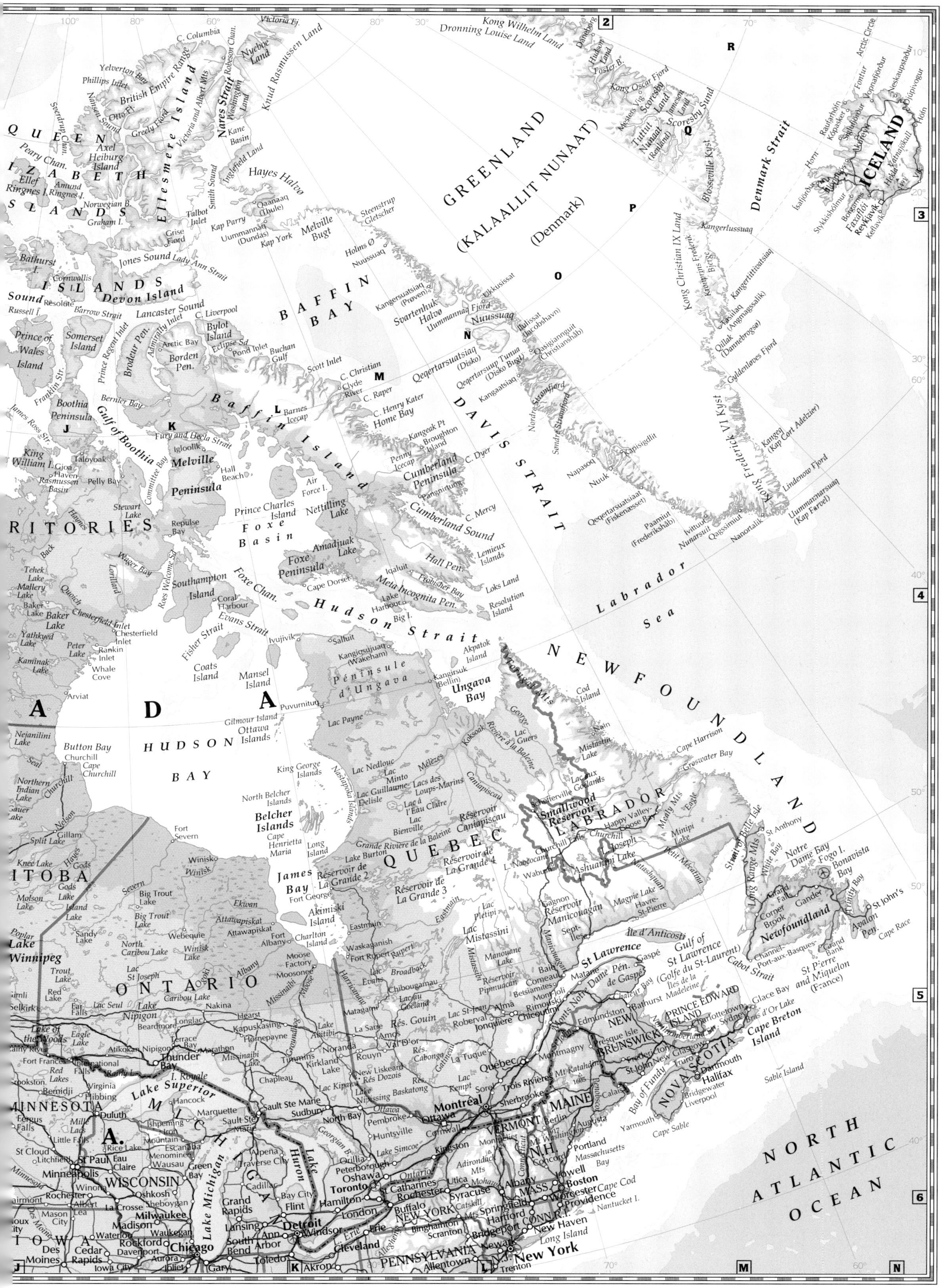

KM / MILES scale bar: 0, 200, 400, 600, 800, 1000 km; 0, 100, 200, 300, 400, 500, 600 miles

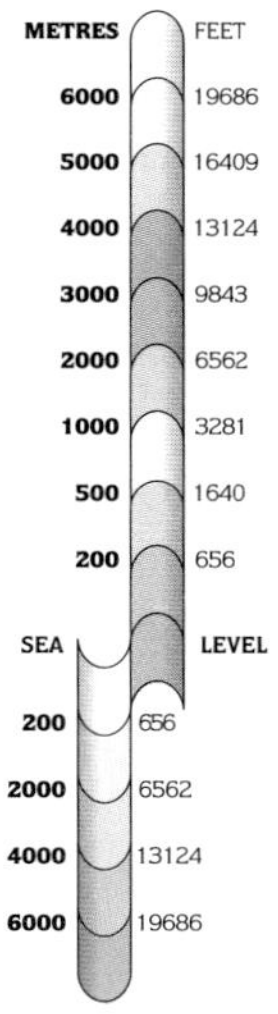

Transverse Mercator Projection

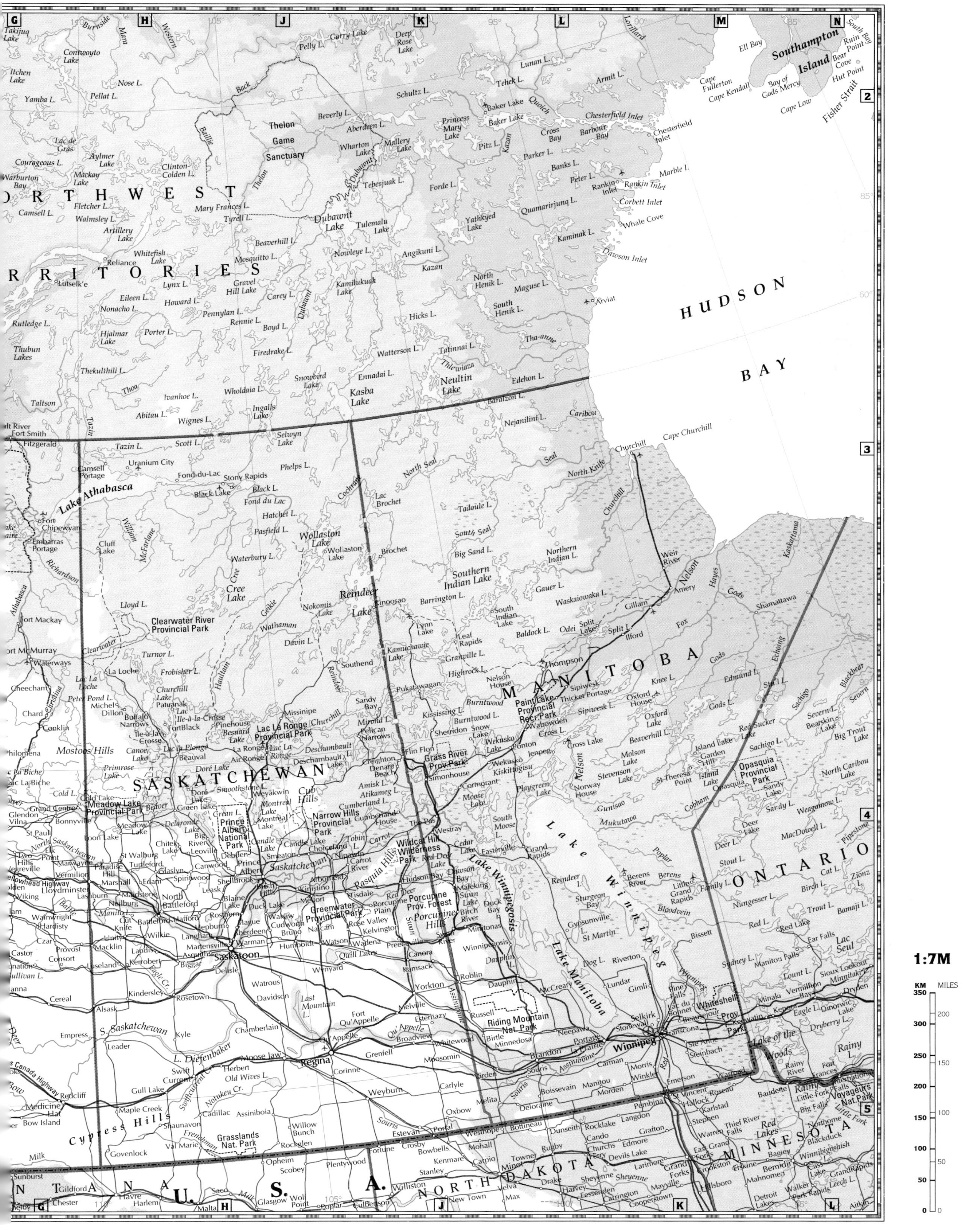
NORTHWEST TERRITORIES
HUDSON BAY
SASKATCHEWAN
MANITOBA
ONTARIO
NORTH DAKOTA
MINNESOTA
U.S.A.
Southampton Island
Thelon Game Sanctuary
Dubawnt Lake
Kasba Lake
Nueltin Lake
Lake Athabasca
Wollaston Lake
Reindeer Lake
Southern Indian Lake
Cree Lake
Clearwater River Provincial Park
Lac La Ronge Provincial Park
Prince Albert National Park
Narrow Hills Provincial Park
Meadow Lake Provincial Park
Greenwater Provincial Park
Grass River Prov. Park
Paint Lake Provincial Recr. Park
Wildcat Hill Wilderness Park
Porcupine Prov. Forest
Opasquia Provincial Park
Riding Mountain Nat. Park
Whiteshell Prov. Park
Grasslands Nat. Park
Voyageurs Nat. Park
Lake Winnipeg
Lake Winnipegosis
Lake Manitoba
Lake of the Woods
Cypress Hills
Churchill
Cape Churchill
Arviat
Rankin Inlet
Chesterfield Inlet
Baker Lake
Whale Cove
Uranium City
Fort Chipewyan
Fort Smith
Fort McMurray
La Loche
Flin Flon
The Pas
Thompson
Gillam
Lynn Lake
Saskatoon
Regina
Moose Jaw
Swift Current
Medicine Hat
Lloydminster
North Battleford
Prince Albert
Yorkton
Brandon
Winnipeg
Dauphin
Selkirk
Portage la Prairie
Estevan
Weyburn
Kenora
Dryden
Sioux Lookout
Red Lake
Minot
Grand Forks
Williston
Fort Frances
1:7M
KM 350 300 250 200 150 100 50 0
MILES 200 150 100 50 0

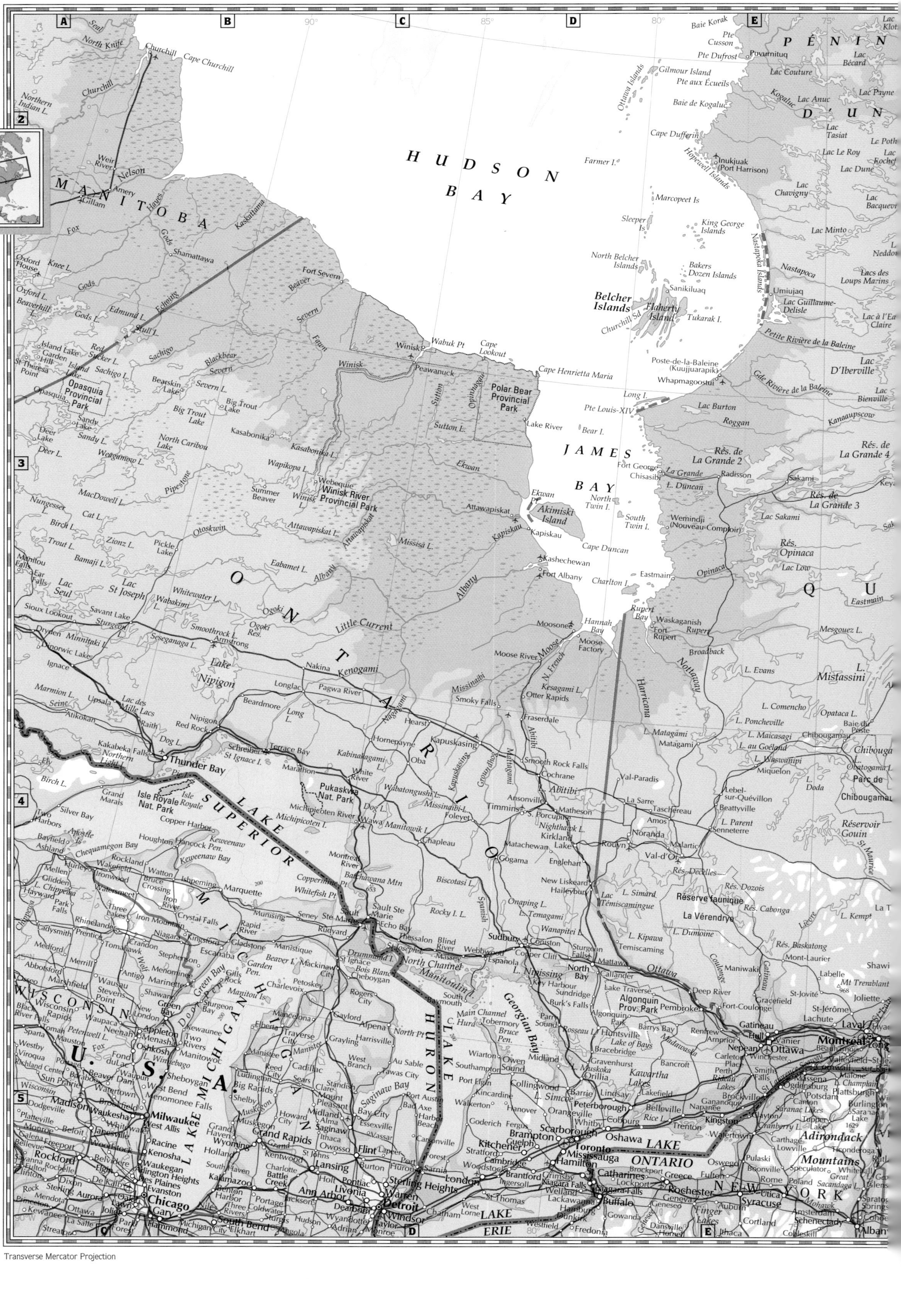

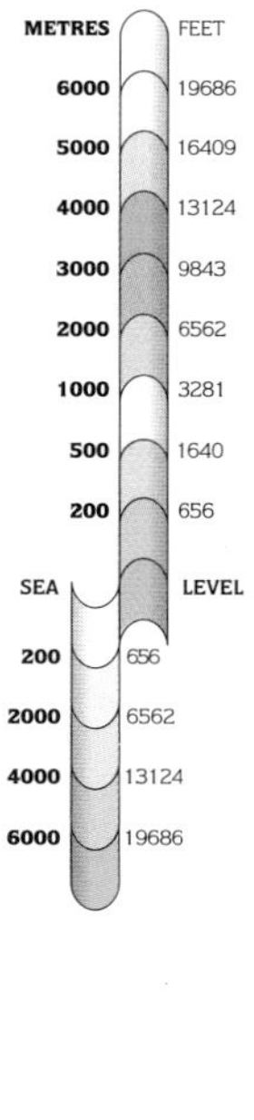

Transverse Mercator Projection

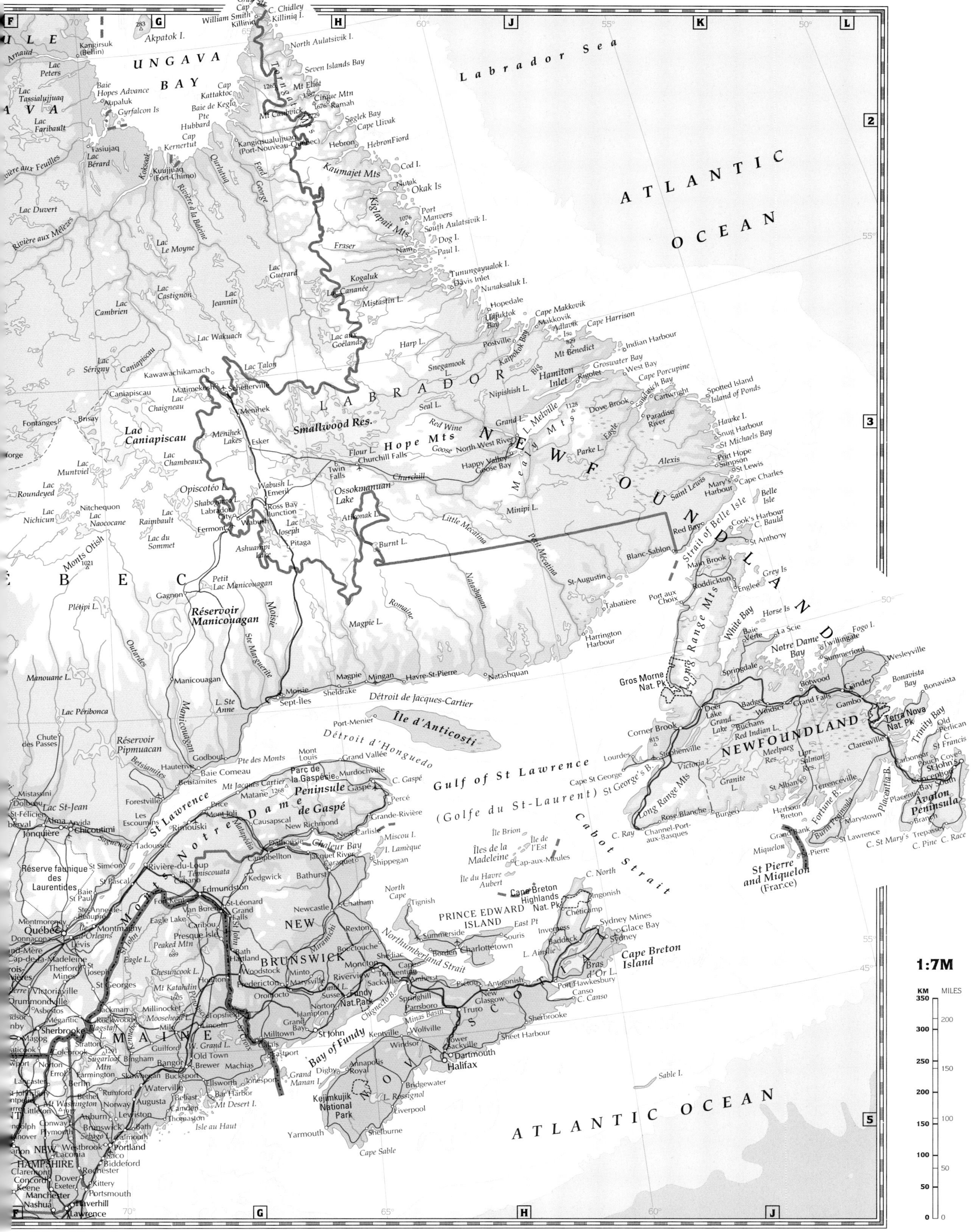
UNGAVA BAY
Akpatok I.
C. Chidley
Killiniq I.
North Aulatsivik I.
Labrador Sea
Seven Islands Bay
Mt Eliot
Cirque Mtn
Ramah
Saglek Bay
Cape Uivak
Hebron
HebronFiord
Kaumajet Mts
Cod I.
Nutak
Okak Is
Kiglapait Mts
Port Manvers
South Aulatsivik I.
Dog I.
Paul I.
Nain
Fraser
Tunungayualok I.
Davis Inlet
Nunaksaluk I.
Kogaluk
Lac Cananée
Mistastin L.
Hopedale
Cape Makkovik
Makkovik
Adlavik Is
Cape Harrison
Postville
Kaipokok Bay
Mt Benedict
Indian Harbour
Groswater Bay
West Bay
Rigolet
Hamilton Inlet
Cape Porcupine
Sandwich Bay
Cartwright
Spotted Island
Island of Ponds
Paradise River
Dove Brook
ATLANTIC OCEAN
LABRADOR
NEWFOUNDLAND
Smallwood Res.
Hope Mts
Mealy Mts
Lake Melville
Goose
North West River
Happy Valley-Goose Bay
Churchill Falls
Churchill
Twin Falls
Grand L.
Harp L.
Snegamook L.
Nipishish L.
Seal L.
Red Wine
Flour L.
Parke L.
Eagle
Alexis
Hawke I.
Snug Harbour
St Michaels Bay
Port Hope Simpson
St Lewis
Mary's Harbour
Cape Charles
Belle Isle
Cook's Harbour
C. Bauld
St Anthony
Strait of Belle Isle
Red Bay
Blanc-Sablon
St-Augustin
Burnt L.
Little Mecatina
Petit Mecatina
Natashquan
Minipi L.
Ossokmanuan Lake
Atikonak L.
Schefferville
Kawawachikamach
Matimekosh
Lac Talon
Menihek
Menihek Lakes
Esker
Lac Caniapiscau
Caniapiscau
Wabush L.
Emeril
Wabush
Labrador City
Shabogamo L.
Fermont
Ross Bay Junction
Lac Joseph
Ashuanipi Lake
Pitaga
Opiscotéo L.
Kujjuaq (Fort-Chimo)
Kangiqsualujjuaq (Port-Nouveau-Québec)
Kangirsuk (Bellin)
Lac Peters
Lac Tassialujjuaq
Lac Faribault
Lac Bérard
Tasiujaq
Lac Duvert
Lac Le Moyne
Lac Guerard
Lac Castignon
Lac Jeannin
Lac Cambrien
Lac Wakuach
Lac Sérigny
Lac Chaigneau
Lac Chambeaux
Lac Muntviel
Lac Roundeyed
Lac Nichicun
Nitchequon
Lac Naococane
Lac Raimbault
Lac du Sommet
Monts Otish
Gagnon
Petit Lac Manicouagan
Réservoir Manicouagan
Plétipi L.
Moisie
Ste Marguerite
Romaine
Magpie L.
Magpie
Mingan
Havre-St-Pierre
Sept-Îles
Sheldrake
Manicouagan
Manouane L.
Lac Péribonca
Réservoir Pipmuacan
Chute des Passes
Outardes
Godbout
L. Ste Anne
Port-Aux Choix
Main Brook
Roddickton
Englee
Grey Is
Tabatière
Harrington Harbour
White Bay
Horse Is
Baie Verte
La Scie
Fogo I.
Notre Dame Bay
Twillingate
Summerford
Wesleyville
Springdale
Botwood
Gander
Bonavista Bay
Bonavista
Gros Morne Nat. Pk
Long Range Mts
Deer Lake
Grand Lake
Badger
Windsor
Grand Falls
Gambo
Terra Nova Nat. Pk
Trinity Bay
Corner Brook
Buchans
Red Indian L.
NEWFOUNDLAND
Meelpaeg Res.
Upper Salmon Res.
Clarenville
Stephenville
Victoria L.
St George's B.
Cape St George
Carbonear
Conception Bay
St John's
Placentia B.
Avalon Peninsula
C. Race
C. Pine
Trepassey
C. St Mary's
St Lawrence
Marystown
Burin Peninsula
Fortune B.
Harbour Breton
Burgeo
Rose Blanche
Channel-Port-aux-Basques
C. Ray
Cabot Strait
St Pierre and Miquelon (France)
Miquelon
St-Pierre
Détroit de Jacques-Cartier
Île d'Anticosti
Port-Menier
Détroit d'Honguedo
Gulf of St Lawrence
(Golfe du St-Laurent)
Îles de la Madeleine
Île Brion
Cap-aux-Meules
Gaspé
C. Gaspé
Péninsule de Gaspé
Parc de la Gaspésie
Mont Jacques Cartier
Matane
Mont Joli
Rimouski
Percé
Chaleur Bay
Campbellton
Bathurst
Caraquet
Shippegan
Miscou I.
Baie Comeau
Forestville
Les Escoumins
Tadoussac
Chicoutimi
Jonquière
Alma
Lac St-Jean
Dolbeau
Mistassini
St-Félicien
Roberval
St Lawrence
Rivière-du-Loup
Edmundston
Québec
Lévis
Thetford Mines
Victoriaville
Drummondville
Sherbrooke
NEW BRUNSWICK
Fredericton
Moncton
Saint John
Bay of Fundy
Fundy Nat. Park
Miramichi
Chatham
Newcastle
Kedgwick
Grand Falls
Woodstock
Oromocto
Sussex
Sackville
Amherst
Northumberland Strait
PRINCE EDWARD ISLAND
Charlottetown
Summerside
Tignish
North Cape
East Pt
Souris
NOVA SCOTIA
Halifax
Dartmouth
Truro
New Glasgow
Antigonish
Pictou
Kentville
Wolfville
Windsor
Annapolis Royal
Digby
Yarmouth
Shelburne
Cape Sable
Liverpool
Bridgewater
Kejimkujik National Park
Sherbrooke
Sheet Harbour
Canso
C. Canso
Port Hawkesbury
Cape Breton Island
Cape Breton Highlands Nat. Pk
Chéticamp
Ingonish
C. North
Sydney
Sydney Mines
Glace Bay
Bras d'Or L.
Inverness
Baddeck
Sable I.
Grand Manan I.
MAINE
Bangor
Augusta
Portland
Lewiston
Waterville
Bar Harbor
Mt Desert I.
Machias
Calais
Eastport
Millinocket
Presque Isle
Caribou
Houlton
Mt Katahdin
Chesuncook L.
Moosehead L.
Belfast
Rockland
Camden
Bath
Brunswick
Biddeford
Saco
Kittery
Portsmouth
Dover
Exeter
Manchester
Concord
Nashua
Haverhill
Lawrence
Keene
NEW HAMPSHIRE
Mt Washington
Berlin
Laconia
ATLANTIC OCEAN
1:7M
KM
MILES
350
300
250
200
150
100
50
0

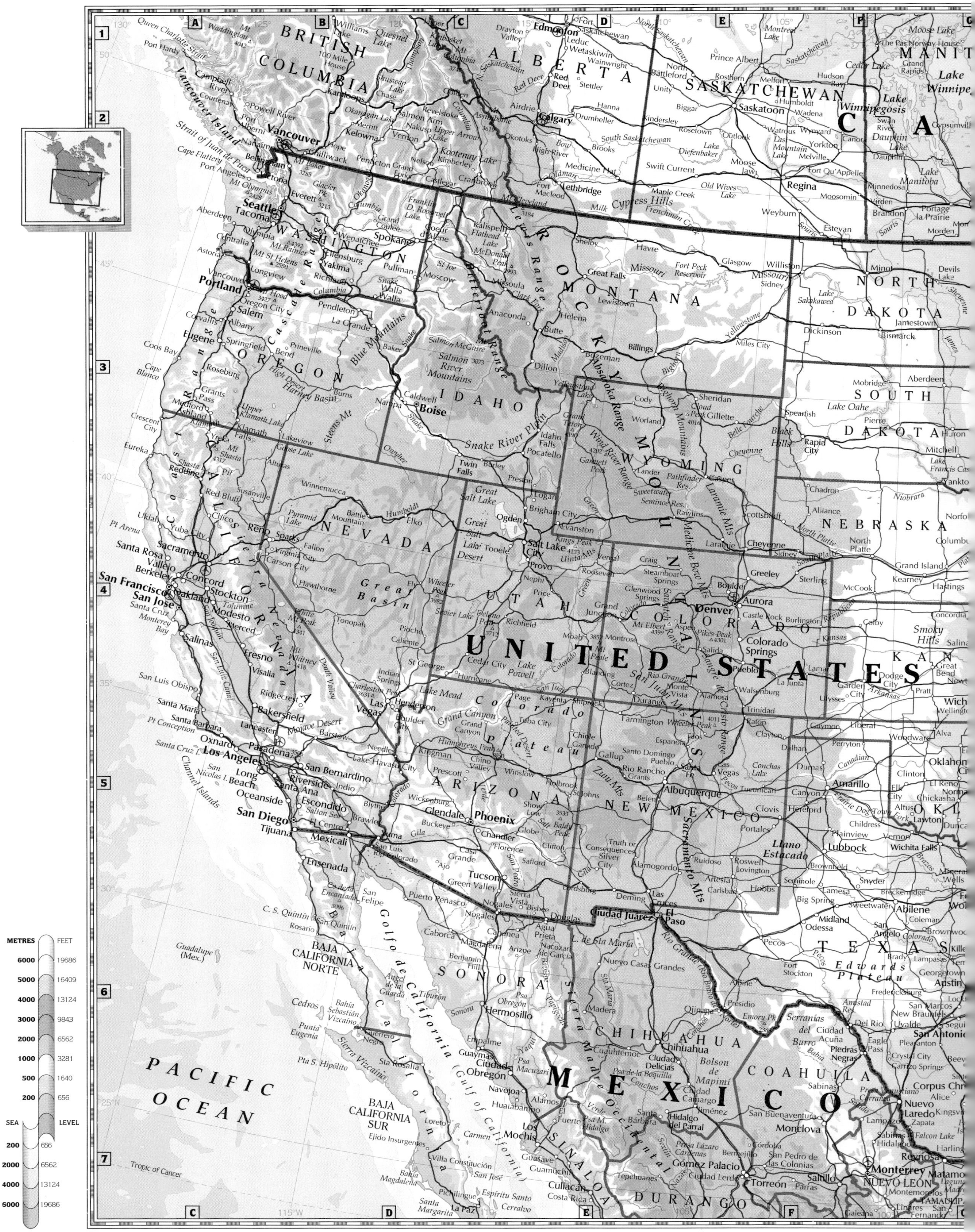

Lambert Conformal Conic Projection

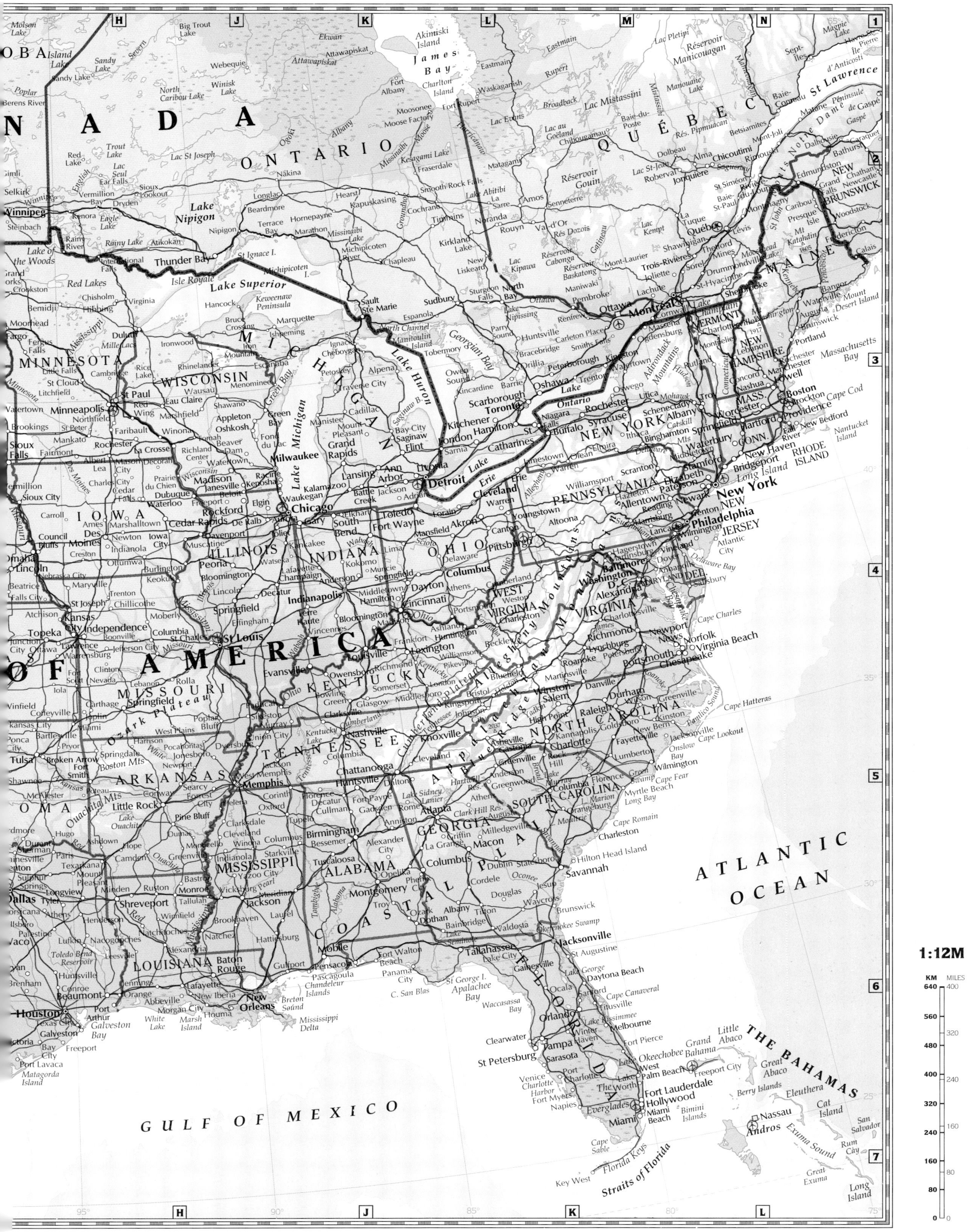
CANADA
ONTARIO
QUÉBEC
NEW BRUNSWICK
UNITED STATES OF AMERICA
MINNESOTA
WISCONSIN
MICHIGAN
IOWA
ILLINOIS
INDIANA
OHIO
MISSOURI
KENTUCKY
TENNESSEE
ARKANSAS
MISSISSIPPI
ALABAMA
GEORGIA
LOUISIANA
FLORIDA
SOUTH CAROLINA
NORTH CAROLINA
VIRGINIA
WEST VIRGINIA
PENNSYLVANIA
NEW YORK
NEW JERSEY
MARYLAND
DEL.
CONN.
RHODE ISLAND
MASS.
VERMONT
NEW HAMPSHIRE
MAINE
THE BAHAMAS
James Bay
Lake Superior
Lake Michigan
Lake Huron
Lake Erie
Lake Ontario
Georgian Bay
Lake Nipigon
ATLANTIC OCEAN
GULF OF MEXICO
Straits of Florida
Appalachian Mountains
Coastal Plain
Ozark Plateau
Cumberland Plateau
Blue Ridge
Winnipeg
Thunder Bay
Duluth
Minneapolis
St Paul
Milwaukee
Chicago
Detroit
Toronto
Ottawa
Montréal
Québec
Boston
New York
Philadelphia
Baltimore
Washington D.C.
Pittsburgh
Cleveland
Columbus
Cincinnati
Indianapolis
Louisville
St Louis
Kansas City
Memphis
Nashville
Atlanta
Charlotte
Jacksonville
Tampa
Miami
New Orleans
Houston
Dallas
Birmingham
Savannah
Charleston
Norfolk
Virginia Beach
Nassau
Key West
1:12M
KM
MILES

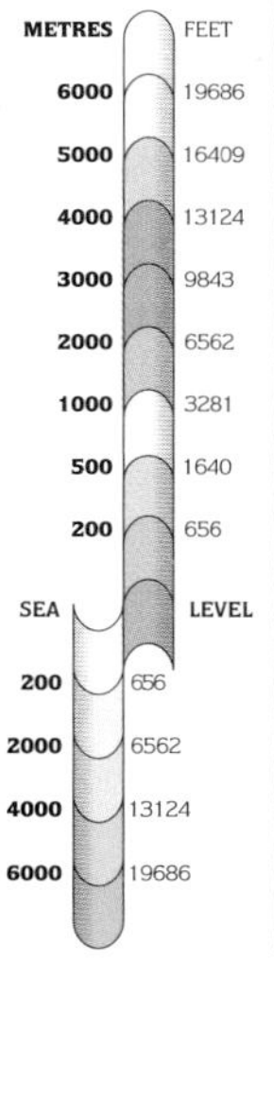

Lambert Conformal Conic Projection

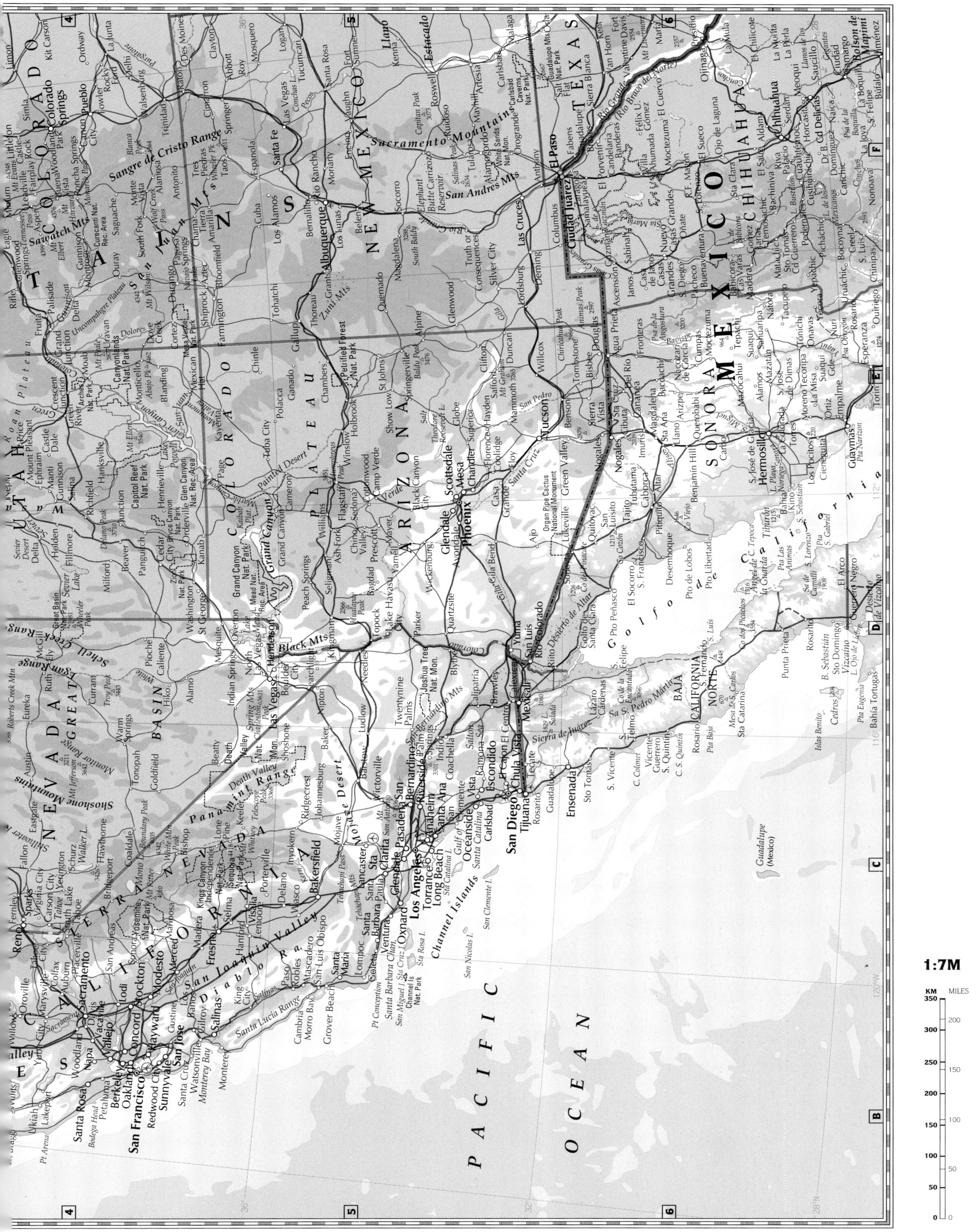
PACIFIC
OCEAN
CALIFORNIA
NEVADA
UTAH
ARIZONA
NEW MEXICO
COLORADO
TEXAS
MEXICO
SONORA
CHIHUAHUA
BAJA CALIFORNIA NORTE
Golfo de California
COLORADO PLATEAU
GREAT BASIN
Mojave Desert
Los Angeles
San Diego
San Francisco
Sacramento
Las Vegas
Phoenix
Tucson
El Paso
Ciudad Juárez
Albuquerque
Santa Fe
Hermosillo
Chihuahua
Tijuana
Mexicali
1:7M
KM
MILES

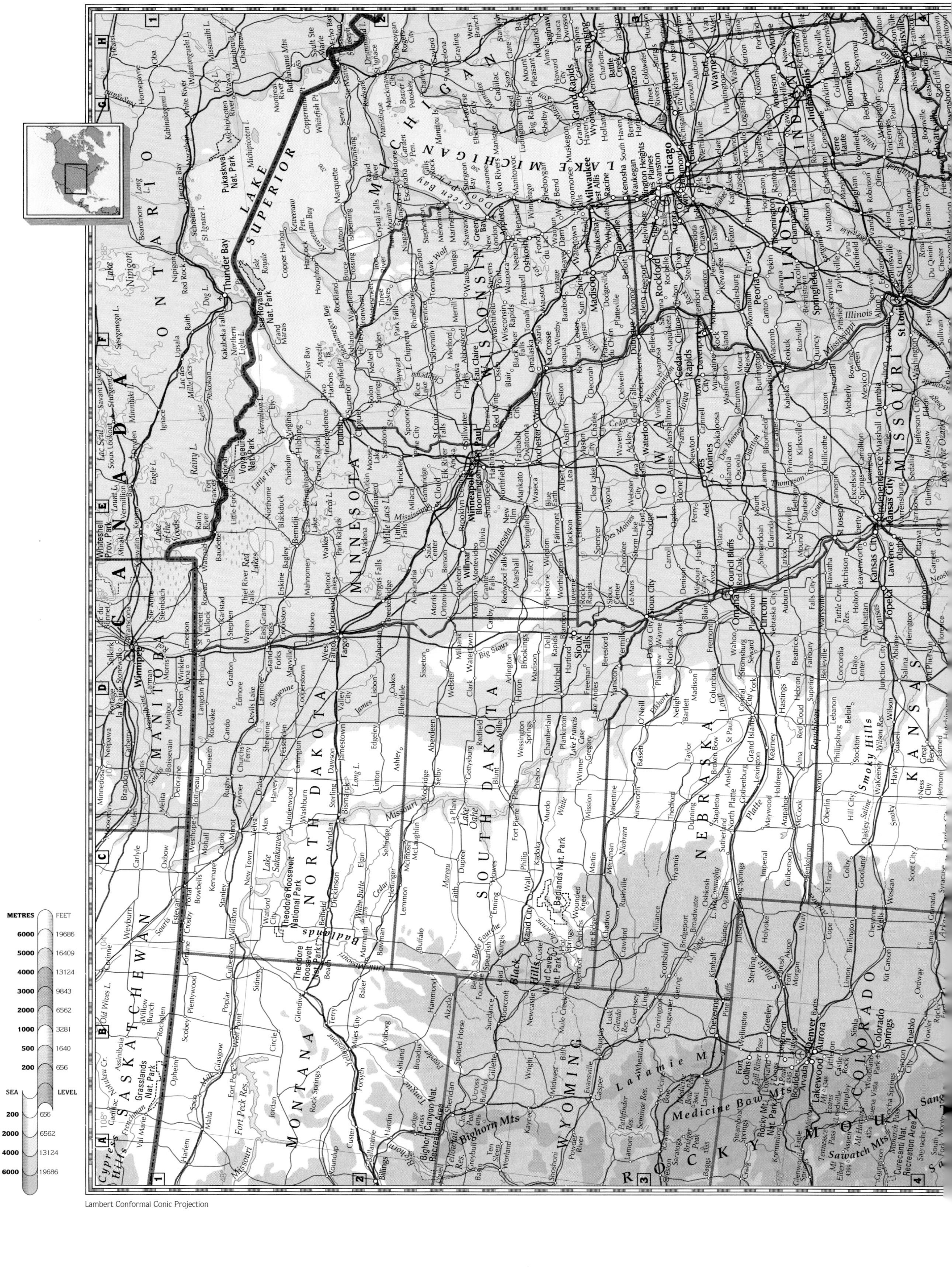

Lambert Conformal Conic Projection

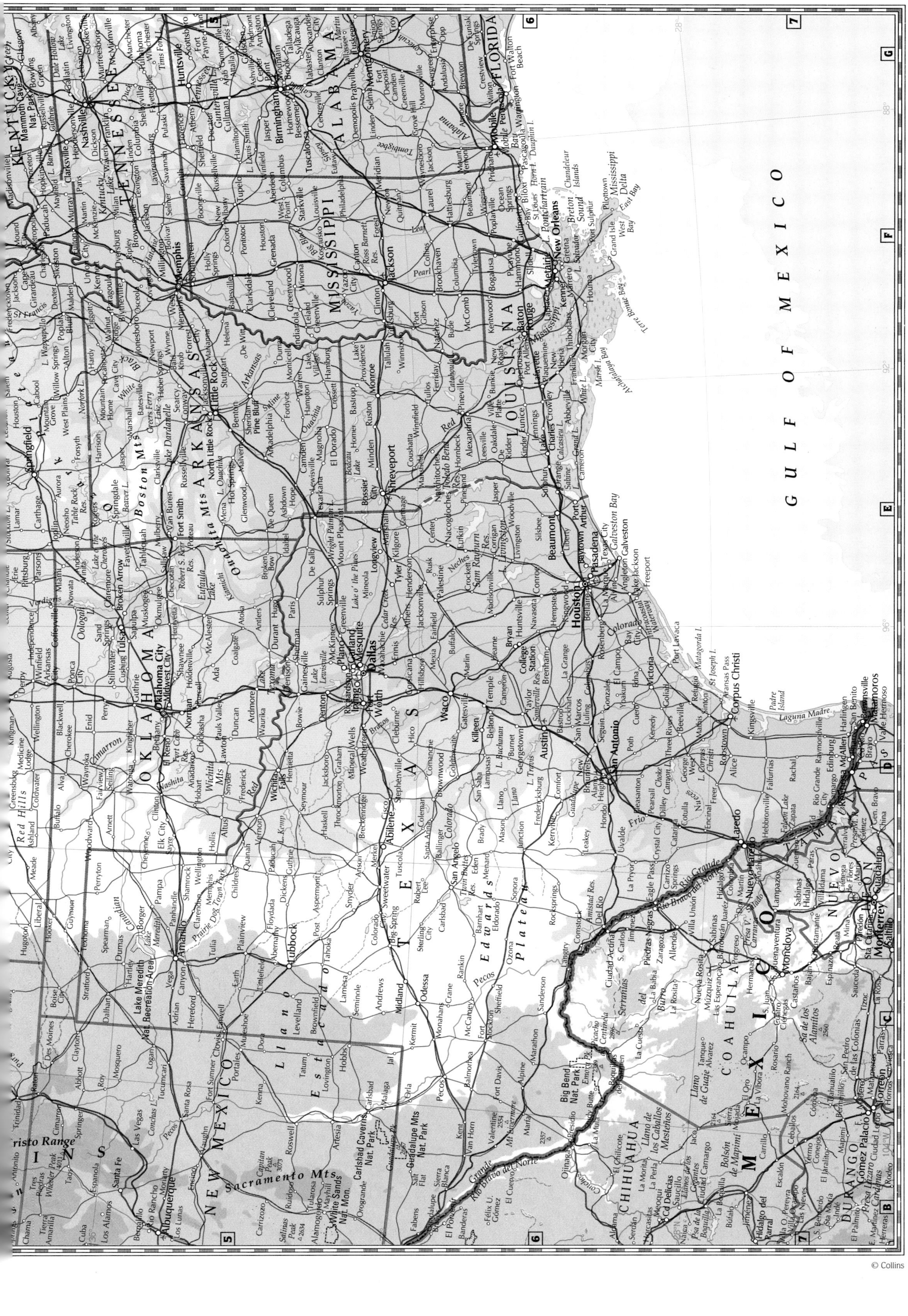

1:7M

KM MILES

350 300 250 200 150 100 50 0

200 150 100 50 0

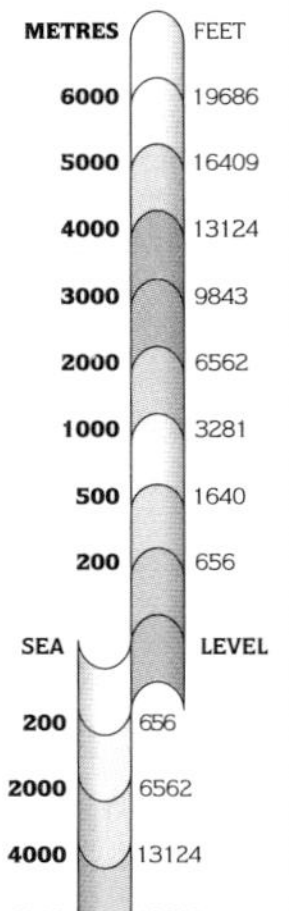

Lambert Conformal Conic Projection

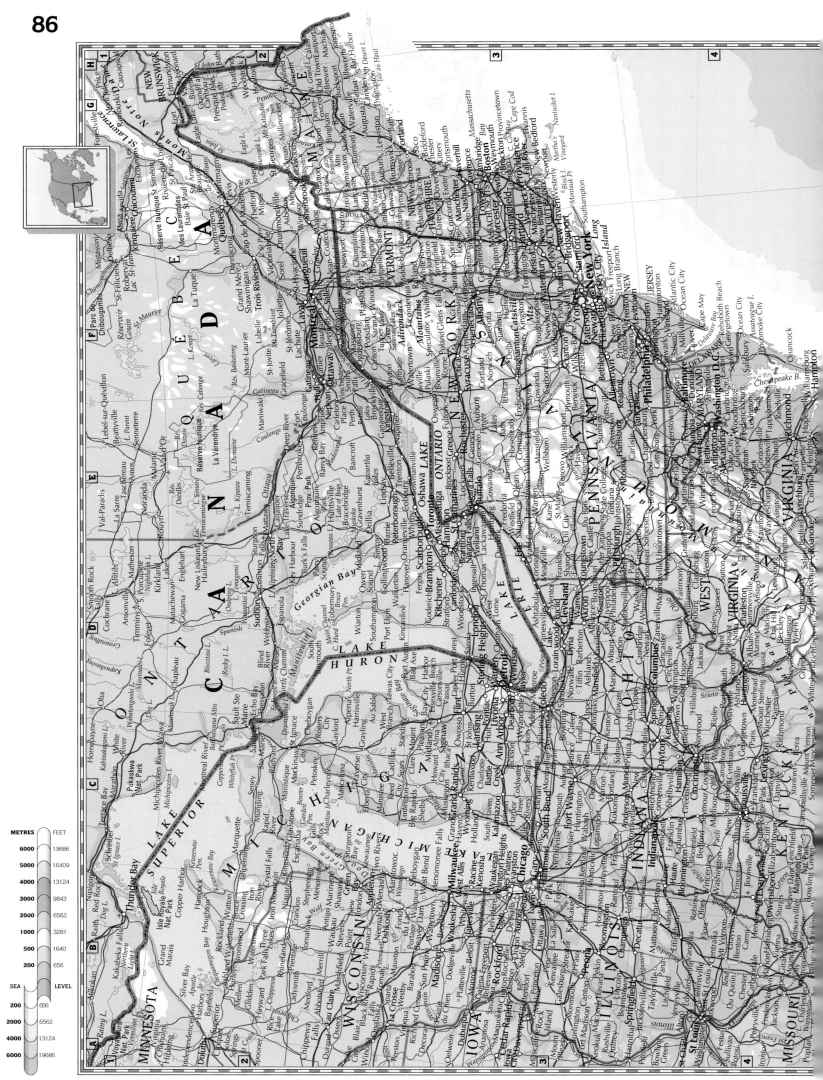

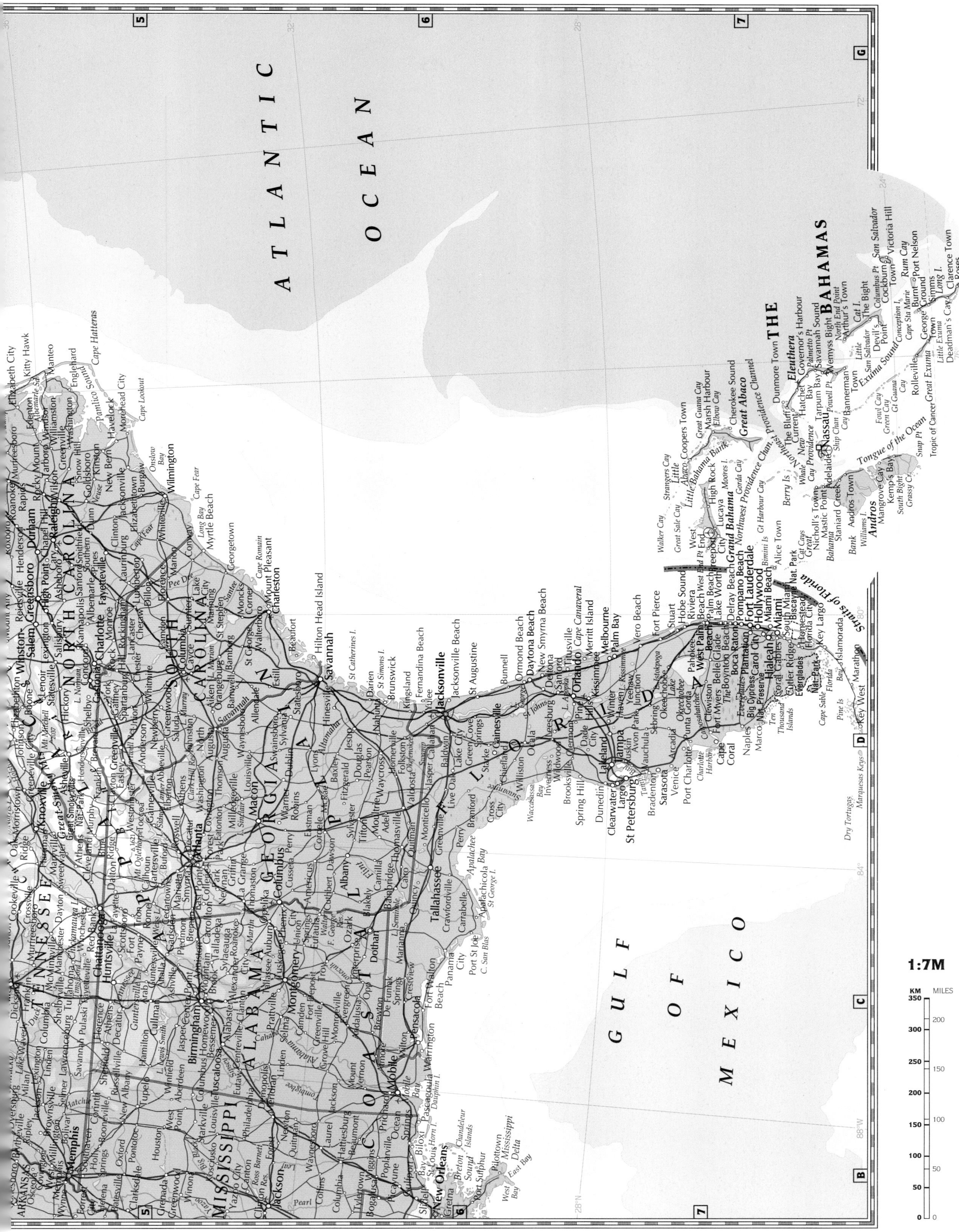
ATLANTIC
OCEAN
GULF
OF
MEXICO
THE BAHAMAS
TENNESSEE
NORTH CAROLINA
SOUTH CAROLINA
GEORGIA
ALABAMA
MISSISSIPPI
FLORIDA
Straits of Florida
Tropic of Cancer
1:7M
KM
MILES
350
300
250
200
150
100
50
0
200
150
100
50
0

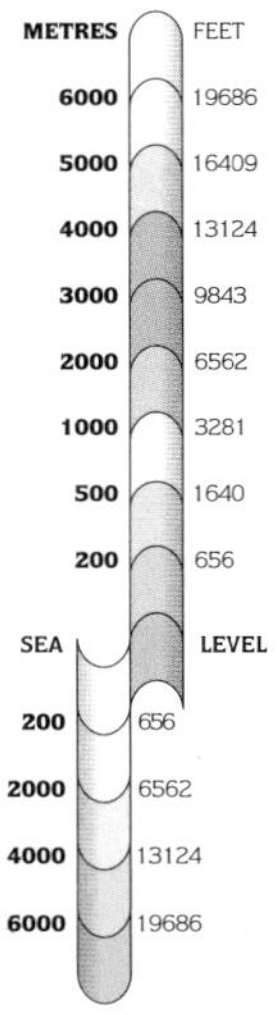

Lambert Conformal Conic Projection

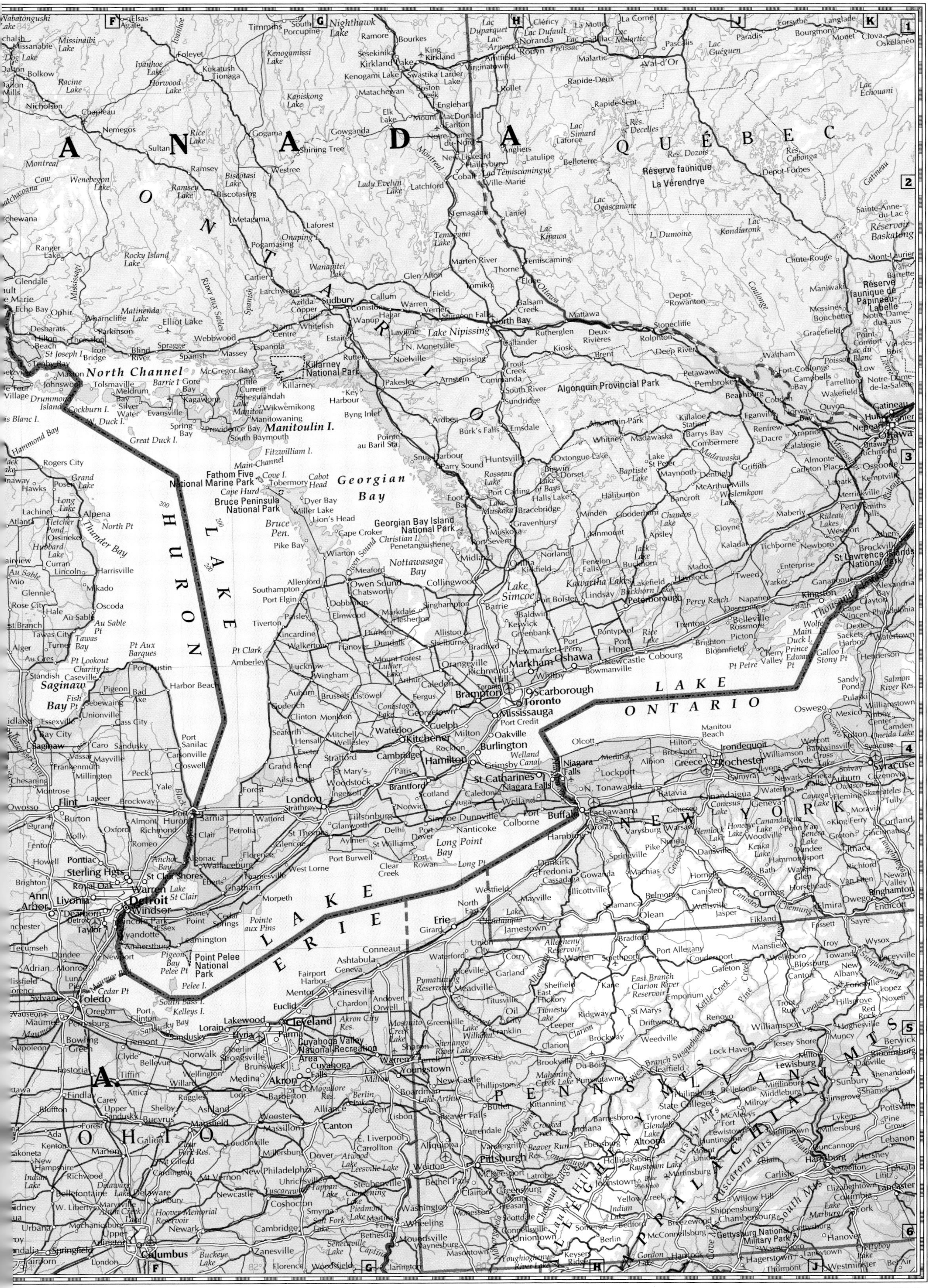

1:3.5M

KM	MILES
175	
150	100
125	75
100	
75	50
50	25
25	
0	0

Lambert Conformal Conic Projection

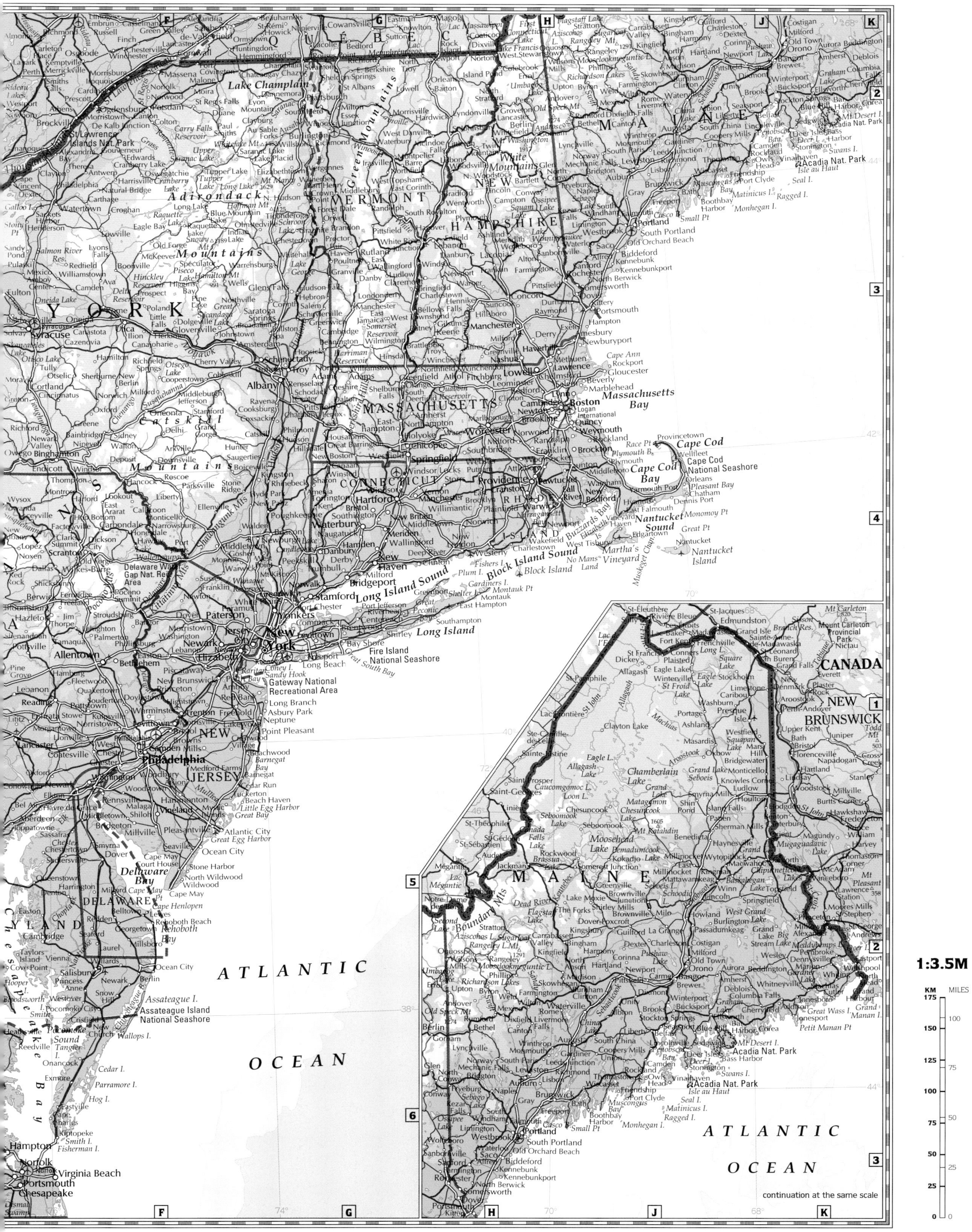
Q U É B E C
Lake Champlain
Adirondack Mountains
Catskill Mountains
Y O R K
VERMONT
NEW HAMPSHIRE
MAINE
MASSACHUSETTS
CONNECTICUT
RHODE ISLAND
NEW JERSEY
DELAWARE
Syracuse
Utica
Albany
Schenectady
Binghamton
Scranton
Allentown
Philadelphia
Wilmington
Trenton
New York
Newark
Paterson
Yonkers
Stamford
Bridgeport
New Haven
Hartford
Springfield
Worcester
Boston
Providence
Manchester
Concord
Portland
Lewiston
Augusta
Bangor
Burlington
Long Island
Long Island Sound
Block Island Sound
Nantucket Sound
Nantucket Island
Martha's Vineyard
Cape Cod
Cape Cod Bay
Massachusetts Bay
Cape Cod National Seashore
Fire Island National Seashore
Gateway National Recreational Area
Delaware Water Gap Nat. Rec. Area
Delaware Bay
Chesapeake Bay
Assateague Island National Seashore
Acadia Nat. Park
Atlantic City
Ocean City
Virginia Beach
Norfolk
Portsmouth
Chesapeake
Hampton
ATLANTIC OCEAN
CANADA
NEW BRUNSWICK
Moosehead Lake
Chamberlain Lake
Mt Katahdin 1605
Boundary Mts
continuation at the same scale
1:3.5M
KM
MILES
175
150
125
100
75
50
25
0

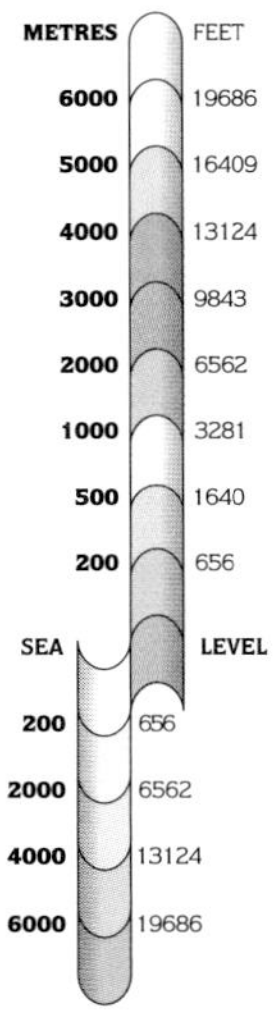

Lambert Conformal Conic Projection

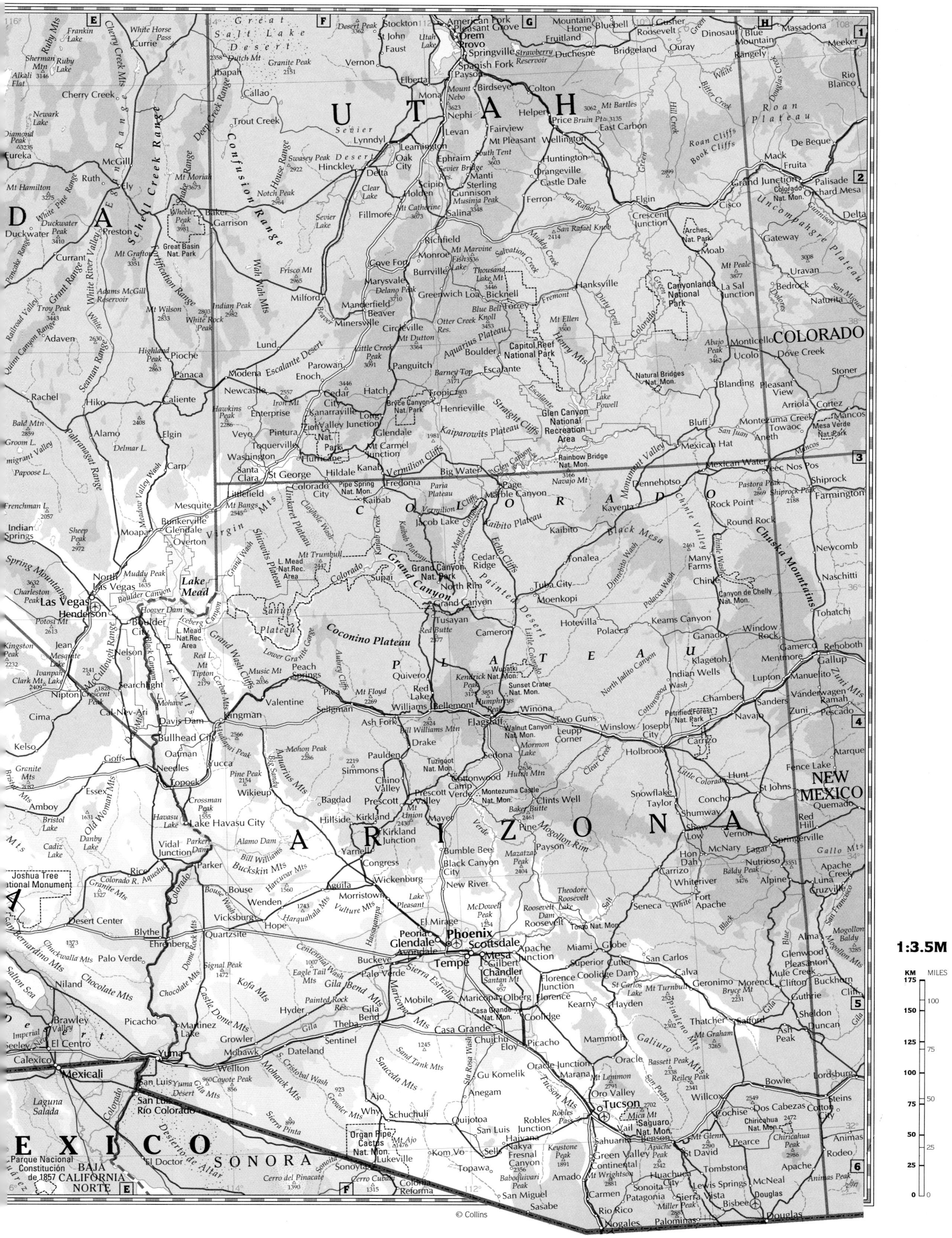
UTAH
COLORADO
NEW MEXICO
ARIZONA
MEXICO
SONORA
BAJA CALIFORNIA NORTE
Great Salt Lake Desert
Colorado Plateau
Grand Canyon
Lake Mead
Lake Powell
Las Vegas
Henderson
Phoenix
Glendale
Scottsdale
Tempe
Mesa
Tucson
Flagstaff
St George
Provo
Orem
Yuma
Mexicali
Grand Junction
Moab
Canyonlands National Park
Capitol Reef National Park
Glen Canyon National Recreation Area
Great Basin Nat. Park
Joshua Tree National Monument
Organ Pipe Cactus Nat. Mon.
Coconino Plateau
Mogollon Rim
Painted Desert
Black Mesa
Chuska Mountains
Uncompahgre Plateau
1:3.5M
KM
MILES
© Collins

Lambert Azimuthal Equal Area Projection

ATLANTIC OCEAN
BERMUDA (U.K.)
Hamilton
Tropic of Cancer
THE BAHAMAS
Grand Bahama
Little Abaco
Great Abaco
Eleuthera
Governor's Harbour
Nassau
Andros
Exuma Sound
Cat Island
San Salvador (Watling)
Rum Cay
Great Exuma
Long Island
Crooked Island Passage
Crooked Island
Acklins Island
Mayaguana
Great Inagua
Matthew Town
TURKS AND CAICOS ISLANDS (U.K.)
Caicos Is
Cockburn Town
Turks Is
Straits of Florida
Great Bahama Bank
NORTH CAROLINA
Raleigh
Durham
Greensboro
Winston Salem
Charlotte
Fayetteville
Wilmington
Cape Fear
Cape Hatteras
Cape Lookout
Jacksonville
New Bern
SOUTH CAROLINA
Columbia
Charleston
Myrtle Beach
Hilton Head Island
GEORGIA
Atlanta
Macon
Augusta
Savannah
Albany
Valdosta
Brunswick
Tallahassee
Jacksonville
St Augustine
Gainesville
Daytona Beach
Ocala
Orlando
Cape Canaveral
Melbourne
Tampa
Clearwater
St Petersburg
Sarasota
Lakeland
Fort Pierce
L. Okeechobee
West Palm Beach
Freeport
Fort Lauderdale
Hollywood
Miami
Miami Beach
Berry Is
Big Cypress Nat. Reserve
Everglades Nat. Park
Cape Sable
Key West
Fort Myers
Port Charlotte
Apalachee Bay
Knoxville
Chattanooga
Greenville
Havana (Habana)
Matanzas
Pinar del Río
Arch. de Sabana
Arch. de Camagüey
Santa Clara
Cienfuegos
Sancti Spíritus
Ciego de Avila
Camagüey
Trinidad
CUBA
Holguín
Bayamo
Manzanillo
Victoria de las Tunas
Santiago de Cuba
Guantánamo
Banes
Baracoa
Isla de la Juventud
G. de Batabanó
Cabo San Antonio
Arch. de los Jardines de la Reina
Golfo de Guacanayabo
Cabo Cruz
Sierra Maestra
Turquino 2005
CAYMAN ISLANDS (U.K.)
Cayman Brac
Little Cayman
Grand Cayman
GREATER ANTILLES
HISPANIOLA
Windward Pass.
Port-de-Paix
Cap-Haïtien
Gonaïves
HAITI
Port-au-Prince
Jérémie
Les Cayes
Jacmel
Île de la Gonâve
Puerto Plata
Santiago
San Francisco de Macorís
DOMINICAN REPUBLIC
Santo Domingo
La Romana
Barahona
Isla Beata
C. Beata
Pico Duarte 3175
Mona Passage
Isla Mona
Aguadilla
San Juan
Mayagüez
Ponce
PUERTO RICO (U.S.A.)
Vieques
St Croix
VIRGIN IS (U.K.)
VIRGIN IS (U.S.A.)
Anegada
ANGUILLA (U.K.)
Saint Martin (Fr.)
Sint Maarten (Neth.)
St Barthélemy (Fr.)
Barbuda
ANTIGUA AND BARBUDA
St John's
St Eustatius
Basseterre
ST KITTS-NEVIS
Plymouth
MONTSERRAT (U.K.)
Pointe-à-Pitre
GUADELOUPE (Fr.)
Basse Terre
Marie Galante
DOMINICA
Roseau
MARTINIQUE (Fr.)
Fort-de-France
Castries
ST LUCIA
Kingstown
ST VINCENT & THE GRENADINES
Bridgetown
BARBADOS
GRENADA
St George's
TRINIDAD AND TOBAGO
Tobago
Scarborough
Trinidad
Port of Spain
Arima
San Fernando
LEEWARD ISLANDS
WINDWARD ISLANDS
Lesser Antilles
JAMAICA
Montego Bay
St Ann's Bay
Savanna la Mar
Mandeville
Kingston
Spanish Town
Jamaica Channel
Swan Islands (Hond.)
CARIBBEAN SEA
NETHERLANDS ANTILLES
ARUBA (Neth.)
Curaçao
Willemstad
Bonaire
Islas Los Roques (Ven.)
I. La Tortuga
I. Orchila (Ven.)
I. Blanquilla (Ven.)
Los Testigos (Ven.)
I. de Margarita
Porlamar
Cumaná
Carúpano
Punta Gallinas
Península de la Guajira
Riohacha
G. de Venezuela
Punto Fijo
Coro
Maracaibo
Cabimas
Mene de Mauroa
L. de Maracaibo
Puerto Cabello
Valencia
Maracay
Caracas
Maiquetía
Los Teques
Barcelona
Barquisimeto
San Felipe
Acarigua
San Carlos
San Juan de los Morros
Valera
Trujillo
Mérida
Barinas
Guanare
San Cristóbal
San Fernando de Apure
Calabozo
Parque Nacional Aguaro-Guariquito
Parque Nacional Sierra Nevada
Cord. de Mérida
Zaraza
Maturín
Tucupita
Ciudad Guayana
Ciudad Bolívar
El Tigre
Upata
El Callao
VENEZUELA
Orinoco
Parque Nacional Cinaruco-Capanaparo
Parque Nacional Canaima
La Gran Sabana
Puerto Carreño
Puerto Ayacucho
Parque Nacional Jaua Sarisariñama
Parque Nacional El Tuparro
Puerto Inírida
Parque Nacional Duida-Marahuaca
Serra Pacaraima
Sierra Parima
Parque Nacional Parima-Tapirapecó
Serranía de la Neblina
Parque Nacional Sierra Nevada de Santa Marta
Santa Marta
Barranquilla
Cartagena
Valledupar
Sincelejo
Montería
Golfo de Morrosquillo
Golfo del Darién
Cúcuta
Bucaramanga
Barrancabermeja
Medellín
Quibdó
Manizales
Pereira
Armenia
Ibagué
Bogotá
Villavicencio
Tunja
Yopal
Sierra Nevada del Cocuy
Cali
Palmira
Buenaventura
Neiva
Popayán
Tumaco
Florencia
San José del Guaviare
Parque Nacional Sumapaz
Parque Nacional La Macarena
Parque Nacional Cord. de los Picachos
Parque Nacional Sanquianga
COLOMBIA
CORDILLERA ORIENTAL
I. de Malpelo (Col.)
Cabo Corrientes
Golfo de Cupica
Parque Nacional de Darién
Parque Nacional Paramillo
Turbo
PANAMA
Panamá
Colón
Panama Canal
G. de Panamá
La Chorrera
David
Santiago
Chitré
Las Tablas
Península de Azuero
Punta Mariato
Golfo de Chiriquí
Isla Coiba
Bocas del Toro
Golfo de los Mosquitos
COSTA RICA
San José
Limón
Alajuela
Cartago
G. de Nicoya
Pen. de Osa
NICARAGUA
Lago de Nicaragua
Bluefields
Puerto Cabezas
Prinzapolca
Costa de Mosquitos
Cayos Miskitos (Nic.)
Isla de Providencia (Col.)
Isla de San Andrés (Col.)
Is del Maíz (Corn Is) (Nic.)
MOSQUITIA
Laguna Caratasca
1:14M
KM
MILES
700
600
500
400
300
200
100
0
400
300
200
100
0

Lambert Conformal Conic Projection

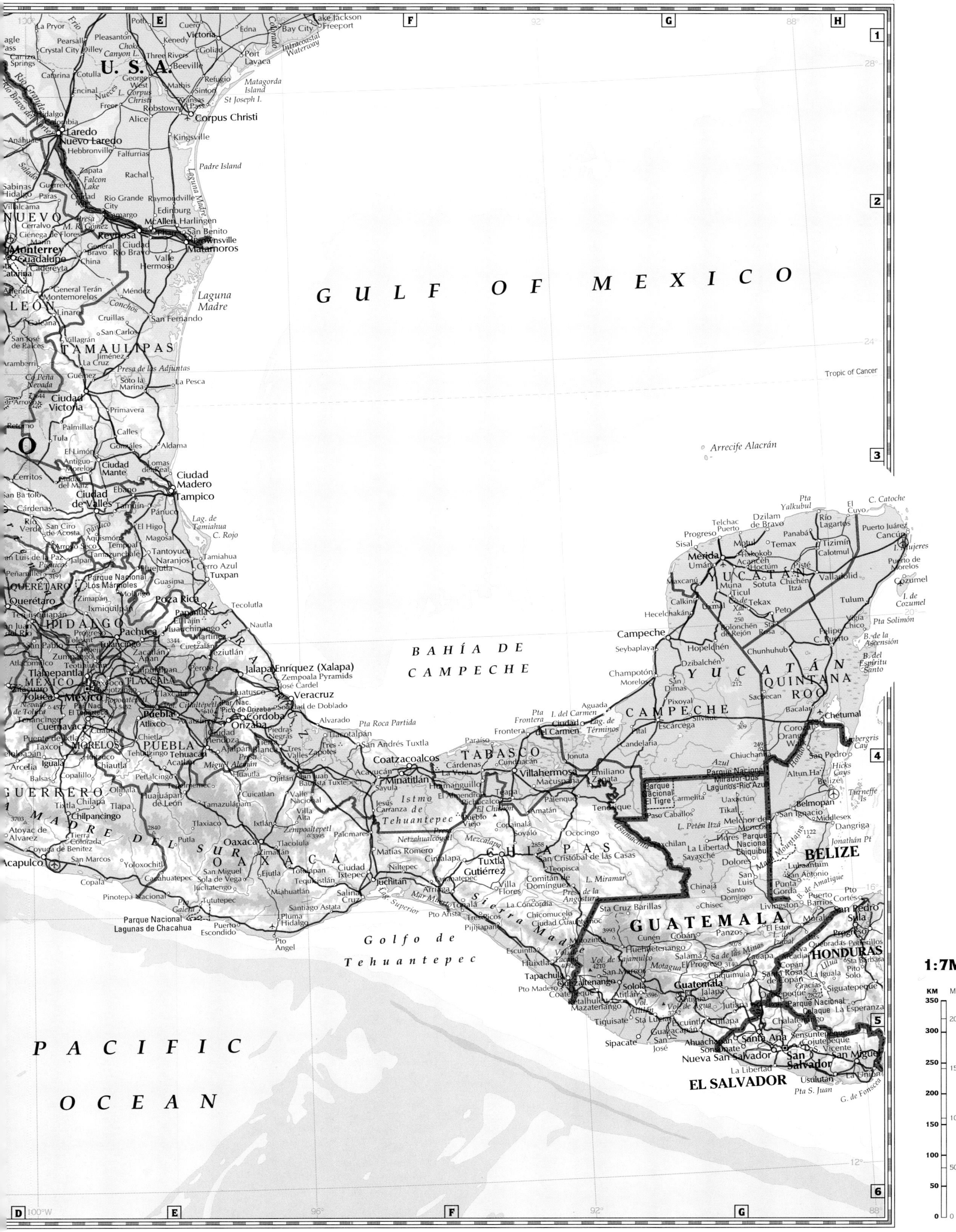
GULF OF MEXICO
PACIFIC OCEAN
BAHÍA DE CAMPECHE
Golfo de Tehuantepec
U. S. A.
MEXICO
GUATEMALA
BELIZE
HONDURAS
EL SALVADOR
TAMAULIPAS
VERACRUZ
HIDALGO
PUEBLA
TLAXCALA
MORELOS
GUERRERO
OAXACA
CHIAPAS
TABASCO
CAMPECHE
YUCATÁN
QUINTANA ROO
QUERÉTARO
Corpus Christi
Laredo
Nuevo Laredo
Monterrey
Matamoros
Brownsville
McAllen
Reynosa
Ciudad Victoria
Tampico
Ciudad Madero
Poza Rica
Pachuca
Tulancingo
Tlalnepantla
México
Toluca
Cuernavaca
Puebla
Orizaba
Córdoba
Jalapa Enríquez (Xalapa)
Veracruz
Acapulco
Chilpancingo
Oaxaca
Coatzacoalcos
Minatitlán
Villahermosa
Tuxtla Gutiérrez
San Cristóbal de las Casas
Tapachula
Campeche
Mérida
Progreso
Cancún
Chetumal
Belmopan
Belize
Guatemala
Quezaltenango
San Pedro Sula
Santa Ana
San Salvador
Nueva San Salvador
Tropic of Cancer
Arrecife Alacrán
Istmo de Tehuantepec
Sierra Madre del Sur
Laguna Madre
Padre Island
1:7M
KM
MILES
350
300
250
200
150
100
50
0
200
150
100
50
0

NORTH
AMERICA
CARIBBEAN SEA
ATLANTIC
OCEAN
Punta Gallinas
G. de Venezuela
Isla de Margarita
Golfo del Darién
L. Maracaibo
Orinoco Delta
Waini Point
Orinoco
Cordillera Occidental
Cordillera Central
Cordillera Oriental
Llanos
Meta
Cerro Yavi 2285
La Gran Sabana
Sa Pacaraima
Essequibo
Maroni
Pointe Isère
Cabo Orange
Cabo Corrientes
I. de Malpelo
Guaviare
Branco
Ilha de Maracá
Mouths of the Amazon
Equator
Cotopaxi 5896
Chimborazo 6310
Caquetá
Japurá
Negro
Represa de Balbina
Amazon
I. de Marajó
Baía de São Marcos
Golfo de Guayaquil
Putumayo
Amazon
Tapajós
Iriri
Tocantins
Parnaíba
Ponta do Calcanhar
Marañón
Represa Tucuruí
SELVAS
Juruá
Purus
Xingu
Madeira
Ucayali
Cordillera Central
ANDES
Teles Pires
Tocantins
6768 Nevado de Huascarán
Juruena
Arinos
Araguaia
Chapada Diamantina
São Francisco
Cordillera Oriental
Jiparaná
Guaporé
Beni
Lago de San Luis
Bahía de Pisco
Cordillera Occidental
Cabo Santo Antônio
San Miguel
Lago Titicaca
Yungas
Altiplano
Bañados del Izozog
L. de Poopó
Velhas
Ponta da Baleia
Paranaíba
Grande
Paraguay
Desierto de Atacama
Gran Chaco
Paraná
Paranapanema
Cabo de São Tomé
Pta Tetas
Tropic of Capricorn
Ilha de São Sebastião
Pilcomayo
Teuco
Iguaçu Falls
Islas de los Desventurados
Pta Ballena
Pta Morro
Cerro Bonete 6872
Salado
Uruguay
Salinas Grandes
Desaguadero
Sierras de Córdoba
Paraná
Lagoa dos Patos
Juan Fernandez Islands
Aconcagua 6960
PAMPAS
Lagoa Mirim
PACIFIC
OCEAN
Río de la Plata
Colorado
ATLANTIC
OCEAN
Pta Galera
Bahía Blanca
Negro
Golfo San Matías
Isla de Chiloé
Península Valdés
Patagonia
Archipiélago de los Chonos
Golfo de San Jorge
Golfo de Penas
Pta Medanosa
L. San Martín
L. Argentino
Bahía Grande
Falkland Islands
West Falkland
East Falkland
Est. de Magallanes
Tierra del Fuego
I.de los Estados
South Georgia
Cape Horn
Sinusoidal Projection
© Collins

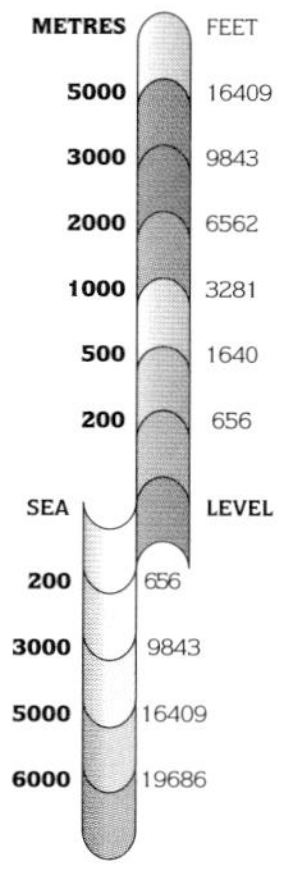
METRES
FEET
5000
16409
3000
9843
2000
6562
1000
3281
500
1640
200
656
SEA
LEVEL
200
656
3000
9843
5000
16409
6000
19686

1:30M

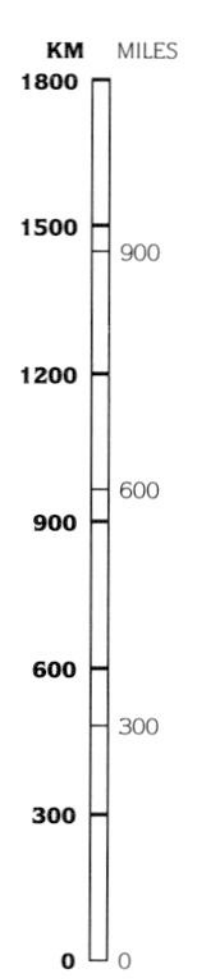
KM
MILES
1800
1500
900
1200
600
900
600
300
300
0
0

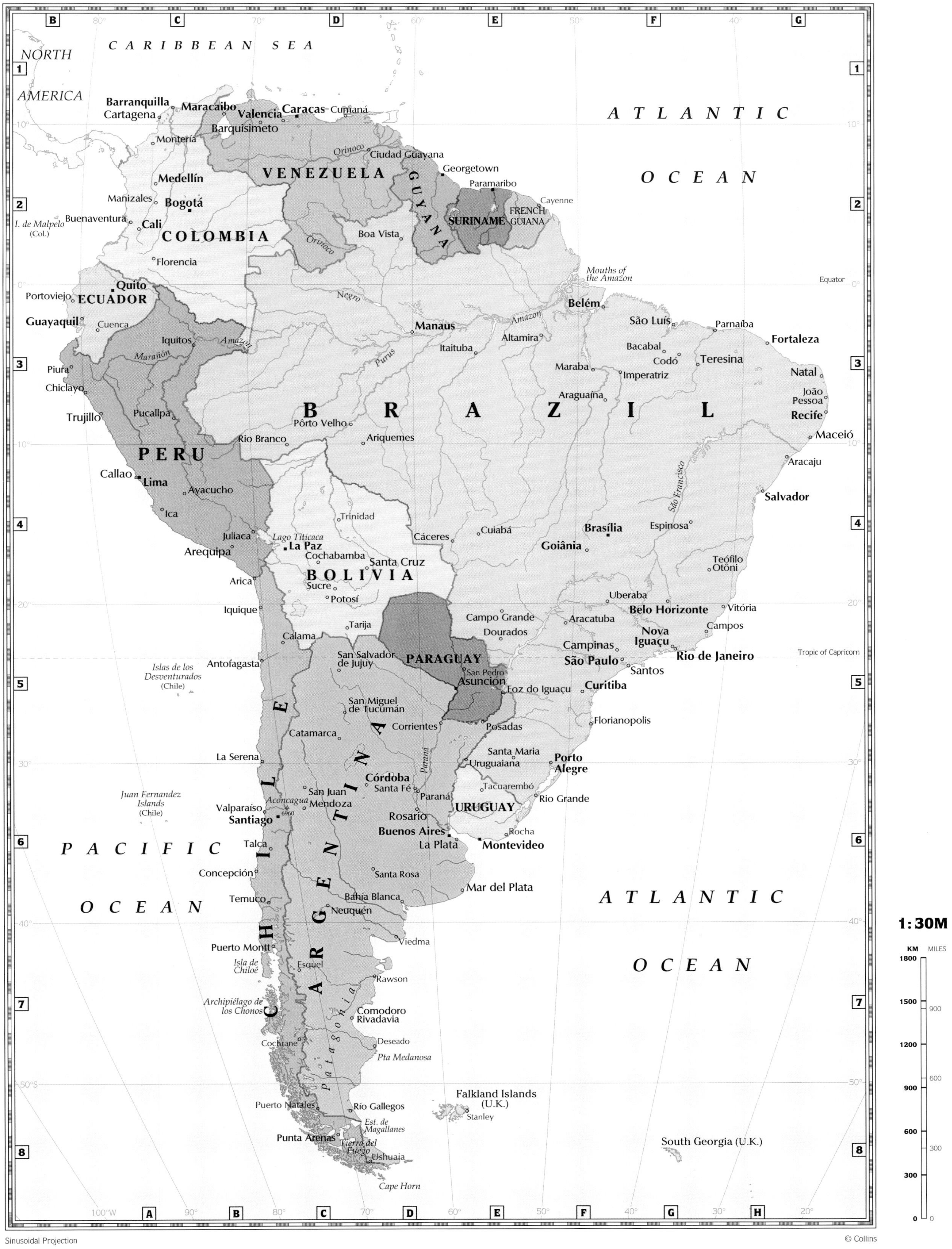
CARIBBEAN SEA
NORTH AMERICA
ATLANTIC OCEAN
PACIFIC OCEAN
VENEZUELA
GUYANA
SURINAME
FRENCH GUIANA
COLOMBIA
ECUADOR
PERU
BRAZIL
BOLIVIA
PARAGUAY
CHILE
ARGENTINA
URUGUAY
Barranquilla
Cartagena
Maracaibo
Valencia
Caracas
Cumaná
Barquisimeto
Montería
Ciudad Guayana
Georgetown
Paramaribo
Cayenne
Medellín
Manizales
Bogotá
Buenaventura
Cali
I. de Malpelo (Col.)
Boa Vista
Florencia
Mouths of the Amazon
Equator
Quito
Portoviejo
Guayaquil
Cuenca
Belém
São Luís
Parnaíba
Fortaleza
Manaus
Altamira
Itaituba
Bacabal
Codó
Teresina
Iquitos
Marañón
Amazon
Negro
Orinoco
Purus
Maraba
Imperatriz
Natal
Piura
Chiclayo
Araguaína
João Pessoa
Recife
Trujillo
Pucallpa
Pôrto Velho
Rio Branco
Ariquemes
Maceió
Aracaju
Callao
Lima
Ayacucho
São Francisco
Salvador
Ica
Trinidad
Juliaca
Lago Titicaca
La Paz
Cáceres
Cuiabá
Brasília
Goiânia
Espinosa
Arequipa
Cochabamba
Santa Cruz
Teófilo Otôni
Arica
Sucre
Potosí
Uberaba
Belo Horizonte
Vitória
Iquique
Campo Grande
Aracatuba
Dourados
Campos
Tarija
Calama
Nova Iguaçu
Campinas
Rio de Janeiro
San Salvador de Jujuy
Tropic of Capricorn
Antofagasta
São Paulo
Santos
Islas de los Desventurados (Chile)
San Pedro
Asunción
Foz do Iguaçu
Curitiba
San Miguel de Tucumán
Corrientes
Florianopolis
Posadas
Catamarca
Santa Maria
Uruguaiana
Porto Alegre
La Serena
Paraná
Córdoba
Santa Fé
Tacuarembó
Juan Fernandez Islands (Chile)
San Juan
Mendoza
Aconcagua 6960
Rio Grande
Valparaíso
Santiago
Rosario
Buenos Aires
La Plata
Rocha
Montevideo
Talca
Concepción
Santa Rosa
Mar del Plata
Temuco
Bahía Blanca
Neuquén
Viedma
Puerto Montt
Isla de Chiloé
Esquel
Rawson
Archipiélago de los Chonos
Comodoro Rivadavia
Patagonia
Cochrane
Deseado
Pta Medanosa
Falkland Islands (U.K.)
Stanley
Puerto Natales
Río Gallegos
Est. de Magallanes
Punta Arenas
Tierra del Fuego
Ushuaia
South Georgia (U.K.)
Cape Horn
1:30M
KM
MILES
Sinusoidal Projection
© Collins

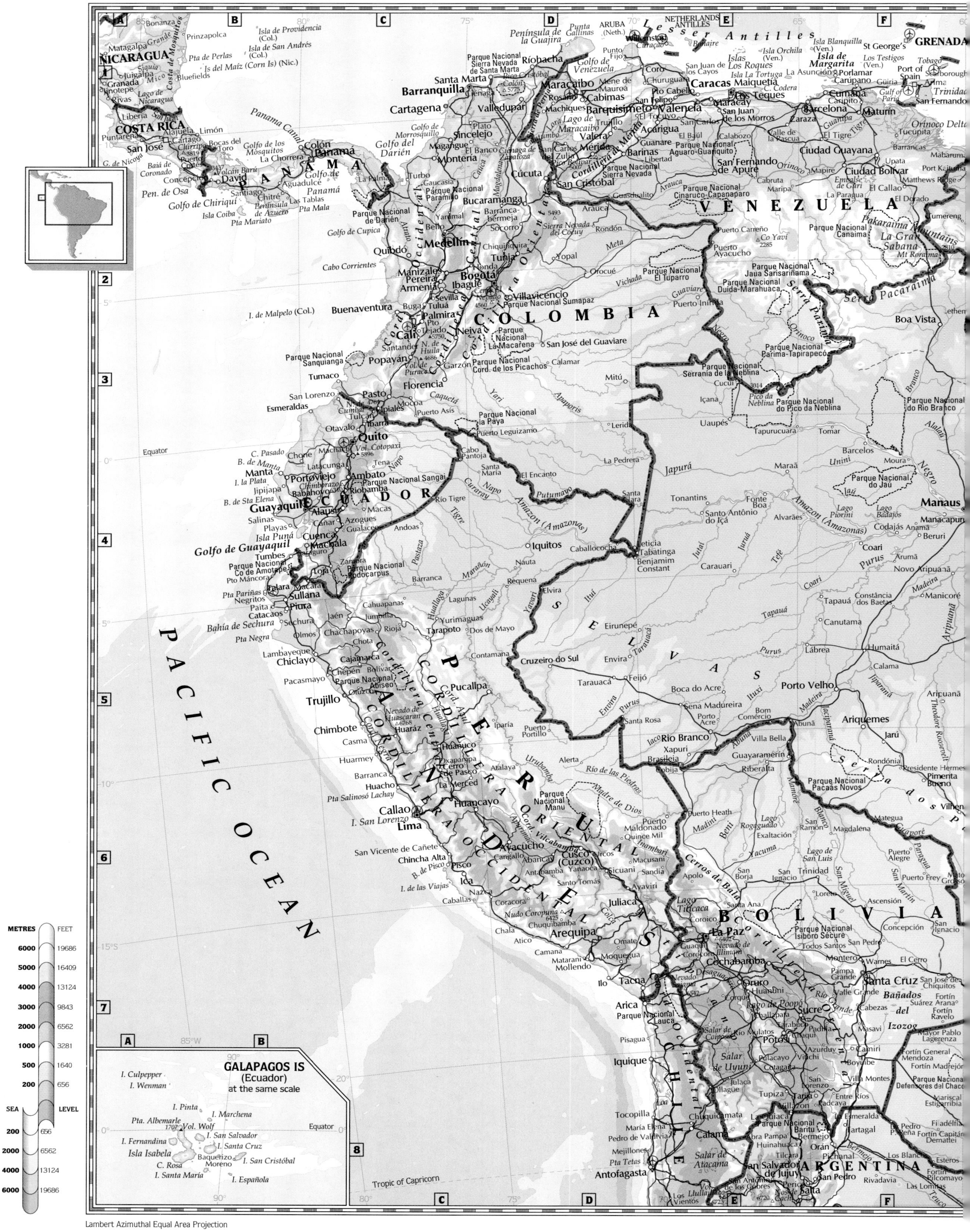

Lambert Azimuthal Equal Area Projection

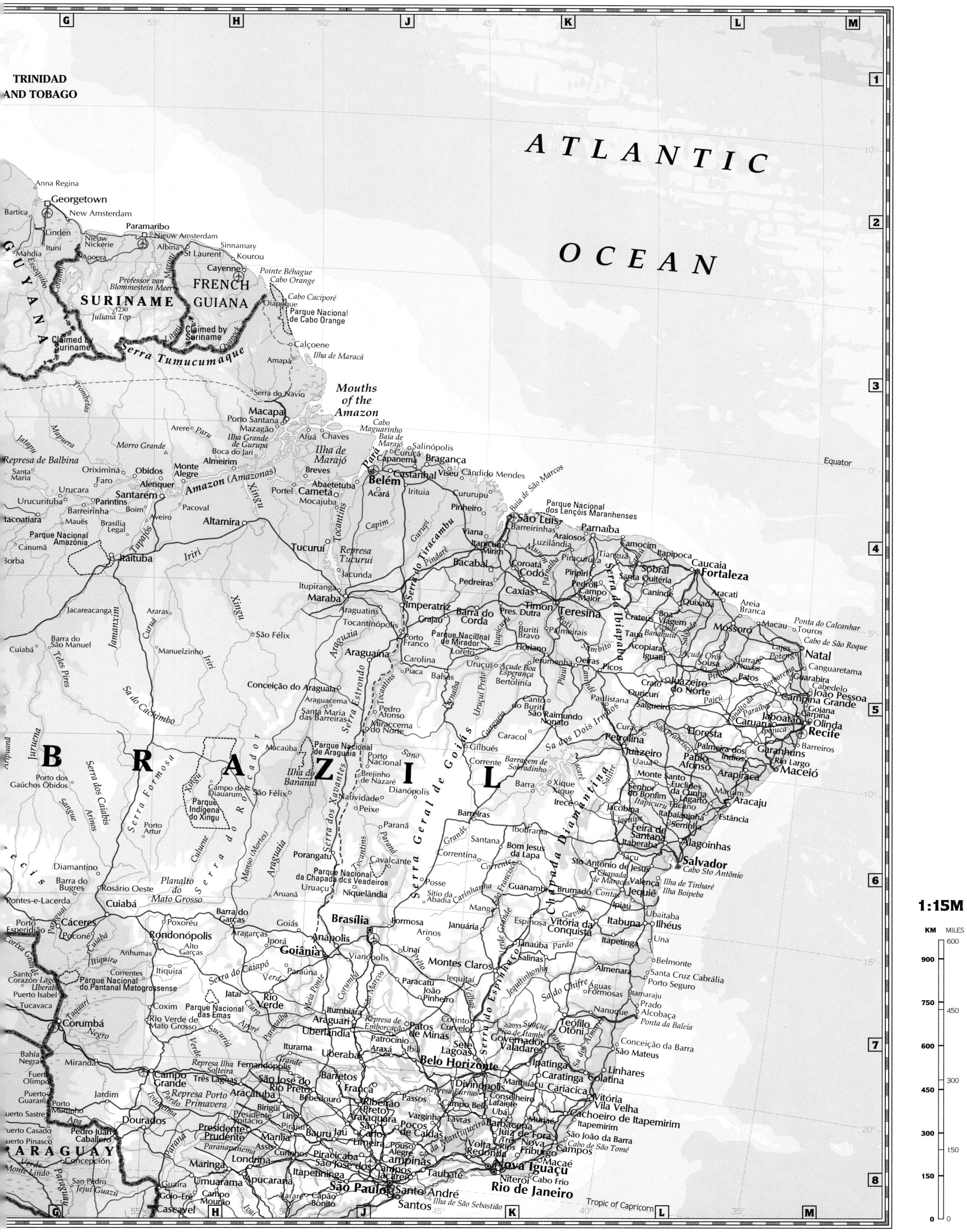
ATLANTIC
OCEAN
TRINIDAD AND TOBAGO
GUYANA
SURINAME
FRENCH GUIANA
BRAZIL
PARAGUAY
Anna Regina
Georgetown
New Amsterdam
Bartica
Linden
Paramaribo
Nieuw Amsterdam
Nieuw Nickerie
Mahdia
Ituni
Albina
St Laurent
Sinnamary
Kourou
Apoera
Cayenne
Pointe Béhague
Cabo Orange
Professor van Blommestein Meer
Juliana Top
1230
Claimed by Suriname
Serra Tumucumaque
Cabo Caciporé
Oiapoque
Parque Nacional de Cabo Orange
Calçoene
Ilha de Maracá
Amapá
Serra do Navio
Mouths of the Amazon
Macapá
Porto Santana
Mazagão
Cabo Maguarinho
Baía de Marajó
Ilha de Marajó
Afuá
Chaves
Arere
Paru
Morro Grande
Ilha Grande de Gurupá
Boca do Jari
Almeirim
Represa de Balbina
Salinópolis
Curuçá
Capanema
Bragança
Viseu
Cândido Mendes
Pará
Belém
Castanhal
Acará
Irituia
Santa Maria
Oriximiná
Óbidos
Monte Alegre
Faro
Alenquer
Amazon (Amazonas)
Breves
Abaetetuba
Urucara
Santarém
Parintins
Portel
Cametá
Mocajuba
Urucurituba
Barreirinha
Boim
Pacoval
Cururupu
Pinheiro
Baía de São Marcos
Parque Nacional dos Lençóis Maranhenses
São Luís
Barreirinhas
Parnaíba
Maués
Brasília Legal
Aveiro
Altamira
Parque Nacional Amazônia
Canumã
Itaituba
Tapajós
Xingu
Iriri
Capim
Gurupi
Viana
Araiosos
Tucuruí
Represa Tucuruí
Jacunda
Serra do Tiracambu
Pindaré
Itapecuru Mirim
Lusilândia
Mearim
Piracuruca
Tianguá
Camocim
Itapipoca
Caucaia
Fortaleza
Bacabal
Coroatá
Codó
Piripiri
Pedro II
Sobral
Santa Quitéria
Canindé
Pedreiras
Caxias
Campo Maior
Aracati
Areia Branca
Itupiranga
Marabá
Araguatins
Tocantinópolis
Imperatriz
Grajaú
Barra do Corda
Pres. Dutra
Timon
Teresina
Serra da Ibiapaba
Crateús
Boa Viagem
Quixadá
Mossoró
Macau
Ponta do Calcanhar
Touros
Cabo de São Roque
Natal
Jacareacanga
Araras
Curuá
Jamanxim
São Félix
Barra do São Manuel
Cuiabá
Teles Pires
Manuelzinho
Araguaia
Porto Franco
Parque Nacional de Mirador
Buriti Bravo
Palmeirais
Tauá
Banabuiú
Lajes
Potengi
Canguaretama
Araguaína
Carolina
Loreto
Floriano
Acopiara
Iguatu
Açude Orós
Sousa
Currais Novos
Patos
Guarabira
Cabedelo
João Pessoa
Campina Grande
Piaca
Balsas
Uruçuí
Açude Boa Esperança
Bertolínia
Jerumenha
Oeiras
Picos
Crato
Juazeiro do Norte
Ouricuri
Salgueiro
Paulistana
Pajeú
Goiana
Carpina
Jaboatão
Olinda
Recife
Caruaru
Ipojuca
Conceição do Araguaia
Araguacema
Santa Maria das Barreiras
Pedro Afonso
Miracema do Norte
Serra Estrondo
Canto do Buriti
São Raimundo Nonato
Serra dos Dois Irmãos
Petrolina
Juazeiro
Curaçá
São Francisco
Floresta
Palmeira dos Índios
Garanhuns
Rio Largo
Maceió
Barreiros
Paulo Afonso
Arapiraca
Caracol
Gilbués
Corrente
Barragem de Sobradinho
Uauá
Monte Santo
Euclides da Cunha
Macaúba
Parque Nacional de Araguaia
Ilha do Bananal
Porto Nacional
Brejinho de Nazaré
Dianópolis
Serra Geral de Goiás
Chapada Diamantina
Xique Xique
Barra
Irecê
Senhor do Bonfim
Itapicuru
Jacobina
Tucano
Lagarto
Maruim
Aracaju
Estância
Juruena
Arinos
Porto dos Gaúchos
Óbidos
Serra dos Caiabis
Sangue
Serra Formosa
Xingu
Campo de Diauarum
Parque Indígena do Xingu
São Félix
Serra do Roncador
Natividade
Peixe
Paraná
Barreiras
Ibotirama
Santana
Bom Jesus da Lapa
Feira de Santana
Serrinha
Itaberaba
Alagoinhas
Salvador
Cabo Sto Antônio
Porto Artur
Culuene
Manso (Mortes)
Porangatu
Cavalcante
Correntina
Santo Antônio de Jesus
Chapada de Maracás
Valença
Ilha de Tinharé
Ilha Boipeba
Jequié
Diamantino
Barra do Bugres
Rosário Oeste
Planalto do Mato Grosso
Aruanã
Parque Nacional da Chapada dos Veadeiros
Uruaçu
Niquelândia
Posse
Sítio da Abadia
Guanambi
Brumado
Contas
Ipiaú
Pontes-e-Lacerda
Cuiabá
Barra do Garças
Brasília
Formosa
Arinos
Januária
Manga
Espinosa
Vitória da Conquista
Gavião
Itabuna
Ilhéus
Ubaitaba
Una
Porto Esperidião
Cáceres
Poconé
Poxoréu
Rondonópolis
Aragarças
Goiás
Iporá
Anápolis
Goiânia
Vianópolis
Unaí
Montes Claros
Janaúba
Pardo
Salinas
Itapetinga
Belmonte
Santa Cruz Cabrália
Porto Seguro
Anhumas
Alto Garças
Itiquira
Correntes
Parque Nacional do Pantanal Matogrossense
Serra do Caiapó
Paraúna
Jataí
Rio Verde
Paracatu
João Pinheiro
Jequitaí
Jequitinhonha
Almenara
Águas Formosas
Itamaraju
Prado
Alcobaça
Ponta da Baleia
Santo Corazón
Laguna Uberaba
Puerto Isabel
Tucavaca
Taquari
Coxim
Parque Nacional das Emas
Rio Verde de Mato Grosso
Itumbiara
Araguari
Uberlândia
Represa de Emborcação
Patos de Minas
Curvelo
Corinto
Pico de Itambé
2033
Teófilo Otôni
Nanuque
Conceição da Barra
São Mateus
Corumbá
Negro
Bahía Negra
Miranda
Represa Ilha Solteira
Fernandópolis
Iturama
Patrocínio
Araxá
Ibiá
Sete Lagoas
Governador Valadares
Serra do Espinhaço
Belo Horizonte
Ipatinga
Caratinga
Colatina
Linhares
Fuerte Olimpo
Puerto Guaraní
Campo Grande
Três Lagoas
Jardim
Represa Porto Primavera
Araçatuba
São José do Rio Preto
Uberaba
Barretos
Franca
Divinópolis
Manhuaçu
Cariacica
Vitória
Vila Velha
Porto Murtinho
Puerto Sastre
Bebedouro
Ribeirão Preto
Passos
Represa Furnas
Conselheiro Lafaiete
Ubá
Cachoeiro de Itapemirim
Itapemirim
Puerto Casado
Puerto Pinasco
Dourados
Birigüi
Presidente Epitácio
Lins
Araraquara
São Carlos
Varginha
Lavras
Campo Belo
Barbacena
Juiz de Fora
Muriaé
São João da Barra
Cabo de São Tomé
Campos
Pedro Juan Caballero
Presidente Prudente
Marília
Pirajuí
Bauru
Jaú
Poços de Caldas
Serra da Mantiqueira
Três Rios
Nova Friburgo
Concepción
Paranapanema
Assis
Piracicaba
Limeira
Pouso Alegre
Volta Redonda
Macaé
Cabo Frio
Monte Lindo
Maringá
Londrina
Itapetininga
São José dos Campos
Campinas
Taubaté
Nova Iguaçu
Niterói
Rio de Janeiro
San Pedro
Umuarama
Apucarana
Jacareí
Guaíra
Goio-Erê
Campo Mourão
Ivaí
Irarará
Capão Bonito
São Paulo
Santo André
Santos
Ilha de São Sebastião
Jejuí Guazú
Cascavel
Tropic of Capricorn
Equator
1:15M
KM
MILES
900
750
600
450
300
150
0
600
450
300
150
0
G
H
J
K
L
M
1
2
3
4
5
6
7
8
55°
50°
45°
40°
35°
10°
5°
0°
5°
10°
15°
20°

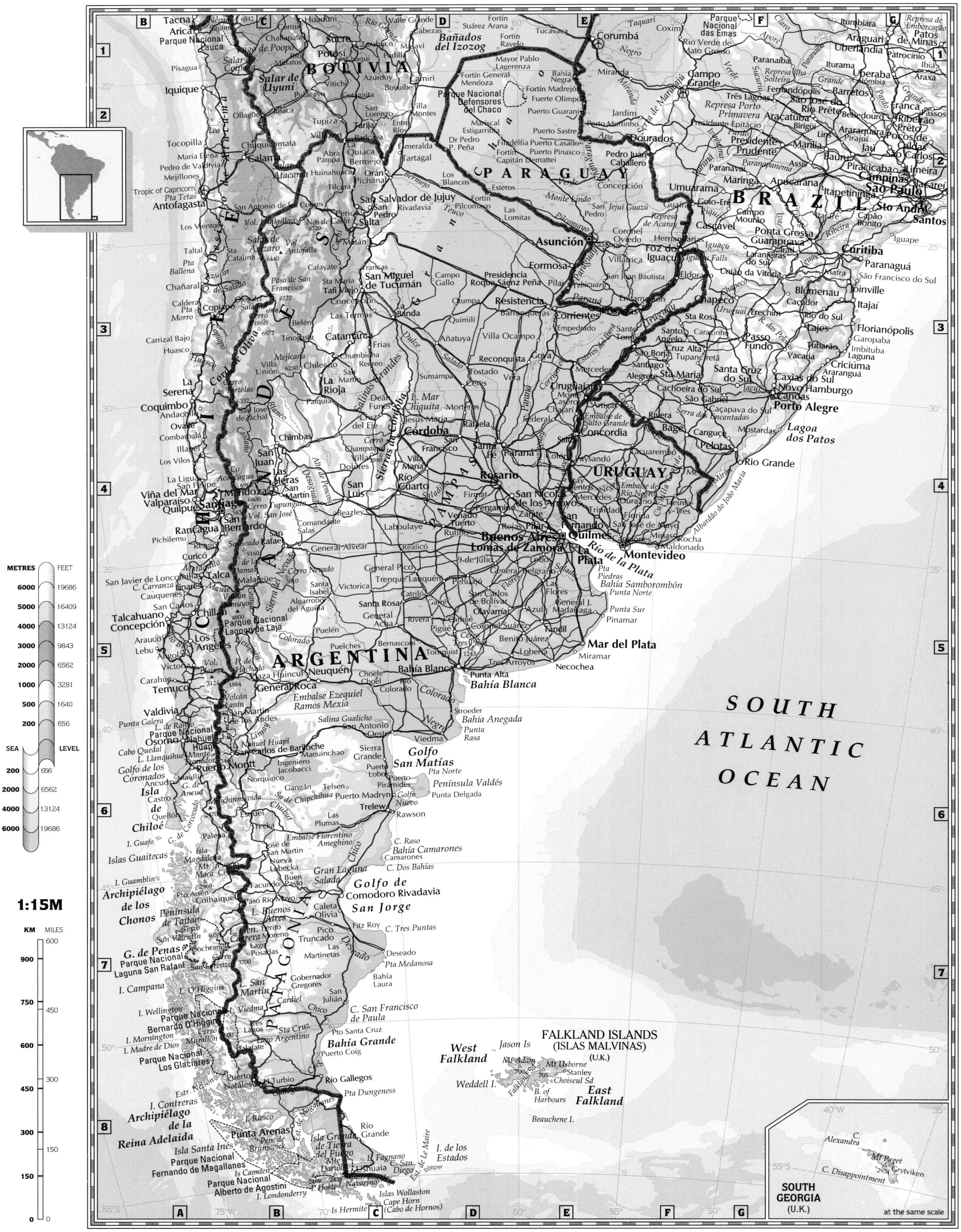

BOLIVIA
PARAGUAY
BRAZIL
URUGUAY
ARGENTINA
CHILE
ANDES
PAMPAS
PATAGONIA
SOUTH ATLANTIC OCEAN
FALKLAND ISLANDS (ISLAS MALVINAS) (U.K.)
West Falkland
East Falkland
Stanley
SOUTH GEORGIA (U.K.)
at the same scale
Grytviken
C. Disappointment
Tacna
Arica
Iquique
Antofagasta
Tropic of Capricorn
Calama
Salar de Uyuni
Potosí
Sucre
Tarija
Asunción
Corumbá
Campo Grande
Curitiba
São Paulo
Santos
Florianópolis
Porto Alegre
Lagoa dos Patos
Rio Grande
Foz do Iguaçu
Posadas
Corrientes
Resistencia
Formosa
Salta
San Salvador de Jujuy
San Miguel de Tucumán
Catamarca
La Rioja
Córdoba
Santa Fe
Paraná
Rosario
Buenos Aires
Lomas de Zamora
La Plata
Quilmes
Montevideo
Río de la Plata
Mar del Plata
Bahía Blanca
Neuquén
Mendoza
San Juan
San Luis
Santiago
Valparaíso
Coquimbo
La Serena
Concepción
Talcahuano
Temuco
Valdivia
Osorno
Puerto Montt
Chiloé
San Carlos de Bariloche
Golfo San Matías
Península Valdés
Trelew
Rawson
Golfo de Comodoro Rivadavia
San Jorge
Bahía Grande
Río Gallegos
Punta Arenas
Isla Grande de Tierra del Fuego
Ushuaia
Cape Horn (Cabo de Hornos)
I. de los Estados
Archipiélago de los Chonos
Archipiélago de la Reina Adelaida
METRES
FEET
SEA LEVEL
1:15M
KM
MILES

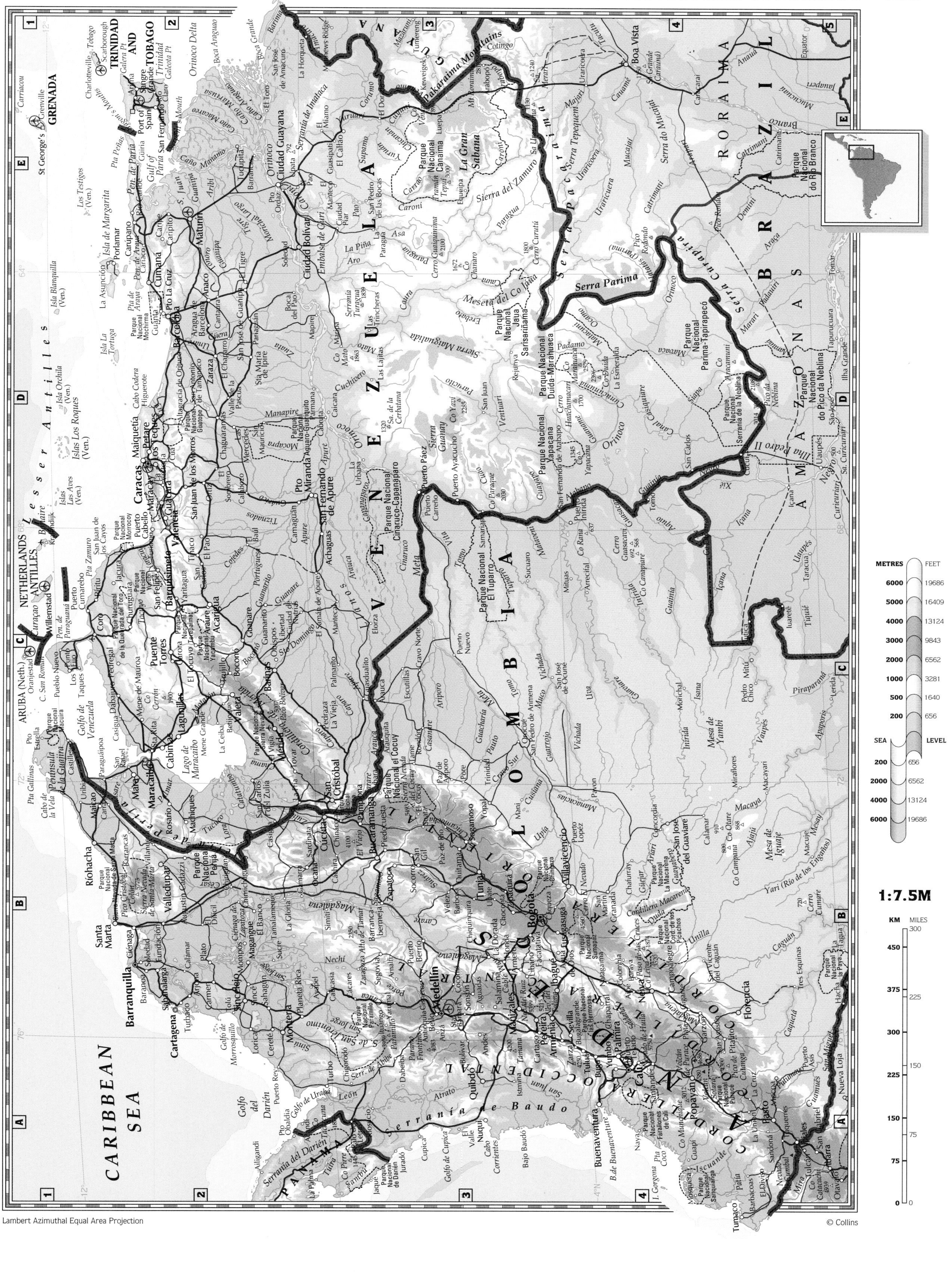

Lambert Azimuthal Equal Area Projection

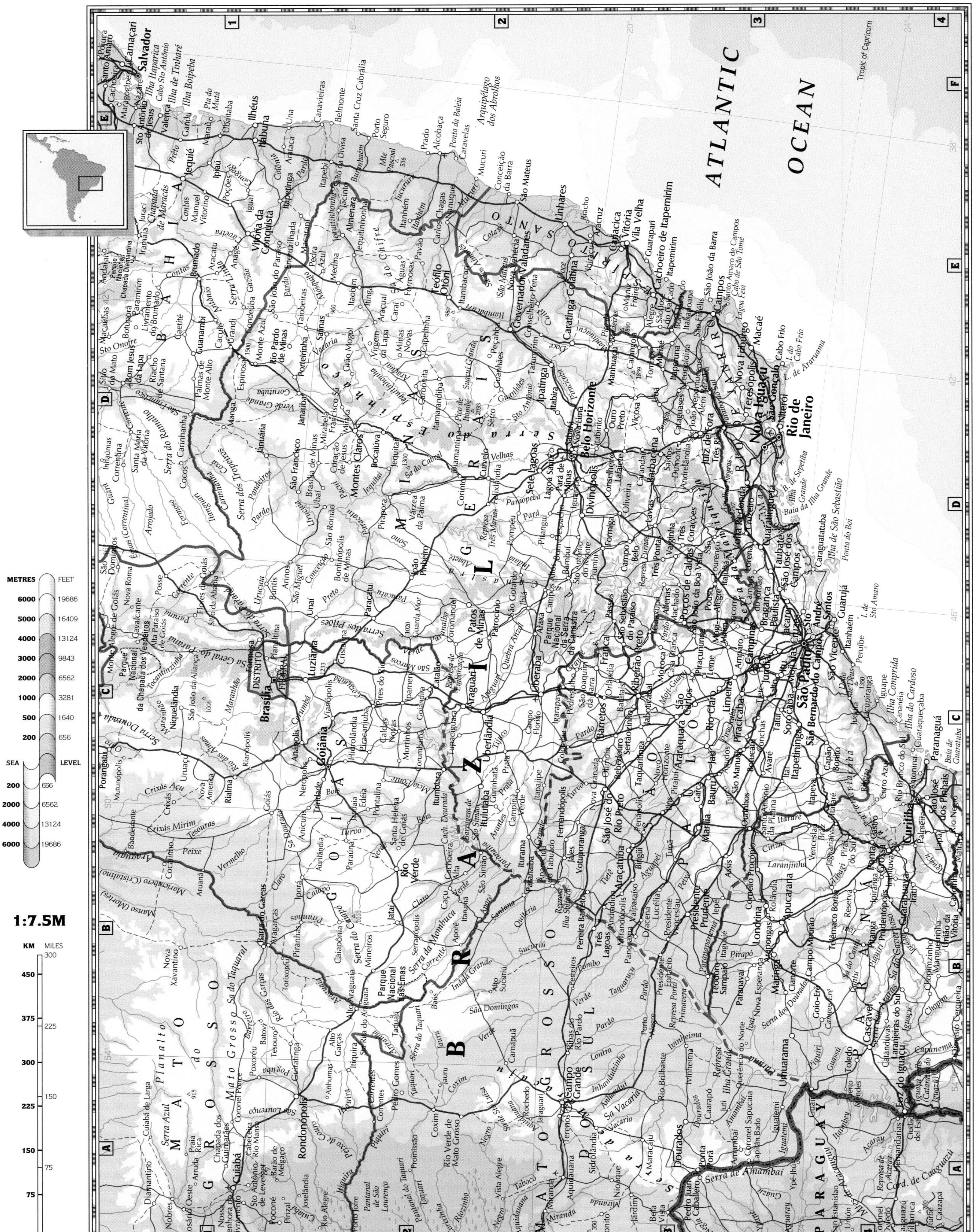
METRES
FEET
6000
19686
5000
16409
4000
13124
3000
9843
2000
6562
1000
3281
500
1640
200
656
SEA
LEVEL
200
656
2000
6562
4000
13124
6000
19686
1:7.5M
KM
MILES
ATLANTIC
OCEAN
Tropic of Capricorn
B R A Z I L
PARAGUAY
Salvador
Belo Horizonte
Rio de Janeiro
São Paulo
Brasília
Goiânia
Curitiba
Campo Grande
Cuiabá
Vitória
Campinas
Santos

Lambert Azimuthal Equal Area Projection

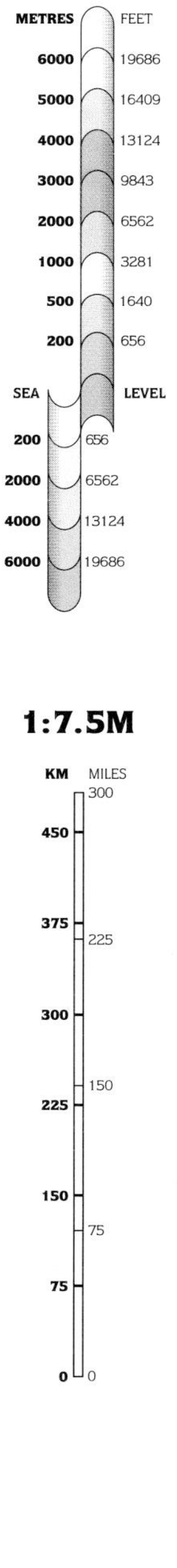

Lambert Azimuthal Equal Area Projection

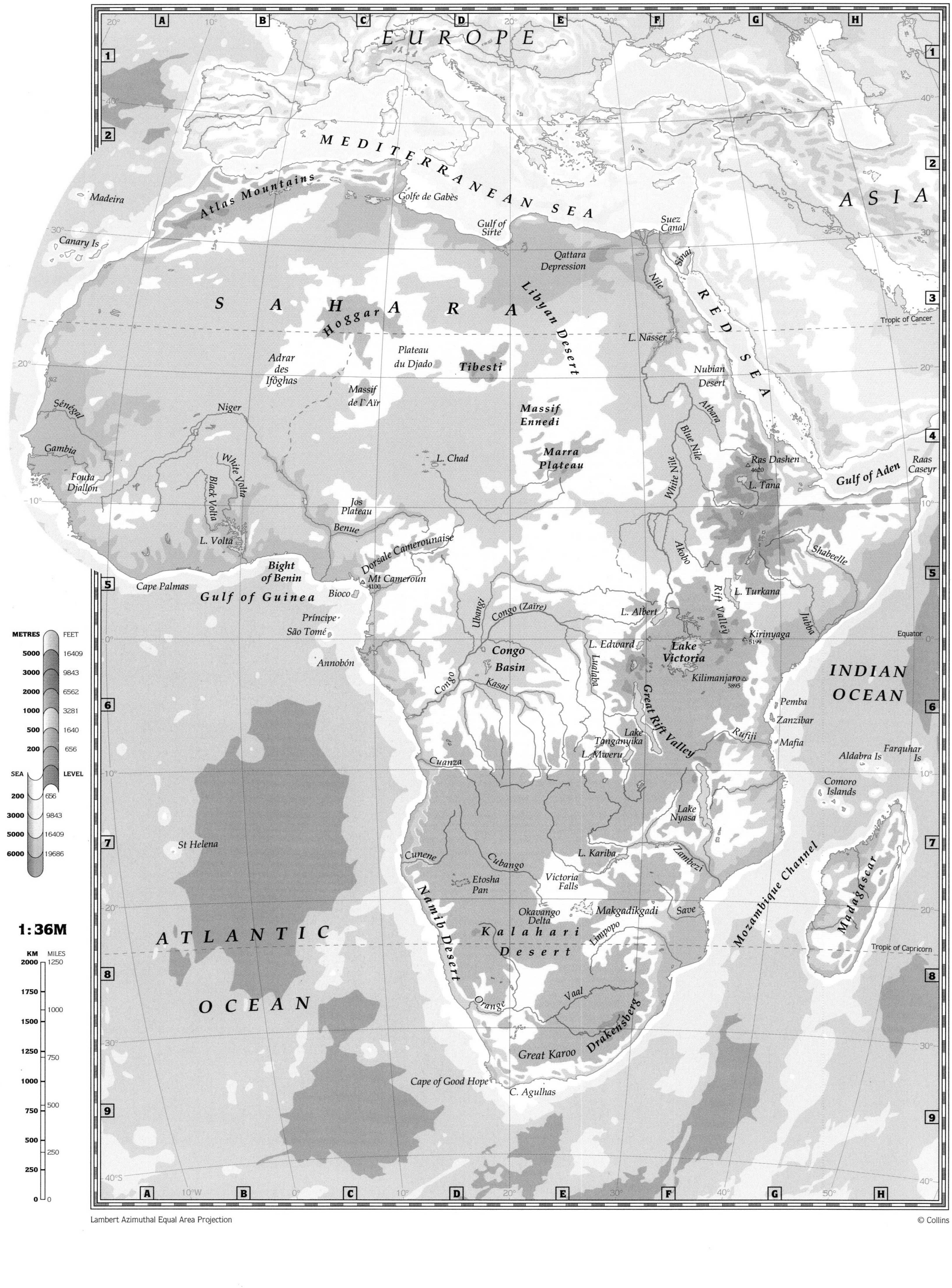
EUROPE
MEDITERRANEAN SEA
ASIA
Atlas Mountains
Madeira
Canary Is
Golfe de Gabès
Gulf of Sirte
Suez Canal
Sinai
Qattara Depression
SAHARA
Hoggar
Libyan Desert
RED SEA
Nile
Tropic of Cancer
L. Nasser
Adrar des Ifôghas
Plateau du Djado
Tibesti
Nubian Desert
Massif de l'Aïr
Massif Ennedi
Atbara
Sénégal
Niger
Blue Nile
Gambia
White Volta
L. Chad
Marra Plateau
Ras Dashen 4620
Raas Caseyr
Fouta Djallon
Black Volta
White Nile
L. Tana
Gulf of Aden
Jos Plateau
Benue
L. Volta
Dorsale Camerounaise
Akobo
Shabeelle
Bight of Benin
Mt Cameroun 4100
Cape Palmas
Gulf of Guinea
Bioco
Rift Valley
L. Turkana
Príncipe
São Tomé
Ubangi
Congo (Zaïre)
L. Albert
Jubba
Equator
L. Edward
Kirinyaga 5199
Congo Basin
Lake Victoria
Annobón
Lualaba
INDIAN OCEAN
Congo
Kasai
Great Rift Valley
Kilimanjaro 5895
Pemba
Zanzibar
Lake Tanganyika
Rufiji
Mafia
L. Mweru
Farquhar Is
Cuanza
Aldabra Is
Comoro Islands
Lake Nyasa
St Helena
Cunene
L. Kariba
Zambezi
Mozambique Channel
Cubango
Etosha Pan
Victoria Falls
Madagascar
Namib Desert
Okavango Delta
Makgadikgadi
Save
ATLANTIC OCEAN
Kalahari Desert
Limpopo
Tropic of Capricorn
Vaal
Orange
Drakensberg
Great Karoo
Cape of Good Hope
C. Agulhas
METRES FEET
5000 16409
3000 9843
2000 6562
1000 3281
500 1640
200 656
SEA LEVEL
200 656
3000 9843
5000 16409
6000 19686
1:36M
KM MILES
2000 1250
1750
1000
1500
1250 750
1000
750 500
500 250
250
0 0

Lambert Azimuthal Equal Area Projection

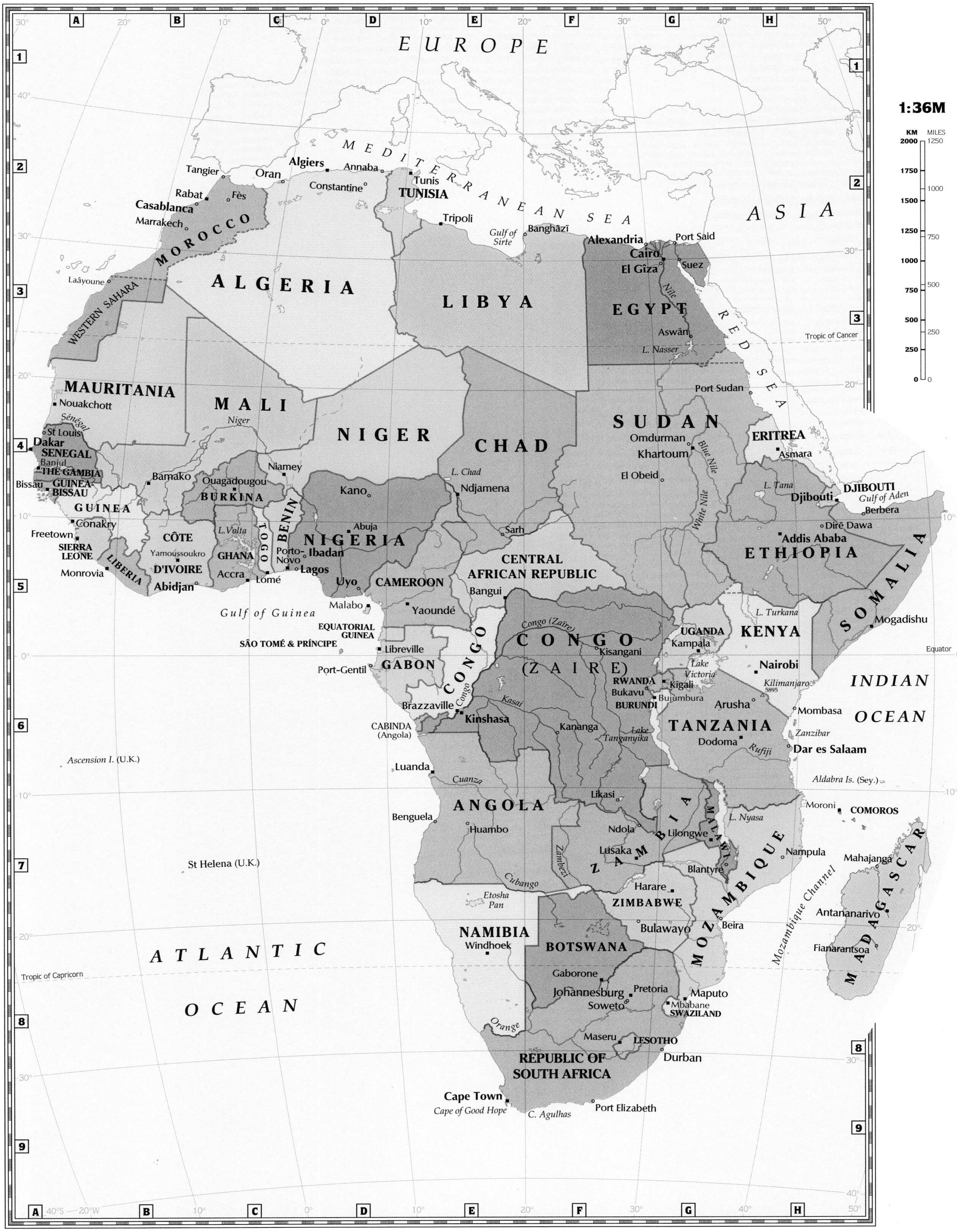
1:36M
KM MILES
2000 1250
1750 1000
1500 750
1250 500
1000 250
750 0
500
250
0
EUROPE
MEDITERRANEAN SEA
ASIA
ATLANTIC OCEAN
INDIAN OCEAN
RED SEA
Gulf of Guinea
Gulf of Aden
Gulf of Sirte
Mozambique Channel
Tropic of Cancer
Tropic of Capricorn
Equator
ALGERIA
MOROCCO
WESTERN SAHARA
TUNISIA
LIBYA
EGYPT
MAURITANIA
MALI
NIGER
CHAD
SUDAN
ERITREA
DJIBOUTI
ETHIOPIA
SOMALIA
SENEGAL
THE GAMBIA
GUINEA-BISSAU
GUINEA
SIERRA LEONE
LIBERIA
CÔTE D'IVOIRE
BURKINA
GHANA
TOGO
BENIN
NIGERIA
CAMEROON
CENTRAL AFRICAN REPUBLIC
EQUATORIAL GUINEA
SÃO TOMÉ & PRÍNCIPE
GABON
CONGO
CONGO (ZAIRE)
CABINDA (Angola)
UGANDA
KENYA
RWANDA
BURUNDI
TANZANIA
ANGOLA
ZAMBIA
MALAWI
MOZAMBIQUE
ZIMBABWE
NAMIBIA
BOTSWANA
SWAZILAND
LESOTHO
REPUBLIC OF SOUTH AFRICA
MADAGASCAR
COMOROS
Tangier
Oran
Algiers
Annaba
Tunis
Constantine
Rabat
Fès
Casablanca
Marrakech
Tripoli
Banghāzī
Alexandria
Port Said
Cairo
El Giza
Suez
Laâyoune
Aswân
L. Nasser
Nile
Port Sudan
Nouakchott
Sénégal
St Louis
Dakar
Banjul
Bissau
Niger
Omdurman
Khartoum
Blue Nile
White Nile
El Obeid
Asmara
Niamey
Bamako
Ouagadougou
L. Chad
Ndjamena
Kano
L. Tana
Djibouti
Berbera
Dire Dawa
Addis Ababa
Abuja
Conakry
Freetown
L. Volta
Yamoussoukro
Porto-Novo
Ibadan
Lagos
Sarh
Monrovia
Accra
Lomé
Abidjan
Uyo
Bangui
Malabo
Yaoundé
L. Turkana
Mogadishu
Congo (Zaïre)
Kampala
Kisangani
Libreville
Port-Gentil
Lake Victoria
Nairobi
Kigali
Bukavu
Bujumbura
Kilimanjaro 5895
Mombasa
Arusha
Brazzaville
Kinshasa
Kasai
Congo
Kananga
Lake Tanganyika
Zanzibar
Dodoma
Rufiji
Dar es Salaam
Ascension I. (U.K.)
Luanda
Cuanza
Aldabra Is. (Sey.)
Likasi
Moroni
Benguela
Huambo
L. Nyasa
Ndola
Lilongwe
Lusaka
Zambezi
Nampula
Mahajanga
St Helena (U.K.)
Blantyre
Cubango
Harare
Etosha Pan
Beira
Antananarivo
Bulawayo
Windhoek
Fianarantsoa
Gaborone
Johannesburg
Pretoria
Soweto
Maputo
Mbabane
Orange
Maseru
Durban
Cape Town
Cape of Good Hope
C. Agulhas
Port Elizabeth
A B C D E F G H
1 2 3 4 5 6 7 8 9
30° 20° 10° 0° 10° 20° 30° 40° 50°
40° 30° 20° 10° 0° 10° 20° 30° 40°
40°S 20°W

Lambert Azimuthal Equal Area Projection

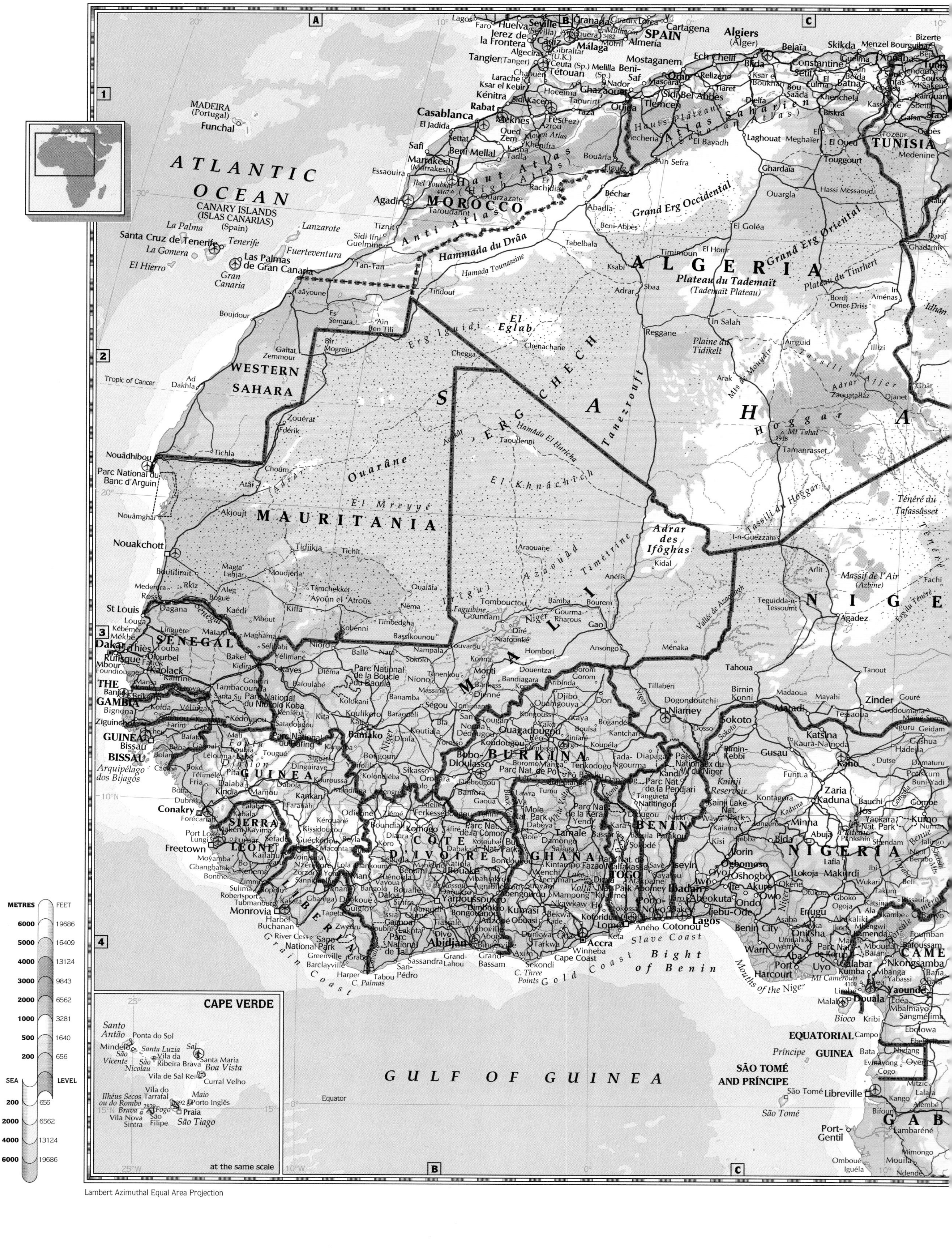

ATLANTIC OCEAN
MADEIRA (Portugal)
Funchal
CANARY ISLANDS (ISLAS CANARIAS) (Spain)
La Palma
Lanzarote
Santa Cruz de Tenerife
Tenerife
La Gomera
Fuerteventura
Las Palmas de Gran Canaria
Gran Canaria
El Hierro
SPAIN
Seville (Sevilla)
Huelva
Faro
Lagos
Jerez de la Frontera
Cádiz
Málaga
Granada
Almería
Cartagena
Algeciras
Gibraltar (U.K.)
Ceuta (Sp.)
Melilla (Sp.)
Tangier (Tanger)
Tétouan
Algiers (Alger)
Oran
Mostaganem
Tlemcen
Sidi Bel Abbès
Constantine
Annaba
Skikda
Bejaïa
Sétif
Batna
Biskra
Tunis
TUNISIA
Kairouan
Sfax
Gabès
Tozeur
Medenine
Rabat
Kénitra
Casablanca
El Jadida
Safi
Meknès
Fès (Fez)
Oujda
Marrakech (Marrakesh)
Essaouira
Agadir
MOROCCO
Haut Atlas (High Atlas)
Anti Atlas
Ouarzazate
Taroudannt
Tiznit
Sidi Ifni
Guelmine
Hammada du Drâa
Tindouf
Tan-Tan
Laâyoune
Boujdour
Es Semara
WESTERN SAHARA
Tropic of Cancer
Ad Dakhla
Béchar
Grand Erg Occidental
Grand Erg Oriental
Ghardaïa
Ouargla
Laghouat
Touggourt
El Goléa
Timimoun
Adrar
Reggane
In Salah
Plateau du Tademaït (Tademaït Plateau)
ALGERIA
Tassili n'Ajjer
Djanet
Hoggar
Mt Tahat 2918
Tamanrasset
Illizi
Tanezrouft
Erg Chech
El Eglab
Erg Iguidi
SAHARA
MAURITANIA
Nouâdhibou
Parc National du Banc d'Arguin
Zouérat
Atâr
Akjoujt
Nouakchott
Tidjikja
Tichit
Kiffa
Néma
Ouarâne
El Mreyyé
El Khnâchîch
Adrar des Ifôghas
Kidal
Tessalit
Tombouctou
Gao
Mopti
Ségou
Bamako
Kayes
Sikasso
MALI
Niger
St Louis
Dakar
Thiès
Rufisque
Kaolack
SENEGAL
Tambacounda
Ziguinchor
Banjul
THE GAMBIA
Bissau
GUINEA-BISSAU
Arquipélago dos Bijagós
Conakry
GUINEA
Kankan
Freetown
SIERRA LEONE
Monrovia
LIBERIA
Yamoussoukro
Abidjan
CÔTE D'IVOIRE
Bouaké
Ouagadougou
BURKINA
Bobo-Dioulasso
Niamey
Zinder
Maradi
Agadez
Arlit
Massif de l'Aïr (Azbine)
Ténéré du Tafassâsset
NIGER
Kumasi
Accra
GHANA
Lomé
TOGO
Porto-Novo
Cotonou
BENIN
Lagos
Ibadan
Abeokuta
Ogbomoso
Ilorin
Abuja
Kaduna
Kano
Sokoto
Zaria
Jos
Makurdi
Enugu
Onitsha
Port Harcourt
Calabar
NIGERIA
Douala
Yaoundé
Malabo
Bioco
EQUATORIAL GUINEA
Bata
SÃO TOMÉ AND PRÍNCIPE
São Tomé
Príncipe
Libreville
Port-Gentil
Lambaréné
GULF OF GUINEA
Bight of Benin
Slave Coast
Gold Coast
Grain Coast
Mouths of the Niger
C. Palmas
C. Three Points
Equator
CAPE VERDE
Santo Antão
São Vicente
Mindelo
Santa Luzia
Sal
Santa Maria
Boa Vista
São Nicolau
Maio
Brava
Fogo
Praia
São Tiago
at the same scale
METRES
FEET
6000
19686
5000
16409
4000
13124
3000
9843
2000
6562
1000
3281
500
1640
200
656
SEA
LEVEL

Lambert Azimuthal Equal Area Projection

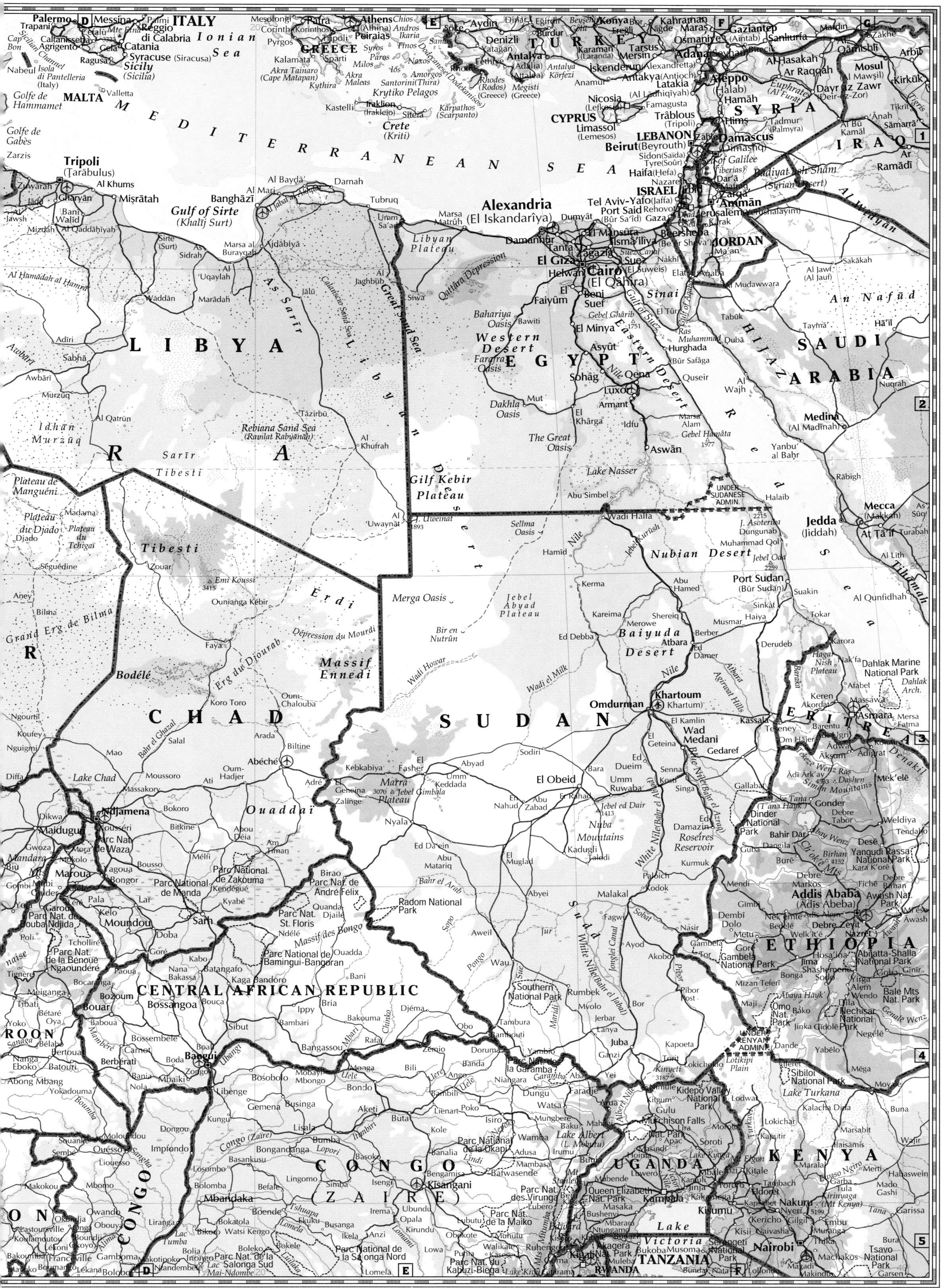
MEDITERRANEAN SEA
Ionian Sea
ITALY
Sicily (Sicilia)
Palermo
Trapani
Messina
Reggio di Calabria
Catania
Syracuse (Siracusa)
Agrigento
Caltanissetta
Ragusa
MALTA
Valletta
Golfe de Hammamet
Golfe de Gabès
Zarzis
GREECE
Athens (Athina)
Peiraias
Patra
Tripoli
Kalamata
Sparti
Crete (Kriti)
Irakleio (Irakleio)
Krytiko Pelagos
Rhodes (Rodos)
Karpathos (Scarpanto)
TURKEY
Aydin
Denizli
Antalya (Adalia)
Konya
Karaman (Laranda)
Tarsus
Mersin
Adana
Iskenderun (Alexandretta)
Antakya (Antioch)
Gaziantep (Aintab)
Sanliurfa
Osmaniye
Kahramanmaras
Nicosia (Lefkosia)
CYPRUS
Limassol (Lemesos)
Famagusta
SYRIA
Aleppo (Halab)
Latakia
Hamah
Hims
Tartous
Tadmur (Palmyra)
Damascus (Dimashq)
Al Hasakah
Ar Raqqah
Dayr az Zawr (Deir-ez-Zor)
Mosul (Al Mawsil)
Arbil
Kirkuk
IRAQ
Ar Ramadi
LEBANON
Beirut (Beyrouth)
Sidon (Saida)
Tyre (Sour)
Haifa (Hefa)
ISRAEL
Tel Aviv-Yafo (Jaffa)
Jerusalem
Gaza
Beersheba (Be'er Sheva)
JORDAN
'Amman
Ma'an
Al 'Aqaba
Elat
Badiyat ash Sham (Syrian Desert)
An Nafud
SAUDI ARABIA
Tabuk
Medina (Al Madinah)
Yanbu' al Bahr
Rabigh
Mecca (Makkah)
Jedda (Jiddah)
At Ta'if
Al Lith
Al Qunfidhah
Al Wajh
Duba
Hijaz
Tihamah
Red Sea
Tripoli (Tarabulus)
Al Khums
Misratah
Gharyan
Bani Walid
Gulf of Sirte (Khalij Surt)
Sirte (Surt)
Banghazi
Al Bayda'
Darnah
Tubruq
Ajdabiya
Marsa al Burayqah
LIBYA
Al Hamadah al Hamra
Waddan
Jalu
Maradah
As Sarir
Calanscio Sand Sea
Sabha
Awbari
Murzuq
Idhan Murzuq
Al Qatrun
Rebiana Sand Sea (Ramlat Rabyanah)
Al Khufrah
Tazirbu
Sarir Tibesti
SAHARA
Alexandria (El Iskandariya)
Marsa Matruh
Damanhur
Tanta
El Mansura
Port Said (Bur Sa'id)
Isma'iliya
Zagazig
Suez (El Suweis)
El Giza
Cairo (El Qahira)
Helwan
Beni Suef
El Faiyum
El Minya
Asyut
Sohag
Qena
Luxor
Armant
Idfu
Aswan
Hurghada
Quseir
Sinai
Siwa
Qattara Depression
Libyan Plateau
Great Sand Sea
Western Desert
Eastern Desert
Bahariya Oasis
Farafra Oasis
Dakhla Oasis
El Kharga
The Great Oasis
EGYPT
Lake Nasser
Abu Simbel
Wadi Halfa
Gilf Kebir Plateau
Libyan Desert
Halaib
UNDER SUDANESE ADMIN.
Nubian Desert
Port Sudan (Bur Sudan)
Suakin
Tokar
Atbara
Berber
Ed Damer
Shereiq
Kareima
Merowe
Kerma
Bayiuda Desert
Khartoum (Khartum)
Omdurman
Wad Medani
Gedaref
Kassala
Sennar
Kosti
El Obeid
En Nahud
Nuba Mountains
Kadugli
Ed Dueim
Roseires Reservoir
Damazin
SUDAN
El Fasher
Geneina
Nyala
Jebel Marra
Wadi Howar
Wadi el Milk
Sudd
Malakal
Nasir
Bor
Juba
Wau
Aweil
Rumbek
Kapoeta
White Nile (Bahr el Jabal)
Blue Nile (Bahr el Azraq)
Tibesti
Emi Koussi 3415
Erdi
Massif Ennedi
Bodélé
Dépression du Mourdi
CHAD
Faya
Abéché
Ati
Mao
Lake Chad
Ndjamena
Moundou
Sarh
Doba
Ouaddaï
Grand Erg de Bilma
Bilma
Maiduguri
Maroua
Garoua
Ngaoundéré
CENTRAL AFRICAN REPUBLIC
Bangui
Bouar
Bossangoa
Bria
Bambari
Berbérati
Bangassou
Mbomou
CONGO (ZAIRE)
Kisangani
Mbandaka
Lisala
Bumba
Buta
Isiro
Bunia
Lake Albert (L. Mobutu)
Lake Edward
Lake Kivu
Lake Victoria
Kampala
UGANDA
Gulu
Mbale
Entebbe
Kisumu
KENYA
Nairobi
Nakuru
Lake Turkana
Lodwar
Marsabit
Moyale
RWANDA
Kigali
TANZANIA
Mwanza
ERITREA
Asmara
Massawa
Keren
Dahlak Marine National Park
ETHIOPIA
Addis Ababa (Adis Abeba)
Gonder
Lake Tana (T'ana Hayk')
Bahir Dar
Dese
Mek'ele
Aksum
Simen Mountains
Ras Dashen
Nazret
Jima
Awash
Gambela
Mizan Teferi
Arba Minch
Nechisar National Park
Bale Mts Nat. Park
Dinder National Park
Parc National de Zakouma
Parc Nat. St. Floris
Parc Nat. de la Garamba
Parc National de l'Okapi
Parc Nat. des Virunga
Queen Elizabeth Nat. Park
Murchison Falls Nat. Park
Kidepo Valley National Park
Sibiloi National Park
Serengeti National Park
Tsavo National Park
Radom National Park
Southern National Park
UNDER KENYAN ADMIN.
Ilemi Triangle
CONGO
Ouesso
Impfondo

1:16M

KM MILES
1000 600
800 500
400
600
300
400
200
200
100
0 0

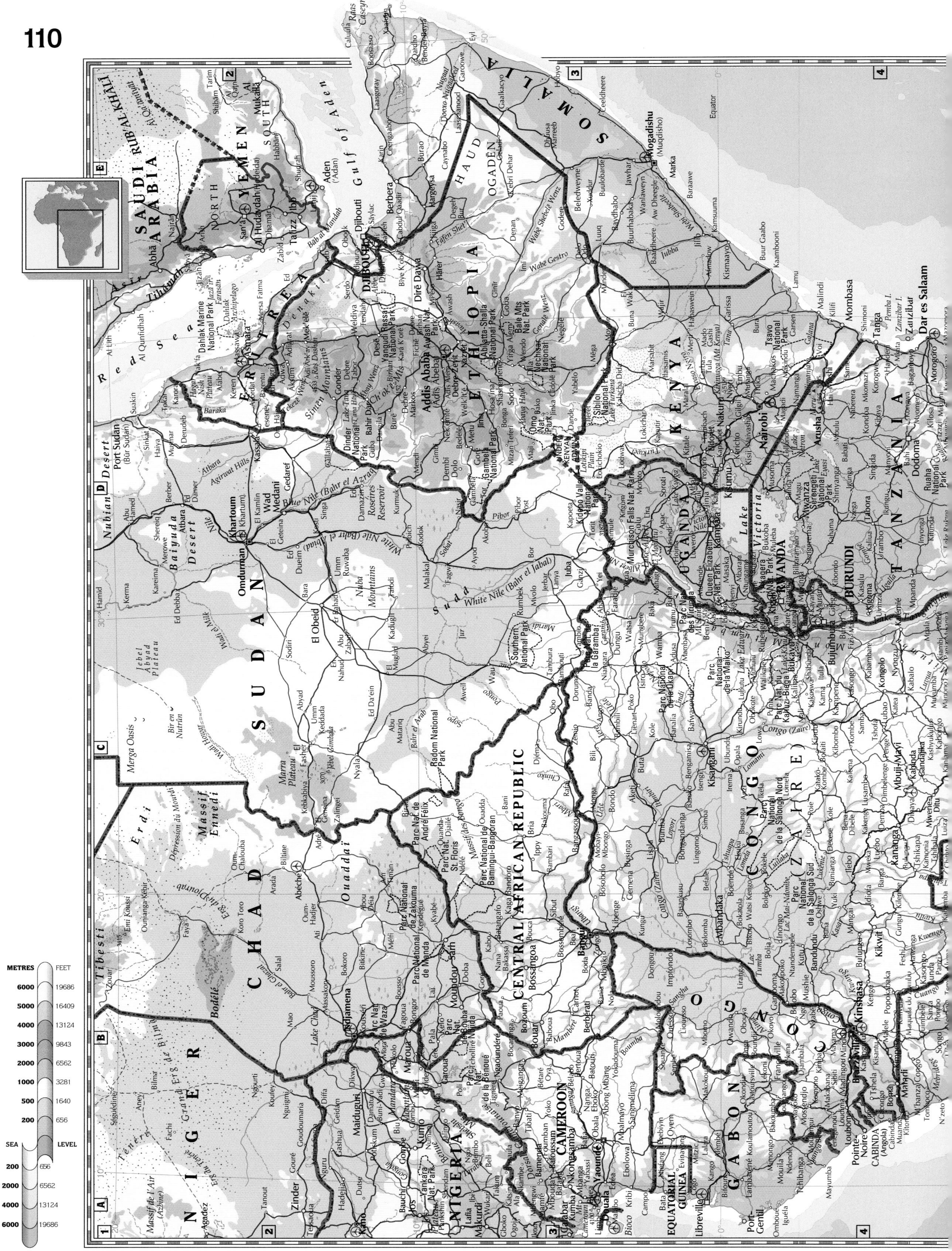

Lambert Azimuthal Equal Area Projection

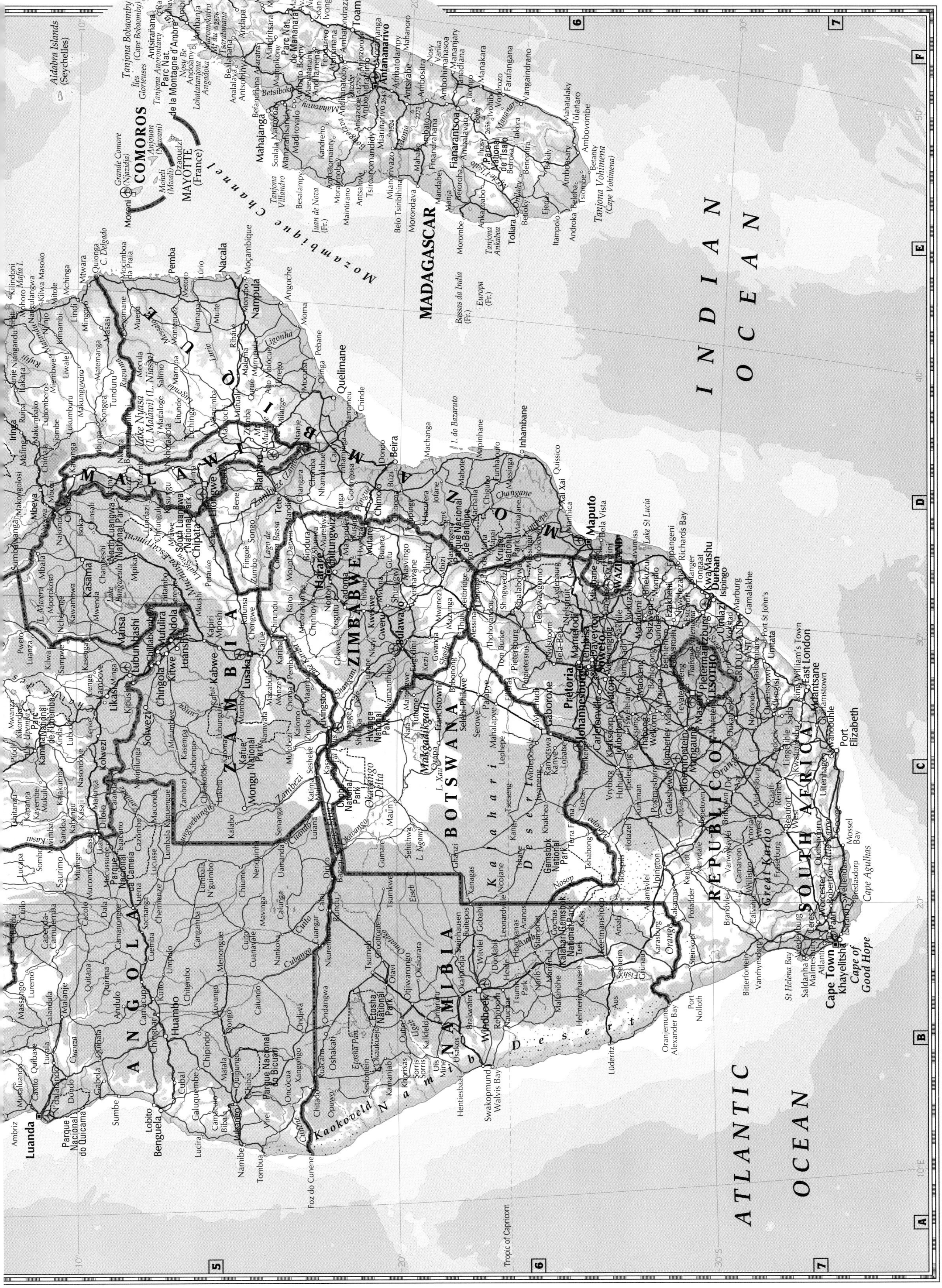

1:16M

KM MILES

0 200 400 600 800 1000

0 100 200 300 400 500 600

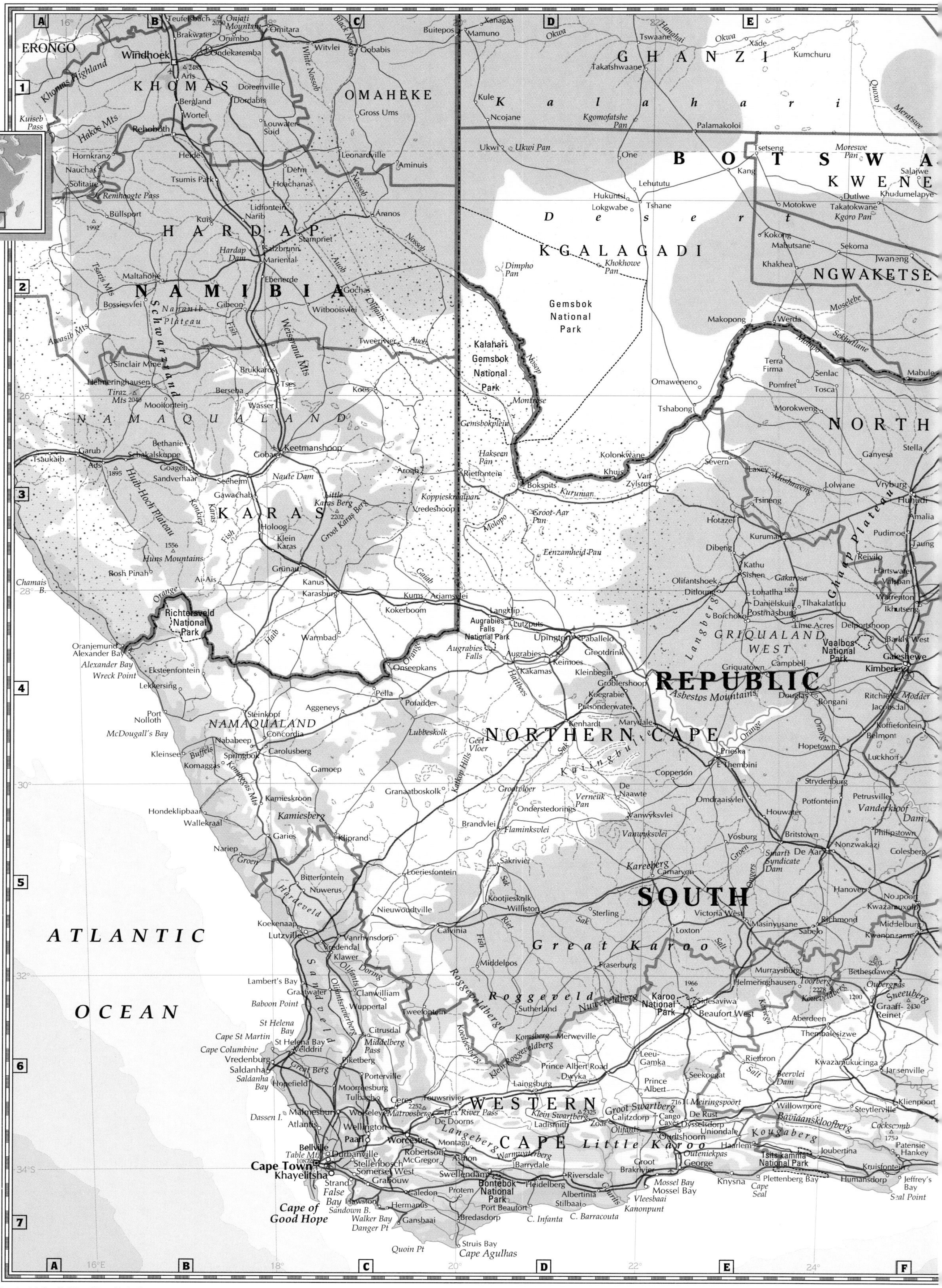

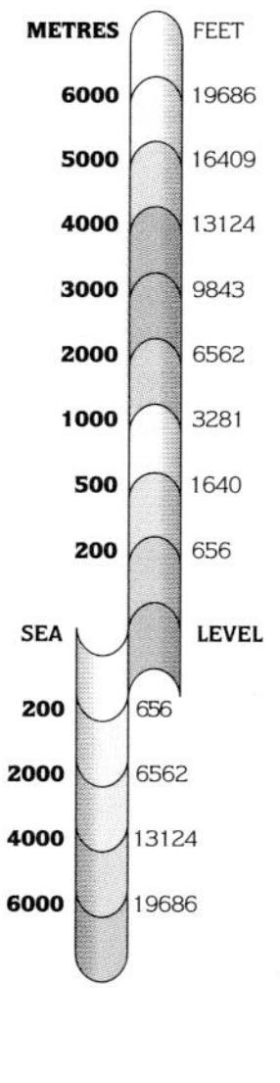

Lambert Azimuthal Equal Area Projection

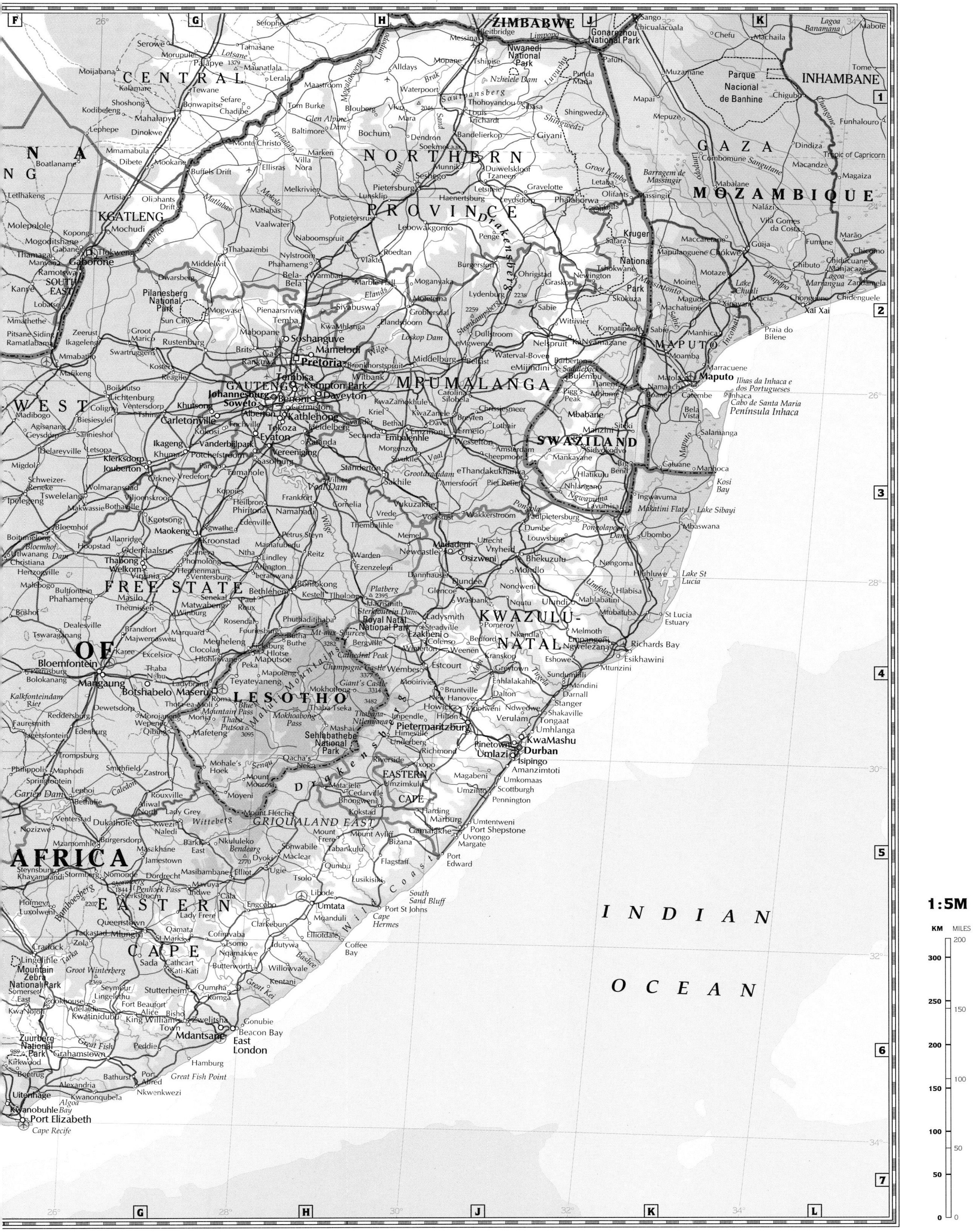
ZIMBABWE
MOZAMBIQUE
GAZA
INHAMBANE
MAPUTO
Maputo
NORTHERN PROVINCE
MPUMALANGA
GAUTENG
Pretoria
Johannesburg
Soweto
SWAZILAND
Mbabane
KWAZULU-NATAL
Durban
Pietermaritzburg
Richards Bay
LESOTHO
Maseru
FREE STATE
Bloemfontein
EASTERN CAPE
East London
Port Elizabeth
Umtata
NORTH WEST
KGATLENG
SOUTH EAST
Gaborone
CENTRAL
Kruger National Park
Drakensberg
Limpopo
Vaal Dam
Lake St Lucia
Wild Coast
INDIAN OCEAN
Tropic of Capricorn
1:5M
KM
MILES
300
250
200
150
100
50
0
200
150
100
50
0

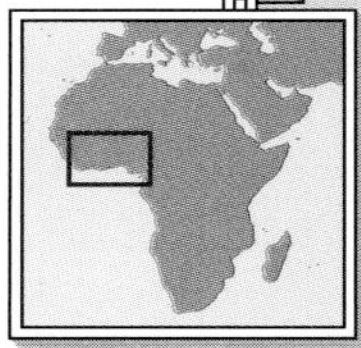

METRES / FEET

METRES	FEET
6000	19686
5000	16409
4000	13124
3000	9843
2000	6562
1000	3281
500	1640
200	656
SEA	LEVEL
200	656
2000	6562
4000	13124
6000	19686

1:10M

KM 600, 500, 400, 300, 200, 100, 0 — MILES 300, 200, 100, 0

Lambert Azimuthal Equal Area Projection

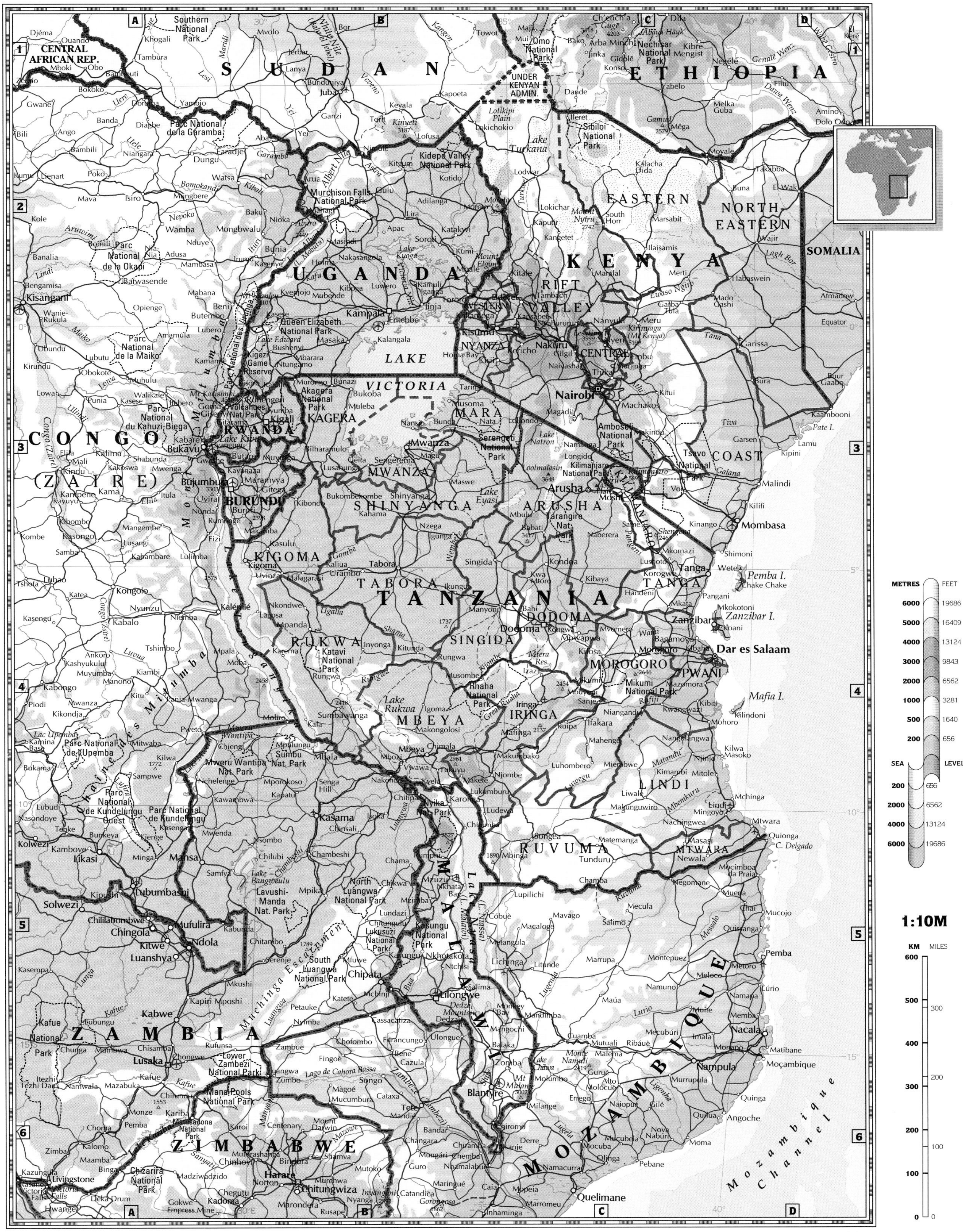

Lambert Azimuthal Equal Area Projection

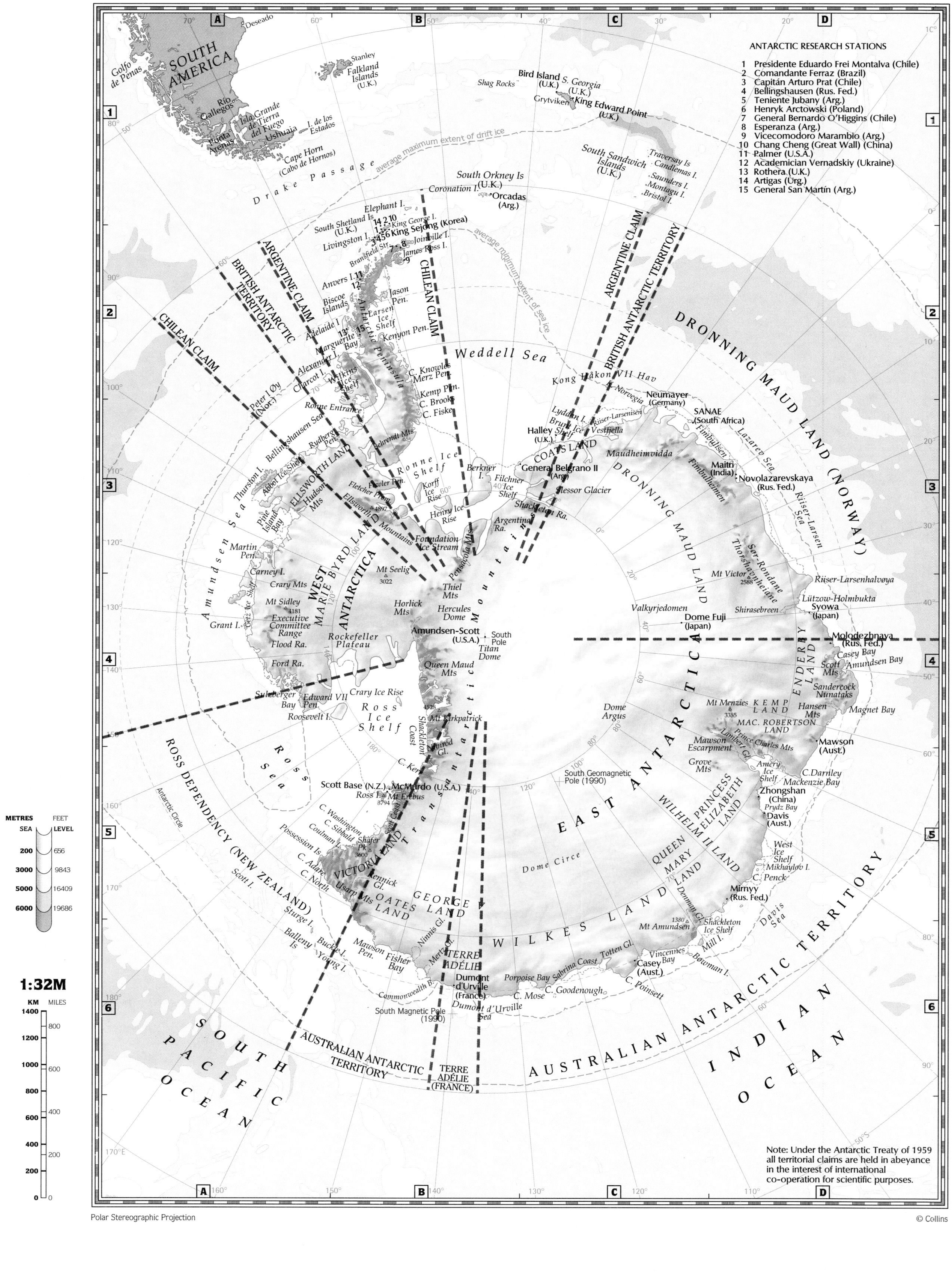
ANTARCTIC RESEARCH STATIONS
1 Presidente Eduardo Frei Montalva (Chile)
2 Comandante Ferraz (Brazil)
3 Capitán Arturo Prat (Chile)
4 Bellingshausen (Rus. Fed.)
5 Teniente Jubany (Arg.)
6 Henryk Arctowski (Poland)
7 General Bernardo O'Higgins (Chile)
8 Esperanza (Arg.)
9 Vicecomodoro Marambio (Arg.)
10 Chang Cheng (Great Wall) (China)
11 Palmer (U.S.A.)
12 Academician Vernadskiy (Ukraine)
13 Rothera (U.K.)
14 Artigas (Urg.)
15 General San Martín (Arg.)
Note: Under the Antarctic Treaty of 1959 all territorial claims are held in abeyance in the interest of international co-operation for scientific purposes.
SOUTH AMERICA
Golfo de Penas
Río Gallegos
Isla Grande de Tierra del Fuego
Punta Arenas
Ushuaia
I. de los Estados
Cape Horn (Cabo de Hornos)
Deseado
Stanley
Falkland Islands (U.K.)
Drake Passage
average maximum extent of drift ice
Shag Rocks
Bird Island (U.K.)
S. Georgia (U.K.)
Grytviken
King Edward Point (U.K.)
South Sandwich Islands (U.K.)
Traversay Is
Candlemas I.
Saunders I.
Montagu I.
Bristol I.
South Orkney Is (U.K.)
Coronation I.
Orcadas (Arg.)
South Shetland Is (U.K.)
Elephant I.
King George I.
Livingston I.
King Sejong (Korea)
Joinville I.
James Ross I.
Bransfield Str.
average minimum extent of sea ice
ARGENTINE CLAIM
BRITISH ANTARCTIC TERRITORY
CHILEAN CLAIM
Anvers I.
Biscoe Islands
Adelaide I.
Marguerite Bay
Jason Pen.
Larsen Ice Shelf
Antarctic Peninsula
Kenyon Pen.
Alexander I.
Charcot I.
Wilkins Ice Shelf
Peter I Øy (Nor.)
Bellingshausen Sea
Ronne Entrance
Rydberg Pen.
Weddell Sea
C. Knowles
Merz Pen.
Kemp Pen.
C. Brooks
C. Fiske
Behrendt Mts
Ronne Ice Shelf
Berkner I.
Filchner Ice Shelf
Korff Ice Rise
Henry Ice Rise
Fowler Pen.
Fletcher Pen.
ELLSWORTH LAND
Ellsworth Mountains
Hudson Mts
Thurston I.
Abbot Ice Shelf
Pine Island Bay
Amundsen Sea
Martin Pen.
Carney I.
Crary Mts
Getz Ice Shelf
Mt Sidley 4181
Executive Committee Range
Grant I.
Flood Ra.
Ford Ra.
WEST ANTARCTICA
MARIE BYRD LAND
Mt Seelig 3022
Foundation Ice Stream
Pensacola Mts
Argentina Ra.
Shackleton Ra.
Slessor Glacier
Thiel Mts
Horlick Mts
Hercules Dome
Rockefeller Plateau
Amundsen-Scott (U.S.A.)
South Pole
Titan Dome
Queen Maud Mts
Transantarctic Mountains
Sulzberger Bay
Edward VII Pen.
Roosevelt I.
Crary Ice Rise
Ross Ice Shelf
Mt Kirkpatrick
Shackleton Coast
Nimrod Gl.
C. Kerr
Ross Sea
Scott Base (N.Z.)
McMurdo (U.S.A.)
Ross I.
Mt Erebus 3794
ROSS DEPENDENCY (NEW ZEALAND)
Antarctic Circle
C. Washington
C. Sibbald
Coulman I.
Possession Is
Shafer Pk 3600
C. Adare
C. North
Scott I.
VICTORIA LAND
Rennick Gl.
Usarp Mts
OATES LAND
GEORGE V LAND
Sturge I.
Balleny Is
Buckle I.
Young I.
Mawson Pen.
Ninnis Gl.
Mertz Gl.
Fisher Bay
TERRE ADÉLIE
Dumont d'Urville (France)
Dumont d'Urville Sea
Commonwealth B.
South Magnetic Pole (1990)
AUSTRALIAN ANTARCTIC TERRITORY
TERRE ADÉLIE (FRANCE)
SOUTH PACIFIC OCEAN
Kong Håkon VII Hav
Neumayer (Germany)
SANAE (South Africa)
Norvegia
Riiser-Larsenisen
Vestfjella
Lyddan I.
Brunt Ice Shelf
Halley (U.K.)
COATS LAND
General Belgrano II (Arg.)
Maudheimvidda
Fimbulisen
Fimbulheimen
Lazarev Sea
Maitri (India)
Novolazarevskaya (Rus. Fed.)
DRONNING MAUD LAND (NORWAY)
DRONNING MAUD LAND
Riiser-Larsen Sea
Sør-Rondane
Thorshavnheiane
Mt Victor 2588
Riiser-Larsenhalvøya
Lützow-Holmbukta
Syowa (Japan)
Shirasebreen
Valkyrjedomen
Dome Fuji (Japan)
Molodezhnaya (Rus. Fed.)
Casey Bay
Amundsen Bay
Scott Mts
ENDERBY LAND
Sandercock Nunataks
Hansen Mts
Magnet Bay
Mt Menzies 3355
KEMP LAND
MAC. ROBERTSON LAND
Dome Argus
Mawson Escarpment
Lambert Gl.
Prince Charles Mts
Mawson (Aust.)
Grove Mts
Amery Ice Shelf
C. Darnley
Mackenzie Bay
Zhongshan (China)
Prydz Bay
Davis (Aust.)
South Geomagnetic Pole (1990)
EAST ANTARCTICA
PRINCESS ELIZABETH LAND
WILHELM II LAND
QUEEN MARY LAND
West Ice Shelf
Mikhaylov I.
C. Penck
Mirnyy (Rus. Fed.)
Dome Circe
Denman Gl.
Mt Amundsen 1380
Shackleton Ice Shelf
Davis Sea
Mill I.
WILKES LAND
Vincennes Bay
Bowman I.
Totten Gl.
Casey (Aust.)
Sabrina Coast
Porpoise Bay
C. Mose
C. Goodenough
C. Poinsett
AUSTRALIAN ANTARCTIC TERRITORY
INDIAN OCEAN
METRES SEA LEVEL FEET
200 656
3000 9843
5000 16409
6000 19686
1:32M
KM MILES
1400 1200 1000 800 600 400 200 0
800 600 400 200 0

Polar Stereographic Projection

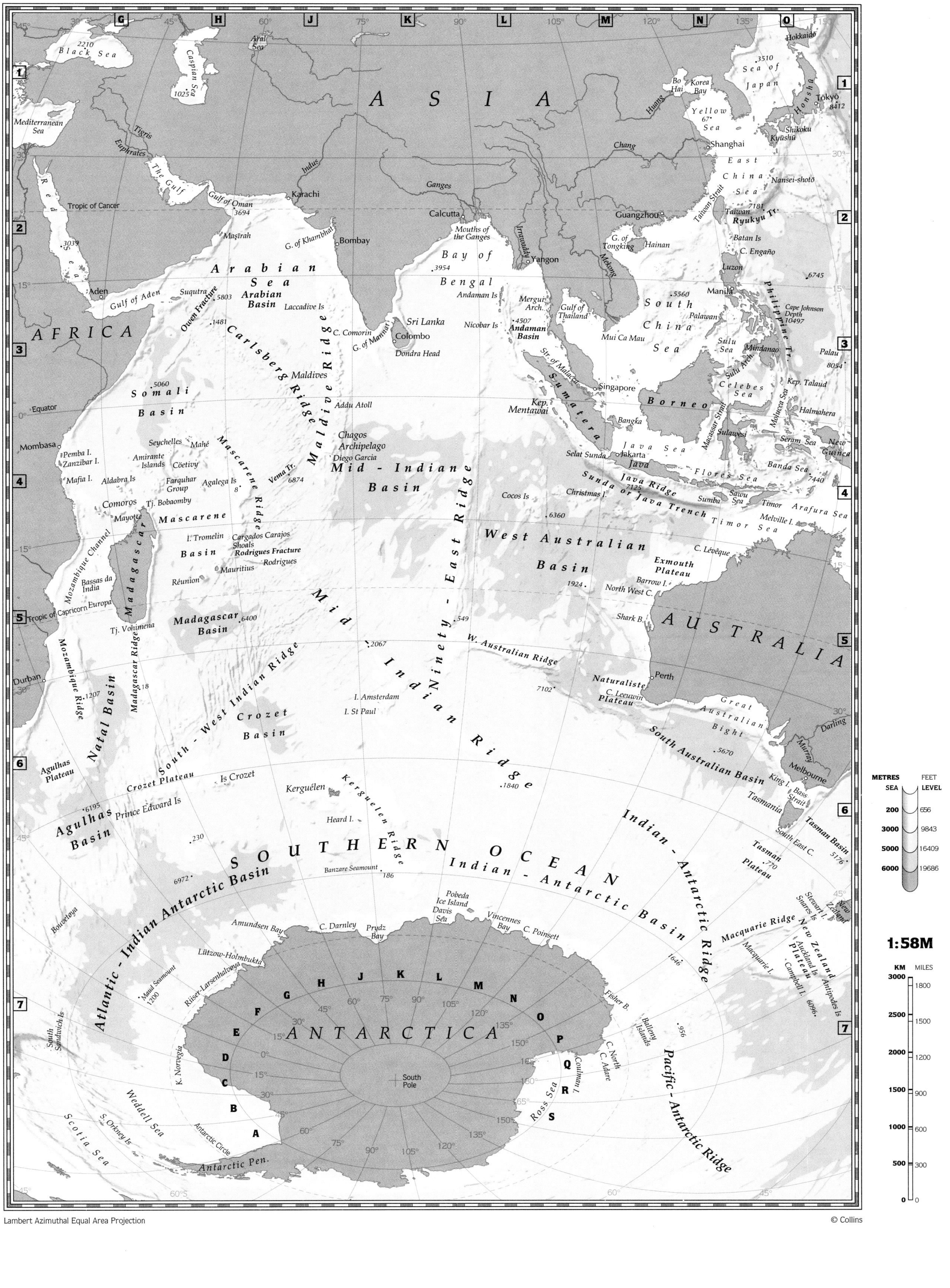
ASIA
AFRICA
AUSTRALIA
ANTARCTICA
SOUTHERN OCEAN
Black Sea
2210
Caspian Sea
1025
Aral Sea
Mediterranean Sea
Tigris
Euphrates
The Gulf
Gulf of Oman
3694
Red Sea
3039
Tropic of Cancer
Karachi
Indus
Ganges
Calcutta
Mouths of the Ganges
Bombay
G. of Khambhat
Masīrah
Arabian Sea
Arabian Basin
Aden
Gulf of Aden
Suqutra
5803
Owen Fracture
1481
Carlsberg Ridge
Laccadive Is
Maldive Ridge
C. Comorin
G. of Mannar
Sri Lanka
Colombo
Dondra Head
Bay of Bengal
3954
Andaman Is
Nicobar Is
4507
Andaman Basin
Mergui Arch.
Gulf of Thailand
Irrawaddy
Yangon
Mekong
G. of Tongking
Hainan
Guangzhou
Chang
Shanghai
Huang
Bo Hai
Korea Bay
Yellow Sea
67
Sea of Japan
3510
Hokkaido
Honshu
Tokyo
8412
Shikoku
Kyushu
East China Sea
Nansei-shotō
Taiwan Strait
Taiwan
7181
Ryukyu Tr.
Batan Is
C. Engaño
Luzon
6745
Manila
Philippine Tr.
Cape Johnson Depth 10497
South China Sea
5560
Palawan
Mui Ca Mau
Sulu Sea
Sulu Arch.
Mindanao
Palau
8054
Str. of Malacca
Singapore
Sumatera
Kep. Mentawai
Borneo
Celebes Sea
Kep. Talaud
Molucca Sea
Halmahera
Makassar Strait
Sulawesi
Bangka
Java Sea
Jakarta
Selat Sunda
Java
Seram Sea
New Guinea
Flores Sea
Banda Sea
7440
Java Ridge
Sunda or Java Trench
7125
Christmas I.
Sumba
Sawu Sea
Timor
Arafura Sea
Timor Sea
Melville I.
Somali Basin
5060
Equator
Maldives
Addu Atoll
Chagos Archipelago
Diego Garcia
Mid - Indian Basin
Mombasa
Pemba I.
Zanzibar I.
Mafia I.
Seychelles
Mahé
Amirante Islands
Cöetivy
Aldabra Is
Farquhar Group
Agalega Is
Mascarene Ridge
Vema Tr.
6874
Comoros
Mayotte
Tj. Bobaomby
Mascarene Basin
I. Tromelin
Cargados Carajos Shoals
Rodrigues Fracture
Rodrigues
Mauritius
Réunion
Mozambique Channel
Madagascar
Bassas da India
Europa
Tropic of Capricorn
Tj. Vohimena
Madagascar Basin
6400
Cocos Is
6360
Ninety - East Ridge
West Australian Basin
1924
C. Lévêque
Exmouth Plateau
Barrow I.
North West C.
Shark B.
549
W. Australian Ridge
Mid Indian Ridge
2067
Perth
Naturaliste Plateau
C. Leeuwin
7102
Durban
Mozambique Ridge
1207
Natal Basin
Madagascar Ridge
South - West Indian Ridge
Crozet Basin
I. Amsterdam
I. St Paul
Great Australian Bight
5670
South Australian Basin
Darling
Murray
Melbourne
King I.
Bass Strait
Tasmania
South East C.
Tasman Basin
5176
Tasman Plateau
770
Agulhas Plateau
Crozet Plateau
Is Crozet
Kerguélen
Kerguelen Ridge
1840
Heard I.
Prince Edward Is
6195
Agulhas Basin
230
Banzare Seamount
186
6972
Atlantic - Indian Antarctic Basin
Indian - Antarctic Basin
Indian - Antarctic Ridge
Pobeda Ice Island
Davis Sea
Vincennes Bay
C. Poinsett
C. Darnley
Prydz Bay
Amundsen Bay
Lützow-Holmbukta
Riiser-Larsenhalvøya
Maud Seamount 1200
Bouvetøya
South Sandwich Is
K. Norvegia
Weddell Sea
S. Orkney Is
Scotia Sea
Antarctic Circle
Antarctic Pen.
South Pole
Ross Sea
Coulman I.
C. North
C. Adare
Balleny Islands
Fisher B.
1646
956
Pacific - Antarctic Ridge
Macquarie Ridge
Macquarie I.
New Zealand Plateau
Auckland Is
Campbell I.
Antipodes Is
6096
Snares Is
Stewart I.
New Zealand
METRES SEA LEVEL FEET
200 656
3000 9843
5000 16409
6000 19686
1:58M
KM MILES
3000 1800
2500 1500
2000 1200
1500 900
1000 600
500 300
0 0

Lambert Azimuthal Equal Area Projection

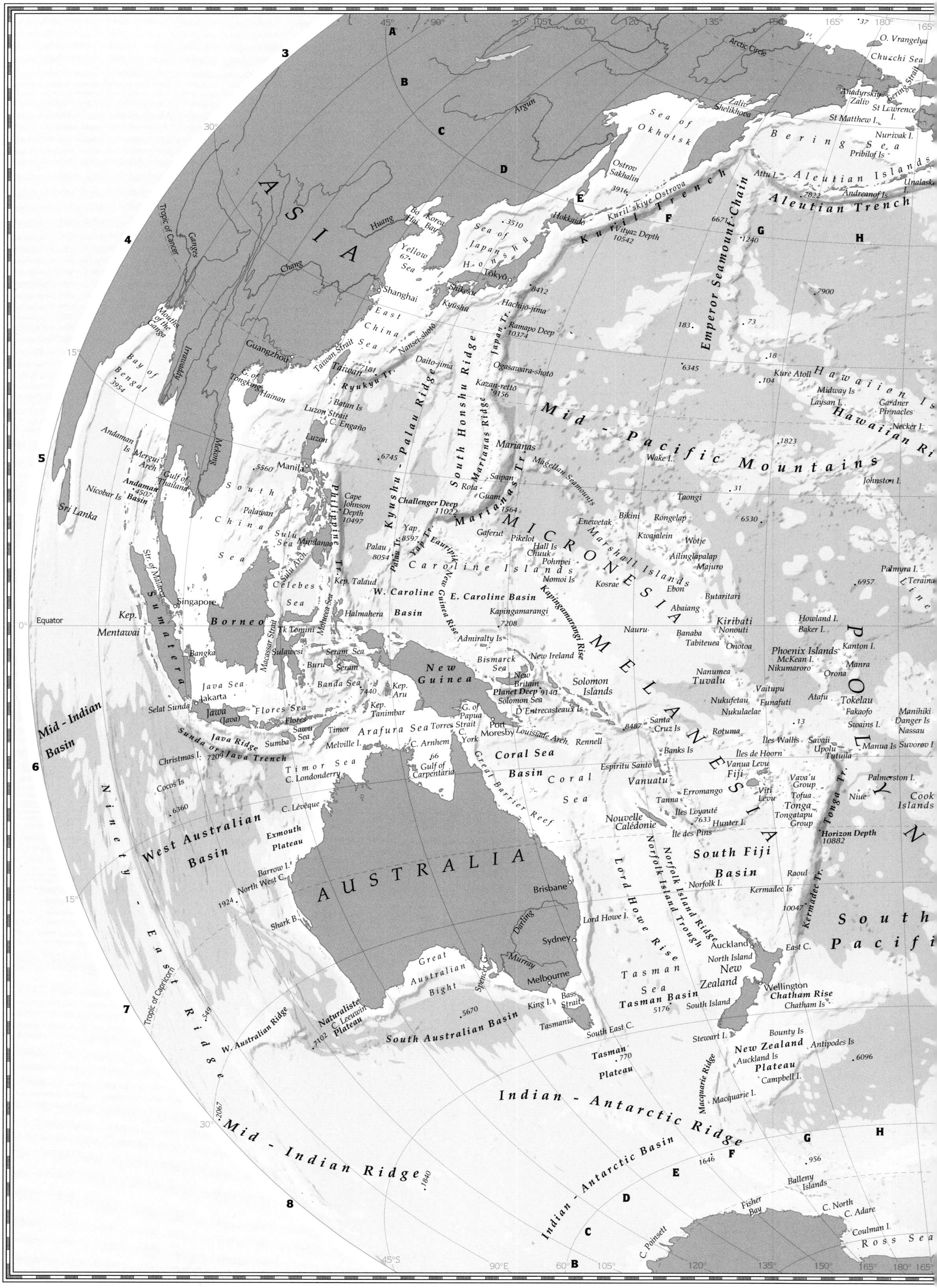
A S I A
Tropic of Cancer
Ganges
Mouths of the Ganga
Bay of Bengal
Irrawaddy
Mekong
Chang
Huang
Argun
Sea of Okhotsk
Zaliv Shelikhova
Anadyrskiy Zaliv
Bering Strait
O. Vrangelya
Chukchi Sea
Arctic Circle
St Lawrence I.
St Matthew I.
Nurivak I.
Bering Sea
Pribilof Is
Ostrov Sakhalin
Attu I.
Aleutian Islands
Unalaska
Andreanof Is
Aleutian Trench
Kuril'skiye Ostrova
Kuril Trench
Vityaz Depth 10542
Emperor Seamount Chain
Hokkaido
Sea of Japan
Honshu
Tokyo
Bo Hai
Korea Bay
Yellow Sea
Shanghai
Shikoku
Kyushu
Hachijo-jima
Ramapo Deep 10374
Japan Tr.
East China Sea
Nansei-shoto
Guangzhou
Taiwan Strait
Taiwan
Ryukyu Tr.
Daito-jima
Ogasawara-shoto
Kazan-retto
G. of Tongking
Hainan
Luzon Strait
Batan Is
C. Engaño
Luzon
Manila
Kyushu - Palau Ridge
South Honshu Ridge
Marianas Ridge
Marianas
Magellan Seamounts
Saipan
Guam
Mariana Tr.
Challenger Deep 11022
Cape Johnson Depth 10497
Philippine Tr.
Mid - Pacific Mountains
Wake I.
Johnston I.
Hawaiian Islands
Kure Atoll
Midway Is
Laysan I.
Gardner Pinnacles
Necker I.
Hawaiian Ridge
Andaman Is
Mergui Arch.
Gulf of Thailand
Andaman Basin
Nicobar Is
Sri Lanka
South China Sea
Palawan
Sulu Sea
Mindanao
Sulu Arch.
Celebes Sea
Kep. Talaud
Palau
Yap
Yap Tr.
Palau Tr.
Eauripik
Caroline Islands
Gaferut
Pikelot
Hall Is
Chuuk
Pohnpei
Nomoi Is
MICRONESIA
Marshall Islands
Enewetak
Bikini
Rongelap
Kwajalein
Wotje
Ailinglapalap
Majuro
Taongi
Kosrae
Ebon
Butaritari
W. Caroline Basin
E. Caroline Basin
New Guinea Rise
Kapingamarangi Rise
Kapingamarangi
Str. of Malacca
Sumatera
Singapore
Borneo
Kep. Mentawai
Equator
Halmahera
Tk. Tomini
Molucca Sea
Makassar Strait
Sulawesi
Bangka
Seram Sea
Seram
Buru
Banda Sea
Kep. Aru
Kep. Tanimbar
Java Sea
Jakarta
Selat Sunda
Jawa (Java)
Flores Sea
Flores
Sawu Sea
Sumba
Timor
Java Ridge
Sunda or Java Trench
Mid - Indian Basin
Christmas I.
Cocos Is
New Guinea
Admiralty Is
Bismarck Sea
New Ireland
New Britain
Planet Deep 9140
Solomon Sea
Solomon Islands
D'Entrecasteaux Is
G. of Papua
Port Moresby
Louisiade Arch.
Rennell
Torres Strait
C. York
Arafura Sea
C. Arnhem
Melville I.
Gulf of Carpentaria
Timor Sea
C. Londonderry
Coral Sea
Coral Sea Basin
Great Barrier Reef
Nauru
Banaba
Tabiteuea
Abaiang
Kiribati
Nonouti
Onotoa
MELANESIA
Howland I.
Baker I.
Phoenix Islands
McKean I.
Nikumaroro
Kanton I.
Manra
Orona
POLYNESIA
Palmyra I.
Teraina
Line
Nanumea
Tuvalu
Nukufetau
Funafuti
Nukulaelae
Vaitupu
Atafu
Tokelau
Fakaofo
Swains I.
Manihiki
Danger Is
Nassau
Santa Cruz Is
Rotuma
Îles Wallis
Savaii
Upolu
Tutuila
Manua Is
Suwarrow I.
Banks Is
Îles de Hoorn
Espiritu Santo
Vanua Levu
Fiji
Viti Levu
Vanuatu
Erromango
Tanna
Îles Loyauté
Nouvelle Calédonie
Île des Pins
Hunter I.
Vava'u Group
Tofua
Tonga
Tongatapu Group
Tonga Tr.
Palmerston I.
Niue
Cook Islands
Horizon Depth 10882
Ninety - East Ridge
West Australian Basin
C. Lévêque
Exmouth Plateau
Barrow I.
North West C.
Shark B.
AUSTRALIA
Brisbane
Darling
Sydney
Murray
Melbourne
Great Australian Bight
Spencer G.
Bass Strait
King I.
Tasmania
South East C.
Lord Howe I.
Lord Howe Rise
Norfolk Island Ridge
Norfolk Island Trough
Norfolk I.
South Fiji Basin
Raoul
Kermadec Is
Kermadec Tr.
South Pacific
Auckland
North Island
New Zealand
East C.
Wellington
Chatham Rise
Chatham Is
Tasman Sea
Tasman Basin
South Island
Stewart I.
Bounty Is
Antipodes Is
New Zealand Plateau
Auckland Is
Campbell I.
Macquarie Ridge
Macquarie I.
Tasman Plateau
South Australian Basin
Naturaliste Plateau
C. Leeuwin
W. Australian Ridge
Tropic of Capricorn
Indian - Antarctic Ridge
Indian - Antarctic Basin
Mid - Indian Ridge
Balleny Islands
C. North
C. Adare
Coulman I.
Ross Sea
Fisher Bay
C. Poinsett

METRES FEET
SEA LEVEL
200 656
3000 9843
5000 16409
6000 19686

Lambert Azimuthal Equal Area Projection

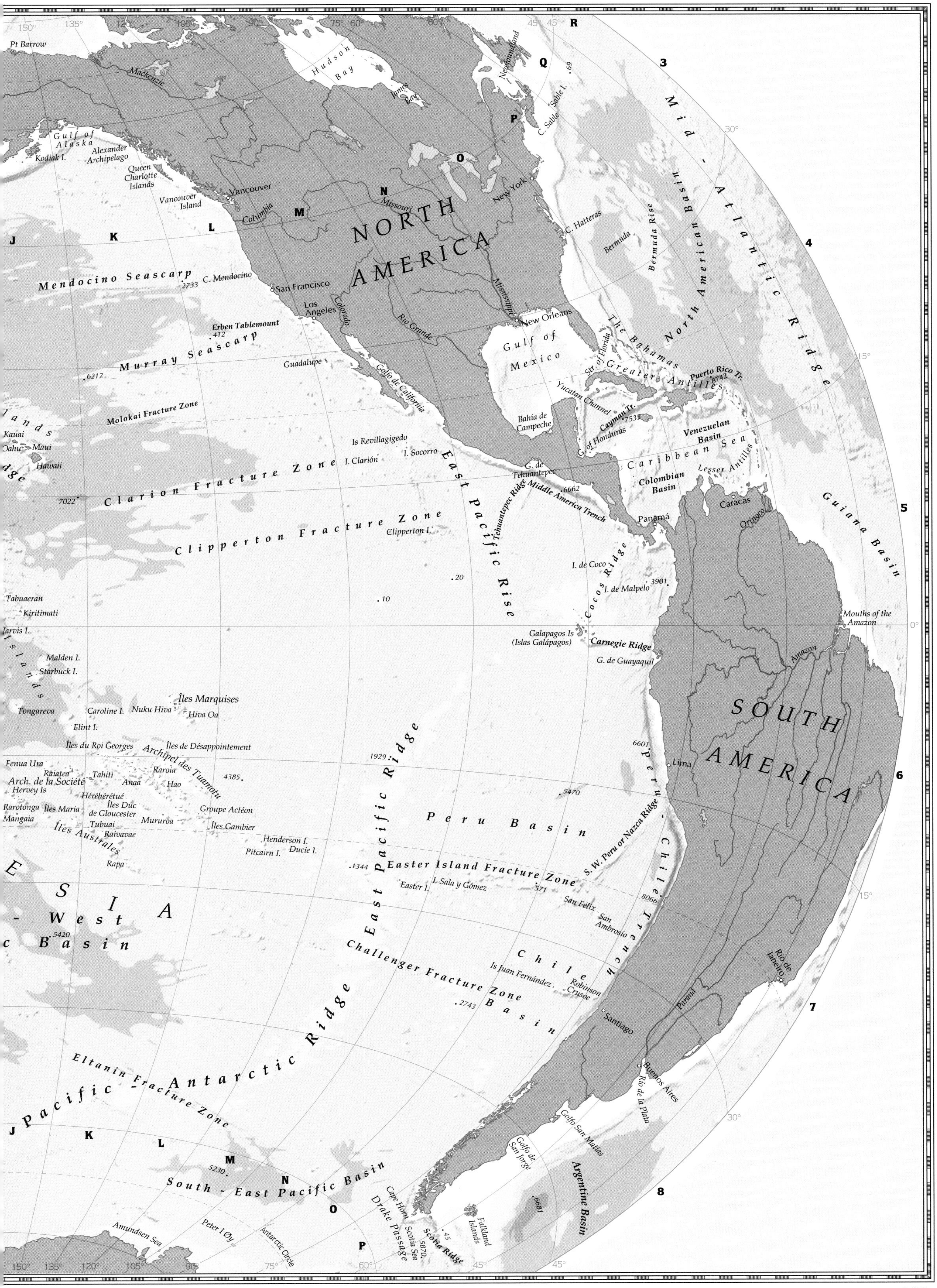

1:58M

KM | MILES
3000 | 1800
2500 | 1500
2000 | 1200
1500 | 900
1000 | 600
500 | 300
0 | 0

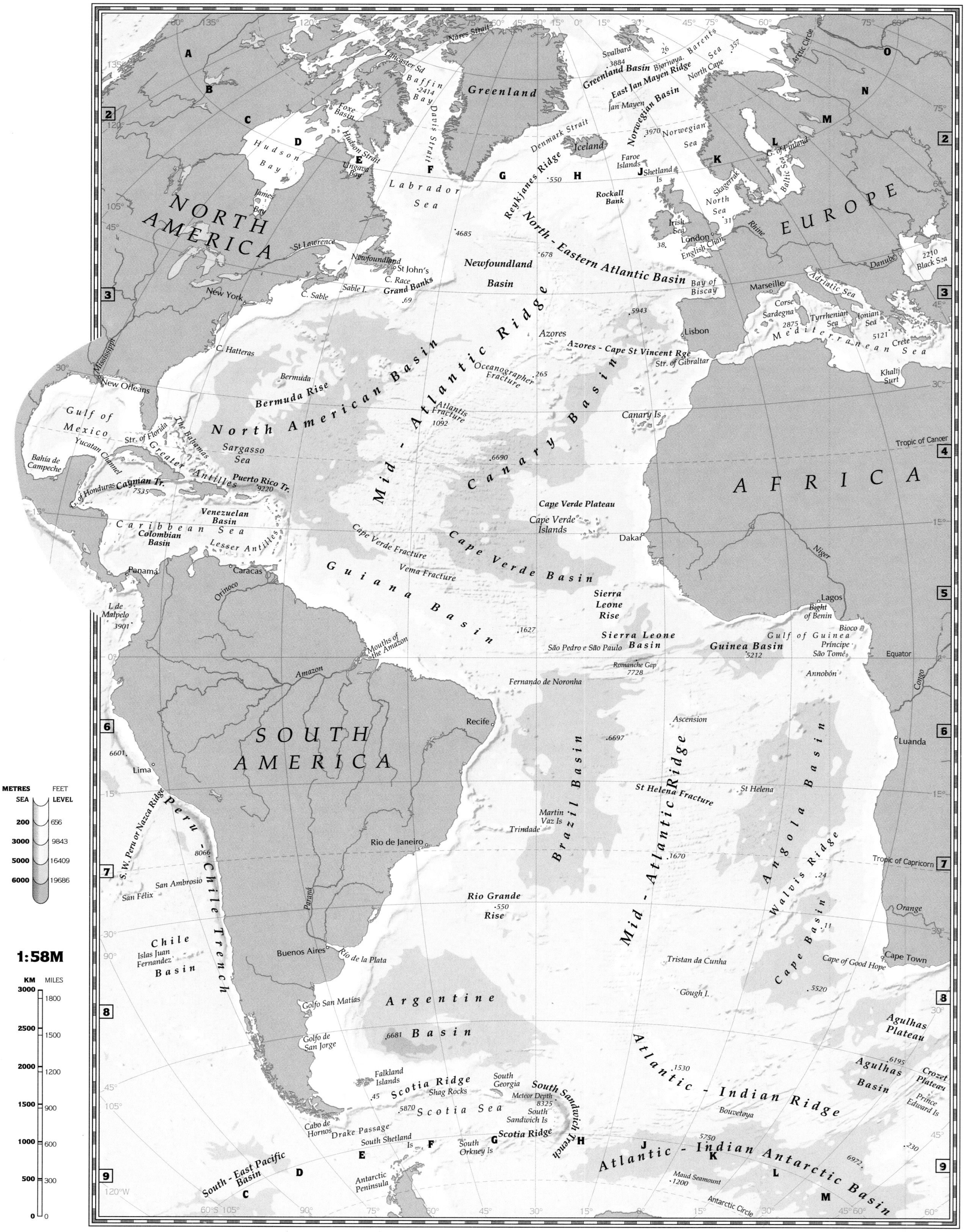

Lambert Azimuthal Equal Area Projection

THE INDEX includes the names on the maps in the main map section of the ATLAS. The names are generally indexed to the largest scale map on which they appear, and can be located using the grid reference letters and numbers around the map frame. Names on insets have a symbol: □, followed by the inset number where more than one inset appears on the page.

A

40 B5 Al Jilh esc. S. Arabia
40 C5 Al Jishshah S. Arabia
36 E6 Al Jīzah Jordan
42 C4 Al Jubayl S. Arabia
40 B5 Al Jubaylah S. Arabia
40 B5 Al Jufayr S. Arabia
40 C5 Al Jurayd i. S. Arabia
40 B5 Al Jurayfah S. Arabia
25 B4 Aljustrel Port.
42 E5 Al Khābūrah Oman
37 K5 Al Khālis Iraq
42 E4 Al Khaşab Oman
40 A6 Al Khāşirah S. Arabia
40 D6 Al Khatam reg. U.A.E.
40 C5 Al Khawr Qatar
40 C5 Al Khīşah well S. Arabia
40 C5 Al Khobar S. Arabia
40 B5 Al Khuff reg. S. Arabia
109 E2 Al Khufrah Libya
109 D1 Al Khums Libya
37 K5 Al Kifl Iraq
40 C5 Al Kir'ānah Qatar
18 B4 Alkmaar Neth.
37 K5 Al Kūfah Iraq
37 L5 Al Kumayt Iraq
37 K5 Al Kūt Iraq
Al Kuwayt see Kuwait
37 H7 Al Labbah plain S. Arabia
Al Lādhiqīyah see Latakia
91 J1 Allagash ME U.S.A.
91 J1 Allagash r. ME U.S.A.
91 J1 Allagash Lake U.S.A.
45 E4 Allahabad India
36 F5 Al Lajā lava Syria
33 P3 Allakh-Yun' Rus. Fed.
113 G3 Allanridge S. Africa
113 H1 Alldays S. Africa
88 E4 Allegan U.S.A.
90 D4 Allegheny r. U.S.A.
90 C6 Allegheny Mountains U.S.A.
90 D4 Allegheny Reservoir U.S.A.
87 D5 Allendale U.S.A.
12 E3 Allendale Town U.K.
96 D1 Allende Coahuila Mex.
97 D2 Allende Nuevo Leon Mex.
18 F5 Allendorf (Eder) Ger.
20 B2 Allendorf (Lumda) Ger.
89 G3 Allenford Can.
14 C3 Allen, Lough l. Rep. of Ireland
91 F4 Allentown U.S.A.
43 B4 Alleppey India
19 G4 Aller r. Ger.
22 B3 Allerey-sur-Saône France
20 D5 Allgäu reg. Ger.
20 D5 Allgäuer Alpen mts Austria
84 C3 Alliance NE U.S.A.
90 C4 Alliance OH U.S.A.
37 J6 Al Līfīyah well Iraq
68 B2 Alligator Creek Austr.
9 O9 Allinge-Sandvig Denmark
89 H3 Alliston Can.
42 B5 Al Līth S. Arabia
11 E4 Alloa U.K.
67 K2 Allora Austr.
22 E2 Allschwil Switz.
43 C3 Allur India
43 C3 Alluru Kottapatnam India
37 J6 Al Lussuf well Iraq
79 F4 Alma Can.
88 E4 Alma MI U.S.A.
84 D3 Alma NE U.S.A.
93 H5 Alma NM U.S.A.
37 J6 Al Ma'ānīyah Iraq
Alma-Ata see Almaty
25 B3 Almada Port.
37 K7 Al Ma'danīyāt well Iraq
68 A1 Almaden Austr.
25 D3 Almadén Spain
Al Madīnah see Medina
37 K5 Al Maḩmūdiyah Iraq
40 B5 Al Majma'ah S. Arabia
37 L1 Almalı Azer.
40 C5 Al Malsūnīyah reg. S. Arabia
39 G4 Almalyk Uzbek.
40 C5 Al Manāmah Bahrain
92 B1 Almanor, Lake U.S.A.
25 F3 Almansa Spain
25 D2 Almanzor mt Spain
37 L6 Al Ma'qil Iraq
40 D6 Al Mariyyah U.A.E.
109 E1 Al Marj Libya
104 C1 Almas, Rio das r. Brazil
39 H3 Almatinskaya Oblast' div. Kazak.
39 J4 Almaty Kazak.
Al Mawşil see Mosul
37 H4 Al Mayādīn Syria
40 B5 Al Mazāḩimīyah S. Arabia
25 E2 Almazán Spain
33 N3 Almaznyy Rus. Fed.
101 H4 Almeirim Brazil
25 B3 Almeirim Port.
18 D4 Almelo Neth.
104 E2 Almenara Brazil
25 C2 Almendra, Embalse de resr Spain
25 C3 Almendralejo Spain
18 C4 Almere Neth.
25 E4 Almería Spain
25 E4 Almería, Golfo de b. Spain
32 G4 Al'met'yevsk Rus. Fed.
9 O8 Älmhult Sweden
40 B5 Al Midhnab S. Arabia
25 D5 Almina, Pta pt Morocco
96 J6 Almirante Panama
42 C4 Al Mish'āb S. Arabia
36 F5 Al Mismīyah Syria
25 B4 Almodôvar Port.
89 F4 Almont U.S.A.
89 J3 Almonte Can.
25 C4 Almonte Spain
48 C6 Almora India
42 C4 Al Mubarrez S. Arabia
36 E7 Al Mudawwara Jordan
40 C5 Al Muharraq Bahrain
42 C7 Al Mukallā Yemen
42 B7 Al Mukhā Yemen
25 E4 Almuñécar Spain
21 G5 Almünster Austria
37 K5 Al Muqdādīyah Iraq
40 B5 Al Murabba S. Arabia
36 F1 Almus Turkey
40 B4 Al Musannāh ridge S. Arabia
37 K5 Al Musayyib Iraq
27 L7 Almyrou, Ormos b. Greece
92 □1 Alna Haina U.S.A.
12 F2 Alnwick U.K.
45 H5 Alon Myanmar
45 H3 Along India
27 K5 Alonnisos i. Greece
57 G8 Alor i. Indon.
57 G8 Alor, Kepulauan is Indon.
59 B1 Alor Setar Malaysia
Alost see Aalst
44 C5 Alot India
22 B2 Aloxe-Corton France
8 W4 Alozero Rus. Fed.
92 C4 Alpaugh U.S.A.
18 D5 Alpen Ger.
89 F3 Alpena U.S.A.
68 B4 Alpha Austr.
68 B4 Alpha Cr. r. Austr.
22 B6 Alpilles h. France
93 H5 Alpine AZ U.S.A.
85 C6 Alpine TX U.S.A.
82 E3 Alpine WY U.S.A.
6 D4 Alps mts Europe
42 C6 Al Qa'āmīyāt reg. S. Arabia
109 D1 Al Qaddāḩīyah Libya
36 F4 Al Qadmūs Syria
40 B5 Al Qā'īyah well S. Arabia
40 C6 Al Qalībah S. Arabia
37 H3 Al Qāmishlī Syria
40 B5 Al Qar'ah well S. Arabia
36 F4 Al Qaryatayn Syria
40 B5 Al Qaşab S. Arabia
42 C6 Al Qaţn Yemen
109 D2 Al Qaţrūn Libya
40 B4 Al Qayşūmah S. Arabia
36 E5 Al Qunayţirah Syria
42 B6 Al Qunfidhah S. Arabia
40 A5 Al Qurayn S. Arabia
37 L6 Al Qurnah Iraq
37 K6 Al Qusayr Iraq
40 B6 Al Qūşūrīyah S. Arabia
36 F5 Al Quţayfah Syria
40 A5 Al Quwārah S. Arabia
40 B5 Al Quwayīyah S. Arabia
19 G1 Als i. Denmark
15 H4 Alsace div. France
24 H2 Alsace reg. France
15 H4 Alsace, Plaine d' v. France
13 E4 Alsager U.K.
37 J6 Al Samīt well Iraq
77 H4 Alsask Can.
19 G6 Alsfeld Ger.
19 J5 Alsleben (Saale) Ger.
12 E3 Alston U.K.
67 K2 Alstonville Austr.
9 R8 Alsunga Latvia
8 S2 Alta Norway
8 S2 Altaelva r. Norway
105 D1 Alta Gracia Arg.
103 D2 Altagracía de Orituco Venez.
30 F5 Altai Mountains China/Mongolia
87 D6 Altamaha r. U.S.A.
101 H4 Altamira Brazil
96 H6 Altamira Costa Rica
70 B6 Alta, Mt N.Z.
26 G4 Altamura Italy
104 C1 Alta Paraiso de Goiás Brazil
96 C2 Altata Mex.
90 D6 Altavista U.S.A.
35 G2 Altay China
48 G2 Altay Mongolia
39 L2 Altay, Respublika div. Rus. Fed.
39 L2 Altayskiy Rus. Fed.
39 K1 Altayskiy Kray div. Rus. Fed.
22 F3 Altdorf Switz.
25 F3 Altea Spain
8 S1 Alteidet Norway
15 G2 Altenahr Ger.
19 L6 Altenberg Ger.
18 E4 Altenberge Ger.
19 K6 Altenburg Ger.
19 H2 Altenholz Ger.
20 F4 Altenmarkt an der Alz Ger.
21 H5 Altenmarkt bei St Gallen Austria
45 H1 Altenqoke China
20 B2 Altenstadt Ger.
20 D4 Altenstadt Ger.
20 B4 Altensteig Ger.
19 L3 Altentreptow Ger.
20 G4 Altheim Austria
37 M1 Altıağaç Azer.
41 H3 Altimur Pass Afgh.
37 K4 Altin Köprü Iraq
27 M5 Altınoluk Turkey
36 C2 Altıntaş Turkey
100 E7 Altiplano plain Bol.
19 J4 Altmark reg. Ger.
20 E4 Altmühl r. Ger.
104 B2 Alto Araguaia Brazil
105 C2 Alto de Pencoso h. Arg.
103 B3 Alto de Tamar mt Col.
104 B2 Alto Garças Brazil
115 C6 Alto Molócuè Moz.
23 J4 Alto, Monte h. Italy
20 E4 Altomünster Ger.
86 B4 Alton IL U.S.A.
85 F4 Alton MO U.S.A.
91 H3 Alton NH U.S.A.
84 D1 Altona Can.
90 D4 Altoona U.S.A.
23 H6 Altopascio Italy
104 B2 Alto Sucuriú Brazil
20 F4 Altötting Ger.
13 E4 Altrincham U.K.
19 K3 Alt Schwerin Ger.
97 G4 Altun Ha tourist site Belize
46 A3 Altun Shan mts China
82 B3 Alturas U.S.A.
85 D5 Altus U.S.A.
20 D5 Altusried Ger.
38 E3 Altynasar tourist site Kazak.
36 G1 Alucra Turkey
9 U8 Alūksne Latvia
37 M5 Alūm Iran
90 B4 Alum Creek Lake U.S.A.
105 B3 Aluminé r. Arg.
105 B3 Aluminé, L. Arg.
29 E6 Alupka Ukr.
109 D1 Al 'Uqaylah Libya
40 C5 Al 'Uqayr S. Arabia
29 E6 Alushta Ukr.
37 K4 'Alut Iran
42 C4 Al 'Uthmānīyah S. Arabia
109 E2 Al 'Uwaynāt Libya
37 J6 Al 'Uwayqīlah S. Arabia
40 A5 Al 'Uyūn S. Arabia
37 L6 Al 'Uzayr Iraq
85 D4 Alva U.S.A.
97 F4 Alvarado Mex.
105 C2 Alvarado, P. de pass Chile
100 F4 Alvarães Brazil
9 M5 Alvdal Norway
9 O6 Älvdalen Sweden
9 O8 Alvesta Sweden
9 K6 Ålvik Norway
85 E6 Alvin U.S.A.
8 R4 Älvsbyn Sweden
34 B4 Al Wajh S. Arabia
40 C5 Al Wakrah Qatar
40 C5 Al Wannān S. Arabia
44 D4 Alwar India
40 B5 Al Warī'ah S. Arabia
43 B4 Alwaye India
37 H5 Al Widyān plat. Iraq/S. Arabia
40 B4 'Al Wusayţ well S. Arabia
50 A2 Alxa Youqi China
50 B2 Alxa Zuoqi China
65 G2 Alyangula Austr.
11 E4 Alyth U.K.
9 T9 Alytus Lith.
20 F4 Alz r. Ger.
82 F2 Alzada U.S.A.
15 G3 Alzette r. Lux.
20 B3 Alzey Ger.
103 E3 Amacuro r. Guyana/Venez.
64 F4 Amadeus, Lake salt flat Austr.
75 L3 Amadjuak Lake Can.
93 G6 Amado U.S.A.
25 B3 Amadora Port.
53 A8 Amakusa-nada b. Japan
9 N7 Åmal Sweden
43 C2 Amalapuram India
49 L1 Amalat r. Rus. Fed.
103 B3 Amalfi Col.
112 F3 Amalia S. Africa
27 J6 Amaliada Greece
44 C5 Amalner India
57 K7 Amamapare Indon.
104 A3 Amambaí Brazil
104 A3 Amambaí r. Brazil
104 A3 Amambaí, Serra de h. Brazil/Para.
49 N6 Amami-guntō is Japan
49 N6 Amami-Ōshima i. Japan
115 A3 Amamula Congo(Zaire)
22 C2 Amance r. France
22 D2 Amancey France
38 D2 Amangel'dy Aktyubinsk. Kazak.
38 F2 Amangel'dy Turgaysk. Kazak.
39 K4 Amankol China
38 E3 Amanotkel' Kazak.
26 G5 Amantea Italy
71 □2 Amanu i. Fr. Polynesia Pac. Oc.
113 J5 Amanzimtoti S. Africa
101 H3 Amapá Brazil
25 C3 Amareleja Port.
92 D3 Amargosa Desert U.S.A.
92 D3 Amargosa Range mts U.S.A.
92 D3 Amargosa Valley U.S.A.
85 C5 Amarillo U.S.A.
26 F3 Amaro, Monte mt Italy
44 E4 Amarpatan India
36 E1 Amasya Turkey
97 F4 Amatán Mex.
96 G5 Amatique, Bahía de b. Guatemala
96 C3 Amatlán de Cañas Mex.
15 F2 Amay Belgium
101 H4 Amazon r. S. America
103 D4 Amazonas div. Brazil
Amazonas r. see Amazon
101 G4 Amazónia, Parque Nacional nat. park Brazil
101 J3 Amazon, Mouths of the est. Brazil
44 C6 Ambad India
43 B2 Ambajogai India
44 D3 Ambala India
43 C5 Ambalangoda Sri Lanka
111 E6 Ambalavao Madag.
114 F4 Ambam Cameroon
111 E5 Ambanja Madag.
41 E4 Ambar Iran
33 S3 Ambarchik Rus. Fed.
43 B4 Ambasamudram India
68 A5 Ambathala Austr.
100 C4 Ambato Ecuador
111 E5 Ambato Boeny Madag.
111 E6 Ambato Finandrahana Madag.
111 E5 Ambatolampy Madag.
111 E5 Ambatomainty Madag.
111 E5 Ambatondrazaka Madag.
20 E3 Amberg Ger.
97 H4 Ambergris Cay i. Belize
22 C4 Ambérieu-en-Bugey France
89 G3 Amberley Can.
22 A3 Ambierle France
45 E5 Ambikapur India
111 E5 Ambilobe Madag.
76 C3 Ambition, Mt Can.
12 F2 Amble U.K.
12 E3 Ambleside U.K.
15 F2 Amblève r. Belgium
111 E6 Amboasary Madag.
111 E5 Ambohidratrimo Madag.
111 E6 Ambohimahasoa Madag.
Amboina see Ambon
57 H7 Ambon Indon.
57 H7 Ambon i. Indon.
115 C3 Amboseli National Park Kenya
111 E6 Ambositra Madag.
111 E6 Ambovombe Madag.
93 E4 Amboy CA U.S.A.
88 C5 Amboy IL U.S.A.
91 F3 Amboy Center U.S.A.
111 B4 Ambriz Angola
71 □9 Ambrym i. Vanuatu
56 □ Ambunten Indon.
43 B3 Ambur India
68 C6 Amby Austr.
45 G2 Amdo China
96 C3 Ameca Mex.
18 C3 Ameland i. Neth.
90 E6 Amelia Court House U.S.A.
91 G4 Amenia U.S.A.
82 D3 American Falls U.S.A.
82 D3 American Falls Resr U.S.A.
93 G1 American Fork U.S.A.
61 K5 American Samoa terr. Pac. Oc.
87 C5 Americus U.S.A.
21 H5 Ameringkogel mt Austria
18 C4 Amersfoort Neth.
113 H3 Amersfoort S. Africa
13 G6 Amersham U.K.
77 L3 Amery Can.
116 D5 Amery Ice Shelf ice feature Ant.
84 E3 Ames U.S.A.
13 F6 Amesbury U.K.
91 H3 Amesbury U.S.A.
45 E4 Amethi India
27 K5 Amfissa Greece
33 P3 Amga Rus. Fed.
49 P2 Amgu Rus. Fed.
108 C2 Amguid Alg.
46 F1 Amgun' r. Rus. Fed.
79 H4 Amherst Can.
91 G3 Amherst MA U.S.A.
91 J2 Amherst ME U.S.A.
90 D6 Amherst VA U.S.A.
89 F4 Amherstburg Can.
26 D3 Amiata, Monte mt Italy
15 C3 Amiens France
37 H5 Amij, Wādī watercourse Iraq
43 A4 Amindivi Islands India
115 D2 Amino Eth.
53 D7 Amino Japan
112 C1 Aminuis Namibia
40 B3 Amīrābād Iran
Amirabad see Fūlād Maïalleh
117 H4 Amirante Islands Seychelles
41 F4 Amir Chah Pak.
77 J4 Amisk L. Can.
96 J6 Amistad, Parque Internacional La nat. park Costa Rica/Panama
96 D1 Amistad Reservoir Mex./U.S.A.
67 K1 Amity Point Austr.
44 D5 Amla Madhya Pradesh India
9 L7 Åmli Norway
13 C4 Amlwch U.K.
36 E6 'Ammān Jordan
13 D6 Ammanford U.K.
8 V4 Ämmänsaari Fin.
8 P4 Ammarnäs Sweden
20 E5 Ammergebirge res. Ger.
18 E3 Ammerland reg. Ger.
20 E4 Ammersee l. Ger.
Ammochostos see Famagusta
54 D4 Amnyong-dan hd N. Korea
44 C5 Amod India
51 B6 Amo Jiang r. China
40 D2 Amol Iran
20 C3 Amorbach Ger.
27 L6 Amorgos i. Greece
78 E4 Amos Can.
19 J1 Åmose r. Denmark
114 B1 Amourj Maur.
Amoy see Xiamen
43 C5 Amparai India
104 C3 Amparo Brazil
20 E4 Amper r. Ger.
20 F4 Ampfing Ger.
25 G2 Amposta Spain
44 D5 Amravati India
44 B5 Amreli India
44 B4 Amri Pak.
36 E4 'Amrit Syria
44 C3 Amritsar India
44 D3 Amroha India
18 F2 Amrum i. Ger.
8 Q4 Åmsele Sweden
18 B4 Amstelveen Neth.
18 B4 Amsterdam Neth.
113 J3 Amsterdam S. Africa
91 F3 Amsterdam U.S.A.
117 K6 Amsterdam, Île i. Ind. Ocean
21 H4 Amstetten Austria
20 C4 Amstetten Ger.
109 E3 Am Timan Chad
38 D3 Amudar'ya r. Turkm./Uzbek.
75 J2 Amund Ringnes I. Can.
116 D4 Amundsen Bay Ant.
116 B4 Amundsen Gl. Ant.
74 F2 Amundsen Gulf g. Can.
116 C5 Amundsen, Mt Ant.
116 B4 Amundsen-Scott U.S.A. Base Ant.
116 A3 Amundsen Sea Ant.
59 E3 Amuntai Indon.
49 P1 Amur r. China/Rus. Fed.
49 P1 Amursk Rus. Fed.
29 F6 Amvrosiyivka Ukr.
88 E1 Amyot Can.
45 H6 An Myanmar
71 □2 Anaa i. Fr. Polynesia Pac. Oc.
57 G7 Anabanua Indon.
33 N2 Anabar r. Rus. Fed.
33 N2 Anabarskiy Zaliv b. Rus. Fed.
66 D4 Ana Branch r. Austr.
92 C4 Anacapa Is U.S.A.
103 D2 Anaco Venez.
82 D2 Anaconda U.S.A.
82 B1 Anacortes U.S.A.
85 D5 Anadarko U.S.A.
36 F1 Anadolu Dağları mts Turkey
33 T3 Anadyr' r. Rus. Fed.
33 U3 Anadyrskiy Zaliv b. Rus. Fed.
27 L6 Anafi i. Greece
104 E1 Anagé Brazil
37 H4 'Ānah Iraq
92 D5 Anaheim U.S.A.
76 D4 Anahim Lake Can.
97 D2 Anáhuac Mex.
43 B4 Anaimalai Hills India
43 B4 Anai Mudi Pk India
43 C2 Anakapalle India
68 B4 Anakie Austr.
111 E5 Analalava Madag.
100 F4 Anamã Brazil
59 C2 Anambas, Kepulauan is Indon.
88 B4 Anamosa U.S.A.
36 D3 Anamur Turkey
36 D3 Anamur Burnu pt Turkey
53 D8 Anan Japan
44 C5 Anand India
45 F5 Ānandapur India
45 F5 Anandpur r. India
39 J4 Anan'ev Kyrg.
43 B3 Anantapur India
44 C2 Anantnag Jammu and Kashmir
29 D6 Anan'yiv Ukr.
29 F6 Anapa Rus. Fed.
104 C2 Anápolis Brazil
40 D4 Anār Iran
40 D3 Anārak Iran
40 C3 Anarbar r. Iran
41 F3 Anardara Afgh.
57 L3 Anatahan i. N. Mariana Is
36 D2 Anatolia reg. Turkey
71 □9 Anatom i. Vanuatu
102 D3 Añatuya Arg.
103 E4 Anauá r. Brazil
40 C2 Anbūh Iran
54 D4 Anbyon N. Korea
24 D3 Ancenis France
74 D3 Anchorage U.S.A.
89 F4 Anchor Bay U.S.A.
23 L6 Ancona Italy
102 B6 Ancud Chile
105 B4 Ancud, Golfo de g. Chile
105 B1 Andacollo Chile
45 F5 Andal India
8 K5 Åndalsnes Norway
25 D4 Andalucía div. Spain
85 G6 Andalusia U.S.A.
56 A5 Andaman and Nicobar Islands div. India
117 L3 Andaman Basin sea feature Ind. Ocean
35 H5 Andaman Islands Andaman and Nicobar Is
56 B4 Andaman Sea Asia
66 B3 Andamooka Austr.
22 B4 Andance France
111 E5 Andapa Madag.
104 E1 Andaraí Brazil
15 B3 Andelle r. France
15 F4 Andelot-Blancheville France
8 P2 Andenes Norway
15 F2 Andenne Belgium
15 E2 Anderlecht Belgium
22 F3 Andermatt Switz.
24 D4 Andernos-les-Bains France
19 L1 Anderslöv Sweden
74 F3 Anderson r. N.W.T. Can.
74 D3 Anderson AK U.S.A.
88 E5 Anderson IN U.S.A.
85 E4 Anderson MO U.S.A.
87 D5 Anderson SC U.S.A.
67 G8 Anderson B. Austr.
103 B3 Andes Col.
98 C6 Andes mts S. America
84 D3 Andes, Lake U.S.A.
8 P2 Andfjorden chan. Norway
43 B2 Andhra Pradesh div. India
111 E5 Andilamena Madag.
111 E5 Andilanatoby Madag.
40 C3 Andīmeshk Iran
36 F3 Andırın Turkey
29 H7 Andiyskoye Koysu r. Rus. Fed.
39 H4 Andizhan Uzbek.
41 G2 Andkhui r. Afgh.
41 G2 Andkhvoy Afgh.
111 E5 Andoany Madag.
100 C4 Andoas Peru
43 B2 Andol India
54 E5 Andong S. Korea
54 E5 Andong-ho l. S. Korea
7 D4 Andorra country Europe
25 G1 Andorra la Vella Andorra
13 F6 Andover U.K.
91 H2 Andover ME U.S.A.
90 C4 Andover OH U.S.A.
8 O2 Andøya i. Norway
104 B3 Andradina Brazil
28 E3 Andreapol' Rus. Fed.
12 C3 Andreas U.K.
110 C3 André Félix, Parc National de nat. park C.A.R.
104 D3 Andrelândia Brazil
85 C5 Andrews U.S.A.
39 F1 Andreyevka Kokshet. Kazak.
39 K3 Andreyevka Taldykorg. Kazak.
38 B1 Andreyevka Rus. Fed.
26 G4 Andria Italy
111 E6 Androka Madag.
87 E7 Andros i. Bahamas
27 L6 Andros i. Greece
91 H2 Androscoggin r. U.S.A.
87 E7 Andros Town Bahamas
43 A4 Āndrott i. India
29 D5 Andrushivka Ukr.
8 Q2 Andselv Norway
25 D3 Andújar Spain
111 B5 Andulo Angola
22 A5 Anduze France
108 C3 Anéfis Mali
95 M5 Anegada i. Virgin Is
105 D4 Anegada, Bahía b. Arg.
93 F5 Anegam U.S.A.
114 D3 Aného Togo
'Aneiza, Jabal h. see 'Unayzah, Jabal
15 B4 Anet France
93 H3 Aneth U.S.A.
25 G1 Aneto mt Spain
109 D3 Aney Niger
51 E5 Anfu China
111 E5 Angadoka, Lohatanjona hd Madag.
48 H1 Angara r. Rus. Fed.
48 H1 Angarsk Rus. Fed.
55 B3 Angat Phil.
9 O5 Ånge Sweden
94 B3 Angel de la Guarda i. Mex.
55 B3 Angeles Phil.
71 □9 Angelghowhat Vanuatu
9 N8 Ängelholm Sweden
68 B6 Angellala Cr. r. Austr.
69 D1 Angelo r. Austr.
92 B2 Angels Camp U.S.A.
8 P4 Ångermanälven r. Sweden
19 M3 Angermünde Ger.
24 D3 Angers France
15 C4 Angerville France
22 E6 Anges, Baie des b. France
77 K2 Angikuni Lake Can.
58 B2 Angkor Cambodia
13 C4 Anglesey i. U.K.
85 E6 Angleton U.S.A.
89 H2 Angliers Can.
Angmagssalik see Tasiilaq
58 □ Ang Mo Kio Sing.
115 A2 Ango Congo(Zaire)
115 C6 Angoche Moz.
41 E5 Angohrān Iran
105 B3 Angol Chile
107 E7 Angola country Africa
88 E5 Angola U.S.A.
120 K7 Angola Basin sea feature Atl. Ocean
97 F5 Angostura, Presa de la resr Mex.
24 E4 Angoulême France
39 G4 Angren Uzbek.
58 B2 Ang Thong Thai.
73 K8 Anguilla terr. Caribbean Sea
50 E1 Anguli Nur l. China
50 E2 Anguo China
104 A3 Anhanduí r. Brazil
9 M8 Anholt i. Denmark
51 D4 Anhua China
50 E3 Anhui div. China
104 A2 Anhumas Brazil
54 D5 Anhŭng S. Korea
104 C2 Anicuns Brazil
38 E2 Anikhovka Rus. Fed.
28 G3 Anikovo Rus. Fed.
93 H6 Animas U.S.A.
93 H6 Animas Peak U.S.A.
96 A1 Animas, Pta Las pt Mex.
58 A2 Anin Myanmar
52 H1 Aniva Rus. Fed.
52 H1 Aniva, Mys c. Rus. Fed.
49 Q2 Aniva, Zaliv b. Rus. Fed.
71 □9 Aniwa i. Vanuatu
15 D3 Anizy-le-Château France
9 U6 Anjalankoski Fin.
43 B4 Anjengo India
51 F4 Anji China
44 D5 Anji India
41 E3 Anjoman Iran
24 D3 Anjou reg. France
111 E5 Anjouan i. Comoros
111 E5 Anjozorobe Madag.
54 C4 Anjŭ N. Korea
111 E6 Ankaboa, Tanjona pt Madag.
50 C3 Ankang China
36 D2 Ankara Turkey
111 E6 Ankazoabo Madag.
111 E5 Ankazobe Madag.
58 D2 An Khê Vietnam
19 L3 Anklam Ger.
44 C5 Ankleshwar India
20 G5 Ankogel mt Austria
115 A4 Ankoro Congo(Zaire)
114 D3 Anloga Ghana
51 B5 Anlong China
58 C2 Ânlong Vêng Cambodia
50 D4 Anlu China
54 D5 Anmyŏn Do i. S. Korea
29 G5 Anna Rus. Fed.
108 C1 Annaba Alg.
20 G2 Annaberg-Buchholtz Ger.
36 F4 An Nabk Syria
42 B4 An Nafūd des. S. Arabia
12 A3 Annahilt U.K.
37 K6 An Najaf Iraq
90 E5 Anna, Lake U.S.A.
14 D3 Annalee r. Rep. of Ireland
14 F3 Annalong U.K.
11 E6 Annan Scot. U.K.
11 E6 Annan r. U.K.
11 E5 Annandale v. U.K.
90 E5 Annapolis U.S.A.
79 G5 Annapolis Royal Can.
45 E3 Annapurna mt Nepal
40 C5 An Naqīrah well S. Arabia
89 F4 Ann Arbor U.S.A.
101 G2 Anna Regina Guyana
37 L6 An Nāşirīyah Iraq
91 H3 Ann, Cape hd U.S.A.
69 D3 Annean, L. salt flat Austr.
22 D4 Annecy France
22 D4 Annecy, Lac d' l. France
22 D4 Annecy-le-Vieux France
22 D3 Annemasse France
18 D3 Annen Neth.
76 C4 Annette I. U.S.A.
51 B5 Anning China
87 C5 Anniston U.S.A.
22 E3 Anniviers, Val d' v. Switz.
106 C6 Annobón i. Equatorial Guinea
22 B4 Annonay France
22 D6 Annot France
42 C4 An Nu'ayrīyah S. Arabia
37 K5 An Nu'mānīyah Iraq
84 E2 Anoka U.S.A.
111 E5 Anorontany, Tanjona hd Madag.
27 L7 Ano Viannos Greece
51 D6 Anpu China
51 C6 Anpu Gang b. China
51 E4 Anqing China
50 F2 Anqiu China
51 D5 Anren China
15 F2 Ans Belgium
50 C2 Ansai China
36 F4 Ansariye, J. el mts Syria
20 D3 Ansbach Ger.
22 B4 Anse France
67 G7 Anser Group is Austr.
21 H4 Ansfelden Austria
54 B3 Anshan China
51 B5 Anshun China
105 C1 Ansilta mt Arg.
105 F1 Ansina Uru.
84 D3 Ansley U.S.A.
85 D5 Anson U.S.A.
114 D1 Ansongo Mali
78 D4 Ansonville Can.
90 C5 Ansted U.S.A.
44 D4 Anta India
100 D6 Antabamba Peru
36 F3 Antakya Turkey
111 F5 Antalaha Madag.
36 C3 Antalya Turkey
36 C3 Antalya Körfezi g. Turkey
111 E5 Antananarivo Madag.
116 B2 Antarctic Peninsula Ant.
11 C3 An Teallach mt U.K.
92 D2 Antelope Range mts U.S.A.
25 D4 Antequera Spain
22 D6 Anthéor France
83 F5 Anthony U.S.A.
108 B2 Anti Atlas mts Morocco
22 E6 Antibes France
79 H4 Anticosti, Île d' i. Can.
88 C3 Antigo U.S.A.
79 H4 Antigonish Can.
95 M5 Antigua i. Antigua
94 F6 Antigua Guatemala
97 G5 Antigua Guatemala
73 K8 Antigua and Barbuda country Caribbean Sea
97 E3 Antiguo-Morelos Mex.
27 K7 Antikythira i. Greece
27 K7 Antikythiro, Steno chan. Greece
Anti Lebanon mts see Sharqi, Jebel esh
Antioch see Antakya
92 B3 Antioch CA U.S.A.
88 C4 Antioch IL U.S.A.
103 B3 Antioquia Col.
60 J8 Antipodes Islands N.Z.
27 L5 Antipsara i. Greece
85 E5 Antlers U.S.A.
102 B2 Antofagasta Chile
102 C3 Antofalla, Vol. volc. Arg.
15 D2 Antoing Belgium
96 J6 Antón Panama
104 C4 Antonina Brazil
104 E1 Antônio r. Brazil
92 B3 Antonio r. U.S.A.
83 F4 Antonito U.S.A.
14 E3 Antrim U.K.
14 E2 Antrim Hills U.K.
111 E5 Antsalova Madag.
111 E5 Antsirabe Madag.
111 E5 Antsirañana Madag.
111 E5 Antsohihy Madag.
8 S3 Anttis Sweden
9 U6 Anttola Fin.
54 E2 Antu China
105 B3 Antuco Chile
105 B3 Antuco, Volcán volc. Chile
15 E1 Antwerp Belgium
91 F2 Antwerp U.S.A.
71 □2 Anuanu Raro i. Fr. Polynesia Pac. Oc.
52 C3 Anuchino Rus. Fed.
78 E2 Anuc, Lac l. Can.
45 F5 Anugul India
54 D6 Anŭi S. Korea
44 C3 Anupgarh India
43 C4 Anuradhapura Sri Lanka
40 D5 Anveh Iran
Anvers see Antwerp
116 B2 Anvers I. Ant.
51 F5 Anxi Fujian China
46 B2 Anxi Gansu China
50 B4 An Xian China
51 D4 Anxiang China
50 E2 Anxin China
66 A4 Anxious Bay Austr.
114 C3 Anyama Côte d'Ivoire
50 E2 Anyang China
54 D5 Anyang S. Korea
27 L6 Anydro i. Greece
46 B3 A'nyêmaqên Shan mts China
51 E4 Anyi China
51 E5 Anyuan China
51 B4 Anyue China
33 S3 Anyuysk Rus. Fed.
103 B3 Anzá Col.
50 D2 Anze China
46 A1 Anzhero-Sudzhensk Rus. Fed.
110 C4 Anzi Congo(Zaire)
26 E4 Anzio Italy
71 □9 Aoba i. Vanuatu
58 A3 Ao Ban Don b. Thai.
50 F1 Aohan Qi China
52 G4 Aomori Japan
44 D3 Aonla India
71 □3 Aorai mt Fr. Polynesia Pac. Oc.
Aoraki, Mt see Cook, Mt
70 D4 Aorere r. N.Z.
58 A3 Ao Sawi b. Thai.
22 E4 Aosta Italy
22 C4 Aoste France
108 B2 Aoukâr reg. Mali/Maur.
101 G8 Apa r. Brazil
115 B2 Apac Uganda
93 H6 Apache U.S.A.
93 H5 Apache Creek U.S.A.
93 G5 Apache Junction U.S.A.
93 G6 Apache Peak U.S.A.
87 C6 Apalachee Bay U.S.A.
87 C6 Apalachicola U.S.A.
97 E4 Apan Mex.
103 C4 Apaporis r. Col.
37 K1 Aparan Armenia
104 B3 Aparecida do Tabuado Brazil
55 B2 Aparri Phil.
71 □2 Apataki i. Fr. Polynesia Pac. Oc.
8 X3 Apatity Rus. Fed.
96 D4 Apatzingán Mex.
9 U8 Ape Latvia
18 C4 Apeldoorn Neth.
19 G4 Apelern Ger.
19 G3 Apensen Ger.
44 E3 Api mt Nepal
71 □4 Apia Western Samoa
23 L6 Apiro Italy
70 E3 Apiti N.Z.
101 G2 Apoera Suriname
19 J5 Apolda Ger.
66 E7 Apollo Bay Austr.
100 E6 Apolo Bol.
55 C5 Apo, Mt volc. Phil.
87 D6 Apopka, L. U.S.A.
104 B2 Aporé Brazil
104 B2 Aporé r. Brazil
86 B2 Apostle Islands U.S.A.
88 B2 Apostle Islands National Lakeshore res. U.S.A.
36 E4 Apostolos Andreas, Cape Cyprus
90 B6 Appalachia U.S.A.
90 C6 Appalachian Mountains U.S.A.
26 E3 Appennino Abruzzese mts Italy
23 H5 Appennino Tosco-Emiliano mts Italy
23 K6 Appennino Umbro-Marchigiano mts Italy
23 G2 Appenzell Switz.
23 J3 Appiano sulla Strada del Vino Italy
67 J5 Appin Austr.
18 D3 Appingedam Neth.
11 C3 Applecross U.K.
84 D2 Appleton MN U.S.A.
88 C3 Appleton WV U.S.A.
92 D4 Apple Valley U.S.A.
90 D6 Appomattox U.S.A.
22 C4 Apprieu France
26 E4 Aprilia Italy
29 F6 Apsheronsk Rus. Fed.
66 D6 Apsley Austr.
89 H3 Apsley Can.
22 C6 Apt France
104 B3 Apucarana Brazil
55 A4 Apurahuan Phil.
103 D3 Apure r. Venez.
100 D6 Apurimac r. Peru
36 E7 'Aqaba Jordan
34 B4 Aqaba, Gulf of g. Asia
39 J4 Aqal China
40 D2 Aqbana Iran
41 G2 Āqchah Afgh.
40 B2 Aq Chai r. Iran
40 D3 Āqdā Iran
40 B2 Aqdoghmish r. Iran
37 K3 Āq Kān Dāgh, Kūh-e mt Iran
48 E4 Aqqikkol Hu salt l. China
37 J3 'Aqrah Iraq
93 F4 Aquarius Mts U.S.A.
93 G3 Aquarius Plateau plat. U.S.A.
26 G4 Aquaviva delle Fonti Italy
104 A3 Aquidauana Brazil
104 A2 Aquidauana r. Brazil
96 D4 Aquila Mex.
103 D4 Aquio r. Col.
97 E3 Aquismón Mex.
24 D4 Aquitaine reg. France
45 F4 Ara India
41 G4 Arab Afgh.
87 C5 Arab U.S.A.
41 E3 'Arabābād Iran
109 E3 Arab, Bahr el watercourse Sudan
117 J3 Arabian Basin sea feature Ind. Ocean
30 D8 Arabian Sea Ind. Ocean
103 E3 Arabopó Venez.
103 E3 Arabopó r. Venez.
36 D1 Araç Turkey
103 E4 Araça r. Brazil
101 L6 Aracaju Brazil
103 D4 Aracamuni, Co summit Venez.
104 A4 Aracanguy, Mtes de h. Para.
101 L4 Aracati Brazil
104 E1 Aracatu Brazil
104 B3 Araçatuba Brazil
25 C4 Aracena Spain
104 E2 Aracruz Brazil
104 D2 Araçuaí Brazil
104 D2 Araçuaí r. Brazil
27 J1 Arad Romania
109 E3 Arada Chad
37 L6 Aradah Iraq
62 D2 Arafura Sea Austr./Indon.
104 B1 Aragarças Brazil
37 J1 Aragats Armenia
37 K1 Aragats Lerr mt Armenia
25 F2 Aragón div. Spain
25 F1 Aragón r. Spain
101 J5 Araguacema Brazil
103 D2 Aragua de Barcelona Venez.
101 J5 Araguaia r. Brazil
101 H6 Araguaia, Parque Nacional de nat. park Brazil
101 J5 Araguaína Brazil
104 C2 Araguari r. Minas Gerais Brazil
104 C2 Araguari Brazil
101 J5 Araguatins Brazil
29 H7 Aragvi r. Georgia
53 F6 Arai Japan
101 K4 Araiosos Brazil
108 C2 Arak Alg.
40 C3 Arāk Iran
56 A2 Arakan Yoma mts Myanmar
43 B3 Arakkonam India
39 K4 Aral China
37 K2 Aralık Turkey
38 D3 Aral Sea salt l. Kazak./Uzbek.
38 E3 Aral'sk Kazak.
Aral'skoye More salt l. see Aral Sea
38 B2 Aralsor, Ozero l. Zapadnyy Kaz. Kazak.
38 C2 Aralsor, Ozero salt l. Zapadnyy Kaz. Kazak.
38 E3 Aralsul'fat Kazak.
68 A4 Aramac Austr.
68 A4 Aramac Cr. watercourse Austr.
40 B5 Aramah plat. S. Arabia
97 E2 Aramberri Mex.
57 L8 Aramia r. P.N.G.
44 D6 Aran r. India
25 E2 Aranda de Duero Spain
37 L4 Arandān Iran
27 J2 Aranđelovac Yugo.
43 B3 Arani India
14 C3 Aran Island Rep. of Ireland
14 B4 Aran Islands Rep. of Ireland
25 E2 Aranjuez Spain
111 B6 Aranos Namibia
85 D7 Aransas Pass U.S.A.
104 B2 Arantes r. Brazil
63 H1 Aranuka i. Gilbert Is
53 B8 Arao Japan
108 B3 Araouane Mali
84 D3 Arapahoe U.S.A.
103 E4 Arapari r. Brazil
105 F1 Arapey Grande r. Uru.
101 L5 Arapiraca Brazil
27 L4 Arapis, Akra pt Greece
36 G2 Arapkir Turkey
104 B3 Arapongas Brazil
45 F4 A Rapti Doon r. Nepal
42 B3 'Ar'ar S. Arabia
102 G3 Araranguá Brazil
104 C3 Araraquara Brazil
101 H5 Araras Brazil
104 B4 Araras, Serra das mts Brazil
37 K2 Ararat Armenia
66 E6 Ararat Austr.
37 K2 Ararat, Mt Turkey
45 F4 Araria India
104 D3 Araruama, Lago de lag. Brazil
37 J6 'Ar'ar, W. watercourse Iraq/S. Arabia
37 J2 Aras Turkey
37 J1 Aras r. Turkey
104 E1 Arataca Brazil
103 C3 Arauca Col.
103 C3 Arauca r. Venez.
105 B3 Arauco Chile
103 C3 Arauquita Col.
103 C2 Araure Venez.
44 C4 Aravalli Range mts India
9 T7 Aravete Estonia
22 D4 Aravis mts France
63 F2 Arawa P.N.G.
104 C2 Araxá Brazil
103 D2 Araya, Pen. de Venez.
103 D2 Araya, Pta de pt Venez.
36 C2 Arayıt Dağı mt Turkey
37 M2 Araz r. Asia
115 C1 Ārba Minch Eth.
37 K4 Arbat Iraq
28 J3 Arbazh Rus. Fed.
37 K3 Arbīl Iraq
9 O7 Arboga Sweden
22 C3 Arbois France
23 G2 Arbon Switz.
77 J4 Arborfield Can.
11 F4 Arbroath U.K.
92 A2 Arbuckle U.S.A.
41 F4 Arbu Lut, Dasht-e des. Afgh.
22 D4 Arc r. France
24 D4 Arcachon France
87 D7 Arcadia U.S.A.
82 A3 Arcata U.S.A.
92 D2 Arc Dome summit U.S.A.
97 D4 Arcelia Mex.
15 D4 Arces-Dilo France
23 K6 Arcevia Italy
28 G1 Archangel Rus. Fed.
65 H2 Archer r. Austr.
65 H2 Archer Bend National Park Austr.
93 H2 Arches Nat. Park U.S.A.
38 D5 Archman Turkm.
15 E4 Arcis-sur-Aube France
37 M2 Ärçivan Azer.
66 A2 Arckaringa watercourse Austr.
23 H4 Arco Italy
82 D3 Arco U.S.A.
25 D4 Arcos de la Frontera Spain
75 K2 Arctic Bay Can.
74 C2 Arctic Ocean
74 C3 Arctic Plains U.S.A.
74 E3 Arctic Red r. Can.
116 B2 Arctowski Poland Base Ant.
23 G5 Arda r. Italy
40 C2 Ardabīl Iran
37 J1 Ardahan Turkey
40 D3 Ardakān Iran
40 C4 Ardal Iran
9 K6 Årdalstangen Norway
14 C3 Ardara Rep. of Ireland
27 L4 Ardas r. Bulg.
28 H4 Ardatov Mordov. Rus. Fed.
28 G4 Ardatov Nizheg. Rus. Fed.
89 G3 Ardbeg Can.
22 B5 Ardèche r. France
14 E4 Ardee Rep. of Ireland
66 B4 Arden, Mount h. Austr.
15 E3 Ardennes, Canal des France
40 D3 Ardestān Iran
14 F3 Ardglass U.K.
25 C3 Ardila r. Port.
96 D3 Ardilla, Cerro la mt Mex.
67 G5 Ardlethan Austr.
85 D5 Ardmore U.S.A.
11 B4 Ardnamurchan, Point of U.K.
15 B2 Ardres France
11 C4 Ardrishaig U.K.
66 B5 Ardrossan Austr.
11 D5 Ardrossan U.K.
11 C3 Ardvasar U.K.
105 E2 Areco r. Arg.
101 L4 Areia Branca Brazil
15 G2 Aremberg h. Ger.

55 B4 Arena rf Phil.
96 H5 Arenal Honduras
92 A2 Arena, Pt U.S.A.
96 B3 Arena, Pta pt Mex.
25 D2 Arenas de San Pedro Spain
9 L7 Arendal Norway
19 J4 Arendsee (Altmark) Ger.
13 D5 Arenig Fawr h. U.K.
22 F5 Arenzano Italy
27 K6 Areopoli Greece
100 D7 Arequipa Peru
101 H4 Arere Brazil
25 D2 Arévalo Spain
23 J6 Arezzo Italy
36 G6 'Arfajah well S. Arabia
50 D1 Argalant Mongolia
25 E2 Arganda Spain
55 B4 Argao Phil.
22 D6 Argens r. France
23 J5 Argenta Italy
24 D2 Argentan France
26 D3 Argentario, Monte h. Italy
22 E5 Argentera, Cima dell' mt Italy
15 H3 Argenthal Ger.
99 D7 Argentina country S. America
116 B3 Argentina Ra. mts Ant.
120 F8 Argentine Basin sea feature Atl. Ocean
116 A2 Argentine Claim reg. Ant.
102 B8 Argentino, Lago l. Arg.
27 L2 Argeș r. Romania
41 G4 Arghandab r. Afgh.
41 G4 Arghastan r. Afgh.
36 C2 Argıthanı Turkey
27 K6 Argolikos Kolpos b. Greece
71 □14 Argo Reefs Fiji
27 K6 Argos Greece
27 J5 Argostoli Greece
15 B3 Argueil France
25 F1 Arguis Spain
49 M1 Argun' r. China/Rus. Fed.
29 H7 Argun Rus. Fed.
114 D2 Argungu Nigeria
92 D4 Argus Range mts U.S.A.
88 C4 Argyle U.S.A.
64 E3 Argyle, Lake Austr.
11 C4 Argyll reg. U.K.
49 M3 Ar Horqin Qi China
9 M8 Århus Denmark
70 E3 Aria N.Z.
67 G5 Ariah Park Austr.
53 B8 Ariake-kai b. Japan
111 B6 Ariamsvlei Namibia
26 F4 Ariano Irpino Italy
103 B4 Ariari r. Col.
105 D2 Arias Arg.
35 F6 Ari Atoll Maldives
103 E2 Aribi r. Venez.
114 C2 Aribinda Burkina
102 B1 Arica Chile
64 D6 Arid, C. Austr.
11 C4 Arienas, Loch l. U.K.
36 F4 Arīḩā Syria
82 G4 Arikaree r. U.S.A.
103 E2 Arima Trinidad and Tobago
104 C1 Arinos Brazil
101 G6 Arinos r. Brazil
22 C3 Arinthod France
96 D4 Ario de Rosáles Mex.
103 C3 Ariporo r. Col.
100 F5 Aripuanã Brazil
100 F5 Aripuanã r. Brazil
100 F5 Ariquemes Brazil
104 B2 Ariranhá r. Brazil
112 B1 Aris Namibia
11 C4 Arisaig U.K.
11 C4 Arisaig, Sound of chan. U.K.
76 D4 Aristazabal I. Can.
93 G4 Arizona div. U.S.A.
80 D5 Arizpe Mex.
40 B5 'Arjah S. Arabia
8 P3 Arjeplog Sweden
103 B2 Arjona Col.
59 D4 Arjuna, G. volc. Indon.
29 G5 Arkadak Rus. Fed.
85 E5 Arkadelphia U.S.A.
11 C4 Arkaig, Loch l. U.K.
39 F2 Arkalyk Kazak.
85 E5 Arkansas div. U.S.A.
85 F5 Arkansas r. U.S.A.
85 D4 Arkansas City U.S.A.
45 G1 Arkatag Shan mts China
Arkhangel'sk see Archangel
28 G2 Arkhangel'skaya Oblast' div. Rus. Fed.
28 F4 Arkhangel'skoye Rus. Fed.
52 C3 Arkhipovka Rus. Fed.
14 E5 Arklow Rep. of Ireland
27 M6 Arkoi i. Greece
19 L2 Arkona, Kap hd Ger.
32 K2 Arkticheskogo Instituta, Ostrova is Rus. Fed.
91 F3 Arkville U.S.A.
22 B6 Arles France
113 G4 Arlington S. Africa
82 B2 Arlington OR U.S.A.
84 D2 Arlington SD U.S.A.
90 E5 Arlington VA U.S.A.
88 D4 Arlington Heights U.S.A.
108 C3 Arlit Niger
15 F3 Arlon Belgium
22 D4 Arly r. France
69 B6 Armadale Austr.
55 C5 Armadores i. Indon.
14 E3 Armagh U.K.
22 B2 Armançon r. France
109 F2 Armant Egypt
29 G6 Armavir Rus. Fed.
31 D5 Armenia country Asia
103 B3 Armenia Col.
103 B3 Armero Col.
67 J3 Armidale Austr.
77 L2 Armit Lake Can.
44 E5 Armori India
76 B3 Armour, Mt Can./U.S.A.
14 E2 Armoy U.K.
76 F4 Armstrong B.C. Can.
78 C3 Armstrong Ont. Can.
52 E1 Armu r. Rus. Fed.
43 B2 Armur India
29 E6 Armyans'k Ukr.
36 D4 Arnaoutis, Cape Cyprus
79 F1 Arnaud r. Can.
22 B2 Arnay-le-Duc France
9 M6 Årnes Norway
85 D4 Arnett U.S.A.
18 C5 Arnhem Neth.
65 G2 Arnhem Bay Austr.
65 G2 Arnhem, C. Austr.
65 F2 Arnhem Land reg. Austr.
23 H6 Arno r. Italy
66 B4 Arno Bay Austr.
13 F4 Arnold U.K.
88 D2 Arnold U.S.A.
89 H1 Arnoux, Lac l. Can.
89 J3 Arnprior Can.
18 F5 Arnsberg Ger.
19 H6 Arnstadt Ger.
89 H3 Arnstein Can.
20 C3 Arnstein Ger.
20 F4 Arnstorf Ger.
89 H1 Arntfield Can.
103 E3 Aro r. Venez.
111 B6 Aroab Namibia
71 □3 Aroa, Pte pt Fr. Polynesia Pac. Oc.
19 G5 Arolsen Ger.
22 F4 Arona Italy
91 K1 Aroostook Can.
91 J1 Aroostook r. Can./U.S.A.
63 H2 Arorae i. Kiribati
55 B3 Aroroy Phil.
29 G7 Arpa r. Armenia/Turkey
37 J1 Arpaçay Turkey
15 B3 Arques-la-Bataille France
41 G5 Arra r. Pak.
37 J5 Ar Ramādī Iraq
36 E7 Ar Ramlah Jordan
11 C5 Arran i. U.K.
37 G4 Ar Raqqah Syria
15 C2 Arras France
40 A5 Ar-Rass S. Arabia
36 F4 Ar Rastan Syria
37 J7 Ar Rawḑ well S. Arabia
40 C5 Ar Rayyān Qatar
103 C4 Arrecifal Col.
105 E2 Arrecifes Arg.
15 E4 Arrentières France
97 F4 Arriaga Chiapas Mex.
96 D3 Arriaga San Luis Potosí Mex.
105 E2 Arribeños Arg.
37 L6 Ar Rifā'ī Iraq
37 K6 Ar Rihāb salt flat Iraq
42 D5 Ar Rimāl reg. S. Arabia
69 B4 Arrino Austr.
93 H3 Arriola U.S.A.
Ar Riyāḑ see Riyadh
11 D4 Arrochar U.K.
105 G2 Arroio Grande Brazil
104 D1 Arrojado r. Brazil
22 E5 Arroscia r. Italy
22 B3 Arroux r. France
88 B1 Arrow Lake Can.
14 C3 Arrow, Lough l. Rep. of Ireland
82 D3 Arrowrock Resr U.S.A.
70 C5 Arrowsmith, Mt N.Z.
105 F3 Arroyo Grande r. Arg.
92 B4 Arroyo Grande U.S.A.
105 E2 Arroyo Seco Arg.
97 E3 Arroyo Seco Mex.
40 B5 Ar Rubay'iyah S. Arabia
104 A1 Arruda Brazil
37 K6 Ar Rumaythah Iraq
36 G4 Ar Ruşāfah Syria
41 E6 Ar Rustāq Oman
37 H5 Ar Ruţba Iraq
40 B6 Ar Ruwayḑah S. Arabia
9 L8 Ārs Denmark
40 B2 Ars Iran
40 D4 Arsenajān Iran
52 C2 Arsen'yev Rus. Fed.
43 B3 Arsikere India
28 J3 Arsk Rus. Fed.
36 E3 Arslanköy Turkey
27 J5 Arta Greece
37 K2 Artashat Armenia
96 D4 Arteaga Mex.
52 C3 Artem Rus. Fed.
29 F5 Artemivs'k Ukr.
52 C3 Artemovskiy Rus. Fed.
15 B4 Artenay France
83 F5 Artesia U.S.A.
69 B2 Arthur r. Austr.
89 G4 Arthur Can.
67 G8 Arthur L. Austr.
66 B3 Arthur, Lake salt flat Austr.
90 C4 Arthur, Lake U.S.A.
68 D4 Arthur Pt Austr.
70 C5 Arthur's Pass N.Z.
70 C5 Arthur's Pass National Park N.Z.
87 F7 Arthur's Town Bahamas
116 B1 Artigas Uru. Base Ant.
105 F1 Artigas Uru.
37 J1 Art'ik Armenia
71 □9 Art, Île i. New Caledonia Pac. Oc.
77 H2 Artillery Lake Can.
113 G2 Artisia Botswana
15 B2 Artois reg. France
15 B2 Artois, Collines d' h. France
37 J2 Artos D. mt Turkey
36 F1 Artova Turkey
21 J4 Artstetten tourist site Austria
29 D6 Artsyz Ukr.
22 D6 Artuby r. France
39 J5 Artux China
37 H1 Artvin Turkey
115 B2 Arua Uganda
104 B1 Aruanã Brazil
73 J8 Aruba terr. Caribbean Sea
57 J8 Aru, Kepulauan is Indon.
100 F4 Arumã Brazil
45 F4 Arun r. Nepal
45 H3 Arunachal Pradesh div. India
13 G7 Arundel U.K.
71 □3 Aruó Fr. Polynesia Pac. Oc.
19 H1 Ārup Denmark
115 C3 Arusha Tanz.
115 C3 Arusha div. Tanz.
71 □2 Arutua i. Fr. Polynesia Pac. Oc.
115 A2 Aruwimi r. Congo(Zaire)
84 B4 Arvada U.S.A.
14 D4 Arvagh Rep. of Ireland
48 H2 Arvayheer Mongolia
22 D3 Arve r. France
44 D5 Arvi India
77 L2 Arviat Can.
79 F4 Arvida Can.
8 Q4 Arvidsjaur Sweden
9 N7 Arvika Sweden
92 C4 Arvin U.S.A.
40 B6 Arwā' S. Arabia
33 O2 Ary Rus. Fed.
39 G1 Arykbalyk Kazak.
39 G4 Arys' Kazak.
39 G4 Arys' r. Kazak.
38 F3 Arys, Oz. salt l. Kazak.
28 G4 Arzamas Rus. Fed.
25 F5 Arzew Alg.
29 H6 Arzgir Rus. Fed.
23 J4 Arzignano Italy
22 A4 Arzon r. France
20 F2 Aš Czech Rep.
103 E3 Asa r. Venez.
114 E3 Asaba Nigeria
41 H3 Asadābād Afgh.
40 C3 Asadābād Iran
41 E3 Asadābād Iran
36 G3 Asad, Buḩayrat al resr Syria
58 A5 Asahan r. Indon.
52 H3 Asahi-dake volc. Japan
53 C7 Asahi-gawa r. Japan
52 H3 Asahikawa Japan
39 H4 Asake Uzbek.
40 C2 Asālem Iran
54 D5 Asan Man b. S. Korea
45 F5 Asansol India
71 □4 Asau Western Samoa
15 H2 Asbach Ger.
79 F4 Asbestos Can.
112 E4 Asbestos Mountains S. Africa
91 F4 Asbury Park U.S.A.
26 F4 Ascea Italy
100 F7 Ascensión Bol.
97 H4 Ascensión, Bahía de la b. Mex.
107 B6 Ascension Island Atl. Ocean
20 C3 Aschaffenburg Ger.
18 E5 Ascheberg Ger.
19 J5 Aschersleben Ger.
23 J6 Asciano Italy
26 E3 Ascoli Piceno Italy
9 O8 Åseda Sweden
8 P4 Åsele Sweden
40 D4 Asemānjerd Iran
13 E6 Ashbourne U.K.
69 C1 Ashburton watercourse Austr.
70 C5 Ashburton N.Z.
88 D1 Ashburton Bay Can.
39 G4 Ashchikol', Oz. salt l. Kazak.
39 F3 Ashchikol', Ozero salt l. Kazak.
76 E4 Ashcroft Can.
36 E6 Ashdod Israel
85 E5 Ashdown U.S.A.
87 E5 Asheboro U.S.A.
87 D5 Asheville U.S.A.
67 J2 Ashford Austr.
13 H6 Ashford U.K.
93 F4 Ash Fork U.S.A.
38 D5 Ashgabat Turkm.
52 H3 Ashibetsu Japan
53 F6 Ashikaga Japan
12 F2 Ashington U.K.
28 J3 Ashit r. Rus. Fed.
53 C8 Ashizuri-misaki pt Japan
40 D4 Ashkazar Iran
85 D4 Ashland KS U.S.A.
90 B5 Ashland KY U.S.A.
91 J1 Ashland ME U.S.A.
82 F2 Ashland MT U.S.A.
91 H3 Ashland NH U.S.A.
90 B4 Ashland OH U.S.A.
82 B3 Ashland OR U.S.A.
90 E6 Ashland VA U.S.A.
88 B2 Ashland WV U.S.A.
67 H2 Ashley Austr.
84 D2 Ashley U.S.A.
64 D2 Ashmore Reef Ashmore and Cartier Is
28 C4 Ashmyany Belarus
93 H5 Ash Peak U.S.A.
36 E6 Ashqelon Israel
44 B4 Ashraf, Md. Pak.
37 J6 Ash Shabakah Iraq
37 H3 Ash Shaddādah Syria
37 H7 Ash Shaqīq well S. Arabia
40 B5 Ash Sha'rā' S. Arabia
37 J4 Ash Sharqāţ Iraq
37 L6 Ash Shaţrah Iraq
42 C7 Ash Shiḩr Yemen
37 K6 Ash Shināfīyah Iraq
40 E5 Ash Shināş Oman
40 B4 Ash Shu'bah S. Arabia
40 B5 Ash Shumlūl S. Arabia
90 C4 Ashtabula U.S.A.
37 K1 Ashtarak Armenia
44 D5 Ashti Maharashtra India
43 A2 Ashti Maharashtra India
40 C3 Ashtiān Iran
112 D6 Ashton S. Africa
82 E2 Ashton U.S.A.
12 E4 Ashton-under-Lyne U.K.
75 M4 Ashuanipi Lake Can.
87 C5 Ashville U.S.A.
23 J4 Asiago Italy
96 D3 Asientos Mex.
43 B2 Asifabad India
45 F6 Āsika India
36 F4 'Āşī, Nahr al r. Asia
26 C4 Asinara, Golfo dell' b. Sardinia Italy
32 K4 Asino Rus. Fed.
28 D4 Asipovichy Belarus
42 B5 'Asīr reg. S. Arabia
37 H2 Aşkale Turkey
38 D1 Askarovo Rus. Fed.
9 M7 Asker Norway
9 O7 Askersund Sweden
9 M7 Askim Norway
48 F1 Askiz Rus. Fed.
19 J2 Askø i. Denmark
37 L2 Aşlāndūz Iran
110 D2 Asmara Eritrea
19 J1 Asnæs pen. Denmark
9 O8 Åsnen l. Sweden
27 L3 Asneovgrad Bulg.
23 H4 Asola Italy
44 C4 Asop India
21 K5 Aspang-Markt Austria
37 M3 Aspar Iran
39 H4 Aspara Kazak.
12 D3 Aspatria U.K.
83 F4 Aspen U.S.A.
20 C4 Asperg Ger.
85 C5 Aspermont U.S.A.
70 B6 Aspiring, Mt N.Z.
77 H4 Asquith Can.
23 J4 Assa r. Italy
39 G4 Assa Kazak.
36 F4 As Sa'an Syria
110 E2 Assab Eritrea
40 C5 As Sabsab well S. Arabia
36 F3 As Safīrah Syria
40 C5 As Saji well S. Arabia
38 D4 Assake-Audan, Vpadina basin Uzbek.
40 B5 As Salamiyah S. Arabia
37 K6 As Salmān Iraq
45 H4 Assam div. India
37 K6 As Samāwah Iraq
36 F5 Aş Şanamayn Syria
109 E2 As Sarīr reg. Libya
91 F5 Assateague I. U.S.A.
91 F6 Assateague Island National Seashore res. U.S.A.
22 D6 Asse r. France
18 D3 Assen Neth.
15 D1 Assenede Belgium
19 G1 Assens Denmark
15 F2 Assesse Belgium
45 F5 Assia Hills India
109 D1 As Sidrah Libya
77 H5 Assiniboia Can.
77 K5 Assiniboine r. Can.
76 F4 Assiniboine, Mt Can.
104 B3 Assis Brazil
26 E3 Assisi Italy
20 B2 Aßlar Ger.
37 L7 Aş Şubayḩīyah Kuwait
40 B4 Aş Şufayrī well S. Arabia
37 G4 As Sukhnah Syria
37 K4 As Sulaymānīyah Iraq
42 C5 As Sulayyil S. Arabia
40 C5 Aş Şulb reg. S. Arabia
40 B5 Aş Şummān plat. S. Arabia
42 B5 As Sūq S. Arabia
37 H4 Aş Şuwār Syria
36 F5 As Suwaydā' Syria
41 E6 As Suwayq Oman
37 K5 Aş Şuwayrah Iraq
11 C2 Assynt, Loch l. U.K.
23 J3 Asta, Cima d' mt Italy
27 M7 Astakida i. Greece
Astalu Island see Astola Island
37 M3 Astaneh Iran
37 M2 Astara Azer.
Asterabad see Gorgān
22 F5 Asti Italy
92 A2 Asti U.S.A.
105 C1 Astica Arg.
23 J4 Astico r. Italy
41 F5 Astola Island Pak.
44 C2 Astor Jammu and Kashmir
44 C2 Astor r. Pak.
25 C1 Astorga Spain
82 B2 Astoria U.S.A.
9 N8 Åstorp Sweden
34 C2 Astrakhan' Rus. Fed.
Astrakhan' Bazar see Călilabad
29 H6 Astrakhanskaya Oblast' div. Rus. Fed.
28 C4 Astravyets Belarus
116 D3 Astrid Ridge sea feature Ant.
71 □9 Astrolabe, Récifs de l' rf New Caledonia Pac. Oc.
25 C1 Asturias div. Spain
27 M6 Astypalaia i. Greece
39 K2 Asubulak Kazak.
71 □4 Asuisui, C. Western Samoa
57 L3 Asuncion i. N. Mariana Is
102 E3 Asunción Para.
115 B2 Aswa r. Uganda
109 F2 Aswân Egypt
109 F2 Asyût Egypt
71 □11 Ata i. Tonga
103 D4 Atabapo r. Col./Venez.
39 G4 Atabay Kazak.
102 C2 Atacama, Desierto de des. Chile
63 J2 Atafu i. Tokelau
71 □3 Ataiti Fr. Polynesia Pac. Oc.
114 D3 Atakpamé Togo
27 K5 Atalanti Greece
96 J7 Atalaya Panama
100 D6 Atalaya Peru
39 G1 Atansor, Ozero salt l. Kazak.
108 A2 Atâr Maur.
58 A1 Ataran r. Myanmar
93 H4 Atarque U.S.A.
92 B4 Atascadero U.S.A.
39 G2 Atasu Kazak.
71 □11 Atata i. Tonga
57 H8 Atauro i. Indon.
109 F3 Atbara Sudan
109 F3 Atbara r. Sudan
39 G2 Atbasar Kazak.
39 H4 At-Bashy Kyrg.
85 F6 Atchafalaya Bay U.S.A.
84 E4 Atchison U.S.A.
114 C3 Atebubu Ghana
96 C3 Atenguillo Mex.
26 E3 Aterno r. Italy
26 F3 Atessa Italy
15 D2 Ath Belgium
76 G4 Athabasca Can.
74 G4 Athabasca r. U.S.A.
77 G3 Athabasca, Lake Can.
14 E4 Athboy Rep. of Ireland
14 C4 Athenry Rep. of Ireland
89 K3 Athens Can.
27 K6 Athens Greece
87 C5 Athens AL U.S.A.
87 D5 Athens GA U.S.A.
90 B5 Athens OH U.S.A.
87 C5 Athens TN U.S.A.
85 E5 Athens TX U.S.A.
13 F5 Atherstone U.K.
68 A1 Atherton Austr.
115 C3 Athi r. Kenya
15 C3 Athies France
Athina see Athens
14 C4 Athleague Rep. of Ireland
14 D4 Athlone Rep. of Ireland
43 A2 Athni India
70 B6 Athol N.Z.
91 G3 Athol U.S.A.
11 E4 Atholl, Forest of reg. U.K.
27 L4 Athos mt Greece
37 J4 Ath Tharthār, Wādī r. Iraq
14 E5 Athy Rep. of Ireland
109 D3 Ati Chad
100 D7 Atico Peru
77 J4 Atikameg L. Can.
78 B4 Atikokan Can.
79 H3 Atikonak L. Can.
71 □3 Atimaono Fr. Polynesia Pac. Oc.
55 B3 Atimonan Phil.
43 B4 Atirampattinam India
97 G5 Atitlán Guatemala
97 G5 Atitlán, Volcán volc. Guatemala
60 M6 Atiu Mauke i. Cook Is Pac. Oc.
33 R3 Atka Rus. Fed.
29 H5 Atkarsk Rus. Fed.
40 B5 'Atk, W. al watercourse S. Arabia
97 E4 Atlacomulco Mex.
87 C5 Atlanta GA U.S.A.
88 C5 Atlanta IL U.S.A.
89 E3 Atlanta MI U.S.A.
36 D2 Atlantı Turkey
84 E3 Atlantic U.S.A.
91 F5 Atlantic City U.S.A.
120 J9 Atlantic-Indian Antarctic Basin sea feature Atl. Ocean
120 J9 Atlantic-Indian Ridge sea feature Ind. Ocean
112 C6 Atlantis S. Africa
120 G3 Atlantis Fracture sea feature Atl. Ocean
106 B2 Atlas Mountains Alg./Morocco
108 C1 Atlas Saharien mts Alg.
76 C3 Atlin Can.
76 C3 Atlin Lake Can.
76 C3 Atlin Prov. Park Can.
36 E5 'Atlit Israel
97 E4 Atlixco Mex.
43 B3 Atmakur Orissa India
43 B3 Atmakur Orissa India
85 G6 Atmore U.S.A.
39 J4 At-Moyen Kyrg.
85 D5 Atoka U.S.A.
96 D3 Atotonilco el Alto Mex.
58 C1 Atouat mt Laos
97 D4 Atoyac de Alvarez Mex.
45 G4 Atrai r. India
40 E2 Atrak r. Iran
103 A3 Atrato r. Col.
40 D2 Atrek r. Iran/Turkm.
38 C5 Atrek r. Iran/Turkm.
91 F5 Atsion U.S.A.
42 B5 Aţ Ţā'if S. Arabia
87 C5 Attalla U.S.A.
58 C2 Attapu Laos
27 M6 Attavyros mt Greece
78 D3 Attawapiskat Can.
78 C3 Attawapiskat r. Can.
78 C3 Attawapiskat L. Can.
37 G7 Aţ Ţawīl mts S. Arabia
40 A4 At Taysīyah plat. S. Arabia
18 E5 Attendorn Ger.
21 G5 Attersee l. Austria
88 D5 Attica IN U.S.A.
90 B4 Attica OH U.S.A.
15 E3 Attigny France
91 H3 Attleboro U.S.A.
13 J5 Attleborough U.K.
36 F7 Aţ Ţubayq reg. S. Arabia
118 G2 Attu Island U.S.A.
40 B5 Aţ Ţulayḩī well S. Arabia
43 B4 Attur India
11 B2 a' Tuath, Loch b. U.K.
105 C2 Atuel r. Arg.
9 O7 Åtvidaberg Sweden
90 C4 Atwood Lake U.S.A.
38 B3 Atyrau Kazak.
38 B3 Atyrauskaya Oblast' div. Kazak.
20 C5 Au Austria
20 D3 Aub Ger.
22 C6 Aubagne France
22 B6 Aubais France
15 F3 Aubange Belgium
55 B2 Aubarede Point Phil.
15 E4 Aube r. France
22 B5 Aubenas France
15 B4 Aubergenville France
22 C2 Auberive France
15 B3 Aubevoye France
22 D3 Aubonne Switz.
15 F3 Auboué France
93 F4 Aubrey Cliffs cliff U.S.A.
74 F3 Aubry Lake Can.
66 C5 Auburn Austr.
68 D5 Auburn r. Austr.
89 G4 Auburn Can.
87 C5 Auburn AL U.S.A.
92 B2 Auburn CA U.S.A.
88 E5 Auburn IN U.S.A.
91 H2 Auburn ME U.S.A.
84 E3 Auburn NE U.S.A.
90 E3 Auburn NY U.S.A.
82 B2 Auburn WA U.S.A.
68 D5 Auburn Ra. h. Austr.
24 F4 Aubusson France
105 C3 Auca Mahuida, Sa de mt Arg.
24 E5 Auch France
11 E4 Auchterarder U.K.
70 E2 Auckland N.Z.
63 G7 Auckland Islands N.Z.
91 H2 Audet Can.
22 C2 Audeux France
22 D2 Audincourt France
15 B2 Audresselles France
15 C2 Audruicq France
20 F2 Aue Ger.
19 G4 Aue r. Ger.
20 F2 Auerbach Ger.
20 E3 Auerbach in der Oberpfalz Ger.
20 F2 Auersberg mt Ger.
15 B3 Auffay France
68 B5 Augathella Austr.
14 D3 Augher U.K.
14 E3 Aughnacloy U.K.
14 E5 Aughrim Rep. of Ireland
112 D4 Augrabies S. Africa
112 D4 Augrabies Falls waterfall S. Africa
112 D4 Augrabies Falls National Park S. Africa
89 F3 Au Gres U.S.A.
20 D4 Augsburg Ger.
69 B7 Augusta Austr.
26 F6 Augusta Sicily Italy
87 D5 Augusta GA U.S.A.
85 D4 Augusta KS U.S.A.
91 J2 Augusta ME U.S.A.
88 B3 Augusta WV U.S.A.
103 B2 Augustin Cadazzi Col.
69 C2 Augustus, Mt Austr.
20 C5 Aulendorf Ger.
23 G5 Aulla Italy
15 D2 Aulnoye-Aymeries France
15 B2 Ault France
15 B3 Aumale France
38 E4 Auminzatau, Gory h. Uzbek.
15 B4 Auneau France
15 C3 Auneuil France
71 □5 Aunuu i. American Samoa Pac. Oc.
111 B6 Auob r. Namibia
79 G2 Aupaluk Can.
22 D6 Aups France
58 C5 Aur i. Malaysia
9 S6 Aura Fin.
20 D3 Aurach Ger.
44 D4 Auraiya India
44 C6 Aurangābād India
22 C5 Aurel France
18 E3 Aurich Ger.
104 B2 Aurilândia Brazil
24 F4 Aurillac France
22 C6 Auriol France
59 D3 Aurkuning Indon.
23 K3 Auronzo di Cadore Italy
55 B5 Aurora Phil.
82 F4 Aurora CO U.S.A.
88 C5 Aurora IL U.S.A.
91 J2 Aurora ME U.S.A.
85 E4 Aurora MO U.S.A.
65 H2 Aurukun Austr.
111 B6 Aus Namibia
89 F3 Au Sable U.S.A.
91 G2 Ausable r. U.S.A.
89 E3 Au Sable r. U.S.A.
91 G2 Ausable Forks U.S.A.
89 F3 Au Sable Pt MI U.S.A.
88 D2 Au Sable Pt MI U.S.A.
11 F1 Auskerry i. U.K.
8 D4 Austari-Jökulsá r. Iceland
84 E3 Austin MN U.S.A.
92 D2 Austin NV U.S.A.
85 D6 Austin TX U.S.A.
69 D3 Austin, L. salt flat Austr.
60 M6 Australes, Îles is Fr. Polynesia Pac. Oc.
61 D6 Australia country Oceania
116 B6 Australian Antarctic Territory reg. Ant.
67 H5 Australian Capital Territory div. Austr.
7 E4 Austria country Europe
8 O2 Austvågøy i. Norway
15 B2 Authie r. France
15 A4 Authon-du-Perche France
96 C4 Autlán Mex.
22 C2 Autrey-lès-Gray France
8 U3 Autti Fin.
22 B3 Autun France
24 F4 Auvergne reg. France
22 A2 Auxerre France
22 A2 Auxerrois reg. France
15 C2 Auxi-le-Château France
22 B2 Auxois reg. France
22 C2 Auxonne France
22 B3 Auxy France
39 K2 Auzzov Kazak.
91 F3 Ava U.S.A.
22 A2 Avallon France
92 C5 Avalon U.S.A.
79 K4 Avalon Peninsula Can.
96 D2 Avalos Mex.
40 B2 Avān Iran
36 E2 Avanos Turkey
104 C3 Avaré Brazil
37 L2 Āvārsīn Iran
92 D4 Avawatz Mts U.S.A.
41 F3 Avaz Iran
101 G4 Aveiro Brazil
25 B2 Aveiro Port.
25 B2 Aveiro, Ria de est. Port.
37 M4 Āvej Iran
105 E2 Avellaneda Arg.
26 F4 Avellino Italy
92 B3 Avenal U.S.A.
66 F6 Avenel Austr.
18 B4 Avenhorn Neth.
26 F4 Aversa Italy
15 D2 Avesnes-sur-Helpe France
9 P6 Avesta Sweden
24 F4 Aveyron r. France
26 E3 Avezzano Italy
23 K3 Aviano Italy
11 E3 Aviemore U.K.
22 E4 Avigliana Italy
26 F4 Avigliano Italy
22 B6 Avignon France
25 D2 Ávila Spain
25 D1 Avilés Spain
15 C2 Avion France
23 J3 Avisio r. Italy
43 C5 Avissawella Sri Lanka
28 H2 Avnyugskiy Rus. Fed.
67 G8 Avoca Tas. Austr.
66 E6 Avoca Vic. Austr.
66 E6 Avoca r. Vic. Austr.
14 E5 Avoca Rep. of Ireland
84 E3 Avoca U.S.A.
26 F6 Avola Sicily Italy
69 C5 Avon r. Austr.
15 C4 Avon France
13 F5 Avon r. Eng. U.K.
13 E6 Avon r. Eng. U.K.
13 F7 Avon r. Eng. U.K.
88 B5 Avon U.S.A.
93 F5 Avondale U.S.A.
68 B3 Avon Downs Austr.
13 E6 Avonmouth U.K.
87 D7 Avon Park U.S.A.
24 D2 Avranches France
15 C3 Avre r. France
63 G2 Avuavu Solomon Is
53 D7 Awaji-shima i. Japan
70 E3 Awakino N.Z.
40 C5 Awālī Bahrain
70 D1 Awanui N.Z.
70 B6 Awarua Pt N.Z.
110 E3 Āwash Eth.
110 D3 Āwash r. Eth.
52 F5 Awa-shima i. Japan
110 D3 Awash National Park Eth.
112 A2 Awasib Mts Namibia
39 K4 Awat China
70 D4 Awatere r. N.Z.
109 D2 Awbārī Libya
14 C5 Awbeg r. Rep. of Ireland
37 L6 'Awdah, Hawr al l. Iraq
110 E3 Aw Dheegle Somalia
109 E4 Aweil Sudan
11 C4 Awe, Loch l. U.K.
114 E3 Awka Nigeria
55 C6 Awu mt Indon.
66 F6 Axedale Austr.
75 J2 Axel Heiburg I. Can.
114 C4 Axim Ghana
13 E7 Axminster U.K.
15 E3 Ay France
39 K3 Ay Kazak.
53 D7 Ayabe Japan
105 E3 Ayacucho Arg.
100 D6 Ayacucho Peru
39 K3 Ayaguz Kazak.
39 J3 Ayaguz watercourse Kazak.
38 F4 Ayakagytma, Vpadina depression Uzbek.
38 F4 Ayakkuduk Uzbek.
46 A3 Ayakkum Hu l. China
25 C4 Ayamonte Spain
46 F1 Ayan Rus. Fed.
29 E7 Ayancık Turkey
54 C4 Ayang N. Korea
103 B2 Ayapel Col.
36 D1 Ayaş Turkey
100 D6 Ayaviri Peru
41 H2 Āybak Afgh.
39 G1 Aydabul Kazak.
29 F5 Aydar r. Ukr.
39 F4 Aydarkul', Ozero l. Uzbek.
38 F3 Aydarly Kazak.
36 A3 Aydın Turkey
36 A2 Aydın Dağları mts Turkey
38 D2 Aydyrlinskiy Rus. Fed.
39 G3 Ayeat, Gora h. Kazak.
58 □ Ayer Chawan, P. i. Sing.
58 □ Ayer Merbau, P. i. Sing.
64 F5 Ayers Rock h. Austr.
39 K2 Aygyrzhal Kazak.
38 E2 Ayke, Oz. l. Kazak.
33 N3 Aykhal Rus. Fed.
28 J2 Aykino Rus. Fed.
39 K4 Aykol China
70 D5 Aylesbury N.Z.
13 G6 Aylesbury U.K.
90 E6 Aylett U.S.A.
25 E2 Ayllón Spain
89 G4 Aylmer Can.
77 H2 Aylmer Lake Can.
38 E2 Aymagambetov Kazak.
39 J3 Aynabulak Kazak.
40 C4 'Ayn al 'Abd well S. Arabia
39 G5 Ayní Tajik.
36 G3 'Ayn 'Īsá Syria
109 F4 Ayod Sudan
33 S3 Ayon, O. i. Rus. Fed.
108 B3 'Ayoûn el 'Atroûs Maur.
68 B2 Ayr Austr.
11 D5 Ayr U.K.
11 D5 Ayr r. U.K.
36 D3 Ayrancı Turkey
12 C3 Ayre, Point of Isle of Man
39 G3 Ayshirak Kazak.
27 M3 Aytos Bulg.
58 B2 Ayutthaya Thai.
27 M5 Ayvacık Turkey
36 F2 Ayvalı Turkey
27 M5 Ayvalık Turkey
96 H5 Azacualpa Honduras
45 E4 Azamgarh India
108 B3 Azaouâd reg. Mali
108 C3 Azaouagh, Vallée de watercourse Mali/Niger
40 B2 Āzarān Iran
114 F2 Azare Nigeria
Azbine mts see Aïr, Massif de l'
36 D1 Azdavay Turkey
31 D5 Azerbaijan country Asia
22 B3 Azergues r. France
40 B2 Āzghān Iran
38 E3 Azhar Kazak.
89 G2 Azilda Can.
91 H2 Aziscohos Lake U.S.A.
100 C4 Azogues Ecuador
32 F3 Azopol'ye Rus. Fed.
4 Azores terr. Europe
120 H3 Azores - Cape St Vincent Ridge sea feature Atl. Ocean
29 F6 Azov Rus. Fed.
29 F6 Azov, Sea of Rus. Fed./Ukr.
108 B1 Azrou Morocco
83 F4 Aztec U.S.A.
25 D3 Azuaga Spain
102 B3 Azucar r. Chile
96 J7 Azuero, Península de pen. Panama
105 E3 Azul Arg.
97 G4 Azul r. Mex.
105 B4 Azul, Cerro mt Arg.
100 C5 Azul, Cordillera mts Peru
104 A1 Azul, Serra h. Brazil
53 G6 Azuma-san volc. Japan
100 F8 Azurduy Bol.
26 B6 Azzaba Alg.
36 F5 Az Zabadānī Syria
37 J6 Aẓ Ẓafīrī reg. Iraq
Aẓ Ẓahrān see Dhahran
23 K4 Azzano Decimo Italy
40 B5 Az Zilfī S. Arabia
37 L6 Az Zubayr Iraq

B

71 □15 Ba Fiji
71 □15 Ba r. Fiji
71 □9 Baâba i. New Caledonia Pac. Oc.
36 E5 Ba'abda Lebanon
36 F4 Ba'albek Lebanon
67 H3 Baan Baa Austr.
110 E3 Baardheere Somalia
27 M5 Baba Burnu pt Turkey
37 M1 Babadağ mt Azer.
27 N2 Babadag Romania
38 D5 Babadurmaz Turkm.
29 C7 Babaeski Turkey
100 C4 Babahoyo Ecuador
43 B1 Babai India
45 E3 Babai r. Nepal
50 B1 Babai Gaxun China
37 L2 Bābā Jān Iran
55 C5 Babak Phil.
41 H3 Bābā, Kūh-e mts Afgh.
42 B7 Bāb al Mandab str. Africa/Asia
57 H8 Babar i. Indon.
57 H8 Babar, Kepulauan is Indon.
115 C3 Babati Tanz.
28 E3 Babayevo Rus. Fed.
29 H7 Babayurt Rus. Fed.
88 B2 Babbitt U.S.A.
67 H7 Babel I. Austr.
80 F6 Babia r. Mex.
68 A1 Babinda Austr.
74 F4 Babine r. Can.
76 D4 Babine Lake Can.
57 J7 Babo Indon.
40 D2 Bābol Iran
38 C5 Bābol Sar Iran
114 F3 Babongo Cameroon
112 C6 Baboon Point S. Africa
93 G6 Baboquivari Peak U.S.A.
110 B3 Baboua C.A.R.
28 D4 Babruysk Belarus
44 B4 Babuhri India
44 C2 Babusar Pass Pak.
48 J1 Babushkin Rus. Fed.
55 A4 Babuyan Phil.
55 B2 Babuyan i. Phil.
51 F7 Babuyan Channel Phil.
55 B2 Babuyan Islands Phil.
37 K5 Babylon Iraq
101 K4 Bacabal Brazil
36 C1 Bacakliyayla T. mt Turkey
97 G4 Bacalar Mex.
57 H7 Bacan i. Indon.
55 B2 Bacarra Phil.
17 N7 Bacău Romania
15 G4 Baccarat France
66 F6 Bacchus Marsh Austr.
51 C6 Băc Giang Vietnam
20 D5 Bach Austria
96 C1 Bachiniva Mex.
39 J5 Bachu China
39 J5 Bachu Liuchang China
75 J3 Back r. N.W.T. Can.
77 J1 Back r. Can.
27 H2 Bačka Palanka Yugo.
76 D2 Backbone Ranges mts Can.
8 P5 Backe Sweden
20 C4 Backnang Ger.
66 B5 Backstairs Pass. chan. Austr.
11 E4 Backwater Reservoir U.K.
51 B6 Bac Lac Vietnam
58 C3 Bac Liêu Vietnam
51 C6 Băc Ninh Vietnam
55 B4 Bacolod Phil.
55 B3 Baco, Mt Phil.
51 B6 Băc Quang Vietnam
15 A3 Bacqueville-en-Caux France
78 F2 Bacqueville, Lac l. Can.
20 F4 Bad Abbach Ger.
43 A4 Badagara India
20 F5 Bad Aibling Ger.
50 A1 Badain Jaran Shamo des. China
100 F4 Badajós, Lago l. Brazil
25 C3 Badajoz Spain
43 A3 Badami India
37 H6 Badanah S. Arabia
45 H4 Badarpur India
21 G5 Bad Aussee Austria
89 F4 Bad Axe U.S.A.
20 A3 Bad Bergzabern Ger.
19 J6 Bad Berka Ger.
18 F5 Bad Berleburg Ger.
19 H3 Bad Bevensen Ger.
20 E2 Bad Blankenburg Ger.
20 F2 Bad Brambach Ger.
19 G3 Bad Bramstedt Ger.
20 C2 Bad Brückenau Ger.
15 J2 Bad Camberg Ger.
79 H4 Baddeck Can.
41 G4 Baddo r. Pak.
19 J2 Bad Doberan Ger.
19 G5 Bad Driburg Ger.
19 K5 Bad Düben Ger.
15 J3 Bad Dürkheim Ger.
19 K5 Bad Dürrenberg Ger.
114 E2 Badéguichéri Niger
36 C3 Bademli Geçidi pass Turkey
15 H2 Bad Ems Ger.
21 K4 Baden Austria
22 F2 Baden Switz.
20 B4 Baden-Baden Ger.
11 D4 Badenoch reg. U.K.
20 B4 Baden-Württemberg div. Ger.
23 L4 Baderna Croatia
18 F4 Bad Essen Ger.
19 M4 Bad Freienwalde Ger.
20 C3 Bad Friedrichshall Ger.
20 G5 Badgastein Austria
79 J4 Badger Can.
21 J6 Bad Gleichenberg Austria
21 G5 Bad Goisern Austria
19 H5 Bad Grund (Harz) Ger.
19 H5 Bad Harzburg Ger.
20 E5 Bad Heilbrunn Ger.
19 G6 Bad Hersfeld Ger.
20 G5 Bad Hofgastein Austria
20 B2 Bad Homburg vor der Höhe Ger.
23 J4 Badia Polesine Italy
44 B4 Badin Pak.
96 C2 Badiraguato Mex.
21 G5 Bad Ischl Austria
Bādiyat ash Shām des. see Syrian Desert
20 D2 Bad Kissingen Ger.
20 C3 Bad König Ger.
19 J5 Bad Kösen Ger.
15 H3 Bad Kreuznach Ger.
22 E2 Bad Krozingen Ger.
18 F6 Bad Laasphe Ger.
84 C2 Badlands reg. U.S.A.
84 C3 Badlands Nat. Park U.S.A.
19 H5 Bad Langensalza Ger.
19 H5 Bad Lauterberg im Harz Ger.
19 L5 Bad Liebenwerda Ger.
18 F5 Bad Lippspringe Ger.
15 H2 Bad Marienberg (Westerwald) Ger.
20 C3 Bad Mergentheim Ger.
20 B2 Bad Nauheim Ger.
15 H2 Bad Neuenahr-Ahrweiler Ger.
20 D2 Bad Neustadt an der Saale Ger.
19 H3 Bad Oldesloe Ger.
50 D4 Badong China
58 C3 Ba Đông Vietnam
15 G4 Badonviller France
19 G5 Bad Pyrmont Ger.
21 J6 Bad Radkersburg Austria
23 G2 Bad Ragaz Switz.
37 K5 Badrah Iraq
20 F5 Bad Reichenhall Ger.
44 D3 Badrinath Peaks mts India
19 H5 Bad Sachsa Ger.
22 E2 Bad Säckingen Ger.
21 H6 Bad St Leonhard im Lavanttal Austria
19 H4 Bad Salzdetfurth Ger.
18 F4 Bad Salzuflen Ger.
19 H6 Bad Salzungen Ger.
20 C4 Bad Schussenried Ger.
20 B2 Bad Schwalbach Ger.
19 H3 Bad Schwartau Ger.
19 H3 Bad Segeberg Ger.
65 H2 Badu I. Austr.
43 C5 Badulla Sri Lanka
20 C4 Bad Urach Ger.
21 K5 Bad Vöslau Austria
20 C5 Bad Waldsee Ger.
19 J4 Bad Wilsnack Ger.
20 D3 Bad Windsheim Ger.
20 D5 Bad Wörishofen Ger.
18 F3 Bad Zwischenahn Ger.
8 B3 Bæir Iceland
67 J4 Baerami Austr.
18 D6 Baesweiler Ger.
25 E4 Baeza Spain
114 F3 Bafang Cameroon
108 A3 Bafatá Guinea-Bissau
75 M2 Baffin Bay Can./Greenland
75 L2 Baffin Island Can.
68 D5 Baffle Cr. r. Austr.
114 F4 Bafia Cameroon
114 D3 Bafilo Togo
114 A2 Bafing r. Guinea/Mali
114 A2 Bafing, Parc National du nat. park Mali
114 A2 Bafoulabé Mali
114 F3 Bafoussam Cameroon
40 D4 Bāfq Iran
36 E1 Bafra Turkey
29 E7 Bafra Burnu pt Turkey
40 E4 Bāft Iran
115 A2 Bafwasende Congo(Zaire)
45 F4 Bagaha India
55 A5 Bagahak, Mt h. Malaysia
43 A2 Bagalkot India
115 C4 Bagamoyo Tanz.
59 B2 Bagan Datuk Malaysia
111 C5 Bagani Namibia
58 B4 Bagan Serai Malaysia
58 B5 Bagansiapiapi Indon.
39 K3 Bagar watercourse Kazak.
93 F4 Bagdad U.S.A.
105 F1 Bagé Brazil
44 D3 Bageshwar India
82 F3 Baggs U.S.A.
13 C6 Baggy Point U.K.
44 C5 Bagh India
41 F2 Baghbaghū Iran

70 D1 **Broadwood** N.Z.
9 S8 **Broceni** Latvia
77 J3 **Brochet** Can.
77 J3 **Brochet, Lac** *l.* Can.
19 H5 **Brocken** *mt* Ger.
74 G2 **Brock I.** Can.
90 E3 **Brockport** U.S.A.
91 H3 **Brockton** U.S.A.
89 K3 **Brockville** Can.
89 F4 **Brockway** *MI* U.S.A.
90 D4 **Brockway** *PA* U.S.A.
21 L3 **Brodek u Přerova** Czech Rep.
19 K2 **Broderstorf** Ger.
75 K2 **Brodeur Peninsula** Can.
88 C4 **Brodhead** U.S.A.
11 C5 **Brodick** U.K.
17 J4 **Brodnica** Pol.
29 C5 **Brody** Ukr.
85 E4 **Broken Arrow** U.S.A.
67 J4 **Broken B.** Austr.
84 D3 **Broken Bow** *NE* U.S.A.
85 E5 **Broken Bow** *OK* U.S.A.
66 D3 **Broken Hill** Austr.
19 G3 **Brokstedt** Ger.
19 H4 **Brome** Ger.
13 G6 **Bromley** U.K.
13 E5 **Bromsgrove** U.K.
9 L8 **Brønderslev** Denmark
113 H2 **Bronkhorstspruit** S. Africa
8 N4 **Brønnøysund** Norway
88 E5 **Bronson** U.S.A.
13 J5 **Brooke** U.K.
55 A4 **Brooke's Point** Phil.
88 C4 **Brookfield** U.S.A.
85 F6 **Brookhaven** U.S.A.
82 A3 **Brookings** *OR* U.S.A.
84 D2 **Brookings** *SD* U.S.A.
91 H3 **Brookline** U.S.A.
88 A5 **Brooklyn** *IA* U.S.A.
88 B5 **Brooklyn** *IL* U.S.A.
84 E2 **Brooklyn Center** U.S.A.
90 D6 **Brookneal** U.S.A.
77 G4 **Brooks** Can.
92 A2 **Brooks** *CA* U.S.A.
91 J2 **Brooks** *ME* U.S.A.
116 B3 **Brooks, C.** Ant.
74 D3 **Brooks Range** *mts* U.S.A.
87 D6 **Brooksville** U.S.A.
69 C6 **Brookton** Austr.
90 D4 **Brookville** U.S.A.
68 E6 **Brooloo** Austr.
64 D3 **Broome** Austr.
69 C6 **Broomehill** Austr.
11 C3 **Broom, Loch** *in.* U.K.
11 E2 **Brora** U.K.
9 O9 **Brösarp** Sweden
14 D4 **Brosna** *r.* Rep. of Ireland
82 B3 **Brothers** U.S.A.
51 □ **Brothers, The** *is H.K.* China
15 B4 **Brou** France
12 E3 **Brough** U.K.
11 E1 **Brough Head** *hd* U.K.
14 E3 **Broughshane** U.K.
66 C4 **Broughton** *r.* Austr.
67 K4 **Broughton Is** Austr.
75 M3 **Broughton Island** Can.
17 P5 **Brovary** Ukr.
9 L8 **Brovst** Denmark
68 A1 **Brown Cr.** *r.* Austr.
85 C5 **Brownfield** U.S.A.
82 D1 **Browning** U.S.A.
69 D5 **Brown, L.** *salt flat* Austr.
66 C4 **Brown, Mt** Austr.
88 D6 **Brownsburg** U.S.A.
91 F5 **Browns Mills** U.S.A.
87 B5 **Brownsville** *TN* U.S.A.
85 D7 **Brownsville** *TX* U.S.A.
91 J2 **Brownville** U.S.A.
91 J2 **Brownville Junction** U.S.A.
85 D6 **Brownwood** U.S.A.
22 D3 **Broye** *r.* Switz.
17 O4 **Brozha** Belarus
15 C2 **Bruay-en-Artois** France
88 C2 **Bruce Crossing** U.S.A.
64 C4 **Bruce, Mt** Austr.
78 D4 **Bruce Pen.** Can.
89 G3 **Bruce Peninsula National Park** Can.
69 D5 **Bruce Rock** Austr.
20 B3 **Bruchsal** Ger.
19 K4 **Brück** Ger.
20 E5 **Bruck am Ziller** Austria
21 K4 **Bruck an der Leitha** Austria
21 J5 **Bruck an der Mur** Austria
21 H6 **Brückl** Austria
20 E5 **Bruckmühl** Ger.
13 E6 **Brue** *r.* U.K.
15 D1 **Bruges** Belgium
Brugge *see* **Bruges**
20 B3 **Brühl** Ger.
18 D6 **Brühl** Ger.
93 G2 **Bruin Pt** *summit* U.S.A.
45 J3 **Bruint** India
112 C2 **Brukkaros** Namibia
88 B2 **Brule** U.S.A.
15 E3 **Brûly** Belgium
104 E1 **Brumado** Brazil
15 H4 **Brumath** France
9 M6 **Brumunddal** Norway
19 J4 **Brunau** Ger.
82 D3 **Bruneau** U.S.A.
82 D3 **Bruneau** *r.* U.S.A.
31 L9 **Brunei** *country* Asia
8 O5 **Brunflo** Sweden
23 J3 **Brunico** Italy
22 F2 **Brunnen** Switz.
70 C5 **Brunner, L.** N.Z.
77 H4 **Bruno** Can.
19 G3 **Brunsbüttel** Ger.
87 D6 **Brunswick** *GA* U.S.A.
91 J3 **Brunswick** *ME* U.S.A.
90 C4 **Brunswick** *OH* U.S.A.
67 K2 **Brunswick Head** Austr.
69 B6 **Brunswick Junction** Austr.
102 B8 **Brunswick, Península de** *pen.* Chile
21 L3 **Bruntál** Czech Rep.
116 C3 **Brunt Ice Shelf** *ice feature* Ant.
113 J4 **Bruntville** S. Africa
67 G9 **Bruny I.** Austr.
82 G3 **Brush** U.S.A.
15 E2 **Brussels** Belgium
89 G4 **Brussels** Can.
88 D3 **Brussels** U.S.A.
17 O5 **Brusyliv** Ukr.
67 G6 **Bruthen** Austr.
Bruxelles *see* **Brussels**
15 G4 **Bruyères** France
90 A4 **Bryan** *OH* U.S.A.
85 D6 **Bryan** *TX* U.S.A.
116 A3 **Bryan Coast** Ant.
66 C4 **Bryan, Mt** *h.* Austr.
28 E4 **Bryansk** Rus. Fed.
28 E4 **Bryanskaya Oblast'** *div.* Rus. Fed.
29 H6 **Bryanskoye** Rus. Fed.
93 F3 **Bryce Canyon Nat. Park** U.S.A.
93 H5 **Bryce Mt** U.S.A.
9 J7 **Bryne** Norway
29 F6 **Bryukhovetskaya** Rus. Fed.
16 H5 **Brzeg** Pol.
21 K1 **Brzeg Dolny** Pol.
71 □13 **Bua** Fiji
115 B5 **Bua** *r.* Malawi
63 F2 **Buala** Solomon Is
108 A3 **Buba** Guinea-Bissau
37 M7 **Būbīyān I.** Kuwait
55 B5 **Bubuan** *i.* Phil.
71 □13 **Buca** Fiji
36 C3 **Bucak** Turkey
103 B3 **Bucaramanga** Col.
55 C4 **Bucas Grande** *i.* Phil.
68 C3 **Bucasia** Austr.
67 H6 **Buchan** Austr.
114 A3 **Buchanan** Liberia
88 D5 **Buchanan** *MI* U.S.A.
90 D6 **Buchanan** *VA* U.S.A.
68 A3 **Buchanan, L.** *salt flat* Austr.
85 D6 **Buchanan, L.** U.S.A.
75 L2 **Buchan Gulf** Can.
79 J4 **Buchans** Can.
27 M2 **Bucharest** Romania
19 H3 **Büchen** Ger.
20 C3 **Buchen (Odenwald)** Ger.
19 K3 **Buchholz** Ger.
19 G3 **Bucholz in der Nordheide** Ger.
92 B4 **Buchon, Point** U.S.A.
15 B3 **Buchy** France
17 M7 **Bucin, Pasul** *pass* Romania
67 F3 **Buckambool Mt** *h.* Austr.
19 G4 **Bückeburg** Ger.
19 G4 **Bücken** Ger.
93 F5 **Buckeye** U.S.A.
90 B5 **Buckeye Lake** U.S.A.
90 C5 **Buckhannon** U.S.A.
90 C5 **Buckhannon** *r.* U.S.A.
11 E4 **Buckhaven** U.K.
89 H3 **Buckhorn** Can.
93 H5 **Buckhorn** U.S.A.
89 H3 **Buckhorn Lake** Can.
90 B6 **Buckhorn Lake** U.S.A.
11 F3 **Buckie** U.K.
89 K3 **Buckingham** Can.
13 G5 **Buckingham** U.K.
90 D6 **Buckingham** U.S.A.
65 G2 **Buckingham Bay** Austr.
68 B5 **Buckland Tableland** *reg.* Austr.
66 B4 **Buckleboo** Austr.
116 A6 **Buckle I.** Ant.
93 F4 **Buckskin Mts** U.S.A.
92 B2 **Bucks Mt** U.S.A.
91 J2 **Bucksport** U.S.A.
19 K4 **Bückwitz** Ger.
21 L3 **Bučovice** Czech Rep.
București *see* **Bucharest**
90 B4 **Bucyrus** U.S.A.
17 P4 **Buda-Kashalyova** Belarus
17 J7 **Budapest** Hungary
44 D3 **Budaun** India
66 F3 **Budda** Austr.
116 C6 **Budd Coast** Ant.
69 B4 **Budd, Mt** *h.* Austr.
11 F4 **Buddon Ness** *pt* U.K.
26 C4 **Buddusò** *Sardinia* Italy
13 C7 **Bude** U.K.
85 F6 **Bude** U.S.A.
19 G2 **Büdelsdorf** Ger.
29 H6 **Budennovsk** Rus. Fed.
67 K1 **Buderim** Austr.
20 C2 **Büdingen** Ger.
44 D5 **Budni** India
28 E3 **Budogoshch'** Rus. Fed.
45 H2 **Budongquan** China
26 C4 **Budoni** *Sardinia* Italy
114 E4 **Buea** Cameroon
22 C5 **Buëch** *r.* France
92 B4 **Buellton** U.S.A.
105 D2 **Buena Esperanza** Arg.
103 A4 **Buenaventura** Col.
94 C3 **Buenaventura** Mex.
103 A4 **Buenaventure, B. de** Col.
83 F4 **Buena Vista** *CO* U.S.A.
90 D6 **Buena Vista** *VA* U.S.A.
25 E2 **Buendia, Embalse de** *resr* Spain
105 B4 **Bueno** *r.* Chile
105 E2 **Buenos Aires** Arg.
105 E3 **Buenos Aires** *div.* Arg.
102 B7 **Buenos Aires, L.** Arg./Chile
102 C7 **Buen Pasto** Arg.
96 C2 **Búfalo** Mex.
76 G3 **Buffalo** *r.* Can.
90 D3 **Buffalo** *NY* U.S.A.
85 D4 **Buffalo** *OK* U.S.A.
84 C2 **Buffalo** *SD* U.S.A.
85 D6 **Buffalo** *TX* U.S.A.
88 B3 **Buffalo** *WV* U.S.A.
82 F2 **Buffalo** *WY* U.S.A.
88 B3 **Buffalo** *r.* U.S.A.
76 F3 **Buffalo Head Hills** Can.
76 F2 **Buffalo Lake** Can.
67 G6 **Buffalo, Mt** Austr.
77 H3 **Buffalo Narrows** Can.
112 B4 **Buffels** *watercourse* S. Africa
113 G1 **Buffels Drift** S. Africa
87 D5 **Buford** U.S.A.
27 L2 **Buftea** Romania
17 K4 **Bug** *r.* Pol.
103 A4 **Buga** Col.
103 A3 **Bugalagrande** Col.
67 H3 **Bugaldie** Austr.
114 E3 **Bugana** Nigeria
38 C5 **Bugdayli** Turkm.
59 D4 **Bugel, Tanjung** *pt* Indon.
15 E1 **Buggenhout** Belgium
26 G2 **Bugojno** Bos.-Herz.
55 A4 **Bugsuk** *i.* Phil.
55 B2 **Buguey** Phil.
38 E3 **Bugun'** Kazak.
38 C1 **Buguruslan** Rus. Fed.
40 D4 **Būhābād** Iran
37 J5 **Buḩayrat ath Tharthār** *l.* Iraq
37 K4 **Buḩayrat Shārī** *l.* Iraq
111 D5 **Buhera** Zimbabwe
55 B3 **Buhi** Phil.
20 B4 **Bühl** Ger.
82 D3 **Buhl** *ID* U.S.A.
88 A2 **Buhl** *MN* U.S.A.
20 C3 **Bühlertann** Ger.
37 J3 **Bühtan** *r.* Turkey
17 N7 **Buhuși** Romania
23 L3 **Buia** Italy
13 D5 **Builth Wells** U.K.
114 C3 **Bui National Park** Ghana
28 J4 **Buinsk** Rus. Fed.
37 L4 **Bu'in Soflā** Iran
49 L2 **Buir Nur** *l.* Mongolia
22 C5 **Buis-les-Baronnies** France
111 B6 **Buitepos** Namibia
27 J3 **Bujanovac** Yugo.
23 L4 **Buje** Croatia
115 A3 **Bujumbura** Burundi
21 K5 **Bük** Hungary
49 L1 **Bukachacha** Rus. Fed.
63 F2 **Buka I.** P.N.G.
115 A4 **Bukama** Congo(Zaire)
40 D4 **Būkand** Iran
39 K1 **Bukanskoye** Rus. Fed.
38 E4 **Bukantau, Gory** *h.* Uzbek.
115 A3 **Bukavu** Congo(Zaire)
38 F5 **Bukhara** Uzbek.
39 L2 **Bukhtarminskoye Vdkhr.** *resr* Kazak.
55 C6 **Bukide** *i.* Indon.
58 □ **Bukit Batok** Sing.
58 B5 **Bukit Fraser** Malaysia
58 □ **Bukit Panjang** Sing.
58 □ **Bukit Timah** Sing.
59 B3 **Bukittinggi** Indon.
115 B3 **Bukoba** Tanz.
115 B3 **Bukombekombe** Tanz.
58 □ **Bukum, P.** *i.* Sing.
57 J7 **Bula** Indon.
28 J4 **Bula** *r.* Rus. Fed.
22 F2 **Bülach** Switz.
67 K4 **Bulahdelal** Austr.
55 B3 **Bulan** Phil.
36 G1 **Bulancak** Turkey
44 D3 **Bulandshahr** India
37 J2 **Bulanık** Turkey
111 C6 **Bulawayo** Zimbabwe
39 G1 **Bulayevo** Kazak.
36 F3 **Bulbul** Syria
36 B2 **Buldan** Turkey
44 D5 **Buldana** India
113 J2 **Bulembu** Swaziland
48 H2 **Bulgan** *Hövsgöl* Mongolia
50 B1 **Bulgan** Mongolia
7 F4 **Bulgaria** *country* Europe
15 F4 **Bulgnéville** France
66 E1 **Bullawarra, Lake** *salt flat* Austr.
22 E3 **Bulle** Switz.
66 D3 **Bullea, Lake** *salt flat* Austr.
70 D4 **Buller** *r.* N.Z.
67 G6 **Buller, Mt** Austr.
69 D5 **Bullfinch** Austr.
93 E4 **Bullhead City** U.S.A.
92 D4 **Bullion Mts** U.S.A.
66 E2 **Bulloo** *watercourse* Austr.
66 E2 **Bulloo Downs** Austr.
66 E2 **Bulloo L.** *salt flat* Austr.
112 B2 **Büllsport** Namibia
58 □ **Buloh, P.** *i.* Sing.
66 E6 **Buloke, Lake** Austr.
113 G4 **Bultfontein** S. Africa
55 C5 **Buluan** Phil.
57 G8 **Bulukumba** Indon.
33 O2 **Bulun** Rus. Fed.
110 B4 **Bulungu** *Bandundu* Congo(Zaire)
110 C4 **Bulungu** *Kasai-Occidental* Congo(Zaire)
39 F5 **Bulungur** Uzbek.
55 C3 **Bulusan** Phil.
110 C3 **Bumba** Congo(Zaire)
50 B1 **Bumbat Sum** China
93 F4 **Bumble Bee** U.S.A.
55 A5 **Bum-Bum** *i.* Malaysia
110 B4 **Buna** Congo(Zaire)
115 C2 **Buna** Kenya
115 B3 **Bunazi** Tanz.
14 C2 **Bunbeg** Rep. of Ireland
69 B6 **Bunbury** Austr.
14 E5 **Bunclody** Rep. of Ireland
14 D2 **Buncrana** Rep. of Ireland
115 B3 **Bunda** Tanz.
68 E5 **Bundaberg** Austr.
67 G2 **Bundaleer** Austr.
67 J3 **Bundarra** Austr.
44 C4 **Bundi** India
14 C3 **Bundoran** Rep. of Ireland
45 F5 **Bundu** India
115 B1 **Bunduqiya** Sudan
13 J5 **Bungay** U.K.
58 B2 **Bung Boraphet** *l.* Thai.
67 H5 **Bungendore** Austr.
116 C6 **Bunger Hills** Ant.
68 C6 **Bungil Cr.** *r.* Austr.
Bungle Bungle National Park *see* **Purnululu National Park**
53 C8 **Bungo-suidō** *chan.* Japan
115 B2 **Bunia** Congo(Zaire)
110 C4 **Bunianga** Congo(Zaire)
66 E6 **Buninyong** Austr.
114 F2 **Buni-Yadi** Nigeria
44 C2 **Bunji** Jammu and Kashmir
68 E4 **Bunker Group** *atolls* Austr.
93 E3 **Bunkerville** U.S.A.
115 A5 **Bunkeya** Congo(Zaire)
85 E6 **Bunkie** U.S.A.
87 D6 **Bunnell** U.S.A.
69 C4 **Buntine** Austr.
36 E2 **Bünyan** Turkey
55 A6 **Bunyu** *i.* Indon.
40 C4 **Bu ol Kheyr** Iran
58 D2 **Buôn Hô** Vietnam
58 D2 **Buôn Mê Thuôt** Vietnam
33 P2 **Buorkhaya, Guba** *b.* Rus. Fed.
42 C4 **Buqayq** S. Arabia
115 C3 **Bura** Kenya
69 C5 **Burakin** Austr.
39 L2 **Buran** Kaz.
44 E3 **Burang** China
104 E2 **Buranhaém** *r.* Brazil
38 C2 **Burannoye** Rus. Fed.
23 K6 **Burano** *r.* Italy
110 E3 **Burao** Somalia
55 C4 **Burauen** Phil.
42 B4 **Buraydah** S. Arabia
18 F6 **Burbach** Ger.
92 C4 **Burbank** U.S.A.
67 G4 **Burcher** Austr.
38 F5 **Burdalyk** Turkm.
68 B3 **Burdekin** *r.* Austr.
68 B3 **Burdekin Falls** *waterfall* Austr.
36 C3 **Burdur** Turkey
110 D2 **Burē** Eth.
13 J5 **Bure** *r.* U.K.
8 R4 **Bureå** Sweden
49 O1 **Bureinskiy Khrebet** *mts* Rus. Fed.
36 D6 **Bûr Fu'ad** Egypt
27 M3 **Burgas** Bulg.
20 D4 **Burgau** Ger.
19 J2 **Burg auf Fehmarn** Ger.
87 E5 **Burgaw** U.S.A.
19 J4 **Burg bei Magdeburg** Ger.
20 D3 **Burgbernheim** Ger.
19 H4 **Burgdorf** Ger.
22 E2 **Burgdorf** Switz.
21 K5 **Burgenland** *div.* Austria
79 J4 **Burgeo** Can.
113 G5 **Burgersdorp** S. Africa
113 J2 **Burgersfort** S. Africa
69 E5 **Burges, Mt** *h.* Austr.
13 G7 **Burgess Hill** U.K.
20 C2 **Burghaun** Ger.
20 F4 **Burghausen** Ger.
11 E3 **Burghead** U.K.
18 A5 **Burgh-Haamstede** Neth.
26 F6 **Burgio, Serra di** *h. Sicily* Italy
20 F3 **Burglengenfeld** Ger.
25 E1 **Burgos** Spain
20 C2 **Burgsinn** Ger.
19 K6 **Burgstädt** Ger.
9 Q8 **Burgsvik** Sweden
Burgundy *reg. see* **Bourgogne**
46 B3 **Burhan Budai Shan** *mts* China
27 M5 **Burhaniye** Turkey
44 D5 **Burhanpur** India
45 E5 **Burhar-Dhanpuri** India
45 F4 **Burhi Gandak** *r.* India
55 B3 **Burias** *i.* Phil.
38 D2 **Buribay** Rus. Fed.
96 J6 **Burica, Pta** *pt* Costa Rica
45 H4 **Buri Dihing** *r.* India
45 E4 **Buri Gandak** *r.* Nepal
79 J4 **Burin Peninsula** Can.
58 B2 **Buriram** Thai.
101 K5 **Buriti Bravo** Brazil
104 C1 **Buritis** Brazil
41 G4 **Burj** Pak.
116 A3 **Burke I.** Ant.
70 C6 **Burke Pass** N.Z.
65 G3 **Burketown** Austr.
107 C4 **Burkina** *country* Africa
89 H3 **Burk's Falls** Can.
39 J1 **Burla** Rus. Fed.
39 J1 **Burla** *r.* Rus. Fed.
82 D3 **Burley** U.S.A.
38 E1 **Burli** Kazak.
38 C2 **Burlin** Kazak.
89 H4 **Burlington** Can.
84 C4 **Burlington** *CO* U.S.A.
88 B5 **Burlington** *IA* U.S.A.
88 D5 **Burlington** *IN* U.S.A.
91 J2 **Burlington** *ME* U.S.A.
91 G2 **Burlington** *VT* U.S.A.
88 C4 **Burlington** *WV* U.S.A.
38 D1 **Burly** Rus. Fed.
Burma *country see* **Myanmar**
85 D6 **Burnet** U.S.A.
68 E5 **Burnett** *r.* Austr.
68 E5 **Burnett Heads** Austr.
82 B3 **Burney** U.S.A.
91 J2 **Burnham** U.S.A.
67 F8 **Burnie** Austr.
12 G3 **Burniston** U.K.
12 E4 **Burnley** U.K.
39 G4 **Burnoye** Kazak.
82 C3 **Burns** U.S.A.
77 H1 **Burnside** *r.* Can.
74 F4 **Burns Lake** Can.
90 C5 **Burnsville Lake** U.S.A.
87 F7 **Burnt Ground** Bahamas
11 E4 **Burntisland** U.K.
79 H3 **Burnt Lake** Can.
77 K3 **Burntwood** *r.* Can.
77 J3 **Burnt Wood Lake** Can.
66 E5 **Buronga** Austr.
38 E4 **Burovoy** Uzbek.
32 K5 **Burqin** China
36 G5 **Burqu'** Jordan
66 C4 **Burra** Austr.
11 □ **Burravoe** U.K.
11 F2 **Burray** *i.* U.K.
27 J4 **Burrel** Albania
67 H4 **Burrendong Reservoir** Austr.
67 H3 **Burren Jct.** Austr.
67 J5 **Burrewarra Pt** Austr.
25 F3 **Burriana** Spain
67 H5 **Burrinjuck** Austr.
67 H5 **Burrinjuck Reservoir** Austr.
90 B5 **Burr Oak Reservoir** U.S.A.
96 D1 **Burro, Serranías del** *mts* Mex.
11 D6 **Burrow Head** *hd* U.K.
68 E5 **Burrum Heads** Austr.
93 G2 **Burrville** U.S.A.
36 B1 **Bursa** Turkey
109 F2 **Bûr Safâga** Egypt
Bûr Sa'îd *see* **Port Said**
20 B3 **Bürstadt** Ger.
Būr Sudan *see* **Port Sudan**
66 D4 **Burta** Austr.
88 E3 **Burt Lake** U.S.A.
89 F4 **Burton** U.S.A.
78 E3 **Burton, Lac** *l.* Can.
14 C3 **Burtonport** Rep. of Ireland
13 F5 **Burton upon Trent** U.K.
8 R4 **Burträsk** Sweden
91 K1 **Burtts Corner** Can.
66 E4 **Burtundy** Austr.
57 H7 **Buru** *i.* Indon.
39 H3 **Burubaytal** Kazak.
36 C6 **Burullus, Bahra el** *lag.* Egypt
107 F6 **Burundi** *country* Africa
115 A3 **Bururi** Burundi
76 B2 **Burwash Landing** Can.
11 F2 **Burwick** U.K.
29 E5 **Buryn'** Ukr.
38 B3 **Burynshyk** Kazak.
13 H5 **Bury St Edmunds** U.K.
44 C2 **Burzil Pass** Jammu and Kashmir
23 F5 **Busalla** Italy
110 C4 **Busanga** Congo(Zaire)
22 E5 **Busca** Italy
19 G2 **Busdorf** Ger.
14 E2 **Bush** *r.* U.K.
40 C4 **Būshehr** Iran
45 E2 **Bushêngcaka** China
115 B3 **Bushenyi** Uganda
Bushire *see* **Būshehr**
14 E2 **Bushmills** U.K.
88 B5 **Bushnell** U.S.A.
110 C3 **Businga** Congo(Zaire)
58 □ **Busing, P.** *i.* Sing.
38 E1 **Buskul'** Kazak.
36 F5 **Buşrá ash Shām** Syria
69 B6 **Busselton** Austr.
18 C4 **Bussum** Neth.
96 D2 **Bustamante** Mex.
96 C1 **Bustillos, L.** Mex.
23 F4 **Busto Arsizio** Italy
55 A3 **Busuanga** Phil.
55 A3 **Busuanga** *i.* Phil.
18 F2 **Büsum** Ger.
110 C3 **Buta** Congo(Zaire)
58 A4 **Butang Group** *is* Thai.
105 C3 **Buta Ranquil** Arg.
115 A3 **Butare** Rwanda
118 G5 **Butaritari** *i.* Pac. Oc.
66 B4 **Bute** Austr.
11 C5 **Bute** *i.* U.K.
76 D4 **Butedale** Can.
76 D4 **Bute In.** *in.* Can.
115 A2 **Butembo** Congo(Zaire)
11 C5 **Bute, Sound of** *chan.* U.K.
113 H4 **Butha Buthe** Lesotho
18 F3 **Butjadingen** *reg.* Ger.
88 E5 **Butler** *IN* U.S.A.
90 D4 **Butler** *PA* U.S.A.
14 D3 **Butlers Bridge** Rep. of Ireland
57 G7 **Buton** *i.* Indon.
19 K3 **Bütow** Ger.
82 D2 **Butte** U.S.A.
92 B1 **Butte Meadows** U.S.A.
20 E3 **Buttenheim** Ger.
59 B1 **Butterworth** Malaysia
113 H6 **Butterworth** S. Africa
14 C5 **Buttevant** Rep. of Ireland
76 D5 **Buttle L.** Can.
11 B2 **Butt of Lewis** *hd* U.K.
75 J4 **Button Bay** Can.
92 C4 **Buttonwillow** U.S.A.
19 J5 **Buttstädt** Ger.
69 E6 **Butty Hd** *pt* Austr.
55 C4 **Butuan** Phil.
51 B5 **Butuo** China
29 G5 **Buturlinovka** Rus. Fed.
45 E4 **Butwal** Nepal
20 B2 **Butzbach** Ger.
19 J3 **Bützow** Ger.
110 E3 **Buulobarde** Somalia
110 E4 **Buur Gaabo** Somalia
110 E3 **Buurhabaka** Somalia
45 F4 **Buxar** India
19 G3 **Buxtehude** Ger.
13 F4 **Buxton** U.K.
28 G3 **Buy** Rus. Fed.
88 A1 **Buyck** U.S.A.
29 H7 **Buynaksk** Rus. Fed.
114 B3 **Buyo, Lac de** *l.* Côte d'Ivoire
Büyük Ağri *mt see* **Ararat, Mt**
36 A3 **Büyükmenderes** *r.* Turkey
54 B3 **Buyun Shan** *mt* China
38 B3 **Buzachi, Poluostrov** *pen.* Kazak.
39 H5 **Buzai Gumbad** Afgh.
15 E3 **Buzancy** France
27 M2 **Buzău** Romania
111 D5 **Búzi** Moz.
38 F2 **Buzuluk** Kazak.
38 C1 **Buzuluk** Rus. Fed.
29 G5 **Buzuluk** *r.* Rus. Fed.
91 H4 **Buzzards Bay** U.S.A.
65 K2 **Bwagaoia** P.N.G.
45 G4 **Byakar** Bhutan
27 L3 **Byala** Bulg.
27 K3 **Byala Slatina** Bulg.
17 O4 **Byalynichy** Belarus
74 H2 **Byam Martin I.** Can.
28 D4 **Byarezina** *r.* Belarus
28 C4 **Byaroza** Belarus
36 E4 **Byblos** Lebanon
21 M1 **Byczyna** Pol.
16 J4 **Bydgoszcz** Pol.
28 D4 **Byerazino** Belarus
82 F4 **Byers** U.S.A.
17 O3 **Byeshankovichy** Belarus
9 K7 **Bygland** Norway
19 M5 **Byhleguhre** Ger.
28 D4 **Bykhaw** Belarus
9 K7 **Bykle** Norway
39 H2 **Bylkyldak** Kazak.
75 L2 **Bylot Island** Can.
89 G3 **Byng Inlet** Can.
116 B5 **Byrd Gl.** Ant.
9 K6 **Byrkjelo** Norway
67 G3 **Byrock** Austr.
88 C4 **Byron** *IL* U.S.A.
91 H2 **Byron** *ME* U.S.A.
67 K2 **Byron Bay** Austr.
67 K2 **Byron, C.** *hd* Austr.
33 M2 **Byrranga, Gory** *mts* Rus. Fed.
8 R4 **Byske** Sweden
21 K3 **Bystřice nad Pernštejnem** Czech Rep.
21 L3 **Bystřice pod Hostýnem** Czech Rep.
39 H4 **Bystrovka** Kyrg.
39 L1 **Bystryy Istok** Rus. Fed.
21 K2 **Bystrzyca Kłodzka** Pol.
21 K2 **Bystrzyckie, Góry** *mts* Czech Rep./Pol.
33 P3 **Bytantay** *r.* Rus. Fed.
21 M3 **Bytča** Slovakia
17 J5 **Bytom** Pol.
16 H3 **Bytów** Pol.
115 B3 **Byumba** Rwanda
38 D5 **Byuzmeyin** Turkm.

C

102 E3 **Caacupé** Para.
104 A4 **Caagazú, Cordillera de** *h.* Para.
104 A4 **Caaguazú** Para.
104 A3 **Caarapó** Brazil
104 A4 **Caazapá** Para.
100 C6 **Caballas** Peru
100 D4 **Caballococha** Peru
96 D1 **Caballos Mesteños, Llano de los** *plain* Mex.
55 B3 **Cabanatuan** Phil.
79 G4 **Cabano** Can.
110 E2 **Cabdul Qaadir** Somalia
104 A1 **Cabeceira Rio Manso** Brazil
101 M5 **Cabedelo** Brazil
25 D3 **Cabeza del Buey** Spain
100 F7 **Cabezas** Bol.
105 E3 **Cabildo** Arg.
103 C2 **Cabimas** Venez.
110 B4 **Cabinda** Angola
110 B4 **Cabinda** *div.* Angola
82 C1 **Cabinet Mts** U.S.A.
103 B3 **Cable Way** *pass* Col.
104 D3 **Cabo Frio** Brazil
104 E3 **Cabo Frio, Ilha do** *i.* Brazil
78 E4 **Cabonga, Réservoir** *resr* Can.
85 E4 **Cabool** U.S.A.
67 K1 **Caboolture** Austr.
101 H3 **Cabo Orange, Parque Nacional de** *nat. park* Brazil
100 C4 **Cabo Pantoja** Peru
94 B2 **Caborca** Mex.
89 G3 **Cabot Head** *pt* Can.
79 J4 **Cabot Strait** Can.
104 D2 **Cabral, Serra do** *mts* Brazil
37 L2 **Cäbrayıl** Azer.
25 H3 **Cabrera** *i.* Spain
25 C1 **Cabrera, Sierra de la** *mts* Spain
25 F3 **Cabriel** *r.* Spain
103 D3 **Cabruta** Venez.
55 B2 **Cabugao** Phil.
102 F3 **Caçador** Brazil
97 E4 **Cacahuatepec** Mex.
27 J3 **Čačak** Yugo.
105 G1 **Caçapava do Sul** Brazil
90 D5 **Cacapon** *r.* U.S.A.
103 B3 **Cácares** Col.
26 C4 **Caccia, Capo** *pt Sardinia* Italy
101 G7 **Cáceres** Brazil
25 C3 **Cáceres** Spain
82 D3 **Cache Peak** U.S.A.
108 A3 **Cacheu** Guinea-Bissau
102 C3 **Cachi** *r.* Arg.
101 H5 **Cachimbo, Serra do** *h.* Brazil
103 B3 **Cáchira** Col.
104 E1 **Cachoeira** Brazil
104 B2 **Cachoeira Alta** Brazil
105 G1 **Cachoeira do Sul** Brazil
104 E3 **Cachoeiro de Itapemirim** Brazil
108 A3 **Cacine** Guinea-Bissau
101 H3 **Caciporé, Cabo** *pt* Brazil
111 B5 **Cacolo** Angola
110 B4 **Cacongo** Angola
92 D3 **Cactus Range** *mts* U.S.A.
104 B2 **Caçu** Brazil
104 D1 **Caculé** Brazil
17 J6 **Čadca** Slovakia
19 G3 **Cadenberge** Ger.
97 D2 **Cadereyta** Mex.
66 A2 **Cadibarrawirracanna, L.** *salt flat* Austr.
55 B3 **Cadig Mountains** Phil.
89 H1 **Cadillac** *Que.* Can.
77 H5 **Cadillac** *Sask.* Can.
88 E3 **Cadillac** U.S.A.
55 B4 **Cadiz** Phil.
25 C4 **Cádiz** Spain
25 C4 **Cádiz, Golfo de** *g.* Spain
93 E4 **Cadiz Lake** U.S.A.
23 K3 **Cadore** *reg.* Italy
69 C5 **Cadoux** Austr.
24 D2 **Caen** France
13 C4 **Caernarfon** U.K.
13 C4 **Caernarfon Bay** U.K.
13 D6 **Caerphilly** U.K.
90 B5 **Caesar Creek Lake** U.S.A.
36 E5 **Caesarea** Israel
104 D1 **Caetité** Brazil
102 C3 **Cafayate** Arg.
55 B4 **Cagayan** *i.* Phil.
55 B2 **Cagayan** *r.* Phil.
55 C4 **Cagayan de Oro** Phil.
55 B4 **Cagayan Islands** Phil.
23 K6 **Cagli** Italy
26 C5 **Cagliari** *Sardinia* Italy
26 C5 **Cagliari, Golfo di** *b. Sardinia* Italy
22 E6 **Cagnes-sur-Mer** France
103 B4 **Caguán** *r.* Col.
14 B6 **Caha** *h.* Rep. of Ireland
87 C5 **Cahaba** *r.* U.S.A.
14 B6 **Caha Mts** *h.* Rep. of Ireland
14 A6 **Cahermore** Rep. of Ireland
14 D5 **Cahir** Rep. of Ireland
14 A6 **Cahirciveen** Rep. of Ireland
115 B6 **Cahora Bassa, Lago de** *resr* Moz.
14 E5 **Cahore Point** Rep. of Ireland
24 E4 **Cahors** France
100 C5 **Cahuapanas** Peru
29 D6 **Cahul** Moldova
115 C6 **Caia** Moz.
101 G6 **Caiabis, Serra dos** *h.* Brazil
111 C5 **Caianda** Angola
104 B2 **Caiapó** *r.* Brazil
104 B2 **Caiapônia** Brazil
104 B2 **Caiapó, Serra do** *mts* Brazil
95 J4 **Caibarién** Cuba
58 C3 **Cai Be** Vietnam
103 D3 **Caicara** Venez.
95 K4 **Caicos Is** Turks and Caicos Is
105 B1 **Caimanes** Chile
55 A3 **Caiman Point** Phil.
25 F2 **Caimodorro** *mt* Spain
58 C3 **Cai Nước** Vietnam
11 E3 **Cairn Gorm** *mt* U.K.
11 E3 **Cairngorm Mountains** U.K.
11 C6 **Cairnryan** U.K.
68 A1 **Cairns** Austr.
11 E3 **Cairn Toul** *mt* U.K.
109 F1 **Cairo** Egypt
87 C6 **Cairo** U.S.A.
22 F5 **Cairo Montenotte** Italy
111 B5 **Caiundo** Angola
67 F2 **Caiwarro** Austr.
100 C5 **Cajamarca** Peru
55 B3 **Cajidiocan** Phil.
26 G1 **Čakovec** Croatia
36 B2 **Çal** Turkey
113 G5 **Cala** S. Africa
114 E4 **Calabar** Nigeria
89 J3 **Calabogie** Can.
103 D2 **Calabozo** Venez.
27 K3 **Calafat** Romania
102 B8 **Calafate** Arg.
55 B3 **Calagua Islands** Phil.
25 F1 **Calahorra** Spain
111 B5 **Calai** Angola
15 B2 **Calais** France
91 K2 **Calais** U.S.A.
100 F5 **Calama** Brazil
102 C2 **Calama** Chile
103 B2 **Calamar** *Bolívar* Col.
103 B4 **Calamar** *Guaviare* Col.
55 A4 **Calamian Group** *is* Phil.
25 F2 **Calamocha** Spain
23 G3 **Calancasca** *r.* Switz.
111 B4 **Calandula** Angola
109 E2 **Calanscio Sand Sea** *des.* Libya
55 B3 **Calapan** Phil.
27 M2 **Călărași** Romania
25 F2 **Calatayud** Spain
19 L5 **Calau** Ger.
55 B3 **Calauag** Phil.
55 B3 **Calavite, Cape** *pt* Phil.
55 A3 **Calawit** *i.* Phil.
55 B2 **Calayan** *i.* Phil.
55 C3 **Calbayog** Phil.
19 J5 **Calbe (Saale)** Ger.
55 C4 **Calbiga** Phil.
105 B4 **Calbuco** Chile
101 L5 **Calcanhar, Ponta do** *pt* Brazil
85 E6 **Calcasieu L.** U.S.A.
101 H3 **Calçoene** Brazil
45 G5 **Calcutta** India
25 B3 **Caldas da Rainha** Port.
104 C2 **Caldas Novas** Brazil
19 G5 **Calden** Ger.
102 B3 **Caldera** Chile
68 B5 **Caldervale** Austr.
37 J2 **Çaldıran** Turkey
82 C3 **Caldwell** U.S.A.
90 D3 **Caledon** Can.
113 G5 **Caledon** *r.* Lesotho/S. Africa
112 C7 **Caledon** S. Africa
89 H4 **Caledonia** Can.
88 B4 **Caledonia** U.S.A.
68 C3 **Calen** Austr.
96 D3 **Calera** Mex.
102 C7 **Caleta Olivia** Arg.
93 E5 **Calexico** U.S.A.
12 C3 **Calf of Man** *i.* U.K.
76 G4 **Calgary** Can.
87 C5 **Calhoun** U.S.A.
103 A4 **Cali** Col.
55 C4 **Calicoan** *i.* Phil.
43 A4 **Calicut** India
92 C4 **Caliente** *CA* U.S.A.
93 E3 **Caliente** *NV* U.S.A.
92 B3 **California** *div.* U.S.A.
92 B3 **California Aqueduct** *canal* U.S.A.
96 B2 **California, Golfo de** *g.* Mex.
California, Gulf of *g. see* **California, Golfo de**
92 C4 **California Hot Springs** U.S.A.
37 M2 **Cälilabad** Azer.
83 D5 **Calipatria** U.S.A.
92 A2 **Calistoga** U.S.A.
112 D6 **Calitzdorp** S. Africa
97 G3 **Calkiní** Mex.
66 D2 **Callabonna, L.** *salt flat* Austr.
92 D2 **Callaghan, Mt** U.S.A.
87 D6 **Callahan** U.S.A.
14 D5 **Callan** Rep. of Ireland
89 H2 **Callander** Can.
11 D4 **Callander** U.K.
100 C6 **Callao** Peru
93 F2 **Callao** U.S.A.
97 E3 **Calles** Mex.
91 F4 **Callicoon** U.S.A.
68 D5 **Callide** Austr.
13 C7 **Callington** U.K.
68 D5 **Calliope** Austr.
89 G2 **Callum** Can.
69 B2 **Callyharra Springs** Austr.
76 G4 **Calmar** Can.
88 B4 **Calmar** U.S.A.
93 E4 **Cal-Nev-Ari** U.S.A.
23 G4 **Calolziocorte** Italy
87 D7 **Caloosahatchee** *r.* U.S.A.
97 G3 **Calotmul** Mex.
67 K1 **Caloundra** Austr.
92 B2 **Calpine** U.S.A.
97 E4 **Calpulálpan** Mex.
26 F6 **Caltanissetta** *Sicily* Italy
22 B4 **Caluire-et-Cuire** France
88 C2 **Calumet** U.S.A.
111 B5 **Calunga** Angola
111 B5 **Caluquembe** Angola
55 B4 **Calusa** *i.* Phil.
110 F2 **Caluula** Somalia
93 G5 **Calva** U.S.A.
65 G3 **Calvert Hills** Austr.
76 D4 **Calvert I.** Can.
26 C3 **Calvi** *Corsica* France
25 H3 **Calvià** Spain
96 D3 **Calvillo** Mex.
112 C5 **Calvinia** S. Africa
26 F4 **Calvo, Monte** *mt* Italy
20 B4 **Calw** Ger.
13 H5 **Cam** *r.* U.K.
104 E1 **Camaçari** Brazil
92 B2 **Camache Reservoir** U.S.A.
96 D2 **Camacho** Mex.
111 B5 **Camacuio** Angola
111 B5 **Camacupa** Angola
103 D2 **Camaguán** Venez.
95 J4 **Camagüey** Cuba
95 J4 **Camagüey, Arch. de** *is* Cuba
59 B1 **Camah, Gunung** *mt* Malaysia
Çamalan *see* **Gülek**
96 A1 **Camalli, Sa de** *mt* Mex.
100 D7 **Camana** Peru
111 C5 **Camanongue** Angola
104 B2 **Camapuã** Brazil
105 G1 **Camaquã** Brazil
105 G1 **Camaquã** *r.* Brazil
36 E3 **Çamardı** Turkey
97 E2 **Camargo** Mex.
22 B6 **Camargue** *reg.* France
96 C3 **Camaronero, L. del** *lag.* Mex.
102 C6 **Camarones** Arg.
102 C6 **Camarones, Bahía** *b.* Arg.
82 B2 **Camas** U.S.A.
58 C3 **Ca Mau** Vietnam
Cambay *see* **Khambhat**
Cambay, Gulf of *g. see* **Khambhat, Gulf of**
13 G6 **Camberley** U.K.
31 K8 **Cambodia** *country* Asia
13 B7 **Camborne** U.K.
15 D2 **Cambrai** France
92 B4 **Cambria** U.S.A.
13 D5 **Cambrian Mountains** *reg.* U.K.
89 G4 **Cambridge** Can.
70 E2 **Cambridge** N.Z.
13 H5 **Cambridge** U.K.
88 B5 **Cambridge** *IL* U.S.A.
91 H3 **Cambridge** *MA* U.S.A.
91 E5 **Cambridge** *MD* U.S.A.
84 E2 **Cambridge** *MN* U.S.A.
91 G3 **Cambridge** *NY* U.S.A.
90 C4 **Cambridge** *OH* U.S.A.
79 G2 **Cambrien, Lac** *l.* Can.
19 J5 **Camburg** Ger.
67 J5 **Camden** Austr.
87 C5 **Camden** *AL* U.S.A.
85 E5 **Camden** *AR* U.S.A.
91 J2 **Camden** *ME* U.S.A.
91 F5 **Camden** *NJ* U.S.A.
91 F3 **Camden** *NY* U.S.A.
87 D5 **Camden** *SC* U.S.A.
102 B8 **Camden, Isla** *i.* Chile
111 C5 **Cameia, Parque Nacional da** *nat. park* Angola
93 G4 **Cameron** *AZ* U.S.A.
85 E6 **Cameron** *LA* U.S.A.
84 E4 **Cameron** *MO* U.S.A.
85 D6 **Cameron** *TX* U.S.A.
88 B3 **Cameron** *WV* U.S.A.
58 B4 **Cameron Highlands** Malaysia
76 F3 **Cameron Hills** Can.
92 B2 **Cameron Park** U.S.A.
107 E5 **Cameroon** *country* Africa
114 E4 **Cameroun, Mont** *mt* Cameroon
101 J4 **Cametá** Brazil
55 C4 **Camiguin** *i.* Phil.
55 B2 **Camiguin** *i.* Phil.
55 B3 **Camiling** Phil.
87 C6 **Camilla** U.S.A.
100 F8 **Camiri** Bol.
96 B2 **Camoa** Mex.
101 K4 **Camocim** Brazil
65 G3 **Camooweal** Austr.
55 C4 **Camotes Sea** *g.* Phil.
105 E2 **Campana** Arg.
103 B4 **Campana, Co** *h.* Col.
102 A7 **Campana, I.** Chile
105 B2 **Campanario** *mt* Arg./Chile
76 D4 **Campania I.** Can.
68 B3 **Campaspe** *r.* Austr.
112 E4 **Campbell** S. Africa
70 E4 **Campbell, Cape** N.Z.
60 H9 **Campbell Island** N.Z.
76 D4 **Campbell River** Can.
89 J3 **Campbells Bay** Can.
86 C4 **Campbellsville** U.S.A.
79 G4 **Campbellton** Can.
67 G8 **Campbell Town** Austr.
11 C5 **Campbeltown** U.K.
97 G4 **Campeche** Mex.
97 G4 **Campeche** *div.* Mex.
97 F4 **Campeche, Bahía de** *g.* Mex.
66 E7 **Camperdown** Austr.
23 J6 **Campi Bisenzio** Italy
27 L2 **Câmpina** Romania
101 L5 **Campina Grande** Brazil
104 C3 **Campinas** Brazil
104 C2 **Campina Verde** Brazil
114 E4 **Campo** Cameroon
103 B4 **Campoalegre** Col.
26 F4 **Campobasso** Italy
104 D3 **Campo Belo** Brazil
101 H6 **Campo de Diauarum** Brazil
23 G3 **Campodolcino** Italy
104 C2 **Campo Florido** Brazil
102 D3 **Campo Gallo** Arg.
104 A3 **Campo Grande** Brazil
101 K4 **Campo Maior** Brazil
25 C3 **Campo Maior** Port.
104 B4 **Campo Mourão** Brazil
104 E3 **Campos** Brazil
104 C2 **Campos Altos** Brazil
104 D3 **Campos do Jordão** Brazil
104 B4 **Campos Eré** *reg.* Brazil
11 D4 **Campsie Fells** *h.* U.K.
90 B6 **Campton** *KY* U.S.A.
91 H3 **Campton** *NH* U.S.A.
27 L2 **Câmpulung** Romania
17 M7 **Câmpulung Moldovenesc** Romania
93 G4 **Camp Verde** U.S.A.
58 D3 **Cam Ranh** Vietnam
76 G4 **Camrose** Can.
13 B6 **Camrose** U.K.
77 G2 **Camsell Lake** Can.
77 H3 **Camsell Portage** Can.
29 C7 **Çan** Turkey
91 G3 **Canaan** U.S.A.
73 E4 **Canada** N. America
105 E2 **Cañada de Gómez** Arg.
91 H2 **Canada Falls Lake** U.S.A.
85 C5 **Canadian** *r.* U.S.A.
103 E3 **Canaima, Parque Nacional** *nat. park* Venez.
91 F3 **Canajoharie** U.S.A.
29 C7 **Çanakkale** Turkey
Çanakkale Boğazı *str. see* **Dardanelles**
71 □9 **Canala** *New Caledonia* Pac. Oc.
105 C2 **Canalejas** Arg.
90 E3 **Canandaigua** U.S.A.
90 E3 **Canandaigua Lake** U.S.A.
94 B2 **Cananea** Mex.
79 H2 **Cananée, Lac** *l.* Can.
104 C4 **Cananéia** Brazil
103 C4 **Canapiare, Co** *h.* Col.
100 C4 **Cañar** Ecuador
Canarias, Islas *is see* **Canary Islands**
120 G4 **Canary Basin** *sea feature* Atl. Ocean
106 A3 **Canary Islands** Atl. Ocean
91 F3 **Canastota** U.S.A.
104 C2 **Canastra, Serra da** *mts* Brazil
96 C2 **Canatlán** Mex.
87 D6 **Canaveral, Cape** U.S.A.
25 E2 **Cañaveras** Spain
104 E1 **Canavieiras** Brazil
23 J3 **Canazei** Italy
67 G3 **Canbelego** Austr.
67 H5 **Canberra** Austr.
82 B3 **Canby** *CA* U.S.A.
84 D2 **Canby** *MN* U.S.A.
15 B2 **Canche** *r.* France
97 H3 **Cancún** Mex.
83 F6 **Candelaria** *Chihuahua* Mex.
97 G4 **Candelaria** Mex.
25 D2 **Candeleda** Spain
67 H6 **Candelo** Austr.
101 J4 **Cândido Mendes** Brazil
36 D1 **Çandır** Turkey
77 H4 **Candle Lake** Can.
116 C1 **Candlemas I.** Atl. Ocean
91 G4 **Candlewood, Lake** U.S.A.
84 D1 **Cando** U.S.A.
55 B2 **Candon** Phil.
105 B1 **Canela Baja** Chile
22 F5 **Canelli** Italy
105 F2 **Canelones** Uru.
105 B3 **Cañete** Chile
25 F2 **Cañete** Spain
100 D6 **Cangallo** Peru
111 B5 **Cangamba** Angola
25 C1 **Cangas del Narcea** Spain
112 E6 **Cango Caves** *caves* S. Africa
101 L5 **Canguaretama** Brazil
105 G1 **Canguçu** Brazil
105 G1 **Canguçu, Sa do** *h.* Brazil
51 D6 **Cangwu** China
50 E2 **Cangzhou** China
79 G3 **Caniapiscau** Can.
79 G2 **Caniapiscau** *r.* Can.
79 G3 **Caniapiscau, Lac** *l.* Can.
26 E6 **Canicattì** *Sicily* Italy
76 E4 **Canim Lake** Can.
76 E4 **Canim Lake** Can.
101 L4 **Canindé** Brazil
101 K5 **Canindé** *r.* Brazil
11 C2 **Canisp** *h.* U.K.
90 E3 **Canisteo** U.S.A.
90 E3 **Canisteo** *r.* U.S.A.
96 D3 **Cañitas de Felipe Pescador** Mex.
36 D1 **Çankırı** Turkey
55 B4 **Canlaon** Phil.
76 F4 **Canmore** Can.
69 B4 **Canna** Austr.
11 B3 **Canna** *i.* U.K.
43 A4 **Cannanore** India
43 A4 **Cannanore Islands** India
22 E6 **Cannes** France
23 H6 **Canneto** Italy
69 C4 **Canning Hill** Austr.
22 F3 **Cannobio** Italy
13 E5 **Cannock** U.K.
68 C3 **Cannonvale** Austr.
67 H6 **Cann River** Austr.
103 E2 **Caño Araguao** *r.* Venez.
102 F3 **Canõas** Brazil
77 H3 **Canoe L.** Can.
104 B4 **Canoinhas** Brazil
103 E2 **Caño Macareo** *r.* Venez.

103 E2 Caño Manamo r. Venez.
103 E2 Caño Mariusa r. Venez.
83 F4 Canon City U.S.A.
68 D4 Canoona Austr.
66 D4 Canopus Austr.
77 J4 Canora Can.
67 H4 Canowindra Austr.
79 H4 Canso, C. hd Can.
25 D1 Cantábrica, Cordillera mts Spain
Cantábrico, Mar sea see Biscay, Bay of
105 C2 Cantantal Arg.
103 D2 Cantaura Venez.
91 K2 Canterbury Can.
13 J6 Canterbury U.K.
70 C6 Canterbury Bight b. N.Z.
70 C5 Canterbury Plains N.Z.
58 C3 Cân Thơ Vietnam
55 C4 Cantilan Phil.
101 K5 Canto do Buriti Brazil
Canton see Guangzhou
88 B5 Canton IL U.S.A.
91 H2 Canton ME U.S.A.
88 B5 Canton MO U.S.A.
85 F5 Canton MS U.S.A.
91 F2 Canton NY U.S.A.
90 C4 Canton OH U.S.A.
90 E4 Canton PA U.S.A.
104 B4 Cantu r. Brazil
23 G4 Cantù Italy
104 B4 Cantu, Serra do h. Brazil
105 E2 Cañuelas Arg.
101 G4 Canumã Brazil
67 K2 Canungra Austr.
100 F5 Canutama Brazil
70 D4 Canvastown N.Z.
13 H6 Canvey Island U.K.
15 A3 Cany-Barville France
85 C5 Canyon U.S.A.
82 C2 Canyon City U.S.A.
93 H3 Canyon de Chelly National Monument res. U.S.A.
82 D2 Canyon Ferry L. U.S.A.
93 H2 Canyonlands National Park U.S.A.
76 D2 Canyon Ranges mts Can.
82 B3 Canyonville U.S.A.
54 C3 Cao r. China
51 C6 Cao Bằng Vietnam
58 D2 Cao Nguyên Đắc Lắc plat. Vietnam
23 K4 Caorle Italy
54 C2 Caoshi China
50 E3 Cao Xian China
55 B5 Cap i. Phil.
103 D3 Capanaparo r. Venez.
101 J4 Capanema Brazil
104 B4 Capanema r. Brazil
104 C4 Capão Bonito Brazil
103 C3 Caparo r. Venez.
103 C4 Caparro, Co h. Brazil
55 B3 Capas Phil.
79 H4 Cap-aux-Meules Can.
79 F4 Cap-de-la-Madeleine Can.
68 B3 Cape r. Austr.
64 D6 Cape Arid National Park Austr.
67 H8 Cape Barren Island Austr.
120 K8 Cape Basin sea feature Atl. Ocean
79 H4 Cape Breton Highlands Nat. Park Can.
79 H4 Cape Breton Island Can.
79 J3 Cape Charles Can.
91 E6 Cape Charles U.S.A.
114 C3 Cape Coast Ghana
91 H4 Cape Cod Bay U.S.A.
91 J4 Cape Cod National Seashore res. U.S.A.
87 D7 Cape Coral U.S.A.
89 G3 Cape Croker Can.
75 L3 Cape Dorset Can.
87 E5 Cape Fear r. U.S.A.
85 F4 Cape Girardeau U.S.A.
118 D5 Cape Johnson Depth depth Pac. Oc.
69 B6 Capel Austr.
104 D2 Capelinha Brazil
68 C4 Capella Austr.
18 B5 Capelle aan de IJssel Neth.
91 F5 Cape May U.S.A.
91 F5 Cape May Court House U.S.A.
91 F5 Cape May Pt U.S.A.
111 B4 Capenda-Camulemba Angola
64 B4 Cape Range National Park Austr.
75 M5 Cape Sable c. Can.
79 J4 Cape St George Can.
79 H4 Cape Tormentine Can.
112 C6 Cape Town S. Africa
5 Cape Verde country Africa
120 G5 Cape Verde Basin sea feature Atl. Ocean
120 F5 Cape Verde Fracture sea feature Atl. Ocean
120 H4 Cape Verde Plateau sea feature Atl. Ocean
91 E2 Cape Vincent U.S.A.
65 H2 Cape York Peninsula Austr.
95 K5 Cap-Haïtien Haiti
101 J4 Capim r. Brazil
116 B2 Capitán Arturo Prat Chile Base Ant.
104 A3 Capitán Bado Para.
83 F5 Capitan Peak U.S.A.
93 G2 Capitol Reef National Park U.S.A.
26 G3 Čapljina Bos.-Herz.
26 F5 Capo d'Orlando Sicily Italy
14 D5 Cappoquin Rep. of Ireland
26 C3 Capraia, Isola di i. Italy
68 D4 Capricorn, C. pt Austr.
68 D4 Capricorn Channel Austr.
68 E4 Capricorn Group atolls Austr.
26 F4 Capri, Isola di i. Italy
111 C5 Caprivi Strip reg. Namibia
Cap St Jacques see Vung Tau
92 □2 Captain Cook U.S.A.
67 H5 Captain's Flat Austr.
90 C5 Captina r. U.S.A.
55 C3 Capul i. Phil.
100 C3 Caquetá r. Col.
103 B3 Cáqueza Col.
55 B3 Carabao i. Phil.
27 L2 Caracal Romania
103 E4 Caracarai Brazil
103 D2 Caracas Venez.
101 K5 Caracol Brazil
96 D4 Carácuaro Mex.
55 C5 Caraga Phil.
105 F2 Caraguatá r. Uru.
104 D3 Caraguatatuba Brazil
105 B3 Carahue Chile
104 E2 Caraí Brazil
104 D3 Carandaí Brazil
104 D3 Carangola Brazil
27 K2 Caransebeş Romania
66 B4 Carappee Hill Austr.
79 H4 Caraquet Can.
103 B3 Carare r. Col.
96 J5 Caratasca Honduras
96 J5 Caratasca, Laguna lag. Honduras
104 D2 Caratinga Brazil
100 E4 Carauari Brazil
Carauná mt see Grande, Serra
25 F3 Caravaca de la Cruz Spain
23 G4 Caravaggio Italy
104 E2 Caravelas Brazil
102 F3 Carazinho Brazil
84 D1 Carberry Can.
26 C5 Carbonara, Capo pt Sardinia Italy
86 B4 Carbondale IL U.S.A.
91 F4 Carbondale PA U.S.A.
79 K4 Carbonear Can.
26 C5 Carbonia Sardinia Italy
104 D2 Carbonita Brazil
25 F3 Carcaixent Spain
55 B4 Carcar Phil.
105 E2 Carcaraña r. Arg.
22 F5 Carcare Italy
24 F5 Carcassonne France
22 D6 Carcès France
76 C2 Carcross Can.
69 A1 Cardabia Austr.
43 B4 Cardamon Hills India
97 E3 Cárdenas San Luis Potosi Mex.
97 F4 Cárdenas Tabasco Mex.
66 E2 Cardenyabba watercourse Austr.
102 B7 Cardiel, L. Arg.
13 D6 Cardiff U.K.
13 C5 Cardigan U.K.
13 C5 Cardigan Bay U.K.
91 F2 Cardinal Can.
90 B4 Cardington U.S.A.
105 F2 Cardona Uru.
104 C4 Cardoso, Ilha do i. Brazil
70 B6 Cardrona N.Z.
76 G5 Cardston Can.
68 B2 Cardwell Austr.
17 L7 Carei Romania
24 D2 Carentan France
90 B4 Carey U.S.A.
69 B2 Carey Downs Austr.
69 E4 Carey, L. salt flat Austr.
77 J2 Carey Lake Can.
117 J5 Cargados Carajos is Mauritius
24 C2 Carhaix-Plouguer France
105 D3 Carhué Arg.
104 E3 Cariacica Brazil
103 E2 Cariaco Venez.
72 J8 Caribbean Sea Atl. Ocean
76 E4 Cariboo Mts Can.
77 K3 Caribou r. Man. Can.
76 D2 Caribou r. N.W.T. Can.
91 K1 Caribou U.S.A.
88 E2 Caribou I. Can.
75 K4 Caribou Lake Can.
76 F3 Caribou Mountains Can.
96 C2 Carichic Mex.
55 C4 Carigara Phil.
15 F3 Carignan France
67 G3 Carinda Austr.
25 F2 Cariñena Spain
104 D1 Carinhanha Brazil
104 D1 Carinhanha r. Brazil
103 E2 Caripe Venez.
103 E2 Caripito Venez.
14 D3 Cark Mountain h. Rep. of Ireland
89 J3 Carleton Place Can.
113 G3 Carletonville S. Africa
82 C3 Carlin U.S.A.
14 E3 Carlingford Lough in. Rep. of Ireland/U.K.
12 E3 Carlisle U.K.
90 A5 Carlisle KY U.S.A.
90 E4 Carlisle PA U.S.A.
24 E5 Carlit, Pic mt France
105 E2 Carlos Casares Arg.
104 E2 Carlos Chagas Brazil
14 E5 Carlow Rep. of Ireland
11 B2 Carloway U.K.
92 D5 Carlsbad CA U.S.A.
83 F5 Carlsbad NM U.S.A.
85 C6 Carlsbad TX U.S.A.
83 F5 Carlsbad Caverns Nat. Park U.S.A.
117 J3 Carlsberg Ridge sea feature Ind. Ocean
116 B3 Carlson In. in. Ant.
11 E5 Carluke U.K.
77 J5 Carlyle Can.
76 B2 Carmacks Can.
22 E5 Carmagnola Italy
77 K5 Carman Can.
13 C6 Carmarthen U.K.
13 C6 Carmarthen Bay U.K.
24 F4 Carmaux France
91 J2 Carmel U.S.A.
13 C4 Carmel Head hd U.K.
97 G4 Carmelita Guatemala
105 E2 Carmelo Uru.
103 B2 Carmen Col.
96 B1 Carmen i. Mex.
55 C4 Carmen Phil.
93 G6 Carmen U.S.A.
105 D4 Carmen de Patagones Arg.
97 G4 Carmen, Isla del i. Mex.
105 C2 Carmensa Arg.
86 B4 Carmi U.S.A.
92 B2 Carmichael U.S.A.
68 C3 Carmila Austr.
69 D6 Carmody, Lake salt flat Austr.
25 D4 Carmona Spain
24 C3 Carnac France
69 B4 Carnamah Austr.
69 A2 Carnarvon Austr.
112 E5 Carnarvon S. Africa
68 C5 Carnarvon Nat. Park Austr.
68 C5 Carnarvon Ra. mts Austr.
69 E2 Carnarvon Range h. Austr.
14 D2 Carndonagh Rep. of Ireland
13 D4 Carnedd Llywelyn mt U.K.
64 D5 Carnegie, L. salt flat Austr.
119 O6 Carnegie Ridge sea feature Pac. Oc.
11 C3 Carn Eighe mt U.K.
88 D3 Carney U.S.A.
116 A3 Carney I. Ant.
12 E3 Carnforth U.K.
56 A5 Car Nicobar i. Andaman and Nicobar Is
14 F3 Carnlough U.K.
11 E4 Carn nan Gabhar mt U.K.
110 B3 Carnot C.A.R.
66 A5 Carnot, C. hd Austr.
11 F4 Carnoustie U.K.
22 C6 Carnoux-en-Provence France
14 E5 Carnsore Point Rep. of Ireland
11 E5 Carnwath U.K.
77 H4 Carnwood Can.
89 F4 Caro U.S.A.
68 E2 Carola Cay rf Coral Sea Is Terr.
87 D7 Carol City U.S.A.
101 J5 Carolina Brazil
113 J3 Carolina S. Africa
60 M4 Caroline Island Kiribati
60 E3 Caroline Islands Pac. Oc.
70 A6 Caroline Pk N.Z.
112 B4 Carolusberg S. Africa
23 G3 Carona Italy
103 E2 Caroní r. Venez.
103 C2 Carora Venez.
93 E3 Carp U.S.A.
23 G5 Carpaneto Piacentino Italy
6 F4 Carpathian Mountains Romania/Ukr.
Carpații Meridionali mts see Transylvanian Alps
68 A2 Carpentaria Downs Austr.
65 G2 Carpentaria, Gulf of g. Austr.
22 C5 Carpentras France
23 H5 Carpi Italy
101 L5 Carpina Brazil
92 C4 Carpinteria U.S.A.
71 J5 Carpio U.S.A.
76 E4 Carp Lake Prov. Park Can.
91 H2 Carrabassett Valley U.S.A.
87 C6 Carrabelle U.S.A.
103 B2 Carraipía Col.
14 B4 Carra, Lough l. Rep. of Ireland
14 B6 Carran h. Rep. of Ireland
14 B6 Carrantuohill mt Rep. of Ireland
105 B2 Carranza, C. pt Chile
96 D2 Carranza, Presa Venustiano resr Mex.
103 E3 Carrao r. Venez.
23 H5 Carrara Italy
67 F5 Carrathool Austr.
105 C3 Carrero, Co mt Arg.
103 E1 Carriacou i. Grenada
11 D5 Carrick reg. U.K.
14 F3 Carrickfergus U.K.
14 E4 Carrickmacross Rep. of Ireland
14 C4 Carrick-on-Shannon Rep. of Ireland
14 D5 Carrick-on-Suir Rep. of Ireland
66 C4 Carrieton Austr.
14 D4 Carrigallen Rep. of Ireland
14 C6 Carrigtwohill Rep. of Ireland
105 C4 Carri Lafquén, L. Arg.
96 D2 Carrillo Mex.
84 D2 Carrington U.S.A.
102 B3 Carrizal Bajo Chile
93 H4 Carrizo AZ U.S.A.
93 G4 Carrizo AZ U.S.A.
92 D5 Carrizo Cr. r. U.S.A.
85 D6 Carrizo Springs U.S.A.
83 F5 Carrizozo U.S.A.
84 E3 Carroll U.S.A.
87 C5 Carrollton GA U.S.A.
86 C4 Carrollton KY U.S.A.
84 E4 Carrollton MO U.S.A.
90 C4 Carrollton OH U.S.A.
77 J4 Carrot r. Can.
77 J4 Carrot River Can.
12 B3 Carrowdore U.K.
14 B3 Carrowmore Lake Rep. of Ireland
91 F2 Carry Falls Reservoir U.S.A.
36 F1 Çarşamba Turkey
88 E4 Carson City MI U.S.A.
92 C2 Carson City NV U.S.A.
92 C2 Carson Lake U.S.A.
92 C2 Carson Sink l. U.S.A.
89 F4 Carsonville U.S.A.
105 B2 Cartagena Chile
103 B2 Cartagena Col.
25 F4 Cartagena Spain
103 B3 Cartago Col.
96 J6 Cartago Costa Rica
22 D6 Cartaya, Cap c. France
87 C5 Cartersville U.S.A.
88 B5 Carthage IL U.S.A.
85 E4 Carthage MO U.S.A.
91 F2 Carthage NY U.S.A.
85 E5 Carthage TX U.S.A.
89 G2 Cartier Can.
64 D2 Cartier Island Ashmore and Cartier Is
12 E3 Cartmel U.K.
79 J3 Cartwright Can.
101 L5 Caruarú Brazil
103 E2 Carúpano Venez.
92 D2 Carvers U.S.A.
15 C2 Carvin France
68 B5 Carwell Austr.
87 E5 Cary U.S.A.
66 E2 Caryapundy Swamp swamp Austr.
108 B1 Casablanca Morocco
104 C3 Casa Branca Brazil
83 E6 Casa de Janos Mex.
93 G5 Casa Grande U.S.A.
93 G5 Casa Grande National Monument res. U.S.A.
22 F4 Casale Monferrato Italy
23 H5 Casalmaggiore Italy
23 G4 Casalpusterlengo Italy
103 C3 Casanare r. Col.
27 H4 Casarano Italy
23 G4 Casatenovo Italy
70 B6 Cascade r. N.Z.
88 B4 Cascade IA U.S.A.
82 C2 Cascade ID U.S.A.
82 E2 Cascade MT U.S.A.
70 B6 Cascade Pt N.Z.
82 B3 Cascade Range mts U.S.A.
82 D2 Cascade Resr U.S.A.
25 B3 Cascais Port.
96 H6 Cascal, Paso del summit Nic.
104 B4 Cascavel Brazil
91 J3 Casco Bay U.S.A.
26 F4 Caserta Italy
89 F4 Caseville U.S.A.
116 C6 Casey Austr. Base Ant.
116 D4 Casey Bay Ant.
14 D5 Cashel Rep. of Ireland
67 H1 Cashmere Austr.
88 B4 Cashton U.S.A.
103 C2 Casigua Falcón Venez.
103 B2 Casigua Zulia Venez.
55 B2 Casiguran Phil.
105 E2 Casilda Arg.
67 K2 Casino Austr.
103 D4 Casiquiare, Canal r. Venez.
21 J3 Čáslav Czech Rep.
100 C5 Casma Peru
88 E4 Casnovia U.S.A.
92 A2 Caspar U.S.A.
25 F2 Caspe Spain
82 F3 Casper U.S.A.
Caspian Lowland see Prikaspiyskaya Nizmennost'
30 B5 Caspian Sea Asia/Europe
90 D5 Cass U.S.A.
89 F4 Cass r. U.S.A.
115 B5 Cassacatiza Moz.
90 D3 Cassadaga U.S.A.
111 C5 Cassai Angola
89 F4 Cass City U.S.A.
15 C2 Cassel France
91 F2 Casselman Can.
76 D3 Cassiar Can.
76 C3 Cassiar Mountains Can.
67 H4 Cassilis Austr.
26 E4 Cassino Italy
22 C6 Cassis France
84 E2 Cass Lake U.S.A.
101 J4 Castanhal Brazil
105 C1 Castaño r. Arg.
96 D2 Castaños Mex.
105 C1 Castaño Viejo Arg.
23 H6 Castelfiorentino Italy
23 J5 Castelfranco Emilia Italy
23 J4 Castelfranco Veneto Italy
24 E4 Casteljaloux France
26 F4 Castellammare di Stabia Italy
22 E4 Castellamonte Italy
22 D6 Castellane France
105 F3 Castelli Arg.
25 F3 Castelló de la Plana Spain
23 H5 Castelnovo ne'Monti Italy
23 H5 Castelnuovo di Garfagnana Italy
25 C3 Castelo Branco Port.
25 C3 Castelo de Vide Port.
23 J3 Castelrotto Italy
23 G4 Castèl San Giovanni Italy
23 J5 Castèl San Pietro Terme Italy
26 C4 Castelsardo Sardinia Italy
26 E6 Casteltermini Sicily Italy
26 E6 Castelvetrano Sicily Italy
66 D6 Casterton Austr.
23 H6 Castiglioncello Italy
23 J5 Castiglione dei Pepoli Italy
23 H4 Castiglione della Stiviere Italy
23 J6 Castiglion Fiorentino Italy
79 G2 Castignon, Lac l. Can.
25 E3 Castilla - La Mancha div. Spain
25 D2 Castilla y León div. Spain
103 C2 Castilletes Col.
105 G2 Castillos Uru.
14 B4 Castlebar Rep. of Ireland
11 A4 Castlebay U.K.
14 E4 Castlebellingham Rep. of Ireland
14 E3 Castleblayney Rep. of Ireland
14 E5 Castlebridge Rep. of Ireland
12 E3 Castle Carrock U.K.
13 E6 Castle Cary U.K.
93 G2 Castle Dale U.S.A.
14 D3 Castlederg U.K.
14 E5 Castledermot Rep. of Ireland
93 E5 Castle Dome Mts U.S.A.
13 F5 Castle Donnington U.K.
11 E6 Castle Douglas U.K.
12 F4 Castleford U.K.
76 F5 Castlegar Can.
14 A5 Castlegregory Rep. of Ireland
14 B5 Castleisland Rep. of Ireland
66 F6 Castlemaine Austr.
14 B5 Castlemaine Rep. of Ireland
14 C6 Castlemartyr Rep. of Ireland
92 B4 Castle Mt U.S.A.
51 □ Castle Peak h. H.K. China
51 □ Castle Peak Bay H.K. China
70 F4 Castlepoint N.Z.
14 D4 Castlepollard Rep. of Ireland
14 C4 Castlerea Rep. of Ireland
67 H3 Castlereagh r. Austr.
83 F4 Castle Rock U.S.A.
88 B4 Castle Rock Lake U.S.A.
12 C3 Castletown Isle of Man
14 D5 Castletown Rep. of Ireland
77 G4 Castor Can.
24 F5 Castres France
18 B4 Castricum Neth.
95 M6 Castries St Lucia
104 C4 Castro Brazil
102 B6 Castro Chile
23 J5 Castrocaro Terme Italy
25 D4 Castro del Río Spain
25 E1 Castro-Urdiales Spain
25 B4 Castro Verde Port.
26 G5 Castrovillari Italy
92 B3 Castroville U.S.A.
70 A6 Caswell Sd in. N.Z.
37 H2 Çat Turkey
100 B5 Catacaos Peru
104 D3 Cataguases Brazil
85 E6 Catahoula L. U.S.A.
55 B3 Cataiñgan Phil.
37 J3 Çatak Turkey
104 C2 Catalão Brazil
25 G2 Cataluña div. Spain
102 C3 Catamarca Arg.
115 B6 Catandica Moz.
55 C3 Catanduanes i. Phil.
104 B4 Catanduvas Brazil
26 F6 Catania Sicily Italy
26 G5 Catanzaro Italy
85 D6 Catarina U.S.A.
55 C3 Catarman Phil.
25 F3 Catarroja Spain
66 A5 Catastrophe, C. hd Austr.
103 B2 Catatumbo r. Venez.
115 B6 Cataxa Moz.
55 C4 Catbalogan Phil.
87 E7 Cat Cays is Bahamas
55 C5 Cateel Phil.
55 C5 Cateel Bay Phil.
113 K3 Catembe Moz.
67 H6 Cathcart Austr.
113 G6 Cathcart S. Africa
113 H4 Cathedral Peak S. Africa
14 A6 Catherdaniel Rep. of Ireland
93 F2 Catherine, Mt U.S.A.
105 B3 Catillo Chile
87 E7 Cat Island Bahamas
78 B3 Cat L. Can.
97 H3 Catoche, C. Mex.
90 E5 Catonsville U.S.A.
96 D3 Catorce Mex.
105 D3 Catriló Arg.
103 E4 Catrimani Brazil
103 E4 Catrimani r. Brazil
91 G3 Catskill U.S.A.
91 F4 Catskill Mts U.S.A.
15 C2 Cats, Mont des h. France
23 K6 Cattolica Italy
113 K3 Catuane Moz.
103 E4 Cauamé r. Brazil
55 B4 Cauayan Phil.
79 H2 Caubvick, Mount Can.
103 B3 Cauca r. Col.
101 L4 Caucaia Brazil
103 B3 Caucasia Col.
6 H4 Caucasus mts Asia/Europe
105 C1 Caucete Arg.
91 J1 Caucomgomoc Lake U.S.A.
15 D2 Caudry France
55 C4 Cauit Point Phil.
105 B2 Cauquenes Chile
103 D3 Caura r. Venez.
79 G4 Causapscal Can.
22 F4 Cavaglià Italy
22 C6 Cavaillon France
104 C1 Cavalcante Brazil
114 B3 Cavally r. Côte d'Ivoire
14 D4 Cavan Rep. of Ireland
23 K4 Cavarzere Italy
85 F4 Cave City U.S.A.
104 E1 Caveira r. Brazil
66 E6 Cavendish Austr.
104 B4 Cavernoso, Serra do mts Brazil
90 B5 Cave Run Lake U.S.A.
55 B4 Cavili rf Phil.
55 B3 Cavite Phil.
22 F4 Cavour, Canale canal Italy
11 E3 Cawdor U.K.
66 D4 Cawndilla Lake Austr.
13 J5 Cawston U.K.
101 K4 Caxias Brazil
102 F3 Caxias do Sul Brazil
111 B4 Caxito Angola
36 C2 Çay Turkey
87 D5 Cayce U.S.A.
36 D1 Çaycuma Turkey
37 H1 Çayeli Turkey
101 H3 Cayenne Fr. Guiana
36 E3 Çayhan Turkey
36 C1 Çayırhan Turkey
95 J5 Cayman Brac i. Cayman Is
73 H8 Cayman Islands terr. Caribbean Sea
120 D4 Cayman Trench sea feature Atl. Ocean
110 E3 Caynabo Somalia
89 H4 Cayuga Can.
90 E3 Cayuga Lake U.S.A.
91 F3 Cazenovia U.S.A.
111 C5 Cazombo Angola
115 B6 Cazula Moz.
96 J7 Cebaco, Isla i. Panama
96 C2 Ceballos Mex.
105 F2 Cebollati r. Uru.
55 B4 Cebu Phil.
55 B4 Cebu i. Phil.
21 J3 Čechtice Czech Rep.
88 C3 Cecil U.S.A.
67 J1 Cecil Plains Austr.
69 E2 Cecil Rhodes, Mt h. Austr.
23 H6 Cecina Italy
23 H6 Cecina r. Italy
88 A4 Cedar r. IA U.S.A.
84 C2 Cedar r. ND U.S.A.
88 D4 Cedarburg U.S.A.
93 F3 Cedar City U.S.A.
85 D5 Cedar Creek Resr U.S.A.
88 A4 Cedar Falls U.S.A.
90 C5 Cedar Grove WV U.S.A.
88 D4 Cedar Grove WV U.S.A.
91 F6 Cedar I. U.S.A.
77 J4 Cedar L. Can.
88 D5 Cedar Lake U.S.A.
90 B4 Cedar Pt U.S.A.
88 B5 Cedar Rapids U.S.A.
93 G3 Cedar Ridge U.S.A.
91 F5 Cedar Run U.S.A.
89 F4 Cedar Springs Can.
88 E4 Cedar Springs U.S.A.
87 C5 Cedartown U.S.A.
113 H5 Cedarville S. Africa
88 E3 Cedarville U.S.A.
23 H3 Cedegolo Italy
96 A1 Cedros i. Mex.
65 F6 Ceduna Austr.
110 E3 Ceeldheere Somalia
110 E2 Ceerigaabo Somalia
26 F5 Cefalù Sicily Italy
17 J7 Cegléd Hungary
51 B5 Ceheng China
21 K4 Čejč Czech Rep.
36 E1 Çekerek Turkey
58 B4 Celah, Gunung mt Malaysia
21 H2 Čelákovice Czech Rep.
96 G5 Celaque, Parque Nacional nat. park Honduras
96 D3 Celaya Mex.
14 E4 Celbridge Rep. of Ireland
Celebes i. see Sulawesi
57 G6 Celebes Sea Indon./Phil.
90 A4 Celina U.S.A.
26 F1 Celje Slovenia
21 L5 Celldömölk Hungary
19 H4 Celle Ger.
10 C6 Celtic Sea Rep. of Ireland/U.K.
36 E1 Cemilbey Turkey
36 G2 Çemişgezek Turkey
23 J3 Čencenighe Agordino Italy
57 K7 Cenderawasih, Teluk b. Indon.
51 C5 Cengong China
23 H5 Ceno r. Italy
115 B6 Centenary Zimbabwe
93 F5 Centennial Wash r. U.S.A.
85 E6 Center U.S.A.
91 G4 Centereach U.S.A.
87 C5 Center Point U.S.A.
90 B5 Centerville U.S.A.
23 J5 Cento Italy
113 G1 Central div. Botswana
115 C3 Central div. Kenya
107 E5 Central African Republic country Africa
41 G4 Central Brahui Range mts Pak.
88 B4 Central City IA U.S.A.
84 D3 Central City NE U.S.A.
103 A4 Central, Cordillera mts Col.
96 J6 Central, Cordillera mts Panama
100 C5 Central, Cordillera mts Peru
55 B2 Central, Cordillera mts Phil.
51 □ Central District H.K. China
86 B4 Centralia IL U.S.A.
82 B2 Centralia WA U.S.A.
88 A2 Central Lakes U.S.A.
41 G5 Central Makran Range mts Pak.
82 B3 Central Point U.S.A.
62 E2 Central Ra. mts P.N.G.
15 B4 Centre div. France
87 C5 Centreville U.S.A.
51 D6 Cenxi China
Cephalonia i. see Kefallonia
Ceram i. see Seram
Ceram Sea g. see Seram Sea
103 D3 Cerbatana, Sa de la mt Venez.
93 E4 Cerbat Mts U.S.A.
23 J4 Cerea Italy
77 G4 Cereal Can.
102 D3 Ceres Arg.
112 C6 Ceres S. Africa
103 B2 Cereté Col.
25 E2 Cerezo de Abajo Spain
26 F4 Cerignola Italy
36 D2 Çerikli Turkey
15 E4 Cerisiers France
36 D1 Çerkeş Turkey
36 E3 Çermelik r. Syria
37 G2 Çermik Turkey
27 N2 Cernavodă Romania
22 E2 Cernay France
21 H3 Černovice Czech Rep.
97 E2 Cerralvo Mex.
96 B2 Cerralvo i. Mex.
27 H4 Cërrik Albania
97 D3 Cerritos Mex.
104 C4 Cerro Azul Brazil
97 E3 Cerro Azul Mex.
100 B4 Cerro de Amotape, Parque Nacional nat. park Peru
100 C6 Cerro de Pasco Peru
96 J7 Cerro Hoya, Parque Nacional nat. park Panama
103 D3 Cerro Jáua, Meseta del plat. Venez.
103 C2 Cerrón, Co mt Venez.
96 C2 Cerro Prieto Mex.
105 C3 Cerros Colorados, Embalse resr Arg.
23 J6 Certaldo Italy
69 B5 Cervantes Austr.
26 F4 Cervati, Monte mt Italy
21 K2 Červená Voda Czech Rep.
23 K5 Cervia Italy
23 L4 Cervignano del Friuli Italy
21 J3 Červiná Řečice Czech Rep.
26 C3 Cervione Corsica France
22 F4 Cervo r. Italy
25 C1 Cervo Spain
23 G4 Cesano Boscone Italy
103 B2 César r. Col.
23 K5 Cesena Italy
23 K5 Cesenatico Italy
9 T8 Cēsis Latvia
21 H2 Česká Kamenice Czech Rep.
21 H2 Česká Lípa Czech Rep.
21 K2 Česká Skalice Czech Rep.
21 H4 České Budějovice Czech Rep.
21 G2 České Středohoří h. Czech Rep.
21 J3 Českomoravská Vysočina reg. Czech Rep.
21 H2 Český Brod Czech Rep.
21 H4 Český Krumlov Czech Rep.
20 F3 Český Les mts Czech Rep./Ger.
21 M3 Český Těšín Czech Rep.
27 M5 Çeşme Turkey
67 J4 Cessnock Austr.
114 B3 Cestos r. Liberia
27 H3 Cetinje Yugo.
15 A4 Ceton France
26 F5 Cetraro Italy
25 D5 Ceuta Spain
23 H3 Cevedale, Monte mt Italy
24 F4 Cévennes mts France
22 A5 Cévennes, Parc National des nat. park France
22 D4 Cevins France
36 E3 Ceyhan Turkey
36 E3 Ceyhan r. Turkey
37 H3 Ceylanpınar Turkey
22 C3 Ceyzériat France
41 F2 Chaacha Turkm.
22 D3 Chablais mts France
22 A2 Chablis France
22 C5 Chabre ridge France
105 E2 Chacabuco Arg.
105 B4 Chacao Chile
105 C3 Chachahuén, Sa mt Arg.
100 C5 Chachapoyas Peru
28 D4 Chachersk Belarus
58 B2 Chachoengsao Thai.
44 B4 Chachro Pak.
76 C4 Chacon, C. U.S.A.
107 E4 Chad country Africa
48 H2 Chadaasan Mongolia
48 F1 Chadan Rus. Fed.
113 G1 Chadibe Botswana
105 C3 Chadileo r. Arg.
109 D3 Chad, Lake Africa
84 C3 Chadron U.S.A.
58 A1 Chae Hom Thai.
39 H4 Chaek Kyrg.
54 D4 Chaeryŏng N. Korea
103 B4 Chafurray Col.
41 G4 Chagai Pak.
41 F4 Chagai Hills Afgh./Pak.
38 F3 Chagan Kzyl-Ordinsk. Kazak.
39 J2 Chagan Semipal. Kazak.
45 F2 Chagdo Kangri reg. China
41 G3 Chaghcharān Afgh.
39 G1 Chaglinka r. Kazak.
22 B3 Chagny France
117 J4 Chagos Archipelago is British Indian Ocean Terr.
28 J4 Chagra r. Rus. Fed.
38 D3 Chagrayskoye Plato plat. Kazak.
96 K6 Chagres, Parque Nacional nat. park Panama
103 D2 Chaguaramas Venez.
38 C4 Chagyl Turkm.
45 G3 Cha'gyüngoinba China
41 F4 Chahah Burjal Afgh.
41 E3 Chāh Ākhvor Iran
41 E3 Chahār Takāb Iran
40 D3 Chāh Badam Iran
41 F5 Chāh Bahār Iran
40 D4 Chāh-e Bāgh well Iran
40 E3 Chāh-e Kavīr well Iran
40 D3 Chāh-e Khorāsān well Iran
41 E3 Chāh-e Khoshāb Iran
41 E4 Chāh-e Malek Iran
40 D3 Chāh-e Mirzā well Iran
40 D3 Chāh-e Mūjān well Iran
40 D3 Chāh-e Nūklok Iran
40 E3 Chāh-e Nūklok well Iran
40 D4 Chāh-e Qeyşar well Iran
40 D3 Chāh-e Qobād well Iran
40 D4 Chāh-e Rāh Iran
41 E4 Chāh-e-Raḥmān well Iran
40 D3 Chāh-e Shūr Iran
40 D3 Chāh-e Shur well Iran
40 D3 Chah Haji Abdulla well Iran
40 D4 Chāh Ḩaqq Iran
41 H2 Chah-i-Ab Afgh.
37 K4 Chah-i-Shurkh Iraq
40 D3 Chāh Pās well Iran
40 C4 Chāh Rūstā'ī Iran
41 F4 Chah Sandan Pak.
54 C2 Chai r. China
115 D5 Chai Moz.
45 F5 Chāībāsa India
79 G3 Chaigneau, Lac l. Can.
58 B2 Chai Si r. Thai.
51 □ Chai Wan H.K. China
58 A3 Chaiya Thai.
58 B2 Chaiyaphum Thai.
105 F1 Chajari Arg.
44 B3 Chakar r. Pak.
45 H5 Chakaria Bangl.
115 C4 Chake Chake Tanz.
41 F4 Chakhānsūr Afgh.
45 E4 Chakia India
41 G5 Chakku Pak.
100 D7 Chala Peru
22 C3 Chalain, Lac de l. France
41 G3 Chalap Dalan mts Afgh.
22 B3 Chalaronne r. France
96 G5 Chalatenango El Salvador
79 G4 Chaleur Bay in. Can.
51 D5 Chaling China
44 C5 Chalisgaon India
27 K5 Chalkida Greece
39 J4 Chalkudysu Kazak.
70 A7 Chalky Inlet in. N.Z.
24 D3 Challans France
100 E7 Challapata Bol.
118 E5 Challenger Deep depth Pac. Oc.
119 M8 Challenger Fracture Zone sea feature Pac. Oc.
82 D2 Challis U.S.A.
22 B2 Chaloire r. France
15 E4 Châlons-en-Champagne France
22 B3 Chalon-sur-Saône France
54 C2 Chaluhe China
40 C2 Chālūs Iran
20 F3 Cham Ger.
83 F4 Chama U.S.A.
103 C2 Chama r. Venez.
115 B5 Chama Zambia
105 D2 Chamaico Arg.
112 A3 Chamais Bay Namibia
41 G4 Chaman Pak.
115 C5 Chamba Tanz.
44 D4 Chambal r. India
22 C4 Chambaran, Plateau de plat. France
79 G3 Chambeaux, Lac l. Can.
22 C3 Chambéria France
77 H4 Chamberlain Can.
84 D3 Chamberlain U.S.A.
91 J1 Chamberlain Lake U.S.A.
93 H4 Chambers U.S.A.
90 E5 Chambersburg U.S.A.
22 C4 Chambéry France
115 B5 Chambeshi Zambia
115 A5 Chambeshi r. Zambia
26 C7 Chambi, Jebel mt Tunisia
22 C4 Chamechaude mt France
40 C3 Cham-e Ḩannā Iran
40 C3 Chameshk Iran
105 C1 Chamical Arg.
45 F4 Chamlang mt Nepal
58 B3 Châmnar Cambodia
22 D4 Chamonix-Mont-Blanc France
78 F4 Chamouchouane r. Can.
22 C5 Chamouse, Montagne de mt France
45 E5 Champa India
76 B2 Champagne Can.
24 G2 Champagne reg. France
15 E4 Champagne-Ardenne div. France
113 H4 Champagne Castle mt S. Africa
22 C3 Champagnole France
88 C5 Champaign U.S.A.
105 D1 Champaqui, Cerro mt Arg.
58 C2 Champasak Laos
15 H4 Champ du Feu mt France
22 E3 Champex Switz.
45 H5 Champhai India
88 D2 Champion U.S.A.
91 G2 Champlain U.S.A.
91 G2 Champlain, L. Can./U.S.A.
22 C2 Champlitte France
97 G4 Champotón Mex.
43 B4 Chamrajnagar India
28 H4 Chamzinka Rus. Fed.
58 B4 Chana Thai.
102 B3 Chañaral Chile
103 E3 Chanaro, Co mt Venez.
22 B4 Chanas France
105 B2 Chanco Chile
Chanda see Chandrapur
74 D3 Chandalar r. U.S.A.
45 E5 Chandarpur India
44 D3 Chandausi India
85 F6 Chandeleur Islands U.S.A.
44 E5 Chandia India
44 D3 Chandigarh India
93 G5 Chandler U.S.A.
89 J3 Chandos Lake Can.
45 G5 Chandpur Bangl.
44 D3 Chandpur India
45 H5 Chandraghona Bangl.
44 D6 Chandrapur India
44 D5 Chandur India
111 D6 Changane r. Moz.
115 B6 Changara Moz.
54 E3 Changbai China
54 D3 Changbai Shan mts China/N. Korea
51 C7 Changcheng China
116 B1 Chang Cheng (Great Wall) China Base Ant.
54 C2 Changchun China
54 C1 Changchunling China
50 F2 Changdao China
51 D4 Changde China
54 D4 Changdo N. Korea
50 E3 Changfeng China
54 E5 Changgi Gap pt S. Korea
54 B4 Changhai China
54 D6 Changhang S. Korea
54 D5 Changhowan S. Korea
51 F5 Chang-hua Taiwan
58 D1 Changhua Jiang r. China
54 D6 Changhung S. Korea
58 □ Changi Sing.
51 C7 Changjiang China
Chang Jiang r. see Yangtze
Changjiang Kou est. see Yangtze, Mouth of the
54 D3 Changjin N. Korea
54 D3 Changjin Reservoir N. Korea
51 F5 Changle China
50 F2 Changli China
54 B1 Changling China
51 D5 Changning China
54 C4 Changnyŏn N. Korea
50 E1 Changping China
54 E5 Changp'yŏng S. Korea
54 C4 Changsan-got pt N. Korea
51 D4 Changsha China
51 F4 Changshan China
54 B4 Changshan Qundao is China
51 C4 Changshou China
51 D4 Changshoujie China
50 F4 Changshu China
51 C5 Changshun China
54 D6 Changsŏng S. Korea
51 E5 Changtai China
51 E5 Changting Fujian China
54 E1 Changting Heilongjiang China
54 C2 Changtu China
96 J6 Changuinola Panama
54 E6 Ch'angwŏn S. Korea
50 C3 Changwu China
54 A4 Changxing Dao i. China
51 D4 Changyang China
50 F2 Changyi China
54 C4 Changyŏn N. Korea
50 E3 Changyuan China
50 D3 Changzhi China
50 F4 Changzhou China
27 L7 Chania Greece
54 D3 Chanjin r. N. Korea
50 B3 Chankou China
43 B3 Channapatna India
10 E7 Channel Islands English Channel
92 C5 Channel Islands U.S.A.
92 B5 Channel Is Nat. Park U.S.A.
79 J4 Channel-Port-aux-Basques Can.
69 D7 Channel Pt Austr.
15 B1 Channel Tunnel tunnel France/U.K.
88 C2 Channing U.S.A.
25 C1 Chantada Spain
58 B2 Chanthaburi Thai.
15 C3 Chantilly France
85 E4 Chanute U.S.A.
32 J4 Chany, Ozero salt l. Rus. Fed.
50 E2 Chaobai Xinhe r. China
50 E4 Chao Hu l. China
58 B2 Chao Phraya r. Thai.
108 B1 Chaouèn Morocco
15 E4 Chaource France
45 H2 Chaowula Shan mts China
50 E4 Chao Xian China
51 E6 Chaoyang Guangdong China
50 F1 Chaoyang Liaoning China
51 E6 Chaozhou China
104 E1 Chapada Diamantina, Parque Nacional nat. park Brazil
104 A1 Chapada dos Guimarães Brazil
104 C1 Chapada dos Veadeiros, Parque Nacional da nat. park Brazil
96 D3 Chapala Mex.
96 D3 Chapala, L. de Mex.
103 B4 Chaparral Col.
38 B2 Chapayev Kazak.
28 J4 Chapayevsk Rus. Fed.
39 H1 Chapayevskoye Kazak.
102 F3 Chapecó Brazil
102 F3 Chapecó r. Brazil
13 F4 Chapel-en-le-Frith U.K.
87 E5 Chapel Hill U.S.A.
15 E2 Chapelle-lez-Herlaimont Belgium
13 F4 Chapeltown U.K.
88 D5 Chapin, Lake U.S.A.
89 F2 Chapleau Can.
28 F4 Chaplygin Rus. Fed.
29 E6 Chaplynka Ukr.
90 B6 Chapmanville U.S.A.
67 G8 Chappell Is Austr.
41 G3 Chapri Pass Afgh.
100 E7 Chaqui Bol.
44 D2 Char Jammu and Kashmir
22 B3 Charbonnat France
96 D3 Charcas Mex.
22 C3 Charchilla France
45 H3 Char Chu r. China
116 A2 Charcot I. Ant.
77 G3 Chard Can.
13 E7 Chard U.K.
37 L3 Chārdagh Iran
39 F4 Chardara Kazak.
39 G4 Chardarinskoye Vdkhr. resr Kazak./Uzbek.
37 L5 Chardāvol Iran
90 C4 Chardon U.S.A.
38 E5 Chardzhev Turkm.
24 E3 Charente r. France
41 H3 Chārīkār Afgh.
84 E3 Chariton r. U.S.A.
89 F3 Charity Is i. U.S.A.
32 G3 Charkayuvom Rus. Fed.
44 D4 Charkhari India
15 E2 Charleroi Belgium
81 L4 Charles, Cape pt U.S.A.
88 A4 Charles City U.S.A.
85 E6 Charles, Lake U.S.A.
69 C3 Charles, Mt h. Austr.
69 E6 Charles Peak h. Austr.
70 C4 Charleston N.Z.
86 B4 Charleston IL U.S.A.
91 J2 Charleston ME U.S.A.
85 F4 Charleston MO U.S.A.
87 E5 Charleston SC U.S.A.
90 C5 Charleston WV U.S.A.
93 E3 Charleston Peak U.S.A.
14 C4 Charlestown Rep. of Ireland
91 G3 Charlestown NH U.S.A.
91 H4 Charlestown RI U.S.A.
90 E5 Charles Town U.S.A.
68 B6 Charleville Austr.
15 E3 Charleville-Mézières France
88 E3 Charlevoix U.S.A.
76 E3 Charlie Lake Can.
22 B3 Charlieu France
88 E4 Charlotte MI U.S.A.
87 D5 Charlotte NC U.S.A.
87 D7 Charlotte Harbor b. U.S.A.
90 D5 Charlottesville U.S.A.
79 H4 Charlottetown Can.
103 E2 Charlotteville Trinidad and Tobago
66 E6 Charlton Austr.
78 E3 Charlton I. Can.
15 G4 Charmes France
28 F2 Charozero Rus. Fed.
44 B2 Charsadda Pak.
39 K2 Charsk Kazak.
68 B3 Charters Towers Austr.
15 B4 Chartres France
22 C4 Chartreuse, Massif de la mts France
71 □15 Charybdis Reef Fiji
39 J4 Charyn Kazak.
39 J4 Charyn r. Kazak.
39 K1 Charysh r. Rus. Fed.
39 K2 Charyshskoye Rus. Fed.
105 E2 Chascomús Arg.

50 C3 Daba Shan mts China
114 A2 Dabatou Guinea
103 A3 Dabeiba Col.
19 J3 Dabel Ger.
44 C5 Dabhoi India
43 A2 Dabhol India
19 M3 Dąbie, Jezioro l. Pol.
50 E4 Dabie Shan mts China
44 D4 Daboh India
108 A3 Dabola Guinea
114 C3 Daboya Ghana
44 D4 Dabra India
19 M4 Dąbroszyn Pol.
17 J5 Dąbrowa Górnicza Pol.
51 E5 Dabu China
54 B1 Dabusu Pao l. China
Dacca see Dhaka
20 E4 Dachau Ger.
50 A2 Dachechang China
43 B2 Dachepalle India
21 J3 Dačice Czech Rep.
89 J3 Dacre Can.
36 D1 Daday Turkey
87 D6 Dade City U.S.A.
44 C5 Dadra India
44 C5 Dadra and Nagar Haveli div. India
44 A4 Dadu Pak.
51 B4 Dadu He r. China
58 C3 Đa Dung r. Vietnam
55 B3 Daet Phil.
51 B5 Dafang China
50 F3 Dafeng China
54 D2 Dafengman China
45 H4 Dafla Hills India
108 A3 Dagana Senegal
50 B3 Dagcanglhamo China
29 H7 Dagestan, Respublika div. Rus. Fed.
22 E2 Dagmersellen Switz.
50 F2 Dagu r. China
51 B5 Daguan China
51 D6 D'Aguilar Peak h. H.K. China
55 B2 Dagupan Phil.
45 H1 Dagur China
45 G3 Dagzê China
45 F3 Dagzê Co salt l. China
44 C6 Dahanu India
50 D1 Dahei r. China
50 C2 Dahei Shan mts China
52 C1 Dahezhen China
46 D2 Da Hinggan Ling mts China
110 E2 Dahlak Archipelago is Eritrea
110 E2 Dahlak Marine National Park Eritrea
15 G2 Dahlem Ger.
19 H3 Dahlenburg Ger.
26 C7 Dahmani Tunisia
19 L5 Dahme Ger.
19 J2 Dahme Ger.
19 L5 Dahme r. Ger.
19 J2 Dahmeshöved hd Ger.
15 H3 Dahn Ger.
44 C5 Dāhod India
44 D2 Dahongliutan China/Jammu and Kashmir
19 H4 Dähre Ger.
37 J3 Dahūk Iraq
54 C3 Dahuofang Shuiku resr China
54 B3 Dahushan China
50 D1 Dai Hai l. China
59 B3 Daik Indon.
11 D5 Dailly U.K.
41 E3 Daim Iran
53 C6 Daimanji-san h. Japan
25 E3 Daimiel Spain
105 E3 Daireaux Arg.
88 A2 Dairyland U.S.A.
53 C7 Daisen volc. Japan
51 G4 Daishan China
51 F5 Daiyun Shan mts China
65 G4 Dajarra Austr.
54 A4 Dajin Chuan r. China
50 B2 Dajing China
45 H1 Da Juh China
108 A3 Dakar Senegal
45 G3 Dakelangsi China
34 C6 Daketa Shet' r. Eth.
45 G5 Dakhin Shahbaz-pur I. Bangl.
109 E2 Dakhla Oasis oasis Egypt
58 C2 Dak Kon Vietnam
28 D4 Dakol'ka r. Belarus
114 E2 Dakoro Niger
84 D3 Dakota City U.S.A.
27 J3 Đakovica Yugo.
27 H2 Đakovo Croatia
111 C5 Dala Angola
20 D5 Dalaas Austria
108 A3 Dalaba Guinea
50 D1 Dalad Qi China
50 E1 Dalai Nur l. China
40 C4 Dalaki, Rud-e r. Iran
50 D1 Dalamamiao China
36 B3 Dalaman Turkey
36 B3 Dalaman r. Turkey
48 H3 Dalandzadgad Mongolia
55 B4 Dalanganem Islands Phil.
61 J3 Dalap-Uliga-Darrit Marshall Is
58 D3 Đa Lat Vietnam
50 B1 Dalay Mongolia
44 A4 Dalbandin Pak.
11 E6 Dalbeattie U.K.
67 J1 Dalby Austr.
19 L1 Dalby Sweden
12 C3 Dalby U.K.
9 J6 Dale Hordaland Norway
9 J6 Dale Sogn og Fjordane Norway
90 E5 Dale City U.S.A.
85 G4 Dale Hollow Lake U.S.A.
18 D4 Dalen Neth.
45 H6 Dalet Myanmar
45 H5 Daletme Myanmar
9 O6 Dalfors Sweden
41 E5 Dalgān Iran
69 C3 Dalgaranger, Mt h. Austr.
67 H6 Dalgety Austr.
69 B2 Dalgety r. Austr.
85 C4 Dalhart U.S.A.
79 G4 Dalhousie Can.
50 D3 Dali Shaanxi China
46 C4 Dali Yunnan China
54 A4 Dalian China
51 B4 Daliang Shan mts China
54 B2 Dalin China
50 F1 Daling r. China
54 D3 Dalizi China
11 E5 Dalkeith U.K.
45 F4 Dālkola India
68 E5 Dalkarnil Austr.
91 F4 Dallas PA U.S.A.
85 D5 Dallas TX U.S.A.
88 B5 Dallas City U.S.A.
82 B2 Dalles, The U.S.A.
76 C4 Dall I. U.S.A.
114 D1 Dallol Bosso watercourse Mali/Niger
69 C5 Dallwallinu Austr.
40 D5 Dalmā i. U.A.E.
105 D2 Dalmacio Vélez Sarsfield Arg.
44 E4 Dalman India
26 G3 Dalmatia reg. Croatia
11 D5 Dalmellington U.K.
52 D2 Dal'negorsk Rus. Fed.
52 C2 Dal'nerechensk Rus. Fed.
114 B3 Daloa Côte d'Ivoire
51 C5 Dalou Shan mts China
40 B5 Dalqān well S. Arabia
11 D5 Dalry U.K.
11 D5 Dalrymple U.K.
68 B3 Dalrymple, L. Austr.
68 C3 Dalrymple, Mt Austr.
45 F4 Daltenganj India
89 E1 Dalton Can.
113 J4 Dalton S. Africa
87 C5 Dalton GA U.S.A.
91 G3 Dalton MA U.S.A.
12 D3 Dalton-in-Furness U.K.
89 E1 Dalton Mills Can.
14 C5 Dalua r. Rep. of Ireland
55 C3 Dalupiri i. Phil.
55 B2 Dalupiri i. Phil.
8 D4 Dalvík Iceland
64 F2 Daly r. Austr.
92 A3 Daly City U.S.A.
65 F3 Daly Waters Austr.
114 E2 Damagaram Takaya Niger
44 C5 Daman India
44 C5 Daman and Diu div. India
109 F1 Damanhûr Egypt
40 C3 Damaq Iran
50 E1 Damaqun Shan mts China
55 C6 Damar Indon.
57 H8 Damar i. Indon.
36 F5 Damascus Syria
105 B2 Damas, P. de las pass Arg./Chile
114 F2 Damaturu Nigeria
40 D3 Damavand Iran
114 F2 Damboa Nigeria
43 C5 Dambulla Sri Lanka
40 D2 Damghan Iran
Damietta see Dumyât
50 E2 Daming China
51 C6 Daming Shan mt China
45 H2 Damjong China
55 B5 Dammai i. Phil.
15 B4 Dammarie France
22 F3 Dammastock mt Switz.
15 D1 Damme Belgium
18 F4 Damme Ger.
44 D5 Damoh India
114 C3 Damongo Ghana
64 C4 Dampier Austr.
64 D3 Dampier Land reg. Austr.
22 C2 Dampierre France
22 C2 Dampierre-sur-Salon France
62 E2 Dampier Strait P.N.G.
57 J7 Dampir, Selat chan. Indon.
Damqoq Kanbab r. see Maquan He
45 H2 Dam Qu r. China
45 H3 Damroh India
15 B4 Damville France
18 D3 Damwoude Neth.
48 F5 Damxung China
114 B3 Danané Côte d'Ivoire
58 D1 Đa Năng Vietnam
55 C4 Danao Phil.
50 A4 Danba China
91 G4 Danbury CT U.S.A.
91 H3 Danbury NH U.S.A.
93 E4 Danby Lake U.S.A.
50 E3 Dancheng China
85 B5 Dandaragan Austr.
110 D3 Dande Eth.
43 A3 Dandeli India
66 F6 Dandenong Austr.
54 C3 Dandong China
13 E4 Dane r. U.K.
75 O2 Daneborg Greenland
91 K2 Danforth U.S.A.
51 E1 Dangan Liedao is China
45 G3 Dangbe La pass China
52 B2 Dangbizhen Rus. Fed.
50 E3 Dangchang China
60 L5 Danger Islands Cook Is Pac. Oc.
112 C7 Danger Pt S. Africa
110 D2 Dangila Eth.
Dangla mts see Tanggula Shan
45 G3 Dangqên China
97 G4 Dangriga Belize
50 E3 Dangshan China
50 F3 Dangtu China
51 D4 Dangyang China
82 E3 Daniel U.S.A.
112 E4 Daniëlskuil S. Africa
28 G3 Danilov Rus. Fed.
39 G1 Danilovka Kazak.
29 H5 Danilovka Rus. Fed.
28 G3 Danilovskaya Vozvyshennost' reg. Rus. Fed.
50 D2 Daning China
19 H2 Dänischenhagen Ger.
19 H3 Dänizkänari Azer.
50 D3 Danjiangkou Sk. resr China
40 E6 Dank Oman
44 D2 Dankhar India
28 F4 Dankov Rus. Fed.
39 J4 Dankova, Pik mt Kyrg.
51 B4 Danleng China
96 H5 Danlí Honduras
Dannebrogsø i. see Qillak
91 G2 Dannemora U.S.A.
19 J3 Dannenberg (Elbe) Ger.
19 J3 Dannenwalde Ger.
70 F4 Dannevirke N.Z.
113 J4 Dannhauser S. Africa
58 B1 Dan Sai Thai.
90 E3 Dansville U.S.A.
44 C4 Danta India
43 C2 Dantewara India
27 M3 Danube r. Europe
88 D5 Danville IL U.S.A.
88 D6 Danville IN U.S.A.
86 C4 Danville KY U.S.A.
89 J5 Danville PA U.S.A.
90 D6 Danville VA U.S.A.
51 C7 Dan Xian China
50 F4 Danyang China
51 C5 Danzhai China
55 B4 Dao Phil.
58 C1 Đao Bach Long Vi i. Vietnam
51 C6 Đao Cai Bâu i. Vietnam
51 C6 Đao Cat Ba i. Vietnam
58 C3 Đao Phu Quôc i. Vietnam
58 B3 Đao Thô Chu i. Vietnam
114 C3 Daoukro Côte d'Ivoire
58 B3 Đao Vây i. Vietnam
51 D5 Dao Xian China
51 C4 Daozhen China
55 C4 Dapa Phil.
114 D2 Dapaong Togo
45 J4 Daphabum mt India
55 B4 Dapiak, Mt Phil.
55 B4 Dapitan Phil.
35 H3 Da Qaidam China
42 E2 Daqing China
50 D1 Daqing Shan mts China
40 D3 Daqq-e Dombūn Iran
41 F3 Daqq-e-Tundi, Dasht-e l. Afgh.
37 K4 Dāqūq Iraq
51 G4 Daqu Shan i. China
36 F5 Dar'ā Syria
40 D4 Dārāb Iran
55 B3 Daraga Phil.
17 O4 Darahanava Belarus
108 D1 Daraj Libya
40 D4 Dārākūyeh Iran
55 D4 Daram i. Phil.
40 D3 Darang, Kūh-e h. Iran
49 K1 Darasun Rus. Fed.
39 H3 Darat Kazak.
40 E4 Darband Iran
45 F4 Darbhanga India
42 C2 Dardanelle, Lake U.S.A.
85 E5 Dardanelle, Lake U.S.A.
27 M4 Dardanelles str. Turkey
19 H5 Dardesheim Ger.
56 F2 Darende Turkey
115 C4 Dar es Salaam Tanz.
23 H4 Darfo Boario Terme Italy
44 B2 Dargai Pak.
38 E4 Dargan-Ata Turkm.
70 D1 Dargaville N.Z.
67 G6 Dargo Austr.
19 K3 Dargun Ger.
48 J2 Darhan Mongolia
50 D1 Darhan Muminggan Lianheqi China
87 D6 Darien U.S.A.
103 A2 Darién, Golfo del g. Col.
96 K7 Darién, Parque Nacional de nat. park Panama
96 K6 Darién, Serranía del mts Panama
39 H2 Dar'inskiy Kazak.
38 B2 Dar'inskoye Kazak.
96 H5 Dario Nic.
45 G4 Dārjiling India
69 C6 Darkan Austr.
66 B4 Darke Peak Austr.
40 C4 Darkhazīneh Iran
50 B3 Darlag China
66 E4 Darling r. Austr.
67 H1 Darling Downs reg. Austr.
69 B5 Darling Range h. Austr.
12 F3 Darlington U.K.
88 B4 Darlington U.S.A.
67 G5 Darlington Point Austr.
69 E3 Darlot, L. salt flat Austr.
16 H3 Darłowo Pol.
40 B5 Darmā S. Arabia
44 E3 Darma Pass China/India
43 B2 Darmaraopet India
40 E4 Dar Mazār Iran
20 B3 Darmstadt Ger.
44 C5 Darna r. India
109 E1 Darnah Libya
113 J4 Darnall S. Africa
15 B3 Darnétal France
15 G4 Darney France
66 E4 Darnick Austr.
74 F3 Darnley Bay Can.
116 D5 Darnley, C. Ant.
25 F2 Daroca Spain
39 H5 Daroot-Korgan Kyrg.
28 G3 Darovka Rus. Fed.
28 H3 Darovskoy Rus. Fed.
105 D3 Darregueira Arg.
40 E3 Darreh Bīd Iran
41 E2 Darreh Gaz Iran
37 L4 Darreh Gozaru r. Iran
41 H3 Darreh-ye Shekārī r. Afgh.
43 B3 Darsi India
37 M6 Darsīyeh Iran
19 K2 Darß pen. Ger.
19 K2 Darßer Ort c. Ger.
13 D7 Dart r. U.K.
13 H6 Dartford U.K.
66 D6 Dartmoor Austr.
13 C7 Dartmoor reg. U.K.
13 D7 Dartmoor National Park U.K.
79 H5 Dartmouth Can.
13 D7 Dartmouth U.K.
68 A6 Dartmouth, L. salt flat Austr.
12 F4 Darton U.K.
14 C3 Dartry Mts h. Rep. of Ireland
62 E2 Daru P.N.G.
108 A4 Daru Sierra Leone
45 G3 Darum Tso l. China
26 G2 Daruvar Croatia
38 D4 Darvaza Turkm.
40 C4 Darvīsla Iran
41 G4 Darwazgai Afgh.
12 E4 Darwen U.K.
41 G4 Darweshan Afgh.
64 F2 Darwin Austr.
102 C8 Darwin, Mte Chile
44 B3 Darya Khan Pak.
41 E4 Dārzīn Iran
40 D5 Dās i. U.A.E.
50 D4 Dashennongjia mt China
38 D4 Dashkhovuz Turkm.
40 E5 Dasht Iran
41 F5 Dasht r. Pak.
40 E4 Dasht Āb Iran
40 C4 Dasht-e Palang r. Iran
41 F5 Dashtiari Iran
39 G5 Dashtidzhum Tajik.
50 C5 Dashuikeng China
50 B2 Dashuitou China
20 E4 Dasing Ger.
44 C2 Daska Pak.
37 L1 Daşkäsän Azer.
44 C1 Daspar mt Pak.
114 D3 Dassa Benin
19 G5 Dassel Ger.
112 C6 Dassen Island S. Africa
19 J3 Dassow Ger.
37 L2 Dastakert Armenia
40 E3 Dastgardān Iran
54 F2 Da Suifen r. China
27 M6 Datça Turkey
52 G3 Date Japan
93 F5 Dateland U.S.A.
44 D4 Datia India
51 E5 Datian China
50 A2 Datong Qinghai China
50 D1 Datong Shanxi China
50 B2 Datong He r. China
50 A2 Datong Shan mts China
55 C5 Datu Piang Phil.
59 C2 Datu, Tanjung c. Indon./Malaysia
28 C3 Daugava r. Belarus/Latvia
9 U9 Daugavpils Latvia
44 A2 Daulatabad Afgh.
44 C6 Daulatabad India
Daulatabad see Malāyer
15 G2 Daun Ger.
43 A2 Daund India
58 A2 Daung Kyun i. Myanmar
77 J4 Dauphin Can.
22 C4 Dauphiné reg. France
22 C5 Dauphiné, Alpes du mts France
88 C5 Dauphin I. U.S.A.
77 K4 Dauphin L. Can.
69 B2 Daurie Creek r. Austr.
44 D4 Dausa India
11 E3 Dava U.K.
37 L1 Däväçi Azer.
37 M1 Däväçi Azer.
43 A3 Davangere India
55 C5 Davao Phil.
55 C5 Davao Gulf Phil.
41 F5 Dāvar Panāh Iran
113 H3 Davel S. Africa
92 A3 Davenport CA U.S.A.
88 B5 Davenport IA U.S.A.
13 F5 Daventry U.K.
113 H3 Daveyton S. Africa
96 J6 David Panama
77 H4 Davidson Can.
77 J3 Davin Lake Can.
116 D5 Davis Austr. Base Ant.
92 B2 Davis U.S.A.
93 E4 Davis Dam U.S.A.
79 H2 Davis Inlet Can.
116 D5 Davis Sea Ant.
75 N3 Davis Strait Can./Greenland
23 G3 Davos Switz.
54 B3 Dawa China
50 A1 Dawan China
110 E3 Dawa Wenz r. Eth.
47 F5 Dawaxung China
55 B4 Dawe China
Dawei see Tavoy
50 E3 Dawen r. China
40 C5 Dawhat Salwah b. Qatar/S. Arabia
58 A1 Dawna Range mts Myanmar/Thai.
42 D6 Dawqah Oman
68 C5 Dawson r. Qld. Austr.
74 E3 Dawson Y.T. Can.
87 C6 Dawson GA U.S.A.
84 D2 Dawson ND U.S.A.
77 J4 Dawson Bay Can.
76 E3 Dawson Creek Can.
77 L2 Dawson Inlet in. Can.
68 C3 Dawson Range mts Can.
50 E4 Dawu Hubei China
46 C3 Dawu Sichuan China
24 D5 Dax France
50 C4 Daxian China
51 C6 Daxin China
50 E2 Daxing China
51 A4 Daxue Shan mts China
54 B4 Dayang r. China
45 H4 Dayang r. India
51 D6 Dayao Shan mts China
51 E4 Daye China
51 B4 Dayi China
66 F6 Daylesford Austr.
92 D3 Daylight Pass U.S.A.
105 F1 Daymán r. Uru.
105 F1 Daymán, Cuchilla del h. Uru.
51 D4 Dayong China
37 H4 Dayr az Zawr Syria
90 A5 Dayton OH U.S.A.
87 C5 Dayton TN U.S.A.
82 C2 Dayton WA U.S.A.
87 D6 Daytona Beach U.S.A.
51 E5 Dayu China
51 D5 Dayu Ling mts China
50 F3 Da Yunhe r. China
82 C2 Dayville U.S.A.
51 D7 Dazhou Dao i. China
51 C4 Dazhu China
51 B4 Dazu China
112 F5 De Aar S. Africa
88 D2 Dead r. U.S.A.
87 F7 Deadman's Cay Bahamas
93 E4 Dead Mts U.S.A.
91 H2 Dead River r. U.S.A.
36 E6 Dead Sea salt l. Asia
13 J6 Deal U.K.
113 F4 Dealesville S. Africa
76 D4 Dean r. Can.
51 E4 De'an China
13 E6 Dean, Forest of U.K.
105 D1 Deán Funes Arg.
89 F4 Dearborn U.S.A.
76 D3 Dease r. Can.
76 C3 Dease Lake Can.
74 H3 Dease Strait Can.
92 D3 Death Valley v. U.S.A.
92 D3 Death Valley Junction U.S.A.
92 D3 Death Valley National Monument res. U.S.A.
24 E2 Deauville France
59 D2 Debak Malaysia
51 C6 Debao China
27 J4 Debar Macedonia
77 H4 Debden Can.
13 J5 Debenham U.K.
93 H2 De Beque U.S.A.
91 J2 Deblois U.S.A.
19 M4 Dębno Pol.
114 C1 Débo, Lac l. Mali
69 D5 Deborah, L. salt flat Austr.
110 D3 Debre Birhan Eth.
17 K7 Debrecen Hungary
110 D2 Debre Markos Eth.
110 D2 Debre Tabor Eth.
110 D3 Debre Zeyit Eth.
87 C5 Decatur AL U.S.A.
87 C5 Decatur GA U.S.A.
88 C6 Decatur IL U.S.A.
88 E5 Decatur IN U.S.A.
88 E4 Decatur MI U.S.A.
43 B2 Deccan plat. India
89 H2 Decelles, Réservoir resr Can.
67 K1 Deception Bay Austr.
21 H2 Děčín Czech Rep.
88 B4 Decorah U.S.A.
13 F6 Deddington U.K.
19 H4 Dedeleben Ger.
19 L3 Dedelow Ger.
19 H4 Dedelstorf Ger.
18 D4 Dedemsvaart Neth.
104 C4 Dedo de Deus mt Brazil
112 C6 De Doorns S. Africa
37 L1 Dedoplis Tsqaro Georgia
114 C2 Dédougou Burkina
28 D3 Dedovichi Rus. Fed.
115 B5 Dedza Malawi
115 B5 Dedza Mountain Malawi
13 E4 Dee r. Eng./Wales U.K.
11 F3 Dee r. Scot. U.K.
13 D4 Dee est. Wales U.K.
14 C5 Deel r. Rep. of Ireland
14 D3 Deele r. Rep. of Ireland
67 C7 Deep r. Austr.
51 Deep Bay H.K. China
90 D5 Deep Creek Lake U.S.A.
93 F2 Deep Creek Range mts U.S.A.
89 J2 Deep River Can.
91 G4 Deep River U.S.A.
77 K1 Deep Rose Lake Can.
92 D3 Deep Springs U.S.A.
67 J2 Deepwater Austr.
90 B5 Deer Creek Lake U.S.A.
79 K4 Deer I. Can.
91 J2 Deer I. U.S.A.
91 J2 Deer Isle U.S.A.
78 B3 Deer L. Can.
79 J4 Deer Lake Nfld Can.
78 B3 Deer Lake Ont. Can.
82 D2 Deer Lodge U.S.A.
102 D2 Defensores del Chaco, Parque Nacional nat. park Para.
90 A4 Defiance U.S.A.
87 C6 De Funiak Springs U.S.A.
46 B3 Dêgê China
110 E3 Degeh Bur Eth.
114 E4 Degema Nigeria
45 G3 Dêgên China
20 F4 Deggendorf Ger.
44 C3 Degh r. Pak.
15 D1 De Haan Belgium
40 D4 Dehaj Iran
41 F4 Dehak Iran
41 F5 Dehak Iran
40 D4 Deh Bīd Iran
40 C4 Deh-e Khalīfeh Iran
40 C3 Deheq Iran
40 C2 Dehgāh Iran
40 B3 Dehgolān Iran
43 B5 Dehiwala-Mount Lavinia Sri Lanka
40 D5 Dehkūyeh Iran
40 B3 Dehlonān Iran
44 D3 Dehra Dun India
45 E4 Dehri India
41 E4 Deh Salm Iran
41 F4 Deh Sard Iran
47 K4 Deh Sheykh Iran
41 F4 Deh Shū Afgh.
51 F5 Dehua China
54 C1 Dehui China
15 D2 Deinze Belgium
36 E5 Deir el Qamer Lebanon
Deir-ez-Zor see Dayr az Zawr
17 L7 Dej Romania
51 C4 Dejiang China
115 A6 Deka Drum Zimbabwe
88 C5 De Kalb IL U.S.A.
85 E5 De Kalb TX U.S.A.
91 F2 De Kalb Junction U.S.A.
49 Q1 De-Kastri Rus. Fed.
42 A6 Dekemhare Eritrea
110 C4 Dekese Congo(Zaire)
39 F5 Dekhkanabad Uzbek.
114 E3 Dekina Nigeria
18 B3 De Koog Neth.
18 B4 De Kooy Neth.
92 C3 Delano U.S.A.
93 F2 Delano Peak U.S.A.
41 F3 Delārām Afgh.
113 F3 Delareyville S. Africa
77 H4 Delaronde Lake Can.
88 C5 Delavan IL U.S.A.
88 C4 Delavan WV U.S.A.
90 B4 Delaware U.S.A.
91 F5 Delaware div. U.S.A.
91 F4 Delaware r. U.S.A.
91 F5 Delaware Bay U.S.A.
90 B4 Delaware Lake U.S.A.
91 F4 Delaware Water Gap National Recreational Area res. U.S.A.
18 F5 Delbrück Ger.
67 H6 Delegate Austr.
22 E2 Delémont Switz.
18 B4 Delft Neth.
43 B4 Delft I. Sri Lanka
18 D3 Delfzijl Neth.
115 D5 Delgado, Cabo pt Moz.
89 G4 Delhi Can.
44 D3 Delhi India
83 F4 Delhi CO U.S.A.
91 F3 Delhi NY U.S.A.
56 Deli i. Indon.
37 J2 Deli r. Turkey
36 E2 Delice Turkey
36 E1 Delice r. Turkey
40 C3 Delījān Iran
76 E1 Délįne Can.
46 B3 Delingha China
77 H4 Delisle Can.
19 K5 Delitzsch Ger.
19 G5 Delligsen Ger.
84 D3 Dell Rapids U.S.A.
25 H4 Dellys Alg.
92 D5 Del Mar U.S.A.
93 E3 Delmar L. U.S.A.
15 G4 Delme France
18 F3 Delmenhorst Ger.
33 R2 De-Longa, O-va is Rus. Fed.
74 B3 De Long Mts U.S.A.
77 J5 Deloraine Can.
88 D5 Delphi U.S.A.
90 A4 Delphos U.S.A.
112 F4 Delportshoop S. Africa
87 D7 Delray Beach U.S.A.
83 E6 Del Río Mex.
85 C6 Del Rio U.S.A.
9 P6 Delsbo Sweden
93 H2 Delta CO U.S.A.
88 A5 Delta IA U.S.A.
93 F2 Delta UT U.S.A.
74 D3 Delta Junction U.S.A.
91 F3 Delta Reservoir U.S.A.
87 D6 Deltona U.S.A.
67 J2 Delungra Austr.
14 D4 Delvin Rep. of Ireland
27 J5 Delvinë Albania
38 C1 Dema r. Rus. Fed.
25 E1 Demanda, Sierra de la mts Spain
110 C4 Demba Congo(Zaire)
110 D3 Dembī Dolo Eth.
28 D4 Demidov Rus. Fed.
83 F5 Deming U.S.A.
103 E4 Demini r. Brazil
36 B2 Demirci Turkey
27 M4 Demirköy Turkey
19 L3 Demmin Ger.
87 C5 Demopolis U.S.A.
88 D5 Demotte U.S.A.
59 B3 Dempo, G. volc. Indon.
44 D2 Dêmqog China/India
28 H2 Dem'yanovo Rus. Fed.
17 Q2 Demyansk Rus. Fed.
112 D5 De Naawte S. Africa
110 E2 Denakil reg. Eritrea
74 C3 Denali U.S.A.
77 J4 Denare Beach Can.
39 F5 Denau Uzbek.
89 J3 Denbigh Can.
13 D4 Denbigh U.K.
18 B3 Den Burg Neth.
58 B1 Den Chai Thai.
59 C3 Dendang Indon.
114 B1 Dendâra Maur.
15 E1 Dendermonde Belgium
113 H1 Dendron S. Africa
50 C1 Dengkou China
45 H3 Dêngqên China
50 D3 Deng Xian China
Den Haag see The Hague
69 A2 Denham Austr.
18 D4 Den Ham Neth.
68 B3 Denham Ra. mts Austr.
69 A2 Denham Sound chan. Austr.
18 B4 Den Helder Neth.
25 G3 Denia Spain
66 F5 Deniliquin Austr.
82 C3 Denio U.S.A.
84 E3 Denison IA U.S.A.
85 D5 Denison TX U.S.A.
36 B3 Denizli Turkey
67 J4 Denman Austr.
116 C5 Denman Glacier Ant.
69 C7 Denmark Austr.
7 D3 Denmark country Europe
75 Q3 Denmark Strait Greenland/Iceland
93 H3 Dennehotso U.S.A.
91 H4 Dennis Port U.S.A.
11 E4 Denny U.K.
91 K2 Dennysville U.S.A.
59 E4 Denpasar Indon.
91 F5 Denton MD U.S.A.
85 D5 Denton TX U.S.A.
22 D4 Dent Parrachée mt France
62 F2 D'Entrecasteaux Islands P.N.G.
69 B7 d'Entrecasteaux National Park Austr.
69 B7 d'Entrecasteaux, Pt Austr.
71 D9 D'Entrecasteaux, Récifs rf New Caledonia Pac. Oc.
22 D3 Dents du Midi mt Switz.
82 F4 Denver U.S.A.
15 H4 Denzlingen Ger.
45 F4 Deo India
44 D3 Deoband India
45 F5 Deogarh India
44 C4 Deogarh mt India
45 F4 Deoghar India
44 D5 Deori India
45 E4 Deoria India
44 C2 Deosai, Plains of Pak.
45 G5 Deosil India
15 C1 De Panne Belgium
18 C5 De Peel reg. Neth.
88 C3 De Pere U.S.A.
91 F3 Deposit U.S.A.
89 J2 Depot-Forbes Can.
89 J2 Depot-Rowanton Can.
88 C5 Depue U.S.A.
33 P3 Deputatskiy Rus. Fed.
46 B4 Dêqên China
51 D6 Deqing Guangdong China
51 F4 Deqing Zhejiang China
51 F5 Deqiu China
85 E5 De Queen U.S.A.
44 B3 Dera Bugti Pak.
44 B3 Dera Ghazi Khan Pak.
44 B3 Dera Ismail Khan Pak.
44 B3 Derawar Fort Pak.
29 J7 Derbent Rus. Fed.
39 F5 Derbent Uzbek.
67 G8 Derby Tas. Austr.
64 D3 Derby W.A. Austr.
13 F5 Derby U.K.
91 G4 Derby CT U.S.A.
85 D4 Derby KS U.S.A.
14 D3 Derg r. Rep. of Ireland/U.K.
29 J5 Dergachi Rus. Fed.
14 C5 Derg, Lough l. Rep. of Ireland
29 F5 Derhachi Ukr.
85 E6 De Ridder U.S.A.
37 H3 Derik Turkey
36 E2 Derinkuyu Turkey
29 F5 Derkul r. Rus. Fed./Ukr.
15 E4 Der, Lac du l. France
112 C1 Derm Namibia
20 D2 Dermbach Ger.
14 D4 Derravaragh, Lough l. Rep. of Ireland
115 C6 Derre Moz.
14 E5 Derry r. Rep. of Ireland
91 H3 Derry U.S.A.
14 C3 Derryveagh Mts h. Rep. of Ireland
50 A1 Derstei China
109 F3 Derudeb Sudan
112 E6 De Rust S. Africa
26 G2 Derventa Bos.-Herz.
67 G9 Derwent r. Austr.
12 G4 Derwent r. U.K.
11 G6 Derwent Reservoir U.K.
12 D3 Derwent Water l. U.K.
38 C1 Derzhavino Rus. Fed.
39 F2 Derzhavinsk Kazak.
105 C2 Desaguadero r. Arg.
100 E7 Desaguadero r. Bol.
71 Désappointement, Îles de is Fr. Polynesia Pac. Oc.
92 D2 Desatoya Mts U.S.A.
89 F2 Desbarats Can.
74 F3 Des Bois, Lac l. Can.
77 J3 Deschambault L. Can.
77 J4 Deschambault Lake Can.
82 B2 Deschutes r. U.S.A.
110 D2 Desē Eth.
102 C7 Deseado Arg.
102 C7 Deseado r. Arg.
23 H4 Desenzano del Garda Italy
89 J3 Deseronto Can.
44 B3 Desert Canal Pak.
93 E5 Desert Center U.S.A.
93 F1 Desert Peak U.S.A.
76 G3 Desmarais Can.
84 E3 Des Moines IA U.S.A.
88 A5 Des Moines r. IA U.S.A.
83 G4 Des Moines NM U.S.A.
28 E5 Desna r. Rus. Fed.
29 D5 Desna Ukr.
28 E4 Desnogorsk Rus. Fed.
55 C4 Desolation Point Phil.
88 D4 Des Plaines U.S.A.
19 K5 Dessau Ger.
15 D1 Destelbergen Belgium
89 H1 Destor Can.
66 B5 D'Estrees B. Austr.
76 B2 Destruction Bay Can.
98 C5 Desventurados, Islas de los is Chile
15 B2 Desvres France
27 J2 Deta Romania
76 G2 Detah Can.
111 C5 Dete Zimbabwe
18 F5 Detmold Ger.
88 D3 Detour, Pt U.S.A.
89 F3 De Tour Village U.S.A.
89 F4 Detroit U.S.A.
89 F4 Detroit airport MI U.S.A.
84 E2 Detroit Lakes U.S.A.
67 H5 Deua Nat. Park Austr.
19 K5 Deuben Ger.
18 C5 Deurne Neth.
21 K5 Deutschkreutz Austria
21 J6 Deutschlandsberg Austria
89 H2 Deux-Rivières Can.
27 K2 Deva Romania
43 B2 Devarkonda India
21 L5 Devecser Hungary
36 E2 Develi Turkey
18 D4 Deventer Neth.
12 E3 Devét Skal h. Czech Rep.
44 B4 Devikot India
14 D5 Devils Bit Mountain h. Rep. of Ireland
13 D5 Devil's Bridge U.K.
92 C4 Devils Den U.S.A.
92 C2 Devils Gate pass U.S.A.
88 B2 Devils I. U.S.A.
84 D1 Devils Lake U.S.A.
92 C3 Devils Peak U.S.A.
92 C3 Devils Postpile National Monument res. U.S.A.
87 F7 Devil's Pt Bahamas
13 F6 Devizes U.K.
44 C4 Devli India
27 M3 Devnya Bulg.
22 C5 Dévoluy mts France
76 G4 Devon Can.
13 G5 Devon r. U.K.
75 J2 Devon Island Can.
67 G8 Devonport Austr.
36 C1 Devrek Turkey
36 D1 Devrekâni Turkey
36 E1 Devrez r. Turkey
43 A2 Devrukh India
44 D5 Dewas India
59 A2 Dewa, Tanjung pt Indon.
113 G4 Dewetsdorp S. Africa
90 B6 Dewey Lake U.S.A.
85 F5 De Witt AR U.S.A.
88 B5 De Witt IA U.S.A.
12 F4 Dewsbury U.K.
51 E4 Dexing China
91 J2 Dexter ME U.S.A.
85 F4 Dexter MO U.S.A.
91 E2 Dexter NY U.S.A.
50 B4 Deyang China
40 E3 Deyhuk Iran
37 M3 Deylaman Iran
38 E5 Deynau Turkm.
59 E3 Deyong, Tanjung pt Indon.
40 C3 Deyyer Iran
37 M6 Dez r. Iran
40 C3 Dezfūl Iran
40 C4 Dez Gerd Iran
50 E2 Dezhou China
40 B5 Dhahlān, J. h. S. Arabia
42 D4 Dhahran S. Arabia
45 G5 Dhaka Bangl.
45 G5 Dhaleswari r. Bangl.
45 H4 Dhaleswari r. India
42 B7 Dhamār Yemen
45 F5 Dhamara India
44 C5 Dhamnod India
45 E5 Dhamtari India
44 B3 Dhana Sar Pak.
45 F5 Dhanbad India
44 C5 Dhandhuka India
45 E3 Dhang Ra. mts Nepal
44 C5 Dhar India
45 F4 Dharan Bazar Nepal
43 B4 Dharapuram India
44 B5 Dhari India
43 B3 Dharmapuri India
43 B3 Dharmavaram India
44 D2 Dharmshala India
43 A3 Dhārwād India
44 D4 Dhasan r. India
45 E3 Dhaulagiri mt Nepal
44 D4 Dhaulpur India
44 C4 Dhebar L. India
45 H4 Dhekiajuli India
36 E6 Dhībān Jordan
45 H4 Dhing India
43 B3 Dhone India
44 B5 Dhoraji India
44 B5 Dhrangadhra India
45 F4 Dhule India
45 F4 Dhunche Nepal
44 D4 Dhund r. India
110 E3 Dhuusa Marreeb Somalia
27 L7 Dia i. Greece
22 E3 Diablerets, Les mts Switz.
92 B3 Diablo, Mt U.S.A.
92 B3 Diablo Range mts U.S.A.
114 B3 Diagbe Congo(Zaire)
114 C1 Diaka r. Mali
105 E2 Diamante Arg.
105 C2 Diamante r. Arg.
65 H4 Diamantina watercourse Austr.
104 D2 Diamantina Brazil
101 K6 Diamantina, Chapada plat. Brazil
104 A1 Diamantino Brazil
92 Diamond Head hd U.S.A.
68 D1 Diamond Islets Coral Sea Is Terr.
93 E2 Diamond Peak U.S.A.
51 D6 Dianbai China
51 B5 Dian Chi l. China
114 B1 Diandioumé Mali
51 C4 Dianjiang China
22 F6 Diano Marina Italy
101 J6 Dianópolis Brazil
114 B3 Dianra Côte d'Ivoire
52 B2 Diaoling China
114 D2 Diapaga Burkina
41 E6 Đibab Oman
45 H4 Dibang r. India
110 C4 Dibaya Congo(Zaire)
114 F1 Dibella well Niger
112 E3 Dibeng S. Africa
78 F2 D'Iberville, Lac l. Can.
113 G1 Dibete Botswana
45 H4 Dibrugarh India
85 C5 Dickens U.S.A.
91 J1 Dickey U.S.A.
84 C2 Dickinson U.S.A.
87 C4 Dickson U.S.A.
91 F4 Dickson City U.S.A.
Dicle r. see Tigris
55 B2 Didicas i. Phil.
44 C4 Didwana India
27 M4 Didymoteicho Greece
22 C5 Die France
15 H2 Dieblich Ger.
114 C2 Diébougou Burkina
20 B3 Dieburg Ger.
77 H4 Diefenbaker, L. Can.
117 J4 Diego Garcia i. British Indian Ocean Terr.
15 G3 Diekirch Lux.
114 B2 Diéma Mali
19 G5 Diemel r. Ger.
51 B6 Điện Biên Vietnam
58 C1 Diễn Châu Vietnam
58 D2 Diễn Khanh Vietnam
18 F4 Diepholz Ger.
15 B3 Dieppe France
50 C2 Di'er Nonchang Qu r. China
54 C1 Di'er Songhua Jiang r. China
18 C5 Diessen Neth.
20 E5 Dießen am Ammersee Ger.
15 F2 Diest Belgium
20 E3 Dietfurt an der Altmühl Ger.
22 F2 Dietikon Switz.
22 C5 Dieulefit France
15 G4 Dieulouard France
15 G4 Dieuze France
20 B2 Diez Ger.
114 F2 Diffa Niger
43 D2 Digapahandi India
79 G5 Digby Can.
45 F5 Digha India
22 D5 Digne-les-Bains France
15 B4 Digny France
22 A3 Digoin France
55 C5 Digos Phil.
44 D5 Digras India
44 B4 Digri Pak.
57 K8 Digul r. Indon.
114 C3 Digya National Park Ghana
22 C2 Dijon France
110 E2 Dikhil Djibouti
27 M5 Dikili Turkey
15 C1 Diksmuide Belgium
32 K2 Dikson Rus. Fed.
114 F2 Dikwa Nigeria
110 D3 Dīla Eth.
41 E4 Dilaram Iran
57 H8 Dili Indon.
37 K1 Dilijan Armenia
58 D3 Di Linh Vietnam
20 B2 Dillenburg Ger.
85 D6 Dilley U.S.A.
20 D4 Dillingen an der Donau Ger.
15 G3 Dillingen (Saar) Ger.
74 C4 Dillingham U.S.A.
77 H3 Dillon Can.
82 D2 Dillon MT U.S.A.
87 E5 Dillon SC U.S.A.
111 C5 Dilolo Congo(Zaire)
15 F1 Dilsen Belgium
37 K5 Diltāwa Iraq
45 H4 Dimapur India
36 F5 Dimashq Syria
110 C4 Dimbelenge Congo(Zaire)
114 C3 Dimbokro Côte d'Ivoire
66 E6 Dimboola Austr.
68 A1 Dimbulah Austr.
27 L3 Dimitrovgrad Bulg.
28 J4 Dimitrovgrad Rus. Fed.
36 E6 Dīmona Israel
112 D2 Dimpho Pan salt pan Botswana
55 C4 Dinagat i. Phil.
45 G4 Dinajpur Bangl.
24 C2 Dinan France
44 C2 Dinanagar India
114 C2 Dinangourou Mali
15 E2 Dinant Belgium
45 F4 Dinapur India
36 C2 Dinar Turkey
26 G2 Dinara mts Croatia
40 C4 Dīnār, Kūh-e mt Iran
109 F3 Dinder National Park Sudan
43 B4 Dindigul India
114 F2 Dindima Nigeria
113 K1 Dindiza Moz.
44 E5 Dindori India
36 D3 Dinek Turkey
50 C2 Dingbian China
110 B4 Dinge Angola
19 H5 Dingelstädt Ger.
51 G4 Dinghai China
45 F4 Dingla Nepal
14 A5 Dingle Rep. of Ireland
14 A5 Dingle Bay Rep. of Ireland
51 E5 Dingnan China
68 C4 Dingo Austr.
20 F4 Dingolfing Ger.
55 B2 Dingras Phil.
51 D3 Dingtao China
114 A2 Dinguiraye Guinea
11 D3 Dingwall U.K.
50 B3 Dingxi China
50 E2 Ding Xian China
50 E2 Dingxing China
50 E3 Dingyuan China
50 F2 Dingzi Gang harbour China
51 C6 Dinh Lập Vietnam
20 D3 Dinkelsbühl Ger.
93 G3 Dinnebito Wash r. U.S.A.
45 G3 Dinngyê China
113 G1 Dinokwe Botswana
77 L5 Dinorwic Lake U.S.A.
93 H1 Dinosaur U.S.A.
82 E3 Dinosaur Nat. Mon. res. U.S.A.
18 D5 Dinslaken Ger.
114 B2 Dioïla Mali
22 C5 Diois, Massif du mts France
104 B4 Dionísio Cerqueira Brazil
22 A3 Diou France
108 A3 Diourbel Senegal
45 H4 Diphu India
44 B4 Diplo Pak.
55 B4 Dipolog Phil.
70 B6 Dipton N.Z.
36 F2 Dirckli Turkey
114 C1 Diré Mali
65 H2 Direction, C. Austr.
110 E3 Dirē Dawa Eth.
111 C5 Dirico Angola
69 A2 Dirk Hartog Island Austr.
67 H2 Dirranbandi Austr.
93 G2 Dirty Devil r. U.S.A.
44 C4 Dīsa India
102 Disappointment, C. Atl. Ocean
82 A2 Disappointment, C. U.S.A.

Disappointment Is see Désappointement, Îles de
64 D4 Disappointment, L. salt flat Austr.
67 H6 Disaster B. Austr.
66 D7 Discovery Bay Austr.
51 □ Discovery Bay H.K. China
23 F3 Disentis Muster Switz.
23 G3 Disgrazia, Monte mt Italy
Disko i. see Qeqertarsuatsiaq
Disko Bugt b. see Qeqertarsuup Tunua
91 E6 Dismal Swamp swamp U.S.A.
45 G4 Dispur India
13 J5 Diss U.K.
104 C1 Distrito Federal div. Brazil
36 C6 Disûq Egypt
55 B4 Dit i. Phil.
112 E4 Ditloung S. Africa
26 F6 Dittaino r. Sicily Italy
20 D2 Dittelbrunn Ger.
20 C4 Ditzingen Ger.
55 C4 Diuata Mountains Phil.
55 C4 Diuata Pt Phil.
40 B3 Dīvāndarreh Iran
28 G4 Diveyevo Rus. Fed.
55 B2 Divilacan Bay Phil.
104 D3 Divinópolis Brazil
29 G6 Divnoye Rus. Fed.
114 B3 Divo Côte d'Ivoire
36 G2 Divriği Turkey
41 G5 Diwana Pak.
91 H2 Dixfield U.S.A.
91 J2 Dixmont U.S.A.
92 B2 Dixon CA U.S.A.
88 C5 Dixon IL U.S.A.
76 C4 Dixon Entrance chan. Can./U.S.A.
76 F3 Dixonville Can.
91 H2 Dixville Can.
37 J2 Diyadin Turkey
37 K5 Diyālā r. Iraq
37 H3 Diyarbakır Turkey
44 B4 Diyodar India
Dizak see Dāvar Panāh
40 D3 Diz Chah Iran
114 F4 Dja r. Cameroon
109 D2 Djado Niger
109 D2 Djado, Plateau du plat. Niger
110 B4 Djambala Congo
108 C2 Djanet Alg.
108 C1 Djelfa Alg.
110 C3 Djéma C.A.R.
114 C2 Djenné Mali
114 C2 Djibo Burkina
107 H4 Djibouti country Africa
110 E2 Djibouti Djibouti
14 E4 Djouce Mountain h. Rep. of Ireland
114 D3 Djougou Benin
114 F4 Djoum Cameroon
8 F4 Djúpivogur Iceland
9 O6 Djurås Sweden
37 K1 Dmanisi Georgia
33 Q2 Dmitriya Lapteva, Proliv chan. Rus. Fed.
52 C2 Dmitriyevka Primorskiy Kray Rus. Fed.
28 G4 Dmitriyevka Tambov. Rus. Fed.
39 L1 Dmitriyevka Rus. Fed.
29 E4 Dmitriyev-L'govskiy Rus. Fed.
28 F3 Dmitrov Rus. Fed.
17 P5 Dnieper r. Europe
17 N6 Dniester r. Ukr.
17 Q7 Dnipro r. Ukr.
29 E5 Dniprodzerzhyns'k Ukr.
29 E5 Dnipropetrovs'k Ukr.
29 E6 Dniprorudne Ukr.
28 D3 Dno Rus. Fed.
109 D4 Doba Chad
45 G3 Doba China
89 G3 Dobbinton Can.
9 S8 Dobele Latvia
19 L5 Döbeln Ger.
57 J7 Doberai Peninsula Indon.
19 L5 Doberlug-Kirchhain Ger.
19 M5 Döbern Ger.
105 D3 Doblas Arg.
57 J8 Dobo Indon.
27 H2 Doboj Bos.-Herz.
20 G3 Dobřany Czech Rep.
27 M3 Dobrich Bulg.
29 G4 Dobrinka Rus. Fed.
21 H3 Dobříš Czech Rep.
21 M2 Dobrodzień Pol.
29 D4 Dobrush Belarus
21 K2 Dobruška Czech Rep.
21 L2 Dobrzeń Wielki Pol.
55 A5 Doc Can rf Phil.
104 E2 Doce r. Brazil
13 H5 Docking U.K.
97 D3 Doctor Arroyo Mex.
96 C1 Doctor Belisario Domínguez Mex.
43 B3 Dod Ballapur India
27 M6 Dodecanese is Greece
Dodekanisos is see Dodecanese
82 C2 Dodge U.S.A.
88 A3 Dodge Center U.S.A.
85 C4 Dodge City U.S.A.
67 G9 Dodges Ferry Austr.
88 B4 Dodgeville U.S.A.
23 J4 Dodici, Cima mt Italy
13 C7 Dodman Point U.K.
115 C4 Dodoma Tanz.
115 C4 Dodoma div. Tanz.
18 D5 Doetinchem Neth.
57 H7 Dofa Indon.
45 G2 Dogai Coring salt l. China
36 F2 Doğanşehir Turkey
76 E4 Dog Creek Can.
45 G3 Dogên Co l. China
79 H2 Dog Island Can.
77 K4 Dog L. Can.
89 E1 Dog Lake Can.
53 C6 Dōgo i. Japan
114 D2 Dogondoutchi Niger
53 C7 Dōgo-yama mt Japan
37 K2 Doğubeyazıt Turkey
45 G3 Do'gyaling China
40 C5 Doha Qatar
Dohad see Dāhod
45 H5 Dohazar Bangl.
45 G3 Doilungdêqên China
58 A1 Doi Saket Thai.
101 K5 Dois Irmãos, Serra dos h. Brazil
27 K4 Dojran, Lake Greece/Macedonia
9 M6 Dokka Norway
18 C3 Dokkum Neth.
44 B4 Dokri Pak.
17 N3 Dokshytsy Belarus
21 H2 Doksy Czech Rep.
38 F2 Dokuchayevka Kazak.
29 F6 Dokuchayevs'k Ukr.
114 C1 Do, Lac l. Mali
57 K8 Dolak, Pulau i. Indon.
77 G4 Dolbeau Can.
13 C5 Dolbenmaen U.K.
24 D2 Dol-de-Bretagne France
22 C2 Dole France
13 D5 Dolgellau U.K.
19 L3 Dolgen Ger.
91 F3 Dolgeville U.S.A.
29 F4 Dolgorukovo Rus. Fed.
29 F4 Dolgoye Rus. Fed.
26 C5 Dolianova Sardinia Italy
49 Q2 Dolinsk Rus. Fed.
19 M3 Doliny Dolnej Odry, Park Krajobrazowy res. Pol.
116 B2 Dolleman I. Ant.
20 E4 Dollnstein Ger.
39 G1 Dolmatovo Kazak.
21 H4 Dolní Dvořiště Czech Rep.
21 L4 Dolní Němčí Czech Rep.
23 K3 Dolomiti Bellunesi, Parco Nazionale delle nat. park Italy
23 J3 Dolomitiche, Alpi mts Italy
110 E3 Dolo Odo Eth.
105 F3 Dolores Arg.
97 G4 Dolores Guatemala
105 E2 Dolores Uru.
93 H2 Dolores r. U.S.A.
96 D3 Dolores Hidalgo Mex.
74 G3 Dolphin and Union Str. Can.
51 B7 Đô Lương Vietnam
29 B5 Dolyna Ukr.
36 B2 Domaniç Turkey
44 E2 Domar China
23 G3 Domat Ems Switz.
20 F3 Domažlice Czech Rep.
45 H2 Domba China
38 D2 Dombarovskiy Rus. Fed.
9 L5 Dombås Norway
16 J7 Dombóvár Hungary
116 C4 Dome Argus ice feature Ant.
116 C5 Dome Circe ice feature Ant.
76 E4 Dome Creek Can.
76 D2 Dome Pk Can.
93 E5 Dome Rock Mts U.S.A.
24 D2 Domfront France
73 K8 Dominica country Caribbean Sea
96 J6 Dominical Costa Rica
73 J8 Dominican Republic country Caribbean Sea
19 J3 Dömitz Ger.
58 C2 Dom Noi, L. r. Thai.
22 F3 Domodossola Italy
27 K5 Domokos Greece
105 F1 Dom Pedrito Brazil
59 E4 Dompu Indon.
105 B3 Domuyo, Volcán volc. Arg.
67 J2 Domville, Mt h. Austr.
68 C3 Don r. Austr.
43 B2 Don r. India
96 B2 Don Mex.
29 G5 Don r. Rus. Fed.
11 F3 Don r. U.K.
14 F3 Donaghadee U.K.
14 E3 Donaghmore U.K.
66 E6 Donald Austr.
Donau r. Austria/Ger. see Danube
20 B5 Donaueschingen Ger.
20 D4 Donauwörth Ger.
25 D3 Don Benito Spain
12 F4 Doncaster U.K.
111 B4 Dondo Angola
111 D5 Dondo Moz.
55 B4 Dondonay i. Phil.
43 C5 Dondra Head c. Sri Lanka
14 C3 Donegal Rep. of Ireland
14 C3 Donegal Bay g. Rep. of Ireland
29 F6 Donets'k Ukr.
29 F5 Donets'kyy Kryazh h. Rus. Fed./Ukr.
51 D5 Dong'an China
69 B4 Dongara Austr.
44 E5 Dongargarh India
51 B5 Dongchuan China
45 F2 Dongco China
51 C7 Dongfang China
52 C1 Dongfanghong China
54 C2 Dongfeng China
59 E3 Donggala Indon.
54 C4 Donggou China
51 E5 Donggu China
51 D6 Dongguan China
58 C1 Đông Hà Vietnam
50 F3 Donghai China
51 D6 Donghai Dao i. China
50 A1 Dong He watercourse Nei Monggol China
50 C3 Dong He r. Sichuan China
58 C1 Đông Hới Vietnam
54 E1 Dongjingcheng China
45 H3 Dongjug Xizang China
45 H3 Dongjug Xizang China
51 D5 Dongkou China
45 G4 Dongkya La pass India
51 C5 Donglan China
50 A2 Dongle China
54 C2 Dongliao r. China
54 B1 Dongminzhutun China
54 F2 Dongning China
111 B5 Dongo Angola
23 G3 Dongo Italy
110 B3 Dongou Congo
58 B1 Dong Phraya Fai mts Thai.
58 B2 Dong Phraya Yen esc. Thai.
51 D6 Dongping Guangdong China
50 E3 Dongping Shandong China
45 G3 Dongqiao China
51 E5 Dongshan China
51 E6 Dongshan Dao i. China
50 D2 Dongsheng China
50 F3 Dongtai Jiangsu China
50 F3 Dongtai r. China
51 D4 Dongting Hu l. China
51 F5 Dongtou China
46 D2 Dong Ujimqin Qi China
51 E4 Dongxiang China
50 B3 Dongxiangzu China
51 F4 Dongyang China
50 F2 Dongying China
50 B2 Dongzhen China
51 E5 Dongzhi China
76 B2 Donjek r. Can.
18 D3 Donkerbroek Neth.
45 G5 Donmanick Islands Bangl.
96 D2 Don Martín Mex.
79 F4 Donnacona Can.
76 F3 Donnelly Can.
70 D1 Donnellys Crossing N.Z.
92 B2 Donner Pass U.S.A.
69 B6 Donnybrook Austr.
Donostia–San Sebastián see San Sebastián
27 L6 Donoussa i. Greece
28 F4 Donskoy Rus. Fed.
29 G6 Donskoye Rus. Fed.
55 B3 Donsol Phil.
58 C2 Don, Xé r. Laos
23 K5 Donzella, Isola della i. Italy
14 A4 Dooagh Rep. of Ireland
14 B5 Doonbeg r. Rep. of Ireland
11 D5 Doon, Loch l. U.K.
18 C4 Doorn Neth.
88 D3 Door Peninsula U.S.A.
18 C5 Doorwerth Neth.
110 E3 Dooxo Nugaaleed v. Somalia
41 F4 Dor watercourse Afgh.
85 C5 Dora U.S.A.
22 F4 Dora Baltea r. Italy
13 E7 Dorchester U.K.
111 B6 Dordabis Namibia
15 C4 Dordives France
24 E4 Dordogne r. France
18 B5 Dordrecht Neth.
113 G5 Dordrecht S. Africa
112 C1 Doreenville Namibia
77 H4 Doré L. Can.
77 H4 Doré Lake Can.
20 F4 Dorfen Ger.
26 C4 Dorgali Sardinia Italy
41 G4 Dori r. Afgh.
114 C2 Dori Burkina
112 C5 Doring r. S. Africa
13 G6 Dorking U.K.
18 D5 Dormagen Ger.
15 D3 Dormans France
23 C5 Dornbirn Austria
50 C1 Dornogovi div. Mongolia
11 D3 Dornoch Firth est. U.K.
18 E3 Dornum Ger.
114 C1 Doro Mali
28 E4 Dorogobuzh Rus. Fed.
17 N7 Dorohoi Romania
48 F2 Döröö Nuur salt l. Mongolia
8 P4 Dorotea Sweden
69 A2 Dorre Island Austr.
67 K3 Dorrigo Austr.
82 B3 Dorris U.S.A.
108 D4 Dorsale Camerounaise slope Cameroon/Nigeria
89 H3 Dorset Can.
22 C3 Dortan France
18 E5 Dortmund Ger.
90 B6 Dorton U.S.A.
36 F3 Dörtyol Turkey
18 F3 Dorum Ger.
115 A2 Doruma Congo(Zaire)
40 E3 Dorüneh Iran
91 G2 Dorval Can.
19 G4 Dörverden Ger.
102 C6 Dos Bahías, C. pt Arg.
93 H5 Dos Cabezas U.S.A.
100 C5 Dos de Mayo Peru
51 C6 Đo Son Vietnam
92 B3 Dos Palos U.S.A.
19 K4 Dosse r. Ger.
114 D2 Dosso Niger
38 C3 Dossor Kazak.
39 K3 Dostyk Kazak.
87 C6 Dothan U.S.A.
15 D2 Douai France
114 E4 Douala Cameroon
24 B2 Douarnenez France
51 □ Double I. H.K. China
68 E5 Double Island Pt Austr.
92 C4 Double Peak U.S.A.
68 B1 Double Pt Austr.
21 J3 Doubrava r. Czech Rep.
22 C3 Doubs r. France/Switz.
70 A6 Doubtful Sound in. N.Z.
70 D1 Doubtless Bay N.Z.
22 C3 Doucier France
114 C2 Douentza Mali
12 C3 Douglas Isle of Man
112 E4 Douglas S. Africa
11 E5 Douglas Scot. U.K.
76 C3 Douglas AK U.S.A.
93 H6 Douglas AZ U.S.A.
93 H6 Douglas airport AZ U.S.A.
87 D6 Douglas GA U.S.A.
82 F3 Douglas WY U.S.A.
76 D4 Douglas Chan. Can.
93 H2 Douglas Creek r. U.S.A.
15 C2 Doullens France
114 D3 Doumè Benin
114 F4 Doumé Cameroon
114 C2 Douna Mali
11 D4 Doune U.K.
104 C2 Dourada, Cach. waterfall Brazil
104 B2 Dourada, Serra h. Brazil
104 C1 Dourada, Serra mts Brazil
104 A3 Dourados Brazil
104 A3 Dourados r. Brazil
104 B3 Dourados, Serra dos h. Brazil
25 C2 Douro r. Port.
22 D4 Doussard France
15 F3 Douzy France
13 F4 Dove r. Eng. U.K.
13 J5 Dove r. Eng. U.K.
79 J3 Dove Brook Can.
76 B3 Dove Creek U.S.A.
67 G9 Dover Austr.
13 J6 Dover U.K.
91 F5 Dover DE U.S.A.
91 H3 Dover NH U.S.A.
91 F4 Dover NJ U.S.A.
90 C4 Dover OH U.S.A.
91 J2 Dover-Foxcroft U.S.A.
13 J7 Dover, Strait of France/U.K.
37 L5 Doveyrīch r. Iran/Iraq
19 H2 Dovnsklint cliff Denmark
88 D5 Dowagiac U.S.A.
40 D4 Dow Chāhī Iran
41 E2 Dowgha'ī Iran
58 A5 Dowi, Tg pt Indon.
41 G3 Dowlatābād Afgh.
41 F2 Dowlatābād Afgh.
41 G2 Dowlatābād Afgh.
41 F4 Dowlatābād Iran
40 D4 Dowlatābād Iran
40 E4 Dowlatābād Iran
41 G3 Dowl at Yār Afgh.
92 B2 Downieville U.S.A.
14 F3 Downpatrick U.K.
91 F3 Downsville U.S.A.
40 C3 Dow Rūd Iran
37 L4 Dow Sar Iran
40 D4 Dowsārī Iran
41 H3 Dowshī Afgh.
92 B1 Doyle U.S.A.
91 F4 Doylestown U.S.A.
53 C6 Dōzen is Japan
89 J2 Dozois, Réservoir resr Can.
104 B3 Dracena Brazil
18 D3 Drachten Neth.
27 L2 Drăgănești-Olt Romania
27 L2 Drăgășani Romania
103 E2 Dragon's Mouths str. Trinidad/Venez.
19 K1 Dragør Denmark
9 S6 Dragsfjärd Fin.
22 D6 Draguignan France
29 C4 Drahichyn Belarus
67 K2 Drake Austr.
93 F4 Drake AZ U.S.A.
77 J5 Drake ND U.S.A.
113 H5 Drakensberg mts Lesotho/S. Africa
113 J2 Drakensberg mts S. Africa
120 E9 Drake Passage str. Ant.
27 L4 Drama Greece
9 M7 Drammen Norway
9 L7 Drangedal Norway
41 G5 Dran juk h. Pak.
19 G5 Dransfeld Ger.
19 L2 Dranske Ger.
13 E3 Draperstown U.K.
44 C2 Dras Jammu and Kashmir
21 H6 Drau r. Austria
21 J6 Dravograd Slovenia
76 G4 Drayton Valley Can.
26 B6 Dréan Alg.
19 M5 Drebkau Ger.
22 B2 Drée r. France
22 D2 Dreieich Ger.
20 E5 Dreitorspitze mt Ger.
19 H2 Drejø i. Denmark
19 L5 Dresden Ger.
28 D4 Dretun' Belarus
15 B4 Dreux France
9 N6 Drevsjø Norway
90 D4 Driftwood U.S.A.
68 D6 Drillham Austr.
14 B6 Drimoleague Rep. of Ireland
26 G3 Drniš Croatia
27 K2 Drobeta-Turnu Severin Romania
19 G3 Drochtersen Ger.
14 E4 Drogheda Rep. of Ireland
29 B5 Drohobych Ukr.
Droichead Átha see Drogheda
13 E5 Droitwich U.K.
45 G4 Drokung India
22 C5 Drôme r. France
19 H4 Drömling reg. Ger.
14 D4 Dromod Rep. of Ireland
14 E3 Dromore Co. Down U.K.
14 D3 Dromore Co. Tyrone U.K.
22 E5 Dronero Italy
13 F4 Dronfield U.K.
75 Q2 Dronning Louise Land reg. Greenland
116 C3 Dronning Maud Land reg. Ant.
18 C4 Dronten Neth.
21 J4 Drosendorf Austria
42 B2 Drosh Pak.
29 F4 Droskovo Rus. Fed.
15 B4 Droué France
67 F7 Drouin Austr.
102 D2 Dr Pedro P. Peña Para.
71 □13 Drua Drua i. Fiji
15 H4 Drulingen France
76 G4 Drumheller Can.
82 D2 Drummond MT U.S.A.
88 B2 Drummond WV U.S.A.
89 F3 Drummond Island U.S.A.
68 B5 Drummond Range h. Austr.
79 F4 Drummondville Can.
11 D6 Drummore U.K.
11 D4 Drumochter Pass U.K.
23 F2 Drusberg mt Switz.
9 T10 Druskininkai Lith.
33 Q3 Druzhina Rus. Fed.
27 L3 Dryanovo Bulg.
76 B3 Dry Bay U.S.A.
77 L5 Dryberry L. Can.
88 E2 Dryburg U.S.A.
78 B4 Dryden Can.
116 D5 Drygalski I. Ant.
116 B5 Drygalski Ice Tongue ice feature Ant.
92 D2 Dry Lake U.S.A.
11 D4 Drymen U.K.
64 E3 Drysdale r. Austr.
64 E3 Drysdale River National Park Austr.
40 C3 Dūāb r. Iran
51 C6 Du'an China
91 F2 Duane U.S.A.
68 C4 Duaringa Austr.
45 G4 Duars reg. India
21 H2 Dubá Czech Rep.
34 B4 Dubā S. Arabia
42 E4 Dubai U.A.E.
77 J2 Dubawnt r. Can.
74 H3 Dubawnt Lake N.W.T. Can.
77 J2 Dubawnt Lake Can.
Dubayy see Dubai
34 B4 Dubbagh, J. ad mt S. Arabia
67 H4 Dubbo Austr.
22 F2 Dübendorf Switz.
21 G2 Dubí Czech Rep.
88 D1 Dublin Can.
14 E4 Dublin Rep. of Ireland
87 D5 Dublin U.S.A.
28 F3 Dubna Rus. Fed.
21 H4 Dubné Czech Rep.
21 M4 Dubnica nad Váhom Slovakia
29 C5 Dubno Ukr.
82 D2 Dubois ID U.S.A.
82 E3 Dubois WY U.S.A.
90 D4 Du Bois U.S.A.
29 H5 Dubovka Rus. Fed.
29 G6 Dubovskoye Rus. Fed.
37 M1 Dübrar P. pass Azer.
108 A4 Dubréka Guinea
27 H3 Dubrovnik Croatia
29 C5 Dubrovytsya Ukr.
28 D4 Dubrowna Belarus
39 K4 Dubun Kazak.
88 B4 Dubuque U.S.A.
9 S9 Dubysa r. Lith.
71 □2 Duc de Gloucester, Îles is Fr. Polynesia Pac. Oc.
51 E4 Duchang China
19 L3 Ducherow Ger.
93 G1 Duchesne U.S.A.
60 P6 Ducie Island Pitcairn Is Pac. Oc.
87 C5 Duck r. U.S.A.
77 J4 Duck Bay Can.
77 H4 Duck Lake Can.
88 E3 Duck Lake U.S.A.
93 E2 Duckwater U.S.A.
93 E2 Duckwater Peak U.S.A.
58 D2 Đưc Phô Vietnam
58 D3 Đưc Trong Vietnam
103 B4 Duda r. Col.
15 G3 Dudelange Lux.
19 H5 Duderstadt Ger.
45 E4 Dudhi India
45 G4 Dudhnai India
32 K3 Dudinka Rus. Fed.
13 E5 Dudley U.K.
44 D6 Dudna r. India
21 L4 Dudváh r. Slovakia
11 F3 Dudwick, Hill of U.K.
114 B3 Duékoué Côte d'Ivoire
25 C2 Duero r. Spain
89 H1 Dufault, Lac l. Can.
116 B4 Dufek Coast Ant.
15 E1 Duffel Belgium
78 E2 Dufferin, Cape hd Can.
90 B6 Duffield U.S.A.
63 G2 Duff Is Solomon Is
22 C5 Duffre, Le mt France
11 E3 Dufftown U.K.
78 E1 Dufrost, Pte pt Can.
39 F5 Dugab Uzbek.
26 F3 Dugi Otok i. Croatia
50 C2 Dugui Qarag China
50 D3 Du He r. China
103 D4 Duida, Co mt Venez.
100 E3 Duida-Marahuaca, Parque Nacional nat. park Venez.
23 L4 Duino Italy
18 D5 Duisburg Ger.
103 B3 Duitama Col.
113 J1 Duiwelskloof S. Africa
37 K4 Dukān Dam dam Iraq
113 G5 Dukathole S. Africa
76 C4 Duke I. U.S.A.
Duke of Gloucester Is see Duc de Gloucester, Îles
40 C5 Dukhān Qatar
17 Q3 Dukhovshchina Rus. Fed.
44 B3 Duki Pak.
51 A5 Dukou China
9 U9 Dūkštas Lith.
46 B3 Dulan China
102 D3 Dulce r. Arg.
96 J6 Dulce, Golfo b. Costa Rica
96 H5 Dulce Nombre de Culmí Honduras
45 E2 Dulishi Hu salt l. China
113 J2 Dullstroom S. Africa
18 E5 Dülmen Ger.
27 M3 Dulovo Bulg.
88 A2 Duluth U.S.A.
88 A2 Duluth/Superior airport U.S.A.
13 D6 Dulverton U.K.
36 F5 Dūmā Syria
55 B4 Dumaguete Phil.
59 B2 Dumai Indon.
55 B4 Dumaran i. Phil.
85 F5 Dumas AR U.S.A.
85 C5 Dumas TX U.S.A.
36 F5 Dumayr Syria
40 B3 Dumbakh Iran
11 D5 Dumbarton U.K.
113 J3 Dumbe S. Africa
71 □9 Dumbéa New Caledonia Pac. Oc.
17 J6 Ďumbier mt Slovakia
69 C6 Dumbleyung Austr.
69 C6 Dumbleyung, Lake salt flat Austr.
44 D2 Dumchele Jammu and Kashmir
45 H4 Dum Duma India
11 E5 Dumfries U.K.
45 F4 Dumka India
18 F4 Dümmer l. Ger.
19 K3 Dummerstorf Ger.
78 E4 Dumoine, L. Can.
116 B6 Dumont d'Urville France base Ant.
116 B6 Dumont d'Urville Sea Ant.
15 G2 Dümpelfeld Ger.
109 F1 Dumyāṭ Egypt
19 H5 Dün ridge Ger.
Duna r. Hungary see Danube
Dunaj r. Slovakia see Danube
16 H7 Dunajská Streda Slovakia
17 J7 Dunakeszi Hungary
67 G9 Dunalley Austr.
14 E4 Dunany Point Rep. of Ireland
Dunărea r. Romania see Danube
27 N2 Dunării, Delta Romania
17 J7 Dunaújváros Hungary
Dunav r. Yugo. see Danube
29 C5 Dunayivtsi Ukr.
70 C6 Dunback N.Z.
11 F4 Dunbar U.K.
11 E4 Dunblane U.K.
14 E4 Dunboyne Rep. of Ireland
76 E5 Duncan Can.
93 H5 Duncan AZ U.S.A.
85 D5 Duncan OK U.S.A.
78 D3 Duncan, Cape Can.
78 E3 Duncan, L. Can.
90 E4 Duncannon U.S.A.
11 E2 Duncansby Head hd U.K.
88 B5 Duncans Mills U.S.A.
14 E5 Duncormick Rep. of Ireland
9 S8 Dundaga Latvia
89 G3 Dundalk Can.
14 E3 Dundalk Rep. of Ireland
90 E5 Dundalk U.S.A.
14 E4 Dundalk Bay Rep. of Ireland
Dundas see Uummannaq
76 C4 Dundas I. Can.
69 E6 Dundas, L. salt flat Austr.
Dun Dealgan see Dundalk
113 J4 Dundee S. Africa
11 F4 Dundee U.K.
89 F5 Dundee MI U.S.A.
90 E3 Dundee NY U.S.A.
14 F3 Dundonald U.K.
67 F1 Dundoo Austr.
11 E6 Dundrennan U.K.
14 F3 Dundrum U.K.
14 F3 Dundrum Bay U.K.
45 E4 Dundwa Range mts India/Nepal
70 C6 Dunedin N.Z.
87 D6 Dunedin U.S.A.
67 H4 Dunedoo Austr.
78 F2 Dune, Lac l. Can.
39 K2 Dunenbay Kazak.
11 E4 Dunfermline U.K.
14 E3 Dungannon U.K.
44 C5 Dungarpur India
14 D5 Dungarvan Rep. of Ireland
13 H7 Dungeness hd U.K.
102 C8 Dungeness, Pta pt Arg.
15 H2 Düngenheim Ger.
14 C3 Dungiven U.K.
14 C3 Dungloe Rep. of Ireland
67 J4 Dungog Austr.
115 A2 Dungu Congo(Zaire)
59 B2 Dungun Malaysia
109 F2 Dungunab Sudan
54 E2 Dunhua China
46 B2 Dunhuang China
66 E6 Dunkeld Austr.
11 E4 Dunkeld U.K.
Dunkerque see Dunkirk
13 D6 Dunkery Beacon h. U.K.
15 C1 Dunkirk France
90 D3 Dunkirk U.S.A.
114 C3 Dunkwa Ghana
14 E4 Dún Laoghaire Rep. of Ireland
14 E4 Dunlavin Rep. of Ireland
14 E4 Dunleer Rep. of Ireland
14 E2 Dunloy U.K.
14 B6 Dunmanus Bay Rep. of Ireland
14 B6 Dunmanway Rep. of Ireland
14 C4 Dunmore Rep. of Ireland
87 E7 Dunmore Town Bahamas
92 D3 Dunmovin U.S.A.
14 F3 Dunmurry U.K.
87 E5 Dunn U.S.A.
11 E2 Dunnet Bay U.K.
11 E2 Dunnet Head hd U.K.
92 B2 Dunnigan U.S.A.
84 C3 Dunning U.S.A.
69 D2 Dunns Range h. Austr.
89 H4 Dunnville Can.
66 E6 Dunolly Austr.
11 D5 Dunoon U.K.
11 F5 Duns U.K.
69 B6 Dunsborough Austr.
84 C1 Dunseith U.S.A.
82 B3 Dunsmuir U.S.A.
13 G6 Dunstable U.K.
70 B6 Dunstan Mts N.Z.
15 F3 Dun-sur-Meuse France
70 C6 Duntroon N.Z.
11 B3 Dunvegan, Loch in. U.K.
44 B3 Dunyapur Pak.
50 E1 Duolun China
51 D5 Dupang Ling mts China
89 H1 Duparquet, Lac l. Can.
84 C2 Dupree U.S.A.
86 B4 Du Quoin U.S.A.
64 E3 Durack r. Austr.
Dura Europos see Qal'at as Ṣāliḥīyah
36 E1 Durağan Turkey
22 B6 Durance r. France
89 F4 Durand MI U.S.A.
88 B3 Durand WV U.S.A.
71 □9 Durand, Récif rf New Caledonia Pac. Oc.
96 C2 Durango Mex.
96 C2 Durango div. Mex.
25 E1 Durango Spain
83 F4 Durango U.S.A.
85 D5 Durant U.S.A.
105 F2 Durazno Uru.
105 F1 Durazno, Cuchilla Grande del h. Uru.
113 J4 Durban S. Africa
24 F5 Durban-Corbières France
112 C6 Durbanville S. Africa
90 D5 Durbin U.S.A.
15 G4 Durbion r. France
15 F2 Durbuy Belgium
18 D6 Düren Ger.
44 E5 Durg India
45 F5 Durgapur India
89 G3 Durham Can.
12 F3 Durham U.K.
92 B2 Durham CA U.S.A.
87 E4 Durham NC U.S.A.
91 H3 Durham NH U.S.A.
29 D6 Durlești Moldova
27 H3 Durmitor mt Yugo.
11 D2 Durness U.K.
21 K4 Dürnkrut Austria
68 D6 Durong South Austr.
27 H4 Durrës Albania
13 E6 Durrington U.K.
14 A6 Dursey Island Rep. of Ireland
36 B2 Dursunbey Turkey
41 F3 Dūruḩ Iran
36 F5 Durūz, Jabal ad mt Syria
70 D4 D'Urville Island N.Z.
57 K7 D'Urville, Tanjung pt Indon.
41 G3 Dushai Afgh.
41 G4 Dushai Pak.
38 D5 Dushak Turkm.
51 C5 Dushan China
39 G5 Dushanbe Tajik.
29 H7 Dusheti Georgia
70 A6 Dusky Sound in. N.Z.
18 D5 Düsseldorf Ger.
39 G4 Dustlik Uzbek.
21 K2 Duszniki-Zdrój Pol.
93 F1 Dutch Mt U.S.A.
112 E1 Dutlwe Botswana
114 E2 Dutse Nigeria
66 B3 Dutton, Lake salt flat Austr.
89 F2 Dutton, Mt mt U.S.A.
28 H3 Duvannoye Rus. Fed.
79 F2 Duvert, Lac l. Can.
39 K5 Duxanbibazar China
51 C5 Duyun China
41 F5 Duzab Pak.
36 C1 Düzce Turkey
Duzdab see Zāhedān
28 D4 Dvina, Western r. Rus. Fed.
21 L3 Dvorce Czech Rep.
29 F5 Dvorichna Ukr.
52 B2 Dvoryanka Rus. Fed.
21 M5 Dvory nad Žitavou Slovakia
21 J2 Dvůr Králové Czech Rep.
44 B5 Dwarka India
113 G2 Dwarsberg S. Africa
69 C6 Dwellingup Austr.
88 C5 Dwight U.S.A.
18 D4 Dwingelderveld, Nationaal Park nat. park Neth.
82 C2 Dworshak Resr U.S.A.
112 D6 Dwyka S. Africa
28 E4 Dyat'kovo Rus. Fed.
11 F3 Dyce U.K.
88 D5 Dyer IN U.S.A.
92 C3 Dyer NV U.S.A.
89 G3 Dyer Bay Can.
75 M3 Dyer, C. Can.
87 B4 Dyersburg U.S.A.
88 B4 Dyersville U.S.A.
13 D5 Dyfi r. U.K.
21 K4 Dyje r. Austria/Czech Rep.
11 E3 Dyke U.K.
29 G7 Dykh Tau mt Georgia/Rus. Fed.
20 F3 Dyleň h. Czech Rep.
17 J4 Dylewska Góra h. Pol.
66 F2 Dynevor Downs Austr.
113 H5 Dyoki S. Africa
68 C4 Dysart Austr.
88 A4 Dysart U.S.A.
112 E6 Dysselsdorp S. Africa
38 E3 Dyurmen'tyube Kazak.
49 K3 Dzamïn Üüd Mongolia
111 E5 Dzaoudzi Mayotte Africa
21 G2 Džbán mts Czech Rep.
28 G3 Dzerzhinsk Rus. Fed.
39 K3 Dzerzhinskoye Kazak.
17 N5 Dzerzhyns'k Ukr.
38 F3 Dzhalagash Kazak.
Dzhalal-Abad see Jalal-Abad
38 C2 Dzhambeyty Kazak.
38 C4 Dzhanga Turkm.
38 B2 Dzhangala Kazak.
38 E4 Dzhankel'dy Uzbek.
29 E6 Dzhankoy Ukr.
39 J3 Dzhansugurov Kazak.
38 A2 Dzharbyek Kazak.
39 F5 Dzharkurgan Uzbek.
38 D5 Dzhebel Turkm.
38 E1 Dzhetygara Kazak.
39 F2 Dzhezkazgan Kazak.
38 E4 Dzhigirbent Turkm.
33 R3 Dzhigudzhak Rus. Fed.
39 F4 Dzhizak Uzbek.
46 F1 Dzhugdzhur, Khrebet mts Rus. Fed.
Dzhul'fa see Culfa
39 F5 Dzhuma Uzbek.
39 J3 Dzhungarskiy Alatau, Khrebet mts China/Kazak.
38 D2 Dzhurun Kazak.
38 F3 Dzhusaly Kazak.
17 K4 Działdowo Pol.
97 G4 Dzibalchén Mex.
21 K2 Dzierżoniów Pol.
97 G3 Dzilam de Bravo Mex.
39 K3 Dzungarian Gate pass China/Kazak.
48 J2 Dzuunmod Mongolia
28 C4 Dzyaniskavichy Belarus
28 C4 Dzyarzhynsk Belarus
17 N4 Dzyatlavichy Belarus

E

78 C3 Eabamet L. Can.
93 H4 Eagar U.S.A.
79 J3 Eagle r. Can.
83 F4 Eagle U.S.A.
91 F3 Eagle Bay U.S.A.
77 H4 Eagle Cr. r. Can.
92 D4 Eagle Crags summit U.S.A.
77 L5 Eagle L. Can.
82 B3 Eagle L. CA U.S.A.
91 J1 Eagle Lake ME U.S.A.
91 J1 Eagle Lake ME U.S.A.
88 B2 Eagle Mtn h. U.S.A.
85 C6 Eagle Pass U.S.A.
74 D3 Eagle Plain Can.
88 C2 Eagle River MI U.S.A.
88 C3 Eagle River WV U.S.A.
76 F3 Eaglesham Can.
93 F5 Eagle Tail Mts U.S.A.
69 E2 Earaheedy Austr.
78 B3 Ear Falls Can.
92 C4 Earlimart U.S.A.
11 F5 Earlston U.K.
89 H2 Earlton Can.
11 E4 Earn r. U.K.
11 D4 Earn, L. U.K.
85 C5 Earth U.S.A.
12 H4 Easington U.K.
87 D5 Easley U.S.A.
116 C5 East Antarctica reg. Ant.
91 F4 East Ararat U.S.A.
90 D3 East Aurora U.S.A.
85 F6 East Bay U.S.A.
91 G3 East Berkshire U.S.A.
13 H7 Eastbourne U.K.
90 D4 East Branch Clarion River Reservoir U.S.A.
91 H4 East Brooklyn U.S.A.
70 G2 East Cape N.Z.
93 G2 East Carbon U.S.A.
118 E5 East Caroline Basin sea feature Pac. Oc.
88 D5 East Chicago U.S.A.
46 E3 East China Sea Asia
70 E2 East Coast Bays N.Z.
91 G2 East Corinth U.S.A.
13 H5 East Dereham U.K.
69 A4 Easter Group is Austr.
71 □8 Easter I. Chile
119 M7 Easter Island Pac. Oc.
119 M7 Easter Island Fracture Zone sea feature Pac. Oc.
115 C2 Eastern div. Kenya
113 G5 Eastern Cape div. S. Africa
109 F2 Eastern Desert Egypt
45 E6 Eastern Ghats mts India
Eastern Group is see Lau Group
44 B4 Eastern Nara canal Pak.
Eastern Transvaal div. see Mpumalanga
77 K4 Easterville Can.
102 E8 East Falkland i. Falkland Is
91 H4 East Falmouth U.S.A.
18 D3 East Frisian Islands Ger.
92 D2 Eastgate U.S.A.
84 D2 East Grand Forks U.S.A.
13 G6 East Grinstead U.K.
91 G4 East Hampton U.S.A.
91 G3 Easthampton U.S.A.
90 D4 East Hickory U.S.A.
91 G3 East Jamaica U.S.A.
120 K1 East Jan Mayen Ridge sea feature Atl. Ocean
88 E3 East Jordan U.S.A.
11 D5 East Kilbride U.K.
88 D3 Eastlake U.S.A.
51 □ East Lamma Channel H.K. China
13 F7 Eastleigh U.K.
90 C4 East Liverpool U.S.A.
11 B3 East Loch Tarbert b. U.K.
113 G6 East London S. Africa
90 B5 East Lynn Lake U.S.A.
78 E3 Eastmain Que. Can.
78 F3 Eastmain r. Que. Can.
91 G2 Eastman Can.
87 D5 Eastman U.S.A.
68 A4 Eastmere Austr.
91 J2 East Millinocket U.S.A.
88 B5 East Moline U.S.A.
88 C5 Easton IL U.S.A.
91 E5 Easton MD U.S.A.
91 F4 Easton PA U.S.A.
119 M8 East Pacific Ridge sea feature Pac. Oc.
119 N5 East Pacific Rise sea feature Pac. Oc.
92 A2 East Park Resr U.S.A.
79 H4 East Point P.E.I. Can.
87 C5 East Point U.S.A.
91 K2 Eastport ME U.S.A.
88 E3 Eastport MI U.S.A.
92 D1 East Range mts U.S.A.
East Retford see Retford
86 B4 East St Louis U.S.A.
33 R2 East Siberian Sea Rus. Fed.
67 H7 East Sister I. Austr.
31 M10 East Timor reg. Asia
45 F4 East Tons r. India
67 F3 East Toorale Austr.
88 C4 East Troy U.S.A.
91 F6 Eastville U.S.A.
92 C2 East Walker r. U.S.A.
91 G3 East Wallingford U.S.A.
87 D5 Eatonton U.S.A.
88 B3 Eau Claire U.S.A.
88 B3 Eau Claire r. U.S.A.
78 F2 Eau Claire, Lac à l' l. Can.
47 G6 Eauripik Atoll Micronesia
118 E5 Eauripik–New Guinea Rise sea feature Pac. Oc.
97 E3 Ebano Mex.
13 D6 Ebbw Vale U.K.
114 F4 Ebebiyin Equatorial Guinea
112 B2 Ebenerde Namibia
90 D4 Ebensburg U.S.A.
21 G5 Ebensee Austria
20 B3 Eberbach Ger.
36 C2 Eber Gölü l. Turkey
19 H5 Ebergötzen Ger.
20 D2 Ebern Ger.
21 H6 Eberndorf Austria
19 M5 Ebersbach Ger.
19 L4 Eberswalde-Finow Ger.
89 F4 Eberts Can.
52 G3 Ebetsu Japan
51 B4 Ebian China
39 K3 Ebinur Hu salt l. China
26 F4 Eboli Italy
114 F4 Ebolowa Cameroon
118 G5 Ebon i. Pac. Oc.
20 D3 Ebrach Ger.
37 K3 Ebrāhim Ḩeṣār Iran
25 G2 Ebro r. Spain
19 H3 Ebstorf Ger.
27 M4 Eceabat Turkey
55 B2 Echague Phil.
25 E1 Echarri-Aranaz Spain
108 C1 Ech Chélif Alg.
25 E1 Echegárate, Puerto pass Spain
51 E4 Echeng China
76 F1 Echo Bay N.W.T. Can.
89 E2 Echo Bay Ont. Can.
93 G3 Echo Cliffs cliff U.S.A.
78 B3 Echoing r. Can.
67 G6 Echo, L. Austr.
89 K2 Échouani, Lac l. Can.
18 C5 Echt Neth.
15 G3 Echternach Lux.
66 F6 Echuca Austr.
25 D4 Écija Spain
20 E3 Eckental Ger.
88 E2 Eckerman U.S.A.
19 G2 Eckernförde Ger.
19 H2 Eckernförder Bucht b. Ger.
75 L2 Eclipse Sound chan. Can.
15 B3 Écos France
22 D5 Écrins, Parc National des nat. park France
99 C3 Ecuador country S. America
78 E2 Écueils, Pte aux pt Can.
110 E2 Ed Eritrea
9 M7 Ed Sweden
77 H4 Edam Can.
18 C4 Edam Neth.
11 F1 Eday i. U.K.
109 E3 Ed Da'ein Sudan
109 F3 Ed Damazin Sudan
109 F3 Ed Damer Sudan
109 F3 Ed Debba Sudan
109 F3 Ed Dueim Sudan
67 H8 Eddystone Pt Austr.
18 C4 Ede Neth.
114 F4 Edéa Cameroon
77 K2 Edehon Lake Can.
104 C2 Edéia Brazil
67 H6 Eden Austr.
12 E3 Eden r. U.K.
85 D6 Eden TX U.S.A.
113 F4 Edenburg S. Africa
70 B7 Edendale N.Z.
14 D4 Edenderry Rep. of Ireland
66 D6 Edenhope Austr.
87 E4 Edenton U.S.A.
113 G3 Edenville S. Africa
20 B1 Eder r. Ger.
27 K4 Edessa Greece
18 E3 Edewecht Ger.
91 H4 Edgartown U.S.A.
84 D2 Edgeley U.S.A.
84 C3 Edgemont U.S.A.
88 C4 Edgerton U.S.A.
14 D4 Edgeworthstown Rep. of Ireland
39 L2 Edigan Rus. Fed.
88 A5 Edina U.S.A.
85 D7 Edinburg U.S.A.
11 E5 Edinburgh U.K.
29 C7 Edirne Turkey
76 F4 Edith Cavell, Mt Can.
69 E2 Edith Withnell, Lake salt flat Austr.
21 K5 Edlitz Austria
82 B2 Edmonds U.S.A.
68 A1 Edmonton Austr.
76 G4 Edmonton Can.
77 K5 Edmore U.S.A.
88 B4 Edmund U.S.A.
77 L4 Edmund L. Can.
79 G4 Edmundston Can.
85 D6 Edna U.S.A.
76 C3 Edna Bay U.S.A.
23 H3 Edolo Italy
27 M5 Edremit Turkey
9 O6 Edsbyn Sweden
76 F4 Edson Can.
105 D2 Eduardo Castex Arg.
66 F5 Edward r. Austr.
88 C1 Edward I. Can.
110 C4 Edward, Lake Congo(Zaire)/Uganda
91 F2 Edwards U.S.A.
66 A2 Edward's Cr. Austr.
85 C6 Edwards Plateau plat. U.S.A.
86 B4 Edwardsville U.S.A.
116 D4 Edward VIII Ice Shelf ice feature Ant.
116 A4 Edward VII Pen. Ant.
76 C3 Edziza Pk Can.
15 D1 Eeklo Belgium
92 A1 Eel r. U.S.A.
18 D3 Eemshaven pt Neth.
18 D3 Eenrum Neth.
112 D3 Eenzamheid Pan salt pan S. Africa
71 □9 Éfaté i. Vanuatu
86 B4 Effingham U.S.A.
36 D1 Eflâni Turkey
93 E2 Egan Range mts U.S.A.

89 J3 Eganville Can.
20 F2 Eger r. Ger.
17 K7 Eger Hungary
9 K7 Egersund Norway
69 C2 Egerton, Mt h. Austr.
18 F5 Eggegebirge h. Ger.
20 F4 Eggenfelden Ger.
19 M3 Eggesin Ger.
15 E2 Eghezée Belgium
8 F4 Egilsstaðir Iceland
36 C3 Eğirdir Turkey
36 C3 Eğirdir Gölü l. Turkey
24 F4 Égletons France
14 D2 Eglinton U.K.
74 F2 Eglinton I. Can.
22 F2 Eglisau Switz.
18 B4 Egmond aan Zee Neth.
70 D3 Egmont, Cape N.Z.
70 E3 Egmont, Mt volc. N.Z.
70 E3 Egmont National Park N.Z.
36 B2 Eğriğöz Dağı mts Turkey
12 G3 Egton U.K.
19 G1 Egtved Denmark
104 D1 Éguas r. Brazil
22 C6 Éguilles France
33 V3 Egvekinot Rus. Fed.
107 F3 Egypt country Africa
20 C4 Ehingen (Donau) Ger.
19 H4 Ehra-Lessien Ger.
93 E5 Ehrenberg U.S.A.
20 D3 Eibelstadt Ger.
18 D4 Eibergen Neth.
21 J6 Eibiswald Austria
20 C2 Eichenzell Ger.
20 E4 Eichstätt Ger.
19 L4 Eichwalde Ger.
19 G2 Eider r. Ger.
9 K6 Eidfjord Norway
68 D5 Eidsvold Austr.
9 M6 Eidsvoll Norway
15 G2 Eifel reg. Ger.
22 F3 Eiger mt Switz.
11 B4 Eigg i. U.K.
43 A4 Eight Degree Chan. India/Maldives
116 A3 Eights Coast Ant.
64 D3 Eighty Mile Beach Austr.
67 F6 Eildon Austr.
67 F6 Eildon, Lake Austr.
11 C4 Eilean Shona i. U.K.
77 H2 Eileen Lake Can.
19 K5 Eilenburg Ger.
19 H4 Eimke Ger.
19 G5 Einbeck Ger.
18 C5 Eindhoven Neth.
22 F2 Einsiedeln Switz.
100 E5 Eirunepé Brazil
19 G5 Eisberg h. Ger.
111 C5 Eiseb watercourse Namibia
19 H6 Eisenach Ger.
19 J6 Eisenberg Ger.
21 H5 Eisenerz Austria
21 H5 Eisenerzer Alpen mts Austria
19 M4 Eisenhüttenstadt Ger.
21 H6 Eisenkappel Austria
21 K5 Eisenstadt Austria
20 D2 Eisfeld Ger.
11 C3 Eishort, Loch in. U.K.
19 J5 Eisleben Lutherstadt Ger.
19 G6 Eiterfeld Ger.
Eivissa see Ibiza
25 F1 Ejea de los Caballeros Spain
111 E6 Ejeda Madag.
96 B2 Ejido Insurgentes Mex.
50 C2 Ejin Horo Qi China
50 A1 Ejin Qi China
37 K1 Ejmiatsin Armenia
97 E4 Ejutla Mex.
9 S7 Ekenäs Fin.
15 E1 Ekeren Belgium
114 E4 Eket Nigeria
70 E4 Eketahuna N.Z.
39 H2 Ekibastuz Kazak.
33 M3 Ekonda Rus. Fed.
9 N6 Ekshärad Sweden
9 O8 Eksjö Sweden
112 B4 Eksteenfontein S. Africa
110 C4 Ekuku Congo(Zaire)
78 D3 Ekwan r. Can.
78 D3 Ekwan Point Can.
27 K6 Elafonisou, Steno chan. Greece
36 B6 El 'Alamein Egypt
97 F4 El Almendro Mex.
36 B6 El 'Amirîya Egypt
113 H2 Elands r. S. Africa
113 H2 Elandsdoorn S. Africa
26 B7 El Aouinet Alg.
36 B6 El 'Arab, Khalîg b. Egypt
96 A1 El Arco Mex.
68 B1 El Arish Austr.
36 D6 El 'Arîsh Egypt
27 K5 Elassona Greece
36 E7 Elat Israel
37 G2 Elazığ Turkey
26 D3 Elba, Isola d' i. Italy
49 P1 El'ban Rus. Fed.
103 B2 El Banco Col.
36 D6 El Bardawîl, Sabkhet lag. Egypt
27 J4 Elbasan Albania
36 E2 Elbaşı Turkey
103 C2 El Baúl Venez.
108 C1 El Bayadh Alg.
19 H3 Elbe r. Ger.
88 D3 Elberta MI U.S.A.
93 G2 Elberta UT U.S.A.
93 F4 Elbert, Mount U.S.A.
87 D5 Elberton U.S.A.
15 B3 Elbeuf France
36 F2 Elbistan Turkey
17 J3 Elbląg Pol.
105 B4 El Bolsón Arg.
87 E7 Elbow Cay i. Bahamas
29 G7 El'brus mt Rus. Fed.
18 C4 Elburg Neth.
25 E2 El Burgo de Osma Spain
105 C4 El Cain Arg.
92 D5 El Cajon U.S.A.
103 E3 El Callao Venez.
85 D6 El Campo U.S.A.
92 D5 El Centro U.S.A.
100 F7 El Cerro Bol.
103 D2 El Chaparro Venez.
25 F3 Elche Spain
97 F4 El Chichón volc. Mex.
96 C1 El Chilicote Mex.
20 D4 Elchingen Ger.
65 G2 Elcho I. Austr.
103 B3 El Cocuy Col.
97 H3 El Cuyo Mex.
25 F3 Elda Spain
19 J3 Elde r. Ger.
89 H2 Eldee Can.
66 D3 Elder, Lake salt flat Austr.
92 D5 El Descanso Mex.
103 B2 El Difícil Col.
33 P3 El'dikan Rus. Fed.
103 A4 El Diviso Col.
93 E6 El Doctor Mex.
85 A5 Eldon IA U.S.A.
84 E4 Eldon MO U.S.A.
102 F3 Eldorado Arg.
96 C2 El Dorado Mex.
85 E5 El Dorado AR U.S.A.
85 D4 El Dorado KS U.S.A.
85 C6 Eldorado U.S.A.
103 E3 El Dorado Venez.
115 C2 Eldoret Kenya
82 E2 Electric Peak U.S.A.
108 B2 El Eglab plat. Alg.
25 E4 El Ejido Spain
28 F4 Elektrostal' Rus. Fed.
100 D4 El Encanto Col.
19 H5 Elend Ger.
83 F5 Elephant Butte Resr U.S.A.
116 B1 Elephant I. Ant.
45 H5 Elephant Point Bangl.
37 J2 Eleşkirt Turkey
97 G5 El Estor Guatemala
108 C1 El Eulma Alg.
87 E7 Eleuthera i. Bahamas
26 C6 El Fahs Tunisia
109 F2 El Faiyûm Egypt
109 E3 El Fasher Sudan
96 B2 El Fuerte Mex.
109 E3 El Geneina Sudan
109 F3 El Geteina Sudan
11 E3 Elgin U.K.
88 C4 Elgin IL U.S.A.
84 C2 Elgin ND U.S.A.
93 E3 Elgin NV U.S.A.
93 G2 Elgin UT U.S.A.
68 B4 Elgin Down Austr.
33 Q3 El'ginskiy Rus. Fed.
109 F2 El Gîza Egypt
96 D3 El Gogorrón, Parque Nacional nat. park Mex.
108 C1 El Goléa Alg.
109 F4 Elgon, Mount Uganda
26 B6 El Hadjar Alg.
36 B6 El Hammâm Egypt
96 J6 El Hato del Volcán Panama
36 F6 El Hazim Jordan
108 A2 El Hierro i. Canary Is
97 E3 El Higo Mex.
108 C2 El Homr Alg.
11 F4 Elie U.K.
70 C5 Elie de Beaumont mt N.Z.
115 A3 Elila Congo(Zaire)
115 A3 Elila r. Congo(Zaire)
74 B3 Elim U.S.A.
79 H2 Eliot, Mount Can.
25 F1 Eliozondo Spain
El Iskandarîya see Alexandria
29 H6 Elista Rus. Fed.
88 B4 Elizabeth IL U.S.A.
91 F4 Elizabeth NJ U.S.A.
90 C5 Elizabeth WV U.S.A.
87 E4 Elizabeth City U.S.A.
91 H4 Elizabeth Is U.S.A.
87 D4 Elizabethton U.S.A.
86 C4 Elizabethtown KY U.S.A.
87 E5 Elizabethtown NC U.S.A.
91 G2 Elizabethtown NY U.S.A.
90 E4 Elizabethtown PA U.S.A.
108 B1 El Jadida Morocco
36 F6 El Jafr Jordan
96 C2 El Jaralito Mex.
26 D7 El Jem Tunisia
96 H5 El Jicaral Nic.
76 G4 Elk r. Can.
17 L4 Ełk Pol.
92 A2 Elk U.S.A.
90 C5 Elk r. U.S.A.
36 F4 El Kaa Lebanon
26 C6 El Kala Alg.
109 F3 El Kamlin Sudan
85 D5 Elk City U.S.A.
92 B2 Elk Creek U.S.A.
115 D1 Ēl Kerē Eth.
92 B2 Elk Grove U.S.A.
109 F2 El Khârga Egypt
88 E5 Elkhart U.S.A.
El Khartûm see Khartoum
108 B2 El Khnâchîch esc. Mali
88 C4 Elkhorn U.S.A.
84 D3 Elkhorn r. U.S.A.
42 B1 El'khotovo Rus. Fed.
27 M3 Elkhovo Bulg.
90 D5 Elkins U.S.A.
76 G4 Elk Island Nat. Park Can.
89 G2 Elk Lake Can.
88 E3 Elk Lake U.S.A.
90 E4 Elkland U.S.A.
76 F5 Elko Can.
82 D3 Elko U.S.A.
77 G4 Elk Point Can.
84 E2 Elk River U.S.A.
91 F5 Elkton MD U.S.A.
90 D5 Elkton VA U.S.A.
37 G4 El Kubar Syria
69 B2 Ellavalla Austr.
77 M2 Ell Bay Can.
75 H2 Ellef Ringnes I. Can.
69 C7 Elleker Austr.
44 C3 Ellenabad India
84 D2 Ellendale U.S.A.
93 G2 Ellen, Mt U.S.A.
82 B2 Ellensburg U.S.A.
91 F4 Ellenville U.S.A.
67 H6 Ellery, Mt Austr.
75 K2 Ellesmere Island Can.
70 D5 Ellesmere, Lake N.Z.
13 E4 Ellesmere Port U.K.
74 H3 Ellice r. Can.
90 D3 Ellicottville U.S.A.
97 E3 El Limón Mex.
71 ▫15 Ellington Fiji
113 G5 Elliot S. Africa
113 H5 Elliotdale S. Africa
89 F2 Elliot Lake Can.
68 B2 Elliot, Mt Austr.
82 D2 Ellis U.S.A.
113 G1 Ellisras S. Africa
66 A4 Elliston Austr.
11 F3 Ellon U.K.
91 J2 Ellsworth ME U.S.A.
88 A3 Ellsworth WI U.S.A.
116 A3 Ellsworth Land reg. Ant.
116 B3 Ellsworth Mountains Ant.
20 D4 Ellwangen (Jagst) Ger.
37 H3 Elmalı Turkey
109 F1 El Manṣûra Egypt
103 E3 El Manteco Venez.
108 C1 El Meghaïer Alg.
20 D5 Elmen Austria
103 E3 El Miamo Venez.
36 E4 El Mîna Lebanon
109 F2 El Minya Egypt
88 E3 Elmira MI U.S.A.
90 E3 Elmira NY U.S.A.
93 F5 El Mirage U.S.A.
25 E4 El Moral Spain
66 F6 Elmore Austr.
105 D2 El Morro mt Arg.
108 B2 El Mreyyé reg. Maur.
19 G3 Elmshorn Ger.
109 E3 El Muglad Sudan
89 G3 Elmwood Can.
88 C5 Elmwood IL U.S.A.
88 A3 Elmwood WV U.S.A.
8 K5 Elnesvågen Norway
103 B3 El Nevado, Cerro mt Col.
55 A4 El Nido Phil.
109 F3 El Obeid Sudan
96 D2 El Oro Mex.
103 C3 Elorza Venez.
108 C1 El Oued Alg.
93 G5 Eloy U.S.A.
96 C2 El Palmito Mex.
103 E2 El Pao Bolívar Venez.
103 C2 El Pao Cojedes Venez.
88 C5 El Paso IL U.S.A.
83 F6 El Paso TX U.S.A.
11 C2 Elphin U.K.
92 C3 El Portal U.S.A.
96 K6 El Porvenir Panama
25 H2 El Prat de Llobregat Spain
97 G5 El Progreso Guatemala
96 B2 El Puerto, Cerro mt Mex.
25 C4 El Puerto de Santa María Spain
El Qâhira see Cairo
36 D6 El Qantara Egypt
36 F7 El Quweira Jordan
96 K6 El Real Panama
85 D5 El Reno U.S.A.
96 D3 El Retorno Mex.
88 B4 Elroy U.S.A.
96 D3 El Rucio Mex.
76 B2 Elsa Can.
23 H6 Elsa r. Italy
36 C7 El Saff Egypt
96 D2 El Salado Mex.
36 D6 El Sâlhîya Egypt
96 C3 El Salto Mex.
73 H8 El Salvador country Central America
96 D2 El Salvador Mex.
55 C4 El Salvador Phil.
103 C3 El Samán de Apure Venez.
89 F1 Elsas Can.
96 C1 El Sauz Mex.
18 F4 Else r. Ger.
45 H2 Elsen Nur l. China
36 D7 El Shatt Egypt
103 D2 El Sombrero Venez.
105 C2 El Sosneado Arg.
19 L5 Elsterwerda Ger.
El Suweis see Suez
97 E3 El Tajín tourist site Mex.
119 L9 Eltanin Fracture Zone sea feature Pac. Oc.
26 C6 El Tarf Alg.
25 C1 El Teleno mt Spain
97 E4 El Tepozteco, Parque Nacional nat. park Mex.
36 E7 El Thamad Egypt
103 D2 El Tigre Venez.
97 G4 El Tigre, Parque Nacional nat. park Guatemala
20 D3 Eltmann Ger.
103 C2 El Tocuyo Venez.
29 H5 El'ton Rus. Fed.
29 H5 El'ton, Ozero l. Rus. Fed.
82 C2 Eltopia U.S.A.
103 E2 El Toro Venez.
105 E2 El Trébol Arg.
96 B3 El Triunfo Mex.
103 C3 El Tuparro, Parque Nacional nat. park Col.
109 F2 El Tûr Egypt
102 B8 El Turbio Chile
43 C2 Eluru India
9 U7 Elva Estonia
103 A3 El Valle Col.
11 E5 Elvanfoot U.K.
25 C3 Elvas Port.
9 M6 Elverum Norway
103 B3 El Viejo mt Col.
103 C2 El Vigía Venez.
100 D5 Elvira Brazil
22 F4 Elvo r. Italy
115 D2 El Wak Kenya
88 E5 Elwood U.S.A.
19 H5 Elxleben Ger.
13 H5 Ely U.K.
88 B2 Ely MN U.S.A.
93 E2 Ely NV U.S.A.
90 B4 Elyria U.S.A.
71 ▫7 El Yunque h. Juan Fernandez Is Chile
20 B2 Elz Ger.
20 B4 Elz r. Ger.
20 B4 Elzach Ger.
19 G4 Elze Ger.
71 ▫9 Émaé i. Vanuatu
40 D2 Emāmrūd Iran
41 H2 Emām Şaḩēb Afgh.
37 L5 Emāmzadeh Naṣrod Dīn Iran
9 O8 Emån r. Sweden
71 ▫9 Emao i. Vanuatu
104 B2 Emas, Parque Nacional das nat. park Brazil
39 K3 Emazar Kazak.
38 D2 Emba Kazak.
38 C3 Emba r. Kazak.
113 H3 Embalenhle S. Africa
77 G3 Embarras Portage Can.
104 C2 Emborcação, Represa de resr Brazil
91 F2 Embrun Can.
22 D5 Embrun France
115 C3 Embu Kenya
18 E3 Emden Ger.
51 B4 Emei China
51 B4 Emei Shan mt China
39 K3 Emel' r. Kazak.
68 C4 Emerald Qld. Austr.
66 F6 Emerald Vic. Austr.
79 G3 Emeril Can.
77 K5 Emerson Can.
36 B2 Emet Turkey
113 J2 eMgwenya S. Africa
93 E3 Emigrant Valley v. U.S.A.
113 J3 eMjindini S. Africa
109 D3 Emi Koussi mt Chad
97 G4 Emiliano Zapata Mex.
23 H5 Emilia-Romagna div. Italy
22 E4 Emilius, Monte mt Italy
39 K3 Emin China
39 K3 Emin He r. China
27 M3 Eminska Planina h. Bulg.
36 C2 Emir D. mt Turkey
36 C2 Emirdağ Turkey
67 H3 Emita Austr.
9 O8 Emmaboda Sweden
9 S7 Emmaste Estonia
67 J2 Emmaville Austr.
22 E2 Emme r. Switz.
18 C4 Emmeloord Neth.
15 H2 Emmelshausen Ger.
18 D4 Emmen Neth.
22 F2 Emmen Switz.
15 H4 Emmendingen Ger.
15 G1 Emmerich Ger.
68 A5 Emmet Austr.
43 B3 Emmiganuru India
85 C6 Emory Pk U.S.A.
96 B2 Empalme Mex.
113 J4 Empangeni S. Africa
102 E3 Empedrado Arg.
118 G3 Emperor Seamount Chain sea feature Pac. Oc.
23 H6 Empoli Italy
84 D4 Emporia KS U.S.A.
90 E6 Emporia VA U.S.A.
90 D4 Emporium U.S.A.
76 G4 Empress Can.
115 A6 Empress Mine Zimbabwe
41 E3 'Emrānī Iran
18 E3 Ems r. Ger.
89 H3 Emsdale Can.
18 E4 Emsdetten Ger.
18 E3 Ems-Jade-Kanal canal Ger.
18 E4 Emsland reg. Ger.
68 D4 Emu Park Austr.
113 G3 Emzinoni S. Africa
8 N5 Enafors Sweden
47 F7 Enarotali Indon.
53 D4 Ena-san mt Japan
39 J2 Enbek Kazak.
94 A2 Encantada, Co de la mt Mex.
105 G1 Encantadas, Serra das h. Brazil
96 A2 Encantado, Cerro mt Mex.
55 B3 Encanto, Cape pt Phil.
105 D3 Encarnación Mex.
102 E3 Encarnación Para.
114 C3 Enchi Ghana
85 D6 Encinal U.S.A.
92 D5 Encinitas U.S.A.
83 F5 Encino U.S.A.
66 C5 Encounter Bay Austr.
104 E1 Encruzilhada Brazil
105 G1 Encruzilhada do Sul Brazil
76 D4 Endako Can.
58 B5 Endau Malaysia
65 H2 Endeavour Strait Austr.
57 G8 Endeh Indon.
19 H1 Endelave i. Denmark
116 D4 Enderby Land reg. Ant.
91 E3 Endicott U.S.A.
76 C3 Endicott Arm in. U.S.A.
74 C3 Endicott Mts U.S.A.
69 B4 Eneabba Austr.
39 J4 Energeticheskiy Kazak.
38 D2 Energetik Rus. Fed.
105 E3 Energía Arg.
29 E6 Enerhodar Ukr.
118 F5 Enewetak i. Pac. Oc.
26 D6 Enfidaville Tunisia
91 G3 Enfield U.S.A.
88 E2 Engadine U.S.A.
8 L5 Engan Norway
55 B2 Engaño, Cape Phil.
Engaños, Río de los r. see Yari
52 H2 Engaru Japan
113 G5 Engcobo S. Africa
21 G4 Engelhartszell Austria
29 H5 Engel's Rus. Fed.
18 B3 Engelschmangat chan. Neth.
66 A2 Engenina watercourse Austr.
59 B4 Enggano i. Indon.
15 E2 Enghien Belgium
13 E5 England div. U.K.
79 J3 Englee Can.
87 F5 Englehard U.S.A.
89 H2 Englehart Can.
81 H1 English r. Can.
10 E6 English Channel France/U.K.
20 C4 Engstingen Ger.
29 G7 Enguri r. Georgia
113 J4 Enhlalakahle S. Africa
85 D4 Enid U.S.A.
52 G3 Eniwa Japan
20 A3 Enkenbach Ger.
18 C4 Enkhuizen Neth.
9 P7 Enköping Sweden
26 F6 Enna Sicily Italy
77 J2 Ennadai Lake Can.
109 E3 En Nahud Sudan
109 E3 Ennedi, Massif mts Chad
14 D4 Ennell, Lough l. Rep. of Ireland
67 F2 Enngonia Austr.
84 C2 Enning U.S.A.
14 C5 Ennis Rep. of Ireland
82 E2 Ennis MT U.S.A.
85 D5 Ennis TX U.S.A.
14 E5 Enniscorthy Rep. of Ireland
14 D3 Enniskillen U.K.
14 B5 Ennistymon Rep. of Ireland
21 H4 Enns Austria
21 H5 Enns r. Austria
8 W5 Eno Fin.
93 F3 Enoch U.S.A.
8 S2 Enontekiö Fin.
51 D6 Enping China
55 B2 Enrile Phil.
67 G6 Ensay Austr.
18 D4 Enschede Neth.
18 F5 Ense Ger.
105 F2 Ensenada Arg.
94 A2 Ensenada Mex.
51 C4 Enshi China
115 B2 Entebbe Uganda
76 F2 Enterprise N.W.T. Can.
89 J3 Enterprise Ont. Can.
87 C6 Enterprise AL U.S.A.
82 C2 Enterprise OR U.S.A.
93 F3 Enterprise UT U.S.A.
22 B5 Entraigues-sur-Sorgues France
76 F4 Entrance Can.
22 C4 Entre-Deux-Guiers France
105 E2 Entre Ríos div. Arg.
100 F8 Entre Ríos Bol.
25 B3 Entroncamento Port.
114 E3 Enugu Nigeria
33 V3 Enurmino Rus. Fed.
100 D5 Envira Brazil
100 D5 Envira r. Brazil
21 M6 Enying Hungary
70 C5 Enys, Mt N.Z.
23 H5 Enza r. Italy
53 F7 Enzan Japan
20 B4 Enzklösterle Ger.
71 ▫9 Eo i. New Caledonia Pac. Oc.
18 C4 Epe Neth.
15 C3 Épernay France
15 B4 Épernon France
93 G2 Ephraim U.S.A.
91 E4 Ephrata PA U.S.A.
82 C2 Ephrata WA U.S.A.
71 ▫9 Epi i. Vanuatu
22 B3 Épinac France
15 G4 Épinal France
36 D4 Episkopi Cyprus
26 E4 Epomeo, Monte h. Italy
13 H6 Epping U.K.
20 B3 Eppingen Ger.
13 G6 Epsom U.K.
105 D3 Epu-pel Arg.
40 D4 Eqlīd Iran
107 D5 Equatorial Guinea country Africa
103 E3 Equeipa Venez.
68 A6 Erac Cr. watercourse Austr.
23 K4 Eraclea Italy
55 A4 Eran Phil.
55 A4 Eran Bay Phil.
36 F1 Erbaa Turkey
20 C3 Erbach Ger.
20 C4 Erbach Ger.
20 F3 Erbendorf Ger.
119 L3 Erben Tablemount depth Pac. Oc.
15 H3 Erbeskopf h. Ger.
37 J2 Erçek Turkey
37 J2 Erciş Turkey
36 E2 Erciyes Dağı mt Turkey
17 J7 Érd Hungary
45 H2 Erdaogou China
42 B2 Erdao Jiang r. China
36 A1 Erdek Turkey
36 E3 Erdemli Turkey
50 C1 Erdenetsogt Mongolia
109 E3 Erdi reg. Chad
20 E4 Erding Ger.
29 H6 Erdniyevskiy Rus. Fed.
103 D3 Erebato r. Venez.
116 B5 Erebus, Mt Ant.
37 K6 Erech Iraq
102 F3 Erechim Brazil
50 D1 Ereentsav Mongolia
36 E3 Ereğli Konya Turkey
36 C1 Ereğli Zonguldak Turkey
26 F6 Erei, Monti mts Sicily Italy
50 D1 Erenhot China
40 E3 Eresk Iran
25 D2 Eresma r. Spain
27 K5 Eretria Greece
Erevan see Yerevan
19 G2 Erfde Ger.
19 J6 Erfurt Ger.
37 G2 Ergani Turkey
108 B2 'Erg Chech sand dunes Alg./Mali
109 D3 Erg du Djourab sand dunes Chad
108 D3 Erg du Ténéré des. Niger
50 C1 Ergel Mongolia
27 M4 Ergene r. Turkey
108 B2 Erg Iguidi sand dunes Alg./Maur.
9 T8 Ērgļi Latvia
20 F4 Ergoldsbach Ger.
52 A1 Ergu China
Ergun He r. see Argun'
54 C3 Erhulai China
11 D2 Eriboll, Loch in. U.K.
11 D4 Ericht, Loch l. U.K.
88 B5 Erie IL U.S.A.
85 E4 Erie KS U.S.A.
90 C3 Erie PA U.S.A.
89 G4 Erie, Lake Can./U.S.A.
52 H3 Erimo Japan
52 H4 Erimo-misaki c. Japan
11 A3 Eriskay i. U.K.
107 G4 Eritrea country Africa
36 E2 Erkilet Turkey
19 L4 Erkner Ger.
20 E3 Erlangen Ger.
21 J5 Erlauf r. Austria
65 F5 Erldunda Austr.
69 E3 Erlistoun watercourse Austr.
54 E2 Erlong Shan mt China
54 C2 Erlongshan Sk. resr China
18 C4 Ermelo Neth.
113 H3 Ermelo S. Africa
36 D3 Ermenek Turkey
27 L6 Ermoupoli Greece
43 B4 Ernakulam India
43 B4 Erode India
112 A1 Erongo div. Namibia
18 C5 Erp Neth.
108 B1 Er Rachidia Morocco
109 F3 Er Rahad Sudan
115 C6 Errego Moz.
26 D7 Er Remla Tunisia
14 C2 Errigal h. Rep. of Ireland
14 A3 Erris Head hd Rep. of Ireland
91 H2 Errol U.S.A.
71 ▫9 Erromango i. Vanuatu
Erronan i. see Futuna
27 J4 Ersekë Albania
84 D2 Erskine U.S.A.
8 R5 Ersmark Sweden
29 G5 Ertil' Rus. Fed.
66 C3 Erudina Austr.
37 J3 Eruh Turkey
105 G2 Erval Brazil
22 A1 Ervy-le-Châtel France
90 D5 Erwin U.S.A.
18 F5 Erwitte Ger.
19 J4 Erxleben Ger.
19 J4 Erxleben Ger.
20 G2 Erzgebirge mts Czech Rep./Ger.
36 F3 Erzin Turkey
37 G2 Erzincan Turkey
37 H2 Erzurum Turkey
52 G4 Esan-misaki pt Japan
52 H2 Esashi Japan
52 G4 Esashi Japan
9 L9 Esbjerg Denmark
93 G3 Escalante U.S.A.
93 G3 Escalante r. U.S.A.
93 F3 Escalante Desert U.S.A.
96 C2 Escalón Mex.
88 D3 Escanaba U.S.A.
97 G4 Escárcega Mex.
55 B2 Escarpada Point Phil.
25 F2 Escatrón Spain
15 D2 Escaut r. Belgium
15 C5 Esch Neth.
18 D4 Esche Ger.
19 H4 Eschede Ger.
15 F3 Esch-sur-Alzette Lux.
19 H5 Eschwege Ger.
18 D6 Eschweiler Ger.
92 D5 Escondido U.S.A.
96 C3 Escuinapa Mex.
97 G5 Escuintla Guatemala
97 F5 Escuintla Mex.
103 C3 Escutillas Col.
114 F4 Eséka Cameroon
36 B3 Eşen Turkey
38 C5 Esenguly Turkm.
18 E3 Esens Ger.
40 C3 Eşfahan Iran
40 D4 Esfandāran Iran
40 C4 Esfarjan Iran
41 E3 Esfīdeh Iran
40 D2 Esfahāb Iran
113 J4 Eshowe S. Africa
36 E6 Esh Sharā mts Jordan
40 C3 Eshtehārd Iran
111 C6 Esigodini Zimbabwe
113 K4 Esikhawini S. Africa
23 L6 Esino r. Italy
67 K1 Esk Austr.
67 G8 Esk r. Austr.
12 D2 Esk r. U.K.
11 E5 Eskdalemuir U.K.
79 G3 Esker Can.
8 F4 Eskifjörður Iceland
9 P7 Eskilstuna Sweden
74 E3 Eskimo Lakes Can.
37 J3 Eski Mosul Iraq
39 H4 Eski-Nookat Kyrg.
36 D1 Eskipazar Turkey
36 C2 Eskişehir Turkey
25 D1 Esla r. Spain
40 B3 Eslamābād e Gharb Iran
36 B3 Esler D. mt Turkey
18 F5 Eslohe (Sauerland) Ger.
9 N9 Eslöv Sweden
36 B2 Eşme Turkey
100 C3 Esmeraldas Ecuador
88 E1 Esnagi Lake Can.
15 D2 Esnes France
41 F5 Espakeh Iran
24 F4 Espalion France
89 G2 Espanola Can.
83 F4 Espanola U.S.A.
100 ▫ Española, Isla i. Galapagos Is Ecuador
92 A2 Esparto U.S.A.
39 H4 Espe Kazak.
18 F4 Espelkamp Ger.
69 E6 Esperance Austr.
69 E7 Esperance Bay Austr.
116 B2 Esperanza Arg. Base Ant.
105 E1 Esperanza Arg.
96 B2 Esperanza Mex.
55 C4 Esperanza Phil.
96 H5 Esperanza, Sa de la mts Honduras
25 B3 Espichel, Cabo hd Port.
25 D1 Espigüete mt Spain
96 D2 Espinazo Mex.
104 D2 Espinhaço, Serra do mts Brazil
104 D1 Espinosa Brazil
104 E2 Espírito Santo div. Brazil
55 B2 Espiritu Phil.
96 B2 Espíritu Santo i. Mex.
71 ▫9 Espiritu Santo i. Vanuatu
97 H4 Espíritu Santo, Bahía del b. Mex.
9 T6 Espoo Fin.
25 F4 Espuña mt Spain
102 B6 Esquel Arg.
76 E5 Esquimalt Can.
55 C5 Essang Indon.
108 B1 Essaouira Morocco
108 A2 Es Semara Western Sahara
15 E1 Essen Belgium
18 E5 Essen Ger.
69 E2 Essendon, Mt h. Austr.
18 E3 Essen (Oldenburg) Ger.
101 G3 Essequibo r. Guyana
89 F4 Essex Can.
93 E4 Essex U.S.A.
91 G2 Essex Junction U.S.A.
89 F4 Essexville U.S.A.
20 D4 Essingen Ger.
20 C4 Esslingen am Neckar Ger.
33 R4 Esso Rus. Fed.
15 C4 Essonne r. France
15 E4 Essoyes France
102 D8 Estados, I. de los Arg.
40 D4 Eşṭahbānāt Iran
89 G2 Estaire Can.
101 L6 Estância Brazil
25 G1 Estats, Pic d' mt France/Spain
113 H4 Estcourt S. Africa
19 G3 Este r. Ger.
23 J4 Este Italy
96 H5 Estelí Nic.
25 E1 Estella Spain
25 D4 Estepa Spain
25 D4 Estepona Spain
22 D6 Esterel reg. France
77 J4 Esterhazy Can.
15 D4 Esternay France
92 B4 Estero Bay U.S.A.
102 D2 Esteros Para.
102 E3 Esteros del Iberá marsh Arg.
77 J5 Estevan Can.
84 E3 Estherville U.S.A.
79 H4 Est, Île de l' i. Can.
87 D5 Estill U.S.A.
15 D4 Estissac France
91 J1 Est, Lac de l' l. Can.
7 F3 Estonia country Europe
15 C3 Estrées-St-Denis France
25 C2 Estrela, Serra da mts Port.
25 E3 Estrella mt Spain
25 C3 Estremoz Port.
101 J5 Estrondo, Serra h. Brazil
37 M4 Estūh Iran
66 C2 Etadunna Austr.
44 D4 Etah India
15 F3 Étain France
15 C4 Étampes France
15 B2 Étaples France
44 D4 Etawah India
22 D2 Éternoz France
113 J3 eThandakukhanya S. Africa
69 D2 Ethel watercourse Austr.
71 ▫15 Ethel Reefs Fiji
112 E4 E'Thembini S. Africa
107 G5 Ethiopia country Africa
36 D2 Etimesğut Turkey
11 C4 Etive, Loch in. U.K.
26 F6 Etna, Monte volc. Sicily Italy
9 J7 Etne Norway
76 C3 Etolin I. U.S.A.
68 C3 Eton Austr.
111 B5 Etosha National Park Namibia
111 B5 Etosha Pan salt pan Namibia
15 B3 Étrépagny France
27 L3 Etropole Bulg.
43 B4 Ettaiyapuram India
15 G3 Ettelbruck Lux.
15 H4 Ettenheim Ger.
18 B5 Etten-Leur Neth.
20 B4 Ettlingen Ger.
11 E5 Ettrick Forest reg. U.K.
20 D4 Ettringen Ger.
22 C2 Étuz France
96 C3 Etzatlán Mex.
15 B2 Eu France
71 ▫11 'Eua i. Tonga
67 G4 Euabalong Austr.
71 ▫11 'Eua Iki i. Tonga
71 ▫10 Euakafa i. Tonga
Euboea i. see Evvoia
64 E6 Eucla Austr.
90 C4 Euclid U.S.A.
101 L6 Euclides da Cunha Brazil
67 H6 Eucumbene, L. Austr.
66 C5 Eudunda Austr.
87 C6 Eufaula U.S.A.
85 E5 Eufaula Lake resr U.S.A.
82 B2 Eugene U.S.A.
96 A2 Eugenia, Pta c. Mex.
67 H4 Eugowra Austr.
67 F2 Eulo Austr.
67 H3 Eumungerie Austr.
68 C3 Eungella Austr.
68 C3 Eungella Nat. Park Austr.
85 E6 Eunice U.S.A.
15 G2 Eupen Belgium
37 K6 Euphrates r. Asia
9 S6 Eura Fin.
15 B3 Eure r. France
82 A3 Eureka CA U.S.A.
82 D1 Eureka MT U.S.A.
93 E2 Eureka NV U.S.A.
66 D3 Eurinilla watercourse Austr.
66 D3 Euriowie Austr.
67 F6 Euroa Austr.
68 C5 Eurombah Austr.
68 C6 Eurombah Cr. r. Austr.
111 E6 Europa, Île i. Ind. Ocean
25 D4 Europa Point Gibraltar
18 A5 Europoort reg. Neth.
15 F4 Eurville-Bienville France
15 G2 Euskirchen Ger.
66 E5 Euston Austr.
76 D4 Eustuk Lake Can.
87 C5 Eutaw U.S.A.
19 H2 Eutin Ger.
19 K5 Eutzsch Ger.
113 H3 Evander S. Africa
76 F4 Evansburg Can.
67 K2 Evans Head Austr.
116 B3 Evans Ice Stream ice feature Ant.
78 E3 Evans, L. Can.
83 F4 Evans, Mt CO U.S.A.
82 D2 Evans, Mt MT U.S.A.
75 K3 Evans Strait Can.
88 D4 Evanston IL U.S.A.
82 E3 Evanston WY U.S.A.
89 F3 Evansville Can.
86 C4 Evansville IN U.S.A.
88 C4 Evansville WV U.S.A.
82 F3 Evansville WY U.S.A.
88 E4 Evart U.S.A.
113 G3 Evaton S. Africa
40 D5 Evaz Iran
88 A2 Eveleth U.S.A.
33 R3 Evensk Rus. Fed.
66 A3 Everard, L. salt flat Austr.
64 F5 Everard Range h. Austr.
18 C5 Everdingen Neth.
45 F4 Everest, Mt China
91 K1 Everett Can.
82 B1 Everett U.S.A.
15 D1 Evergem Belgium
87 D7 Everglades Nat. Park U.S.A.
87 D7 Everglades, The swamp U.S.A.
87 C6 Evergreen U.S.A.
13 F5 Evesham U.K.
13 F5 Evesham, Vale of reg. U.K.
22 D3 Évian-les-Bains France
8 S5 Evijärvi Fin.
114 F4 Evinayong Equatorial Guinea
9 K7 Evje Norway
22 E3 Evolène Switz.
25 C3 Évora Port.
29 P1 Evoron, Ozero l. Rus. Fed.
37 K2 Evowghlī Iran
15 B3 Évreux France
27 K6 Evrotas r. Greece
15 C4 Évry France
36 D4 Evrychou Cyprus
27 L5 Evvoia i. Greece
92 ▫1 Ewa Beach U.S.A.
115 C2 Ewaso Ngiro r. Kenya
11 C3 Ewe, Loch in. U.K.
116 B2 Ewing I. Ant.
100 E6 Exaltación Bol.
113 G4 Excelsior S. Africa
92 C2 Excelsior Mtn U.S.A.
92 C2 Excelsior Mts U.S.A.
84 E4 Excelsior Springs U.S.A.
13 D6 Exe r. U.K.
116 A4 Executive Committee Range mts Ant.
67 J5 Exeter Austr.
89 G4 Exeter Can.
13 D7 Exeter U.K.
92 C3 Exeter CA U.S.A.
91 H3 Exeter NH U.S.A.
18 D4 Exloo Neth.
13 D7 Exminster U.K.
13 D6 Exmoor Forest reg. U.K.
13 D6 Exmoor National Park U.K.
91 F6 Exmore U.S.A.
64 B4 Exmouth Austr.
13 D7 Exmouth U.K.
64 B4 Exmouth Gulf Austr.
67 H3 Exmouth, Mt Austr.
117 M5 Exmouth Plateau sea feature Ind. Ocean
68 C5 Expedition Range mts Austr.
71 ▫14 Exploring Is Fiji
25 D3 Extremadura div. Spain
87 E7 Exuma Sound chan. Bahamas
115 B3 Eyasi, Lake salt l. Tanz.
13 J5 Eye U.K.
11 F5 Eyemouth U.K.
11 B2 Eye Peninsula U.K.
22 C5 Eygues r. France
8 D5 Eyjafjallajökull ice cap Iceland
8 D3 Eyjafjörður in. Iceland
110 E3 Eyl Somalia
13 F6 Eynsham U.K.
66 B2 Eyre, Lake (North) salt flat Austr.
66 B2 Eyre, Lake (South) salt flat Austr.
70 B6 Eyre Mountains N.Z.
66 A4 Eyre Peninsula Austr.
22 B5 Eyrieux r. France
19 G4 Eystrup Ger.
8 ▫ Eysturoy i. Faroe Is
114 E3 Eyumojok Cameroon
113 J4 Ezakheni S. Africa
113 H3 Ezenzeleni S. Africa
105 C3 Ezequiel Ramos Mexía, Embalse resr Arg.
28 J2 Ezhva Rus. Fed.
27 M5 Ezine Turkey
36 F1 Ezinepazar Turkey
37 L6 Ezra's Tomb Iraq
15 B4 Ézy-sur-Eure France

F

71 ▫3 Faaa Fr. Polynesia Pac. Oc.
43 A5 Faadhippolhu Atoll Maldives
71 ▫2 Faaite i. Fr. Polynesia Pac. Oc.
71 ▫3 Faaone Fr. Polynesia Pac. Oc.
85 B6 Fabens U.S.A.
76 F2 Faber Lake Can.
58 ▫ Faber, Mt h. Sing.
9 M9 Fåborg Denmark
23 K6 Fabriano Italy
103 B3 Facatativá Col.
15 D2 Faches-Thumesnil France
108 D3 Fachi Niger
68 D4 Facing I. Austr.
91 F4 Factoryville U.S.A.
102 B7 Facundo Arg.
114 D2 Fada-Ngourma Burkina
23 J5 Faenza Italy
Faeroes terr. see Faroe Islands
71 ▫11 Fafa i. Tonga
57 J7 Fafanlap Indon.
110 E3 Fafen Shet' watercourse Eth.
71 ▫4 Fagaloa Bay Western Samoa
71 ▫4 Fagamalo Western Samoa
27 L2 Făgăraș Romania
9 L6 Fagernes Norway
9 O7 Fagersta Sweden
102 C8 Fagnano, L. Arg./Chile
15 E2 Fagne reg. Belgium
114 C1 Faguibine, Lac l. Mali
8 E5 Fagurhólsmýri Iceland
109 F4 Fagwir Sudan
41 E4 Fahraj Iran
74 D3 Fairbanks U.S.A.
90 B5 Fairborn U.S.A.
84 D3 Fairbury U.S.A.
90 E5 Fairfax U.S.A.
92 A2 Fairfield CA U.S.A.
88 B5 Fairfield IA U.S.A.
86 C4 Fairfield OH U.S.A.
85 D6 Fairfield TX U.S.A.
91 G3 Fair Haven U.S.A.
14 E2 Fair Head hd U.K.
55 A4 Fairie Queen sand bank Phil.
11 G1 Fair Isle i. U.K.
84 E3 Fairmont MN U.S.A.
90 C5 Fairmont WV U.S.A.
83 F4 Fairplay U.S.A.
88 D3 Fairport U.S.A.
90 C4 Fairport Harbor U.S.A.
76 F3 Fairview Can.
89 E3 Fairview MI U.S.A.
85 D4 Fairview OK U.S.A.
93 G2 Fairview UT U.S.A.
51 ▫ Fairview Park H.K. China
76 B3 Fairweather, Cape U.S.A.
76 B3 Fairweather, Mt Can./U.S.A.
47 G6 Fais i. Micronesia
44 C3 Faisalabad Pak.
15 E3 Faissault France
84 C2 Faith U.S.A.
11 ▫ Faither, The pt U.K.
45 E4 Faizabad India
71 ▫2 Fakaina i. Fr. Polynesia Pac. Oc.
63 J2 Fakaofo i. Tokelau
71 ▫2 Fakarava i. Fr. Polynesia Pac. Oc.
13 H5 Fakenham U.K.
8 O5 Fåker Sweden
57 J7 Fakfak Indon.
40 D4 Fakhrabad Iran
19 K1 Fakse Denmark
19 K1 Fakse Bugt b. Denmark
54 B2 Faku China
13 C7 Fal r. U.K.
108 A4 Falaba Sierra Leone
24 D2 Falaise France
45 G4 Falakata India
45 H5 Falam Myanmar
40 C3 Falavarjan Iran
Falcon i. see Fonuafo'ou
23 L6 Falconara Marittima Italy
97 E2 Falcon Lake Mex./U.S.A.
71 ▫4 Falealupo Western Samoa
71 ▫4 Faleasiu Western Samoa
71 ▫4 Faleula Western Samoa
71 ▫4 Falelima Western Samoa
85 D7 Falfurrias U.S.A.
76 F3 Falher Can.
19 L5 Falkenberg Ger.
9 N8 Falkenberg Sweden
19 K3 Falkenhagen Ger.
19 K5 Falkenhain Ger.
19 L4 Falkensee Ger.
11 E5 Falkirk U.K.
11 E4 Falkland U.K.
99 E8 Falkland Islands Ant.
102 D8 Falkland Sound chan. Falkland Is
9 N7 Falköping Sweden
92 D5 Fallbrook U.S.A.
19 G4 Fallingbostel Ger.
92 C2 Fallon U.S.A.
91 H4 Fall River U.S.A.
82 F3 Fall River Pass U.S.A.
84 E3 Falls City U.S.A.
13 B7 Falmouth U.K.
90 A5 Falmouth KY U.S.A.
91 H3 Falmouth ME U.S.A.
88 E3 Falmouth MI U.S.A.
112 C7 False Bay S. Africa
96 B3 Falso, Cabo c. Mex.
9 M9 Falster i. Denmark
17 N7 Fălticeni Romania
9 O6 Falun Sweden
36 D4 Famagusta Cyprus
15 G3 Fameck France
15 E2 Famenne v. Belgium
77 K4 Family L. Can.
114 B2 Fana Mali
50 F4 Fanchang China
14 E4 Fane r. Rep. of Ireland
58 A1 Fang Thai.
71 ▫2 Fangatau i. Fr. Polynesia Pac. Oc.
71 ▫11 Fanga Uta lag. Tonga
50 D3 Fangcheng China

51 C4 Fangdou Shan mts China
51 F6 Fang-liao Taiwan
50 E2 Fangshan Beijing China
50 D2 Fangshan Shanxi China
51 F6 Fangshan Taiwan
50 D3 Fang Xian China
52 A2 Fangzheng China
51 □ Fanling H.K. China
11 C3 Fannich, Loch l. U.K.
41 E5 Fannūj Iran
18 F1 Fanø i. Denmark
23 L6 Fano Italy
18 F1 Fanø Bugt b. Denmark
51 F5 Fanshan China
50 D2 Fanshi China
51 B6 Fan Si Pan mt Vietnam
38 E5 Farab Turkm.
114 A2 Faraba Mali
115 A2 Faradje Congo(Zaire)
111 E6 Farafangana Madag.
109 E2 Farafra Oasis oasis Egypt
41 F3 Farāh Afgh.
41 F4 Farah Rūd r. Afgh.
57 L2 Farallon de Pajaros i.
N. Mariana Is
103 A4 Farallones de Cali, Parque
Nacional nat. park Col.
114 A2 Faranah Guinea
47 G6 Faraulep Atoll Micronesia
13 F7 Fareham U.K.
70 D4 Farewell, Cape N.Z.
70 D4 Farewell Spit spit N.Z.
9 N7 Färgelanda Sweden
84 D2 Fargo U.S.A.
84 E2 Faribault U.S.A.
79 F2 Faribault, Lac l. Can.
44 D3 Faridabad India
44 C3 Faridkot India
45 G5 Faridpur Bangl.
38 D5 Fārīğ Iran
108 A3 Farim Guinea-Bissau
41 E3 Farīmān Iran
9 P8 Färjestaden Sweden
39 G5 Farkhor Tajik.
37 M4 Farmahin Iran
88 C5 Farmer City U.S.A.
78 D2 Farmer Island Can.
76 E3 Farmington Can.
88 B5 Farmington IA U.S.A.
88 B5 Farmington IL U.S.A.
91 H2 Farmington ME U.S.A.
91 H3 Farmington NH U.S.A.
93 H3 Farmington NM U.S.A.
82 E3 Farmington UT U.S.A.
76 D4 Far Mt. Can.
90 D6 Farmville U.S.A.
21 M4 Farná Slovakia
13 G6 Farnborough U.K.
12 F2 Farne Islands U.K.
13 G6 Farnham U.K.
76 F4 Farnham, Mt Can.
101 G4 Faro Brazil
76 C2 Faro Can.
25 C4 Faro Port.
9 Q8 Fårö i. Sweden
6 C2 Faroe Islands terr. Atl. Ocean
9 Q8 Fårösund Sweden
106 H6 Farquhar Islands Seychelles
40 D4 Farrāshband Iran
90 C4 Farrell U.S.A.
89 K3 Farrellton Can.
41 E3 Farrokhī Iran
Farrukhabad see Fatehgarh
40 D3 Farsakh Iran
27 K5 Farsala Greece
41 F3 Fārsī Afgh.
82 E3 Farson U.S.A.
9 K7 Farsund Norway
Farvel, Kap c. see
Uummannarsuaq
85 C5 Farwell U.S.A.
40 D4 Fasā Iran
26 G4 Fasano Italy
19 H4 Faßberg Ger.
90 E4 Fassett U.S.A.
29 D5 Fastiv Ukr.
44 D4 Fatehgarh India
44 C4 Fatehpur Rajasthan India
44 E4 Fatehpur Uttar Pradesh India
40 D4 Fatḩābād Iran
89 G3 Fathom Five National Marine
Park Can.
108 A3 Fatick Senegal
71 □10 Fatumanga i. Tonga
15 G3 Faulquemont France
69 A2 Faure Island Austr.
113 F4 Fauresmith S. Africa
8 O3 Fauske Norway
93 F1 Faust U.S.A.
15 A3 Fauville-en-Caux France
23 K6 Favalto, Monte mt Italy
22 D2 Faverney France
26 E6 Favignana, Isola i. Sicily Italy
76 G4 Fawcett Can.
13 F7 Fawley U.K.
78 C3 Fawn r. Can.
8 B4 Faxaflói b. Iceland
8 P5 Faxälven r. Sweden
51 B5 Faxian Hu l. China
109 D3 Faya Chad
22 D6 Fayence France
88 D3 Fayette U.S.A.
85 E4 Fayetteville AR U.S.A.
87 E5 Fayetteville NC U.S.A.
87 C5 Fayetteville TN U.S.A.
36 D6 Fâyid Egypt
37 M7 Faylakah i. Kuwait
22 C2 Fayl-la-Forêt France
23 L5 Fažana Croatia
114 D3 Fazao Malfakassa, Parc
National de nat. park Togo
44 C3 Fazilka India
40 C5 Fazrān, J. h. S. Arabia
108 A2 Fdérik Maur.
14 B5 Feale r. Rep. of Ireland
87 E5 Fear, Cape U.S.A.
92 B2 Feather Falls U.S.A.
70 E4 Featherston N.Z.
67 G6 Feathertop, Mt Austr.
24 E2 Fécamp France
22 E1 Fecht r. France
105 F1 Federación Arg.
102 E4 Federal Arg.
38 E1 Fedorovka Kustanay. Kazak.
39 J1 Fedorovka Pavlodarsk. Kazak.
38 B2 Fedorovka Zapadnyy Kaz.
Kazak.
38 C1 Fedorovka Rus. Fed.
19 J2 Fehmarn i. Ger.
19 H2 Fehmarnsund chan. Ger.
19 K4 Fehrbellin Ger.
104 E3 Feia, Lagoa lag. Brazil
50 E4 Feidong China
50 F3 Feihuanghe Kou est. China
100 D5 Feijó Brazil
70 E4 Feilding N.Z.
101 L6 Feira de Santana Brazil
50 E4 Feixi China
21 M5 Fejér div. Hungary
19 J2 Fejø i. Denmark
36 E3 Feke Turkey
25 H3 Felanitx Spain
88 D3 Felch U.S.A.
21 M5 Felcsút Hungary
20 D2 Felda r. Ger.
21 J6 Feldbach Austria
19 L3 Feldberg Ger.
22 F2 Feldberg mt Ger.
20 C5 Feldkirch Austria
21 H6 Feldkirchen in Kärnten Austria
20 E5 Feldkirchen-Westerham Ger.
71 □10 Feletoa Tonga
105 E1 Feliciano r. Arg.
97 G4 Felipe C. Puerto Mex.
104 D2 Felixlândia Brazil
13 J6 Felixstowe U.K.
23 L3 Fella r. Italy
23 J3 Feltre Italy
19 J2 Femer Bælt str. Denmark/Ger.
19 J2 Femø i. Denmark
9 M5 Femunden l. Norway
9 N5 Femundsmarka Nasjonalpark
nat. park Norway
50 D2 Fen r. China
26 D3 Fenaio, Punta del pt Italy
93 H4 Fence Lake U.S.A.
89 H3 Fenelon Falls Can.
27 L4 Fengari mt Greece
51 E4 Fengcheng Jiangxi China
54 C3 Fengcheng Liaoning China
51 C4 Fengdu China
51 C5 Fenggang China
54 D1 Fengguang China
51 F4 Fenghua China
51 C5 Fenghuang China
50 C4 Fengjie China
51 D6 Fengkai China
51 F6 Fenglin Taiwan
50 F2 Fengnan China
50 E1 Fengning China
50 E3 Fengqiu China
51 C5 Fengshan China
51 E6 Fengshun China
50 E3 Fengtai China
51 E4 Fengxin China
50 E3 Fengyang China
50 D1 Fengzhen China
45 G5 Feni Bangl.
63 F2 Feni Is P.N.G.
24 F5 Fenille, Col de la pass France
88 B4 Fennimore U.S.A.
111 E5 Fenoarivo Atsinanana Madag.
13 G5 Fens, The reg. U.K.
89 F4 Fenton U.S.A.
60 M5 Fenua Ura i. Fr. Polynesia
Pac. Oc.
50 D2 Fenxi China
50 D2 Fenyang China
51 E5 Fenyi China
29 E6 Feodosiya Ukr.
26 B6 Fer, Cap de hd Alg.
19 L3 Ferdinandshof Ger.
41 E3 Ferdows Iran
15 D4 Fère-Champenoise France
39 G4 Fergana Uzbek.
39 H4 Fergana Too Tizmegi mts Kyrg.
89 G4 Fergus Can.
84 D2 Fergus Falls U.S.A.
62 F2 Fergusson I. P.N.G.
26 C7 Fériana Tunisia
114 B3 Ferkessédougou Côte d'Ivoire
21 H6 Ferlach Austria
26 E3 Fermo Italy
79 G3 Fermont Can.
25 C2 Fermoselle Spain
14 C5 Fermoy Rep. of Ireland
87 D6 Fernandina Beach U.S.A.
100 □ Fernandina, Isla i.
Galapagos Is Ecuador
102 B8 Fernando de Magallanes,
Parque Nacional nat. park
Chile
120 G6 Fernando de Noronha i.
Atl. Ocean
104 B3 Fernandópolis Brazil
82 B1 Ferndale U.S.A.
13 F7 Ferndown U.K.
76 F5 Fernie Can.
67 G2 Fernlee Austr.
92 C2 Fernley U.S.A.
91 F4 Fernridge U.S.A.
14 E5 Ferns Rep. of Ireland
82 C2 Fernwood U.S.A.
23 J5 Ferrara Italy
23 J5 Ferrarese r. Italy
104 B3 Ferreiros Brazil
85 F6 Ferriday U.S.A.
26 C4 Ferro, Capo pt Sardinia Italy
25 B1 Ferrol Spain
93 G2 Ferron U.S.A.
38 D1 Fershampenuaz Rus. Fed.
21 K5 Fertő l. Austria/Hungary
21 K5 Fertő-tavi nat. park Hungary
18 C3 Ferwerd Neth.
108 B1 Fès Morocco
110 B4 Feshi Congo(Zaire)
77 K5 Fessenden U.S.A.
84 F4 Festus U.S.A.
11 □ Fethaland, Point of U.K.
14 D5 Fethard Rep. of Ireland
36 B3 Fethiye Turkey
38 C4 Fetisovo Kazak.
11 □ Fetlar i. U.K.
11 F4 Fettercairn U.K.
20 D3 Feuchtwangen Ger.
79 F2 Feuilles, Rivière aux r. Can.
15 B2 Feuquières-en-Vimeu France
22 B4 Feurs France
36 F3 Fevzipaşa Turkey
41 H2 Feyzābād Afgh.
41 E3 Feyzābād Iran
Fez see Fès
13 D5 Ffestiniog U.K.
111 E6 Fianarantsoa Madag.
110 D3 Fichē Eth.
20 E2 Fichtelgebirge reg. Ger.
113 G4 Ficksburg S. Africa
23 H5 Fidenza Italy
76 F4 Field B.C. Can.
89 G2 Field Ont. Can.
27 H4 Fier Albania
88 E3 Fife Lake U.S.A.
11 F4 Fife Ness pt U.K.
67 G4 Fifield Austr.
88 B3 Fifield U.S.A.
22 D6 Figaniéres France
24 F4 Figeac France
25 B2 Figueira da Foz Port.
25 H1 Figueres Spain
108 B1 Figuig Morocco
114 F3 Figuil Cameroon
61 J5 Fiji country Pac. Oc.
96 H6 Filadelfia Costa Rica
102 D2 Filadélfia Para.
116 B3 Filchner Ice Shelf ice feature
Ant.
12 G3 Filey U.K.
27 J5 Filippiada Greece
9 O7 Filipstad Sweden
8 L5 Fillan Norway
92 C4 Fillmore CA U.S.A.
93 F2 Fillmore UT U.S.A.
115 D1 Fīltu Eth.
116 C3 Fimbulheimen mts Ant.
116 C3 Fimbulisen ice feature Ant.
23 J5 Finale Emilia Italy
22 F5 Finale Ligure Italy
91 F2 Finch Can.
68 C3 Finch Hatton Austr.
11 E3 Findhorn r. U.K.
37 H3 Fındık Turkey
90 B4 Findlay U.S.A.
67 H8 Fingal Austr.
78 E5 Finger Lakes U.S.A.
115 B6 Fingoè Moz.
36 C3 Finike Turkey
36 C3 Finike Körfezi b. Turkey
Finisterre, Cape see
Fisterra, Cabo
64 F4 Finke Gorge National Park
Austr.
7 F2 Finland country Europe
9 S7 Finland, Gulf of g. Europe
76 D3 Finlay r. Can.
76 D3 Finlay, Mt Can.
67 F5 Finley Austr.
19 J5 Finne ridge Ger.
66 A4 Finniss, C. pt Austr.
8 P2 Finnsnes Norway
9 O7 Finspång Sweden
22 F3 Finsteraarhorn mt Switz.
19 L5 Finsterwalde Ger.
14 D3 Fintona U.K.
14 C3 Fintown Rep. of Ireland
11 C3 Fionn Loch l. U.K.
11 B4 Fionnphort U.K.
70 A6 Fiordland National Park N.Z.
23 G5 Fiorenzuola d'Arda Italy
Fırat r. see Euphrates
92 B3 Firebaugh U.S.A.
77 J2 Firedrake Lake Can.
91 G4 Fire Island National Seashore
res. U.S.A.
Firenze see Florence
37 K6 Firk, Sha'īb watercourse Iraq
105 E2 Firmat Arg.
22 B4 Firminy France
17 Q2 Firovo Rus. Fed.
44 B3 Firoza Pak.
44 D4 Firozabad India
41 G3 Firozkoh reg. Afgh.
44 C3 Firozpur India
91 H2 First Connecticut L. U.S.A.
Firuzabad see Rāsk
40 D4 Fīrūzābād Iran
15 H3 Fischbach Ger.
111 B6 Fish r. Namibia
112 D5 Fish r. S. Africa
116 B6 Fisher Bay Ant.
91 F6 Fisherman I. U.S.A.
91 H4 Fishers I. U.S.A.
77 N2 Fisher Strait Can.
13 C6 Fishguard U.K.
76 E2 Fish Lake Can.
88 A2 Fish Lake MN U.S.A.
93 G2 Fish Lake UT U.S.A.
51 □ Fish Ponds l. H.K. China
89 F4 Fish Pt U.S.A.
116 B3 Fiske, C. Ant.
Fiskenæsset see
Qeqertarsuatsiaat
15 D3 Fismes France
25 B1 Fisterra Spain
25 B1 Fisterra, Cabo c. Spain
91 H3 Fitchburg U.S.A.
71 □4 Fito mt Western Samoa
77 G3 Fitzgerald Can.
87 D6 Fitzgerald U.S.A.
66 B4 Fitzgerald Bay Austr.
69 D6 Fitzgerald River Nat. Park
Austr.
102 C7 Fitz Roy Arg.
68 D4 Fitzroy r. Austr.
64 E3 Fitzroy Crossing Austr.
89 G3 Fitzwilliam I. Can.
14 D3 Fivemiletown U.K.
23 H5 Fivizzano Italy
115 A3 Fizi Congo(Zaire)
9 L6 Flå Norway
113 H5 Flagstaff S. Africa
93 G4 Flagstaff U.S.A.
91 H2 Flagstaff Lake U.S.A.
78 E2 Flaherty Island Can.
88 B3 Flambeau r. U.S.A.
12 G3 Flamborough Head hd U.K.
19 K4 Fläming h. Ger.
82 E3 Flaming Gorge Resr l. U.S.A.
112 D5 Flaminksvlei salt pan S. Africa
15 C2 Flandre reg. France
11 A2 Flannan Isles U.K.
8 O4 Flåsjön l. Sweden
88 E4 Flat r. U.S.A.
82 D2 Flathead L. U.S.A.
80 D2 Flathead Lake U.S.A.
68 C4 Flat Is Austr.
70 E4 Flat Point N.Z.
65 J2 Flattery, C. pt Austr.
82 A1 Flattery, C. U.S.A.
19 J4 Fleetmark Ger.
68 A4 Fleetwood Austr.
12 D4 Fleetwood U.K.
91 F4 Fleetwood U.S.A.
9 K7 Flekkefjord Norway
90 E3 Fleming U.S.A.
90 B5 Flemingsburg U.S.A.
9 P7 Flen Sweden
19 G2 Flensborg Fjord in.
Denmark/Ger.
19 G2 Flensburg Ger.
24 D2 Flers France
89 G3 Flesherton Can.
77 H2 Fletcher Lake Can.
89 F3 Fletcher Pond l. U.S.A.
116 B3 Fletcher Prom. hd Ant.
15 B3 Fleury-sur-Andelle France
20 C2 Flieden Ger.
65 H3 Flinders r. Austr.
69 B7 Flinders Bay Austr.
66 B5 Flinders Chase Nat. Park Austr.
67 H7 Flinders I. Austr.
66 A4 Flinders Island Austr.
68 C2 Flinders Passage Austr.
66 C3 Flinders Ranges mts Austr.
66 C3 Flinders Ranges Nat. Park
Austr.
68 C1 Flinders Reefs Coral Sea Is Terr.
77 J4 Flin Flon Can.
13 D4 Flint U.K.
87 C6 Flint r. GA U.S.A.
89 F4 Flint r. MI U.S.A.
89 F4 Flint U.S.A.
19 H2 Flintbek Ger.
60 M5 Flint Island Kiribati
67 H1 Flinton Austr.
20 D5 Flirsch Austria
9 N6 Flisa Norway
15 C2 Flixecourt France
12 E2 Flodden U.K.
20 D2 Floh Ger.
19 L6 Flöha Ger.
19 L6 Flöha r. Ger.
20 G2 Flöha r. Ger.
116 A4 Flood Ra. mts Ant.
88 A2 Floodwood U.S.A.
86 B4 Flora U.S.A.
24 F4 Florac France
15 G3 Florange France
68 B1 Flora Reef Coral Sea Is Terr.
89 F4 Florence Can.
23 J6 Florence Italy
87 C5 Florence AL U.S.A.
93 G5 Florence AZ U.S.A.
84 D4 Florence KS U.S.A.
90 C5 Florence OH U.S.A.
82 A3 Florence OR U.S.A.
87 E5 Florence SC U.S.A.
93 G5 Florence Junction U.S.A.
91 K1 Florenceville Can.
103 B4 Florencia Col.
15 E2 Florennes Belgium
102 C6 Florentino Ameghino, Embalse
resr Arg.
105 E2 Flores r. Arg.
97 G4 Flores Guatemala
57 G8 Flores i. Indon.
104 C1 Flores de Goiás Brazil
57 F8 Flores Sea Indon.
101 L5 Floresta Brazil
101 K5 Floriano Brazil
102 G3 Florianópolis Brazil
105 F2 Florida Uru.
87 D6 Florida div. U.S.A.
87 D7 Florida Bay U.S.A.
87 D7 Florida City U.S.A.
63 G2 Florida Is Solomon Is
81 K7 Florida Keys is U.S.A.
95 H4 Florida, Straits of
Bahamas/U.S.A.
27 J4 Florina Greece
9 J6 Florø Norway
20 B2 Flörsheim am Main Ger.
79 H3 Flour Lake Can.
88 A4 Floyd IA U.S.A.
90 C6 Floyd VA U.S.A.
85 C5 Floydada U.S.A.
93 F4 Floyd, Mt U.S.A.
23 G2 Flums Switz.
62 E2 Fly r. P.N.G.
90 C5 Fly U.S.A.
71 □12 Foa i. Tonga
27 H3 Foča Bos.-Herz.
11 E3 Fochabers U.K.
113 G3 Fochville S. Africa
27 M2 Focşani Romania
51 D6 Fogang China
26 F4 Foggia Italy
108 □ Fogo i. Cape Verde
79 K4 Fogo I. Can.
39 G4 Fogoleva Kazak.
18 F2 Föhr i. Ger.
23 J6 Foiano della Chiana Italy
11 D2 Foinaven h. U.K.
24 E5 Foix France
8 O3 Folda chan. Norway
8 N4 Foldereid Norway
8 M4 Foldfjorden chan. Norway
27 L6 Folegandros i. Greece
89 F1 Foleyet Can.
26 E3 Foligno Italy
13 J6 Folkestone U.K.
13 G5 Folkingham U.K.
87 D6 Folkston U.S.A.
9 M5 Folldal Norway
26 D3 Follonica Italy
92 B2 Folsom Lake U.S.A.
29 G6 Fomin Rus. Fed.
28 J2 Fominskiy Rus. Fed.
77 H3 Fond-du-Lac Can.
77 J3 Fond du Lac r. Can.
88 C4 Fond du Lac U.S.A.
25 B2 Fondevila Spain
26 E4 Fondi Italy
63 H2 Fongafale Tuvalu
26 C4 Fonni Sardinia Italy
96 H5 Fonseca, Golfo do b.
Central America
22 C4 Fontaine France
15 C4 Fontainebleau France
15 C4 Fontainebleau, Forêt de forest
France
22 C2 Fontaine-Française France
22 C2 Fontaine-lès-Dijon France
79 F3 Fontanges Can.
76 E3 Fontas Can.
76 E3 Fontas r. Can.
100 E4 Fonte Boa Brazil
24 D3 Fontenay-le-Comte France
23 J4 Fonti, Cima mt Italy
8 F3 Fontur pt Iceland
71 □12 Fonuafo'ou i. Tonga
71 □10 Fonua Unga i. Tonga
21 L6 Fonyód Hungary
89 H3 Foot's Bay Can.
50 C3 Foping China
67 H4 Forbes Austr.
82 C1 Forbes, Mt Can.
20 E3 Forchheim Ger.
79 G2 Ford r. Can.
88 D2 Ford r. U.S.A.
9 J6 Førde Norway
77 K2 Forde Lake Can.
13 H5 Fordham U.K.
13 F7 Fordingbridge U.K.
116 A4 Ford Ra. mts Ant.
67 F2 Fords Bridge Austr.
85 E5 Fordyce U.S.A.
108 A4 Forécariah Guinea
13 F7 Foreland hd U.K.
13 D6 Foreland Point U.K.
76 D4 Foresight Mtn Can.
89 G4 Forest Can.
85 F5 Forest MS U.S.A.
90 B4 Forest OH U.S.A.
91 G3 Forest Dale U.S.A.
67 G5 Forest Hill N.S.W. Austr.
68 A5 Forest Hill Qld. Austr.
92 B2 Foresthill U.S.A.
67 H9 Forestier, C. hd Austr.
67 H9 Forestier Pen. Austr.
88 A3 Forest Lake U.S.A.
87 C5 Forest Park U.S.A.
79 G4 Forestville Can.
22 B4 Forez, Plaine du plain France
11 F4 Forfar U.K.
15 B3 Forges-les-Eaux France
82 A2 Forks U.S.A.
91 J2 Forks, The U.S.A.
90 E4 Forksville U.S.A.
23 K5 Forlì Italy
12 D4 Formby U.K.
25 G3 Formentera i. Spain
25 H3 Formentor, Cap de pt Spain
15 B3 Formerie France
104 D3 Formiga Brazil
23 H5 Formigine Italy
102 E3 Formosa Arg.
Formosa country see Taiwan
104 C1 Formosa Brazil
101 G6 Formosa, Serra h. Brazil
104 D1 Formoso r. Brazil
23 H5 Fornovo di Taro Italy
11 E3 Forres U.K.
66 E7 Forrest Vic. Austr.
64 E6 Forrest W.A. Austr.
88 C5 Forrest U.S.A.
85 F5 Forrest City U.S.A.
88 C4 Forreston U.S.A.
8 P5 Fors Sweden
65 H3 Forsayth Austr.
8 S3 Forsnäs Sweden
9 S6 Forssa Fin.
19 M5 Forst Ger.
67 K4 Forster Austr.
85 E4 Forsyth MO U.S.A.
82 F2 Forsyth MT U.S.A.
89 J1 Forsythe Can.
44 C3 Fort Abbas Pak.
78 D3 Fort Albany Can.
101 L4 Fortaleza Brazil
93 H5 Fort Apache U.S.A.
76 G4 Fort Assiniboine Can.
88 C4 Fort Atkinson U.S.A.
11 D3 Fort Augustus U.K.
113 G6 Fort Beaufort S. Africa
82 E2 Fort Benton U.S.A.
77 H3 Fort Black Can.
92 A2 Fort Bragg U.S.A.
Fort-Chimo see Kuujjuaq
77 G3 Fort Chipewyan Can.
85 D5 Fort Cobb Resr U.S.A.
82 F3 Fort Collins U.S.A.
89 J3 Fort-Coulonge Can.
91 F2 Fort Covington U.S.A.
85 C6 Fort Davis U.S.A.
95 M6 Fort-de-France Martinique
87 C5 Fort Deposit U.S.A.
84 E3 Fort Dodge U.S.A.
64 C4 Fortescue r. Austr.
84 E1 Fort Frances U.S.A.
78 E3 Fort George Can.
74 F3 Fort Good Hope Can.
11 D4 Forth r. U.K.
11 F4 Forth, Firth of est. U.K.
93 E2 Fortification Range mts U.S.A.
102 D2 Fortín Capitán Demattei Para.
102 D2 Fortín General Mendoza Para.
102 E2 Fortín Madrejón Para.
102 D2 Fortín Pilcomayo Arg.
100 F7 Fortín Ravelo Bol.
100 F7 Fortín Suárez Arana Bol.
91 J1 Fort Kent U.S.A.
87 D7 Fort Lauderdale U.S.A.
76 E2 Fort Liard Can.
88 B3 Fort McCoy U.S.A.
77 G3 Fort Mackay Can.
76 G5 Fort Macleod Can.
77 G3 Fort McMurray Can.
74 E3 Fort McPherson Can.
88 B5 Fort Madison U.S.A.
15 B2 Fort-Mahon-Plage France
82 G3 Fort Morgan U.S.A.
87 D7 Fort Myers U.S.A.
76 E3 Fort Nelson Can.
76 E3 Fort Nelson r. Can.
76 D2 Fort Norman Can.
87 C5 Fort Payne U.S.A.
82 F1 Fort Peck U.S.A.
82 F2 Fort Peck Resr U.S.A.
87 D7 Fort Pierce U.S.A.
84 C2 Fort Pierre U.S.A.
76 F2 Fort Providence Can.
77 J4 Fort Qu'Appelle Can.
76 G2 Fort Resolution Can.
70 B7 Fortrose N.Z.
11 D3 Fortrose U.K.
92 A2 Fort Ross U.S.A.
78 E3 Fort Rupert Can.
76 E4 Fort St James Can.
76 E3 Fort St John Can.
76 G4 Fort Saskatchewan Can.
85 E4 Fort Scott U.S.A.
78 C2 Fort Severn Can.
38 B3 Fort-Shevchenko Kazak.
76 E2 Fort Simpson Can.
77 G2 Fort Smith Can.
85 E5 Fort Smith U.S.A.
85 C6 Fort Stockton U.S.A.
83 F5 Fort Sumner U.S.A.
82 A3 Fortuna U.S.A.
84 C1 Fortune U.S.A.
79 J4 Fortune B. Can.
76 F3 Fort Vermilion Can.
87 C6 Fort Walton Beach U.S.A.
88 E5 Fort Wayne U.S.A.
11 C4 Fort William U.K.
85 D5 Fort Worth U.S.A.
74 D3 Fort Yukon U.S.A.
40 D5 Forūr, Jazīreh-ye i. Iran
8 N4 Forvik Norway
51 D6 Foshan China
114 C3 Foso Ghana
22 E5 Fossano Italy
23 K6 Fossombrone Italy
22 B6 Fos-sur-Mer France
67 G7 Foster Austr.
75 Q2 Foster B. Greenland
76 B3 Foster, Mt Can./U.S.A.
90 B4 Fostoria U.S.A.
13 G4 Fotherby U.K.
114 F2 Fotokol Cameroon
24 D2 Fougères France
11 □ Foula i. U.K.
22 C1 Foulain France
13 H6 Foulness Point U.K.
43 C4 Foul Pt Sri Lanka
70 C4 Foulwind, Cape N.Z.
114 F3 Foumban Cameroon
116 B3 Foundation Ice Stream
ice feature Ant.
108 A3 Foundiougne Senegal
88 A4 Fountain U.S.A.
15 F4 Fourches, Mont des h. France
92 D4 Four Corners U.S.A.
113 H4 Fouriesburg S. Africa
15 E2 Fourmies France
27 M6 Fournoi i. Greece
88 C2 Fourteen Mile Pt U.S.A.
108 A3 Fouta Djallon reg. Guinea
70 A7 Foveaux Strait N.Z.
87 E7 Fowl Cay i. Bahamas
83 F4 Fowler CO U.S.A.
88 D5 Fowler IN U.S.A.
88 E4 Fowler MI U.S.A.
116 B3 Fowler Pen. Ant.
64 F6 Fowlers Bay Austr.
37 M3 Fowman Iran
77 L3 Fox r. Can.
88 C4 Fox r. U.S.A.
68 B4 Fox Cr. r. Austr.
76 F4 Fox Creek Can.
12 C3 Foxdale U.K.
75 K3 Foxe Basin g. Can.
75 K3 Foxe Channel Can.
75 L3 Foxe Peninsula Can.
70 C5 Fox Glacier N.Z.
76 G3 Fox Lake Can.
88 C4 Fox Lake U.S.A.
70 E4 Foxton N.Z.
11 D3 Foyers U.K.
14 D3 Foyle r. Rep. of Ireland/U.K.
14 D2 Foyle, Lough b.
Rep. of Ireland/U.K.
14 B5 Foynes Rep. of Ireland
111 B5 Foz do Cunene Angola
104 A4 Foz do Iguaçu Brazil
25 G2 Fraga Spain
15 H4 Fraize France
116 D4 Framnes Mts Ant.
104 C3 Franca Brazil
63 G3 Français, Récif des rf
New Caledonia
71 □9 Français, Récif des rf
New Caledonia Pac. Oc.
7 D4 France country Europe
66 D6 Frances Austr.
76 D2 Frances r. Can.
76 D2 Frances Lake Can.
76 D2 Frances Lake Can.
88 D5 Francesville U.S.A.
110 B4 Franceville Gabon
22 D2 Franche-Comté div. France
84 D3 Francis Case, Lake U.S.A.
96 D2 Francisco I. Madero Coahuila
Mex.
96 C2 Francisco I. Madero Durango
Mex.
104 D2 Francisco Sá Brazil
91 H2 Francis, Lake U.S.A.
111 C6 Francistown Botswana
76 D4 François Lake Can.
82 E3 Francs Peak U.S.A.
18 C3 Franeker Neth.
19 L6 Frankenberg Ger.
18 F5 Frankenberg (Eder) Ger.
89 F4 Frankenmuth U.S.A.
20 B3 Frankenthal (Pfalz) Ger.
20 E2 Frankenwald forest Ger.
113 H3 Frankfort S. Africa
88 D5 Frankfort IN U.S.A.
86 C4 Frankfort KY U.S.A.
88 D3 Frankfort MI U.S.A.
20 B2 Frankfurt am Main Ger.
19 M4 Frankfurt an der Oder Ger.
93 E1 Frankin Lake U.S.A.
20 D4 Fränkische Alb reg. Ger.
20 C2 Fränkische Saale r. Ger.
20 E3 Fränkische Schweiz reg. Ger.
69 C7 Frankland r. Austr.
82 E3 Franklin ID U.S.A.
86 C4 Franklin IN U.S.A.
85 F6 Franklin LA U.S.A.
91 H3 Franklin MA U.S.A.
87 D5 Franklin NC U.S.A.
91 H3 Franklin NH U.S.A.
91 F4 Franklin NJ U.S.A.
90 D4 Franklin PA U.S.A.
87 C5 Franklin TN U.S.A.
90 E6 Franklin VA U.S.A.
90 D5 Franklin WV U.S.A.
74 F3 Franklin Bay Can.
82 C1 Franklin D. Roosevelt Lake
U.S.A.
66 B4 Franklin Harbour b. Austr.
116 B5 Franklin I. Ant.
76 E2 Franklin Mountains Can.
70 A6 Franklin Mts N.Z.
67 G8 Franklin Sd chan. Austr.
75 J2 Franklin Str. Can.
66 F7 Frankston Austr.
9 P5 Fränsta Sweden
88 E1 Franz Can.
70 C5 Franz Josef Glacier N.Z.
32 G2 Franz Josef Land is Rus. Fed.
26 C5 Frasca, Capo della pt Sardinia
Italy
26 E4 Frascati Italy
76 E4 Fraser r. B.C. Can.
79 H2 Fraser r. Nfld Can.
112 D5 Fraserburg S. Africa
11 F3 Fraserburgh U.K.
78 D4 Fraserdale Can.
68 E5 Fraser Island Austr.
68 E5 Fraser Island National Park
Austr.
76 E4 Fraser Lake Can.
69 D2 Fraser, Mt h. Austr.
76 E4 Fraser Plateau plat. Can.
70 F3 Frasertown N.Z.
22 D3 Frasne France
20 C5 Frastanz Austria
88 E2 Frater Can.
23 F2 Frauenfeld Switz.
105 E2 Fray Bentos Uru.
18 D6 Frechen Ger.
12 E4 Freckleton U.K.
88 E3 Frederic MI U.S.A.
88 A3 Frederic WV U.S.A.
9 L9 Fredericia Denmark
69 C2 Frederick r. Austr.
90 E5 Frederick MD U.S.A.
85 D5 Frederick OK U.S.A.
85 D6 Fredericksburg TX U.S.A.
90 E5 Fredericksburg VA U.S.A.
76 C3 Frederick Sound chan. U.S.A.
85 F4 Fredericktown U.S.A.
79 G4 Fredericton Can.
19 K1 Frederiksberg Denmark
Frederikshåb see Paamiut
9 M8 Frederikshavn Denmark
9 N9 Frederiksværk Denmark
93 F3 Fredonia AZ U.S.A.
90 D3 Fredonia NY U.S.A.
8 Q4 Fredrika Sweden
9 M7 Fredrikstad Norway
91 F4 Freehold U.S.A.
91 F4 Freeland U.S.A.
66 C3 Freeling Heights mt Austr.
92 C2 Freel Peak U.S.A.
84 D3 Freeman U.S.A.
88 D5 Freeman, Lake U.S.A.
88 C4 Freeport IL U.S.A.
91 H3 Freeport ME U.S.A.
91 G4 Freeport NY U.S.A.
85 E6 Freeport TX U.S.A.
87 E7 Freeport City Bahamas
85 D7 Freer U.S.A.
113 G4 Free State div. S. Africa
108 A4 Freetown Sierra Leone
25 C3 Fregenal de la Sierra Spain
24 C2 Fréhel, Cap pt France
19 L6 Freiberg Ger.
22 E1 Freiburg im Breisgau Ger.
20 C2 Freiensteinau Ger.
20 F5 Freilassing Ger.
15 H3 Freisen Ger.
20 E4 Freising Ger.
21 H4 Freistadt Austria
19 L6 Freital Ger.
22 D6 Fréjus France
22 D4 Fréjus Tunnel tunnel
France/Italy
69 B6 Fremantle Austr.
88 E4 Fremont MI U.S.A.
84 D3 Fremont NE U.S.A.
90 B4 Fremont OH U.S.A.
93 G2 Fremont r. U.S.A.
90 B6 Frenchburg U.S.A.
90 C4 French Creek r. U.S.A.
99 E2 French Guiana terr. S. America
66 F7 French I. Austr.
77 H5 Frenchman r. Can./U.S.A.
92 C2 Frenchman U.S.A.
67 F9 Frenchman Cap mt Austr.
92 B2 Frenchman L. CA U.S.A.
93 E3 Frenchman L. NV U.S.A.
14 C4 Frenchpark Rep. of Ireland
70 D4 French Pass N.Z.
61 N5 French Polynesia terr. Pac. Oc.
5 French Southern and Antarctic
Lands terr.
Southern Ocean
91 J1 Frenchville U.S.A.
21 M3 Frenštát pod Radhoštěm
Czech Rep.
18 E4 Freren Ger.
14 D5 Freshford Rep. of Ireland
93 G6 Fresnal Canyon U.S.A.
96 D3 Fresnillo Mex.
92 C3 Fresno U.S.A.
92 C3 Fresno r. U.S.A.
22 C2 Fretigney-et-Velloreille France
25 H3 Freu, Cap des pt Spain
18 E6 Freudenberg Ger.
20 B4 Freudenstadt Ger.
15 C2 Frévent France
69 B7 Freycinet, Cape pt Austr.
69 A3 Freycinet Estuary in. Austr.
67 H9 Freycinet Nat. Park Austr.
67 H9 Freycinet Peninsula Austr.
19 K3 Freyenstein Ger.
15 G3 Freyming-Merlebach France
105 D1 Freyre Arg.
21 G4 Freyung Ger.
108 A3 Fria Guinea
92 C3 Friant U.S.A.
102 C3 Frias Arg.
22 E3 Fribourg Switz.
22 F2 Frick Switz.
21 K5 Friedberg Austria
20 D4 Friedberg Ger.
20 B2 Friedberg (Hessen) Ger.
18 E3 Friedeburg Ger.
19 L3 Friedland Ger.
19 M4 Friedland Ger.
20 C5 Friedrichshafen Ger.
18 F2 Friedrichskoog Ger.
19 G2 Friedrichstadt Ger.
91 J3 Friendship U.S.A.
21 H6 Friesach Austria
19 K4 Friesack Ger.
18 C3 Friese Wad tidal flats Neth.
18 E3 Friesoythe Ger.
13 J6 Frinton-on-Sea U.K.
85 D6 Frio r. U.S.A.
11 B4 Frisa, Loch l. U.K.
93 F2 Frisco Mt U.S.A.
19 G5 Fritzlar Ger.
23 K3 Friuli - Venezia Giulia div.
Italy
75 M3 Frobisher Bay Can.
77 H3 Frobisher Lake Can.
19 H1 Frøbjerg Bavnehøj h. Denmark
8 L5 Frohavet b. Norway
19 K5 Frohburg Ger.
21 J5 Frohnleiten Austria
15 C3 Froissy France
29 G5 Frolovo Rus. Fed.
28 K2 Frolovskaya Rus. Fed.
66 C2 Frome watercourse Austr.
13 E6 Frome U.K.
66 C3 Frome Downs Austr.
66 C3 Frome, Lake salt flat Austr.
18 E5 Fröndenberg Ger.
97 F4 Frontera Mex.
97 F4 Frontera, Pta pt Mex.
90 D5 Front Royal U.S.A.
26 E4 Frosinone Italy
90 D5 Frostburg U.S.A.
8 L5 Frøya i. Norway
15 C2 Fruges France
93 H2 Fruita U.S.A.
93 G1 Fruitland U.S.A.
Frunze see Bishkek
39 G4 Frunze Kyrg.
22 E3 Frutigen Switz.
21 M3 Frýdek-Místek Czech Rep.
21 J2 Frýdlant Czech Rep.
91 H2 Fryeburg U.S.A.
71 □11 Fua'amotu Tonga
51 F5 Fu'an China
23 H6 Fucecchio Italy
51 D5 Fuchuan China
51 F4 Fuchun Jiang r. China
11 A3 Fuday i. U.K.
51 E5 Fude China
51 F5 Fuding China
25 E2 Fuenlabrada Spain
25 D3 Fuente Obejuna Spain
102 E2 Fuerte Olimpo Para.
108 A2 Fuerteventura i. Canary Is
55 B2 Fuga i. Phil.
50 E3 Fugou China
50 D2 Fugu China
48 E2 Fuhai China
37 J4 Fuḩaymī Iraq
42 E4 Fujairah U.A.E.
53 F7 Fuji Japan
51 E5 Fujian div. China
50 C4 Fu Jiang r. China
53 F7 Fuji-Hakone-Izu National Park
Japan
52 B1 Fujin China
53 F7 Fujinomiya Japan
53 F7 Fuji-san volc. Japan
52 H3 Fukagawa Japan
53 D7 Fukuchiyama Japan
53 A8 Fukue Japan
53 A8 Fukue-jima i. Japan
53 E6 Fukui Japan
53 B8 Fukuoka Japan
53 G6 Fukushima Japan
53 B9 Fukuyama Japan
40 D2 Fūlād Maialleh Iran
71 □14 Fulaga i. Fiji
20 C2 Fulda Ger.
19 G5 Fulda r. Ger.
13 G6 Fulham U.K.
50 E3 Fuliji China
51 C4 Fuling China
77 M2 Fullerton, Cape hd Can.
21 L3 Fulnek Czech Rep.
20 E5 Fulpmes Austria
88 B5 Fulton IL U.S.A.
86 B4 Fulton KY U.S.A.
84 F4 Fulton MO U.S.A.
91 E3 Fulton NY U.S.A.
113 K2 Fumane Moz.
15 E3 Fumay France
53 F7 Funabashi Japan
63 H2 Funafuti i. Tuvalu
108 A1 Funchal Port.
103 B2 Fundación Col.
25 C2 Fundão Port.
96 B2 Fundición Mex.
79 G5 Fundy, Bay of g. Can.
79 G4 Fundy Nat. Park Can.
92 D3 Funeral Peak U.S.A.
111 D6 Funhalouro Moz.
50 F3 Funing Jiangsu China
51 B6 Funing Yunnan China
50 D3 Funiu Shan mts China
68 C4 Funnel Cr. r. Austr.
114 E2 Funtua Nigeria
11 □ Funzie U.K.
51 F5 Fuqing China
54 D2 Fur r. China
115 B5 Furancungo Moz.
52 H3 Furano Japan
40 E5 Fūrgun, Kūh-e mt Iran
28 G3 Furmanov Rus. Fed.
39 H3 Furmanovka Kazak.
38 B2 Furmanovo Kazak.
52 D3 Furmanovo Rus. Fed.
92 D3 Furnace Creek U.S.A.
104 C3 Furnas, Represa resr Brazil
65 J7 Furneaux Group is Austr.
18 E4 Fürstenau Ger.
19 L3 Fürstenberg Ger.
21 K5 Fürstenfeld Austria
20 E4 Fürstenfeldbruck Ger.
19 M4 Fürstenwalde Ger.
20 D3 Fürth Ger.
20 B3 Fürth Ger.
20 F3 Furth im Wald Ger.
52 G3 Furubira Japan
52 G5 Furukawa Japan
75 K3 Fury and Hecla Strait Can.
103 B3 Fusagasugá Col.
51 C7 Fushan Hainan China
54 A5 Fushan Shandong China
54 B3 Fushun Liaoning China
51 B4 Fushun Sichuan China
54 B3 Fushuncheng China
54 D2 Fusong China
51 C6 Fusui China
53 B8 Futago-san volc. Japan
71 □9 Futuna i. Vanuatu
51 E5 Futun Xi r. China
50 C3 Fu Xian China
54 A2 Fuxin Liaoning China
52 F5 Fuya Japan
50 E3 Fuyang Anhui China
51 F4 Fuyang Zhejiang China
50 E2 Fuyang r. China
46 E2 Fuyu Heilongjiang China
54 C1 Fuyu China
51 B5 Fuyuan China
48 E2 Fuyun China
51 F5 Fuzhou Fujian China
51 E5 Fuzhou Jiangxi China
54 A4 Fuzhou Wan b. China
37 L2 Füzuli Azer.
9 M9 Fyn i. Denmark
11 C5 Fyne, Loch in. U.K.
19 H1 Fyns Hoved hd Denmark
F.Y.R.O.M. country see
Macedonia

G

26 C6 Gaâfour Tunisia
110 E3 Gaalkacyo Somalia
113 F2 Gabane Botswana
92 D2 Gabbs U.S.A.
92 C2 Gabbs Valley Range mts U.S.A.
111 B5 Gabela Angola
108 D1 Gabès Tunisia
109 D1 Gabès, Golfe de g. Tunisia
67 H6 Gabo I. Austr.
107 E6 Gabon country Africa
111 C6 Gaborone Botswana
41 E5 Gābrīk Iran
41 E5 Gābrīk watercourse Iran
27 L3 Gabrovo Bulg.
108 A3 Gabú Guinea-Bissau
40 C2 Gach Sār Iran
40 C4 Gach Sārān Iran
43 A3 Gadag India
8 O4 Gäddede Sweden
19 J3 Gadebusch Ger.
44 B5 Gadhra India
44 B4 Gadra Pak.
87 C5 Gadsden U.S.A.
43 B2 Gadwal India
38 E5 Gadyn Turkm.
8 S2 Gædnovuoppe Norway
13 D6 Gaer U.K.
27 L2 Găeşti Romania
26 E4 Gaeta Italy
26 E4 Gaeta, Golfo di g. Italy
60 F3 Gaferut i. Micronesia
87 D5 Gaffney U.S.A.
108 C1 Gafsa Tunisia
28 E4 Gagarin Rus. Fed.
39 G4 Gagarin Uzbek.
28 H4 Gagino Rus. Fed.
114 B3 Gagnoa Côte d'Ivoire
79 G3 Gagnon Can.
29 G7 Gagra Georgia
37 L4 Gahvāreh Iran
112 C3 Gaiab watercourse Namibia
45 G4 Gaibandha Bangl.

23 K5 Gorino Italy
37 L2 Goris Armenia
19 L3 Göritz Ger.
23 L4 Gorizia Italy
Gor'kiy see Nizhniy Novgorod
29 H5 Gor'ko-Solenoye, Ozero l. Rus. Fed.
28 G3 Gor'kovskoye Vdkhr. resr Rus. Fed.
39 K1 Gor'koye, Ozero salt l. Rus. Fed.
19 J1 Gørlev Denmark
17 K6 Gorlice Pol.
21 H1 Görlitz Ger.
44 D4 Gormi India
27 L3 Gorna Oryakhovitsa Bulg.
21 J6 Gornja Radgona Slovenia
27 J2 Gornji Milanovac Yugo.
26 G3 Gornji Vakuf Bos.-Herz.
39 L2 Gorno-Altaysk Rus. Fed.
39 L2 Gornoye Kazak.
52 G1 Gornozavodsk Rus. Fed.
52 K2 Gornyak Rus. Fed.
52 C2 Gornye Klyuchi Rus. Fed.
52 C2 Gornyy Primorskiy Kray Rus. Fed.
29 J5 Gornyy Saratov. Obl. Rus. Fed.
29 H5 Gornyy Balykley Rus. Fed.
28 G3 Gorodets Rus. Fed.
29 H5 Gorodishche Rus. Fed.
29 G6 Gorodovikovsk Rus. Fed.
62 E2 Goroka P.N.G.
66 D6 Goroke Austr.
28 G3 Gorokhovets Rus. Fed.
114 C2 Gorom Gorom Burkina
111 D5 Gorongosa Moz.
115 B6 Gorongosa mt Moz.
57 G6 Gorontalo Indon.
29 G5 Gorshechnoye Rus. Fed.
21 K2 Gór Stołowych, Park Narodowy nat. park Pol.
14 C4 Gort Rep. of Ireland
14 C2 Gortahork Rep. of Ireland
104 D1 Gorutuba r. Brazil
29 F6 Goryachiy Klyuch Rus. Fed.
19 K4 Görzke Ger.
21 M1 Gorzów Śląski Pol.
16 G4 Gorzów Wielkopolski Pol.
21 K6 Gősfai Hegy h. Hungary
67 J4 Gosford Austr.
12 F2 Gosforth U.K.
88 E5 Goshen IN U.S.A.
91 F4 Goshen NY U.S.A.
52 G4 Goshogawara Japan
19 H5 Goslar Ger.
26 F2 Gospić Croatia
13 F7 Gosport U.K.
114 C1 Gossi Mali
20 E3 Gößweinstein Ger.
27 J4 Gostivar Macedonia
21 H5 Göstling an der Ybbs Austria
Göteborg see Gothenburg
9 N7 Götene Sweden
19 H6 Gotha Ger.
9 M8 Gothenburg Sweden
84 C3 Gothenburg U.S.A.
9 Q8 Gotland i. Sweden
27 K4 Gotse Delchev Bulg.
9 Q7 Gotska Sandön i. Sweden
53 C7 Götsu Japan
23 G5 Gottero, Monte mt Italy
19 G5 Göttingen Ger.
76 E4 Gott Peak Can.
Gottwaldow see Zlín
54 A3 Goubangzi China
18 B4 Gouda Neth.
108 A3 Goudiri Senegal
114 F2 Goudoumaria Niger
88 E1 Goudreau Can.
120 J8 Gough Island Atl. Ocean
78 F4 Gouin, Réservoir resr Can.
88 E2 Goulais River Can.
67 G4 Goulburn r. N.S.W. Austr.
66 F6 Goulburn r. Vic. Austr.
67 H5 Goulburn Austr.
65 F2 Goulburn Is Austr.
88 E2 Gould City U.S.A.
116 B4 Gould Coast Ant.
69 C2 Gould, Mt h. Austr.
22 D2 Goumois France
114 C1 Goundam Mali
22 G4 Gouraya Alg.
22 D6 Gourdan, Sommet de mt France
114 F2 Gouré Niger
15 E4 Gourgançon France
112 D7 Gourits r. S. Africa
114 C1 Gourma-Rharous Mali
15 B3 Gournay-en-Bray France
67 H6 Gourock Range mts Austr.
15 C3 Goussainville France
91 F2 Gouverneur U.S.A.
77 H5 Govenlock Can.
104 E2 Governador Valadares Brazil
55 C5 Governor Generoso Phil.
87 E7 Governor's Harbour Bahamas
48 G3 Govĭ Altayn Nuruu mts Mongolia
45 E4 Govind Ballash Pant Sāgar resr India
44 D3 Govind Sagar resr India
38 F5 Govurdak Turkm.
90 D3 Gowanda U.S.A.
68 A5 Gowan Ra. h. Austr.
41 G4 Gowārān Afgh.
40 D3 Gowd-e Aḥmad Iran
40 E3 Gowd-e Hasht Tekkeh waterhole Iran
40 D4 Gowd-e Mokh l. Iran
13 C6 Gower pen. U.K.
89 G2 Gowganda Can.
41 E4 Gowk Iran
14 D4 Gowna, Lough l. Rep. of Ireland
102 E3 Goya Arg.
37 L1 Göyçay Azer.
37 H2 Göynük Turkey
114 F3 Goyoum Cameroon
52 G5 Goyō-zan mt Japan
37 M2 Göytäpä Azer.
41 F3 Gōzareh Afgh.
36 G2 Gözene Turkey
44 E2 Gozha Co salt l. China
26 F6 Gozo i. Malta
112 F6 Graaff-Reinet S. Africa
112 C6 Graafwater S. Africa
20 B3 Graben-Neudorf Ger.
20 D2 Grabfeld plain Ger.
114 B4 Grabo Côte d'Ivoire
112 C7 Grabouw S. Africa
39 H1 Grabovo Kazak.
19 J3 Grabow Ger.
26 F2 Gračac Croatia
89 J2 Gracefield Can.
69 D6 Grace, L. salt flat Austr.
38 C1 Grachevka Rus. Fed.
39 J2 Grachi Kazak.
96 G5 Gracias Honduras
23 L4 Grado Italy
69 C5 Grady, Lake salt flat Austr.
21 G4 Grafenau Ger.
20 E3 Gräfenberg Ger.
19 K5 Gräfenhainichen Ger.
67 K2 Grafton Austr.
84 D1 Grafton ND U.S.A.
90 C5 Grafton WV U.S.A.
88 D4 Grafton WV U.S.A.
68 A1 Grafton, C. pt Austr.
63 E2 Grafton, Mt U.S.A.
68 B1 Grafton Passage Austr.
85 D5 Graham U.S.A.
Graham Bell Island see Greem-Bell, Ostrov
75 J2 Graham I. Can.
76 C4 Graham Island Can.
91 J2 Graham Lake U.S.A.
116 B2 Graham Land reg. Ant.
93 H5 Graham, Mt U.S.A.
113 G6 Grahamstown S. Africa
14 E5 Graigue Rep. of Ireland
114 A3 Grain Coast Liberia
88 C5 Grajaú Brazil
11 B1 Gralisgeir i. U.K.
19 G1 Gram Denmark
19 M3 Grambow Ger.
19 J2 Grammendorf Ger.
22 E6 Grammont, Mont mt Italy
27 J4 Grámmos mt Greece
11 D4 Grampian Mountains U.K.
66 E6 Grampians mts Austr.
19 M3 Gramzow Ger.
22 F4 Grana r. Italy
22 E5 Grana r. Italy
112 C5 Granaatboskolk S. Africa
103 B4 Granada Col.
96 H6 Granada Nic.
25 E4 Granada Spain
84 C4 Granada U.S.A.
14 D4 Granard Rep. of Ireland
105 C3 Gran Bajo Salitroso salt flat Arg.
76 F4 Granby Can.
108 A2 Gran Canaria i. Canary Is
102 D3 Gran Chaco reg. Arg./Para.
86 C3 Grand r. MI U.S.A.
84 E3 Grand r. MO U.S.A.
87 E7 Grand Bahama i. Bahamas
22 E4 Grand Ballon mt France
79 J4 Grand Bank Can.
120 F2 Grand Banks sea feature Atl. Ocean
114 C3 Grand-Bassam Côte d'Ivoire
79 G4 Grand Bay Can.
89 G4 Grand Bend Can.
14 D4 Grand Canal Rep. of Ireland
93 F3 Grand Canyon U.S.A.
93 F3 Grand Canyon gorge U.S.A.
93 F3 Grand Canyon Nat. Park U.S.A.
95 H5 Grand Cayman i. Cayman Is
77 G4 Grand Centre Can.
92 C2 Grand Coulee U.S.A.
105 C3 Grande r. Arg.
101 J6 Grande r. Bahia Brazil
104 B2 Grande r. São Paulo Brazil
102 C3 Grande, Bahía b. Arg.
76 E4 Grande Cache Can.
22 D4 Grande Casse, Pointe de la mt France
111 E5 Grande Comore i. Comoros
105 F1 Grande, Cuchilla h. Uru.
104 D3 Grande, Ilha i. Brazil
76 F3 Grande Prairie Can.
109 D3 Grand Erg de Bilma sand dunes Niger
108 B1 Grand Erg Occidental des. Alg.
108 C2 Grand Erg Oriental des. Alg.
79 H4 Grande-Rivière Can.
78 F3 Grande Rivière de la Baleine r. Can.
82 C2 Grande Ronde r. U.S.A.
103 E4 Grande, Serra mt Brazil
22 C5 Grande Tête de l'Obiou mt France
79 G4 Grand Falls N.B. Can.
79 J4 Grand Falls Nfld Can.
76 F5 Grand Forks Can.
84 D2 Grand Forks U.S.A.
91 F3 Grand Gorge U.S.A.
91 K2 Grand Harbour Can.
88 D4 Grand Haven U.S.A.
76 F2 Grandin, Lac l. Can.
84 D3 Grand Island MI U.S.A.
88 D2 Grand Island NE U.S.A.
85 F6 Grand Isle LA U.S.A.
91 J1 Grand Isle ME U.S.A.
93 H2 Grand Junction U.S.A.
114 B3 Grand-Lahou Côte d'Ivoire
79 G4 Grand Lake N.B. Can.
79 J4 Grand Lake Nfld Can.
79 H3 Grand Lake Nfld Can.
85 E6 Grand Lake LA U.S.A.
91 K2 Grand Lake ME U.S.A.
89 F3 Grand Lake MI U.S.A.
91 J1 Grand Lake Matagamon U.S.A.
90 A4 Grand Lake St Marys U.S.A.
91 J1 Grand Lake Seboeis U.S.A.
91 K2 Grand Lake Stream U.S.A.
88 E4 Grand Ledge U.S.A.
79 G5 Grand Manan I. Can.
88 E2 Grand Marais MI U.S.A.
88 B2 Grand Marais MN U.S.A.
79 F4 Grand-Mère Can.
22 D4 Grand Mont, Le mt France
25 B3 Grândola Port.
63 G3 Grand Passage New Caledonia
88 C2 Grand Portage U.S.A.
77 K4 Grand Rapids Can.
88 E4 Grand Rapids MI U.S.A.
84 E2 Grand Rapids MN U.S.A.
63 G3 Grand Récif de Cook rf New Caledonia
71 □9 Grand Récif de Cook rf New Caledonia Pac. Oc.
63 G4 Grand Récif du Sud rf New Caledonia
71 □9 Grand Récif du Sud rf New Caledonia Pac. Oc.
82 E3 Grand Teton mt U.S.A.
82 E3 Grand Teton Nat. Park U.S.A.
88 E3 Grand Traverse Bay U.S.A.
79 G4 Grand Vallée Can.
82 C2 Grandview U.S.A.
22 D2 Grandvillars France
15 B3 Grandvilliers France
93 F3 Grand Wash r. U.S.A.
93 E4 Grand Wash Cliffs cliff U.S.A.
22 B5 Grane France
105 B2 Graneros Chile
14 D6 Grange Rep. of Ireland
22 D3 Grange, Mont de mt France
82 E3 Granger U.S.A.
9 O6 Grängesberg Sweden
82 C2 Grangeville U.S.A.
76 D3 Granisle Can.
84 E2 Granite Falls U.S.A.
79 J4 Granite Lake Can.
93 E4 Granite Mts U.S.A.
82 E2 Granite Peak MT U.S.A.
82 D3 Granite Peak UT U.S.A.
39 H4 Granitogorsk Kazak.
26 E6 Granitola, Capo c. Sicily Italy
100 C6 Gran Laguna Salada l. Arg.
9 O7 Gränna Sweden
22 E4 Gran Paradiso mt Italy
22 E4 Gran Paradiso, Parco Nazionale del nat. park Italy
23 J3 Gran Pilastro mt Austria/Italy
19 L3 Gransee Ger.
13 G5 Grantham U.K.
116 A4 Grant I. Ant.
92 C2 Grant, Mt NV U.S.A.
92 D2 Grant, Mt NV U.S.A.
11 E3 Grantown-on-Spey U.K.
93 E2 Grant Range mts U.S.A.
83 F5 Grants U.S.A.
82 B3 Grants Pass U.S.A.
24 D2 Granville France
88 C5 Granville IL U.S.A.
91 G3 Granville NY U.S.A.
77 J3 Granville Lake Can.
104 D2 Grão Mogol Brazil
92 C4 Grapevine U.S.A.
92 D3 Grapevine Mts U.S.A.
91 G3 Graphite U.S.A.
113 J2 Graskop S. Africa
77 G2 Gras, Lac de l. Can.
91 F2 Grass r. U.S.A.
22 D6 Grasse France
12 F3 Grassington U.K.
77 H5 Grasslands Nat. Park Can.
69 E6 Grass Patch Austr.
82 E2 Grassrange U.S.A.
77 J4 Grass River Prov. Park Can.
92 B2 Grass Valley U.S.A.
67 F8 Grassy Austr.
87 E7 Grassy Cr. r. Bahamas
9 N7 Grästorp Sweden
88 B4 Gratiot U.S.A.
21 J5 Gratkorn Austria
25 G1 Graus Spain
77 J2 Gravel Hill Lake Can.
15 C2 Gravelines France
22 F4 Gravellona Toce Italy
113 J1 Gravelotte S. Africa
89 H3 Gravenhurst Can.
20 B2 Grävenwiesbach Ger.
67 J2 Gravesend Austr.
13 H6 Gravesend U.K.
26 G4 Gravina in Puglia Italy
88 E3 Grawn U.S.A.
22 C2 Gray France
91 H3 Gray U.S.A.
88 E3 Grayling U.S.A.
13 H6 Grays U.K.
82 A2 Grays Harbor in. U.S.A.
82 E3 Grays L. U.S.A.
90 B5 Grayson U.S.A.
79 H1 Gray Strait Can.
86 B4 Grayville U.S.A.
21 J5 Graz Austria
87 E7 Great Abaco i. Bahamas
71 □14 Great Astrolabe Reef Fiji
64 E6 Great Australian Bight g. Austr.
13 H6 Great Baddow U.K.
95 J3 Great Bahama Bank sea feature Bahamas
70 E2 Great Barrier Island N.Z.
65 J3 Great Barrier Reef Austr.
65 J3 Great Barrier Reef Marine Park (Cairns Section) Austr.
65 K4 Great Barrier Reef Marine Park (Capricorn Section) Austr.
65 J3 Great Barrier Reef Marine Park (Central Section) Austr.
65 H2 Great Barrier Reef Marine Park (Far North Section) Austr.
91 G3 Great Barrington U.S.A.
83 B4 Great Basin U.S.A.
83 E2 Great Basin Nat. Park U.S.A.
91 F5 Great Bay U.S.A.
76 E1 Great Bear r. Can.
76 E1 Great Bear Lake Can.
84 D4 Great Bend U.S.A.
112 C6 Great Berg r. S. Africa
14 A4 Great Bernera i. U.K.
14 A5 Great Blasket I. Rep. of Ireland
12 D3 Great Clifton U.K.
13 D5 Great Cumbrae i. U.K.
65 J4 Great Dividing Range mts Austr.
12 G3 Great Driffield U.K.
89 F3 Great Duck I. Can.
91 F5 Great Egg Harbor in. U.S.A.
95 H4 Greater Antilles is Caribbean Sea
95 J4 Great Exuma i. Bahamas
82 E2 Great Falls U.S.A.
113 G6 Great Fish r. S. Africa
113 G6 Great Fish Point S. Africa
45 F4 Great Gandak r. India
87 E7 Great Guana Cay i. Bahamas
87 E7 Great Harbour Cay i. Bahamas
95 K4 Great Inagua i. Bahamas
112 D5 Great Karoo plat. S. Africa
113 H6 Great Kei r. S. Africa
68 D4 Great. Keppel I. Austr.
67 G8 Great Lake Austr.
13 E5 Great Malvern U.K.
90 A5 Great Miami r. U.S.A.
56 A5 Great Nicobar i. Andaman and Nicobar Is
109 F2 Great Oasis, The oasis Egypt
13 D4 Great Ormes Head hd U.K.
13 H5 Great Ouse r. U.K.
68 B2 Great Palm Islands Austr.
91 G4 Great Peconic Bay U.S.A.
91 H4 Great Pt U.S.A.
13 D5 Great Rhos h. U.K.
115 C4 Great Ruaha r. Tanz.
91 F3 Great Sacandaga L. U.S.A.
22 E4 Great St Bernard Pass Italy/Switz.
87 E7 Great Sale Cay i. Bahamas
82 D3 Great Salt Lake U.S.A.
82 D3 Great Salt Lake Desert U.S.A.
109 E2 Great Sand Sea des. Egypt/Libya
64 D4 Great Sandy Desert Austr.
71 □13 Great Sea Reef Fiji
74 G3 Great Slave Lake N.W.T. Can.
87 D5 Great Smoky Mts U.S.A.
87 D5 Great Smoky Mts Nat. Park U.S.A.
76 E3 Great Snow Mtn Can.
91 G4 Great South Bay U.S.A.
13 H7 Greatstone-on-Sea U.K.
13 J6 Great Stour r. U.K.
13 C7 Great Torrington U.K.
64 E5 Great Victoria Desert Austr.
50 F1 Great Wall China
13 H6 Great Waltham U.K.
91 K2 Great Wass I. U.S.A.
62 B3 Great Western Tiers mts Austr.
12 G3 Great Whernside h. U.K.
13 J5 Great Yarmouth U.K.
37 J3 Great Zab r. Iraq
26 E4 Greco, Monte mt Italy
25 D2 Gredos, Sa de mts Spain
7 F5 Greece country Europe
82 F3 Greeley U.S.A.
75 K1 Greely Fiord in. Can.
32 H1 Greem-Bell, Ostrov i. Rus. Fed.
86 C4 Green r. KY U.S.A.
82 E3 Green r. UT/WY U.S.A.
89 H3 Greenbank Can.
88 D3 Green Bay U.S.A.
88 C3 Green Bay b. U.S.A.
69 B6 Greenbushes Austr.
67 J6 Green C. hd Austr.
13 E6 Greencastle U.K.
86 C4 Greencastle U.S.A.
87 E7 Green Cay i. Bahamas
87 D6 Green Cove Springs U.S.A.
84 A4 Greene IA U.S.A.
91 F3 Greene NY U.S.A.
87 D4 Greeneville U.S.A.
92 B3 Greenfield CA U.S.A.
88 E6 Greenfield IN U.S.A.
91 G3 Greenfield MA U.S.A.
90 B5 Greenfield OH U.S.A.
88 C4 Greenfield WI U.S.A.
69 B5 Green Head Austr.
69 B5 Green Head hd Austr.
68 B1 Green I. Austr.
55 A4 Green Island Bay Phil.
77 H4 Green Lake Can.
88 C4 Green Lake U.S.A.
73 M2 Greenland terr. Arctic Ocean
120 J1 Greenland Basin sea feature Arctic Ocean
11 F5 Greenlaw U.K.
66 A5 Greenly Island Austr.
91 G2 Green Mountains U.S.A.
11 D5 Greenock U.K.
14 E3 Greenore Rep. of Ireland
69 B4 Greenough Austr.
69 B4 Greenough r. Austr.
91 G4 Greenport U.S.A.
83 E4 Green River UT U.S.A.
82 E3 Green River WY U.S.A.
87 E4 Greensboro U.S.A.
86 C4 Greensburg IN U.S.A.
84 D4 Greensburg KS U.S.A.
90 D4 Greensburg PA U.S.A.
11 C3 Greenstone Point U.K.
81 L5 Green Swamp swamp U.S.A.
90 B5 Greenup U.S.A.
68 A2 Greenvale Austr.
91 F2 Green Valley Can.
93 G6 Green Valley U.S.A.
88 C5 Greenview U.S.A.
114 B3 Greenville Liberia
87 C6 Greenville AL U.S.A.
92 B1 Greenville CA U.S.A.
87 D6 Greenville FL U.S.A.
91 J2 Greenville ME U.S.A.
88 E4 Greenville MI U.S.A.
85 F5 Greenville MS U.S.A.
87 E5 Greenville NC U.S.A.
91 H3 Greenville NH U.S.A.
90 A4 Greenville OH U.S.A.
90 C4 Greenville PA U.S.A.
87 D5 Greenville SC U.S.A.
85 D5 Greenville TX U.S.A.
77 J4 Greenwater Provincial Park Can.
67 J5 Greenwell Point Austr.
91 G4 Greenwich CT U.S.A.
91 G3 Greenwich NY U.S.A.
93 G2 Greenwich UT U.S.A.
85 F5 Greenwood MS U.S.A.
87 D5 Greenwood SC U.S.A.
85 E5 Greers Ferry Lake U.S.A.
84 D3 Gregory U.S.A.
65 G3 Gregory Downs Austr.
69 D2 Gregory, L. salt flat Austr.
66 C2 Gregory, L. salt flat Austr.
64 E4 Gregory Lake salt flat Austr.
64 F3 Gregory National Park Austr.
65 H3 Gregory Range h. Austr.
19 L2 Greifswald Ger.
19 L2 Greifswalder Bodden b. Ger.
19 L2 Greifswalder Oie i. Ger.
20 F2 Greiz Ger.
36 E4 Greko, Cape Cyprus
9 M8 Grenå Denmark
73 K8 Grenada country Caribbean Sea
85 F5 Grenada U.S.A.
24 E5 Grenade France
22 E2 Grenchen Switz.
9 M8 Grenen spit Denmark
67 H4 Grenfell Austr.
77 J4 Grenfell Can.
22 C4 Grenoble France
103 E1 Grenville Grenada
65 H2 Grenville, C. hd Austr.
19 K2 Gresenhorst Ger.
82 B2 Gresham U.S.A.
56 □ Gresik Indon.
12 F3 Greta r. U.K.
11 E6 Gretna U.K.
85 F6 Gretna U.S.A.
19 H5 Greußen Ger.
15 C4 Greux France
23 J6 Greve in Chianti Italy
18 A5 Grevelingen chan. Neth.
27 J4 Grevena Greece
18 C5 Grevenbicht Neth.
18 D5 Grevenbroich Ger.
15 G3 Grevenmacher Lux.
19 J3 Grevesmühlen Ger.
70 C5 Grey r. N.Z.
82 E2 Greybull U.S.A.
76 B2 Grey Hunter Pk Can.
79 J3 Grey Is Can.
70 C5 Greymouth N.Z.
66 E2 Grey Range h. Austr.
113 J4 Greytown S. Africa
15 E2 Grez-Doiceau Belgium
29 G5 Gribanovskiy Rus. Fed.
92 B2 Gridley CA U.S.A.
88 C5 Gridley IL U.S.A.
20 G4 Griesbach im Rottal Ger.
21 J5 Grieskirchen Austria
87 C5 Griffin U.S.A.
67 G5 Griffith Austr.
89 J3 Griffith Can.
74 F2 Griffiths Point Can.
67 F8 Grim, C. Austr.
19 K5 Grimma Ger.
19 L2 Grimmen Ger.
19 L2 Grimnitzsee l. Ger.
89 H4 Grimsby Can.
12 G4 Grimsby U.K.
8 E3 Grimsey i. Iceland
76 F3 Grimshaw Can.
8 E4 Grímsstaðir Iceland
9 L7 Grimstad Norway
8 B5 Grindavík Iceland
9 L9 Grindsted Denmark
27 N2 Grindul Chituc spit Romania
83 E4 Grinnell U.S.A.
20 E5 Grinzens Austria
113 H5 Griqualand East reg. S. Africa
112 E4 Griqualand West reg. S. Africa
112 E4 Griquatown S. Africa
75 K2 Grise Fiord Can.
59 B3 Grisik Indon.
15 B2 Gris Nez, Cap pt France
11 F2 Gritley U.K.
26 G2 Grmeč mts Bos.-Herz.
15 E1 Grobbendonk Belgium
113 H2 Groblersdal S. Africa
112 E4 Groblershoop S. Africa
21 G5 Gröbming Austria
32 E2 Grodekovo Rus. Fed.
19 L5 Gröditz Ger.
21 L2 Grodków Pol.
Grodno see Hrodna
112 B5 Groen watercourse Northern Cape S. Africa
112 E5 Groen watercourse Northern Cape S. Africa
24 C3 Groix, Île de i. France
21 J1 Gromadka Pol.
26 B6 Gromballa Tunisia
19 H2 Grömitz Ger.
18 E4 Gronau (Westfalen) Ger.
8 N4 Grong Norway
18 D3 Groningen Neth.
18 D3 Groninger Wad tidal flats Neth.
19 K2 Grønsund chan. Denmark
93 E3 Groom L. U.S.A.
112 D3 Groot-Aar Pan salt pan S. Africa
112 E7 Groot Brakrivier S. Africa
113 H3 Grootdraaidam dam S. Africa
112 E6 Grootdrink S. Africa
65 G2 Groote Eylandt i. Austr.
111 B5 Grootfontein Namibia
112 C3 Groot Karas Berg plat. Namibia
113 J1 Groot Letaba r. S. Africa
113 G2 Groot Marico S. Africa
112 D6 Groot Swartberg mts S. Africa
112 D5 Grootvloer salt pan S. Africa
113 G6 Groot Winterberg mt S. Africa
88 E3 Gros Cap U.S.A.
79 J4 Gros Morne Nat. Pk Can.
22 B3 Grosne r. France
19 G3 Großenaspe Ger.
19 J2 Großenbrode Ger.
19 L5 Großenhain Ger.
18 F4 Großenkneten Ger.
20 C2 Großenlüder Ger.
20 D2 Großer Beerberg h. Ger.
21 H5 Grosser Bösenstein mt Austria
20 G3 Großer Osser mt Czech Rep./Ger.
19 H2 Großer Plöner See l. Ger.
21 H5 Grosser Priel mt Austria
21 G4 Großer Rachel mt Ger.
21 G6 Grosser Speikkofel mt Austria
21 G6 Grosser Speikkogel mt Austria
20 E2 Großer Waldstein h. Ger.
26 D3 Grosseto Italy
20 B3 Groß-Gerau Ger.
20 F5 Großglockner mt Austria
19 L4 Groß Köris Ger.
19 M4 Groß Leine Ger.
19 M4 Groß Lindow Ger.
19 H4 Groß Oesingen Ger.
20 C3 Großostheim Ger.
21 K5 Großpetersdorf Austria
21 H5 Großraming Austria
19 M5 Großräschen Ger.
19 L4 Groß Schönebeck Ger.
112 C1 Gross Ums Namibia
20 F5 Großvenediger mt Austria
82 E3 Gros Ventre Range mts U.S.A.
79 J3 Groswater Bay Can.
91 E3 Groton U.S.A.
90 D5 Grottoes U.S.A.
76 F3 Grouard Can.
114 C3 Groumania Côte d'Ivoire
78 D4 Groundhog r. Can.
18 C3 Grouw Neth.
90 C4 Grove City U.S.A.
87 C6 Grove Hill U.S.A.
92 B3 Groveland U.S.A.
116 D5 Grove Mts Ant.
92 B4 Grover Beach U.S.A.
91 H2 Groveton U.S.A.
93 F5 Growler U.S.A.
93 F5 Growler Mts U.S.A.
29 H7 Groznyy Rus. Fed.
27 M3 Grudovo Bulg.
17 J4 Grudziądz Pol.
11 C3 Gruinard Bay U.K.
21 G5 Grünau Austria
111 B6 Grünau Namibia
8 B4 Grundarfjörður Iceland
90 B6 Grundy U.S.A.
88 A4 Grundy Center U.S.A.
19 M4 Grunow Ger.
20 B3 Grünstadt Ger.
22 E3 Gruyère, Lac de la l. Switz.
29 F4 Gryazi Rus. Fed.
28 G3 Gryazovets Rus. Fed.
16 G4 Gryfice Pol.
16 G4 Gryfino Pol.
16 G5 Gryfów Śląski Pol.
8 P2 Gryllefjord Norway
102 □ Grytviken Atl. Ocean
20 C4 Gschwend Ger.
22 E3 Gstaad Switz.
45 F5 Gua India
23 J4 Guà r. Italy
95 J4 Guacanayabo, Golfo de b. Cuba
103 D2 Guacara Venez.
103 C3 Guacharía r. Col.
25 D4 Guadajoz r. Spain
96 D3 Guadalajara Mex.
25 E2 Guadalajara Spain
25 F2 Guadalaviar r. Spain
63 G2 Guadalcanal i. Solomon Is
25 C4 Guadalete r. Spain
25 C4 Guadalope r. Spain
25 D4 Guadalquivir r. Spain
97 D2 Guadalupe Nuevo León Mex.
96 D3 Guadalupe Zacatecas Mex.
80 C6 Guadalupe i. Mex.
92 B4 Guadalupe U.S.A.
85 D6 Guadalupe r. U.S.A.
96 C2 Guadalupe Aguilera Mex.
85 B6 Guadalupe Mts Nat. Park U.S.A.
85 B6 Guadalupe Pk U.S.A.
25 D3 Guadalupe, Sierra de mts Spain
96 C2 Guadalupe Victoria Mex.
96 C2 Guadalupe y Calvo Mex.
25 D2 Guadarrama, Sierra de mts Spain
73 K8 Guadeloupe terr. Caribbean Sea
105 C2 Guadel, Sa de mts Arg.
25 C4 Guadiana r. Port./Spain
25 E4 Guadix Spain
102 B6 Guafo, I. Chile
96 H5 Guaimaca Honduras
103 D4 Guainía r. Col./Venez.
103 E3 Guaiquinima, Cerro mt Venez.
104 A4 Guaíra Brazil
102 B6 Guaitecas, Islas Chile
96 D2 Guaje, Llano de plain Mex.
103 C1 Guajira, Península de pen. Col.
100 C4 Gualaceo Ecuador
92 A2 Gualala U.S.A.
23 K6 Gualdo Tadino Italy
105 E2 Gualeguay Arg.
105 E2 Gualeguay r. Arg.
105 E2 Gualeguaychu Arg.
105 B4 Gualjaina Arg.
57 L4 Guam terr. Pac. Oc.
102 A6 Guamblin, I. Chile
105 D3 Guamini Arg.
96 B2 Guamúchil Mex.
103 A4 Guamués r. Col.
58 B4 Gua Musang Malaysia
96 H6 Guanacaste, Parque Nacional nat. park Costa Rica
96 C2 Guanacevi Mex.
105 D3 Guanaco, Co h. Arg.
96 H4 Guanaja Honduras
96 D3 Guanajuato Mex.
96 D3 Guanajuato div. Mex.
104 D1 Guanambi Brazil
103 D4 Guanare r. Venez.
103 C2 Guanare Venez.
103 C2 Guanare r. Venez.
103 C2 Guanarito Venez.
103 C2 Guanarito r. Venez.
103 D3 Guanay, Sierra mts Venez.
50 D2 Guandi Shan mt China
95 H4 Guane Cuba
51 C4 Guang'an China
51 D6 Guangdong div. China
51 F4 Guangfeng China
51 D6 Guanghai China
50 B4 Guanghan China
51 E5 Guangchang China
50 E3 Guanghe China
51 D5 Guangji China
50 E2 Guangling China
54 B4 Guanglu Dao i. China
51 B5 Guangnan China
51 D6 Guangning China
50 F2 Guangrao China
51 E4 Guangshan China
50 D4 Guangshui China
51 C6 Guangxi div. China
50 B3 Guangyuan China
51 E5 Guangze China
51 D6 Guangzhou China
104 D2 Guanhães Brazil
104 D2 Guanhães r. Brazil
103 E2 Guanipa r. Venez.
51 B5 Guanling China
50 C4 Guanmian Shan mts China
50 D3 Guanpo China
54 C3 Guanshui China
103 D2 Guanta Venez.
95 J4 Guantánamo Cuba
50 E1 Guanting Sk. resr China
50 B4 Guan Xian China
51 D5 Guanyang China
50 F3 Guanyun China
103 A4 Guapí Col.
96 J6 Guápiles Costa Rica
100 F6 Guaporé r. Bol./Brazil
100 E7 Guaquí Bol.
104 D1 Guará r. Brazil
101 L5 Guarabira Brazil
104 E2 Guarapari Brazil
104 B4 Guarapuava Brazil
104 C4 Guaraqueçaba Brazil
104 C3 Guaratinguetá Brazil
104 C4 Guaratuba, Baía de b. Brazil
25 C2 Guarda Port.
104 C2 Guarda Mor Brazil
25 D1 Guardo Spain
103 D2 Guárico r. Venez.
103 C3 Guarrojo r. Col.
104 C4 Guarujá Brazil
103 D4 Guasacavi r. Col.
103 C4 Guasacavi, Cerro h. Col.
103 B2 Guasare r. Venez.
96 B2 Guasave Mex.
103 C3 Guasdualito Venez.
97 E3 Guasima Mex.
103 E3 Guasipati Venez.
104 B4 Guassú r. Brazil
23 H5 Guastalla Italy
73 G8 Guatemala country Central America
97 G5 Guatemala Guatemala
103 D2 Guatopo, Parque Nacional nat. park Venez.
105 D3 Guatrache Arg.
103 C4 Guaviare r. Col.
104 C3 Guaxupé Brazil
103 B4 Guayabero r. Col.
103 D3 Guayapo r. Venez.
100 C4 Guayaquil Ecuador
100 B4 Guayaquil, Golfo de g. Ecuador
100 E6 Guayaramerín Bol.
96 B2 Guaymas Mex.
97 G5 Guazacapán Guatemala
110 D2 Guba Eth.
43 B3 Gubbi India
23 K6 Gubbio Italy
19 M5 Guben Ger.
114 F2 Gubio Nigeria
29 F5 Gubkin Rus. Fed.
50 D3 Gucheng China
29 G7 Gudaut'a Georgia
9 M6 Gudbrandsdalen v. Norway
29 H7 Gudermes Rus. Fed.
43 C2 Gudivada India
43 B3 Gudiyattam India
19 H1 Gudme Denmark
15 F4 Gudmont-Villiers France
54 E2 Gudong r. China
41 G5 Gudri r. Pak.
36 D1 Güdül Turkey
43 B3 Gudur Andhra Pradesh India
43 B3 Gudur Andhra Pradesh India
9 K6 Gudvangen Norway
22 E2 Guebwiller France
114 A3 Guéckédou Guinea
89 J1 Guéguen, Lac l. Can.
103 B4 Guéjar r. Col.
108 C1 Guelma Alg.
108 A2 Guelmine Morocco
89 G4 Guelph Can.
97 E3 Guémez Mex.
15 G3 Guénange France
103 D2 Güera r. Venez.
79 G2 Guérard, Lac l. Can.
24 E3 Guéret France
10 E7 Guernsey i. U.K.
82 F3 Guernsey U.S.A.
114 A1 Guérou Maur.
97 E2 Guerrero Mex.
97 D4 Guerrero div. Mex.
96 A2 Guerrero Negro Mex.
75 M4 Guers, Lac l. Can.
22 B3 Gueugnon France
115 C1 Gugē mt Eth.
40 D3 Gügerd, Küh-e mts Iran
120 F5 Guiana Basin sea feature Atl. Ocean
66 C6 Guichen B. Austr.
51 E4 Guichi China
105 F2 Guichón Uru.
50 A3 Guide China
114 F3 Guider Cameroon
51 C5 Guiding China
26 E4 Guidonia-Montecelio Italy
114 B3 Guiglo Côte d'Ivoire
15 D3 Guignicourt France
113 K2 Guija Moz.
51 D6 Gui Jiang r. China
51 F4 Guiji Shan mts China
22 D5 Guil r. France
13 G6 Guildford U.K.
91 J2 Guilford U.S.A.
22 B5 Guilherand France
51 D5 Guilin China
78 E2 Guillaume-Delisle, Lac l. Can.
22 D5 Guillaumes France
22 D5 Guillestre France
22 B2 Guillon France
25 B2 Guimarães Port.
55 B4 Guimaras Str. Phil.
50 E3 Guimeng Ding mt China
50 A3 Guinan China
92 A2 Guinda U.S.A.
55 C4 Guindulman Phil.
107 B4 Guinea country Africa
120 J5 Guinea Basin sea feature Atl. Ocean
107 B4 Guinea-Bissau country Africa
106 B5 Guinea, Gulf of g. Africa
95 H4 Güines Cuba
15 B2 Guînes France
24 C2 Guingamp France
24 B2 Guipavas France
51 D6 Guiping China
104 B2 Guiratinga Brazil
103 E2 Güiria Venez.
22 D5 Guisane r. France
15 D3 Guiscard France
15 D3 Guise France
55 C4 Guiuan Phil.
51 E4 Guixi China
51 C6 Gui Xian China
51 C5 Guiyang Guizhou China
51 D5 Guiyang Hunan China
51 C5 Guizhou div. China
44 B5 Gujarat div. India
44 C2 Gujar Khan Pak.
44 C2 Gujranwala Pak.
44 C2 Gujrat Pak.
93 G5 Gu Komelik U.S.A.
29 F5 Gukovo Rus. Fed.
37 K3 Gük Tappeh Iran
44 D2 Gulabgarh Jammu and Kashmir
50 B2 Gulang China
67 H3 Gulargambone Austr.
43 B2 Gulbarga India
9 U8 Gulbene Latvia
39 H4 Gul'chö Kyrg.
19 J2 Guldborg Sund str. Denmark
36 E3 Gülek Turkey
85 F6 Gulfport U.S.A.
40 C4 Gulf, The g. Asia
67 H4 Gulgong Austr.
46 E1 Gulian China
51 B5 Gulin China
41 G4 Gulistan Pak.
39 G4 Gulistan Uzbek.
19 J3 Gülitz Ger.
88 E3 Gull I. U.S.A.
77 H4 Gull Lake Can.
8 R3 Gullträsk Sweden
36 D3 Gülnar Turkey
41 F3 Gulran Afgh.
29 G7 Gul'rip'shi Georgia
36 E2 Gülşehir Turkey
39 H3 Gul'shad Kazak.
115 B2 Gulu Uganda
68 D6 Guluguba Austr.
44 B3 Gumal r. Pak.
111 C5 Gumare Botswana
38 C5 Gumdag Turkm.
114 E2 Gumel Nigeria
45 F5 Gumia India
45 F5 Gumla India
18 E5 Gummersbach Ger.
36 E1 Gümüşhacıköy Turkey
37 G1 Gümüşhane Turkey
44 D4 Guna India
67 F5 Gunbar Austr.
39 G5 Gund r. Tajik.
67 H5 Gundagai Austr.
36 D3 Gündoğmuş Turkey
110 B4 Gungu Congo(Zaire)
29 H7 Gunib Rus. Fed.
77 K4 Gunisao r. Can.
68 A1 Gunnawarra Austr.
67 J3 Gunnedah Austr.
116 D3 Gunnerus Ridge sea feature Ant.
67 H5 Gunning Austr.
83 F4 Gunnison CO U.S.A.
93 G2 Gunnison UT U.S.A.
83 E4 Gunnison r. U.S.A.
43 B3 Guntakal India
19 H5 Güntersberge Ger.
87 C5 Guntersville U.S.A.
87 C5 Guntersville L. U.S.A.
43 C2 Guntur India
65 G3 Gununa Austr.
59 A2 Gunungsitoli Indon.
58 A5 Gunungtua Indon.
43 C2 Gunupur India
20 D4 Günzburg Ger.
20 D3 Gunzenhausen Ger.
50 E1 Guojiatun China
50 E3 Guoyang China
50 A2 Gurban Hudag China
50 D1 Gurban Obo China
41 F5 Gurdim Iran
36 B2 Güre Turkey
44 D3 Gurgaon India
101 K5 Gurgueia r. Brazil
44 B4 Gurha India
103 E3 Guri, Embalse de resr Venez.
64 F2 Gurig National Park Austr.
104 C2 Gurinhatã Brazil
29 H7 Gurjaani Georgia
21 H6 Gurk r. Austria
41 E4 Gur Khar Iran
38 E4 Gurlen Uzbek.
115 B6 Guro Moz.
37 J2 Gürpınar Turkey
45 G3 Guru China
115 C6 Guruè Moz.
36 F2 Gürün Turkey
101 J4 Gurupi r. Brazil
44 C4 Guru Sikhar mt India
28 B4 Gur'yevsk Rus. Fed.
114 E2 Gusau Nigeria
19 J4 Güsen Ger.
28 B4 Gusev Rus. Fed.
54 B4 Gushan China
93 H1 Gusher U.S.A.
41 F3 Gushgy Turkm.
50 E3 Gushi China
55 A5 Gusi Malaysia
33 M2 Gusikha Rus. Fed.
28 D4 Gusino Rus. Fed.
48 J1 Gusinoozersk Rus. Fed.
45 F5 Guskara India
28 G4 Gus'-Khrustal'nyy Rus. Fed.
19 M4 Gusow Ger.
26 C5 Guspini Sardinia Italy
21 K5 Güssing Austria
21 J5 Gußwerk Austria
76 B3 Gustavus U.S.A.
19 J5 Güsten Ger.
92 B3 Gustine U.S.A.
19 K3 Güstrow Ger.
45 H3 Gutang China
21 J5 Gutenstein Austria
19 L4 Güterfelde Ger.
18 F5 Gütersloh Ger.
93 H5 Guthrie AZ U.S.A.
86 C4 Guthrie KY U.S.A.
85 D5 Guthrie OK U.S.A.
85 C5 Guthrie TX U.S.A.
51 F5 Gutian Fujian China
51 E5 Gutian Fujian China
15 G3 Gutland reg. Ger./Lux.
45 F3 Gutsuo China
88 B4 Guttenberg U.S.A.
111 D5 Gutu Zimbabwe
45 G4 Guwahati India
37 J3 Guwēr Iraq
19 G5 Guxhagen Ger.
99 E2 Guyana country S. America
50 D1 Guyang China
85 C4 Guymon U.S.A.
40 D4 Güyom Iran
67 J3 Guyra Austr.
50 E1 Guyuan Hebei China
50 C3 Guyuan Ningxia China
38 F5 Guzar Uzbek.
51 C4 Guzhang China
50 E3 Guzhen China
17 K3 Gvardeysk Rus. Fed.
67 H3 Gwabegar Austr.
114 E2 Gwadabawa Nigeria
41 F5 Gwadar Pak.
41 F5 Gwadar West Bay Pak.
44 D4 Gwalior India
68 C5 Gwambegwine Austr.
111 C6 Gwanda Zimbabwe
115 A2 Gwane Congo(Zaire)
41 G4 Gwash Pak.
41 F5 Gwatar Bay Pak.
14 C3 Gweebarra Bay Rep. of Ireland
14 C2 Gweedore Rep. of Ireland
111 C5 Gweru Zimbabwe
115 A5 Gweshe Congo(Zaire)
88 D2 Gwinn U.S.A.
114 F4 Gwoza Nigeria
67 H2 Gwydir r. Austr.
22 C2 Gy France
45 H3 Gyaca China
50 B3 Gyagartang China
45 F3 Gyangrang China
45 G3 Gyangzê China
45 G3 Gyaring Co l. China
46 B3 Gyaring Hu l. China
27 L6 Gyaros i. Greece
45 H3 Gyarubtang China
32 J2 Gydanskiy Poluostrov pen. Rus. Fed.
45 H3 Gyimda China
45 F3 Gyirong Xizang China
45 F3 Gyirong Xizang China
45 H2 Gyiza China
75 O3 Gyldenløves Fjord in. Greenland
19 J1 Gyldenløveshøj h. Denmark
68 E6 Gympie Austr.
17 J7 Gyöngyös Hungary
16 H7 Győr Hungary
21 L5 Győr-Moson-Sopron div. Hungary
77 K4 Gypsumville Can.
79 G2 Gyrfalcon Is i. Can.
27 K6 Gytheio Greece
17 K7 Gyula Hungary
37 J1 Gyumri Armenia
38 D5 Gyzylarbat Turkm.

H

21 H4 Haag Austria
71 □10 Ha'alaufuli Tonga
71 □12 Ha'ano i. Tonga
71 □12 Ha'apai Group is Tonga
8 T5 Haapajärvi Fin.
8 T4 Haapavesi Fin.
71 □3 Haapiti Fr. Polynesia Pac. Oc.
9 S7 Haapsalu Estonia
18 B4 Haarlem Neth.
112 E6 Haarlem S. Africa
18 F5 Haarstrang ridge Ger.
70 B5 Haast N.Z.
71 □11 Ha'atua Tonga
41 G5 Hab r. Pak.

48 E2 Habahe China
Habana see Havana
43 C4 Habarane Sri Lanka
115 C2 Habaswein Kenya
76 F3 Habay Can.
42 C7 Habbān Yemen
37 J5 Habbānīyah Iraq
37 J5 Ḩabbānīyah, Hawr al l. Iraq
41 G5 Hab Chauki Pak.
45 G4 Habiganj Bangl.
50 E1 Habirag China
45 G5 Habra India
103 B5 Hacha Col.
105 B3 Hachado, P. de pass Arg./Chile
53 F8 Hachijō-jima i. Japan
52 G4 Hachinohe Japan
53 F7 Hachiōji Japan
36 E2 Hacıbektaş Turkey
37 H2 Hacıömer Turkey
66 C3 Hack, Mt Austr.
111 D6 Hacufera Moz.
40 C6 Hadabat al Budū plain S. Arabia
43 A3 Hadagalli India
11 F5 Haddington U.K.
114 F2 Hadejia Nigeria
114 E2 Hadejia watercourse Nigeria
36 E5 Hadera Israel
9 L9 Haderslev Denmark
42 C6 Ḩaḑramawt reg. Yemen
36 D3 Hadım Turkey
13 H5 Hadleigh U.K.
74 H2 Hadley Bay Can.
54 D6 Hadong S. Korea
36 F6 Ḩadraj, Wādī watercourse S. Arabia
9 M8 Hadsund Denmark
29 E5 Hadyach Ukr.
105 F1 Haedo, Cuchilla de h. Uru.
54 C4 Haeju N. Korea
54 C5 Haeju-man b. N. Korea
54 D6 Haenam S. Korea
113 H1 Haenertsburg S. Africa
40 B4 Hafar al Bāţin S. Arabia
77 H4 Hafford Can.
36 F2 Hafik Turkey
44 C2 Hafizabad Pak.
45 H4 Hāflong India
8 C4 Hafnarfjörður Iceland
40 C4 Haft Gel Iran
8 B4 Hafursfjörður b. Iceland
89 G2 Hagar Can.
43 B3 Hagari r. India
110 D2 Hagar Nish Plateau plat. Eritrea
15 E2 Hageland reg. Belgium
18 E5 Hagen Ger.
62 E2 Hagen, Mount P.N.G.
19 J3 Hagenow Ger.
90 E5 Hagerstown U.S.A.
24 D5 Hagetmau France
9 N6 Hagfors Sweden
53 B7 Hagi Japan
51 B6 Ha Giang Vietnam
13 E5 Hagley U.K.
14 B5 Hag's Head hd Rep. of Ireland
77 H4 Hague Can.
24 D2 Hague, Cap de la pt France
18 B4 Hague, The Neth.
15 H4 Haguenau France
46 G4 Hahajima-rettō is Japan
50 E2 Hai r. China
115 C3 Hai Tanz.
50 F3 Hai'an China
112 B4 Haib watercourse Namibia
54 B3 Haicheng China
51 C6 Hai Dương Vietnam
36 E5 Haifa Israel
36 E5 Haifa, Bay of Israel
51 E6 Haifeng China
18 F6 Haiger Ger.
51 D6 Haikang China
51 D6 Haikou China
42 B4 Hā'il S. Arabia
46 D2 Hailar China
89 H2 Haileybury Can.
54 E1 Hailin China
54 C2 Hailong China
13 H7 Hailsham U.K.
8 T4 Hailuoto Fin.
50 F4 Haimen China
51 C7 Hainan div. China
51 D7 Hainan i. China
21 K4 Hainburg an der Donau Austria
114 A3 Haindi Liberia
76 B3 Haines U.S.A.
76 B2 Haines Junction Can.
21 J4 Hainfeld Austria
19 H5 Hainich ridge Ger.
19 L6 Hainichen Ger.
19 H5 Hainleite ridge Ger.
51 C6 Hai Phong Vietnam
50 A2 Hairag China
50 B1 Hairhan Namag China
51 F5 Haitan Dao i. China
73 J8 Haiti country Caribbean Sea
51 C7 Haitou China
93 G5 Haivana Nakya U.S.A.
92 D3 Haiwee Reservoir U.S.A.
50 E2 Haixing China
109 F3 Haiya Sudan
50 A2 Haiyan Qinghai China
51 F4 Haiyan Zhejiang China
54 A5 Haiyang China
54 B4 Haiyang Dao i. China
50 B2 Haiyuan China
50 F3 Haizhou Wan b. China
21 L5 Hajagos r. Hungary
17 K7 Hajdúböszörmény Hungary
26 C7 Hajeb El Ayoun Tunisia
42 D7 Ḩajhir mt Yemen
52 F5 Hajiki-zaki pt Japan
45 F4 Hajipur India
40 D4 Ḩājjīābād Iran
40 D4 Ḩājjīābād Iran
42 E6 Hajmah Oman
45 H5 Haka Myanmar
92 2 Hakalau U.S.A.
71 12 Hakau Fusi rf Tonga
105 C4 Hakelhuincul, Altiplanicie de plat. Arg.
Hakha see Haka
37 J3 Hakkârı Turkey
8 R3 Hakkas Sweden
53 D7 Hakken-zan mt Japan
52 H2 Hako-dake mt Japan
52 G4 Hakodate Japan
112 B1 Hakos Mts Namibia
112 D3 Hakseen Pan salt pan S. Africa
53 E6 Hakui Japan
53 E6 Haku-san volc. Japan
53 E6 Haku-san National Park Japan
44 B4 Hala Pak.
Ḩalab see Aleppo
40 B6 Halabān S. Arabia
37 K4 Halabja Iraq
54 C1 Halaha China
54 C1 Halahai China
109 F2 Halaib Sudan
42 E6 Ḩalāniyāt, Juzur al is Oman
92 2 Halawa U.S.A.
36 F4 Halba Lebanon
48 G2 Halban Mongolia
19 J5 Halberstadt Ger.
55 B3 Halcon, Mt Phil.
8 Haldarsvik Faroe Is
9 M7 Halden Norway
19 J4 Haldensleben Ger.
45 G5 Haldi r. India
45 G5 Haldia India
45 G4 Haldibari India
44 D3 Haldwani India
89 F3 Hale U.S.A.
37 G4 Halebiye Syria
92 1 Haleiwa U.S.A.
69 C3 Hale, Mt h. Austr.
13 E5 Halesowen U.K.
13 J5 Halesworth U.K.
114 C3 Half Assini Ghana
36 F3 Halfeti Turkey
70 B7 Halfmoon Bay N.Z.
76 E3 Halfway r. Can.
14 C6 Halfway Rep. of Ireland
18 B4 Halfweg Neth.
45 E4 Halia India
89 H3 Haliburton Can.
68 B2 Halifax Austr.
79 H5 Halifax Can.
12 F4 Halifax U.K.
90 D6 Halifax U.S.A.
68 B2 Halifax Bay Austr.
68 B2 Halifax, Mt Austr.
11 E2 Halkirk U.K.
8 P5 Hälla Sweden
54 D7 Halla-san mt S. Korea
66 A5 Hall Bay Austr.
75 K3 Hall Beach Can.
15 E2 Halle Belgium
18 D5 Halle Neth.
9 O7 Hällefors Sweden
20 G5 Hallein Austria
15 B3 Hallencourt France
19 J5 Halle-Neustadt Ger.
19 J5 Halle (Saale) Ger.
116 A5 Hallett, C. Ant.
116 C3 Halley U.K. Base Ant.
18 F2 Halligen is Ger.
60 G3 Hall Islands Micronesia
8 Q4 Hällnäs Sweden
84 D1 Hallock U.S.A.
75 M3 Hall Peninsula Can.
9 O7 Hallsberg Sweden
64 E3 Halls Creek Austr.
89 H3 Halls Lake Can.
15 D2 Halluin Belgium
8 O5 Hallviken Sweden
22 F2 Hallwiler See l. Switz.
57 H6 Halmahera i. Indon.
9 N8 Halmstad Sweden
44 C5 Halol India
9 M8 Hals Denmark
19 J1 Halsskov Denmark
8 T5 Halsua Fin.
18 E5 Haltern Ger.
12 E3 Haltwhistle U.K.
40 D5 Ḩālūl i. Qatar
18 E5 Halver Ger.
68 D6 Haly, Mount h. Austr.
15 D3 Ham France
53 C7 Hamada Japan
108 B2 Ḩamâda El Ḩaricha des. Mali
40 C3 Hamadān Iran
108 B2 Hamada Tounassine des. Alg.
36 F4 Ḩamāh Syria
52 G3 Hamamasu Japan
53 E7 Hamamatsu Japan
9 M6 Hamar Norway
8 O2 Hamarøy Norway
52 H2 Hamatonbetsu Japan
43 C5 Hambantota Sri Lanka
18 F3 Hambergen Ger.
12 F3 Hambleton Hills U.K.
19 G3 Hamburg Ger.
113 G6 Hamburg S. Africa
85 F5 Hamburg AR U.S.A.
90 D3 Hamburg NY U.S.A.
91 F4 Hamburg PA U.S.A.
18 F3 Hamburgisches Wattenmeer, Nationalpark nat. park Ger.
91 G4 Hamden U.S.A.
9 T6 Hämeenlinna Fin.
69 B3 Hamelin Austr.
69 B3 Hamelin Pool b. Austr.
19 G4 Hameln Ger.
64 C4 Hamersley Range mts Austr.
69 D1 Hamersley Range Nat. Park Austr.
54 D4 Hamhŭng N. Korea
46 B2 Hami China
40 C4 Ḩamīd Iran
109 F2 Ḩamīd Sudan
66 E6 Hamilton Austr.
95 M2 Hamilton Bermuda
89 H4 Hamilton Can.
70 E2 Hamilton N.Z.
11 D5 Hamilton U.K.
87 C5 Hamilton AL U.S.A.
88 B5 Hamilton IL U.S.A.
82 D2 Hamilton MT U.S.A.
91 F3 Hamilton NY U.S.A.
90 A5 Hamilton OH U.S.A.
92 A2 Hamilton City U.S.A.
92 B3 Hamilton, Mt CA U.S.A.
93 E2 Hamilton, Mt NV U.S.A.
9 U6 Hamina Fin.
44 D3 Hamirpur India
37 H6 Ḩāmir, W. watercourse S. Arabia
79 J3 Hamiton Inlet in. Can.
54 D4 Hamju N. Korea
66 C5 Hamley Bridge Austr.
88 D3 Hamlin Lake U.S.A.
18 E5 Hamm Ger.
108 B2 Hammada du Drâa plat. Alg.
37 J3 Hammam Ali Iraq
26 D6 Hammamet Tunisia
109 D1 Hammamet, Golfe de b. Tunisia
37 L6 Ḩammār, Hawr al l. Iraq
8 P5 Hammarstrand Sweden
20 C2 Hammelburg Ger.
19 M1 Hammenhög Sweden
8 O5 Hammerdal Sweden
8 S1 Hammerfest Norway
18 D5 Hamminkeln Ger.
66 C4 Hammond Austr.
88 D5 Hammond IN U.S.A.
85 F6 Hammond LA U.S.A.
82 F2 Hammond MT U.S.A.
89 E3 Hammond Bay U.S.A.
90 E3 Hammondsport U.S.A.
91 F5 Hammonton U.S.A.
15 F2 Hamoir Belgium
70 C6 Hampden N.Z.
13 F6 Hampshire Downs h. U.K.
79 G4 Hampton Can.
85 E5 Hampton AR U.S.A.
91 H3 Hampton NH U.S.A.
91 E6 Hampton VA U.S.A.
37 K4 Hamrīn, Jabal h. Iraq
58 C3 Ham Tân Vietnam
44 D2 Hamta Pass India
41 E5 Hāmūn-e Jaz Mūrīān salt marsh Iran
41 F4 Hāmūn Helmand salt flat Afgh./Iran
41 G4 Hamun-i-Lora l. Pak.
41 F4 Hāmūn Pu marsh Afgh.
37 J2 Hamur Turkey
21 L3 Haná r. Czech Rep.
92 2 Hana U.S.A.
112 E1 Hanahai watercourse Botswana/Namibia
92 2 Hanalei U.S.A.
52 G5 Hanamaki Japan
20 B2 Hanau Ger.
50 D3 Hancheng China
90 D5 Hancock MD U.S.A.
88 C2 Hancock MI U.S.A.
91 F4 Hancock NY U.S.A.
11 C2 Handa Island U.K.
50 E2 Handan China
115 C4 Handeni Tanz.
19 G2 Handewitt Ger.
19 G2 Hanerau-Hademarschen Ger.
92 C3 Hanford U.S.A.
43 A3 Hangal India
71 8 Hanga Roa Easter I. Chile
48 G2 Hangayn Nuruu mts Mongolia
50 C1 Hanggin Houqi China
50 C2 Hanggin Qi China
50 E2 Hangu China
44 B2 Hangu Pak.
51 D5 Hanguang China
51 F4 Hangzhou China
51 F4 Hangzhou Wan b. China
37 H2 Hani Turkey
40 C5 Ḩanīdh S. Arabia
50 B2 Hanjiaoshui China
19 H4 Hankensbüttel Ger.
112 F6 Hankey S. Africa
9 S7 Hanko Fin.
93 G2 Hanksville U.S.A.
44 D2 Hanle Jammu and Kashmir
70 D5 Hanmer Springs N.Z.
41 F4 Hanmni Mashkel salt flat Pak.
77 G4 Hanna Can.
78 D3 Hannah Bay Can.
88 B6 Hannibal U.S.A.
19 G4 Hannover Ger.
19 G5 Hannoversch Münden Ger.
15 F2 Hannut Belgium
9 O9 Hanöbukten b. Sweden
Hanoi see Ha Nôi
51 B6 Ha Nôi Vietnam
89 G3 Hanover Can.
112 F5 Hanover S. Africa
91 G3 Hanover NH U.S.A.
90 E5 Hanover PA U.S.A.
21 L5 Hanság h. Hungary
116 D4 Hansen Mts Ant.
51 D4 Hanshou China
50 E4 Han Shui r. China
44 D3 Hansi India
8 Q2 Hansnes Norway
66 B3 Hanson, L. salt flat Austr.
90 B6 Hansonville U.S.A.
9 L8 Hanstholm Denmark
15 G4 Han-sur-Nied France
28 C4 Hantsavichy Belarus
44 C3 Hanumangarh India
21 K2 Hanušovice Czech Rep.
67 G5 Hanwwod Austr.
51 E4 Hanyang China
50 C3 Hanyin China
51 B4 Hanyuan China
50 C3 Hanzhong China
60 N5 Hao atoll Fr. Polynesia Pac. Oc.
71 2 Hao i. Fr. Polynesia Pac. Oc.
45 G5 Hāora India
8 T4 Haparanda Sweden
45 H4 Hāpoli India
79 H3 Happy Valley-Goose Bay Can.
54 E3 Hapsu N. Korea
44 D3 Hapur India
43 C5 Haputale Sri Lanka
40 C5 Ḩaraḑ well S. Arabia
40 C5 Ḩaraḑh S. Arabia
28 D4 Haradok Belarus
71 2 Haraiki i. Fr. Polynesia Pac. Oc.
53 G6 Haramachi Japan
44 C2 Haramukh mt India
44 C3 Harappa Road Pak.
111 D5 Harare Zimbabwe
49 J2 Har-Ayrag Mongolia
114 A3 Harbel Liberia
46 E2 Harbin China
89 F4 Harbor Beach U.S.A.
88 E3 Harbor Springs U.S.A.
79 J4 Harbour Breton Can.
102 E8 Harbours, B. of Falkland Is
93 F5 Harcuvar Mts U.S.A.
44 D5 Harda Khās India
9 K6 Hardangervidda plat. Norway
9 K6 Hardangervidda Nasjonalpark nat. park Norway
112 B2 Hardap div. Namibia
112 B2 Hardap Dam dam Namibia
18 D4 Hardenberg Neth.
59 E2 Harden, Bukit mt Indon.
18 C4 Harderwijk Neth.
112 C5 Hardeveld mts S. Africa
82 F2 Hardin U.S.A.
113 H5 Harding S. Africa
77 G4 Hardisty Can.
76 F2 Hardisty Lake Can.
44 E4 Hardoi India
91 G2 Hardwick U.S.A.
66 B5 Hardwicke B. Austr.
85 F4 Hardy U.S.A.
88 E4 Hardy Reservoir U.S.A.
36 D6 Hareidîn, W. watercourse Egypt
15 D2 Harelbeke Belgium
18 D3 Haren Neth.
18 E4 Haren (Ems) Ger.
110 E3 Härer Eth.
91 F4 Harford U.S.A.
110 E3 Hargeysa Somalia
17 M7 Harghita-Mădăraș, Vârful mt Romania
37 H2 Harhal D. mts Turkey
50 C2 Harhatan China
46 B3 Har Hu l. China
44 D3 Haridwar India
43 A3 Harihar India
70 C5 Harihari N.Z.
53 D7 Harima-nada b. Japan
45 G5 Haringhat r. Bangl.
18 B5 Haringvliet est. Neth.
41 G3 Hari Rūd r. Afgh./Iran
9 S6 Harjavalta Fin.
84 E3 Harlan IA U.S.A.
90 B6 Harlan KY U.S.A.
13 C5 Harlech U.K.
82 E1 Harlem U.S.A.
13 J5 Harleston U.K.
18 C3 Harlingen Neth.
85 D7 Harlingen U.S.A.
13 H6 Harlow U.K.
82 E2 Harlowtown U.S.A.
15 D3 Harly France
91 J2 Harmony ME U.S.A.
88 A4 Harmony MN U.S.A.
19 H3 Harmsdorf Ger.
44 A3 Harnai Pak.
15 C2 Harnes France
82 B3 Harney Basin U.S.A.
82 C3 Harney L. U.S.A.
9 P5 Härnösand Sweden
46 E2 Har Nur China
48 F2 Har Nuur l. Mongolia
11 Haroldswick U.K.
15 G4 Haroué France
114 B4 Harper Liberia
92 D4 Harper Lake U.S.A.
90 E5 Harpers Ferry U.S.A.
79 H2 Harp Lake Can.
18 F4 Harpstedt Ger.
37 G2 Harput Turkey
50 F1 Harqin China
50 F1 Harqin Qi China
93 F5 Harquahala Mts U.S.A.
37 G3 Harran Turkey
36 F5 Harrat er Rujeila lava Jordan
78 E3 Harricana r. Can.
87 C5 Harriman U.S.A.
91 G3 Harriman Reservoir U.S.A.
67 K3 Harrington Austr.
91 F5 Harrington U.S.A.
79 J3 Harrington Harbour Can.
11 B3 Harris U.K.
86 B4 Harrisburg IL U.S.A.
90 E4 Harrisburg PA U.S.A.
66 A3 Harris, Lake salt flat Austr.
19 G2 Harrislee Ger.
69 C6 Harrismith Austr.
113 H4 Harrismith S. Africa
85 E4 Harrison AR U.S.A.
88 E3 Harrison MI U.S.A.
74 C2 Harrison Bay U.S.A.
90 D5 Harrisonburg U.S.A.
79 J3 Harrison, Cape Can.
76 E5 Harrison L. Can.
84 E4 Harrisonville U.S.A.
11 A3 Harris, Sound of chan. U.K.
89 F3 Harrisville MI U.S.A.
91 F2 Harrisville NY U.S.A.
90 C5 Harrisville WV U.S.A.
12 F4 Harrogate U.K.
19 G3 Harsefeld Ger.
40 B3 Harsin Iran
36 G1 Harşit r. Turkey
27 M2 Hârşova Romania
8 P2 Harstad Norway
19 G4 Harsum Ger.
88 D4 Hart U.S.A.
54 B2 Hartao China
112 D4 Hartbees watercourse S. Africa
21 J5 Hartberg Austria
9 K6 Harteigan mt Norway
11 E5 Hart Fell h. U.K.
91 G4 Hartford CT U.S.A.
88 D4 Hartford MI U.S.A.
84 D3 Hartford SD U.S.A.
88 C4 Hartford WV U.S.A.
76 E3 Hart Highway Can.
66 B3 Hart, L. salt flat Austr.
79 G4 Hartland Can.
13 C7 Hartland U.K.
91 J2 Hartland U.S.A.
13 C6 Hartland Point U.K.
12 F3 Hartlepool U.K.
85 C5 Hartley U.S.A.
76 D4 Hartley Bay Can.
9 U6 Hartola Fin.
76 E4 Hart Ranges mts Can.
20 D4 Härtsfeld h. Ger.
112 F3 Hartswater S. Africa
87 D5 Hartwell Resr U.S.A.
48 F2 Har Us Nuur l. Mongolia
41 F3 Harut watercourse Afgh.
88 C4 Harvard U.S.A.
83 F4 Harvard, Mt U.S.A.
69 B6 Harvey Austr.
91 K2 Harvey Can.
88 D2 Harvey MI U.S.A.
84 C2 Harvey ND U.S.A.
13 J6 Harwich U.K.
67 K2 Harwood Austr.
44 C3 Haryana div. India
36 F6 Ḩaşāh, Wādī al watercourse Jordan
39 J5 Hasalbag China
40 B2 Hasan Iran
44 C2 Hasan Abdal Pak.
36 E2 Hasan Dağı mts Turkey
37 H3 Hasankeyf Turkey
40 E5 Ḩasan Langī Iran
43 B2 Hasanparti India
36 E5 Hasbani r. Lebanon
36 E2 Hasbek Turkey
37 K6 Ḩasb, Sha'īb watercourse Iraq
45 E5 Hasdo r. India
18 E4 Hase r. Ger.
18 E4 Haselünne Ger.
22 E2 Hasenmatt mt Switz.
40 C3 Hashtgerd Iran
40 C2 Hashtpar Iran
85 D5 Haskell U.S.A.
19 M1 Hasle Denmark
13 G6 Haslemere U.K.
19 J1 Haslev Denmark
17 M7 Hăşmaşul Mare mt Romania
43 B3 Hassan India
37 K4 Hassan Iraq
93 F5 Hassayampa r. U.S.A.
20 D2 Haßberge reg. Ger.
15 F2 Hasselt Belgium
18 D4 Hasselt Neth.
108 C1 Hassi Messaoud Alg.
9 N8 Hässleholm Sweden
66 F7 Hastings Austr.
67 K3 Hastings r. Austr.
70 F3 Hastings N.Z.
13 H7 Hastings U.K.
88 E4 Hastings MI U.S.A.
88 A3 Hastings MN U.S.A.
84 D3 Hastings NE U.S.A.
Hatay see Antakya
93 F3 Hatch U.S.A.
87 E7 Hatchet Bay Bahamas
77 J3 Hatchet Lake Can.
87 B5 Hatchie r. U.S.A.
66 E4 Hatfield Austr.
12 G4 Hatfield U.K.
48 H1 Hatgal Mongolia
44 D4 Hathras India
45 F4 Hatia Nepal
58 C3 Ha Tiên Vietnam
58 C1 Ha Tinh Vietnam
66 E5 Hattah Austr.
87 F5 Hatteras, Cape U.S.A.
8 N4 Hattfjelldal Norway
45 E6 Hatti r. India
87 B6 Hattiesburg U.S.A.
18 E5 Hattingen Ger.
19 G2 Hattstedt Ger.
110 E3 Haud reg. Eth.
9 K7 Hauge Norway
9 J7 Haugesund Norway
70 E3 Hauhungaroa mt N.Z.
9 K7 Haukeligrend Norway
8 T4 Haukipudas Fin.
9 V5 Haukivesi l. Fin.
77 H3 Haultain r. Can.
70 E2 Hauraki Gulf g. N.Z.
70 A7 Hauroko, L. N.Z.
71 3 Hauru, Pte pt Fr. Polynesia Pac. Oc.
15 J4 Hausach Ger.
21 G4 Hausruck mts Austria
23 G3 Hausstock mt Switz.
108 B1 Haut Atlas mts Morocco
15 B3 Haute-Normandie div. France
79 G4 Hauterive Can.
22 B2 Haut-Folin h. France
91 J3 Haut, Isle au i. U.S.A.
108 B1 Hauts Plateaux plat. Alg.
92 1 Hauula U.S.A.
21 G4 Hauzenberg Ger.
95 H4 Havana Cuba
88 B5 Havana U.S.A.
13 G7 Havant U.K.
93 E4 Havasu Lake U.S.A.
19 K4 Havel r. Ger.
15 F2 Havelange Belgium
19 K4 Havelberg Ger.
19 K4 Havelländisches Luch marsh Ger.
89 J3 Havelock Can.
87 E5 Havelock U.S.A.
70 F3 Havelock North N.Z.
13 C6 Haverfordwest U.K.
91 H3 Haverhill U.S.A.
43 A3 Haveri India
15 F2 Haversin Belgium
18 E5 Havixbeck Ger.
21 J3 Havlíčkův Brod Czech Rep.
8 T1 Havøysund Norway
27 M5 Havran Turkey
82 E1 Havre U.S.A.
79 H4 Havre Aubert, Île du i. Can.
91 E5 Havre de Grace U.S.A.
79 H3 Havre-St-Pierre Can.
27 M4 Havsa Turkey
36 E1 Havza Turkey
60 M2 Hawaii is Pac. Oc.
92 2 Hawaii i. U.S.A.
118 H4 Hawaiian Ridge sea feature Pac. Oc.
92 2 Hawaii Volcanoes National Park U.S.A.
37 L7 Ḩawallī Kuwait
13 D4 Hawarden U.K.
70 B6 Hawea, L. N.Z.
70 E3 Hawera N.Z.
12 E3 Hawes U.K.
92 2 Hawi U.S.A.
11 F5 Hawick U.K.
37 L6 Ḩawīzah, Hawr al l. Iraq
70 B6 Hawkdun Range mts N.Z.
70 F3 Hawke Bay N.Z.
79 J3 Hawke Island Can.
66 C3 Hawker Austr.
66 D2 Hawkers Gate Austr.
91 F2 Hawkesbury Can.
93 F3 Hawkins Peak U.S.A.
89 F3 Hawks U.S.A.
91 K2 Hawkshaw Can.
68 D5 Hawkwood Austr.
91 F4 Hawley U.S.A.
37 J5 Ḩawrān, Wādī watercourse Iraq
112 C7 Hawston S. Africa
92 C2 Hawthorne U.S.A.
54 C1 Haxat China
12 F3 Haxby U.K.
66 F5 Hay Austr.
76 F2 Hay r. Can.
88 B3 Hay r. U.S.A.
50 B1 Haya China
52 G5 Hayachine-san mt Japan
40 B2 Haydarābad Iran
93 G5 Hayden AZ U.S.A.
82 C2 Hayden ID U.S.A.
77 L3 Hayes r. Can.
75 M2 Hayes Halvø pen. Greenland
13 B7 Hayle U.K.
36 D2 Haymana Turkey
68 C3 Hayman I. Austr.
90 E5 Haymarket U.S.A.
75 J3 Haynes r. Can.
91 J2 Haynesville U.S.A.
13 D5 Hay-on-Wye U.K.
29 C7 Hayrabolu Turkey
76 F2 Hay River Can.
84 D4 Hays U.S.A.
29 D5 Haysyn Ukr.
92 A3 Hayward CA U.S.A.
88 B2 Hayward WV U.S.A.
13 G7 Haywards Heath U.K.
41 G3 Hazarajat reg. Afgh.
90 B6 Hazard U.S.A.
45 F5 Hazārībāg India
45 E5 Hazaribagh Range mts India
15 C2 Hazebrouck France
76 D3 Hazelton Can.
91 F4 Hazelton U.S.A.
74 G2 Hazen Strait Can.
18 B4 Hazerswoude-Rijndijk Neth.
37 G6 Ḩazm al Jalāmid ridge S. Arabia
41 G2 Hazrat Sultan Afgh.
37 H2 Hazro Turkey
14 B4 Headford Rep. of Ireland
64 F6 Head of Bight b. Austr.
92 A2 Healdsburg U.S.A.
66 F6 Healesville Austr.
13 F4 Heanor U.K.
117 J7 Heard Island Ind. Ocean
85 D6 Hearne U.S.A.
78 D4 Hearst Can.
116 B2 Hearst I. Ant.
13 H7 Heathfield U.K.
91 E6 Heathsville U.S.A.
85 D7 Hebbronville U.S.A.
50 E2 Hebei div. China
67 G2 Hebel Austr.
85 E5 Heber Springs U.S.A.
50 E3 Hebi China
79 H2 Hebron Can.
88 D5 Hebron IN U.S.A.
84 D3 Hebron NE U.S.A.
91 G3 Hebron NY U.S.A.
36 E6 Hebron West Bank
79 H2 Hebron Fiord in. Can.
74 E4 Hecate Strait B.C. Can.
76 C4 Hecate Strait Can.
97 G3 Hecelchakán Mex.
76 C3 Heceta I. U.S.A.
51 C5 Hechi China
20 B4 Hechingen Ger.
51 C4 Hechuan China
9 N5 Hede Sweden
9 O6 Hedemora Sweden
51 D6 Hede Sk. resr China
82 C2 He Devil Mt. U.S.A.
88 A5 Hedrick U.S.A.
18 C4 Heeg Neth.
18 E4 Heek Ger.
15 E2 Heer Belgium
18 D4 Heerde Neth.
18 C4 Heerenveen Neth.
18 B4 Heerhugowaard Neth.
18 C6 Heerlen Neth.
Hefa see Haifa
50 E4 Hefei China
51 D4 Hefeng China
52 B1 Hegang China
20 B5 Hegau reg. Ger.
53 E6 Hegura-jima i. Japan
19 J5 Heidberg h. Ger.
19 G2 Heide Ger.
111 B6 Heide Namibia
20 B3 Heidelberg Ger.
113 H3 Heidelberg Gauteng S. Africa
112 D7 Heidelberg Western Cape S. Africa
19 L6 Heidenau Ger.
20 D4 Heidenheim an der Brenz Ger.
21 J4 Heidenreichstein Austria
Heihe see Aihui
113 G3 Heilbron S. Africa
20 C3 Heilbronn Ger.
19 H2 Heiligenhafen Ger.
21 K6 Heiligenkreuz im Lafnitztal Austria
51 Hei Ling Chau i. H.K. China
54 E1 Heilongjiang div. China
49 P1 Heilong Jiang r. China/Rus. Fed.
20 D3 Heilsbronn Ger.
8 M5 Heimdal Norway
20 C5 Heimenkirch Ger.
9 U6 Heinola Fin.
58 A2 Heinze Is Myanmar
54 B3 Heishan China
15 E1 Heist-op-den-Berg Belgium
50 E2 Hejian China
51 B4 Hejiang China
51 D6 He Jiang r. China
50 D3 Hejin China
36 F2 Hekimhan Turkey
8 D5 Hekla volc. Iceland
50 B2 Hekou Gansu China
51 B6 Hekou Yunnan China
8 N5 Helagsfjället mt Sweden
50 B2 Helan Shan mts China
19 J5 Helbra Ger.
45 H4 Helem India
85 F5 Helena AR U.S.A.
82 E2 Helena MT U.S.A.
92 D3 Helen, Mt U.S.A.
11 D4 Helensburgh U.K.
36 E6 Helez Israel
18 E2 Helgoland i. Ger.
18 F2 Helgoländer Bucht b. Ger.
67 K1 Helidon Austr.
8 C5 Hella Iceland
8 P2 Helland Norway
40 C4 Helleh r. Iran
18 B5 Hellevoetsluis Neth.
8 R2 Helligskogen Norway
25 F3 Hellín Spain
82 C2 Hells Canyon gorge U.S.A.
41 F4 Helmand r. Afgh.
20 E2 Helmbrechts Ger.
19 J5 Helme r. Ger.
111 B6 Helmeringhausen Namibia
18 C5 Helmond Neth.
11 E2 Helmsdale U.K.
11 E2 Helmsdale r. U.K.
19 J4 Helmstedt Ger.
54 E2 Helong China
93 G2 Helper U.S.A.
19 L3 Helpter Berge h. Ger.
9 N8 Helsingborg Sweden
9 N8 Helsingør Denmark
9 T6 Helsinki Fin.
13 B7 Helston U.K.
12 D3 Helvellyn mt U.K.
14 D5 Helvick Head hd Rep. of Ireland
109 F2 Helwân Egypt
13 G6 Hemel Hempstead U.K.
92 D5 Hemet U.S.A.
90 E3 Hemlock Lake U.S.A.
19 G4 Hemmingen Ger.
91 G2 Hemmingford Can.
19 G3 Hemmoor Ger.
85 D6 Hempstead U.S.A.
13 J5 Hemsby U.K.
9 Q8 Hemse Sweden
50 A3 Henan Qinghai China
50 D3 Henan div. China
25 E2 Henares r. Spain
52 F4 Henashi-zaki pt Japan
36 C1 Hendek Turkey
105 E3 Henderson Arg.
86 C4 Henderson KY U.S.A.
87 E4 Henderson NC U.S.A.
93 E3 Henderson NV U.S.A.
91 E3 Henderson NY U.S.A.
85 E5 Henderson TX U.S.A.
60 P6 Henderson Island Pitcairn Is Pac. Oc.
87 D5 Hendersonville NC U.S.A.
87 C4 Hendersonville TN U.S.A.
40 C4 Hendījān Iran
13 G6 Hendon U.K.
40 D5 Hendorābī i. Iran
46 B4 Hengduan Shan mts China
18 D4 Hengelo Neth.
51 D5 Hengshan Hunan China
51 D5 Heng Shan mt Hunan China
50 C2 Hengshan Shaanxi China
54 F1 Hengshan China
50 D2 Heng Shan mt China
50 E2 Hengshui China
51 C6 Heng Xian China
51 D5 Hengyang Hunan China
29 E6 Heniches'k Ukr.
70 C6 Henley N.Z.
13 G6 Henley-on-Thames U.K.
91 F5 Henlopen, Cape pt U.S.A.
18 E6 Hennef (Sieg) Ger.
113 G3 Hennenman S. Africa
22 D1 Hennezel France
19 L4 Hennigsdorf Berlin Ger.
91 H3 Henniker U.S.A.
85 D5 Henrietta U.S.A.
78 D2 Henrietta Maria, Cape Can.
93 G3 Henrieville U.S.A.
69 B1 Henry r. Austr.
88 C5 Henry U.S.A.
116 B3 Henry Ice Rise ice feature Ant.
75 M3 Henry Kater, C. hd Can.
93 G2 Henry Mts U.S.A.
89 G4 Hensall Can.
19 G3 Henstedt-Ulzburg Ger.
111 B6 Hentiesbaai Namibia
67 G5 Henty Austr.
56 B3 Henzada Myanmar
77 H4 Hepburn Can.
51 E5 Heping China
51 C6 Hepu China
50 D2 Hequ China
68 C1 Herald Cays atolls Coral Sea Is Terr.
41 F3 Herāt Afgh.
24 F5 Hérault r. France
22 C4 Herbasse r. France
68 A2 Herbert r. Austr.
77 H4 Herbert Can.
20 C4 Herbertingen Ger.
68 A1 Herberton Austr.
20 B2 Herborn Ger.
20 D4 Herbrechtingen Ger.
20 C2 Herbstein Ger.
116 B4 Hercules Dome ice feature Ant.
18 E5 Herdecke Ger.
18 E6 Herdorf Ger.
96 H6 Heredia Costa Rica
13 E5 Hereford U.K.
85 C5 Hereford U.S.A.
71 2 Héréhérétué i. Fr. Polynesia Pac. Oc.
60 N5 Héréhérétué i. Pac. Oc.
15 E2 Herent Belgium
18 F4 Herford Ger.
22 D2 Héricourt France
22 D2 Hérimoncourt France
19 H6 Heringen (Werra) Ger.
19 J2 Heringsdorf Ger.
84 D4 Herington U.S.A.
40 B2 Herīs Iran
23 G2 Herisau Switz.
116 B3 Heritage Ra. mts Ant.
91 F3 Herkimer U.S.A.
21 G6 Hermagor Austria
96 D2 Hermanas Mex.
11 Herma Ness hd U.K.
19 H4 Hermannsburg Ger.
112 C7 Hermanus S. Africa
21 J3 Heřmanův Městec Czech Rep.
113 H5 Hermes, Cape pt S. Africa
67 G3 Hermidale Austr.
82 C2 Hermiston U.S.A.
102 C9 Hermite, Is is Chile
62 E2 Hermit Is P.N.G.
Hermon, Mount see Sheikh, Jebel esh
105 B2 Hermosa, P. de V. pass Chile
103 B4 Hermosas, Parque Nacional las nat. park Col.
96 B1 Hermosillo Mex.
19 J6 Hermsdorf Ger.
102 F3 Hernandarias Para.
18 E5 Herne Ger.
13 J6 Herne Bay U.K.
9 L8 Herning Denmark
88 D1 Heron Bay Can.
68 D4 Heron I. Austr.
21 K5 Herpenyő r. Hungary
96 D3 Herradura Mex.
20 B4 Herrenberg Ger.
25 D3 Herrera del Duque Spain
96 C2 Herreras Mex.
67 G8 Herrick Austr.
20 D3 Herrieden Ger.
21 H1 Herrnhut Ger.
90 E4 Hershey U.S.A.
13 G6 Hertford U.K.
113 F4 Hertzogville S. Africa
15 F2 Herve Belgium
68 E5 Hervey Bay Austr.
68 E5 Hervey Bay Austr.
60 M5 Hervey Is Pac. Oc.
68 B2 Hervey Ra. mts Austr.
19 K4 Herzberg Ger.
19 L5 Herzberg Ger.
18 E4 Herzlake Ger.
20 D3 Herzogenaurach Ger.
21 J4 Herzogenburg Austria
19 K3 Herzsprung Ger.
37 M4 Heşar Iran
15 E2 Hesbaye reg. Belgium
15 C2 Hesdin France
18 E3 Hesel Ger.
51 C6 Heshan China
50 C3 Heshui China
50 D2 Heshun China
92 D4 Hesperia U.S.A.
76 C2 Hess r. Can.
20 D3 Heßdorf Ger.
19 G6 Hessen div. Ger.
19 G5 Hessisch Lichtenau Ger.
51 B6 Het r. Laos
92 B3 Hetch Hetchy Aqueduct canal U.S.A.
18 C5 Heteren Neth.
21 K6 Hetés h. Hungary
20 C4 Hettingen Ger.
84 C2 Hettinger U.S.A.
12 E3 Hetton U.K.
19 J5 Hettstedt Ger.
12 E3 Hexham U.K.
50 F4 He Xian Anhui China
51 D5 He Xian Guangxi China
50 B2 Hexibao China
50 E1 Hexigten Qi China
112 C6 Hex River Pass S. Africa
50 D3 Heyang China
40 D4 Ḩeydarābād Iran
41 F4 Ḩeydarābād Iran
12 E3 Heysham U.K.
51 E6 Heyuan China
66 D7 Heywood Austr.
12 E4 Heywood U.K.
88 C5 Heyworth U.S.A.
50 E3 Heze China
51 B5 Hezhang China
50 B3 Hezheng China
50 B3 Hezuozhen China
87 D7 Hialeah U.S.A.
84 E4 Hiawatha U.S.A.
88 A2 Hibbing U.S.A.
67 F9 Hibbs, Pt hd Austr.
87 D5 Hickory U.S.A.
70 G2 Hicks Bay N.Z.
97 G4 Hicks Cays is Belize
77 K2 Hicks L. Can.
90 A4 Hicksville U.S.A.
85 D5 Hico U.S.A.
52 H3 Hidaka-sanmyaku mts Japan
97 E2 Hidalgo Mex.
97 E3 Hidalgo div. Mex.
96 C2 Hidalgo del Parral Mex.
96 B2 Hidalgo, Presa M. resr Mex.
19 L2 Hiddensee i. Ger.
104 C2 Hidrolândia Brazil
21 H5 Hieflau Austria
71 9 Hienghène New Caledonia Pac. Oc.
53 C7 Higashi-Hiroshima Japan
52 G5 Higashine Japan
53 D7 Higashi-ōsaka Japan
53 A8 Higashi-suidō chan. Japan
91 F3 Higgins Bay U.S.A.
88 E3 Higgins Lake U.S.A.
High Atlas mts see Haut Atlas
82 B3 High Desert U.S.A.
88 C3 High Falls Reservoir U.S.A.
88 E3 High I. U.S.A.
51 High Island Resr H.K. China
88 D4 Highland Park U.S.A.
92 C2 Highland Peak CA U.S.A.
93 E3 Highland Peak NV U.S.A.
76 F3 High Level Can.
45 F5 High Level Canal India
87 E5 High Point U.S.A.
76 F3 High Prairie Can.
76 G4 High River Can.
87 E7 High Rock Bahamas
77 J3 Highrock Lake Can.
67 F9 High Rocky Pt hd Austr.
12 E3 High Seat h. U.K.
91 F4 Hightstown U.S.A.
13 G6 High Wycombe U.K.
96 B2 Higuera de Zaragoza Mex.
103 D2 Higuerote Venez.
9 S7 Hiiumaa i. Estonia
42 A4 Hijaz reg. S. Arabia
93 E3 Hiko U.S.A.
53 E7 Hikone Japan
71 2 Hikuéru i. Fr. Polynesia Pac. Oc.
70 G2 Hikurangi mt N.Z.
93 F3 Hildale U.S.A.
20 D2 Hildburghausen Ger.
20 D2 Hilders Ger.
19 G4 Hildesheim Ger.
45 G4 Hili Bangl.
69 B5 Hill r. Austr.
116 B5 Hillary Coast Ant.
84 D4 Hill City U.S.A.
93 H2 Hill Creek r. U.S.A.
18 B4 Hillegom Neth.
9 N9 Hillerød Denmark
68 A2 Hillgrove Austr.
84 D2 Hillsboro ND U.S.A.
91 H3 Hillsboro NH U.S.A.
90 B5 Hillsboro OH U.S.A.
85 D5 Hillsboro TX U.S.A.
90 C5 Hillsboro WV U.S.A.
88 B4 Hillsboro WV U.S.A.
68 C3 Hillsborough, C. pt Austr.
88 E5 Hillsdale MI U.S.A.
91 G3 Hillsdale NY U.S.A.
90 E4 Hillsgrove U.S.A.
11 F4 Hillside U.K.
93 F4 Hillside U.S.A.
67 F4 Hillston Austr.
90 C6 Hillsville U.S.A.
67 J5 Hilltop Austr.
92 2 Hilo U.S.A.
113 J4 Hilton S. Africa
90 E3 Hilton U.S.A.
89 F2 Hilton Beach Can.
87 D5 Hilton Head Island U.S.A.
37 G3 Hilvan Turkey
18 C4 Hilversum Neth.
44 D3 Himachal Pradesh div. India
30 F6 Himalaya mts Asia
45 F3 Himalchul mt Nepal
8 S4 Himanka Fin.
27 H4 Himarë Albania
44 C5 Himatnagar India
21 K4 Himberg Austria
53 D7 Himeji Japan
52 G5 Himekami-dake mt Japan
113 H4 Himeville S. Africa
53 E6 Himi Japan
36 F4 Ḩimş Syria
36 F4 Ḩimş, Baḩrat resr Syria
55 C4 Hinatuan Phil.
68 B2 Hinchinbrook I. Austr.
13 F5 Hinckley U.K.
88 A2 Hinckley MN U.S.A.
93 F2 Hinckley UT U.S.A.
91 F3 Hinckley Reservoir U.S.A.
44 D3 Hindan r. India
44 D4 Hindaun India
12 G3 Hinderwell U.K.
37 K5 Hindīyah Barrage Iraq
12 E4 Hindley U.K.
90 B6 Hindman U.S.A.
66 D6 Hindmarsh, L. Austr.
45 F5 Hindola India
19 H1 Hindsholm pen. Denmark
41 G3 Hindu Kush mts Afgh./Pak.
43 B3 Hindupur India
76 F3 Hines Creek Can.
87 D6 Hinesville U.S.A.
44 D5 Hinganghat India
41 G5 Hinglaj Pak.
41 G5 Hingol r. Pak.
44 D6 Hingoli India
37 H2 Hınıs Turkey
92 D4 Hinkley U.S.A.
8 O2 Hinnøya i. Norway
55 B4 Hinobaan Phil.
25 D3 Hinojosa del Duque Spain
53 C7 Hino-misaki pt Japan
91 G3 Hinsdale U.S.A.
18 E3 Hinte Ger.
20 D2 Hinternah Ger.
76 F4 Hinton Can.
90 C6 Hinton U.S.A.
18 B4 Hippolytushoef Neth.
37 K2 Hirabit Dağ mt Turkey
53 A8 Hirado Japan
53 A8 Hirado-shima i. Japan
45 E5 Hirakud Reservoir India
52 H3 Hiroo Japan
52 G4 Hirosaki Japan
53 C7 Hiroshima Japan

20 E3 **Hirschaid** Ger.
20 E2 **Hirschberg** Ger.
20 E5 **Hirschberg** *mt* Ger.
22 E2 **Hirsingue** France
15 E3 **Hirson** France
9 L8 **Hirtshals** Denmark
37 M3 **Hisar** Iran
41 G3 **Hisar, Koh-i-** *mts* Afgh.
36 D1 **Hisarönü** Turkey
36 E6 **Hisban** Jordan
39 G5 **Hisor** Tajik.
95 K4 **Hispaniola** *i.* Caribbean Sea
44 C3 **Hissar** India
45 F4 **Hisua** India
37 J5 **Hīt** Iraq
53 G6 **Hitachi** Japan
53 G6 **Hitachi-ōta** Japan
71 □3 **Hitiaa** *Fr. Polynesia* Pac. Oc.
53 B8 **Hitoyoshi** Japan
8 L5 **Hitra** *i.* Norway
19 J3 **Hitzacker** Ger.
71 □9 **Hiu** *i.* Vanuatu
53 C7 **Hiuchi-nada** *b.* Japan
60 O4 **Hiva Oa** *i. Fr. Polynesia* Pac. Oc.
76 E4 **Hixon** Can.
68 E4 **Hixson Cay** *rf* Austr.
37 J2 **Hizan** Turkey
9 O7 **Hjälmaren** *l.* Sweden
77 H2 **Hjalmar Lake** Can.
19 K2 **Hjelm Bugt** *b.* Denmark
9 L5 **Hjerkinn** Norway
9 O7 **Hjo** Sweden
9 M8 **Hjørring** Denmark
113 J4 **Hlabisa** S. Africa
45 F3 **Hlako Kangri** *mt* China
113 J3 **Hlatikulu** Swaziland
21 J3 **Hlinsko** Czech Rep.
29 E5 **Hlobyne** Ukr.
113 G4 **Hlohlowane** S. Africa
21 L4 **Hlohovec** Slovakia
113 H4 **Hlotse** Lesotho
21 M3 **Hlučín** Czech Rep.
113 K4 **Hluhluwe** S. Africa
29 E5 **Hlukhiv** Ukr.
17 O4 **Hlusha** Belarus
28 C4 **Hlybokaye** Belarus
114 D3 **Ho** Ghana
111 B6 **Hoachanas** Namibia
67 G9 **Hobart** Austr.
85 D5 **Hobart** U.S.A.
85 C5 **Hobbs** U.S.A.
116 A4 **Hobbs Coast** Ant.
87 D7 **Hobe Sound** U.S.A.
9 L8 **Hobro** Denmark
110 E3 **Hobyo** Somalia
20 C3 **Höchberg** Ger.
19 G2 **Hochdonn** Ger.
15 H4 **Hochfelden** France
58 C3 **Hô Chi Minh** Vietnam
21 J5 **Hochschwab** *mt* Austria
21 H5 **Hochschwab** *mts* Austria
20 D3 **Höchstadt an der Aisch** Ger.
21 H5 **Hochtor** *mt* Austria
20 B3 **Hockenheim** Ger.
90 B5 **Hocking** *r.* U.S.A.
97 G3 **Hoctúm** Mex.
44 D4 **Hodal** India
12 E4 **Hodder** *r.* U.K.
13 G6 **Hoddesdon** U.K.
Hodeida *see* **Al Hudaydah**
91 K1 **Hodgdon** U.S.A.
17 K7 **Hódmezővásárhely** Hungary
25 J5 **Hodna, Chott el** *salt l.* Alg.
54 D4 **Hodo dan** *pt* N. Korea
21 L4 **Hodonín** Czech Rep.
18 B5 **Hoek van Holland** Neth.
18 C6 **Hoensbroek** Neth.
54 E2 **Hoeryŏng** N. Korea
54 D4 **Hoeyang** N. Korea
20 E2 **Hof** Ger.
20 D2 **Hofheim in Unterfranken** Ger.
113 F5 **Hofmeyr** S. Africa
8 F4 **Höfn** Iceland
9 P6 **Hofors** Sweden
8 D4 **Hofsjökull** *ice cap* Iceland
53 B7 **Hōfu** Japan
9 N8 **Höganäs** Sweden
67 G7 **Hogan Group** *is* Austr.
68 B5 **Hoganthulla Cr.** *r.* Austr.
108 C2 **Hoggar** *plat.* Alg.
91 F6 **Hog I.** U.S.A.
9 P8 **Högsby** Sweden
21 J5 **Hohenberg** Austria
19 L5 **Hohenbucko** Ger.
20 E3 **Hohenburg** Ger.
20 E3 **Hohenfels** Ger.
20 D5 **Hohenfurch** Ger.
19 G3 **Hohenlockstedt** Ger.
20 C3 **Hohenloher Ebene** *plain* Ger.
19 K5 **Hohenmölsen** Ger.
19 K4 **Hohennauen** Ger.
21 H5 **Hohe Nock** *mt* Austria
21 G6 **Hohenthurn** Austria
20 C5 **Hohenweiler** Austria
19 G2 **Hohenwestedt** Ger.
21 G5 **Hoher Dachstein** *mt* Austria
20 G5 **Hoher Göll** *mt* Austria/Ger.
20 C2 **Hohe Rhön** *mts* Ger.
20 F5 **Hohe Tauern** *mts* Austria
20 F5 **Hohe Tauern, Nationalpark** *nat. park* Austria
15 G2 **Hohe Venn** *moorland* Belgium
50 D1 **Hohhot** China
22 E1 **Hohneck** *mt* France
45 G2 **Hoh Xil Hu** *salt l.* China
45 G2 **Hoh Xil Shan** *mts* China
58 D2 **Hôi An** Vietnam
115 B2 **Hoima** Uganda
51 B6 **Hôi Xuân** Vietnam
45 H4 **Hojai** India
53 C8 **Hōjo** Japan
70 D1 **Hokianga Harbour** *in.* N.Z.
70 C5 **Hokitika** N.Z.
52 H3 **Hokkaidō** *i.* Japan
9 L7 **Hokksund** Norway
37 K1 **Hoktemberyan** Armenia
9 L6 **Hol** Norway
43 B3 **Holalkere** India
9 M9 **Holbæk** Denmark
13 H5 **Holbeach** U.K.
68 C2 **Holborne I.** Austr.
93 G4 **Holbrook** U.S.A.
88 B3 **Holcombe Flowage** *resr* U.S.A.
77 G4 **Holden** Can.
93 F2 **Holden** U.S.A.
85 D5 **Holdenville** U.S.A.
84 D3 **Holdrege** U.S.A.
43 B3 **Hole Narsipur** India
21 L3 **Holešov** Czech Rep.
95 J4 **Holguín** Cuba
21 L4 **Holíč** Slovakia
9 N6 **Höljes** Sweden
21 K4 **Hollabrunn** Austria
88 D4 **Holland** U.S.A.
20 E4 **Hollenbach** Ger.
90 D4 **Hollidaysburg** U.S.A.
76 C3 **Hollis** *AK* U.S.A.
85 D5 **Hollis** *OK* U.S.A.
92 B3 **Hollister** U.S.A.
89 F4 **Holly** U.S.A.
85 F5 **Holly Springs** U.S.A.
87 D7 **Hollywood** U.S.A.
8 N4 **Holm** Norway
74 G2 **Holman** Can.
68 B1 **Holmes Reef** Coral Sea Is Terr.
8 T2 **Holmestrand** *Finnmark* Norway
9 M7 **Holmestrand** *Vestfold* Norway
8 R5 **Holmön** *i.* Sweden
75 M2 **Holms Ø** *i.* Greenland
8 R5 **Holmsund** Sweden
71 □10 **Holonga** Tonga
112 B3 **Holoog** Namibia
9 L8 **Holstebro** Denmark
19 F1 **Holsted** Denmark
87 D4 **Holston** *r.* U.S.A.
90 C6 **Holston Lake** U.S.A.
13 C7 **Holsworthy** U.K.
13 J5 **Holt** U.K.
88 E4 **Holt** U.S.A.
84 E4 **Holton** U.S.A.
18 C3 **Holwerd** Neth.
14 D5 **Holycross** Rep. of Ireland
13 C4 **Holyhead** U.K.
13 C4 **Holyhead Bay** U.K.
12 F2 **Holy Island** *Eng.* U.K.
13 C4 **Holy Island** *Wales* U.K.
91 G3 **Holyoke** U.S.A.
13 D4 **Holywell** U.K.
19 K5 **Holzhausen** Ger.
20 E5 **Holzkirchen** Ger.
19 G5 **Holzminden** Ger.
115 B3 **Homa Bay** Kenya
40 C3 **Homâyunshahr** Iran
19 G5 **Homberg (Efze)** Ger.
114 C1 **Hombori** Mali
15 H3 **Homburg** Ger.
75 M3 **Home Bay** Can.
15 F3 **Homécourt** France
68 B2 **Home Hill** Austr.
85 E5 **Homer** U.S.A.
87 D6 **Homerville** U.S.A.
68 A3 **Homestead** Austr.
87 D7 **Homestead** U.S.A.
87 C5 **Homewood** U.S.A.
43 B2 **Homnabad** India
55 C4 **Homonhon** *pt* Phil.
Homs *see* **Ḩimş**
29 D4 **Homyel'** Belarus
43 A3 **Honavar** India
103 B3 **Honda** Col.
55 A4 **Honda Bay** Phil.
93 H4 **Hon Dah** U.S.A.
112 B5 **Hondeklipbaai** S. Africa
50 C1 **Hondlon Ju** China
97 G4 **Hondo** *r.* Belize/Mex.
85 D6 **Hondo** U.S.A.
18 D3 **Hondsrug** *reg.* Neth.
73 H8 **Honduras** *country* Central America
96 H4 **Honduras, Gulf of** *g.* Belize/Honduras
9 M6 **Hønefoss** Norway
91 F4 **Honesdale** U.S.A.
92 B1 **Honey Lake** U.S.A.
91 E3 **Honeyoye Lake** U.S.A.
24 E2 **Honfleur** France
50 E4 **Hong'an** China
54 D5 **Hongch'ŏn** S. Korea
51 C6 **Hông Gai** Vietnam
51 E6 **Honghai Wan** *b.* China
51 B6 **Honghe** China
50 E3 **Hong He** *r.* China
51 D4 **Honghu** China
51 C5 **Hongjiang** China
51 E6 **Hong Kong** China
51 E6 **Hong Kong** *div.* China
51 □ **Hong Kong Island** *H.K.* China
50 C2 **Hongliu** *r.* China
50 B2 **Hongliuyuan** China
58 C3 **Hông Ngư** Vietnam
51 C6 **Hong or Red River, Mouths of the** *est.* Vietnam
51 C7 **Hongqizhen** China
50 B2 **Hongshansi** China
54 D2 **Hongshi** China
51 D6 **Hongshui He** *r.* China
51 C6 **Hông, Sông** *r.* Vietnam
50 D2 **Hongtong** China
79 G4 **Honguedo, Détroit d'** *chan.* Can.
54 D3 **Hongwŏn** N. Korea
54 B1 **Hongxing** China
50 B3 **Hongyuan** China
50 F3 **Hongze** China
50 F3 **Hongze Hu** *l.* China
63 F2 **Honiara** Solomon Is
13 D7 **Honiton** U.K.
52 G5 **Honjō** Japan
9 S6 **Honkajoki** Fin.
58 C3 **Hon Khoai** *i.* Vietnam
58 D2 **Hon Lơn** *i.* Vietnam
58 C1 **Hon Mê** *i.* Vietnam
43 A3 **Honnali** India
8 T1 **Honningsvåg** Norway
92 □2 **Honokaa** U.S.A.
92 □1 **Honolulu** U.S.A.
58 C3 **Hon Rai** *i.* Vietnam
53 C7 **Honshū** *i.* Japan
82 B2 **Hood, Mt** *volc.* U.S.A.
69 D7 **Hood Pt** Austr.
18 F2 **Hooge** Ger.
18 F2 **Hooge** *i.* Ger.
18 D4 **Hoogeveen** Neth.
18 D3 **Hoogezand-Sappemeer** Neth.
85 C4 **Hooker** U.S.A.
14 E5 **Hook Head** *hd* Rep. of Ireland
68 C3 **Hook Island** Austr.
Hook of Holland *see* **Hoek van Holland**
68 E5 **Hook Point** Austr.
68 C2 **Hook Reef** Austr.
76 B3 **Hoonah** U.S.A.
74 B3 **Hooper Bay** U.S.A.
91 E5 **Hooper I.** U.S.A.
88 D5 **Hoopeston** U.S.A.
113 F3 **Hoopstad** S. Africa
9 N9 **Höör** Sweden
18 C4 **Hoorn** Neth.
91 G3 **Hoosick** U.S.A.
93 E3 **Hoover Dam** *dam* U.S.A.
90 B4 **Hoover Memorial Reservoir** U.S.A.
37 H1 **Hopa** Turkey
91 F4 **Hop Bottom** U.S.A.
76 E5 **Hope** *B.C.* Can.
70 D5 **Hope** *r.* N.Z.
85 E5 **Hope** *AR* U.S.A.
93 F5 **Hope** *AZ* U.S.A.
79 J2 **Hopedale** Can.
112 C6 **Hopefield** S. Africa
69 E6 **Hope, L.** *salt flat S.A.* Austr.
66 C2 **Hope, L.** *salt flat W.A.* Austr.
97 G4 **Hopelchén** Mex.
79 H3 **Hope Mountains** Can.
32 D2 **Hopen** *i.* Svalbard
74 B3 **Hope, Point** *c.* U.S.A.
70 D4 **Hope Saddle** *pass* N.Z.
79 G2 **Hopes Advance, Baie** *b.* Can.
66 E5 **Hopetoun** Austr.
69 E6 **Hopetoun** Austr.
112 F4 **Hopetown** S. Africa
90 E6 **Hopewell** U.S.A.
78 E2 **Hopewell Islands** Can.
64 E4 **Hopkins, L.** *salt flat* Austr.
86 C4 **Hopkinsville** U.S.A.
92 A2 **Hopland** U.S.A.
82 B2 **Hoquiam** U.S.A.
50 A3 **Hor** China
37 L2 **Horadiz** Azer.
37 J1 **Horasan** Turkey
21 G3 **Horažďovice** Czech Rep.
20 B4 **Horb am Neckar** Ger.
9 N9 **Hörby** Sweden
96 C1 **Horcasitas** Mex.
88 C4 **Horeb, Mount** U.S.A.
50 B1 **Hörh Uul** *mts* Mongolia
88 C4 **Horicon** U.S.A.
50 D1 **Horinger** China
118 H6 **Horizon Depth** *depth* Pac. Oc.
28 D4 **Horki** Belarus
116 B4 **Horlick Mts** Ant.
29 F5 **Horlivka** Ukr.
41 F4 **Hormak** Iran
40 E5 **Hormoz** *i.* Iran
40 E5 **Hormuz, Strait of** Iran/Oman
21 J4 **Horn** Austria
76 F2 **Horn** *r.* Can.
8 B3 **Horn** *c.* Iceland
8 P3 **Hornavan** *l.* Sweden
85 E6 **Hornbeck** U.S.A.
19 H4 **Hornburg** Ger.
102 C9 **Horn, Cape** Chile
13 G4 **Horncastle** U.K.
9 P6 **Horndal** Sweden
19 G3 **Horneburg** Ger.
8 Q5 **Hörnefors** Sweden
90 E3 **Hornell** U.S.A.
78 D4 **Hornepayne** Can.
87 B6 **Horn I.** U.S.A.
21 H4 **Horní Planá** Czech Rep.
20 B4 **Hornisgrinde** *mt* Ger.
20 F2 **Horní Slavkov** Czech Rep.
63 J3 **Horn, Îsles de** *is* Wallis and Futuna Is
112 B1 **Hornkranz** Namibia
105 B4 **Hornopiren, V.** *volc.* Chile
96 D2 **Hornos** Mex.
Hornos, Cabo de *c. see* **Horn, Cape**
15 B3 **Hornoy-le-Bourg** France
67 J4 **Hornsby** Austr.
12 G4 **Hornsea** U.K.
9 P6 **Hornslandet** *pen.* Sweden
18 F2 **Hörnum** Ger.
17 M6 **Horodenka** Ukr.
29 D5 **Horodnya** Ukr.
29 C5 **Horodok** *Khmel'nyts'kyy* Ukr.
29 B5 **Horodok** *L'viv* Ukr.
52 H2 **Horokanai** Japan
17 M5 **Horokhiv** Ukr.
52 H3 **Horoshiri-dake** *mt* Japan
21 G3 **Hořovice** Czech Rep.
54 A2 **Horqin Shadi** *reg.* China
46 E2 **Horqin Youyi Qianqi** China
54 A1 **Horqin Youyi Zhongqi** China
54 B2 **Horqin Zuoyi Houqi** China
54 B1 **Horqin Zuoyi Zhongqi** China
13 C7 **Horrabridge** U.K.
69 B4 **Horrocks** Austr.
45 G3 **Horru** China
76 E4 **Horsefly** Can.
90 E3 **Horseheads** U.S.A.
79 J3 **Horse Is** Can.
14 C4 **Horseleap** Rep. of Ireland
9 L9 **Horsens** Denmark
82 C3 **Horseshoe Bend** U.S.A.
66 E6 **Horsham** Austr.
13 G6 **Horsham** U.K.
20 F3 **Horšovský Týn** Czech Rep.
20 C2 **Horst** *h.* Ger.
18 E4 **Hörstel** Ger.
9 M7 **Horten** Norway
74 F3 **Horton** *r.* Can.
89 F1 **Horwood Lake** Can.
17 N5 **Horyn'** *r.* Ukr.
45 H2 **Ho Sai Hu** *l.* China
110 D3 **Hosa'ina** Eth.
43 B3 **Hosdurga** India
37 L4 **Hoseynābād** Iran
40 C4 **Ḩoseynīyeh** Iran
41 F5 **Hoshab** Pak.
44 D5 **Hoshangabad** India
44 C3 **Hoshiarpur** India
43 B3 **Hospet** India
14 C5 **Hospital** Rep. of Ireland
105 F1 **Hospital, Cuchilla del** *h.* Uru.
114 F3 **Hosséré Vokre** *mt* Cameroon
102 C9 **Hoste, I.** Chile
21 K4 **Hostěradice** Czech Rep.
21 J2 **Hostinné** Czech Rep.
8 O5 **Hotagen** *l.* Sweden
39 K5 **Hotan** China
39 K5 **Hotan He** *watercourse* China
112 E3 **Hotazel** S. Africa
93 G4 **Hotevilla** U.S.A.
69 C6 **Hotham** *r.* Austr.
67 G6 **Hotham, Mt** Austr.
8 P4 **Hoting** Sweden
85 E5 **Hot Springs** *AR* U.S.A.
84 C3 **Hot Springs** *SD* U.S.A.
76 F1 **Hottah Lake** Can.
95 K5 **Hotte, Massif de la** *mts* Haiti
71 □9 **Houaïlu** *New Caledonia* Pac. Oc.
15 B4 **Houdan** France
15 F4 **Houdelaincourt** France
15 F2 **Houffalize** Belgium
58 □ **Hougang** Sing.
68 B2 **Houghton** *r.* Austr.
88 C2 **Houghton** U.S.A.
88 E3 **Houghton Lake** U.S.A.
88 E3 **Houghton Lake** U.S.A.
12 F3 **Houghton le Spring** U.K.
91 K1 **Houlton** U.S.A.
50 D3 **Houma** China
71 □11 **Houma** Tonga
71 □11 **Houma** Tonga
85 F6 **Houma** U.S.A.
71 □11 **Houma Toloa** *pt* Tonga
114 C2 **Houndé** Burkina
15 B2 **Hourdel, Pte du** *pt* France
11 C3 **Hourn, Loch** *in.* U.K.
91 G3 **Housatonic** U.S.A.
93 F2 **House Range** *mts* U.S.A.
76 D4 **Houston** Can.
85 F4 **Houston** *MO* U.S.A.
85 F5 **Houston** *MS* U.S.A.
85 E6 **Houston** *TX* U.S.A.
113 H1 **Hout** *r.* S. Africa
69 A4 **Houtman Abrolhos** *is* Austr.
11 E2 **Houton** U.K.
112 E5 **Houwater** S. Africa
48 F2 **Hovd** Mongolia
13 G7 **Hove** U.K.
13 J5 **Hoveton** U.K.
40 C4 **Hoveyzeh** Iran
9 O8 **Hovmantorp** Sweden
50 C1 **Hövsgöl** Mongolia
48 H1 **Hövsgöl Nuur** *l.* Mongolia
48 H3 **Hövüün** Mongolia
68 E5 **Howard** Austr.
88 E4 **Howard City** U.S.A.
77 H2 **Howard Lake** Can.
12 G4 **Howden** U.K.
67 H6 **Howe, C.** *hd* Austr.
89 F4 **Howell** U.S.A.
84 C2 **Howes** U.S.A.
91 G2 **Howick** Can.
113 J4 **Howick** S. Africa
66 C1 **Howitt, L.** *salt flat* Austr.
67 G6 **Howitt, Mt** Austr.
91 J2 **Howland** U.S.A.
63 J1 **Howland Island** Pac. Oc.
67 G5 **Howlong** Austr.
14 E4 **Howth** Rep. of Ireland
40 D3 **Ḩowz-e Dūmatu** Iran
40 E4 **Ḩowz-e Panj** Iran
19 G5 **Höxter** Ger.
11 E2 **Hoy** *i.* U.K.
19 G4 **Hoya** Ger.
9 K6 **Hoyanger** Norway
19 M5 **Hoyerswerda** Ger.
8 N4 **Høylandet** Norway
19 J5 **Hoym** Ger.
8 V5 **Höytiäinen** *l.* Fin.
37 G2 **Hozat** Turkey
21 J2 **Hradec Králové** Czech Rep.
21 L4 **Hradište p. Vrátnom** Slovakia
27 H3 **Hrasnica** Bos.-Herz.
37 K1 **Hrazdan** Armenia
29 E5 **Hrebinka** Ukr.
28 B4 **Hrodna** Belarus
21 M4 **Hron** *r.* Slovakia
21 K3 **Hrotovice** Czech Rep.
21 K4 **Hrušovany nad Jevišovkou** Czech Rep.
51 F6 **Hsi-hsu-p'ing Hsü** *i.* Taiwan
51 F5 **Hsin-chu** Taiwan
56 B2 **Hsipaw** Myanmar
51 F5 **Hsueh Shan** *mt* Taiwan
51 E5 **Hua'an** China
103 D4 **Huachamacari, Cerro** *mt* Venez.
50 C2 **Huachi** China
100 C6 **Huacho** Peru
52 B1 **Huachuan** China
93 G6 **Huachuca City** U.S.A.
105 C1 **Huaco** Arg.
50 D1 **Huade** China
54 D2 **Huadian** China
71 □2 **Huahine** *i. Fr. Polynesia* Pac. Oc.
50 E1 **Huai'an** *Hebei* China
50 F3 **Huai'an** *Jiangsu* China
50 E3 **Huaibei** China
50 E3 **Huaibin** China
54 C2 **Huaide** China
54 C2 **Huaidezhen** China
50 F3 **Huai He** *r.* China
51 C5 **Huaihua** China
51 D6 **Huaiji** China
50 E1 **Huailai** China
51 B7 **Huai Luang** *r.* Thai.
50 E3 **Huainan** China
51 E4 **Huaining** China
50 D2 **Huairen** China
50 E3 **Huaiyang** China
50 F3 **Huaiyin** China
50 E3 **Huaiyuan** *Anhui* China
51 C5 **Huaiyuan** *Guangxi* China
50 B3 **Huajialing** China
97 E4 **Huajuápan de León** Mex.
57 H8 **Huaki** Indon.
93 F4 **Hualapai Peak** U.S.A.
51 F5 **Hua-lien** Taiwan
100 C5 **Huallaga** *r.* Peru
50 B2 **Hualong** China
111 B5 **Huambo** Angola
52 B1 **Huanan** China
105 C4 **Huancache, Sa** *mts* Arg.
100 C6 **Huancayo** Peru
50 E2 **Huangbizhuang Sk.** *resr* China
50 A2 **Huangcheng** China
50 E3 **Huangchuan** China
51 E4 **Huanggang** China
Huang Hai *sea see* **Yellow Sea**
50 E2 **Huang He** *r.* China
50 F2 **Huanghe Kou** *est.* China
45 H2 **Huanghetan** China
50 E2 **Huanghua** China
50 C3 **Huangling** China
51 C7 **Huangliu** China
51 E4 **Huangmei** China
54 D2 **Huangnihe** China
51 E4 **Huangpi** China
51 C5 **Huangping** China
50 D1 **Huangqi Hai** *l.* China
51 F4 **Huangshan** China
51 F4 **Huang Shan** *mt* China
51 E4 **Huangshi** China
50 B2 **Huang Shui** *r.* China
50 C2 **Huangtu Gaoyuan** *plat.* China
50 F2 **Huang Xian** China
51 F4 **Huangyan** China
50 A2 **Huangyuan** China
51 C5 **Huanjiang** China
50 C2 **Huan Jiang** *r.* China
54 C3 **Huanren** China
50 F2 **Huantai** China
100 C5 **Huanuco** Peru
100 E7 **Huanuni** Bol.
50 C2 **Huan Xian** China
51 G5 **Hua-p'ing Hsü** *i.* Taiwan
100 C5 **Huaráz** Peru
100 C6 **Huarmey** Peru
51 D4 **Huarong** China
100 C5 **Huascaran, Nevado de** *mt* Peru
102 B3 **Huasco** Chile
102 B3 **Huasco** *r.* Chile
54 D2 **Huashulinzi** China
96 B2 **Huatabampo** Mex.
50 C3 **Huating** China
54 A3 **Huatong** China
97 E4 **Huatusco** Mex.
97 E3 **Huauchinango** Mex.
97 E4 **Huautla** Mex.
51 D6 **Hua Xian** *Guangdong* China
50 E3 **Hua Xian** *Henan* China
50 D4 **Huayuan** *Hubei* China
51 C4 **Huayuan** *Hunan* China
51 C4 **Huayun** China
51 D6 **Huazhou** China
89 F3 **Hubbard Lake** U.S.A.
76 B2 **Hubbard, Mt** Can./U.S.A.
79 G2 **Hubbard, Pointe** *hd* Can.
50 D4 **Hubei** *div.* China
43 A3 **Hubli** India
54 D3 **Huch'ang** N. Korea
18 D5 **Hückelhoven** Ger.
13 F4 **Hucknall** U.K.
15 B2 **Hucqueliers** France
12 F4 **Huddersfield** U.K.
90 B6 **Huddy** U.S.A.
9 P6 **Hudiksvall** Sweden
88 E5 **Hudson** *MI* U.S.A.
91 G3 **Hudson** *NY* U.S.A.
88 A3 **Hudson** *WV* U.S.A.
86 F3 **Hudson** *r.* U.S.A.
77 J4 **Hudson Bay** *Sask.* Can.
75 K4 **Hudson Bay** Can.
91 G3 **Hudson Falls** U.S.A.
75 Q2 **Hudson Land** *reg.* Greenland
116 A3 **Hudson Mts** Ant.
76 E3 **Hudson's Hope** Can.
75 L3 **Hudson Strait** Can.
58 C1 **Huê** Vietnam
105 A4 **Huechucuicui, Pta** *pt* Chile
97 G5 **Huehuetenango** Guatemala
96 C2 **Huehueto, Cerro** *mt* Mex.
97 E4 **Huejotzingo** Mex.
97 E3 **Huejutla** Mex.
25 C4 **Huelva** Spain
105 B1 **Huentelauquén** Chile
105 B4 **Huequi, Volcán** *volc.* Chile
25 F4 **Huércal-Overa** Spain
25 F1 **Huesca** Spain
25 E4 **Huéscar** Spain
96 D4 **Huétamo** Mex.
68 A3 **Hughenden** Austr.
90 E4 **Hughesville** U.S.A.
45 F5 **Hugli** *est.* India
45 G5 **Hugli-Chunchura** India
85 E5 **Hugo** U.S.A.
85 C4 **Hugoton** U.S.A.
50 D2 **Huguan** China
112 F3 **Huhudi** S. Africa
51 F5 **Hui'an** China
50 C2 **Hui'anbu** China
70 F3 **Huiarau Range** *mts* N.Z.
112 B3 **Huib-Hoch Plateau** *plat.* Namibia
51 E5 **Huichang** China
54 D3 **Huich'ŏn** N. Korea
51 E6 **Huidong** *Guangdong* China
51 B5 **Huidong** *Sichuan* China
54 D2 **Huifa** *r.* China
18 B5 **Huijbergen** Neth.
50 E3 **Huiji** *r.* China
51 E6 **Huilai** China
103 B4 **Huila, Nevado de** *mt* Col.
51 B5 **Huili** China
97 F4 **Huimanguillo** Mex.
50 E2 **Huimin** China
102 C2 **Huinahuaca** Arg.
54 D2 **Huinan** China
105 D2 **Huinca Renancó** Arg.
50 B3 **Huining** China
51 C5 **Huishui** China
45 G2 **Huiten Nur** *l.* China
51 C5 **Huitong** China
9 S6 **Huittinen** Fin.
97 E4 **Huitzuco** Mex.
50 C3 **Hui Xian** *Gansu* China
50 D3 **Hui Xian** *Henan* China
97 F5 **Huixtla** Mex.
51 B5 **Huize** China
51 E6 **Huizhou** China
48 H2 **Hujirt** Mongolia
51 E4 **Hukou** China
112 D1 **Hukuntsi** Botswana
88 E2 **Hulbert Lake** U.S.A.
40 B3 **Hulilan** Iran
52 C2 **Hulin** China
21 L3 **Hulín** Czech Rep.
89 K3 **Hull** Can.
9 O8 **Hultsfred** Sweden
54 A3 **Huludao** China
46 D2 **Hulun Nur** *l.* China
29 F6 **Hulyaypole** Ukr.
46 E1 **Huma** China
100 F5 **Humaitá** Brazil
112 F7 **Humansdorp** S. Africa
40 B5 **Humayyān, J.** *h.* S. Arabia
12 H4 **Humber, Mouth of the** *est.* U.K.
77 H4 **Humboldt** Can.
82 C3 **Humboldt** *r.* U.S.A.
82 A3 **Humboldt Bay** U.S.A.
92 C1 **Humboldt Lake** U.S.A.
71 □9 **Humboldt, Mt** *New Caledonia* Pac. Oc.
92 C1 **Humboldt Range** *mts* U.S.A.
92 D2 **Humbolt Salt Marsh** U.S.A.
67 F1 **Humeburn** Austr.
41 E5 **Hūmedān** Iran
51 D6 **Hu Men** *chan.* China
21 L3 **Humenec** *h.* Czech Rep.
17 K6 **Humenné** Slovakia
67 G6 **Hume Reservoir** Austr.
92 C3 **Humphreys, Mt** U.S.A.
93 G4 **Humphreys Peak** U.S.A.
21 J3 **Humpolec** Czech Rep.
54 B3 **Hun** *r.* China
8 C4 **Húnaflói** *b.* Iceland
51 D5 **Hunan** *div.* China
54 F2 **Hunchun** China
54 F2 **Hunchun** *r.* China
19 K5 **Hundeluft** Ger.
9 M9 **Hundested** Denmark
27 K2 **Hunedoara** Romania
20 C2 **Hünfeld** Ger.
71 □10 **Hunga** *i.* Tonga
71 □12 **Hunga Ha'apai** *i.* Tonga
7 E4 **Hungary** *country* Europe
71 □12 **Hunga Tonga** *i.* Tonga
66 F2 **Hungerford** Austr.
54 D4 **Hŭngnam** N. Korea
82 D1 **Hungry Horse Resr** U.S.A.
51 □ **Hung Shui Kiu** *H.K.* China
51 C6 **Hưng Yên** Vietnam
54 D3 **Hunjiang** China
54 C3 **Hun Jiang** *r.* China
112 B3 **Huns Mountains** Namibia
15 H3 **Hunsrück** *reg.* Ger.
13 H5 **Hunstanton** U.K.
43 B3 **Hunsur** India
93 H4 **Hunt** U.S.A.
18 F3 **Hunte** *r.* Ger.
67 J4 **Hunter** *r.* Austr.
91 F3 **Hunter** U.S.A.
67 F8 **Hunter I.** *Tas.* Austr.
76 D4 **Hunter I.** Can.
63 H4 **Hunter I.** New Caledonia
67 F8 **Hunter Is** Austr.
45 H6 **Hunter's Bay** Myanmar
70 C6 **Hunters Hills, The** N.Z.
91 F2 **Huntingdon** Can.
13 G5 **Huntingdon** U.K.
90 E4 **Huntingdon** U.S.A.
88 E5 **Huntington** *IN* U.S.A.
93 G2 **Huntington** *UT* U.S.A.
90 B5 **Huntington** *WV* U.S.A.
92 D5 **Huntington Beach** U.S.A.
70 E2 **Huntly** N.Z.
11 F3 **Huntly** U.K.
66 B2 **Hunt Pen.** *ridge* Austr.
89 H3 **Huntsville** Can.
87 C5 **Huntsville** *AL* U.S.A.
85 E6 **Huntsville** *TX* U.S.A.
50 D2 **Hunyuan** China
44 C1 **Hunza** Pak.
44 C2 **Hunza** *r.* Pak.
39 K3 **Huocheng** China
54 B1 **Huolin** *r.* China
50 E2 **Huolu** China
71 □9 **Huon** *i. New Caledonia* Pac. Oc.
58 C1 **Hương Khê** Vietnam
58 C1 **Hương Thuy** Vietnam
62 E2 **Huon Peninsula** P.N.G.
67 G9 **Huonville** Austr.
50 E3 **Huoqiu** China
50 E4 **Huoshan** China
50 E4 **Huo Shan** *mt* China
51 F6 **Huo-shao Tao** *i.* Taiwan
50 D2 **Huo Xian** China
40 E4 **Ḩūr** Iran
21 M5 **Hurbanovo** Slovakia
89 G3 **Hurd, Cape** *hd* Can.
50 C1 **Hure Jadgai** China
54 A2 **Hure Qi** China
109 F2 **Hurghada** Egypt
88 C1 **Hurkett** Can.
14 C5 **Hurler's Cross** Rep. of Ireland
88 B2 **Hurley** U.S.A.
84 D2 **Huron** U.S.A.
88 C2 **Huron Bay** U.S.A.
89 F3 **Huron, Lake** Can./U.S.A.
88 D2 **Huron Mts** *h.* U.S.A.
93 F3 **Hurricane** U.S.A.
13 F6 **Hursley** U.K.
13 H6 **Hurst Green** U.K.
70 D5 **Hurunui** *r.* N.Z.
8 E3 **Húsavík** *Norðurland eystra* Iceland
8 C4 **Húsavík** *Vestfirðir* Iceland
19 G2 **Husby** Ger.
17 O7 **Huși** Romania
21 G3 **Husinec** Czech Rep.
9 O8 **Huskvarna** Sweden
74 C3 **Huslia** U.S.A.
9 J7 **Husnes** Norway
45 F4 **Hussainabad** India
21 K4 **Hustopeče** Czech Rep.
19 G2 **Husum** Ger.
8 Q5 **Husum** Sweden
48 H2 **Hutag** Mongolia
84 D4 **Hutchinson** U.S.A.
93 G4 **Hutch Mtn** U.S.A.
58 A1 **Huthi** Myanmar
52 C2 **Hutou** China
77 N2 **Hut Point** Can.
69 B4 **Hutt** *r.* Austr.
21 H6 **Hüttenberg** Austria
21 G4 **Hutthurm** Ger.
68 C5 **Hutton, Mt** *h.* Austr.
90 D5 **Huttonsville** U.S.A.
50 D2 **Hutuo** *r.* China
36 G3 **Hüvek** Turkey
50 C3 **Hu Xian** China
50 F4 **Huzhou** China
50 A2 **Huzhu** China
8 E4 **Hvannadalshnúkur** *mt* Iceland
26 G3 **Hvar** *i.* Croatia
29 E6 **Hvardiys'ke** Ukr.
8 C4 **Hveragerði** Iceland
9 L8 **Hvide Sande** Denmark
8 C4 **Hvíta** *r.* Iceland
54 E3 **Hwadae** N. Korea
111 C5 **Hwange** Zimbabwe
111 C5 **Hwange National Park** Zimbabwe
54 C4 **Hwangju** N. Korea
111 D5 **Hwedza** Zimbabwe
91 H4 **Hyannis** *MA* U.S.A.
84 C3 **Hyannis** *NE* U.S.A.
48 F2 **Hyargas Nuur** *salt l.* Mongolia
76 C3 **Hydaburg** U.S.A.
70 C6 **Hyde** N.Z.
69 D6 **Hyden** Austr.
90 B6 **Hyden** U.S.A.
91 G4 **Hyde Park** U.S.A.
93 F5 **Hyder** U.S.A.
43 B2 **Hyderabad** India
44 B4 **Hyderabad** Pak.
22 D6 **Hyères** France
22 D6 **Hyères, Îles d'** *is* France
54 E3 **Hyesan** N. Korea
76 D2 **Hyland** *r.* Can.
19 J2 **Hyllekrog** *i.* Denmark
9 J6 **Hyllestad** Norway
9 N8 **Hyltebruk** Sweden
66 D6 **Hynam** Austr.
67 K3 **Hyndland, Mt** Austr.
53 D7 **Hyōnosen** *mt* Japan
8 V4 **Hyrynsalmi** Fin.
76 F3 **Hythe** Can.
13 J6 **Hythe** U.K.
53 B8 **Hyūga** Japan
9 T6 **Hyvinkää** Fin.

I

100 E6 **Iaco** *r.* Brazil
101 K6 **Iaçu** Brazil
111 E6 **Iakora** Madag.
27 M2 **Ialomița** *r.* Romania
27 M2 **Ianca** Romania
17 N7 **Iași** Romania
55 A3 **Iba** Phil.
114 D3 **Ibadan** Nigeria
103 B3 **Ibagué** Col.
93 F1 **Ibapah** U.S.A.
100 C3 **Ibarra** Ecuador
42 B7 **Ibb** Yemen
18 E4 **Ibbenbüren** Ger.
114 E2 **Ibeto** Nigeria
58 A4 **Ibi** Indon.
114 E3 **Ibi** Nigeria
104 C2 **Ibiá** Brazil
101 K4 **Ibiapaba, Serra da** *h.* Brazil
105 F1 **Ibicuí da Cruz** *r.* Brazil
104 E2 **Ibiraçu** Brazil
25 G3 **Ibiza** *i. Balearic Is* Spain
25 G3 **Ibiza** Spain
26 F6 **Iblei, Monti** *mts Sicily* Italy
40 B5 **Ibn Buşayyiş** *well* S. Arabia
101 K6 **Ibotirama** Brazil
42 E5 **Ibrā'** Oman
42 E5 **Ibrī** Oman
55 B1 **Ibuhos** *i.* Phil.
53 B9 **Ibusuki** Japan
100 C6 **Ica** Peru
103 D4 **Içana** Brazil
103 D4 **Içana** *r.* Brazil
93 E3 **Iceberg Canyon** U.S.A.
36 E3 **İçel** Turkey
7 B2 **Iceland** Europe
43 A2 **Ichalkaranji** India
43 D2 **Ichchapuram** India
53 B8 **Ichifusa-yama** *mt* Japan
52 G5 **Ichinoseki** Japan
33 R4 **Ichinskaya Sopka** *mt* Rus. Fed.
29 E5 **Ichnya** Ukr.
54 D4 **Ich'ŏn** N. Korea
54 D5 **Ich'ŏn** S. Korea
15 D1 **Ichtegem** Belgium
19 H6 **Ichtershausen** Ger.
76 B3 **Icy Pt** U.S.A.
76 B3 **Icy Strait** U.S.A.
85 E5 **Idabel** U.S.A.
114 E3 **Idah** Nigeria
82 D2 **Idaho** *div.* U.S.A.
82 D3 **Idaho City** U.S.A.
82 D3 **Idaho Falls** U.S.A.
15 H3 **Idar-Oberstein** Ger.
109 F2 **Idfu** Egypt
108 D2 **Idhān Awbārī** *des.* Libya
109 D2 **Idhān Murzūq** *des.* Libya
23 J5 **Idice** *r.* Italy
110 B4 **Idiofa** Congo(Zaire)
74 C3 **Iditarod** U.S.A.
8 S2 **Idivuoma** Sweden
36 C6 **Idku** Egypt
36 F4 **Idlib** Syria
9 N6 **Idre** Sweden
23 H4 **Idro, Lago d'** *l.* Italy
20 B2 **Idstein** Ger.
113 H6 **Idutywa** S. Africa
9 T8 **Iecava** Latvia
104 B3 **Iepê** Brazil
15 C2 **Ieper** Belgium
27 L7 **Ierapetra** Greece
115 C4 **Ifakara** Tanz.
111 E6 **Ifanadiana** Madag.
114 D3 **Ife** Nigeria
8 U1 **Ifjord** Norway
59 D2 **Igan** Malaysia
115 B2 **Iganga** Uganda
104 C3 **Igarapava** Brazil
32 K3 **Igarka** Rus. Fed.
44 C6 **Igatpuri** India
114 D3 **Igbeti** Nigeria
37 K2 **Iğdır** Turkey
9 P6 **Iggesund** Sweden
26 C5 **Iglesias** *Sardinia* Italy
75 K3 **Igloolik** Can.
78 B4 **Ignace** Can.
9 U9 **Ignalina** Lith.
29 C7 **İğneada** Turkey
27 N4 **İğneada Burnu** *pt* Turkey
22 B2 **Ignon** *r.* France
115 B4 **Igoma** Tanz.
17 Q3 **Igorevskaya** Rus. Fed.
27 J5 **Igoumenitsa** Greece
32 H3 **Igrim** Rus. Fed.
104 B4 **Iguaçu** *r.* Brazil
104 A4 **Iguaçu Falls** *waterfall* Arg./Brazil
104 E1 **Iguaí** Brazil
103 B4 **Iguaje, Mesa de** *h.* Col.
97 E4 **Iguala** Mex.
25 G2 **Igualada** Spain
104 C4 **Iguape** Brazil
104 D3 **Iguarapé** Brazil
104 A3 **Iguatemi** Brazil
104 A3 **Iguatemi** *r.* Brazil
101 L5 **Iguatu** Brazil
Iguazú, Cataratas do *waterfall see* **Iguaçu Falls**
110 A4 **Iguéla** Gabon
115 B3 **Igunga** Tanz.
111 E5 **Iharaña** Madag.
48 J3 **Ihbulag** Mongolia
111 E6 **Ihosy** Madag.
54 B2 **Ih Tal** China
53 F6 **Iide-san** *mt* Japan
8 U2 **Iijärvi** *l.* Fin.
8 T4 **Iijoki** *r.* Fin.
8 U5 **Iisalmi** Fin.
53 B8 **Iizuka** Japan
114 D3 **Ijebu-Ode** Nigeria
37 K1 **Ijevan** Armenia
18 B4 **IJmuiden** Neth.
18 C4 **IJssel** *r.* Neth.
18 C4 **IJsselmeer** *l.* Neth.
18 A5 **IJzendijke** Neth.
9 S6 **Ikaalinen** Fin.
113 G2 **Ikageleng** S. Africa
113 G3 **Ikageng** S. Africa
27 M6 **Ikaria** *i.* Greece
9 L8 **Ikast** Denmark
52 H3 **Ikeda** Japan
110 C4 **Ikela** Congo(Zaire)
27 K3 **Ikhtiman** Bulg.
112 F4 **Ikhutseng** S. Africa
53 A8 **Iki** *i.* Japan
29 H6 **Iki-Burul** Rus. Fed.
114 D3 **Ikire** Nigeria
114 E3 **Ikom** Nigeria
111 E6 **Ikongo** Madag.
29 H6 **Ikryanoye** Rus. Fed.
115 B4 **Ikungu** Tanz.
55 B2 **Ilagan** Phil.
115 C2 **Ilaisamis** Kenya
40 B3 **Īlām** Iran
45 F4 **Ilam** Nepal
23 G3 **Ilanz** Switz.
114 D3 **Ilaro** Nigeria
21 M4 **Ilava** Slovakia
17 J4 **Iława** Pol.
77 H3 **Île-à-la-Crosse** Can.
77 H3 **Île-à-la-Crosse, Lac** *l.* Can.
110 C4 **Ilebo** Congo(Zaire)
15 C4 **Île-de-France** *div.* France
38 C2 **Ilek** Kazak.
38 C2 **Ilek** *r.* Rus. Fed.
115 C2 **Ileret** Kenya
28 G2 **Ileza** Rus. Fed.
77 K3 **Ilford** Can.
13 H6 **Ilford** U.K.
68 A4 **Ilfracombe** Austr.
13 C6 **Ilfracombe** U.K.
36 D1 **Ilgaz** Turkey
36 D1 **Ilgaz D.** *mts* Turkey
36 C2 **Ilgın** Turkey
103 D5 **Ilha Grande** Brazil
104 D3 **Ilha Grande, Baía da** *b.* Brazil
104 B3 **Ilha Grande, Represa** *resr* Brazil
104 B3 **Ilha Solteíra, Represa** *resr* Brazil
25 B2 **Ílhavo** Port.
104 E1 **Ilhéus** Brazil
108 □ **Ilhéus Secos ou do Rombo** *i.* Cape Verde
74 C4 **Iliamna Lake** U.S.A.
36 G2 **İliç** Turkey
39 G4 **Il'ich** Kazak.
55 C4 **Iligan** Phil.
55 C4 **Iligan Bay** Phil.
39 K4 **Ili He** *r.* China
39 J3 **Ili He** *r.* Kazak.
38 D2 **Il'inka** Kazak.
28 H2 **Il'insko-Podomskoye** Rus. Fed.
91 F3 **Ilion** U.S.A.
43 B3 **Ilkal** India
13 F5 **Ilkeston** U.K.
12 F4 **Ilkley** U.K.
15 H4 **Ill** *r.* France
55 B5 **Illana Bay** Phil.
105 B1 **Illapel** Chile
105 B1 **Illapel** *r.* Chile
23 J4 **Illasi** *r.* Italy
20 D4 **Iller** *r.* Ger.
20 D4 **Illertissen** Ger.
29 D6 **Illichivs'k** Ukr.
15 B4 **Illiers-Combray** France
100 E7 **Illimani, Nevado de** *mt* Bol.
20 B4 **Illingen** Ger.
88 C5 **Illinois** *div.* U.S.A.
88 C5 **Illinois** *r.* U.S.A.
88 B5 **Illinois and Mississippi Canal** U.S.A.
29 D5 **Illintsi** Ukr.
108 C2 **Illizi** Alg.
20 E4 **Ilm** *r.* Ger.
19 J5 **Ilm** *r.* Ger.
8 S5 **Ilmajoki** Fin.
20 D2 **Ilmenau** Ger.
19 H3 **Ilmenau** *r.* Ger.
28 D3 **Il'men', Ozero** *l.* Rus. Fed.
13 E7 **Ilminster** U.K.
100 D7 **Ilo** Peru
55 A4 **Iloc** *i.* Phil.
55 B4 **Iloilo** Phil.
8 W5 **Ilomantsi** Fin.
114 D3 **Ilorin** Nigeria
29 F6 **Ilovays'k** Ukr.
29 G5 **Ilovlya** Rus. Fed.
29 H5 **Ilovlya** *r.* Rus. Fed.
19 H4 **Ilsede** Ger.
20 C3 **Ilshofen** Ger.
67 K2 **Iluka** Austr.
75 N3 **Ilulissat** Greenland
114 E3 **Ilushi** Nigeria
38 D4 **Il'yaly** Turkm.
21 G4 **Ilz** *r.* Ger.
53 C7 **Imabari** Japan
53 F6 **Imaichi** Japan
115 C5 **Imala** Moz.
37 K6 **Imām al Ḩamzah** Iraq
36 E3 **İmamoğlu** Turkey
37 K5 **Imām Ḩamīd** Iraq
52 D2 **Iman** *r.* Rus. Fed.
53 A8 **Imari** Japan
103 E3 **Imataca, Serranía de** *mts* Venez.
9 V6 **Imatra** Fin.
53 E7 **Imazu** Japan
102 G3 **Imbituba** Brazil
104 B4 **Imbituva** Brazil
28 G3 **imeni Babushkina** Rus. Fed.
41 F2 **imeni Chapayeva** Turkm.
39 H5 **imeni Kalinina** Tajik.
110 E3 **Īmī** Eth.
37 M2 **İmişli** Azer.
54 D6 **Imja-do** *i.* S. Korea
54 D4 **Imjin** *r.* N. Korea
20 D5 **Immenstadt im Allgäu** Ger.
23 J5 **Imola** Italy
113 H4 **Impendle** S. Africa
101 J5 **Imperatriz** Brazil
22 F6 **Imperia** Italy
84 C3 **Imperial** U.S.A.
92 D5 **Imperial Beach** U.S.A.
93 E5 **Imperial Valley** *v.* U.S.A.
110 B3 **Impfondo** Congo
45 H4 **Imphal** India
27 L4 **İmroz** Turkey
20 D5 **Imst** Austria
36 F5 **Imtān** Syria
55 A4 **Imuruan Bay** Phil.
53 E7 **Ina** Japan
19 M3 **Ina** *r.* Pol.
100 E6 **Inambari** *r.* Peru
108 C2 **In Aménas** Alg.
70 C4 **Inangahua Junction** N.Z.
57 J7 **Inanwatan** Indon.
8 U2 **Inari** Fin.
8 U2 **Inarijärvi** *l.* Fin.
8 T2 **Inarijoki** *r.* Fin./Norway
25 H3 **Inca** Spain
29 C7 **İnce Burnu** *pt* Turkey
29 E7 **İnce Burnu** *pt* Turkey
36 D3 **İncekum Burnu** *pt* Turkey
36 E2 **İncesu** Turkey
14 E5 **Inch** Rep. of Ireland
11 C2 **Inchard, Loch** *b.* U.K.
15 B2 **Incheville** France
11 E4 **Inchkeith** *i.* U.K.
54 D5 **Inch'ŏn** S. Korea
113 K2 **Incomati** *r.* Moz.
11 B5 **Indaal, Loch** *in.* U.K.
104 D2 **Indaiá** *r.* Brazil
104 B2 **Indaiá Grande** *r.* Brazil
8 P5 **Indalsälven** *r.* Sweden
9 J6 **Indalstø** Norway
96 C2 **Indé** Mex.
92 C3 **Independence** *CA* U.S.A.
88 B4 **Independence** *IA* U.S.A.
85 E4 **Independence** *KS* U.S.A.
88 A2 **Independence** *MN* U.S.A.
84 E4 **Independence** *MO* U.S.A.
90 C6 **Independence** *VA* U.S.A.
88 B3 **Independence** *WV* U.S.A.
82 C3 **Independence Mts** U.S.A.
38 B2 **Inderborskiy** Kazak.
38 B2 **Inder, Oz.** *salt l.* Kazak.
43 B2 **Indi** India
31 G7 **India** *country* Asia
88 D2 **Indian** *r.* U.S.A.
90 D4 **Indiana** U.S.A.
88 D5 **Indiana** *div.* U.S.A.
88 D5 **Indiana Dunes National Lakeshore** *res.* U.S.A.
117 M7 **Indian-Antarctic Basin** *sea feature* Ind. Ocean
117 O7 **Indian-Antarctic Ridge** *sea feature* Pac. Oc.
88 D6 **Indianapolis** U.S.A.
Indian Desert *see* **Thar Desert**
79 J3 **Indian Harbour** Can.

J

9 U7 **Jõgeva** Estonia
9 U7 **Jõgua** Estonia
113 G3 **Johannesburg** S. Africa
92 D4 **Johannesburg** U.S.A.
44 E5 **Johilla** *r.* India
82 C2 **John Day** U.S.A.
82 B2 **John Day** *r.* U.S.A.
76 F3 **John d'Or Prairie** Can.
90 D6 **John H. Kerr Resr** U.S.A.
11 E2 **John o'Groats** U.K.
87 D4 **Johnson City** U.S.A.
76 C2 **Johnson's Crossing** Can.
87 D5 **Johnston** U.S.A.
11 D5 **Johnstone** U.K.
60 L2 **Johnston I.** Pac. Oc.
69 E6 **Johnston, L.** *salt flat* Austr.
69 D4 **Johnston Range** *h.* Austr.
14 D5 **Johnstown** Rep. of Ireland
91 F3 **Johnstown** *NY* U.S.A.
90 D4 **Johnstown** *PA* U.S.A.
89 F3 **Johnswood** U.S.A.
59 B2 **Johor Bahru** Malaysia
9 U7 **Jõhvi** Estonia
102 G3 **Joinville** Brazil
15 F4 **Joinville** France
116 B2 **Joinville I.** Ant.
8 Q3 **Jokkmokk** Sweden
8 F4 **Jökulsá á Brú** *r.* Iceland
8 E3 **Jökulsá á Fjöllum** *r.* Iceland
8 F4 **Jökulsá í Fljótsdal** *r.* Iceland
40 B2 **Jolfa** Iran
88 C5 **Joliet** U.S.A.
78 F4 **Joliette** Can.
55 B5 **Jolo** Phil.
55 B5 **Jolo** *i.* Phil.
55 B3 **Jomalig** *i.* Phil.
59 D4 **Jombang** Indon.
48 G5 **Jomda** China
97 G4 **Jonathán Pt** Belize
9 T9 **Jonava** Lith.
50 B3 **Jonê** China
85 F5 **Jonesboro** *AR* U.S.A.
91 K2 **Jonesboro** *ME* U.S.A.
116 A3 **Jones Mts** Ant.
91 K2 **Jonesport** U.S.A.
75 K2 **Jones Sound** *chan.* Can.
90 B6 **Jonesville** U.S.A.
109 F4 **Jonglei Canal** Sudan
45 E5 **Jonk** *r.* India
9 O8 **Jönköping** Sweden
79 F4 **Jonquière** Can.
97 F4 **Jonuta** Mex.
85 E4 **Joplin** U.S.A.
91 E5 **Joppatowne** U.S.A.
44 D4 **Jora** India
31 C6 **Jordan** *country* Asia
36 E6 **Jordan** *r.* Asia
82 F2 **Jordan** U.S.A.
82 E3 **Jordan** *r.* U.S.A.
68 B4 **Jordan Cr.** *watercourse* Austr.
82 C3 **Jordan Valley** U.S.A.
104 B4 **Jordão** *r.* Brazil
9 N6 **Jordet** Norway
19 H1 **Jordløse** Denmark
45 H4 **Jorhat** India
39 J5 **Jor Hu** *l.* China
19 G3 **Jork** Ger.
8 R4 **Jörn** Sweden
9 U5 **Joroinen** Fin.
9 K7 **Jørpeland** Norway
114 E3 **Jos** Nigeria
55 C5 **Jose Abad Santos** Phil.
97 E4 **José Cardel** Mex.
102 B6 **José de San Martin** Arg.
104 A2 **Joselândia** Brazil
105 F2 **José Pedro Varela** Uru.
64 E2 **Joseph Bonaparte Gulf** *g.* Austr.
93 G4 **Joseph City** U.S.A.
79 G3 **Joseph, Lac** *l.* Can.
75 M4 **Joseph, Lake** Can.
53 F6 **Jōshinetsu-kōgen National Park** Japan
93 E5 **Joshua Tree National Monument** *res.* U.S.A.
114 E3 **Jos Plateau** *plat.* Nigeria
9 K6 **Jostedalsbreen Nasjonalpark** *nat. park* Norway
9 L6 **Jotunheimen Nasjonalpark** *nat. park* Norway
112 E6 **Joubertina** S. Africa
113 G3 **Jouberton** S. Africa
22 C6 **Jouques** France
18 C4 **Joure** Neth.
9 U6 **Joutsa** Fin.
9 V6 **Joutseno** Fin.
22 D3 **Joux, Lac de** *l.* Switz.
15 B4 **Jouy** France
15 G3 **Jouy-aux-Arches** France
45 H4 **Jowai** India
14 B4 **Joyce's Country** *reg.* Rep. of Ireland
96 D2 **Juan Aldama** Mex.
82 A1 **Juan de Fuca, Str. of** U.S.A.
111 E5 **Juan de Nova** *i.* Ind. Ocean
98 B6 **Juan Fernández, Islas** *is* Chile
8 V5 **Juankoski** Fin.
96 H6 **Juan Santamaria** *airport* Costa Rica
96 D2 **Juaréz** Mex.
101 K5 **Juàzeiro** Brazil
101 L5 **Juàzeiro do Norte** Brazil
114 B3 **Juazohn** Liberia
109 F4 **Juba** Sudan
110 E3 **Jubba** *r.* Somalia
19 G2 **Jübek** Ger.
92 D4 **Jubilee Pass** U.S.A.
25 F3 **Júcar** *r.* Spain
97 E4 **Juchatengo** Mex.
96 D3 **Juchipila** Mex.
97 F4 **Juchitán** Mex.
96 C3 **Juchitlán** Mex.
104 E2 **Jucururu** *r.* Brazil
9 J7 **Judaberg** Norway
37 H6 **Judaidat al Hamir** Iraq
37 H6 **Judayyidat 'Ar'ar** *well* Iraq
21 H5 **Judenburg** Austria
9 M9 **Juelsminde** Denmark
50 C2 **Juh** China
50 F1 **Juhua Dao** *i.* China
96 H5 **Juigalpa** Nic.
18 E3 **Juist** *i.* Ger.
104 D3 **Juiz de Fora** Brazil
100 E8 **Julaca** Bol.
84 C3 **Julesburg** U.S.A.
100 D7 **Juliaca** Peru
65 H4 **Julia Creek** Austr.
18 B4 **Julianadorp** Neth.
101 G3 **Juliana Top** *summit* Suriname
18 D6 **Jülich** Ger.
26 E1 **Julijske Alpe** *mts* Slovenia
105 E2 **Julio, 9 de** Arg.
100 C5 **Jumbilla** Peru
25 F3 **Jumilla** Spain
45 E3 **Jumla** Nepal
44 B5 **Junagadh** India
45 E6 **Junagarh** India
50 F3 **Junan** China
105 B2 **Juncal** *mt* Chile
105 D4 **Juncal, L.** Arg.
85 D6 **Junction** *TX* U.S.A.
83 D4 **Junction** *UT* U.S.A.
84 D4 **Junction City** U.S.A.
104 C3 **Jundiaí** Brazil
76 C3 **Juneau** U.S.A.
67 G5 **Junee** Austr.
22 E3 **Jungfrau** *mt* Switz.
46 A2 **Junggar Pendi** *basin* China
44 A4 **Jungshahi** Pak.
90 E4 **Juniata** *r.* U.S.A.
105 E2 **Junín** Arg.
105 B3 **Junín de los Andes** Arg.
91 K1 **Juniper** Can.
92 B3 **Junipero Serro Peak** U.S.A.
51 B4 **Junlian** China
43 A2 **Junnar** India
8 P5 **Junsele** Sweden
82 C3 **Juntura** U.S.A.
50 D3 **Jun Xian** China
9 T8 **Juodupė** Lith.
104 C4 **Juquiá** Brazil
109 E4 **Jur** *r.* Sudan
22 D3 **Jura** *mts* France/Switz.
11 C4 **Jura** *i.* U.K.
104 E1 **Jurací** Brazil
103 A3 **Juradó** Col.
11 C5 **Jura, Sound of** *chan.* U.K.
9 S9 **Jurbarkas** Lith.
36 E6 **Jurf ed Darāwīsh** Jordan
19 K3 **Jürgenstorf** Ger.
54 A1 **Jurh** China
54 A1 **Jurhe** China
45 G2 **Jurhen Ul Shan** *mts* China
69 B5 **Jurien** Austr.
9 S8 **Jūrmala** Latvia
8 U4 **Jurmu** Fin.
50 F4 **Jurong** China
58 □ **Jurong** Sing.
100 E4 **Juruá** *r.* Brazil
101 G6 **Juruena** *r.* Brazil
8 R5 **Jurva** Fin.
40 E2 **Jūshqān** Iran
22 C2 **Jussey** France
105 D2 **Justo Daract** Arg.
100 E4 **Jutaí** *r.* Brazil
19 L5 **Jüterbog** Ger.
104 A3 **Juti** Brazil
97 G5 **Jutiapa** Guatemala
96 H5 **Juticalpa** Honduras
8 P3 **Jutis** Sweden
19 G1 **Jutland** *pen.* Denmark
8 V5 **Juuka** Fin.
9 U6 **Juva** Fin.
95 H4 **Juventud, Isla de la** *i.* Cuba
41 F4 **Juwain** Afgh.
50 F3 **Ju Xian** China
50 A1 **Juyan** China
50 E3 **Juye** China
41 E3 **Jūymand** Iran
40 D4 **Jūyom** Iran
111 C6 **Jwaneng** Botswana
19 J1 **Jyderup** Denmark
Jylland *pen.* see **Jutland**
39 J4 **Jyrgalang** Kyrg.
9 T5 **Jyväskylä** Fin.

K

44 D2 **K2** *mt* China/Jammu and Kashmir
54 C4 **Ka** *i.* N. Korea
38 D5 **Kaakhka** Turkm.
92 □1 **Kaala** *mt* U.S.A.
71 □9 **Kaala-Gomen** *New Caledonia* Pac. Oc.
115 D3 **Kaambooni** Kenya
9 S6 **Kaarina** Fin.
19 J3 **Kaarßen** Ger.
18 D5 **Kaarst** Ger.
8 V5 **Kaavi** Fin.
57 G8 **Kabaena** *i.* Indon.
38 E5 **Kabakly** Turkm.
108 A4 **Kabala** Sierra Leone
115 A3 **Kabale** Uganda
115 A4 **Kabalo** Congo(Zaire)
115 A3 **Kabambare** Congo(Zaire)
111 C5 **Kabangu** Congo(Zaire)
58 A5 **Kabanjahe** Indon.
38 B1 **Kabanovka** Rus. Fed.
71 □14 **Kabara** *i.* Fiji
29 G7 **Kabardino-Balkarskaya Respublika** *div.* Rus. Fed.
115 A3 **Kabare** Congo(Zaire)
8 R3 **Kåbdalis** Sweden
88 E1 **Kabenung Lake** Can.
78 D4 **Kabinakagami Lake** Can.
110 C4 **Kabinda** Congo(Zaire)
40 B3 **Kabīrkūh** *mts* Iran
44 B3 **Kabirwala** Pak.
110 B3 **Kabo** C.A.R.
111 C5 **Kabompo** Zambia
115 A4 **Kabongo** Congo(Zaire)
41 F3 **Kabūdeh** Iran
41 E2 **Kabūd Gonbad** Iran
40 C3 **Kabūd Rāhang** Iran
55 B2 **Kabugao** Phil.
41 H3 **Kābul** Afgh.
41 H3 **Kabul** *r.* Afgh.
115 A5 **Kabunda** Congo(Zaire)
55 C6 **Kaburuang** *i.* Indon.
115 A5 **Kabwe** Zambia
38 E2 **Kabyrga** *r.* Kazak.
38 F1 **Kabzn'** Kazak.
41 F4 **Kacha Kuh** *mts* Iran/Pak.
29 H5 **Kachalinskaya** Rus. Fed.
44 B5 **Kachchh, Gulf of** *g.* India
39 J1 **Kachiry** Kazak.
44 C1 **Kach Pass** Afgh.
48 J1 **Kachug** Rus. Fed.
37 H1 **Kaçkar Dağı** *mt* Turkey
43 B4 **Kadaiyanallur** India
20 G2 **Kadaň** Czech Rep.
44 A3 **Kadanai** *r.* Afgh./Pak.
58 A2 **Kadan Kyun** *i.* Myanmar
71 □14 **Kadavu** *i.* Fiji
71 □15 **Kadavu Passage** Fiji
114 C3 **Kade** Ghana
37 K5 **Kādhimain** Iraq
44 C5 **Kadi** India
36 B1 **Kadıköy** Turkey
66 B4 **Kadina** Austr.
36 D2 **Kadınhanı** Turkey
114 B2 **Kadiolo** Mali
43 B3 **Kadiri** India
36 F3 **Kadirli** Turkey
43 A4 **Kadmat** *i.* India
84 C3 **Kadoka** U.S.A.
111 C5 **Kadoma** Zimbabwe
109 E3 **Kadugli** Sudan
114 E2 **Kaduna** Nigeria
114 E3 **Kaduna** *r.* Nigeria
45 J3 **Kadusam** *mt* China
28 F3 **Kaduy** Rus. Fed.
43 A2 **Kadwa** *r.* India
28 G3 **Kadyy** Rus. Fed.
32 G3 **Kadzherom** Rus. Fed.
Kadzhi-Say *see* **Kajy-Say**
54 C4 **Kaechon** N. Korea
108 A3 **Kaédi** Maur.
114 F2 **Kaélé** Cameroon
92 □1 **Kaena Pt** U.S.A.
70 D1 **Kaeo** N.Z.
54 D5 **Kaesŏng** N. Korea
36 F6 **Kāf** S. Arabia
111 C4 **Kafakumba** Congo(Zaire)
108 A3 **Kaffrine** Senegal
27 L5 **Kafireas, Akra** *pt* Greece
36 C6 **Kafr el Sheik** Egypt
115 B2 **Kafu** *r.* Uganda
115 A6 **Kafue** Zambia
115 A6 **Kafue** *r.* Zambia
115 A5 **Kafue National Park** Zambia
53 E6 **Kaga** Japan
110 B3 **Kaga Bandoro** C.A.R.
29 G6 **Kagal'nitskaya** Rus. Fed.
38 F5 **Kagan** Uzbek.
89 F3 **Kagawong** Can.
8 R4 **Kåge** Sweden
115 B3 **Kagera** *div.* Tanz.
115 B3 **Kagera, Parc National de la** *nat. park* Rwanda
37 J1 **Kağızman** Turkey
59 A3 **Kagologolo** Indon.
53 B9 **Kagoshima** Japan
40 C2 **Kahak** Iran
92 □1 **Kahaluu** U.S.A.
115 B3 **Kahama** Tanz.
92 □1 **Kahana** U.S.A.
29 D5 **Kaharlyk** Ukr.
59 D3 **Kahayan** *r.* Indon.
110 B4 **Kahemba** Congo(Zaire)
70 A6 **Kaherekoau Mts** N.Z.
19 J6 **Kahla** Ger.
Kahnu *see* **Kahnūj**
41 E5 **Kahnūj** Iran
88 B5 **Kahoka** U.S.A.
92 □2 **Kahoolawe** *i.* U.S.A.
36 F3 **Kahraman Maraş** Turkey
44 B3 **Kahror** Pak.
36 G3 **Kahta** Turkey
92 □1 **Kahuku** U.S.A.
92 □1 **Kahuku Pt** U.S.A.
92 □2 **Kahului** U.S.A.
70 D4 **Kahurangi Point** N.Z.
44 C2 **Kahuta** Pak.
115 A3 **Kahuzi-Biega, Parc National du** *nat. park* Congo(Zaire)
114 D3 **Kaiama** Nigeria
70 D5 **Kaiapoi** N.Z.
93 F3 **Kaibab** U.S.A.
83 D4 **Kaibab Plat.** *plat.* U.S.A.
57 J8 **Kai Besar** *i.* Indon.
93 G3 **Kaibito** U.S.A.
93 G3 **Kaibito Plateau** *plat.* U.S.A.
50 E3 **Kaifeng** *Henan* China
50 E3 **Kaifeng** *Henan* China
51 F4 **Kaihua** China
112 D4 **Kaiingveld** *reg.* S. Africa
50 C4 **Kaijiang** China
57 J8 **Kai Kecil** *i.* Indon.
57 J8 **Kai, Kepulauan** *is* Indon.
70 D5 **Kaikoura** N.Z.
70 D5 **Kaikoura Peninsula** N.Z.
51 □ **Kai Kung Leng** *h.* *H.K.* China
108 A4 **Kailahun** Sierra Leone
Kailas *mt* *see* **Kangrinboqê Feng**
45 G4 **Kailāshahar** India
Kailas Range *mts* *see* **Gangdisê Shan**
51 C5 **Kaili** China
54 A2 **Kailu** China
92 □1 **Kailua** U.S.A.
92 □2 **Kailua Kona** U.S.A.
70 E2 **Kaimai Range** *h.* N.Z.
57 J7 **Kaimana** Indon.
70 E3 **Kaimanawa Mountains** N.Z.
45 H2 **Kaimar** China
44 E4 **Kaimur Range** *h.* India
9 S7 **Käina** Estonia
53 D8 **Kainan** Japan
53 D7 **Kainan** Japan
39 H4 **Kaindy** Kyrg.
114 D2 **Kainji Lake National Park** Nigeria
114 D2 **Kainji Reservoir** Nigeria
70 E2 **Kaipara Harbour** *in.* N.Z.
93 G3 **Kaiparowits Plateau** *plat.* U.S.A.
51 D6 **Kaiping** China
79 J3 **Kaipokok Bay** *in.* Can.
44 D3 **Kairana** India
108 D1 **Kairouan** Tunisia
20 F5 **Kaisergebirge** *mts* Austria
15 H3 **Kaiserslautern** Ger.
54 E2 **Kaishantun** China
70 D1 **Kaitaia** N.Z.
70 B7 **Kaitangata** N.Z.
70 F3 **Kaitawa** N.Z.
44 D3 **Kaithal** India
8 R3 **Kaitum** Sweden
57 H8 **Kaiwatu** Indon.
92 □2 **Kaiwi Channel** U.S.A.
50 C4 **Kai Xian** China
51 C5 **Kaiyang** China
51 B6 **Kaiyuan** *Yunnan* China
54 C2 **Kaiyuan** China
8 U4 **Kajaani** Fin.
65 H4 **Kajabbi** Austr.
41 G3 **Kajaki** Afgh.
58 B5 **Kajang** Malaysia
44 B3 **Kajanpur** Pak.
37 L2 **K'ajaran** Armenia
21 L5 **Kajárpéc** Hungary
41 G3 **Kajrān** Afgh.
37 L3 **Kaju** Iran
39 J4 **Kajy-Say** Kyrg.
78 C4 **Kakabeka Falls** Can.
64 F2 **Kakadu National Park** Austr.
112 D4 **Kakamas** S. Africa
115 B2 **Kakamega** Kenya
70 C6 **Kakanui Mts** N.Z.
Kakar Range *mts* *see* **Toba & Kakar Ranges**
114 A3 **Kakata** Liberia
70 E3 **Kakatahi** N.Z.
45 H4 **Kakching** India
53 C7 **Kake** Japan
76 C3 **Kake** U.S.A.
110 C4 **Kakenge** Congo(Zaire)
19 J4 **Kakerbeck** Ger.
29 E6 **Kakhovka** Ukr.
29 E6 **Kakhovs'ke Vodoskhovyshche** *resr* Ukr.
40 C4 **Kākī** Iran
43 C2 **Kākināda** India
76 F2 **Kakisa** Can.
76 F2 **Kakisa** *r.* Can.
76 F2 **Kakisa Lake** Can.
53 D7 **Kakogawa** Japan
115 A3 **Kakoswa** Congo(Zaire)
114 C3 **Kakpin** Côte d'Ivoire
44 D4 **Kakrala** India
74 D2 **Kaktovik** U.S.A.
53 G6 **Kakuda** Japan
76 F4 **Kakwa** *r.* Can.
44 B3 **Kala** Pak.
115 B4 **Kala** Tanz.
26 D7 **Kalaâ Kebira** Tunisia
44 B2 **Kalabagh** Pak.
57 G8 **Kalabahi** Indon.
55 A5 **Kalabakan** Malaysia
66 D3 **Kalabity** Austr.
111 C5 **Kalabo** Zambia
29 G5 **Kalach** Rus. Fed.
115 C2 **Kalacha Dida** Kenya
29 G5 **Kalach-na-Donu** Rus. Fed.
45 H5 **Kaladan** *r.* India/Myanmar
89 J3 **Kaladar** Can.
92 □2 **Ka Lae** *c.* U.S.A.
106 D8 **Kalahari Desert** Africa
111 B6 **Kalahari Gemsbok National Park** S. Africa
41 F3 **Kala-I-Mor** Turkm.
8 S4 **Kalajoki** Fin.
8 T4 **Kalajoki** *r.* Fin.
114 D2 **Kalalé** Benin
44 C2 **Kalam** Pak.
113 G1 **Kalamare** Botswana
27 K4 **Kalamaria** Greece
27 K6 **Kalamata** Greece
88 E4 **Kalamazoo** U.S.A.
88 D4 **Kalamazoo** *r.* U.S.A.
27 J5 **Kalampaka** Greece
66 C1 **Kalamurra, Lake** *salt flat* Austr.
44 C3 **Kalanaur** India
29 E6 **Kalanchak** Ukr.
41 F4 **Kalandi** Pak.
66 D6 **Kalangadoo** Austr.
115 B3 **Kalangala** Uganda
69 C5 **Kalannie** Austr.
44 C3 **Kalanwali** India
55 C5 **Kalaong** Phil.
43 C4 **Kala Oya** *r.* Sri Lanka
41 F5 **Kalar** *watercourse* Iran
37 K4 **Kalār** Iraq
Kalāt *see* **Kabūd Gonbad**
41 G4 **Kalat** Pak.
71 □11 **Kalau** *i.* Tonga
92 □2 **Kalaupapa** U.S.A.
29 G6 **Kalaus** *r.* Rus. Fed.
37 L1 **Kälbäcär** Azer.
67 K1 **Kalbar** Austr.
69 B3 **Kalbarri** Austr.
69 B3 **Kalbarri Nat. Park** Austr.
19 J4 **Kalbe (Milde)** Ger.
39 K2 **Kalbinskiy Khr.** *mts* Kazak.
41 E3 **Kalbū** Iran
38 C2 **Kaldygayty** *r.* Kazak.
36 B3 **Kale** *Denizli* Turkey
37 G1 **Kale** Turkey
36 D1 **Kalecik** Turkey
19 H5 **Kalefeld** Ger.
37 M3 **Kaleh Sarai** Iran
110 C4 **Kalema** Congo(Zaire)
115 A4 **Kalémié** Congo(Zaire)
38 B2 **Kalenoye** Kazak.
88 D3 **Kaleva** U.S.A.
8 W4 **Kalevala** Rus. Fed.
45 H5 **Kalewa** Myanmar
69 C7 **Kalgan** *r.* Austr.
69 E5 **Kalgoorlie** Austr.
114 F2 **Kalguéri** Niger
26 F2 **Kali** Croatia
44 E3 **Kali** *r.* India/Nepal
55 B4 **Kalibo** Phil.
45 F4 **Kali Gadaki** *r.* Nepal
115 A3 **Kalima** Congo(Zaire)
59 D3 **Kalimantan** *reg.* Indon.
44 E4 **Kali Nadi** *r.* India
43 A3 **Kalinadi** *r.* India
38 D4 **Kalinin** Turkm.
28 B4 **Kaliningrad** Rus. Fed.
28 B4 **Kaliningradskaya Oblast'** *div.* Rus. Fed.
28 G3 **Kalinino** Rus. Fed.
39 G5 **Kalininobod** Tajik.
29 H5 **Kalininsk** Rus. Fed.
29 F6 **Kalininskaya** Rus. Fed.
29 D4 **Kalinkavichy** Belarus
38 C2 **Kalinovka** Kazak.
44 D4 **Kali Sindh** *r.* India
82 D1 **Kalispell** U.S.A.
16 J5 **Kalisz** Pol.
21 L1 **Kalisz** *div.* Pol.
29 G5 **Kalitva** *r.* Rus. Fed.
115 B4 **Kaliua** Tanz.
8 S4 **Kalix** Sweden
8 S3 **Kalixälven** *r.* Sweden
45 H4 **Kalkalighat** India
36 B3 **Kalkan** Turkey
64 F3 **Kalkaringi** Austr.
88 E3 **Kalkaska** U.S.A.
111 B6 **Kalkfeld** Namibia
113 F4 **Kalkfonteindam** *dam* S. Africa
15 G2 **Kall** Ger.
58 □ **Kallang** Sing.
9 U7 **Kallaste** Estonia
8 U5 **Kallavesi** *l.* Fin.
8 N5 **Kallsedet** Sweden
8 N5 **Kallsjön** *l.* Sweden
38 F3 **Kalmakkyrgan** *watercourse* Kazak.
39 K1 **Kalmanka** Rus. Fed.
9 P8 **Kalmar** Sweden
9 P8 **Kalmarsund** *chan.* Sweden
29 F6 **Kal'mius** *r.* Ukr.
43 C5 **Kalmunai** Sri Lanka
29 H6 **Kalmykiya, Respublika** *div.* Rus. Fed.
38 B2 **Kalmykovo** Kazak.
45 G4 **Kalni** *r.* Bangl.
17 N5 **Kalodnaye** Belarus
44 C5 **Kalol** India
55 C6 **Kaloma** *i.* Indon.
115 A6 **Kalomo** Zambia
76 D4 **Kalone Pk** Can.
40 B2 **Kalow** *r.* Iran
44 D3 **Kalpa** India
43 A4 **Kalpeni** *i.* India
44 D4 **Kalpi** India
39 J4 **Kalpin** China
37 L4 **Kal Safīd** Iran
74 C3 **Kaltag** U.S.A.
19 G3 **Kaltenkirchen** Ger.
20 D2 **Kaltensundheim** Ger.
44 C3 **Kalu** India
28 F4 **Kaluga** Rus. Fed.
9 M9 **Kalundborg** Denmark
44 B2 **Kalur Kot** Pak.
29 C5 **Kalush** Ukr.
43 B5 **Kalutara** Sri Lanka
28 E4 **Kaluzhskaya Oblast'** *div.* Rus. Fed.
8 S5 **Kälviä** Fin.
21 H5 **Kalwang** Austria
43 A2 **Kalyan** India
28 F3 **Kalyazin** Rus. Fed.
27 M6 **Kalymnos** *i.* Greece
17 M6 **Kalynivka** Ukr.
115 A3 **Kama** Congo(Zaire)
52 G5 **Kamaishi** Japan
44 C3 **Kamalia** Pak.
36 D2 **Kaman** Turkey
115 A3 **Kamande** Congo(Zaire)
111 B5 **Kamanjab** Namibia
43 B2 **Kamareddi** India
41 F5 **Kamarod** Pak.
114 A3 **Kamaron** Sierra Leone
39 F5 **Kamashi** Uzbek.
69 E5 **Kambalda** Austr.
43 B4 **Kambam** India
39 K4 **Kamberdi** China
54 E3 **Kambo Ho** *mt* N. Korea
115 A5 **Kambove** Congo(Zaire)
33 S4 **Kamchatka** *r.* Rus. Fed.
33 R4 **Kamchatka Peninsula** Rus. Fed.
27 M3 **Kamchiya** *r.* Bulg.
38 B2 **Kamelik** *r.* Rus. Fed.
18 E5 **Kamen** Ger.
21 J3 **Kamenice nad Lipou** Czech Rep.
27 K4 **Kamenitsa** *mt* Bulg.
38 B2 **Kamenka** Kazak.
28 H4 **Kamenka** *Penzen.* Rus. Fed.
52 E2 **Kamenka** *Primorskiy Kray* Rus. Fed.
32 K4 **Kamen'-na-Obi** Rus. Fed.
28 F3 **Kamenniki** Rus. Fed.
28 D2 **Kamennogorsk** Rus. Fed.
29 G6 **Kamennomostskiy** Rus. Fed.
29 G6 **Kamenolomni** Rus. Fed.
52 C2 **Kamen'-Rybolov** Rus. Fed.
33 S3 **Kamenskoye** Rus. Fed.
29 G5 **Kamensk-Shakhtinskiy** Rus. Fed.
32 H4 **Kamensk-Ural'skiy** Rus. Fed.
19 M5 **Kamenz** Ger.
28 G3 **Kameshkovo** Rus. Fed.
44 D3 **Kamet** *mt* China
21 K2 **Kamienna Gora** Pol.
19 M3 **Kamień Pomorski** Pol.
112 C5 **Kamiesberge** *mts* S. Africa
112 B5 **Kamieskroon** S. Africa
77 J2 **Kamilukuak Lake** Can.
111 C4 **Kamina** Congo(Zaire)
115 A4 **Kamina Base** Congo(Zaire)
77 L2 **Kaminak Lake** Can.
17 M5 **Kamin'-Kashyrs'kyy** Ukr.
52 H3 **Kamishihoro** Japan
45 H4 **Kamjong** India
39 G3 **Kamkaly** Kazak.
45 F4 **Kamla** *r.* India
39 L2 **Kamlak** Rus. Fed.
76 E4 **Kamloops** Can.
37 K1 **Kamo** Armenia
53 G7 **Kamogawa** Japan
44 C3 **Kamoke** Pak.
110 C4 **Kamonia** Congo(Zaire)
58 C2 **Kamon, Xé** *r.* Laos
21 J4 **Kamp** *r.* Austria
115 B2 **Kampala** Uganda
59 B2 **Kampar** *r.* Indon.
58 B4 **Kampar** Malaysia
18 C4 **Kampen** Neth.
115 A3 **Kampene** Congo(Zaire)
58 A1 **Kamphaeng Phet** Thai.
43 B3 **Kampli** India
58 C3 **Kâmpóng Cham** Cambodia
58 C2 **Kâmpóng Chhnăng** Cambodia
58 C2 **Kâmpóng Khleăng** Cambodia
58 C3 **Kâmpóng Spoe** Cambodia
58 C2 **Kâmpóng Thum** Cambodia
58 C3 **Kâmpôt** Cambodia
114 C2 **Kampti** Burkina
Kampuchea *country* *see* **Cambodia**
57 J7 **Kamrau, Teluk** *b.* Indon.
77 J4 **Kamsack** Can.
32 G4 **Kamskoye Vdkhr.** *resr* Rus. Fed.
110 E3 **Kamsuuma** Somalia
77 J3 **Kamuchawie Lake** Can.
115 B2 **Kamuli** Uganda
29 C5 **Kam"yane** Ukr.
29 C5 **Kam"yanets'-Podil's'kyy** Ukr.
29 C5 **Kam"yanka-Buz'ka** Ukr.
17 L4 **Kamyanyets** Belarus
40 B3 **Kāmyārān** Iran
29 F6 **Kamyshevatskaya** Rus. Fed.
29 H5 **Kamyshin** Rus. Fed.
38 E3 **Kamyshlybash** Kazak.
38 E3 **Kamyslybas, Oz.** *l.* Kazak.
29 J6 **Kamyzyak** Rus. Fed.
40 E5 **Kamzar** Oman
78 F3 **Kanaaupscow** *r.* Can.
93 F3 **Kanab** U.S.A.
93 F3 **Kanab Creek** *r.* U.S.A.
71 □13 **Kanacea** *i.* Fiji
41 G4 **Kanak** Pak.
23 L3 **Kanal** Slovenia
37 K5 **Kan'ān** Iraq
110 C4 **Kananga** Congo(Zaire)
67 J4 **Kanangra Nat. Park** Austr.
38 D1 **Kananikol'skoye** Rus. Fed.
93 F3 **Kanarraville** U.S.A.
28 H4 **Kanash** Rus. Fed.
90 C5 **Kanawha** *r.* U.S.A.
53 E7 **Kanayama** Japan
53 E6 **Kanazawa** Japan
58 A2 **Kanchanaburi** Thai.
43 B3 **Kanchipuram** India
44 A3 **Kand** *mt* Pak.
41 G4 **Kandahār** Afgh.
8 X3 **Kandalaksha** Rus. Fed.
58 A5 **Kandang** Indon.
15 J4 **Kandel** *mt* Ger.
22 E3 **Kandersteg** Switz.
44 B2 **Kandhura** Pak.
114 D2 **Kandi** Benin
41 G4 **Kandi** Pak.
44 B4 **Kandiaro** Pak.
36 C1 **Kandıra** Turkey
67 H4 **Kandos** Austr.
111 E5 **Kandreho** Madag.
43 B3 **Kandukur** India
43 C5 **Kandy** Sri Lanka
90 D4 **Kane** U.S.A.
75 M2 **Kane Basin** *b.* Can./Greenland
40 D5 **Kaneh** *watercourse* Iran
92 □1 **Kaneohe** U.S.A.
92 □1 **Kaneohe Bay** U.S.A.
29 F6 **Kanevskaya** Rus. Fed.
111 C6 **Kang** Botswana
45 G5 **Kanga** *r.* Bangl.
75 N3 **Kangaatsiaq** Greenland
114 B2 **Kangaba** Mali
36 F2 **Kangal** Turkey
41 E5 **Kangān** Iran
40 D5 **Kangan** Iran
59 B1 **Kangar** Malaysia
66 B5 **Kangaroo I.** Austr.
8 V5 **Kangaslampi** Fin.
9 U6 **Kangasniemi** Fin.
40 B3 **Kangāvar** Iran
50 E1 **Kangbao** China
45 G4 **Kangchenjunga** *mt* Nepal
51 A4 **Kangding** China
54 D4 **Kangdong** N. Korea
75 M3 **Kangeak Pt** Can.
59 E4 **Kangean, Kepulauan** *is* Indon.
115 B1 **Kangen** *r.* Sudan
75 O3 **Kangeq** *hd* Greenland
75 O4 **Kangerlussuaq** *in.* Greenland
75 N2 **Kangersuatsiaq** Greenland
75 P3 **Kangertittivatsiaq** *in.* Greenland
115 C2 **Kangetet** Kenya
54 D3 **Kanggye** N. Korea
54 D5 **Kanghwa** S. Korea
54 D5 **Kanghwa Do** *i.* S. Korea
79 G2 **Kangiqsualujjuaq** Can.
75 L3 **Kangiqsujuaq** Can.
79 G1 **Kangirsuk** Can.
50 B3 **Kangle** China
45 G3 **Kangmar** *Xizang* China
45 F3 **Kangmar** *Xizang* China
54 E5 **Kangnŭng** S. Korea
110 B3 **Kango** Gabon
54 B2 **Kangping** China
45 J3 **Kangri Karpo Pass** India
44 E3 **Kangrinboqê Feng** *mt* China
45 H4 **Kangto** *mt* China
45 F2 **Kangtog** China
50 B3 **Kang Xian** China
44 D5 **Kanhan** *r.* India
45 E4 **Kanhar** *r.* India
114 B3 **Kani** Côte d'Ivoire
110 C4 **Kaniama** Congo(Zaire)
39 G4 **Kanibadam** Tajik.
70 C5 **Kaniere, L.** N.Z.
43 B3 **Kanigiri** India
32 F3 **Kanin, Poluostrov** *pen.* Rus. Fed.
37 K3 **Kānī Rash** Iraq
29 D5 **Kaniv** Ukr.
66 D6 **Kaniva** Austr.
9 S6 **Kankaanpää** Fin.
88 D5 **Kankakee** U.S.A.
88 C5 **Kankakee** *r.* U.S.A.
114 B2 **Kankan** Guinea
45 E5 **Kanker** India
43 C4 **Kankesanturai** Sri Lanka
58 A3 **Kanmaw Kyun** *i.* Myanmar
44 D4 **Kannanj** India
87 D5 **Kannapolis** U.S.A.
43 B4 **Kanniyakumari** India
Kanniya Kumari *c.* *see* **Comorin, Cape**
44 D5 **Kannod** India
8 T5 **Kannonkoski** Fin.
Kannur *see* **Cannanore**
8 S5 **Kannus** Fin.
114 E2 **Kano** Nigeria
39 J2 **Kanonerka** Kazak.
112 D7 **Kanonpunt** *pt* S. Africa
44 C4 **Kanor** India
53 B9 **Kanoya** Japan
44 E4 **Kanpur** India
44 B3 **Kanpur** Pak.
41 G5 **Kanrach** *reg.* Pak.
84 D4 **Kansas** *div.* U.S.A.
84 D4 **Kansas** *r.* U.S.A.
84 E4 **Kansas City** *KS* U.S.A.
84 E4 **Kansas City** *MO* U.S.A.
46 B1 **Kansk** Rus. Fed.
54 E4 **Kansŏng** S. Korea
39 H5 **Kansu** China
39 H4 **Kant** Kyrg.
45 G4 **Kantanagar** Bangl.
58 C2 **Kantaralak** Thai.
114 D2 **Kantchari** Burkina
29 F5 **Kantemirovka** Rus. Fed.
45 F5 **Kānthi** India
45 F4 **Kanti** India
44 C3 **Kantli** *r.* India
63 J2 **Kanton Island** Kiribati
14 C5 **Kanturk** Rep. of Ireland
112 C3 **Kanus** Namibia
113 J2 **KaNyamazane** S. Africa
111 C6 **Kanye** Botswana
71 □12 **Kao** *i.* Tonga
58 B3 **Kaôh Kŏng** *i.* Cambodia
58 B3 **Kaôh Rŭng** *i.* Cambodia
58 B3 **Kaôh Rŭng Sânlŏem** *i.* Cambodia
51 F6 **Kao-hsiung** Taiwan
58 B3 **Kaôh Smăch** *i.* Cambodia
111 B5 **Kaokoveld** *plat.* Namibia
108 A3 **Kaolack** Senegal
111 C5 **Kaoma** Zambia
92 □2 **Kapaa** U.S.A.
92 □2 **Kapaau** U.S.A.
39 J3 **Kapal** Kazak.
37 L2 **Kapan** Armenia
111 C4 **Kapanga** Congo(Zaire)
115 B4 **Kapatu** Zambia
39 J4 **Kapchagay** Kazak.
39 J4 **Kapchagayskoye Vdkhr.** *resr* Kazak.
15 E1 **Kapellen** Belgium
27 K6 **Kapello, Akra** *pt* Greece
9 Q7 **Kapellskär** Sweden
21 J5 **Kapfenberg** Austria
36 A1 **Kapıdağı Yarımadası** *pen.* Turkey
45 G4 **Kapili** *r.* India
118 F5 **Kapingamarangi Rise** *sea feature* Pac. Oc.
118 F5 **Kapingaramarangi** *i.* Pac. Oc.
44 B3 **Kapip** Pak.
115 A5 **Kapiri Mposhi** Zambia
75 N3 **Kapisigdlit** Greenland
78 D3 **Kapiskau** Can.
78 D3 **Kapiskau** *r.* Can.
89 G2 **Kapiskong Lake** Can.
70 E4 **Kapiti I.** N.Z.
21 H4 **Kaplice** Czech Rep.
58 A3 **Kapoe** Thai.
109 F4 **Kapoeta** Sudan
21 M6 **Kapos** *r.* Romania
16 H7 **Kaposvár** Hungary
41 F5 **Kappar** Pak.
15 H3 **Kappel** Ger.
21 H6 **Kappel am Krappfeld** Austria
19 G2 **Kappeln** Ger.
20 D5 **Kappl** Austria
44 D4 **Kapran** India
115 C2 **Kapsabet** Kenya
54 E3 **Kapsan** N. Korea
59 D3 **Kapuas** *r.* Indon.
59 D2 **Kapuas** *r.* Indon.
66 C5 **Kapunda** Austr.
44 C4 **Kapūriya** India
44 C3 **Kapurthala** India
78 D4 **Kapuskasing** Can.
78 D4 **Kapuskasing** *r.* Can.
29 H5 **Kapustin Yar** Rus. Fed.
67 J3 **Kaputar** *mt* Austr.
115 C2 **Kaputir** Kenya
16 H7 **Kapuvár** Hungary
28 C4 **Kapyl'** Belarus
54 D5 **Kap'yŏng** S. Korea
114 D3 **Kara** Togo
37 H2 **Kara** *r.* Turkey
27 M5 **Kara Ada** *i.* Turkey
36 D2 **Karaali** Turkey
39 H4 **Kara-Balta** Kyrg.
39 H2 **Karabas** Kazak.
38 C2 **Karabau** Kazak.
38 F5 **Karabekaul** Turkm.
41 F2 **Karabil', Vozvyshennost'** *reg.* Turkm.
38 C4 **Kara-Bogaz-Gol** Turkm.
38 C4 **Kara-Bogaz Gol, Zaliv** *b.* Turkm.
36 D1 **Karabük** Turkey
39 J3 **Karabulak** *Taldykorg.* Kazak.
39 L3 **Karabulak** *Vostochnyy Kaz.* Kazak.
39 H2 **Karabulakskaya** Kazak.
38 E2 **Karabutak** Kazak.
36 B1 **Karacabey** Turkey
37 G3 **Karacadağ** Turkey
36 D3 **Karacadağ** *mts* Turkey
36 B1 **Karacaköy** Turkey
37 G3 **Karacalı Dağ** *mt* Turkey
36 B3 **Karacasu** Turkey
36 C3 **Karaca Yarımadası** *pen.* Turkey
29 G7 **Karachayevo-Cherkesskaya Respublika** *div.* Rus. Fed.
29 G7 **Karachayevsk** Rus. Fed.
28 E4 **Karachev** Rus. Fed.
41 G5 **Karachi** Pak.
37 J2 **Karaçoban** Turkey
43 A2 **Karad** India
37 J3 **Kara Dağ** *mt* Turkey
36 D3 **Kara Dağ** *mt* Turkey
39 H4 **Kara-Darya** *r.* Kyrg.
Kara Deniz *sea* *see* **Black Sea**
39 H2 **Karaganda** Kazak.
39 G2 **Karagandinskaya Oblast'** *div.* Kazak.
39 H2 **Karagayly** Kazak.
33 S4 **Karaginskiy Zaliv** *b.* Rus. Fed.
38 B4 **Karagiye, Vpadina** *depression* Kazak.
39 K2 **Karaguzhikha** Kazak.
36 B2 **Karahallı** Turkey
36 E2 **Karahasanlı** Turkey
43 B4 **Kāraikāl** India
43 B4 **Karaikkudi** India
39 L3 **Kara Irtysh** *r.* Kazak.
36 E3 **Karaisalı** Turkey
40 C3 **Karaj** Iran
40 C3 **Karaj** *r.* Iran
36 E6 **Karak** Jordan
38 D5 **Kara Kala** Turkm.
38 D3 **Karakalpakiya** Uzbek.
38 F4 **Karakatinskaya, Vpadina** *depression* Uzbek.
44 E1 **Karakax He** *r.* China
37 G3 **Karakeçi** Turkey
36 D2 **Karakeçili** Turkey
47 E6 **Karakelong** *i.* Indon.
38 F3 **Karaketken** Kazak.
69 B5 **Karakin Lakes** *salt flat* Austr.
37 H2 **Karakoçan** Turkey
39 G3 **Karakoin, Oz.** *salt l.* Kazak.
38 C2 **Karakol'** Kazak.
39 J4 **Karakol** Kyrg.
39 J4 **Karakol** Kyrg.
39 H4 **Kara-Köl** Kyrg.
44 D2 **Karakoram Pass** China/Jammu and Kashmir
35 F3 **Karakoram Range** *mts* Asia
110 D2 **Kara K'orē** Eth.
38 E5 **Karakul'** Uzbek.
38 E5 **Karakul'** Uzbek.
38 E1 **Karakul'skoye** Rus. Fed.
39 J3 **Karakum** Kazak.
38 D5 **Karakum Desert** Turkm.
38 C3 **Karakum, Peski** *des.* Kazak.
38 E5 **Karakumskiy Kanal** *canal* Turkm.
38 D5 **Kara Kumy** *des.* Turkm.
37 J1 **Karakurt** Turkey
9 R7 **Karala** Estonia
69 D3 **Karalundi** Austr.
36 D3 **Karaman** Turkey
36 B3 **Karamanlı** Turkey
35 G2 **Karamay** China
44 C1 **Karambar Pass** Afgh./Pak.
70 D4 **Karamea** N.Z.
70 C4 **Karamea Bight** *b.* N.Z.
38 F5 **Karamet-Niyaz** Turkm.
45 F1 **Karamiran** China
45 F1 **Karamiran Shankou** *pass* China
36 B1 **Karamürsel** Turkey
28 D3 **Karamyshevo** Rus. Fed.
40 C5 **Karān** *i.* S. Arabia
40 B3 **Karand** Iran
44 D5 **Karanja** India
43 B2 **Karanja** *r.* India
45 F5 **Karanjia** India
44 C3 **Karanpura** India
39 J1 **Karaoba** Kazak.
39 H3 **Karaoy** *Almatinsk.* Kazak.
38 F3 **Karaozek** Kazak.
36 D3 **Karapınar** Turkey
39 J4 **Karaqi** China
112 B3 **Karas** *div.* Namibia
112 B3 **Karas** *watercourse* Namibia
39 J4 **Kara-Say** Kyrg.
111 B6 **Karasburg** Namibia
32 J2 **Kara Sea** Rus. Fed.
39 H2 **Karashoky** Kazak.
8 T2 **Karasjok** Norway
39 H2 **Karasor** Kazak.
39 H2 **Karasor, Ozero** *salt l.* *Karagan.* Kazak.
39 H1 **Karasor, Ozero** *salt l.* *Pavlodarsk.* Kazak.
38 F1 **Karasu** *Kustanay.* Kazak.
39 H3 **Karasu** *Zhezkaz.* Kazak.
39 H1 **Karasu** *r.* Kazak.
36 C1 **Karasu** Turkey
37 J2 **Karasu** *r.* Turkey
39 J1 **Karasuk** Rus. Fed.
39 H4 **Kara-Suu** Kyrg.
39 L3 **Karatan** Kazak.
36 E3 **Karataş** Turkey
36 E3 **Karataş Burun** *pt* Turkey
39 G4 **Karatau** Kazak.
39 F3 **Karatau, Khrebet** *mts* Kazak.
44 E2 **Karatax Shan** *mts* China
58 A3 **Karathuri** Myanmar
43 B4 **Karativu** *i.* Sri Lanka
38 C2 **Karatobe** Kazak.
38 D3 **Karatobe, Mys** *pt* Kazak.
38 D2 **Karatogay** Kazak.
39 J3 **Karatol** *r.* Kazak.
38 E1 **Karatomarskoye Vdkhr.** *resr* Kazak.
38 C3 **Karaton** Kazak.
45 G4 **Karatoya** *r.* Bangl.
53 A8 **Karatsu** Japan
55 C5 **Karatung** *i.* Indon.
38 F2 **Karaturgay** *r.* Kazak.
39 J2 **Karaul** Kazak.
38 F5 **Karaulbazar** Uzbek.
44 D4 **Karauli** India
37 J1 **Karaurgan** Turkey
39 G4 **Karavan** Kyrg.
59 C4 **Karawang** Indon.
39 K4 **Karayulgan** China
39 G2 **Karazhal** Kazak.
38 D4 **Karazhar** Uzbek.
39 H3 **Karazhingil** Kazak.
37 K5 **Karbalā'** Iraq
39 H2 **Karbushevka** Kazak.
17 K7 **Karcag** Hungary
15 H2 **Karden** Ger.
27 J5 **Karditsa** Greece
9 S7 **Kärdla** Estonia
113 G4 **Karee** S. Africa
112 D5 **Kareeberge** *mts* S. Africa
109 F3 **Kareima** Sudan
29 G7 **K'areli** Georgia
44 D5 **Kareli** India
28 E2 **Kareliya, Respublika** *div.* Rus. Fed.
115 B4 **Karema** Tanz.
49 L1 **Karenga** *r.* Rus. Fed.
44 D4 **Karera** India
8 S2 **Karesuando** Sweden
41 F5 **Kārevāndar** Iran
38 C2 **Kargala** Rus. Fed.
29 H7 **Kargalinskaya** Rus. Fed.
38 D2 **Kargalinskoye** Kazak.
39 H2 **Kargaly** Kazak.
37 H2 **Kargapazarı Dağları** *mts* Turkey
36 E1 **Kargı** Turkey
44 D2 **Kargil** Jammu and Kashmir
28 F2 **Kargopol'** Rus. Fed.
9 P6 **Karholmsbruk** Sweden
114 F2 **Kari** Nigeria
111 C5 **Kariba** Zimbabwe
111 C5 **Kariba, Lake** *resr* Zambia/Zimbabwe
52 F3 **Kariba-yama** *volc.* Japan
112 E6 **Kariega** *r.* S. Africa
8 T2 **Karigasniemi** Fin.
9 R5 **Karijoki** Fin.
70 D1 **Karikari, Cape** N.Z.
40 D3 **Karīmābād** Iran
59 C3 **Karimata, Pulau Pulau** *is* Indon.
59 C3 **Karimata, Selat** *str.* Indon.
43 B2 **Karimnagar** India
59 D4 **Karimunjawa, Pulau Pulau** *is* Indon.
110 E2 **Karin** Somalia
115 A3 **Karisimbi, Mt** *volc.* Rwanda
40 E3 **Karit** Iran
43 A2 **Karjat** India
45 F5 **Karkai** *r.* India
43 A3 **Karkal** India
39 H2 **Karkaralinsk** Kazak.
55 C5 **Karkaralong, Kepulauan** *is* Indon.
62 E2 **Karkar I.** P.N.G.
40 C4 **Karkheh** *r.* Iran
41 E5 **Kārkīn Dar** Iran
29 E6 **Karkinits'ka Zatoka** *g.* Ukr.
9 T6 **Kärkölä** Fin.
9 T7 **Karksi-Nuia** Estonia
37 H2 **Karlıova** Turkey
29 E5 **Karlivka** Ukr.
Karl-Marx-Stadt *see* **Chemnitz**
26 F2 **Karlovac** Croatia
27 L3 **Karlovo** Bulg.
20 F2 **Karlovy Vary** Czech Rep.
20 B4 **Karlsbad** Ger.
9 O7 **Karlsborg** Sweden
19 L3 **Karlsburg** Ger.
20 E4 **Karlsfeld** Ger.
9 O8 **Karlshamn** Sweden
9 O7 **Karlskoga** Sweden
9 O8 **Karlskrona** Sweden
20 B3 **Karlsruhe** Ger.
9 N7 **Karlstad** Sweden
84 D1 **Karlstad** U.S.A.
20 C3 **Karlstadt** Ger.
28 D4 **Karma** Belarus
43 A2 **Karmala** India
38 B2 **Karmanovka** Kazak.
9 J7 **Karmøy** *i.* Norway
45 H5 **Karnafuli Reservoir** Bangl.
44 D3 **Karnal** India
45 E3 **Karnali** *r.* Nepal
43 A3 **Karnataka** *div.* India
27 M3 **Karnobat** Bulg.
21 G6 **Kärnten** *div.* Austria
41 G5 **Karodi** Pak.
111 C5 **Karoi** Zimbabwe
71 □13 **Karoko** Fiji
45 G3 **Karo La** *pass* China
45 H4 **Karong** India
115 B4 **Karonga** Malawi
39 J4 **Karool-Döbö** Kyrg.
112 E6 **Karoo National Park** S. Africa
66 C5 **Karoonda** Austr.
44 B3 **Karor** Pak.
110 D2 **Karora** Eritrea
19 K3 **Karow** Ger.
27 M7 **Karpathos** *i.* Greece
27 M6 **Karpathou, Steno** *chan.* Greece
Karpaty *see* **Carpathian Mountains**
27 J5 **Karpenisi** Greece
28 H1 **Karpogory** Rus. Fed.
64 C4 **Karratha** Austr.
41 F3 **Karrukh** Afgh.
37 J1 **Kars** Turkey
8 T5 **Kärsämäki** Fin.
9 U8 **Kārsava** Latvia
38 C4 **Karshi** Turkm.
38 F5 **Karshi** Uzbek.
45 G4 **Kārsiyāng** India

13 E4 Kirkby U.K.
13 F4 Kirkby in Ashfield U.K.
12 E3 Kirkby Lonsdale U.K.
12 E3 Kirkby Stephen U.K.
11 E4 Kirkcaldy U.K.
11 C6 Kirkcolm U.K.
14 F3 Kirkcubbin U.K.
11 D6 Kirkcudbright U.K.
9 N6 Kirkenær Norway
8 W2 Kirkenes Norway
89 H3 Kirkfield Can.
11 D5 Kirkintilloch U.K.
9 T6 Kirkkonummi Fin.
93 F4 Kirkland U.S.A.
93 F4 Kirkland Junction U.S.A.
89 G1 Kirkland Lake Can.
29 C7 Kırklareli Turkey
12 C3 Kirk Michael U.K.
12 E3 Kirkoswald U.K.
84 E3 Kirksville U.S.A.
37 K4 Kirkūk Iraq
11 F2 Kirkwall U.K.
113 F6 Kirkwood S. Africa
92 B2 Kirkwood CA U.S.A.
84 F4 Kirkwood MO U.S.A.
36 C1 Kırmır r. Turkey
15 H3 Kirn Ger.
39 G4 Kirov Kyrg.
28 E4 Kirov Kaluzh. Obl. Rus. Fed.
Kirov see Vyatka
Kirovabad see Gäncä
38 C2 Kirovo Kazak.
28 J3 Kirovo-Chepetsk Rus. Fed.
29 E5 Kirovohrad Ukr.
37 M2 Kirovsk Azer.
28 D3 Kirovsk Leningrad. Rus. Fed.
8 X3 Kirovsk Murmansk. Rus. Fed.
38 E5 Kirovsk Turkm.
28 J3 Kirovskaya Oblast' div. Rus. Fed.
39 J3 Kirovskiy Kazak.
52 C2 Kirovskiy Rus. Fed.
116 B4 Kirpatrick, Mt Ant.
38 D5 Kirpili Turkm.
11 E4 Kirriemuir U.K.
28 K3 Kirs Rus. Fed.
28 G4 Kirsanov Rus. Fed.
36 E2 Kırşehir Turkey
41 G5 Kirthar Range mts Pak.
19 G6 Kirtorf Ger.
8 R3 Kiruna Sweden
115 A3 Kirundu Congo(Zaire)
28 H4 Kirya Rus. Fed.
53 F6 Kiryū Japan
9 O8 Kisa Sweden
115 A2 Kisangani Congo(Zaire)
110 B4 Kisantu Congo(Zaire)
59 A2 Kisaran Indon.
21 M5 Kisbér Hungary
46 A1 Kiselevsk Rus. Fed.
45 F4 Kishanganj India
44 B4 Kishangarh Rajasthan India
44 C4 Kishangarh Rajasthan India
44 C2 Kishen Ganga r. India/Pak.
53 B9 Kishika-zaki pt Japan
Kishinev see Chişinău
53 D6 Kishiwada Japan
45 G4 Kishorganj Bangl.
44 C2 Kishtwar Jammu and Kashmir
114 D3 Kisi Nigeria
115 B3 Kisii Kenya
77 K4 Kiskittogisu L. Can.
17 J7 Kiskunfélegyháza Hungary
17 J7 Kiskunhalas Hungary
29 G7 Kislovodsk Rus. Fed.
110 E4 Kismaayo Somalia
115 A3 Kisoro Uganda
52 F3 Kiso-sanmyaku mts Japan
114 A3 Kissidougou Guinea
87 D6 Kissimmee U.S.A.
87 D7 Kissimmee, L. U.S.A.
20 D4 Kissing Ger.
77 J3 Kississing L. Can.
20 C5 Kißlegg Ger.
Kistna r. see Krishna
115 B3 Kisumu Kenya
114 B2 Kita Mali
39 F5 Kitab Uzbek.
52 G6 Kitaibaraki Japan
52 G5 Kitakami Japan
52 G5 Kitakami-gawa r. Japan
53 F6 Kitakata Japan
53 B8 Kita-Kyūshū Japan
115 C2 Kitale Kenya
46 G2 Kitami Japan
83 G4 Kit Carson U.S.A.
89 G4 Kitchener Can.
8 W5 Kitee Fin.
115 B2 Kitgum Uganda
79 D4 Kitimat Can.
8 U3 Kitinen r. Fin.
110 B4 Kitona Congo(Zaire)
53 B8 Kitsuki Japan
66 C2 Kittakittaooloo, L. salt flat Austr.
90 D4 Kittanning U.S.A.
91 F4 Kittatinny Mts h. U.S.A.
91 H3 Kittery U.S.A.
8 T3 Kittilä Fin.
87 F4 Kitty Hawk U.S.A.
115 A4 Kitu Congo(Zaire)
115 C3 Kitui Kenya
115 B4 Kitunda Tanz.
76 D3 Kitwanga Can.
115 A5 Kitwe Zambia
20 F5 Kitzbühel Austria
20 F5 Kitzbüheler Alpen mts Austria
20 D3 Kitzingen Ger.
19 K5 Kitzscher Ger.
8 U5 Kiuruvesi Fin.
8 T5 Kivijärvi Fin.
9 U7 Kiviõli Estonia
110 C4 Kivu, Lake Congo(Zaire)/Rwanda
39 G2 Kiyakty, Ozero salt l. Kazak.
39 G2 Kiyevka Kazak.
52 C3 Kiyevka Rus. Fed.
27 N4 Kıyıköy Turkey
39 F2 Kiyma Kazak.
32 G4 Kizel Rus. Fed.
28 H2 Kizema Rus. Fed.
39 K4 Kizil China
38 B3 Kızılca D. mt Turkey
36 D1 Kızılcahamam Turkey
36 G2 Kızıl D. mt Turkey
36 D1 Kızılırmak Turkey
36 D2 Kızılırmak r. Turkey
36 C3 Kızılkaya Turkey
36 D3 Kızılören Turkey
38 D1 Kizil'skoye Rus. Fed.
37 H3 Kızıltepe Turkey
29 H7 Kizil"yurt Rus. Fed.
29 H7 Kizlyar Rus. Fed.
38 C5 Kizyl-Atrek Turkm.
38 F5 Kizylayak Turkm.
8 U1 Kjøllefjord Norway
8 P2 Kjøpsvik Norway
21 H2 Kladno Czech Rep.
21 H6 Klagenfurt Austria
93 H4 Klagetoh U.S.A.
9 R9 Klaipėda Lith.
8 □ Klaksvík Faroe Is
82 B3 Klamath r. U.S.A.
82 B3 Klamath Falls U.S.A.
82 B3 Klamath Mts U.S.A.
9 N6 Klarälven r. Sweden
20 G2 Klášterec nad Ohří Czech Rep.
20 G3 Klatovy Czech Rep.
19 H2 Klausdorf Ger.
112 C5 Klawer S. Africa
76 C3 Klawock U.S.A.
18 D4 Klazienaveen Neth.
76 E4 Kleena Kleene Can.
112 D4 Kleinbegin S. Africa
19 L3 Klein Bünzow Ger.
112 C3 Klein Karas Namibia
112 D6 Klein Roggeveldberg mts S. Africa
112 B4 Kleinsee S. Africa
112 D6 Klein Swartberg mts S. Africa
76 D4 Klemtu Can.
113 G3 Klerksdorp S. Africa
28 E4 Kletnya Rus. Fed.
29 G5 Kletskiy Rus. Fed.
22 F2 Klettgau reg. Ger./Switz.
18 D5 Kleve Ger.
112 F6 Klienpoort S. Africa
28 D4 Klimavichy Belarus
29 E4 Klimovo Rus. Fed.
28 F4 Klimovsk Rus. Fed.
28 F3 Klin Rus. Fed.
76 D4 Klinaklini r. Can.
20 C3 Klingenberg am Main Ger.
19 K3 Klink Ger.
20 G2 Klínovec mt Czech Rep.
9 Q8 Klintehamn Sweden
29 J5 Klintsovka Rus. Fed.
28 E4 Klintsy Rus. Fed.
112 C5 Kliprand S. Africa
18 F2 Klixbüll Ger.
26 G2 Ključ Bos.-Herz.
21 M2 Kłobuck Pol.
16 H5 Kłodzko Pol.
18 D4 Kloosterhaar Neth.
20 D5 Klösterle Austria
21 K4 Klosterneuburg Austria
23 G3 Klosters Switz.
19 J4 Klötze (Altmark) Ger.
24 F1 Klotz, Lac l. Can.
76 A2 Kluane Game Sanctuary Can.
76 B2 Kluane Lake Can.
76 B2 Kluane National Park Can.
16 J5 Kluczbork Pol.
44 B4 Klupro Pak.
19 J3 Klütz Ger.
28 C4 Klyetsk Belarus
33 S4 Klyuchevskaya Sopka volc. Rus. Fed.
39 J1 Klyuchi Rus. Fed.
9 O6 Knåda Sweden
12 F3 Knaresborough U.K.
77 L3 Knee Lake Can.
88 B1 Knife Lake Can./U.S.A.
76 D4 Knight In. in. Can.
13 D5 Knighton U.K.
88 E6 Knightstown U.S.A.
26 G2 Knin Croatia
21 H5 Knittelfeld Austria
27 K3 Knjaževac Yugo.
69 B4 Knobby Head hd Austr.
14 C4 Knock Rep. of Ireland
14 B6 Knockaboy h. Rep. of Ireland
14 B5 Knockacummer h. Rep. of Ireland
14 C3 Knockalongy h. Rep. of Ireland
14 B5 Knockalough Rep. of Ireland
11 F3 Knock Hill U.K.
14 E2 Knocklayd h. U.K.
15 D1 Knokke-Heist Belgium
19 L3 Knorrendorf Ger.
13 F5 Knowle U.K.
116 B2 Knowles, C. Ant.
91 J1 Knowles Corner U.S.A.
91 G2 Knowlton Can.
88 D5 Knox U.S.A.
76 C4 Knox, C. Can.
116 C6 Knox Coast Ant.
92 A2 Knoxville CA U.S.A.
88 B5 Knoxville IL U.S.A.
87 D4 Knoxville TN U.S.A.
11 C3 Knoydart U.K.
75 N1 Knud Rasmussen Land reg. Greenland
19 G6 Knüllgebirge h. Ger.
112 E7 Knysna S. Africa
115 C4 Koani Tanz.
53 B9 Kobayashi Japan
8 V2 Kobbfoss Norway
53 D7 Kōbe Japan
København see Copenhagen
108 B3 Kobenni Maur.
15 H2 Koblenz Ger.
28 J3 Kobra Rus. Fed.
57 J8 Kobroör i. Indon.
28 C4 Kobryn Belarus
29 G7 K'obulet'i Georgia
27 K4 Kočani Macedonia
36 B1 Kocasu r. Turkey
26 F2 Kočevje Slovenia
58 A3 Ko Chan i. Thai.
54 D6 Kŏch'ang S. Korea
54 D5 Koch'ang S. Korea
58 B2 Ko Chang i. Thai.
45 G4 Koch Bihār India
20 E5 Kochelsee l. Ger.
20 C3 Kocher r. Ger.
Kochi see Cochin
53 C8 Kōchi Japan
39 H4 Kochkor Kyrg.
28 H4 Kochkurovo Rus. Fed.
29 H6 Kochubey Rus. Fed.
29 G6 Kochubeyevskoye Rus. Fed.
21 M5 Kocs Hungary
43 B4 Kodaikanal India
43 D2 Kodala India
74 C4 Kodiak U.S.A.
74 C4 Kodiak Island U.S.A.
114 C3 Kodibeleng Botswana
28 F2 Kodino Rus. Fed.
109 F4 Kodok Sudan
29 G7 Kodori r. Georgia
29 D5 Kodyma Ukr.
27 L4 Kodzhaele mt Bulg./Greece
112 D6 Koedoesberg mts S. Africa
112 D4 Koegrabie S. Africa
112 C5 Koekenaap S. Africa
45 E4 Koel r. India
15 C1 Koersel Belgium
111 B6 Koës Namibia
93 F5 Kofa Mts U.S.A.
112 F4 Koffiefontein S. Africa
21 J5 Köflach Austria
114 C3 Koforidua Ghana
53 F7 Kōfu Japan
78 E2 Kogaluc r. Can.
78 E2 Kogaluc, Baie de b. Can.
79 H2 Kogaluk r. Can.
67 J1 Kogan Austr.
9 N9 Køge Denmark
19 K1 Køge Bugt b. Denmark
44 B5 Kohan Pak.
44 B2 Kohat Pak.
9 T7 Kohila Estonia
45 H4 Kohima India
44 B3 Kohlu Pak.
41 F3 Kohsan Afgh.
9 U7 Kohtla-Järve Estonia
70 E2 Kohukohunui h. N.Z.
54 D6 Kohŭng S. Korea
76 A2 Koidern Can.
43 B3 Koilkuntla India
54 D3 Koindong N. Korea
37 K3 Koi Sanjaq Iraq
54 E6 Kŏje do i. S. Korea
21 L3 Kojetín Czech Rep.
53 F8 Ko-jima i. Japan
52 F4 Ko-jima i. Japan
69 C6 Kojonup Austr.
58 A1 Kok r. Thai.
91 J2 Kokadjo U.S.A.
38 F2 Kokalaat Kazak.
39 G4 Kokand Uzbek.
9 R7 Kökar Fin.
38 E3 Kokaral Kazak.
39 H4 Kök-Aygyr Kyrg.
41 H2 Kokcha r. Afgh.
9 R6 Kokemäenjoki r. Fin.
112 C4 Kokerboom Namibia
17 O3 Kokhanava Belarus
28 G3 Kokhma Rus. Fed.
39 H4 Kök-Janggak Kyrg.
43 C4 Kokkilai Sri Lanka
8 S5 Kokkola Fin.
92 □1 Koko Hd hd U.S.A.
88 D5 Kokomo U.S.A.
112 E2 Kokong Botswana
113 G3 Kokosi S. Africa
39 K2 Kokpekty Kazak.
54 D4 Koksan N. Korea
39 G4 Koksaray Kazak.
39 J4 Kokshaal-Tau mts China/Kyrg.
28 H3 Koksharka Rus. Fed.
39 G1 Kokshetau Kazak.
39 G1 Kokshetauskaya Oblast' div. Kazak.
79 G2 Koksoak r. Can.
113 H5 Kokstad S. Africa
39 J3 Koksu Taldykorg. Kazak.
39 G4 Koksu Yuzhnyy Kaz. Kazak.
39 J3 Koktal Kazak.
39 J3 Kokterek Taldykorg. Kazak.
38 B2 Kokterek Zapadnyy Kaz. Kazak.
38 D2 Koktubek Kazak.
39 K3 Koktuma Kazak.
58 B3 Ko Kut i. Thai.
39 J5 Kokyar China
8 X2 Kola Rus. Fed.
Kolab r. see Sābari
41 G5 Kolachi r. Pak.
42 C2 Kolahoi mt India
57 G7 Kolaka Indon.
58 A4 Ko Lanta Thai.
58 A4 Ko Lanta i. Thai.
32 E3 Kola Peninsula Rus. Fed.
43 B3 Kolar Karnataka India
44 E6 Kolar Madhya Pradesh India
44 D4 Kolaras India
43 B3 Kolar Gold Fields India
8 S3 Kolari Fin.
44 C4 Kolayat India
28 F3 Kol'chugino Rus. Fed.
108 A3 Kolda Senegal
9 L9 Kolding Denmark
110 C4 Kole Kasai-Oriental Congo(Zaire)
115 A2 Kole Congo(Zaire)
24 D5 Koléa Alg.
8 R4 Koler Sweden
32 F3 Kolguyev, O. i. Rus. Fed.
45 F5 Kolhan reg. India
43 A2 Kolhapur India
58 A4 Ko Libong i. Thai.
114 A1 Kolimbiné watercourse Mali/Maur.
21 J2 Kolín Czech Rep.
9 S7 Kõljala Estonia
9 S8 Kolkasrags pt Latvia
39 G5 Kolkhozobod Tajik.
Kollam see Quilon
20 F4 Kollbach r. Ger.
20 E1 Kölleda Ger.
43 B3 Kollegal India
43 C2 Kolleru L. India
18 D3 Kollum Neth.
Köln see Cologne
71 □11 Koloa i. Tonga
16 G3 Kołobrzeg Pol.
28 H3 Kologriv Rus. Fed.
114 B2 Kolokani Mali
63 F2 Kolombangara i. Solomon Is
28 F4 Kolomna Rus. Fed.
29 C5 Kolomyya Ukr.
114 B2 Kolondiéba Mali
57 G7 Kolonedale Indon.
71 □11 Kolonga Tonga
112 D3 Kolonkwane Botswana
71 □11 Kolovai Tonga
38 B2 Kolovertnoye Kazak.
32 K4 Kolpashevo Rus. Fed.
29 F4 Kolpny Rus. Fed.
39 K4 Kol'shat Kazak.
Kol'skiy Poluostrov pen. see Kola Peninsula
38 B1 Koltubanovskiy Rus. Fed.
42 B7 Koluli Eritrea
39 G2 Koluton Kazak.
43 A2 Kolvan India
8 M4 Kolvereid Norway
8 T1 Kolvik Norway
41 G5 Kolwa reg. Pak.
115 A5 Kolwezi Congo(Zaire)
33 R3 Kolyma r. Rus. Fed.
33 R3 Kolymskaya Nizmennost' lowland Rus. Fed.
33 R3 Kolymskiy, Khrebet mts Rus. Fed.
28 H4 Kolyshley Rus. Fed.
27 K3 Kom mt Bulg.
114 F2 Komadugu-gana watercourse Nigeria
52 G3 Komaga-take volc. Japan
112 B4 Komaggas S. Africa
112 B4 Komaggas Mts S. Africa
33 S4 Komandorskiye Ostrova is Rus. Fed.
16 J7 Komárno Slovakia
21 M5 Komárom-Esztergom div. Hungary
113 J2 Komatipoort S. Africa
53 E6 Komatsu Japan
53 D7 Komatsushima Japan
115 A3 Kombe Congo(Zaire)
114 C2 Kombissiri Burkina
59 B3 Komering r. Indon.
113 G6 Komga S. Africa
29 D6 Kominternivs'ke Ukr.
28 J2 Komi, Respublika div. Rus. Fed.
26 G3 Komiža Croatia
27 H1 Komló Hungary
38 E5 Kommuna Turkm.
114 B3 Komodou Guinea
114 C3 Komoé r. Côte d'Ivoire
110 B4 Komono Congo
53 F6 Komoro Japan
21 M1 Komorzno Pol.
27 L4 Komotini Greece
112 D6 Komsberg mts S. Africa
38 E1 Komsomolets Kazak.
33 L1 Komsomolets, O. i. Rus. Fed.
38 C3 Komsomolets, Zaliv b. Kazak.
28 G3 Komsomol'sk Rus. Fed.
38 E5 Komsomol'sk Turkm.
29 E5 Komsomol's'k Ukr.
38 C3 Komsomol'skiy Kazak.
29 H6 Komsomol'skiy Kalmykiya Rus. Fed.
28 H4 Komsomol'skiy Mordov. Rus. Fed.
49 P1 Komsomol'sk-na-Amure Rus. Fed.
38 D4 Komsomol'sk-na-Ustyurte Uzbek.
38 E2 Komsomol'skoye Kazak.
32 H3 Komsonol'skiy Rus. Fed.
37 J1 Kömürlü Turkey
93 F6 Kom Vo U.S.A.
28 F3 Konakovo Rus. Fed.
45 F5 Konar Resr India
44 D4 Konch India
45 E6 Kondagaon India
89 J2 Kondiaronk, Lac l. Can.
69 D6 Kondinin Austr.
115 C3 Kondoa Tanz.
28 E2 Kondopoga Rus. Fed.
28 E4 Kondrovo Rus. Fed.
71 □9 Kone New Caledonia Pac. Oc.
114 F3 Kong Cameroon
75 P3 Kong Christian IX Land reg. Greenland
72 N2 Kong Christian X Land reg. Greenland
72 O2 Kong Frederik VIII Land reg. Greenland
75 O3 Kong Frederik VI Kyst reg. Greenland
116 C2 Kong Håkon VII Hav sea Ant.
54 D5 Kongju S. Korea
32 D2 Kong Karl's Land is Svalbard
59 E2 Kongkemul mt Indon.
115 A4 Kongolo Congo(Zaire)
75 Q2 Kong Oscar Fjord in. Greenland
114 C2 Kongoussi Burkina
9 L7 Kongsberg Norway
9 N6 Kongsvinger Norway
58 C2 Kông, T. r. Cambodia
39 H5 Kongur Shan mt China
115 C4 Kongwa Tanz.
75 Q2 Kong Wilhelm Land reg. Greenland
58 C2 Kong, Xé r. Laos
20 D4 Königsbrunn Ger.
19 M5 Königswartha Ger.
15 H2 Königswinter Ger.
19 L4 Königs Wusterhausen Ger.
16 J4 Konin Pol.
46 F1 Konin r. Rus. Fed.
22 E3 Köniz Switz.
26 G3 Konjic Bos.-Herz.
112 B3 Konkiep watercourse Namibia
114 C2 Konna Mali
19 J5 Könnern Ger.
8 U5 Konnevesi Fin.
28 G2 Konosha Rus. Fed.
53 F6 Kōnosu Japan
29 E5 Konotop Ukr.
58 D2 Kon Plong Vietnam
115 C1 Konso Eth.
20 F3 Konstantinovy Lázně Czech Rep.
20 C5 Konstanz Ger.
114 E2 Kontagora Nigeria
8 V5 Kontiolahti Fin.
8 U4 Konttila Fin.
58 C2 Kon Tum Vietnam
58 D2 Kontum, Plateau du plat. Vietnam
36 D3 Konya Turkey
39 J3 Konyrolen Kazak.
38 C2 Konystanu Kazak.
15 G3 Konz Ger.
69 E4 Kookynie Austr.
92 □1 Koolau Range mts U.S.A.
69 D5 Koolyanobbing Austr.
66 F5 Koondrook Austr.
90 D5 Koon Lake U.S.A.
67 H5 Koorawatha Austr.
69 C5 Koorda Austr.
82 C2 Kooskia U.S.A.
76 F5 Kootenay r. Can./U.S.A.
76 F5 Kootenay L. Can.
76 F4 Kootenay Nat. Park Can.
112 D5 Kootjieskolk S. Africa
39 J3 Kopa Kazak.
29 H6 Kopanovka Rus. Fed.
44 C6 Kopargaon India
8 E3 Kópasker Iceland
39 J3 Kopbirlik Kazak.
26 E2 Koper Slovenia
38 D5 Kopet Dag, Khrebet mts Turkm.
58 B3 Ko Phangan i. Thai.
58 A3 Ko Phra Thong i. Thai.
58 A4 Ko Phuket i. Thai.
21 J2 Kopidlno Czech Rep.
9 P7 Köping Sweden
8 Q5 Köpmanholmen Sweden
113 F2 Kopong Botswana
43 B3 Koppal India
9 M6 Koppang Norway
21 L6 Koppány r. Romania
9 O7 Kopparberg Sweden
113 G3 Koppies S. Africa
112 D3 Koppieskraalpan salt pan S. Africa
26 G1 Koprivnica Croatia
36 C3 Köprü r. Turkey
40 D4 Kor watercourse Iran
28 G4 Korablino Rus. Fed.
41 G5 Korak Pak.
78 E1 Korak, Baie b. Can.
43 B2 Korangal India
41 G5 Korangi Pak.
43 C2 Koraput India
Korat see Nakhon Ratchasima
45 E5 Korba India
26 D6 Korba Tunisia
18 F5 Korbach Ger.
58 B4 Korbu, Gunung mt Malaysia
27 J4 Korçë Albania
39 K1 Korchino Rus. Fed.
26 G3 Korčula Croatia
26 G3 Korčula i. Croatia
26 G3 Korčulanski Kanal chan. Croatia
37 M4 Kord Khvord Iran
40 D2 Kord Küy Iran
41 F5 Kords reg. Iran
40 D4 Kord Sheykh Iran
54 B4 Korea Bay g. China/N. Korea
31 M5 Korea, North country Asia
31 M6 Korea, South country Asia
53 A7 Korea Strait Japan/S. Korea
43 C2 Koregaon India
29 F6 Korenovsk Rus. Fed.
29 C5 Korets' Ukr.
36 B1 Körfez Turkey
116 B3 Korff Ice Rise ice feature Ant.
39 K3 Korgas China
8 N3 Korgen Norway
114 B3 Korhogo Côte d'Ivoire
44 B5 Kori Creek in. India
27 K5 Korinthiakos Kolpos chan. Greece
27 K6 Korinthos Greece
16 H7 Kőris-hegy mt Hungary
27 J3 Koritnik mt Albania
53 G6 Kōriyama Japan
36 C3 Korkuteli Turkey
36 D4 Kormakitis, Cape Cyprus
16 H7 Körmend Hungary
39 H2 Korneyevka Karagan. Kazak.
39 G1 Korneyevka Severnyy Kaz. Kazak.
39 K1 Kornilovo Rus. Fed.
20 C4 Kornwestheim Ger.
114 B3 Koro Côte d'Ivoire
71 □15 Koro i. Fiji
114 C2 Koro Mali
71 □15 Koroba mt Fiji
29 F5 Korocha Rus. Fed.
36 D1 Köroğlu Dağları mts Turkey
36 D1 Köroğlu Tepesi mt Turkey
115 C4 Korogwe Tanz.
66 E7 Koroit Austr.
71 □15 Korolevu Fiji
66 E6 Korong Vale Austr.
27 K4 Koronia, L. Greece
57 J5 Koror Palau
71 □13 Koro Sea Fiji
29 D5 Korosten' Ukr.
29 D5 Korostyshiv Ukr.
71 □13 Korotasere Fiji
109 D3 Koro Toro Chad
71 □15 Koroyanitu mt Fiji
9 T5 Korpilahti Fin.
9 R6 Korpo Fin.
38 C3 Korsak Kazak.
49 Q2 Korsakov Rus. Fed.
28 J3 Korshik Rus. Fed.
8 R5 Korsnäs Fin.
9 M9 Korsør Denmark
29 D5 Korsun'-Shevchenkivs'kyy Ukr.
17 K3 Korsze Pol.
8 S5 Kortesjärvi Fin.
28 J2 Kortkeros Rus. Fed.
15 D2 Kortrijk Belgium
28 G3 Kortsovo Rus. Fed.
67 F7 Korumburra Austr.
114 E3 Korup, Parc National de nat. park Cameroon
8 U3 Korvala Fin.
44 D4 Korwai India
46 H1 Koryakskaya Sopka volc. Rus. Fed.
33 S3 Koryakskiy Khrebet mts Rus. Fed.
28 H2 Koryazhma Rus. Fed.
54 E6 Koryŏng S. Korea
29 E5 Koryukivka Ukr.
38 C2 Korzhun Kazak.
27 M6 Kos i. Greece
38 E2 Kosagal Kazak.
58 B3 Ko Samui i. Thai.
54 D4 Kosan N. Korea
38 F2 Kosay Kazak.
38 C3 Koschagyl Kazak.
20 E4 Kösching Ger.
16 H4 Kościan Pol.
85 F5 Kosciusko U.S.A.
76 C3 Kosciusko I. U.S.A.
67 H6 Kosciusko, Mt Austr.
67 H6 Kosciusko National Park Austr.
37 G1 Köse Turkey
36 F1 Köse Dağı mt Turkey
43 B2 Kosgi India
48 E2 Kosh-Agach Rus. Fed.
38 B2 Koshankol' Kazak.
39 H4 Kosh-Döbö Kyrg.
53 A9 Koshikijima-rettō is Japan
41 F3 Koshkak Iran
41 F3 Koshk-e-Kohneh Afgh.
88 C4 Koshkoning, Lake U.S.A.
38 E4 Koshkupyr Uzbek.
38 C4 Koshoba Turkm.
39 F4 Koshrabad Uzbek.
44 D4 Kosi India
44 D3 Kosi r. India
113 K3 Kosi Bay S. Africa
17 K6 Košice Slovakia
43 B3 Kosigi India
38 D2 Kos-Istek Kazak.
39 F2 Koskol' Kazak.
39 J3 Koskuduk Kazak.
8 R3 Koskullskule Sweden
28 J2 Koslan Rus. Fed.
54 E4 Kosŏng N. Korea
54 E3 Kosŏng-ni N. Korea
27 J3 Kosovo div. Yugo.
27 J3 Kosovska Mitrovica Yugo.
60 H3 Kosrae i. Micronesia
39 J5 Kosrap China
114 B3 Kossou, Lac de l. Côte d'Ivoire
27 K3 Kostenets Bulg.
113 G2 Koster S. Africa
109 F3 Kosti Sudan
27 K3 Kostinbrod Bulg.
32 K3 Kostino Rus. Fed.
28 D1 Kostomuksha Rus. Fed.
29 C5 Kostopil' Ukr.
28 G3 Kostroma Rus. Fed.
28 G3 Kostroma r. Rus. Fed.
28 G3 Kostromskaya Oblast' div. Rus. Fed.
16 G4 Kostrzyn Pol.
29 F5 Kostyantynivka Ukr.
16 H3 Koszalin Pol.
16 H7 Kőszeg Hungary
21 K5 Kőszegi-hegység h. Hungary
45 E5 Kota Madhya Pradesh India
44 C4 Kota Rajasthan India
59 B4 Kotaagung Indon.
44 C4 Kota Barrage barrage India
59 E3 Kotabaru Indon.
59 D3 Kotabaru Indon.
59 B1 Kota Bharu Malaysia
59 B3 Kotabumi Indon.
44 C4 Kota Dam dam India
59 E1 Kota Kinabalu Malaysia
39 J3 Kotanemel', Gora mt Kazak.
58 A3 Ko Tao i. Thai.
43 C2 Kotapārh India
58 B5 Kotapinang Indon.
44 C4 Kotari r. India
59 B2 Kota Tinggi Malaysia
28 J3 Kotel'nich Rus. Fed.
29 G6 Kotel'nikovo Rus. Fed.
33 P2 Kotel'nyy, O. i. Rus. Fed.
44 D3 Kotgarh India
19 J5 Köthen (Anhalt) Ger.
44 E4 Kothi India
115 B2 Kotido Uganda
9 U6 Kotka Fin.
44 C3 Kot Kapura India
28 H2 Kotlas Rus. Fed.
44 C2 Kotli Pak.
74 B3 Kotlik U.S.A.
8 D5 Kötlutangi pt Iceland
9 V7 Kotly Rus. Fed.
114 E3 Koton-Karifi Nigeria
114 E2 Kotorkoshi Nigeria
26 G2 Kotor Varoš Bos.-Herz.
114 C3 Kotouba Côte d'Ivoire
29 H5 Kotovo Rus. Fed.
29 G4 Kotovsk Rus. Fed.
29 D6 Kotovs'k Ukr.
44 C4 Kotra India
44 E6 Kotri r. India
44 B4 Kotri Pak.
38 D2 Kotr-Tas Kazak.
44 A5 Kot Sarae Pak.
20 G6 Kötschach Austria
43 C2 Kottagudem India
43 B4 Kottarakara India
43 B4 Kottayam India
43 B5 Kotte Sri Lanka
43 B3 Kotturu India
71 □12 Kotu i. Tonga
71 □12 Kotu Group is Tonga
38 C5 Koturdepe Turkm.
33 M2 Kotuy r. Rus. Fed.
74 B3 Kotzebue U.S.A.
74 B3 Kotzebue Sound b. U.S.A.
20 F3 Kötzting Ger.
108 A3 Koubia Guinea
114 C2 Koudougou Burkina
112 E6 Koueveldberg mts S. Africa
114 F2 Koufey Niger
27 M7 Koufonisi i. Greece
112 E6 Kougaberg mts S. Africa
36 D4 Kouklia Cyprus
110 B4 Koulamoutou Gabon
114 B2 Koulikoro Mali
63 G4 Koumac New Caledonia
71 □9 Koumac New Caledonia Pac. Oc.
68 C3 Koumala Austr.
108 A3 Koundâra Guinea
39 H3 Kounradskiy Zhezkaz. Kazak.
39 H3 Kounradskiy Zhezkaz. Kazak.
114 C2 Koupéla Burkina
21 H3 Kouřim Czech Rep.
101 H2 Kourou Fr. Guiana
114 B2 Kouroussa Guinea
109 D3 Kousséri Cameroon
114 B2 Koutiala Mali
71 □9 Koutoumo i. New Caledonia Pac. Oc.
9 U6 Kouvola Fin.
8 W3 Kovdor Rus. Fed.
8 W3 Kovdozero, Oz. l. Rus. Fed.
29 C5 Kovel' Ukr.
28 G3 Kovernino Rus. Fed.
43 B4 Kovilpatti India
28 G3 Kovrov Rus. Fed.
28 G4 Kovylkino Rus. Fed.
28 F2 Kovzhskoye, Ozero l. Rus. Fed.
21 J2 Kowary Pol.
70 C5 Kowhitirangi N.Z.
51 □ Kowloon Peninsula H.K. China
51 □ Kowloon Pk h. H.K. China
54 D4 Kowŏn N. Korea
39 K5 Koxlax China
39 J5 Koxtag China
53 B7 Kōyama-misaki pt Japan
58 A3 Ko Yao Yai i. Thai.
36 B3 Köyceğiz Turkey
28 J2 Koygorodok Rus. Fed.
38 C4 Koymatdag, Gory h. Turkm.
43 A2 Koyna Resr India
28 H1 Koynas Rus. Fed.
74 C3 Koyukuk r. U.S.A.
36 F1 Koyulhisar Turkey
28 F3 Koza Rus. Fed.
53 A7 Kō-zaki pt Japan
36 E3 Kozan Turkey
27 J4 Kozani Greece
26 G2 Kozara mts Bos.-Herz.
21 M4 Kozárovce Slovakia
29 D5 Kozelets' Ukr.
28 E4 Kozel'sk Rus. Fed.
Kozhikode see Calicut
36 C1 Kozlu Turkey
28 H3 Koz'modem'yansk Rus. Fed.
39 G4 Kozmoldak Kazak.
27 K4 Kožuf mts Greece/Macedonia
53 F7 Kōzu-shima i. Japan
29 D5 Kozyatyn Ukr.
114 D3 Kpalimé Togo
114 D3 Kpandu Ghana
58 A3 Krabi Thai.
58 A3 Kra Buri Thai.
58 C2 Krâcheh Cambodia
19 M3 Krackow Ger.
8 P4 Kraddsele Sweden
9 L7 Kragerø Norway
18 C4 Kraggenburg Neth.
27 J2 Kragujevac Yugo.
58 A3 Kra, Isthmus of isth. Thai.
19 M3 Krajnik Dolny Pol.
59 C4 Krakatau i. Indon.
58 C2 Krâkôr Cambodia
17 J5 Kraków Pol.
19 K3 Krakower See l. Ger.
58 B2 Krâlănh Cambodia
103 C1 Kralendijk Neth. Ant.
21 K2 Králíky Czech Rep.
21 G3 Kralovice Czech Rep.
21 H2 Kralupy nad Vltavou Czech Rep.
21 H3 Králův Dvůr Czech Rep.
29 F5 Kramators'k Ukr.
8 P5 Kramfors Sweden
18 B5 Krammer est. Neth.
27 K6 Kranidi Greece
26 F1 Kranj Slovenia
58 □ Kranji Resr Sing.
21 G6 Kranjska Gora Slovenia
113 J4 Kranskop S. Africa
21 L2 Krapkowice Pol.
28 H2 Krasavino Rus. Fed.
32 G2 Krasino Rus. Fed.
52 B3 Kraskino Rus. Fed.
9 U9 Krāslava Latvia
20 F2 Kraslice Czech Rep.
17 P4 Krasnapollye Belarus
28 D4 Krasnaya Gora Rus. Fed.
39 H2 Krasnaya Polyana Kazak.
39 G1 Krasnoarmeysk Kazak.
29 H5 Krasnoarmeysk Rus. Fed.
29 F5 Krasnoarmiys'k Ukr.
28 H2 Krasnoborsk Rus. Fed.
29 F6 Krasnodar Rus. Fed.
29 F6 Krasnodarskiy Kray div. Rus. Fed.
29 F5 Krasnodon Ukr.
39 H4 Krasnogorka Kazak.
28 D3 Krasnogorodskoye Rus. Fed.
49 Q2 Krasnogorsk Rus. Fed.
29 G6 Krasnogvardeyskoye Rus. Fed.
29 E6 Krasnohrad Ukr.
29 E6 Krasnohvardiys'ke Ukr.
38 C2 Krasnokholm Rus. Fed.
39 H1 Krasnokutsk Kazak.
17 R2 Krasnomayskiy Rus. Fed.
29 E6 Krasnoperekops'k Ukr.
52 D2 Krasnorechenskiy Rus. Fed.
9 V6 Krasnosel'skoye Rus. Fed.
28 G4 Krasnoslobodsk Rus. Fed.
38 D1 Krasnousol'skiy Rus. Fed.
38 C4 Krasnovodskiy Zaliv b. Turkm.
38 C4 Krasnovodsk, Mys pt Turkm.
38 C4 Krasnovodskoye Plato plat. Turkm.
38 C2 Krasnoyar Kazak.
46 B1 Krasnoyarsk Rus. Fed.
39 G2 Krasnoznamenskoye Kazak.
17 P3 Krasnyy Rus. Fed.
28 H3 Krasnyye Baki Rus. Fed.
29 H6 Krasnyye Barrikady Rus. Fed.
28 F3 Krasnyy Kholm Rus. Fed.
29 H5 Krasnyy Kut Rus. Fed.
28 D3 Krasnyy Luch Rus. Fed.
29 F5 Krasnyy Lyman Ukr.
39 J3 Krasnyy Oktyabr' Kazak.
39 G1 Krasnyy Yar Kazak.
29 J6 Krasnyy Yar Astrak. Rus. Fed.
29 H5 Krasnyy Yar Volgograd. Rus. Fed.
38 B1 Krasnyy Yar Rus. Fed.
29 C5 Krasyliv Ukr.
29 H7 Krayinovka Rus. Fed.
18 D5 Krefeld Ger.
29 E5 Kremenchuk Ukr.
29 E5 Kremenchuts'ka Vodoskhovshche resr Ukr.
29 G5 Kremenskaya Rus. Fed.
21 J3 Křemešník h. Czech Rep.
82 F3 Kremmling U.S.A.
21 J4 Krems r. Austria
21 J4 Krems an der Donau Austria
21 H4 Kremsmünster Austria
21 H4 Křemže Czech Rep.
21 K4 Křepice Czech Rep.
33 U3 Kresta, Zaliv b. Rus. Fed.
28 E3 Kresttsy Rus. Fed.
9 R9 Kretinga Lith.
20 E5 Kreuth Ger.
18 D6 Kreuzau Ger.
20 G6 Kreuzeck mt Austria
23 G2 Kreuzlingen Switz.
18 E6 Kreuztal Ger.
17 N3 Kreva Belarus
114 E4 Kribi Cameroon
21 J5 Krieglach Austria
113 H3 Kriel S. Africa
27 J5 Krikellos Greece
52 H2 Kril'on, Mys c. Rus. Fed.
21 G5 Krippenstein mt Austria
35 F5 Krishna r. India
43 B3 Krishnagiri India
43 C3 Krishna, Mouths of the river mouth India
45 G5 Krishnanagar India
43 B3 Krishnaraja Sagara l. India
9 K7 Kristiansand Norway
9 O8 Kristianstad Sweden
8 K5 Kristiansund Norway
9 O7 Kristinehamn Sweden
9 R5 Kristinestad Fin.
Kriti i. see Crete
39 K2 Kriushi Kazak.
21 G3 Křivoklátská Vrchovina reg. Czech Rep.
Krivoy Rog see Kryvyy Rih
21 K3 Křižanov Czech Rep.
26 G1 Križevci Croatia
26 F2 Krk i. Croatia
21 J2 Krkonoše mts Czech Rep.
21 J2 Krkonošský narodní park nat. park Czech Rep./Pol.
21 L2 Krnov Czech Rep.
21 K2 Krobielowice tourist site Pol.
8 O5 Krokom Sweden
8 L5 Krokstadøra Norway
8 O3 Krokstranda Norway
29 E5 Krolevets' Ukr.
20 E2 Kronach Ger.
58 B3 Krŏng Kaôh Kŏng Cambodia
8 S5 Kronoby Fin.
75 P3 Kronprins Frederik Bjerge mt Greenland
19 H2 Kronshagen Ger.
58 A2 Kronwa Myanmar
113 G3 Kroonstad S. Africa
19 J2 Kröpelin Ger.
29 G6 Kropotkin Rus. Fed.
19 G2 Kropp Ger.
19 K5 Kropstädt Ger.
17 K6 Krosno Pol.
16 H5 Krotoszyn Pol.
113 J2 Kruger National Park S. Africa
17 O3 Kruhlaye Belarus
59 B4 Krui Indon.
112 F7 Kruisfontein S. Africa
27 H4 Krujë Albania
20 C5 Krumbach Austria
20 D4 Krumbach (Schwaben) Ger.
27 L4 Krumovgrad Bulg.
Krungkao see Ayutthaya
Krung Thep see Bangkok
21 G2 Krupka Czech Rep.
17 O3 Krupki Belarus
27 J3 Kruševac Yugo.
20 F2 Krušné Hory mts Czech Rep.
76 B3 Kruzof I. U.S.A.
28 D4 Krychaw Belarus
29 F6 Krymsk Rus. Fed.
39 H3 Krypsalo Kazak.
27 L6 Krytiko Pelagos sea Greece
29 E6 Kryvyy Rih Ukr.
21 M2 Krzepice Pol.
19 M4 Krzymów Pol.
108 B2 Ksabi Alg.
108 C1 Ksar el Boukhari Alg.
108 B1 Ksar el Kebir Morocco
29 F5 Kshenskiy Rus. Fed.
26 D7 Ksour Essaf Tunisia
28 H3 Kstovo Rus. Fed.
58 A4 Kuah Malaysia
58 B4 Kuala Kangsar Malaysia
58 B4 Kuala Kerai Malaysia
58 B5 Kuala Kubu Baharu Malaysia
59 B2 Kuala Lipis Malaysia
59 B2 Kuala Lumpur Malaysia
58 B4 Kuala Nerang Malaysia
58 B5 Kuala Pilah Malaysia
58 B5 Kuala Rompin Malaysia
59 D3 Kualasampit Indon.
58 A4 Kualasimpang Indon.
59 B1 Kuala Terengganu Malaysia
55 A5 Kuamut Malaysia
54 C3 Kuandian China
51 F6 Kuanshan Taiwan
59 B2 Kuantan Malaysia
29 G6 Kuban' r. Rus. Fed.
37 J5 Kubaysah Iraq
28 F3 Kubenskoye, Ozero l. Rus. Fed.
19 L2 Kubitzer Bodden b. Ger.
28 H4 Kubnya r. Rus. Fed.
27 M3 Kubrat Bulg.
71 □13 Kubulau Pt Fiji
44 C4 Kuchāman India
44 C4 Kuchera India
59 D2 Kuching Malaysia
53 A10 Kuchino-shima i. Japan
20 G5 Kuchl Austria
39 J1 Kuchukskoye, Oz. salt l. Rus. Fed.
Kucing see Kuching
27 H4 Kuçovë Albania
43 A3 Kudal India
59 E1 Kudat Malaysia
43 B3 Kudligi India
21 K2 Kudowa-Zdrój Pol.
43 A2 Kudremukh mt India
59 D4 Kudus Indon.
20 F5 Kufstein Austria
39 J3 Kugaly Kazak.
28 H3 Kugesi Rus. Fed.
74 E3 Kugmallit Bay Can.
41 F5 Kūhak Iran
45 E3 Kuhanbokano mt China
20 E4 Kühbach Ger.
19 K3 Kuhbier Ger.
40 E4 Kūhbonān Iran
40 B3 Kūhdasht Iran
37 L2 Kūhhaye Sabalan mts Iran
37 M3 Kūhīn Iran
9 V4 Kuhmo Fin.
9 T6 Kuhmoinen Fin.
40 D3 Kūhpāyeh Iran
40 E5 Kūh, Ra's-al pt Iran
112 B2 Kuis Namibia
112 A1 Kuiseb Pass Namibia
111 B5 Kuito Angola
76 C3 Kuiu Island U.S.A.
8 T4 Kuivaniemi Fin.
40 B5 Kū', J. al h. S. Arabia
45 F5 Kujang India
54 C4 Kujang-Dong N. Korea
52 G4 Kuji Japan
53 B8 Kujū-san volc. Japan
89 F1 Kukatush Can.
69 D6 Kukerin Austr.
27 J3 Kukës Albania
28 J3 Kukmor Rus. Fed.
58 B5 Kukup Malaysia
40 D5 Kūl r. Iran
36 B2 Kula Turkey
38 B2 Kulagino Kazak.
45 G3 Kula Kangri mt Bhutan
38 B3 Kulaly, O. i. Kazak.
39 H4 Kulanak Kyrg.
38 D3 Kulandy Kazak.
38 D3 Kulandy, Poluostrov pen. Kazak.
41 F5 Kulaneh reg. Pak.
39 J4 Kulansarak China
39 G2 Kulanutpes watercourse Kazak.
41 G4 Kulao r. Pak.
33 P2 Kular Rus. Fed.
55 B5 Kulassein i. Phil.
45 H4 Kulaura Bangl.
9 R8 Kuldīga Latvia
38 E4 Kul'dzhuktau, Gory h. Uzbek.
112 D1 Kule Botswana
28 G4 Kulebaki Rus. Fed.
58 C2 Kulen Cambodia
38 E1 Kulevchinskoye Rus. Fed.
65 F5 Kulgera Austr.
58 B4 Kulim Malaysia
69 D6 Kulin Austr.
39 H5 Kuli Sarez l. Tajik.
69 C5 Kulja Austr.
38 E4 Kulkuduk Uzbek.
66 F3 Kulkyne watercourse Austr.
44 D3 Kullu India
20 E2 Kulmbach Ger.
39 G5 Kŭlob Tajik.
37 H2 Kulp Turkey
44 D4 Kulpahar India
91 F4 Kulpsville U.S.A.
38 C3 Kul'sary Kazak.
20 C3 Külsheim Ger.
36 D2 Kulu Turkey
36 C3 Kulübe Tepe mt Turkey
39 J1 Kulunda Rus. Fed.
39 K1 Kulunda r. Rus. Fed.
39 H1 Kulundinskaya Step' plain Kazak./Rus. Fed.
39 J1 Kulundinskoye, Ozero salt l. Rus. Fed.
40 D4 Kūlvand Iran
66 E5 Kulwin Austr.
54 D5 Kŭm r. S. Korea
29 H6 Kuma r. Rus. Fed.
53 F6 Kumagaya Japan
52 F3 Kumaishi Japan
59 D3 Kumai, Teluk b. Indon.
38 E2 Kumak Rus. Fed.
38 D2 Kumak r. Rus. Fed.

53 B8 **Kumamoto** Japan
53 E8 **Kumano** Japan
27 J3 **Kumanovo** Macedonia
114 C3 **Kumasi** Ghana
Kumayri *see* **Gyumri**
114 E4 **Kumba** Cameroon
43 B4 **Kumbakonam** India
21 J5 **Kumberg** Austria
36 C2 **Kümbet** Turkey
112 E1 **Kumchuru** Botswana
40 D3 **Kumel** *well* Iran
38 C1 **Kumertau** Rus. Fed.
54 E4 **Kumgang-san** *mt* N. Korea
54 E6 **Kŭmho** *r.* S. Korea
54 D4 **Kumhwa** S. Korea
54 E5 **Kumi** S. Korea
115 B2 **Kumi** Uganda
9 O7 **Kumla** Sweden
19 K3 **Kummerower See** *l.* Ger.
19 L4 **Kummersdorf-Alexanderdorf** Ger.
114 F2 **Kumo** Nigeria
54 D6 **Kŭmō-do** *i.* S. Korea
39 F3 **Kumola** *watercourse* Kazak.
23 L5 **Kumpare, Rt** *pt* Croatia
58 B1 **Kumphawapi** Thai.
112 C4 **Kums** Namibia
43 A3 **Kumta** India
115 A2 **Kumu** Congo(Zaire)
29 H7 **Kumukh** Rus. Fed.
41 H3 **Kunar** *r.* Afgh.
39 K4 **Künas Chang** China
49 R3 **Kunashir, Ostrov** *i.* Rus. Fed.
45 E2 **Kunchuk Tso** *salt l.* China
9 U7 **Kunda** Estonia
45 E4 **Kunda** India
43 A3 **Kundāpura** India
44 B2 **Kundar** *r.* Afgh./Pak.
115 A5 **Kundelungu Ouest, Parc National de** *nat. park* Congo(Zaire)
115 A5 **Kundelungu, Parc National de** *nat. park* Congo(Zaire)
41 H2 **Kunduz** Afgh.
41 H2 **Kunduz** *r.* Afgh.
Künes *see* **Xinyuan**
39 K4 **Künes He** *r.* China
9 M8 **Kungälv** Sweden
39 J4 **Kungei Alatau** *mts* Kazak./Kyrg.
76 C4 **Kunghit I.** Can.
38 D4 **Kungrad** Uzbek.
38 E4 **Kungradkol'** Uzbek.
9 N8 **Kungsbacka** Sweden
9 M7 **Kungshamn** Sweden
110 B3 **Kungu** Congo(Zaire)
44 D6 **Kuni** *r.* India
53 B8 **Kunimi-dake** *mt* Japan
45 F5 **Kunjabar** India
45 G4 **Kunlui** *r.* India/Nepal
35 F3 **Kunlun Shan** *mts* China
45 H2 **Kunlun Shankou** *pass* China
51 B5 **Kunming** China
44 D4 **Kuno** *r.* India
54 D6 **Kunsan** S. Korea
50 F4 **Kunshan** China
64 E3 **Kununurra** Austr.
44 D4 **Kunwari** *r.* India
28 D3 **Kun'ya** Rus. Fed.
54 A5 **Kunyu Shan** *h.* China
20 C2 **Künzell** Ger.
20 C3 **Künzelsau** Ger.
19 J5 **Künzels Berg** *h.* Ger.
51 F4 **Kuocang Shan** *mts* China
9 T6 **Kuohijärvi** *l.* Fin.
8 V3 **Kuolayarvi** Rus. Fed.
8 U5 **Kuopio** Fin.
8 S5 **Kuortane** Fin.
26 F2 **Kupa** *r.* Croatia/Slovenia
57 G9 **Kupang** Indon.
9 T9 **Kupiškis** Lith.
76 C3 **Kupreanof Island** U.S.A.
29 F5 **Kup''yans'k** Ukr.
39 K4 **Kuqa** China
37 M2 **Kür** *r.* Azer.
37 K1 **Kura** *r.* Azer./Georgia
29 G7 **Kura** *r.* Georgia/Rus. Fed.
69 C1 **Kurabuka** *r.* Austr.
39 H4 **Kuragaty** Kazak.
29 H7 **Kurakh** Rus. Fed.
68 A1 **Kuranda** Austr.
39 G2 **Kurchasayskiy** Kazak.
53 C7 **Kurashiki** Japan
45 E5 **Kurasia** India
53 C7 **Kurayoshi** Japan
36 B1 **Kurban Dağı** *mt* Turkey
29 E5 **Kurchatov** Rus. Fed.
39 K2 **Kurchum** Kazak.
37 M1 **Kürdämir** Azer.
39 H4 **Kurday** Kazak.
37 M2 **Kür Dili** *pt* Azer.
43 B2 **Kurduvadi** India
27 L4 **Kŭrdzhali** Bulg.
53 C7 **Kure** Japan
36 D1 **Küre** Turkey
118 H4 **Kure Atoll** *atoll HI* U.S.A.
9 S7 **Kuressaare** Estonia
39 G2 **Kurgal'dzhinskiy** Kazak.
32 H4 **Kurgan** Rus. Fed.
29 G6 **Kurganinsk** Rus. Fed.
39 F2 **Kurgasyn** Kazak.
41 H2 **Kuri** Afgh.
44 B4 **Kuri** India
Kuria Muria Islands *see* **Ḥalāniyāt, Juzur al**
8 S5 **Kurikka** Fin.
52 G5 **Kurikoma-yama** *volc.* Japan
52 G4 **Kuril Islands** Rus. Fed.
38 B2 **Kurilovka** Rus. Fed.
46 G2 **Kuril'sk** Japan
Kuril'skiye Ostrova *is* *see* **Kuril Islands**
118 F2 **Kuril Trench** *sea feature* Pac. Oc.
21 K3 **Kuřim** Czech Rep.
38 B1 **Kurmanayevka** Rus. Fed.
110 F3 **Kurmuk** Sudan
43 B3 **Kurnool** India
36 E6 **Kurnub** Israel
53 E6 **Kurobe** Japan
52 G4 **Kuroishi** Japan
53 G6 **Kuroiso** Japan
53 A9 **Kuro-shima** *i.* Japan
28 F4 **Kurovskoye** Rus. Fed.
70 C6 **Kurow** N.Z.
44 B2 **Kurram** *r.* Afgh./Pak.
67 J4 **Kurri Kurri** Austr.
Kuršių Marios *lag.* *see* **Courland Lagoon**
29 F5 **Kursk** Rus. Fed.
29 H6 **Kurskaya** Rus. Fed.
29 F5 **Kurskaya Oblast'** *div.* Rus. Fed.
Kurskiy Zaliv *lag.* *see* **Courland Lagoon**
36 D1 **Kurşunlu** Turkey
37 H3 **Kurtalan** Turkey
39 J3 **Kurty** *r.* Kazak.
45 G4 **Kuru** *r.* Bhutan
36 F2 **Kuruçay** Turkey
44 D3 **Kurukshetra** India
46 A2 **Kuruktag** *mts* China
112 E3 **Kuruman** S. Africa
112 D3 **Kuruman** *watercourse* S. Africa
53 B8 **Kurume** Japan
49 K1 **Kurumkan** Rus. Fed.
43 C5 **Kurunegala** Sri Lanka
109 F2 **Kurush, Jebel** *reg.* Sudan
39 K2 **Kur'ya** Rus. Fed.
38 B4 **Kuryk** Kazak.
27 M6 **Kuşadası** Turkey
27 M6 **Kuşadası Körfezi** *b.* Turkey
76 B2 **Kusawa Lake** Can.
15 H3 **Kusel** Ger.
36 A1 **Kuş Gölü** *l.* Turkey
29 F6 **Kushchevskaya** Rus. Fed.
53 B9 **Kushikino** Japan
53 D8 **Kushimoto** Japan
52 J3 **Kushiro** Japan
52 J3 **Kushiro-Shitsugen National Park** Japan
41 F3 **Kushka** *r.* Turkm.
37 M5 **Kūshkak** Iran
38 F1 **Kushmurun** Kazak.
43 B3 **Kushtagi** India
45 G5 **Kushtia** Bangl.
50 C2 **Kushui** *r.* China
38 B2 **Kushum** Kazak.
38 B2 **Kushum** *r.* Kazak.
74 C3 **Kuskokwim** *r.* U.S.A.
74 B4 **Kuskokwim Bay** U.S.A.
74 C3 **Kuskokwim Mts** U.S.A.
54 C4 **Kusŏng** N. Korea
52 J3 **Kussharo-ko** *l.* Japan
38 E1 **Kustanay** Kazak.
38 E1 **Kustanayskaya Oblast'** *div.* Kazak.
18 E3 **Küstenkanal** *canal* Ger.
40 C4 **Kut** Iran
37 M6 **Kūt Abdollāh** Iran
55 A5 **Kutacane** Indon.
36 B2 **Kütahya** Turkey
29 G7 **K'ut'aisi** Georgia
Kut-al-Imara *see* **Al Kūt**
29 H6 **Kutan** Rus. Fed.
52 G3 **Kutchan** Japan
37 M5 **Kūt-e Gapu** Iran
26 G2 **Kutina** Croatia
26 G2 **Kutjevo** Croatia
21 J3 **Kutná Hora** Czech Rep.
17 J4 **Kutno** Pol.
110 B4 **Kutu** Congo(Zaire)
45 G5 **Kutubdia I.** Bangl.
74 G2 **Kuujjua** *r.* Can.
79 G2 **Kuujjuaq** Can.
Kuujjuarapik *see* **Poste-de-la-Baleine**
38 C4 **Kuuli-Mayak** Turkm.
8 V4 **Kuusamo** Fin.
9 U6 **Kuusankoski** Fin.
38 D2 **Kuvandyk** Rus. Fed.
111 B5 **Kuvango** Angola
28 E3 **Kuvshinovo** Rus. Fed.
Kuwaé *i.* *see* **Tongoa**
31 D7 **Kuwait** *country* Asia
37 L7 **Kuwait** Kuwait
37 L7 **Kuwait Jun** *b.* Kuwait
53 E7 **Kuwana** Japan
28 G1 **Kuya** Rus. Fed.
32 J4 **Kuybyshev** *Novosibirsk* Rus. Fed.
Kuybyshev *see* **Samara**
38 F1 **Kuybyshevskiy** Kazak.
38 B1 **Kuybyshevskoye Vdkhr.** *resr* Rus. Fed.
50 D2 **Kuye** *r.* China
39 H3 **Kuygan** Kazak.
35 G2 **Kuytun** China
39 K3 **Kuytun He** *r.* China
27 N6 **Kuyucak** Turkey
38 F2 **Kuyukkol', Ozero** *salt l.* Kazak.
39 L2 **Kuyus** Rus. Fed.
9 V6 **Kuznechnoye** Rus. Fed.
28 H4 **Kuznetsk** Rus. Fed.
52 F1 **Kuznetsovo** Rus. Fed.
29 C5 **Kuznetsovs'k** Ukr.
8 R1 **Kvænangen** *chan.* Norway
19 H1 **Kværndrup** Denmark
8 Q2 **Kvaløya** *i.* Norway
8 S1 **Kvalsund** Norway
Kvareli *see* **Qvareli**
38 D1 **Kvarkeno** Rus. Fed.
26 F2 **Kvarnerić** *chan.* Croatia
74 C4 **Kvichak Bay** U.S.A.
76 D3 **Kwadacha Wilderness Prov. Park** Can.
51 **Kwai Tau Leng** *h.* *H.K.* China
118 G5 **Kwajalein** *i.* Pac. Oc.
58 A5 **Kwala** Indon.
114 E3 **Kwale** Nigeria
113 J4 **KwaMashu** S. Africa
113 H2 **KwaMhlanga** S. Africa
115 C4 **Kwa Mtoro** Tanz.
54 D5 **Kwangch'ŏn** S. Korea
54 D6 **Kwangju** S. Korea
110 B4 **Kwango** *r.* Congo(Zaire)
115 C4 **Kwangwazi** Tanz.
54 D6 **Kwangyang** S. Korea
113 F6 **Kwanobuhle** S. Africa
113 F6 **KwaNojoli** S. Africa
113 G6 **Kwanonqubela** S. Africa
112 F5 **Kwanonzame** S. Africa
113 G6 **Kwatinidubu** S. Africa
113 H3 **KwaZamokhule** S. Africa
112 F6 **Kwazamukucinga** S. Africa
113 F5 **Kwazamuxolo** S. Africa
113 H3 **KwaZanele** S. Africa
113 J4 **Kwazulu-Natal** *div.* S. Africa
111 C5 **Kwekwe** Zimbabwe
112 F1 **Kweneng** *div.* Botswana
110 B4 **Kwenge** *r.* Congo(Zaire)
113 G5 **Kwezi-Naledi** S. Africa
17 J4 **Kwidzyn** Pol.
74 B4 **Kwigillingok** U.S.A.
62 E2 **Kwikila** P.N.G.
110 B4 **Kwilu** *r.* Angola/Congo(Zaire)
57 F7 **Kwoka** *mt* Indon.
51 **Kwun Tong** *H.K.* China
109 D4 **Kyabé** Chad
66 F6 **Kyabram** Austr.
56 B3 **Kyaikto** Myanmar
58 A1 **Kya-in Seikkyi** Myanmar
48 J1 **Kyakhta** Rus. Fed.
65 F5 **Kyalite** Austr.
64 A4 **Kyancutta** Austr.
28 F1 **Kyanda** Rus. Fed.
58 A1 **Kyaukhnyat** Myanmar
56 A2 **Kyaukpyu** Myanmar
45 H5 **Kyauktaw** Myanmar
9 S9 **Kybartai** Lith.
66 D6 **Kybybolite** Austr.
115 B4 **Kyela** Tanz.
44 D2 **Kyelang** India
115 B2 **Kyenjojo** Uganda
39 H3 **Kyeylbelen, Gora** *h.* Kazak.
50 A2 **Kyikug** China
Kyiv *see* **Kiev**
21 L3 **Kyjov** Czech Rep.
Kyklades *is* *see* **Cyclades**
77 H4 **Kyle** Can.
11 C3 **Kyle of Lochalsh** U.K.
53 G3 **Kyll** *r.* Ger.
27 K6 **Kyllini** *mt* Greece
66 F6 **Kyneton** Austr.
115 B2 **Kyoga, Lake** Uganda
53 C7 **Kyōga-misaki** *pt* Japan
67 K2 **Kyogle** Austr.
58 A1 **Kyondo** Myanmar
68 A3 **Kyong** Austr.
54 E6 **Kyŏngju** S. Korea
53 D7 **Kyōto** Japan
27 J6 **Kyparissia** Greece
27 K6 **Kyparissiakos Kolpos** *b.* Greece
39 G2 **Kypshak, Ozero** *salt l.* Kazak.
27 L5 **Kyra Panagia** *i.* Greece
31 G5 **Kyrgyzstan** *country* Asia
19 K4 **Kyritz** Ger.
8 L5 **Kyrksæterøra** Norway
32 G3 **Kyrta** Rus. Fed.
28 H1 **Kyssa** Rus. Fed.
21 M3 **Kysucké Nové Mesto** Slovakia
33 P3 **Kytalyktakh** Rus. Fed.
27 K6 **Kythira** *i.* Greece
27 K6 **Kythnos** *i.* Greece
57 A2 **Kyungyaung** Myanmar
38 D3 **Kyushe** Kazak.
53 B8 **Kyūshū** *i.* Japan
118 D5 **Kyushu - Palau Ridge** *sea feature* Pac. Oc.
27 K3 **Kyustendil** Bulg.
67 G5 **Kywong** Austr.
29 D5 **Kyyivs'ke Vdskh.** *resr* Ukr.
8 T5 **Kyyjärvi** Fin.
38 C3 **Kyzan** Kazak.
48 F1 **Kyzyl** Rus. Fed.
39 H5 **Kyzylart Pass** Kyrg.
39 G2 **Kyzyldyykan** Kazak.
39 K3 **Kyzylkesek** Kazak.
38 C2 **Kyzylkoga** Kazak.
38 E2 **Kyzylkol', Ozero** *l.* Kazak.
38 E4 **Kyzylkum Desert** Uzbek.
39 H4 **Kyzyl-Kyya** Kyrg.
38 F3 **Kyzyl-Mazhalyk** Rus. Fed.
39 H5 **Kyzyl-Suu** *r.* Kyrg.
39 H2 **Kyzyltas** Kazak.
39 H3 **Kyzyltau** Kazak.
38 B3 **Kyzylysor** Kazak.
39 G2 **Kzyl-Dzhar** Kazak.
38 F3 **Kzyl-Orda** Kazak.
38 E3 **Kzyl-Ordinskaya Oblast'** *div.* Kazak.
39 H1 **Kzyltu** Kazak.
38 F2 **Kzyluy** Kazak.

L

21 K4 **Laa an der Thaya** Austria
15 H2 **Laacher See** *l.* Ger.
19 K3 **Laage** Ger.
9 T7 **Laagri** Estonia
8 U2 **Laanila** Fin.
105 B3 **La Araucania** *div.* Chile
110 E3 **Laascaanood** Somalia
110 E2 **Laasgoray** Somalia
103 E2 **La Asunción** Venez.
108 A2 **Laâyoune** Western Sahara
29 G6 **Laba** *r.* Rus. Fed.
96 D1 **La Babia** Mex.
30 D3 **La Banda** Arg.
82 E3 **La Barge** U.S.A.
71 13 **Labasa** Fiji
24 C3 **La Baule-Escoublac** France
15 A4 **La Bazoche-Gouet** France
21 H2 **Labe** *r.* Czech Rep.
108 A3 **Labé** Guinea
78 F4 **Labelle** Can.
88 B5 **La Belle** U.S.A.
78 B2 **Laberge, Lake** Can.
55 A5 **Labian, Tg** *pt* Malaysia
76 E2 **La Biche** *r.* Can.
29 G6 **Labinsk** Rus. Fed.
58 B5 **Labis** Malaysia
55 B3 **Labo** Phil.
96 C2 **La Boquilla** Mex.
69 D2 **Labouchere, Mt** *h.* Austr.
74 F4 **Laboué** Lebanon
24 D4 **Labouheyre** France
105 D2 **Laboulaye** Arg.
79 H3 **Labrador** Can.
79 G3 **Labrador City** Can.
75 N3 **Labrador Sea** Can./Greenland
100 F5 **Lábrea** Brazil
96 B2 **La Brecha** Mex.
59 E1 **Labuan** Malaysia
59 C4 **Labuhan** Indon.
59 B2 **Labuhanbilik** Indon.
58 A5 **Labuhanruku** Indon.
55 A5 **Labuk** *r.* Malaysia
59 E1 **Labuk, Telukan** *b.* Malaysia
57 H7 **Labuna** Indon.
66 A3 **Labyrinth, L.** *salt flat* Austr.
32 H3 **Labytnangi** Rus. Fed.
27 H4 **Laç** Albania
105 D1 **La Calera** Arg.
105 B2 **La Calera** Chile
15 D3 **La Capelle** France
105 D4 **La Carlota** Arg.
25 E3 **La Carolina** Spain
27 M2 **Lăcăuți, Vârful** *mt* Romania
91 J1 **Lac-Baker** Can.
34 F5 **Laccadive Islands** India
77 K4 **Lac du Bonnet** Can.
96 H5 **La Ceiba** Honduras
103 C2 **La Ceiba** Venez.
66 C6 **Lacepede B.** Austr.
91 E4 **Laceyville** U.S.A.
28 F2 **Lacha, Ozero** *l.* Rus. Fed.
22 D2 **La Chaux-de-Fonds** Switz.
19 H4 **Lachendorf** Ger.
22 D6 **Lachens, Montagne de** *mt* France
89 F3 **Lachine** U.S.A.
66 F5 **Lachlan** *r.* Austr.
96 K6 **La Chorrera** Panama
78 F4 **Lachute** Can.
37 M1 **Laçın** Azer.
22 C6 **La Ciotat** France
96 C3 **La Ciudad** Mex.
90 D3 **Lackawanna** U.S.A.
77 G4 **Lac La Biche** Can.
76 E3 **Lac La Hache** Can.
76 F2 **Lac la Martre** Can.
77 H3 **Lac La Ronge Provincial Park** Can.
22 C5 **La Cluse** France
79 F4 **Lac Mégantic** Can.
76 F4 **Lacolle** Can.
96 B1 **La Colorada** Mex.
76 G4 **Lacombe** Can.
97 F4 **La Concordia** Mex.
26 C5 **Laconi** *Sardinia* Italy
91 H3 **Laconia** U.S.A.
89 J1 **La Corne** Can.
88 B4 **La Crescent** U.S.A.
88 B4 **La Crosse** U.S.A.
103 A4 **La Cruz** Col.
96 C3 **La Cruz** *Sinaloa* Mex.
97 E3 **La Cruz** *Tamaulipas* Mex.
96 H5 **La Cruz** Nic.
96 D1 **La Cuesta** Mex.
84 C4 **La Cygne** U.S.A.
44 D2 **Ladakh Range** *mts* India
58 A4 **Ladang** *i.* Thai.
21 K2 **Lądek-Zdrój** Pol.
36 E1 **Lâdik** Turkey
112 D6 **Ladismith** S. Africa
41 F4 **Lādīz** Iran
44 C4 **Ladnun** India
103 B3 **La Dorada** Col.
Ladozhskoye Ozero *l.* *see* **Ladoga, Lake**
45 H4 **Ladu** *mt* India
28 E2 **Ladva** Rus. Fed.
28 E2 **Ladva-Vetka** Rus. Fed.
75 K2 **Lady Ann Strait** Can.
11 E4 **Ladybank** U.K.
113 G4 **Ladybrand** S. Africa
89 G2 **Lady Evelyn Lake** Can.
113 G5 **Lady Frere** S. Africa
113 G5 **Lady Grey** S. Africa
76 E5 **Ladysmith** Can.
113 H4 **Ladysmith** S. Africa
88 B3 **Ladysmith** U.S.A.
39 G2 **Ladyzhenka** Kazak.
62 E2 **Lae** P.N.G.
58 B2 **Laem Ngop** Thai.
58 B4 **Laem Pho** *pt* Thai.
9 K6 **Lærdalsøyri** Norway
100 F8 **La Esmeralda** Bol.
103 D4 **La Esmeralda** Venez.
9 M8 **Læsø** *i.* Denmark
96 G5 **La Esperanza** Honduras
50 E2 **Lafa** China
105 D1 **La Falda** Arg.
22 C5 **La Faurie** France
82 F4 **Lafayette** *CO* U.S.A.
88 D5 **Lafayette** *IN* U.S.A.
85 E6 **Lafayette** *LA* U.S.A.
87 C5 **La Fayette** U.S.A.
15 D3 **La Fère** France
15 D4 **La Ferté-Gaucher** France
15 D3 **La Ferté-Milon** France
15 D4 **La Ferté-sous-Jouarre** France
40 C5 **Laffān, Ra's** *pt* Qatar
114 E3 **Lafia** Nigeria
114 E3 **Lafiagi** Nigeria
24 D3 **La Flèche** France
21 K5 **Lafnitz** Austria
71 9 **La Foa** *New Caledonia* Pac. Oc.
90 A6 **La Follette** U.S.A.
89 H2 **Laforce** Can.
89 G2 **Laforest** Can.
79 F3 **Laforge** Can.
103 B2 **La Fría** Venez.
40 D5 **Laft** Iran
26 C6 **La Galite** *i.* Tunisia
29 H6 **Lagan'** Rus. Fed.
14 E3 **Lagan** *r.* U.K.
101 L6 **Lagarto** Brazil
114 F3 **Lagdo, Lac de** *l.* Cameroon
18 F5 **Lage** Ger.
9 L7 **Lågen** *r.* Norway
11 C5 **Lagg** U.K.
11 D3 **Laggan** U.K.
11 D4 **Laggan, Loch** *l.* U.K.
115 D2 **Lagh Bor** *watercourse* Kenya/Somalia
108 C1 **Laghouat** Alg.
45 F2 **Lagkor Co** *salt l.* China
103 B2 **La Gloria** Col.
104 D2 **Lagoa Santa** Brazil
28 D2 **Lagoda, Lake** Rus. Fed.
37 L1 **Lagodekhi** Georgia
58 D5 **Lagong** *i.* Indon.
55 B3 **Lagonoy Gulf** Phil.
102 B7 **Lago Posadas** Arg.
105 B4 **Lago Ranco** Chile
114 D3 **Lagos** Nigeria
25 B4 **Lagos** Port.
115 A4 **Lagosa** Tanz.
96 D3 **Lagos de Moreno** Mex.
22 B4 **La Grand-Croix** France
78 E3 **La Grande** *r.* Can.
82 C2 **La Grande** U.S.A.
22 B5 **La Grande-Combe** France
22 B6 **La Grande-Motte** France
78 F3 **La Grande 4, Réservoir de** *resr* Can.
78 E3 **La Grande 3, Réservoir de** *resr* Can.
78 E3 **La Grande 2, Réservoir de** *resr* Can.
64 D3 **Lagrange** Austr.
87 C5 **La Grange** *GA* U.S.A.
91 J2 **La Grange** *ME* U.S.A.
88 D5 **La Grange** *MI* U.S.A.
88 B5 **La Grange** *MO* U.S.A.
85 D6 **La Grange** *TX* U.S.A.
88 E5 **Lagrange** U.S.A.
103 E3 **La Gran Sabana** *plat.* Venez.
22 D4 **La Grave** France
102 G3 **Laguna** Brazil
92 D5 **Laguna Beach** U.S.A.
105 B3 **Laguna de Laja, Parque Nacional** *nat. park* Chile
96 J5 **Laguna de Perlas** Nic.
92 D5 **Laguna Mts** U.S.A.
100 C5 **Lagunas** Peru
102 A7 **Laguna San Rafael, Parque Nacional** *nat. park* Chile
97 E5 **Lagunas de Chacahua, Parque Nacional** *nat. park* Mex.
103 C2 **Lagunillas** Venez.
59 E1 **Lahad Datu** Malaysia
55 A5 **Lahad Datu, Telukan** *b.* Malaysia
92 2 **Lahaina** U.S.A.
37 M3 **Lahargin** Iran
59 B3 **Lahat** Indon.
58 A5 **Lahewa** Indon.
42 B7 **Laḥij** Yemen
40 C2 **Lāhījān** Iran
92 1 **Lahilahi Pt** U.S.A.
18 F6 **Lahn** *r.* Ger.
15 H2 **Lahnstein** Ger.
9 N8 **Laholm** Sweden
92 C2 **Lahontan Resr** U.S.A.
44 C3 **Lahore** Pak.
103 E3 **La Horqueta** Venez.
44 B3 **Lahri** Pak.
20 E3 **Lahr (Schwarzwald)** Ger.
9 T6 **Lahti** Fin.
96 C4 **La Huerta** Mex.
109 D4 **Laï** Chad
50 F3 **Lai'an** China
51 C6 **Laibin** China
40 E4 **Laidāru** Iran
67 K1 **Laidley** Austr.
92 1 **Laie** U.S.A.
92 1 **Laie Pt** U.S.A.
51 C4 **Laifeng** China
22 C5 **L'Aigle** France
22 B2 **Laignes** France
96 G5 **La Iguala** Honduras
8 S5 **Laihia** Fin.
45 H4 **Laimakuri** India
71 9 **Laimbélé, Mt** *h.* Vanuatu
15 F4 **Laimont** France
112 D6 **Laingsburg** S. Africa
8 S3 **Lainioälven** *r.* Sweden
21 H4 **Lainsitz** *r.* Austria
11 D2 **Lairg** U.K.
55 C5 **Lais** Phil.
9 R6 **Laitila** Fin.
23 H3 **Laives** Italy
50 E2 **Laiwu** China
50 F2 **Laiyang** China
50 E2 **Laiyuan** China
50 F2 **Laizhou Wan** *b.* China
105 B3 **Laja** *r.* Chile
105 B3 **Laja, Lago de la** *l.* Chile
67 F3 **Lajamanu** Austr.
101 L5 **Lajes** *Rio Grande do Norte* Brazil
102 F3 **Lajes** *Santa Catarina* Brazil
96 C2 **La Joya** Mex.
21 L5 **Lajta** *r.* Austria/Hungary
83 G4 **La Junta** U.S.A.
82 E2 **Lake** U.S.A.
71 14 **Lakeba** *i.* Fiji
71 14 **Lakeba Passage** Fiji
36 D6 **Lake Bardawil Reserve** Egypt
66 E6 **Lake Bolac** Austr.
67 G4 **Lake Cargelligo** Austr.
67 K3 **Lake Cathie** Austr.
82 B1 **Lake Chelan Nat. Recreation Area** *res.* U.S.A.
87 D6 **Lake City** *FL* U.S.A.
88 E3 **Lake City** *MI* U.S.A.
88 A3 **Lake City** *MN* U.S.A.
87 E5 **Lake City** *SC* U.S.A.
23 G2 **Lake Constance** *l.* Ger./Switz.
12 D3 **Lake District National Park** U.K.
92 D5 **Lake Elsinore** U.S.A.
66 B2 **Lake Eyre Nat. Park** Austr.
89 H3 **Lakefield** Can.
65 H2 **Lakefield National Park** Austr.
88 C4 **Lake Geneva** U.S.A.
69 D6 **Lake Grace** Austr.
75 M3 **Lake Harbour** Can.
93 E4 **Lake Havasu City** U.S.A.
92 C4 **Lake Isabella** U.S.A.
69 D6 **Lake King** Austr.
87 D6 **Lakeland** U.S.A.
88 C2 **Lake Linden** U.S.A.
76 F4 **Lake Louise** Can.
93 E4 **Lake Mead National Recreation Area** *res.* U.S.A.
91 J2 **Lake Moxie** U.S.A.
69 E2 **Lake Nabberu** *salt l.* Austr.
82 B2 **Lake Oswego** U.S.A.
70 B5 **Lake Paringa** N.Z.
91 G2 **Lake Placid** U.S.A.
92 A2 **Lakeport** U.S.A.
70 C6 **Lake Pukaki** N.Z.
70 C6 **Lake Pukaki** N.Z.
78 D3 **Lake River** Can.
89 H3 **Lake St Peter** Can.
67 H6 **Lakes Entrance** Austr.
88 E2 **Lake Superior National Park** Can.
67 J5 **Lake Tabourie** Austr.
70 C6 **Lake Tekapo** N.Z.
78 E4 **Lake Traverse** Can.
82 B3 **Lakeview** U.S.A.
82 F4 **Lakewood** *CO* U.S.A.
91 F4 **Lakewood** *NJ* U.S.A.
90 C4 **Lakewood** *OH* U.S.A.
87 D7 **Lake Worth** U.S.A.
28 D2 **Lakhdenpokh'ya** Rus. Fed.
44 E4 **Lakhimpur** India
44 D5 **Lakhnadon** India
44 B5 **Lakhpat** India
27 K6 **Lakonikos Kolpos** *b.* Greece
114 B3 **Lakota** Côte d'Ivoire
8 U1 **Laksefjorden** *chan.* Norway
8 T1 **Lakselv** Norway
43 A4 **Lakshadweep** *div.* India
45 G5 **Laksham** Bangl.
43 B2 **Lakshettipet** India
45 G5 **Lakshmikantapur** India
55 B5 **Lala** Phil.
105 D2 **La Laguna** Arg.
105 B3 **La Laja** Chile
71 10 **Lalalomei Bank** *rf* Tonga
105 B3 **Lalara** Gabon
68 C4 **Laleham** Austr.
19 K3 **Lalendorf** Ger.
40 C3 **Lālī** Iran
105 B4 **La Libertad** El Salvador
97 G4 **La Libertad** Guatemala
105 B2 **La Ligua** Chile
54 D1 **Lalin** China
54 C1 **Lalin** *r.* China
25 B1 **Lalín** Spain
25 D4 **La Línea de la Concepción** Spain
44 D4 **Lalitpur** India
22 C5 **Lalley** France
55 B2 **Lal-Lo** Phil.
77 H3 **La Loche** Can.
77 H3 **La Loche, Lac** *l.* Can.
71 4 **Lalomanu** Western Samoa
15 B4 **La Loupe** France
15 E2 **La Louvière** Belgium
28 H2 **Lal'sk** Rus. Fed.
45 H5 **Lama** Bangl.
26 C4 **La Maddalena** *Sardinia* Italy
55 A5 **Lamag** Malaysia
58 A2 **Lamaing** Myanmar
La Manche *str.* *see* **English Channel**
84 C4 **Lamar** *CO* U.S.A.
85 E4 **Lamar** *MO* U.S.A.
22 C1 **Lamarche** France
40 D5 **Lamard** Iran
26 C5 **La Marmora, Punta** *mt* *Sardinia* Italy
105 D3 **Lamarque** Arg.
85 E6 **La Marque** U.S.A.
76 F2 **La Martre, Lac** *l.* Can.
22 B5 **Lamastre** France
110 B4 **Lambaréné** Gabon
100 C5 **Lambayeque** Peru
14 F4 **Lambay Island** Rep. of Ireland
116 D4 **Lambert Gl.** Ant.
112 C6 **Lambert's Bay** S. Africa
22 C6 **Lambesc** France
44 C3 **Lambi** India
21 L2 **Łambinowice** Pol.
13 F6 **Lambourn Downs** *h.* U.K.
23 G4 **Lambro** *r.* Italy
58 C2 **Lam Chi** *r.* Thai.
25 C2 **Lamego** Port.
71 9 **Lamen** Vanuatu
79 H4 **Lamèque, I.** Can.
100 C6 **La Merced** Peru
66 D5 **Lameroo** Austr.
92 D5 **La Mesa** U.S.A.
85 C5 **Lamesa** U.S.A.
27 K5 **Lamia** Greece
21 J5 **Laming** *r.* Austria
67 K2 **Lamington Nat. Park** Austr.
96 B1 **La Misa** Mex.
96 D2 **La Misión** Mex.
55 B5 **Lamitan** Phil.
51 **Lamma I.** *H.K.* China
70 B6 **Lammerlaw Ra.** *mts* N.Z.
11 F5 **Lammermuir Hills** U.K.
9 O8 **Lammhult** Sweden
9 T6 **Lammi** Fin.
88 C5 **La Moille** U.S.A.
91 G2 **Lamoille** *r.* U.S.A.
88 B5 **La Moine** *r.* U.S.A.
22 D6 **La Môle** France
55 B3 **Lamon Bay** Phil.
23 J5 **Lamone** *r.* Italy
84 E3 **Lamoni** U.S.A.
82 F3 **Lamont** U.S.A.
96 C1 **La Morita** Mex.
89 H1 **La Motte** Can.
22 C4 **La Motte-Servolex** France
58 B1 **Lam Pao Resr** Thai.
85 D6 **Lampasas** U.S.A.
96 D2 **Lampazos** Mex.
26 E7 **Lampedusa, Isola di** *i.* *Sicily* Italy
20 B3 **Lampertheim** Ger.
13 C5 **Lampeter** U.K.
58 B2 **Lam Plai Mat** *r.* Thai.
28 F4 **Lamskoye** Rus. Fed.
51 **Lam Tin** *H.K.* China
115 D3 **Lamu** Kenya
45 H6 **Lamu** Myanmar
22 C5 **La Mure** France
23 J3 **Lana** Italy
92 2 **Lanai** *i.* U.S.A.
92 2 **Lanai City** U.S.A.
55 C5 **Lanao, Lake** Phil.
89 J3 **Lanark** Can.
11 E5 **Lanark** U.K.
88 C4 **Lanark** U.S.A.
55 A5 **Lanas** Malaysia
58 A3 **Lanbi Kyun** *i.* Myanmar
Lancang Jiang *r.* *see* **Mekong**
91 F2 **Lancaster** Can.
12 E3 **Lancaster** U.K.
92 C4 **Lancaster** *CA* U.S.A.
88 A5 **Lancaster** *MO* U.S.A.
91 H2 **Lancaster** *NH* U.S.A.
90 B5 **Lancaster** *OH* U.S.A.
91 E4 **Lancaster** *PA* U.S.A.
87 D5 **Lancaster** *SC* U.S.A.
88 B4 **Lancaster** *WI* U.S.A.
12 E4 **Lancaster Canal** U.K.
75 K2 **Lancaster Sound** *str.* Can.
69 B5 **Lancelin** Austr.
26 F3 **Lanciano** Italy
105 B3 **Lanco** Chile
50 F2 **Lancun** China
22 D3 **Lancy** Switz.
20 F4 **Landau an der Isar** Ger.
20 B3 **Landau in der Pfalz** Ger.
20 D5 **Landeck** Austria
82 E3 **Lander** U.S.A.
19 G4 **Landesbergen** Ger.
77 H4 **Landis** Can.
69 C2 **Landor** Austr.
23 G3 **Landquart** Switz.
20 D4 **Landsberg am Lech** Ger.
68 A4 **Landsborough Cr.** *watercourse* Austr.
13 B7 **Land's End** *pt* U.K.
20 F4 **Landshut** Ger.
9 N9 **Landskrona** Sweden
15 H3 **Landstuhl** Ger.
18 F3 **Land Wursten** *reg.* Ger.
14 D4 **Lanesborough** Rep. of Ireland
58 C3 **La Nga** *r.* Vietnam
44 E3 **La'nga Co** *l.* China
50 C3 **Langao** China
39 H5 **Langar** Afgh.
41 F3 **Langar** Iran
11 B2 **Langavat, Loch** *l.* U.K.
112 E4 **Langberg** *mts* S. Africa
84 D1 **Langdon** U.S.A.
112 C6 **Langeberg** *mts* S. Africa
19 H2 **Langelands Bælt** *str.* Denmark
9 T6 **Längelmäki** Fin.
9 T6 **Längelmävesi** *l.* Fin.
19 H5 **Langelsheim** Ger.
18 D3 **Langen** Ger.
20 C3 **Langenburg** Ger.
20 D5 **Langenfeld** Austria
19 G4 **Langenhagen** Ger.
15 H2 **Langenhahn** Ger.
21 J4 **Langenlois** Austria
15 H3 **Langenlonsheim** Ger.
22 E2 **Langenthal** Switz.
19 J4 **Langenweddingen** Ger.
18 E3 **Langeoog** Ger.
18 E3 **Langeoog** *i.* Ger.
19 H1 **Langeskov** Denmark
9 L7 **Langesund** Norway
58 B5 **Langgapayung** Indon.
20 B2 **Langgöns** Ger.
77 H4 **Langham** Can.
23 H5 **Langhirano** Italy
8 C4 **Langjökull** *ice cap* Iceland
59 A2 **Langka** Indon.
59 A1 **Langkawi** *i.* Malaysia
58 A3 **Lang Kha Toek, Khao** *mt* Thai.
112 D4 **Langklip** S. Africa
55 A5 **Langkon** Malaysia
89 K1 **Langlade** Can.
88 C3 **Langlade** U.S.A.
68 A5 **Langlo** *watercourse* Austr.
68 A6 **Langlo Crossing** Austr.
22 A5 **Langogne** France
8 O2 **Langøya** *i.* Norway
45 F3 **Langphu** *mt* China
13 E6 **Langport** U.K.
51 F5 **Langqi** China
22 C2 **Langres** France
22 B2 **Langres, Plateau de** *plat.* France
44 D1 **Langru** China
59 A2 **Langsa** Indon.
58 A4 **Langsa, Teluk** *b.* Indon.
8 P5 **Långsele** Sweden
50 C1 **Langshan** China
50 C1 **Lang Shan** *mts* China
51 C6 **Lang Sơn** Vietnam
12 G3 **Langtoft** U.K.
85 C6 **Langtry** U.S.A.
24 F5 **Languedoc** *reg.* France
22 B5 **Languedoc-Roussillon** *div.* France
8 R4 **Långvattnet** Sweden
19 G4 **Langwedel** Ger.
50 F4 **Langxi** China
50 C4 **Langzhong** China
89 H2 **Laniel** Can.
77 H4 **Lanigan** Can.
92 1 **Lanikai** U.S.A.
105 B3 **Lanín, Parque Nacional** *nat. park* Arg.
105 B3 **Lanín, Volcán** *volc.* Arg.
50 E3 **Lankao** China
37 M2 **Länkäran** Azer.
21 J6 **Lannach** Austria
24 C2 **Lannion** France
96 C3 **La Noria** Mex.
8 S3 **Lansån** Sweden
88 C2 **L'Anse** U.S.A.
76 C2 **Lansing** *r.* Can.
88 A4 **Lansing** *IA* U.S.A.
88 E4 **Lansing** *MI* U.S.A.
21 K3 **Lanškroun** Czech Rep.
51 D6 **Lantau I.** *H.K.* China
51 **Lantau Island** *H.K.* China
51 **Lantau Peak** *h.* *H.K.* China
22 D2 **Lanterne** *r.* France
22 E6 **Lantosque** France
22 A3 **Lanty** France
55 C4 **Lanuza Bay** Phil.
51 F4 **Lanxi** China
109 F4 **Lanya** Sudan
51 F6 **Lan Yü** *i.* Taiwan
108 A2 **Lanzarote** *i.* Canary Is
50 B2 **Lanzhou** China
54 B1 **Lanzijing** China
22 E4 **Lanzo Torinese** Italy
55 B2 **Laoag** Phil.
55 C3 **Laoang** Phil.
51 B6 **Lao Cai** Vietnam
39 K3 **Laofengkou** China
50 F1 **Laoha** *r.* China
50 D3 **Laohekou** China
54 A4 **Laohutun** China
54 D3 **Laoling** China
54 D3 **Lao Ling** *mts* China
15 D3 **Laon** France
88 C3 **Laona** U.S.A.
31 K8 **Laos** *country* Asia
50 F2 **Lao Shan** *mt* China
54 E2 **Laotougou** China
54 C3 **Laotuding Shan** *h.* China
Laowohi *pass* *see* **Khardung La**
50 A1 **Laoximiao** China
54 E2 **Laoye Ling** *mts* China
104 C4 **Lapa** Brazil
55 B5 **Lapac** *i.* Phil.
22 A3 **La Pacaudière** France
96 K6 **La Palma** Panama
25 C4 **La Palma del Condado** Spain
105 F2 **La Paloma** Uru.
105 D3 **La Pampa** *div.* Arg.
92 B4 **La Panza Range** *mts* U.S.A.
103 E3 **La Paragua** Venez.
55 A5 **Laparan** *i.* Phil.
105 E1 **La Paz** *Entre Ríos* Arg.
105 C2 **La Paz** *Mendoza* Arg.
100 E7 **La Paz** Bol.
96 H5 **La Paz** Honduras
96 B2 **La Paz** Mex.
88 D5 **Lapaz** U.S.A.
100 E4 **La Pedrera** Col.
89 F4 **Lapeer** U.S.A.
96 C1 **La Perla** Mex.
71 8 **La Pérouse, Bahía** *b.* *Easter I.* Chile
52 G2 **La Pérouse Strait** Japan/Rus. Fed.
97 E3 **La Pesca** Mex.
96 D3 **La Piedad** Mex.
103 E3 **La Piña** *r.* Venez.
82 B3 **La Pine** U.S.A.
55 C3 **Lapinig** Phil.
55 C4 **Lapinin** *i.* Phil.
8 U5 **Lapinlahti** Fin.
36 D4 **Lapithos** Cyprus
85 F6 **Laplace** U.S.A.
84 C2 **La Plant** U.S.A.
105 F2 **La Plata** Arg.
103 B4 **La Plata** Col.
28 G1 **Lapominka** Rus. Fed.
88 D5 **La Porte** U.S.A.
88 A4 **La Porte City** U.S.A.
15 H4 **Lapoutroie** France
8 S5 **Lappajärvi** Fin.
8 S5 **Lappajärvi** *l.* Fin.
9 V6 **Lappeenranta** Fin.
20 F3 **Lappersdorf** Ger.
8 S2 **Lappland** *reg.* Europe
91 G2 **La Prairie** Can.
85 D6 **La Pryor** U.S.A.
27 M4 **Lâpseki** Turkey
33 N2 **Laptev Sea** Rus. Fed.
8 S5 **Lapua** Fin.
55 B4 **Lapu-Lapu** Phil.
26 E3 **L'Aquila** Italy
92 D5 **La Quinta** U.S.A.
40 D5 **Lār** Iran
108 B1 **Larache** Morocco
22 C5 **Laragne-Montéglin** France
40 E5 **Lārak** *i.* Iran
82 F3 **Laramie** U.S.A.
82 F3 **Laramie Mts** U.S.A.
Laranda *see* **Karaman**
104 B4 **Laranjeiras do Sul** Brazil
104 B3 **Laranjinha** *r.* Brazil
57 G8 **Larantuka** Indon.
57 J8 **Larat** *i.* Indon.
25 H4 **Larba** Alg.
22 B4 **L'Arbresle** France
9 Q8 **Lärbro** Sweden
89 G2 **Larchwood** Can.
89 H1 **Larder Lake** Can.
25 E1 **Laredo** Spain
85 D7 **Laredo** U.S.A.
22 D5 **L'Argentière-la-Bessée** France
87 D7 **Largo** U.S.A.
11 D5 **Largs** U.K.
40 B2 **Lārī** Iran
26 D6 **L'Ariana** Tunisia
77 K5 **Larimore** U.S.A.
102 C3 **La Rioja** Arg.
105 C1 **La Rioja** *div.* Arg.
25 E1 **La Rioja** *div.* Spain
27 K5 **Larisa** Greece
40 E5 **Laristan** *reg.* Iran
44 B4 **Larkana** Pak.
22 D3 **Larmont** *mt* France/Switz.
36 D4 **Larnaka** Cyprus
14 F3 **Larne** U.K.
84 D4 **Larned** U.S.A.
11 C6 **Larne Lough** *in.* U.K.
25 D1 **La Robla** Spain
15 F2 **La Roche-en-Ardenne** Belgium
24 D3 **La Rochelle** France
22 D3 **La Roche-sur-Foron** France
24 D3 **La Roche-sur-Yon** France
25 E3 **La Roda** Spain
95 L5 **La Romana** Dom. Rep.
77 H3 **La Ronge** Can.
96 D2 **La Rosa** Mex.
96 D1 **La Rosita** Mex.
64 F3 **Larrimah** Austr.
116 B2 **Larsen Ice Shelf** *ice feature* Ant.
8 S5 **Larsmo** Fin.
9 M7 **Larvik** Norway
93 H2 **La Sal Junction** U.S.A.
91 G2 **La Salle** Can.
88 C5 **La Salle** U.S.A.
78 E4 **La Sarre** Can.
103 D2 **Las Aves, Islas** *is* Venez.
105 B2 **Las Cabras** Chile
79 J4 **La Scie** Can.
83 F5 **Las Cruces** U.S.A.
95 K5 **La Selle** *mt* Haiti
105 B1 **La Serena** Chile
96 D2 **Las Esperanças** Mex.
22 C6 **La Seyne-sur-Mer** France
105 E3 **Las Flores** Arg.
41 F5 **Lāshār** *r.* Iran
77 H4 **Lashburn** Can.
105 C2 **Las Heras** Arg.
56 B2 **Lashio** Myanmar
41 G4 **Lashkar Gāh** Afgh.
105 D3 **Las Lajas** Arg.
103 D3 **Las Lajitas** Venez.
102 D2 **Las Lomitas** Arg.
25 C4 **Las Marismas** *marsh* Spain
102 C7 **Las Martinetas** Arg.
103 D2 **Las Mercedes** Venez.
96 C2 **Las Nieves** Mex.
92 D5 **Las Palmas** *r.* Mex.
108 A2 **Las Palmas de Gran Canaria** Canary Is
23 G5 **La Spezia** Italy
105 F2 **Las Piedras** Uru.
102 C6 **Las Plumas** Arg.
96 G5 **Las Quebradas** Guatemala
105 E2 **Las Rosas** Arg.
82 B3 **Lassen Pk** *volc.* U.S.A.
82 B3 **Lassen Volcanic Nat. Park** U.S.A.
116 B2 **Lassiter Coast** Ant.
96 J7 **Las Tablas** Panama
102 D3 **Las Termas** Arg.
77 H4 **Last Mountain L.** Can.
110 B4 **Lastoursville** Gabon
26 G3 **Lastovo** *i.* Croatia
103 D3 **Las Trincheras** Venez.
18 E4 **Lastrup** Ger.
83 F6 **Las Varas** *Chihuahua* Mex.
96 C3 **Las Varas** Mex.
105 D1 **Las Varillas** Arg.
83 F5 **Las Vegas** *NM* U.S.A.
93 E3 **Las Vegas** *NV* U.S.A.
25 D3 **Las Villuercas** *mt* Spain
100 C4 **Latacunga** Ecuador
116 A2 **Latady I.** Ant.
103 B5 **La Tagua** Col.
36 E4 **Latakia** Syria
89 H2 **Latchford** Can.
71 12 **Late** *i.* Tonga
45 F5 **Latehar** India
23 J3 **Latemar** *mt* Italy
24 D4 **La Teste** France
69 C4 **Latham** Austr.
18 E4 **Lathen** Ger.
11 E2 **Latheron** U.K.
26 E4 **Latina** Italy
23 L4 **Latisana** Italy
105 D2 **La Toma** Arg.
103 D2 **La Tortuga, Isla** *i.* Venez.
22 C4 **La Tour-du-Pin** France
22 E6 **La Trinité** France
67 G8 **Latrobe** Austr.
90 D4 **Latrobe** U.S.A.
18 D4 **Lattrop** Neth.
89 H2 **Latulipe** Can.
78 F4 **La Tuque** Can.
43 B2 **Latur** India
7 F3 **Latvia** *country* Europe
71 13 **Laucala** *i.* Fiji
102 C1 **Lauca, Parque Nacional** *nat. park* Chile
19 L5 **Lauchhammer** Ger.
20 B5 **Lauchringen** Ger.
20 C3 **Lauda-Königshofen** Ger.
11 F5 **Lauder** U.K.
19 G3 **Lauenbrück** Ger.
19 H3 **Lauenburg (Elbe)** Ger.
20 E3 **Lauf an der Pegnitz** Ger.
20 F5 **Laufen** Ger.
22 E2 **Laufen** Switz.
20 C3 **Lauffen am Neckar** Ger.
88 D2 **Laughing Fish Pt** U.S.A.
71 14 **Lau Group** *is* Fiji
9 S7 **Lauka** Estonia
8 V1 **Laukvik** Norway
71 4 **Laulii** Western Samoa
52 D2 **Laulyu** Rus. Fed.
58 A3 **Laun** Thai.
67 G8 **Launceston** Austr.
13 C7 **Launceston** U.K.
14 B5 **Laune** *r.* Rep. of Ireland
58 A2 **Launglon Bok Is** Myanmar
105 B4 **La Unión** Chile
103 A4 **La Union** Col.
97 H5 **La Unión** El Salvador
96 D4 **La Unión** Mex.
20 C4 **Laupheim** Ger.
55 B3 **Laur** Phil.
65 H3 **Laura** *Qld.* Austr.
66 C4 **Laura** *S.A.* Austr.
103 D3 **La Urbana** Venez.
91 F5 **Laurel** *DE* U.S.A.
85 F6 **Laurel** *MS* U.S.A.
82 E2 **Laurel** *MT* U.S.A.
90 D4 **Laurel Hill** U.S.A.
90 A6 **Laurel River Lake** U.S.A.
11 F4 **Laurencekirk** U.K.

M

40 B2 Mahābād Iran
43 A2 Mahabaleshwar India
Mahabalipuram see Māmallapuram
45 F4 Mahabharat Range mts Nepal
111 E6 Mahabo Madag.
43 A2 Mahad India
44 D5 Mahadeo Hills India
71 □3 Mahaena Fr. Polynesia Pac. Oc.
110 D3 Mahagi Congo(Zaire)
115 B2 Mahagi Port Congo(Zaire)
44 C3 Mahajan India
111 E5 Mahajanga Madag.
59 D2 Mahakam r. Indon.
111 C6 Mahalapye Botswana
111 E5 Mahalevona Madag.
40 C3 Mahallāt Iran
44 D3 Maham India
59 D4 Mahameru, Gunung volc. Indon.
40 E4 Mahān Iran
45 F5 Mahanadi r. India
111 E5 Mahanoro Madag.
44 C6 Maharashtra div. India
45 E5 Mahasamund India
58 B1 Maha Sarakham Thai.
111 E6 Mahatalaky Madag.
111 E5 Mahavanona Madag.
111 E5 Mahavavy r. Madag.
43 C5 Mahaweli Ganga r. Sri Lanka
58 C1 Mahaxai Laos
43 C2 Mahbubabad India
43 B2 Mahbubnagar India
40 D5 Maḩḑah Oman
101 G2 Mahdia Guyana
26 D7 Mahdia Tunisia
117 H4 Mahé i. Seychelles
43 D2 Mahendragiri mt India
115 C4 Mahenge Tanz.
44 C5 Mahesāna India
44 C5 Maheshwar India
44 C5 Mahi r. India
41 E4 Māhī watercourse Iran
70 F3 Mahia Peninsula N.Z.
28 D4 Mahilyow Belarus
71 □3 Mahina Fr. Polynesia Pac. Oc.
43 C5 Mahiyangana Sri Lanka
113 J4 Mahlabatini S. Africa
19 J4 Mahlsdorf Ger.
41 H3 Maḩmūd-e 'Erāqī Afgh.
37 M4 Mahniān Iran
84 D2 Mahnomen U.S.A.
44 D4 Mahoba India
25 J3 Mahón Spain
90 D4 Mahoning Creek Lake U.S.A.
45 H5 Mahudaung Hgts mts Myanmar
44 B5 Mahuva India
27 M4 Mahya Dağı mt Turkey
40 C3 Mahyār Iran
71 □5 Maia American Samoa Pac. Oc.
45 H4 Maibang India
103 B2 Maicao Col.
78 E4 Maicasagi, Lac l. Can.
22 D2 Maîche France
51 C6 Maichen China
13 G6 Maidenhead U.K.
77 H4 Maidstone Can.
13 H6 Maidstone U.K.
114 F2 Maiduguri Nigeria
103 D3 Maigualida, Sierra mts Venez.
14 C5 Maigue r. Rep. of Ireland
44 E4 Maihar India
50 C3 Maiji Shan mt China
44 E5 Maikala Range h. India
115 A3 Maiko r. Congo(Zaire)
115 A3 Maiko, Parc National de la nat. park Congo(Zaire)
44 E3 Mailani India
15 E4 Mailly-le-Camp France
20 C3 Main r. Ger.
14 E3 Main r. U.K.
79 J3 Main Brook Can.
20 E4 Mainburg Ger.
89 F3 Main Channel Can.
110 B4 Mai-Ndombe, Lac l. Congo(Zaire)
20 E3 Main-Donau-Kanal canal Ger.
89 J4 Main Duck I. Can.
91 J2 Maine div. U.S.A.
114 F2 Maïné-Soroa Niger
58 A2 Maingy I. Myanmar
20 C3 Mainhardt Ger.
55 C4 Mainit Phil.
55 C4 Mainit, Lake Phil.
11 E1 Mainland i. Orkney U.K.
11 □ Mainland i. Shetland U.K.
20 E2 Mainleus Ger.
45 E5 Mainpat reg. India
44 D4 Mainpuri India
15 B4 Maintenon France
111 E5 Maintirano Madag.
20 B2 Mainz Ger.
108 □ Maio i. Cape Verde
105 C2 Maipó, Vol. volc. Chile
105 F3 Maipú Buenos Aires Arg.
105 C2 Maipú Mendoza Arg.
103 D2 Maiquetía Venez.
22 E5 Maira r. Italy
45 G5 Maiskhal I. Bangl.
15 C4 Maisons-Laffitte France
111 C6 Maitengwe Botswana
67 J4 Maitland N.S.W. Austr.
66 B5 Maitland S.A. Austr.
69 E3 Maitland, Lake salt flat Austr.
69 D2 Maitland, Mt h. Austr.
116 D3 Maitri India Base Ant.
45 G3 Maizhokunggar China
96 J5 Maíz, Islas del is Nic.
53 D7 Maizuru Japan
27 H3 Maja Jezercë mt Albania
43 B2 Mājalgaon India
103 E4 Majari r. Brazil
59 E3 Majene Indon.
40 C6 Majḩūd well S. Arabia
110 D3 Majī Eth.
50 E2 Majia r. China
51 D6 Majiang China
Majorca i. see Mallorca
45 H4 Majuli I. India
118 G5 Majuro i. Pac. Oc.
113 G4 Majwemasweu S. Africa
110 B4 Makabana Congo
92 □1 Makaha U.S.A.
57 F7 Makale Indon.
45 F4 Makalu, Mt China
115 A3 Makamba Burundi
39 K3 Makanchi Kazak.
92 □1 Makapuu Hd hd U.S.A.
28 J2 Makar-Ib Rus. Fed.
49 Q2 Makarov Rus. Fed.
26 G3 Makarska Croatia
38 C3 Makat Kazak.
71 □2 Makatéa i. Fr. Polynesia Pac. Oc.
113 K3 Makatini Flats lowland S. Africa
71 □11 Makau Mama'o i. Tonga
71 □2 Makemo i. Fr. Polynesia Pac. Oc.
108 A4 Makeni Sierra Leone
115 B4 Makete Tanz.
111 C6 Makgadikgadi salt pan Botswana
29 H7 Makhachkala Rus. Fed.
38 B3 Makhambet Kazak.
36 G4 Makhfar al Ḩammām Syria
37 J4 Makhmūr Iraq
39 G1 Makhorovka Kazak.
115 C3 Makindu Kenya
39 G1 Makinsk Kazak.
29 F5 Makiyivka Ukr.
Makkah see Mecca
79 J2 Makkovik Can.
79 J2 Makkovik, Cape Can.
18 C3 Makkum Neth.
27 J1 Makó Hungary
71 □15 Makodroga i. Fiji
71 □15 Makogai i. Fiji
110 B3 Makokou Gabon
115 B4 Makongolosi Tanz.
112 E2 Makopong Botswana
110 B4 Makotipoko Congo
21 M3 Makov Slovakia
41 F5 Makran reg. Iran/Pak.
44 C4 Makrana India
Makran Coast Range mts see Talar-i-Band
45 E6 Makri India
27 L6 Makronisi i. Greece
28 E3 Maksatikha Rus. Fed.
52 E1 Maksimovka Rus. Fed.
41 F4 Maksotag Iran
40 B2 Mākū Iran
45 H4 Makum India
115 B4 Makumbako Tanz.
115 C5 Makunguwiro Tanz.
53 B9 Makurazaki Japan
114 E3 Makurdi Nigeria
40 D4 Makūyeh Iran
113 F3 Makwassie S. Africa
8 Q4 Malå Sweden
55 C5 Malabang Phil.
43 A3 Malabar Coast India
114 E4 Malabo Equatorial Guinea
55 A4 Malabuñgan Phil.
59 A2 Malacca, Strait of Indon./Malaysia
21 L4 Malacky Slovakia
82 D3 Malad City U.S.A.
28 C4 Maladzyechna Belarus
25 D4 Málaga Spain
91 F5 Malaga NJ U.S.A.
83 F5 Malaga NM U.S.A.
115 B4 Malagarasi Tanz.
63 G2 Malaita i. Solomon Is
109 F4 Malakal Sudan
43 C2 Malakanagiri India
71 □15 Malake i. Fiji
71 □9 Malakula i. Vanuatu
44 C2 Malakwal Pak.
57 G7 Malamala Indon.
68 A1 Malanda Austr.
59 D4 Malang Indon.
111 B4 Malanje Angola
41 G5 Malan, Ras pt Pak.
105 C1 Malanzán, Sa de mts Arg.
43 B4 Malappuram India
96 J7 Mala, Pta pt Panama
9 P7 Mälaren l. Sweden
105 C2 Malargüe Arg.
89 H1 Malartic Can.
89 H1 Malartic, Lac l. Can.
76 A3 Malaspina Glacier U.S.A.
36 G2 Malatya Turkey
71 □13 Malau Fiji
22 C5 Malaucène France
44 C3 Malaut India
37 L5 Malāvi Iran
55 A5 Malawali i. Malaysia
107 G7 Malawi country Africa
Malawi, Lake see Nyasa, Lake
58 B5 Malaya reg. Malaysia
28 E3 Malaya Vishera Rus. Fed.
55 C4 Malaybalay Phil.
40 C3 Malāyer Iran
68 C1 Malay Reef Coral Sea Is Terr.
39 J3 Malay Sary Kazak.
31 K9 Malaysia country Asia
37 J2 Malazgirt Turkey
17 J3 Malbork Pol.
15 G3 Malborn Ger.
19 K3 Malchin Ger.
19 K3 Malchiner See l. Ger.
69 E4 Malcolm Austr.
45 G4 Māldah India
15 D1 Maldegem Belgium
85 F4 Malden U.S.A.
60 M4 Malden Island Kiribati
117 J4 Maldive Ridge sea feature Ind. Ocean
31 G9 Maldives country Ind. Ocean
13 H6 Maldon U.K.
105 F2 Maldonado Uru.
27 K6 Maleas, Akra i. Greece
35 F6 Male Atoll Maldives
113 F4 Malebogo S. Africa
43 B2 Malegaon Maharashtra India
44 C5 Malegaon Maharashtra India
16 H6 Malé Karpaty h. Slovakia
37 L3 Malek Kandī Iran
110 B4 Malele Congo(Zaire)
115 C5 Malema Moz.
19 H2 Malente Ger.
44 C3 Maler Kōtla India
15 C4 Malesherbes France
41 G3 Mālestān Afgh.
29 H7 Malgobek Rus. Fed.
8 P4 Malgomaj l. Sweden
40 B5 Malham S. Arabia
82 C3 Malheur L. U.S.A.
107 C4 Mali country Africa
71 □13 Mali i. Fiji
108 A3 Mali Guinea
115 A3 Mali Congo(Zaire)
50 C3 Malian r. China
44 E4 Malihabad India
22 D5 Malijai France
41 F4 Malik Naro mt Pak.
58 A2 Mali Kyun i. Myanmar
57 G7 Malili Indon.
71 □13 Malima i. Fiji
59 C4 Malimping Indon.
115 D3 Malindi Kenya
14 D2 Malin Head hd Rep. of Ireland
14 C3 Malin More Rep. of Ireland
71 □11 Malinoa i. Tonga
39 G2 Malinovka Kazak.
52 D2 Malinovka r. Rus. Fed.
39 J2 Malinovoye Ozero Rus. Fed.
51 B6 Malipo China
26 F2 Mali Raginac mt Croatia
55 C5 Malita Phil.
58 A3 Maliwun Myanmar
44 B5 Maliya India
37 L5 Malkaili Iran
44 D5 Malkapur India
29 C7 Malkara Turkey
17 N4 Mal'kavichy Belarus
27 M4 Malko Tŭrnovo Bulg.
67 H6 Mallacoota Austr.
67 H6 Mallacoota Inlet b. Austr.
11 C4 Mallaig U.K.
66 C5 Mallala Austr.
66 E5 Mallee Cliffs Nat. Park Austr.
77 K2 Mallery Lake Can.
25 H3 Mallorca i. Spain
14 C5 Mallow Rep. of Ireland
13 D5 Mallwyd U.K.
8 M4 Malm Norway
8 R3 Malmberget Sweden
15 G2 Malmédy Belgium
112 C6 Malmesbury S. Africa
13 E6 Malmesbury U.K.
9 N9 Malmö Sweden
19 L1 Malmö (Sturup) airport Sweden
28 J3 Malmyzh Rus. Fed.
23 F4 Malnate Italy
71 □9 Malo i. Vanuatu
71 □15 Malolo i. Fiji
71 □15 Malolo Barrier Reef Fiji
55 B3 Malolos Phil.
91 F2 Malone U.S.A.
51 B5 Malong China
111 C5 Malonga Congo(Zaire)
28 F2 Maloshuyka Rus. Fed.
9 J6 Måløy Norway
28 F4 Maloyaroslavets Rus. Fed.
100 B3 Malpelo, Isla de i. Col.
43 A3 Malprabha r. India
21 H4 Malše r. Czech Rep.
7 E5 Malta country Europe
9 U8 Malta Latvia
82 F1 Malta U.S.A.
26 F6 Malta Channel Italy/Malta
111 B6 Maltahöhe Namibia
13 F4 Maltby U.K.
13 H4 Maltby le Marsh U.K.
12 G3 Malton U.K.
71 □4 Malua Western Samoa
Maluku is see Moluccas
9 N6 Malung Sweden
113 H4 Maluti Mountains Lesotho
63 G2 Malu'u Solomon Is
43 A2 Malvan India
85 E5 Malvern U.S.A.
Malvinas, Islas terr. see Falkland Islands
21 L4 Malý Dunaj r. Slovakia
29 D5 Malyn Ukr.
33 S3 Malyy Anyuy r. Rus. Fed.
38 C5 Malyy Balkhan, Khrebet h. Turkm.
29 H6 Malyye Derbety Rus. Fed.
38 B1 Malyy Irgiz r. Rus. Fed.
29 G7 Malyy Kavkaz mts Asia
33 Q2 Malyy Lyakhovskiy, Ostrov i. Rus. Fed.
38 B2 Malyy Uzen' r. Kazak./Rus. Fed.
33 Q3 Mama r. Rus. Fed.
113 H3 Mamafubedu S. Africa
43 C3 Māmallapuram India
71 □15 Mamanuca Group is Fiji
55 A5 Mambahenauhan i. Phil.
55 C4 Mambajao Phil.
115 A2 Mambasa Congo(Zaire)
110 B3 Mambéré r. C.A.R.
55 B3 Mamburao Phil.
113 H2 Mamelodi S. Africa
114 E3 Mamfé Cameroon
22 D2 Mamirolle France
39 F1 Mamlyutka Kazak.
93 G5 Mammoth U.S.A.
86 C4 Mammoth Cave Nat. Park U.S.A.
92 C3 Mammoth Lakes U.S.A.
100 E6 Mamoré r. Bol./Brazil
108 A3 Mamou Guinea
111 E5 Mampikony Madag.
114 C3 Mampong Ghana
105 B3 Mamuil Malal, P. pass Arg./Chile
59 E3 Mamuju Indon.
112 D1 Mamuno Botswana
38 D2 Mamyt Kazak.
114 B3 Man Côte d'Ivoire
71 □15 Mana i. Fiji
103 B3 Manacacias r. Col.
100 F4 Manacapuru Brazil
25 H3 Manacor Spain
57 G6 Manadao Indon.
96 H5 Managua Nic.
96 H5 Managua, L. de Nic.
111 E6 Manakara Madag.
70 D5 Manakau mt N.Z.
62 E2 Manam I. P.N.G.
92 □1 Manana i. U.S.A.
111 E6 Mananara r. Madag.
111 E5 Mananara Avaratra Madag.
111 E5 Mananara, Parc National de nat. park Madag.
66 E5 Manangatang Austr.
111 E6 Mananjary Madag.
114 A2 Manantali, Lac de l. Mali
43 B4 Mānantavādi India
103 D2 Manapire r. Venez.
115 A6 Mana Pools National Park Zimbabwe
70 A6 Manapouri, L. N.Z.
111 E5 Manarantsandry Madag.
45 G4 Manas r. Bhutan
44 D3 Mana Shankou pass India
46 A2 Manas Hu l. China
45 F3 Manaslu mt Nepal
90 E5 Manassas U.S.A.
57 H8 Manatuto Indon.
100 F4 Manaus Brazil
36 C3 Manavgat Turkey
70 E4 Manawatu r. N.Z.
55 C5 Manay Phil.
36 F3 Manbij Syria
13 H4 Manby U.K.
88 E3 Mancelona U.S.A.
13 E4 Manchester U.K.
92 A2 Manchester CA U.S.A.
91 G4 Manchester CT U.S.A.
88 B4 Manchester IA U.S.A.
90 B6 Manchester KY U.S.A.
89 E4 Manchester MI U.S.A.
91 H3 Manchester NH U.S.A.
90 B5 Manchester OH U.S.A.
87 C5 Manchester TN U.S.A.
91 G3 Manchester VT U.S.A.
44 A4 Manchhar L. Pak.
20 E4 Manching Ger.
36 F2 Mancılık Turkey
93 H3 Mancos U.S.A.
93 H3 Mancos r. U.S.A.
40 D4 Mand r. Iran
41 F5 Mand Pak.
111 E6 Mandabe Madag.
59 B2 Mandah Indon.
58 □ Mandai Sing.
41 F3 Mandal Afgh.
44 C4 Mandal India
9 K7 Mandal Norway
57 L7 Mandala, Pk Indon.
56 B2 Mandalay Myanmar
48 J2 Mandalgovĭ Mongolia
37 K5 Mandalī Iraq
50 D1 Mandalt Sum China
84 C2 Mandan U.S.A.
55 B3 Mandaon Phil.
109 D4 Manda, Parc National de nat. park Chad
114 F2 Mandara Mountains Cameroon/Nigeria
26 C5 Mandas Sardinia Italy
22 D6 Mandelieu-la-Napoule France
23 G4 Mandello del Lario Italy
110 E3 Mandera Kenya
93 F2 Manderfield U.S.A.
15 G2 Manderscheid Ger.
95 J5 Mandeville Jamaica
70 B6 Mandeville N.Z.
44 B4 Mandha India
48 C5 Mandi India
114 B2 Mandiana Guinea
44 C3 Mandi Burewala Pak.
115 B6 Mandié Moz.
115 C5 Mandimba Moz.
113 J4 Mandini S. Africa
45 F5 Mandira Dam dam India
44 E5 Mandla India
18 F1 Mandø i. Denmark
111 E5 Mandritsara Madag.
44 C4 Mandsaur India
55 A6 Mandul i. Indon.
69 B6 Mandurah Austr.
26 G4 Manduria Italy
44 B5 Mandvi Gujarat India
44 C5 Mandvi Gujarat India
43 B3 Mandya India
43 B2 Maner r. India
23 H4 Manerbio Italy
17 M5 Manevychi Ukr.
26 F4 Manfredonia Italy
26 G4 Manfredonia, Golfo di g. Italy
104 D1 Manga Brazil
114 C2 Manga Burkina
110 B4 Mangai Congo(Zaire)
60 M6 Mangaia i. Cook Is Pac. Oc.
70 E3 Mangakino N.Z.
43 C2 Mangalagiri India
45 H4 Mangaldai India
27 N3 Mangalia Romania
43 A3 Mangalore India
43 A2 Mangalvedha India
45 G4 Mangan India
43 C2 Mangapet India
55 C6 Mangarang Indon.
113 G4 Mangaung S. Africa
70 E3 Mangaweka N.Z.
45 G4 Mangde r. Bhutan
115 A3 Mangembe Congo(Zaire)
14 B6 Mangerton Mt h. Rep. of Ireland
20 E5 Mangfall r. Ger.
59 C3 Manggar Indon.
38 C3 Mangistauskaya Oblast' div. Kazak.
38 E4 Mangit Uzbek.
46 B3 Mangnai China
115 C5 Mangochi Malawi
57 H7 Mangole i. Indon.
13 E6 Mangotsfield U.K.
44 B5 Māngral India
87 E7 Mangrove Cay Bahamas
25 C2 Mangualde Port.
41 G4 Manguchar Pak.
105 G2 Mangueira, L. Brazil
104 B4 Mangueirinha Brazil
109 D2 Manguéni, Plateau de plat. Niger
46 E1 Mangui China
55 C5 Mangupung i. Indon.
38 B4 Mangyshlak Kazak.
38 B3 Mangyshlak, Poluostrov pen. Kazak.
38 B3 Mangyshlakskiy Zaliv b. Kazak.
84 D4 Manhattan KS U.S.A.
92 D2 Manhattan NV U.S.A.
111 D6 Manhica Moz.
113 K3 Manhoca Moz.
104 D3 Manhuaçu Brazil
104 E2 Manhuaçu r. Brazil
103 B3 Mani Col.
111 E5 Mania r. Madag.
23 K3 Maniago Italy
100 F5 Manicoré Brazil
79 G3 Manicouagan Can.
79 G3 Manicouagan r. Can.
79 G3 Manicougan, Réservoir resr Can.
40 C5 Manīfah S. Arabia
68 D4 Manifold, Cape Austr.
71 □2 Manihi i. Fr. Polynesia Pac. Oc.
60 L5 Manihiki i. Cook Is Pac. Oc.
Manikgarh see Rajura
44 E4 Manikpur India
55 B3 Manila Phil.
82 E3 Manila U.S.A.
67 H4 Manildra Austr.
67 J3 Manilla Austr.
65 F2 Maningrida Austr.
71 □10 Maninita i. Tonga
Manipur see Imphal
45 H4 Manipur div. India
27 M5 Manisa Turkey
37 L5 Manisht Kūh mt Iran
12 C3 Man, Isle of terr. Europe
88 D3 Manistee U.S.A.
88 E3 Manistee r. U.S.A.
88 D3 Manistique U.S.A.
88 E2 Manistique Lake U.S.A.
78 B2 Manitoba div. Can.
77 K4 Manitoba, Lake Can.
77 H4 Manito L. Can.
77 K5 Manitou Can.
90 E3 Manitou Beach U.S.A.
78 B3 Manitou Falls Can.
88 D2 Manitou Island U.S.A.
86 C2 Manitou Islands U.S.A.
89 G3 Manitou, Lake Can.
89 F3 Manitoulin I. Can.
89 G3 Manitowaning Can.
78 D4 Manitowik Lake Can.
88 D3 Manitowoc U.S.A.
89 K2 Maniwaki Can.
103 B3 Manizales Col.
111 E6 Manja Madag.
113 K2 Manjacaze Moz.
43 B4 Manjeri India
54 D3 Man Jiang r. China
37 M3 Manjil Iran
69 C7 Manjimup Austr.
43 B2 Manjra r. India
84 E2 Mankato U.S.A.
113 J3 Mankayane Swaziland
114 B3 Mankono Côte d'Ivoire
43 C4 Mankulam Sri Lanka
67 J4 Manly Austr.
44 C5 Manmad India
59 B3 Manna Indon.
66 C4 Mannahill Austr.
43 B4 Mannar Sri Lanka
35 F6 Mannar, Gulf of India/Sri Lanka
43 B4 Mannar, Gulf of g. India/Sri Lanka
43 B3 Manneru r. India
20 B3 Mannheim Ger.
14 A4 Mannin Bay Rep. of Ireland
76 F3 Manning Can.
87 D5 Manning U.S.A.
13 J6 Manningtree U.K.
26 C4 Mannu, Capo pt Sardinia Italy
66 C5 Mannum Austr.
57 J7 Manokwari Indon.
115 A4 Manono Congo(Zaire)
58 A3 Manoron Myanmar
22 C6 Manosque France
75 L4 Manouane Lake Can.
54 D3 Manp'o N. Korea
63 J2 Manra i. Kiribati
25 G2 Manresa Spain
44 C3 Mānsa India
115 A5 Mansa Zambia
108 A3 Mansa Konko The Gambia
44 C2 Mansehra Pak.
75 L3 Mansel Island Can.
67 G6 Mansfield Austr.
13 F4 Mansfield U.K.
85 E5 Mansfield LA U.S.A.
90 B4 Mansfield OH U.S.A.
90 E4 Mansfield PA U.S.A.
101 H6 Manso r. Brazil
76 E3 Manson Creek Can.
37 M6 Manşūrī Iran
36 E3 Mansurlu Turkey
100 B4 Manta Ecuador
100 B4 Manta, B. de Ecuador
55 A4 Mantalingajan, Mount Phil.
54 E3 Mantapsan mt N. Korea
92 B3 Manteca U.S.A.
103 C3 Mantecal Venez.
87 F5 Manteo U.S.A.
15 B4 Mantes-la-Jolie France
43 B2 Manthani India
93 G2 Manti U.S.A.
104 D3 Mantiqueira, Serra da mts Brazil
22 C2 Mantoche France
88 E3 Manton U.S.A.
23 H4 Mantova Italy
9 T6 Mäntsälä Fin.
9 T5 Mänttä Fin.
Mantua see Mantova
68 B5 Mantuan Downs Austr.
28 H3 Manturovo Rus. Fed.
9 U6 Mäntyharju Fin.
8 U3 Mäntyjärvi Fin.
71 □5 Manua Is American Samoa Pac. Oc.
93 H4 Manuelito U.S.A.
105 F2 Manuel J. Cobo Arg.
104 E1 Manuel Vitorino Brazil
101 H5 Manuelzinho Brazil
71 □2 Manuhangi i. Fr. Polynesia Pac. Oc.
57 G7 Manui i. Indon.
41 E5 Manūjān Iran
55 B4 Manukan Phil.
70 E2 Manukau N.Z.
70 E2 Manukau Harbour in. N.Z.
55 A5 Manuk Manka i. Phil.
66 C4 Manunda watercourse Austr.
100 D6 Manu, Parque Nacional nat. park Peru
62 E2 Manus I. P.N.G.
43 B3 Manvi India
115 B6 Manyame r. Moz./Zimbabwe
113 F2 Manyana Botswana
29 G6 Manych-Gudilo, Ozero l. Rus. Fed.
93 H3 Many Farms U.S.A.
115 B4 Manyoni Tanz.
68 D5 Many Peaks Austr.
69 D7 Many Peaks, Mt h. Austr.
36 D6 Manzala, Bahra el l. Egypt
25 E3 Manzanares Spain
95 J4 Manzanillo Cuba
96 C4 Manzanillo Mex.
96 K6 Manzanillo, Pta pt Panama
40 C3 Manzariyeh Iran
46 D2 Manzhouli China
113 J3 Manzini Swaziland
109 D3 Mao Chad
Maó see Mahón
50 D4 Maocifan China
50 C2 Maojiachuan China
113 G3 Maokeng S. Africa
57 K7 Maoke, Pegunungan mts Indon.
54 B3 Maokui Shan h. China
54 B2 Maolin China
50 B2 Maomao Shan mt China
51 D6 Maoming China
51 □ Ma On Shan h. H.K. China
111 D6 Mapai Moz.
44 E3 Mapam Yumco l. China
57 G7 Mapane Indon.
113 F5 Maphodi S. Africa
96 D2 Mapimí Mex.
55 A5 Mapin i. Phil.
111 D6 Mapinhane Moz.
103 D3 Mapire Venez.
88 E4 Maple r. U.S.A.
77 H5 Maple Creek Can.
113 G4 Mapoteng Lesotho
57 L7 Maprik P.N.G.
101 G4 Mapuera r. Brazil
113 K2 Mapulanguene Moz.
111 D6 Maputo Moz.
113 K2 Maputo div. Moz.
113 K3 Maputo r. Moz.
113 G4 Maputsoe Lesotho
37 H6 Maqar an Na'am well Iraq
50 B3 Maqu China
45 E3 Maquan He r. China
110 B4 Maquela do Zombo Angola
105 C4 Maquinchao Arg.
105 C4 Maquinchao r. Arg.
88 B4 Maquoketa U.S.A.
88 B4 Maquoketa r. U.S.A.
41 G5 Mar r. Pak.
77 H1 Mara r. Can.
45 E5 Māra India
113 H1 Mara S. Africa
115 B3 Mara div. Tanz.
103 C2 Mara Venez.
100 E4 Maraã Brazil
71 □3 Maraa Fr. Polynesia Pac. Oc.
101 J5 Marabá Brazil
68 C4 Maraboon, L. resr Austr.
103 C2 Maracaibo Venez.
103 C2 Maracaibo, Lago de l. Venez.
101 H3 Maracá, Ilha de i. Brazil
104 A3 Maracaju Brazil
104 A3 Maracajú, Serra de h. Brazil
104 E1 Maracás, Chapada de reg. Brazil
103 D2 Maracay Venez.
109 D2 Marādah Libya
114 E2 Maradi Niger
40 B2 Marāgheh Iran
104 E1 Maragogipe Brazil
55 B3 Maragondon Phil.
103 D4 Marahuaca, Co mt Venez.
101 J4 Marajó, Baía de est. Brazil
101 J3 Marajó, Ilha de i. Brazil
43 B3 Marakkanam India
115 C2 Maralal Kenya
44 C2 Marala Weir barrage Pak.
37 J1 Maralik Armenia
64 F6 Maralinga Austr.
63 G2 Maramasike i. Solomon Is
55 C5 Marampit i. Indon.
41 G4 Maran mt Pak.
37 K4 Marāna Iraq
93 G5 Marana U.S.A.
40 B2 Marand Iran
23 H5 Maranello Italy
58 B4 Marang Malaysia
58 A3 Marang Myanmar
104 C1 Maranhão r. Brazil
68 C6 Maranoa r. Austr.
100 D4 Marañón r. Peru
113 L2 Marão Moz.
25 C2 Marão mt Port.
114 B3 Maraoué r. Côte d'Ivoire
114 B3 Maraoué, Parc National de la nat. park Côte d'Ivoire
103 D4 Marari r. Brazil
70 A6 Mararoa r. N.Z.
88 D1 Marathon Can.
87 D7 Marathon FL U.S.A.
85 C6 Marathon TX U.S.A.
104 E1 Maraú Brazil
59 D3 Marau Indon.
103 D4 Marauiá r. Brazil
55 C4 Marawi Phil.
37 M1 Märäzä Azer.
25 D4 Marbella Spain
64 C4 Marble Bar Austr.
93 G3 Marble Canyon U.S.A.
93 G3 Marble Canyon gorge U.S.A.
113 H2 Marble Hall S. Africa
91 H3 Marblehead U.S.A.
77 L2 Marble I. Can.
22 C3 Marboz France
113 J5 Marburg S. Africa
18 F6 Marburg an der Lahn Ger.
90 E5 Marburg, Lake U.S.A.
16 H7 Marcali Hungary
23 H4 Marcaria Italy
21 K4 March r. Austria/Slovakia
13 H5 March U.K.
66 C4 Marchant Hill Austr.
22 D2 Marchaux France
23 K6 Marche div. Italy
15 F2 Marche-en-Famenne Belgium
25 D4 Marchena Spain
100 □ Marchena, Isla i. Galapagos Is Ecuador
105 D1 Mar Chiquita, L. Arg.
21 H4 Marchtrenk Austria
15 D4 Marcilly-le-Hayer France
15 B2 Marck France
87 D7 Marco U.S.A.
15 D2 Marcoing France
78 E2 Marcopeet Islands Can.
105 D2 Marcos Juárez Arg.
22 D5 Marcoux France
91 G2 Marcy, Mt U.S.A.
44 C2 Mardan Pak.
105 F3 Mar del Plata Arg.
37 H3 Mardın Turkey
22 B4 Mare r. France
63 G4 Maré i. New Caledonia
71 □9 Maré i. New Caledonia Pac. Oc.
23 K6 Marecchia r. Italy
68 A1 Mareeba Austr.
11 C3 Maree, Loch l. U.K.
88 A5 Marengo IA U.S.A.
88 C4 Marengo IL U.S.A.
26 E6 Marettimo, Isola i. Sicily Italy
28 E3 Marevo Rus. Fed.
85 B6 Marfa U.S.A.
45 F2 Margai Caka salt l. China
66 B2 Margaret watercourse Austr.
69 B6 Margaret River Austr.
103 E2 Margarita, Isla de i. Venez.
52 D3 Margaritovo Rus. Fed.
67 G9 Margate Austr.
113 J5 Margate S. Africa
13 J6 Margate U.K.
39 G4 Margilan Uzbek.
41 F4 Margo, Dasht-i des. Afgh.
55 B5 Margosatubig Phil.
18 C6 Margraten Neth.
88 E3 Margrethe, Lake U.S.A.
76 E4 Marguerite Can.
116 B2 Marguerite Bay Ant.
45 G3 Margyang China
37 L5 Marhaj Khalīl Iraq
37 J3 Marhan D. h. Iraq
29 E6 Marhanets' Ukr.
102 C2 María Elena Chile
65 G2 Maria I. N.T. Austr.
67 H9 Maria I. Tas. Austr.
105 E3 María Ignacia Arg.
60 M6 Maria, Îles i. Fr. Polynesia Pac. Oc.
119 J7 Maria, Îles is Pac. Oc.
20 F6 Maria Luggau Austria
68 C3 Marian Austr.
118 E4 Marianas Ridge sea feature Pac. Oc.
118 E5 Marianas Tr. sea feature Pac. Oc.
45 H4 Mariani India
76 F2 Marian Lake Can.
85 F5 Marianna AR U.S.A.
87 C6 Marianna FL U.S.A.
20 F3 Mariánské Lázně Czech Rep.
96 C3 Marías, Islas is Mex.
96 J7 Mariato, Pta pt Panama
70 D1 Maria van Diemen, Cape N.Z.
21 J5 Mariazell Austria
19 J2 Maribo Denmark
26 F1 Maribor Slovenia
93 F5 Maricopa AZ U.S.A.
92 C4 Maricopa CA U.S.A.
93 F5 Maricopa Mts U.S.A.
109 E4 Maridi watercourse Sudan
116 A4 Marie Byrd Land reg. Ant.
95 M5 Marie Galante i. Guadeloupe
9 Q6 Mariehamn Fin.
104 B1 Mariembero r. Brazil
18 E3 Marienhafe Ger.
111 B6 Mariental Namibia
9 N7 Mariestad Sweden
87 C5 Marietta GA U.S.A.
90 C5 Marietta OH U.S.A.
22 C6 Marignane France
15 D4 Marigny-le-Châtel France
46 G1 Marii, Mys pt Rus. Fed.
46 A1 Mariinsk Rus. Fed.
38 E1 Mariinskoye Rus. Fed.
9 S9 Marijampolė Lith.
104 C3 Marília Brazil
96 D2 Marin Mex.
25 B1 Marín Spain
23 H6 Marina di Castagneto Donoratico Italy
26 G5 Marina di Gioiosa Ionica Italy
23 K5 Marina di Ravenna Italy
28 D4 Mar"ina Horka Belarus
23 K5 Marina Romea Italy
55 B3 Marinduque i. Phil.
88 D3 Marinette U.S.A.
104 B3 Maringá Brazil
115 B6 Maringué Moz.
25 B3 Marinha Grande Port.
86 B4 Marion IL U.S.A.
88 E5 Marion IN U.S.A.
91 K2 Marion ME U.S.A.
90 B4 Marion OH U.S.A.
87 E5 Marion SC U.S.A.
90 C6 Marion VA U.S.A.
87 D5 Marion, L. U.S.A.
68 E2 Marion Reef Coral Sea Is Terr.
103 D3 Maripa Venez.
92 C3 Mariposa U.S.A.
102 D2 Mariscal Estigarribia Para.
22 D5 Maritime Alps mts France/Italy
27 L3 Maritsa r. Bulg.
28 J3 Mari-Turek Rus. Fed.
29 F6 Mariupol' Ukr.
40 B3 Marīvān Iran
28 J3 Mariy El, Respublika div. Rus. Fed.
110 E3 Marka Somalia
39 L2 Markakol', Ozero Kazak.
114 B2 Markala Mali
48 G6 Markam China
37 K2 Mārkān Iran
43 B3 Markapur India
9 N8 Markaryd Sweden
89 G3 Markdale Can.
19 J2 Markelsdorfer Huk pt Ger.
113 H1 Marken S. Africa
18 C4 Markermeer l. Neth.
13 G5 Market Deeping U.K.
13 E5 Market Drayton U.K.
13 G5 Market Harborough U.K.
14 E3 Markethill U.K.
12 G4 Market Weighton U.K.
33 N3 Markha r. Rus. Fed.
89 H4 Markham Can.
39 H4 Markhamet Uzbek.
116 B4 Markham, Mt Ant.
39 J5 Markit China
29 F5 Markivka Ukr.
19 K5 Markkleeberg Ger.
19 G4 Marklohe Ger.
33 T3 Markovo Rus. Fed.
114 D2 Markoye Burkina
19 K5 Markranstädt Ger.
29 H5 Marks Rus. Fed.
19 H6 Marksuhl Ger.
20 C3 Marktheidenfeld Ger.
20 F2 Marktleuthen Ger.
20 D5 Marktoberdorf Ger.
20 F2 Marktredwitz Ger.
21 K5 Markt St Martin Austria
88 B6 Mark Twain Lake U.S.A.
18 E5 Marl Ger.
65 F5 Marla Austr.
69 C4 Marlandy Hill Austr.
68 C4 Marlborough Qld. Austr.
91 H3 Marlborough U.S.A.
13 F6 Marlborough Downs h. U.K.
15 D3 Marle France
22 C3 Marlieux France
85 D6 Marlin U.S.A.
90 C5 Marlinton U.S.A.
67 H6 Marlo Austr.
24 E4 Marmande France
Marmara Denizi g. see Marmara, Sea of
36 B2 Marmara Gölü l. Turkey
36 B1 Marmara, Sea of g. Turkey
36 B3 Marmaris Turkey
84 C2 Marmarth U.S.A.
90 C5 Marmet U.S.A.
78 B4 Marmion L. Can.
23 J3 Marmolada mt Italy
22 C2 Marnay France
15 E4 Marne r. France
19 G3 Marne Ger.
15 G4 Marne au Rhin, Canal de la France
15 C4 Marne-la-Vallée France
37 K1 Marneuli Georgia
19 J3 Marnitz Ger.
66 E6 Marnoo Austr.
111 E5 Maroantsetra Madag.
71 □2 Marokau i. Fr. Polynesia Pac. Oc.
20 D2 Maroldsweisach Ger.
15 B3 Maromme France
111 E5 Maromokotro mt Madag.
111 D5 Marondera Zimbabwe
101 H2 Maroni r. Fr. Guiana
67 K1 Maroochydore Austr.
69 B1 Maroonah Austr.
60 N6 Marotiri is Fr. Polynesia Pac. Oc.
114 F2 Maroua Cameroon
111 E5 Marovoay Madag.
37 H4 Marqādah Syria
50 A3 Mar Qu r. China
113 G4 Marquard S. Africa
87 D7 Marquesas Keys is U.S.A.
88 D2 Marquette U.S.A.
15 D2 Marquion France
15 B2 Marquise France
60 O4 Marquises, Îles is Fr. Polynesia Pac. Oc.
66 E3 Marra Austr.
67 G3 Marra r. Austr.
113 K2 Marracuene Moz.
23 J5 Marradi Italy
108 B1 Marrakech Morocco
Marrakesh see Marrakech
113 L2 Marrangua, Lagoa l. Moz.
109 E3 Marra Plateau plat. Sudan
67 G5 Marrar Austr.
67 F8 Marrawah Austr.
66 C2 Marree Austr.
85 F6 Marrero U.S.A.
115 C6 Marromeu Moz.
115 C5 Marrupa Moz.
109 F2 Marsa Alam Egypt
109 D1 Marsa al Burayqah Libya
115 C2 Marsabit Kenya
23 G5 Marsaglia Italy
26 E6 Marsala Sicily Italy
109 E1 Marsa Matrûh Egypt
18 F5 Marsberg Ger.
26 E3 Marsciano Italy
67 G4 Marsden Austr.
18 B4 Marsdiep chan. Neth.
22 C6 Marseille France
15 B3 Marseille-en-Beauvaisis France
88 C5 Marseilles U.S.A.
104 D3 Mar, Serra do mts Brazil
8 O4 Marsfjället mt Sweden
77 H4 Marshall Can.
85 E5 Marshall AR U.S.A.
86 C4 Marshall IL U.S.A.
88 E4 Marshall MI U.S.A.
84 E2 Marshall MN U.S.A.
84 E4 Marshall MO U.S.A.
85 E5 Marshall TX U.S.A.
67 G7 Marshall B. Austr.
61 H2 Marshall Islands country Pac. Oc.
84 E3 Marshalltown U.S.A.
88 B3 Marshfield U.S.A.
87 E7 Marsh Harbour Bahamas
91 K1 Mars Hill U.S.A.
85 F6 Marsh Island U.S.A.
76 C2 Marsh Lake Can.
37 M3 Marshūn Iran
82 C3 Marsing U.S.A.
9 P7 Märsta Sweden
19 H2 Marstal Denmark
19 H2 Marstal Bugt b. Denmark
45 F4 Marsyangdi r. Nepal
58 A1 Martaban Myanmar
56 B3 Martaban, Gulf of g. Myanmar
59 D3 Martapura Kalimantan Indon.
59 B3 Martapura Sumatera Indon.
89 H2 Marten River Can.
77 H4 Martensville Can.
91 H4 Martha's Vineyard i. U.S.A.
22 E3 Martigny Switz.
15 F4 Martigny-les-Gerbonvaux France
17 J6 Martin Slovakia
84 C3 Martin SD U.S.A.
87 B4 Martin TN U.S.A.
97 E3 Martínez Mex.
96 C2 Martínez, E. Mex.
93 E5 Martinez Lake U.S.A.
73 K8 Martinique terr. Caribbean Sea
87 C5 Martin, L. U.S.A.
116 A3 Martin Pen. Ant.
21 J4 Martinsberg Austria
90 D4 Martinsburg PA U.S.A.
90 E5 Martinsburg WV U.S.A.
90 C4 Martins Ferry U.S.A.
90 D6 Martinsville U.S.A.
120 H7 Martin Vas, Is Atl. Ocean
70 E4 Marton N.Z.
25 G2 Martorell Spain
25 E4 Martos Spain
38 D2 Martuk Kazak.
37 K1 Martuni Armenia
41 F3 Maruchak Afgh.
53 C7 Marugame Japan
70 D5 Maruia r. N.Z.
101 L6 Maruim Brazil
29 G7 Marukhis Ugheltekhili pass Georgia/Rus. Fed.
67 H5 Marulan Austr.
71 □9 Marum, Mt Vanuatu
39 L1 Marushka Rus. Fed.
71 □2 Marutéa i. Fr. Polynesia Pac. Oc.
40 D4 Marvast Iran
24 F4 Marvejols France
93 G2 Marvine, Mt U.S.A.
77 G4 Marwayne Can.
68 E6 Mary r. Austr.
38 E5 Mary Turkm.
68 E5 Maryborough Qld. Austr.
66 E6 Maryborough Vic. Austr.
112 E4 Marydale S. Africa
28 J4 Mar'yevka Rus. Fed.
77 H2 Mary Frances Lake Can.
91 E5 Maryland div. U.S.A.
12 D3 Maryport U.K.
79 J3 Mary's Harbour Can.
79 K4 Marystown Can.
93 F2 Marysvale U.S.A.
79 G4 Marysville Can.
92 B2 Marysville CA U.S.A.
84 D4 Marysville KS U.S.A.
90 B4 Marysville OH U.S.A.
68 A2 Maryvale Austr.
84 E3 Maryville U.S.A.
87 C5 Maryville U.S.A.
19 K4 Marzahna Ger.
96 H6 Masachapa Nic.
40 D4 Masāhūn, Kūh-e mt Iran
115 B3 Masaka Uganda
113 G5 Masakhane S. Africa
37 M2 Masallı Azer.
57 G7 Masamba Indon.
54 E6 Masan S. Korea
91 J1 Masardis U.S.A.
115 C5 Masasi Tanz.
100 F7 Masavi Bol.
96 H6 Masaya Nic.
55 B3 Masbate Phil.
55 B4 Masbate i. Phil.
108 C1 Mascara Alg.
117 H4 Mascarene Basin sea feature Ind. Ocean
117 J4 Mascarene Ridge sea feature Ind. Ocean
91 G2 Mascouche Can.
113 G4 Maseru Lesotho
22 D2 Masevaux France
113 H4 Mashai Lesotho
51 C6 Mashan China
44 D2 Masherbrum mt Pak.
41 E2 Mashhad Iran
44 C4 Mashi r. India
37 L2 Mashīrān Iran

40 E4 Mashīz Iran
41 F5 Mashket r. Pak.
41 F4 Mashki Chah Pak.
41 F5 Māshkīd r. Iran
8 S2 Masi Norway
96 B2 Masiáca Mex.
113 G5 Masibambane S. Africa
113 G4 Masilo S. Africa
115 B2 Masindi Uganda
55 A3 Masinloc Phil.
112 E5 Masinyusane S. Africa
42 E5 Maṣīrah i. Oman
42 E6 Maṣīrah, Gulf of Oman
37 K1 Masis Armenia
40 C4 Masjed Soleymān Iran
36 G3 Maskanah Syria
14 B4 Mask, Lough l. Rep. of Ireland
41 E5 Maskūtān Iran
41 G4 Maslti Pak.
111 F5 Masoala, Tanjona c. Madag.
88 E4 Mason MI U.S.A.
92 C2 Mason NV U.S.A.
85 D6 Mason TX U.S.A.
70 A7 Mason Bay N.Z.
84 E3 Mason City IA U.S.A.
85 C5 Mason City IL U.S.A.
69 D3 Mason, Lake salt flat Austr.
90 D5 Masontown U.S.A.
Masqaṭ see Muscat
23 H5 Massa Italy
91 G3 Massachusetts div. U.S.A.
91 H3 Massachusetts Bay U.S.A.
23 H6 Massaciuccoli, Lago di l. Italy
91 H2 Massadona U.S.A.
26 G4 Massafra Italy
109 D3 Massakory Chad
26 D3 Massa Marittimo Italy
111 D6 Massangena Moz.
111 B4 Massango Angola
23 H6 Massarosa Italy
110 D2 Massawa Eritrea
91 G2 Massawippi, Lac l. Can.
91 F2 Massena U.S.A.
76 C4 Masset Can.
89 F2 Massey Can.
24 F4 Massif Central mts France
90 C4 Massillon U.S.A.
114 B2 Massina Mali
111 D6 Massinga Moz.
111 D6 Massingir Moz.
113 K1 Massingir, Barragem de resr Moz.
113 K2 Massintonto r. Moz./S. Africa
116 D3 Massivet mts Ant.
89 K3 Masson Can.
116 D5 Masson I. Ant.
37 M1 Maştağa Azer.
39 G5 Mastchoh Tajik.
70 E4 Masterton N.Z.
27 M5 Masticho, Akra pt Greece
87 E7 Mastic Point Bahamas
44 C1 Mastuj Pak.
41 G4 Mastung Pak.
28 C4 Masty Belarus
53 B7 Masuda Japan
37 M3 Masuleh Iran
Masulipatam see Machilipatnam
111 D6 Masvingo Zimbabwe
115 B3 Maswe Tanz.
36 F4 Maşyāf Syria
71 □15 Matacawa Levu i. Fiji
89 G2 Matachewan Can.
96 C1 Matachic Mex.
103 D4 Matacuni r. Venez.
110 B4 Matadi Congo(Zaire)
96 H5 Matagalpa Nic.
78 E4 Matagami Can.
78 E4 Matagami, Lac l. Can.
85 D6 Matagorda I. U.S.A.
71 □3 Mataiea Fr. Polynesia Pac. Oc.
71 □2 Mataiva i. Fr. Polynesia Pac. Oc.
58 C5 Matak i. Indon.
39 H2 Matak Kazak.
70 F2 Matakana Island N.Z.
111 B5 Matala Angola
43 C5 Matale Sri Lanka
108 A3 Matam Senegal
114 E2 Matameye Niger
96 D2 Matamoros Coahuila Mex.
97 E2 Matamoros Tamaulipas Mex.
55 B5 Matanal Point Phil.
115 C4 Matandu r. Tanz.
79 G4 Matane Can.
44 B2 Matanui Pak.
95 H4 Matanzas Cuba
Matapan, Cape pt see Tainaro, Akra
21 M2 Mata Panew r. Pol.
79 G4 Matapédia r. Can.
113 B2 Mataquito r. Chile
43 C5 Matara Sri Lanka
59 E4 Mataram Indon.
100 D7 Matarani Peru
64 F2 Mataranka Austr.
25 H2 Mataró Spain
113 H5 Matatiele S. Africa
71 □5 Matatula, C. American Samoa Pac. Oc.
70 B7 Mataura N.Z.
70 B7 Mataura r. N.Z.
71 □4 Matautu Western Samoa
71 □10 Mata'utuliki i. Tonga
71 □4 Matavanu Crater crater Western Samoa
103 C3 Mataveni r. Col.
71 □8 Mataveri Easter I. Chile
70 F3 Matawai N.Z.
100 F6 Mategua Bol.
96 D3 Matehuala Mex.
23 L6 Matelica Italy
115 C5 Matemanga Tanz.
26 G4 Matera Italy
26 C6 Mateur Tunisia
78 D4 Matheson Can.
85 D6 Mathis U.S.A.
66 F5 Mathoura Austr.
44 D4 Mathura India
55 C5 Mati Phil.
45 G4 Matiali India
51 D5 Matianxu China
44 B4 Matiari Pak.
97 F4 Matías Romero Mex.
115 D5 Matibane Moz.
79 G3 Matimekosh Can.
89 F2 Matinenda Lake Can.
91 J3 Matinicus I. U.S.A.
71 □3 Matiti Fr. Polynesia Pac. Oc.
45 G5 Matla r. India
113 G2 Matlabas S. Africa
113 G2 Matlabas r. S. Africa
44 B4 Matli Pak.
13 F4 Matlock U.K.
103 D3 Mato r. Venez.
103 D3 Mato, Co mt Venez.
100 G7 Mato Grosso Brazil
104 A1 Mato Grosso div. Brazil
104 A3 Mato Grosso do Sul div. Brazil
104 A1 Mato Grosso, Planalto do plat. Brazil
113 K2 Matola Moz.
25 B2 Matosinhos Port.
42 E5 Maṭraḥ Oman
20 F5 Matrei in Osttirol Austria
112 C6 Matroosberg mt S. Africa
53 C7 Matsue Japan
52 G4 Matsumae Japan
53 E6 Matsumoto Japan
53 E7 Matsusaka Japan
51 F5 Matsu Tao i. Taiwan
53 C8 Matsuyama Japan
78 D4 Mattagami r. Can.
79 H2 Mattawa Can.
91 J2 Mattawamkeag U.S.A.
22 E4 Matterhorn mt Italy/Switz.
82 D3 Matterhorn mt U.S.A.
21 K5 Mattersburg Austria
63 H4 Matthew i. New Caledonia
103 E3 Matthews Ridge Guyana
95 K4 Matthew Town Bahamas
40 D6 Maṭṭī, Sabkhat salt pan S. Arabia
86 B4 Mattoon U.S.A.
Matturai see Matara
43 C5 Matugama Sri Lanka
71 □14 Matuku i. Fiji
Matun see Khowst
103 E2 Maturín Venez.
115 A6 Matusadona National Park Zimbabwe
55 C5 Matutuang i. Indon.
113 G4 Matwabeng S. Africa
45 E4 Mau Uttar Pradesh India
44 E4 Mau Uttar Pradesh India
115 C5 Maúa Moz.
45 E4 Mau Aimma India
15 D2 Maubeuge France
24 E5 Maubourguet France
11 D5 Mauchline U.K.
70 A7 Maudheimvidda mts Ant.
69 A1 Maud, Pt Austr.
117 E7 Maud Seamount depth Ind. Ocean
101 G4 Maués Brazil
45 E4 Mauganj India
57 L2 Maug Islands N. Mariana Is
22 B6 Mauguio France
92 □2 Maui i. U.S.A.
119 J7 Mauke i. Pac. Oc.
20 B4 Maulbronn Ger.
105 B2 Maule div. Chile
105 B2 Maule r. Chile
105 B4 Maullín Chile
14 B3 Maumakeogh h. Rep. of Ireland
90 B4 Maumee U.S.A.
90 B4 Maumee r. U.S.A.
89 F5 Maumee Bay U.S.A.
57 G8 Maumere Indon.
14 B4 Maumturk Mts h. Rep. of Ireland
111 C5 Maun Botswana
92 □2 Mauna Kea volc. U.S.A.
92 □2 Mauna Loa volc. U.S.A.
92 □1 Maunalua B. U.S.A.
113 G1 Maunatlala Botswana
72 D2 Maungaturoto N.Z.
45 H5 Maungdaw Myanmar
58 A2 Maungmagan Is Myanmar
74 F3 Maunoir, Lac l. Can.
66 B5 Maupertuis B. Austr.
22 D6 Maures, Massif des reg. France
64 F5 Maurice, L. salt flat Austr.
12 C5 Maurik Neth.
107 B4 Mauritania country Africa
5 Mauritius country Indian Ocean
88 B4 Mauston U.S.A.
21 G5 Mauterndorf Austria
115 A2 Mava Congo(Zaire)
103 D4 Mavaca r. Venez.
115 C5 Mavago Moz.
111 C5 Mavinga Angola
113 G5 Mavuya S. Africa
44 D3 Mawana India
114 B4 Mawanga Congo(Zaire)
51 D4 Ma Wang Dui China
58 A3 Mawdaung Pass Myanmar/Thai.
70 G3 Mawhai Pt N.Z.
56 A2 Mawlaik Myanmar
116 D4 Mawson Aust. Base Ant.
116 D4 Mawson Coast Ant.
116 D4 Mawson Escarpment esc. Ant.
116 B6 Mawson Pen. Ant.
58 A3 Maw Taung mt Myanmar
42 D7 Max U.S.A.
97 G3 Maxcanú Mex.
26 C5 Maxia, Punta mt Sardinia Italy
88 D5 Maxinkuckee, Lake U.S.A.
8 S5 Maxmo Fin.
89 F2 Maxton U.S.A.
92 A2 Maxwell U.S.A.
55 B3 Maya i. Indon.
46 F1 Maya r. Rus. Fed.
95 K4 Mayaguana i. Bahamas
95 L5 Mayagüez Puerto Rico
114 E2 Mayahi Niger
41 H2 Mayakovskogo mt Tajik.
39 G4 Mayakum Kazak.
110 B4 Mayama Congo
40 D2 Mayamey Iran
97 G4 Maya Mountains Belize
50 B3 Mayan China
51 C5 Mayang China
52 F5 Maya-san mt Japan
11 D5 Maybole U.K.
37 K4 Maydān Iraq
41 H3 Maydā Shahr Afgh.
67 G3 Maydena Austr.
15 H2 Mayen Ger.
24 D2 Mayenne France
24 D2 Mayenne r. France
73 G4 Mayer U.S.A.
76 F4 Mayerthorpe Can.
70 C5 Mayfield N.Z.
86 B4 Mayfield U.S.A.
93 F5 Mayhill U.S.A.
54 E1 Mayi r. China
11 F4 May, Isle of i. U.K.
39 H2 Maykain Kazak.
39 J3 Maykamys Kazak.
39 G5 Maykhura Tajik.
29 G6 Maykop Rus. Fed.
53 B3 Maylibash Kazak.
29 H4 Mayly-Say Kyrg.
39 L1 Mayma Rus. Fed.
39 G4 Maymak Kazak.
50 B2 Maymyo Myanmar
48 F1 Mayna Rus. Fed.
43 A2 Mayni India
89 J3 Maynooth Can.
70 B2 Mayo Can.
105 E2 Mayo, 25 de Buenos Aires Arg.
105 C3 Mayo, 25 de La Pampa Arg.
114 F3 Mayo Alim Cameroon
55 C5 Mayo Bay Phil.
114 F3 Mayo-Belwa Nigeria
110 B4 Mayoko Congo
76 B2 Mayo Lake Can.
55 B3 Mayon volc. Phil.
105 D3 Mayor Buratovich Arg.
70 F2 Mayor I. N.Z.
102 D1 Mayor Pablo Lagerenza Para.
111 E5 Mayotte terr. Africa
55 B2 Mayraira Point Phil.
20 E5 Mayrhofen Austria
49 N1 Mayskiy Rus. Fed.
90 B5 Maysville U.S.A.
110 B4 Mayumba Gabon
43 E3 Mayum La pass China
43 B4 Mayuram India
89 F4 Mayville MI U.S.A.
84 D2 Mayville ND U.S.A.
90 D3 Mayville NY U.S.A.
88 C4 Mayville WV U.S.A.
84 C3 Maywood U.S.A.
105 D3 Maza Arg.
28 F3 Maza Rus. Fed.
115 A6 Mazabuka Zambia
101 H4 Mazagão Brazil
24 F5 Mazamet France
44 D1 Mazar China
26 E6 Mazara del Vallo Sicily Italy
41 G2 Mazār-e Sharīf Afgh.
41 G3 Mazar, Koh-i- mt Afgh.
103 E3 Mazaruni r. Guyana
96 B1 Mazatán Mex.
97 G5 Mazatenango Guatemala
96 C3 Mazatlán Mex.
93 G4 Mazatzal Peak U.S.A.
40 C3 Mazdaj Iran
9 S8 Mažeikiai Lith.
37 G2 Mazgirt Turkey
9 S8 Mazirbe Latvia
115 C4 Mazomora Tanz.
115 B6 Mazowe r. Zimbabwe
37 M3 Mazr'eh Iran
37 M5 Māzū Iran
111 C6 Mazunga Zimbabwe
29 D4 Mazyr Belarus
113 J3 Mbabane Swaziland
114 C3 Mbahiakro Côte d'Ivoire
110 B3 Mbaïki C.A.R.
110 B3 Mbakaou, Lac de l. Cameroon
115 B4 Mbala Zambia
115 B2 Mbale Uganda
114 F4 Mbalmayo Cameroon
114 F3 Mbam r. Cameroon
110 B4 Mbandaka Congo(Zaire)
114 F4 Mbang Cameroon
114 E4 Mbanga Cameroon
110 B4 M'banza Congo Angola
115 B3 Mbarara Uganda
110 C3 Mbari r. C.A.R.
113 K3 Mbaswana S. Africa
115 C4 Mbemkuru r. Tanz.
114 F4 Mbengwi Cameroon
115 B4 Mbeya Tanz.
115 B4 Mbeya div. Tanz.
115 C5 Mbinga Tanz.
114 E4 Mbini Equatorial Guinea
114 E4 Mbini r. Equatorial Guinea
111 D6 Mbizi Zimbabwe
115 A1 Mboki C.A.R.
110 B3 Mbomo Congo
110 F3 Mbouda Cameroon
108 A3 Mbour Senegal
108 A3 Mbout Maur.
115 B4 Mbozi Tanz.
110 C4 Mbuji-Mayi Congo(Zaire)
115 C3 Mbulu Tanz.
115 C4 Mbuyuni Tanz.
115 C4 Mchinga Tanz.
115 B5 Mchinji Malawi
113 G6 Mdantsane S. Africa
26 B6 M'Daourouch Alg.
85 C4 Meade U.S.A.
93 E3 Mead, Lake U.S.A.
69 B3 Meadow Austr.
77 H4 Meadow Lake Can.
77 H4 Meadow Lake Provincial Park Can.
93 E3 Meadow Valley Wash r. U.S.A.
90 C4 Meadville U.S.A.
89 G3 Meaford Can.
52 J3 Meaken-dake volc. Japan
11 A2 Mealasta Island U.K.
25 B2 Mealhada Port.
11 D4 Meall a'Bhuiridh mt U.K.
79 J3 Mealy Mountains Can.
38 E5 Meana Turkm.
67 H1 Meandarra Austr.
76 F3 Meander River Can.
55 C5 Meares i. Indon.
15 C4 Meaux France
110 B4 Mebridege r. Angola
42 A5 Mecca S. Arabia
91 H2 Mechanic Falls U.S.A.
90 B4 Mechanicsburg U.S.A.
88 B5 Mechanicsville U.S.A.
15 E1 Mechelen Belgium
18 C6 Mechelen Neth.
108 B1 Mecheria Alg.
15 G2 Mechernich Ger.
36 E1 Mecitözü Turkey
23 G2 Meckenbeuren Ger.
15 H2 Meckenheim Ger.
19 J2 Mecklenburger Bucht b. Ger.
19 J3 Mecklenburgische Seenplatte reg. Ger.
19 J3 Mecklenburg-Vorpommern div. Ger.
115 C5 Mecubúri Moz.
115 C5 Mecula Moz.
25 C2 Meda Port.
43 B2 Medak India
43 B2 Medan Indon.
105 D3 Médanos Arg.
102 C7 Medanosa, Pta pt Arg.
43 C4 Medawachchiya Sri Lanka
43 B2 Medchal India
91 K2 Meddybemps L. U.S.A.
25 H4 Médéa Alg.
18 F5 Medebach Ger.
103 B3 Medellín Col.
13 F4 Meden r. U.K.
108 D1 Medenine Tunisia
108 A3 Mederdra Maur.
82 B3 Medford OR U.S.A.
88 B3 Medford WV U.S.A.
91 F5 Medford Farms U.S.A.
27 N2 Medgidia Romania
37 L4 Medhīkhan Iran
88 B5 Media U.S.A.
105 C2 Media Luna Arg.
17 M7 Mediaş Romania
82 C2 Medical Lake U.S.A.
82 F3 Medicine Bow U.S.A.
82 F3 Medicine Bow Mts U.S.A.
82 F3 Medicine Bow Peak U.S.A.
77 G4 Medicine Hat Can.
85 D4 Medicine Lodge U.S.A.
104 E2 Medina Brazil
42 A5 Medina S. Arabia
90 D3 Medina NY U.S.A.
90 C4 Medina OH U.S.A.
25 E2 Medinaceli Spain
25 D2 Medina del Campo Spain
25 D2 Medina de Rioseco Spain
45 F5 Medinīpur India
6 C5 Mediterranean Sea Africa/Europe
26 B6 Medjerda, Monts de la mts Alg.
38 D2 Mednogorsk Rus. Fed.
24 D4 Médoc reg. France
28 H3 Medvedevo Rus. Fed.
29 H5 Medveditsa r. Rus. Fed.
26 F2 Medvednica mts Croatia
33 S2 Medvezh'i, O-va is Rus. Fed.
49 P2 Medvezh'ya, Gora mt China/Rus. Fed.
28 E2 Medvezh'yegorsk Rus. Fed.
13 H6 Medway r. U.K.
69 B3 Meeberrie Austr.
69 D3 Meekatharra Austr.
93 H1 Meeker U.S.A.
92 B2 Meeks Bay U.S.A.
79 J4 Meelpaeg Resr Can.
19 K6 Meerane Ger.
18 D5 Meerlo Neth.
44 D3 Meerut India
82 E2 Meeteetse U.S.A.
110 D3 Mēga Eth.
59 B3 Mega i. Indon.
45 G4 Meghalaya div. India
45 F5 Meghāsani mt India
45 G5 Meghna r. Bangl.
37 L2 Meghri Armenia
36 B3 Megisti i. Greece
8 U1 Mehamn Norway
50 B3 Mehar Pak.
64 C4 Meharry, Mt Austr.
44 D5 Mehekar India
45 G5 Meherpur Bangl.
90 E6 Meherrin r. U.S.A.
71 □2 Mehetia i. Fr. Polynesia Pac. Oc.
60 N5 Méhétia i. Pac. Oc.
37 L2 Mehrābān Iran
40 D5 Mehrān watercourse Iran
37 L5 Mehrān Iraq
15 G2 Mehren Ger.
40 D4 Mehrīz Iran
41 H3 Mehtar Lām Afgh.
104 C2 Meia Ponte r. Brazil
114 F3 Meiganga Cameroon
51 B4 Meigu China
51 E5 Mei Jiang r. China
18 D5 Meijnweg, Nationaal Park De nat. park Neth.
11 D5 Meikle Millyea h. U.K.
56 B2 Meiktila Myanmar
22 F2 Meilen Switz.
19 H4 Meine Ger.
19 H4 Meinersen Ger.
20 C5 Meiningen Austria
20 D2 Meiningen Ger.
22 F3 Meiringen Switz.
112 E6 Meiringspoort pass S. Africa
51 B4 Meishan China
19 L5 Meißen Ger.
51 C5 Meitan China
20 D4 Meitingen Ger.
50 C3 Mei Xian China
51 E5 Meizhou China
44 D4 Mej r. India
22 B5 Méjan, Sommet de mt France
102 C3 Mejicana mt Arg.
102 B2 Mejillones Chile
114 F4 Mékambo Gabon
110 D2 Mek'elē Eth.
108 A3 Mékhé Senegal
44 B3 Mekhtar Pak.
26 C7 Meknassy Tunisia
108 B1 Meknès Morocco
58 C2 Mekong r. Asia
46 B3 Mekong r. China
58 C3 Mekong, Mouths of the est. Vietnam
59 B2 Melaka Malaysia
118 G6 Melanesia is Pac. Oc.
55 A5 Melaut r. Malaysia
59 D3 Melawi r. Indon.
66 F6 Melbourne Austr.
87 D6 Melbourne U.S.A.
11 □ Melby U.K.
97 G4 Melchor de Mencos Guatemala
23 K5 Meldola Italy
19 G2 Meldorf Ger.
89 F3 Meldrum Bay Can.
22 F6 Mele, Capo pt Italy
23 G4 Melegnano Italy
36 E2 Melendiz Dağı mt Turkey
28 G4 Melenki Rus. Fed.
38 C1 Meleuz Rus. Fed.
79 F2 Mélèzes, Rivière aux r. Can.
109 D3 Mélfi Chad
26 F4 Melfi Italy
77 J4 Melfort Can.
8 M5 Melhus Norway
25 C1 Melide Spain
108 B1 Melilla Spain
105 E2 Melincué Arg.
59 E3 Melintang, Danau l. Indon.
105 B2 Melipilla Chile
18 A5 Meliskerke Neth.
77 J5 Melita Can.
29 E6 Melitopol' Ukr.
21 J4 Melk Austria
115 C2 Melka Guba Eth.
113 H1 Melkrivier S. Africa
13 E6 Melksham U.K.
23 H4 Mella r. Italy
8 T3 Mellakoski Fin.
8 Q5 Mellansel Sweden
20 C5 Mellau Austria
18 F4 Melle Ger.
88 B2 Mellen U.S.A.
9 N7 Mellerud Sweden
20 D2 Mellrichstadt Ger.
18 F3 Mellum i. Ger.
113 J4 Melmoth S. Africa
21 H2 Mělník Czech Rep.
105 F2 Melo Uru.
115 C5 Meloco Moz.
69 E3 Melrose Austr.
66 C4 Melrose Austr.
11 F5 Melrose U.K.
19 G5 Melsungen Ger.
55 A5 Melta, Mt Malaysia
13 G5 Melton Mowbray U.K.
15 C4 Melun France
77 J4 Melville Can.
65 G2 Melville Bay Austr.
75 M2 Melville Bugt b. Greenland
65 H2 Melville, C. Austr.
55 A5 Melville, C. Phil.
64 F2 Melville Island Austr.
74 G2 Melville Island N.W.T. Can.
79 J3 Melville, Lake Can.
75 K3 Melville Peninsula Can.
14 C3 Melvin, Lough l. Rep. of Ireland/U.K.
33 T3 Melyuveyem Rus. Fed.
45 E2 Mêmar Co salt l. China
115 D5 Memba Moz.
57 K7 Memberamo r. Indon.
57 F8 Memboro Indon.
113 H3 Memel S. Africa
20 D3 Memmelsdorf Ger.
20 D5 Memmingen Ger.
15 E3 Mémorial Américain h. France
59 C2 Mempawah Indon.
36 C7 Memphis Egypt
88 A5 Memphis MO U.S.A.
87 B5 Memphis TN U.S.A.
85 C5 Memphis TX U.S.A.
91 G2 Memphrémagog, Lac l. Can.
52 H3 Memuro-dake mt Japan
29 E5 Mena Ukr.
85 E5 Mena U.S.A.
114 D1 Ménaka Mali
Ménam Khong r. see Mekong
85 D6 Menard U.S.A.
88 C3 Menasha U.S.A.
15 F4 Menaucourt France
59 D3 Mendawai r. Indon.
24 F4 Mende France
37 M3 Mendejīn Iran
74 B4 Mendenhall, C. pt U.S.A.
76 C3 Mendenhall Glacier U.S.A.
39 F2 Mendesh Kazak.
97 E2 Méndez Mex.
110 D3 Mendī Eth.
62 E2 Mendi P.N.G.
13 E6 Mendip Hills U.K.
92 A2 Mendocino U.S.A.
82 A3 Mendocino, C. U.S.A.
119 K3 Mendocino Seascarp sea feature Pac. Oc.
88 E4 Mendon U.S.A.
67 H3 Mendooran Austr.
92 B3 Mendota CA U.S.A.
88 C5 Mendota IL U.S.A.
88 C4 Mendota, Lake U.S.A.
105 C2 Mendoza Arg.
105 C2 Mendoza div. Arg.
105 C2 Mendoza r. Arg.
103 C2 Mene de Mauroa Venez.
23 G5 Menegosa, Monte mt Italy
103 C2 Mene Grande Venez.
27 M5 Menemen Turkey
50 E3 Mengcheng China
20 C4 Mengen Ger.
36 D1 Mengen Turkey
59 C3 Menggala Indon.
20 F4 Mengkofen Ger.
51 D5 Mengshan China
50 F3 Meng Shan mts China
50 E3 Mengyin China
51 B6 Mengzi China
79 G3 Menihek Can.
79 G3 Menihek Lakes Can.
66 E4 Menindee Austr.
66 E4 Menindee Lake Austr.
66 C5 Meningie Austr.
37 M4 Menjān Iran
33 O3 Menkere Rus. Fed.
15 C4 Mennecy France
88 D3 Menominee U.S.A.
88 D3 Menominee r. U.S.A.
88 C4 Menomonee Falls U.S.A.
88 B3 Menomonie U.S.A.
111 B5 Menongue Angola
25 J2 Menorca i. Spain
22 C5 Mens France
55 A6 Mensalong Indon.
59 A3 Mentawai, Kepulauan is Indon.
58 B5 Mentekab Malaysia
19 H5 Menteroda Ger.
93 H4 Mentmore U.S.A.
59 C3 Mentok Indon.
22 E6 Menton France
90 C4 Mentor U.S.A.
108 C1 Menzel Bourguiba Tunisia
26 D6 Menzel Temime Tunisia
69 E4 Menzies Austr.
116 D4 Menzies, Mt Ant.
96 C1 Meoqui Mex.
18 D4 Meppel Neth.
18 E4 Meppen Ger.
113 K1 Mepuze Moz.
113 G4 Meqheleng S. Africa
23 G3 Mera r. Italy
28 G3 Mera r. Rus. Fed.
59 C4 Merak Indon.
8 M5 Meråker Norway
84 F4 Meramec r. U.S.A.
23 J3 Merano Italy
103 E3 Merari, Sa. mt Brazil
112 F1 Meratswe r. Botswana
59 E3 Meratus, Pegunungan mts Indon.
57 L8 Merauke Indon.
66 E5 Merbein Austr.
22 D5 Mercantour, Parc National du nat. park France
92 B3 Merced U.S.A.
105 B1 Mercedario, Cerro mt Arg.
105 E2 Mercedes Buenos Aires Arg.
102 E3 Mercedes Corrientes Arg.
105 D2 Mercedes San Luis Arg.
105 E2 Mercedes Uru.
90 A4 Mercer OH U.S.A.
88 B2 Mercer WV U.S.A.
76 F4 Mercoal Can.
70 E2 Mercury Islands N.Z.
75 M3 Mercy, C. hd Can.
15 D2 Mere Belgium
13 E6 Mere U.K.
91 H3 Meredith U.S.A.
85 C5 Meredith, Lake U.S.A.
85 C5 Meredith Nat. Recreation Area, Lake res. U.S.A.
88 B6 Meredosia U.S.A.
29 F5 Merefa Ukr.
71 □9 Mere Lava i. Vanuatu
109 E3 Merga Oasis oasis Sudan
38 B2 Mergenevo Kazak.
58 A2 Mergui Myanmar
58 A3 Mergui Archipelago is Myanmar
66 D5 Meribah Austr.
27 M4 Meriç r. Greece/Turkey
97 G3 Mérida Mex.
25 C3 Mérida Spain
103 C2 Mérida Venez.
103 C2 Mérida, Cordillera de mts Venez.
91 G4 Meriden U.S.A.
92 B2 Meridian CA U.S.A.
85 F5 Meridian MS U.S.A.
71 □9 Merig i. Vanuatu
24 D4 Mérignac France
8 T4 Merijärvi Fin.
9 R6 Merikarvia Fin.
67 H6 Merimbula Austr.
66 C3 Merinda Austr.
20 D4 Mering Ger.
66 D5 Meringur Austr.
66 D6 Merino Austr.
68 B5 Merivale r. Austr.
39 H4 Merke Kazak.
85 C5 Merkel U.S.A.
15 B2 Merl-les-Bains France
58 □ Merlimau, P. i. Sing.
22 C5 Merlu Rocher, Le mt France
109 F3 Merowe Sudan
69 B5 Merredin Austr.
11 D5 Merrick h. U.K.
89 K3 Merrickville Can.
88 C3 Merrill U.S.A.
88 D5 Merrillville U.S.A.
84 C3 Merriman U.S.A.
76 E4 Merritt Can.
87 D6 Merritt Island U.S.A.
67 J4 Merriwa Austr.
67 H3 Merrygoen Austr.
110 E2 Mersa Fatma Eritrea
15 G3 Mersch Lux.
19 J5 Merseburg (Saale) Ger.
13 E4 Mersey est. U.K.
Mersin see İçel
59 B2 Mersing Malaysia
9 S8 Mērsrags Latvia
44 C4 Merta India
13 D6 Merthyr Tydfil U.K.
115 C2 Merti Kenya
25 C4 Mértola Port.
116 B6 Mertz Gl. Ant.
15 C3 Méru France
115 C2 Meru Kenya
115 C3 Meru volc. Tanz.
41 F4 Merui Pak.
Merv see Mary
112 D6 Merweville S. Africa
36 E1 Merzifon Turkey
15 G3 Merzig Ger.
116 B2 Merz Pen. Ant.
93 G5 Mesa U.S.A.
88 A2 Mesabi Range h. U.S.A.
26 G4 Mesagne Italy
27 L7 Mesara, Ormos b. Greece
93 H3 Mesa Verde Nat. Park U.S.A.
103 B4 Mesay r. Col.
18 F5 Meschede Ger.
8 P4 Meselefors Sweden
78 F3 Mesgouez L. Can.
28 J2 Meshchura Rus. Fed.
Meshed see Mashhad
38 E4 Meshekli Uzbek.
41 E2 Meshkān Iran
29 G5 Meshkovskaya Rus. Fed.
83 E3 Mesick U.S.A.
27 K4 Mesimeri Greece
23 K5 Mesola Italy
27 J5 Mesolongi Greece
37 J4 Mesopotamia reg. Iraq
93 E3 Mesquite NV U.S.A.
85 D5 Mesquite TX U.S.A.
93 E4 Mesquite Lake U.S.A.
115 C5 Messalo r. Moz.
26 F5 Messina Sicily Italy
113 J1 Messina S. Africa
26 F5 Messina, Stretta di str. Italy
89 J2 Messines Can.
27 K6 Messini Greece
27 K6 Messiniakos Kolpos b. Greece
20 C5 Meßkirch Ger.
75 Q2 Mesters Vig Greenland
15 J3 Mestlin Ger.
21 J2 Město Albrechtice Czech Rep.
27 L5 Meston, Akra pt Greece
23 K4 Mestre Italy
36 F1 Mesudiye Turkey
103 C3 Meta r. Col./Venez.
89 G2 Metagama Can.
75 L3 Meta Incognita Pen. Can.
85 F6 Metairie U.S.A.
23 H6 Metallifere, Colline mts Italy
88 C5 Metamora U.S.A.
102 C3 Metán Arg.
115 B5 Metangula Moz.
23 K6 Metauro r. Italy
68 C5 Meteor Creek r. Austr.
120 H9 Meteor Depth depth Atl. Ocean
27 J6 Methoni Greece
91 H3 Methuen U.S.A.
11 E4 Methven U.K.
69 E2 Methwin, Mt h. Austr.
26 G3 Metković Croatia
76 C3 Metlakatla U.S.A.
21 H6 Metnitz Austria
21 H6 Metnitz r. Austria
115 C5 Metoro Moz.
59 C4 Metro Indon.
86 B4 Metropolis U.S.A.
15 E2 Mettet Belgium
18 E4 Mettingen Ger.
92 C4 Mettler U.S.A.
43 B4 Mettur India
110 D3 Metu Eth.
15 G3 Metz France
20 C4 Metzingen Ger.
15 G4 Meurthe r. France
15 F4 Meuse r. Belgium/France
19 K5 Meuselwitz Ger.
13 C7 Mevagissey U.K.
50 B3 Mêwa China
85 D6 Mexia U.S.A.
94 A2 Mexicali Mex.
93 H3 Mexican Hat U.S.A.
96 C1 Mexicanos, L. de los Mex.
93 H3 Mexican Water U.S.A.
73 F7 Mexico country Central America
97 E4 México Mex.
97 E4 México div. Mex.
91 H2 Mexico ME U.S.A.
84 F4 Mexico MO U.S.A.
91 E3 Mexico NY U.S.A.
72 G7 Mexico, Gulf of Mex./U.S.A.
40 D3 Meybod Iran
19 K3 Meyenburg Ger.
22 C4 Meylan France
41 G3 Meymaneh Afgh.
40 C3 Meymeh Iran
40 B3 Meymeh r. Iran
22 D4 Meythet France
97 F4 Mezcalapa r. Mex.
27 K3 Mezdra Bulg.
32 F3 Mezen' Rus. Fed.
22 B5 Mézenc, Mont mt France
48 E1 Mezhdurechensk Kemerovsk. Rus. Fed.
28 J2 Mezhdurechensk Komi Rus. Fed.
32 G2 Mezhdusharskiy, O. i. Rus. Fed.
38 D1 Mezhozernyy Rus. Fed.
17 K7 Mezőtúr Hungary
96 C3 Mezquital Mex.
96 C3 Mezquital r. Mex.
96 D3 Mezquitic Mex.
9 U8 Mežvidi Latvia
23 J3 Mezzano Italy
23 G3 Mezzola, Lago di l. Italy
23 J3 Mezzolombardo Italy
115 B5 Mfuwe Zambia
43 A2 Mhasvad India
113 J3 Mhlume Swaziland
44 C5 Mhow India
45 H5 Mi r. Myanmar
97 E4 Miahuatlán Mex.
25 D3 Miajadas Spain
93 G5 Miami AZ U.S.A.
87 D7 Miami FL U.S.A.
85 E4 Miami OK U.S.A.
87 D7 Miami Beach U.S.A.
40 C4 Mīān Āb Iran
41 F5 Mianaz Pak.
40 D3 Mīāndarreh Iran
40 B2 Miandowāb Iran
111 E5 Miandrivazo Madag.
40 B2 Mīāneh Iran
55 C5 Miangas i. Phil.
41 G5 Miani Hor b. Pak.
41 G3 Mianjoi Afgh.
51 B4 Mianning China
44 B2 Mianwali Pak.
50 C3 Mian Xian China
51 D4 Mianyang Hubei China
50 B4 Mianyang Sichuan China
50 B4 Mianzhu China
50 F2 Miao Dao i. China
50 F2 Miaodao Qundao is China
39 K3 Miao'ergou China
51 F5 Miaoli Taiwan
111 E5 Miarinarivo Madag.
32 H4 Miass Rus. Fed.
93 G5 Mica Mt U.S.A.
50 C3 Micang Shan mts China
17 K6 Michalovce Slovakia
77 H3 Michel Can.
19 L4 Michendorf Ger.
88 C2 Michigamme Lake U.S.A.
88 C2 Michigamme Reservoir U.S.A.
88 D2 Michigan div. U.S.A.
88 D5 Michigan City U.S.A.
88 D4 Michigan, Lake U.S.A.
88 E2 Michipicoten Bay Can.
88 E2 Michipicoten I. Can.
88 E1 Michipicoten River Can.
96 D4 Michoacán div. Mex.
27 M3 Michurin Bulg.
28 G4 Michurinsk Rus. Fed.
96 H5 Mico r. Nic.
118 E5 Micronesia is Pac. Oc.
61 G3 Micronesia, Federated States of country Pac. Oc.
59 C2 Midai i. Indon.
120 F4 Mid-Atlantic Ridge sea feature Atl. Ocean
112 C6 Middelberg Pass S. Africa
18 A5 Middelburg Neth.
112 F5 Middelburg Eastern Cape S. Africa
113 H2 Middelburg Mpumalanga S. Africa
9 L9 Middelfart Denmark
18 B5 Middelharnis Neth.
112 D5 Middelpos S. Africa
113 G2 Middelwit S. Africa
82 C3 Middle Alkali Lake U.S.A.
119 N5 Middle America Trench sea feature Pac. Oc.
56 A4 Middle Andaman i. Andaman and Nicobar Is
91 H4 Middleboro U.S.A.
90 E4 Middleburg U.S.A.
91 F3 Middleburgh U.S.A.
91 G2 Middlebury U.S.A.
70 C6 Middlemarch N.Z.
68 C4 Middlemount Austr.
90 B6 Middlesboro U.S.A.
13 F3 Middlesbrough U.K.
97 G4 Middlesex Belize
92 A2 Middletown CA U.S.A.
91 G4 Middletown CT U.S.A.
91 F5 Middletown DE U.S.A.
91 F4 Middletown NY U.S.A.
90 A5 Middletown OH U.S.A.
88 E4 Middleville U.S.A.
13 G7 Midhurst U.K.
117 K4 Mid-Indian Basin sea feature Ind. Ocean
117 J6 Mid-Indian Ridge sea feature Ind. Ocean
89 H3 Midland Can.
89 E4 Midland MI U.S.A.
85 C5 Midland TX U.S.A.
69 B5 Midland Junction Austr.
14 C6 Midleton Rep. of Ireland
118 F4 Mid-Pacific Mountains sea feature Pac. Oc.
8 □ Miðvágur Faroe Is
Midway see Thamarīt
118 H4 Midway Islands Pac. Oc.
82 F3 Midwest U.S.A.
85 D5 Midwest City U.S.A.
18 C4 Midwoud Neth.
37 H3 Midyat Turkey
11 □ Mid Yell U.K.
27 K3 Midzhur mt Bulg./Yugo.
20 E5 Mieders Austria
21 K2 Międzylesie Pol.
19 M3 Międzyzdroje Pol.
9 U6 Miehikkälä Fin.
8 T3 Miekojärvi l. Fin.
17 K5 Mielec Pol.
115 C4 Miembwe Tanz.
20 E5 Mieming Austria
67 G8 Miena Austr.
8 U2 Mieraslompolo Fin.
17 M7 Miercurea-Ciuc Romania
25 D1 Mieres Spain
21 K2 Mieroszów Pol.
20 E5 Miesbach Ger.
110 E3 Mī'ēso Eth.
19 J4 Mieste Ger.
19 M4 Mieszkowice Pol.
90 E4 Mifflinburg U.S.A.
90 E4 Mifflintown U.S.A.
50 C3 Migang Shan mt China
113 F3 Migdol S. Africa
41 E4 Mīghān Iran
45 H3 Miging India
96 D2 Miguel Auza Mex.
36 C2 Mihalıççık Turkey
53 C7 Mihara Japan
53 F7 Mihara-yama volc. Japan
25 F2 Mijares r. Spain
18 B4 Mijdrecht Neth.
89 F3 Mikado U.S.A.
17 N4 Mikashevichy Belarus
28 F4 Mikhaylov Rus. Fed.
116 D5 Mikhaylov I. Ant.
39 G4 Mikhaylovka Kazak.
39 J1 Mikhaylovka Kazak.
52 C3 Mikhaylovka Primorskiy Kray Rus. Fed.
29 G5 Mikhaylovka Volgograd. Rus. Fed.
39 J2 Mikhaylovskiy Rus. Fed.
45 H4 Mikir Hills India
9 U6 Mikkeli Fin.
9 U6 Mikkelin mlk Fin.
76 G3 Mikkwa r. Can.
21 K4 Mikulov Czech Rep.
115 C4 Mikumi Tanz.
115 C4 Mikumi National Park Tanz.
28 J2 Mikun' Rus. Fed.
52 F6 Mikuni-sammyaku mts Japan
53 F8 Mikura-jima i. Japan
84 E2 Milaca U.S.A.
43 A5 Miladhunmadulu Atoll atoll Maldives
23 G4 Milan Italy
87 B5 Milan U.S.A.
66 C5 Milang Austr.
115 C6 Milange Moz.
Milano see Milan
36 A3 Milas Turkey
26 F5 Milazzo Sicily Italy
84 D2 Milbank U.S.A.
13 H5 Mildenhall U.K.
66 E5 Mildura Austr.
51 B5 Mile China
68 D6 Miles Austr.
82 F2 Miles City U.S.A.
14 C5 Milestone Rep. of Ireland
26 F4 Miletto, Monte mt Italy
69 C3 Mileura Austr.
21 H3 Milevsko Czech Rep.
14 D2 Milford Rep. of Ireland
92 B1 Milford CA U.S.A.
91 G4 Milford CT U.S.A.
91 F5 Milford DE U.S.A.
88 D5 Milford IL U.S.A.
91 H3 Milford MA U.S.A.
91 J2 Milford ME U.S.A.
91 H3 Milford NH U.S.A.
91 F3 Milford NY U.S.A.
93 F2 Milford UT U.S.A.
13 B6 Milford Haven U.K.
70 A6 Milford Sound N.Z.
70 A6 Milford Sound in. N.Z.
69 D4 Milgoo, Mt h. Austr.
69 D2 Milgun Austr.
25 H4 Miliana Alg.
69 C5 Miling Austr.
74 G5 Milk r. Alta. Can.
82 F1 Milk r. Can./U.S.A.
33 R4 Mil'kovo Rus. Fed.
68 A1 Millaa Millaa Austr.
25 F2 Millārs r. Spain
24 F4 Millau France
92 B1 Mill Creek r. U.S.A.
87 D5 Milledgeville GA U.S.A.
88 C5 Milledgeville IL U.S.A.
84 E2 Mille Lacs L. U.S.A.
78 B4 Mille Lacs, Lac des l. Can.
66 A2 Miller watercourse Austr.
84 D2 Miller U.S.A.
88 B3 Miller Dam Flowage resr U.S.A.
89 G3 Miller Lake Can.
29 G5 Millerovo Rus. Fed.
93 G6 Miller Peak U.S.A.
90 C4 Millersburg OH U.S.A.
90 E4 Millersburg PA U.S.A.
66 B3 Millers Creek Austr.
90 E6 Millers Tavern U.S.A.
92 C3 Millerton Lake U.S.A.
11 C5 Milleur Point U.K.
116 C6 Mill I. Ant.
66 D6 Millicent Austr.
89 F4 Millington MI U.S.A.
87 B5 Millington TN U.S.A.
91 J2 Millinocket U.S.A.
67 J1 Millmerran Austr.
12 D3 Millom U.K.
11 D5 Millport U.K.
91 F5 Millsboro U.S.A.
76 F2 Mills Lake Can.
90 C5 Millstone U.S.A.
79 G4 Milltown Can.
14 B5 Milltown Malbay Rep. of Ireland
91 K1 Millville Can.
91 F5 Millville U.S.A.
15 C4 Milly-la-Forêt France
69 C3 Milly Milly Austr.
19 L3 Milmersdorf Ger.
114 B2 Milo r. Guinea
91 J2 Milo U.S.A.
52 D3 Milogradovo Rus. Fed.
27 L6 Milos i. Greece
28 F4 Miloslavskoye Rus. Fed.
66 D2 Milparinka Austr.
90 E4 Milroy U.S.A.
20 C3 Miltenberg Ger.
89 H4 Milton Can.
70 B7 Milton N.Z.
87 C6 Milton FL U.S.A.
88 A5 Milton IA U.S.A.
90 E4 Milton PA U.S.A.
91 G2 Milton VT U.S.A.
82 C2 Milton-Freewater U.S.A.
13 G5 Milton Keynes U.K.
90 C4 Milton, Lake U.S.A.
19 L2 Miltzow Ger.
51 D4 Miluo China
88 D4 Milwaukee U.S.A.
38 E2 Mily Kazak.
29 G5 Milyutinskaya Rus. Fed.
24 D4 Mimizan France
21 H2 Mimoň Czech Rep.
110 B4 Mimongo Gabon
96 D2 Mina Mex.
92 C2 Mina U.S.A.
40 E5 Mīnāb Iran
57 G6 Minahassa Peninsula Indon.
40 D5 Mina Jebel Ali U.A.E.
77 L4 Minaki Can.

53 B8 Minamata Japan
53 E7 Minami Alps National Park Japan
59 B2 Minas Indon.
105 F2 Minas Uru.
37 M7 Minā Sa'ūd Kuwait
79 H4 Minas Basin b. Can.
105 F1 Minas de Corrales Uru.
104 D2 Minas Gerais div. Brazil
104 D2 Minas Novas Brazil
97 G5 Minas, Sa de las mts Guatemala
97 F4 Minatitlán Mex.
45 H5 Minbu Myanmar
45 H5 Minbya Myanmar
102 B6 Minchinmávida volc. Chile
11 C2 Minch, The str. U.K.
26 D2 Mincio r. Italy
37 L2 Mincivan Azer.
55 C5 Mindanao i. Phil.
66 D5 Mindarie Austr.
20 D4 Mindel r. Ger.
20 D4 Mindelheim Ger.
108 □ Mindelo Cape Verde
89 H3 Minden Can.
18 F4 Minden Ger.
85 E5 Minden LA U.S.A.
92 C2 Minden NV U.S.A.
45 H6 Mindon Myanmar
66 E4 Mindona L. Austr.
55 B3 Mindoro i. Phil.
55 A3 Mindoro Strait Phil.
110 B4 Mindouli Congo
38 D1 Mindyak Rus. Fed.
14 D6 Mine Head hd Rep. of Ireland
13 D6 Minehead U.K.
104 B2 Mineiros Brazil
85 E5 Mineola U.S.A.
92 B1 Mineral U.S.A.
92 C3 Mineral King U.S.A.
29 G6 Mineral'nyye Vody Rus. Fed.
88 B4 Mineral Point U.S.A.
85 D5 Mineral Wells U.S.A.
93 F2 Minersville U.S.A.
26 G4 Minervino Murge Italy
45 E1 Minfeng China
115 A5 Minga Congo(Zaire)
37 L1 Mingäçevir Azer.
37 L1 Mingäçevir Su Anbarı resr Azer.
79 H3 Mingan Can.
66 D4 Mingary Austr.
38 E4 Mingbulak Uzbek.
68 B2 Mingela Austr.
69 B4 Mingenew Austr.
50 E3 Minggang China
39 H4 Ming-Kush Kyrg.
25 F3 Minglanilla Spain
115 C5 Mingoyo Tanz.
51 B4 Ming-shan China
46 E2 Mingshui China
39 H5 Mingteke China
11 A4 Mingulay i. U.K.
51 E5 Mingxi China
50 B2 Minhe China
51 F5 Minhou China
43 A4 Minicoy i. India
69 B1 Minilya Austr.
69 B1 Minilya r. Austr.
114 B2 Mininian Côte d'Ivoire
79 H3 Minipi Lake Can.
77 J4 Minitonas Can.
51 F5 Min Jiang r. Fujian China
50 B4 Min Jiang r. Sichuan China
50 A2 Minle China
114 E3 Minna Nigeria
9 O5 Minne Sweden
84 E2 Minneapolis U.S.A.
77 K4 Minnedosa Can.
81 G2 Minnesota r. MN U.S.A.
88 A2 Minnesota div. U.S.A.
69 B1 Minnie Creek Austr.
66 A4 Minnipa Austr.
78 B4 Minnitaki L. Can.
25 B2 Miño r. Port./Spain
88 C3 Minocqua U.S.A.
88 B2 Minong U.S.A.
88 C5 Minonk U.S.A.
Minorca i. see Menorca
84 C1 Minot U.S.A.
50 B2 Minqin China
51 F5 Minqing China
50 B3 Min Shan mts China
45 H4 Minsin Myanmar
28 C4 Minsk Belarus
17 K4 Mińsk Mazowiecki Pol.
13 E5 Minsterley U.K.
114 F4 Minta Cameroon
44 C1 Mintaka Pass China/Jammu and Kashmir
79 G4 Minto Can.
74 G2 Minto Inlet in. Can.
79 F2 Minto, Lac l. Can.
84 F3 Minturn U.S.A.
36 C6 Minûf Egypt
48 F1 Minusinsk Rus. Fed.
45 J3 Minutang India
50 B3 Min Xian China
66 E6 Minyip Austr.
89 E3 Mio U.S.A.
22 B4 Mionnay France
78 E4 Miquelon Can.
79 J4 Miquelon i. N. America
103 A4 Mira r. Col.
23 K4 Mira Italy
41 F4 Mirabad Afgh.
91 F2 Mirabel Can.
104 D2 Mirabela Brazil
101 J5 Miracema do Norte Brazil
97 G4 Mirador-Dos Lagunos-Río Azul, Parque Nacional nat. park Guatemala
101 J5 Mirador, Parque Nacional de nat. park Brazil
103 B4 Miraflores Col.
104 D2 Miralta Brazil
105 F3 Miramar Arg.
23 K5 Miramare Italy
97 G4 Miramar, L. Mex.
22 C6 Miramas France
79 G4 Miramichi r. Can.
27 L7 Mirampelou, Kolpos b. Greece
44 B2 Miram Shah Pak.
104 A3 Miranda Brazil
104 A3 Miranda r. Brazil
92 A1 Miranda U.S.A.
25 E1 Miranda de Ebro Spain
25 C2 Mirandela Port.
23 J5 Mirandola Italy
104 B3 Mirandópolis Brazil
23 K4 Mirano Italy
37 H5 Mirā', Wādī al watercourse Iraq/S. Arabia
42 D6 Mirbāt Oman
15 G4 Mirecourt France
24 E5 Mirepoix France
59 D2 Miri Malaysia
41 F4 Miri mt Pak.
43 B2 Mirialguda India
105 G2 Mirim, Lagoa l. Brazil
41 F4 Mīrjāveh Iran
23 L4 Mirna r. Croatia
116 D5 Mirnyy Rus. Fed. Base Ant.
28 G2 Mirnyy Archangel. Rus. Fed.
33 N3 Mirnyy Resp. Sakha Rus. Fed.
77 J3 Mirond L. Can.
19 K3 Mirow Ger.
44 C2 Mirpur Pak.
44 B4 Mirpur Batoro Pak.
44 B4 Mirpur Khas Pak.
44 A4 Mirpur Sakro Pak.
76 G4 Mirror Can.
51 □ Mirs Bay H.K. China
41 E5 Mīr Shahdād Iran
68 B3 Mirtna Austr.
27 K6 Mirtoö Pelagos sea Greece
54 E6 Miryang S. Korea
38 E5 Mirzachirla Turkm.
45 E4 Mirzapur India
53 C8 Misaki Japan
39 K5 Misalay China
52 G4 Misawa Japan
79 H4 Miscou I. Can.
44 C1 Misgar Pak.
52 B2 Mishan China
40 C5 Mishāsh al Hādī well S. Arabia
88 D5 Mishawaka U.S.A.
88 E1 Mishibishu Lake Can.
53 B7 Mi-shima i. Japan
45 H3 Mishmi Hills India
62 F3 Misima I. P.N.G.
96 J5 Miskitos, Cayos is Nic.
17 K6 Miskolc Hungary
22 C5 Mison France
57 J7 Misoöl i. Indon.
88 B2 Misquah Hills U.S.A.
109 D1 Mişrātah Libya
44 E4 Misrikh India
89 E1 Missanabie Can.
78 D3 Missinaibi r. Can.
78 D4 Missinaibi Lake Can.
77 J3 Missinipe Can.
76 E5 Mission Can.
84 C3 Mission SD U.S.A.
68 B1 Mission Beach Austr.
78 D3 Missisa L. Can.
78 E3 Missisagi r. Can.
89 H4 Mississauga Can.
88 E5 Mississinewa Lake U.S.A.
89 J3 Mississippi r. Can.
85 F5 Mississippi div. U.S.A.
85 F6 Mississippi r. U.S.A.
85 F6 Mississippi Delta U.S.A.
82 D2 Missoula U.S.A.
88 A6 Missouri div. U.S.A.
84 E4 Missouri r. U.S.A.
84 E3 Missouri Valley U.S.A.
68 B4 Mistake Cr. r. Austr.
75 L4 Mistassibi r. Can.
79 F4 Mistassini Can.
79 F4 Mistassini r. Can.
78 F3 Mistassini, L. Can.
79 H2 Mistastin Lake Can.
21 K4 Mistelbach Austria
76 C3 Misty Fjords National Monument res. U.S.A.
96 C3 Mita, Pta de hd Mex.
67 K2 Mitchell r. N.S.W. Austr.
68 B6 Mitchell Qld. Austr.
65 H3 Mitchell r. Qld. Austr.
67 G6 Mitchell r. Vic. Austr.
89 G4 Mitchell Can.
84 D3 Mitchell U.S.A.
88 E3 Mitchell, Lake U.S.A.
87 D5 Mitchell, Mt U.S.A.
14 C5 Mitchelstown Rep. of Ireland
36 C6 Mīt Ghamr Egypt
44 B3 Mithankot Pak.
44 B4 Mithi Pak.
44 B4 Mithrani Canal Pak.
27 M5 Mithymna Greece
76 C3 Mitkof I. U.S.A.
53 G6 Mito Japan
115 C4 Mitole Tanz.
70 E4 Mitre mt N.Z.
63 H3 Mitre Island Solomon Is
67 J5 Mittagong Austr.
67 F6 Mitta Mitta Austr.
20 D5 Mittelberg Austria
20 D6 Mittelberg Austria
18 F4 Mittellandkanal canal Ger.
20 E5 Mittenwald Ger.
21 G4 Mitterding Austria
15 G4 Mittersheim France
20 F5 Mittersill Austria
20 F3 Mitterteich Ger.
19 K6 Mittweida Ger.
103 C4 Mitú Col.
103 C4 Mituas Col.
115 A4 Mitumba, Chaîne des mts Congo(Zaire)
115 A3 Mitumba, Monts mts Congo(Zaire)
115 A4 Mitwaba Congo(Zaire)
110 B3 Mitzic Gabon
53 F7 Miura Japan
37 G4 Miyah, Wādī el watercourse Syria
53 F7 Miyake-jima i. Japan
52 G5 Miyako Japan
53 B9 Miyakonojō Japan
53 B9 Miyaluo China
38 C2 Miyaly Kazak.
45 H4 Miyāni India
53 B9 Miyazaki Japan
53 D7 Miyazu Japan
51 B5 Miyi China
53 C7 Miyoshi Japan
50 E1 Miyun China
50 E1 Miyun Sk. resr China
41 G3 Mīzāni Afgh.
110 D3 Mīzan Teferī Eth.
109 D1 Mizdah Libya
14 B6 Mizen Head hd Rep. of Ireland
29 B5 Mizhhir"ya Ukr.
50 D2 Mizhi China
45 H5 Mizoram div. India
52 G5 Mizusawa Japan
9 O7 Mjölby Sweden
115 C4 Mkata Tanz.
115 C4 Mkokotoni Tanz.
115 C3 Mkomazi Tanz.
115 A5 Mkushi Zambia
21 H2 Mladá Boleslav Czech Rep.
27 J2 Mladenovac Yugo.
17 K4 Mława Pol.
26 G3 Mljet i. Croatia
115 C5 Mlungisi S. Africa
17 M5 Mlyniv Ukr.
113 F2 Mmabatho S. Africa
113 G1 Mmamabula Botswana
113 F2 Mmathethe Botswana
21 H2 Mnichovo Hradiště Czech Rep.
9 J6 Mo Norway
93 H2 Moab U.S.A.
65 H2 Moa I. Austr.
71 □14 Moala i. Fiji
40 D3 Mo'alla Iran
112 A2 Moamba Moz.
66 D2 Moanba, Lake salt flat Austr.
93 E3 Moapa U.S.A.
14 D4 Moate Rep. of Ireland
115 A4 Moba Congo(Zaire)
53 G7 Mobara Japan
40 C3 Mobārakeh Iran
110 C3 Mobayi-Mbongo Congo(Zaire)
84 E4 Moberly U.S.A.
87 B6 Mobile AL U.S.A.
93 F5 Mobile AZ U.S.A.
87 B6 Mobile Bay U.S.A.
84 C2 Mobridge U.S.A.
Mobutu, Lake see Albert, Lake
104 C1 Mocajuba Brazil
115 D6 Moçambique Moz.
103 D2 Mocapra r. Venez.
51 B6 Môc Châu Vietnam
71 □14 Moce i. Fiji
103 D2 Mochirma, Parque Nacional nat. park Venez.
111 C6 Mochudi Botswana
115 D5 Mocimboa da Praia Moz.
19 J4 Möckern Ger.
20 C3 Möckmühl Ger.
9 N8 Mockträsk Sweden
103 A4 Mocoa Col.
104 C3 Mococa Brazil
96 C2 Mocorito Mex.
96 D3 Moctezuma Mex.
115 C6 Mocuba Moz.
22 D4 Modane France
44 C5 Modasa India
112 F4 Modder r. S. Africa
23 H5 Modena Italy
93 F3 Modena U.S.A.
92 B3 Modesto U.S.A.
21 K4 Mödling Austria
21 L4 Modra Slovakia
21 J6 Modriach Austria
67 G7 Moe Austr.
13 D5 Moel Sych h. U.K.
9 M6 Moely Norway
8 Q2 Moen Norway
93 G3 Moenkopi U.S.A.
70 C6 Moeraki Pt N.Z.
18 D5 Moers Ger.
11 E5 Moffat U.K.
44 C3 Moga India
Mogadishu see Muqdisho
90 C4 Mogadore Reservoir U.S.A.
113 H1 Mogalakwena r. S. Africa
113 H2 Moganyaka S. Africa
19 K4 Mögelin Ger.
21 K6 Mogersdorf Austria
39 F5 Moghiyon Tajik.
104 C3 Mogi-Mirim Brazil
23 K4 Mogliano Veneto Italy
49 L1 Mogocha Rus. Fed.
26 C6 Mogod mts Tunisia
113 F2 Mogoditshane Botswana
56 B2 Mogok Myanmar
93 H5 Mogollon Baldy mt U.S.A.
93 H5 Mogollon Mts U.S.A.
93 G4 Mogollon Rim plat. U.S.A.
113 G2 Mogwase S. Africa
27 H2 Mohács Hungary
70 F3 Mohaka r. N.Z.
113 G5 Mohale's Hoek Lesotho
77 J5 Mohall U.S.A.
41 E3 Mohammad Iran
Mohammadābād see Darreh Gaz
25 G5 Mohammadia Alg.
44 E3 Mohan r. India/Nepal
93 E4 Mohave, L. U.S.A.
93 F5 Mohawk U.S.A.
91 F3 Mohawk r. U.S.A.
93 F5 Mohawk Mts U.S.A.
111 E5 Moheli i. Comoros
21 K3 Mohelnice Czech Rep.
14 D4 Mohill Rep. of Ireland
18 F5 Möhne r. Ger.
93 F4 Mohon Peak U.S.A.
115 C4 Mohoro Tanz.
96 D2 Mohovano Ranch Mex.
37 M5 Moh Reza Shah Pahlavi resr Iran
29 C5 Mohyliv Podil's'kyy Ukr.
9 K7 Moi Norway
113 G1 Moijabana Botswana
71 □9 Moindou New Caledonia Pac. Oc.
113 K2 Moine Moz.
17 N7 Moineşti Romania
22 B4 Moingt France
39 F3 Moinkum, Peski des. Kazak.
91 F2 Moira U.S.A.
8 O3 Mo i Rana Norway
45 H4 Moirang India
9 T7 Mõisaküla Estonia
105 E1 Moisés Ville Arg.
79 G3 Moisie Can.
79 G3 Moisie r. Can.
24 E4 Moissac France
92 C4 Mojave U.S.A.
92 D4 Mojave r. U.S.A.
92 D4 Mojave Desert U.S.A.
104 C3 Moji das Cruzes Brazil
104 C3 Moji-Guaçu r. Brazil
53 B8 Mojikō Japan
45 F4 Mokāma India
92 □1 Mokapu Pen. U.S.A.
70 E3 Mokau N.Z.
70 E3 Mokau r. N.Z.
92 B2 Mokelumne r. U.S.A.
113 H4 Mokhoabong Pass Lesotho
113 H4 Mokhotlong Lesotho
26 D7 Moknine Tunisia
70 E1 Mokohinau Is N.Z.
114 F2 Mokolo Cameroon
113 G2 Mokolo r. S. Africa
54 D6 Mokp'o S. Korea
28 G4 Moksha r. Rus. Fed.
28 H4 Mokshan Rus. Fed.
92 □1 Mokuauia I. U.S.A.
92 □1 Mokulua Is U.S.A.
114 E3 Mokwa Nigeria
112 E3 Molango Mex.
25 F3 Molatón mt Spain
Moldavia country see Moldova
8 K5 Molde Norway
8 O3 Moldjord Norway
7 F4 Moldova country Europe
27 L2 Moldoveanu, Vârful mt Romania
18 D7 Mole r. U.K.
114 C3 Mole National Park Ghana
111 C6 Molepolole Botswana
9 T9 Molėtai Lith.
26 G4 Molfetta Italy
19 H2 Molfsee Ger.
54 C2 Molihong Shan h. China
25 F2 Molina de Aragón Spain
88 B5 Moline U.S.A.
23 J5 Molinella Italy
22 C3 Molinges France
115 B4 Moliro Congo(Zaire)
9 N7 Molkom Sweden
37 M4 Molla Bodāgh Iran
19 G1 Møllbjerg h. Ger.
45 H4 Mol Len mt India
19 L3 Möllenbeck Ger.
100 D7 Mollendo Peru
69 C5 Mollerin, L. salt flat Austr.
15 C3 Molliens-Dreuil France
19 J3 Mölln Ger.
9 N8 Mölnlycke Sweden
28 F3 Molochnoye Rus. Fed.
8 X2 Molochnyy Rus. Fed.
116 D4 Molodezhnaya Rus. Fed. Base Ant.
39 G1 Molodogvardeyskoye Kazak.
28 E3 Molodoy Tud Rus. Fed.
92 □2 Molokai i. U.S.A.
119 K4 Molokai Fracture Zone sea feature Pac. Oc.
28 J3 Moloma r. Rus. Fed.
67 H4 Molong Austr.
112 F2 Molopo watercourse Botswana/S. Africa
109 D4 Moloundou Cameroon
15 H4 Molsheim France
77 K4 Molson L. Can.
57 H7 Moluccas is Indon.
57 H6 Molucca Sea g. Indon.
115 C6 Molumbo Moz.
23 H3 Molveno, Lago di l. Italy
115 C6 Moma Moz.
66 E3 Momba Austr.
115 C3 Mombasa Kenya
45 H4 Mombi New India
20 C2 Mömbris Ger.
104 B2 Mombuca, Serra da h. Brazil
29 C7 Momchilgrad Bulg.
88 D5 Momence U.S.A.
22 F4 Momo Italy
103 B2 Mompós Col.
9 N9 Møn i. Denmark
92 G2 Mona U.S.A.
11 A3 Monach Islands U.K.
11 A3 Monach, Sound of chan. U.K.
7 D4 Monaco country Europe
11 D3 Monadhliath Mountains U.K.
14 E3 Monaghan Rep. of Ireland
85 C6 Monahans U.S.A.
95 L5 Mona, I. Puerto Rico
95 L5 Mona Passage Dom. Rep./Puerto Rico
115 D5 Monapo Moz.
76 D4 Monarch Mt. Can.
83 F4 Monarch Pass U.S.A.
11 C3 Monar, Loch l. U.K.
76 F4 Monashee Mts Can.
26 D7 Monastir Tunisia
17 P3 Monastyrshchina Rus. Fed.
29 D5 Monastyryshche Ukr.
71 □15 Monavatu mt Fiji
52 H2 Monbetsu Japan
52 H3 Monbetsu Japan
22 E4 Moncalieri Italy
25 F2 Moncayo mt Spain
8 X3 Monchegorsk Rus. Fed.
18 D5 Mönchengladbach Ger.
21 K5 Mönchhof Austria
25 B4 Monchique Port.
87 E5 Moncks Corner U.S.A.
96 D2 Monclova Mex.
79 H4 Moncton Can.
25 C2 Mondego r. Port.
113 J3 Mondlo S. Africa
23 L6 Mondolfo Italy
22 E5 Mondovì Italy
88 B3 Mondovi U.S.A.
22 B5 Mondragon France
26 E4 Mondragone Italy
27 K6 Monemvasia Greece
52 G1 Moneron, Ostrov i. Rus. Fed.
90 D4 Monessen U.S.A.
89 K1 Monet Can.
22 A2 Monéteau France
14 D5 Moneygall Rep. of Ireland
14 E3 Moneymore U.K.
23 L4 Monfalcone Italy
25 C1 Monforte Spain
110 C3 Monga Congo(Zaire)
115 B2 Mongbwalu Congo(Zaire)
51 C6 Mông Cai Vietnam
69 C4 Mongers Lake salt flat Austr.
Monggolküre see Zhaosu
54 C4 Monggŭmp'o-ri N. Korea
23 J5 Monghidoro Italy
58 A1 Mong Mau Myanmar
31 J5 Mongolia country Asia
114 F2 Mongonu Nigeria
44 C2 Mongora Pak.
111 C5 Mongu Zambia
91 J3 Monhegan I. U.S.A.
20 D4 Monheim Ger.
11 E5 Moniaive U.K.
84 B4 Monistrol-sur-Loire France
92 D2 Monitor Mt U.S.A.
92 D2 Monitor Range mts U.S.A.
14 C4 Monivea Rep. of Ireland
115 C5 Monkey Bay Malawi
89 G4 Monkton Can.
45 F3 Mon La pass China
13 E6 Monmouth U.K.
88 B5 Monmouth IL U.S.A.
91 H2 Monmouth ME U.S.A.
76 E4 Monmouth Mt. Can.
13 E6 Monnow r. U.K.
114 D3 Mono r. Togo
92 C3 Mono Lake U.S.A.
91 H4 Monomoy Pt U.S.A.
88 D5 Monon U.S.A.
88 B4 Monona U.S.A.
26 G4 Monopoli Italy
90 C5 Monongahela r. U.S.A.
25 F2 Monreal del Campo Spain
26 E5 Monreale Sicily Italy
85 E5 Monroe LA U.S.A.
89 F5 Monroe MI U.S.A.
87 D5 Monroe NC U.S.A.
91 F4 Monroe NY U.S.A.
93 F2 Monroe UT U.S.A.
88 C4 Monroe WV U.S.A.
88 B6 Monroe City U.S.A.
87 C6 Monroeville U.S.A.
114 B4 Monrovia Liberia
15 D2 Mons Belgium
15 G2 Monschau Ger.
23 J4 Monselice Italy
15 H2 Montabaur Ger.
111 E5 Montagne d'Ambre, Parc National de la nat. park Madag.
112 D6 Montagu S. Africa
78 D4 Montague U.S.A.
63 F2 Montague Range h. Austr.
116 C1 Montagu I. Atl. Ocean
22 C3 Montalieu-Vercieu France
26 F5 Montalto mt Italy
26 G5 Montalto Uffugo Italy
27 K3 Montana Bulg.
82 E2 Montana div. U.S.A.
96 H5 Montaña de Yoro nat. park Honduras
24 F3 Montargis France
24 E4 Montauban France
91 G4 Montauk U.S.A.
91 H4 Montauk Pt U.S.A.
22 B2 Montbard France
22 C2 Montbarrey France
22 D2 Montbéliard France
22 D4 Mont Blanc mt France/Italy
25 G2 Montblanc Spain
22 D4 Mont Blanc Tunnel tunnel France/Italy
22 B3 Montbrison France
22 B3 Montceau-les-Mines France
22 B3 Montcenis France
22 D4 Mont Cenis, Lac du l. France
22 B3 Montchanin France
15 E3 Montcornet France
24 D5 Mont-de-Marsan France
15 C3 Montdidier France
101 H4 Monte Alegre Brazil
104 C1 Monte Alegre de Goiás Brazil
104 D1 Monte Azul Brazil
78 E4 Montebello Can.
26 F6 Montebello Ionico Italy
23 K4 Montebelluna Italy
105 D2 Monte Buey Arg.
22 E6 Monte-Carlo Monaco
105 F1 Monte Caseros Arg.
23 H6 Montecatini Terme Italy
23 K6 Montecchio Italy
23 J4 Montecchio Maggiore Italy
113 G1 Monte Christo S. Africa
105 C2 Monte Comán Arg.
95 K5 Monte Cristi Dom. Rep.
26 C3 Montecristo, Isola di i. Italy
23 L6 Montegiorgio Italy
95 J5 Montego Bay Jamaica
22 B5 Montélimar France
102 E2 Monte Lindo r. Para.
26 F4 Montella Italy
88 C4 Montello U.S.A.
97 E2 Montemorelos Mex.
25 B3 Montemor-o-Novo Port.
27 H3 Montenegro div. Yugo.
115 C5 Montepuez Moz.
26 D3 Montepulciano Italy
15 C4 Montereau-faut-Yonne France
92 B3 Monterey CA U.S.A.
90 D5 Monterey VA U.S.A.
92 B3 Monterey Bay U.S.A.
103 B2 Montería Col.
23 J6 Monteriggioni Italy
100 F7 Montero Bol.
22 E4 Monte Rosa summit Italy/Switz.
97 D2 Monterrey Mex.
26 F4 Montesano sulla Marcellana Italy
101 L6 Monte Santo Brazil
104 D2 Montes Claros Brazil
26 F3 Montesilvano Italy
23 J6 Montevarchi Italy
105 F2 Montevideo Uru.
84 E2 Montevideo U.S.A.
83 F4 Monte Vista U.S.A.
88 A5 Montezuma U.S.A.
93 G4 Montezuma Castle National Monument res. U.S.A.
93 H3 Montezuma Creek U.S.A.
92 D3 Montezuma Peak U.S.A.
22 D6 Montferrat France
18 C5 Montfort Neth.
22 D5 Montgenèvre France
15 C4 Montgeron France
13 D5 Montgomery U.K.
87 C5 Montgomery U.S.A.
22 D3 Monthey Switz.
23 K4 Monticano r. Italy
85 F5 Monticello AR U.S.A.
87 D6 Monticello FL U.S.A.
88 B4 Monticello IA U.S.A.
88 D5 Monticello IN U.S.A.
91 K1 Monticello ME U.S.A.
88 B5 Monticello MO U.S.A.
91 F4 Monticello NY U.S.A.
93 H3 Monticello UT U.S.A.
88 C4 Monticello WV U.S.A.
23 H4 Montichiari Italy
105 E1 Montiel, Cuchilla de h. Arg.
15 E4 Montier-en-Der France
24 E4 Montignac France
15 E2 Montignies-le-Tilleul Belgium
15 G3 Montigny-lès-Metz France
22 B2 Montigny-sur-Aube France
96 J7 Montijo, G. de Panama
25 D4 Montilla Spain
79 G4 Mont Joli Can.
89 K2 Mont-Laurier Can.
79 G4 Mont Louis Can.
24 F3 Montluçon France
79 F4 Montmagny Can.
15 F3 Montmédy France
22 D4 Montmélian France
22 D6 Montmeyan France
15 D4 Montmirail France
15 A4 Montmirail France
88 D5 Montmorenci U.S.A.
79 F4 Montmorency Can.
24 E3 Montmorillon France
15 D4 Montmort-Lucy France
68 D5 Monto Austr.
82 E3 Montpelier ID U.S.A.
88 E5 Montpelier IN U.S.A.
90 A4 Montpelier OH U.S.A.
91 G2 Montpelier VT U.S.A.
24 F5 Montpellier France
78 F4 Montréal Can.
89 G2 Montreal r. Can.
89 F2 Montreal r. Can.
22 B2 Montréal France
88 E2 Montreal I. Can.
77 H4 Montreal L. Can.
77 H4 Montreal Lake Can.
91 F2 Montréal-Mirabel Can.
88 E2 Montreal River Can.
15 B2 Montreuil France
22 D3 Montreux Switz.
112 D3 Montrose well S. Africa
11 F4 Montrose U.K.
83 F4 Montrose CO U.S.A.
89 F4 Montrose MI U.S.A.
91 F4 Montrose PA U.S.A.
15 B3 Mont-St-Aignan France
73 K8 Montserrat terr. Caribbean Sea
22 C3 Mont-sous-Vaudrey France
79 G4 Monts, Pte des pt Can.
15 B3 Montville France
93 G3 Monument Valley reg. U.S.A.
56 B2 Monywa Myanmar
23 G4 Monza Italy
115 A6 Monze Zambia
25 G2 Monzón Spain
113 J4 Mooi r. S. Africa
112 B3 Mooifontein Namibia
113 H4 Mooirivier S. Africa
113 G1 Mookane Botswana
66 C2 Moolawatana Austr.
67 H2 Moomin Cr. r. Austr.
66 A3 Moonaree Austr.
67 J3 Moonbi Ra. mts Austr.
67 J1 Moonie Qld. Austr.
67 H2 Moonie r. Austr.
66 B5 Moonta Austr.
69 C5 Moora Austr.
82 F2 Moorcroft U.S.A.
69 B5 Moore r. Austr.
71 □3 Moorea i. Fr. Polynesia Pac. Oc.
90 D5 Moorefield U.S.A.
69 C4 Moore, Lake salt flat Austr.
87 E7 Moores I. Bahamas
91 K2 Moores Mills Can.
11 E5 Moorfoot Hills U.K.
84 D2 Moorhead U.S.A.
66 E4 Moornanyah Lake Austr.
66 D5 Mooroopna Austr.
66 F6 Mooroopna Austr.
112 C6 Moorreesburg S. Africa
20 F4 Moos Ger.
20 B5 Moos Ger.
20 E4 Moosburg an der Isar Ger.
78 D3 Moose r. Can.
78 D3 Moose Factory Can.
91 J2 Moosehead Lake U.S.A.
77 H4 Moose Jaw Can.
77 J4 Moose Lake Can.
88 A2 Moose Lake U.S.A.
91 H2 Mooselookmeguntic Lake U.S.A.
78 D3 Moose River Can.
77 J4 Moosomin Can.
78 D3 Moosonee Can.
66 E3 Mootwingee Austr.
113 H1 Mopane S. Africa
115 C6 Mopeia Moz.
114 C2 Mopti Mali
41 G3 Moqor Afgh.
100 D7 Moquegua Peru
21 M5 Mór Hungary
114 F2 Mora Cameroon
25 E3 Mora Spain
9 O6 Mora Sweden
105 B2 Mora, Cerro mt Arg./Chile
44 A3 Morad r. Pak.
44 D3 Moradabad India
111 E5 Morafenobe Madag.
96 G5 Morales Guatemala
43 B2 Moram India
111 E5 Moramanga Madag.
88 E3 Moran MI U.S.A.
82 E3 Moran WY U.S.A.
68 C3 Moranbah Austr.
11 C4 Morar, Loch l. U.K.
43 B5 Moratuwa Sri Lanka
21 K4 Morava r. Austria/Slovakia
40 D2 Moraveh Tappeh Iran
21 K3 Moravia reg. Czech Rep.
91 E3 Moravia U.S.A.
21 L3 Moravice r. Czech Rep.
21 J4 Moravská Dyje r. Czech Rep.
21 K3 Moravská Třebová Czech Rep.
21 J3 Moravské Budějovice Czech Rep.
21 M3 Moravskoslezské Beskydy mts Czech Rep.
21 L3 Moravský Beroun Czech Rep.
69 B4 Morawa Austr.
68 B3 Moray Downs Austr.
11 E3 Moray Firth est. U.K.
15 H3 Morbach Ger.
23 G3 Morbegno Italy
44 B5 Morbi India
24 D4 Morcenx France
46 E1 Mordaga China
37 K3 Mor Dağı mt Turkey
77 K5 Morden Can.
66 F7 Mordialloc Austr.
28 H4 Mordoviya, Respublika div. Rus. Fed.
29 G4 Mordovo Rus. Fed.
84 C2 Moreau r. U.S.A.
12 E3 Morecambe U.K.
12 D3 Morecambe Bay U.K.
67 H2 Moree Austr.
62 E2 Morehead P.N.G.
90 B5 Morehead U.S.A.
87 E5 Morehead City U.S.A.
44 D4 Morel r. India
More Laptevykh sea see Laptev Sea
96 D4 Morelia Mex.
25 F2 Morella Spain
11 D2 More, Loch l. U.K.
97 E4 Morelos div. Mex.
44 D4 Morena India
25 D3 Morena, Sierra mts Spain
93 H5 Morenci AZ U.S.A.
89 E5 Morenci MI U.S.A.
27 L2 Moreni Romania
105 E2 Moreno Arg.
96 B1 Moreno Mex.
92 D5 Moreno Valley U.S.A.
76 C4 Moresby Island Can.
22 C4 Morestel France
112 F1 Moreswe Pan salt pan Botswana
67 K1 Moreton B. Austr.
67 K1 Moreton I. Austr.
13 F6 Moreton-in-Marsh U.K.
15 C3 Moreuil France
22 D3 Morez France
36 D4 Morfou Cyprus
36 D4 Morfou Bay Cyprus
66 C5 Morgan Austr.
85 F6 Morgan City U.S.A.
92 B3 Morgan Hill U.S.A.
92 C3 Morgan, Mt U.S.A.
91 F4 Morgantown PA U.S.A.
90 D5 Morgantown WV U.S.A.
113 H3 Morgenzon S. Africa
22 D3 Morges Switz.
22 E4 Morgex Italy
45 F4 Morhar r. India
48 F3 Mori China
52 G3 Mori Japan
93 E2 Moriah, Mt U.S.A.
83 F5 Moriarty U.S.A.
67 F2 Moriarty's Ra. h. Austr.
103 C4 Morichal Col.
103 E2 Morichal Largo r. Venez.
113 G4 Morija Lesotho
19 G5 Moringen Ger.
28 D3 Morino Rus. Fed.
52 G5 Morioka Japan
67 J4 Morisset Austr.
52 G5 Moriyoshi-zan volc. Japan
8 S3 Morjärv Sweden
41 F4 Morjen r. Pak.
28 J3 Morki Rus. Fed.
24 C2 Morlaix France
12 F4 Morley U.K.
15 C4 Mormant France
93 G4 Mormon Lake U.S.A.
22 B5 Mornas France
65 G3 Mornington I. Austr.
102 A7 Mornington, I. Chile
44 A4 Moro Pak.
62 E2 Morobe P.N.G.
107 C2 Morocco country Africa
88 D5 Morocco U.S.A.
115 C4 Morogoro Tanz.
115 C4 Morogoro div. Tanz.
55 B5 Moro Gulf g. Phil.
113 G4 Morojaneng S. Africa
112 E3 Morokweng S. Africa
96 D3 Moroleón Mex.
111 E6 Morombe Madag.
95 J4 Morón Cuba
48 H2 Mörön Mongolia
111 E6 Morondava Madag.
25 D4 Morón de la Frontera Spain
114 B3 Morondo Côte d'Ivoire
111 E5 Moroni Comoros
57 H6 Morotai i. Indon.
115 B2 Moroto Uganda
115 B2 Moroto, Mt Uganda
29 G5 Morozovsk Rus. Fed.
89 G4 Morpeth Can.
12 F2 Morpeth U.K.
104 C2 Morrinhos Brazil
77 K5 Morris Can.
88 C5 Morris IL U.S.A.
84 E2 Morris MN U.S.A.
91 F2 Morrisburg Can.
88 C5 Morrison U.S.A.
93 F5 Morristown AZ U.S.A.
91 F4 Morristown NJ U.S.A.
91 F2 Morristown NY U.S.A.
87 D4 Morristown TN U.S.A.
91 F4 Morrisville PA U.S.A.
91 G2 Morrisville VT U.S.A.
92 B4 Morro Bay U.S.A.
103 C2 Morrocoy, Parque Nacional nat. park Venez.
96 D4 Morro de Petatlán hd Mex.
101 H4 Morro Grande h. Brazil
102 B3 Morro, Pta pt Chile
103 B2 Morrosquillo, Golfo de b. Col.
19 G5 Morschen Ger.
88 D5 Morse Reservoir U.S.A.
28 G4 Morshansk Rus. Fed.
28 H2 Morskaya Masel'ga Rus. Fed.
26 C7 Morsott Alg.
24 E2 Mortagne-au-Perche France
24 D3 Mortagne-sur-Sèvre France
22 F4 Mortara Italy
22 D2 Morteau France
13 C6 Mortehoe U.K.
105 E1 Morteros Arg.
Mortes r. see Manso
66 E7 Mortlake Austr.
Mortlock Is see Tauu
13 G5 Morton U.K.
88 C5 Morton IL U.S.A.
82 B2 Morton WA U.S.A.
67 J5 Morton Nat. Park Austr.
20 F6 Mörtschach Austria
67 G5 Morundah Austr.
113 G1 Morupule Botswana
67 J5 Moruya Austr.
22 B3 Morvan reg. France
68 B6 Morven Austr.
11 C4 Morvern reg. U.K.
Morvi see Morbi
67 G7 Morwell Austr.
20 C3 Mosbach Ger.
13 F4 Mosborough U.K.
28 F4 Moscow Rus. Fed.
82 C2 Moscow U.S.A.
116 C6 Moscow Univ. Ice Shelf ice feature Ant.
116 B6 Mose, C. Ant.
15 H2 Mosel r. Ger.
112 E2 Moselebe watercourse Botswana
15 G3 Moselle r. France
19 J4 Möser Ger.
82 C2 Moses Lake U.S.A.
92 D1 Moses, Mt U.S.A.
70 C6 Mosgiel N.Z.
112 E3 Moshaweng watercourse S. Africa
115 C3 Moshi Tanz.
88 C3 Mosinee U.S.A.
8 N4 Mosjøen Norway
8 N3 Moskenesøy i. Norway
28 F4 Moskovskaya Oblast' div. Rus. Fed.
Moskva see Moscow
71 □9 Moso i. Vanuatu
21 L5 Mosoni-Duna r. Hungary
16 H7 Mosonmagyaróvár Hungary
103 A4 Mosquera Col.
83 F5 Mosquero U.S.A.
96 H5 Mosquitia reg. Honduras
104 E1 Mosquito r. Brazil
90 C4 Mosquito Creek Lake U.S.A.
96 J6 Mosquitos, Golfo de los b. Panama
77 J2 Mosquitto Lake Can.
9 M7 Moss Norway
11 F3 Mossat U.K.
70 B6 Mossburn N.Z.
112 E7 Mossel Bay S. Africa
112 E7 Mossel Bay b. S. Africa
110 B4 Mossendjo Congo
66 F4 Mossgiel Austr.
68 A1 Mossman Austr.
101 L5 Mossoró Brazil
67 J5 Moss Vale Austr.
21 G2 Most Czech Rep.
40 D3 Moşţafaabad Iran
108 C1 Mostaganem Alg.
26 G3 Mostar Bos.-Herz.
102 F4 Mostardas Brazil
77 G3 Mostoos Hills Can.
29 G6 Mostovskoy Rus. Fed.
59 E2 Mostyn Malaysia
37 J3 Mosul Iraq
9 L7 Møsvatnet l. Norway
71 □9 Mota i. Vanuatu
97 G5 Motagua r. Guatemala
9 O7 Motala Sweden
71 □9 Mota Lava i. Vanuatu
103 C2 Motatán r. Venez.
113 K2 Motaze Moz.
113 H2 Motetema S. Africa
44 D4 Moth India
11 E5 Motherwell U.K.
45 F4 Motihari India
25 F3 Motilla del Palancar Spain
70 F2 Motiti I. N.Z.
54 B3 Motian Ling h. China
112 E2 Motokwe Botswana
97 F5 Motozintla Mex.
25 E4 Motril Spain
27 K2 Motru Romania
20 D4 Möttingen Ger.
71 □8 Motu Iti i. Easter I. Chile
71 □2 Motu Iti i. Fr. Polynesia Pac. Oc.
97 G3 Motul Mex.
71 □8 Motu Nui i. Easter I. Chile
60 M5 Motu One i. Pac. Oc.
71 □15 Moturiki i. Fiji
71 □8 Motu Tautara i. Easter I. Chile
114 E4 Mouanko Cameroon
22 C3 Mouchard France
51 A5 Mouding China
108 A3 Moudjéria Maur.
22 D3 Moudon Switz.
27 L5 Moudros Greece
22 E6 Mougins France
9 S6 Mouhijärvi Fin.
Mouhoun r. see Black Volta
110 B4 Mouila Gabon
66 F5 Moulamein Austr.
66 F5 Moulamein r. Austr.
110 B4 Moulèngui Binza Gabon
24 F3 Moulins France
22 A3 Moulins-Engilbert France
58 A1 Moulmein Myanmar
87 D6 Moultrie U.S.A.
81 L5 Moultrie, Lake U.S.A.
86 B4 Mound City IL U.S.A.
84 E3 Mound City MO U.S.A.
109 D4 Moundou Chad
90 C5 Moundsville U.S.A.
44 C4 Mount Abu India
87 C5 Mountain Brook U.S.A.
90 C6 Mountain City U.S.A.
85 E4 Mountain Grove U.S.A.
85 E4 Mountain Home AR U.S.A.
82 D3 Mountain Home ID U.S.A.
113 F6 Mountain Zebra National Park S. Africa
90 C6 Mount Airy U.S.A.
70 B6 Mount Aspiring National Park N.Z.
69 C2 Mount Augustus Austr.
113 H4 Mt-aux-Sources mt Lesotho
113 H5 Mount Ayliff S. Africa
84 E3 Mount Ayr U.S.A.
66 C5 Mount Barker S.A. Austr.
69 C7 Mount Barker Austr.
67 G6 Mount Beauty Austr.
14 C4 Mount Bellew Rep. of Ireland
65 J7 Mount Bogong National Park Austr.
67 G6 Mount Buffalo National Park Austr.
68 A1 Mount Carbine Austr.
91 K1 Mount Carleton Provincial Park Can.
93 F3 Mount Carmel Junction U.S.A.
88 C4 Mount Carroll U.S.A.
69 C2 Mount Clere Austr.
70 C5 Mount Cook N.Z.
70 C5 Mount Cook National Park N.Z.
68 B3 Mount Coolon Austr.
111 D5 Mount Darwin Zimbabwe
91 J2 Mount Desert Island U.S.A.
66 A1 Mount Dutton Austr.
66 A3 Mount Eba Austr.
67 G9 Mount Field Nat. Park Austr.
113 H5 Mount Fletcher S. Africa
89 G4 Mount Forest Can.
113 H5 Mount Frere S. Africa
66 D6 Mount Gambier Austr.
68 A1 Mount Garnet Austr.
90 B4 Mount Gilead U.S.A.
62 E2 Mount Hagen P.N.G.
67 F4 Mount Hope N.S.W. Austr.
66 A5 Mount Hope S.A. Austr.
90 C6 Mount Hope U.S.A.
65 G4 Mount Isa Austr.
91 G4 Mount Kisco U.S.A.
68 D4 Mount Larcom Austr.
66 C5 Mount Lofty Range mts Austr.
89 G2 Mount MacDonald Can.
69 C4 Mount Magnet Austr.
66 E4 Mount Manara Austr.
92 B1 Mount Meadows Reservoir U.S.A.
14 D4 Mountmellick Rep. of Ireland
68 A1 Mount Molloy Austr.
113 G5 Mount Moorosi Lesotho
68 D4 Mount Morgan Austr.
66 E3 Mount Murchison Austr.
68 D5 Mount Perry Austr.
88 B5 Mount Pleasant IA U.S.A.
88 E4 Mount Pleasant MI U.S.A.
86 C3 Mount Pleasant MI U.S.A.
90 D4 Mount Pleasant PA U.S.A.
87 E5 Mount Pleasant SC U.S.A.
85 E5 Mount Pleasant TX U.S.A.
93 G2 Mount Pleasant UT U.S.A.
88 C5 Mount Pulaski U.S.A.
82 B2 Mount Rainier Nat. Park U.S.A.
76 F4 Mount Robson Prov. Park Can.
90 C6 Mount Rogers National Recreation Area res. U.S.A.
13 B7 Mount's Bay U.K.
13 F5 Mountsorrel U.K.
88 B6 Mount Sterling IL U.S.A.
90 B5 Mount Sterling KY U.S.A.
90 D5 Mount Storm U.S.A.
68 A2 Mount Surprise Austr.
90 E4 Mount Union U.S.A.
69 D2 Mount Vernon Austr.
87 B6 Mount Vernon AL U.S.A.
88 B5 Mount Vernon IA U.S.A.
86 B4 Mount Vernon IL U.S.A.
90 A6 Mount Vernon KY U.S.A.
90 B4 Mount Vernon OH U.S.A.
82 B1 Mount Vernon WA U.S.A.
116 C3 Mt. Victor mt Ant.

66 A4 Mount Wedge Austr.
67 H8 Mount William National Park Austr.
68 C5 Moura Austr.
100 F4 Moura Brazil
114 B2 Mourdiah Mali
109 E3 Mourdi, Dépression du depression Chad
68 B1 Mourilyan Harbour Austr.
14 D3 Mourne r. U.K.
14 E3 Mourne Mountains h. U.K.
15 D2 Mouscron Belgium
109 D3 Moussoro Chad
22 D4 Moûtiers France
57 G6 Moutong Indon.
15 C3 Mouy France
108 C2 Mouydir, Mts de plat. Alg.
15 F3 Mouzon France
67 J1 Mowbullan, Mt Austr.
14 C4 Moy r. Rep. of Ireland
110 D3 Moyale Eth.
108 A4 Moyamba Sierra Leone
43 B4 Moyar r. India
108 B1 Moyen Atlas mts Morocco
113 G5 Moyeni Lesotho
14 E4 Moyer h. Rep. of Ireland
79 G2 Moyne, Lac Le l. Can.
39 J5 Moyu China
39 H3 Moyynty Kazak.
107 G8 Mozambique country Africa
111 E5 Mozambique Channel Africa
117 G5 Mozambique Ridge sea feature Ind. Ocean
29 H7 Mozdok Rus. Fed.
41 F2 Mozdūrān Iran
28 F4 Mozhaysk Rus. Fed.
41 F3 Mozhnābād Iran
45 H5 Mozo Myanmar
115 A4 Mpala Congo(Zaire)
115 B4 Mpanda Tanz.
115 B5 Mpika Zambia
113 J4 Mpolweni S. Africa
115 B4 Mporokoso Zambia
115 B4 Mpulungu Zambia
113 H2 Mpumalanga div. S. Africa
115 C4 Mpwapwa Tanz.
113 H5 Mqanduli S. Africa
26 G2 Mrkonjić-Grad Bos.-Herz.
108 D1 M'Saken Tunisia
28 D3 Mshinskaya Rus. Fed.
25 J5 M'Sila Alg.
28 E3 Msta r. Rus. Fed.
28 D4 Mstsislaw Belarus
115 C4 Mtera Reservoir Tanz.
28 F4 Mtsensk Rus. Fed.
113 K4 Mtubatuba S. Africa
113 J4 Mtunzini S. Africa
115 D5 Mtwara Tanz.
115 C5 Mtwara div. Tanz.
71 11 Mu'a Tonga
110 B4 Muanda Congo(Zaire)
58 B2 Muang Chainat Thai.
58 A1 Muang Chiang Rai Thai.
51 B6 Muang Hiam Laos
58 B1 Muang Hôngsa Laos
58 B1 Muang Kalasin Thai.
58 C1 Muang Khammouan Laos
58 C2 Muang Không Laos
58 C2 Muang Khôngxédôn Laos
58 B1 Muang Khon Kaen Thai.
58 C1 Muang Khoua Laos
58 A3 Muang Kirirath r. Thai.
58 A1 Muang Lampang Thai.
58 A1 Muang Lamphun Thai.
58 B1 Muang Loei Thai.
58 B1 Muang Lom Sak Thai.
58 A1 Muang Long Thai.
58 A3 Muang Luang r. Thai.
58 C2 Muang Mai Thai.
58 C1 Muang Mok Laos
58 C1 Muang Nakhon Phanom Thai.
58 B2 Muang Nakhon Sawan Thai.
58 B1 Muang Nan Thai.
51 B6 Muang Ngoy Laos
58 C1 Muang Nong Laos
51 A6 Muang Ou Nua Laos
58 B1 Muang Pakxan Laos
58 C1 Muang Phalan Laos
58 A1 Muang Phan Thai.
58 A1 Muang Phayao Thai.
58 B1 Muang Phetchabun Thai.
58 B1 Muang Phiang Laos
58 B1 Muang Phichai Thai.
58 B1 Muang Phichit Thai.
58 C1 Muang Phin Laos
58 B1 Muang Phitsanulok Thai.
58 B1 Muang Phôn-Hông Laos
58 B1 Muang Phrae Thai.
58 B1 Muang Roi Et Thai.
58 C1 Muang Sakon Nakhon Thai.
58 B2 Muang Samut Prakan Thai.
58 B1 Muang Souy Laos
58 B2 Muang Uthai Thani Thai.
51 B6 Muang Va Laos
58 B1 Muang Vangviang Laos
58 B1 Muang Xaignabouri Laos
51 B6 Muang Xon Laos
58 C2 Muang Yasothon Thai.
59 B2 Muar Malaysia
59 B5 Muar r. Malaysia
59 B3 Muarabungo Indon.
59 B3 Muaradua Indon.
59 A3 Muarasiberut Indon.
59 A2 Muarasipongi Indon.
59 B3 Muaratembesi Indon.
45 E4 Mubarakpur India
41 F5 Mubarek Uzbek.
37 H7 Mubarraz well S. Arabia
115 B2 Mubende Uganda
114 F2 Mubi Nigeria
103 E4 Mucajaí r. Brazil
103 E4 Mucajaí, Serra do mts Brazil
18 E6 Much Ger.
69 B5 Muchea Austr.
115 B5 Muchinga Escarpment esc. Zambia
51 B4 Muchuan China
11 B4 Muck i. U.K.
68 C6 Muckadilla Austr.
11 Muckle Roe i. U.K.
103 C3 Muco r. Col.
115 D5 Mucojo Moz.
111 C5 Muconda Angola
115 C6 Mucubela Moz.
103 E4 Mucucuaú r. Brazil
115 B6 Mucumbura Moz.
36 E2 Mucur Turkey
104 E2 Mucuri Brazil
104 E2 Mucuri r. Brazil
111 C5 Mucussueje Angola
43 B3 Mūdabidri India
54 E1 Mudanjiang China
54 E1 Mudan Jiang r. China
36 B1 Mudanya Turkey
37 L7 Mudayrah Kuwait
90 C5 Muddlety U.S.A.
8 R3 Muddus Nationalpark nat. park Sweden
93 G2 Muddy Creek r. U.S.A.
93 E3 Muddy Peak U.S.A.
41 E3 Mūd-e-Dahanāb Iran
20 A2 Mudersbach Ger.
67 H4 Mudgee Austr.
43 A2 Mudhol India
44 C3 Mudki India
92 D3 Mud Lake U.S.A.
58 A1 Mudon Myanmar
36 C1 Mudurnu Turkey
28 F2 Mud'yuga Rus. Fed.
115 C5 Mueda Moz.
71 9 Muéo New Caledonia Pac. Oc.
97 F4 Muerto, Mar l. Mex.
28 H1 Muftyuga Rus. Fed.
115 A5 Mufulira Zambia
111 C5 Mufumbwe Zambia
51 E4 Mufu Shan mts China
91 K2 Mugaguadavic Lake Can.
37 M2 Muğan Düzü lowland Azer.
45 F2 Mugarripug China
Mughalbhin see Jati
45 E4 Mughal Sarai India
40 D3 Mūghār Iran
36 F7 Mughayrā' S. Arabia
39 G5 Mughsu r. Tajik.
36 B3 Muğla Turkey
38 D3 Mugodzhary ridge Kazak.
45 H2 Mug Qu r. China
45 E3 Mugu Karnali r. Nepal
45 H2 Mugxung China
109 F2 Muhammad Qol Sudan
40 B5 Muḩayriqah S. Arabia
20 B4 Mühlacker Ger.
19 K5 Mühlanger Ger.
19 L5 Mühlberg Ger.
20 G6 Mühldorf Austria
20 F4 Mühldorf am Inn Ger.
19 H5 Mühlhausen (Thüringen) Ger.
20 B2 Mühlheim am Main Ger.
21 H4 Mühlviertel reg. Austria
8 T4 Muhos Fin.
115 A3 Muhulu Congo(Zaire)
115 C1 Mui Eth.
58 C3 Mui Ca Mau c. Vietnam
58 D3 Mui Dinh hd Vietnam
71 11 Mui Hopohoponga pt Tonga
58 D2 Mui Nây pt Vietnam
14 E5 Muine Bheag Rep. of Ireland
11 D5 Muirkirk U.K.
11 B2 Muirneag h. U.K.
11 D3 Muir of Ord U.K.
92 A3 Muir Woods National Monument res. U.S.A.
115 C5 Muite Moz.
97 H3 Mujeres, Isla i. Mex.
54 D6 Muju S. Korea
29 B5 Mukacheve Ukr.
59 D2 Mukah Malaysia
45 F3 Mükangsar China
21 H3 Mukařov Czech Rep.
58 C1 Mukdahan Thai.
69 D5 Mukinbudin Austr.
59 B3 Mukomuko Indon.
38 F5 Mukry Turkm.
44 C3 Muktsar India
38 C3 Mukur Atyrausk. Kazak.
39 K2 Mukur Semipal. Kazak.
77 K4 Mukutawa r. Can.
88 C4 Mukwonago U.S.A.
44 D5 Mul India
43 A2 Mula r. India
44 A3 Mula r. Pak.
39 J3 Mulaly Kazak.
52 A2 Mulan China
55 B3 Mulanay Phil.
115 C6 Mulanje, Mt Malawi
40 B5 Mulayḩ S. Arabia
85 E5 Mulberry U.S.A.
105 B3 Mulchén Chile
19 K5 Mulde r. Ger.
115 B3 Muleba Tanz.
93 H5 Mule Creek NM U.S.A.
82 F3 Mule Creek WY U.S.A.
96 A2 Mulegé Mex.
85 C5 Muleshoe U.S.A.
25 E4 Mulhacén mt Spain
18 D5 Mülheim an der Ruhr Ger.
22 E2 Mulhouse France
51 A5 Muli China
71 4 Mulifanua Western Samoa
54 F1 Muling China
54 F1 Muling China
54 F1 Muling r. China
71 4 Mulitapuili, C. Western Samoa
11 C4 Mull i. U.K.
37 M3 Mulla Ali Iran
14 B5 Mullaghareirk Mts h. Rep. of Ireland
43 C4 Mullaittivu Sri Lanka
67 H3 Mullaley Austr.
67 G3 Mullengudgery Austr.
59 D2 Muller, Pegunungan mts Indon.
88 E3 Mullett Lake U.S.A.
69 B4 Mullewa Austr.
11 F1 Mull Head hd U.K.
22 E2 Müllheim Ger.
91 F5 Mullica r. U.S.A.
14 D4 Mullingar Rep. of Ireland
67 H4 Mullion Cr. Austr.
19 M4 Müllrose Ger.
11 B4 Mull, Sound of chan. U.K.
111 C5 Mulobezi Zambia
43 A2 Mulshi L. India
44 D5 Multai India
44 B3 Multan Pak.
9 T5 Multia Fin.
15 C4 Multien reg. France
41 F5 Mūmān Iran
Mumbai see Bombay
67 H4 Mumbil Austr.
115 A5 Mumbwa Zambia
39 G5 Mū'minobod Tajik.
29 H6 Mumra Rus. Fed.
57 G8 Muna i. Indon.
97 G3 Muna Mex.
33 N3 Muna r. Rus. Fed.
8 C3 Munaðarnes Iceland
38 C3 Munayly Kazak.
38 C4 Munayshy Kazak.
20 E2 Münchberg Ger.
19 M4 Müncheberg Ger.
20 E4 München Ger.
18 F6 Münchhausen Ger.
103 A4 Munchique, Co mt Col.
76 D3 Muncho Lake Can.
76 D3 Muncho Lake Provincial Park Can.
54 D4 Munch'ŏn N. Korea
88 E5 Muncie U.S.A.
90 E4 Muncy U.S.A.
43 C5 Mundel L. Sri Lanka
13 J5 Mundesley U.K.
13 H5 Mundford U.K.
69 E1 Mundiwindi Austr.
64 E6 Mundrabilla Austr.
68 D5 Mundubbera Austr.
44 C4 Mundwa India
43 C2 Muneru r. India
68 B6 Mungallala Austr.
67 G2 Mungallala Cr. r. Austr.
68 A1 Mungana Austr.
44 D4 Mungaoli India
115 B4 Mungári Moz.
115 A2 Mungbere Congo(Zaire)
45 E5 Mungeli India
45 F4 Munger India
66 C2 Mungeranie Austr.
58 D5 Munggureak, Tanjung pt Indon.
67 H2 Mungindi Austr.
Munich see München
101 K4 Munim r. Brazil
88 D2 Munising U.S.A.
104 E3 Muniz Freire Brazil
9 V2 Munkelva Norway
9 N7 Munkfors Sweden
113 H1 Munnik S. Africa
67 H8 Munro, Mt Austr.
54 D5 Munsan S. Korea
20 C4 Münsingen Ger.
22 E3 Münsingen Switz.
20 B3 Münster Ger.
19 H4 Münster Ger.
18 E5 Münster Ger.
18 E5 Münsterland reg. Ger.
69 D5 Muntadgin Austr.
79 F3 Muntviel, Lac l. Can.
8 V4 Muojärvi l. Fin.
58 C1 Mương Lam Vietnam
51 B6 Mương Nhie Vietnam
8 S3 Muonio Fin.
8 S2 Muonioälven r. Fin./Sweden
54 A5 Muping China
110 E3 Muqdisho Somalia
37 M1 Müqtädır Azer.
21 J6 Mur r. Austria
21 K6 Mura r. Croatia/Slovenia
37 J2 Muradiye Turkey
52 F5 Murakami Japan
102 B7 Murallón, Cerro mt Chile
115 A3 Muramvya Burundi
115 C3 Muranga Kenya
58 Mura Resr Sing.
28 J3 Murashi Rus. Fed.
37 H2 Murat r. Turkey
36 B2 Murat Dağı mts Turkey
36 A1 Muratlı Turkey
21 H5 Murau Austria
52 G5 Murayama Japan
40 C3 Murcheh Khvort Iran
19 L3 Murchin Ger.
66 F6 Murchison Austr.
69 B3 Murchison watercourse Austr.
115 B2 Murchison Falls National Park Uganda
69 C3 Murchison, Mt h. Austr.
25 F4 Murcia Spain
25 F4 Murcia div. Spain
84 C3 Murdo U.S.A.
79 G4 Murdochville Can.
21 J6 Mureck Austria
111 D5 Murehwa Zimbabwe
17 M7 Mureş r. Romania
22 E4 Muret France
87 E4 Murfreesboro NC U.S.A.
87 C5 Murfreesboro TN U.S.A.
39 H5 Murgab r. Tajik.
38 F5 Murgab Turkm.
41 F2 Murgab r. Turkm.
41 G3 Murghab r. Afgh.
44 B3 Murgha Kibzai Pak.
39 H5 Murghob Tajik.
41 H3 Murgh Pass Afgh.
68 D6 Murgon Austr.
69 C3 Murgoo Austr.
50 A2 Muri China
45 F5 Muri India
40 E2 Mūrī Iran
104 D3 Muriaé Brazil
111 C4 Muriege Angola
19 K3 Müritz l. Ger.
19 L3 Müritz, Nationalpark nat. park Ger.
19 K3 Müritz Seenpark res. Ger.
8 X2 Murmansk Rus. Fed.
26 C4 Muro, Capo di pt Corsica France
28 G4 Murom Rus. Fed.
115 B3 Murongo Tanz.
52 G3 Muroran Japan
25 B1 Muros Spain
53 D8 Muroto Japan
53 D8 Muroto-zaki pt Japan
88 D5 Murphey Lake, J. C. U.S.A.
82 C3 Murphy ID U.S.A.
87 D5 Murphy NC U.S.A.
92 B2 Murphys U.S.A.
67 G2 Murra Murra Austr.
66 D5 Murray S.A. Austr.
69 B6 Murray r. Austr.
76 E3 Murray r. Can.
86 B4 Murray KY U.S.A.
82 E3 Murray UT U.S.A.
66 C5 Murray Bridge Austr.
87 D5 Murray, L. U.S.A.
62 E2 Murray, Lake P.N.G.
112 E5 Murraysburg S. Africa
119 L3 Murray Seascarp sea feature Pac. Oc.
66 D5 Murrayville Austr.
67 H5 Murringo Austr.
14 B4 Murrisk reg. Rep. of Ireland
14 B4 Murroogh Rep. of Ireland
67 H5 Murrumbateman Austr.
66 F5 Murrumbidgee r. Austr.
115 C6 Murrupula Moz.
67 J3 Murrurundi Austr.
26 G1 Murska Sobota Slovenia
22 E3 Murten Switz.
22 E3 Murtensee l. Switz.
66 D2 Murteree, Lake salt flat Austr.
66 E6 Murtoa Austr.
43 A2 Murud India
59 D2 Murud, Gunung mt Indon.
39 E4 Murun Sum Sk. resr China
43 C4 Murunkan Sri Lanka
70 F3 Murupara N.Z.
60 N6 Mururoa atoll Fr. Polynesia Pac. Oc.
60 N6 Mururoa i. Fr. Polynesia Pac. Oc.
45 E5 Murwara India
67 K2 Murwillumbah Austr.
21 J5 Mürz r. Austria
109 D2 Murzūq Libya
21 J5 Mürzzuschlag Austria
37 H2 Muş Turkey
44 B3 Musa Khel Bazar Pak.
27 K3 Musala mt Bulg.
59 A2 Musala i. Indon.
54 E2 Musan N. Korea
40 E5 Musandam Peninsula Oman
41 G3 Musa Qala Afgh.
41 G3 Musa Qala, Rūd-i r. Afgh.
Musay'īd see Umm Sa'id
42 E5 Muscat Oman
88 B5 Muscatine U.S.A.
88 B4 Muscoda U.S.A.
91 J3 Muscongus Bay U.S.A.
65 H2 Musgrave Austr.
64 F5 Musgrave Ranges mts Austr.
14 C5 Musheramore h. Rep. of Ireland
110 B4 Mushie Congo(Zaire)
43 B2 Musi r. India
59 B3 Musi r. Indon.
93 F2 Music Mt U.S.A.
93 G2 Musinia Peak U.S.A.
76 E2 Muskeg r. Can.
91 H4 Muskeget Channel U.S.A.
88 D4 Muskegon U.S.A.
88 D4 Muskegon r. U.S.A.
90 C5 Muskingum r. U.S.A.
85 E5 Muskogee U.S.A.
89 H3 Muskoka Can.
89 H3 Muskoka, Lake Can.
76 E3 Muskwa r. Can.
36 F3 Muslimīyah Syria
109 F3 Musmar Sudan
115 B3 Musoma Tanz.
115 B4 Musombe Tanz.
62 E2 Mussau I. P.N.G.
11 E5 Musselburgh U.K.
18 E4 Musselkanaal Neth.
82 E2 Musselshell r. U.S.A.
36 B1 Mustafakemalpaşa Turkey
38 C2 Mustayevo Rus. Fed.
9 S7 Mustjala Estonia
54 E3 Musu-dan pt N. Korea
67 J4 Muswellbrook Austr.
109 E2 Mut Egypt
36 D3 Mut Turkey
104 E1 Mutá, Pta do pt Brazil
111 D5 Mutare Zimbabwe
57 G8 Mutis, G. mt Indon.
115 B6 Mutoko Zimbabwe
66 D4 Mutooroo Austr.
111 D5 Mutorashanga Zimbabwe
52 G4 Mutsu Japan
52 G4 Mutsu-wan b. Japan
68 A4 Muttaburra Austr.
70 B7 Muttonbird Is N.Z.
70 A7 Muttonbird Islands N.Z.
14 B5 Mutton Island Rep. of Ireland
115 C5 Mutuali Moz.
104 C1 Mutunópolis Brazil
43 C4 Mutur Sri Lanka
8 U2 Mutusjärvi r. Fin.
8 T3 Muurola Fin.
50 C2 Mu Us Shamo des. China
111 B4 Muxaluando Angola
28 E2 Muyezerskiy Rus. Fed.
115 B3 Muyinga Burundi
38 D4 Muynak Uzbek.
115 A4 Muyumba Congo(Zaire)
39 G3 Muyunkum, Peski des. Kazak.
50 D4 Muyuping China
44 C2 Muzaffarabad Pak.
44 B3 Muzaffargarh Pak.
44 D3 Muzaffarnagar India
45 F4 Muzaffarpur India
113 K1 Muzamane Moz.
39 K4 Muzat He r. China
41 F5 Mūzīn Iran
76 C4 Muzon, C. U.S.A.
96 D2 Múzquiz Mex.
45 F1 Muztag mt China
44 E2 Muztag mt China
39 H5 Muztagata mt China
114 F4 Mvadi Gabon
114 F4 Mvangan Cameroon
109 E4 Mvolo Sudan
115 C4 Mvomero Tanz.
111 D5 Mvuma Zimbabwe
Mwali i. see Moheli
115 B3 Mwanza Tanz.
115 B3 Mwanza div. Tanz.
115 A4 Mwanza Congo(Zaire)
14 B4 Mweelrea h. Rep. of Ireland
110 C4 Mweka Congo(Zaire)
115 A5 Mwenda Zambia
110 C4 Mwene-Ditu Congo(Zaire)
111 D6 Mwenezi Zimbabwe
115 A3 Mwenga Congo(Zaire)
115 A4 Mweru, Lake Congo(Zaire)/Zambia
115 A4 Mweru Wantipa,Lake l. Zambia
115 A4 Mweru Wantipa Nat. Park Zambia
111 C4 Mwimba Congo(Zaire)
111 C5 Mwinilunga Zambia
28 C4 Myadzyel Belarus
45 H5 Myaing Myanmar
44 B4 Myājlār India
67 K4 Myall L. Austr.
56 B3 Myanaung Myanmar
31 J7 Myanmar country Asia
53 B9 Myanoura-dake mt Japan
11 E2 Mybster U.K.
45 H5 Myebon Myanmar
56 B2 Myingyan Myanmar
58 A2 Myinmoletkat mt Myanmar
56 B1 Myitkyina Myanmar
58 A2 Myitta Myanmar
45 H5 Myittha r. Myanmar
21 L4 Myjava Slovakia
21 L4 Myjava r. Slovakia
29 E6 Mykolayiv Ukr.
27 L6 Mykonos Greece
27 L6 Mykonos i. Greece
32 G3 Myla Rus. Fed.
45 G4 Mymensingh Bangl.
9 S6 Mynämäki Fin.
39 H3 Mynaral Kazak.
13 D5 Mynydd Eppynt h. U.K.
13 C6 Mynydd Preseli h. U.K.
45 H5 Myohaung Myanmar
53 F6 Myōkō-san volc. Japan
54 E3 Myonggan N. Korea
28 C4 Myory Belarus
8 D5 Mýrdalsjökull ice cap Iceland
8 O2 Myre Norway
8 R4 Myrheden Sweden
29 E5 Myrhorod Ukr.
29 D5 Myronivka Ukr.
87 E5 Myrtle Beach U.S.A.
67 G6 Myrtleford Austr.
82 A3 Myrtle Point U.S.A.
16 G4 Myślibórz Pol.
43 B3 Mysore India
33 U3 Mys Shmidta Rus. Fed.
91 F5 Mystic Islands U.S.A.
58 C3 My Tho Vietnam
27 M5 Mytilini Greece
28 F4 Mytishchi Rus. Fed.
21 G3 Mýto Czech Rep.
39 H2 Myylybulak Kazak.
113 G5 Mzamomhle S. Africa
111 D5 Mzimba Malawi
115 B5 Mzuzu Malawi

N

20 E3 Naab r. Ger.
92 2 Naalehu U.S.A.
9 S6 Naantali Fin.
14 E4 Naas Rep. of Ireland
112 B4 Nababeep S. Africa
43 C2 Nabarangapur India
53 E7 Nabari Japan
55 B4 Nabas Phil.
36 E5 Nabatîyet et Tahta Lebanon
71 13 Nabavatu Fiji
115 C3 Naberera Tanz.
28 J4 Naberezhnyye Chelny Rus. Fed.
109 D1 Nabeul Tunisia
44 D3 Nabha India
67 K4 Nabiac Austr.
41 E4 Nabīd Iran
57 K7 Nabire Indon.
36 E5 Nablus West Bank
113 H2 Naboomspruit S. Africa
71 15 Naboutini Fiji
71 13 Nabouwalu Fiji
58 A2 Nabule Myanmar
115 D5 Nacala Moz.
96 H5 Nacaome Honduras
82 B2 Naches U.S.A.
115 C5 Nachingwea Tanz.
44 B4 Nāchna India
21 K2 Náchod Czech Rep.
71 15 Nacilau Pt Fiji
92 B4 Nacimiento Reservoir U.S.A.
85 E6 Nacogdoches U.S.A.
94 C2 Nacozari de García Mex.
71 15 Nadarivatu Fiji
71 15 Nadi Fiji
44 C5 Nadiad India
71 15 Nadi B. Fiji
40 D4 Nādīk Iran
108 B1 Nador Morocco
71 15 Nadrau Plateau plat. Fiji
71 13 Naduri Fiji
40 D3 Nadūshan Iran
29 C5 Nadvirna Ukr.
32 E3 Nadvoitsy Rus. Fed.
32 J3 Nadym Rus. Fed.
9 M9 Næstved Denmark
27 J5 Nafpaktos Greece
27 K6 Nafplio Greece
37 K5 Naft r. Iraq
40 C4 Naft-e Safīd Iran
37 K5 Naft Khaneh Iraq
40 B3 Naft Shahr Iran
40 B5 Nafūd al Jur'ā sand dunes S. Arabia
40 A6 Nafūd as Surrah sand dunes S. Arabia
40 B5 Nafūd Qunayfidhah sand dunes S. Arabia
40 A5 Nafy S. Arabia
55 B3 Naga Phil.
76 D4 Nagagami r. Can.
53 C8 Nagahama Japan
45 H4 Naga Hills India
53 G5 Nagai Japan
45 H4 Nagaland div. India
66 F6 Nagambie Austr.
53 F6 Nagano Japan
53 F6 Nagaoka Japan
45 H4 Nagaon India
43 B4 Nagappattinam India
44 D2 Nagar India
43 B2 Nāgārjuna Sāgar Reservoir India
44 B4 Nagar Parkar Pak.
45 G3 Nagarzê China
53 A8 Nagasaki Japan
53 B7 Nagato Japan
44 C4 Nagaur India
43 C2 Nagavali r. India
45 G2 Nag, Co l. China
44 C5 Nagda India
43 B4 Nagercoil India
41 G5 Nagha Kalat Pak.
44 D3 Nagina India
45 E3 Nagma Nepal
20 B4 Nagold Ger.
28 J3 Nagorsk Rus. Fed.
53 E7 Nagoya Japan
44 D5 Nagpur India
45 H3 Nagqu China
55 C3 Nagumbuaya Point Phil.
32 F1 Nagurskoye Rus. Fed.
26 G1 Nagyatád Hungary
21 L6 Nagybajom Hungary
16 H7 Nagykanizsa Hungary
49 N6 Naha Japan
44 D3 Nahan India
41 F5 Nahang r. Iran/Pak.
76 E2 Nahanni Butte Can.
76 D2 Nahanni National Park Can.
36 E5 Nahariyya Israel
40 C3 Nahāvand Iran
15 H3 Nahe r. Ger.
71 9 Nahoï, Cap c. Vanuatu
37 K5 Nahrawān canal Iraq
37 L6 Nahr 'Umr Iraq
105 B3 Nahuelbuta, Parque Nacional nat. park Chile
105 B4 Nahuel Huapí, L. Arg.
105 B4 Nahuel Huapí, Parque Nacional nat. park Arg.
87 D6 Nahunta U.S.A.
96 C2 Naica Mex.
71 13 Naidi Fiji
71 15 Naigani i. Fiji
45 H2 Naij Tal China
71 13 Nailotha Pk h. Fiji
54 A2 Naiman Qi China
79 H2 Nain Can.
40 D3 Nā'īn Iran
44 D3 Naini Tal India
44 E5 Nainpur India
115 C6 Naiopué Moz.
71 15 Nairai i. Fiji
11 E3 Nairn U.K.
89 G2 Nairn Centre Can.
66 C2 Nairn, Mt h. Austr.
115 C3 Nairobi Kenya
71 13 Naituaba i. Fiji
115 C3 Naivasha Kenya
54 D2 Naizishan China
40 B5 Na'jān S. Arabia
42 B4 Najd reg. S. Arabia
25 E1 Nájera Spain
44 D3 Najibabad India
54 F2 Najin N. Korea
42 B6 Najrān S. Arabia
53 A8 Nakadōri-shima i. Japan
53 B8 Nakama Japan
Nakambé watercourse see White Volta
53 C8 Nakamura Japan
33 M3 Nakanno Rus. Fed.
53 F6 Nakano Japan
53 C6 Nakano-shima i. Japan
41 H3 Naka Pass Afgh.
115 B2 Nakasongola Uganda
53 B8 Nakatsu Japan
53 E7 Nakatsugawa Japan
110 D2 Nak'fa Eritrea
109 F1 Nakhl Egypt
49 O3 Nakhodka Rus. Fed.
58 B2 Nakhon Nayok Thai.
58 B2 Nakhon Pathom Thai.
58 B2 Nakhon Ratchasima Thai.
58 A3 Nakhon Si Thammarat Thai.
44 B5 Nakhtarana India
76 C3 Nakina B.C. Can.
78 C3 Nakina Ont. Can.
74 C4 Naknek U.S.A.
115 B4 Nakonde Zambia
9 M9 Nakskov Denmark
54 E6 Naktong r. S. Korea
115 C3 Nakuru Kenya
76 F4 Nakusp Can.
41 G5 Nal Pak.
41 G5 Nal r. Pak.
113 K2 Nalázi Moz.
45 G4 Nalbari India
29 G7 Nal'chik Rus. Fed.
43 B2 Naldurg India
43 B2 Nalgonda India
43 B3 Nallamala Hills India
36 C1 Nallıhan Turkey
39 G1 Nalobino Kazak.
108 D1 Nālūt Libya
21 G3 Nalžovské Hory Czech Rep.
54 D4 Nam r. N. Korea
113 K2 Namaacha Moz.
71 15 Namacu Fiji
115 C6 Namacurra Moz.
113 H3 Namahadi S. Africa
40 C3 Namak, Daryācheh-ye salt flat Iran
41 E4 Namakzar-e Shadad salt flat Iran
115 C3 Namanga Kenya
39 G4 Namangan Uzbek.
71 15 Namanu-i-Ra i. Fiji
115 C5 Namapa Moz.
112 B3 Namaqualand reg. Namibia
112 B4 Namaqualand reg. S. Africa
62 F2 Namatanai P.N.G.
67 K1 Nambour Austr.
67 K3 Nambucca Heads Austr.
58 C3 Năm Căn Vietnam
45 H3 Namcha Barwa mt China
54 D4 Namch'ŏn N. Korea
46 B3 Nam Co l. China
8 N4 Namdalen v. Norway
8 M4 Namdalseid Norway
51 C6 Nam Định Vietnam
88 B3 Namekagon r. U.S.A.
71 15 Namena Barrier Reef Fiji
71 15 Namenalala i. Fiji
21 K3 Náměšť nad Oslavou Czech Rep.
54 E6 Namhae-do i. S. Korea
111 B6 Namib Desert Namibia
111 B5 Namibe Angola
107 E8 Namibia country Africa
53 G6 Namie Japan
58 B1 Nam Khan r. Laos
57 H7 Namlea Indon.
45 E3 Namling China
58 A1 Nammekon Myanmar
51 B6 Nam Na r. China/Vietnam
58 B1 Nam Ngum r. Laos
67 H3 Namoi r. Austr.
71 15 Namotu Pks mts Fiji
51 B6 Nam Ou r. Laos
76 F3 Nampa Can.
44 E3 Nampa mt Nepal
82 C3 Nampa U.S.A.
114 B1 Nampala Mali
58 B1 Nam Pat Thai.
58 B1 Nam Phong Thai.
54 C4 Namp'o N. Korea
115 C6 Nampula Moz.
45 G2 Namru Co l. China
48 G6 Namrup India
51 B7 Nam Sam r. Laos/Vietnam
45 E3 Namsê La pass Nepal
8 N4 Namsen r. Norway
37 K3 Namshir Iran
45 H4 Namsi India
45 G3 Namsi La pass Bhutan
8 M4 Namsos Norway
56 B4 Nam Tok Thai.
33 O3 Namtsy Rus. Fed.
56 B2 Namtu Myanmar
71 14 Namuka-i-lau i. Fiji
115 C6 Namuli, Monte mt Moz.
115 C5 Namuno Moz.
15 E2 Namur Belgium
115 A6 Namwala Zambia
54 D6 Namwŏn S. Korea
21 L1 Namysłów Pol.
110 B3 Nana Bakassa C.A.R.
76 E5 Nanaimo Can.
92 1 Nanakuli U.S.A.
54 E3 Nanam N. Korea
51 F5 Nan'an China
68 E6 Nanango Austr.
112 B2 Nananib Plateau plat. Namibia
51 E6 Nan'ao China
53 E6 Nanao Japan
53 E6 Nanatsu-shima i. Japan
50 C4 Nanbu China
52 A1 Nancha China
51 E4 Nanchang Jiangxi China
51 E4 Nanchang Jiangxi China
51 E5 Nancheng China
50 C4 Nanchong China
51 C4 Nanchuan China
15 G4 Nancy France
44 E3 Nanda Devi mt India
44 E3 Nanda Kot mt India
51 C5 Nandan China
43 B2 Nānded India
67 J3 Nandewar Range mts Austr.
44 C5 Nandgaon India
Nandi see Nadi
51 D6 Nandu Jiang r. China
44 C5 Nandurbar India
43 B3 Nandyal India
51 D6 Nanfeng Guangdong China
51 E5 Nanfeng Jiangxi China
114 F4 Nanga Eboko Cameroon
59 D3 Nangahpinoh Indon.
54 E2 Nangang Shan mts China/N. Korea
44 C2 Nanga Parbat mt Jammu and Kashmir
59 D3 Nangatayap Indon.
114 D3 Nangbéto, Retenue de resr Togo
58 A3 Nangin Myanmar
15 D4 Nangis France
54 D3 Nangnim N. Korea
54 D3 Nangnim Sanmaek mts N. Korea
50 E2 Nangong China
115 C4 Nangulangwa Tanz.
45 H3 Nang Xian China
50 A2 Nanhua China
50 F4 Nanhui China
43 B3 Nanjangud India
50 C3 Nanjiang China
51 E5 Nanjing Fujian China
50 F3 Nanjing Jiangsu China
51 E5 Nankang China
Nanking see Nanjing
53 C8 Nankoku Japan
111 B5 Nankova Angola
50 E2 Nanle China
50 F4 Nanling China
51 D5 Nan Ling mts China
51 C6 Nanliu Jiang r. China
69 D3 Nannine Austr.
51 C6 Nanning China
69 B6 Nannup Austr.
51 E6 Nan'oa Dao i. China
75 O3 Nanortalik Greenland
51 C5 Nanpan Jiang r. China
48 D6 Nanpara India
54 A3 Nanpiao China
51 F5 Nanping Fujian China
50 B3 Nanping Sichuan China
51 F5 Nanri Dao i. China
49 N7 Nansei-shotō is Japan
75 J1 Nansen Sound chan. Can.
115 B3 Nansio Tanz.
24 D3 Nantes France
15 C3 Nanteuil-le-Haudouin France
43 C4 Nanthi Kadal lag. Sri Lanka
89 G4 Nanticoke Can.
91 F5 Nanticoke r. U.S.A.
76 G4 Nanton Can.
50 F4 Nantong Jiangsu China
50 F3 Nantong Jiangsu China
51 F6 Nant'ou Taiwan
91 H4 Nantucket U.S.A.
91 H4 Nantucket I. U.S.A.
91 H4 Nantucket Sound g. U.S.A.
13 E4 Nantwich U.K.
71 15 Nanukuloa Fiji
71 13 Nanuku Passage Fiji
71 13 Nanuku Reef Fiji
63 H2 Nanumanga i. Tuvalu
63 H2 Nanumea i. Tuvalu
104 E2 Nanuque Brazil
55 C5 Nanusa, Kepulauan is Indon.
64 C4 Nanutarra Roadhouse Austr.
51 B4 Nanxi China
51 D4 Nan Xian China
51 E5 Nanxiong China
50 D3 Nanyang China
115 C3 Nanyuki Kenya
54 C3 Nanzamu China
50 D4 Nanzhang China
50 D3 Nanzhao China
25 G3 Nao, Cabo de la hd Spain
79 F3 Naococane, Lac l. Can.
45 G4 Naogaon Bangl.
44 B4 Naokot Pak.
52 C1 Naoli r. China
41 F3 Naomid, Dasht-e des. Afgh./Iran
44 C2 Naoshera Jammu and Kashmir
51 D6 Naozhou Dao i. China
92 A2 Napa U.S.A.
91 K1 Napadogan Can.
89 J3 Napanee Can.
44 C4 Napasar India
75 N3 Napasoq Greenland
88 C5 Naperville U.S.A.
70 F3 Napier N.Z.
116 D4 Napier Mts Ant.
91 G2 Napierville Can.
26 F4 Naples Italy
87 D7 Naples FL U.S.A.
91 H3 Naples ME U.S.A.
51 B6 Napo China
100 D4 Napo r. Ecuador/Peru
90 A4 Napoleon U.S.A.
Napoli see Naples
105 D3 Naposta Arg.
105 D3 Naposta r. Arg.
88 E5 Nappanee U.S.A.
71 2 Napuka i. Fr. Polynesia Pac. Oc.
37 K3 Naqadeh Iran
36 E6 Naqb Ashtar Jordan
37 M4 Naqqash Iran
53 D7 Nara Japan
114 B1 Nara Mali
17 N3 Narach Belarus
66 D6 Naracoorte Austr.
67 G4 Naradhan Austr.
44 C4 Naraina India
44 E6 Narainpur India
96 B2 Naranjo Mex.
97 E3 Naranjos Mex.
43 D2 Narasannapeta India
43 C2 Narasapatnam, Pt India
43 C2 Narasapur India
43 C2 Narasaraopet India
45 F5 Narasinghapur India
58 B4 Narathiwat Thai.
43 A2 Narayangaon India
Narbada r. see Narmada
13 C6 Narberth U.K.
24 F5 Narbonne France
25 C1 Narcea r. Spain
40 D2 Nardin Iran
27 H4 Nardò Italy
105 E1 Nare Arg.
44 B3 Narechi r. Pak.
69 D6 Narembeen Austr.
75 M1 Nares Strait Can./Greenland
17 K4 Narew r. Pol.
54 D2 Narhong China
44 A3 Nari r. Pak.
111 B6 Narib Namibia
112 B5 Nariep S. Africa
29 H6 Narimanov Rus. Fed.
39 L2 Narimskiy Khr. mts Kazak.
41 H2 Narin Afgh.
41 H3 Narin reg. Afgh.
36 G3 Narince Turkey
45 H1 Narin Gol watercourse China
53 G7 Narita Japan
96 B2 Narizon, Pta pt Mex.
44 C5 Narmada r. India
37 H1 Narman Turkey
44 D3 Narnaul India
26 E3 Narni Italy
17 O5 Narodychi Ukr.
28 F4 Naro-Fominsk Rus. Fed.
67 J6 Narooma Austr.
28 G4 Narovchat Rus. Fed.
29 D5 Narowlya Belarus
9 R5 Närpes Fin.
67 H3 Narrabri Austr.
91 H4 Narragansett Bay U.S.A.
67 G2 Narran r. Austr.
67 G5 Narrandera Austr.
67 G2 Narran L. Austr.
69 C6 Narrogin Austr.
67 H4 Narromine Austr.
77 J4 Narrow Hills Provincial Park Can.
90 C6 Narrows U.S.A.
91 F4 Narrowsburg U.S.A.
44 D5 Narsimhapur India
45 G5 Narsingdi Bangl.
44 D5 Narsinghgarh India
43 C2 Narsipatnam India
50 E1 Nart China
53 D7 Naruto Japan
9 V7 Narva Estonia
9 U7 Narva Bay Estonia/Rus. Fed.
55 B2 Narvacan Phil.
8 P2 Narvik Norway
9 V7 Narvskoye Vdkhr. resr Estonia/Rus. Fed.
44 D3 Narwana India
44 D4 Narwar India
32 G3 Nar'yan-Mar Rus. Fed.
39 H4 Naryn Kyrg.
39 H4 Naryn r. Kyrg.
39 K4 Narynkol Kazak.
8 P5 Näsåker Sweden
21 J3 Nasavrky Czech Rep.
93 H3 Naschitti U.S.A.
70 C6 Naseby N.Z.
88 A4 Nashua IA U.S.A.
91 H3 Nashua NH U.S.A.
87 C4 Nashville U.S.A.
36 F5 Nasib Syria
9 S6 Näsijärvi l. Fin.
44 C5 Nasik India
109 F4 Nasir Sudan
Nasirabad see Mymensingh
44 B3 Nasirabad Pak.
115 A5 Nasondoye Congo(Zaire)
71 13 Nasorolevu mt Fiji
36 C6 Nasr Egypt
41 E3 Naşrābād Iran
40 C3 Naşrābād Iran
Nasratabad see Zābol
37 L5 Nasrīān-e-Pā'īn Iran
76 D3 Nass r. Can.
87 E7 Nassau Bahamas
20 A2 Nassau Ger.
60 L5 Nassau i. Cook Is Pac. Oc.
109 F2 Nasser, Lake resr Egypt
9 O8 Nässjö Sweden
78 E2 Nastapoca r. Can.
78 E2 Nastapoka Islands Can.
53 F6 Nasu-dake volc. Japan
55 B3 Nasugbu Phil.
17 P2 Nasva Rus. Fed.
111 C6 Nata Botswana
115 B3 Nata Tanz.
103 B4 Natagaima Col.
101 L5 Natal Brazil
Natal div. see Kwazulu-Natal
117 G6 Natal Basin sea feature Ind. Ocean
40 C3 Naṭanz Iran
79 H3 Natashquan Can.
79 H3 Natashquan r. Can.
85 F6 Natchez U.S.A.
85 E6 Natchitoches U.S.A.
71 13 Natewa Bay Fiji
66 F6 Nathalia Austr.
44 C4 Nathdwara India
66 D6 Natimuk Austr.
92 D5 National City U.S.A.
25 H2 Nati, Pta pt Spain
114 D2 Natitingou Benin
68 B4 Native Companion Cr. r. Austr.
101 J6 Natividade Brazil
96 B1 Nátora Mex.
52 G5 Natori Japan
115 C3 Natron, Lake salt l. Tanz.
58 A1 Nattaung mt Myanmar
20 E5 Natters Austria
59 C2 Natuna Besar i. Indon.
59 C2 Natuna, Kepulauan is Indon.
91 F2 Natural Bridge U.S.A.
93 G3 Natural Bridges National Monument res. U.S.A.
69 B6 Naturaliste, C. Austr.
69 A2 Naturaliste Channel Austr.
117 M6 Naturaliste Plateau sea feature Ind. Ocean
93 H2 Naturita U.S.A.
88 E2 Naubinway U.S.A.
111 B6 Nauchas Namibia
20 D6 Nauders Austria
19 K4 Nauen Ger.
91 G4 Naugatuck U.S.A.
55 B3 Naujan Phil.
55 B3 Naujan, L. Phil.
9 S8 Naujoji Akmenė Lith.
44 C4 Naukh India
19 G5 Naumburg (Hessen) Ger.
19 J5 Naumburg (Saale) Ger.
19 L6 Naundorf Ger.
58 A1 Naungpale Myanmar
36 E6 Na'ūr Jordan
41 G4 Nauroz Kalat Pak.
61 H4 Nauru country Pac. Oc.
44 B4 Naushara Pak.
71 15 Nausori Fiji
9 J6 Naustdal Norway
100 D4 Nauta Peru
112 C3 Naute Dam dam Namibia
97 E3 Nautla Mex.
41 G3 Nauzad Afgh.
45 G5 Navadwīp India
28 C4 Navahrudak Belarus
93 H4 Navajo U.S.A.
83 F4 Navajo Lake U.S.A.
93 G3 Navajo Mt U.S.A.
55 C4 Naval Phil.

91 H2 **North Woodstock** U.S.A.
12 G3 **North York Moors** *reg.* U.K.
12 G3 **North York Moors National Park** U.K.
79 G4 **Norton** Can.
12 G3 **Norton** U.K.
84 D4 **Norton** *KS* U.S.A.
90 B6 **Norton** *VA* U.S.A.
91 H2 **Norton** *VT* U.S.A.
111 D5 **Norton** Zimbabwe
74 B3 **Norton Sound** *b.* U.S.A.
19 G2 **Nortorf** Ger.
116 C2 **Norvegia, K.** *c.* Ant.
91 G4 **Norwalk** *CT* U.S.A.
90 B4 **Norwalk** *OH* U.S.A.
7 D2 **Norway** *country* Europe
91 H2 **Norway** U.S.A.
89 J3 **Norway Bay** Can.
77 K4 **Norway House** Can.
120 J1 **Norwegian Basin** *sea feature* Atl. Ocean
75 J2 **Norwegian Bay** Can.
6 C2 **Norwegian Sea** Atl. Ocean
89 G4 **Norwich** Can.
13 J5 **Norwich** U.K.
91 G4 **Norwich** *CT* U.S.A.
91 F3 **Norwich** *NY* U.S.A.
91 H3 **Norwood** *MA* U.S.A.
91 F2 **Norwood** *NY* U.S.A.
90 A5 **Norwood** *OH* U.S.A.
102 C2 **Nos de Cachi** *mt* Arg.
77 H1 **Nose Lake** Can.
27 M3 **Nos Emine** *pt* Bulg.
27 M3 **Nos Galata** *pt* Bulg.
52 G2 **Noshappu-misaki** *hd* Japan
52 G4 **Noshiro** Japan
29 D5 **Nosivka** Ukr.
27 N3 **Nos Kaliakra** *pt* Bulg.
28 H3 **Noskovo** Rus. Fed.
112 D2 **Nosop** *r.* Botswana/S. Africa
32 G3 **Nosovaya** Rus. Fed.
41 E4 **Noşratābād** Iran
104 A1 **Nossa Senhora do Livramento** Brazil
9 N7 **Nossebro** Sweden
19 L5 **Nossen** Ger.
27 N3 **Nos Shabla** *pt* Bulg.
11 **Noss, Isle of** *i.* U.K.
112 C2 **Nossob** *r.* Namibia
111 E5 **Nosy Be** *i.* Madag.
111 F5 **Nosy Boraha** *i.* Madag.
111 E6 **Nosy Varika** Madag.
93 F2 **Notch Peak** U.S.A.
16 H4 **Noteć** *r.* Pol.
9 L7 **Notodden** Norway
26 F6 **Noto, Golfo di** *g. Sicily* Italy
53 E6 **Noto-hantō** *pen.* Japan
79 K4 **Notre Dame Bay** Can.
89 K3 **Notre-Dame-de-la-Salette** Can.
91 H2 **Notre-Dame-des-Bois** Can.
89 K2 **Notre-Dame-du-Laus** Can.
89 H2 **Notre-Dame-du-Nord** Can.
79 G4 **Notre Dame, Monts** *mts* Can.
89 G3 **Nottawasaga Bay** Can.
78 E3 **Nottaway** *r.* Can.
13 F5 **Nottingham** U.K.
90 E6 **Nottoway** *r.* U.S.A.
18 E5 **Nottuln** Ger.
77 H5 **Notukeu Cr.** *r.* Can.
108 A2 **Nouâdhibou** Maur.
108 A3 **Nouakchott** Maur.
108 A3 **Nouâmghâr** Maur.
58 C2 **Nouei** Vietnam
71 9 **Nouméa** *New Caledonia* Pac. Oc.
114 C2 **Nouna** Burkina
112 F5 **Noupoort** S. Africa
8 V3 **Nousu** Fin.
Nouveau-Comptoir *see* **Wemindji**
Nouvelle Calédonie *is see* **New Caledonia**
15 B2 **Nouvion** France
39 G4 **Nov** Tajik.
104 C1 **Nova América** Brazil
21 M4 **Nová Baňa** Slovakia
21 J3 **Nová Bystřice** Czech Rep.
104 B3 **Nova Esperança** Brazil
104 D3 **Nova Friburgo** Brazil
23 L4 **Nova Gorica** Slovenia
26 G2 **Nova Gradiška** Croatia
104 C3 **Nova Granada** Brazil
104 D3 **Nova Iguaçu** Brazil
29 E6 **Nova Kakhovka** Ukr.
21 M4 **Nováky** Slovakia
104 D2 **Nova Lima** Brazil
28 D4 **Novalukoml'** Belarus
115 C6 **Nova Nabúri** Moz.
29 D6 **Nova Odesa** Ukr.
21 J2 **Nová Paka** Czech Rep.
22 F4 **Novara** Italy
104 C1 **Nova Roma** Brazil
79 H5 **Nova Scotia** *div.* Can.
23 G3 **Novate Mezzola** Italy
92 A2 **Novato** U.S.A.
104 C2 **Nova Venécia** Brazil
104 B1 **Nova Xavantino** Brazil
38 B2 **Novaya Kazanka** Kazak.
33 R2 **Novaya Sibir', Ostrov** *i.* Rus. Fed.
32 G2 **Novaya Zemlya** *is* Rus. Fed.
27 M3 **Nova Zagora** Bulg.
21 H4 **Nové Hrady** Czech Rep.
25 F3 **Novelda** Spain
23 H5 **Novellara** Italy
21 K2 **Nové Město nad Metuj** Czech Rep.
16 J7 **Nové Zámky** Slovakia
28 D3 **Novgorod** Rus. Fed.
28 E3 **Novgorodskaya Oblast'** *div.* Rus. Fed.
29 E5 **Novhorod-Sivers'kyy** Ukr.
39 K1 **Novichikha** Rus. Fed.
23 H5 **Novi di Modena** Italy
27 K3 **Novi Iskŭr** Bulg.
22 F5 **Novi Ligure** Italy
27 M3 **Novi Pazar** Bulg.
27 J3 **Novi Pazar** Yugo.
27 H2 **Novi Sad** Yugo.
29 G6 **Novoaleksandrovsk** Rus. Fed.
38 C2 **Novoalekseyevka** Kazak.
29 G5 **Novoanninskiy** Rus. Fed.
100 F5 **Novo Aripuanã** Brazil
29 F6 **Novoazovs'k** Ukr.
29 G5 **Novobod** Tajik.
28 H3 **Novocheboksarsk** Rus. Fed.
29 G6 **Novocherkassk** Rus. Fed.
39 H2 **Novodolinka** Kazak.
28 G1 **Novodvinsk** Rus. Fed.
102 F3 **Novo Hamburgo** Brazil
104 C3 **Novo Horizonte** Brazil
21 H4 **Novohradské Hory** *mts* Czech Rep.
29 C5 **Novohrad-Volyns'kyy** Ukr.
38 D1 **Novokaolinovyy** Rus. Fed.
38 E3 **Novokazalinsk** Kazak.
29 G6 **Novokubansk** Rus. Fed.
38 B1 **Novokuybyshevsk** Rus. Fed.
46 A1 **Novokuznetsk** Rus. Fed.
116 D3 **Novolazarevskaya** *Rus. Fed. Base* Ant.
33 M2 **Novoletov'ye** Rus. Fed.
39 H2 **Novomarkovka** Kazak.
26 F2 **Novo Mesto** Slovenia
28 F4 **Novomichurinsk** Rus. Fed.
29 G6 **Novomikhaylovskiy** Rus. Fed.
28 F4 **Novomoskovsk** Rus. Fed.
29 E5 **Novomoskovs'k** Ukr.
29 D5 **Novomyrhorod** Ukr.
29 C5 **Novonikolayevskiy** Rus. Fed.
39 G1 **Novonikol'skoye** Kazak.
29 E6 **Novooleksiyivka** Ukr.
28 D2 **Novoorsk** Rus. Fed.
38 F1 **Novopokrovka** *Kustanay.* Kazak.
39 K2 **Novopokrovka** *Semipal.* Kazak.
39 F1 **Novopokrovka** *Severnyy Kaz.* Kazak.
52 D2 **Novopokrovka** Rus. Fed.
29 G6 **Novopokrovskaya** Rus. Fed.
29 J5 **Novorepnoye** Rus. Fed.
29 F6 **Novorossiysk** Rus. Fed.
38 D2 **Novorossiyskoye** Kazak.
33 M2 **Novorybnoye** Rus. Fed.
17 O2 **Novorzhev** Rus. Fed.
29 E6 **Novoselivs'ke** Ukr.
17 O1 **Novosel'ye** Rus. Fed.
38 C1 **Novosergiyevka** Rus. Fed.
29 G5 **Novoshakhtinsk** Rus. Fed.
52 C2 **Novoshakhtinskiy** Rus. Fed.
32 K4 **Novosibirsk** Rus. Fed.
33 Q2 **Novosibirskiye Ostrova** *is* Rus. Fed.
28 D3 **Novosokol'niki** Rus. Fed.
28 H4 **Novospasskoye** Rus. Fed.
38 D2 **Novotroitsk** Rus. Fed.
39 H4 **Novotroitskoye** Kazak.
29 E6 **Novotroyits'ke** Ukr.
29 D5 **Novoukrayinka** Ukr.
38 D2 **Novoural'sk** Rus. Fed.
29 J5 **Novouzensk** Rus. Fed.
39 H1 **Novovarshavka** Rus. Fed.
29 C5 **Novovolyns'k** Ukr.
29 F5 **Novovoronezh** Rus. Fed.
39 K2 **Novoyegor'yevskoye** Rus. Fed.
29 D4 **Novozybkov** Rus. Fed.
21 H2 **Nový Bor** Czech Rep.
21 J2 **Nový Bydžov** Czech Rep.
21 L3 **Nový Jičín** Czech Rep.
33 M2 **Novyy** Rus. Fed.
29 F5 **Novyy Oskol** Rus. Fed.
32 J3 **Novyy Port** Rus. Fed.
28 J3 **Novyy Tor'yal** Rus. Fed.
32 J3 **Novyy Urengoy** Rus. Fed.
49 O1 **Novyy Urgal** Rus. Fed.
38 C4 **Novyy Uzen'** Kazak.
40 D4 **Now** Iran
21 K2 **Nowa Ruda** Pol.
84 E4 **Nowata** U.S.A.
40 C3 **Nowbarān** Iran
41 E3 **Now Deh** Iran
37 M3 **Nowdī** Iran
19 M3 **Nowe Czarnowo** Pol.
Nowgong *see* **Nagaon**
44 D4 **Nowgong** *Madhya Pradesh* India
77 J2 **Nowleye Lake** Can.
16 G4 **Nowogard** Pol.
67 J5 **Nowra** Austr.
40 C2 **Now Shahr** Iran
40 C2 **Nowshahr** Iran
44 C2 **Nowshera** Pak.
17 K6 **Nowy Sącz** Pol.
17 K6 **Nowy Targ** Pol.
87 E4 **Noxen** U.S.A.
32 J3 **Noyabr'sk** *r.* Rus. Fed.
22 C5 **Noyers-sur-Jabron** France
76 C3 **Noyes I.** U.S.A.
15 C3 **Noyon** France
58 C1 **Noy, Xé** *r.* Laos
113 F5 **Nozizwe** S. Africa
114 B4 **Nozza** *r.* Italy
113 G6 **Nqamakwe** S. Africa
113 J4 **Nqutu** S. Africa
115 C6 **Nsanje** Malawi
115 A5 **Nsombo** Zambia
113 A4 **Ntandembele** Congo(Zaire)
115 B5 **Ntchisi** Malawi
114 F4 **Ntem** *r.* Cameroon
113 G3 **Ntha** S. Africa
114 F4 **Ntui** Cameroon
115 B3 **Ntungamo** Uganda
71 10 **Nuapapu** *i.* Tonga
109 F3 **Nuba Mountains** Sudan
37 K1 **Nubarashen** Armenia
109 F2 **Nubian Desert** Sudan
105 B3 **Nuble** *r.* Chile
50 D1 **Nüden** Mongolia
100 D7 **Nudo Coropuna** *mt* Peru
85 D6 **Nueces** *r.* U.S.A.
96 G5 **Nueva Arcadia** Honduras
96 H5 **Nueva Armenia** Honduras
105 F2 **Nueva Helvecia** Uru.
105 B3 **Nueva Imperial** Chile
103 A4 **Nueva Loja** Ecuador
102 B6 **Nueva Lubecka** Arg.
96 D2 **Nueva Rosita** Mex.
97 G5 **Nueva San Salvador** El Salvador
95 J4 **Nuevitas** Cuba
94 C2 **Nuevo Casas Grandes** Mex.
105 D4 **Nuevo, Golfo** *g.* Arg.
96 C2 **Nuevo Ideal** Mex.
97 E2 **Nuevo Laredo** Mex.
97 E2 **Nuevo León** *div.* Mex.
110 E3 **Nugaal** *watercourse* Somalia
68 C5 **Nuga Nuga, L.** Austr.
70 B7 **Nugget Pt** N.Z.
63 F2 **Nuguria Is** P.N.G.
70 F3 **Nuhaka** N.Z.
63 H2 **Nui** *i.* Tuvalu
58 C2 **Nui Ti On** *mt* Vietnam
22 B2 **Nuits-St-Georges** France
46 B3 **Nu Jiang** *r.* China
66 A4 **Nukey Bluff** *h.* Austr.
71 11 **Nuku** *i.* Tonga
71 11 **Nuku'alofa** Tonga
71 13 **Nukubasaga** *rf* Fiji
63 H2 **Nukufetau** *i.* Tuvalu
60 O4 **Nuku Hiva** *i.* Pac. Oc.
63 H2 **Nukulaelae** *i.* Tuvalu
63 F2 **Nukumanu Is** P.N.G.
63 J2 **Nukunono** *i.* Pac. Oc.
38 D4 **Nukus** Uzbek.
64 D4 **Nullagine** Austr.
64 F6 **Nullarbor** Austr.
64 F6 **Nullarbor National Park** Austr.
64 E6 **Nullarbor Plain** Austr.
50 F1 **Nulu'erhu Shan** *mts* China
66 F2 **Numalla, Lake** *salt flat* Austr.
114 F3 **Numan** Nigeria
53 F6 **Numata** Japan
53 F7 **Numazu** Japan
65 G2 **Numbulwar** Austr.
9 L6 **Numedal** *v.* Norway
57 J7 **Numfor** *i.* Indon.
67 F6 **Numurkah** Austr.
79 H2 **Nunaksaluk Island** Can.
75 O3 **Nunarsuit** *i.* Greenland
15 J5 **Nünchritz** Ger.
90 E3 **Nunda** U.S.A.
115 A3 **Nunda** Congo(Zaire)
67 J3 **Nundle** Austr.
13 F5 **Nuneaton** U.K.
69 D5 **Nungarin** Austr.
78 B3 **Nungesser L.** Can.
74 B4 **Nunivak I.** U.S.A.
44 D2 **Nunkun** *mt* India
33 U3 **Nunligran** Rus. Fed.
25 C2 **Nuñomoral** Spain
18 C4 **Nunspeet** Neth.
26 C4 **Nuoro** *Sardinia* Italy
63 G3 **Nupani** *i.* Solomon Is
42 B4 **Nuqrah** S. Arabia
103 A3 **Nuquí** Col.
44 E1 **Nur** China
40 D2 **Nur** *r.* Iran
39 H2 **Nura** Kazak.
39 G2 **Nura** *r.* Kazak.
40 C4 **Nūrābād** Iran
39 B4 **Nurata** Uzbek.
39 F4 **Nuratau, Khrebet** *mts* Uzbek.
23 G4 **Nure** *r.* Italy
Nuremberg *see* **Nürnberg**
37 J2 **Nurettin** Turkey
41 G4 **Nūr Gamma** Pak.
96 B1 **Nuri** Mex.
66 C5 **Nuriootpa** Austr.
41 H3 **Nuristan** *reg.* Afgh.
28 J4 **Nurlaty** Rus. Fed.
8 V5 **Nurmes** Fin.
8 S5 **Nurmo** Fin.
20 E3 **Nürnberg** Ger.
67 G3 **Nurri, Mt** *h.* Austr.
20 C4 **Nürtingen** Ger.
45 H1 **Nur Turu** China
37 H3 **Nusaybin** Turkey
41 G4 **Nushki** Pak.
79 H2 **Nutak** Can.
93 H5 **Nutrioso** U.S.A.
44 B3 **Nuttal** Pak.
8 U3 **Nuupas** Fin.
71 3 **Nuupéré, Pte** *pt Fr. Polynesia* Pac. Oc.
75 N2 **Nuussuaq** Greenland
75 N2 **Nuussuaq** *pen.* Greenland
71 5 **Nuuuli** *American Samoa* Pac. Oc.
43 C5 **Nuwara Eliya** Sri Lanka
112 C5 **Nuwerus** S. Africa
112 D6 **Nuweveldberg** *mts* S. Africa
69 C7 **Nuyts, Pt** Austr.
37 K4 **Nuzi** Iraq
113 J1 **Nwanedi National Park** S. Africa
69 D6 **Nyabing** Austr.
32 H3 **Nyagan'** Rus. Fed.
115 C2 **Nyahururu** Kenya
66 E5 **Nyah West** Austr.
45 G3 **Nyainqêntanglha Feng** *mt* China
45 G3 **Nyainqêntanglha Shan** *mts* China
45 H2 **Nyainrong** China
8 Q5 **Nyaker** Sweden
109 E3 **Nyala** Sudan
45 F3 **Nyalam** China
111 C5 **Nyamandhiovu** Zimbabwe
28 G2 **Nyandoma** Rus. Fed.
28 F2 **Nyandomskiy Vozvyshennost'** *reg.* Rus. Fed.
110 B4 **Nyanga** *r.* Gabon
111 D5 **Nyanga** Zimbabwe
45 H3 **Nyang Qu** *r. Xizang* China
45 G3 **Nyang Qu** *r. Xizang* China
115 B3 **Nyanza** *div.* Kenya
111 D5 **Nyasa, Lake** Africa
28 C4 **Nyasvizh** Belarus
9 M9 **Nyborg** Denmark
8 V1 **Nyborg** Norway
9 O8 **Nybro** Sweden
75 N1 **Nyeboe Land** *reg.* Greenland
45 G3 **Nyêmo** China
115 C3 **Nyeri** Kenya
115 B6 **Nyika National Park** Zambia
45 F3 **Nyima** China
115 B5 **Nyimba** Zambia
46 B4 **Nyingchi** China
17 K7 **Nyíregyháza** Hungary
115 C2 **Nyiru, Mount** Kenya
8 S5 **Nykarleby** Fin.
9 M9 **Nykøbing** Denmark
9 M9 **Nykøbing Sjælland** Denmark
9 P7 **Nyköping** Sweden
8 P5 **Nyland** Sweden
113 H2 **Nylstroom** S. Africa
67 G4 **Nymagee** Austr.
67 K2 **Nymboida** Austr.
67 K2 **Nymboida** *r.* Austr.
21 J2 **Nymburk** Czech Rep.
9 P7 **Nynäshamn** Sweden
67 G3 **Nyngan** Austr.
17 L4 **Nyoman** *r.* Belarus/Lith.
22 D3 **Nyon** Switz.
114 F4 **Nyong** *r.* Cameroon
54 E5 **Nyongwol** S. Korea
45 F3 **Nyonni Ri** *mt* China
22 C5 **Nyons** France
20 E2 **Nýřany** Czech Rep.
32 G3 **Nyrob** Rus. Fed.
20 E3 **Nýrsko** Czech Rep.
16 H5 **Nysa** Pol.
21 L2 **Nysa Kłodzka** *r.* Pol.
Nysa Łużycka *r. see* **Neiß**
19 J2 **Nysted** Denmark
28 H2 **Nyuchpas** Rus. Fed.
52 F5 **Nyūdō-zaki** *pt* Japan
115 A4 **Nyunzu** Congo(Zaire)
33 N3 **Nyurba** Rus. Fed.
28 J2 **Nyuvchim** Rus. Fed.
29 E6 **Nyzhn'ohirs'kyy** Ukr.
115 B3 **Nzega** Tanz.
114 B3 **Nzérékoré** Guinea
110 B4 **N'zeto** Angola
113 J1 **Nzhelele Dam** *dam* S. Africa
114 C3 **Nzi** *r.* Côte d'Ivoire
Nzwani *i. see* **Anjouan**

O

84 C2 **Oahe, Lake** U.S.A.
92 1 **Oahu** *i.* U.S.A.
66 D4 **Oakbank** Austr.
93 F2 **Oak City** U.S.A.
85 E6 **Oakdale** U.S.A.
84 D2 **Oakes** U.S.A.
67 J1 **Oakey** Austr.
13 G5 **Oakham** U.K.
82 B1 **Oak Harbor** U.S.A.
90 C6 **Oak Hill** U.S.A.
92 C3 **Oakhurst** U.S.A.
88 B2 **Oak I.** U.S.A.
92 A3 **Oakland** *CA* U.S.A.
92 A3 **Oakland** *airport CA* U.S.A.
90 D5 **Oakland** *MD* U.S.A.
84 D3 **Oakland** *NE* U.S.A.
82 B3 **Oakland** *OR* U.S.A.
67 G5 **Oaklands** Austr.
88 D5 **Oak Lawn** U.S.A.
84 C4 **Oakley** U.S.A.
64 D4 **Oakover** *r.* Austr.
82 B3 **Oakridge** U.S.A.
87 C4 **Oak Ridge** U.S.A.
66 D4 **Oakvale** Austr.
89 H4 **Oakville** Can.
70 C6 **Oamaru** N.Z.
11 B5 **Oa, Mull of** *hd* U.K.
70 B5 **Oaro** N.Z.
55 B3 **Oas** Phil.
115 B5 **Oates Land** *reg.* Ant.
67 G9 **Oatlands** Austr.
93 E4 **Oatman** U.S.A.
97 E4 **Oaxaca** Mex.
97 E4 **Oaxaca** *div.* Mex.
32 H3 **Ob'** *r.* Rus. Fed.
114 F4 **Obala** Cameroon
37 D7 **Obama** Japan
11 C4 **Oban** U.K.
52 G5 **Obanazawa** Japan
25 C1 **O Barco** Spain
78 D4 **Obatogama L.** Can.
76 F4 **Obed** Can.
70 B6 **Obelisk** *mt* N.Z.
22 F3 **Oberalpstock** *mt* Switz.
20 E5 **Oberammergau** Ger.
20 E5 **Oberau** Ger.
19 G6 **Oberaula** Ger.
19 H5 **Oberdorla** Ger.
20 F6 **Oberdrauburg** Austria
20 D5 **Obergünzburg** Ger.
20 C6 **Obergurgl** Austria
20 J6 **Oberhaag** Austria
19 H5 **Oberharz** *nat. park* Ger.
18 D5 **Oberhausen** Ger.
20 B4 **Oberkirch** Ger.
84 C4 **Oberlin** *KS* U.S.A.
90 B4 **Oberlin** *OH* U.S.A.
15 H3 **Obermoschel** Ger.
20 C4 **Obernai** France
20 B4 **Oberndorf am Neckar** Ger.
67 H4 **Oberon** Austr.
21 G4 **Oberösterreich** *div.* Austria
20 F3 **Oberpfälzer Wald** *mts* Ger.
20 E4 **Oberpframmern** Ger.
21 K5 **Oberpullendorf** Austria
20 D5 **Oberstdorf** Ger.
20 F6 **Obertilliach** Austria
20 F3 **Oberviechtach** Ger.
19 G5 **Oberwälder Land** *reg.* Ger.
21 K5 **Oberwart** Austria
57 H7 **Obi** *i.* Indon.
101 G4 **Óbidos** Brazil
39 G5 **Obigarm** Tajik.
52 H3 **Obihiro** Japan
29 H6 **Obil'noye** Rus. Fed.
103 C2 **Obispos** Venez.
49 O2 **Obluch'ye** Rus. Fed.
28 F4 **Obninsk** Rus. Fed.
110 C3 **Obo** C.A.R.
50 A2 **Obo** China
110 E2 **Obock** Djibouti
115 A3 **Obokote** Congo(Zaire)
54 E3 **Obŏk-tong** N. Korea
110 B4 **Obouya** Congo
29 F5 **Oboyan'** Rus. Fed.
28 G2 **Obozerskiy** Rus. Fed.
45 E4 **Obra** India
45 E4 **Obra Dam** *dam* India
96 B1 **Obregón, Presa** *resr* Mex.
27 J2 **Obrenovac** Yugo.
36 D2 **Obruk** Turkey
38 B2 **Obshchiy Syrt** *h.* Rus. Fed.
32 J2 **Obskaya Guba** *chan.* Rus. Fed.
114 C3 **Obuasi** Ghana
29 D5 **Obukhiv** Ukr.
32 H2 **Ob''yachevo** Rus. Fed.
87 D6 **Ocala** U.S.A.
103 D4 **Ocamo** *r.* Venez.
96 D2 **Ocampo** *Coahuila* Mex.
103 B2 **Ocaña** Col.
25 E3 **Ocaña** Spain
23 J5 **Occhiobello** Italy
100 E7 **Occidental, Cordillera** *mts* Chile
103 A4 **Occidental, Cordillera** *mts* Col.
100 C6 **Occidental, Cordillera** *mts* Peru
76 B3 **Ocean Cape** *pt* U.S.A.
91 F5 **Ocean City** *MD* U.S.A.
91 F5 **Ocean City** *NJ* U.S.A.
76 D4 **Ocean Falls** Can.
120 G3 **Oceanographer Fracture** *sea feature* Atl. Ocean
92 D5 **Oceanside** U.S.A.
85 F6 **Ocean Springs** U.S.A.
29 D6 **Ochakiv** Ukr.
29 G7 **Och'amch'ire** Georgia
11 E4 **Ochil Hills** U.K.
44 C1 **Ochili Pass** Afgh.
20 D3 **Ochsenfurt** Ger.
20 C4 **Ochsenhausen** Ger.
18 E4 **Ochtrup** Ger.
9 P6 **Ockelbo** Sweden
18 F2 **Ockholm** Ger.
97 F4 **Ococingo** Mex.
17 M7 **Ocolaşul Mare, Vârful** *mt* Romania
81 K5 **Oconee** *r.* U.S.A.
88 C4 **Oconomowoc** U.S.A.
88 D3 **Oconto** U.S.A.
96 H5 **Ocotal** Nic.
97 G5 **Ocotepeque** Honduras
93 E5 **Ocotillo Wells** U.S.A.
96 D3 **Ocotlán** Mex.
114 C3 **Oda** Ghana
53 C7 **Ōda** Japan
8 E4 **Ódáðahraun** *lava* Iceland
54 E3 **Ŏdaejin** N. Korea
52 G4 **Ōdate** Japan
53 F7 **Odawara** Japan
9 K6 **Odda** Norway
77 K3 **Odei** *r.* Can.
88 C5 **Odell** U.S.A.
25 B4 **Odemira** Port.
36 A2 **Ödemiş** Turkey
113 G3 **Odendaalsrus** S. Africa
9 M9 **Odense** Denmark
19 H1 **Odense Fjord** *b.* Denmark
20 B3 **Odenwald** *reg.* Ger.
19 H5 **Oder** *r.* Ger./Pol.
19 M4 **Oderberg** Ger.
19 M4 **Oderbruch** *reg.* Ger.
19 L2 **Oderbucht** *b.* Ger.
19 M3 **Oderhaff** *b.* Ger.
23 K4 **Oderzo** Italy
29 D6 **Odesa** Ukr.
9 O7 **Ödeshog** Sweden
85 C6 **Odessa** U.S.A.
32 H3 **Odesskoye** Rus. Fed.
39 H1 **Odesskoye** Rus. Fed.
25 C4 **Odiel** *r.* Spain
114 B3 **Odienné** Côte d'Ivoire
28 F4 **Odintsovo** Rus. Fed.
58 C3 **Ŏdôngk** Cambodia
21 L3 **Odra** *r.* Ger./Pol.
101 K5 **Oeiras** Brazil
84 C3 **Oelrichs** U.S.A.
20 F2 **Oelsnitz** Ger.
88 B4 **Oelwein** U.S.A.
18 C3 **Oenkerk** Neth.
60 O6 **Oeno** *i. Pitcairn Is* Pac. Oc.
20 A2 **Oestrich-Winkel** Ger.
20 D4 **Oettingen in Bayern** Ger.
19 E2 **Oetz** Austria
19 D2 **Oeversee** Ger.
37 H1 **Of** Turkey
26 G4 **Ofanto** *r.* Italy
114 D3 **Offa** Nigeria
20 B2 **Offenbach am Main** Ger.
15 H4 **Offenburg** Ger.
20 D4 **Offingen** Ger.
27 M6 **Ofidoussa** *i.* Greece
71 12 **Ofolanga** *i.* Tonga
71 5 **Ofu** *i. American Samoa* Pac. Oc.
71 10 **Ofu** *i.* Tonga
52 G5 **Ōfunato** Japan
52 F5 **Oga** Japan
110 E3 **Ogadēn** *reg.* Eth.
52 F5 **Oga-hantō** *pen.* Japan
53 E7 **Ōgaki** Japan
84 C3 **Ogallala** U.S.A.
46 G4 **Ogasawara-shotō** *is* Japan
89 H2 **Ogascanane, Lac** *l.* Can.
114 D3 **Ogbomoso** Nigeria
84 E3 **Ogden** *IA* U.S.A.
82 E3 **Ogden** *UT* U.S.A.
76 C3 **Ogden, Mt** Can.
91 F2 **Ogdensburg** U.S.A.
71 14 **Ogea** *i.* Fiji
74 E3 **Ogilvie** *r.* Can.
74 E3 **Ogilvie Mts** Can.
38 C5 **Oglanly** Turkm.
87 C5 **Oglethorpe, Mt** U.S.A.
23 H4 **Oglio** *r.* Italy
68 C4 **Ogmore** Austr.
22 C2 **Ognon** *r.* France
114 E3 **Ogoja** Nigeria
73 C3 **Ogoki** *r.* Can.
78 C3 **Ogoki Resr** Can.
27 K3 **Ogosta** *r.* Bulg.
9 T8 **Ogre** Latvia
26 F2 **Ogulin** Croatia
114 D3 **Ogun** *r.* Nigeria
38 C5 **Ogurchinskiy, Ostrov** *i.* Turkm.
37 L1 **Oğuz** Azer.
70 A6 **Ohai** N.Z.
70 E3 **Ohakune** N.Z.
52 G3 **Ohata** Japan
70 B6 **Ohau, L.** N.Z.
105 B7 **O'Higgins** *div.* Chile
71 8 **O'Higgins, C.** *Easter I.* Chile
105 B7 **O'Higgins, L.** Chile
90 B4 **Ohio** *div.* U.S.A.
86 C4 **Ohio** *r.* U.S.A.
20 B2 **Ohm** *r.* Ger.
71 11 **Ohonua** Tonga
19 H6 **Ohrdruf** Ger.
21 H2 **Ohře** *r.* Czech Rep.
19 J4 **Ohre** *r.* Ger.
27 J4 **Ohrid** Macedonia
27 J4 **Ohrid, Lake** Albania/Macedonia
113 J2 **Ohrigstad** S. Africa
20 C3 **Öhringen** Ger.
70 E3 **Ohura** N.Z.
101 H3 **Oiapoque** Brazil
11 D3 **Oich, Loch** *l.* U.K.
45 H3 **Oiga** China
15 C2 **Oignies** France
22 C3 **Oignin** *r.* France
90 D4 **Oil City** U.S.A.
92 C4 **Oildale** U.S.A.
15 C3 **Oise** *r.* France
15 D3 **Oise à l'Aisne, Canal de l'** France
15 B3 **Oisemont** France
53 B8 **Ōita** Japan
27 K5 **Oiti** *mt* Greece
92 C4 **Ojai** U.S.A.
105 D2 **Ojeda** Arg.
88 B3 **Ojibwa** U.S.A.
94 D3 **Ojinaga** Mex.
97 E4 **Ojitlán** Mex.
53 F6 **Ojiya** Japan
96 A2 **Ojo de Liebre, L.** *b.* Mex.
102 C3 **Ojos del Salado** *mt* Arg.
114 E3 **Oju** Nigeria
28 G4 **Oka** *r.* Rus. Fed.
111 B6 **Okahandja** Namibia
70 E3 **Okahukura** N.Z.
111 B6 **Okakarara** Namibia
79 H2 **Okak Islands** Can.
76 F5 **Okanagan Falls** Can.
76 F4 **Okanagan Lake** Can.
82 C1 **Okanogan** *r.* Can./U.S.A.
76 F5 **Okanogan** U.S.A.
82 B1 **Okanogan Range** *mts* U.S.A.
115 A2 **Okapi, Parc National de la** *nat. park* Congo(Zaire)
44 C3 **Okara** Pak.
38 C5 **Okarem** Turkm.
111 B5 **Okaukuejo** Namibia
111 C5 **Okavango** *r.* Botswana/Namibia
111 C5 **Okavango Delta** *swamp* Botswana
53 F6 **Okaya** Japan
53 C7 **Okayama** Japan
53 E7 **Okazaki** Japan
87 D7 **Okeechobee** U.S.A.
87 D7 **Okeechobee, L.** U.S.A.
87 D6 **Okefenokee Swamp** *swamp* U.S.A.
13 C7 **Okehampton** U.K.
114 E3 **Okene** Nigeria
19 H4 **Oker** *r.* Ger.
44 B5 **Okha** India
46 G1 **Okha** Rus. Fed.
45 F4 **Okhaldhunga** Nepal
44 B5 **Okha Rann** *marsh* India
33 Q3 **Okhotka** *r.* Rus. Fed.
33 Q4 **Okhotsk** Rus. Fed.
46 G2 **Okhotsk, Sea of** *g.* Rus. Fed.
29 E5 **Okhtyrka** Ukr.
49 N6 **Okinawa** *i.* Japan
49 N6 **Okinawa-guntō** *is* Japan
53 B7 **Okino-shima** *i.* Japan
49 O4 **Oki-shotō** *i.* Japan
53 C6 **Oki-shotō** *is* Japan
85 D5 **Oklahoma** *div.* U.S.A.
85 D5 **Oklahoma City** U.S.A.
85 D5 **Okmulgee** U.S.A.
110 B4 **Okondja** Gabon
76 G4 **Okotoks** Can.
28 E4 **Okovskiy Les** *forest* Rus. Fed.
110 B4 **Okoyo** Congo
39 K3 **Okpety, Gora** *mt* Kazak.
8 S1 **Øksfjord** Norway
28 F2 **Oksovskiy** Rus. Fed.
38 H5 **Oksu** *r.* Tajik.
39 G5 **Oktyabr'** Tajik.
38 E2 **Oktyabr'sk** Kazak.
28 J4 **Oktyabr'sk** Rus. Fed.
28 G2 **Oktyabr'skiy** *Archangel.* Rus. Fed.
46 H1 **Oktyabr'skiy** *Kamchatsk.* Rus. Fed.
32 G4 **Oktyabr'skiy** *Resp. Bashkor.* Rus. Fed.
29 G6 **Oktyabr'skiy** *Volgograd.* Rus. Fed.
39 F5 **Oktyabr'skiy** Uzbek.
38 F1 **Oktyabr'skoye** Kazak.
38 E1 **Oktyabr'skoye** *Chelyabinsk.* Rus. Fed.
32 H3 **Oktyabr'skoye** *Khanty.* Rus. Fed.
38 C1 **Oktyabr'skoye** *Orenburg.* Rus. Fed.
33 L2 **Oktyabr'skoy Revolyutsii, Ostrov** *i.* Rus. Fed.
38 D4 **Oktyah'sk** Turkm.
28 E3 **Okulovka** Rus. Fed.
52 F3 **Okushiri-tō** *i.* Japan
114 D3 **Okuta** Nigeria
112 E1 **Okwa** *watercourse* Botswana
8 B4 **Ólafsvík** Iceland
92 C3 **Olancha** U.S.A.
92 C3 **Olancha Peak** U.S.A.
96 H5 **Olanchito** Honduras
9 P8 **Öland** *i.* Sweden
8 W3 **Olanga** Rus. Fed.
66 D4 **Olary** Austr.
66 D4 **Olary** *watercourse* Austr.
84 E4 **Olathe** U.S.A.
105 E3 **Olavarría** Arg.
16 H5 **Oława** Pol.
93 G5 **Olberg** U.S.A.
20 G2 **Olbernhau** Ger.
26 C4 **Olbia** *Sardinia* Italy
90 D3 **Olcott** U.S.A.
67 K3 **Old Bar** Austr.
43 C2 **Old Bastar** India
14 D4 **Oldcastle** Rep. of Ireland
74 E3 **Old Crow** Can.
18 C3 **Oldeboorn** Neth.
18 F3 **Oldenburg** Ger.
19 H2 **Oldenburg in Holstein** Ger.
18 D4 **Oldenzaal** Neth.
8 R2 **Olderdalen** Norway
69 E6 **Oldfield** *r.* Austr.
91 F3 **Old Forge** *NY* U.S.A.
91 F4 **Old Forge** *PA* U.S.A.
69 D3 **Old Gidgee** Austr.
12 E4 **Oldham** U.K.
14 C6 **Old Head of Kinsale** *hd* Rep. of Ireland
76 G4 **Oldman** *r.* Can.
11 F3 **Oldmeldrum** U.K.
91 H3 **Old Orchard Beach** U.S.A.
79 K4 **Old Perlican** Can.
76 G4 **Olds** Can.
91 J2 **Old Town** U.S.A.
77 H4 **Old Wives L.** Can.
93 E4 **Old Woman Mts** U.S.A.
90 D3 **Olean** U.S.A.
17 L3 **Olecko** Pol.
22 F4 **Oleggio** Italy
33 O4 **Olekma** *r.* Rus. Fed.
33 O3 **Olekminsk** Rus. Fed.
29 E5 **Oleksandriya** Ukr.
28 H1 **Olema** Rus. Fed.
9 J7 **Ølen** Norway
8 X2 **Olenegorsk** Rus. Fed.
33 N3 **Olenek** Rus. Fed.
33 O2 **Olenek** *r.* Rus. Fed.
33 O2 **Olenek B.** Rus. Fed.
28 E3 **Olenino** Rus. Fed.
39 H1 **Olenty** *r. Pavlodarsk.* Kazak.
38 C2 **Olenty** *r. Zapadnyy Kaz.* Kazak.
21 L1 **Oleśnica** Pol.
21 M2 **Olesno** Pol.
29 C5 **Olevs'k** Ukr.
52 D3 **Ol'ga** Rus. Fed.
25 C4 **Olhão** Port.
112 C2 **Olifants** *watercourse* Namibia
113 J2 **Olifants** S. Africa
112 C5 **Olifants** *r.* S. Africa
112 E3 **Olifantshoek** S. Africa
112 C6 **Olifantsrivierberg** *mts* S. Africa
105 F2 **Olimar Grande** *r.* Uru.
104 C3 **Olímpia** Brazil
97 E4 **Olinalá** Mex.
101 M5 **Olinda** Brazil
115 C6 **Olinga** Moz.
113 G2 **Oliphants Drift** Botswana
105 D2 **Oliva** Arg.
25 F3 **Oliva** Spain
102 C3 **Oliva, Cordillera de** *mts* Arg./Chile
105 C1 **Olivares, Co del** *mt* Chile
90 B5 **Olive Hill** U.S.A.
104 D3 **Oliveira** Brazil
25 C3 **Olivenza** Spain
84 E2 **Olivia** U.S.A.
28 G4 **Ol'khi** Rus. Fed.
102 C2 **Ollagüe** Chile
22 C6 **Ollioules** France
105 B1 **Ollita, Cordillera de** *mts* Arg./Chile
105 B1 **Ollitas** *mt* Arg.
22 D3 **Ollon** Switz.
100 C5 **Olmos** Peru
91 G3 **Olmstedville** U.S.A.
13 G5 **Olney** U.K.
86 C4 **Olney** U.S.A.
9 O8 **Olofström** Sweden
21 L3 **Olomouc** Czech Rep.
28 E2 **Olonets** Rus. Fed.
55 B3 **Olongapo** Phil.
24 D5 **Oloron-Ste-Marie** France
71 5 **Olosega** *i. American Samoa* Pac. Oc.
25 H1 **Olot** Spain
49 L1 **Olovyannaya** Rus. Fed.
44 C5 **Olpad** India
18 E5 **Olpe** Ger.
21 L3 **Olšava** *r.* Czech Rep.
21 M3 **Olše** *r.* Czech Rep.
17 K4 **Olsztyn** Pol.
22 E2 **Olten** Switz.
27 M2 **Olteniţa** Romania
37 H1 **Oltu** Turkey
55 B5 **Olutanga** *i.* Phil.
82 B2 **Olympia** U.S.A.
82 A2 **Olympic Nat. Park** *WA* U.S.A.
Olympus *mt see* **Troödos, Mount**
27 K4 **Olympus** *mt* Greece
82 B2 **Olympus, Mt** U.S.A.
33 S3 **Olyutorskiy** Rus. Fed.
33 T4 **Olyutorskiy, Mys** *c.* Rus. Fed.
33 S4 **Olyutorskiy Zaliv** *b.* Rus. Fed.
45 E2 **Oma** China
52 G4 **Ōma** Japan
53 E6 **Ōmachi** Japan
53 F7 **Omae-zaki** *pt* Japan
14 D3 **Omagh** U.K.
84 E3 **Omaha** U.S.A.
112 C1 **Omaheke** *div.* Namibia
82 C1 **Omak** U.S.A.
31 E8 **Oman** *country* Asia
41 E5 **Oman, Gulf of** *g.* Asia
70 B6 **Omarama** N.Z.
111 B6 **Omaruru** Namibia
111 B5 **Omatako** *watercourse* Namibia
100 D7 **Omate** Peru
112 E2 **Omaweneno** Botswana
52 G4 **Ōma-zaki** *c.* Japan
Omba *i. see* **Aoba**
110 A4 **Omboué** Gabon
26 D3 **Ombrone** *r.* Italy
45 G3 **Ombu** China
112 E5 **Omdraaisvlei** S. Africa
109 F3 **Omdurman** Sudan
22 F4 **Omegna** Italy
67 G6 **Omeo** Austr.
96 H6 **Ometepe, Isla de** *i.* Nic.
110 D2 **Om Hājer** Eritrea
40 C4 **Omīdīyeh** Iran
76 D3 **Omineca Mountains** Can.
112 C1 **Omitara** Namibia
53 F7 **Ōmiya** Japan
76 C3 **Ommaney, Cape** *hd* U.S.A.
18 D4 **Ommen** Neth.
50 B1 **Ömnögovĭ** *div.* Mongolia
19 J1 **Omø** *i.* Denmark
33 R3 **Omolon** *r.* Rus. Fed.
110 D3 **Omo National Park** Eth.
52 G5 **Omono-gawa** *r.* Japan
32 J4 **Omsk** Rus. Fed.
33 R3 **Omsukchan** Rus. Fed.
52 H2 **Ōmū** Japan
53 A8 **Ōmura** Japan
27 L2 **Omu, Vârful** *mt* Romania
88 B4 **Onalaska** U.S.A.
91 F6 **Onancock** U.S.A.
78 D4 **Onaping Lake** Can.
96 B1 **Onavas** Mex.
89 E3 **Onaway** U.S.A.
58 A2 **Onbingwin** Myanmar
105 D1 **Oncativo** Arg.
23 C1 **Onchan** U.K.
111 B5 **Oncócua** Angola
111 B5 **Ondangwa** Namibia
112 B1 **Ondekaremba** Namibia
112 D5 **Onderstedorings** S. Africa
111 B5 **Ondjiva** Angola
114 D3 **Ondo** Nigeria
49 K2 **Öndörhaan** Mongolia
54 A1 **Ondor Had** China
50 B1 **Ondor Mod** China
50 D1 **Ondor Sum** China
28 E2 **Ondozero** Rus. Fed.
112 D1 **One** Botswana
71 14 **Oneata** *i.* Fiji
28 F2 **Onega** Rus. Fed.
28 F2 **Onega** *r.* Rus. Fed.
28 E2 **Onega, Lake** Rus. Fed.
76 E4 **100 Mile House** Can.
91 F3 **Oneida** U.S.A.
91 F3 **Oneida Lake** U.S.A.
84 D3 **O'Neill** U.S.A.
46 H2 **Onekotan, O.** *i.* Rus. Fed.
91 F3 **Oneonta** U.S.A.
70 E2 **Oneroa** N.Z.
17 N7 **Oneşti** Romania
71 11 **Onevai** *i.* Tonga
28 E1 **Onezhskaya Guba** *g.* Rus. Fed.
Onezhskoye Ozero *l. see* **Onega, Lake**
45 E5 **Ong** *r.* India
110 B4 **Onga** Gabon
70 F3 **Ongaonga** N.Z.
112 E4 **Ongers** *watercourse* S. Africa
69 D6 **Ongerup** Austr.
54 C5 **Ongjin** N. Korea
50 F1 **Ongniud Qi** China
43 C3 **Ongole** India
39 L2 **Onguday** Rus. Fed.
29 G7 **Oni** Georgia
111 E6 **Onilahy** *r.* Madag.
114 E3 **Onitsha** Nigeria
112 B1 **Onjati Mountain** Namibia
71 14 **Ono** *i.* Fiji
53 E7 **Ōno** Japan
63 J4 **Ono-i-Lau** *i.* Fiji
53 C7 **Onomichi** Japan
63 H2 **Onotoa** *i.* Kiribati
76 G4 **Onoway** Can.
112 C4 **Onseepkans** S. Africa
64 C4 **Onslow** Austr.
87 E5 **Onslow Bay** U.S.A.
54 F2 **Onsong** N. Korea
18 E3 **Onstwedde** Neth.
53 E7 **Ontake-san** *volc.* Japan
78 B3 **Ontario** *div.* Can.
82 C2 **Ontario** U.S.A.
89 H4 **Ontario, Lake** Can./U.S.A.
88 C2 **Ontonagon** U.S.A.
63 F2 **Ontong Java Atoll** *atoll* Solomon Is
65 G5 **Oodnadatta** Austr.
64 F6 **Ooldea** Austr.
85 E4 **Oologah L.** *resr* U.S.A.
18 A5 **Oostburg** Neth.
Oostende *see* **Ostend**
18 C4 **Oostendorp** Neth.
18 B5 **Oosterhout** Neth.
18 A5 **Oosterschelde** *est.* Neth.
18 D4 **Oosterwolde** Neth.
15 C2 **Oostvleteren** Belgium
18 C3 **Oost-Vlieland** Neth.
76 D4 **Ootsa Lake** Can.
76 D4 **Ootsa Lake** Can.
90 E5 **Opal** U.S.A.
110 C4 **Opala** Congo(Zaire)
28 J3 **Oparino** Rus. Fed.
78 B3 **Opasquia** Can.
78 B3 **Opasquia Provincial Park** Can.
78 E3 **Opataca, L.** Can.
21 J2 **Opatovice nad Labem** Czech Rep.
21 M1 **Opatów** Pol.
21 L3 **Opava** Czech Rep.
21 L2 **Opava** *r.* Czech Rep.
87 C5 **Opelika** U.S.A.
85 E6 **Opelousas** U.S.A.
82 F1 **Opheim** U.S.A.
89 F2 **Ophir** Can.
59 B2 **Ophir, Gunung** *volc.* Indon.
115 A2 **Opienge** Congo(Zaire)
70 C6 **Opihi** *r.* N.Z.
78 E3 **Opinaca** *r.* Can.
78 E3 **Opinaca, Réservoir** *resr* Can.
78 D3 **Opinnagau** *r.* Can.
37 K5 **Opis** Iraq
79 G3 **Opiscotéo L.** Can.
18 B4 **Opmeer** Neth.
28 D3 **Opochka** Rus. Fed.
16 H5 **Opole** Pol.
21 L2 **Opole** *div.* Pol.
38 C3 **Opornyy** Kazak.
25 B2 **Oporto** Port.
70 F3 **Opotiki** N.Z.
87 C6 **Opp** U.S.A.
8 L5 **Oppdal** Norway
20 B4 **Oppenau** Ger.
23 L4 **Oprtalj** Croatia
70 D3 **Opunake** N.Z.
111 B5 **Opuwo** Namibia
38 B3 **Opytnoye** Kazak.
88 B5 **Oquawka** U.S.A.
91 H2 **Oquossoc** U.S.A.
38 D2 **Or'** *r.* Rus. Fed.
93 G5 **Oracle** U.S.A.
93 G5 **Oracle Junction** U.S.A.
17 K7 **Oradea** Romania
8 E4 **Öræfajökull** *gl.* Iceland
27 J3 **Orahovac** Yugo.
44 D4 **Orai** India
22 C3 **Orain** *r.* France
108 B1 **Oran** Alg.
102 D2 **Orán** Arg.
58 C2 **O Rang** Cambodia
54 E3 **Orang** N. Korea
67 H4 **Orange** Austr.
22 B5 **Orange** France
111 B6 **Orange** *r.* Namibia/S. Africa
91 G3 **Orange** *MA* U.S.A.
85 E6 **Orange** *TX* U.S.A.
90 D5 **Orange** *VA* U.S.A.
87 D5 **Orangeburg** U.S.A.
101 H3 **Orange, Cabo** *c.* Brazil
Orange Free State *div. see* **Free State**
89 G3 **Orangeville** Can.
93 G2 **Orangeville** U.S.A.
97 G4 **Orange Walk** Belize
55 B3 **Orani** Phil.
19 L4 **Oranienburg** Ger.
111 B6 **Oranjemund** Namibia
103 C1 **Oranjestad** Aruba
14 C4 **Oranmore** Rep. of Ireland
111 C6 **Orapa** Botswana
55 C3 **Oras** Phil.
27 K2 **Orăştie** Romania
8 S5 **Oravais** Fin.
27 J2 **Oravita** Romania
22 F5 **Orba** *r.* Italy
44 E2 **Orba Co** *l.* China
19 H1 **Ørbæk** Denmark
22 E4 **Orbassano** Italy
22 D3 **Orbe** *r.* Switz.
26 D3 **Orbetello** Italy
25 D1 **Orbigo** *r.* Spain
67 H6 **Orbost** Austr.
116 B1 **Orcadas** *Arg. Base* Ant.
93 H2 **Orchard Mesa** U.S.A.
103 D2 **Orchila, Isla** *i.* Venez.
22 E4 **Orco** *r.* Italy
92 B4 **Orcutt** U.S.A.
64 E3 **Ord** *r.* Austr.
83 D4 **Orderville** U.S.A.
25 B1 **Ordes** Spain
64 E3 **Ord, Mt** *h.* Austr.
92 D4 **Ord Mt** U.S.A.
36 F1 **Ordu** Turkey
37 L2 **Ordubad** Azer.
83 G4 **Ordway** U.S.A.
38 E1 **Ordzhonikidze** Kazak.
Ordzhonikidze *see* **Vladikavkaz**
29 E6 **Ordzhonikidze** Ukr.
92 C1 **Orenna** U.S.A.
9 O7 **Örebro** Sweden
88 C4 **Oregon** *IL* U.S.A.
90 B4 **Oregon** *OH* U.S.A.
88 C4 **Oregon** *WV* U.S.A.
82 B3 **Oregon** *div.* U.S.A.
82 B2 **Oregon City** U.S.A.
28 F4 **Orekhovo-Zuyevo** Rus. Fed.
28 F4 **Orel** Rus. Fed.
46 F1 **Orel', Ozero** *l.* Rus. Fed.
93 G1 **Orem** U.S.A.
36 A2 **Ören** Turkey
27 M6 **Ören** Turkey
38 C2 **Orenburg** Rus. Fed.
38 C2 **Orenburgskaya Oblast'** *div.* Rus. Fed.
105 E3 **Orense** Arg.
70 A7 **Orepuki** N.Z.
9 N9 **Øresund** *str.* Denmark
70 B7 **Oreti** *r.* N.Z.
70 E2 **Orewa** N.Z.
15 F2 **Oreye** Belgium
27 K4 **Orfanou, Kolpos** *b.* Greece
67 G9 **Orford** Austr.
13 J5 **Orford** U.K.
13 J5 **Orford Ness** *spit* U.K.
93 F5 **Organ Pipe Cactus National Monument** *res.* U.S.A.
22 C3 **Orgelet** France
41 H3 **Orgūn** Afgh.
36 B2 **Orhaneli** Turkey
29 D7 **Orhangazi** Turkey
28 J3 **Orichi** Rus. Fed.
91 J1 **Orient** U.S.A.
100 E7 **Oriental, Cordillera** *mts* Bol.
103 B3 **Oriental, Cordillera** *mts* Col.
100 D6 **Oriental, Cordillera** *mts* Peru
105 E3 **Oriente** Arg.
25 F3 **Orihuela** Spain
29 E6 **Orikhiv** Ukr.
89 H3 **Orillia** Can.
9 T6 **Orimattila** Fin.
103 E2 **Orinoco** *r.* Col./Venez.
103 E2 **Orinoco Delta** Venez.
45 F5 **Orissa** *div.* India
9 S7 **Orissaare** Estonia
26 C5 **Oristano** *Sardinia* Italy
9 T6 **Orivesi** Fin.

8 V5 Orivesi *l.* Fin.
101 G4 Oriximiná Brazil
97 E4 Orizaba Mex.
8 L5 Orkanger Norway
9 N8 Örkelljunga Sweden
8 L5 Orkla *r.* Norway
113 G3 Orkney S. Africa
11 E1 Orkney Islands U.K.
85 C6 Orla U.S.A.
92 A2 Orland U.S.A.
104 C3 Orlândia Brazil
87 D6 Orlando U.S.A.
24 E3 Orléans France
19 J4 Orleans *MA* U.S.A.
91 G2 Orleans *VT* U.S.A.
22 D3 Or, Le Mont d' *mt* France
21 J2 Orlice *r.* Czech Rep.
21 K2 Orlické Hory *mts* Czech Rep.
21 L2 Orlík *mt* Czech Rep.
28 J3 Orlov Rus. Fed.
21 M3 Orlová Czech Rep.
28 F4 Orlovskaya Oblast' *div.* Rus. Fed.
29 G6 Orlovskiy Rus. Fed.
41 G5 Ormara Pak.
41 G5 Ormara, Ras *hd* Pak.
55 C4 Ormoc Phil.
87 D6 Ormond Beach U.S.A.
21 K6 Ormož Slovenia
12 E4 Ormskirk U.K.
91 G2 Ormstown Can.
15 F4 Ornain *r.* France
22 D2 Ornans France
24 D2 Orne *r.* France
8 N3 Ørnes Norway
8 Q5 Örnsköldsvik Sweden
54 D4 Oro N. Korea
23 G3 Orobie, Alpi *mts* Italy
103 C3 Orocué Col.
114 C2 Orodara Burkina
82 C2 Orofino U.S.A.
83 F5 Orogrande U.S.A.
71 ▫3 Orohena *mt* *Fr. Polynesia* Pac. Oc.
23 J4 Orolo *r.* Italy
79 G4 Oromocto Can.
36 E6 Oron Israel
63 J2 Orona *i.* Kiribati
91 J2 Orono U.S.A.
11 B4 Oronsay *i.* U.K.
Orontes *r.* see 'Āṣī, Nahr al
22 E4 Oropa Italy
46 E1 Oroqen Zizhiqi China
55 B4 Oroquieta Phil.
101 L5 Orós, Açude *resr* Brazil
26 C4 Orosei *Sardinia* Italy
26 C4 Orosei, Golfo di *b.* *Sardinia* Italy
17 K7 Orosháza Hungary
21 M5 Oroszlány Hungary
93 G5 Oro Valley U.S.A.
92 B2 Oroville *CA* U.S.A.
82 C1 Oroville *WA* U.S.A.
92 B2 Oroville, Lake U.S.A.
68 B2 Orpheus I. Austr.
66 C4 Orroroo Austr.
9 Q6 Orsa Sweden
15 C4 Orsay France
28 D4 Orsha Belarus
22 D4 Orsiera, Monte *mt* Italy
38 D2 Orsk Rus. Fed.
9 K5 Ørsta Norway
22 F4 Orta, Lago d' *l.* Italy
25 C1 Ortegal, Cabo *c.* Spain
15 H4 Ortenau *reg.* Ger.
20 C2 Ortenberg Ger.
20 G4 Ortenburg Ger.
24 D5 Orthez France
25 C1 Ortigueira Spain
96 B1 Ortiz Mex.
103 D2 Ortiz Venez.
23 H3 Ortles *mt* Italy
12 E3 Orton U.K.
26 F3 Ortona Italy
84 D2 Ortonville U.S.A.
19 L5 Ortrand Ger.
33 O3 Orulgan, Khrebet *mts* Rus. Fed.
112 B1 Orumbo Namibia
40 B2 Orūmīyeh Iran
40 B2 Orūmīyeh, Daryācheh-ye *salt l.* Iran
100 E7 Oruro Bol.
26 E3 Orvieto Italy
116 B3 Orville Coast Ant.
90 C4 Orwell *OH* U.S.A.
91 G3 Orwell *VT* U.S.A.
23 G4 Orzinuovi Italy
9 M5 Os Norway
88 A4 Osage U.S.A.
84 E4 Osage *r.* U.S.A.
53 D7 Ōsaka Japan
39 H2 Osakarovka Kazak.
96 J6 Osa, Pen. de Costa Rica
19 G1 Øsby Denmark
9 N8 Osby Sweden
85 F5 Osceola *AR* U.S.A.
84 E3 Osceola *IA* U.S.A.
15 L5 Oschatz Ger.
19 J4 Oschersleben (Bode) Ger.
26 C4 Oschiri *Sardinia* Italy
89 F3 Oscoda U.S.A.
28 F4 Osetr *r.* Rus. Fed.
53 A8 Ōse-zaki *pt* Japan
89 K3 Osgoode Can.
39 H4 Osh Kyrg.
111 B5 Oshakati Namibia
52 G3 Oshamanbe Japan
89 H4 Oshawa Can.
52 G5 Oshika-hantō *pen.* Japan
53 F7 Ō-shima *i.* Japan
53 F6 Ō-shima *i.* Japan
88 C3 Oshkosh *NE* U.S.A.
88 C3 Oshkosh *WV* U.S.A.
40 B2 Oshnovīyeh Iran
114 D3 Oshogbo Nigeria
37 M5 Oshtorān Kūh *mt* Iran
37 M5 Oshtorīnān Iran
110 B4 Oshwe Congo(Zaire)
27 H2 Osijek Croatia
23 L6 Osimo Italy
44 C4 Osiyān India
113 J3 Osizweni S. Africa
26 G2 Osječenica *mt* Bos.-Herz.
8 O5 Ösjön *l.* Sweden
84 E3 Oskaloosa U.S.A.
9 P8 Oskarshamn Sweden
89 K1 Oskélanéo Can.
29 F5 Oskol *r.* Rus. Fed.
21 K3 Oslava *r.* Czech Rep.
9 M7 Oslo Norway
55 B4 Oslob Phil.
9 M7 Oslofjorden *chan.* Norway
43 B2 Osmānābād India
36 E1 Osmancık Turkey
36 B1 Osmaneli Turkey
36 F3 Osmaniye Turkey
9 V7 Os'mino Rus. Fed.
18 F4 Osnabrück Ger.
27 K3 Osogovske Planine *mts* Bulg./Macedonia
105 B4 Osorno Chile
25 D1 Osorno Spain
105 B4 Osorno, Vol. *volc.* Chile
76 F5 Osoyoos Can.
9 J6 Osøyri Norway
65 J2 Osprey Reef Coral Sea Is Terr.
18 C5 Oss Neth.
67 G8 Ossa, Mt Austr.
114 E3 Osse *r.* Nigeria
88 B3 Osseo U.S.A.
21 G6 Ossiach Austria
21 G6 Ossiacher See *l.* Austria
89 F3 Ossineke U.S.A.
91 H3 Ossipee Lake U.S.A.
19 J5 Oßmannstedt Ger.
79 H3 Ossokmanuan Lake Can.
28 E3 Ostashkov Rus. Fed.
18 E4 Ostbevern Ger.
19 G3 Oste *r.* Ger.
15 C1 Ostend Belgium
19 G2 Ostenfeld (Husum) Ger.
17 P5 Oster Ukr.
21 G4 Osterbach *r.* Ger.
19 J4 Osterburg (Altmark) Ger.
20 C3 Osterburken Ger.
9 O8 Österbymo Sweden
9 N6 Österdalälven *l.* Sweden
9 M5 Østerdalen *v.* Norway
19 J5 Osterfeld Ger.
18 F2 Osterhever Ger.
20 G4 Osterhofen Ger.
18 F3 Osterholz-Scharmbeck Ger.
22 E3 Ostermundigen Switz.
19 H5 Osterode am Harz Ger.
8 O5 Östersund Sweden
19 H5 Osterwieck Ger.
Ostfriesische Inseln *is* see East Frisian Islands
18 E3 Ostfriesland *reg.* Ger.
9 Q6 Östhammar Sweden
23 J4 Ostiglia Italy
21 M3 Ostrava Czech Rep.
17 J4 Ostróda Pol.
29 F5 Ostrogozhsk Rus. Fed.
21 J2 Ostroměř Czech Rep.
20 F2 Ostrov Czech Rep.
28 D3 Ostrov Rus. Fed.
17 K5 Ostrowiec Świętokrzyski Pol.
17 K4 Ostrów Mazowiecka Pol.
16 H5 Ostrów Wielkopolski Pol.
19 L2 Ostseebad Binz Ger.
19 K2 Ostseebad Dierhagen Ger.
19 L2 Ostseebad Göhren Ger.
19 K2 Ostseebad Graal-Müritz Ger.
19 J2 Ostseebad Kühlungsborn Ger.
19 K2 Ostseebad Prerow am Darß Ger.
19 J2 Ostseebad Rerik Ger.
19 K2 Ostseebad Wustrow Ger.
20 F5 Osttirol *reg.* Austria
27 L3 Osŭm *r.* Bulg.
53 B9 Ōsumi-Kaikyō *chan.* Japan
53 B9 Ōsumi-shotō *is* Japan
25 D4 Osuna Spain
91 F2 Oswegatchie U.S.A.
88 C5 Oswego *IL* U.S.A.
90 E3 Oswego *NY* U.S.A.
91 E3 Oswego *r.* U.S.A.
13 D5 Oswestry U.K.
53 F6 Ota Japan
70 C6 Otago Peninsula N.Z.
70 E4 Otaki N.Z.
8 U4 Otanmäki Fin.
39 H4 Otar Kazak.
103 B4 Otare, Co *h.* Col.
52 G3 Otaru Japan
70 B7 Otatara N.Z.
21 G3 Otava *r.* Czech Rep.
100 C3 Otavalo Ecuador
111 B5 Otavi Namibia
53 G6 Ōtawara Japan
70 C6 Otematata N.Z.
9 U7 Otepää Estonia
15 D4 Othe, Forêt d' *forest* France
82 C2 Othello U.S.A.
114 D3 Oti *r.* Ghana/Togo
96 C2 Otinapa Mex.
70 C5 Otira N.Z.
91 E3 Otisco Lake U.S.A.
79 F3 Otish, Monts *mts* Can.
111 B6 Otjiwarongo Namibia
12 F4 Otley U.K.
21 L2 Otmuchów Pol.
50 C2 Otog Qi China
52 H2 Otoineppu Japan
70 E3 Otorohanga N.Z.
78 C3 Otoskwin *r.* Can.
38 B3 Otpan, Gora *h.* Kazak.
38 B1 Otradnyy Rus. Fed.
27 H4 Otranto Italy
27 H4 Otranto, Strait of Albania/Italy
21 L3 Otrokovice Czech Rep.
29 H5 Otrozhnyy Rus. Fed.
88 E4 Otsego U.S.A.
88 E3 Otsego Lake *MI* U.S.A.
91 F3 Otsego Lake *NY* U.S.A.
91 F3 Otselic U.S.A.
53 D7 Ōtsu Japan
9 L6 Otta Norway
89 K3 Ottawa Can.
89 H2 Ottawa *r.* Can.
88 C5 Ottawa *IL* U.S.A.
84 E4 Ottawa *KS* U.S.A.
90 A4 Ottawa *OH* U.S.A.
78 D2 Ottawa Islands Can.
21 H4 Ottensheim Austria
21 J4 Ottenstein Stausee *resr* Austria
12 E2 Otterburn U.K.
93 G2 Otter Creek Reservoir U.S.A.
88 D1 Otter I. Can.
78 D3 Otter Rapids Can.
19 G3 Ottersberg Ger.
13 C7 Ottery *r.* U.K.
15 E2 Ottignies Belgium
20 E4 Ottobrunn Ger.
75 K1 Otto Fjord *in.* Can.
39 H4 Ottuk Kyrg.
88 A5 Ottumwa U.S.A.
15 H3 Ottweiler Ger.
114 E3 Otukpa Nigeria
114 E3 Oturkpo Nigeria
102 D3 Otumpa Arg.
71 ▫12 'Otu Tolu Group *is* Tonga
100 C5 Otuzco Peru
66 E7 Otway, C. Austr.
20 D6 Ötztaler Alpen *mts* Austria
85 E5 Ouachita *r.* U.S.A.
85 E5 Ouachita, L. U.S.A.
85 E5 Ouachita Mts U.S.A.
110 C3 Ouadda C.A.R.
109 E3 Ouaddaï *reg.* Chad
114 C2 Ouagadougou Burkina
114 C2 Ouahigouya Burkina
108 B3 Oualâta Maur.
114 D2 Oualé *r.* Burkina
114 D2 Ouallam Niger
115 A1 Ouando C.A.R.
108 D2 Ouanoum *reg.* Maur.
108 C1 Ouargla Alg.
114 B4 Ouarville France
108 B1 Ouarzazate Morocco
112 F6 Oubergpas *pass* S. Africa
22 B2 Ouche *r.* France
15 D2 Oudenaarde Belgium
18 E2 Oude Pekela Neth.
112 E6 Oudtshoorn S. Africa
25 F5 Oued Tlélat Alg.
108 B1 Oued Zem Morocco
26 B6 Oued Zénati Alg.
71 ▫9 Ouegoa *New Caledonia* Pac. Oc.
114 D3 Ouémé *r.* Benin
71 ▫9 Ouen *i.* *New Caledonia* Pac. Oc.
24 B2 Ouessant, Île d' *i.* France
110 B3 Ouesso Congo
114 D3 Ouidah Benin
108 B2 Ouirieg Mex.
108 B1 Oujda Morocco
8 T4 Oulainen Fin.
25 G4 Ouled Farès Alg.
22 B4 Oullins France
8 T4 Oulu Fin.
8 U4 Oulujärvi *l.* Fin.
8 U4 Oulujoki *r.* Fin.
8 T4 Oulunsalo Fin.
22 D4 Oulx Italy
109 E3 Oum-Chalouba Chad
108 B3 Oumé Côte d'Ivoire
109 D3 Oum-Hadjer Chad
8 T3 Ounasjoki *r.* Fin.
13 G5 Oundle U.K.
109 E3 Ounianga Kébir Chad
15 F2 Oupeye Belgium
15 G3 Our *r.* Lux.
83 F4 Ouray *CO* U.S.A.
93 H1 Ouray *UT* U.S.A.
25 C1 Ourense Spain
101 K5 Ouricurí Brazil
104 C3 Ourinhos Brazil
101 H4 Ouro *r.* Brazil
104 D3 Ouro Preto Brazil
82 B3 Ouroux-sur-Saône France
15 F2 Ourthe *r.* Belgium
15 G3 Our, Vallée de l' *v.* Ger./Lux.
15 A3 Ourville-en-Caux France
13 H7 Ouse *r.* *Eng.* U.K.
12 G4 Ouse *r.* *Eng.* U.K.
79 G3 Outardes *r.* Can.
112 E6 Outeniekpas *pass* S. Africa
11 A2 Outer Hebrides *is* U.K.
88 B2 Outer I. U.S.A.
92 C5 Outer Santa Barbara Channel U.S.A.
111 B6 Outjo Namibia
74 H4 Outlook Can.
8 V5 Outokumpu Fin.
15 B2 Outreau France
11 ▫ Out Skerries *is* U.K.
71 ▫9 Ouvéa *i.* *New Caledonia* Pac. Oc.
51 D5 Ouyang Hai Sk. *resr* China
66 E5 Ouyen Austr.
13 G5 Ouzel *r.* U.K.
26 C4 Ovace, Pte d' *mt* *Corsica* France
37 G2 Ovacık Turkey
22 F5 Ovada Italy
71 ▫15 Ovalau *i.* Fiji
105 B1 Ovalle Chile
25 B2 Ovar Port.
105 D2 Oveja *mt* Arg.
67 G6 Ovens *r.* Austr.
18 E6 Overath Ger.
8 S3 Överkalix Sweden
69 B3 Overlander Roadhouse Austr.
93 E3 Overton U.S.A.
8 S3 Övertorneå Sweden
9 P8 Överum Sweden
18 B4 Overveen Neth.
88 E4 Ovid U.S.A.
25 D1 Oviedo Spain
8 T2 Øvre Anarjåkka Nasjonalpark *nat. park* Norway
8 Q2 Øvre Dividal Nasjonalpark *nat. park* Norway
9 M6 Øvre Rendal Norway
29 D5 Ovruch Ukr.
70 B7 Owaka N.Z.
110 B4 Owando Congo
53 E7 Owase Japan
84 E2 Owatonna U.S.A.
41 F3 Owbeh Afgh.
91 E3 Owego U.S.A.
117 H3 Owen Fracture *sea feature* Ind. Ocean
14 B3 Owenmore *r.* Rep. of Ireland
70 D4 Owen River N.Z.
92 C3 Owens *r.* U.S.A.
86 C4 Owensboro U.S.A.
92 D3 Owens Lake U.S.A.
89 G3 Owen Sound Can.
89 G3 Owen Sound *in.* Can.
62 E2 Owen Stanley Range *mts* P.N.G.
114 E3 Owerri Nigeria
76 D4 Owikeno L. Can.
90 B5 Owingsville U.S.A.
91 J2 Owls Head U.S.A.
114 E3 Owo Nigeria
89 E4 Owosso U.S.A.
37 L4 Owrāmān, Kūh-e *mts* Iran/Iraq
92 C3 Owyhee U.S.A.
82 C3 Owyhee *r.* U.S.A.
82 C3 Owyhee Mts U.S.A.
100 C5 Oxapampa Peru
8 C4 Öxarfjörður *b.* Iceland
77 J5 Oxbow Can.
91 J1 Oxbow U.S.A.
9 P7 Oxelösund Sweden
70 D5 Oxford N.Z.
13 F6 Oxford U.K.
89 F4 Oxford *MI* U.S.A.
85 F5 Oxford *MS* U.S.A.
91 F3 Oxford *NY* U.S.A.
91 F5 Oxford *PA* U.S.A.
68 C3 Oxford Downs Austr.
77 K4 Oxford House Can.
77 K4 Oxford L. Can.
66 F5 Oxley Austr.
67 J3 Oxleys Pk Austr.
92 C4 Oxnard U.S.A.
89 H3 Oxtongue Lake Can.
8 N3 Øya Norway
53 F6 Oyama Japan
101 H3 Oyapock *r.* Brazil/Fr. Guiana
110 B3 Oyem Gabon
11 D3 Oykel *r.* U.K.
114 D3 Oyo Nigeria
22 C3 Oyonnax France
39 K3 Oyshilik Kazak.
67 H9 Oyster B. Austr.
45 H5 Oyster I. Myanmar
39 H4 Oy-Tal Kyrg.
19 G2 Oyten Ger.
114 D2 Ozalp Turkey
55 B4 Ozamiz Phil.
15 B4 Ozanne *r.* France
57 C6 Ozark *AL* U.S.A.
88 E2 Ozark *MI* U.S.A.
85 E4 Ozark Plateau *plat.* U.S.A.
84 E4 Ozarks, Lake of the U.S.A.
40 E3 Ozbagū Iran
29 G7 Ozerget'i Georgia
46 H1 Ozernovskiy Rus. Fed.
38 B2 Ozernoye Rus. Fed.
38 E1 Ozernyy Kazak.
28 E2 Ozernyy *Orenburg.* Rus. Fed.
28 E4 Ozernyy *Smolensk.* Rus. Fed.
17 L3 Ozersk Rus. Fed.
28 F4 Ozery Rus. Fed.
39 H4 Özgön Kyrg.
33 Q3 Ozhogino Rus. Fed.
26 C4 Ozieri *Sardinia* Italy
21 M2 Ozimek Pol.
38 B2 Ozinki Rus. Fed.
15 C4 Ozoir-le-Ferrière France
85 C6 Ozona U.S.A.
53 B7 Ozuki Japan

P

71 ▫9 Paama *i.* Vanuatu
75 O3 Paamiut Greenland
58 A1 Pa-an Myanmar
112 C6 Paarl S. Africa
112 D4 Paballelo S. Africa
54 E3 Pabal-ri N. Korea
11 A3 Pabbay *i.* *Scot.* U.K.
11 A4 Pabbay *i.* *Scot.* U.K.
17 J5 Pabianice Pol.
45 G4 Pabna Bangl.
9 T9 Pabradė Lith.
41 G5 Pab Range *mts* Pak.
101 F6 Pacaás Novos, Parque Nacional *nat. park* Brazil
103 E4 Pacaraima, Serra *mts* Brazil
100 C5 Pacasmayo Peru
83 E6 Pacheco *Chihuahua* Mex.
96 D2 Pacheco Mex.
28 H2 Pachikha Rus. Fed.
26 F6 Pachino *Sicily* Italy
43 B1 Pachmarhi India
44 D5 Pachore India
97 E3 Pachuca Mex.
92 B2 Pacific U.S.A.
119 L9 Pacific-Antarctic Ridge *sea feature* Pac. Oc.
55 C4 Pacijan *i.* Phil.
59 D4 Pacitan Indon.
21 H6 Pack Austria
21 J3 Pacov Czech Rep.
101 H4 Pacoval Brazil
104 D2 Pacuí *r.* Brazil
16 H5 Paczków Pol.
55 C5 Padada Phil.
103 D4 Padamo *r.* Venez.
59 B3 Padang Indon.
59 B3 Padangpanjang Indon.
59 A2 Padangsidimpuan Indon.
59 C3 Padangtikar *i.* Indon.
28 E2 Padany Rus. Fed.
37 M5 Padatha, Kūh-e *mt* Iran
103 D4 Padauiri *r.* Brazil
100 F8 Padcaya Bol.
76 F3 Paddle Prairie Can.
90 C5 Paden City U.S.A.
18 F5 Paderborn Ger.
27 K2 Padeşu, Vârful *mt* Romania
100 F7 Padilla Bol.
8 P3 Padjelanta Nationalpark *nat. park* Sweden
45 G5 Padma *r.* Bangl.
Padova see Padua
85 D7 Padre Island U.S.A.
26 C3 Padro, Monte *mt* *Corsica* France
13 C7 Padstow U.K.
17 N3 Padsvillye Belarus
66 D6 Padthaway Austr.
43 C2 Pādua India
23 J4 Padua Italy
86 B4 Paducah *KY* U.S.A.
85 C5 Paducah *TX* U.S.A.
44 D2 Padum Jammu and Kashmir
71 ▫3 Paea *Fr. Polynesia* Pac. Oc.
54 E3 Paegam N. Korea
54 C5 Paengnyŏng-do *i.* N. Korea
70 E2 Paeroa N.Z.
23 K4 Paese Italy
55 B3 Paete Phil.
36 D4 Pafos Cyprus
113 J1 Pafúri Moz.
26 F2 Pag Croatia
26 F2 Pag *i.* Croatia
55 B5 Pagadian Phil.
59 B3 Pagai Selatan *i.* Indon.
59 B3 Pagai Utara *i.* Indon.
57 L3 Pagan *i.* N. Mariana Is
23 J3 Paganella *mt* Italy
59 E3 Pagatan Indon.
93 G3 Page U.S.A.
9 R9 Pagėgiai Lith.
68 E2 Paget Cay *rf* Coral Sea Is Terr.
102 ▫ Paget, Mt Atl. Ocean
71 ▫5 Pago Pago *American Samoa* Pac. Oc.
83 F4 Pagosa Springs U.S.A.
45 G4 Pagri China
78 C3 Pagwa River Can.
92 ▫2 Pahala U.S.A.
44 B2 Paharpur Pak.
70 A7 Pahia Pt N.Z.
38 C5 Pahlavī Dezh Iran
92 ▫2 Pahoa U.S.A.
87 D7 Pahokee U.S.A.
41 F3 Pahra Kariz Afgh.
93 E3 Pahranagat Range *mts* U.S.A.
44 D4 Pahuj *r.* India
92 D3 Pahute Mesa *plat.* U.S.A.
58 A1 Pai Thai.
9 T7 Paide Estonia
13 D7 Paignton U.K.
9 T6 Päijänne *l.* Fin.
63 F5 Paiku Co *l.* China
58 B2 Pailin Cambodia
105 B4 Paillaco Chile
92 ▫1 Pailolo Chan. U.S.A.
9 S6 Paimio Fin.
105 B2 Paine Chile
90 C4 Painesville U.S.A.
93 G3 Painted Desert U.S.A.
93 F5 Painted Rock Reservoir U.S.A.
66 C3 Painter, Mount Austr.
77 K3 Paint Lake Provincial Recr. Park Can.
90 B6 Paintsville U.S.A.
89 G3 Paisley Can.
11 D5 Paisley U.K.
71 ▫9 Païta *New Caledonia* Pac. Oc.
100 B5 Paita Peru
55 A5 Paitan, Teluk *b.* Malaysia
51 D4 Paizhou China
8 S3 Pajala Sweden
101 L5 Pajeú *r.* Brazil
58 B4 Paka Malaysia
100 F2 Pakaraima Mountains Guyana
54 C4 Pakch'ŏn N. Korea
89 G3 Pakesley Can.
53 S3 Pakhacha Rus. Fed.
38 D2 Pakhar' Kazak.
39 H4 Pakhtaabad Uzbek.
114 E2 Paki Nigeria
31 F7 Pakistan *country* Asia
Paknampho see Muang Nakhon Sawan
56 B2 Pakokku Myanmar
70 D2 Pakotai N.Z.
44 C3 Pakpattan Pak.
58 B4 Pak Phayun Thai.
9 S9 Pakruojis Lith.
17 J7 Paks Hungary
41 H3 Paktīkā *reg.* Afgh.
58 C2 Pakxé Laos
109 D4 Pala Chad
58 A2 Pala Myanmar
59 C4 Palabuhanratu Indon.
59 C4 Palabuhanratu, Teluk *b.* Indon.
26 G3 Palagruža *i.* Croatia
27 K7 Palaiochora Greece
15 C4 Palaiseau France
Palakkat see Palghat
43 D1 Pāla Laharha India
112 E1 Palamakoloi Botswana
25 H2 Palamós Spain
44 C4 Palana India
33 R4 Palana Rus. Fed.
55 B2 Palanan Phil.
55 B2 Palanan Point Phil.
41 F4 Palangān, Kūh-e *mts* Iran
59 D3 Palangkaraya Indon.
43 B4 Palani India
44 C4 Palanpur India
41 G5 Palantak Pak.
55 C3 Palapag Phil.
111 C6 Palapye Botswana
43 B3 Palar *r.* India
45 G4 Palasbari India
33 R3 Palatka Rus. Fed.
31 N9 Palau *country* Pac. Oc.
55 B2 Palaui *i.* Phil.
55 A3 Palauig Phil.
58 A2 Palauk Myanmar
71 ▫4 Palauli B. Western Samoa
118 E5 Palau Tr. *sea feature* Pac. Oc.
58 A2 Palaw Myanmar
55 A4 Palawan *i.* Phil.
55 B3 Palayan Phil.
23 G4 Palazzolo sull'Oglio Italy
9 T7 Paldiski Estonia
45 H5 Pale Myanmar
59 B3 Palembang Indon.
102 B6 Palena Chile
25 D1 Palencia Spain
97 G4 Palenque Mex.
26 E5 Palermo *Sicily* Italy
85 E6 Palestine U.S.A.
45 H5 Paletwa Myanmar
43 B4 Palghat India
69 B1 Palgrave, Mt *h.* Austr.
44 C4 Pali India
61 G3 Palikir Micronesia
55 C5 Palimbang Phil.
26 F4 Palinuro, Capo *c.* Italy
93 H2 Palisade U.S.A.
15 F3 Paliseul Belgium
44 B5 Palitana India
9 S7 Palivere Estonia
43 B4 Palk Bay Sri Lanka
28 D3 Palkino Rus. Fed.
43 C2 Pālkohda India
43 B3 Palkonda Range *mts* India
43 B4 Palk Strait India/Sri Lanka
67 J2 Pallamallawa Austr.
14 C5 Pallas Green Rep. of Ireland
8 S2 Pallas-ja Ounastunturin Kansallispuisto *nat. park* Fin.
29 H5 Pallasovka Rus. Fed.
43 C2 Pallavaram India
43 B2 Palleru *r.* India
69 D7 Pallinup *r.* Austr.
70 E4 Palliser Bay N.Z.
70 E4 Palliser, Cape N.Z.
71 ▫2 Palliser, Îles *is* *Fr. Polynesia* Pac. Oc.
44 C3 Pallu India
25 D4 Palma del Río Spain
25 H3 Palma de Mallorca Spain
108 A2 Palma, La *i.* Canary Is
23 L4 Palmanova Italy
103 B2 Palmar *r.* Venez.
103 C3 Palmarito Venez.
114 B4 Palmas, Cape Liberia
104 D1 Palmas de Monte Alto Brazil
87 D7 Palm Bay U.S.A.
87 D7 Palm Beach U.S.A.
92 C4 Palmdale U.S.A.
104 B4 Palmeira Brazil
101 L5 Palmeira dos Índios Brazil
101 K5 Palmeirais Brazil
116 B2 Palmer *U.S.A. Base* Ant.
74 D3 Palmer *AK* U.S.A.
116 B2 Palmer Land *reg.* Ant.
70 C6 Palmerston N.Z.
68 C3 Palmerston, C. *pt* Austr.
60 L5 Palmerston Island *Cook Is* Pac. Oc.
70 E4 Palmerston North N.Z.
91 F4 Palmerton U.S.A.
87 E7 Palmetto Pt Bahamas
26 F5 Palmi Italy
97 E3 Palmillas Mex.
103 A4 Palmira Col.
96 C3 Palmito del Verde, Isla *i.* Mex.
92 D5 Palm Springs U.S.A.
68 C5 Palm Tree Cr. *r.* Austr.
Palmyra see Tadmur
88 B6 Palmyra *MO* U.S.A.
90 E3 Palmyra *NY* U.S.A.
88 C4 Palmyra *WV* U.S.A.
60 L3 Palmyra I. Pac. Oc.
45 F5 Palmyras Point India
92 A3 Palo Alto U.S.A.
103 A3 Palo de las Letras Col.
109 F3 Paloich Sudan
8 S2 Palojärvi Fin.
8 U2 Palomaa Fin.
97 F4 Palomares Mex.
92 D5 Palomar Mt U.S.A.
93 G6 Palominas U.S.A.
43 C2 Paloncha India
57 G7 Palopo Indon.
25 F4 Palos, Cabo de *c.* Spain
93 F5 Palo Verde *AZ* U.S.A.
93 E5 Palo Verde *CA* U.S.A.
8 U4 Paltamo Fin.
57 F7 Palu Indon.
37 G2 Palu Turkey
55 B3 Paluan Phil.
15 A3 Paluel France
38 F5 Pal'vart Turkm.
44 D3 Palwal India
33 T3 Palyavaam *r.* Rus. Fed.
21 L4 Pama Austria
43 B4 Pamban Channel India
47 H6 Pambula Austr.
56 ▫ Pamekasan Indon.
59 C4 Pameungpeuk Indon.
21 K5 Pamhagen Austria
43 B3 Pamidi India
24 E5 Pamiers France
39 H5 Pamir *r.* Afgh./Tajik.
39 H5 Pamir *mts* Asia
87 E5 Pamlico Sound *chan.* U.S.A.
85 C5 Pampa U.S.A.
105 C2 Pampa de la Salinas *salt pan* Arg.
100 F7 Pampa Grande Bol.
105 D2 Pampas *reg.* Arg.
103 B3 Pamplona Col.
55 B4 Pamplona Phil.
25 F1 Pamplona Spain
19 J3 Pampow Ger.
36 C1 Pamukova Turkey
90 E6 Pamunkey *r.* U.S.A.
44 D2 Pamzal Jammu and Kashmir
86 B4 Pana U.S.A.
97 G3 Panabá Mex.
55 C5 Panabo Phil.
93 E3 Panaca U.S.A.
55 A4 Panagtaran Point Phil.
59 C4 Panaitan *i.* Indon.
43 A3 Panaji India
96 J7 Panamá Panama
73 H9 Panama *country* S. America
96 K6 Panamá, Bahía de *b.* Panama
96 K6 Panama Canal Panama
87 C6 Panama City U.S.A.
96 K7 Panamá, Golfo de *g.* Panama
92 D3 Panamint Range *mts* U.S.A.
92 D3 Panamint Springs U.S.A.
92 D3 Panamint Valley U.S.A.
55 C4 Panaon *i.* Phil.
45 G4 Panar *r.* India
26 F5 Panarea, Isola *i.* Italy
23 H5 Panaro *r.* Italy
55 C3 Panay *i.* Phil.
55 B4 Panay *i.* Phil.
55 B4 Panay Gulf Phil.
93 E2 Pancake Range *mts* U.S.A.
27 J2 Pančevo Yugo.
55 B4 Pandan Phil.
55 C3 Pandan Phil.
55 B4 Pandan B. Phil.
58 ▫ Pandan Resr Sing.
44 E5 Pandaria India
104 D1 Pandeiros *r.* Brazil
43 A2 Pandharpur India
44 D5 Pandhurna India
105 F2 Pando Uru.
13 E6 Pandy U.K.
9 T9 Panevėžys Lith.
71 ▫12 Pangai Tonga
71 ▫10 Pangaimotu *i.* Tonga
56 ▫ Pangandaran Indon.
115 C4 Pangani Tanz.
115 B4 Pangani *r.* Tanz.
44 C2 Pangi Range *mts* Pak.
59 D3 Pangkalanbuun Indon.
59 A2 Pangkalansusu Indon.
59 C3 Pangkalpinang Indon.
57 G7 Pangkalsiang, Tanjung *pt* Indon.
55 B4 Panglao *i.* Phil.
75 M3 Pangnirtung Can.
32 J3 Pangody Rus. Fed.
44 D2 Pangong Tso *l.* India
56 ▫ Pangrango *volc.* Indon.
105 B3 Panguipulli Chile
105 B3 Panguipulli, L. Chile
93 F3 Panguitch U.S.A.
58 A5 Panguruan Indon.
55 B5 Pangutaran *i.* Phil.
55 B5 Pangutaran Group *is* Phil.
85 C5 Panhandle U.S.A.
115 A4 Pania-Mwanga Congo(Zaire)
71 ▫9 Panié, Mt *New Caledonia* Pac. Oc.
44 B5 Pānikoita *i.* India
29 G5 Panino Rus. Fed.
44 D3 Panipat India
55 A4 Panitan Phil.
39 G5 Panj Tajik.
41 G3 Panjāb Afgh.
39 F5 Panjakent Tajik.
58 D5 Panjang *i.* Indon.
37 L5 Panjbarār Iran
41 G5 Panjgur Pak.
44 C5 Panjhra *r.* India
Panjim see Panaji
44 B2 Panjkora *r.* Pak.
44 B3 Panjnad *r.* Pak.
8 W5 Pankakoski Fin.
114 E3 Pankshin Nigeria
54 F2 Pan Ling *mts* China
48 D7 Panna India
44 D4 Panna *reg.* India
64 C4 Pannawonica Austr.
21 L5 Pannonhalmo Hungary
104 B3 Panorama Brazil
43 B4 Panruti India
54 B3 Panshan China
54 D2 Panshi China
104 A2 Pantanal de São Lourenço *marsh* Brazil
104 A2 Pantanal do Taquari *marsh* Brazil
101 G7 Pantanal Matogrossense, Parque Nacional do *nat. park* Brazil
26 D6 Pantelleria *Sicily* Italy
26 E6 Pantelleria, Isola di *i.* *Sicily* Italy
57 G8 Pantemakassar Indon.
55 C5 Pantukan Phil.
97 E3 Pánuco Mex.
97 E3 Pánuco *r.* Mex.
43 A2 Panvel India
51 B5 Pan Xian China
51 D6 Panyu China
110 B4 Panzi Congo(Zaire)
97 G5 Panzos Guatemala
103 E3 Pao *r.* Venez.
26 G5 Paola Italy
86 C4 Paoli U.S.A.
71 ▫3 Paopao *Fr. Polynesia* Pac. Oc.
110 B3 Paoua C.A.R.
39 G4 Pap Uzbek.
16 H7 Pápa Hungary
70 E2 Papakura N.Z.
26 F4 Papa, Monte del *mt* Italy
97 E3 Papantla Mex.
71 ▫3 Papara *Fr. Polynesia* Pac. Oc.
43 C2 Pāparhāhandi India
70 C5 Paparoa Range *mts* N.Z.
11 ▫ Papa Stour *i.* U.K.
21 L5 Pápateszér Hungary
70 E2 Papatoetoe N.Z.
70 B7 Papatowai N.Z.
71 ▫3 Papava *Fr. Polynesia* Pac. Oc.
11 F1 Papa Westray *i.* U.K.
71 ▫3 Papeete *Fr. Polynesia* Pac. Oc.
18 E3 Papenburg Ger.
71 ▫3 Papenoo *Fr. Polynesia* Pac. Oc.
71 ▫3 Papenoo *r.* *Fr. Polynesia* Pac. Oc.
71 ▫3 Papetoai *Fr. Polynesia* Pac. Oc.
80 E6 Papigochic *r.* Mex.
89 K2 Papineau-Labelle, Réserve faunique de *res.* Can.
93 E3 Papoose L. U.S.A.
11 B5 Paps of Jura *h.* U.K.
14 B5 Paps, The *h.* Rep. of Ireland
62 E2 Papua, Gulf of *g.* P.N.G.
61 F4 Papua New Guinea *country* Oceania
48 G8 Papun Myanmar
13 C7 Par U.K.
104 D2 Pará *r.* Brazil
101 J4 Pará *r.* Brazil
28 G4 Para *r.* Rus. Fed.
69 C1 Paraburdoo Austr.
55 B3 Paracale Phil.
104 C2 Paracatu *Minas Gerais* Brazil
104 D2 Paracatu *r.* Brazil
66 C3 Parachilna Austr.
27 J3 Paraćin Yugo.
104 D2 Pará de Minas Brazil
89 J1 Paradis Can.
92 B2 Paradise *CA* U.S.A.
88 E2 Paradise *MI* U.S.A.
77 H4 Paradise Hill Can.
92 D2 Paradise Peak U.S.A.
79 J3 Paradise River Can.
85 F4 Paragould U.S.A.
100 F6 Paraguá *r.* Bol.
103 E3 Paragua *r.* Venez.
101 G7 Paraguai *r.* Brazil
103 C2 Paraguáipoa Venez.
103 C1 Paraguaná, Pen. de Venez.
102 E3 Paraguay *r.* Arg./Para.
99 E5 Paraguay *country* S. America
103 E3 Paraíba *r.* Brazil
104 D3 Paraíba do Sul *r.* Brazil
97 F4 Paraíso Mex.
114 D3 Parakou Benin
66 B3 Parakylia Austr.
43 D2 Parālakhemundi India
44 E6 Paralkot India
43 B4 Paramakkudi India
101 G2 Paramaribo Suriname
103 B3 Paramillo *mt* Col.
103 A3 Paramillo, Parque Nacional *nat. park* Col.
104 D1 Paramirim Brazil
103 A3 Paramo Frontino *mt* Col.
91 F4 Paramus U.S.A.
46 H1 Paramushir, O. *i.* Rus. Fed.
105 E1 Paraná Arg.
101 J6 Paranã Brazil
104 B4 Paraná *div.* Brazil
104 C1 Paraná *r.* Brazil
105 E2 Paraná *r.* S. America
104 C4 Paranaguá Brazil
104 B2 Paranaíba Brazil
104 B2 Paranaíba *r.* Brazil
105 E2 Paraná Ibicuy *r.* Arg.
104 B3 Paranapanema *r.* Brazil
104 C4 Paranapiacaba, Serra *mts* Brazil
104 C1 Paranã, Sa do *h.* Brazil
104 B3 Paranavaí Brazil
55 B5 Parang Phil.
43 B4 Parangipettai India
27 K2 Parângul Mare, Vârful *mt* Romania
44 C5 Parantij India
71 ▫2 Paraoa *i.* *Fr. Polynesia* Pac. Oc.
104 D2 Paraopeba *r.* Brazil
37 K4 Pārāpāra Iraq
70 E4 Paraparaumu N.Z.
103 D3 Paraque, Co *mt* Venez.
97 E2 Paras Mex.
44 D5 Paratwada India
104 B2 Paraúna Brazil
22 B3 Paray-le-Monial France
44 D4 Parbati *r.* India
43 B2 Parbhani India
39 F1 Parchevka Kazak.
19 J3 Parchim Ger.
22 J6 Parco Nazionale del Monte *nat. park* Italy
88 C4 Pardeeville U.S.A.
45 G2 Parding China
104 E1 Pardo *r.* *Bahia/Minas Gerais* Brazil
104 B3 Pardo *r.* *Mato Grosso do Sul* Brazil
104 D1 Pardo *r.* *Minas Gerais* Brazil
104 C3 Pardo *r.* *São Paulo* Brazil
21 J2 Pardubice Czech Rep.
56 ▫ Pare Indon.
44 D2 Pare Chu *r.* China
45 G4 Pare Chu *r.* India
100 F6 Parecis, Serra dos *h.* Brazil
96 D2 Paredón Mex.
70 D1 Parengarenga Harbour *in.* N.Z.
78 E4 Parent, Lac *l.* Can.
70 C6 Pareora N.Z.
59 E3 Parepare Indon.
105 D2 Parera Arg.
28 G3 Parfen'yevo Rus. Fed.
17 P2 Parfino Rus. Fed.
27 J5 Parga Greece
9 S6 Pargas Fin.
103 D2 Pariaguán Venez.
103 E2 Paria, Gulf of *g.* Trinidad/Venez.
103 E2 Paria, Península de *pen.* Venez.
93 F3 Paria Plateau *plat.* U.S.A.
9 V6 Parikkala Fin.
Parima *r.* see Uatatás
103 D4 Parima, Serra *mts* Brazil
103 D4 Parima-Tapirapecó, Parque Nacional *nat. park* Venez.
100 B4 Pariñas, Pta *pt* Peru
66 D5 Paringa Austr.
101 G4 Parintins Brazil
89 G4 Paris Can.
15 C4 Paris France
90 A5 Paris *KY* U.S.A.
87 B4 Paris *TN* U.S.A.
85 E5 Paris *TX* U.S.A.
88 E2 Parisienne, Île *i.* Can.
96 J7 Parita Panama
40 D4 Pārīz Iran
81 E1 Parizh Rus. Fed.
14 D3 Park U.K.
41 E5 Parkā Bandar Iran
43 B2 Parkal India
9 S5 Parkano Fin.
79 J3 Parke Lake Can.
39 G4 Parkent Uzbek.
93 E4 Parker U.S.A.
93 E4 Parker Dam *dam* U.S.A.
68 F5 Parker Hill Austr.
77 K2 Parker Lake Can.
51 ▫ Parker, Mt *h.* *H.K.* China
69 D5 Parker Range *h.* Austr.
88 A4 Parkersburg *IA* U.S.A.
90 C5 Parkersburg *WV* U.S.A.
67 H4 Parkes Austr.
88 B3 Park Falls U.S.A.
88 D5 Park Forest U.S.A.
89 F2 Parkinson Can.
84 E2 Park Rapids U.S.A.
20 F4 Parkstetten Ger.
76 E5 Parksville Can.
91 F4 Parksville U.S.A.
43 D2 Parla Kimedi India
40 B2 Parli Vaijnath India
23 H5 Parma Italy
23 H5 Parma *r.* Italy
82 C3 Parma *ID* U.S.A.
90 C4 Parma *OH* U.S.A.
103 D3 Parmana Venez.
101 K4 Parnaíba Brazil
101 K4 Parnaíba *r.* Brazil
27 K5 Parnassos *mt* Greece
70 D5 Parnassus N.Z.
66 B5 Parndana Austr.
21 K5 Parndorf Austria
88 A5 Parnell U.S.A.
27 K6 Parnon *mts* Greece
9 T7 Pärnu Estonia
9 T7 Pärnu-Jaagupi Estonia
15 D4 Paron France
66 F2 Paroo *watercourse* Austr.
41 F3 Paropamisus *mts* Afgh.
27 L6 Paros Greece
27 L6 Paros *i.* Greece
93 F3 Parowan U.S.A.
105 B3 Parral Chile
91 F6 Parramore I. U.S.A.
96 D2 Parras Mex.
105 F3 Parravicini Arg.
13 E6 Parrett *r.* U.K.
96 H6 Parrita Costa Rica
79 H4 Parrsboro Can.
74 F2 Parry, Cape *pt* Can.
74 G2 Parry Islands Can.
75 L2 Parry, Kap *c.* Greenland
89 G3 Parry Sound Can.
85 E4 Parsons *KS* U.S.A.
90 D5 Parsons *WV* U.S.A.
19 L4 Parsteiner See *l.* Ger.
24 D3 Parthenay France
52 C3 Partizansk Rus. Fed.
21 M4 Partizánske Slovakia
13 H4 Partney U.K.
14 B4 Partry Rep. of Ireland
14 B4 Partry Mts *h.* Rep. of Ireland
101 H4 Paru *r.* Brazil
103 D3 Parucito *r.* Venez.
43 C2 Parvatipuram India
44 C4 Parvatsar India
44 D4 Parwan *r.* India
45 E3 Paryang China
113 G3 Parys S. Africa
92 C4 Pasadena *CA* U.S.A.
85 E6 Pasadena *TX* U.S.A.
100 B4 Pasado, C. *pt* Ecuador
40 D4 Pasargadae Iran
59 B3 Pasarseblat Indon.
58 A1 Pasawng Myanmar
85 F6 Pascagoula U.S.A.
89 J1 Pascalis Can.
17 N7 Paşcani Romania
82 C2 Pasco U.S.A.
104 E2 Pascoal, Monte *h.* Brazil
Pascua, Isla de *i.* see Easter I.
55 B3 Pasculai Phil.
Pas de Calais *str.* see Dover, Strait of
19 M3 Pasewalk Ger.
77 H3 Pasfield Lake Can.
28 E2 Pasha Rus. Fed.
55 B3 Pasig Phil.
37 H2 Pasinler Turkey
58 ▫ Pasir Gudang Malaysia
58 ▫ Pasir Panjang Sing.
59 B1 Pasir Putih Malaysia
92 A2 Paskenta U.S.A.
38 E3 Paskevicha, Zaliv *l.* Kazak.
41 F5 Paskūh Iran
41 F5 Pasni Pak.
97 G4 Paso Caballos Guatemala
103 B4 Paso de las Cruces *mt* Col.
105 F2 Paso de los Toros Uru.
102 B7 Paso Río Mayo Arg.
92 B4 Paso Robles U.S.A.
77 J4 Pasquia Hills Can.
40 D4 Pasrūdak Iran
91 J2 Passadumkeag U.S.A.
88 C1 Passage I. U.S.A.
21 G4 Passau Ger.
55 B4 Passi Phil.
23 J3 Passirio *r.* Italy
102 F3 Passo Fundo Brazil
104 C3 Passos Brazil
19 M3 Passow Ger.
28 C4 Pastavy Belarus
100 C4 Pastaza *r.* Peru
77 J4 Pas, The Can.
103 A4 Pasto Col.
93 H3 Pastora Peak U.S.A.
55 B2 Pasuquin Phil.
59 D4 Pasuruan Indon.

105 B4 Purranque Chile
96 D3 Puruandiro Mex.
45 F5 Puruliya India
100 F4 Purus r. Brazil
9 V6 Puruvesi l. Fin.
59 C4 Purwakarta Indon.
59 D4 Purwodadi Indon.
59 C4 Purwokerto Indon.
54 E2 Puryŏng N. Korea
44 D6 Pus r. India
44 D6 Pusad India
54 E6 Pusan S. Korea
91 J2 Pushaw Lake U.S.A.
28 H2 Pushemskiy Rus. Fed.
44 C4 Pushkar India
28 D3 Pushkin Rus. Fed.
29 H5 Pushkino Rus. Fed.
28 D3 Pushkinskiye Gory Rus. Fed.
41 F4 Pusht-i-Rud reg. Afgh.
23 G4 Pusiano, Lago di l. Italy
23 J3 Pusteria, Val v. Italy
17 O2 Pustoshka Rus. Fed.
17 L4 Puszcza Augustowska forest Pol.
16 G4 Puszcza Natecka forest Pol.
46 B4 Putao Myanmar
19 L2 Putbus Ger.
51 F5 Putian China
59 D3 Puting, Tanjung pt Indon.
39 L2 Putintsevo Kazak.
97 E4 Putla Mex.
41 G4 Putla Khan Afgh.
19 K3 Putlitz Ger.
27 M2 Putna r. Romania
91 H4 Putnam U.S.A.
91 G3 Putney U.S.A.
45 H3 Putrang La pass China
112 D4 Putsonderwater S. Africa
43 B4 Puttalam Sri Lanka
43 B4 Puttalam Lagoon Sri Lanka
15 G3 Puttelange-aux-Lacs France
18 C4 Putten Neth.
18 B5 Puttershoek Neth.
19 J2 Puttgarden Ger.
100 D4 Putumayo r. Col.
36 G2 Pütürge Turkey
59 D2 Putusibau Indon.
28 G4 Putyatino Rus. Fed.
29 E5 Putyvl' Ukr.
9 V6 Puumala Fin.
92 □2 Puuwai U.S.A.
78 E1 Puvurnituq Can.
82 B2 Puyallup U.S.A.
50 E3 Puyang China
105 B4 Puyehue Chile
105 B4 Puyehue, Parque Nacional nat. park Chile
24 F5 Puylaurens France
70 A7 Puysegur Pt N.Z.
115 C4 Pwani div. Tanz.
115 A4 Pweto Congo(Zaire)
13 C5 Pwllheli U.K.
28 E2 Pyal'ma Rus. Fed.
28 H4 P'yana r. Rus. Fed.
44 B1 Pyandzh r. Afgh./Tajik.
8 W3 Pyaozero, Ozero l. Rus. Fed.
8 W4 Pyaozerskiy Rus. Fed.
56 B3 Pyapon Myanmar
32 K2 Pyasina r. Rus. Fed.
29 G6 Pyatigorsk Rus. Fed.
38 B2 Pyatimarskoye Kazak.
29 E5 P"yatykhatky Ukr.
56 B3 Pyè Myanmar
70 B7 Pye, Mt h. N.Z.
29 D4 Pyetrykaw Belarus
8 T4 Pyhäjoki Fin.
8 T4 Pyhäjoki r. Fin.
8 U4 Pyhäntä Fin.
8 T5 Pyhäsalmi Fin.
8 V5 Pyhäselkä l. Fin.
45 H5 Pyingaing Myanmar
56 B3 Pyinmana Myanmar
13 E6 Pyle U.K.
32 K3 Pyl'karamo Rus. Fed.
27 J6 Pylos Greece
90 C4 Pymatuning Reservoir U.S.A.
54 C5 Pyŏksŏng N. Korea
54 C3 Pyŏktong N. Korea
54 D4 P'yŏnggang N. Korea
54 E5 P'yŏnghae S. Korea
54 C4 P'yŏngsong N. Korea
54 D5 P'yŏngt'aek S. Korea
54 C4 P'yŏngyang N. Korea
66 F6 Pyramid Hill Austr.
92 C2 Pyramid Lake U.S.A.
88 E3 Pyramid Pt U.S.A.
92 C2 Pyramid Range mts U.S.A.
6 C4 Pyrenees mts France/Spain
Pyrénées mts see Pyrenees
27 J6 Pyrgos Greece
29 E5 Pyryatyn Ukr.
16 G4 Pyrzyce Pol.
28 H3 Pyshchug Rus. Fed.
17 N2 Pytalovo Rus. Fed.
27 K5 Pyxaria mt Greece

Q

75 M2 Qaanaaq Greenland
40 D6 Qābil Oman
37 J6 Qabr Bandar Iraq
113 H5 Qacha's Nek Lesotho
37 K4 Qādir Karam Iraq
37 J4 Qadissiya Dam dam Iraq
50 C1 Qagan Ders China
54 C1 Qagan Nur l. Jilin China
50 E1 Qagan Nur l. Nei Monggol China
50 D1 Qagan Nur China
50 C2 Qagan Nur China
50 E1 Qagan Nur resr China
50 D1 Qagan Teg China
50 E1 Qagan Us China
45 H3 Qagbasêrag China
45 E2 Qagcaka China
75 O3 Qagssimiut Greenland
50 D1 Qahar Youyi Qianqi China
50 D1 Qahar Youyi Zhongqi China
46 B3 Qaidam Pendi basin China
41 G3 Qaisar Afgh.
41 G3 Qaisar, Koh-i- mt Afgh.
37 K3 Qaļā Diza Iraq
39 G5 Qal'aikhum Tajik.
41 G3 Qala Shinia Takht Afgh.
41 G3 Qalāt Afgh.
40 D4 Qalat Iran
37 H4 Qal'at as Sālihīyah Syria
36 E6 Qal'at el Hasal Jordan
37 L6 Qal'at Sālih Iraq
37 L6 Qal'at Sukkar Iraq
41 F3 Qala Vali Afgh.
40 B2 Qal'eh D. mt Iran
41 F3 Qal 'eh-ye Now Afgh.
41 G4 Qal 'eh-ye Bost Afgh.
37 M5 Qal'eh-ye-Now Iran
37 K7 Qalīb Bāqūr well Iraq
36 C6 Qalyūb Egypt
113 G5 Qamata S. Africa
44 B4 Qambar Pak.
48 G5 Qamdo China
71 □13 Qamea i. Fiji
44 B3 Qamruddin Karez Pak.
40 C3 Qamşar Iran
40 B2 Qandaranbashi mt Iran
50 E1 Qangdin Sum China
39 K4 Qapqal China
37 M2 Qaraçala Azer.
37 J4 Qarachōq, J. mts Iraq
37 K4 Qara D. r. Iraq
37 H7 Qārah S. Arabia
41 H3 Qarah Bāgh Afgh.
40 A5 Qa'rah, J. al h. S. Arabia
37 L3 Qaranqu r. Iran
110 E3 Qardho Somalia
37 L3 Qar'eh Aqāj Iran
37 L2 Qareh D. mts Iran
37 L3 Qareh Dāsh, Kūh-e mt Iran
40 B2 Qareh Sū r. Iran
37 L3 Qareh Urgān, Kūh-e mt Iran
45 H1 Qarhan China
39 H5 Qarokŭl l. Tajik.
41 G3 Qarqīn Afgh.
37 K6 Qaryat al Gharab Iraq
40 B5 Qaryat al Ulyā S. Arabia
40 E3 Qasamī Iran
41 F3 Qasa Murg mts Afgh.
40 C3 Qash Qai reg. Iran
75 N3 Qasigiannguit Greenland
37 J5 Qasr al Khubbāz Iraq
40 C4 Qaşr aş Şabīyah Kuwait
36 F6 Qasr el Azraq Jordan
41 F5 Qasr-e-Qand Iran
40 B3 Qasr-e-Shirin Iran
37 L6 Qaşr Shaqrah Iraq
36 F4 Qaţanā Syria
31 E7 Qatar country Asia
36 F6 Qatrāna Jordan
22 E7 Qaţrānī, Gebel esc. Egypt
109 E2 Qattâra Depression depression Egypt
37 L1 Qax Azer.
39 K4 Qaxi China
41 E3 Qāyen Iran
45 H3 Qayü China
37 J4 Qayyarah Iraq
37 L2 Qazangöldağ mt Azer.
37 K1 Qazax Azer.
44 B4 Qazi Ahmad Pak.
37 M1 Qazımämmäd Azer.
40 C2 Qazvīn Iran
50 A1 Qeh China
71 □13 Qeleni Fiji
109 F2 Qena Egypt
75 N3 Qeqertarsuatsiaat Greenland
75 N3 Qeqertarsuatsiaq i. Greenland
75 N3 Qeqertarsuup Tunua b. Greenland
40 B3 Qeshlag r. Iran
37 L4 Qeshlaq Iran
40 E5 Qeshm Iran
40 C2 Qeydār Iran
40 D5 Qeys i. Iran
40 C2 Qezel Owzan r. Iran
36 E6 Qeẕi'ot Israel
50 C3 Qian r. China
50 F2 Qian'an China
54 C1 Qian'an China
51 C5 Qiancheng China
39 K4 Qianfodong China
50 F3 Qiang r. China
54 C1 Qian Gorlos China
51 D4 Qianjiang Hubei China
51 C4 Qianjiang Sichuan China
54 E1 Qianjin China
50 A4 Qianning China
54 B1 Qianqihao China
53 B4 Qian Shan mts China
51 C5 Qianxi China
50 C3 Qian Xian China
51 D5 Qianyang Hunan China
50 C3 Qianyang Shaanxi China
51 F4 Qianyang Zhejiang China
50 D2 Qiaocun China
51 B5 Qiaojia China
37 K2 Qīās Iran
40 B5 Qibā' S. Arabia
113 G4 Qibing S. Africa
51 D5 Qidong Hunan China
50 F4 Qidong Jiangsu China
45 H2 Qidukou China
35 G3 Qiemo China
50 E2 Qihe China
51 C4 Qijiang China
32 L5 Qijiaojing China
41 G4 Qila Abdullah Pak.
41 F5 Qila Ladgasht Pak.
51 E2 Qilaotu Shan mts China
41 F4 Qila Safed Pak.
41 G4 Qila Saifullah Pak.
46 B3 Qilian Shan mts China
75 P3 Qillak i. Greenland
45 G1 Qimantag mts China
51 E4 Qimen China
50 D3 Qin r. China
50 D3 Qin'an China
54 C2 Qing r. China
54 B3 Qingchengzi China
50 F2 Qingdao China
50 A2 Qinghai div. China
50 A2 Qinghai Hu salt l. China
46 B3 Qinghai Nanshan mts China
52 A1 Qinghe China
54 C3 Qinghecheng China
50 D2 Qingjian China
50 F3 Qingjiang Jiangsu China
51 E4 Qingjiang Jiangxi China
51 D4 Qing Jiang r. China
51 E5 Qingliu China
51 B5 Qinglong Guizhou China
50 F1 Qinglong Hebei China
50 F2 Qinglong r. China
51 C6 Qingping China
51 F4 Qingpu China
50 C3 Qingshui China
50 D2 Qingshuihe China
39 K3 Qingshuihezi China
51 E2 Qing Xian China
50 D2 Qingxu China
51 E4 Qingyang Anhui China
50 C2 Qingyang Gansu China
51 D6 Qingyuan Guangdong China
51 F5 Qingyuan Zhejiang China
54 C2 Qingyuan China
51 C5 Qingzhen China
50 F2 Qinhuangdao China
50 C3 Qin Ling mts China
50 C3 Qintongxia China
50 D3 Qin Xian China
50 D3 Qinyang China
50 D2 Qinyuan China
51 C6 Qinzhou China
51 C6 Qinzhou Wan b. China
51 D7 Qionghai China
51 B4 Qionglai China
50 B4 Qionglai Shan mts China
51 D7 Qiongshan China
51 C6 Qiongzhou Haixia str. China
46 E2 Qiqihar China
40 D4 Qīr Iran
37 M5 Qīr Iran
44 E1 Qira China
36 E6 Qiryat Gat Israel
36 F6 Qitab ash Shāmah crater S. Arabia
52 B2 Qitaihe China
51 B5 Qiubei China
50 F2 Qixia China
50 E3 Qi Xian Henan China
50 D2 Qi Xian Shanxi China
51 E5 Qixing r. China
51 D5 Qiyang China
50 C2 Qiying China
51 D7 Qizhou Liedao i. China
37 M2 Qızılağac Körfäzi b. Azer.
39 H5 Qizilrabot Tajik.
Qogir Feng mt see K2
50 C1 Qog Qi China
40 B2 Qojūr Iran
40 C3 Qom Iran
45 H3 Qomdo China
40 C3 Qomisheh Iran
Qomolangma Feng mt see Everest, Mt
37 M1 Qonaqkänd Azer.
40 C3 Qonāq, Kūh-e h. Iran
45 G3 Qonggyai China
50 C1 Qongi China
36 F4 Qornet es Saouda mt Lebanon
40 B3 Qorveh Iran
40 E5 Qoţbābād Iran
37 K2 Qoţūr Iran
91 G3 Quabbin Reservoir U.S.A.
92 D4 Quail Mts U.S.A.
69 C6 Quairading Austr.
18 E4 Quakenbrück Ger.
91 F4 Quakertown U.S.A.
77 K2 Quamarirjung Lake Can.
66 E5 Quambatook Austr.
67 G3 Quambone Austr.
85 D5 Quanah U.S.A.
50 D3 Quanbao Shan mt China
58 D2 Quang Ngai Vietnam
58 C1 Quang Tri Vietnam
51 C6 Quang Yen Vietnam
51 E5 Quannan China
51 F5 Quanzhou Fujian China
51 D5 Quanzhou Guangxi China
77 J4 Qu'Appelle Can.
77 J4 Qu'Appelle r. Can.
105 F1 Quaraí Brazil
105 F1 Quaraí r. Brazil
24 E2 Quarré-les-Tombes France
51 Quarry Bay H.K. China
26 C5 Quartu Sant'Elena Sardinia Italy
93 D4 Quartzite Mt U.S.A.
93 E5 Quartzsite U.S.A.
76 D4 Quatsino Sound in. Can.
38 A4 Quba Azer.
41 E2 Quchan Iran
67 H5 Queanbeyan Austr.
79 F4 Québec Can.
75 L4 Quebec div. Can.
104 C2 Quebra Anzol r. Brazil
103 C2 Quebrada del Toro, Parque Nacional de la nat. park Venez.
105 B4 Quedal, C. hd Chile
19 J5 Quedlinburg Ger.
76 E4 Queen Bess, Mt Can.
76 C4 Queen Charlotte Can.
76 C4 Queen Charlotte Islands Can.
76 D4 Queen Charlotte Sound chan. Can.
76 D4 Queen Charlotte Str. Can.
75 H1 Queen Elizabeth Islands Can.
115 B2 Queen Elizabeth National Park Uganda
116 C5 Queen Mary Land reg. Ant.
74 H3 Queen Maud Gulf Can.
Queen Maud Land reg. see Dronning Maud Land
116 B4 Queen Maud Mts Ant.
65 H4 Queensland div. Austr.
67 F9 Queenstown Austr.
70 B6 Queenstown N.Z.
113 G5 Queenstown S. Africa
58 Queenstown Sing.
91 E5 Queenstown U.S.A.
82 A2 Queets U.S.A.
105 F2 Queguay Grande r. Uru.
105 D3 Quehué Arg.
101 H4 Queimada ou Serraria, Ilha i. Brazil
115 C6 Quelimane Moz.
102 B6 Quellón Chile
Quelpart Island see Cheju-do
93 H4 Quemado U.S.A.
105 B4 Quemchi Chile
105 D3 Quemú-Quemú Arg.
105 E3 Quequén Grande r. Arg.
104 B3 Querência do Norte Brazil
97 D3 Querétaro Mex.
97 E3 Querétaro div. Mex.
19 J5 Querfurt Ger.
50 E3 Queshan China
76 E4 Quesnel Can.
76 E4 Quesnel r. Can.
76 E4 Quesnel L. Can.
88 B1 Quetico Provincial Park Can.
44 A3 Quetta Pak.
105 B3 Queuco Chile
105 B3 Queule Chile
97 G5 Quezaltenango Guatemala
55 A4 Quezon Phil.
55 B3 Quezon City Phil.
50 E3 Qufu China
115 B5 Quibala Angola
111 B4 Quibaxe Angola
103 A3 Quibdó Col.
24 C3 Quiberon France
111 B4 Quiçama, Parque Nacional do nat. park Angola
58 C1 Qui Châu Vietnam
93 F5 Quijotoa U.S.A.
105 D1 Quilino Arg.
24 F5 Quillan France
77 J4 Quill Lakes Can.
105 B2 Quillota Chile
105 E2 Quilmes Arg.
43 B4 Quilon India
65 H5 Quilpie Austr.
105 B2 Quilpué Chile
115 C6 Quilua Moz.
110 B4 Quimbele Angola
102 D3 Quimilí Arg.
24 B3 Quimper France
24 C3 Quimperlé France
100 D6 Quince Mil Peru
22 E4 Quincinetto Italy
92 B2 Quincy CA U.S.A.
87 C6 Quincy FL U.S.A.
88 B6 Quincy IL U.S.A.
91 H3 Quincy MA U.S.A.
105 D2 Quines Arg.
115 D6 Quinga Moz.
22 C2 Quingey France
58 D2 Qui Nhơn Vietnam
93 E3 Quinn Canyon Range mts U.S.A.
25 E3 Quintanar de la Orden Spain
97 G4 Quintana Roo div. Mex.
105 B2 Quintero Chile
105 D2 Quinto r. Arg.
25 F2 Quinto Spain
23 H4 Quinzano d'Oglio Italy
115 D5 Quionga Moz.
111 B5 Quipungo Angola
105 B3 Quirihue Chile
111 B5 Quirima Angola
67 J3 Quirindi Austr.
105 E2 Quiroga Arg.
22 B6 Quissac France
115 D5 Quissanga Moz.
111 D6 Quissico Moz.
111 B5 Quitapa Angola
104 B2 Quitéria r. Brazil
87 D6 Quitman GA U.S.A.
87 B5 Quitman MS U.S.A.
100 C4 Quito Ecuador
83 D6 Quitovac Mex.
93 F4 Quivero U.S.A.
101 L4 Quixadá Brazil
51 D5 Qujiang China
51 C4 Qu Jiang r. China
51 D6 Qujie China
51 B5 Qujing China
37 L7 Qulbān Layyah well Iraq
39 G5 Qullai Garmo mt Tajik.
45 H2 Qumar He r. China
45 H2 Qumarlêb China
45 H2 Qumarrabdün China
45 H2 Qumaryan China
113 H5 Qumbu S. Africa
113 G6 Qumrha S. Africa
40 B6 Qunayy well S. Arabia
77 L2 Quoich r. Can.
11 C3 Quoich, Loch l. U.K.
14 F3 Quoile r. U.K.
112 C7 Quoin Pt S. Africa
45 E2 Quong Muztag mt China
65 G6 Quorn Austr.
112 F1 Quoxo r. Botswana
37 L3 Qurābeh Iran
41 E6 Qurayat Oman
39 G5 Qŭrghonteppa Tajik.
Qurlurtuuq see Coppermine
79 G2 Qurlutu r. Can.
37 K2 Qūrū Gol pass Iran
37 M1 Qusar Azer.
38 A4 Qusar Azer.
109 F2 Quseir Egypt
40 B2 Qūshchī Iran
37 L2 Qūsheh D. mts Iran
40 C3 Qūtīābād Iran
50 B2 Quwu Shan mts China
50 C4 Qu Xian China
45 G3 Qūxü China
58 C1 Quynh Luu Vietnam
51 B6 Quynh Nhai Vietnam
89 J3 Quyon Can.
50 E2 Quzhou Hebei China
51 F4 Quzhou Zhejiang China
29 H7 Qvareli Georgia
Qyteti Stalin see Kuçovë

R

21 J6 Raab r. Austria
8 T4 Raahe Fin.
8 V5 Rääkkylä Fin.
18 D4 Raalte Neth.
8 T3 Raanujärvi Fin.
59 D4 Raas i. Indon.
11 B3 Raasay i. U.K.
11 B3 Raasay, Sound of chan. U.K.
110 F2 Raas Caseyr c. Somalia
21 L5 Rába r. Hungary
59 E4 Raba Indon.
44 E2 Rabang China
39 G5 Rabat Kyrg.
26 F7 Rabat Malta
108 B1 Rabat Morocco
41 E3 Rabāt-e Kamah Iran
62 F2 Rabaul P.N.G.
23 J6 Rabbi r. Italy
76 D3 Rabbit r. Can.
21 L5 Rábca r. Hungary
71 □13 Rabi i. Fiji
42 A5 Rābigh S. Arabia
45 G5 Rabnabad Islands Bangl.
29 D6 Râbniţa Moldova
40 E4 Rābor Iran
22 E5 Racconigi Italy
90 B5 Raccoon Creek r. U.S.A.
79 K4 Race, C. Can.
91 H3 Race Pt U.S.A.
36 E5 Rachaïya Lebanon
85 D7 Rachal U.S.A.
93 E3 Rachel U.S.A.
58 C3 Rach Gia Vietnam
16 J5 Racibórz Pol.
88 D4 Racine U.S.A.
89 F1 Racine Lake Can.
88 E2 Raco U.S.A.
17 M7 Rădăuţi Romania
20 F3 Radbuza r. Czech Rep.
86 C4 Radcliff U.S.A.
19 L5 Radeberg Ger.
19 L5 Radebeul Ger.
19 L5 Radeburg Ger.
21 K6 Radenci Slovenia
21 G6 Radenthein Austria
20 E5 Radfeld Austria
90 C6 Radford U.S.A.
44 B5 Radhanpur India
38 A1 Radishchevo Rus. Fed.
78 E3 Radisson Can.
76 F4 Radium Hot Springs Can.
21 K2 Radków Pol.
21 J6 Radlje ob Dravi Slovenia
27 L3 Radnevo Bulg.
20 B5 Radolfzell am Bodensee Ger.
17 K5 Radom Pol.
27 K3 Radomir Bulg.
109 E4 Radom National Park Sudan
17 J5 Radomsko Pol.
29 D5 Radomyshl' Ukr.
27 K4 Radoviš Macedonia
13 E6 Radstock U.K.
28 C4 Radun' Belarus
9 S9 Radviliškis Lith.
17 M5 Radyvyliv Ukr.
44 E4 Rae Bareli India
76 F2 Rae-Edzo Can.
71 □2 Raeffsky, Îles is Fr. Polynesia Pac. Oc.
76 F2 Rae Lakes Can.
69 E4 Raeside, Lake salt flat Austr.
70 E3 Raetihi N.Z.
105 E1 Rafaela Arg.
36 E6 Rafaḩ Gaza
110 C3 Rafaï C.A.R.
42 B4 Rafḩā S. Arabia
40 E4 Rafsanjān Iran
55 C5 Ragang, Mt volc. Phil.
55 B3 Ragay Gulf Phil.
19 K3 Rägelin Ger.
91 J3 Ragged I. U.S.A.
40 B4 Raghwah S. Arabia
19 K4 Ragösen Ger.
19 K5 Raguhn Ger.
26 F6 Ragusa Sicily Italy
50 A3 Ra'gyagoinba China
57 G7 Raha Indon.
28 D4 Rahachow Belarus
Rahaeng see Tak
18 F4 Rahden Ger.
37 J5 Raḩḩālīyah Iraq
43 A2 Rahimatpur India
44 B3 Rahimyar Khan Pak.
40 C3 Rāhjerd Iran
105 B3 Rahue mt Chile
43 A2 Rahuri India
41 F3 Rahzanak Afgh.
71 □2 Raiatea i. Fr. Polynesia Pac. Oc.
43 B2 Raichur India
45 G4 Raiganj India
45 E5 Raigarh India
71 □10 Raihifahifa i. Tonga
93 E2 Railroad Valley v. U.S.A.
79 G3 Raimbault, Lac l. Can.
20 D4 Rain Ger.
66 E5 Rainbow Austr.
68 E5 Rainbow Beach Austr.
93 G3 Rainbow Bridge Nat. Mon. res. U.S.A.
76 F3 Rainbow Lake Can.
90 C6 Rainelle U.S.A.
82 B2 Rainier, Mt volc. U.S.A.
44 B3 Raini N. r. Pak.
78 B4 Rainy r. U.S.A.
75 J5 Rainy Lake Can.
77 L5 Rainy River Can.
45 E5 Raipur Madhya Pradesh India
44 C4 Raipur Rajasthan India
9 S6 Raisio Fin.
15 D2 Raismes France
60 N6 Raivavae i. Fr. Polynesia Pac. Oc.
43 C2 Rajahmundry India
8 V2 Raja-Jooseppi Fin.
43 B3 Rajampet India
59 D2 Rajang r. Malaysia
43 B4 Rajapalaiyam India
43 A2 Rajapur India
44 C4 Rajasthan div. India
44 C3 Rajasthan Canal India
45 F4 Rajauli India
45 G5 Rajbari Bangl.
21 M3 Rajec Slovakia
44 D4 Rajgarh Rajasthan India
44 C3 Rajgarh Rajasthan India
21 K3 Rajhradice Czech Rep.
36 F6 Rajil, W. watercourse Jordan
45 E5 Rajim India
44 B5 Rajkot India
45 F4 Rajmahal India
45 F4 Rajmahal Hills India
44 E5 Raj Nandgaon India
44 D3 Rajpura India
45 G4 Rajshahi Bangl.
43 B2 Rajura India
45 F3 Raka China
60 L5 Rakahanga i. Pac. Oc.
70 C5 Rakaia r. N.Z.
44 C1 Rakaposhi mt Pak.
45 F3 Raka Zangbo r. China
29 C5 Rakhiv Ukr.
44 B3 Rakhni Pak.
41 G5 Rakhshan r. Pak.
71 □15 Rakiraki Fiji
56 □ Rakit i. Indon.
29 E5 Rakitnoye Belgorod. Obl. Rus. Fed.
52 D2 Rakitnoye Primorskiy Kray Rus. Fed.
9 U7 Rakke Estonia
9 M7 Rakkestad Norway
44 B3 Rakni r. Pak.
21 G2 Rakovník Czech Rep.
38 C3 Rakusha Kazak.
38 B4 Rakushechnyy, Mys pt Kazak.
9 U7 Rakvere Estonia
87 E5 Raleigh U.S.A.
60 H3 Ralik Chain is Marshall Is
88 D2 Ralph U.S.A.
76 E2 Ram r. Can.
79 H2 Ramah Can.
93 H4 Ramah U.S.A.
104 D1 Ramalho, Serra do h. Brazil
36 E6 Ramallah West Bank
43 B3 Ramanagaram India
43 B4 Ramanathapuram India
118 E3 Ramapo Deep depth Pac. Oc.
43 A3 Ramas, C. India
113 F2 Ramatlabama S. Africa
15 G4 Rambervillers France
15 B4 Rambouillet France
15 F4 Rambucourt France
62 E2 Rambutyo I. P.N.G.
43 A3 Ramdurg India
13 C7 Rame Head hd U.K.
111 E5 Ramena Madag.
28 F3 Rameshki Rus. Fed.
43 B4 Rameswaram India
44 D4 Ramganga r. India
45 G5 Ramgarh Bangl.
45 F5 Ramgarh Bihar India
44 B4 Ramgarh Rajasthan India
40 C4 Rāmhormoz Iran
36 E7 Ram, Jebel mt Jordan
36 E6 Ramla Israel
Ramlat Rabyānah des. see Rebiana Sand Sea
Ramnad see Ramanathapuram
44 D3 Ramnagar India
27 M2 Râmnicu Sărat Romania
27 L2 Râmnicu Vâlcea Romania
92 D5 Ramona U.S.A.
89 G1 Ramore Can.
111 C6 Ramotswa Botswana
44 D3 Rampur India
44 C4 Rampura India
Rampur Boalia see Rajshahi
45 F4 Rampur Hat India
45 H6 Ramree I. Myanmar
8 P5 Ramsele Sweden
89 F2 Ramsey Can.
12 C3 Ramsey Isle of Man
13 G5 Ramsey Eng. U.K.
13 B6 Ramsey Island U.K.
89 F2 Ramsey Lake Can.
13 J6 Ramsgate U.K.
44 D5 Ramtek India
9 T9 Ramygala Lith.
103 C4 Rana, Co h. Col.
45 G5 Ranaghat India
44 C5 Ranapur India
59 E1 Ranau Malaysia
105 B2 Rancagua Chile
45 F5 Ranchi India
105 B4 Ranco, L. de Chile
67 G5 Rand Austr.
14 E3 Randalstown U.K.
26 F6 Randazzo Sicily Italy
9 M8 Randers Denmark
91 H3 Randolph MA U.S.A.
91 G3 Randolph VT U.S.A.
19 M3 Randow r. Ger.
9 N5 Randsjö Sweden
8 S4 Råneå Sweden
70 C6 Ranfurly N.Z.
58 B4 Rangae Thai.
45 H5 Rangamati Bangl.
70 D1 Rangaunu Bay N.Z.
91 H2 Rangeley U.S.A.
91 H2 Rangeley Lake U.S.A.
93 H1 Rangely U.S.A.
89 F2 Ranger Lake Can.
70 D5 Rangiora N.Z.
71 □2 Rangiroa i. Fr. Polynesia Pac. Oc.
70 F3 Rangitaiki r. N.Z.
70 C5 Rangitata r. N.Z.
70 E4 Rangitikei r. N.Z.
56 Rangkasbitung Indon.
39 H5 Rangkŭl Tajik.
Rangoon see Yangon
45 G4 Rangpur Bangl.
43 A3 Ranibennur India
45 F5 Raniganj India
45 E5 Ranijula Peak India
44 B4 Ranipur Pak.
85 C6 Rankin U.S.A.
77 L2 Rankin Inlet Can.
77 L2 Rankin Inlet in. Can.
67 G4 Rankin's Springs Austr.
9 U7 Ranna Estonia
68 D5 Rannes Austr.
38 C2 Ranneye Rus. Fed.
11 D4 Rannoch, L. U.K.
11 D4 Rannoch Moor moorland U.K.
44 B4 Rann of Kachchh marsh India
71 □8 Rano Kao h. Easter I. Chile
71 □9 Ranon Vanuatu
58 A3 Ranong Thai.
71 □8 Rano Raraku h. Easter I. Chile
58 B4 Ranot Thai.
28 G4 Ranova r. Rus. Fed.
37 M5 Rānsa Iran
9 N6 Ransby Sweden
57 J7 Ransiki Indon.
9 V5 Rantasalmi Fin.
59 A2 Rantauprapat Indon.
57 F7 Rantepao Indon.
88 C5 Rantoul U.S.A.
17 R2 Rantsevo Rus. Fed.
8 T4 Rantsila Fin.
8 U4 Ranua Fin.
37 K3 Rānya Iraq
52 C1 Raohe China
15 G4 Raon-l'Étape France
51 E6 Raoping China
60 K6 Raoul i. N.Z.
60 N6 Rapa i. Fr. Polynesia Pac. Oc.
23 G5 Rapallo Italy
44 B5 Rapar India
41 E5 Rapch watercourse Iran
105 B2 Rapel r. Chile
75 M3 Raper, C. pt Can.
14 D3 Raphoe Rep. of Ireland
90 E5 Rapidan r. U.S.A.
66 C5 Rapid Bay Austr.
84 C2 Rapid City U.S.A.
89 H2 Rapide-Deux Can.
89 H2 Rapide-Sept Can.
88 D3 Rapid River U.S.A.
9 T7 Rapla Estonia
90 E5 Rappahannock r. U.S.A.
45 E4 Rapti r. India
44 B5 Rapur India
55 C3 Rapurapu i. Phil.
91 F2 Raquette r. U.S.A.
91 F3 Raquette Lake U.S.A.
91 F3 Raquette Lake U.S.A.
71 □2 Raraka i. Fr. Polynesia Pac. Oc.
91 F4 Raritan Bay U.S.A.
22 E3 Raron Switz.
60 M6 Rarotonga i. Cook Is Pac. Oc.
55 A4 Rasa i. Phil.
42 E5 Ra's al Ḩadd pt Oman
40 D5 Ra's al Khaymah U.A.E.
105 D4 Rasa, Pta pt Arg.
110 D2 Ras Dashen mt Eth.
9 S9 Raseiniai Lith.
114 C1 Râs el Mâ Mali
36 C6 Rashîd Egypt
41 G4 Rashid Qala Afgh.
40 D3 Rashm Iran
40 C2 Rasht Iran
43 B4 Rasipuram India
41 F5 Rāsk Iran
44 C1 Raskam mts China
41 G4 Raskoh mts Pak.
109 F2 Ras Muhammad c. Egypt
75 J3 Rasmussen Basin b. Can.
102 C6 Raso, C. pt Arg.
28 D4 Rasony Belarus
45 E4 Rasra India
26 D6 Rass Jebel Tunisia
28 G4 Rasskazovo Rus. Fed.
42 D4 Ras Tannūrah S. Arabia
20 B4 Rastatt Ger.
18 F3 Rastede Ger.
19 J3 Rastow Ger.
40 D5 Rasūl watercourse Iran
60 J2 Ratak Chain is Marshall Is
9 O5 Rätan Sweden
113 H3 Ratanda S. Africa
44 C3 Ratangarh India
45 E5 Ratanpur India
9 O5 Rätansbyn Sweden
58 A2 Rat Buri Thai.
38 C1 Ratchino Rus. Fed.
19 H3 Ratekau Ger.
44 D4 Rath India
14 E4 Rathangan Rep. of Ireland
14 D5 Rathdowney Rep. of Ireland
14 E5 Rathdrum Rep. of Ireland
45 H5 Rathedaung Myanmar
19 K4 Rathenow Ger.
14 E3 Rathfriland U.K.
14 C5 Rathkeale Rep. of Ireland
14 E2 Rathlin Island U.K.
14 C5 Rathluirc Rep. of Ireland
18 D5 Ratingen Ger.
44 C3 Ratiya India
44 C5 Ratlam India
43 A2 Ratnagiri India
43 C5 Ratnapura Sri Lanka
29 C5 Ratne Ukr.
44 B4 Rato Dero Pak.
83 F4 Raton U.S.A.
22 C6 Ratonneau, Île i. France
21 J5 Ratten Austria
11 G3 Rattray Head hd U.K.
9 O6 Rättvik Sweden
19 H3 Ratzeburg Ger.
76 C3 Ratz, Mt Can.
58 B5 Raub Malaysia
20 E5 Raubling Ger.
105 E3 Rauch Arg.
37 L7 Raudhatain Kuwait
20 E2 Rauenstein Ger.
8 F3 Raufarhöfn Iceland
70 G2 Raukumara mt N.Z.
70 F3 Raukumara Range mts N.Z.
9 R6 Rauma Fin.
56 E8 Raung, G. volc. Indon.
45 F5 Raurkela India
52 J2 Rausu Japan
8 V5 Rautavaara Fin.
9 V6 Rautjärvi Fin.
71 □2 Ravahéré i. Fr. Polynesia Pac. Oc.
82 D2 Ravalli U.S.A.
37 L4 Ravānsar Iran
40 E4 Rāvar Iran
15 E1 Ravels Belgium
91 G3 Ravena U.S.A.
12 D3 Ravenglass U.K.
23 K5 Ravenna Italy
68 A5 Ravensbourne Cr. watercourse Austr.
20 C5 Ravensburg Ger.
68 A1 Ravenshoe Austr.
69 E6 Ravensthorpe Austr.
68 B3 Ravenswood Austr.
90 C5 Ravenswood U.S.A.
44 C3 Ravi r. Pak.
21 H6 Ravne na Koroškem Slovenia
38 E5 Ravnina Turkm.
41 F2 Ravnina Turkm.
38 F3 Ravnina Dar'yalyktakyr plain Kazak.
38 D4 Ravshan Uzbek.
37 H4 Rāwah Iraq
60 K4 Rawaki i. Kiribati
44 C2 Rawalpindi Pak.
37 K3 Rawāndiz Iraq
44 C3 Rawatsar India
16 H5 Rawicz Pol.
90 D5 Rawley Springs U.S.A.
66 E4 Rawlinna Austr.
82 F3 Rawlins U.S.A.
102 C6 Rawson Arg.
45 F4 Raxaul India
43 B3 Rayachoti India
43 B3 Rayadurg India
43 C2 Rāyagarha India
36 F5 Rayak Lebanon
79 J4 Ray, C. hd Can.
49 N2 Raychikhinsk Rus. Fed.
22 C5 Raye, La mt France
38 C1 Rayevskiy Rus. Fed.
13 H6 Rayleigh U.K.
76 G5 Raymond Can.
91 H3 Raymond NH U.S.A.
82 B2 Raymond WA U.S.A.
67 J4 Raymond Terrace Austr.
85 D7 Raymondville U.S.A.
58 B2 Rayong Thai.
90 D4 Raystown Lake U.S.A.
40 C3 Razan Iran
37 M5 Rāzān Iran
Razdan see Hrazdan
52 B3 Razdol'noye Rus. Fed.
40 C3 Razeh Iran
27 M3 Razgrad Bulg.
27 N2 Razim, Lacul lag. Romania
27 K4 Razlog Bulg.
24 B2 Raz, Pte du pt France
13 G6 Reading U.K.
91 F4 Reading U.S.A.
88 B4 Readstown U.S.A.
113 G2 Reagile S. Africa
105 D2 Realicó Arg.
24 F5 Réalmont France
58 B2 Reăng Kesei Cambodia
96 D2 Reata Mex.
15 D4 Rebais France
109 E2 Rebiana Sand Sea des. Libya
28 D2 Reboly Rus. Fed.
39 K1 Rebrikha Rus. Fed.
52 G2 Rebun-tō i. Japan
23 L6 Recanati Italy
22 B2 Recey-sur-Ource France
64 D6 Recherche, Archipelago of the is Austr.
44 C3 Rechna Doab lowland Pak.
29 D4 Rechytsa Belarus
101 M5 Recife Brazil
113 F7 Recife, Cape c. S. Africa
18 E5 Recklinghausen Ger.
19 K2 Recknitz r. Ger.
23 J4 Recoaro Terme Italy
102 E3 Reconquista Arg.
102 C3 Recreo Arg.
77 K5 Red r. Can./U.S.A.
85 E6 Red r. U.S.A.
58 B4 Redang i. Malaysia
91 F4 Red Bank NJ U.S.A.
87 C5 Red Bank TN U.S.A.
79 J3 Red Bay Can.
69 D2 Red Bluff h. Austr.
92 A1 Red Bluff U.S.A.
93 F4 Red Butte summit U.S.A.
12 F3 Redcar U.K.
77 G4 Redcliff Can.
69 E4 Redcliffe, Mt h. Austr.
66 E5 Red Cliffs Austr.
84 D3 Red Cloud U.S.A.
76 G4 Red Deer Alta. Can.
77 G4 Red Deer r. Alta. Can.
77 J4 Red Deer r. Sask. Can.
77 J4 Red Deer L. Can.
91 F5 Redden U.S.A.
113 G4 Reddersburg S. Africa
82 B3 Redding U.S.A.
13 F5 Redditch U.K.
91 F3 Redfield NY U.S.A.
84 D2 Redfield SD U.S.A.
66 C4 Redhill Austr.
93 H4 Red Hill U.S.A.
85 D4 Red Hills U.S.A.
79 J4 Red Indian L. Can.
88 E5 Redkey U.S.A.
77 L4 Red L. Can.
93 E4 Red L. U.S.A.
77 L4 Red Lake Can.
84 E1 Red Lakes U.S.A.
82 E2 Red Lodge U.S.A.
82 B2 Redmond U.S.A.
84 E3 Red Oak U.S.A.
25 C3 Redondo Port.
88 C1 Red Rock Can.
91 E4 Red Rock U.S.A.
106 F3 Red Sea Africa/Asia
76 E4 Redstone Can.
76 D2 Redstone r. Can.
77 L4 Red Sucker L. Can.
114 C2 Red Volta r. Burkina/Ghana
76 G4 Redwater Can.
79 H3 Red Wine r. Can.
88 A3 Red Wing U.S.A.
92 A3 Redwood City U.S.A.
84 E2 Redwood Falls U.S.A.
82 B3 Redwood Nat. Park U.S.A.
92 A2 Redwood Valley U.S.A.
88 E4 Reed City U.S.A.
92 C3 Reedley U.S.A.
88 C4 Reedsburg U.S.A.
82 A3 Reedsport U.S.A.
91 E6 Reedville U.S.A.
68 A4 Reedy Cr. watercourse Austr.
116 B4 Reedy Gl. Ant.
Reef Islands see Rowa
70 C5 Reefton N.Z.
14 D4 Ree, Lough l. Rep. of Ireland
18 D5 Rees Ger.
36 G2 Refahiye Turkey
85 D6 Refugio U.S.A.
20 G4 Regen Ger.
20 F3 Regen r. Ger.
20 F3 Regensburg Ger.
20 F3 Regenstauf Ger.
108 C2 Reggane Alg.
23 J6 Reggello Italy
26 F5 Reggio di Calabria Italy
23 H5 Reggio nell'Emilia Italy
17 M7 Reghin Romania
77 J4 Regina Can.
41 G4 Registan reg. Afgh.
8 W4 Regozero Rus. Fed.
20 F2 Rehau Ger.
19 G4 Rehburg (Rehburg-Loccum) Ger.
44 D5 Rehli India
19 J3 Rehna Ger.
111 B6 Rehoboth Namibia
93 H4 Rehoboth U.S.A.
91 F5 Rehoboth Bay U.S.A.
91 F5 Rehoboth Beach U.S.A.
36 E6 Rehovot Israel
19 K5 Reibitz Ger.
21 J5 Reichenau an der Rax Austria
20 F2 Reichenbach Ger.
21 H5 Reichenfels Austria
20 F5 Reichenspitze mt Austria
15 H4 Reichshoffen France
71 □14 Reid Reef Fiji
87 E4 Reidsville U.S.A.
13 G6 Reigate U.K.
24 D3 Ré, Île de i. France
93 G5 Reiley Peak U.S.A.
22 C6 Reillanne France
15 E3 Reims France
102 B8 Reina Adelaida, Archipiélago de la is Chile
22 E2 Reinach Switz.
88 A4 Reinbeck U.S.A.
19 H3 Reinbek Ger.
77 J3 Reindeer r. Can.
77 K4 Reindeer I. Can.
77 J3 Reindeer Lake Can.
8 N3 Reine Norway
19 H3 Reinfeld (Holstein) Ger.
70 D1 Reinga, Cape N.Z.
20 B3 Reinheim Ger.
25 D1 Reinosa Spain
15 G3 Reinsfeld Ger.
8 B4 Reiphólsfjöll mt Iceland
8 R2 Reisaelva r. Norway
8 S2 Reisa Nasjonalpark nat. park Norway
20 F4 Reisbach Ger.
8 T5 Reisjärvi Fin.
20 E5 Reith bei Seefeld Austria
113 H3 Reitz S. Africa
112 F3 Reivilo S. Africa
103 D3 Rejunya Venez.
18 E5 Reken Ger.
77 H2 Reliance Can.
108 C1 Relizane Alg.
19 G3 Rellingen Ger.
15 H2 Remagen Ger.
66 C4 Remarkable, Mt Austr.
41 E5 Remeshk Iran
112 B1 Remhoogte Pass Namibia
22 D1 Remiremont France
44 D2 Remo Gl. India
29 G6 Remontnoye Rus. Fed.
18 E5 Remscheid Ger.
88 E4 Remus U.S.A.
9 M6 Rena Norway
43 B2 Renapur India
86 B4 Rend L. U.S.A.
63 F2 Rendova i. Solomon Is
19 G2 Rendsburg Ger.
22 D3 Renens Switz.
89 J3 Renfrew Can.
11 D5 Renfrew U.K.
105 B2 Rengo Chile
50 C3 Ren He r. China
50 E4 Renheji China
51 D5 Renhua China
51 C5 Renhuai China
29 D6 Reni Ukr.
Renland reg. see Tuttut Nunaat
66 D5 Renmark Austr.
63 G3 Rennell i. Solomon Is
20 B2 Rennerod Ger.
24 D2 Rennes France
116 B5 Rennick Gl. Ant.
77 H2 Rennie Lake Can.

S

36 D1 Şabanözü Turkey
104 D2 Šabará Brazil
43 C2 Sābari r. India
44 C5 Sabarmati r. India
26 E4 Sabaudia Italy
41 E3 Sabeh Iran
112 E5 Sabelo S. Africa
109 D2 Sabhā Libya
40 B6 Şabḩā' S. Arabia
44 D3 Sabi r. India
113 K2 Sabie Moz.
113 K2 Sabie r. Moz./S. Africa
113 J2 Sabie S. Africa
96 D2 Sabinas Mex.
97 D2 Sabinas Hidalgo Mex.
85 E6 Sabine L. U.S.A.
37 M1 Sabirabad Azer.
55 B3 Sablayan Phil.
87 D7 Sable, Cape U.S.A.
71 9 Sable, Î. de i. New Caledonia Pac. Oc.
63 F3 Sable, Île de i. New Caledonia
75 N5 Sable Island Can.
89 F2 Sables, River aux r. Can.
22 B6 Sablon, Pointe du pt France
114 E2 Sabon Kafi Niger
116 C6 Sabrina Coast Ant.
55 B1 Sabtang i. Phil.
25 C2 Sabugal Port.
88 B4 Sabula U.S.A.
42 B6 Şabyā S. Arabia
Sabzawar see Shīndand
41 E2 Sabzevār Iran
27 N2 Sacalinul Mare, Insula i. Romania
97 G4 Sacbecan Mex.
22 E5 Saccarel, Mt France/Italy
27 L2 Săcele Romania
111 B5 Sachanga Angola
78 B3 Sachigo r. Can.
78 B3 Sachigo L. Can.
44 C5 Sachin India
54 E6 Sach'ŏn S. Korea
44 D2 Sach Pass India
19 L5 Sachsen div. Ger.
19 J5 Sachsen-Anhalt div. Ger.
74 F2 Sachs Harbour Can.
19 M6 Sächsische Schweiz, Nationalpark nat. park Ger.
91 E3 Sackets Harbor U.S.A.
18 F6 Sackpfeife h. Ger.
79 H4 Sackville Can.
91 H3 Saco ME U.S.A.
82 F1 Saco MT U.S.A.
55 B5 Sacol i. Phil.
92 B2 Sacramento airport CA U.S.A.
92 B2 Sacramento U.S.A.
92 B2 Sacramento r. U.S.A.
83 F5 Sacramento Mts U.S.A.
82 B3 Sacramento Valley v. U.S.A.
113 G6 Sada S. Africa
25 F1 Sádaba Spain
40 C4 Sa'dabad Iran
36 F4 Şadad Syria
58 B4 Sadao Thai.
113 J2 Saddleback pass S. Africa
Saddle I. see Mota Lava
58 C3 Sa Đec Vietnam
45 H3 Sadêng China
41 E5 Sadīj watercourse Iran
44 B3 Sadiqabad Pak.
44 C1 Sad Istragh mt Afgh./Pak.
48 G6 Sadiya India
37 L5 Sa'dīyah, Hawr as l. Iraq
40 D5 Sa'diyyat i. U.A.E.
40 E2 Sad-Kharv Iran
25 B3 Sado r. Port.
53 F6 Sadoga-shima i. Japan
46 F3 Sado-Shima i. Japan
25 H3 Sa Dragonera i. Spain
9 M8 Sæby Denmark
Safad see Zefat
71 4 Safata B. Western Samoa
37 L6 Safayal Maqūf well Iraq
41 H2 Safed Khirs mts Afgh.
41 G3 Safed Koh mts Afgh.
9 N7 Säffle Sweden
93 H5 Safford U.S.A.
13 H5 Saffron Walden U.K.
36 E6 Safi Jordan
108 B1 Safi Morocco
40 C2 Safīd r. Iran
40 D3 Safid Ab Iran
41 F4 Safidabeh Iran
37 M5 Safid Dasht Iran
36 F4 Şāfītā Syria
8 X2 Safonovo Murmansk. Rus. Fed.
28 E4 Safonovo Smolensk. Rus. Fed.
32 F3 Safonovo Rus. Fed.
71 4 Safotu Western Samoa
40 A5 Safrā' al Asyāḩ esc. S. Arabia
36 D1 Safranbolu Turkey
71 4 Safune Western Samoa
37 L6 Safwān Iraq
45 F3 Saga China
53 B8 Saga Japan
38 E2 Saga Turgaysk. Kazak.
38 F2 Saga Turgaysk. Kazak.
53 F7 Sagamihara Japan
53 F7 Sagami-nada g. Japan
53 F7 Sagami-wan b. Japan
103 B3 Sagamoso r. Col.
114 D3 Sagamu Nigeria
39 J4 Sagankuduk China
58 A2 Saganthit Kyun i. Myanmar
43 B2 Sagar Karnataka India
43 A3 Sagar Karnataka India
44 D5 Sagar Madhya Pradesh India
19 L2 Sagard Ger.
29 H7 Sagarejo Georgia
45 G5 Sagar I. India
33 O2 Sagastyr Rus. Fed.
40 D3 Saghand Iran
41 F3 Saghar Afgh.
43 B3 Sagileru r. India
89 F4 Saginaw U.S.A.
89 F4 Saginaw Bay U.S.A.
38 C2 Sagiz Kazak.
38 C2 Sagiz r. Kazak.
79 H2 Saglek Bay Can.
26 C3 Sagone, Golfe de b. Corsica France
25 B4 Sagres Port.
45 H5 Sagu Myanmar
83 F4 Saguache U.S.A.
95 H4 Sagua la Grande Cuba
93 G5 Saguaro National Monument res. U.S.A.
79 F4 Saguenay r. Can.
25 F3 Sagunto-Sagunt Spain
44 C5 Sagwara India
103 B2 Sahagún Col.
25 D1 Sahagún Spain
37 L3 Sahand, Kūh-e mt Iran
106 B3 Sahara des. Africa
Saharan Atlas mts see Atlas Saharien
44 D3 Saharanpur India
45 F4 Saharsa India
44 D3 Sahaswan India
40 C6 Sahba', W. as watercourse S. Arabia
44 C3 Sahiwal Pak.
41 E3 Sahlābād Iran
37 L4 Şaḩneh Iran
37 K6 Şaḩrā al Ḩijārah reg. Iraq
96 B1 Sahuaripa Mex.
93 G6 Sahuarita U.S.A.
96 D3 Sahuayo Mex.
58 D2 Sa Huynh Vietnam
Sahyadri mts see Western Ghats
44 C5 Sahyadriparvat Range h. India
44 E4 Sai r. India
58 B4 Sai Buri Thai.
58 B4 Sai Buri r. Thai.
Saïda see Sidon
40 D4 Sa'īdābād Iran
58 B2 Sai Dao Tai, Khao mt Thai.
41 F5 Sa'īdī Iran
45 G4 Saidpur Bangl.
44 C2 Saidu Pak.
53 C6 Saigō Japan
Saigon see Hô Chi Minh
45 H5 Saiha India
50 A1 Saihan Toroi China
53 C8 Saijō Japan
53 B8 Saiki Japan
51 Sai Kung H.K. China
9 V6 Saimaa l. Fin.
36 F2 Saimbeyli Turkey
40 D4 Sā'īn Iran
96 D3 Sain Alto Mex.
41 F4 Saindak Pak.
40 B2 Sa'īndezh Iran
11 F5 St Abb's Head hd U.K.
21 J5 St Aegyd am Neuwalde Austria
13 B7 St Agnes U.K.
13 A8 St Agnes i. U.K.
22 B4 St-Agrève France
79 J4 St Alban's Can.
13 G6 St Albans U.K.
91 G2 St Albans VT U.S.A.
90 C5 St Albans WV U.S.A.
13 E7 St Alban's Head hd U.K.
76 G4 St Albert Can.
15 D2 St-Amand-les-Eaux France
24 F3 St-Amand-Montrond France
22 B5 St-Ambroix France
22 C3 St-Amour France
21 H6 St Andrä Austria
22 C4 St-André-le-Gaz France
22 D6 St-André-les-Alpes France
91 K2 St Andrews Can.
11 F4 St Andrews U.K.
95 J5 St Ann's Bay Jamaica
14 F6 St Ann's Head hd U.K.
79 J3 St Anthony Can.
82 E3 St Anthony U.S.A.
66 E6 St Arnaud Austr.
70 D5 St Arnaud Range mts N.Z.
15 B4 St-Arnoult-en-Yvelines France
79 J3 St-Augustin Can.
18 E6 St Augustin Ger.
87 D6 St Augustine U.S.A.
13 C7 St Austell U.K.
24 E3 St-Avertin France
15 G3 St-Avold France
95 M5 St Barthélémy i. Guadeloupe
12 D3 St Bees U.K.
12 D3 St Bees Head hd U.K.
22 D2 St-Blaise Switz.
15 F4 St-Blin-Semilly France
22 B4 St-Bonnet-le-Froid France
13 B6 St Bride's Bay U.K.
24 C2 St-Brieuc France
22 B2 St-Brisson France
93 G5 St Carlos Lake U.S.A.
89 H4 St Catharines Can.
87 D6 St Catherines I. U.S.A.
13 F7 St Catherine's Point U.K.
24 E4 St-Céré France
91 G2 St-Césaire Can.
22 B4 St-Chamond France
82 E3 St Charles ID U.S.A.
90 E5 St Charles MD U.S.A.
88 A4 St Charles MN U.S.A.
84 F4 St Charles MO U.S.A.
22 B5 St-Christol-lès-Alès France
89 F4 St Clair U.S.A.
89 F4 St Clair Shores U.S.A.
22 C3 St-Claude France
13 C6 St Clears U.K.
15 G4 St-Clément France
84 E2 St Cloud U.S.A.
79 G4 St Croix r. Can.
88 A2 St Croix r. U.S.A.
95 M5 St Croix i. Virgin Is
88 A3 St Croix Falls U.S.A.
15 C4 St-Cyr-l'École France
93 G6 St David U.S.A.
14 F6 St David's U.K.
13 B6 St David's Head hd U.K.
15 C4 St-Denis France
15 G4 St-Dié France
15 E4 St-Dizier France
77 K5 Ste Anne Can.
79 F4 Ste-Anne-de-Beaupré Can.
91 J1 Sainte-Anne-de-Madawaska Can.
89 K2 Sainte-Anne-du-Lac Can.
79 G3 Ste Anne, L. Can.
91 H1 Ste-Camille-de-Lellis Can.
22 C5 Ste-Croix France
22 D6 Ste-Croix, Lac de l. France
22 C4 St-Égrève France
91 H1 Sainte-Justine Can.
91 J1 St-Éleuthère Can.
76 B2 St Elias Mountains Can.
79 G3 Ste Marguerite r. Can.
22 D6 Ste-Maxime France
24 D4 Saintes France
22 B6 Stes-Maries-de-la-Mer France
91 G2 Ste-Thérèse Can.
22 B4 St-Étienne France
22 D5 St-Étienne-de-Tinée France
15 B3 St-Étienne-du-Rouvray France
22 C5 St-Étienne-les-Orgues France
91 F2 St Eugene Can.
91 G2 St-Eustache Can.
95 M5 St Eustatius i. Neth. Ant.
79 F4 St-Félicien Can.
22 B4 St-Félicien France
14 F3 Saintfield U.K.
26 C3 St-Florent Corsica France
22 A1 St-Florentin France
24 F3 St-Florent-sur-Cher France
110 C3 St. Floris, Parc National nat. park C.A.R.
22 B4 St-Fons France
91 J1 St-Francis r. Can./U.S.A.
84 C4 St Francis KS U.S.A.
91 J1 St Francis ME U.S.A.
85 F4 St Francis r. U.S.A.
79 K4 St Francis, C. Can.
91 J1 St Froid Lake U.S.A.
23 G2 St Gallen Switz.
22 B4 St-Galmier France
24 E5 St-Gaudens France
91 H2 St-Gédéon Can.
22 B4 St-Genest-Lerpt France
67 H2 St George Austr.
91 K2 St George Can.
87 D5 St George SC U.S.A.
93 F3 St George UT U.S.A.
63 F2 St George, C. pt P.N.G.
87 C6 St George I. U.S.A.
21 H4 St Georgen an der Gusen Austria
20 B4 St Georgen im Schwarzwald Ger.
82 A3 St George, Pt U.S.A.
79 F4 St Georges Can.
95 M6 St George's Grenada
79 J4 St George's B. Can.
62 F2 St George's Channel P.N.G.
10 C6 St George's Channel Rep. of Ireland/U.K.
22 D4 St-Gervais-les-Bains France
22 B6 St-Gilles France
22 D3 St-Gingolph France
13 C6 St Govan's Head hd U.K.
88 E3 St Helen U.S.A.
107 C7 St Helena terr. Atl. Ocean
92 A2 St Helena U.S.A.
112 C6 St Helena Bay S. Africa
112 C6 St Helena Bay S. Africa
120 J7 St Helena Fracture sea feature Atl. Ocean
67 H8 St Helens Austr.
13 E4 St Helens U.K.
82 B2 St Helens U.S.A.
82 B2 St Helens, Mt volc. U.S.A.
67 H8 St Helens Pt Austr.
10 E7 St Helier U.K.
22 A3 St-Hilaire-Fontaine France
22 D5 St-Honorat, Mont mt France
15 F2 St-Hubert Belgium
78 F4 St-Hyacinthe Can.
88 E3 St Ignace U.S.A.
88 C1 St Ignace I. Can.
22 D2 St-Imier Switz.
13 C6 St Ishmael U.K.
13 B7 St Ives Eng. U.K.
13 G5 St Ives Eng. U.K.
91 J1 St-Jacques Can.
88 E3 St James U.S.A.
76 C4 St James, Cape pt Can.
24 D4 St-Jean-d'Angély France
22 C2 St-Jean-de-Losne France
22 D4 St-Jean-de-Maurienne France
24 C3 St-Jean-de-Monts France
79 F4 St-Jean, Lac l. Can.
78 F4 St-Jean-sur-Richelieu Can.
78 F4 St-Jérôme Can.
82 C2 St Joe r. U.S.A.
20 G5 St Johann im Pongau Austria
20 F6 St Johann im Walde Austria
20 F5 St Johann in Tirol Austria
79 G4 St John Can.
91 K2 St John r. Can./U.S.A.
114 B3 St John r. Liberia
93 F1 St John U.S.A.
95 M5 St John i. Virgin Is
95 M5 St John's Antigua
79 K4 St John's Can.
93 H4 St Johns AZ U.S.A.
88 E4 St Johns MI U.S.A.
87 D6 St Johns r. U.S.A.
91 H2 St Johnsbury U.S.A.
12 E3 St John's Chapel U.K.
79 F4 St Joseph Can.
88 D4 St Joseph MI U.S.A.
84 E4 St Joseph MO U.S.A.
88 E5 St Joseph r. U.S.A.
89 F2 St Joseph I. Can.
85 D7 St Joseph I. U.S.A.
78 B3 St Joseph, Lac l. Can.
78 F4 St Jovité Can.
22 B4 St-Julien-Chapteuil France
24 E4 St-Junien France
22 B5 St-Just France
13 B7 St Just U.K.
15 C3 St-Just-en-Chaussée France
22 B4 St-Just-St-Rambert France
13 B7 St Keverne U.K.
10 B3 St Kilda is U.K.
73 K8 St Kitts-Nevis country Caribbean Sea
101 H2 St Laurent Fr. Guiana
St-Laurent, Golfe du g. see St Lawrence, Gulf of
68 C4 St Lawrence Austr.
79 K4 St Lawrence Nfld Can.
79 G4 St Lawrence in. Que. Can.
79 H4 St Lawrence, Gulf of g. Can./U.S.A.
74 B3 St Lawrence I. U.S.A.
89 K3 St Lawrence Islands National Park Can.
91 F2 St Lawrence Seaway chan. Can./U.S.A.
79 G4 St-Léonard Can.
15 G4 St-Léonard France
79 J3 St Lewis Can.
79 J3 Saint Lewis r. Can.
24 D2 St-Lô France
22 C3 St-Lothian France
22 E2 St-Louis France
108 A3 St Louis Senegal
88 E4 St Louis MI U.S.A.
84 F4 St Louis MO U.S.A.
88 A2 St Louis r. U.S.A.
22 D2 St-Loup-sur-Semouse France
15 B4 St-Lubin-des-Joncherets France
73 K8 St Lucia country Caribbean Sea
113 K4 St Lucia Estuary S. Africa
113 K3 St Lucia, Lake S. Africa
15 E4 St-Lyé France
95 M5 St Maarten i. Neth. Ant.
11 St Magnus Bay U.K.
24 D3 St-Maixent-l'École France
24 C2 St-Malo France
24 C2 St-Malo, Golfe de g. France
22 C4 St-Marcellin France
22 B2 St-Marc-sur-Seine France
19 G3 St Margarethen Ger.
113 G6 St Marks S. Africa
95 M5 Saint Martin i. Guadeloupe
112 B6 St Martin, Cape hd S. Africa
22 B6 St-Martin-de-Crau France
22 C4 St-Martin-d'Hères France
88 D3 St Martin I. U.S.A.
77 K4 St Martin, L. Can.
13 A8 St Martin's i. U.K.
45 H5 St Martin's I. Bangl.
66 C3 St Mary Pk Austr.
67 H8 St Marys Austr.
89 G4 St Mary's Can.
11 F2 St Mary's U.K.
13 A8 St Mary's i. U.K.
90 A4 St Marys OH U.S.A.
90 D4 St Marys PA U.S.A.
90 C5 Saint Marys U.S.A.
90 A4 St Marys r. U.S.A.
79 K4 St Mary's, C. hd Can.
74 A3 St Matthew I. U.S.A.
62 E2 St Matthias Group is P.N.G.
78 F4 St Maurice r. Can.
13 B7 St Mawes U.K.
22 C6 St-Maximin-la-Ste-Baume France
24 D4 St-Médard-en-Jalles France
21 G5 St Michael im Lungau Austria
19 G3 St Michaelisdonn Ger.
79 J3 St Michaels Bay Can.
22 D4 St-Michel-de-Maurienne France
23 G3 St Moritz Switz.
15 G4 St-Nabord France
24 C3 St-Nazaire France
13 G5 St Neots U.K.
15 G4 St-Nicolas-de-Port France
15 E1 St-Niklaas Belgium
22 E3 St Niklaus Switz.
15 C2 St-Omer France
91 J1 St-Pamphile Can.
79 G4 St Pascal Can.
77 G4 St Paul Can.
114 A3 St Paul r. Liberia
88 A3 St Paul MN U.S.A.
84 D3 St Paul NE U.S.A.
90 B6 St Paul VA U.S.A.
117 K6 St Paul, Île i. Ind. Ocean
22 B5 St-Paul-le-Jeune France
71 6 St Paul's Point Pitcairn I. Pac. Oc.
22 B5 St-Péray France
22 A2 St-Père France
81 H3 St Peter 950723 U.S.A.
21 H5 St Peter am Kammersberg Austria
18 F2 St Peter-Ording Ger.
10 E7 St Peter Port U.K.
28 D3 St Petersburg Rus. Fed.
87 D7 St Petersburg U.S.A.
22 C6 St-Pierre mt France
79 J4 St-Pierre St Pierre and Miquelon N. America
75 N5 St Pierre and Miquelon terr. N. America
24 D4 St-Pierre-d'Oléron France
78 F4 St Pierre, Lac l. Can.
24 F3 St-Pierre-le-Moûtier France
15 C2 St-Pol-sur-Ternoise France
21 J4 St Pölten Austria
24 F3 St-Pourçain-sur-Sioule France
22 B4 St-Priest France
91 H1 Saint-Prosper Can.
15 D3 St-Quentin France
22 B4 St-Rambert-d'Albon France
22 D6 St-Raphaël France
91 F2 St Regis r. U.S.A.
91 F2 St Regis Falls U.S.A.
91 G2 St-Rémi Can.
22 B3 St-Rémy France
22 B3 St-Rigaud, Mont mt France
15 B2 St-Riquier France
15 B3 St-Saëns France
91 H2 St-Sébastien Can.
79 G4 St Siméon Can.
87 D6 St Simons I. U.S.A.
91 K2 St Stephen Can.
87 E5 St Stephen U.S.A.
91 H2 St-Théophile Can.
77 L4 St Theresa Point Can.
89 G4 St Thomas Can.
22 B3 St-Trivier-sur-Moignans France
22 D6 St-Tropez France
22 D6 St-Tropez, Cap de c. France
15 F2 St-Truiden Belgium
15 A3 St-Valery-en-Caux France
15 B2 St-Valery-sur-Somme France
22 B3 St-Vallier France
22 B4 St-Vallier France
22 D6 St-Vallier-de-Thiey France
21 H6 St Veit an der Glan Austria
22 D5 St-Véran France
22 E4 St-Vincent Italy
77 K5 St Vincent U.S.A.
73 K8 St Vincent and the Grenadines country Caribbean Sea
67 F9 St Vincent, C. hd Austr.
St Vincent, Cape see São Vicente, Cabo de
66 B5 St. Vincent, Gulf Austr.
15 G2 St-Vith Belgium
77 H4 St Walburg Can.
15 H3 St Wendel Ger.
89 G4 St Williams Can.
24 E4 St-Yrieix-la-Perche France
44 E3 Saipal mt Nepal
57 L3 Saipan i. N. Mariana Is
45 H5 Saitlai Myanmar
8 T3 Saittanulkki h. Fin.
51 Sai Wan H.K. China
100 E7 Sajama, Nevado mt Bol.
40 B5 Sājir S. Arabia
112 D5 Sak watercourse S. Africa
53 D7 Sakai Japan
53 C7 Sakaide Japan
53 C7 Sakaiminato Japan
42 B3 Sakākah S. Arabia
41 G5 Saka Kalat Pak.
84 C2 Sakakawea, Lake U.S.A.
78 E3 Sakami Can.
78 F3 Sakami r. Can.
78 E3 Sakami, Lac l. Can.
27 M4 Sakar mts Bulg.
36 C1 Sakarya Turkey
36 C1 Sakarya r. Turkey
114 B3 Sakassou Côte d'Ivoire
52 F5 Sakata Japan
54 C3 Sakchu N. Korea
58 B2 Sa Keo r. Thai.
114 D3 Sakété Benin
49 Q1 Sakhalin i. Rus. Fed.
46 G1 Sakhalinskiy Zaliv b. Rus. Fed.
44 C3 Sakhi India
113 H3 Sakhile S. Africa
40 C2 Sakht-Sar Iran
37 L1 Şäki Azer.
114 D3 Şaki Nigeria
9 S9 Šakiai Lith.
44 A3 Sakir mt Pak.
49 M7 Sakishima-Guntō is Japan
44 B4 Sakrand Pak.
58 Sakra, P. i. Sing.
112 D5 Sakrivier S. Africa
38 E3 Saksaul'skiy Kazak.
53 B9 Sakura-jima volc. Japan
29 E6 Saky Ukr.
9 S6 Säkylä Fin.
108 Sal i. Cape Verde
29 G6 Sal r. Rus. Fed.
9 S8 Sala Latvia
21 L4 Šaľa Slovakia
9 P7 Sala Sweden
78 F4 Salaberry-de-Valleyfield Can.
9 T8 Salacgrīva Latvia
26 F4 Sala Consilina Italy
93 E5 Salada, Laguna salt l. Mex.
105 E2 Saladillo Buenos Aires Arg.
105 D2 Saladillo r. Córdoba Arg.
105 E2 Salado r. Buenos Aires Arg.
105 C2 Salado r. Mendoza/San Luis Arg.
105 D4 Salado r. Río Negro Arg.
105 E1 Salado r. Santa Fé Arg.
97 E2 Salado r. Mex.
102 B3 Salado, Quebrada de r. Chile
114 C3 Salaga Ghana
71 4 Salailua Western Samoa
112 F1 Salajwe Botswana
109 D3 Salal Chad
42 D6 Şalālah Oman
97 G5 Salamá Guatemala
96 H5 Salamá Honduras
105 B1 Salamanca Chile
96 D3 Salamanca Mex.
25 D2 Salamanca Spain
90 D3 Salamanca U.S.A.
113 K3 Salamanga Moz.
40 B3 Salamatabad Iran
103 B3 Salamina Col.
36 F4 Salamīyah Syria
88 E5 Salamonie r. U.S.A.
88 E5 Salamonie Lake U.S.A.
45 F5 Salandi r. India
71 4 Salani Western Samoa
9 R8 Salantai Lith.
102 C2 Salar de Arizaro salt flat Arg.
102 C2 Salar de Atacama salt flat Chile
25 C1 Salas Spain
9 T8 Salaspils Latvia
56 Salatiga Indon.
38 C1 Salavat Rus. Fed.
57 J7 Salawati i. Indon.
44 B5 Salaya India
57 G8 Salayar i. Indon.
119 N7 Sala y Gómez, Isla i. Chile
105 D3 Salazar Arg.
9 T9 Šalčininkai Lith.
13 D7 Salcombe U.K.
103 B4 Saldaña r. Col.
25 D1 Saldaña Spain
112 B6 Saldanha S. Africa
112 B6 Saldanha Bay S. Africa
105 E3 Saldungaray Arg.
9 S8 Saldus Latvia
67 G7 Sale Austr.
40 C3 Sāleḩābād Iran
37 L5 Şālehābād Iran
32 H3 Salekhard Rus. Fed.
71 4 Salelologa Western Samoa
43 B4 Salem India
91 H3 Salem MA U.S.A.
85 F4 Salem MO U.S.A.
86 F4 Salem NJ U.S.A.
91 G3 Salem NY U.S.A.
90 C4 Salem OH U.S.A.
82 B2 Salem OR U.S.A.
86 D4 Salem VA U.S.A.
11 C4 Salen U.K.
11 C4 Salen U.K.
26 F4 Salerno Italy
26 F4 Salerno, Golfo di g. Italy
13 E4 Salford U.K.
101 L5 Salgado r. Brazil
17 J6 Salgótarján Hungary
101 L5 Salgueiro Brazil
41 F4 Salian Afgh.
55 C6 Salibabu i. Indon.
80 E4 Salida U.S.A.
24 D5 Salies-de-Béarn France
36 B2 Salihli Turkey
28 C4 Salihorsk Belarus
115 B5 Salima Malawi
115 C5 Salimo Moz.
84 D4 Salina KS U.S.A.
93 G2 Salina UT U.S.A.
97 F4 Salina Cruz Mex.
105 D4 Salina Gualicho salt flat Arg.
26 F5 Salina, Isola i. Italy
105 C2 Salina Llancanelo salt flat Arg.
104 D2 Salinas Brazil
100 B4 Salinas Ecuador
96 D3 Salinas Mex.
92 B3 Salinas CA U.S.A.
92 B3 Salinas r. CA U.S.A.
102 C4 Salinas Grandes salt flat Arg.
83 F5 Salinas Peak U.S.A.
71 7 Salinas, Punta pt Juan Fernandez Is Chile
85 E5 Saline r. AR U.S.A.
84 C4 Saline r. KS U.S.A.
25 H3 Salines, Cap de ses pt Spain
92 D3 Saline Valley v. U.S.A.
101 J4 Salinópolis Brazil
100 C6 Salinosó Lachay, Pta pt Peru
13 F6 Salisbury U.K.
91 F5 Salisbury MD U.S.A.
87 D5 Salisbury NC U.S.A.
13 E6 Salisbury Plain U.K.
101 K6 Salitre r. Brazil
36 F5 Şalkhad Syria
45 F5 Salki r. India
8 V3 Salla Fin.
22 D4 Sallanches France
105 D3 Salliqueló Arg.
85 E5 Sallisaw U.S.A.
75 L3 Salluit Can.
45 E3 Sallyana Nepal
40 B2 Salmās Iran
28 D2 Salmi Rus. Fed.
76 F5 Salmo Can.
82 D2 Salmon U.S.A.
82 D2 Salmon r. U.S.A.
76 F4 Salmon Arm Can.
69 E6 Salmon Gums Austr.
91 F3 Salmon Reservoir U.S.A.
82 D2 Salmon River Mountains U.S.A.
15 G3 Salmtal Ger.
9 S6 Salo Fin.
23 H4 Salò Italy
45 E4 Salon India
22 C6 Salon-de-Provence France
110 C4 Salonga Nord, Parc National de la nat. park Congo(Zaire)
110 C4 Salonga Sud, Parc National de la nat. park Congo(Zaire)
17 K7 Salonta Romania
96 H5 Sal, Pta pt Honduras
105 D1 Salsacate Arg.
29 G6 Sal'sk Rus. Fed.
23 G5 Salsomaggiore Terme Italy
36 E5 Salt Jordan
112 E5 Salt watercourse S. Africa
93 G5 Salt r. AZ U.S.A.
88 B6 Salt r. MO U.S.A.
102 C2 Salta Arg.
13 C7 Saltash U.K.
11 D5 Saltcoats U.K.
90 B5 Salt Creek r. U.S.A.
14 E5 Saltee Islands Rep. of Ireland
8 O3 Saltfjellet Svartisen Nasjonalpark nat. park Norway
85 B6 Salt Flat U.S.A.
90 C4 Salt Fork Lake U.S.A.
96 D2 Saltillo Mex.
82 E3 Salt Lake City U.S.A.
66 E3 Salt L., The salt flat Austr.
105 E2 Salto Arg.
104 C3 Salto Brazil
105 F1 Salto Uru.
104 E2 Salto da Divisa Brazil
102 E4 Salto Grande, Embalse de resr Uru.
93 E5 Salton Sea salt l. U.S.A.
44 C2 Salt Ra. h. Pak.
77 G2 Salt River Can.
90 B5 Salt Rock U.S.A.
71 4 Saluafata Western Samoa
87 D5 Saluda SC U.S.A.
90 E6 Saluda VA U.S.A.
44 C4 Salumbar India
43 C2 Salur India
22 E5 Saluzzo Italy
105 D1 Salvador Arg.
104 E1 Salvador Brazil
85 F6 Salvador, L. U.S.A.
96 D3 Salvatierra Mex.
93 G2 Salvation Creek r. U.S.A.
40 C5 Salwah Qatar
56 B3 Salween r. Myanmar
37 M2 Salyan Azer.
90 B6 Salyersville U.S.A.
21 J5 Salza r. Austria
20 F4 Salzach r. Austria/Ger.
112 B2 Salzbrunn Namibia
20 G5 Salzburg Austria
20 F5 Salzburg div. Austria
19 H4 Salzgitter Ger.
19 H3 Salzhausen Ger.
21 G5 Salzkammergut reg. Austria
18 F5 Salzkotten Ger.
19 J5 Salzmünde Ger.
19 J4 Salzwedel Ger.
44 B4 Sam India
40 B4 Samāh well S. Arabia
40 B3 Samaida Iran
55 C5 Samal i. Phil.
71 4 Samalaeulu Western Samoa
55 B5 Samales Group is Phil.
43 C2 Samalkot India
36 E3 Samandağı Turkey
52 H3 Samani Japan
36 C6 Samannûd Egypt
55 C4 Samar i. Phil.
38 B1 Samara Rus. Fed.
38 B1 Samara r. Rus. Fed.
103 D3 Samariapo Venez.
59 E3 Samarinda Indon.
52 D2 Samarka Rus. Fed.
39 F5 Samarkand Uzbek.
39 G5 Samarkand, Pik mt Tajik.
37 J4 Sāmarrā' Iraq
55 C4 Samar Sea g. Phil.
28 J4 Samarskaya Oblast' div. Rus. Fed.
39 H2 Samarskoye Karagan. Kazak.
39 K2 Samarskoye Vostochnyy Kaz. Kazak.
71 4 Samatau Western Samoa
37 M1 Şamaxı Azer.
115 A3 Samba Congo(Zaire)
59 E2 Sambaliung mts Indon.
45 F5 Sambalpur India
59 D3 Sambar, Tanjung pt Indon.
59 C2 Sambas Indon.
111 F5 Sambava Madag.
45 G4 Sambha India
44 D3 Sambhal India
44 C4 Sambhar L. India
29 B5 Sambir Ukr.
101 K5 Sambito r. Brazil
105 F2 Samborombón, Bahía b. Arg.
15 D2 Sambre r. Belgium/France
103 A3 Sambú r. Panama
54 E5 Samch'ŏk S. Korea
54 E6 Samch'ŏnp'o S. Korea
37 K3 Samdi Dag mt Turkey
115 C3 Same Tanz.
15 B2 Samer France
115 A5 Samfya Zambia
Samirum see Yazd-e Khvāst
54 E3 Samjiyŏn N. Korea
37 L1 Şämkir Azer.
40 D3 Samnan va Damghan reg. Iran
26 F2 Samobor Croatia
28 G2 Samoded Rus. Fed.
27 K3 Samokov Bulg.
16 H6 Šamorín Slovakia
27 M6 Samos i. Greece
59 A2 Samosir i. Indon.
27 L4 Samothraki Greece
27 L4 Samothraki i. Greece
55 B3 Sampaloc Point Phil.
22 E5 Sampeyre Italy
15 F4 Sampigny France
59 D3 Sampit Indon.
59 D3 Sampit, Teluk b. Indon.
115 A4 Sampwe Congo(Zaire)
54 E6 Samrangjin S. Korea
85 E6 Sam Rayburn Resr U.S.A.
45 E3 Samsang China
58 C1 Sâm Sơn Vietnam
36 F1 Samsun Turkey
19 L2 Samtens Ger.
29 G7 Samtredia Georgia
29 J7 Samur r. Azer./Rus. Fed.
58 B2 Samut Sakhon Thai.
58 B2 Samut Songkhram Thai.
45 G3 Samyai China
114 C2 San Mali
42 B6 Şan'ā Yemen
116 C3 Sanae S. Africa Base Ant.
114 F4 Sanaga r. Cameroon
103 A4 San Agustín Col.
55 C5 San Agustin, Cape Phil.
40 B6 Sanām S. Arabia
40 B3 Sanandaj Iran
92 B2 San Andreas U.S.A.
55 C3 San Andres Phil.
96 J5 San Andrés, Isla de i. Col.
100 B1 San Andrés, Isla de i. Col.
83 F5 San Andres Mts U.S.A.
97 F4 San Andrés Tuxtla Mex.
85 C6 San Angelo U.S.A.
96 G4 San Antonio Belize
105 B2 San Antonio Chile
55 B3 San Antonio Phil.
85 D6 San Antonio U.S.A.
25 G3 San Antonio Abad Spain
95 H4 San Antonio, C. pt Cuba
105 F3 San Antonio, Cabo pt Arg.
102 C2 San Antonio de los Cobres Arg.
103 D2 San Antonio de Tamanaco Venez.
92 D4 San Antonio, Mt U.S.A.
105 D4 San Antonio Oeste Arg.
92 B4 San Antonio Reservoir U.S.A.
92 B3 San Ardo U.S.A.
105 E3 San Augustín Arg.
105 C1 San Augustín de Valle Fértil Arg.
44 D5 Sanawad India
97 D3 San Bartolo Mex.
26 E3 San Benedetto del Tronto Italy
23 H4 San Benedetto Po Italy
94 B5 San Benedicto, I. Mex.
85 D7 San Benito U.S.A.
92 B3 San Benito r. U.S.A.
92 B3 San Benito Mt U.S.A.
92 D4 San Bernardino U.S.A.
83 C5 San Bernardino Mts U.S.A.
105 B2 San Bernardo Chile
96 C2 San Bernardo Mex.
53 C7 Sanbe-san volc. Japan
96 C3 San Blas Nayarit Mex.
96 B2 San Blas Sinaloa Mex.
96 K6 San Blas, Archipiélago de is Panama
87 C6 San Blas, C. U.S.A.
96 K6 San Blas, Cordillera de mts Panama
100 E6 San Borja Bol.
91 H3 Sanbornville U.S.A.
96 D2 San Buenaventura Mex.
23 K3 San Candido Italy
105 C2 San Carlos Arg.
105 B3 San Carlos Chile
96 D1 San Carlos Coahuila Mex.
97 E2 San Carlos Tamaulipas Mex.
96 H6 San Carlos Nic.
55 B3 San Carlos Luzon Phil.
55 B4 San Carlos Negros Phil.
105 F2 San Carlos Uru.
93 G5 San Carlos U.S.A.
103 D4 San Carlos Amazonas Venez.
103 C2 San Carlos Cojedes Venez.
105 E1 San Carlos Centro Arg.
105 B4 San Carlos de Bariloche Arg.
105 E3 San Carlos de Bolívar Arg.
103 C2 San Carlos del Zulia Venez.
83 D6 San Carlos, Mesa de h. Mex.
50 C2 Sancha Gansu China
50 D2 Sancha Shanxi China
54 D1 Sanchahe China
51 C5 Sancha He r. China
39 J5 Sanchakou China
15 B4 Sancheville France
51 B6 San Chien Pau mt Laos
44 B4 Sanchor India
50 D2 Sanchuan r. China
28 H3 Sanchursk Rus. Fed.
97 E3 San Ciro de Acosta Mex.
105 B2 San Clemente Chile
92 D5 San Clemente U.S.A.
92 C5 San Clemente I. U.S.A.
24 F3 Sancoins France
23 L6 San Costanzo Italy
105 E1 San Cristóbal Arg.
63 G3 San Cristobal i. Solomon Is
103 B3 San Cristóbal Venez.
97 F4 San Cristóbal de las Casas Mex.
100 San Cristóbal, Isla i. Galapagos Is Ecuador
93 F5 San Cristobal Wash r. U.S.A.
95 J4 Sancti Spíritus Cuba
113 H1 Sand r. S. Africa
52 D3 Sandagou Rus. Fed.
11 C5 Sanda Island U.K.
59 E1 Sandakan Malaysia
9 K6 Sandane Norway
27 K4 Sandanski Bulg.
114 A2 Sandaré Mali
19 K4 Sandau Ger.
11 F1 Sanday i. U.K.
11 F1 Sanday Sound chan. U.K.
13 E4 Sandbach U.K.
9 M7 Sandefjord Norway
116 D4 Sandercock Nunataks nunatak Ant.
93 H4 Sanders U.S.A.
19 J5 Sandersleben Ger.
85 C6 Sanderson U.S.A.
64 D3 Sandfire Roadhouse Austr.
67 K1 Sandgate Austr.
11 D6 Sandhead U.K.
88 B2 Sand I. U.S.A.
100 E6 Sandia Peru
92 D5 San Diego U.S.A.
102 C8 San Diego, C. Arg.
36 C2 Sandıklı Turkey
44 E4 Sandila India
69 B2 Sandiman, Mt h. Austr.
97 G4 San Dimas Campeche Mex.
96 C2 San Dimas Durango Mex.
19 L4 Sandkrug Ger.
21 H4 Sandl Austria
88 E2 Sand Lake Can.
9 J7 Sandnes Norway
8 N3 Sandnessjøen Norway
111 C4 Sandoa Congo(Zaire)
17 K5 Sandomierz Pol.
103 A4 Sandoná Col.
23 K4 San Donà di Piave Italy
56 A3 Sandoway Myanmar
13 F7 Sandown U.K.
112 C7 Sandown Bay S. Africa
8 Sandoy i. Faroe Is
82 C1 Sandpoint U.S.A.
11 A4 Sandray i. U.K.
17 N7 Şandrul Mare, Vârful mt Romania
9 O6 Sandsjö Sweden
76 C4 Sandspit Can.
85 D4 Sand Springs U.S.A.
92 C2 Sand Springs Salt Flat salt flat U.S.A.
69 D3 Sandstone Austr.
88 A2 Sandstone U.S.A.
93 F5 Sand Tank Mts U.S.A.
51 C5 Sandu Guizhou China
51 D5 Sandu Hunan China
89 F4 Sandusky MI U.S.A.
90 B4 Sandusky OH U.S.A.
90 B4 Sandusky Bay U.S.A.
112 C5 Sandveld mts S. Africa
112 B3 Sandverhaar Namibia
9 M7 Sandvika Norway
8 N5 Sandvika Sweden
9 P6 Sandviken Sweden
79 J3 Sandwich Bay Can.
11 Sandwick U.K.
45 G5 Sandwip Ch. chan. Bangl.
91 H2 Sandy r. U.S.A.
77 J3 Sandy Bay Can.
67 F8 Sandy C. hd Austr.
68 E5 Sandy Cape Austr.
90 B5 Sandy Hook U.S.A.
91 F4 Sandy Hook pt U.S.A.
41 F2 Sandykachi Turkm.
78 B3 Sandy L. Can.
78 B3 Sandy Lake Can.
91 E3 Sandy Pond U.S.A.
22 C3 Sâne r. France
104 A4 San Estanislao Para.
55 B2 San Fabian Phil.
105 B2 San Felipe Chile
94 B2 San Felipe Baja California Norte Mex.
96 C2 San Felipe Chihuahua Mex.
96 D3 San Felipe Guanajuato Mex.
103 C2 San Felipe Venez.
105 E2 San Fernando Arg.
105 B2 San Fernando Chile
97 E2 San Fernando Mex.
55 B2 San Fernando Luzon Phil.
55 B3 San Fernando Luzon Phil.
25 C4 San Fernando Spain
103 E2 San Fernando Trinidad and Tobago
92 C4 San Fernando U.S.A.
103 D3 San Fernando de Apure Venez.
103 D3 San Fernando de Atabapo Venez.
92 D5 San Filipe Creek r. U.S.A.
69 C3 Sanford r. Austr.
87 D6 Sanford FL U.S.A.
91 H3 Sanford ME U.S.A.
87 E5 Sanford NC U.S.A.
88 E4 Sanford Lake U.S.A.
105 D1 San Francisco Arg.
96 J6 San Francisco Panama
92 A3 San Francisco CA U.S.A.
92 A3 San Francisco airport CA U.S.A.
93 H5 San Francisco r. NM U.S.A.
92 A3 San Francisco Bay in. U.S.A.
96 C2 San Francisco del Oro Mex.
96 D3 San Francisco del Rincón Mex.
95 K5 San Francisco de Macorís Dom. Rep.
102 C7 San Francisco de Paula, C. pt Arg.
96 G5 San Francisco Gotera El Salvador
25 G3 San Francisco Javier Spain
102 C3 San Francisco, Paso de pass Arg.
103 A4 San Gabriel Ecuador
92 C4 San Gabriel Mts U.S.A.
96 A1 San Gabriel, Pta pt Mex.
100 C4 Sangai, Parque Nacional nat. park Ecuador
44 C6 Sangamner India
88 C6 Sangamon r. U.S.A.
41 G3 Sangan Afgh.
41 F4 Sangān Iran
41 G3 Sangan, Koh-i- mt Afgh.
44 B3 Sangar r. Pak.
33 O3 Sangar Rus. Fed.
43 B2 Sangāreddi India
26 C5 San Gavino Monreale Sardinia Italy
41 E3 Sang Bast Iran
55 B5 Sangboy Islands Phil.
59 E4 Sangeang i. Indon.
50 B1 Sangejing China
92 C3 Sanger U.S.A.
19 J5 Sangerhausen Ger.
50 E1 Sanggan r. China
50 B3 Sanggarmai China
59 D2 Sanggau Indon.
54 B5 Sanggou Wan b. China
110 B3 Sangha r. Congo
44 B4 Sanghar Pak.
103 B3 San Gil Col.
26 G5 San Giovanni in Fiore Italy
23 J5 San Giovanni in Persiceto Italy
23 J4 San Giovanni Lupatoto Italy
26 F4 San Giovanni Rotondo Italy
23 J6 San Giovanni Valdarno Italy
55 C6 Sangir i. Indon.
57 G6 Sangir, Kepulauan is Indon.
23 H6 San Giuliano Terme Italy
23 K6 San Giustino Italy
56 Sangkapura Indon.
59 E2 Sangkulirang Indon.
43 A2 Sangli India
41 H2 Sanglich Afgh.
114 F4 Sangmélima Cameroon
44 D3 Sangnam India
45 H3 Sangngagqoling China
54 E3 Sang-ni N. Korea
111 D6 Sango Zimbabwe
43 A2 Sāngole India
92 D4 San Gorgonio Mt U.S.A.
22 F3 San Gottardo, Passo del pass Switz.
83 F4 Sangre de Cristo Range mts U.S.A.
103 E2 Sangre Grande Trinidad and Tobago
44 C3 Sangrur India
45 F3 Sangsang China
76 G4 Sangudo Can.
101 G6 Sangue r. Brazil
23 G4 San Guiliano Milanese Italy
113 K1 Sangutane r. Moz.
39 G5 Sangvor Tajik.
51 D4 Sangzhi China
96 A2 San Hipólito, Pta pt Mex.
36 C7 Sanhûr Egypt
97 G4 San Ignacio Belize
100 E6 San Ignacio Beni Bol.
100 F7 San Ignacio Santa Cruz Bol.
96 A2 San Ignacio Mex.
78 E2 Sanikiluaq Can.
55 B2 San Ildefonso, Cape Phil.
55 B2 San Ildefonso Peninsula Phil.
55 C4 San Isidro Phil.
19 K2 Sanitz Ger.
55 B3 San Jacinto Phil.
92 D5 San Jacinto U.S.A.
92 D5 San Jacinto Peak U.S.A.
45 F5 Sanjai, R r. India

58 Selat Pandan *chan.* Sing.
74 B3 Selawik U.S.A.
20 F2 Selb Ger.
8 L5 Selbekken Norway
8 M5 Selbu Norway
12 F4 Selby U.K.
84 C2 Selby U.S.A.
111 C6 Selebi-Phikwe Botswana
46 F1 Selemdzhinsky Khr. *mts* Rus. Fed.
36 B2 Selendi Turkey
19 H2 Selenter See *l.* Ger.
15 H4 Sélestat France
58 Seletar Sing.
58 Seletar, P. *i.* Sing.
58 Seletar Resr Sing.
39 H2 Seletinskoye Kazak.
39 H1 Selety *r.* Kazak.
39 H1 Seletyteniz, Ozero *salt l.* Kazak.
Seleucia Pieria *see* Samandağı
84 C2 Selfridge U.S.A.
28 J2 Selib Rus. Fed.
108 A3 Sélibabi Maur.
20 B2 Seligenstadt Ger.
28 E3 Seliger, Oz. *l.* Rus. Fed.
93 F4 Seligman U.S.A.
109 E2 Selîma Oasis *oasis* Sudan
114 B2 Sélingué, Lac de *l.* Mali
89 J5 Selinsgrove U.S.A.
17 Q2 Selishche Rus. Fed.
29 H6 Selitrennoye Rus. Fed.
17 Q2 Selizharovo Rus. Fed.
9 L7 Seljord Norway
19 J5 Selke *r.* Ger.
77 K4 Selkirk Can.
11 F5 Selkirk U.K.
76 F4 Selkirk Mountains Can.
12 D3 Sellafield U.K.
68 B3 Sellheim *r.* Austr.
22 C3 Sellières France
93 G6 Sells U.S.A.
18 E5 Selm Ger.
87 C5 Selma *AL* U.S.A.
92 C3 Selma *CA* U.S.A.
87 B5 Selmer U.S.A.
19 H3 Selmsdorf Ger.
22 C2 Selongey France
22 D5 Selonnet France
114 A2 Sélouma Guinea
41 F4 Selseleh-ye Pīr Shūrān *mts* Iran
13 G7 Selsey Bill *hd* U.K.
15 J4 Seltz France
54 C4 Seluan *i.* Indon.
100 D5 Selvas *reg.* Brazil
77 J3 Selwyn Lake Can.
76 C2 Selwyn Mountains Can.
74 E3 Selwyn Mts Can.
65 G4 Selwyn Range *h.* Austr.
20 B3 Selz *r.* Ger.
59 D4 Semarang Indon.
59 C2 Sematan Malaysia
59 E3 Semayang, Danau *l.* Indon.
55 A6 Sembakung *r.* Indon.
58 Sembawang Sing.
110 B3 Sembé Congo
37 K3 Şemdinli Turkey
59 D4 Semenanjung Blambangan *pen.* Indon.
29 E4 Semenivka Ukr.
38 C1 Semenkino Rus. Fed.
28 H3 Semenov Rus. Fed.
29 G6 Semikarakorsk Rus. Fed.
29 F5 Semiluki Rus. Fed.
21 J2 Semily Czech Rep.
22 C3 Semine *r.* France
82 F3 Seminoe Resr U.S.A.
85 C5 Seminole U.S.A.
87 C6 Seminole, L. U.S.A.
39 L2 Seminskiy Khrebet *mts* Rus. Fed.
38 F1 Semiozernoye Kazak.
39 K2 Semipalatinsk Kazak.
39 J2 Semipalatinskaya Oblast' *div.* Kazak.
39 F1 Semipolka Kazak.
55 B3 Semirara *i.* Phil.
55 B4 Semirara Islands Phil.
40 C4 Semīrom Iran
39 J2 Semiyarka Kazak.
39 H2 Semiz-Bugu Kazak.
21 J5 Semmering *pass* Austria
40 D3 Semnān Iran
15 F3 Semois *r.* Belgium
15 E3 Semois, Vallée de la *v.* Belgium/France
22 F2 Sempach Switz.
22 F2 Sempacher See *l.* Switz.
59 E2 Semporna Malaysia
59 D4 Sempu *i.* Indon.
22 B2 Semur-en-Auxois France
55 A5 Senaja Malaysia
29 G7 Senaki Georgia
100 E5 Sena Madureira Brazil
43 C5 Senanayake Samudra *l.* Sri Lanka
111 C5 Senanga Zambia
15 B3 Senarpont France
22 C6 Sénas France
53 B9 Sendai Japan
52 G5 Sendai Japan
20 D4 Senden Ger.
45 H3 Sêndo China
58 B5 Senebui, Tanjung *pt* Indon.
21 L4 Senec Slovakia
93 G5 Seneca *AZ* U.S.A.
88 C5 Seneca *IL* U.S.A.
82 C2 Seneca *OR* U.S.A.
90 E3 Seneca Falls U.S.A.
90 E3 Seneca Lake U.S.A.
90 D5 Seneca Rocks U.S.A.
90 C5 Senecaville Lake U.S.A.
107 B4 Senegal *country* Africa
108 A3 Sénégal *r.* Maur./Senegal
113 G4 Senekal S. Africa
88 E2 Seney U.S.A.
19 M5 Senftenberg Ger.
115 B4 Senga Hill Zambia
44 D4 Sengar *r.* India
115 B3 Sengerema Tanz.
28 J4 Sengiley Rus. Fed.
38 C4 Sengiri, Mys *pt* Kazak.
21 H5 Sengsengebirge *mts* Austria
101 K6 Senhor do Bonfim Brazil
21 L4 Senica Slovakia
23 L6 Senigallia Italy
23 J5 Senio *r.* Italy
26 F2 Senj Croatia
8 P2 Senja *i.* Norway
37 J1 Şenkaya Turkey
44 D2 Senku Jammu and Kashmir
112 E2 Senlac S. Africa
54 F2 Senlin Shan *mt* China
15 C3 Senlis France
58 C2 Senmonorom Cambodia
13 B7 Sennen U.K.
89 J1 Senneterre Can.
113 H5 Senqu *r.* Lesotho
15 D4 Sens France
15 D2 Senséé *r.* France
96 G5 Sensuntepeque El Salvador
27 J2 Senta Yugo.
39 K2 Sentas Kazak.
44 D3 Senthal India
21 J6 Šentilj Slovenia
93 F5 Sentinel U.S.A.
76 E3 Sentinel Pk Can.
116 B3 Sentinel Ra. *mts* Ant.
58 Sentosa *i.* Sing.
37 H3 Şenyurt Turkey
44 D5 Seoni India
45 E5 Seorinarayan India
54 D5 Seoul S. Korea
70 D4 Separation Pt N.Z.
37 L4 Separ Shāhābād Iran
104 D3 Sepetiba, Baía de *b.* Brazil
40 C4 Sepīdān Iran
62 E2 Sepik *r.* P.N.G.
54 D4 Sep'o N. Korea
22 C6 Septèmes-les-Vallons France
15 B4 Septeuil France
79 G3 Sept-Îles Can.
92 C3 Sequoia National Park U.S.A.
37 L3 Serā Iran
29 G5 Serafimovich Rus. Fed.
41 F2 Serakhs Turkm.
57 H7 Seram *i.* Indon.
57 J7 Seram Sea *g.* Indon.
59 C4 Serang Indon.
58 Serangoon Harbour *chan.* Sing.
58 D5 Serasan *i.* Indon.
59 C2 Serasan, Selat *chan.* Indon.
23 H6 Seravezza Italy
58 D5 Seraya *i.* Indon.
58 Seraya, P. *i.* Sing.
27 J3 Serbia *div.* Yugo.
23 H5 Serchio *r.* Italy
96 C1 Serdán Mex.
Serdar *see* Kaypak
110 E2 Serdo Eth.
29 H4 Serdoba *r.* Rus. Fed.
29 H4 Serdobsk Rus. Fed.
39 K2 Serebryansk Kazak.
21 L4 Sereď Slovakia
28 D3 Seredka Rus. Fed.
36 D2 Şereflikoçhisar Turkey
22 B2 Serein *r.* France
59 B2 Seremban Malaysia
115 B3 Serengeti National Park Tanz.
115 B5 Serenje Zambia
20 D5 Serfaus Austria
28 H4 Sergach Rus. Fed.
52 C3 Sergeyevka Rus. Fed.
39 G2 Sergeyevka *Akmolinsk.* Kazak.
39 F1 Sergeyevka *Severnyy Kaz.* Kazak.
28 F3 Sergiyev Posad Rus. Fed.
59 D2 Seria Brunei
59 D2 Serian Malaysia
15 B3 Sérifontaine France
27 L6 Serifos *i.* Greece
79 G2 Sérigny, Lac *l.* Can.
36 C3 Serik Turkey
39 J5 Serikbuya China
66 C3 Serle, Mt *h.* Austr.
57 H8 Sermata, Kepulauan *is* Indon.
28 J3 Sernur Rus. Fed.
28 J3 Sernyy Zavod Turkm.
39 D6 Seroglazka Rus. Fed.
32 H4 Serov Rus. Fed.
111 C6 Serowe Botswana
25 C4 Serpa Port.
69 B6 Serpentine *r.* Austr.
103 E2 Serpent's Mouth *chan.* Trinidad/Venez.
28 F4 Serpukhov Rus. Fed.
104 C3 Serra da Canastra, Parque Nacional da *nat. park* Brazil
23 H5 Serramazzoni Italy
103 D4 Serranía de la Neblina, Parque Nacional *nat. park* Venez.
104 B2 Serranópolis Brazil
23 K6 Serravalle San Marino
23 F5 Serravalle Scrivia Italy
15 D3 Serre *r.* France
22 C5 Serres France
27 K4 Serres Greece
105 D1 Serrezuela Arg.
101 L6 Serrinha Brazil
104 D2 Sêrro Brazil
105 B4 Serrucho *mt* Arg.
26 C6 Sers Tunisia
104 C3 Sertãozinho Brazil
28 D2 Sertolovo Rus. Fed.
58 A4 Seruai Indon.
57 K7 Serui Indon.
59 D3 Seruyan *r.* Indon.
46 B3 Sêrxu China
59 E2 Sesayap Indon.
55 A6 Sesayap *r.* Indon.
78 B3 Seseganaga L. Can.
89 G1 Sesekinika Can.
71 13 Seseleka *h.* Fiji
111 B5 Sesfontein Namibia
113 H3 Seshego S. Africa
111 C5 Sesheke Zambia
22 F4 Sesia *r.* Italy
26 E4 Sessa Aurunca Italy
23 J6 Sesto Fiorentino Italy
23 G4 Sesto San Giovanni Italy
22 D5 Sestriere Italy
23 G5 Sestri Levante Italy
28 D2 Sestroretsk Rus. Fed.
52 F3 Setana Japan
24 F5 Sète France
104 D2 Sete Lagoas Brazil
8 Q2 Setermoen Norway
9 K7 Setesdal *v.* Norway
45 F4 Seti *r. Gandakhi* Nepal
44 E3 Seti *r. Seti* Nepal
108 C1 Sétif Alg.
53 E7 Seto Japan
58 C2 Set, P. *mt* Laos
108 B1 Settat Morocco
22 F5 Settepani, Monte *mt* Italy
22 E4 Settimo Torinese Italy
12 E3 Settle U.K.
25 B3 Setúbal Port.
25 B3 Setúbal, Baía de *b.* Port.
88 E3 Seul Choix Pt U.S.A.
78 B3 Seul, Lac *l.* Can.
22 C2 Seurre France
37 K1 Sevan Armenia
Sevana Lich *l. see* Sevan, Lake
37 K1 Sevan, Lake Armenia
29 E6 Sevastopol' Ukr.
79 H2 Seven Islands Bay Can.
69 C1 Seven Mile Creek *r.* Austr.
13 H6 Sevenoaks U.K.
Seven Pagodas *see* Māmallapuram
24 F4 Sévérac-le-Château France
67 J2 Severn *r.* Austr.
78 B3 Severn *r.* Can.
112 E3 Severn S. Africa
13 E6 Severn *r.* U.K.
28 H2 Severnaya Dvina *r.* Rus. Fed.
29 H7 Severnaya Osetiya, Respublika *div.* Rus. Fed.
33 M1 Severnaya Zemlya *is* Rus. Fed.
78 B3 Severn L. Can.
32 H3 Severnyy Rus. Fed.
38 D3 Severnyy Chink Ustyurta *esc.* Kazak.
46 D1 Severo Baykalskoye Nagorye *mts* Rus. Fed.
21 G2 Severočeský *div.* Czech Rep.
28 F1 Severodvinsk Rus. Fed.
33 R4 Severo-Kuril'sk Rus. Fed.
21 L3 Severomoravský *div.* Czech Rep.
8 X2 Severomorsk Rus. Fed.
32 L3 Severo-Yeniseyskiy Rus. Fed.
29 F6 Severskaya Rus. Fed.
23 G4 Seveso Italy
23 G4 Seveso *r.* Italy
83 D4 Sevier *r.* U.S.A.
93 G2 Sevier Bridge Reservoir U.S.A.
93 F2 Sevier Desert U.S.A.
93 F2 Sevier Lake *salt l.* U.S.A.
103 B3 Sevilla Col.
Sevilla *see* Seville
25 D4 Seville Spain
27 L3 Sevlievo Bulg.
22 C3 Sevron *r.* France
43 C3 Sewāni India
74 D3 Seward *AK* U.S.A.
84 D3 Seward *NE* U.S.A.
74 B3 Seward Peninsula U.S.A.
76 F3 Sexsmith Can.
80 E6 Sextín *r.* Mex.
41 F3 Seyah Band Koh *mts* Afgh.
37 K2 Seyah Cheshmeh Iran
32 J2 Seyakha Rus. Fed.
97 G4 Seybaplaya Mex.
5 Seychelles *country* Indian Ocean
38 E5 Seydi Turkm.
36 C3 Seydişehir Turkey
8 F4 Seyðisfjörður Iceland
40 B2 Seydvān Iran
Seyhan *see* Adana
36 E3 Seyhan *r.* Turkey
29 E5 Seym *r.* Rus. Fed.
33 R3 Seymchan Rus. Fed.
66 F6 Seymour Austr.
113 G6 Seymour S. Africa
86 C4 Seymour *IN* U.S.A.
85 D5 Seymour *TX* U.S.A.
22 B5 Seynes France
22 D4 Seynod France
41 H3 Seyyedābād Afgh.
23 L4 Sežana Slovenia
15 D4 Sézanne France
27 L7 Sfakia Greece
27 L2 Sfântu Gheorghe Romania
108 D1 Sfax Tunisia
27 K4 Sfikia, Limni *resr* Greece
99 G8 S. Georgia *i.* Atl. Ocean
18 C4 's-Graveland Neth.
's-Gravenhage *see* The Hague
11 C4 Sgurr Dhomhnuill *h.* U.K.
11 C3 Sgurr Mor *mt* U.K.
50 E2 Sha *r.* China
50 C3 Shaanxi *div.* China
40 B2 Shabestar Iran
79 G3 Shabogamo Lake Can.
115 A3 Shabunda Congo(Zaire)
39 J5 Shache China
116 B4 Shackleton Coast Ant.
116 B4 Shackleton Gl. Ant.
116 D6 Shackleton Ice Shelf *ice feature* Ant.
116 C3 Shackleton Ra. *mts* Ant.
44 A4 Shadadkot Pak.
40 C4 Shādegān Iran
44 A3 Shadikhak Pass Pak.
40 D4 Shādkām *watercourse* Iran
88 D5 Shafer, Lake U.S.A.
116 B5 Shafer Pk Ant.
92 C4 Shafter U.S.A.
13 E6 Shaftesbury U.K.
38 D3 Shagan *watercourse Aktyubinsk.* Kazak.
39 J2 Shagan *watercourse Semipal.* Kazak.
74 C3 Shageluk U.S.A.
39 G1 Shaglyteniz, Oz. *l.* Kazak.
39 G1 Shaglyteniz, Oz. *l.* Kazak.
70 C6 Shag Pt N.Z.
116 B2 Shag Rocks *is* Atl. Ocean
38 D4 Shagyrlyk Uzbek.
40 C2 Shah *r.* Iran
43 B2 Shahabad *Karnataka* India
44 E4 Shahabad *Uttar Pradesh* India
44 C5 Shahada India
59 B2 Shah Alam Malaysia
43 A3 Shahapur India
44 A4 Shahbandar Pak.
41 G5 Shahbaz Kalat Pak.
45 G5 Shahbazpur *chan.* Bangl.
41 E4 Shahdāb Iran
44 E5 Shahdol India
41 G3 Shah Fuladi *mt* Afgh.
44 B4 Shahgarh India
37 K3 Shahīdān Iraq
37 K3 Shahi Pen. Iran
41 G4 Shah Ismail Afgh.
44 D5 Shahjahanpur India
40 E2 Shāh Jehān, Kūh-e *mts* Iran
41 E4 Shāh Kūh *mt* Iran
40 D3 Shahmīrzād Iran
43 B2 Shahpur India
44 B3 Shahpur Pak.
44 E5 Shahpura *Madhya Pradesh* India
44 D4 Shahpura *Rajasthan* India
41 G3 Shahrak Afgh.
41 F3 Shahrakht Iran
40 E4 Shahr-e Bābāk Iran
40 C3 Shahr-e Kord Iran
Shahrezā *see* Qomishēh
40 C3 Shahr Rey Iran
39 G5 Shahrtuz Tajik.
40 D3 Shahrud Bustam *reg.* Iran
39 F4 Shaidara, Step' *plain* Kazak.
41 G4 Shaikh Husain *mt* Pak.
40 C5 Shaj'ah, J. h. S. Arabia
45 C3 Shajianzi China
54 C3 Shakar Bolāghī Iran
113 J4 Shakaville S. Africa
39 G5 Shakhdara *r.* Tajik.
39 G5 Shakhdara *r.* Tajik.
41 E3 Shākhen Iran
28 E3 Shakhovskaya Rus. Fed.
39 F5 Shakhrisabz Uzbek.
39 H2 Shakhtinsk Kazak.
38 D2 Shakhty Kazak.
29 G6 Shakhty Rus. Fed.
28 H3 Shakhun'ya Rus. Fed.
84 E2 Shakopee U.S.A.
52 G3 Shakotan-hantō *pen.* Japan
52 G3 Shakotan-misaki *c.* Japan
28 G2 Shalakusha Rus. Fed.
38 E4 Shalday Kazak.
39 J2 Shalginskiy Kazak.
38 E2 Shalkar Karashatau *salt l.* Kazak.
38 B2 Shalkar, Oz. *salt l.* Kazak.
46 B3 Shaluli Shan *mts* China
45 H3 Shaluni *mt* India
115 B4 Shama *r.* Tanz.
77 L3 Shamattawa Can.
51 Sham Chun *h. H.K.* China
40 E5 Shamil Iran
40 E5 Shamis U.A.E.
90 E4 Shamokin U.S.A.
85 C5 Shamrock U.S.A.
111 D5 Shamva Zimbabwe
14 B6 Shanacrane Rep. of Ireland
41 G3 Shand Afgh.
41 F4 Shāndak Iran
42 D3 Shandan China
37 M3 Shanderman Iran
50 E1 Shandian *r.* China
42 E1 Shandiz Iran
92 B4 Shandon U.S.A.
54 A5 Shandong *div.* China
50 F2 Shandong Bandao *pen.* China
37 K5 Shandrūkh Iraq
41 F3 Shandur Pass Pak.
111 C5 Shangani *r.* Zimbabwe
50 E3 Shangcai China
51 C5 Shangchao China
51 D4 Shangcheng China
45 G3 Shang Chu *r.* China
51 D6 Shangchuan Dao *i.* China
50 D1 Shangdu China
51 E4 Shanggao China
50 E4 Shanghai China
51 F4 Shanghai *div.* China
51 E5 Shanghang China
50 E2 Shanghe China
54 C3 Shanghekou China
50 D3 Shangjin China
51 D6 Shanglin China
50 F3 Shangmian China
50 D3 Shangqiu *Henan* China
51 E4 Shangrao *Jiangxi* China
39 K3 Shangsanshilipu China
50 E3 Shangshui China
51 C6 Shangsi China
51 E4 Shangtang China
50 D1 Shangyi China
51 E5 Shangyou China
39 K4 Shangyou Sk. *salt flat* China
51 F4 Shangyu China
54 D1 Shangzhi China
50 C3 Shangzhou China
54 D1 Shanhetun China
14 C5 Shannon *est.* Rep. of Ireland
14 C4 Shannon *r.* Rep. of Ireland
14 B5 Shannon, Mouth of the *est.* Rep. of Ireland
69 C7 Shannon National Park Austr.
48 F3 Shanshan China
39 K4 Shanshanggang China
45 G5 Shāntipur India
51 E5 Shantou China
51 E6 Shanwei China
50 D2 Shanxi *div.* China
50 E3 Shan Xian China
50 C3 Shanyang China
50 D2 Shanyin China
51 D5 Shaodong China
51 D5 Shaoguan China
51 E5 Shaowu China
51 F4 Shaoxing China
51 D5 Shaoyang *Hunan* China
12 E3 Shap U.K.
51 D6 Shapa China
11 F1 Shapinsay *i.* U.K.
42 C4 Shaqrā' S. Arabia
37 J6 Sharaf *well* Iraq
44 B3 Sharan Jogizai Pak.
39 K1 Sharchino Rus. Fed.
39 F5 Shargun Uzbek.
29 D5 Sharhorod Ukr.
52 J3 Shari-dake *volc.* Japan
42 E4 Sharjah U.A.E.
17 N3 Sharkawshchyna Belarus
69 A2 Shark Bay Austr.
38 C5 Sharlouk Turkm.
38 C1 Sharlyk Rus. Fed.
91 G4 Sharon *CT* U.S.A.
90 C4 Sharon *PA* U.S.A.
51 Sharp Peak *h. H.K.* China
36 E5 Sharqi, Jebel esh *mts* Lebanon/Syria
28 H3 Shar'ya Rus. Fed.
111 C6 Shashe *r.* Botswana/Zimbabwe
110 D3 Shashemenē Eth.
51 D4 Shashi China
39 H3 Shashubay Kazak.
82 B3 Shasta L. U.S.A.
82 B3 Shasta, Mt *volc.* U.S.A.
51 Sha Tin *H.K.* China
28 H4 Shatki Rus. Fed.
37 M7 Shatt al Arab *r.* Iran/Iraq
37 K6 Shatt al Hillah *r.* Iraq
40 C4 Shaṭṭ, Ra's osh *pt* Iran
28 F4 Shatura Rus. Fed.
36 E6 Shaubak Jordan
39 G4 Shaul'der Kazak.
77 H5 Shaunavon Can.
90 D5 Shavers Fork *r.* U.S.A.
91 F4 Shawangunk Mts *h.* U.S.A.
88 C3 Shawano U.S.A.
88 C3 Shawano Lake U.S.A.
78 F4 Shawinigan Can.
36 F4 Shawmarīyah, Jebel ash *mts* Syria
85 D5 Shawnee U.S.A.
51 F5 Sha Xi *r.* China
51 E5 Sha Xian China
51 D4 Shayang China
64 D4 Shay Gap Austr.
37 L5 Shaykh Jūwī Iraq
37 L5 Shaykh Sa'd Iraq
40 D4 Shaytūr Iran
39 H5 Shazud Tajik.
28 F4 Shchekino Rus. Fed.
33 S3 Shcherbakovo Rus. Fed.
39 J1 Shcherbakty Kazak.
29 F5 Shchigry Rus. Fed.
29 D5 Shchors Ukr.
39 G1 Shchuchinsk Kazak.
29 C4 Shchuchyn Belarus
39 L2 Shebalino Rus. Fed.
29 F5 Shebekino Rus. Fed.
41 G2 Sheberghān Afgh.
88 D4 Sheboygan U.S.A.
114 F3 Shebshi Mountains Nigeria
79 H4 Shediac Can.
76 D3 Shedin Pk Can.
14 D4 Sheelin, Lough *l.* Rep. of Ireland
14 D2 Sheep Haven *b.* Rep. of Ireland
113 J3 Sheepmoor S. Africa
93 E3 Sheep Peak U.S.A.
13 H6 Sheerness U.K.
79 H5 Sheet Harbour Can.
67 G8 Sheffield Austr.
70 D5 Sheffield N.Z.
13 F4 Sheffield U.K.
87 C5 Sheffield *AL* U.S.A.
88 C5 Sheffield *IL* U.S.A.
90 D4 Sheffield *PA* U.S.A.
85 C6 Sheffield *TX* U.S.A.
89 G3 Sheguiandah Can.
50 B4 Shehong China
36 E5 Sheikh, Jebel esh *mt* Lebanon/Syria
44 C3 Shekhupura Pak.
51 Shek Kwu Chau *i. H.K.* China
51 Shek Pik Reservoir *H.K.* China
28 F3 Sheksna Rus. Fed.
51 Shek Uk Shan *h. H.K.* China
41 F4 Shelag *watercourse* Afgh./Iran
33 T2 Shelagskiy, Mys *pt* Rus. Fed.
88 A6 Shelbina U.S.A.
79 G5 Shelburne *N.S.* Can.
89 G3 Shelburne *Ont.* Can.
91 G3 Shelburne Falls U.S.A.
88 D4 Shelby *MI* U.S.A.
82 E1 Shelby *MT* U.S.A.
87 D5 Shelby *NC* U.S.A.
90 B4 Shelby *OH* U.S.A.
86 C4 Shelbyville *IN* U.S.A.
88 A6 Shelbyville *MO* U.S.A.
87 C5 Shelbyville *TN* U.S.A.
93 H5 Sheldon *AZ* U.S.A.
88 D5 Sheldon *IL* U.S.A.
91 G2 Sheldon Springs U.S.A.
79 H3 Sheldrake Can.
33 R3 Shelikhova, Zaliv *g.* Rus. Fed.
74 C4 Shelikof Strait U.S.A.
77 H4 Shellbrook Can.
82 D3 Shelley U.S.A.
67 J5 Shellharbour Austr.
92 A1 Shell Mt U.S.A.
76 F4 Shelter Bay Can.
92 A1 Shelter Cove U.S.A.
51 Shelter I. *H.K.* China
91 G4 Shelter I. U.S.A.
70 B7 Shelter Pt N.Z.
39 K2 Shemonaikha Kazak.
84 E3 Shenandoah *IA* U.S.A.
91 E4 Shenandoah *PA* U.S.A.
90 D5 Shenandoah *VA* U.S.A.
90 D5 Shenandoah *r.* U.S.A.
90 D5 Shenandoah Mountains U.S.A.
90 D5 Shenandoah National Park U.S.A.
90 C4 Shenango River Lake U.S.A.
38 E2 Shenbertal Kazak.
114 E3 Shendam Nigeria
52 C1 Shending Shan *h.* China
43 E3 Shengel'dy Kazak.
115 C3 Shengena *mt* Tanz.
50 G4 Shengsi China
51 F4 Sheng Xian China
28 G2 Shenkursk Rus. Fed.
50 D2 Shenmu China
50 D4 Shennongjia China
50 E3 Shenqiu China
52 A1 Shenshu China
54 B3 Shenwo Sk. *resr* China
54 B3 Shenyang China
51 E6 Shenzhen China
29 C5 Shepetivka Ukr.
71 9 Shepherd Is Vanuatu
67 F6 Shepparton Austr.
13 H6 Sheppey, Isle of *i.* U.K.
39 F5 Sherabad Uzbek.
13 E7 Sherborne U.K.
79 H4 Sherbrooke *N.S.* Can.
79 F4 Sherbrooke *Que.* Can.
91 F3 Sherburne U.S.A.
14 E4 Shercock Rep. of Ireland
41 H3 Sher Dahan Pass Afgh.
109 F3 Shereiq Sudan
44 C4 Shergarh India
85 E5 Sheridan *AR* U.S.A.
82 F2 Sheridan *WY* U.S.A.
66 A4 Sheringa Austr.
13 J5 Sheringham U.K.
85 D5 Sherman U.S.A.
91 J2 Sherman Mills U.S.A.
93 E1 Sherman Mtn U.S.A.
45 G4 Sherpur Bangl.
77 J3 Sherridon Can.
18 C5 's-Hertogenbosch Neth.
13 F4 Sherwood Forest *reg.* U.K.
76 C3 Sheslay Can.
10 F1 Shetland Islands U.K.
38 C3 Shetpe Kazak.
51 Sheung Shui *H.K.* China
51 Sheung Sze Mun *chan. H.K.* China
43 B4 Shevaroy Hills India
51 F4 She Xian China
50 F3 Sheyang China
84 D2 Sheyenne U.S.A.
84 D2 Sheyenne *r.* U.S.A.
40 D5 Sheykh Sho'eyb *i.* Iran
11 B3 Shiant Islands U.K.
46 H2 Shiashkotan, O. *i.* Rus. Fed.
89 E4 Shiawassee *r.* U.S.A.
42 C6 Shibām Yemen
41 H3 Shibar Pass Afgh.
53 F6 Shibata Japan
52 J3 Shibetsu Japan
52 H2 Shibetsu Japan
109 F1 Shibîn el Kôm Egypt
53 F6 Shibukawa Japan
51 E5 Shicheng China
54 B4 Shicheng Dao *i.* China
91 E4 Shickshinny U.S.A.
36 G6 Shidād al Mismā' *h.* S. Arabia
54 B5 Shidao China
54 B5 Shidao Wan *b.* China
39 H2 Shiderty *r.* Kazak.
11 C4 Shiel, Loch *l.* U.K.
50 B4 Shifang China
28 J4 Shigony Rus. Fed.
50 D1 Shiguaigou China
35 G2 Shihezi China
50 E2 Shijiazhuang China
41 F4 Shikar *r.* Pak.
43 A3 Shikarpur India
44 B4 Shikarpur Pak.
44 D4 Shikohabad India
53 C8 Shikoku *i.* Japan
53 C8 Shikoku-sanchi *mts* Japan
52 G3 Shikotsu-Tōya National Park Japan
38 D2 Shil'da Rus. Fed.
12 F3 Shildon U.K.
28 H1 Shilega Rus. Fed.
45 G4 Shiliguri India
51 D4 Shilipu China
44 D2 Shilla *mt* India
14 E5 Shillelagh Rep. of Ireland
89 G1 Shillington Can.
45 G4 Shillong India
28 G4 Shilovo Rus. Fed.
28 B2 Shil'naya Balka Kazak.
91 F5 Shiloh U.S.A.
50 D2 Shilou China
28 G4 Shilovo Rus. Fed.
53 B8 Shimabara Japan
53 F7 Shimada Japan
49 N1 Shimanovsk Rus. Fed.
51 D4 Shimen China
51 B4 Shimian China
53 F7 Shimizu Japan
44 D3 Shimla India
53 F7 Shimoda Japan
43 A3 Shimoga India
115 C3 Shimoni Kenya
53 B8 Shimonoseki Japan
44 C1 Shimshal Jammu and Kashmir
28 D3 Shimsk Rus. Fed.
51 C6 Shinan China
41 F3 Shīndand Afgh.
44 B3 Shinghar Pak.
41 C1 Shingshal Pass Pak.
88 D2 Shingleton U.S.A.
51 Shing Mun Resr *H.K.* China
39 K3 Shingozha Kazak.
53 E8 Shingū Japan
113 J1 Shingwedzi S. Africa
113 J1 Shingwedzi *r.* S. Africa
13 E4 Shining Tor *h.* U.K.
89 G2 Shining Tree Can.
52 G5 Shinjō Japan
41 G4 Shinkāy Afgh.
11 D2 Shin, Loch *l.* U.K.
53 E6 Shinminato Japan
91 J1 Shin Pond U.S.A.
115 B3 Shinyanga Tanz.
115 B3 Shinyanga *div.* Tanz.
52 G5 Shiogama Japan
53 D8 Shiono-misaki *c.* Japan
53 G6 Shioya-zaki *pt* Japan
87 E7 Ship Chan Cay *i.* Bahamas
51 B6 Shiping China
44 D3 Shipki Pass China/India
12 F4 Shipley U.K.
79 H4 Shippegan Can.
90 E4 Shippensburg U.S.A.
93 H3 Shiprock U.S.A.
93 H3 Shiprock Peak U.S.A.
51 F4 Shipu China
39 K1 Shipunovo Rus. Fed.
51 D5 Shiqian China
50 C3 Shiquan China
44 E2 Shiquan He *r.* China
50 C3 Shiquan Sk. *resr* China
37 M2 Shīrābād Iran
53 G6 Shirakawa Japan
46 F3 Shirane-san *mt* Japan
53 F6 Shirane-san *volc.* Japan
116 D4 Shirasebreen *ice feature* Ant.
116 A4 Shirase Coast Ant.
40 D4 Shīrāz Iran
36 C6 Shirbîn Egypt
115 B6 Shire *r.* Malawi
52 J2 Shiretoko-misaki *c.* Japan
38 E3 Shirikrabat *tourist site* Kazak.
39 G4 Shirin Uzbek.
41 G4 Shirinab *r.* Pak.
52 G4 Shiriya-zaki *c.* Japan
38 D3 Shirkala *reg.* Kazak.
91 G4 Shirley U.S.A.
91 J2 Shirley Mills U.S.A.
114 E3 Shiroro Reservoir Nigeria
53 E7 Shirotori Japan
44 C5 Shirpur India
41 E2 Shīrvān Iran
37 L5 Shīrvān Iran
51 D4 Shishou China
51 E4 Shitai China
51 F4 Shitang China
37 J5 Shithāthah Iraq
44 B4 Shiv India
86 C4 Shively U.S.A.
44 D4 Shivpuri India
93 F3 Shivwits Plateau *plat.* U.S.A.
41 H2 Shiwal *l.* Afgh.
51 C6 Shiwan Dashan *mts* China
51 E5 Shixing China
50 D3 Shiyan China
51 C4 Shizhu China
51 B5 Shizong China
52 G5 Shizugawa Japan
50 C2 Shizuishan China
53 F7 Shizuoka Japan
28 D4 Shklow Belarus
27 H3 Shkodër Albania
32 K1 Shmidta, Ostrov *i.* Rus. Fed.
68 D4 Shoalwater B. Austr.
53 C7 Shōbara Japan
52 G3 Shokanbetsu-dake *mt* Japan
38 F2 Sholakkorgan Kazak.
38 D3 Sholakkorgan Kazak.
28 J2 Shomvukva Rus. Fed.
38 D3 Shoptykol' *Aktyubinsk.* Kazak.
39 H2 Shoptykol' *Pavlodarsk.* Kazak.
44 D2 Shor India
43 B4 Shoranur India
41 G5 Shorap Pak.
41 G4 Shorawak *reg.* Afgh.
38 D4 Shor Barsa-Kel'mes *salt marsh* Uzbek.
37 K3 Shor Gol Iran
44 C3 Shorkot Pak.
38 C4 Shorkozakhly, Solonchak *depression* Turkm.
39 G2 Shortandy Kazak.
52 G2 Shosanbetsu Japan
92 D4 Shoshone *CA* U.S.A.
82 D3 Shoshone *ID* U.S.A.
82 E2 Shoshone *r.* U.S.A.
82 E2 Shoshone L. U.S.A.
83 C4 Shoshone Mts U.S.A.
113 G1 Shoshong Botswana
82 E3 Shoshoni U.S.A.
29 E5 Shostka Ukr.
50 F2 Shouguang China
51 F5 Shouning China
50 E3 Shou Xian China
50 D2 Shouyang China
50 C3 Shouyang Shan *mt* China
45 E3 Shovo Tso *salt l.* China
93 G4 Show Low U.S.A.
29 G6 Shpakovskoye Rus. Fed.
29 D5 Shpola Ukr.
85 E5 Shreveport U.S.A.
13 E5 Shrewsbury U.K.
43 A2 Shrigonda India
45 G5 Shrirampur India
50 F3 Shu *r.* China
37 L6 Shu'aiba Iraq
51 A5 Shuangbai China
54 D1 Shuangcheng China
50 C4 Shuanghechang China
48 G7 Shuangjiang China
54 B2 Shuangliao China
51 D5 Shuangpai China
54 A3 Shuangtaizihe Kou *b.* China
54 C2 Shuangyang China
52 B1 Shuangyashan China
38 D2 Shubarkuduk Kazak.
38 D2 Shubarshi Kazak.
50 E4 Shucheng China
39 H5 Shufu China
51 F5 Shuiji China
50 A2 Shuiquanzi China
56 B3 Shujaabad Pak.
40 C4 Shūl *watercourse* Iran
54 D1 Shulan China
39 J5 Shule China
50 E2 Shulu China
38 D4 Shumanay Uzbek.
52 H2 Shumarinai-ko *l.* Japan
111 C5 Shumba Zimbabwe
27 M3 Shumen Bulg.
28 H4 Shumerlya Rus. Fed.
17 O3 Shumilina Belarus
93 G4 Shumway U.S.A.
29 D5 Shumyachi Rus. Fed.
39 H3 Shunak, Gora *mt* Kazak.
51 E5 Shunchang China
51 D6 Shunde China
74 C3 Shungnak U.S.A.
50 E1 Shunyi China
51 C6 Shuolong China
50 D2 Shuo Xian China
42 C7 Shuqrah Yemen
41 F3 Shūr *r.* Iran
40 D4 Shūr *r.* Iran
40 D4 Shūr *r.* Iran
41 E3 Shur *watercourse* Iran
40 C4 Shūr *watercourse* Iran
40 D5 Shūr *watercourse* Iran
40 E3 Shūrāb Iran
40 C3 Shūr Āb Iran
40 E3 Shūrāb Iran
40 E4 Shūr Āb *watercourse* Iran
39 F5 Shurchi Uzbek.
40 D3 Shureghestan Iran
41 E4 Shūr Gaz Iran
40 D4 Shūrjestān Iran
39 G4 Shūrob Tajik.
111 D5 Shurugwi Zimbabwe
37 K6 Shuruppak Iraq
41 F4 Shusf Iran
40 C3 Shūsh Iran
40 C3 Shushtar Iran
76 F4 Shuswap L. Can.
41 G3 Shutar Khun Pass Afgh.
28 G3 Shuya Rus. Fed.
51 F5 Shuyang China
40 D4 Shūzū Iran
56 B2 Shwebo Myanmar
58 A1 Shwegun Myanmar
56 B3 Shwegyin Myanmar
39 G4 Shymkent Kazak.
44 D2 Shyok *r.* India
44 D2 Shyok Jammu and Kashmir
29 F5 Shypuvate Ukr.
29 E6 Shyroke Ukr.
57 J8 Sia Indon.
44 D2 Siachen Gl. India
22 D6 Siagne *r.* France
41 F5 Siahan Range *mts* Pak.
41 G3 Siah Koh *mts* Afgh.
40 D3 Siāh Kūh *mts* Iran
41 G4 Siah Sang Pass Afgh.
44 C2 Sialkot Pak.
58 C5 Siantan *i.* Indon.
103 D4 Siapa *r.* Venez.
41 F4 Siāreh Iran
55 C4 Siargao *i.* Phil.
55 B5 Siasi Phil.
55 B5 Siasi *i.* Phil.
55 B4 Siaton Phil.
9 S9 Šiauliai Lith.
41 F5 Sib Iran
40 C3 Sībak Iran
113 J1 Sibasa S. Africa
55 B4 Sibay *i.* Phil.
38 D1 Sibay Rus. Fed.
113 K3 Sibayi, Lake S. Africa
116 B5 Sibbald, C. Ant.
26 F3 Šibenik Croatia
58 A3 Siberut *i.* Indon.
44 A3 Sibi Pak.
115 C2 Sibiloi National Park Kenya
52 C2 Sibirtsevo Rus. Fed.
110 B4 Sibiti Congo
27 L2 Sibiu Romania
59 A2 Sibolga Indon.
58 A5 Siborongborong Indon.
45 H4 Sibsagar India
59 D2 Sibu Malaysia
55 B5 Sibuco Phil.
55 B5 Sibuguey *r.* Phil.
55 B5 Sibuguey Bay Phil.
110 B3 Sibut C.A.R.
55 A5 Sibutu *i.* Phil.
55 A5 Sibutu Passage Phil.
55 B3 Sibuyan *i.* Phil.
55 B3 Sibuyan Sea Phil.
55 B2 Sicapoo *mt* Phil.
51 B4 Sichuan *div.* China
51 B4 Sichuan Pendi *basin* China
22 C6 Sicié, Cap *c.* France
Sicilia *i. see* Sicily
26 E6 Sicilian Channel Italy/Tunisia
26 E6 Sicily *i.* Italy
100 D6 Sicuani Peru
52 D2 Sidatun Rus. Fed.
44 C5 Siddhapur India
43 B2 Siddipet India
27 M7 Sideros, Akra *pt* Greece
112 E6 Sidesaviwa S. Africa
25 H5 Sidi Aïssa Alg.
25 G4 Sidi Ali Alg.
108 B1 Sidi Bel Abbès Alg.
26 C7 Sidi Bouzid Tunisia
26 D7 Sidi El Hani, Sebkhet de *salt pan* Tunisia
108 A2 Sidi Ifni Morocco
108 B1 Sidi Kacem Morocco
58 A5 Sidikalang Indon.
114 B2 Sidikila Guinea
11 E4 Sidlaw Hills U.K.
116 A4 Sidley, Mt Ant.
13 D7 Sidmouth U.K.
76 E5 Sidney Can.
82 F2 Sidney *MT* U.S.A.
84 C3 Sidney *NE* U.S.A.
91 F3 Sidney *NY* U.S.A.
90 A4 Sidney *OH* U.S.A.
87 D5 Sidney Lanier, L. U.S.A.
56 Sidoarjo Indon.
45 H5 Sidoktaya Myanmar
36 E5 Sidon Lebanon
28 G3 Sidorovo Rus. Fed.
104 A3 Sidrolândia Brazil
113 J3 Sidvokodvo Swaziland
39 G4 Sidzhak Uzbek.
24 F5 Sié, Col de *pass* France
17 L4 Siedlce Pol.
18 F6 Siegen Ger.
21 K4 Sieghartskirchen Austria
58 B2 Siĕmréab Cambodia
23 J6 Siena Italy
17 J5 Sieradz Pol.
21 M1 Sieradz *div.* Pol.
21 H4 Sierning Austria
85 B6 Sierra Blanca U.S.A.
105 C4 Sierra Colorada Arg.
93 F5 Sierra Estrella *mts* U.S.A.
105 D4 Sierra Grande Arg.
107 B5 Sierra Leone *country* Africa
120 H5 Sierra Leone Basin *sea feature* Atl. Ocean
120 H5 Sierra Leone Rise *sea feature* Atl. Ocean
92 C4 Sierra Madre Mts U.S.A.
96 D2 Sierra Mojada Mex.
92 B1 Sierra Nevada *mts* U.S.A.
103 B2 Sierra Nevada de Santa Marta, Parque Nacional *nat. park* Col.
103 C2 Sierra Nevada, Parque Nacional *nat. park* Venez.
105 D4 Sierra, Punta *pt* Arg.
92 B2 Sierraville U.S.A.
93 G6 Sierra Vista U.S.A.
22 E3 Sierre Switz.
23 J6 Sieve *r.* Italy
8 T5 Sievi Fin.
51 C6 Sifang Ling *mts* China
27 L6 Sifnos *i.* Greece
25 F5 Sig Alg.
71 15 Sigatoka Fiji
71 15 Sigatoka *r.* Fiji
17 L7 Sighetu Marmaţiei Romania
17 M7 Sighişoara Romania
58 Siglap Sing.
59 A1 Sigli Indon.
8 D3 Siglufjörður Iceland
55 B4 Sigma Phil.
20 C4 Sigmaringen Ger.
15 G2 Signal de Botrange *h.* Belgium
22 C6 Signal de la Ste-Baume *mt* France
22 B4 Signal de St-André *h.* France
93 E5 Signal Peak U.S.A.
116 B1 Signy *U.K. Base* Ant.
15 E3 Signy-l'Abbaye France
88 A5 Sigourney U.S.A.
27 L5 Sigri, Akra *pt* Greece
97 H5 Siguatepeque Honduras
25 E2 Sigüenza Spain
114 B2 Siguiri Guinea
9 T8 Sigulda Latvia
58 B3 Sihanoukville Cambodia
22 F2 Sihlsee *l.* Switz.
50 F3 Sihong China
44 E5 Sihora India
51 D6 Sihui China
8 T4 Siikajoki Fin.
8 U5 Siilinjärvi Fin.
37 H3 Siirt Turkey
59 B3 Sijunjung Indon.
44 B5 Sika India
76 E3 Sikanni Chief Can.
76 E3 Sikanni Chief *r.* Can.
44 C4 Sikar India
41 H3 Sikaram *mt* Afgh.
114 B2 Sikasso Mali
39 L3 Sikeshu China
85 F4 Sikeston U.S.A.
49 P2 Sikhote-Alin' *mts* Rus. Fed.
27 L6 Sikinos *i.* Greece
45 G4 Sikkim *div.* India
8 P4 Sikṣjö Sweden
59 E1 Sikuati Malaysia
25 C1 Sil *r.* Spain
55 C4 Şilago Phil.
9 S9 Šilalė Lith.
23 H3 Silandro Italy
96 D3 Silao Mex.
55 B4 Silay Phil.
19 G3 Silberberg *h.* Ger.
45 H4 Silchar India
36 B1 Şile Turkey
43 C2 Šileru *r.* India
44 E3 Silgarhi Nepal
26 C6 Siliana Tunisia
36 D3 Silifke Turkey
45 G3 Siling Co *salt l.* China
71 4 Silisili *mt* Western Samoa
Silistat *see* Bozkır
27 M2 Silistra Bulg.
36 B1 Silivri Turkey
9 O6 Siljan *l.* Sweden
9 L8 Silkeborg Denmark
68 B1 Silkwood Austr.
9 U7 Sillamäe Estonia
20 F6 Sillian Austria
44 C5 Sillod India
113 J3 Silobela S. Africa
45 G3 Silong China
85 E6 Silsbee U.S.A.
8 U3 Siltaharju Fin.
114 E3 Siluko Nigeria
41 F5 Sīlūp *r.* Iran
9 R9 Šilutė Lith.
37 H2 Silvan Turkey
44 C5 Silvassa India
88 B2 Silver Bay U.S.A.
83 E5 Silver City U.S.A.
88 C1 Silver Islet Can.
92 D4 Silver Lake *CA* U.S.A.
88 D2 Silver Lake *MI* U.S.A.
82 B3 Silver Lake U.S.A.
14 C5 Silvermine Mts *h.* Rep. of Ireland
92 D3 Silver Peak Range *mts* U.S.A.
90 E5 Silver Spring U.S.A.
92 C2 Silver Springs U.S.A.
66 D3 Silverton Austr.
13 D7 Silverton U.K.
89 F3 Silver Water Can.
97 G4 Silvituc Mex.
23 H3 Silvretta Gruppe *mts* Switz.

59 D2 Simanggang Malaysia
55 B3 Simara i. Phil.
89 H2 Simard, Lac l. Can.
37 L5 Sīmareh r. Iran
45 F4 Simaria India
36 B2 Simav Turkey
36 B2 Simav Dağları mts Turkey
110 C3 Simba Congo(Zaire)
20 G4 Simbach am Inn Ger.
Simbirsk see Ul'yanovsk
Simbor i. see Pānikoita
89 G4 Simcoe Can.
89 H3 Simcoe, Lake Can.
45 F5 Simdega India
110 D2 Simēn Mountains Eth.
59 A2 Simeuluë i. Indon.
29 E6 Simferopol' Ukr.
45 E3 Simikot Nepal
23 H3 Similaun mt Austria/Italy
103 B3 Simití Col.
92 C4 Simi Valley U.S.A.
83 F4 Simla U.S.A.
17 L7 Şimleu Silvaniei Romania
22 E3 Simme r. Switz.
15 G2 Simmerath Ger.
15 H3 Simmern (Hunsrück) Ger.
92 C4 Simmler U.S.A.
93 F4 Simmons U.S.A.
87 F7 Simms Bahamas
8 U3 Simojärvi l. Fin.
76 F4 Simonette r. Can.
77 J4 Simonhouse Can.
21 M6 Simontornya Hungary
22 F3 Simplon Pass Switz.
65 G5 Simpson Desert Austr.
65 G5 Simpson Desert National Park Austr.
88 D1 Simpson I. Can.
92 D2 Simpson Park Mts U.S.A.
9 O9 Simrishamn Sweden
55 A5 Simunul i. Phil.
46 H2 Simushir, O. i. Rus. Fed.
43 A2 Sina r. India
59 A2 Sinabang Indon.
58 A5 Sinabung volc. Indon.
109 F2 Sinai reg. Egypt
15 E3 Sinai, Mont h. France
96 B2 Sinaloa div. Mex.
23 J6 Sinalunga Italy
51 C5 Sinan China
54 C4 Sinanju N. Korea
45 H5 Sinbyugyun Myanmar
36 F2 Sincan Turkey
103 B2 Sincé Col.
103 B2 Sincelejo Col.
87 D5 Sinclair, L. U.S.A.
76 E4 Sinclair Mills Can.
112 B2 Sinclair Mine Namibia
11 E2 Sinclair's Bay U.K.
44 D4 Sind r. India
55 B4 Sindañgan Phil.
59 C4 Sindangbarang Indon.
44 B4 Sindari India
20 B4 Sindelfingen Ger.
43 B2 Sindgi India
44 B4 Sindh div. Pak.
43 B3 Sindhnur India
36 B2 Sındırgı Turkey
44 D6 Sindkhed India
44 C5 Sindkheda India
54 C4 Sin-do i. China
28 J2 Sindor Rus. Fed.
45 F5 Sindri India
44 B3 Sind Sagar Doab lowland Pak.
28 J3 Sinegor'ye Rus. Fed.
27 M4 Sinekçi Turkey
25 B4 Sines Port.
25 B4 Sines, Cabo de pt Port.
8 T3 Sinettä Fin.
114 B3 Sinfra Côte d'Ivoire
109 F3 Singa Sudan
44 E3 Singahi India
54 D3 Sin'galp'a China
44 D2 Singa Pass India
31 K9 Singapore country Asia
58 B5 Singapore Sing.
58 B5 Singapore, Strait of Indon./Sing.
59 E4 Singaraja Indon.
58 B2 Sing Buri Thai.
20 B5 Singen (Hohentwiel) Ger.
89 G3 Singhampton Can.
115 B3 Singida Tanz.
115 B4 Singida div. Tanz.
57 G7 Singkang Indon.
59 C2 Singkawang Indon.
58 A5 Singkil Indon.
67 J4 Singleton Austr.
69 C4 Singleton, Mt h. Austr.
Singora see Songkhla
54 D4 Sin'gye N. Korea
54 D3 Sinhŭng N. Korea
26 C4 Siniscola Sardinia Italy
38 E2 Siniy-Shikhan Rus. Fed.
26 G3 Sinj Croatia
57 G8 Sinjai Indon.
37 H3 Sinjār Iraq
37 H3 Sinjār, Jabal mt Iraq
37 K3 Sīnji Iran
109 F3 Sinkat Sudan
Sinkiang Uighur Aut. Region div. see Xinjiang Uygur Zizhiqu
54 C4 Sinmi i. N. Korea
20 B2 Sinn Ger.
101 H2 Sinnamary Fr. Guiana
Sinneh see Sanandaj
27 N2 Sinoie, Lacul lag. Romania
29 E7 Sinop Turkey
54 D3 Sinpa N. Korea
54 E3 Sinp'o N. Korea
54 E3 Sinp'ung-dong N. Korea
54 D4 Sinp'yŏng N. Korea
54 D4 Sinsang N. Korea
20 B3 Sinsheim Ger.
59 D2 Sintang Indon.
85 D6 Sinton U.S.A.
103 A2 Sinú r. Col.
54 C3 Sinŭiju N. Korea
15 H2 Sinzig Ger.
21 M6 Sió r. Romania
55 B5 Siocon Phil.
16 J7 Siófok Hungary
22 E3 Sion Switz.
14 D3 Sion Mills U.K.
84 D3 Sioux Center U.S.A.
84 D3 Sioux City U.S.A.
84 D3 Sioux Falls U.S.A.
78 B3 Sioux Lookout Can.
97 G5 Sipacate Guatemala
55 B4 Sipalay Phil.
54 C2 Siping China
77 K3 Sipiwesk Can.
77 K3 Sipiwesk L. Can.
116 B4 Siple Coast Ant.
116 A4 Siple, Mt Ant.
44 C5 Sipra r. India
87 C5 Sipsey r. U.S.A.
59 A3 Sipura i. Indon.
96 H5 Siquia r. Nic.
55 B4 Siquijor Phil.
55 B4 Siquijor i. Phil.
44 B5 Sir r. Pak.
43 B3 Sira India
9 K7 Sira r. Norway
40 D5 Şīr Abū Nu'āyr i. U.A.E.
Siracusa see Syracuse
76 E4 Sir Alexander, Mt Can.
37 G1 Şiran Turkey
41 G5 Siranda Lake Pak.
40 D5 Şīr Banī Yās i. U.A.E.
37 M3 Sīrdān Iran
65 G3 Sir Edward Pellew Group is Austr.
88 A3 Siren U.S.A.
41 F5 Sīrgān Iran
58 B1 Siri Kit Dam dam Thai.
40 D4 Sīrīz Iran
76 D2 Sir James McBrien, Mt Can.
Sirjan see Sa'īdābād
40 D4 Sīrjan salt flat Iran
66 B5 Sir Joseph Bank's Group is Austr.
40 E5 Sirk Iran
44 E4 Sirmour India
37 J3 Şırnak Turkey
43 C2 Sironcha India
44 D4 Sironj India
43 B2 Sirpur India
92 C4 Sirretta Peak U.S.A.
40 D5 Sīrrī, Jazīreh-ye i. Iran
44 C3 Sirsa Haryana India
45 E4 Sirsa Uttar Pradesh India
76 F4 Sir Sanford, Mt Can.
43 A3 Sirsi Karnataka India
44 D3 Sirsi Uttar Pradesh India
43 B2 Sirsilla India
109 D1 Sirte Libya
109 D1 Sirte, Gulf of g. Libya
43 A2 Sirur India
37 J2 Şirvan Turkey
9 T9 Širvintos Lith.
37 K4 Sīrwān r. Iraq
76 F4 Sir Wilfred Laurier, Mt Can.
26 G2 Sisak Croatia
58 C2 Sisaket Thai.
40 C4 Sīsakht Iran
97 G3 Sisal Mex.
112 E3 Sishen S. Africa
37 L2 Sisian Armenia
88 C2 Siskiwit Bay U.S.A.
58 B2 Sisŏphŏn Cambodia
92 B4 Sisquoc r. U.S.A.
84 D2 Sisseton U.S.A.
91 K1 Sisson Branch Reservoir Can.
41 F4 Sistan reg. Iran
41 F4 Sīstan, Daryācheh-ye marsh Afgh.
22 C5 Sisteron France
67 F8 Sisters Beach Austr.
44 C5 Sitamau India
55 A5 Sitangkai Phil.
44 E4 Sitapur India
27 M7 Siteia Greece
113 J3 Siteki Swaziland
27 K4 Sithonia pen. Greece
104 C1 Sítio da Abadia Brazil
104 D1 Sítio do Mato Brazil
76 B3 Sitka U.S.A.
44 B3 Sitpur Pak.
18 C6 Sittard Neth.
45 H4 Sittaung Myanmar
19 G3 Sittensen Ger.
21 H6 Sittersdorf Austria
13 H6 Sittingbourne U.K.
56 A2 Sittwe Myanmar
51 Siu A Chau i. H.K. China
71 4 Siumu Western Samoa
96 H5 Siuna Nic.
45 F5 Siuri India
43 B4 Sivaganga India
43 B4 Sivakasi India
40 D4 Sivand Iran
36 F2 Sivas Turkey
36 B2 Sivaslı Turkey
37 G3 Siverek Turkey
37 G2 Sivrice Turkey
36 C2 Sivrihisar Turkey
113 H3 Sivukile S. Africa
109 E2 Siwa Egypt
44 D3 Siwalik Range mts India/Nepal
45 F4 Siwan India
44 C4 Siwana India
22 C6 Six-Fours-les-Plages France
50 E3 Si Xian China
88 E4 Six Lakes U.S.A.
14 D3 Sixmilecross U.K.
113 H2 Siyabuswa S. Africa
50 F3 Siyang China
38 A4 Siyäzän Azer.
37 M1 Siyäzän Azer.
50 C1 Siyitang China
40 D3 Sīyunī Iran
50 D1 Siziwang Qi China
Sjælland i. see Zealand
27 J3 Sjenica Yugo.
9 N9 Sjöbo Sweden
8 P2 Sjøvegan Norway
29 E6 Skadovs'k Ukr.
19 J1 Skælsør Denmark
18 F1 Skærbæk Denmark
8 E4 Skaftafell National Park Iceland
8 E5 Skaftárós est. Iceland
8 D3 Skagafjörður in. Iceland
9 M8 Skagen Denmark
9 L8 Skagerrak str. Denmark/Norway
82 B1 Skagit r. Can./U.S.A.
76 B3 Skagway U.S.A.
8 T1 Skaidi Norway
8 P2 Skaland Norway
8 O4 Skalmodal Sweden
9 L8 Skanderborg Denmark
91 E3 Skaneateles Lake U.S.A.
88 C2 Skanee U.S.A.
19 L1 Skåne-Tranås Sweden
27 L5 Skantzoura i. Greece
9 N7 Skara Sweden
9 R7 Skargadshavets Nationalpark nat. park Fin.
9 M6 Skarnes Norway
17 K5 Skarżysko-Kamienna Pol.
8 R3 Skaulo Sweden
17 J6 Skawina Pol.
76 D3 Skeena r. Can.
76 D3 Skeena Mountains Can.
13 H4 Skegness U.K.
8 R4 Skellefteå Sweden
8 Q4 Skellefteälven r. Sweden
8 R4 Skelleftehamn Sweden
14 A6 Skellig Rocks is Rep. of Ireland
12 E4 Skelmersdale U.K.
14 E4 Skerries Rep. of Ireland
9 M7 Ski Norway
27 K5 Skiathos i. Greece
14 B6 Skibbereen Rep. of Ireland
8 R2 Skibotn Norway
12 D3 Skiddaw mt U.K.
9 L7 Skien Norway
17 K5 Skierniewice Pol.
108 C1 Skikda Alg.
12 G4 Skipsea U.K.
66 E6 Skipton Austr.
12 E4 Skipton U.K.
9 L8 Skive Denmark
8 E4 Skjálfandafljót r. Iceland
9 L9 Skjern Denmark
9 K6 Skjolden Norway
39 H5 Skobeleva, Pik mt Kyrg.
21 M3 Skoczów Pol.
8 K5 Skodje Norway
8 T2 Skoganvarre Norway
14 F6 Skokholm Island U.K.
88 D4 Skokie U.S.A.
38 E2 Skol' Kazak.
13 B6 Skomer Island U.K.
27 K5 Skopelos i. Greece
28 F4 Skopin Rus. Fed.
27 J4 Skopje Macedonia
29 F5 Skorodnoye Rus. Fed.
9 N7 Skövde Sweden
49 M1 Skovorodino Rus. Fed.
91 J2 Skowhegan U.S.A.
21 G6 Škrlutica mt Slovenia
9 S8 Skrunda Latvia
76 B2 Skukum, Mt Can.
113 J2 Skukuza S. Africa
92 D3 Skull Peak U.S.A.
88 B5 Skunk r. U.S.A.
9 R8 Skuodas Lith.
9 N9 Skurup Sweden
21 J3 Skuteč Czech Rep.
9 P6 Skutskär Sweden
29 D5 Skvyra Ukr.
11 B3 Skye i. U.K.
27 L5 Skyros Greece
27 L5 Skyros i. Greece
116 B3 Skytrain Ice Rise ice feature Ant.
21 L4 Sládkovičovo Slovakia
9 M9 Slagelse Denmark
8 Q4 Slagnäs Sweden
59 C4 Slamet, Gunung volc. Indon.
14 E4 Slane Rep. of Ireland
14 E5 Slaney r. Rep. of Ireland
28 D3 Slantsy Rus. Fed.
21 H2 Slaný Czech Rep.
21 K3 Šlapanice Czech Rep.
29 G5 Slashchevskaya Rus. Fed.
68 B2 Slashers Reefs Austr.
88 D1 Slate Is Can.
26 G2 Slatina Croatia
27 L2 Slatina Romania
77 G2 Slave r. Can.
108 C4 Slave Coast Africa
76 G3 Slave Lake Can.
39 J1 Slavgorod Rus. Fed.
17 O2 Slavkovichi Rus. Fed.
21 K3 Slavkov u Brna Czech Rep.
21 J4 Slavonice Czech Rep.
27 H2 Slavonija reg. Croatia
27 H2 Slavonski Brod Croatia
29 C5 Slavuta Ukr.
29 D5 Slavutych Ukr.
39 G4 Slavyanka Kazak.
52 B3 Slavyanka Rus. Fed.
29 F6 Slavyansk-na-Kubani Rus. Fed.
28 D4 Slawharad Belarus
16 H3 Sławno Pol.
13 G4 Sleaford U.K.
66 A5 Sleaford B. Austr.
14 A5 Slea Head hd Rep. of Ireland
11 C3 Sleat pen. U.K.
11 C3 Sleat, Sound of chan. U.K.
78 E2 Sleeper Islands Can.
88 D3 Sleeping Bear Dunes National Seashore res. U.S.A.
88 D3 Sleeping Bear Pt U.S.A.
29 H7 Sleptsovskaya Rus. Fed.
116 C3 Slessor Glacier Ant.
21 K2 Ślęża r. Pol.
85 F6 Slidell U.S.A.
14 A5 Slievanea h. Rep. of Ireland
14 D3 Slieve Anierin h. Rep. of Ireland
14 D5 Slieveardagh Hills Rep. of Ireland
14 C4 Slieve Aughty Mts h. Rep. of Ireland
14 D3 Slieve Beagh h. Rep. of Ireland/U.K.
14 C5 Slieve Bernagh h. Rep. of Ireland
14 D4 Slieve Bloom Mts h. Rep. of Ireland
14 B5 Slievecallan h. Rep. of Ireland
14 B3 Slieve Car h. Rep. of Ireland
14 F3 Slieve Donard h. U.K.
14 B4 Slieve Elva h. Rep. of Ireland
14 C3 Slieve Gamph h. Rep. of Ireland
14 C3 Slieve League h. Rep. of Ireland
14 B5 Slieve Mish Mts h. Rep. of Ireland
14 B5 Slieve Miskish Mts h. Rep. of Ireland
14 A3 Slieve More h. Rep. of Ireland
14 D4 Slieve na Calliagh h. Rep. of Ireland
14 D5 Slievenamon h. Rep. of Ireland
14 D2 Slieve Snaght mt Rep. of Ireland
11 B3 Sligachan U.K.
14 C3 Sligo Rep. of Ireland
14 C3 Sligo Bay Rep. of Ireland
9 Q8 Slite Sweden
27 M3 Sliven Bulg.
28 H2 Sloboda Rus. Fed.
28 J2 Slobodchikovo Rus. Fed.
27 M2 Slobozia Romania
76 F5 Slocan Can.
18 D3 Slochteren Neth.
28 C4 Slonim Belarus
18 B4 Slootdorp Neth.
18 C4 Sloten Neth.
18 C4 Slotermeer l. Neth.
63 F2 Slot, The chan. Solomon Is
13 G6 Slough U.K.
7 E4 Slovakia country Europe
7 E4 Slovenia country Europe
26 F1 Slovenj Gradec Slovenia
21 J6 Slovenska Bistrica Slovenia
21 J6 Slovenske Gorice h. Slovenia
29 F5 Slov''yans'k Ukr.
19 M4 Słubice Pol.
21 H1 Šluknov Czech Rep.
21 L3 Slunečná h. Czech Rep.
16 H3 Słupsk Pol.
8 P4 Slussfors Sweden
28 C4 Slutsk Belarus
14 A4 Slyne Head hd Rep. of Ireland
48 H1 Slyudyanka Rus. Fed.
19 J1 Smålandsfarvandet chan. Denmark
91 J3 Small Pt U.S.A.
79 H3 Smallwood Reservoir Can.
28 D4 Smalyavichy Belarus
17 N3 Smarhon' Belarus
112 E5 Smartt Syndicate Dam resr S. Africa
77 J4 Smeaton Can.
27 J2 Smederevo Yugo.
27 J2 Smederevska Palanka Yugo.
90 D4 Smethport U.S.A.
29 D5 Smila Ukr.
18 D4 Smilde Neth.
9 T8 Smiltene Latvia
39 G1 Smirnovo Kazak.
76 G3 Smith Can.
92 C2 Smith U.S.A.
90 C6 Smith r. U.S.A.
74 C2 Smith Bay U.S.A.
76 D4 Smithers Can.
113 G5 Smithfield S. Africa
87 E5 Smithfield NC U.S.A.
82 E3 Smithfield UT U.S.A.
116 A3 Smith Glacier Ant.
116 B2 Smith I. S. Shetland Is Ant.
91 E5 Smith I. MD U.S.A.
91 F6 Smith I. VA U.S.A.
90 D6 Smith Mountain Lake U.S.A.
76 D3 Smith River Can.
89 J3 Smiths Falls Can.
75 L2 Smith Sound str. Can./Greenland
67 F8 Smithton Austr.
92 C1 Smoke Creek Desert U.S.A.
76 F4 Smoky r. Can.
84 C4 Smoky r. U.S.A.
67 K3 Smoky C. hd Austr.
78 D3 Smoky Falls Can.
84 D4 Smoky Hills U.S.A.
76 G4 Smoky Lake Can.
8 K5 Smøla i. Norway
38 B2 Smolenka Rus. Fed.
28 E4 Smolensk Rus. Fed.
28 E4 Smolenskaya Oblast' div. Rus. Fed.
39 L1 Smolenskoye Rus. Fed.
27 L4 Smolyan Bulg.
52 C3 Smolyoninovo Rus. Fed.
78 D4 Smooth Rock Falls Can.
78 C3 Smoothrock L. Can.
77 H4 Smoothstone Lake Can.
8 T1 Smørfjord Norway
19 L1 Smygehamn Sweden
116 B3 Smyley I. Ant.
91 F5 Smyrna DE U.S.A.
87 C5 Smyrna GA U.S.A.
90 C4 Smyrna OH U.S.A.
91 J1 Smyrna Mills U.S.A.
8 F4 Snæfell mt Iceland
12 C3 Snaefell h. U.K.
76 A2 Snag Can.
82 D3 Snake r. U.S.A.
93 E2 Snake Range mts U.S.A.
82 D3 Snake River Plain U.S.A.
87 E7 Snap Pt Bahamas
76 G2 Snare Lake Can.
63 G6 Snares Is N.Z.
8 N4 Snasa Norway
18 C3 Sneek Neth.
14 B6 Sneem Rep. of Ireland
112 F6 Sneeuberg mts S. Africa
79 H3 Snegamook Lake Can.
13 H5 Snettisham U.K.
32 K3 Snezhnogorsk Rus. Fed.
26 F2 Snežnik mt Slovenia
17 K4 Śniardwy, Jezioro l. Pol.
29 E6 Snihurivka Ukr.
11 B3 Snizort, Loch b. U.K.
82 B2 Snohomish U.S.A.
82 B2 Snoqualmie Pass U.S.A.
8 N3 Snøtinden mt Norway
77 J2 Snowbird Lake Can.
13 C4 Snowdon mt U.K.
13 D5 Snowdonia National Park U.K.
93 G4 Snowflake U.S.A.
91 F5 Snow Hill MD U.S.A.
87 E5 Snow Hill NC U.S.A.
77 J4 Snow Lake Can.
66 C4 Snowtown Austr.
82 D3 Snowville U.S.A.
67 H6 Snowy r. Austr.
67 H6 Snowy Mts Austr.
67 G9 Snug Austr.
79 J3 Snug Harbour Nfld Can.
89 G3 Snug Harbour Ont. Can.
58 C2 Snuŏl Cambodia
85 D5 Snyder OK U.S.A.
85 C5 Snyder TX U.S.A.
111 E5 Soalala Madag.
111 E5 Soanierana-Ivongo Madag.
54 D6 Soan kundo i. S. Korea
103 B3 Soata Col.
11 B3 Soay i. U.K.
54 D6 Sobaek Sanmaek mts S. Korea
109 F4 Sobat r. Sudan
15 H3 Sobernheim Ger.
21 H3 Soběslav Czech Rep.
57 L7 Sobger r. Indon.
53 B8 Sobo-san mt Japan
21 J6 Soboth Austria
101 K6 Sobradinho, Barragem de resr Brazil
101 K4 Sobral Brazil
23 L3 Soča r. Slovenia
So-chaoson-man g. see Korea Bay
29 F7 Sochi Rus. Fed.
54 D5 Sŏch'on S. Korea
60 M5 Société, Arch. de la arch. Pac. Oc.
71 2 Société, Archipel de la is Fr. Polynesia Pac. Oc.
Society Islands see Société, Archipel de la
104 C3 Socorro Brazil
103 B3 Socorro Col.
83 F5 Socorro U.S.A.
94 B5 Socorro, I. Mex.
42 D7 Socotra i. Yemen
58 C3 Soc Trăng Vietnam
25 E3 Socuéllamos Spain
92 D4 Soda Lake U.S.A.
8 U3 Sodankylä Fin.
44 D2 Soda Plains China/Jammu and Kashmir
82 E3 Soda Springs U.S.A.
9 P6 Söderhamn Sweden
9 P7 Söderköping Sweden
9 P7 Södertälje Sweden
109 E3 Sodiri Sudan
110 D3 Sodo Eth.
9 Q6 Södra Kvarken str. Fin./Sweden
113 H1 Soekmekaar S. Africa
18 C5 Soerendonk Neth.
18 F5 Soest Ger.
18 C4 Soest Neth.
67 H4 Sofala Austr.
27 K3 Sofia Bulg.
Sofiya see Sofia
8 W4 Sofporog Rus. Fed.
53 G10 Sōfu-gan i. Japan
103 B3 Sogamoso Col.
37 G1 Soğanlı Dağları mts Turkey
18 E4 Sögel Ger.
9 K7 Søgne Norway
9 J6 Sognefjorden in. Norway
55 C4 Sogod Phil.
50 A1 Sogo Nur l. China
28 H2 Sogra Rus. Fed.
50 A3 Sogruma China
36 C1 Söğüt Turkey
54 D7 Sŏgwip'o S. Korea
45 H3 Sog Xian China
109 F2 Sohâg Egypt
44 D5 Sohagpur India
39 G2 Sohalinskiy Kazak.
13 H5 Soham U.K.
44 B2 Sohan r. Pak.
63 F2 Sohano P.N.G.
45 E5 Sohela India
44 D3 Sohna India
54 E3 Sŏho-ri N. Korea
54 C6 Sohŭksan i. S. Korea
15 E2 Soignes, Forêt de forest Belgium
15 E2 Soignies Belgium
8 T5 Soini Fin.
15 D3 Soissons France
44 C4 Sojat India
55 B4 Sojoton Point Phil.
29 C5 Sokal' Ukr.
54 E4 Sokch'o S. Korea
27 M6 Söke Turkey
39 G5 Sokh Tajik.
29 G7 Sokhumi Georgia
114 D3 Sokodé Togo
51 Soko Islands H.K. China
28 G3 Sokol Rus. Fed.
17 L4 Sokółka Pol.
114 B2 Sokolo Mali
20 F2 Sokolov Czech Rep.
52 C3 Sokolovka Rus. Fed.
17 L4 Sokołow Podlaski Pol.
114 E2 Sokoto Nigeria
114 D2 Sokoto r. Nigeria
29 C5 Sokyryany Ukr.
44 D3 Solan India
70 A7 Solander I. N.Z.
43 A2 Solāpur India
20 E6 Sölden Austria
103 B2 Soledad Col.
92 B3 Soledad U.S.A.
103 E2 Soledad Venez.
97 E4 Soledad de Doblado Mex.
29 G6 Solenoye Rus. Fed.
13 F7 Solent, The str. U.K.
8 N3 Solfjellsjøen Norway
37 H2 Solhan Turkey
28 G3 Soligalich Rus. Fed.
13 F5 Solihull U.K.
32 G4 Solikamsk Rus. Fed.
38 C2 Sol'-Iletsk Rus. Fed.
97 H4 Solimón, Pta pt Mex.
18 E5 Solingen Ger.
112 A1 Solitaire Namibia
37 M1 Şollar Azer.
8 P5 Sollefteå Sweden
19 K5 Söllichau Ger.
19 G5 Solling h. Ger.
19 H5 Sollstedt Ger.
28 F3 Solnechnogorsk Rus. Fed.
59 B3 Solok Indon.
97 G5 Sololá Guatemala
61 H4 Solomon Islands country Pac. Oc.
62 F2 Solomon Sea P.N.G./Solomon Is
39 L2 Soloneshnoye Rus. Fed.
88 B2 Solon Springs U.S.A.
57 G8 Solor, Kepulauan is Indon.
22 E2 Solothurn Switz.
28 E1 Solovetskiye Ostrova is Rus. Fed.
28 H3 Solovetskoye Rus. Fed.
26 G3 Šolta i. Croatia
41 E2 Soltānābād Iran
40 C4 Solţānābād Iran
41 E3 Solţānābād Iran
19 G4 Soltau Ger.
28 D3 Sol'tsy Rus. Fed.
91 E3 Solvay U.S.A.
9 O8 Sölvesborg Sweden
11 E6 Solway Firth est. U.K.
115 B5 Solwezi Zambia
53 G6 Sōma Japan
36 A2 Soma Turkey
15 D2 Somain France
107 H5 Somalia country Africa
117 H3 Somali Basin sea feature Ind. Ocean
111 C4 Sombo Angola
27 H2 Sombor Yugo.
96 D3 Sombrerete Mex.
44 C4 Somdari India
91 J2 Somerest Junction U.S.A.
9 S6 Somero Fin.
86 C4 Somerset KY U.S.A.
88 E4 Somerset MI U.S.A.
90 D5 Somerset PA U.S.A.
113 F6 Somerset East S. Africa
75 J2 Somerset Island Can.
68 E6 Somerset, L. Austr.
91 G3 Somerset Reservoir U.S.A.
112 C7 Somerset West S. Africa
91 H3 Somersworth U.S.A.
85 D6 Somerville Resr U.S.A.
22 F4 Somma Lombardo Italy
22 A3 Somme r. France
15 B2 Somme, Baie de la b. France
9 O7 Sommen l. Sweden
19 J5 Sömmerda Ger.
15 E3 Somme-Soude r. France
15 E4 Sommesous France
79 G3 Sommet, Lac du l. Can.
44 B5 Somnath India
21 L6 Somogy div. Hungary
21 L6 Somogyvár Hungary
88 C5 Somonauk U.S.A.
71 13 Somosomo Fiji
71 13 Somosomo Str. Fiji
96 H5 Somotillo Nic.
96 H5 Somoto Nic.
105 C4 Somuncurá, Mesa Volcánica de plat. Arg.
45 F4 Son r. India
96 J7 Soná Panama
39 G2 Sonaly Kazak.
45 F5 Sonamukhi India
45 G5 Sonamura India
45 E5 Sonapur India
44 D4 Sonar r. India
45 H4 Sonari India
54 C4 Sŏnch'ŏn N. Korea
23 H3 Sondalo Italy
28 E2 Sondaly Rus. Fed.
19 G2 Sønderå r. Denmark
9 L9 Sønderborg Denmark
19 G2 Sønderborg airport Denmark
19 H5 Sondershausen Ger.
19 H1 Søndersø Denmark
19 J2 Sondersø l. Denmark
75 N3 Søndre Strømfjord in. Greenland
23 G3 Sondrio Italy
43 B2 Sonepet India
44 B5 Songad India
50 E4 Songbu China
58 D2 Sông Cau Vietnam
58 C1 Song Con r. Vietnam
51 B6 Sông Đa r. Vietnam
58 D2 Sông Đa Răng r. Vietnam
115 C5 Songea Tanz.
15 B3 Songeons France
54 D3 Songgan N. Korea
58 C3 Song Hâu Giang r. Vietnam
54 D2 Songhua Hu resr China
54 C1 Songhua Jiang China
52 B1 Songhua Jiang r. China
50 F4 Songjiang China
54 D2 Songjianghe China
51 C4 Songkan China
58 B4 Songkhla Thai.
39 H4 Songköl l. Kyrg.
51 C6 Sông Ky Cung r. Vietnam
46 E2 Songling China
50 F1 Song Ling mts China
51 B6 Sông Ma r. Laos/Vietnam
54 D5 Sŏngnam S. Korea
58 C1 Sông Ngan Sau r. Vietnam
54 C4 Songnim N. Korea
110 B4 Songo Angola
115 B6 Songo Moz.
50 B3 Songpan China
58 C3 Song Saigon r. Vietnam
45 G4 Songsak India
54 D7 Sŏngsan S. Korea
50 D3 Song Shan mt China
54 D3 Songshuzhen China
51 C4 Songtao China
51 F5 Songxi China
50 D3 Song Xian China
51 D4 Songzi China
58 D2 Sơn Ha Vietnam
50 D1 Sonid Youqi China
50 D1 Sonid Zuoqi China
44 D3 Sonīpat India
8 U5 Sonkajärvi Fin.
51 B6 Sơn La Vietnam
41 G5 Sonmiani Pak.
41 G5 Sonmiani Bay Pak.
20 E2 Sonneberg Ger.
19 L5 Sonnewalde Ger.
104 D2 Sono r. Minas Gerais Brazil
101 J6 Sono r. Tocantins Brazil
93 F6 Sonoita r. Mex.
93 G6 Sonoita U.S.A.
96 B2 Sonora div. Mex.
96 B1 Sonora r. Mex.
92 B3 Sonora CA U.S.A.
85 C6 Sonora TX U.S.A.
93 F6 Sonoyta Mex.
40 B3 Sonqor Iran
103 B3 Sonsón Col.
97 G5 Sonsonate El Salvador
51 B6 Sơn Tây Vietnam
20 D5 Sonthofen Ger.
113 H5 Sonwabile S. Africa
105 F1 Sopas r. Uru.
33 S4 Sopka Shiveluch mt Rus. Fed.
109 E4 Sopo watercourse Sudan
27 L3 Sopot Bulg.
17 J3 Sopot Pol.
16 H7 Sopron Hungary
39 H4 Sopu-Korgon Kyrg.
26 E4 Sora Italy
45 F6 Sorada India
9 P5 Soråker Sweden
54 E4 Sŏraksan mt S. Korea
38 D3 Sor Donyztau l. Kazak.
78 F4 Sorel Can.
67 G9 Sorell Austr.
67 G9 Sorell L. Austr.
22 B5 Sorgues France
36 E2 Sorgun Turkey
25 E2 Soria Spain
32 C2 Sørkappøya i. Svalbard
38 C3 Sor Kaydak l. Kazak.
40 D3 Sorkheh Iran
40 D3 Sorkh, Kūh-e mts Iran
8 N4 Sørli Norway
38 C3 Sor Mertvyy Kultuk l. Kazak.
19 J1 Sorø Denmark
45 F5 Soro India
29 D5 Soroca Moldova
104 C3 Sorocaba Brazil
38 C1 Sorochinsk Rus. Fed.
39 L1 Sorokino Rus. Fed.
47 G6 Sorol i. Micronesia
57 J7 Sorong Indon.
115 B2 Soroti Uganda
8 S1 Sørøya i. Norway
25 B3 Sorraia r. Port.
8 Q2 Sorreisa Norway
66 F7 Sorrento Austr.
111 B6 Sorris Sorris Namibia
116 D3 Sør-Rondane mts Ant.
8 P4 Sorsele Sweden
55 C3 Sorsogon Phil.
28 D2 Sortavala Rus. Fed.
8 O2 Sortland Norway
28 J2 Sortopolovskaya Rus. Fed.
28 J3 Sorvizhi Rus. Fed.
54 D5 Sŏsan S. Korea
113 H2 Soshanguve S. Africa
29 F4 Sosna r. Rus. Fed.
105 C2 Sosneado mt Arg.
28 K2 Sosnogorsk Rus. Fed.
39 J2 Sosnovka Kazak.
28 H2 Sosnovka Archangel. Rus. Fed.
28 G4 Sosnovka Tambov. Rus. Fed.
32 F3 Sosnovka Rus. Fed.
49 K1 Sosnovo-Ozerskoye Rus. Fed.
8 X4 Sosnovyy Rus. Fed.
9 V7 Sosnovyy Bor Rus. Fed.
17 J5 Sosnowice Pol.
29 F6 Sosyka r. Rus. Fed.
103 A4 Sotara, Volcán volc. Col.
8 V4 Sotkamo Fin.
105 D1 Soto Arg.
97 E3 Soto la Marina Mex.
15 B3 Sotteville-lès-Rouen France
97 G3 Sotuta Mex.
110 B3 Souanké Congo
114 B3 Soubré Côte d'Ivoire
91 F4 Souderton U.S.A.
15 H4 Soufflenheim France
27 M4 Soufli Greece
24 E4 Souillac France
15 F3 Souilly France
108 C1 Souk Ahras Alg.
Sŏul see Seoul
15 E4 Soulaines-Dhuys France
24 D5 Soulom France
22 E2 Soultz-Haut-Rhin France
Soûr see Tyre
25 H4 Sour el Ghozlane Alg.
77 J5 Souris Man. Can.
79 H4 Souris P.E.I. Can.
77 J5 Souris r. Can./U.S.A.
101 L5 Sousa Brazil
108 D1 Sousse Tunisia
24 D5 Soustons France
42 C7 South div. Yemen
107 F9 South Africa, Republic of country Africa
89 G3 Southampton Can.
13 F7 Southampton U.K.
91 G4 Southampton U.S.A.
75 K3 Southampton I. Can.
77 M2 Southampton Island Can.
56 A4 South Andaman i. Andaman and Nicobar Is
90 E6 South Anna r. U.S.A.
13 F4 South Anston U.K.
79 H2 South Aulatsivik Island Can.
65 F5 South Australia div. Austr.
117 N6 South Australian Basin sea feature Ind. Ocean
85 F5 Southaven U.S.A.
83 F5 South Baldy mt U.S.A.
12 F3 South Bank U.K.
90 B4 South Bass I. U.S.A.
77 N2 South Bay Can.
89 F3 South Baymouth Can.
88 D5 South Bend IN U.S.A.
82 B2 South Bend WA U.S.A.
87 E7 South Bight chan. Bahamas
90 D6 South Boston U.S.A.
70 D5 Southbridge N.Z.
91 G3 Southbridge U.S.A.
71 13 South Cape Fiji
South Cape see Ka Lae
87 D5 South Carolina div. U.S.A.
91 J2 South China U.S.A.
56 E4 South China Sea Pac. Oc.
84 C2 South Dakota div. U.S.A.
91 G3 South Deerfield U.S.A.
13 G7 South Downs h. U.K.
113 F2 South East div. Botswana
67 G9 South East Cape Tas. Austr.
67 G7 South East Cape Vic. Austr.
119 N10 South-East Pacific Basin sea feature Pac. Oc.
77 J3 Southend Can.
11 C5 Southend U.K.
13 H6 Southend-on-Sea U.K.
88 A5 South English U.S.A.
70 C5 Southern Alps mts N.Z.
69 D5 Southern Cross Austr.
77 K3 Southern Indian Lake Can.
109 E4 Southern National Park Sudan
61 C7 Southern Ocean
87 E5 Southern Pines U.S.A.
116 C1 Southern Thule I. Atl. Ocean
11 D5 Southern Uplands reg. U.K.
11 F4 South Esk r. U.K.
88 B6 South Fabius r. U.S.A.
118 G7 South Fiji Basin sea feature Pac. Oc.
83 F4 South Fork U.S.A.
92 A2 South Fork Eel r. U.S.A.
92 C4 South Fork Kern r. U.S.A.
90 D5 South Fork South Branch r. U.S.A.
88 E3 South Fox I. U.S.A.
116 C5 South Geomagnetic Pole Ant.
11 A3 South Harris i. U.K.
45 G5 South Hatia I. Bangl.
88 D4 South Haven U.S.A.
77 K2 South Henik Lake Can.
91 G2 South Hero U.S.A.
90 D6 South Hill U.S.A.
118 E4 South Honshu Ridge sea feature Pac. Oc.
115 C2 South Horr Kenya
77 K3 South Indian Lake Can.
91 G4 Southington U.S.A.
70 C6 South Island N.Z.
55 A4 South Islet rf Phil.
45 F5 South Koel r. India
92 B2 South Lake Tahoe U.S.A.
115 B5 South Luangwa National Park Zambia
116 B6 South Magnetic Pole Ant.
88 D3 South Manitou I. U.S.A.
87 D7 South Miami U.S.A.
13 H6 Southminster U.K.
77 J4 South Moose L. Can.
90 E5 South Mts h. U.S.A.
76 D2 South Nahanni r. Can.
11 South Nesting Bay U.K.
120 G10 South Orkney Is Atl. Ocean
63 H5 South Pacific Ocean
91 H2 South Paris U.S.A.
82 G3 South Platte r. U.S.A.
116 B4 South Pole Ant.
89 G1 South Porcupine Can.
67 K1 Southport Austr.
12 D4 Southport U.K.
91 H3 South Portland U.S.A.
89 H3 South River Can.
11 F2 South Ronaldsay i. U.K.
91 G3 South Royalton U.S.A.
113 J5 South Sand Bluff pt S. Africa
120 H9 South Sandwich Islands Atl. Ocean
120 H9 South Sandwich Trench sea feature Atl. Ocean
77 H4 South Saskatchewan r. Can.
77 K3 South Seal r. Can.
120 F9 South Shetland Is Ant.
12 F2 South Shields U.K.
12 G4 South Skirlaugh U.K.
88 A5 South Skunk r. U.S.A.
70 E3 South Taranaki Bight b. N.Z.
93 G2 South Tent summit U.S.A.
45 E4 South Tons r. India
78 E3 South Twin I. Can.
12 E3 South Tyne r. U.K.
11 A3 South Uist i. U.K.
67 G9 South West C. hd Austr.
70 A7 South West Cape N.Z.
117 H6 South-West Indian Ridge sea feature Ind. Ocean
68 C1 South West Island Coral Sea Is Terr.
67 G9 South West Nat. Park Austr.
119 J8 South-West Pacific Basin sea feature Pac. Oc.
119 O7 South-West Peru Ridge sea feature Pac. Oc.
67 K3 South West Rocks Austr.
88 E5 South Whitley U.S.A.
91 H3 South Windham U.S.A.
13 J5 Southwold U.K.
113 H1 Soutpansberg mts S. Africa
26 G5 Soverato Italy
28 B4 Sovetsk Kaliningrad. Rus. Fed.
28 J3 Sovetsk Kirovsk. Rus. Fed.
49 Q2 Sovetskaya Gavan' Rus. Fed.
28 D2 Sovetskiy Leningrad. Rus. Fed.
28 J3 Sovetskiy Mariy El. Rus. Fed.
32 H3 Sovetskiy Rus. Fed.
71 15 Sovi B. Fiji
113 G3 Soweto S. Africa
40 E4 Sowghān Iran
97 F4 Soyaló Mex.
52 G2 Sōya-misaki c. Japan
54 D4 Soyang-ho l. S. Korea
17 P4 Sozh r. Belarus
27 M3 Sozopol Bulg.
15 F2 Spa Belgium
116 B3 Spaatz I. Ant.
7 C4 Spain country Europe
13 G5 Spalding U.K.
13 D6 Span Head h. U.K.
89 F2 Spanish Can.
89 G2 Spanish r. Can.
93 G1 Spanish Fork U.S.A.
95 J5 Spanish Town Jamaica
92 C2 Sparks U.S.A.
90 C6 Sparta NC U.S.A.
88 B4 Sparta WV U.S.A.
87 D5 Spartanburg U.S.A.
27 K6 Sparti Greece
26 G6 Spartivento, Capo c. Italy
76 G5 Sparwood Can.
28 E4 Spas-Demensk Rus. Fed.
28 E2 Spasskaya Guba Rus. Fed.
49 O3 Spassk-Dal'niy Rus. Fed.
27 K7 Spatha, Akra pt Greece
76 D3 Spatsizi Plateau Wilderness Provincial Park Can.
84 C2 Spearfish U.S.A.
85 C4 Spearman U.S.A.
91 F3 Speculator U.S.A.
21 J5 Speikkogel mt Austria
84 E3 Spencer IA U.S.A.
82 D2 Spencer ID U.S.A.
90 C5 Spencer WV U.S.A.
66 B5 Spencer, C. hd Austr.
76 B3 Spencer, Cape U.S.A.
66 B5 Spencer Gulf est. Austr.
76 E4 Spences Bridge Can.
12 F3 Spennymoor U.K.
14 D3 Sperrin Mountains h. U.K.
90 D5 Sperryville U.S.A.
20 C3 Spessart reg. Ger.
27 K6 Spetses i. Greece
11 E3 Spey r. U.K.
20 B3 Speyer Ger.
41 G4 Spezand Pak.
18 E3 Spiekeroog i. Ger.
21 J6 Spielfeld Austria
22 E3 Spiez Switz.
18 D3 Spijk Neth.
18 B5 Spijkenisse Neth.
23 K3 Spilimbergo Italy
13 H4 Spilsby U.K.
41 G4 Spīn Būldak Afgh.
44 B3 Spintangi Pak.
76 F3 Spirit River Can.
88 C3 Spirit River Flowage resr U.S.A.
77 H4 Spiritwood Can.
41 G3 Spirsang Pass Afgh.
17 K6 Spišská Nová Ves Slovakia
37 K1 Spitak Armenia
21 H5 Spital am Pyhrn Austria
44 D3 Spiti r. India
32 C2 Spitsbergen i. Svalbard
21 G6 Spittal an der Drau Austria
26 G3 Split Croatia
77 K3 Split Lake Can.
77 K3 Split Lake Can.
19 H2 Spodsbjerg Denmark
82 C2 Spokane U.S.A.
23 H3 Spöl r. Italy
26 E3 Spoleto Italy
23 H3 Spondigna Italy
58 C2 Spong Cambodia
88 B3 Spooner U.S.A.
19 J3 Spornitz Ger.
82 F2 Spotted Horse U.S.A.
79 J3 Spotted Island Can.
89 F2 Spragge Can.
76 E4 Spranger, Mt Can.
82 C2 Spray U.S.A.
19 M5 Spree r. Ger.
19 L5 Spreewald reg. Ger.
19 M5 Spremberg Ger.
15 F2 Sprimont Belgium
89 F3 Spring Bay Can.
112 B4 Springbok S. Africa
79 J4 Springdale Can.
85 E4 Springdale U.S.A.
19 G4 Springe Ger.
83 F4 Springer U.S.A.
93 H4 Springerville U.S.A.
85 C4 Springfield CO U.S.A.
88 C6 Springfield IL U.S.A.
91 G3 Springfield MA U.S.A.
91 J2 Springfield ME U.S.A.
84 E2 Springfield MN U.S.A.
85 E4 Springfield MO U.S.A.
90 B5 Springfield OH U.S.A.
85 C4 Springfield OR U.S.A.
91 G3 Springfield VT U.S.A.
90 D5 Springfield WV U.S.A.
88 C6 Springfield, Lake U.S.A.
113 F5 Springfontein S. Africa
88 B4 Spring Green U.S.A.
88 B4 Spring Grove U.S.A.
79 H4 Springhill Can.
87 D6 Spring Hill U.S.A.
88 D4 Spring Lake U.S.A.
93 E3 Spring Mountains U.S.A.
70 D5 Springs Junction N.Z.
68 C5 Springsure Austr.
88 A4 Spring Valley U.S.A.
90 D3 Springville NY U.S.A.

93 G1 Springville UT U.S.A.
13 J5 Sprowston U.K.
76 G4 Spruce Grove Can.
90 D5 Spruce Knob-Seneca Rocks National Recreation Area res. U.S.A.
82 D3 Spruce Mt. U.S.A.
12 H4 Spurn Head c. U.K.
76 E5 Spuzzum Can.
76 E5 Squamish Can.
91 H3 Squam Lake U.S.A.
91 J1 Squapan Lake U.S.A.
91 J1 Square Lake U.S.A.
26 G5 Squillace, Golfo di g. Italy
Srbija div. see Serbia
58 B3 Srê Âmbĕl Cambodia
33 R4 Sredninyy Khrebet mts Rus. Fed.
27 K3 Sredna Gora mts Bulg.
33 R3 Srednekolymsk Rus. Fed.
32 E4 Sredne-Russkaya Vozvyshennost' reg. Rus. Fed.
33 M3 Sredne-Sibirskoye Ploskogor'ye plat. Rus. Fed.
8 W4 Sredneye Kuyto, Oz. l. Rus. Fed.
27 L3 Srednogorie Bulg.
58 C2 Srêpok, T. r. Cambodia
49 L1 Sretensk Rus. Fed.
43 C3 Sriharikota I. India
43 D2 Srikakulam India
43 B3 Sri Kālahasti India
44 D3 Sri Kanta mt India
31 H9 Sri Lanka country Asia
44 D3 Srinagar India
44 C2 Srinagar Jammu and Kashmir
43 B4 Srirangam India
58 B1 Sri Thep Thai.
43 B4 Srivaikuntam India
43 A2 Srivardhan India
43 B4 Srivilliputtur India
21 K1 Środa Śląska Pol.
39 L1 Srostki Rus. Fed.
43 C2 Srungavarapukota India
65 H3 Staaten River National Park Austr.
21 K4 Staatz Austria
21 G3 Stachy Czech Rep.
19 G3 Stade Ger.
15 D2 Staden Belgium
18 D4 Stadskanaal Neth.
19 G6 Stadtallendorf Ger.
19 G4 Stadthagen Ger.
19 J6 Stadtilm Ger.
18 D5 Stadtlohn Ger.
19 G5 Stadtoldendorf Ger.
19 J6 Stadtroda Ger.
11 B4 Staffa i. U.K.
20 D2 Staffelstein Ger.
23 G4 Staffora r. Italy
13 E5 Stafford U.K.
90 E5 Stafford U.S.A.
9 T8 Staicele Latvia
21 H5 Stainach Austria
13 G6 Staines U.K.
21 J6 Stainz Austria
29 F5 Stakhanov Ukr.
13 E7 Stalbridge U.K.
13 J5 Stalham U.K.
Stalingrad see Volgograd
76 E3 Stalin, Mt Can.
20 G6 Stall Austria
17 L5 Stalowa Wola Pol.
27 L3 Stamboliyski Bulg.
13 G5 Stamford U.K.
91 G4 Stamford CT U.S.A.
91 F3 Stamford NY U.S.A.
Stampalia i. see Astypalaia
111 B6 Stampriet Namibia
8 N2 Stamsund Norway
84 E3 Stanberry U.S.A.
18 B5 Standdaarbuiten Neth.
113 H3 Standerton S. Africa
89 F4 Standish U.S.A.
86 C4 Stanford U.S.A.
113 J4 Stanger S. Africa
87 E7 Staniard Ck Bahamas
27 K3 Stanke Dimitrov Bulg.
20 G3 Staňkov Czech Rep.
67 F8 Stanley Austr.
91 K1 Stanley Can.
51 Stanley H.K. China
102 E8 Stanley Falkland Is
12 F3 Stanley U.K.
82 D2 Stanley ID U.S.A.
84 C1 Stanley ND U.S.A.
88 B3 Stanley WV U.S.A.
110 C3 Stanley, Mount Congo(Zaire)/Uganda
67 F8 Stanley, Mt h. Austr.
43 B4 Stanley Reservoir India
12 F2 Stannington U.K.
33 R3 Stanovaya Rus. Fed.
46 D1 Stanovoye Nagor'ye mts Rus. Fed.
46 E1 Stanovoy Khrebet mts Rus. Fed.
22 F3 Stans Switz.
67 J2 Stanthorpe Austr.
13 H5 Stanton U.K.
90 B6 Stanton KY U.S.A.
88 E4 Stanton MI U.S.A.
84 C3 Stapleton U.S.A.
68 B2 Star r. Austr.
17 K5 Starachowice Pol.
Stara Planina mts see Balkan Mountains
28 H4 Staraya Kulatka Rus. Fed.
28 H5 Staraya Poltavka Rus. Fed.
28 D3 Staraya Russa Rus. Fed.
17 P2 Staraya Toropa Rus. Fed.
28 J4 Staraya Tumba Rus. Fed.
27 L3 Stara Zagora Bulg.
60 M4 Starbuck Island Kiribati
16 G4 Stargard Szczeciński Pol.
28 E3 Staritsa Rus. Fed.
38 D2 Star Karabutak Kazak.
87 D6 Starke U.S.A.
85 F5 Starkville U.S.A.
20 E5 Starnberg Ger.
20 E5 Starnberger See l. Ger.
39 K2 Staroaleyskoye Rus. Fed.
29 F5 Starobil's'k Ukr.
17 Q4 Starodub Rus. Fed.
17 J4 Starogard Gdański Pol.
29 C5 Starokostyantyniv Ukr.
29 F6 Starominskaya Rus. Fed.
29 F6 Staroshcherbinovskaya Rus. Fed.
38 D1 Starosubkhangulovo Rus. Fed.
92 C1 Star Peak U.S.A.
13 D7 Start Point U.K.
17 O4 Staryya Darohi Belarus
17 S3 Staryy Oskol Rus. Fed.
19 J5 Staßfurt Ger.
90 E4 State College U.S.A.
87 D5 Statesboro U.S.A.
87 D5 Statesville U.S.A.
19 L5 Stauchitz Ger.
90 D5 Staunton U.S.A.
9 J7 Stavanger Norway
13 F4 Staveley U.K.
29 G6 Stavropol' Rus. Fed.
38 F1 Stavropolka Kazak.
29 G6 Stavropol'skaya Vozvyshennost' reg. Rus. Fed.
29 G6 Stavropol'skiy Kray div. Rus. Fed.
66 E6 Stawell Austr.
113 H4 Steadville S. Africa
92 C2 Steamboat U.S.A.
82 F3 Steamboat Springs U.S.A.
116 B2 Steele I. Ant.
90 E4 Steelton U.S.A.
18 D4 Steenderen Neth.
113 J2 Steenkampsberge mts S. Africa
76 F3 Steen River Can.
82 C3 Steens Mt. U.S.A.
75 N2 Steenstrup Gletscher gl. Greenland
15 C2 Steenvoorde France
18 D4 Steenwijk Neth.
69 A3 Steep Pt Austr.
116 D4 Stefansson Bay Ant.
74 H2 Stefansson I. Can.
22 E3 Steffisburg Switz.
19 K2 Stege Denmark
21 K5 Stegersbach Austria
21 H5 Steiermark div. Austria
20 D3 Steigerwald forest Ger.
20 E3 Stein Ger.
20 E2 Steinach Ger.
77 K5 Steinbach Can.
22 E2 Steinen Ger.
18 F4 Steinfeld (Oldenburg) Ger.
18 E4 Steinfurt Ger.
20 D5 Steingaden Ger.
19 K2 Steinhagen Ger.
111 B6 Steinhausen Namibia
20 E2 Steinheid Ger.
19 G5 Steinheim Ger.
19 G4 Steinhuder Meer l. Ger.
8 M4 Steinkjer Norway
112 B4 Steinkopf S. Africa
93 H5 Steins U.S.A.
8 M4 Steinsdalen Norway
19 M4 Steinsdorf Ger.
112 F3 Stella S. Africa
112 C6 Stellenbosch S. Africa
26 C3 Stello, Monte mt Corsica France
23 H3 Stelvio, Parco Nazionale dello nat. park Italy
15 F3 Stenay France
19 J4 Stendal Ger.
66 B5 Stenhouse Bay Austr.
51 Stenhouse, Mt h. H.K. China
11 E4 Stenhousemuir U.K.
19 K2 Stensved Denmark
9 M7 Stenungsund Sweden
Stepanakert see Xankändi
29 H7 Step'anavan Armenia
77 K5 Stephen U.S.A.
66 D4 Stephens watercourse Austr.
70 D4 Stephens, Cape N.Z.
66 D3 Stephens Creek Austr.
88 D3 Stephenson U.S.A.
76 C3 Stephens Passage U.S.A.
79 J4 Stephenville Can.
85 D5 Stephenville U.S.A.
39 H1 Stepnogorsk Kazak.
39 H4 Stepnoy Kyrg.
38 E1 Stepnoye Chelyabinsk. Rus. Fed.
29 H5 Stepnoye Saratov. Obl. Rus. Fed.
38 G1 Stepnyak Kazak.
71 Steps Pt American Samoa Pac. Oc.
113 H4 Sterkfontein Dam resr S. Africa
113 G5 Sterkstroom S. Africa
38 C1 Sterlibashevo Rus. Fed.
112 D5 Sterling S. Africa
82 G3 Sterling CO U.S.A.
88 C5 Sterling IL U.S.A.
84 C2 Sterling ND U.S.A.
93 G2 Sterling UT U.S.A.
85 C6 Sterling City U.S.A.
89 F4 Sterling Hgts U.S.A.
38 C1 Sterlitamak Rus. Fed.
19 J3 Sternberg Ger.
21 L3 Šternberk Czech Rep.
19 G2 Sterup Ger.
21 H2 Štětí Czech Rep.
76 G4 Stettler Can.
20 D2 Steuben U.S.A.
90 C4 Steubenville U.S.A.
13 G6 Stevenage U.K.
77 K4 Stevenson L. Can.
88 C3 Stevens Point U.S.A.
74 D3 Stevens Village U.S.A.
76 D3 Stewart Can.
76 B2 Stewart r. Can.
76 B2 Stewart Crossing Can.
70 A7 Stewart Island N.Z.
63 G2 Stewart Islands Solomon Is
75 K3 Stewart Lake Can.
11 D5 Stewarton U.K.
88 A4 Stewartville U.S.A.
113 F5 Steynsburg S. Africa
21 H4 Steyr Austria
112 F6 Steytlerville S. Africa
18 C3 Stiens Neth.
76 C3 Stikine r. Can./U.S.A.
76 C3 Stikine Ranges mts Can.
112 D7 Stilbaai S. Africa
88 A3 Stillwater MN U.S.A.
92 C2 Stillwater NV U.S.A.
85 D4 Stillwater OK U.S.A.
83 C4 Stillwater Ra. mts U.S.A.
13 G5 Stilton U.K.
27 K4 Štip Macedonia
66 C5 Stirling Austr.
11 E4 Stirling U.K.
92 B2 Stirling City U.S.A.
69 A3 Stirling, Mt h. Austr.
66 B4 Stirling North Austr.
69 C7 Stirling Range mts Austr.
69 D7 Stirling Range National Park Austr.
23 H5 Stirone r. Italy
8 M5 Stjørdalshalsen Norway
21 G2 Stochov Czech Rep.
20 C5 Stockach Ger.
19 H3 Stockelsdorf Ger.
21 K4 Stockerau Austria
9 Q7 Stockholm Sweden
91 J1 Stockholm U.S.A.
22 E3 Stockhorn mt Switz.
13 E4 Stockport U.K.
92 B3 Stockton CA U.S.A.
84 D4 Stockton KS U.S.A.
93 F1 Stockton UT U.S.A.
88 B2 Stockton I. U.S.A.
85 E4 Stockton L. U.S.A.
12 F3 Stockton-on-Tees U.K.
91 J2 Stockton Springs U.S.A.
20 G3 Stod Czech Rep.
9 P5 Stöde Sweden
58 B2 Stœng Sângke r. Cambodia
58 C2 Stœng Sên r. Cambodia
58 C2 Stœng Trêng Cambodia
11 C2 Stoer, Point of U.K.
13 E4 Stoke-on-Trent U.K.
69 E6 Stokes Inlet in. Austr.
12 F3 Stokesley U.K.
67 E8 Stokes Pt Austr.
8 C5 Stokkseyri Iceland
8 N3 Stokkvågen Norway
8 O2 Stokmarknes Norway
26 G3 Stolac Bos.-Herz.
19 H5 Stolberg Ger.
18 D6 Stolberg (Rheinland) Ger.
39 L2 Stolboukha Kazak.
29 C5 Stolin Belarus
20 F2 Stollberg Ger.
19 G4 Stolzenau Ger.
13 E5 Stone U.K.
89 J2 Stonecliffe Can.
91 F5 Stone Harbor U.S.A.
11 F4 Stonehaven U.K.
76 E3 Stone Mountain Prov. Park Can.
93 H3 Stoner U.S.A.
91 F4 Stone Ridge U.S.A.
77 K4 Stonewall Can.
90 C5 Stonewall Jackson Lake U.S.A.
89 F4 Stoney Point Can.
91 J2 Stonington U.S.A.
92 A2 Stonyford U.S.A.
91 E3 Stony Pt U.S.A.
77 H3 Stony Rapids Can.
8 Q3 Stora Inlevatten l. Sweden
8 P3 Stora Sjöfallets Nationalpark nat. park Sweden
8 Q4 Storavan l. Sweden
9 M9 Store Bælt chan. Denmark
19 K1 Store Heddinge Denmark
8 M5 Støren Norway
8 O3 Storforshei Norway
8 O3 Storjord Norway
74 H2 Storkerson Peninsula Can.
19 L4 Storkow Ger.
67 G9 Storm Bay Austr.
113 G5 Stormberg S. Africa
113 G5 Stormberg mts S. Africa
84 E3 Storm Lake U.S.A.
9 K6 Stornosa mt Norway
11 B2 Stornoway U.K.
23 H4 Storo Italy
28 K2 Storozhevsk Rus. Fed.
29 C5 Storozhynets' Ukr.
91 G4 Storrs U.S.A.
11 B3 Storr, The h. U.K.
8 P4 Storseleby Sweden
8 O5 Storsjön l. Sweden
8 O5 Storskrymten mt Norway
8 R2 Storslett Norway
19 J2 Storstrømmen chan. Denmark
18 C3 Stortemelk chan. Neth.
8 P4 Storuman Sweden
8 P4 Storuman l. Sweden
9 P6 Storvik Sweden
9 M8 Storvorde Denmark
9 P7 Storvreta Sweden
13 G5 Stotfold U.K.
88 C4 Stoughton U.S.A.
13 E7 Stour r. Eng. U.K.
13 H6 Stour r. Eng. U.K.
13 F5 Stour r. Eng. U.K.
13 J6 Stour r. Eng. U.K.
13 E5 Stourbridge U.K.
13 E5 Stourport-on-Severn U.K.
77 L4 Stout L. Can.
28 C4 Stowbtsy Belarus
91 F4 Stowe U.S.A.
13 H5 Stowmarket U.K.
14 D3 Strabane U.K.
14 D4 Stradbally Rep. of Ireland
13 J5 Stradbroke U.K.
23 G4 Stradella Italy
67 F9 Strahan Austr.
93 G3 Straight Cliffs cliff U.S.A.
21 G3 Strakonice Czech Rep.
19 L2 Stralsund Ger.
112 C7 Strand S. Africa
8 K5 Stranda Norway
87 E7 Strangers Cay i. Bahamas
14 F3 Strangford U.K.
14 F3 Strangford Lough l. U.K.
11 C6 Stranraer U.K.
15 H4 Strasbourg France
19 L3 Strasburg Ger.
90 D5 Strasburg U.S.A.
20 G5 Straßwalchen Austria
67 G6 Stratford Austr.
89 G4 Stratford Can.
70 E3 Stratford N.Z.
85 C4 Stratford TX U.S.A.
88 B3 Stratford WV U.S.A.
13 F5 Stratford-upon-Avon U.K.
66 C5 Strathalbyn Austr.
11 D5 Strathaven U.K.
11 D3 Strathbeg, Loch of l. U.K.
11 D3 Strathcarron v. U.K.
76 D5 Strathcona Prov. Park Can.
11 D3 Strathconon v. U.K.
11 D3 Strath Dearn v. U.K.
11 D2 Strath Fleet v. U.K.
76 G4 Strathmore Can.
11 D3 Strathnaver Can.
76 E4 Strathnaver v. U.K.
11 E2 Strath of Kildonan v. U.K.
89 G4 Strathroy Can.
11 E3 Strathspey v. U.K.
11 E2 Strathy U.K.
11 D2 Strathy Point U.K.
13 C7 Stratton U.K.
91 H2 Stratton U.S.A.
20 F4 Straubing Ger.
8 B3 Straumnes pt Iceland
19 M5 Straupitz Ger.
17 L3 Strausberg Ger.
88 B4 Strawberry Point U.S.A.
93 G1 Strawberry Reservoir U.S.A.
65 F6 Streaky Bay Austr.
65 F6 Streaky Bay b. Austr.
88 C5 Streator U.S.A.
21 H3 Středočeský div. Czech Rep.
21 M3 Středoslovenský div. Slovakia
13 E6 Street U.K.
27 K2 Strehaia Romania
19 L5 Strehla Ger.
20 G2 Střela r. Czech Rep.
33 R3 Strelka Rus. Fed.
9 T8 Strenči Latvia
20 F3 Stříbro Czech Rep.
11 F3 Strichen U.K.
57 L8 Strickland r. P.N.G.
27 K4 Strimonas r. Greece
105 D4 Stroeder Arg.
14 C4 Strokestown Rep. of Ireland
11 E2 Stroma, Island of U.K.
26 F5 Stromboli, Isola i. Italy
11 E2 Stromness U.K.
84 D3 Stromsburg U.S.A.
9 M7 Strömstad Sweden
8 O5 Strömsund Sweden
23 H4 Strone r. Italy
90 C4 Strongsville U.S.A.
11 F1 Stronsay i. U.K.
21 H4 Stropnice r. Czech Rep.
67 J4 Stroud Austr.
13 E6 Stroud U.K.
67 J4 Stroud Road Austr.
91 F4 Stroudsburg U.S.A.
9 L8 Struer Denmark
27 J4 Struga Macedonia
28 D3 Strugi-Krasnyye Rus. Fed.
112 D7 Struis Bay S. Africa
27 K4 Struma r. Bulg.
13 B5 Strumble Head hd U.K.
27 K4 Strumica Macedonia
27 L3 Stryama r. Bulg.
112 D7 Strydenburg S. Africa
9 K6 Stryn Norway
29 B5 Stryy Ukr.
21 K2 Strzegomka r. Pol.
21 M2 Strzelce Opolskie Pol.
66 D2 Strzelecki Cr. watercourse Austr.
67 H8 Strzelecki Pk h. Austr.
21 L2 Strzelin Pol.
68 D6 Stuart r. Austr.
87 D7 Stuart FL U.S.A.
90 C6 Stuart VA U.S.A.
76 E4 Stuart Lake Can.
90 D5 Stuarts Draft U.S.A.
67 H4 Stuart Town Austr.
20 E5 Stubaier Alpen mts Austria
19 L2 Stubbenkammer hd Ger.
70 C6 Studholme Junction N.Z.
8 Q5 Studsviken Sweden
85 C6 Study Butte U.S.A.
77 L4 Stull L. Can.
58 C2 Stung Chinit r. Cambodia
28 F4 Stupino Rus. Fed.
23 F4 Stura r. Italy
22 E5 Stura di Demonte r. Italy
116 A6 Sturge I. Ant.
88 D2 Sturgeon r. U.S.A.
77 K4 Sturgeon Bay Can.
88 D3 Sturgeon Bay WV U.S.A.
88 E3 Sturgeon Bay U.S.A.
88 D3 Sturgeon Bay Canal chan. U.S.A.
89 H2 Sturgeon Falls Can.
78 B3 Sturgeon L. Can.
86 C4 Sturgis KY U.S.A.
88 E5 Sturgis MI U.S.A.
84 C2 Sturgis SD U.S.A.
66 B5 Sturt Bay Austr.
64 E3 Sturt Creek r. Austr.
66 D2 Sturt Desert Austr.
66 D2 Sturt, Mt h. Austr.
66 D2 Sturt Nat. Park Austr.
113 G6 Stutterheim S. Africa
20 C4 Stuttgart Ger.
85 F5 Stuttgart U.S.A.
8 B4 Stykkishólmur Iceland
17 M5 Styr r. Ukr.
104 D2 Suaçuí Grande r. Brazil
109 F3 Suakin Sudan
51 F5 Su'ao Taiwan
95 B1 Suaqui Gde Mex.
103 B3 Suárez r. Col.
17 M3 Subačius Lith.
56 Subang Indon.
45 H4 Subansiri r. India
45 F5 Subarnarekha r. India
37 G6 Şubayhah S. Arabia
59 C2 Subi Besar i. Indon.
27 H1 Subotica Yugo.
23 H5 Succiso, Alpi di mts Italy
17 N7 Suceava Romania
52 C3 Suchan r. Rus. Fed.
21 H4 Suchdol nad Lužnicí Czech Rep.
14 C4 Suck r. Rep. of Ireland
19 J3 Suckow Ger.
100 E7 Sucre Bol.
103 B2 Sucre Col.
103 C3 Sucuaro Col.
104 B2 Sucuriú r. Brazil
29 E6 Sudak Ukr.
107 F4 Sudan country Africa
28 G3 Suday Rus. Fed.
37 K6 Sudayr watercourse Iraq
89 G2 Sudbury Can.
13 H5 Sudbury U.K.
109 E4 Sudd swamp Sudan
19 J3 Sude r. Ger.
19 G2 Süderbrarup Ger.
19 G2 Süderhastedt Ger.
19 F2 Süderlügum Ger.
18 F2 Süderoog i. Ger.
21 K2 Sudety mts Czech Rep./Pol.
91 F5 Sudlersville U.S.A.
28 G4 Sudogda Rus. Fed.
36 D7 Sudr Egypt
8 Suðuroy i. Faroe Is
109 E4 Sue watercourse Sudan
25 F3 Sueca Spain
109 F2 Suez Egypt
109 F1 Suez Canal Egypt
71 F2 Suez, Gulf of g. Egypt
90 E6 Suffolk U.S.A.
40 B2 Sūfiān Iran
88 C4 Sugar r. U.S.A.
91 H2 Sugarloaf Mt. U.S.A.
67 K4 Sugarloaf Pt Austr.
55 C4 Sugbuhan Point Phil.
39 J5 Sugun China
59 E1 Sugut r. Malaysia
55 A5 Sugut, Tg pt Malaysia
50 B2 Suhait China
42 E5 Şuḩār Oman
48 J1 Sühbaatar Mongolia
20 D2 Suhl Ger.
19 H4 Suhlendorf Ger.
36 C2 Şuhut Turkey
44 B3 Sui Pak.
52 B1 Suibin China
54 F4 Suichang China
51 E5 Suichuan China
50 D2 Suide China
54 F1 Suifenhe Sweden
44 B4 Suigam India
46 E2 Suihua China
51 B4 Suijiang China
51 D5 Suining Hunan China
50 E3 Suining Jiangsu China
51 B4 Suining Sichuan China
50 E3 Suiping China
15 E3 Suippes France
14 D5 Suir r. Rep. of Ireland
50 E3 Suixi China
50 E3 Sui Xian China
51 C5 Suiyang China
50 F1 Suizhong China
50 D4 Suizhou China
50 C1 Suj China
44 C4 Sujangarh India
44 D3 Sujanpur India
44 B4 Sujawal Pak.
56 Sukabumi Indon.
53 C3 Sukadana Indon.
53 G6 Sukagawa Japan
55 A5 Sukau Malaysia
54 C4 Sukchŏn N. Korea
52 C1 Sukhanovka Rus. Fed.
28 E4 Sukhinichi Rus. Fed.
28 H2 Sukhona r. Rus. Fed.
58 A1 Sukhothai Thai.
28 E2 Sukkozero Rus. Fed.
44 B4 Sukkur Pak.
43 C2 Sukma India
44 C4 Sukri r. India
28 F3 Sukromny Rus. Fed.
53 C8 Sukumo Japan
9 J6 Sula i. Norway
83 B4 Sulaiman Ranges mts Pak.
29 H7 Sulak r. Rus. Fed.
57 H7 Sula, Kepulauan is Indon.
40 C4 Sūlār Iran
11 B1 Sula Sgeir i. U.K.
59 E3 Sulawesi i. Indon.
37 K4 Sulaymān Beg Iraq
42 C2 Suledeh Iran
11 D1 Sule Skerry i. U.K.
11 D1 Sule Stack i. U.K.
36 F3 Süleymanlı Turkey
108 A4 Sulima Sierra Leone
18 F4 Sulingen Ger.
8 P3 Sulitjelma Norway
9 V6 Sulkava Fin.
100 B4 Sullana Peru
84 F4 Sullivan U.S.A.
77 G4 Sullivan L. Can.
91 J1 Sully Can.
26 E3 Sulmona Italy
85 E6 Sulphur U.S.A.
85 E5 Sulphur Springs U.S.A.
82 F2 Sultan Can.
Sultanabad see Arāk
36 C2 Sultan Dağları mts Turkey
36 D2 Sultanhanı Turkey
41 F4 Sultan, Koh-i- mts Pak.
45 E4 Sultanpur India
55 B5 Sulu Archipelago is Phil.
39 G5 Sülüktü Kyrg.
36 F2 Sulusaray Turkey
55 A4 Sulu Sea Phil.
38 F3 Sulutobe Kazak.
20 B4 Sulz am Neckar Ger.
20 E3 Sulzbach-Rosenberg Ger.
20 C5 Sulzberg Austria
116 A4 Sulzberger Bay Ant.
41 E6 Sumāil Oman
102 D3 Sumampa Arg.
103 B4 Sumapaz, Parque Nacional nat. park Col.
37 K5 Sūmar Iran
59 B3 Sumatera i. Indon.
Sumatra i. see Sumatera
21 G3 Šumava mts Czech Rep.
21 G3 Šumava nat. park Czech Rep.
57 G8 Sumba i. Indon.
38 C5 Sumbar r. Turkm.
57 F8 Sumba, Selat chan. Indon.
59 E4 Sumbawa i. Indon.
59 E4 Sumbawabesar Indon.
115 B4 Sumbawanga Tanz.
111 B5 Sumbe Angola
115 B4 Sumbu National Park Zambia
11 Sumburgh U.K.
11 Sumburgh Head hd U.K.
44 D2 Sumdo China/Jammu and Kashmir
56 Sumedang Indon.
37 M3 Sume'eh Sarā Iran
21 L6 Sümeg Hungary
59 D4 Sumenep Indon.
Sumgait see Sumqayıt
53 F9 Sumisu-jima i. Japan
37 J3 Summēl Iraq
78 C3 Summer Beaver Can.
79 K4 Summerford Can.
88 D3 Summer I. U.S.A.
11 C2 Summer Isles U.K.
79 H4 Summerside Can.
90 C5 Summersville U.S.A.
90 C5 Summersville Lake U.S.A.
76 E4 Summit Lake Can.
88 E5 Summit Lake U.S.A.
92 D2 Summit Mt U.S.A.
44 D2 Sumnal China/India
70 D5 Sumner N.Z.
88 A4 Sumner U.S.A.
70 D5 Sumner, L. N.Z.
76 C3 Sumner Strait U.S.A.
53 F6 Sumon-dake mt Japan
53 D7 Sumoto Japan
21 K3 Šumperk Czech Rep.
38 A4 Sumqayıt Azer.
37 M1 Sumqayıt r. Azer.
44 B4 Sumrahu Pak.
87 D5 Sumter U.S.A.
29 E5 Sumy Ukr.
82 D2 Sun r. U.S.A.
28 J3 Suna Rus. Fed.
52 G3 Sunagawa Japan
45 G4 Sunamganj Bangl.
54 C4 Sunan N. Korea
11 C4 Sunart, Loch in. U.K.
40 D6 Şunaynah Oman
37 K4 Sunbula Kuh mts Iran
82 E1 Sunburst U.S.A.
66 F6 Sunbury Austr.
90 B4 Sunbury OH U.S.A.
90 E4 Sunbury PA U.S.A.
105 E1 Sunchales Arg.
54 C4 Sunch'ŏn N. Korea
54 D6 Sunch'ŏn S. Korea
113 G2 Sun City S. Africa
91 H3 Suncook U.S.A.
82 F2 Sundance U.S.A.
45 F5 Sundargarh India
44 D3 Sundarnagar India
59 C4 Sunda, Selat chan. Indon.
117 M4 Sunda Trench sea feature Ind. Ocean
12 F3 Sunderland U.K.
18 F5 Sundern (Sauerland) Ger.
36 C2 Sündiken Dağları mts Turkey
89 H3 Sundridge Can.
9 P5 Sundsvall Sweden
38 E5 Sundukli, Peski des. Turkm.
113 J4 Sundumbili S. Africa
44 D4 Sunel India
58 B5 Sungaikabung Indon.
59 C3 Sungailiat Indon.
59 B2 Sungai Pahang r. Malaysia
59 B3 Sungaipenuh Indon.
59 B1 Sungei Petani Malaysia
58 Sungei Seletar Resr Sing.
36 E1 Sungurlu Turkey
39 H3 Sunkar, Gora mt Kazak.
45 F4 Sun Kosi r. Nepal
9 K6 Sunndal Norway
8 L5 Sunndalsøra Norway
9 N7 Sunne Sweden
82 C2 Sunnyside U.S.A.
92 A3 Sunnyvale U.S.A.
88 C4 Sun Prairie U.S.A.
92 Sunset Beach U.S.A.
93 G4 Sunset Crater National Monument res. U.S.A.
33 N3 Suntar Rus. Fed.
41 F5 Suntsar Pak.
82 D3 Sun Valley U.S.A.
54 C5 Sunwi Do i. N. Korea
114 C5 Sunyani Ghana
8 U3 Suolijärvet l. Fin.
88 C1 Suomi Can.
8 V4 Suomussalmi Fin.
53 B8 Suō-nada b. Japan
8 U5 Suonenjoki Fin.
58 C3 Suŏng Cambodia
51 B7 Suong r. Laos
28 E2 Suoyarvi Rus. Fed.
43 A3 Supa India
93 F3 Supai U.S.A.
103 E3 Supamo r. Venez.
45 F4 Supaul India
93 G5 Superior AZ U.S.A.
84 D3 Superior NE U.S.A.
88 A2 Superior WV U.S.A.
97 F4 Superior, L. Mex.
88 C2 Superior, Lake Can./U.S.A.
58 B2 Suphan Buri Thai.
37 J2 Süphan Dağı mt Turkey
28 E4 Suponevo Rus. Fed.
116 B3 Support Force Glacier Ant.
54 C3 Supung N. Korea
37 L6 Sūq ash Shuyūkh Iraq
50 F3 Suqian China
Suquţrā i. see Socotra
21 M5 Súr Hungary
42 E5 Şūr Oman
28 H4 Sura Rus. Fed.
28 H4 Sura r. Rus. Fed.
37 M1 Şuraabad Azer.
41 G4 Surab Pak.
59 D4 Surabaya Indon.
55 E5 Sūrak Iran
59 D4 Surakarta Indon.
22 C3 Suran r. France
21 M4 Šurany Slovakia
67 H1 Surat Austr.
44 C5 Surat India
44 C3 Suratgarh India
58 A3 Surat Thani Thai.
28 E4 Surazh Rus. Fed.
68 B4 Surbiton Austr.
71 Sur, Cabo c. Easter I. Chile
37 K4 Sūrdāsh Iraq
27 K3 Surdulica Yugo.
15 F3 Sûre r. Lux.
44 B5 Surendranagar India
96 J6 Suretka Costa Rica
92 B4 Surf U.S.A.
32 J3 Surgut Rus. Fed.
18 D3 Surhuisterveen Neth.
43 B2 Suriapet India
55 C4 Surigao Phil.
55 C4 Surigao Str. Phil.
58 B2 Surin Thai.
99 E2 Suriname country S. America
41 H3 Surkhab r. Afgh.
39 F5 Surkhandar'ya r. Uzbek.
45 E3 Surkhet Nepal
39 G5 Surkhob r. Tajik.
40 D4 Surmāq Iran
37 H1 Sürmene Turkey
29 G5 Surovikino Rus. Fed.
71 Surprise, Î. i. New Caledonia Pac. Oc.
92 B3 Sur, Pt U.S.A.
105 F3 Sur, Pta pt Arg.
90 E6 Surry U.S.A.
23 F3 Surselva reg. Switz.
28 H4 Sursk Rus. Fed.
Surt see Sirte
Surt, Khalīj g. see Sirte, Gulf of
8 C5 Surtsey i. Iceland
36 G3 Sürüç Turkey
53 F7 Suruga-wan b. Japan
59 B3 Surulangun Indon.
55 C5 Surup Phil.
18 E4 Surwold Ger.
40 B2 Suşa Azer.
22 E4 Susa Italy
53 B7 Susa Japan
53 C8 Susaki Japan
40 C4 Süsangerd Iran
28 G3 Susanino Rus. Fed.
92 B1 Susanville U.S.A.
36 G1 Suşehri Turkey
19 H2 Süsel Ger.
21 G3 Sušice Czech Rep.
58 A4 Suso Thai.
51 E4 Susong China
91 E4 Susquehanna r. U.S.A.
79 G4 Sussex Can.
91 F4 Sussex U.S.A.
55 A5 Susul Malaysia
33 Q3 Susuman Rus. Fed.
36 B2 Susurluk Turkey
44 D2 Sutak Jammu and Kashmir
92 C2 Sutcliffe U.S.A.
112 D6 Sutherland S. Africa
84 C3 Sutherland U.S.A.
44 C3 Sutlej r. Pak.
92 B2 Sutter Creek U.S.A.
13 G5 Sutterton U.K.
91 G2 Sutton Can.
78 D3 Sutton r. Can.
13 H5 Sutton U.K.
90 C5 Sutton U.S.A.
13 F5 Sutton Coldfield U.K.
13 F4 Sutton in Ashfield U.K.
78 D3 Sutton L. Can.
90 C5 Sutton Lake U.S.A.
68 B3 Suttor r. Austr.
52 G3 Suttsu Japan
52 D1 Sutungu Rus. Fed.
39 H4 Suusamyr Kyrg.
71 15 Suva Fiji
28 F4 Suvorov Rus. Fed.
60 L5 Suvorov I. Cook Is Pac. Oc.
53 F6 Suwa Japan
17 L3 Suwałki Pol.
58 B2 Suwannaphum Thai.
87 D6 Suwannee r. U.S.A.
37 K5 Suwayqīyah, Hawr as l. Iraq
37 H6 Suwayr well S. Arabia
54 D5 Suwŏn S. Korea
39 G3 Suzak Kazak.
53 F6 Suzaka Japan
28 G3 Suzdal' Rus. Fed.
50 E3 Suzhou Anhui China
50 F4 Suzhou Jiangsu China
54 C3 Suzi r. China
38 C4 Suz, Mys pt Kazak.
53 E6 Suzu Japan
53 E7 Suzuka Japan
53 E6 Suzu-misaki pt Japan
23 H5 Suzzara Italy
8 U1 Sværholthalvøya pen. Norway
120 K1 Svalbard terr. Arctic Ocean
19 J1 Svallerup Denmark
75 N2 Svartenhuk Halvø pen. Greenland
29 F5 Svatove Ukr.
58 C3 Svay Riĕng Cambodia
9 O5 Sveg Sweden
9 U8 Sveki Latvia
9 J6 Svelgen Norway
8 L5 Svellingen Norway
9 T9 Švenčionėliai Lith.
9 U9 Švenčionys Lith.
9 M9 Svendborg Denmark
8 Q2 Svensby Norway
8 O5 Svenstavik Sweden
Sverdlovsk see Yekaterinburg
29 F5 Sverdlovs'k Ukr.
75 J1 Sverdrup Channel Can.
27 J4 Sveti Nikole Macedonia
21 J3 Světlá nad Sázavou Czech Rep.
49 P2 Svetlaya Rus. Fed.
49 Q2 Svetlodarskoye Rus. Fed.
28 B4 Svetlogorsk Kaliningrad. Rus. Fed.
32 K3 Svetlogorsk Krasnoy. Rus. Fed.
29 G6 Svetlograd Rus. Fed.
28 B4 Svetlyy Kaliningrad. Rus. Fed.
38 E2 Svetlyy Orenburg. Rus. Fed.
29 H5 Svetlyy Yar Rus. Fed.
28 D2 Svetogorsk Rus. Fed.
8 E4 Svíahnúkar volc. Iceland
20 G3 Švihov Czech Rep.
27 M4 Svilengrad Bulg.
27 K2 Svinecea Mare, Vârful mt Romania
19 J1 Svinninge Denmark
28 C4 Svir Belarus
28 E2 Svir' r. Rus. Fed.
27 L3 Svishtov Bulg.
21 K3 Svitava r. Czech Rep.
21 K3 Svitávka Czech Rep.
21 K3 Svitavy Czech Rep.
29 E5 Svitlovods'k Ukr.
28 J4 Sviyaga r. Rus. Fed.
49 N1 Svobodnyy Rus. Fed.
8 O2 Svolvær Norway
21 K3 Svratka r. Czech Rep.
27 K3 Svrljiške Planine mts Yugo.
29 D4 Svyetlahorsk Belarus
13 F5 Swadlincote U.K.
13 H5 Swaffham U.K.
68 E3 Swain Reefs Austr.
87 D5 Swainsboro U.S.A.
60 K5 Swains Island Pac. Oc.
111 B6 Swakopmund Namibia
12 F3 Swale r. U.K.
63 G3 Swallow Is Solomon Is
66 E5 Swan r. Austr.
77 J4 Swan r. Can.
13 F7 Swanage U.K.
66 E5 Swan Hill Austr.
76 F4 Swan Hills Can.
95 H5 Swan Islands Honduras
77 J4 Swan L. Can.
13 H6 Swanley U.K.
66 C5 Swan Reach Austr.
77 J4 Swan River Can.
67 J4 Swansea N.S.W. Austr.
67 H9 Swansea Tas. Austr.
13 D6 Swansea U.K.
13 D6 Swansea Bay U.K.
91 J2 Swans I. U.S.A.
91 G2 Swanton U.S.A.
113 G2 Swartruggens S. Africa
93 F2 Swasey Peak U.S.A.
89 G1 Swastika Can.
44 B2 Swat r. Pak.
Swatow see Shantou
107 G8 Swaziland country Africa
7 E3 Sweden country Europe
82 B2 Sweet Home U.S.A.
87 C5 Sweetwater TN U.S.A.
85 C5 Sweetwater TX U.S.A.
82 E3 Sweetwater r. U.S.A.
112 D7 Swellendam S. Africa
16 H5 Świdnica Pol.
16 G4 Świdwin Pol.
21 K2 Świebodzice Pol.
16 G4 Świebodzin Pol.
17 J4 Świecie Pol.
21 J1 Świerzawa Pol.
91 H1 Swift r. U.S.A.
77 H4 Swift Current Can.
77 H5 Swiftcurrent Cr. r. Can.
76 C2 Swift River Can.
14 D2 Swilly, Lough in. Rep. of Ireland
13 F6 Swindon U.K.
14 C4 Swinford Rep. of Ireland
16 G4 Świnoujście Pol.
11 F5 Swinton U.K.
23 H3 Swiss National Park Switz.
7 D4 Switzerland country Europe
14 E4 Swords Rep. of Ireland
52 D1 Syain Rus. Fed.
28 E2 Syamozero, Oz. l. Rus. Fed.
28 G2 Syamzha Rus. Fed.
17 O3 Syanno Belarus
28 E2 Syas'troy Rus. Fed.
28 H3 Syava Rus. Fed.
88 C5 Sycamore U.S.A.
21 L1 Syców Pol.
67 J4 Sydney Austr.
79 H4 Sydney Can.
77 L4 Sydney L. Can.
79 H4 Sydney Mines Can.
29 F5 Syeverodonets'k Ukr.
18 F4 Syke Ger.
28 J2 Syktyvkar Rus. Fed.
87 C5 Sylacauga U.S.A.
8 N5 Sylarna mt Norway/Sweden
45 G4 Sylhet Bangl.
28 G2 Syloga Rus. Fed.
18 F2 Sylt i. Ger.
69 D1 Sylvania Austr.
87 D5 Sylvania GA U.S.A.
90 B4 Sylvania OH U.S.A.
76 G4 Sylvan Lake Can.
87 D6 Sylvester U.S.A.
76 E3 Sylvia, Mt Can.
27 M6 Symi i. Greece
96 D2 Symon Mex.
29 E5 Synel'nykove Ukr.
116 D4 Syowa Japan Base Ant.
26 F6 Syracuse Sicily Italy
84 C4 Syracuse KS U.S.A.
91 E3 Syracuse NY U.S.A.
91 E3 Syracuse airport NY U.S.A.
39 F4 Syrdar'ya r. Kazak.
39 G4 Syrdar'ya Uzbek.
31 C6 Syria country Asia
56 B3 Syriam Myanmar
42 A3 Syrian Desert Asia
27 M6 Syrna i. Greece
27 L6 Syros i. Greece
9 T6 Sysmä Fin.
28 J2 Sysola r. Rus. Fed.
28 J4 Syzran' Rus. Fed.
21 M6 Szakály Hungary
21 L5 Szany Hungary
16 G4 Szczecin Pol.
16 H4 Szczecinek Pol.
17 K4 Szczytno Pol.
17 K7 Szeged Hungary
17 J7 Székesfehérvár Hungary
17 J7 Szekszárd Hungary
17 K7 Szentes Hungary
16 H7 Szentgotthárd Hungary
26 G1 Szigetvár Hungary
17 K7 Szolnok Hungary
16 H7 Szombathely Hungary
21 J2 Szrenica mt Czech Rep.

T

55 B3 Taal, L. Phil.
71 Taapuma Fr. Polynesia Pac. Oc.
21 M6 Tab Hungary
55 B3 Tabaco Phil.
113 H5 Tabankulu S. Africa
36 G4 Ţabaqah Syria
62 F2 Tabar Is P.N.G.
36 E4 Tabarja Lebanon
26 C6 Tabarka Tunisia
40 E3 Ţabas Iran
97 F4 Tabasco div. Mex.
41 E4 Tabāsin Iran
40 C4 Tābask, Kūh-e mt Iran
79 J3 Tabatière Can.
100 E4 Tabatinga Col.
55 B2 Tabayoo, Mt Phil.
67 F5 Tabbita Austr.
108 B2 Tabelbala Alg.
77 G5 Taber Can.
45 F3 Tabia Tsaka salt l. China
63 H2 Tabiteuea i. Kiribati
9 U7 Tabivere Estonia
55 B3 Tablas i. Phil.
55 B3 Tablas Strait Phil.
70 F3 Table Cape N.Z.
112 C6 Table Mountain S. Africa
85 E4 Table Rock Resr U.S.A.
68 A2 Tabletop, Mount h. Austr.
114 D3 Tabligbo Togo
104 A2 Taboco r. Brazil
21 H3 Tábor Czech Rep.
115 B4 Tabora Tanz.
115 B4 Tabora div. Tanz.
39 G4 Taboshar Tajik.
114 B4 Tabou Côte d'Ivoire
40 B2 Tabrīz Iran
60 M3 Tabueran i. Kiribati
42 A4 Tabūk S. Arabia
67 K2 Tabulam Austr.
39 J1 Tabuny Rus. Fed.
71 9 Tabwémasana mt Vanuatu
9 Q7 Täby Sweden
96 D4 Tacámbaro Mex.
97 F5 Tacaná, Volcán de volc. Mex.
96 K6 Tacarcuna, Cerro mt Panama
39 K3 Tacheng China
20 F3 Tachov Czech Rep.
55 C4 Tacloban Phil.
100 D7 Tacna Peru
82 B2 Tacoma U.S.A.
105 F1 Tacuarembó Uru.
105 G2 Tacuarí r. Uru.
96 B1 Tacupeto Mex.
103 E4 Tacutu r. Brazil
12 F4 Tadcaster U.K.
108 C2 Tademaït, Plateau du plat. Alg.
63 G4 Tadine New Caledonia
71 9 Tadine New Caledonia Pac. Oc.
110 E2 Tadjoura Djibouti
36 G4 Tadmur Syria
77 K3 Tadoule Lake Can.
79 G4 Tadoussac Can.
54 D4 T'aebaek Sanmaek mts N. Korea/S. Korea
54 D5 Taech'ŏn S. Korea
54 C5 Taech'ŏngdo i. N. Korea
54 C4 Taedasa-do N. Korea
54 D4 Taedong r. N. Korea
54 C5 Taedong man b. N. Korea
54 E6 Taegu S. Korea
54 C6 Taehŭksan-kundo i. S. Korea
54 D5 Taejŏn S. Korea
54 D7 Taejŏng S. Korea
54 E5 T'aepaek S. Korea
13 C6 Taf r. U.K.
63 J3 Tafahi i. Tonga
25 F1 Tafalla Spain
40 D4 Tafīhān Iran
36 E6 Tafila Jordan
114 B3 Tafiré Côte d'Ivoire
102 C3 Tafí Viejo Arg.
40 C3 Tafresh Iran
40 D4 Taft Iran
92 C4 Taft U.S.A.
41 F4 Taftān, Kūh-e mt Iran
71 4 Taga Western Samoa
29 F6 Taganrog Rus. Fed.
29 F6 Taganrog, Gulf of Rus. Fed./Ukr.
55 C3 Tagapula i. Phil.
55 B3 Tagaytay City Phil.
55 B4 Tagbilaran Phil.

45 E2 Tagchagpu Ri mt China
22 E6 Taggia Italy
14 E5 Taghmon Rep. of Ireland
76 C2 Tagish Can.
23 K3 Tagliamento r. Italy
25 J4 Tagma, Col de pass Alg.
55 C4 Tagoloan r. Phil.
55 B4 Tagolo Point Phil.
63 F3 Tagula I. P.N.G.
55 C5 Tagum Phil.
25 B3 Tagus r. Port./Spain
71 □2 Tahaa i. Fr. Polynesia Pac. Oc.
76 F4 Tahaetkun Mt. Can.
71 □2 Tahanea i. Fr. Polynesia Pac. Oc.
58 B4 Tahan, Gunung mt Malaysia
108 C2 Tahat, Mt Alg.
46 E1 Tahe China
70 D1 Taheke N.Z.
71 □3 Tahiti i. Fr. Polynesia Pac. Oc.
60 N5 Tahiti i. Pac. Oc.
41 F4 Tahlab r. Iran/Pak.
41 F4 Tahlab, Dasht-i plain Pak.
85 E5 Tahlequah U.S.A.
92 B2 Tahoe City U.S.A.
74 H3 Tahoe Lake Can.
92 B2 Tahoe, Lake U.S.A.
85 C5 Tahoka U.S.A.
114 E2 Tahoua Niger
76 D4 Tahtsa Pk Can.
55 C6 Tahuna Indon.
114 B3 Taï Côte d'Ivoire
51 □ Tai a Chau i. H.K. China
54 B3 Tai'an Liaoning China
50 E2 Tai'an Shandong China
71 □3 Taiarapu, Presqu'île de pen. Fr. Polynesia Pac. Oc.
50 C3 Taibai Shan mt China
50 E1 Taibus Qi China
51 F5 T'ai-chung Taiwan
70 C6 Taieri r. N.Z.
50 D2 Taigu China
50 D2 Taihang Shan mts China
70 E3 Taihape N.Z.
50 E3 Taihe Anhui China
51 E5 Taihe Jiangxi China
51 E4 Taihu China
46 E3 Tai Hu l. China
51 C5 Taijiang China
50 E3 Taikang China
51 □ Tai Lam Chung Resr H.K. China
66 C5 Tailem Bend Austr.
22 C4 Taillefer, Le mt France
51 □ Tai Long Bay H.K. China
51 F5 T'ai-lu-ko Taiwan
41 F3 Taimani reg. Afgh.
51 □ Tai Mo Shan h. H.K. China
51 F6 T'ai-nan Taiwan
51 F6 T'ainan Taiwan
27 K6 Tainaro, Akra pt Greece
51 E5 Taining China
22 B4 Tain-l'Hermitage France
51 □ Tai O H.K. China
104 D1 Taiobeiras Brazil
114 B3 Taï, Parc National de nat. park Côte d'Ivoire
51 F5 T'ai-pei Taiwan
51 F4 Taiping Anhui China
51 D6 Taiping Guangxi China
59 B2 Taiping Malaysia
50 A2 Taipingbao China
54 B1 Taipingchuan China
51 □ Tai Po H.K. China
52 H3 Taisetsu-zan National Park Japan
53 C7 Taisha Japan
51 D6 Taishan China
51 F5 Taishun China
15 E3 Taissy France
70 D5 Taitanu N.Z.
102 B7 Taitao, Península de pen. Chile
51 F6 T'ai-tung Taiwan
8 V4 Taivalkoski Fin.
8 T2 Taivaskero h. Fin.
31 M7 Taiwan country Asia
51 F6 Taiwan Shan mts Taiwan
51 F5 Taiwan Strait China/Taiwan
50 F3 Tai Xian China
50 F3 Taixing China
50 D3 Taiyuan China
50 D2 Taiyue Shan mts China
50 F3 Taizhou China
51 F4 Taizhou Wan b. China
54 C3 Taizi r. China
42 B7 Ta'izz Yemen
44 B4 Tajal Pak.
97 G5 Tajamulco, Volcán de volc. Guatemala
26 C7 Tajerouine Tunisia
31 G6 Tajikistan country Asia
Tajo r. see Tagus
58 A1 Tak Thai.
40 B2 Takāb Iran
115 D2 Takabba Kenya
53 C7 Takahashi Japan
70 D4 Takaka N.Z.
44 D5 Takal India
53 D7 Takamatsu Japan
44 D4 Takanpur India
53 E6 Takaoka Japan
70 F4 Takapau N.Z.
71 □2 Takapoto i. Fr. Polynesia Pac. Oc.
70 E2 Takapuna N.Z.
71 □2 Takaroa i. Fr. Polynesia Pac. Oc.
53 F6 Takasaki Japan
112 F2 Takatokwane Botswana
112 D1 Takatshwaane Botswana
53 C8 Takatsuki-yama mt Japan
53 E6 Takayama Japan
58 B4 Tak Bai Thai.
53 E7 Takefu Japan
39 G4 Takelī Tajik.
53 B8 Takeo Japan
Take-shima i. see Tok-tō
53 B9 Take-shima i. Japan
40 C2 Takestān Iran
53 B8 Taketa Japan
58 C3 Takêv Cambodia
37 K7 Takhādīd well Iraq
38 D4 Takhiatash Uzbek.
58 C3 Ta Khmau Cambodia
38 D4 Takhta Turkm.
38 E4 Takhta-Bazar Turkm.
39 F1 Takhtabrod Kazak.
38 E4 Takhtakupyr Uzbek.
37 M5 Takht Apān, Kūh-e mt Iran
41 G4 Takhta Pul Post Afgh.
44 B3 Takht-i-Sulaiman mt Pak.
40 C2 Takht-i-Suleiman mt Iran
77 G1 Takijuq Lake Can.
52 G3 Takikawa Japan
52 H2 Takinoue Japan
70 A6 Takitimu Mts N.Z.
76 D3 Takla Lake Can.
76 D3 Takla Landing Can.
Taklimakan Desert see Taklimakan Shamo
35 G3 Taklimakan Shamo des. China
39 G5 Takob Tajik.
45 H3 Takpa Shiri mt China
76 C3 Taku r. Can.
114 E3 Takum Nigeria
71 □2 Takumé i. Fr. Polynesia Pac. Oc.
105 F2 Tala Uru.
28 D4 Talachyn Belarus
43 B4 Talaimannar Sri Lanka
44 C5 Talaja India
45 H4 Talap India
100 B4 Talara Peru
41 F5 Talar-i-Band mts Pak.
39 G4 Talas r. Kazak./Kyrg.
39 H4 Talas Kyrg.
39 G4 Talas Ala-Too mts Kyrg.
57 H6 Talaud, Kepulauan is Indon.
25 D3 Talavera de la Reina Spain
33 R3 Talaya Rus. Fed.
55 C5 Talayan Phil.
75 L2 Talbot Inlet b. Can.
67 H4 Talbragar r. Austr.
105 B2 Talca Chile
105 B3 Talcahuano Chile
45 F5 Talcher India
38 D4 Taldyk Uzbek.
39 J3 Taldykorgan Kazak.
39 J3 Taldykorganskaya Oblast' div. Kazak.
39 G2 Taldysayskiy Kazak.
39 J4 Taldy-Suu Kyrg.
37 M5 Taleh Zang Iran
71 □12 Taleki Tonga i. Tonga
71 □12 Taleki Vavu'u i. Tonga
40 C2 Tālesh Iran
39 J4 Talgar Kazak.
13 D6 Talgarth U.K.
66 A4 Talia Austr.
57 G7 Taliabu i. Indon.
22 D5 Taliard France
55 C4 Talibon Phil.
43 B2 Talikota India
38 F5 Talimardzhan Uzbek.
37 J1 T'alin Armenia
43 A3 Taliparamba India
55 B4 Talisay Phil.
55 C4 Talisayan Phil.
37 M2 Taliş Dağları mts Azer./Iran
28 H3 Talitsa Rus. Fed.
59 E4 Taliwang Indon.
87 C5 Talladega U.S.A.
37 J3 Tall 'Afar Iraq
87 C6 Tallahassee U.S.A.
67 G6 Tallangatta Austr.
87 C5 Tallassee U.S.A.
36 F6 Tall as Suwaysh h. Jordan
37 H3 Tall Baydar Syria
69 B4 Tallering Pk h. Austr.
37 H4 Tall Fadghāmī Syria
9 T7 Tallinn Estonia
36 F4 Tall Kalakh Syria
37 J3 Tall Kayf Iraq
14 C5 Tallow Rep. of Ireland
85 F5 Tallulah U.S.A.
37 J3 Tall 'Uwaynāt Iraq
24 D3 Talmont-St-Hilaire France
29 D5 Tal'ne Ukr.
109 F3 Talodi Sudan
79 G2 Talon, Lac l. Can.
41 H2 Tāloqān Afgh.
38 B2 Talovaya Kazak.
29 G5 Talovaya Rus. Fed.
75 J3 Taloyoak Can.
44 C2 Tal Pass Pak.
9 S8 Talsi Latvia
102 B3 Taltal Chile
77 G2 Taltson r. Can.
40 B3 Talvar r. Iran
8 S1 Talvik Norway
67 H2 Talwood Austr.
66 E4 Talyawalka r. Austr.
88 A4 Tama U.S.A.
69 A3 Tamala Austr.
103 B2 Tamalameque Col.
114 C3 Tamale Ghana
103 A3 Tamana mt Col.
63 H2 Tamana i. Kiribati
53 C7 Tamano Japan
108 C2 Tamanrasset Alg.
45 H4 Tamanthi Myanmar
103 B3 Tama, Parque Nacional el nat. park Venez.
91 F4 Tamaqua U.S.A.
13 C7 Tamar r. U.K.
113 G1 Tamasane Botswana
21 M6 Tamási Hungary
97 E2 Tamaulipas div. Mex.
96 C2 Tamazula Mex.
97 E4 Tamazulápam Mex.
97 E3 Tamazunchale Mex.
115 C2 Tambach Kenya
114 A3 Tambacounda Senegal
45 F4 Tamba Kosi r. Nepal
69 C7 Tambellup Austr.
55 A5 Tambisan Malaysia
68 B5 Tambo Qld. Austr.
67 G6 Tambo r. Vic. Austr.
59 E4 Tambora, Gunung volc. Indon.
67 G6 Tamboritha mt Austr.
28 G4 Tambov Rus. Fed.
28 G4 Tambovskaya Oblast' div. Rus. Fed.
25 B1 Tambre r. Spain
55 A5 Tambunan, Bukit h. Malaysia
109 E4 Tambura Sudan
55 A5 Tambuyukon, Gunung mt Malaysia
108 A3 Tâmchekkeț Maur.
38 D2 Tamdy Kazak.
38 F4 Tamdybulak Uzbek.
103 C3 Tame Col.
25 C2 Tâmega r. Port.
45 H4 Tamenglong India
26 B7 Tamerza Tunisia
97 E3 Tamiahua Mex.
97 E3 Tamiahua, Lag. de Mex.
58 A4 Tamiang, Ujung pt Indon.
43 B4 Tamil Nadu div. India
28 F1 Tamitsa Rus. Fed.
36 C7 Tâmîya Egypt
38 E2 Tamkamys Kazak.
58 C1 Tam Ky Vietnam
87 D7 Tampa U.S.A.
87 D7 Tampa Bay U.S.A.
41 E5 Tamp-e Gīrān Iran
9 S6 Tampere Fin.
97 E3 Tampico Mex.
58 □ Tampines Sing.
49 L2 Tamsagbulag Mongolia
50 B1 Tamsag Muchang China
21 G5 Tamsweg Austria
45 H4 Tamu Myanmar
97 E3 Tamuín Mex.
45 F4 Tamur r. Nepal
67 J3 Tamworth Austr.
13 F5 Tamworth U.K.
39 J2 Tan Kazak.
115 C3 Tana r. Kenya
53 D8 Tanabe Japan
8 V1 Tana Bru Norway
8 V1 Tanafjorden chan. Norway
T'ana Hāyk' l. see Tana, Lake
59 E3 Tanahgrogot Indon.
57 G8 Tanahjampea i. Indon.
59 A3 Tanahmasa i. Indon.
57 J6 Tanahmerah Indon.
58 B4 Tanah Merah Malaysia
59 C4 Tanah, Tanjung pt Indon.
110 D2 Tana, Lake Eth.
64 F3 Tanami Desert Austr.
58 C3 Tân An Vietnam
74 C3 Tanana U.S.A.
22 F4 Tanaro r. Italy
55 C4 Tanauan Phil.
71 □15 Tanavuso Pt Fiji
50 F3 Tancheng China
54 E3 Tanch'ŏn N. Korea
96 D4 Tancítaro, Cerro de mt Mex.
114 C3 Tanda Côte d'Ivoire
45 E4 Tanda India
55 C4 Tandag Phil.
27 M2 Țăndărei Romania
55 A5 Tandek Malaysia
44 D2 Tandi India
105 E3 Tandil Arg.
105 E3 Tandil, Sa del h. Arg.
41 G5 Tando Adam Pak.
41 G5 Tando Bago Pak.
66 E4 Tandou L. Austr.
14 E3 Tandragee U.K.
43 B2 Tandur India
70 F3 Taneatua N.Z.
53 B9 Tanega-shima i. Japan
58 A1 Tanen Taunggyi mts Thai.
90 E5 Taneytown U.S.A.
108 B2 Tanezrouft reg. Alg./Mali
115 C4 Tanga Tanz.
115 C4 Tanga div. Tanz.
70 E2 Tangaehe N.Z.
45 G4 Tangail Bangl.
63 F2 Tanga Is P.N.G.
43 C5 Tangalla Sri Lanka
110 C4 Tanganyika, Lake Africa
40 D2 Tangar Iran
43 B4 Tangasseri India
51 B5 Tangdan China
38 C5 Tangeli Iran
116 D4 Tange Prom. hd Ant.
Tanger see Tangier
59 C4 Tangerang Indon.
19 J4 Tangerhütte Ger.
19 J4 Tangermünde Ger.
50 B3 Tanggor China
45 G2 Tanggula Shan mts China
45 G2 Tanggula Shankou pass China
50 D3 Tanghe China
44 B2 Tangi Pak.
108 B1 Tangier Morocco
91 E6 Tangier I. U.S.A.
54 D5 Tangjin S. Korea
45 G4 Tangla India
58 □ Tanglin Sing.
45 H3 Tangmai China
45 F3 Tangra Yumco salt l. China
50 F2 Tangshan China
55 B4 Tangub Phil.
114 D2 Tanguieta Benin
52 A1 Tangwang r. China
51 C4 Tangyan He r. China
50 E3 Tangyin China
52 A1 Tangyuan China
8 U3 Tanhua Fin.
58 C3 Tani Cambodia
45 H3 Taniantaweng Shan mts China
57 J8 Tanimbar, Kepulauan is Indon.
55 B4 Tanjay Phil.
Tanjore see Thanjavur
58 A5 Tanjungbalai Indon.
59 C4 Tanjungkarang Telukbetung Indon.
59 C3 Tanjungpandan Indon.
59 B2 Tanjungpinang Indon.
58 A5 Tanjungpura Indon.
59 E2 Tanjungredeb Indon.
59 E2 Tanjungselor Indon.
44 B2 Tank Pak.
44 D2 Tankse Jammu and Kashmir
45 F4 Tankuhi India
71 □9 Tanna i. Vanuatu
11 F4 Tannadice U.K.
9 N5 Tannäs Sweden
20 D5 Tannheim Austria
68 D4 Tannum Sands Austr.
48 F1 Tannu-Ola, Khrebet mts Rus. Fed.
114 C3 Tano r. Ghana
55 B4 Tañon Strait Phil.
44 B4 Tanot India
114 E2 Tanout Niger
19 M3 Tanowo Pol.
96 D2 Tanque Alvarez Mex.
45 E4 Tansen Nepal
51 F5 Tan-shui Taiwan
39 J3 Tansyk Kazak.
109 F1 Tanta Egypt
108 A2 Tan-Tan Morocco
66 D6 Tantanoola Austr.
97 E3 Tantoyuca Mex.
43 C2 Tanuku India
9 M7 Tanumshede Sweden
66 C5 Tanunda Austr.
54 E5 Tanyang S. Korea
107 G6 Tanzania country Africa
54 B1 Tao'an China
54 B1 Tao'er r. China
50 B3 Tao He r. China
51 D4 Taojiang China
50 C2 Taole China
118 G5 Taongi i. Pac. Oc.
58 □ Tao Payoh Sing.
26 F6 Taormina Sicily Italy
83 F4 Taos U.S.A.
108 B2 Taoudenni Mali
108 B1 Taourirt Morocco
51 E5 Taoxi China
51 D4 Taoyuan China
51 F5 T'ao-yuan Taiwan
9 T7 Tapa Estonia
55 B5 Tapaan Passage Phil.
97 F5 Tapachula Mex.
101 G4 Tapajós r. Brazil
59 A2 Tapaktuan Indon.
105 E3 Tapalqué Arg.
97 F4 Tapanatepec Mex.
58 A5 Tapanuli, Teluk b. Indon.
100 F5 Tapauá Brazil
100 E5 Tapauá r. Brazil
114 B3 Tapeta Liberia
44 C5 Tāpi r. India
55 B5 Tapiantana i. Phil.
88 C2 Tapiola U.S.A.
58 B4 Tapis mt Malaysia
45 F4 Taplejung Nepal
51 □ Tap Mun Chau i. H.K. China
21 L6 Tapolca Hungary
90 E6 Tappahannock U.S.A.
52 G4 Tappi-zaki pt Japan
40 C3 Tappeh, Kūh-e h. Iran
70 D4 Tapuaenuku mt N.Z.
55 B5 Tapul Phil.
55 B5 Tapul Group is Phil.
103 D5 Tapurucuara Brazil
37 L4 Tāq-e Bostan mt Iraq
37 K4 Taqtaq Iraq
104 B1 Taquaral, Serra do h. Brazil
104 B2 Taquari Brazil
101 G7 Taquari r. Brazil
104 A2 Taquari, Serra do h. Brazil
104 C3 Taquaritinga Brazil
104 B3 Taquaruçu r. Brazil
14 D5 Tar r. Rep. of Ireland
67 J1 Tara Austr.
114 F3 Taraba r. Nigeria
100 E7 Tarabuco Bol.
Țarābulus see Tripoli
103 C4 Taracua Brazil
14 E4 Tara, Hill of Rep. of Ireland
44 B4 Tar Ahmad Rind Pak.
44 E4 Tarahuwan India
45 G4 Tarai reg. India
59 E2 Tarakan Indon.
55 A6 Tarakan i. Indon.
36 C1 Taraklı Turkey
67 H5 Taralga Austr.
67 H4 Tarana Austr.
44 C3 Tarana India
44 C3 Tārānagar India
Taranaki, Mt volc. see Egmont, Mt
25 E2 Tarancón Spain
115 C3 Tarangire Nat. Park Tanz.
37 J3 Taran, Mys pt Rus. Fed.
11 A3 Taransay i. U.K.
26 G4 Taranto Italy
26 G4 Taranto, Golfo di g. Italy
100 C5 Tarapoto Peru
22 B4 Tarare France
70 E4 Tararua Range mts N.Z.
17 P6 Tarashcha Ukr.
100 D5 Tarauacá Brazil
100 E5 Tarauacá r. Brazil
71 □3 Taravao Fr. Polynesia Pac. Oc.
71 □3 Taravao, Isthme de isth. Fr. Polynesia Pac. Oc.
60 J3 Tarawa i. Kiribati
70 F3 Tarawera N.Z.
70 F3 Tarawera, Mt N.Z.
25 F2 Tarazona Spain
25 F3 Tarazona de la Mancha Spain
39 K3 Tarbagatay Kazak.
39 K3 Tarbagatay, Khrebet mts Kazak.
11 E3 Tarbat Ness pt U.K.
44 C2 Tarbela Dam dam Pak.
14 B5 Tarbert Rep. of Ireland
11 B3 Tarbert Scot. U.K.
11 C5 Tarbert Scot. U.K.
24 E5 Tarbes France
87 E5 Tarboro U.S.A.
23 L3 Tarcento Italy
66 A3 Tarcoola Austr.
67 G3 Tarcoon Austr.
67 G5 Tarcutta Austr.
49 P2 Tardoki-Yani, Gora mt Rus. Fed.
67 K3 Taree Austr.
66 E3 Tarella Austr.
32 L2 Tareya Rus. Fed.
114 E1 Targa well Niger
82 E2 Targhee Pass U.S.A.
27 L2 Târgoviște Romania
27 K2 Târgu Jiu Romania
17 M7 Târgu Mureș Romania
17 N7 Târgu Neamț Romania
17 N7 Târgu Secuiesc Romania
40 B3 Tarhān Iran
57 L8 Tari P.N.G.
50 C2 Tarian Gol China
40 D5 Tarif U.A.E.
25 D4 Tarifa Spain
25 D4 Tarifa o Marroqui, Pta de pt Spain
100 F8 Tarija Bol.
57 K7 Tariku r. Indon.
42 C6 Tarīm Yemen
Tarim Basin see Tarim Pendi
115 B3 Tarime Tanz.
39 K4 Tarim He r. China
35 G3 Tarim Pendi basin China
41 G3 Tarīn Kowt Afgh.
57 K7 Taritatu r. Indon.
37 K4 Tarjīl Iraq
113 F6 Tarka r. S. Africa
113 G6 Tarkastad S. Africa
114 E2 Tarka, Vallée de watercourse Niger
84 E3 Tarkio U.S.A.
32 J3 Tarko-Sale Rus. Fed.
114 C3 Tarkwa Ghana
55 B3 Tarlac Phil.
19 G3 Tarmstedt Ger.
24 F4 Tarn r. France
8 O4 Tärnaby Sweden
41 G3 Tarnak r. Afgh.
17 M7 Târnăveni Romania
17 K5 Tarnobrzeg Pol.
28 G2 Tarnogskiy Gorodok Rus. Fed.
17 K5 Tarnów Pol.
21 M2 Tarnowskie Góry Pol.
23 H5 Taro r. Italy
45 E3 Tarok Tso salt l. China
40 D4 Țārom Iran
68 C5 Taroom Austr.
108 B1 Taroudannt Morocco
19 G2 Tarp Ger.
66 D6 Tarpenna Austr.
87 E7 Tarpum Bay Bahamas
40 C3 Tarq Iran
26 D3 Tarquinia Italy
25 G2 Tarragona Spain
8 Q3 Tärrajaur Sweden
67 G9 Tarraleah Austr.
67 G4 Tarran Hills Austr.
70 B6 Tarras N.Z.
25 G2 Tàrrega Spain
19 J2 Tårs Denmark
36 E3 Tarsus Turkey
19 J2 Tårs Vig b. Denmark
38 C4 Tarta Turkm.
102 D2 Tartagal Arg.
37 L1 Tärtär Azer.
37 L1 Tärtär r. Azer.
23 K4 Tartaro r. Italy
24 D5 Tartas France
9 U7 Tartu Estonia
36 E4 Țarțūs Syria
104 E2 Tarumirim Brazil
29 H6 Tarumovka Rus. Fed.
58 A5 Tarutung Indon.
23 L3 Tarvisio Italy
40 E4 Tarz Iran
39 H3 Tasaral Kazak.
38 E3 Tasbuget Kazak.
78 E4 Taschereau Can.
43 A2 Tasgaon India
54 A3 Tashan China
45 G4 Tashigang Bhutan
Tashio Chho see Thimphu
37 K1 Tashir Armenia
40 D4 Tashk Iran
40 D4 Tashk, Daryācheh-ye l. Iran
39 G4 Tashkent Uzbek.
41 F2 Tashkepri Turkm.
39 H4 Tash-Kömür Kyrg.
38 C2 Tashla Rus. Fed.
78 F2 Tasiat, Lac l. Can.
75 P3 Tasiilaq Greenland
59 C4 Tasikmalaya Indon.
79 G2 Tasiujaq Can.
114 F1 Tasker Niger
39 H2 Taskesken Kazak.
36 E1 Taşköprü Turkey
37 J2 Taşlıçay Turkey
118 F8 Tasman Basin sea feature Pac. Oc.
70 D4 Tasman Bay N.Z.
67 G9 Tasman Head hd Austr.
67 F9 Tasmania div. Austr.
70 D4 Tasman Mountains N.Z.
67 H9 Tasman Pen. Austr.
118 E9 Tasman Plateau sea feature Pac. Oc.
63 F5 Tasman Sea Pac. Oc.
36 F1 Taşova Turkey
92 B3 Tassajara Hot Springs U.S.A.
114 E1 Tassara Niger
79 F2 Tassialujjuaq, Lac l. Can.
108 C2 Tassili du Hoggar plat. Alg.
108 C2 Tassili n'Ajjer plat. Alg.
39 G3 Tasty Kazak.
38 F2 Tasty-Taldy Kazak.
37 K2 Tasūj Iran
33 N3 Tas-Yuryakh Rus. Fed.
21 M5 Tata Hungary
71 □3 Tataa, Pte pt Fr. Polynesia Pac. Oc.
57 G7 Tataba Indon.
17 J7 Tatabánya Hungary
29 D6 Tatarbunary Ukr.
32 J4 Tatarsk Rus. Fed.
49 Q1 Tatarskiy Proliv str. Rus. Fed.
28 J4 Tatarstan, Respublika div. Rus. Fed.
71 □3 Tatatua, Pte pt Fr. Polynesia Pac. Oc.
40 B2 Tatavi r. Iran
68 A1 Tate r. Austr.
53 F7 Tateyama Japan
53 E6 Tate-yama volc. Japan
76 F2 Tathlina Lake Can.
42 B5 Tathlīth, W. watercourse S. Arabia
67 H6 Tathra Austr.
77 K2 Tatinnai Lake Can.
29 H5 Tatishchevo Rus. Fed.
82 A1 Tatla Lake Can.
76 D3 Tatlatui Prov. Park Can.
57 H8 Tat Mailau, G. mt Indon.
67 G6 Tatong Austr.
17 J6 Tatry mts Pol./Slovakia
76 B3 Tatshenshini r. Can.
29 G5 Tatsinskiy Rus. Fed.
53 D7 Tatsuno Japan
44 A4 Tatta Pak.
39 H4 Tatti Kazak.
104 C3 Tatuí Brazil
76 E4 Tatuk Mtn Can.
85 C5 Tatum U.S.A.
66 F6 Tatura Austr.
37 J2 Tatvan Turkey
9 J7 Tau Norway
71 □5 Tau i. American Samoa Pac. Oc.
71 □11 Tau i. Tonga
101 K5 Tauá Brazil
104 D3 Taubaté Brazil
20 C3 Tauber r. Ger.
20 C3 Tauberbischofsheim Ger.
19 K5 Taucha Ger.
38 B3 Tauchik Kazak.
19 M5 Tauer Ger.
71 □2 Tauère i. Fr. Polynesia Pac. Oc.
20 F4 Taufkirchen (Vils) Ger.
20 C2 Taufstein h. Ger.
39 H3 Taukum, Peski des. Kazak.
71 □10 Taula i. Tonga
112 F3 Taung S. Africa
56 B2 Taung-gyi Myanmar
58 A2 Taungnyo Range mts Myanmar
71 □3 Taunoa Fr. Polynesia Pac. Oc.
13 D6 Taunton U.K.
91 H4 Taunton U.S.A.
20 B2 Taunus h. Ger.
70 F3 Taupo N.Z.
70 E3 Taupo, Lake N.Z.
9 S9 Tauragė Lith.
70 F2 Tauranga N.Z.
26 G5 Taurianova Italy
70 D1 Tauroa Pt N.Z.
71 □6 Tautama pt Pitcairn I. Pac. Oc.
71 □3 Tautira Fr. Polynesia Pac. Oc.
63 F2 Tauu is P.N.G.
23 L3 Tavagnacco Italy
36 B3 Tavas Turkey
22 C2 Tavaux France
13 J5 Taverham U.K.
23 G5 Taverone r. Italy
71 □13 Taveuni i. Fiji
25 C4 Tavira Port.
13 C7 Tavistock U.K.
58 A2 Tavoy Myanmar
58 A2 Tavoy Pt Myanmar
52 B3 Tavrichanka Rus. Fed.
36 B2 Tavşanlı Turkey
71 □15 Tavua Fiji
71 □15 Tavua i. Fiji
13 C6 Taw r. U.K.
89 F3 Tawas Bay U.S.A.
89 F3 Tawas City U.S.A.
59 E2 Tawau Malaysia
13 D6 Tawe r. U.K.
44 C2 Tawi r. India
55 A5 Tawitawi i. Phil.
51 F6 T'a-wu Taiwan
97 E4 Taxco Mex.
39 H5 Taxkorgan China
76 C2 Tay r. Can.
11 E4 Tay r. U.K.
55 B3 Tayabas Bay Phil.
8 X2 Taybola Rus. Fed.
11 E4 Tay, Firth of est. U.K.
11 C5 Tayinloan U.K.
69 E6 Tay, L. salt flat Austr.
11 D4 Tay, Loch l. U.K.
76 E3 Taylor Can.
93 G4 Taylor AZ U.S.A.
89 F4 Taylor MI U.S.A.
88 B6 Taylor MO U.S.A.
84 D3 Taylor NE U.S.A.
85 D6 Taylor TX U.S.A.
91 E5 Taylors Island U.S.A.
86 B4 Taylorville U.S.A.
42 A4 Taymā' S. Arabia
33 L3 Taymura r. Rus. Fed.
33 M2 Taymyr, Ozero l. Rus. Fed.
33 L2 Taymyr, Poluostrov pen. Rus. Fed.
39 G1 Tayncha Kazak.
58 C3 Tây Ninh Vietnam
38 C2 Taysoygan, Peski des. Kazak.
55 B3 Taytay Phil.
55 A4 Taytay Phil.
55 A4 Taytay Bay Phil.
56 □ Tayu Indon.
41 F3 Tayyebād Iran
33 K2 Taz r. Rus. Fed.
108 B1 Taza Morocco
37 K4 Țāza Khurmātū Iraq
37 L2 Tazeh Kand Azer.
90 B6 Tazewell TN U.S.A.
90 C6 Tazewell VA U.S.A.
77 H2 Tazin r. Can.
77 H3 Tazin Lake Can.
109 E2 Tāzirbū Libya
25 J4 Tazmalt Alg.
32 J3 Tazovskaya Guba chan. Rus. Fed.
37 J1 Tba Khozap'ini l. Georgia
29 H7 T'bilisi Georgia
29 G6 Tbilisskaya Rus. Fed.
114 F4 Tchabal Mbabo mt Cameroon
114 D3 Tchamba Togo
110 B4 Tchibanga Gabon
109 D2 Tchigaï, Plateau du plat. Niger
114 F3 Tchollire Cameroon
17 J3 Tczew Pol.
96 C3 Teacapán Mex.
69 E2 Teague, L. salt flat Austr.
71 □3 Teahupoo Fr. Polynesia Pac. Oc.
70 A6 Te Anau N.Z.
70 A6 Te Anau, L. N.Z.
69 C2 Teano Range mts Austr.
97 F4 Teapa Mex.
70 G2 Te Araroa N.Z.
70 E2 Te Aroha N.Z.
71 □3 Teavaro Fr. Polynesia Pac. Oc.
70 E2 Te Awamutu N.Z.
12 E3 Tebay U.K.
77 K2 Tebesjuak Lake Can.
108 C1 Tébessa Alg.
26 B7 Tébessa, Monts de mts Alg.
102 E3 Tebicuary r. Para.
59 A2 Tebingtinggi Indon.
59 B3 Tebingtinggi Indon.
26 C6 Tébourba Tunisia
26 C6 Téboursouk Tunisia
29 H7 Tebulos Mt'a Georgia/Rus. Fed.
20 G6 Techendorf Austria
114 C3 Techiman Ghana
102 B6 Tecka Arg.
18 E4 Tecklenburger Land reg. Ger.
97 E3 Tecolutla Mex.
96 D4 Tecomán Mex.
92 D4 Tecopa U.S.A.
96 C2 Tecoríchic Mex.
96 B1 Tecoripa Mex.
96 D4 Técpan Mex.
17 N7 Tecuci Romania
89 F5 Tecumseh U.S.A.
38 E5 Tedzhen Turkm.
41 F2 Tedzhen r. Turkm.
41 F2 Tedzhenstroy Turkm.
93 H3 Teec Nos Pos U.S.A.
48 F1 Teeli Rus. Fed.
12 F3 Tees r. U.K.
12 E3 Teesdale reg. U.K.
55 A4 Teeth, The mt Phil.
100 E4 Tefé r. Brazil
36 B3 Tefenni Turkey
59 C4 Tegal Indon.
13 D5 Tegid, Llyn l. U.K.
114 E2 Tegina Nigeria
23 H3 Teglio Italy
71 □9 Tegua i. Vanuatu
96 H5 Tegucigalpa Honduras
114 E1 Teguidda-n-Tessoumt Niger
92 C4 Tehachapi U.S.A.
83 C5 Tehachapi Mts U.S.A.
92 C4 Tehachapi Pass U.S.A.
77 K2 Tehek Lake Can.
Teheran see Tehrān
114 C3 Téhini Côte d'Ivoire
40 C3 Tehrān Iran
44 D3 Tehri Uttar Pradesh India
Tehri see Tikamgarh
97 E4 Tehuacán Mex.
97 F5 Tehuantepec, Golfo de g. Mex.
97 F4 Tehuantepec, Istmo de isth. Mex.
119 N5 Tehuantepec Ridge sea feature Pac. Oc.
97 E4 Tehuitzingo Mex.
39 L3 Teichanggou China
13 C5 Teifi r. U.K.
13 D7 Teign r. U.K.
13 D7 Teignmouth U.K.
Tejo r. see Tagus
92 C4 Tejon Pass U.S.A.
70 D1 Te Kao N.Z.
70 C5 Tekapo, L. N.Z.
45 F4 Tekari India
97 G3 Tekax Mex.
38 E2 Tekeli Aktyubinsk. Kazak.
39 J3 Tekeli Taldykorg. Kazak.
39 H1 Teke, Oz. salt l. Kazak.
39 K4 Tekes China
39 K4 Tekes Kazak.
39 K3 Tekes He r. China
110 D2 Tekezē Wenz r. Eritrea/Eth.
44 E1 Tekiliktag mt China
36 A1 Tekirdağ Turkey
43 D2 Tekkali India
37 H2 Tekman Turkey
45 H5 Teknaf Bangl.
88 E4 Tekonsha U.S.A.
70 E3 Te Kuiti N.Z.
45 E5 Tel r. India
96 H5 Tela Honduras
114 D1 Télataï Mali
29 H7 T'elavi Georgia
36 E5 Tel Aviv-Yafo Israel
21 J3 Telč Czech Rep.
97 G3 Telchac Puerto Mex.
76 C3 Telegraph Creek Can.
22 B2 Télégraphe, Le h. France
104 B4 Telêmaco Borba Brazil
105 D3 Telén Arg.
59 E2 Telen r. Indon.
27 L2 Teleorman r. Romania
92 D3 Telescope Peak U.S.A.
101 G5 Teles Pires r. Brazil
13 E5 Telford U.K.
20 E5 Telfs Austria
18 E5 Telgte Ger.
96 H5 Telica Nic.
108 A3 Télimélé Guinea
37 J3 Tel Kotchek Syria
76 D4 Telkwa Can.
74 B3 Teller U.S.A.
43 A4 Tellicherry India
15 F2 Tellin Belgium
19 G2 Tellingstedt Ger.
37 L6 Telloh Iraq
38 D4 Tel'mansk Turkm.
58 □ Telok Blangah Sing.
97 E4 Teloloapán Mex.
105 C4 Telsen Arg.
9 S9 Telšiai Lith.
19 L4 Teltow Ger.
59 B2 Teluk Anson Malaysia
59 A2 Telukdalam Indon.
89 H2 Temagami Can.
89 G2 Temagami Lake Can.
59 D4 Temanggung Indon.
97 G3 Temax Mex.
113 H2 Temba S. Africa
59 C2 Tembelan, Kepulauan is Indon.
33 L3 Tembenchi r. Rus. Fed.
59 B3 Tembilahan Indon.
113 H3 Tembisa S. Africa
110 B4 Tembo Aluma Angola
13 E5 Teme r. U.K.
92 D5 Temecula U.S.A.
36 D2 Temelli Turkey
59 B2 Temerloh Malaysia
37 M5 Temīleh Iran
38 D2 Temir Kazak.
39 G4 Temirlanovka Kazak.
39 H2 Temirtau Kazak.
89 H2 Temiscaming Can.
89 H2 Témiscamingue, Lac l. Can.
79 G4 Témiscouata, L. Can.
67 F8 Temma Austr.
8 T4 Temmes Fin.
28 G4 Temnikov Rus. Fed.
67 G5 Temora Austr.
96 C1 Temósachic Mex.
93 G5 Tempe U.S.A.
26 C4 Tempio Pausania Sardinia Italy
88 E3 Temple MI U.S.A.
85 D6 Temple TX U.S.A.
13 C5 Temple Bar U.K.
14 D5 Templemore Rep. of Ireland
55 A4 Templer Bank sand bank Phil.
12 E3 Temple Sowerby U.K.
19 L3 Templin Ger.
97 E3 Tempoal Mex.
29 F6 Temryuk Rus. Fed.
105 B3 Temuco Chile
70 C6 Temuka N.Z.
100 C4 Tena Ecuador
92 D1 Tenabo, Mt U.S.A.
43 C2 Tenali India
97 E4 Tenancingo Mex.
58 A2 Tenasserim Myanmar
58 A2 Tenasserim r. Myanmar
13 E5 Tenbury Wells U.K.
13 C6 Tenby U.K.
89 F2 Tenby Bay Can.
22 B4 Tence France
110 E2 Tendaho Eth.
22 E5 Tende France
35 H6 Ten Degree Chan. Andaman and Nicobar Is
52 G5 Tendō Japan
22 D3 Tendre, Mont mt Switz.
37 J2 Tendürük Dağı mt Turkey
114 C2 Ténenkou Mali
108 D3 Ténéré reg. Niger
108 D2 Ténéré du Tafassâsset des. Niger
108 A2 Tenerife i. Canary Is
25 G4 Ténès Alg.
59 E4 Tengah, Kepulauan is Indon.
38 C4 Tenge Kazak.
58 □ Tengeh Resr Sing.
50 B2 Tengger Shamo des. China
58 B4 Tenggul i. Malaysia
39 G2 Tengiz, Oz. salt l. Kazak.
51 C7 Tengqiao China
114 B2 Tengréla Côte d'Ivoire
51 D6 Teng Xian Guangxi China
50 E3 Teng Xian Shandong China
22 D5 Ténibre, Mont mt France/Italy
116 B2 Teniente Jubany Arg. Base Ant.
116 B2 Teniente Rodolfo Marsh Chile Base Ant.
22 E1 Teningen Ger.
38 F1 Teniz, Oz. l. Kazak.
115 A6 Tenke Congo(Zaire)
33 Q2 Tenkeli Rus. Fed.
114 C2 Tenkodogo Burkina
65 F3 Tennant Creek Austr.
90 B6 Tennessee div. U.S.A.
87 C5 Tennessee r. U.S.A.
83 F4 Tennessee Pass U.S.A.
8 P2 Tennevoll Norway
105 B2 Teno r. Chile
8 U2 Tenojoki r. Fin./Norway
97 G4 Tenosique Mex.
82 F2 Ten Sleep U.S.A.
57 G7 Tenteno Indon.
13 H6 Tenterden U.K.
67 K2 Tenterfield Austr.
87 D7 Ten Thousand Islands U.S.A.
25 C3 Tentudia mt Spain
104 B3 Teodoro Sampaio Brazil
104 E2 Teófilo Otôni Brazil
97 F4 Teopisca Mex.
97 E4 Teotihuacán tourist site Mex.
71 □9 Teouta New Caledonia Pac. Oc.
57 H8 Tepa Indon.
83 E6 Tepachi Mex.
70 D1 Te Paki N.Z.
96 D4 Tepalcatepec Mex.
71 □3 Tepati Fr. Polynesia Pac. Oc.
96 D3 Tepatitlán Mex.
37 H3 Tepe Turkey
37 J3 Tepe Gawra Iraq
96 C2 Tepehuanes Mex.
97 E4 Tepeji Mex.
27 J4 Tepelenë Albania
97 E4 Tepelmemec Mex.
20 F3 Tepelská Vrchovina reg. Czech Rep.
103 E4 Tepequem, Serra mts Brazil
96 C3 Tepic Mex.
70 C5 Te Pirita N.Z.
21 G2 Teplice Czech Rep.
28 K2 Teplogorka Rus. Fed.
28 F4 Teploye Rus. Fed.
71 □2 Tepoto i. Fr. Polynesia Pac. Oc.
70 F2 Te Puke N.Z.
97 F4 Tequisistlán Mex.
97 E3 Tequisquiapán Mex.
25 H1 Ter r. Spain
60 L3 Teraina i. Kiribati
44 D2 Teram Kangri mt China/Jammu and Kashmir
26 E3 Teramo Italy
66 E7 Terang Austr.
18 E4 Ter Apel Neth.
44 B3 Teratani r. Pak.
29 F4 Terbuny Rus. Fed.
37 H2 Tercan Turkey
22 F4 Terdoppio r. Italy
17 M6 Terebovlya Ukr.
29 H7 Terek Rus. Fed.
29 H7 Terek r. Rus. Fed.
39 G2 Terekty Kazak.
28 J4 Teren'ga Rus. Fed.
104 A3 Terenos Brazil
38 F3 Terenozek Kazak.
38 D2 Terensay Rus. Fed.
103 C2 Terepaima, Parque Nacional nat. park Venez.
28 H4 Tereshka r. Rus. Fed.
101 K5 Teresina Brazil
104 D3 Teresópolis Brazil
71 □8 Terevaka h. Easter I. Chile
15 D3 Tergnier France
23 J4 Tergola r. Italy
36 F1 Terme Turkey
39 F5 Termez Uzbek.
69 E7 Termination Island Austr.
26 E6 Termini Imerese Sicily Italy
97 G4 Términos, Lag. de Mex.
114 F1 Termit well Niger
114 F1 Termit-Kaoboul Niger
26 F4 Termoli Italy
13 E5 Tern r. U.K.
57 H6 Ternate Indon.
18 A5 Terneuzen Neth.
52 E2 Terney Rus. Fed.
26 E3 Terni Italy
21 K5 Ternitz Austria
29 C5 Ternopil' Ukr.
66 C4 Terowie Austr.
49 Q2 Terpeniya, Mys c. Rus. Fed.
49 Q2 Terpeniya, Zaliv g. Rus. Fed.
76 D4 Terrace Can.
88 D1 Terrace Bay Can.
69 E4 Terraces, The h. Austr.
112 E2 Terra Firma S. Africa
8 N4 Terråk Norway
26 C5 Terralba Sardinia Italy
79 K4 Terra Nova Nat. Pk Can.
23 J6 Terranuova Bracciolini Italy
116 B6 Terre Adélie reg. Ant.
85 F6 Terre Bonne Bay U.S.A.
86 C4 Terre Haute U.S.A.
79 K4 Terrenceville Can.
22 B2 Terre Plaine plain France
82 F2 Terry U.S.A.
29 G5 Tersa r. Rus. Fed.
39 F2 Tersakkan r. Kazak.
18 C3 Terschelling i. Neth.
39 J4 Terskey Ala-Too mts Kyrg.
26 C5 Tertenia Sardinia Italy
25 F2 Teruel Spain
58 A4 Terutao i. Thai.
8 T3 Tervola Fin.
26 G2 Tešanj Bos.-Herz.
110 D2 Teseney Eritrea
28 G4 Tesha r. Rus. Fed.
74 C2 Teshekpuk Lake U.S.A.
52 J3 Teshikaga Japan
52 G2 Teshio Japan
52 H3 Teshio-dake mt Japan
52 G2 Teshio-gawa r. Japan
76 C2 Teslin Can.
76 C2 Teslin r. Can.
76 C2 Teslin Lake Can.
104 B1 Tesouras r. Brazil
104 B2 Tesouro Brazil
114 E2 Tessaoua Niger
19 K2 Tessin Ger.
13 F6 Test r. U.K.
26 C6 Testour Tunisia
102 B2 Tetas, Pta pt Chile
115 B6 Tete Moz.
22 C6 Tête du Grand Puech h. France
70 F3 Te Teko N.Z.
17 P5 Teteriv r. Ukr.
19 K3 Teterow Ger.
71 □2 Tetiaroa i. Fr. Polynesia Pac. Oc.
17 O6 Tetiyiv Ukr.
12 G4 Tetney U.K.
82 E2 Teton r. U.S.A.
82 E3 Teton Ra. mts U.S.A.
108 B1 Tétouan Morocco
27 J3 Tetovo Macedonia
44 B5 Tetpur India
20 C5 Tettnang Ger.
28 J4 Tetyushi Rus. Fed.
102 D2 Teuco r. Arg.
112 B1 Teufelsbach Namibia
19 G3 Teufels Moor reg. Ger.
52 G2 Teuri-tō i. Japan
18 F4 Teutoburger Wald h. Ger.
9 R5 Teuva Fin.
Tevere r. see Tiber
Teverya see Tiberias
11 F5 Teviot r. U.K.
11 F5 Teviotdale v. U.K.
70 A7 Te Waewae Bay N.Z.
113 G1 Tewane Botswana
68 E6 Tewantin Austr.
70 E4 Te Wharau N.Z.
13 E6 Tewkesbury U.K.
50 B3 Têwo China
76 E5 Texada I. Can.
85 E5 Texarkana U.S.A.
67 J2 Texas Austr.
85 D6 Texas div. U.S.A.
85 E6 Texas City U.S.A.
97 E4 Texcoco Mex.
18 B3 Texel i. Neth.
85 C4 Texhoma U.S.A.
85 D5 Texoma, Lake U.S.A.
113 G4 Teyateyaneng Lesotho
28 G3 Teykovo Rus. Fed.
41 G3 Teyvareh Afgh.
28 G3 Teza r. Rus. Fed.

U

115 B4 Uvinza Tanz.
115 A3 Uvira Congo(Zaire)
113 J5 Uvongo S. Africa
48 F1 Uvs Nuur salt l. Mongolia
53 C8 Uwajima Japan
109 E2 Uweinat, Jebel mt Sudan
13 G6 Uxbridge U.K.
50 C2 Uxin Ju China
50 C2 Uxin Qi China
97 G3 Uxmal tourist site Mex.
38 E1 Uy r. Rus. Fed.
38 E3 Uyaly Kazak.
39 K3 Uyaly, Oz. l. Kazak.
46 B1 Uyar Rus. Fed.
50 C1 Üydzin Mongolia
33 Q3 Uyega Rus. Fed.
114 E3 Uyo Nigeria
39 G4 Uyuk Kazak.
100 E8 Uyuni, Salar de salt flat Bol.
28 H4 Uza r. Rus. Fed.
37 K4 'Uzaym, Nahr al r. Iraq
31 E5 Uzbekistan country Asia
38 C4 Uzen' Kazak.
22 B5 Uzès France
29 B5 Uzhhorod Ukr.
27 H3 Užice Yugo.
28 F4 Uzlovaya Rus. Fed.
28 G3 Uzola r. Rus. Fed.
36 C3 Üzümlü Turkey
39 G5 Uzun Uzbek.
39 J4 Uzunagach Kazak.
39 J3 Uzunbulak China
37 L3 Üzün Darreh r. Iran
29 C7 Uzunköprü Turkey
29 D5 Uzyn Ukr.
38 E3 Uzynkair Kazak.

V

9 T5 Vaajakoski Fin.
113 H3 Vaal r. S. Africa
8 U4 Vaala Fin.
112 F4 Vaalbos National Park S. Africa
113 H3 Vaal Dam dam S. Africa
113 H2 Vaalwater S. Africa
8 R5 Vaasa Fin.
38 F4 Vabkent Uzbek.
17 J7 Vác Hungary
104 A3 Vacaria r. Mato Grosso do Sul Brazil
104 D2 Vacaria r. Minas Gerais Brazil
102 F3 Vacaria Brazil
104 A3 Vacaria, Serra h. Brazil
92 B2 Vacaville U.S.A.
20 D2 Vacha Ger.
28 G4 Vad r. Rus. Fed.
44 C6 Vada India
9 K7 Vadla Norway
44 C5 Vadodara India
8 V1 Vadsø Norway
23 G2 Vaduz Liechtenstein
8 N3 Værøy i. Norway
28 G2 Vaga r. Rus. Fed.
9 L6 Vågåmo Norway
26 F2 Vaganski Vrh mt Croatia
8 Vágar i. Faroe Is
8 Q4 Vägsele Sweden
8 Vágur Faroe Is
21 M5 Váh r. Slovakia
8 S5 Vähäkyrö Fin.
23 J6 Vaiano Italy
71 3 Vaiau, Pte pt Fr. Polynesia Pac. Oc.
9 T7 Vaida Estonia
43 B4 Vaigai r. India
71 8 Vaihu Easter I. Chile
93 G5 Vail U.S.A.
15 D3 Vailly-sur-Aisne France
71 4 Vailoa Western Samoa
71 11 Vaina Tonga
71 3 Vairao Fr. Polynesia Pac. Oc.
22 D6 Vaire r. France
22 C5 Vaison-la-Romaine France
71 5 Vaitogi American Samoa Pac. Oc.
63 H2 Vaitupu i. Tuvalu
71 4 Vaiusu Western Samoa
71 10 Vaka'eitu i. Tonga
39 H5 Vakhan Tajik.
39 G5 Vakhsh Tajik.
23 C5 Valachchenai Sri Lanka
21 L3 Valašská Polanka Czech Rep.
21 M3 Valašské Klobouky Czech Rep.
21 L3 Valašské Meziříčí Czech Rep.
89 K2 Val-Barrette Can.
9 P6 Valbo Sweden
22 D4 Val-Cenis France
105 C4 Valcheta Arg.
23 J4 Valdagno Italy
22 D2 Valdahon France
22 B2 Valday Rus. Fed.
28 E3 Valdayskaya Vozvyshennost' reg. Rus. Fed.
25 D3 Valdecañas, Embalse de resr Spain
9 S8 Valdemārpils Latvia
9 P7 Valdemarsvik Sweden
22 C1 Val-de-Meuse France
25 E3 Valdepeñas Spain
15 B3 Val-de-Reuil France
89 K3 Val-des-Bois Can.
105 D4 Valdés, Península pen. Arg.
74 D3 Valdez U.S.A.
22 D4 Val-d'Isère France
105 B3 Valdivia Chile
23 J4 Valdobbiadene Italy
22 D2 Valdoie France
89 J1 Val-d'Or Can.
87 D6 Valdosta U.S.A.
9 L6 Valdres v. Norway
37 J1 Vale Georgia
82 C2 Vale U.S.A.
76 F4 Valemount Can.
104 E1 Valença Brazil
22 B5 Valence France
25 F3 Valencia Spain
25 F3 Valencia div. Spain
103 D2 Valencia Venez.
25 C3 Valencia de Alcántara Spain
25 D1 Valencia de Don Juan Spain
25 G3 Valencia, Golfo de g. Spain
14 A6 Valencia Island Rep. of Ireland
15 D2 Valenciennes France
22 D6 Valensole, Plateau de plat. France
52 D3 Valentin Rus. Fed.
93 F4 Valentine AZ U.S.A.
84 C3 Valentine NE U.S.A.
85 B6 Valentine TX U.S.A.
22 F4 Valenza Italy
55 B3 Valenzuela Phil.
9 M6 Våler Norway
103 C2 Valera Venez.
22 C5 Valernes France
15 G4 Valfroicourt France
39 G1 Valikhanovo Kazak.
71 13 Valili Pk h. Fiji
27 H2 Valjevo Yugo.
9 T8 Valka Latvia
9 T6 Valkeakoski Fin.
18 C5 Valkenswaard Neth.
29 E5 Valky Ukr.
116 C4 Valkyrjedomen ice feature Ant.
97 G3 Valladolid Mex.
25 D2 Valladolid Spain
15 F4 Vallage reg. France
22 E6 Vallauris France
25 F3 Vall de Uxó Spain
9 K7 Valle Norway
22 E4 Valle d'Aosta div. Italy
103 D2 Valle de la Pascua Venez.
96 D3 Valle de Santiago Mex.
103 B2 Valledupar Col.
105 C1 Valle Fértil, Sa de mts Arg.
100 F7 Valle Grande Bol.
97 E2 Valle Hermoso Mex.
92 A2 Vallejo U.S.A.
97 E4 Valle Nacional Mex.
26 F7 Valletta Malta
13 C4 Valley U.K.
84 D2 Valley City U.S.A.
82 B3 Valley Falls U.S.A.
90 C5 Valley Head U.S.A.
76 F3 Valleyview Can.
23 J6 Vallombrosa Italy
25 G2 Valls Spain
77 H5 Val Marie Can.
9 T8 Valmiera Latvia
25 E1 Valnera mt Spain
29 C4 Valozhyn Belarus
22 E4 Val-Paradis Can.
104 B3 Valparaíso Brazil
105 B2 Valparaíso Chile
105 B2 Valparaíso div. Chile
96 D3 Valparaíso Mex.
88 D5 Valparaiso U.S.A.
22 E4 Valpelline v. Italy
22 B5 Valréas France
44 C5 Valsād India
22 B5 Vals-les-Bains France
112 F3 Valspan S. Africa
57 K8 Vals, Tanjung c. Indon.
22 B2 Val-Suzon France
28 H1 Val'tevo Rus. Fed.
8 V5 Valtimo Fin.
22 E4 Valtournenche Italy
71 15 Valukoula Fiji
29 G6 Valuyevka Rus. Fed.
29 F5 Valuyki Rus. Fed.
25 C4 Valverde del Camino Spain
21 K2 Vamberk Czech Rep.
58 C3 Vam Co Tay r. Vietnam
19 G1 Vamdrup Denmark
9 S6 Vammala Fin.
43 C2 Vamsadhara r. India
37 J2 Van Turkey
37 K1 Vanadzor Armenia
15 E4 Vanault-les-Dames France
85 E5 Van Buren AR U.S.A.
91 K1 Van Buren ME U.S.A.
58 D2 Vân Canh Vietnam
91 K2 Vanceboro U.S.A.
90 B5 Vanceburg U.S.A.
76 E5 Vancouver Can.
82 B2 Vancouver U.S.A.
69 D7 Vancouver, C. pt Austr.
76 D5 Vancouver Island Can.
76 A2 Vancouver, Mt Can./U.S.A.
86 B4 Vandalia IL U.S.A.
90 A5 Vandalia OH U.S.A.
22 A3 Vandenesse France
113 G3 Vanderbijlpark S. Africa
88 E3 Vanderbilt U.S.A.
90 D4 Vandergrift U.S.A.
76 E4 Vanderhoof Can.
112 F5 Vanderkloof Dam resr S. Africa
65 G3 Vanderlin I. Austr.
93 H4 Vanderwagen U.S.A.
64 F2 Van Diemen Gulf Austr.
9 T7 Vändra Estonia
9 N7 Vänern l. Sweden
9 N7 Vänersborg Sweden
90 E3 Van Etten U.S.A.
111 E6 Vangaindrano Madag.
37 J2 Van Gölü salt l. Turkey
85 B6 Van Horn U.S.A.
89 K3 Vanier Can.
63 G3 Vanikoro Is Solomon Is
62 E2 Vanimo P.N.G.
49 Q2 Vanino Rus. Fed.
43 B3 Vanivilasa Sagara resr India
43 B3 Vaniyambadi India
39 G5 Vanj Tajik.
39 G5 Vanj, Qatorkŭhi mts Tajik.
33 V3 Vankarem Rus. Fed.
91 F2 Vankleek Hill Can.
Van, Lake salt l. see Van Gölü
8 R1 Vanna i. Norway
8 Q5 Vännäs Sweden
24 C3 Vannes France
39 G4 Vannovka Kazak.
22 D4 Vanoise, Massif de la mts France
22 D4 Vanoise, Parc National de la nat. park France
57 K7 Van Rees, Pegunungan mts Indon.
112 C5 Vanrhynsdorp S. Africa
9 O6 Vansbro Sweden
9 T6 Vantaa Fin.
71 14 Vanua Balavu i. Fiji
63 G3 Vanua Lava i. Vanuatu
71 9 Vanua Lava i. Vanuatu
71 13 Vanua Levu i. Fiji
71 15 Vanua Levu Barrier Reef Fiji
61 H5 Vanuatu country Pac. Oc.
90 A4 Van Wert U.S.A.
112 D5 Vanwyksvlei S. Africa
112 D5 Vanwyksvlei l. S. Africa
51 B6 Văn Yên Vietnam
112 E3 Van Zylsrus S. Africa
71 9 Vao New Caledonia Pac. Oc.
22 E6 Var r. France
23 G5 Vara r. Italy
43 A3 Varada r. India
22 C6 Varages France
9 U8 Varakļāni Latvia
114 C3 Varalé Côte d'Ivoire
22 F4 Varallo Italy
40 C3 Varāmīn Iran
45 E4 Varanasi India
8 V1 Varangerfjorden chan. Norway
8 V1 Varangerhalvøya pen. Norway
26 G1 Varaždin Croatia
22 F5 Varazze Italy
9 N8 Varberg Sweden
43 B2 Vardannapet India
27 K4 Vardar r. Macedonia
9 L9 Varde Denmark
18 F1 Varde r. Denmark
40 B1 Vardenis Armenia
8 W1 Vardø Norway
18 F3 Varel Ger.
105 C2 Varela Arg.
9 T9 Varėna Lith.
15 B3 Varenne r. France
23 F4 Varese Italy
52 C2 Varfolomeyevka Rus. Fed.
9 N7 Vårgårda Sweden
104 D3 Varginha Brazil
18 C5 Varik Neth.
8 U5 Varkaus Fin.
27 M3 Varna Bulg.
38 E1 Varna Rus. Fed.
9 O8 Värnamo Sweden
9 N6 Värnäs Sweden
28 H3 Varnavino Rus. Fed.
36 D4 Varosia Cyprus
8 U5 Varpaisjärvi Fin.
16 J7 Várpalota Hungary
22 D5 Vars France
41 F2 Varsaj Afgh.
37 H2 Varto Turkey
45 E4 Varuna r. India
90 D3 Varysburg U.S.A.
40 D3 Varzaneh Iran
104 D2 Várzea da Palma Brazil
22 F3 Varzo Italy
21 K5 Vas div. Hungary
28 H2 Vashka r. Rus. Fed.
Vasht see Khāsh
28 H2 Vasilevo Rus. Fed.
9 U7 Vasknarva Estonia
17 N7 Vaslui Romania
89 F4 Vassar U.S.A.
21 K5 Vas-Soproni-síkság h. Hungary
9 P7 Västerås Sweden
9 N6 Västerdalälven r. Sweden
8 P3 Västerfjäll Sweden
9 Q7 Vasterhaninge Sweden
9 P8 Västervik Sweden
26 F3 Vasto Italy
21 K5 Vasvár Hungary
29 D5 Vasyl'kiv Ukr.
24 E3 Vatan France
Vaté i. see Éfaté
11 A4 Vatersay i. U.K.
43 A2 Vathar India
27 M6 Vathy Greece
71 15 Vatia Pt Fiji
7 E4 Vatican City country Europe
8 E4 Vatnajökull ice cap Iceland
17 M7 Vatra Dornei Romania
9 O7 Vättern l. Sweden
71 15 Vatu-i-Ra Channel Fiji
71 15 Vatu-i-Thake i. Fiji
71 15 Vatulele i. Fiji
71 13 Vatu Vara i. Fiji
22 C6 Vaucluse, Monts de mts France
22 C2 Vauconcourt-Nervezain France
15 F4 Vaucouleurs France
83 F5 Vaughn U.S.A.
103 C4 Vaupés r. Col.
22 B6 Vauvert France
15 C4 Vaux-le-Pénil France
71 10 Vava'u Group is Tonga
114 B3 Vavoua Côte d'Ivoire
43 C4 Vavuniya Sri Lanka
28 C4 Vawkavysk Belarus
9 O8 Växjö Sweden
43 B3 Vāyalpād India
28 H1 Vazhgort Rus. Fed.
111 E5 Vazobe mt Madag.
58 B2 Veal Vêng Cambodia
19 L1 Veberöd Sweden
18 F4 Vechta Ger.
18 D4 Vechte r. Ger.
19 G5 Veckerhagen (Reinhardshagen) Ger.
43 B4 Vedaranniyam India
9 N8 Veddige Sweden
27 L3 Vedea r. Romania
29 H7 Vedeno Rus. Fed.
37 K2 Vedi Armenia
105 E2 Vedia Arg.
28 E2 Vedlozero Rus. Fed.
88 D5 Veedersburg U.S.A.
18 D3 Veendam Neth.
18 C4 Veenendaal Neth.
8 M4 Vega i. Norway
85 C5 Vega U.S.A.
77 G4 Vegreville Can.
9 U6 Vehkalahti Fin.
44 B3 Vehoa r. Pak.
44 B4 Veirwaro Pak.
20 C3 Veitshöchheim Ger.
19 G1 Vejen Denmark
25 D4 Vejer de la Frontera Spain
19 G1 Vejle Denmark
19 H2 Vejsnæs Nakke pt Denmark
103 B1 Vela, Cabo de la pt Col.
43 B4 Velanai I. Sri Lanka
18 E5 Velbert Ger.
27 K3 Velbŭzhdki Prokhod pass Macedonia
112 C6 Velddrif S. Africa
21 H6 Velden am Wörther See Austria
26 F2 Velebit mts Croatia
18 D5 Velen Ger.
21 M5 Velencei-tó l. Hungary
26 F1 Velenje Slovenia
27 J4 Veles Macedonia
103 B3 Vélez Col.
25 D4 Vélez-Málaga Spain
25 E4 Vélez-Rubio Spain
104 D2 Velhas r. Brazil
29 H6 Velichayevskoye Rus. Fed.
26 G2 Velika Gorica Croatia
27 J2 Velika Plana Yugo.
33 T3 Velikaya r. Chukotsk. Rus. Fed.
28 D3 Velikaya r. Pskov. Obl. Rus. Fed.
28 J3 Velikaya Rus. Fed.
28 E2 Velikaya Guba Rus. Fed.
52 E2 Velikaya Kema Rus. Fed.
28 D3 Velikiye Luki Rus. Fed.
28 H2 Velikiy Ustyug Rus. Fed.
43 B3 Velikonda Ra. h. India
17 Q2 Velikooktyabr'skiy Rus. Fed.
27 L3 Veliko Tŭrnovo Bulg.
27 B3 Velikoye Rus. Fed.
28 F3 Velikoye, Oz. l. Rus. Fed.
28 G4 Velikoye, Oz. l. Rus. Fed.
26 F2 Veli Lošinj Croatia
108 A3 Vélingara Senegal
17 P3 Velizh Rus. Fed.
21 K3 Velká Bíteš Czech Rep.
21 L4 Velká Javořina h. Czech Rep./Slovakia
21 K3 Velké Meziříčí Czech Rep.
21 M3 Veľký Javorník mt Slovakia
16 H7 Veľký Meder Slovakia
21 H1 Velký Šenov Czech Rep.
63 F2 Vella Lavella i. Solomon Is
43 B4 Vellar r. India
20 C3 Vellberg Ger.
19 L1 Vellinge Sweden
19 G5 Vellmar Ger.
43 B3 Vellore India
19 H4 Velpke Ger.
28 G2 Vel'sk Rus. Fed.
19 L2 Velten Ger.
21 H2 Veltrusy Czech Rep.
18 C4 Veluwe reg. Neth.
18 D4 Veluwezoom, Nationaal Park nat. park Neth.
77 J5 Velva U.S.A.
120 G5 Vema Fracture sea feature Atl. Ocean
117 J4 Vema Trough sea feature Ind. Ocean
43 B4 Vembanād L. India
11 C4 Venachar, Loch l. U.K.
105 E2 Venado Tuerto Arg.
26 F4 Venafro Italy
83 E3 Venamo r. Guyana/Venez.
103 E3 Venamo, Co mt Venez.
22 B2 Venarey-les-Laumes France
22 E4 Venaria Italy
22 E6 Vence France
104 C3 Venceslau Bráz Brazil
15 H4 Vendenheim France
15 E4 Vendeuvre-sur-Barse France
24 E3 Vendôme France
26 C3 Venelles France
26 E3 Veneta, Laguna lag. Italy
23 J4 Veneto div. Italy
28 F4 Venev Rus. Fed.
Venezia see Venice
23 K4 Venezia, Golfo di g. Europe
99 D2 Venezuela country S. America
103 C2 Venezuela, Golfo de g. Venez.
120 E4 Venezuelan Basin sea feature Atl. Ocean
43 A3 Vengurla India
23 K4 Venice Italy
87 D7 Venice U.S.A.
22 B4 Vénissieux France
43 B3 Venkatagiri India
43 C2 Venkatapuram India
18 D5 Venlo Neth.
9 K7 Vennesla Norway
23 H3 Venosta, Val v. Italy
22 A2 Venoy France
18 C5 Venray Neth.
9 R8 Venta r. Latvia/Lith.
9 S8 Venta Lith.
105 D3 Ventana, Serra de la h. Arg.
22 D5 Ventavon France
113 G4 Ventersburg S. Africa
113 G3 Ventersdorp S. Africa
113 F5 Venterstad S. Africa
71 2 Vent, Îles du is Fr. Polynesia Pac. Oc.
71 2 Vent, Îles sous le is Fr. Polynesia Pac. Oc.
13 F7 Ventnor U.K.
22 C5 Ventoux, Mont mt France
9 R8 Ventspils Latvia
103 D3 Ventuari r. Venez.
92 C4 Ventucopa U.S.A.
92 C4 Ventura U.S.A.
66 F7 Venus B. Austr.
71 3 Vénus, Pte pt Fr. Polynesia Pac. Oc.
23 L3 Venzone Italy
102 D3 Vera Arg.
25 F4 Vera Spain
97 E4 Veracruz Mex.
97 E3 Veracruz div. Mex.
44 B5 Veraval India
22 F4 Verbania Italy
22 F4 Vercelli Italy
22 C5 Vercors reg. France
8 M5 Verdalsøra Norway
105 D4 Verde r. Arg.
104 C2 Verde r. Goiás Brazil
104 C2 Verde r. Goiás/Minas Gerais Brazil
104 B2 Verde r. Mato Grosso do Sul Brazil
80 E6 Verde r. Mex.
102 E2 Verde r. Para.
93 G4 Verde r. U.S.A.
104 D1 Verde Grande r. Brazil
55 B3 Verde Island Pass. chan. Phil.
19 G4 Verden (Aller) Ger.
105 D3 Verde, Pen. Arg.
85 E4 Verdigris r. U.S.A.
22 C6 Verdon r. France
15 F3 Verdun France
22 C3 Verdun-sur-le-Doubs France
113 G3 Vereeniging S. Africa
89 J2 Vérendrye, Réserve faunique La res. Can.
105 G2 Vergara Uru.
91 G2 Vergennes U.S.A.
25 C2 Verín Spain
38 D1 Verkhne-Avzyan Rus. Fed.
29 F6 Verkhnebakanskiy Rus. Fed.
17 Q3 Verkhnedneprovskiy Rus. Fed.
32 K3 Verkhneimbatskoye Rus. Fed.
8 W2 Verkhnetulomskiy Rus. Fed.
38 D1 Verkhneural'sk Rus. Fed.
33 O3 Verkhnevilyuysk Rus. Fed.
28 D1 Verkhneye Kuyto, Oz. l. Rus. Fed.
29 H5 Verkhniy Baskunchak Rus. Fed.
29 J5 Verkhniy Kushum Rus. Fed.
8 W3 Verkhnyaya Pirenga, Oz. l. Rus. Fed.
28 H2 Verkhnyaya Toyma Rus. Fed.
28 G2 Verkhovazh'ye Rus. Fed.
29 C5 Verkhovyna Ukr.
33 P3 Verkhoyansk Rus. Fed.
33 O3 Verkhoyanskiy Khrebet mts Rus. Fed.
39 K2 Verkhuba Kazak.
15 D3 Vermand France
104 B1 Vermelho r. Brazil
77 G4 Vermilion Can.
88 C5 Vermilion r. U.S.A.
93 F3 Vermilion Cliffs cliff U.S.A.
88 A2 Vermilion Lake U.S.A.
88 A2 Vermilion Range h. U.S.A.
84 D3 Vermillion U.S.A.
77 L5 Vermillion Bay Can.
91 G3 Vermont div. U.S.A.
116 B2 Vernadsky Ukr. Base Ant.
82 E3 Vernal U.S.A.
89 G2 Verner Can.
15 A4 Verneuil-sur-Avre France
112 D5 Verneuk Pan salt pan S. Africa
22 D3 Vernier Switz.
23 H5 Vernio Italy
76 F4 Vernon Can.
15 B3 Vernon France
93 H4 Vernon AZ U.S.A.
91 G4 Vernon CT U.S.A.
85 D5 Vernon TX U.S.A.
93 F1 Vernon UT U.S.A.
69 C2 Vernon, Mount h. Austr.
15 B4 Vernouillet France
87 D7 Vero Beach U.S.A.
27 K4 Veroia Greece
23 H4 Verona Italy
105 F2 Verónica Arg.
66 B4 Verran Austr.
22 E4 Verres Italy
15 C4 Versailles France
18 F4 Versmold Ger.
22 D3 Versoix Switz.
21 M5 Vértesacsa Hungary
15 B2 Verton France
15 F4 Vertus France
113 J4 Verulam S. Africa
15 F2 Verviers Belgium
15 D3 Vervins France
22 E5 Verzuolo Italy
15 E3 Verzy France
26 C3 Vescovato Corsica France
32 G4 Veselaya, G. mt Rus. Fed.
29 E6 Vesele Ukr.
21 H3 Veselí nad Lužnicí Czech Rep.
21 L4 Veselí nad Moravou Czech Rep.
29 G6 Veselovskoye Vdkhr. resr Rus. Fed.
39 K2 Veseloyarsk Rus. Fed.
38 F1 Veselyy Podol Kazak.
29 G5 Veshenskaya Rus. Fed.
15 D3 Vesle r. France
22 D3 Vesoul France
52 D3 Vesselyy Yar Rus. Fed.
18 C5 Vessem Neth.
8 O2 Vesterålen is Norway
8 N2 Vesterålsfjorden chan. Norway
116 C3 Vestfjella mts Ant.
9 L7 Vestfjorddalen v. Norway
8 N3 Vestfjorden chan. Norway
8 Vestmanna Faroe Is
8 C5 Vestmannaeyjar Iceland
8 C5 Vestmannaeyjar is Iceland
8 K5 Vestnes Norway
8 F4 Vesturhorn hd Iceland
22 E6 Vésubie r. France
Vesuvio volc. see Vesuvius
26 F4 Vesuvius volc. Italy
28 F3 Ves'yegonsk Rus. Fed.
28 F3 Veszprém Hungary
21 L5 Veszprém div. Hungary
71 13 Vetauua i. Fiji
8 S5 Veteli Fin.
9 O8 Vetlanda Sweden
28 H3 Vetluga Rus. Fed.
28 H3 Vetluga r. Rus. Fed.
19 M5 Vetschau Ger.
26 E3 Vettore, Monte mt Italy
15 C1 Veurne Belgium
111 B3 Veveno r. Sudan
22 D3 Vevey Switz.
22 B3 Veyle r. France
22 C5 Veynes France
22 A2 Veynon r. France
93 F3 Veyo U.S.A.
40 C4 Veys Iran
15 G4 Vézelise France
24 E4 Vézère r. France
36 E1 Vezirköprü Turkey
9 P5 Vi Sweden
23 H5 Viadana Italy
100 C6 Viajas, I. de las Peru
101 K4 Viana Brazil
25 B2 Viana do Castelo Port.
18 C5 Vianen Neth.
Viangchan see Vientiane
104 C2 Vianópolis Brazil
23 H6 Viareggio Italy
9 L8 Viborg Denmark
26 G5 Vibo Valentia Italy
15 A4 Vibraye France
25 H2 Vic Spain
116 B2 Vicecomodoro Marambio Arg. Base Ant.
92 C5 Vicente, Pt U.S.A.
23 J4 Vicenza Italy
100 D3 Vichada r. Col.
103 C3 Vichada r. Col.
28 G3 Vichuga Rus. Fed.
105 B2 Vichuquén Chile
24 F3 Vichy France
93 F5 Vicksburg AZ U.S.A.
85 F5 Vicksburg MS U.S.A.
104 D3 Viçosa Brazil
88 A5 Victor U.S.A.
66 C5 Victor Harbour Austr.
105 E2 Victoria Arg.
66 F6 Victoria div. Austr.
64 F3 Victoria r. Austr.
76 E5 Victoria Can.
105 B3 Victoria Chile
96 H5 Victoria Honduras
Victoria see Labuan
26 F6 Victoria Malta
85 D6 Victoria U.S.A.
75 L2 Victoria and Albert Mts Can.
95 J4 Victoria de las Tunas Cuba
111 C5 Victoria Falls waterfall Zambia/Zimbabwe
115 A6 Victoria Falls Zimbabwe
75 O1 Victoria Fjord in. Greenland
51 Victoria Harbour chan. H.K. China
87 F7 Victoria Hill Bahamas
74 G2 Victoria Island Can.
110 D4 Victoria, Lake Africa
66 D4 Victoria, Lake Austr.
67 G6 Victoria, Lake Austr.
79 J4 Victoria Lake Can.
116 B5 Victoria Land reg. Ant.
Victoria, Mt see Tomanivi
45 H5 Victoria, Mt Myanmar
62 E2 Victoria, Mt P.N.G.
110 D3 Victoria Nile r. Sudan/Uganda
70 D5 Victoria Range mts N.Z.
64 F3 Victoria River Austr.
64 F3 Victoria River Downs Austr.
79 F4 Victoriaville Can.
112 E5 Victoria West S. Africa
105 D3 Victorica Arg.
116 D3 Victor, Mt Ant.
92 D4 Victorville U.S.A.
105 D2 Vicuña Mackenna Arg.
93 E4 Vidal Junction U.S.A.
27 L2 Videle Romania
27 K3 Vidin Bulg.
44 D5 Vidisha India
11 Vidlin U.K.
28 E2 Vidlitsa Rus. Fed.
17 N3 Vidzy Belarus
28 J2 Vidz'yuyar Rus. Fed.
20 F3 Viechtach Ger.
105 D4 Viedma Arg.
102 B7 Viedma, L. Arg.
19 J3 Vielank Ger.
15 F2 Vielsalm Belgium
19 H5 Vienenburg Ger.
21 K4 Vienna Austria
86 B4 Vienna IL U.S.A.
91 F5 Vienna MD U.S.A.
90 C5 Vienna WV U.S.A.
22 B4 Vienne France
24 E3 Vienne r. France
58 B1 Vientiane Laos
105 B3 Viento, Cordillera del mts Arg.
95 L5 Vieques i. Puerto Rico
8 U5 Vieremä Fin.
18 D5 Viersen Ger.
22 F2 Vierwaldstätter See l. Switz.
24 F3 Vierzon France
96 D2 Viesca Mex.
9 T8 Viesīte Latvia
26 G4 Vieste Italy
8 Q3 Vietas Sweden
31 K8 Vietnam country Asia
51 B6 Viêt Tri Vietnam
22 C4 Vif France
55 B2 Vigan Phil.
22 F4 Vigevano Italy
97 H4 Vigía Chico Mex.
15 C2 Vignacourt France
24 D5 Vignemale mt France
23 J5 Vignola Italy
15 B3 Vigny France
25 B1 Vigo Spain
22 E5 Vigone Italy
8 T4 Vihanti Fin.
44 C3 Vihari Pak.
9 T6 Vihti Fin.
8 T5 Viitasaari Fin.
43 C3 Vijainagar India
43 A2 Vijayapati India
43 C2 Vijayawada India
8 D5 Vík Iceland
8 U3 Vikajärvi Fin.
77 G4 Viking Can.
8 M4 Vikna i. Norway
9 K6 Vikøyri Norway
108 Vila da Ribeira Brava Cape Verde
108 Vila do Sal Rei Cape Verde
108 Vila do Tarrafal Cape Verde
25 B3 Vila Franca de Xira Port.
25 B1 Vilagarcía de Arousa Spain
113 K2 Vila Gomes da Costa Moz.
25 C1 Vilalba Spain
25 B2 Vila Nova de Gaia Port.
25 G2 Vilanova i la Geltrú Spain
108 Vila Nova Sintra Cape Verde
25 C2 Vila Real Port.
25 C2 Vilar Formoso Port.
43 B4 Vilavankod India
101 H3 Vila Velha Brazil
100 D6 Vilcabamba, Cordillera mts Peru
8 P4 Vilhelmina Sweden
100 F6 Vilhena Brazil
9 T7 Viljandi Estonia
113 G3 Viljoenskroon S. Africa
9 S9 Vilkaviškis Lith.
9 S9 Vilkija Lith.
33 L2 Vil'kitskogo, Proliv str. Rus. Fed.
97 E4 Villa Alta Mex.
100 E6 Villa Bella Bol.
25 C1 Villablino Spain
105 E2 Villa Cañás Arg.
25 E3 Villacañas Spain
21 G6 Villach Austria
26 C5 Villacidro Sardinia Italy
105 E2 Villa Constitución Arg.
96 B2 Villa Constitución Mex.
96 D4 Villa de Alvarez Mex.
96 D3 Villa de Cos Mex.
105 D1 Villa del Rosario Arg.
105 D2 Villa del Totoral Arg.
105 D1 Villa Dolores Arg.
23 F3 Villadossola Italy
97 F4 Villa Flores Mex.
23 H4 Villafranca di Verona Italy
85 F5 Village, Lake U.S.A.
105 F3 Villa Gesell Arg.
97 E2 Villagrán Mex.
105 E1 Villaguay Arg.
97 F4 Villahermosa Mex.
105 D2 Villa Huidobro Arg.
105 D3 Villa Iris Arg.
25 F3 Villajoyosa Spain
97 D2 Villaldama Mex.
105 D3 Villálonga Arg.
105 D2 Villa María Arg.
105 E1 Villa María Grande Arg.
100 F8 Villa Montes Bol.
111 E5 Villandro, Tanjona pt Madag.
113 H1 Villa Nora S. Africa
103 B2 Villanueva Col.
96 D3 Villanueva Mex.
25 D3 Villanueva de la Serena Spain
25 E3 Villanueva de los Infantes Spain
102 E3 Villa Ocampo Arg.
96 C2 Villa Ocampo Mex.
96 C2 Villa Orestes Pereyra Mex.
26 C5 Villaputzu Sardinia Italy
22 C4 Villard-Bonnot France
22 C4 Villard-de-Lans France
105 C3 Villa Regina Arg.
105 B3 Villarrica Chile
102 E3 Villarrica Para.
105 B3 Villarrica, L. Chile
105 B3 Villarrica, Parque Nacional nat. park Chile
105 B3 Villarrica, Volcán volc. Chile
25 E3 Villarrobledo Spain
26 F5 Villa San Giovanni Italy
105 E2 Villa San José Arg.
105 C1 Villa Santa Rita de Catuna Arg.
102 C3 Villa Unión Arg.
96 D1 Villa Unión Coahuila Mex.
96 C3 Villa Unión Durango Mex.
96 C3 Villa Unión Sinaloa Mex.
105 D2 Villa Valeria Arg.
103 B3 Villavicencio Col.
100 E8 Villazon Bol.
15 H4 Villé France
22 C4 Villefontaine France
24 F4 Villefranche-de-Rouergue France
22 B4 Villefranche-sur-Saône France
22 C6 Villelaure France
89 H2 Ville-Marie Can.
25 F3 Villena Spain
15 D4 Villenauxe-la-Grande France
22 C6 Villeneuve France
15 D4 Villeneuve-l'Archevêque France
22 B6 Villeneuve-lès-Avignon France
24 E4 Villeneuve-sur-Lot France
15 D4 Villeneuve-sur-Yonne France
85 E6 Ville Platte U.S.A.
15 D3 Villers-Cotterêts France
15 G4 Villers-lès-Nancy France
15 F3 Villerupt France
15 E4 Villery France
22 B4 Villeurbanne France
113 H3 Villiers S. Africa
15 F4 Villiers-le-Sec France
20 B4 Villingen Ger.
15 F4 Villotte-sur-Aire France
77 G4 Vilna Can.
9 T9 Vilnius Lith.
29 E6 Vil'nyans'k Ukr.
9 T5 Vilppula Fin.
20 F3 Vils r. Ger.
20 F4 Vilsbiburg Ger.
20 G4 Vilshofen Ger.
43 B4 Viluppuram India
15 E2 Vilvoorde Belgium
28 C4 Vilyeyka Belarus
33 O3 Vilyuy r. Rus. Fed.
33 M3 Vilyuyskoye Vdkhr. resr Rus. Fed.
23 G4 Vimercate Italy
9 O8 Vimmerby Sweden
21 G3 Vimperk Czech Rep.
15 C2 Vimy France
114 F3 Vina r. Cameroon
92 A2 Vina U.S.A.
105 B2 Viña del Mar Chile
91 J2 Vinalhaven U.S.A.
25 G2 Vinaròs Spain
22 C4 Vinay France
86 C4 Vincennes U.S.A.
116 C6 Vincennes Bay Ant.
8 Q4 Vindelälven r. Sweden
8 Q4 Vindeln Sweden
44 C5 Vindhya Range h. India
91 F5 Vineland U.S.A.
91 H4 Vineyard Haven U.S.A.
22 C2 Vingeanne r. France
58 C1 Vinh Vietnam
58 C1 Vinh Linh Vietnam
58 C3 Vinh Long Vietnam
58 C3 Vinh Rach Gia b. Vietnam
51 B6 Vinh Yên Vietnam
85 E4 Vinita U.S.A.
27 H2 Vinkovci Croatia
29 D5 Vinnytsya Ukr.
116 B3 Vinson Massif mt Ant.
9 L6 Vinstra Norway
88 A4 Vinton U.S.A.
43 B2 Vinukonda India
19 G2 Viöl Ger.
23 J3 Vipiteno Italy
19 K3 Vipperow Ger.
55 C3 Virac Phil.
44 C5 Viramgam India
37 G3 Viranşehir Turkey
44 B4 Virawah Pak.
77 J5 Virden Can.
24 D2 Vire France
111 B5 Virei Angola
104 D2 Virgem da Lapa Brazil
89 H1 Virginatown Can.
14 D4 Virginia Rep. of Ireland
113 G4 Virginia S. Africa
88 A2 Virginia U.S.A.
90 D6 Virginia div. U.S.A.
91 E6 Virginia Beach U.S.A.
92 C2 Virginia City U.S.A.
73 K8 Virgin Islands (U.K.) terr. Caribbean Sea
73 K8 Virgin Islands (U.S.A.) terr. Caribbean Sea
93 F3 Virgin Mts U.S.A.
9 T6 Virkkala Fin.
58 C2 Virôchey Cambodia
88 B4 Viroqua U.S.A.
26 G2 Virovitica Croatia
9 S5 Virrat Fin.
15 F3 Virton Belgium
9 S7 Virtsu Estonia
43 B4 Virudunagar India
115 A3 Virunga, Parc National des nat. park Congo(Zaire)
22 D3 Viry France
26 G3 Vis i. Croatia
9 U9 Visaginas Lith.
92 C3 Visalia U.S.A.
44 B5 Visavadar India
55 B4 Visayan Sea Phil.
18 F4 Visbek Ger.
9 Q8 Visby Sweden
74 G2 Viscount Melville Sound str. Can.
15 F2 Visé Belgium
27 H3 Višegrad Bos.-Herz.
32 J2 Vise, O. i. Rus. Fed.
23 K5 Viserba Italy
101 J4 Viseu Brazil
25 C2 Viseu Port.
43 C2 Vishakhapatnam India
39 H2 Vishnevka Kazak.
9 U8 Višķi Latvia
44 C5 Visnagar India
27 H3 Visoko Bos.-Herz.
22 E5 Viso, Monte mt Italy
22 E3 Visp Switz.
19 G4 Visselhövede Ger.
92 D5 Vista U.S.A.
104 A2 Vista Alegre Brazil
27 L4 Vistonida, Limni lag. Greece
103 C3 Vita r. Col.
44 B3 Vitakri Pak.
26 E3 Viterbo Italy
26 G2 Vitez Bos.-Herz.
100 E8 Vitichi Bol.
25 C2 Vitigudino Spain
71 15 Viti Levu i. Fiji
49 L1 Vitim r. Rus. Fed.
49 K1 Vitimskoye Ploskogor'ye plat. Rus. Fed.
21 L3 Vítkov Czech Rep.
104 E3 Vitória Brazil
Vitoria see Vitoria-Gasteiz
104 E1 Vitória da Conquista Brazil
25 E1 Vitoria-Gasteiz Spain
24 D2 Vitré France
22 C6 Vitrolles France
15 C2 Vitry-en-Artois France
15 E4 Vitry-le-François France
28 D4 Vitsyebsk Belarus
8 R3 Vittangi Sweden
22 B2 Vitteaux France
15 F4 Vittel France
26 F6 Vittoria Sicily Italy
23 K4 Vittorio Veneto Italy
118 F3 Vityaz Depth depth Pac. Oc.
22 B5 Vivarais, Monts du mts France
25 C1 Viveiro Spain
22 F4 Viverone, Lago di l. Italy
113 H1 Vivo S. Africa
66 B6 Vivonne B. Austr.
71 15 Viwa i. Fiji
Vizagapatam see Vishakhapatnam
96 A2 Vizcaíno, Desierto de des. Mex.
96 A2 Vizcaíno, Sierra mts Mex.
29 C7 Vize Turkey
43 C2 Vizianagaram India
22 C4 Vizille France
28 J2 Vizinga Rus. Fed.
21 L3 Vizovice Czech Rep.
18 B5 Vlaardingen Neth.
17 L7 Vlădeasa, Vârful mt Romania
29 H7 Vladikavkaz Rus. Fed.
52 D3 Vladimir Primorskiy Kray Rus. Fed.
28 G3 Vladimir Vladimir. Obl. Rus. Fed.
52 C3 Vladimiro-Aleksandrovskoye Rus. Fed.
38 F1 Vladimirovka Kustanay. Kazak.
38 B2 Vladimirovka Zapadnyy Kaz. Kazak.
21 J3 Vladislav Czech Rep.
49 O3 Vladivostok Rus. Fed.
28 G4 Vlaimirskaya Oblast' div. Rus. Fed.
113 H2 Vlakte S. Africa
21 H3 Vlašim Czech Rep.
27 K3 Vlasotince Yugo.
112 D7 Vleesbaai b. S. Africa
18 C3 Vlieland i. Neth.
18 A5 Vlissingen Neth.
27 H4 Vlorë Albania
18 F4 Vlotho Ger.
21 H2 Vltava r. Czech Rep.
21 G4 Vöcklabruck Austria
28 F2 Vodlozero, Ozero l. Rus. Fed.
21 H3 Vodňany Czech Rep.
11 Voe U.K.
22 C1 Vôge, La reg. France
20 C2 Vogelsberg h. Ger.
23 G5 Voghera Italy
71 9 Voh New Caledonia Pac. Oc.
20 E4 Vohburg an der Donau Ger.
20 F3 Vohenstrauß Ger.
Vohimena, Cape see Vohimena, Tanjona
111 E6 Vohimena, Tanjona c. Madag.
18 F5 Vöhl Ger.
9 T7 Võhma Estonia
115 C3 Voi Kenya
114 B3 Voinjama Liberia
15 E4 Voire r. France
22 C4 Voiron France
22 C3 Voiteur France
21 J5 Voitsberg Austria
9 L9 Vojens Denmark
27 H2 Vojvodina div. Yugo.
28 H3 Vokhma Rus. Fed.
28 D1 Voknavolok Rus. Fed.
21 G4 Volary Czech Rep.
82 F2 Volborg U.S.A.
105 B1 Volcán, Co del mt Chile
Volcano Bay see Uchiura-wan
Volcano Is see Kazan-rettō
39 K1 Volchikha Rus. Fed.
9 K5 Volda Norway
18 C4 Volendam Neth.
29 H6 Volga Rus. Fed.
88 B4 Volga r. U.S.A.
29 G6 Volgodonsk Rus. Fed.
32 F5 Volgograd Rus. Fed.
29 H5 Volgogradskaya Oblast' div. Rus. Fed.
71 15 Voli Voli Pt Fiji
21 H6 Völkermarkt Austria
28 E3 Volkhov Rus. Fed.
28 D3 Volkhov r. Rus. Fed.
15 G3 Völklingen Ger.
113 H3 Volksrust S. Africa
52 B3 Vol'no-Nadezhdinskoye Rus. Fed.
29 F6 Volnovakha Ukr.
39 G1 Vol'noye Rus. Fed.
33 L2 Volochanka Rus. Fed.
29 C5 Volochys'k Ukr.
29 F6 Volodars'ke Ukr.
29 J6 Volodarskiy Rus. Fed.
39 G1 Volodarskoye Kazak.
17 O5 Volodars'k-Volyns'kyy Ukr.
17 N5 Volodymyrets' Ukr.
29 C5 Volodymyr-Volyns'kyy Ukr.
28 F3 Vologda Rus. Fed.
28 G3 Vologodskaya Oblast' div. Rus. Fed.
29 F5 Volokonovka Rus. Fed.
27 K5 Volos Greece
28 D3 Volosovo Rus. Fed.
17 P2 Volot Rus. Fed.
29 F4 Volovo Rus. Fed.
19 J3 Völschow Ger.
29 H4 Vol'sk Rus. Fed.
114 C3 Volta, Lake Ghana
104 D3 Volta Redonda Brazil
23 H6 Volterra Italy
26 F4 Volturno r. Italy
27 K4 Volvi, L. Greece
21 G3 Volyně Czech Rep.
28 J4 Volzhsk Rus. Fed.
38 B1 Volzhskiy Samarsk. Rus. Fed.
29 H5 Volzhskiy Volgograd. Rus. Fed.
19 L1 Vombsjön l. Sweden
71 15 Vomo i. Fiji
111 E6 Vondrozo Madag.
28 G1 Vonga Rus. Fed.
8 F4 Vopnafjörður Iceland
8 F4 Vopnafjörður b. Iceland
28 B2 Võrå Fin.
17 M3 Voranava Belarus
20 C5 Vorarlberg div. Austria
28 J3 Vorchanka Rus. Fed.
19 J1 Vordingborg Denmark
28 E2 Vorenzha Rus. Fed.
32 H3 Vorkuta Rus. Fed.
9 S7 Vormsi i. Estonia
29 G5 Vorona r. Rus. Fed.
29 F5 Voronezh Rus. Fed.
29 F5 Voronezh r. Rus. Fed.
29 G5 Voronezhskaya Oblast' div. Rus. Fed.
39 K2 Vorontsovka Kazak.
28 G3 Voron'ye Rus. Fed.
Voroshilovgrad see Luhans'k
17 Q3 Vorot'kovo Rus. Fed.
19 K2 Vorpommersche Boddenlandschaft, Nationalpark nat. park Ger.

W

21 K2 Wielka Sowa mt Pol.
17 J5 Wieluń Pol.
Wien see Vienna
21 K4 Wien div. Austria
21 K5 Wiener Neustadt Austria
18 D4 Wierden Neth.
19 H4 Wieren Ger.
18 B4 Wieringermeer Polder reclaimed land Neth.
18 C4 Wieringerwerf Neth.
21 M1 Wieruszów Pol.
21 J6 Wies Austria
20 B2 Wiesbaden Ger.
21 J4 Wieselburg Austria
20 F3 Wiesenfelden Ger.
20 D3 Wiesentheid Ger.
20 E5 Wiesing Austria
20 B3 Wiesloch Ger.
18 E3 Wiesmoor Ger.
19 G4 Wietze Ger.
19 G4 Wietzendorf Ger.
16 J3 Wieżyca h. Pol.
12 E4 Wigan U.K.
85 F6 Wiggins U.S.A.
13 F7 Wight, Isle of i. U.K.
77 H2 Wignes Lake Can.
13 F5 Wigston U.K.
12 D3 Wigton U.K.
11 D6 Wigtown U.K.
11 D6 Wigtown Bay U.K.
18 C5 Wijchen Neth.
18 D4 Wijhe Neth.
15 E1 Wijnegem Belgium
93 F4 Wikieup U.S.A.
89 G3 Wikwemikong Can.
23 G2 Wil Switz.
82 C2 Wilbur U.S.A.
66 E3 Wilcannia Austr.
19 L4 Wildau Ger.
20 B4 Wildbad im Schwarzwald Ger.
19 K4 Wildberg Ger.
77 J4 Wildcat Hill Wilderness Park Can.
92 D2 Wildcat Peak U.S.A.
113 H5 Wild Coast S. Africa
18 F4 Wildeshausen Ger.
88 C1 Wild Goose Can.
76 F4 Wildhay r. Can.
22 E3 Wildhorn mt Switz.
21 J6 Wildon Austria
88 A2 Wild Rite Lake U.S.A.
67 F9 Wild Rivers Nat. Park Austr.
20 D6 Wildspitze mt Austria
87 D6 Wildwood FL U.S.A.
91 F5 Wildwood NJ U.S.A.
113 H3 Wilge r. Free State S. Africa
113 H2 Wilge r. Gauteng/Mpumalanga S. Africa
66 A3 Wilgena Austr.
116 C5 Wilhelm II Land reg. Ant.
90 C4 Wilhelm, Lake U.S.A.
62 E2 Wilhelm, Mt P.N.G.
21 J4 Wilhelmsburg Austria
20 C5 Wilhelmsdorf Ger.
18 F3 Wilhelmshaven Ger.
91 F4 Wilkes-Barre U.S.A.
116 B6 Wilkes Coast Ant.
116 B6 Wilkes Land reg. Ant.
77 H4 Wilkie Can.
116 B2 Wilkins Coast Ant.
116 B2 Wilkins Ice Shelf ice feature Ant.
82 B2 Willamette r. U.S.A.
13 D7 Willand U.K.
66 F4 Willandra Billabong watercourse Austr.
82 B1 Willapa B. U.S.A.
90 B4 Willard U.S.A.
91 F5 Willards U.S.A.
93 H5 Willcox U.S.A.
19 G5 Willebadessen Ger.
15 E1 Willebroek Belgium
95 L6 Willemstad Neth. Ant.
77 H3 William r. Can.
69 B1 Williambury Austr.
66 B2 William Cr. Austr.
66 E6 William, Mt Austr.
69 C6 William, Mt h. Austr.
69 C5 Williams Austr.
69 C6 Williams r. Austr.
93 F4 Williams AZ U.S.A.
92 A2 Williams CA U.S.A.
88 A5 Williamsburg IA U.S.A.
90 A6 Williamsburg KY U.S.A.
88 E3 Williamsburg MI U.S.A.
90 E6 Williamsburg VA U.S.A.
87 E7 Williams I. Bahamas
76 E4 Williams Lake Can.
79 H1 William Smith, Cap c. Can.
90 E3 Williamson NY U.S.A.
90 B6 Williamson WV U.S.A.
88 D5 Williamsport IN U.S.A.
90 E4 Williamsport PA U.S.A.
87 E5 Williamston U.S.A.
91 G3 Williamstown MA U.S.A.
91 F3 Williamstown NY U.S.A.
90 C5 Williamstown WV U.S.A.
91 G4 Willimantic U.S.A.
68 D1 Willis Group atolls Coral Sea Is Terr.
112 D5 Williston S. Africa
87 D6 Williston FL U.S.A.
84 C1 Williston ND U.S.A.
76 E3 Williston Lake Can.
13 D6 Williton U.K.
92 A2 Willits U.S.A.
84 E2 Willmar U.S.A.
76 F4 Willmore Wilderness Prov. Park Can.
76 D3 Will, Mt Can.
66 B3 Willochra watercourse Austr.
76 E4 Willow r. Can.
77 H5 Willow Bunch Can.
90 E4 Willow Hill U.S.A.
76 F2 Willow Lake Can.
112 E6 Willowmore S. Africa
88 C3 Willow Reservoir U.S.A.
92 A2 Willows U.S.A.
85 F4 Willow Springs U.S.A.
67 J3 Willow Tree Austr.
113 H6 Willowvale S. Africa
91 G2 Willsboro U.S.A.
64 E4 Wills, L. salt flat Austr.
66 C5 Willunga Austr.
19 L3 Wilmersdorf Ger.
66 C4 Wilmington Austr.
91 F5 Wilmington DE U.S.A.
87 E5 Wilmington NC U.S.A.
90 B5 Wilmington OH U.S.A.
91 G3 Wilmington VT U.S.A.
13 E4 Wilmslow U.K.
18 F6 Wilnsdorf Ger.
66 C3 Wilpena watercourse Austr.
15 L5 Wilsdruff Ger.
19 G3 Wilseder Berg h. Ger.
84 D4 Wilson KS U.S.A.
87 E5 Wilson NC U.S.A.
116 B5 Wilson Hills Ant.
83 F4 Wilson, Mt CO U.S.A.
93 E2 Wilson, Mt NV U.S.A.
84 D4 Wilson Resr U.S.A.
91 H2 Wilsons Mills U.S.A.
67 G7 Wilson's Promontory pen. Austr.
67 G7 Wilson's Promontory Nat. Park Austr.
18 D4 Wilsum Ger.
21 K4 Wiltersdorf Austria
19 M5 Wilthen Ger.
88 B5 Wilton IA U.S.A.
91 H2 Wilton ME U.S.A.
15 F3 Wiltz Lux.
69 E3 Wiluna Austr.
15 B2 Wimereux France
66 E6 Wimmera r. Austr.
88 D5 Winamac U.S.A.
68 A6 Winbin watercourse Austr.
91 G4 Winburg S. Africa
13 E6 Wincanton U.K.
91 G3 Winchendon U.S.A.
78 E4 Winchester Can.
13 F6 Winchester U.K.
88 B6 Winchester IL U.S.A.
88 E5 Winchester IN U.S.A.
90 A6 Winchester KY U.S.A.
91 G3 Winchester NH U.S.A.
87 C5 Winchester TN U.S.A.
90 D5 Winchester VA U.S.A.
74 E3 Wind r. N.W.T. Can.
82 E3 Wind r. U.S.A.
66 B3 Windabout, L. salt flat Austr.
84 C3 Wind Cave Nat. Park U.S.A.
12 E3 Windermere U.K.
12 E3 Windermere l. U.K.
111 B6 Windhoek Namibia
21 H5 Windischgarsten Austria
84 E3 Windom U.S.A.
65 H5 Windorah Austr.
93 H4 Window Rock U.S.A.
88 D4 Wind Pt U.S.A.
82 E3 Wind River Range mts U.S.A.
13 F6 Windrush r. U.K.
20 D3 Windsbach Ger.
67 J4 Windsor Austr.
79 J4 Windsor Nfld Can.
79 H5 Windsor N.S. Can.
89 F4 Windsor Ont. Can.
79 F4 Windsor Que. Can.
13 G6 Windsor U.K.
91 G4 Windsor CT U.S.A.
87 E5 Windsor NC U.S.A.
91 F3 Windsor NY U.S.A.
90 E6 Windsor VA U.S.A.
91 G3 Windsor VT U.S.A.
91 G4 Windsor Locks U.S.A.
95 M5 Windward Islands Caribbean Sea
95 K5 Windward Passage Cuba/Haiti
87 C5 Winfield AL U.S.A.
88 B5 Winfield IA U.S.A.
85 D4 Winfield KS U.S.A.
12 F3 Wingate U.K.
67 J3 Wingen Austr.
15 D1 Wingene Belgium
15 H4 Wingen-sur-Moder France
67 K3 Wingham Austr.
89 G4 Wingham Can.
78 C2 Winisk Can.
78 C2 Winisk r. Can.
78 C3 Winisk L. Can.
78 C3 Winisk River Provincial Park Can.
58 A2 Winkana Myanmar
77 K5 Winkler Can.
91 J2 Winn U.S.A.
114 C3 Winneba Ghana
88 C3 Winnebago, Lake U.S.A.
88 C3 Winneconne U.S.A.
82 C3 Winnemucca U.S.A.
92 C1 Winnemucca Lake U.S.A.
20 C4 Winnenden Ger.
84 D3 Winner U.S.A.
85 E6 Winnfield U.S.A.
84 E2 Winnibigoshish L. U.S.A.
69 B1 Winning Austr.
77 K5 Winnipeg Can.
77 K4 Winnipeg r. Can.
77 K4 Winnipeg, Lake Can.
77 J4 Winnipegosis Can.
77 J4 Winnipegosis, Lake Can.
91 H3 Winnipesaukee, L. U.S.A.
85 F5 Winnsboro U.S.A.
93 G4 Winona AZ U.S.A.
88 C2 Winona MI U.S.A.
88 B3 Winona MN U.S.A.
85 F5 Winona MS U.S.A.
91 G2 Winooski U.S.A.
91 G2 Winooski r. U.S.A.
18 E3 Winschoten Neth.
19 G4 Winsen (Aller) Ger.
19 H3 Winsen (Luhe) Ger.
13 E4 Winsford U.K.
93 G4 Winslow U.S.A.
91 G4 Winsted U.S.A.
87 D4 Winston-Salem U.S.A.
18 F5 Winterberg Ger.
87 D6 Winter Haven U.S.A.
91 J2 Winterport U.S.A.
92 B2 Winters U.S.A.
18 D5 Winterswijk Neth.
22 F2 Winterthur Switz.
113 H4 Winterton S. Africa
91 J1 Winterville U.S.A.
91 J2 Winthrop U.S.A.
65 H4 Winton Austr.
70 B7 Winton N.Z.
13 G5 Winwick U.K.
66 C4 Wirrabara Austr.
13 D4 Wirral pen. U.K.
66 B3 Wirraminna Austr.
66 D6 Wirrega Austr.
66 A4 Wirrulla Austr.
13 H5 Wisbech U.K.
91 J2 Wiscasset U.S.A.
19 G3 Wischhafen Ger.
88 C3 Wisconsin div. U.S.A.
88 B4 Wisconsin r. U.S.A.
88 C4 Wisconsin Dells U.S.A.
88 C4 Wisconsin, Lake U.S.A.
88 C3 Wisconsin Rapids U.S.A.
90 B6 Wise U.S.A.
11 E5 Wishaw U.K.
17 J4 Wisła r. Pol.
19 J3 Wismar Ger.
19 J2 Wismarbucht b. Ger.
15 B2 Wissant France
88 B3 Wissota L. U.S.A.
76 D4 Wistaria Can.
113 H2 Witbank S. Africa
112 C2 Witbooisvlei Namibia
13 H6 Witham U.K.
13 G4 Witham r. U.K.
91 G2 Witherbee U.S.A.
12 H4 Withernsea U.K.
65 G5 Witjira National Park Austr.
18 C3 Witmarsum Neth.
13 F6 Witney U.K.
19 M4 Witnica Pol.
113 J2 Witrivier S. Africa
15 E3 Witry-lès-Reims France
18 F2 Wittdün Ger.
113 G5 Witteberg mts S. Africa
88 C3 Wittenberg U.S.A.
19 J4 Wittenberge Ger.
19 J3 Wittenburg Ger.
22 E2 Wittenheim France
19 M5 Wittichenau Ger.
19 H4 Wittingen Ger.
15 G3 Wittlich Ger.
18 E3 Wittmund Ger.
19 K3 Wittstock Ger.
62 E2 Witu Is P.N.G.
111 B6 Witvlei Namibia
19 G5 Witzenhausen Ger.
67 K1 Wivenhoe, Lake Austr.
54 D3 Wiwon N. Korea
16 J3 Władysławowo Pol.
17 J4 Włocławek Pol.
91 H2 Woburn Can.
67 G6 Wodonga Austr.
21 M2 Wodzisław Śląski Pol.
15 H4 Wœrth France
15 F3 Woëvre, Plaine de la plain France
18 F6 Wohra r. Ger.
19 G2 Wöhrden Ger.
15 G3 Woippy France
57 J8 Wokam i. Indon.
52 B1 Woken r. China
45 H4 Wokha India
39 H5 Wokhin Pass Afghan./Tajik.
13 G6 Woking U.K.
13 G6 Wokingham U.K.
88 D5 Wolcott IN U.S.A.
90 E3 Wolcott NY U.S.A.
21 M1 Wołczyn Pol.
19 L3 Woldegk Ger.
18 E3 Woldendorp Neth.
76 C2 Wolf r. Can.
88 C3 Wolf r. U.S.A.
20 B4 Wolfach Ger.
82 D2 Wolf Creek U.S.A.
83 F4 Wolf Creek Pass U.S.A.
91 H3 Wolfeboro U.S.A.
89 J3 Wolfe I. Can.
19 K5 Wolfen Ger.
19 H4 Wolfenbüttel Ger.
19 G5 Wolfhagen Ger.
76 C2 Wolf Lake Can.
82 F1 Wolf Point U.S.A.
21 H6 Wolfsberg Austria
19 H4 Wolfsburg Ger.
15 H3 Wolfstein Ger.
79 H4 Wolfville Can.
100 Wolf, Volcán volc. Galapagos Is Ecuador
19 L2 Wolgast Ger.
22 F2 Wolhusen Switz.
16 G4 Wolin Pol.
19 M3 Woliński Park Narodowy nat. park Pol.
21 K4 Wolkersdorf Austria
102 C9 Wollaston, Islas is Chile
77 J3 Wollaston Lake Can.
77 J3 Wollaston Lake Can.
74 G3 Wollaston Peninsula Can.
22 F2 Wollerau Switz.
67 J5 Wollongong Austr.
113 F3 Wolmaransstad S. Africa
19 J4 Wolmirstedt Ger.
20 E4 Wolnzach Ger.
21 K1 Wołów Pol.
66 D6 Wolseley Austr.
112 C6 Wolseley S. Africa
12 F3 Wolsingham U.K.
18 D4 Wolvega Neth.
13 E5 Wolverhampton U.K.
88 E3 Wolverine U.S.A.
15 E1 Wommelgem Belgium
15 H3 Womrather Höhe h. Ger.
68 D6 Wondai Austr.
20 F3 Wondreb r. Ger.
66 F3 Wongalarroo Lake salt l. Austr.
69 C5 Wongan Hills Austr.
67 H4 Wongarbon Austr.
45 G4 Wong Chu r. Bhutan
51 Wong Chuk Hang H.K. China
54 D5 Wŏnju S. Korea
65 J7 Wonnangatta Moroka National Park Austr.
66 E3 Wonominta watercourse Austr.
54 D4 Wŏnsan N. Korea
66 F7 Wonthaggi Austr.
69 D2 Wonyulgunna, Mt h. Austr.
66 B3 Woocalla Austr.
65 G2 Woodah, Isle i. Austr.
13 J5 Woodbridge U.K.
90 E5 Woodbridge U.S.A.
76 G3 Wood Buffalo National Park Can.
67 K2 Woodburn Austr.
82 B2 Woodburn U.S.A.
88 E4 Woodbury MI U.S.A.
91 F5 Woodbury NJ U.S.A.
90 A6 Wood Creek Lake U.S.A.
92 C2 Woodfords U.S.A.
92 C3 Woodlake U.S.A.
92 B2 Woodland U.S.A.
83 F4 Woodland Park U.S.A.
58 Woodlands Sing.
63 F2 Woodlark I. P.N.G.
64 F5 Woodroffe, Mt Austr.
82 E3 Woodruff U.S.A.
90 C5 Woodsfield U.S.A.
67 G7 Woodside Austr.
65 F3 Woods, L. salt flat Austr.
77 L5 Woods, Lake of the Can./U.S.A.
67 G6 Woods Pt Austr.
68 B2 Woodstock Austr.
79 G4 Woodstock N.B. Can.
89 G4 Woodstock Ont. Can.
88 C4 Woodstock IL U.S.A.
90 D5 Woodstock VA U.S.A.
91 F5 Woodstown U.S.A.
91 G2 Woodville U.S.A.
70 E4 Woodville N.Z.
91 E3 Woodville NY U.S.A.
85 E6 Woodville TX U.S.A.
85 D4 Woodward U.S.A.
12 F2 Wooler U.K.
67 K3 Woolgoolga Austr.
67 K2 Wooli Austr.
66 C3 Wooltana Austr.
66 B3 Woomera Austr.
66 A3 Woomera Prohibited Area res. Austr.
91 H4 Woonsocket U.S.A.
68 C5 Woorabinda Austr.
69 B2 Wooramel r. Austr.
90 C4 Wooster U.S.A.
22 E3 Worb Switz.
19 H5 Worbis Ger.
112 C6 Worcester S. Africa
13 E5 Worcester U.K.
91 H3 Worcester U.S.A.
20 F5 Wörgl Austria
12 D3 Workington U.K.
13 F4 Worksop U.K.
18 C4 Workum Neth.
82 F2 Worland U.S.A.
19 K5 Wörlitz Ger.
18 B4 Wormerveer Neth.
20 B3 Worms Ger.
13 C6 Worms Head hd U.K.
20 B3 Wörrstadt Ger.
112 B1 Wortel Namibia
20 B3 Wörth am Rhein Ger.
21 H6 Wörther See l. Austria
13 G7 Worthing U.K.
84 E3 Worthington U.S.A.
118 G5 Wotje i. Pac. Oc.
57 G7 Wotu Indon.
18 B5 Woudrichem Neth.
84 C3 Wounded Knee U.S.A.
15 H3 Woustviller France
68 D4 Wowan Austr.
57 G7 Wowoni i. Indon.
Wrangel Island see Vrangelya, O.
76 C3 Wrangell U.S.A.
74 C3 Wrangell I. U.S.A.
74 C4 Wrangell Mountains U.S.A.
11 C2 Wrath, Cape U.K.
84 C3 Wray U.S.A.
13 F5 Wreake r. U.K.
112 B4 Wreck Point S. Africa
19 H4 Wrestedt Ger.
13 E4 Wrexham U.K.
19 M4 Wriezen Ger.
56 C4 Wright Phil.
82 F3 Wright U.S.A.
85 E5 Wright Patman L. U.S.A.
93 G6 Wrightson, Mt U.S.A.
76 E2 Wrigley Can.
20 E3 Wrist Ger.
16 H5 Wrocław Pol.
21 K1 Wrocław div. Pol.
19 G2 Wrohm Ger.
16 H4 Września Pol.
50 E2 Wu'an China
69 C5 Wubin Austr.
50 D2 Wubu China
54 D1 Wuchang Heilongjiang China
51 E4 Wuchang Hubei China
51 D6 Wuchuan Guangdong China
51 C4 Wuchuan Guizhou China
50 D1 Wuchuan Nei Monggol China
50 C2 Wuda China
50 D3 Wudang Shan mt China
50 C3 Wudang Shan mts China
54 A4 Wudao China
45 H2 Wudaoliang China
50 E2 Wudi China
51 B5 Wuding China
50 D2 Wuding r. China
66 A4 Wudinna Austr.
50 B3 Wudu China
51 D4 Wufeng China
51 D5 Wugang China
50 C3 Wugong China
50 C2 Wuhai China
51 E4 Wuhan China
50 E3 Wuhe China
50 F4 Wuhu China
51 E6 Wuhua China
44 D2 Wüjang China
51 C6 Wujia China
51 C4 Wu Jiang r. China
114 E3 Wukari Nigeria
45 H3 Wulang China
44 C2 Wular L. India
54 B5 Wuleidao Wan l. China
50 E3 Wulian China
51 B4 Wulian Feng mts China
46 C4 Wuliang Shan mts China
57 J8 Wuliaru i. Indon.
51 C4 Wuling Shan mts China
51 C4 Wulong China
51 B5 Wumeng Shan mts China
51 C6 Wuming China
19 G3 Wümme r. Ger.
50 A4 Wungda China
51 E4 Wuning China
18 F5 Wünnenberg Ger.
19 L4 Wünsdorf Ger.
20 F3 Wunsiedel Ger.
19 G4 Wunstorf Ger.
56 B2 Wuntho Myanmar
93 G4 Wupatki National Monument res. U.S.A.
51 E5 Wuping China
18 E5 Wuppertal Ger.
112 C6 Wuppertal S. Africa
50 C2 Wuqi China
50 E2 Wuqiao China
50 E2 Wuqing China
69 C4 Wurarga Austr.
20 E2 Würzbach Ger.
20 C3 Würzburg Ger.
19 K5 Wurzen Ger.
50 B3 Wushan Gansu China
50 C4 Wushan Sichuan China
50 D4 Wu Shan mts China
51 C4 Wusheng China
51 C6 Wushi China
39 J4 Wushi China
Wusuli Jiang r. see Ussuri
50 D2 Wutai China
50 D2 Wutai Shan mt China
62 E2 Wuvulu I. P.N.G.
50 E4 Wuwei Anhui China
50 B2 Wuwei Gansu China
50 F4 Wuxi Jiangsu China
50 C4 Wuxi Sichuan China
Wuxing see Huzhou
51 C6 Wuxu China
51 C6 Wuxuan China
50 D3 Wuyang China
51 F4 Wuyi China
46 E2 Wuyiling China
51 E5 Wuyi Shan mts China
51 E4 Wuyuan Jiangxi China
50 C1 Wuyuan Nei Monggol China
50 D2 Wuzhai China
50 D4 Wuzhen China
50 C2 Wuzhong China
51 D6 Wuzhou China
88 B5 Wyaconda r. U.S.A.
69 C5 Wyalkatchem Austr.
67 G4 Wyalong Austr.
89 F4 Wyandotte U.S.A.
68 A6 Wyandra Austr.
88 C5 Wyanet U.S.A.
67 H4 Wyangala Reservoir Austr.
66 F2 Wyara, Lake salt flat Austr.
66 E6 Wycheproof Austr.
13 E6 Wye r. U.K.
69 D4 Wyemandoo h. Austr.
15 H4 Wyhl Ger.
18 F2 Wyk auf Föhr Ger.
13 F6 Wylye r. U.K.
13 J5 Wymondham U.K.
64 E3 Wyndham Austr.
85 F5 Wynne U.S.A.
74 G2 Wynniatt Bay Can.
67 F8 Wynyard Austr.
77 J4 Wynyard Can.
88 C5 Wyoming IL U.S.A.
88 E4 Wyoming MI U.S.A.
82 E3 Wyoming div. U.S.A.
82 E3 Wyoming Peak U.S.A.
67 J4 Wyong Austr.
66 D5 Wyperfeld Nat. Park Austr.
12 E4 Wyre r. U.K.
21 J2 Wysoka Kopa mt Pol.
91 E4 Wysox U.S.A.
17 K4 Wyszków Pol.
13 F5 Wythall U.K.
90 C6 Wytheville U.S.A.
91 J2 Wytopitlock U.S.A.

X

110 F2 Xaafuun Somalia
38 A4 Xacmaz Azer.
37 M1 Xaçmaz Azer.
112 E1 Xade Botswana
45 H3 Xagquka China
44 D1 Xaidulla China
15 G4 Xaintois reg. France
45 G3 Xainza China
111 D6 Xai Xai Moz.
Xalapa see Jalapa Enríquez
97 G3 Xal, Cerro de h. Mex.
56 C2 Xam Hua Laos
58 B1 Xan r. Laos
111 C6 Xanagas Botswana
50 B1 Xangd China
50 D1 Xangdin Hural China
111 B5 Xangongo Angola
37 L2 Xankändi Azer.
27 L4 Xanthi Greece
58 C2 Xan, Xé r. Vietnam
100 E6 Xapuri Brazil
39 J2 Xaqung China
37 M2 Xaraba Şähär Sayı i. Azer.
37 M2 Xärä Zirä Adası is Azer.
45 F3 Xarba La pass China
50 B1 Xar Burd China
50 F1 Xar Moron r. Nei Monggol China
50 D1 Xar Moron r. Nei Monggol China
25 F3 Xátiva Spain
111 C6 Xau, Lake Botswana
101 J6 Xavantes, Serra dos h. Brazil
58 C3 Xa Vo Đat Vietnam
39 K4 Xayar China
39 K4 Xekar China
90 B5 Xenia U.S.A.
15 G4 Xertigny France
54 A3 Xi r. China
54 F1 Xiachengzi China
51 D6 Xiachuan Dao i. China
50 B3 Xiahe China
51 E5 Xiajiang China
50 E2 Xiajin China
51 F5 Xiamen China
50 C3 Xi'an China
50 E3 Xiancheng China
50 C2 Xianchengbu China
51 C4 Xianfeng China
50 D3 Xiangcheng China
50 D3 Xiangfan China
50 D1 Xianghuang Qi China
58 B1 Xiangkhoang Laos
44 D3 Xiangquan He r. China
51 F4 Xiangshan China
51 D5 Xiangtan China
51 D5 Xiangxiang China
51 D4 Xiangyin China
51 F4 Xianju China
51 E4 Xianning China
51 E5 Xianxia Ling mts China
50 E2 Xian Xian China
50 C3 Xianyang China
51 F5 Xianyou China
51 C6 Xiaodong China
51 D4 Xiaogan China
46 E1 Xiao Hinggan Ling mts China
50 B4 Xiaojin China
45 H2 Xiaonanchuan China
51 F4 Xiaoshan China
51 E5 Xiaotao China
50 E2 Xiaowutai Shan mt China
50 E3 Xiao Xian China
51 B4 Xiaoxiang Ling mts China
50 D2 Xiaoyi China
51 F5 Xiapu China
54 A2 Xiawa China
50 D3 Xiayukou China
51 B5 Xichang China
50 B4 Xichong China
51 B6 Xichou China
50 D3 Xichuan China
51 B4 Xide China
103 D4 Xié r. Brazil
50 C1 Xieyang Dao i. China
50 E3 Xifei He r. China
51 C5 Xifeng Guizhou China
54 C2 Xifeng Liaoning China
50 C3 Xifengzhen China
35 G4 Xigazê China
50 C3 Xihan Shui r. China
50 B3 Xihe China
50 A1 Xi He watercourse China
50 B3 Xiji China
51 D6 Xi Jiang r. China
45 G2 Xijir China
45 G2 Xijir Ulan Hu salt l. China
50 B2 Xijishui China
50 D1 Xil China
54 B2 Xiliao r. China
51 B5 Xilin China
50 E1 Xilinhot China
50 D1 Xilin Qagan Obo Mongolia
50 A1 Ximiao China
51 F4 Xin'anjiang China
51 F4 Xin'anjiang Sk. resr China
113 K2 Xinavane Moz.
54 C3 Xinbin China
50 E3 Xincai China
51 F4 Xinchang China
50 A2 Xincheng Gansu China
51 C5 Xincheng Guangxi China
50 C2 Xincheng Ningxia China
50 C2 Xinchengbu China
51 D6 Xindu Guangxi China
50 B4 Xindu Sichuan China
51 E5 Xinfeng Guangdong China
51 E5 Xinfeng Jiangxi China
51 E6 Xinfengjiang Sk. resr China
51 D5 Xing'an China
51 E5 Xingan China
54 A4 Xingangzhen China
50 F1 Xingcheng China
51 E5 Xingguo China
35 H3 Xinghai China
50 D1 Xinghe China
50 F3 Xinghua China
51 F5 Xinghua Wan b. China
52 C2 Xingkai China
Xingkai Hu l. see Khanka, Lake
51 E5 Xingning China
51 D4 Xingou China
50 C3 Xingping China
51 B5 Xingren China
50 A3 Xingsagoinba China
50 D4 Xingshan China
50 E2 Xingtai China
101 H4 Xingu r. Brazil
101 H6 Xingu, Parque Indígena do nat. park Brazil
51 B4 Xingwen China
50 D2 Xing Xian China
50 D3 Xingyang China
51 B5 Xingyi China
51 E2 Xinhe China
39 K4 Xinhe China
51 D5 Xinhua China
50 B2 Xinhuacun China
51 C5 Xinhuang China
51 D6 Xinhui China
50 A2 Xining China
51 E4 Xinjian China
50 D3 Xinjiang China
46 A2 Xinjiang Uygur Zizhiqu div. China
50 C2 Xinjie China
51 B4 Xinjin Sichuan China
54 B4 Xinjin China
54 B1 Xinkai r. China
54 B3 Xinmin China
51 D5 Xinning China
51 A5 Xinping China
51 E5 Xinquan China
51 D5 Xinshao China
50 E3 Xintai China
51 D5 Xintian China
50 E4 Xin Xian Henan China
50 D2 Xin Xian Shanxi China
50 D3 Xinxiang China
51 D6 Xinxing China
50 E3 Xinyang Henan China
50 D3 Xinye China
50 F3 Xinye r. China
51 D6 Xinyi Guangdong China
50 F3 Xinyi Jiangsu China
51 C7 Xinying China
51 E5 Xinyu China
39 K4 Xinyuan China
50 D2 Xinzhou China
25 C1 Xinzo de Limia Spain
54 B3 Xiongyuecheng China
50 E3 Xiping Henan China
50 D3 Xiping Henan China
50 A3 Xiqing Shan mts China
101 K6 Xique Xique Brazil
51 C5 Xishui Guizhou China
51 E4 Xishui Hubei China
46 D2 Xi Ujimqin Qi China
51 F4 Xiuning China
51 C4 Xiushan China
51 E4 Xiu Shui r. China
51 C5 Xiuwen China
50 D3 Xiuwu China
54 B3 Xiuyan China
51 D6 Xiuying China
45 F3 Xixabangma Feng mt China
50 D3 Xixia China
50 E3 Xi Xian Henan China
50 D2 Xi Xian Shanxi China
50 C3 Xi Xiang China
51 F5 Xiyang Dao i. China
51 B5 Xiyang Jiang r. China
35 G3 Xizang Gaoyuan plat. China
46 B3 Xizang Zizhiqu div. China
54 A4 Xizhong Dao i. China
45 H3 Xoka China
58 C3 Xom An Lôc Vietnam
58 C3 Xom Duc Hanh Vietnam
50 F4 Xuancheng China
51 C4 Xuan'en China
50 C4 Xuanhan China
50 B2 Xuanhepu China
50 E1 Xuanhua China
58 C3 Xuân Lôc Vietnam
51 B5 Xuanwei China
49 K5 Xuchang Henan China
50 D3 Xuchang Henan China
38 A4 Xudat Azer.
37 M1 Xudat Azer.
110 E3 Xuddur Somalia
51 C5 Xuefeng Shan mts China
45 H2 Xugui China
45 E2 Xungba China
45 F3 Xungru China
50 C3 Xun He r. China
51 D6 Xun Jiang r. China
51 E5 Xunwu China
50 E3 Xun Xian China
50 C3 Xunyang China
50 C3 Xunyi China
51 D5 Xupu China
45 F3 Xuru Co salt l. China
50 E2 Xushui China
51 D6 Xuwen China
50 F3 Xuyi China
51 B4 Xuyong China
50 E3 Xuzhou China

Y

68 D4 Yaamba Austr.
51 B4 Ya'an China
66 E5 Yaapeet Austr.
114 E4 Yabassi Cameroon
110 D3 Yabēlo Eth.
49 K1 Yablonovyy Khrebet mts Rus. Fed.
50 B2 Yabrai Shan mts China
50 B2 Yabrai Yanchang China
36 F5 Yabrūd Syria
54 E1 Yabuli China
103 C2 Yacambu, Parque Nacional nat. park Venez.
71 13 Yacata i. Fiji
51 C7 Yacheng China
51 C5 Yachi He r. China
100 E6 Yacuma r. Bol.
44 D5 Yadgir India
81 K4 Yadkin r. U.S.A.
28 H4 Yadrin Rus. Fed.
71 13 Yadua i. Fiji
114 C2 Yagaba Ghana
52 G2 Yagishiri-tō i. Japan
38 C5 Yagman Turkm.
109 D3 Yagoua Cameroon
45 E3 Yagra China
45 H2 Yagradagzê Shan mt China
105 F1 Yaguari r. Uru.
Yaguarón r. see Jaguarão
58 B4 Yaha Thai.
36 D2 Yahşihan Turkey
36 E2 Yahyalı Turkey
41 H4 Yahya Wana Afgh.
53 F6 Yaita Japan
53 F7 Yaizu Japan
51 A4 Yajiang China
36 F3 Yakacık Turkey
41 G4 Yakhehal Afgh.
82 B2 Yakima U.S.A.
82 B2 Yakima r. U.S.A.
40 D3 Yakinish Iran
39 F5 Yakkabag Uzbek.
41 F4 Yakmach Pak.
114 C2 Yako Burkina
76 B3 Yakobi I. U.S.A.
52 C2 Yakovlevka Rus. Fed.
39 K4 Yakrik China
52 G3 Yakumo Japan
53 B9 Yaku-shima i. Japan
76 B3 Yakutat U.S.A.
76 B3 Yakutat Bay U.S.A.
33 O3 Yakutsk Rus. Fed.
29 E6 Yakymivka Ukr.
58 B4 Yala Thai.
89 F4 Yale U.S.A.
69 C4 Yalgoo Austr.
97 G3 Yalkubul, Pta pt Mex.
68 A5 Yalleroi Austr.
68 B6 Yallingup Austr.
67 G7 Yallourn Austr.
51 A5 Yalong Jiang r. China
36 B1 Yalova Turkey
29 F6 Yalta Donets'k Ukr.
29 E6 Yalta Krym Ukr.
54 C3 Yalu Jiang r. China/N. Korea
54 C4 Yalujiang Kou river mouth N. Korea
36 C2 Yalvaç Turkey
52 G5 Yamada Japan
52 G5 Yamagata Japan
53 B9 Yamagawa Japan
53 B7 Yamaguchi Japan
32 H2 Yamal, Poluostrov pen. Rus. Fed.
67 K2 Yamba Austr.
67 E7 Yambacoona Austr.
77 G2 Yamba Lake Can.
103 C4 Yambi, Mesa de h. Col.
109 E4 Yambio Sudan
27 M3 Yambol Bulg.
32 J3 Yamburg Rus. Fed.
50 A2 Yamenzhuang China
56 B2 Yamethin Myanmar
53 G6 Yamizo-san mt Japan
9 V7 Yamm Rus. Fed.
114 B3 Yamoussoukro Côte d'Ivoire
82 E3 Yampa r. U.S.A.
29 D5 Yampil' Ukr.
44 E4 Yamuna r. India
44 D3 Yamunanagar India
39 J2 Yamyshevo Kazak.
45 G3 Yamzho Yumco l. China
45 G4 Yana r. China
33 P3 Yana r. Rus. Fed.
66 D6 Yanac Austr.
43 C2 Yanam India
48 D8 Yanam India
50 C2 Yan'an China
100 D6 Yanaoca Peru
51 A5 Yanbian China
42 A5 Yanbu' al Baḥr S. Arabia
50 F3 Yancheng China
69 B5 Yanchep Austr.
50 C2 Yanchi China
50 D2 Yanchuan China
67 G5 Yanco Austr.
51 C4 Yanco Glen Austr.
66 D3 Yanco Glen Austr.
67 F3 Yanda watercourse Austr.
66 D3 Yandama Cr. watercourse Austr.
69 D3 Yandil Austr.
68 E6 Yandina Austr.
114 B2 Yanfolila Mali
51 E1 Yang r. China
45 H3 Ya'ngamdo China
45 G3 Yangbajain China
50 D3 Yangcheng China
51 D6 Yangchun China
54 D4 Yangdok N. Korea
50 D1 Yanggao China
50 E2 Yanggu China
39 G4 Yangiabad Uzbek.
39 F4 Yangikishlak Uzbek.
38 F5 Yangi-Nishan Uzbek.
38 F4 Yangirabad Uzbek.
39 G4 Yangiyul' Uzbek.
51 D6 Yangjiang China
56 B3 Yangon Myanmar
50 D4 Yangping China
50 D2 Yangquan China
51 D5 Yangshan China
51 D5 Yangshuo China
51 E4 Yangtze r. China
50 F4 Yangtze, Mouth of the est. China
110 E2 Yangudi Nassa National Park Eth.
50 C3 Yang Xian China
54 E5 Yangyang S. Korea
50 E1 Yangyuan China
50 F3 Yangzhou China
51 C4 Yanhe China
45 E2 Yanhuqu China
66 A4 Yaninee, Lake salt flat Austr.
54 F2 Yanji China
51 B4 Yanjin China
114 F3 Yankara Nat. Park Nigeria
84 D3 Yankton U.S.A.
33 P2 Yano-Indigirskaya Nizmennost' lowland Rus. Fed.
43 C4 Yan Oya r. Sri Lanka
48 E3 Yanqi China
50 E1 Yanqing China
50 E2 Yanshan Hebei China
51 E4 Yanshan Jiangxi China
51 B6 Yanshan Yunnan China
50 E1 Yan Shan mts China
45 H2 Yanshiping China
54 E1 Yanshou China
33 P2 Yanskiy Zaliv g. Rus. Fed.
67 F2 Yantabulla Austr.
37 M2 Yantagh Iran
54 A5 Yantai China
66 E2 Yantara Lake salt flat Austr.
17 J3 Yantarnyy Rus. Fed.
54 D2 Yantongshan China
71 15 Yanuca i. Fiji
71 13 Yanuca rf Fiji
71 15 Yanuya i. Fiji
51 A5 Yanyuan China
50 E3 Yanzhou China
114 F4 Yaoundé Cameroon
50 C3 Yao Xian China
47 F6 Yap i. Micronesia
103 D4 Yapacana, Co mt Venez.
57 K7 Yapen i. Indon.
57 K7 Yapen, Selat chan. Indon.
118 E5 Yap Tr. sea feature Pac. Oc.
71 13 Yaqaga i. Fiji
71 15 Yaqeta i. Fiji
96 B1 Yaqui r. Mex.
103 C2 Yaracuy r. Venez.
65 H4 Yaraka Austr.
28 H3 Yaransk Rus. Fed.
66 A4 Yardea Austr.
36 C3 Yardımcı Burnu pt Turkey
37 M2 Yardımlı Azer.
13 J5 Yare r. U.K.
28 K2 Yarega Rus. Fed.
63 G2 Yaren Nauru
28 J2 Yarensk Rus. Fed.
103 B4 Yari r. Col.
53 E6 Yariga-take mt Japan
103 C2 Yaritagua Venez.
39 J5 Yarkant He r. China
89 J3 Yarker Can.
44 C1 Yarkhun r. Pak.
79 G5 Yarmouth Can.
13 F7 Yarmouth U.K.
91 H4 Yarmouth Port U.S.A.
93 F4 Yarnell U.S.A.
28 F3 Yaroslavl' Rus. Fed.
28 F3 Yaroslavskaya Oblast' div. Rus. Fed.
52 C2 Yaroslavskiy Rus. Fed.
67 G7 Yarram Austr.
67 J1 Yarraman Austr.
66 F6 Yarra Yarra r. Austr.
69 B4 Yarra Yarra Lakes salt flat Austr.
68 A6 Yarronvale Austr.
68 A3 Yarrowmere Austr.
45 H3 Yartö Tra La pass China
32 K3 Yartsevo Krasnoy. Rus. Fed.
28 E4 Yartsevo Smolensk. Rus. Fed.
103 B3 Yarumal Col.
45 F5 Yasai r. India
71 14 Yasawa i. Fiji
71 15 Yasawa Group is Fiji
29 G6 Yasenskaya Rus. Fed.
29 G6 Yashalta Rus. Fed.
39 H5 Yashilkŭl l. Tajik.
29 H6 Yashkul' Rus. Fed.
52 E2 Yasnaya Polyana Rus. Fed.
38 D2 Yasnyy Rus. Fed.
67 H5 Yass Austr.
67 H5 Yass r. Austr.
40 C4 Yāsūj Iran
36 B3 Yatağan Turkey
63 G4 Yaté New Caledonia
71 9 Yaté New Caledonia Pac. Oc.
85 E4 Yates Center U.S.A.
77 K2 Yathkyed Lake Can.
53 F7 Yatsuga-take volc. Japan
53 B8 Yatsushiro Japan
13 E6 Yatton U.K.
51 Yau Tong H.K. China
100 D5 Yavari r. Brazil/Peru
96 B2 Yávaros Mex.
44 D5 Yavatmāl India
37 H2 Yavi Turkey
103 D3 Yavi, Co mt Venez.
29 B5 Yavoriv Ukr.
53 C8 Yawatahama Japan
45 E1 Yawatongguz He r. China
45 H5 Yaw Ch. r. Myanmar
97 G4 Yaxchilan tourist site Guatemala
40 D4 Yazd Iran
41 F3 Yazdān Iran
40 D4 Yazd-e Khvāst Iran
36 G2 Yazıhan Turkey
85 F5 Yazoo r. U.S.A.
85 F5 Yazoo City U.S.A.
21 H5 Ybbs r. Austria
21 J4 Ybbs an der Donau Austria
21 H5 Ybbsitz Austria
9 L9 Yding Skovhøj h. Denmark
27 K6 Ydra i. Greece
58 A2 Ye Myanmar
66 F6 Yea Austr.
13 D7 Yealmpton U.K.
39 J5 Yecheng China
96 B1 Yécora Mex.
87 D7 Yeehaw Junction U.S.A.
66 A5 Yeelanna Austr.
28 F4 Yefremov Rus. Fed.
37 K2 Yeghegnadzor Armenia
39 J2 Yegindybulak Kazak.
29 G6 Yegorlyk r. Rus. Fed.
29 G6 Yegorlykskaya Rus. Fed.
28 E3 Yegorova, Mys pt Rus. Fed.
28 F4 Yegor'yevsk Rus. Fed.
109 F4 Yei Sudan
115 B1 Yei r. Sudan
114 C3 Yeji Ghana
50 E4 Yejiaji China
32 H4 Yekaterinburg Rus. Fed.
29 G5 Yelan' Rus. Fed.
29 G5 Yelan' r. Rus. Fed.
67 J2 Yelarbon Austr.
38 E5 Yelbarsli Turkm.
29 F5 Yelets Rus. Fed.
114 A1 Yélimané Mali
11 Yell i. U.K.
43 C2 Yellandu India
43 A3 Yellapur India
88 B3 Yellow r. U.S.A.
90 D4 Yellow Creek U.S.A.
69 D5 Yellowdine Austr.
74 G3 Yellowknife Can.
67 G4 Yellow Mt h. Austr.
Yellow River r. see Huang He